Integrative Production Technology for High-Wage Countries

Christian Brecher
Editor

Integrative Production Technology for High-Wage Countries

Editor
Christian Brecher
Werkzeugmaschinenlabor (WZL)
der RWTH Aachen
Lehrstuhl für Werkzeugmaschinen
Steinbachstraße 19
52074 Aachen Nordrhein-Westfalen
Germany
c.brecher@wzl.rwth-aachen.de

The Cluster of Excellence "Integrative Production Technology for High-Wage Countries" has been funded by the German Research Foundation (DFG) as part of the Excellence Initiative. The authors wish to thank the DFG for their support.

ISBN 978-3-642-21066-2   ISBN 978-3-642-21067-9 (eBook)
DOI 10.1007/978-3-642-21067-9
Springer Heidelberg Dordrecht London New York

Library of Congress Control Number: 2011944180

© Springer-Verlag Berlin Heidelberg 2012
This work is subject to copyright. All rights are reserved, whether the whole or part of the material is concerned, specifically the rights of translation, reprinting, reuse of illustrations, recitation, broadcasting, reproduction on microfilm or in any other way, and storage in data banks. Duplication of this publication or parts thereof is permitted only under the provisions of the German Copyright Law of September 9, 1965, in its current version, and permission for use must always be obtained from Springer. Violations are liable to prosecution under the German Copyright Law.
The use of general descriptive names, registered names, trademarks, etc. in this publication does not imply, even in the absence of a specific statement, that such names are exempt from the relevant protective laws and regulations and therefore free for general use.

Cover design: eStudio Calamar S.L.

Printed on acid-free paper

Springer is part of Springer Science+Business Media (www.springer.com)

Preface

CEO of the Cluster of Excellence "Integrative Production Technology for High-Wage Countries"

This book summarizes the recent findings in basic research fields of the Cluster of Excellence "Integrative Production Technology for High-Wage countries" that provide answers to the question how future production in high-wage countries can turn out to be both sustainable and succesful. As part of a targeted production oriented research the challenging dimensions of a manufacturing company have to be examined in a holistic manner. In the framework of integrated production technology these cannot be examined separately but have to be considered comprehensively.

Even today, the research results show that the integrative approach and working method holds great potential for the future. The implementation of such a research approach is associated with an initial effort that pushes forward the crosslinking of different actors, allowing to identify synergies and fully exploit them. The coming years will therefore stand more than ever under the theme of strategic and structural alignment of material and production research in Aachen: The Excellence Initiative will set the course for a research program until 2017 that will greatly benefit from the initial efforts already taken. The research program defined in the running Cluster of Excellence with focus on individualization, virtualization, integrated technologies and self-optimization of production sets the framework for integrative research approaches to follow in the future.

The aim of this book is to present the Cluster of Excellence as a whole and emphasize the innovative character of the presented research whose detailed results are further described in renown technical publications. They clearly show that integrativity in production technology can only be realised with an integrative approach in research.

I want to thank all the scientists for their extraordinary results and the German Research Foundation (DFG) for the funding of the work described in the Cluster of Excellence "Integrative Production Technology for High-Wage Countries."

Aachen   Christian Brecher

Rector of RWTH Aachen University

Innovation—Production Engineering—Aachen: Since many decades, this triad represents the excellent research at RWTH Aachen University. Together with its production research institutes, RWTH Aachen University provides a sound basis for nationally and internationally acknowledged innovations, which have been transferred into industrial practice numerous times. Amongst others, its production research has significantly contributed to the fact that RWTH Aachen University today has the highest level of third-party funding in Germany (in terms of grantings) and came first in the Excellence Initiative of the German Research Foundation.

The success of RWTH Aachen University is also the success of its institutes and professors, which complement each other to an outstanding extent—especially in the field of production research. One of this strong network's achievements is the Cluster of Excellence "Integrative Production Technology for High-Wage Countries". Due to its structural aim, it has changed and will continue changing the participating institutes while continuously enhancing the collaboration among the involved production research institutes.

The Cluster of Excellence involves more than 20 professors of RWTH Aachen University from the fields of material science and production research as well as numerous affiliated institutes like for example of the Fraunhofer Gesellschaft. Together they contribute to securing employment-relevant industrial production in high-wage countries and thus eventually to securing Germany's wealth as well. The research carried out in the Cluster of Excellence therefore focuses on economically relevant products, which are successfully being produced in Germany. The solution of the issues, which the cluster addresses in a fundamental manner, requires a novel understanding of product and production interrelations. Hence, the Cluster of Excellence aims for developing the fundamentals of a sustainable, production research-oriented strategy and theory as well as the required technologies.

I am happy that with this publication, the extraordinary results of the Cluster of Excellence are made accessible to a broad audience. I personally would like to thank all of the participants, who have contributed to the articles and to the realisation of this book with outstanding commitment. I wish the Cluster of Excellence "Integrative Production Technology for High-Wage Countries" all the best in the future.

Aachen   Ernst Schmachtenberg

Contents

1 **Introduction** 1
   Christian Brecher, Wilhelm Oliver Karmann and Stefan Kozielski

2 **Integrative Production Technology for High-wage Countries** 17
   Christian Brecher, Sabina Jeschke, Günther Schuh, Susanne Aghassi,
   Jens Arnoscht, Fabian Bauhoff, Sascha Fuchs, Claudia Jooß,
   Oliver Karmann, Stefan Kozielski, Simon Orilski, Anja Richert,
   Andreas Roderburg, Michael Schiffer, Johannes Schubert,
   Sebastian Stiller, Stefan Tönissen and Florian Welter

3 **Individualised Production** 77
   Günther Schuh, Marek Behr, Christian Brecher, Andreas Bührig-Polaczek,
   Walter Michaeli, Robert Schmitt, Jens Arnoscht, Arne Bohl,
   Damien Buchbinder, Jan Bültmann, Andrei Diatlov, Stefanie Elgeti,
   Werner Herfs, Christian Hinke, Andreas Karlberger, Daniel Kupke,
   Michael Lenders, Christopher Nußbaum, Markus Probst,
   Yann Queudeville, Jerome Quick, Henrich Schleifenbaum,
   Michael Vorspel-Rüter and Christian Windeck

4 **Virtual Production Systems** 241
   Wolfgang Schulz, Christian Bischof, Kirsten Bobzin, Christian Brecher,
   Thomas Gries, Sabina Jeschke, Achim Kampker, Fritz Klocke,
   Torsten Kuhlen, Günther Schuh, Markus Apel, Tim Arping,
   Nazlim Bagcivan, Markus Bambach, Thomas Baranowski,
   Stephan Bäumler, Thomas Beer, Stefan Benke, Thomas Bergs,
   Peter Burggräf, Gustavo Francisco Cabral, Urs Eppelt, Patrick Fayek,
   Marcel Fey, Bastian Franzkoch, Stephan Freyberger, Lothar Glasmacher,
   Barbara Heesel, Thomas Henke, Werner Herfs, Ulrich Jansen,
   Tatyana Kashko, Sergey Konovalov, Britta Kuckhoff, Gottfried Laschet,
   Markus Linke, Wolfram Lohse, Tobias Meisen, Meysam Minoufekr,
   Jan Nöcker, Ulrich Prahl, Hendrik Quade, Matthias Rasim,
   Marcus Rauhut, Rudolf Reinhard, Jan Rosenbaum, Eduardo Rossiter,
   Daniel Schilberg, Georg J. Schmitz, Johannes Triebs, Hagen Wegner
   and Cathrin Wesch-Potente

5 **Hybrid Production Systems** 435
   Gerhard Hirt, Wolfgang Bleck, Kirsten Bobzin, Christian Brecher,
   Andreas Bührig-Polazcek, Edmund Haberstroh, Fritz Klocke,
   Peter Loosen, Walter Michaeli, Reinhart Poprawe, Uwe Reisgen,
   Kristian Arntz, Nazlim Bagcivan, Markus Bambach, Stephan Bäumler,
   Stefan Beckemper, Georg Bergweiler, Tobias Breitbach, Steffen Buchholz,
   Jan Bültmann, Jörg Diettrich, Dennis Do-Khac, Stephan Eilbracht,
   Michael Emonts, Dustin Flock, Kai Gerhardt, Arnold Gillner,
   Alexander Göttmann, Oliver Grönlund, Claudia Hartmann, Daniel Heinen,
   Werner Herfs, Jan-Patrick Hermani, Jens Holtkamp, Todor Ivanov,
   Matthias Jakob, Andreas Janssen, Andreas Karlberger, Fritz Klaiber,
   Pia Kutschmann, Andreas Neuß, Ulrich Prahl, Andreas Roderburg,
   Chris-Jörg Rosen, Andreas Rösner, Alireza Saeed-Akbari, Micha Scharf,
   Sven Scheik, Markus Schleser, Maximilian Schöngart, Lars Stein,
   Marius Steiners, Jochen Stollenwerk, Babak Taleb Araghi, Sebastian Theiß
   and Johannes Wunderle

6 **Self-optimising Production Systems** 697
   Robert Schmitt, Christian Brecher, Burkhard Corves, Thomas Gries,
   Sabina Jeschke, Fritz Klocke, Peter Loosen, Walter Michaeli,
   Rainer Müller, Reinhard Poprawe, Uwe Reisgen, Christopher M. Schlick,
   Günther Schuh, Thomas Auerbach, Fabian Bauhoff, Marion Beckers,
   Daniel Behnen, Tobias Brosze, Guido Buchholz, Christian Büscher,
   Urs Eppelt, Martin Esser, Daniel Ewert, Kamil Fayzullin,
   Reinhard Freudenberg, Peter Fritz, Sascha Fuchs, Yves-Simon Gloy,
   Sebastian Haag, Eckart Hauck, Werner Herfs, Niklas Hering,
   Mathias Hüsing, Mario Isermann, Markus Janßen, Bernhard Kausch,
   Tobias Kempf, Stephan Kratz, Sinem Kuz, Matthis Laass, Juliane Lose,
   Adam Malik, Marcel Ph. Mayer, Thomas Molitor, Simon Müller,
   Barbara Odenthal, Alberto Pavim, Dirk Petring, Till Potente,
   Nicolas Pyschny, Axel Reßmann, Martin Riedel, Simone Runge,
   Heiko Schenuit, Daniel Schilberg, Wolfgang Schulz, Maik Schürmeyer,
   Jens Schüttler, Ulrich Thombansen, Dražen Veselovac, Matthias Vette,
   Carsten Wagels and Konrad Willms

7 **Integrative Business and Technology Cases** 987
   Christian Brecher, Achim Kampker, Fritz Klocke, Peter Loosen,
   Walter Michaeli, Robert Schmitt, Günther Schuh, Thomas Auerbach,
   Arne Bohl, Peter Burggräf, Sascha Fuchs, Max Funck, Alexander Gatej,
   Lothar Glasmacher, Julio Aguilar, Robert Guntlin, Ulrike Hecht,
   Rick Hilchner, Mario Isermann, Stephan Kratz, Matthis Laass,
   Meysam Minoufekr, Valentin Morasch, Andreas Neuß,
   Christian Niggemann, Jan Noecker, Till Potente, André Schievenbusch,
   Georg J. Schmitz, Stephan Schmitz, Jochen Stollenwerk,
   Dražen Veselovac, Cathrin Wesch-Potente and Johannes Wunderle

Index 1077

List of Contributors

Dipl.-Inform. Susanne Aghassi Fraunhofer-Institut für Produktionstechnologie IPT, Steinbachstr. 17, 52074 Aachen, Germany
e-mail: susanne.aghassi@ipt.fraunhofer.de

Dr.-Ing. Julio Aguilar ACCESS e.V., Intzestr. 5, 52072 Aachen, Germany
e-mail: j.aguilar@access.rwth-aachen.de

Dr. rer. nat. Markus Apel ACCESS e.V., Intzestr. 5, 52072 Aachen,
e-mail: m.apel@access.rwth-aachen.de

Dr.-Ing. Jens Arnoscht Werkzeugmaschinenlabor WZL der RWTH Aachen, Steinbachstr. 19, 52074 Aachen, Germany
e-mail: j.arnoscht@wzl.rwth-aachen.de

Dipl.-Ing. Kristian Arntz Fraunhofer-Institut für Produktionstechnologie IPT, Steinbachstr. 17, 52074 Aachen, Germany
e-mail: kristian.arntz@ipt.fraunhofer.de

Dipl.-Ing. Tim Arping Institut für Kunststoffverarbeitung (IKV) in Industrie und Handwerk an der RWTH Aachen, Pontstr. 49, 52062 Aachen, Germany
e-mail: arping@ikv.rwth-aachen.de

Dipl.-Ing. Thomas Auerbach Werkzeugmaschinenlabor WZL der RWTH Aachen, Steinbachstr. 19, 52074 Aachen, Germany
e-mail: t.auerbach@wzl.rwth-aachen.de

Dr.-Ing. Nazlim Bagcivan Lehrstuhl für Oberflächentechnik im Maschinenbau der RWTH Aachen, Templergraben 55, 52062 Aachen, Germany
e-mail: bagcivan@iot.rwth-aachen.de

Dr.-Ing. Markus Bambach Institut für Bildsame Formgebung der RWTH Aachen, Intzestr. 10, 52056 Aachen, Germany
e-mail: bambach@ibf.rwth-aachen.de

Dipl.-Ing. Thomas Baranowski Institut für Kunststoffverarbeitung (IKV) in Industrie und Handwerk an der RWTH Aachen, Pontstr. 49, 52062 Aachen, Germany
e-mail: baranowski@ikv.rwth-aachen.de

Dipl.-Wirt. Ing. Fabian Bauhoff Forschungsinstitut für Rationalisierung (FIR) e. V. an der RWTH Aachen, Pontdriesch 14-16, 52062 Aachen, Germany
e-mail: fabian.bauhoff@fir.rwth-aachen.de

Dipl.-Ing. Stephan Bäumler Werkzeugmaschinenlabor WZL der RWTH Aachen, Steinbachstr. 19, 52074 Aachen, Germany
e-mail: s.baeumler@wzl.rwth-aachen.de

Dipl.-Phys. Stefan Beckemper Fraunhofer-Institut für Lasertechnik ILT, Steinbachstr. 15, 52074 Aachen, Germany
e-mail: stefan.beckemper@ilt.fraunhofer.de

Dipl.-Inform. Marion Beckers Institut für Schweißtechnik und Fügetechnik der RWTH Aachen, Pontstr. 49, 52062 Aachen, Germany
e-mail: beckers@isf.rwth-aachen.de

M.Sc. Thomas Beer Rechen- und Kommunikationszentrum RWTH Aachen (RZ), Dienstgebäude Seffenter Weg 23, 52074 Aachen, Germany
e-mail: beer@vr.rwth-aachen.de

Dipl.-Inform. Daniel Behnen Werkzeugmaschinenlabor WZL der RWTH Aachen, Steinbachstr. 19, 52074 Aachen, Germany
e-mail: D.Behnen@wzl.rwth-aachen.de

Prof. Ph.D. Marek Behr Lehrstuhl für Computergestützte Analyse Technischer Systeme der RWTH Aachen, Schinkelstr. 2, 52062 Aachen,
e-mail: behr@cats.rwth-aachen.de

Dr.-Ing. Stefan Benke ACCESS e.V., Intzestr. 5, 52072 Aachen, Germany
e-mail: s.benke@access.rwth-aachen.de

Dr.-Ing. Thomas Bergs Fraunhofer-Institut für Produktionstechnologie IPT, Steinbachstr. 17, 52074 Aachen, Germany
e-mail: thomas.bergs@ipt.fraunhofer.de

Dipl.-Ing. Georg Bergweiler Fraunhofer-Institut für Lasertechnik ILT, Steinbachstr. 15, 52074 Aachen, Germany
e-mail: georg.bergweiler@ilt.fraunhofer.de

List of Contributors

Dr.-Ing. Stephan Bichmann Fraunhofer-Institut für Produktionstechnologie IPT, Steinbachstr. 17, 52074 Aachen, Germany
e-mail: stephan.bichmann@ipt.fraunhofer.de

Prof. Ph.D. Christian Bischof Rechen- und Kommunikationszentrum RWTH Aachen (RZ), Dienstgebäude Seffenter Weg 23, 52074 Aachen, Germany
e-mail: bischof@sc.rwth-aachen.de

Prof. Dr.-Ing. Wolfgang Bleck Institut für Eisenhüttenkunde der RWTH Aachen, Intzestr. 1, 52072 Aachen, Germany
e-mail: bleck@iehk.rwth-aachen.de

Prof. Dr.-Ing. Kirsten Bobzin Lehrstuhl für Oberflächentechnik im Maschinenbau der RWTH Aachen, Templergraben 55, 52062 Aachen, Germany
e-mail: info@iot.rwth-aachen.de

Dipl.-Ing. Dipl.-Wirt. Ing. Arne Bohl Werkzeugmaschinenlabor WZL der RWTH Aachen, Steinbachstr. 19, 52074 Aachen, Germany
e-mail: a.bohl@wzl.rwth-aachen.de

Prof. Dr.-Ing. Christian Brecher Werkzeugmaschinenlabor WZL der RWTH Aachen, Steinbachstr. 19, 52074 Aachen, Germany
e-mail: c.brecher@wzl.rwth-aachen.de

Dipl.-Ing. Tobias Breitbach Werkzeugmaschinenlabor WZL der RWTH Aachen, Steinbachstr. 19, 52074 Aachen, Germany
e-mail: t.breitbach@wzl.rwth-aachen.de

Dr.-Ing. Dipl.-Wirt.-Ing. Tobias Brosze Forschungsinstitut für Rationalisierung (FIR) e. V. an der RWTH Aachen, Pontdriesch 14-16, 52062 Aachen, Germany
e-mail: br@fir.rwth-aachen.de

Dipl.-Ing. Damien Buchbinder Fraunhofer-Institut für Lasertechnik ILT, Steinbachstr. 15, 52074 Aachen, Germany
e-mail: damien.buchbinder@ilt.fraunhofer.de

Dipl.-Ing. Guido Buchholz Institut für Schweißtechnik und Fügetechnik der RWTH Aachen, Pontstr. 49, 52062 Aachen, Germany
e-mail: bu@isf.rwth-aachen.de

Dipl.-Ing. Steffen Buchholz Werkzeugmaschinenlabor WZL der RWTH Aachen, Steinbachstr. 19, 52074 Aachen, Germany
e-mail: s.buchholz@wzl.rwth-aachen.de

Prof. Dr.-Ing. Andreas Bührig-Polaczek Gießerei-Institut der RWTH Aachen, Intzestr. 5, 52072 Aachen, Germany
e-mail: sekretariat@gi.rwth-aachen.de

Dipl.-Ing. Jan Bültmann Institut für Eisenhüttenkunde der RWTH Aachen, Intzestr. 1, 52072 Aachen, Germany
e-mail: jan.bueltmann@iehk.rwth-aachen.de

Dipl.-Ing. Peter Burggräf Werkzeugmaschinenlabor WZL der RWTH Aachen, Steinbachstr. 19, 52074 Aachen, Germany
e-mail: p.burggraef@wzl.rwth-aachen.de

Dipl.-Wirt. Ing. Christian Büscher IMA/ZLW & IfU der RWTH Aachen, Dennewartstr. 27, 52068 Aachen, Germany
e-mail: christian.buescher@ima-zlw-ifu.rwth-aachen.de

M.Eng. Gustavo Francisco Cabral Fraunhofer-Institut für Produktionstechnologie IPT, Steinbachstr. 17, 52074 Aachen, Germany
e-mail: gustavo.francisco.cabral@ipt.fraunhofer.de

Prof. Dr.-Ing. Burkhard Corves Institut für Getriebetechnik und Maschinendynamik der RWTH Aachen, Eilfschornsteinstr. 18, 52062 Aachen, Germany
e-mail: corves@igm.rwth-aachen.de

Dipl.-Phys. Andrei Diatlov Fraunhofer-Institut für Lasertechnik ILT, Steinbachstr. 15, 52074 Aachen, Germany
e-mail: andrei.diatlov@ilt.fraunhofer.de

Dipl.-Ing. Jörg Diettrich Fraunhofer-Institut für Lasertechnik ILT, Steinbachstr. 15, 52074 Aachen, Germany
e-mail: joerg.diettrich@ilt.fraunhofer.de

Dipl.-Ing. Dennis Do-Khac Werkzeugmaschinenlabor WZL der RWTH Aachen, Steinbachstr. 19, 52074 Aachen, Germany
e-mail: d.do-khac@wzl.rwth-aachen.de

Dipl.-Ing. Stephan Eilbracht Institut für Kunststoffverarbeitung (IKV) in Industrie und Handwerk an der RWTH Aachen, Pontstr. 49, 52062 Aachen, Germany
e-mail: eilbracht@ikv.rwth-aachen.de

Dipl.-Ing. Stefanie Elgeti Lehrstuhl für Computergestützte Analyse Technischer Systeme der RWTH Aachen, Schinkelstr. 2, 52062 Aachen, Germany
e-mail: elgeti@cats.rwth-aachen.de

Dr.-Ing. Michael Emonts Fraunhofer-Institut für Produktionstechnologie IPT, Steinbachstr. 17, 52074 Aachen, Germany
e-mail: michael.emonts@ipt.fraunhofe.de

Dipl.-Phys. Urs Eppelt Fraunhofer-Institut für Lasertechnik ILT, Steinbachstr. 15, 52074 Aachen, Germany
e-mail: urs.eppelt@ilt.fraunhofer.de

Dr.Ing. Dipl.Kfm. Martin Esser Werkzeugmaschinenlabor WZL der RWTH Aachen, Steinbachstr. 19, 52074 Aachen, Germany
e-mail: m.esser@wzl.rwth-aachen.de

Dipl.-Inform. Daniel Ewert IMA/ZLW & IfU der RWTH Aachen, Dennewartstr. 27, 52068 Aachen, Germany
e-mail: daniel.ewert@ima-zlw-ifu.rwth-aachen.de

Dipl.-Ing. Patrick Fayek Institut für Eisenhüttenkunde der RWTH Aachen, Intzestr. 1, 52072 Aachen, Germany
e-mail: patrick.fayek@iehk.rwth-aachen.de

Dr.-Ing. Kamil Fayzullin, M.O.R. Werkzeugmaschinenlabor WZL der RWTH Aachen, Steinbachstr. 19, 52074 Aachen, Germany
e-mail: k.fayzullin@wzl.rwth-aachen.de

Dipl.-Ing. Marcel Fey Werkzeugmaschinenlabor WZL der RWTH Aachen, Steinbachstr. 19, 52074 Aachen, Germany
e-mail: M.Fey@wzl.rwth-aachen.de

Dipl.-Ing. Dustin Flock Lehr- und Forschungsgebiet Kautschuktechnologie der RWTH Aachen, Templergraben 55, 52056 Aachen, Germany
e-mail: flock@lfk.rwth-aachen.de

Dr.-Ing. Bastian Franzkoch Werkzeugmaschinenlabor WZL der RWTH Aachen, Steinbachstr. 19, 52074 Aachen, Germany
e-mail: b.franzkoch@wzl.rwth-aachen.de

Dr.-Ing. Reinhard Freudenberg Werkzeugmaschinenlabor WZL der RWTH Aachen, Steinbachstr. 19, 52074 Aachen, Germany
e-mail: r.freudenberg@wzl.rwth-aachen.de

Dipl.-Ing. Stephan Freyberger Gießerei-Institut der RWTH Aachen, Intzestr. 5, 52072 Aachen, Germany
e-mail: s.freyberger@gi.rwth-aachen.de

Dipl.-Inform. Peter Fritz Werkzeugmaschinenlabor WZL der RWTH Aachen, Steinbachstr. 19, 52074 Aachen, Germany
e-mail: P.Fritz@wzl.rwth-aachen.de

Dipl.-Ing. Sascha Fuchs Werkzeugmaschinenlabor WZL der RWTH Aachen, Steinbachstr. 19, 52074 Aachen, Germany
e-mail: S.Fuchs@wzl.rwth-aachen.de

Dipl.-Ing. Max Funck Lehrstuhl für Technologie Optischer Systeme der RWTH Aachen, Steinbachstr. 15, 52074 Aachen, Germany
e-mail: max.funck@ilt.fraunhofer.de

Dipl.-Wirt. Ing. Alexander Gatej Lehrstuhl für Technologie Optischer Systeme der RWTH Aachen, Steinbachstr. 15, 52074 Aachen, Germany
e-mail: alexander.gatej@ilt.fraunhofer.de

Dipl.-Ing. Kai Gerhardt Institut für Bildsame Formgebung der RWTH Aachen, Intzestr. 10, 52056 Aachen, Germany
e-mail: gerhardt@ibf.rwth-aachen.de

Dr.-Ing. Arnold Gillner Fraunhofer-Institut für Lasertechnik ILT, Steinbachstr. 15, 52074 Aachen, Germany
e-mail: arnold.gillner@ilt.fraunhofer.de

Dipl.-Ing. Dipl.-Inform. Lothar Glasmacher Fraunhofer-Institut für Produktionstechnologie IPT, Steinbachstr. 17, 52074 Aachen, Germany
e-mail: lothar.glasmacher@ipt.fraunhofer.de

Dipl.-Ing. Yves-Simon Gloy Institut für Textiltechnik der RWTH Aachen, Otto-Blumenthal-Str. 1, 52074 Aachen, Germany
e-mail: yves.gloy@ita.rwth-aachen.de

Dipl.-Ing. Alexander Göttmann Institut für Bildsame Formgebung der RWTH Aachen, Intzestr. 10, 52056 Aachen, Germany
e-mail: Goettmann@ibf.rwth-aachen.de

Prof. Dr.-Ing. Dipl.-Wirt. Ing. Thomas Gries Institut für Textiltechnik der RWTH Aachen, Otto-Blumenthal-Str. 1, 52074 Aachen, Germany
e-mail: thomas.gries@ita.rwth-aachen.de

Dipl.-Ing. Oliver Grönlund Institut für Kunststoffverarbeitung (IKV) in Industrie und Handwerk an der RWTH Aachen, Pontstr. 49, 52062 Aachen, Germany
e-mail: groenlund@ikv.rwth-aachen.de

Robert Guntlin ACCESS e.V., Intzestr. 5, 52072 Aachen, Germany
e-mail: r.guntlin@access.rwth-aachen.de

Dipl.-Ing. Sebastian Haag Fraunhofer-Institut für Produktionstechnologie IPT, Steinbachstr. 17, 52074 Aachen, Germany
e-mail: sebastian.haag@ipt.fraunhofer.de

Prof. Dr.-Ing. Edmund Haberstroh Lehr- und Forschungsgebiet Kautschuktechnologie der RWTH Aachen, Templergraben 55, 52056 Aachen, Germany
e-mail: haberstroh@lfk.rwth-aachen.de

Dipl.-Ing. (FH) Claudia Hartmann Fraunhofer-Institut für Lasertechnik ILT, Steinbachstr. 15, 52074 Aachen, Germany
e-mail: claudia.hartmann@ilt.fraunhofer.de

Dr.-Ing. Eckart Hauck IMA/ZLW & IfU der RWTH Aachen, Dennewartstr. 27, 52068 Aachen, Germany
e-mail: eckart.hauck@ima-zlw-ifu.rwth-aachen.de

Dipl.-Ing. Ulrike Hecht ACCESS e.V., Intzestr. 5, 52072 Aachen, Germany
e-mail: ulrike.hecht@access.rwth-aachen.de

Dipl.-Ing. Barbara Heesel Institut für Kunststoffverarbeitung (IKV) in Industrie und Handwerk an der RWTH Aachen, Pontstr. 49, 52062 Aachen, Germany
e-mail: heesel@ikv.rwth-aachen.de

Dipl.-Ing. Daniel Heinen Fraunhofer-Institut für Produktionstechnologie IPT, Steinbachstr. 17, 52074 Aachen, Germany
e-mail: daniel.heinen@ipt.fraunhofer.de

Dipl.-Ing. Thomas Henke Institut für Bildsame Formgebung der RWTH Aachen, Intzestr. 10, 52056 Aachen, Germany
e-mail: henke@ibf.rwth-aachen.de

Dr.-Ing. Werner Herfs Werkzeugmaschinenlabor WZL der RWTH Aachen, Steinbachstr. 19, 52074 Aachen, Germany
e-mail: w.herfs@wzl.rwth-aachen.de

Dipl.-Wirt. Ing. Niklas Hering Forschungsinstitut für Rationalisierung (FIR) e. V. an der RWTH Aachen, Pontdriesch 14-16, 52062 Aachen, Germany
e-mail: niklas.hering@fir.rwth-aachen.de

Dipl.-Ing. Jan-Patrick Hermani Fraunhofer-Institut für Produktionstechnologie IPT, Steinbachstr. 17, 52074 Aachen, Germany
e-mail: jan-patrick.hermani@ipt.fraunhofer.de

Dipl.-Ing. Rick Hilchner Werkzeugmaschinenlabor WZL der RWTH Aachen, Steinbachstr. 19, 52074 Aachen, Germany
e-mail: R.Hilchner@wzl.rwth-aachen.de

Dipl.-Phys. Christian Hinke Fraunhofer-Institut für Lasertechnik ILT, Steinbachstr. 15, 52074 Aachen, Germany
e-mail: christian.hinke@ilt.fraunhofer.de

Prof. Dr.-Ing. Gerhard Hirt Institut für Bildsame Formgebung der RWTH Aachen, Intzestr. 10, 52056 Aachen, Germany
e-mail: hirt@ibf.rwth-aachen.de

Dr.-Ing. Jens Holtkamp Fraunhofer-Institut für Lasertechnik ILT, Steinbachstr. 15, 52074 Aachen, Germany
e-mail: jens.holtkamp@ilt.fraunhofer.de

Dr.-Ing. Mathias Hüsing Institut für Getriebetechnik und Maschinendynamik der RWTH Aachen, Eilfschornsteinstr. 18, 52062 Aachen, Germany
e-mail: huesing@igm.rwth-aachen.de

Dipl.-Ing. Mario Isermann Fraunhofer Institut für Produktionstechnologie IPT, Steinbachstr. 17, 52074 Aachen, Germany
e-mail: mario.isermann@ipt.fraunhofer.de

Dipl.-Ing. Todor Ivanov Gießerei-Institut der RWTH Aachen, Intzestr. 5, 52072 Aachen, Germany
e-mail: t.ivanov@gi.rwth-aachen.de

Dipl.-Ing. Matthias Jakob Gießerei-Institut der RWTH Aachen, Intzestr. 5, 52072 Aachen, Germany
e-mail: m.jakob@gi.rwth-aachen.de

Dipl.-Phys. Ulrich Jansen Fraunhofer-Institut für Lasertechnik ILT, Steinbachstr. 15, 52074 Aachen, Germany
e-mail: ulrich.jansen@ilt.fraunhofer.de

Dipl.-Ing. Markus Janßen Werkzeugmaschinenlabor WZL der RWTH Aachen, Steinbachstr. 19, 52074 Aachen, Germany
e-mail: M.Janssen@wzl.rwth-aachen.de

Prof. Dr. rer. nat. Sabina Jeschke IMA/ZLW & IfU der RWTH Aachen, Dennewartstr. 27, 52068 Aachen, Germany
e-mail: jeschke.office@ima-zlw-ifu.rwth-aachen.de

Claudia Jooß M.A. IMA/ZLW & IfU der RWTH Aachen, Dennewartstr. 27, 52068 Aachen, Germany
e-mail: claudia.jooss@ima-zlw-ifu.rwth-aachen.de

Prof. Dr.-Ing. Achim Kampker Werkzeugmaschinenlabor WZL der RWTH Aachen, Steinbachstr. 19, 52074 Aachen, Germany
e-mail: a.kampker@wzl.rwth-aachen.de

Dipl.-Ing. Andreas Karlberger Werkzeugmaschinenlabor WZL der RWTH Aachen, Steinbachstr. 19, 52074 Aachen, Germany
e-mail: a.karlberger@wzl.rwth-aachen.de

Dipl.-Ing. Wilhelm Oliver Karmann Werkzeugmaschinenlabor WZL der RWTH Aachen, Steinbachstr. 19, 52074 Aachen, Germany
e-mail: w.karmann@wzl.rwth-aachen.de

Dipl. Math. Tatyana Kashko Lehrstuhl für Oberflächentechnik im Maschinenbau der RWTH Aachen, Templergraben 55, 52062 Aachen, Germany
e-mail: kashko@iot.rwth-aachen.de

Dr.-Ing. Bernhard Kausch Institut für Arbeitswissenschaft der RWTH Aachen, Bergdriesch 27, 52062 Aachen, Germany
e-mail: b.kausch@iaw.rwth-aachen.de

Dr.-Ing. Tobias Kempf Werkzeugmaschinenlabor WZL der RWTH Aachen, Steinbachstr. 19, 52074 Aachen, Germany
e-mail: t.kempf@wzl.rwth-aachen.de

M.Sc. Fritz Klaiber Institut für Kunststoffverarbeitung (IKV) in Industrie und Handwerk an der RWTH Aachen, Pontstr. 49, 52062 Aachen, Germany
e-mail: klaiber@ikv.rwth-aachen.de

Prof. Dr.-Ing. Dr.-Ing. E.h. Dr. h.c. Dr. h.c. Fritz Klocke Werkzeugmaschinenlabor WZL der RWTH Aachen, Steinbachstr. 19, 52074 Aachen, Germany
e-mail: f.klocke@wzl.rwth-aachen.de

M.Sc. Sergey Konovalov Institut für Eisenhüttenkunde der RWTH Aachen, Intzestr. 1, 52072 Aachen, Germany
e-mail: sergey.konovalov@iehk.rwth-aachen.de

Dr.-Ing. Dipl.-Wirt Ing. Stefan Kozielski Werkzeugmaschinenlabor WZL der RWTH Aachen, Steinbachstr. 19, 52074 Aachen, Germany
e-mail: s.kozielski@wzl.rwth-aachen.de

Dipl.-Ing. Stephan Kratz Werkzeugmaschinenlabor WZL der RWTH Aachen, Steinbachstr. 19, 52074 Aachen, Germany
e-mail: s.kratz@wzl.rwth-aachen.de

Dipl.-Ing. Britta Kuckhoff Institut für Textiltechnik der RWTH Aachen, Otto-Blumenthal-Str. 1, 52074 Aachen, Germany
e-mail: britta.kuckhoff@ita.rwth-aachen.de

Prof. Dr. rer. nat. Torsten Kuhlen Rechen- und Kommunikationszentrum RWTH Aachen (RZ), Dienstgebäude Seffenter Weg 23, 52074 Aachen, Germany
e-mail: kuhlen@rz.rwth-aachen.de

Dipl.-Ing. Daniel Kupke Werkzeugmaschinenlabor WZL der RWTH Aachen, Steinbachstr. 19, 52074 Aachen,
e-mail: d.kupke@wzl.rwth-aachen.de

Dipl.-Ing. Pia Kutschmann Lehrstuhl für Oberflächentechnik im Maschinenbau der RWTH Aachen, Templergraben 55, 52062 Aachen, Germany
e-mail: kutschmann@iot.rwth-aachen.de

Dipl.-Inform. Sinem Kuz Institut für Arbeitswissenschaft der RWTH Aachen, Bergdriesch 27, 52062 Aachen, Germany
e-mail: s.kuz@iaw.rwth-aachen.de

Dipl.-Ing. (FH) Matthis Laass Werkzeugmaschinenlabor WZL der RWTH Aachen, Steinbachstr. 19, 52074 Aachen, Germany
e-mail: m.laass@wzl.rwth-aachen.de

Dr.-Ing. Gottfried Laschet ACCESS e.V., Intzestr. 5, 52072 Aachen, Germany
e-mail: g.laschet@access.rwth-aachen.de

Dr.-Ing. Dipl.Kfm. Markus Linke Institut für Textiltechnik der RWTH Aachen, Otto-Blumenthal-Str. 1, 52074 Aachen, Germany
e-mail: markus.linke@ita.rwth-aachen.de

Dipl.-Ing. Wolfram Lohse Werkzeugmaschinenlabor WZL der RWTH Aachen, Steinbachstr. 19, 52074 Aachen, Germany
e-mail: w.lohse@wzl.rwth-aachen.de

Prof. Dr. rer. nat. Peter Loosen Lehrstuhl für Technologie Optischer Systeme der RWTH Aachen und Fraunhofer-Institut für Lasertechnik ILT, Steinbachstr. 15, 52074 Aachen, Germany
e-mail: peter.loosen@ilt.fraunhofer.de

List of Contributors

Dipl.-Ing. Juliane Lose Werkzeugmaschinenlabor WZL der RWTH Aachen, Steinbachstr. 19, 52074 Aachen, Germany
e-mail: j.lose@wzl.rwth-aachen.de

Dipl.-Inform. Adam Malik Werkzeugmaschinenlabor WZL der RWTH Aachen, Steinbachstr. 19, 52074 Aachen, Germany
e-mail: a.malik@wzl.rwth-aachen.de

Dipl.-Ing. Marcel Ph. Mayer Institut für Arbeitswissenschaft der RWTH Aachen, Bergdriesch 27, 52062 Aachen, Germany
e-mail: m.mayer@iaw.rwth-aachen.de

Dipl.-Ing. Alexander Meckelnborg Werkzeugmaschinenlabor WZL der RWTH Aachen, Steinbachstr. 19, 52074 Aachen, Germany
e-mail: a.meckelnborg@wzl.rwth-aachen.de

Dipl.-Inform. Tobias Meisen IMA/ZLW & IfU der RWTH Aachen, Dennewartstr. 27, 52068 Aachen, Germany
e-mail: tobias.meisen@ima-zlw-ifu.rwth-aachen.de

Prof. Dr.-Ing. Dr.-Ing. E.h. Walter Michaeli Institut für Kunststoffverarbeitung (IKV) in Industrie und Handwerk an der RWTH Aachen, Pontstr. 49, 52062 Aachen, Germany
e-mail: zentrale@ikv.rwth-aachen.de

Dipl.-Inform. Meysam Minoufekr Fraunhofer-Institut für Produktionstechnologie IPT, Steinbachstr. 17, 52074 Aachen, Germany
e-mail: meysam.minoufekr@ipt.fraunhofer.de

M.Sc. Thomas Molitor Fraunhofer-Institut für Lasertechnik ILT, Steinbachstr. 15, 52074 Aachen, Germany
e-mail: thomas.molitor@ilt.fraunhofer.de

Dipl.-Ing. Valentin Morasch Lehrstuhl für Technologie Optischer Systeme der RWTH Aachen, Steinbachstr. 15, 52074 Aachen, Germany
e-mail: valentin.morasch@ilt.fraunhofer.de

Prof. Dr.-Ing. Rainer Müller Werkzeugmaschinenlabor WZL der RWTH Aachen, Steinbachstr. 19, 52074 Aachen, Germany
e-mail: r.mueller@wzl.rwth-aachen.de

Dipl.-Ing. Simon Müller Werkzeugmaschinenlabor WZL der RWTH Aachen, Steinbachstr. 19, 52074 Aachen, Germany
e-mail: S.Mueller@wzl.rwth-aachen.de

Dipl.-Ing. Andreas Neuß Institut für Kunststoffverarbeitung (IKV) in Industrie und Handwerk an der RWTH Aachen, Pontstr. 49, 52062 Aachen, Germany
e-mail: neuss@ikv.rwth-aachen.de

Dipl.-Ing. Christian Niggemann Werkzeugmaschinenlabor WZL der RWTH Aachen, Steinbachstr. 19, 52074 Aachen, Germany
e-mail: c.niggemann@wzl.rwth-aachen.de

Dipl.-Ing. Jan Nöcker Werkzeugmaschinenlabor WZL der RWTH Aachen, Steinbachstr. 19, 52074 Aachen, Germany
e-mail: j.noecker@wzl.rwth-aachen.de

Dr.-Ing. Christopher Nußbaum Werkzeugmaschinenlabor WZL der RWTH Aachen, Steinbachstr. 19, 52074 Aachen, Germany
e-mail: c.nussbaum@wzl.rwth-aachen.de

Dipl.-Ing. Barbara Odenthal Institut für Arbeitswissenschaft der RWTH Aachen, Bergdriesch 27, 52062 Aachen, Germany
e-mail: b.odenthal@iaw.rwth-aachen.de

Dipl.-Ing. Dipl.-Wirt. Ing. Simon Orilski Fraunhofer-Institut für Produktionstechnologie IPT, Steinbachstr. 17, 52074 Aachen, Germany
e-mail: simon.orilski@googlemail.com

M.Eng. Alberto Pavim Werkzeugmaschinenlabor WZL der RWTH Aachen, Steinbachstr. 19, 52074 Aachen, Germany
e-mail: a.pavim@wzl.rwth-aachen.de

Dr. Dirk Petring Fraunhofer-Institut für Lasertechnik ILT, Steinbachstr. 15, 52074 Aachen, Germany
e-mail: dirk.petring@ilt.fraunhofer.de

Prof. Dr. rer. nat. Reinhart Poprawe M.A. Fraunhofer-Institut für Lasertechnik ILT, Steinbachstr. 15, 52074 Aachen, Germany
e-mail: reinhart.poprawe@ilt.fraunhofer.de

Dipl.-Ing. Till Potente Werkzeugmaschinenlabor WZL der RWTH Aachen, Steinbachstr. 19, 52074 Aachen, Germany
e-mail: t.potente@wzl.rwth-aachen.de

Dr.-Ing. Ulrich Prahl Institut für Eisenhüttenkunde der RWTH Aachen, Intzestr. 1, 52072 Aachen, Germany
e-mail: ulrich.prahl@iehk.rwth-aachen.de

Dipl. Math. Markus Probst Lehrstuhl für Computergestützte Analyse Technischer Systeme der RWTH Aachen, Schinkelstr. 2, 52062 Aachen, Germany
e-mail: probst@cats.rwth-aachen.de

Dipl.-Ing. Dipl.-Wirt. Ing. Nicolas Pyschny Fraunhofer-Institut für Produktionstechnologie IPT, Steinbachstr. 17, 52074 Aachen, Germany
e-mail: nicolas.pyschny@ipt.fraunhofer.de

Dipl.-Ing. Hendrik Quade Institut für Eisenhüttenkunde der RWTH Aachen, Intzestr. 1, 52072 Aachen, Germany
e-mail: hendrik.quade@iehk.rwth-aachen.de

Dipl.-Ing. Yann Queudeville Gießerei-Institut der RWTH Aachen, Intzestr. 5, 52072 Aachen, Germany
e-mail: y.queudeville@gi.rwth-aachen.de

Dipl.-Ing. oec. Jerome Quick Forschungsinstitut für Rationalisierung (FIR) e. V. an der RWTH Aachen, Pontdriesch 14-16, 52062 Aachen, Germany
e-mail: qi@fir.rwth-aachen.de

Dipl.-Ing. Matthias Rasim Werkzeugmaschinenlabor WZL der RWTH Aachen, Steinbachstr. 19, 52074 Aachen, Germany
e-mail: m.rasim@wzl.rwth-aachen.de

Dr.-Ing. Dipl.-Phys. oec. Marcus Rauhut Werkzeugmaschinenlabor WZL der RWTH Aachen, Steinbachstr. 19, 52074 Aachen, Germany
e-mail: m.rauhut@wzl.rwth-aachen.de

Dipl. Math. Rudolf Reinhard IMA/ZLW & IfU der RWTH Aachen, Dennewartstr. 27, 52068 Aachen, Germany
e-mail: rudolf.reinhard@ima-zlw-ifu.rwth-aachen.de

Prof. Dr.-Ing. Uwe Reisgen Institut für Schweißtechnik und Fügetechnik der RWTH Aachen, Pontstr. 49, 52062 Aachen, Germany
e-mail: reisgen@isf.rwth-aachen.de

Dipl.-Ing. Axel Reßmann Institut für Kunststoffverarbeitung (IKV) in Industrie und Handwerk an der RWTH Aachen, Pontstr. 49, 52062 Aachen, Germany
e-mail: ressmann@ikv.rwth-aachen.de

Dr. Anja Richert IMA/ZLW & IfU der RWTH Aachen, Dennewartstr. 27, 52068 Aachen, Germany
e-mail: anja.richert@ima-zlw-ifu.rwth-aachen.de

Dipl.-Ing. Martin Riedel Institut für Getriebetechnik und Maschinendynamik der RWTH Aachen, Eilfschornsteinstr. 18, 52062 Aachen, Germany
e-mail: riedel@igm.rwth-aachen.de

Dipl.-Ing. Andreas Roderburg Werkzeugmaschinenlabor WZL der RWTH Aachen, Steinbachstr. 19, 52074 Aachen, Germany
e-mail: a.roderburg@wzl.rwth-aachen.de

Dipl.-Ing. M.Eng. Dipl.-Kfm. Chris- Jörg Rosen Fraunhofer-Institut für Produktionstechnologie IPT, Steinbachstr. 17, 52074 Aachen, Germany
e-mail: chris-joerg.rosen@ipt.fraunhofer.de

Jan Rosenbaum ACCESS e.V., Intzestr. 5, 52072 Aachen, Germany
e-mail: j.rosenbaum@access.rwth-aachen.de

Dipl.-Ing. Andreas Rösner Fraunhofer-Institut für Lasertechnik ILT, Steinbachstr. 15, 52074 Aachen, Germany
e-mail: andreas.roesner@ilt.fraunhofer.de

M.Sc. Eduardo Rossiter Institut für Schweißtechnik und Fügetechnik der RWTH Aachen, Pontstr. 49, 52062 Aachen, Germany
e-mail: ro@isf.rwth-aachen.de

Dipl.-Math. Simone Runge Forschungsinstitut für Rationalisierung (FIR) e. V. an der RWTH Aachen, Pontdriesch 14-16, 52062 Aachen, Germany
e-mail: simone.runge@fir.rwth-aachen.de

M.Sc. Alireza Saeed-Akbari Institut für Eisenhüttenkunde der RWTH Aachen, Intzestr. 1, 52072 Aachen, Germany
e-mail: alireza.saeed-akbari@iehk.rwth-aachen.de

Dipl.-Ing. Micha Scharf Institut für Kunststoffverarbeitung (IKV) in Industrie und Handwerk an der RWTH Aachen, Pontstr. 49, 52062 Aachen, Germany
e-mail: scharf@ikv.rwth-aachen.de

Dipl.-Ing. Sven Scheik Institut für Schweißtechnik und Fügetechnik der RWTH Aachen, Pontstr. 49, 52062 Aachen, Germany
e-mail: scheik@isf.rwth-aachen.de

Dipl.-Ing. Heiko Schenuit Institut für Textiltechnik der RWTH Aachen, Otto-Blumenthal-Str. 1, 52074 Aachen, Germany
e-mail: heiko.schenuit@ita.rwth-aachen.de

Dr.-Ing. André Schievenbusch ACCESS e.V., Intzestr. 5, 52072 Aachen, Germany
e-mail: sekretariat@access.rwth-aachen.de

List of Contributors

Dipl.-Wirt. Ing. Michael Schiffer Werkzeugmaschinenlabor WZL der RWTH Aachen, Steinbachstr. 19, 52074 Aachen, Germany
e-mail: m.schiffer@wzl.rwth-aachen.de

Dr.-Ing. Daniel Schilberg IMA/ZLW & IfU der RWTH Aachen, Dennewartstr. 27, 52068 Aachen, Germany
e-mail: daniel.schilberg@ima-zlw-ifu.rwth-aachen.de

Dipl.-Ing. Dipl.-Wirt. Ing. Henrich Schleifenbaum Fraunhofer-Institut für Lasertechnik ILT, Steinbachstr. 15, 52074 Aachen, Germany
e-mail: henrich.schleifenbaum@ilt.fraunhofer.de

Dipl.-Ing. Markus Schleser Institut für Schweißtechnik und Fügetechnik der RWTH Aachen, Pontstr. 49, 52062 Aachen, Germany
e-mail: sr@isf.rwth-aachen.de

Prof. Dr.-Ing. Dipl.-Wirt. Ing. Christopher M. Schlick Institut für Arbeitswissenschaft der RWTH Aachen, Bergdriesch 27, 52062 Aachen, Germany
e-mail: c.schlick@iaw.rwth-aachen.de

Prof. Dr.-Ing. Robert Schmitt Werkzeugmaschinenlabor WZL der RWTH Aachen, Steinbachstr. 19, 52074 Aachen, Germany
e-mail: r.schmitt@wzl.rwth-aachen.de

Dr. rer. nat. Georg J. Schmitz ACCESS e.V., Intzestr. 5, 52072 Aachen, Germany
e-mail: g.j.schmitz@access.rwth-aachen.de

Dipl.-Ing. Stephan Schmitz Werkzeugmaschinenlabor WZL der RWTH Aachen, Steinbachstr. 19, 52074 Aachen, Germany
e-mail: st.schmitz@wzl.rwth-aachen.de

Dipl.-Ing. Maximilian Schöngart Institut für Kunststoffverarbeitung (IKV) in Industrie und Handwerk an der RWTH Aachen, Pontstr. 49, 52062 Aachen, Germany
e-mail: schoengart@ikv.rwth-aachen.de

Dipl.-Phys. Dipl.-Wirt. Phys. Johannes Schubert Fraunhofer-Institut für Produktionstechnologie IPT, Steinbachstr. 17, 52074 Aachen, Germany
e-mail: johannes.schubert@ipt.fraunhofer.de

Prof. Dr.-Ing. Dipl.-Wirt. Ing. Günther Schuh Werkzeugmaschinenlabor WZL der RWTH Aachen, Steinbachstr. 19, 52074 Aachen, Germany
e-mail: g.schuh@wzl.rwth-aachen.de

Prof. Dr. rer. nat. Wolfgang Schulz Fraunhofer-Institut für Lasertechnik ILT, Steinbachstr. 15, 52074 Aachen, Germany
e-mail: wolfgang.schulz@ilt.fraunhofer.de

Dipl.-Ing. Maik Schürmeyer Forschungsinstitut für Rationalisierung (FIR) e. V. an der RWTH Aachen, Pontdriesch 14-16, 52062 Aachen, Germany
e-mail: maik.schuermeyer@fir.rwth-aachen.de

Dr. rer. nat. Jens Schüttler Fraunhofer-Institut für Lasertechnik ILT, Steinbachstr. 15, 52074 Aachen, Germany
e-mail: jens.schuettler@ilt.fraunhofer.de

Dr.-Ing. Lars Stein Institut für Schweißtechnik und Fügetechnik der RWTH Aachen, Pontstr. 49, 52062 Aachen,
e-mail: stein@isf.rwth-aachen.de

Dipl.-Ing. Marius Steiners Institut für Schweißtechnik und Fügetechnik der RWTH Aachen, Pontstr. 49, 52062 Aachen, Germany
e-mail: steiners@isf.rwth-aachen.de

Dipl.-Wirt. Ing. Sebastian Stiller Werkzeugmaschinenlabor WZL der RWTH Aachen, Steinbachstr. 19, 52074 Aachen, Germany
e-mail: s.stiller@wzl.rwth-aachen.de

Dr.-Ing. Jochen Stollenwerk Lehrstuhl für Technologie Optischer Systeme der RWTH Aachen und Fraunhofer-Institut für Lasertechnik ILT, Steinbachstr. 15, 52074 Aachen, Germany
e-mail: jochen.stollenwerk@ilt.fraunhofer.de

Dipl.-Ing. Babak Taleb Araghi Institut für Bildsame Formgebung der RWTH Aachen, Intzestr. 10, 52056 Aachen, Germany
e-mail: taleb@ibf.rwth-aachen.de

Dipl.-Ing. Sebastian Theiß Lehrstuhl für Oberflächentechnik im Maschinenbau der RWTH Aachen, Templergraben 55, 52062 Aachen, Germany
e-mail: theiss@iot.rwth-aachen.de

M.Sc. Dipl.-Ing. (FH) B. Eng. (hon) Ulrich Thombansen Fraunhofer-Institut für Lasertechnik ILT, Steinbachstr. 15, 52074 Aachen, Germany
e-mail: ulrich.thombansen@ilt.fraunhofer.de

Dipl.-Ing. Dipl.-Wirt.Ing. Stefan Tönissen Werkzeugmaschinenlabor WZL der RWTH Aachen, Steinbachstr. 19, 52064 Aachen, Germany
e-mail: s.toenissen@wzl.rwth-aachen.de

Dipl.-Ing. Johannes Triebs Werkzeugmaschinenlabor WZL der RWTH Aachen, Steinbachstr. 19, 52074 Aachen, Germany
e-mail: j.triebs@wzl.rwth-aachen.de

Dipl.-Ing. Dražen Veselovac Werkzeugmaschinenlabor WZL der RWTH Aachen, Steinbachstr. 19, 52074 Aachen, Germany
e-mail: d.veselovac@wzl.rwth-aachen.de

Dipl.-Wirt. Ing. Matthias Vette Werkzeugmaschinenlabor WZL der RWTH Aachen, Steinbachstr. 19, 52074 Aachen, Germany
e-mail: m.vette@wzl.rwth-aachen.de

Dipl.-Ing. Michael Vorspel-Rüter Werkzeugmaschinenlabor WZL der RWTH Aachen, Steinbachstr. 19, 52074 Aachen, Germany
e-mail: m.vorspel-rueter@wzl.rwth-aachen.de

Dipl.-Inform. Carsten Wagels Werkzeugmaschinenlabor WZL der RWTH Aachen, Steinbachstr. 19, 52074 Aachen, Germany
e-mail: c.wagels@wzl.rwth-aachen.de

Dr.-Ing. Hagen Wegner Werkzeugmaschinenlabor WZL der RWTH Aachen, Steinbachstr. 19, 52074 Aachen, Germany
e-mail: h.wegener@wzl.rwth-aachen.de

Florian Welter M.A. IMA/ZLW & IfU der RWTH Aachen, Dennewartstr. 27, 52068 Aachen, Germany
e-mail: florian.welter@ima-zlw-ifu.rwth-aachen.de

Dipl.-Ing. Cathrin Wesch-Potente Werkzeugmaschinenlabor WZL der RWTH Aachen, Steinbachstr. 19, 52074 Aachen, Germany
e-mail: c.wesch@wzl.rwth-aachen.de

Dipl.-Ing. Konrad Willms Institut für Schweißtechnik und Fügetechnik der RWTH Aachen, Pontstr. 49, 52062 Aachen, Germany
e-mail: wls@isf.rwth-aachen.de

Dipl.-Ing. Christian Windeck Institut für Kunststoffverarbeitung (IKV) in Industrie und Handwerk an der RWTH Aachen, Pontstr. 49, 52062 Aachen, Germany
e-mail: windeck@ikv.rwth-aachen.de

Dipl.-Ing. Johannes Wunderle Institut für Kunststoffverarbeitung (IKV) in Industrie und Handwerk an der RWTH Aachen, Pontstr. 49, 52062 Aachen, Germany
e-mail: Wunderle@ikv.rwth-aachen.de

Chapter 1
Introduction

Christian Brecher, Wilhelm Oliver Karmann and Stefan Kozielski

Contents

1.1 The Cluster of Excellence "Integrative Production Technology for High-Wage Countries" 1
1.2 Contents and Authors 2

1.1 The Cluster of Excellence "Integrative Production Technology for High-Wage Countries"

More than ever, German companies have to compete in an environment of increasing global competition. The recent years in particular have seen increasing relocation of production. It therefore has to be the concern of high-wage countries to evaluate and define the conditions under which domestic businesses can successfully develop and produce corresponding products. In terms of economic relevance, these products have to address not only niche markets but also markets of adequate volume. How can these companies sustainably increase the level of value creation while minimising necessary planning activities, which by definition are not value adding? And how can they solve the dichotomy between individualisation and mass production? Delivering answers to these questions is crucial to strengthen domestic industrial production and has the potential to sustainably lower unemployment.

In order to meet these challenges, a fundamentally new understanding of product and production interrelations is required. The Cluster of Excellence "Integrative Production Technology for High-Wage Countries" therefore aims at developing a viable, production-scientific strategy and theory of production including necessary technology approaches.

C. Brecher (✉)
Werkzeugmaschinenlabor WZL der RWTH Aachen, Steinbachstr. 19,
52074 Aachen, Deutschland
e-mail: c.brecher@wzl.rwth-aachen.de

The Cluster of Excellence (CoE) brings together over 100 scientists of RWTH Aachen University and is funded by the German Research Foundation (DFG) with nearly 40 million Euros. This enables the CoE to address the key future questions of production technology and deliver sustainable solutions.

The result of the first funding period is a comprehensive technology roadmap for the future development of industrial production in high-wage countries. It addresses various issues of individualisation, virtualisation and hybridisation of industrial production as well as its self-optimisation and contributes to organisational and technological innovations. In order to succesfully put the corresponding research into industrial practice, the cluster comprises a number of business and technology cases which were defined and executed in close cooperation with its partner companies. These business and technology cases involve projects from a large spectrum of industrial production such as an automobile axle, a machine tool modularised with regard to the concept of design-to-cast or an integrated micro-laser.

With the CoE, RWTH Aachen University has set up the "Aachen House of Integrative Production Technology", which bundles the university's production engineering competencies and allows for industrial partners to be integrated into the research activities of the cluster. In order to support companies to work more efficiently and sustainably, the cluster qualifies companies of all sizes, particularly medium-sized businesses, to identify future potentials and to define future competencies necessary to secure their business under high-wage country conditions.

The authors would like to thank the German Research Foundation (DFG) for funding the work done in the Cluster of Excellence "Integrative Production Technology for High-Wage Countries."

1.2 Contents and Authors

This compendium comprises the main results of the Cluster of Excellence "Integrative Production Technology for High-Wage Countries", which have been developed within the funding period from 2006 to 2011. The following abstracts provide the reader with an overview of the results and corresponding authors.

Chapter 2 "Integrative Production Technology for High-Wage Countries"

Section 2.5 "Development of a Production Theory"

Günther Schuh, Stefan Tönissen, Fabian Bauhoff, Sascha Fuchs,
Andreas Roderburg, Michael Schiffer, Johannes Schubert and Sebastian Stiller

The planning, design and controlling of a production system necessarily contains technical as well as economical aspects, which cannot be separated completely from each other. While economic research focuses on the economic analysis and thus concentrates on a general view, production engineering often specifically focuses on the fundamental analysis of aspects of the production itself. As a consequence, the technical scientific disciplines lack a comprehensive understanding of the interdependencies of industrial production, which together with the results of the economic analysis can be combined to a profitability analysis. In this chapter, existing production theories are evaluated against their potential to address this issue. Subsequently, new approaches for the development of a technically based production theory are identified, which contribute to a comprehensive analysis of production systems.

Section 2.6 "Technology Roadmapping for Production in High-Wage Countries"

Günther Schuh, Johannes Schubert, Simon Orilski and Susanne Aghassi

Being used as an integrative planning tool, technology roadmapping is increasingly employed as a vital instrument to support the forecast of future trends and developments as well as identification of consistent strategies. This chapter contains both the advancement and implementation of technology roadmapping in scientific research and results in the four dimensions of the Cluster of Excellence, in particular the corresponding technology radars, which structure the relevant technologies and provide information on their stage of development. Furthermore, the text provides information on technological challenges as well as developments and current research and presents the results of a survey on the identification of relevant trends.

Section 2.7 "Organisation and Management of Integrative Research"

Florian Welter, Claudia Jooß, Anja Richert, Sabina Jeschke and Christian Brecher

The initiation of Clusters of Excellence is a milestone in the development of integrative and interdisciplinary research. The composition of nearly 20 university institutes and various non-university institutes as well as several partnering companies poses structural and cultural challenges to its management. In order to address the overall issues with regards to the high complexity of this research cooperation, the Cluster of Excellence features several cross sectional processes, which are designed to provide efficient crosslinking of its protagonists. This chapter develops an empiric model for the management of these cross sectional processes, which contains specific activities to support interdisciplinary cooperation and contributes to the quality of the research results in all stages of the network development.

Chapter 3 "Individualised Production"

Section 3.2 "Integrative Assessment and Configuration Logic for Production Systems"

Günther Schuh, Robert Schmitt, Jens Arnoscht, Arne Bohl, Daniel Kupke, Michael Lenders, Christopher Nußbaum, Jerome Quick and Michael Vorspel-Rüter

In contradiction to the incompatibility of economies of scale and economies of scope, as described by Porter, manufacturing companies in high-wage countries are facing increasing challenges to satisfy individual customer needs and withstand the cost pressure enforced by the globalisation of the markets. This challenge corresponds to a resolution of the scale-scope-dichotomy. Due to the high degree of interdependency, structure-forming elements of product production systems have to be adjusted to one another regarding their specific degree of standardisation in order to resolve the dichotomy.

This adjustment corresponds to the task of the integrative assessment and configuration logic which will be presented in the following. Based on an integrated assessment model, that classifies product production systems into four quantifiable fields of tension, the current operating point of a production system can be analysed. A configuration logic derived from the results of the analysis then provides the means to control a production system's configuration process.

Section 3.3 "Tool-less Production Technologies for Individualised Products"

Damien Buchbinder, Jan Bültmann, Andrei Diatlov, Christian Hinke
and Henrich Schleifenbaum

The resolution of the dilemma between scale and scope, expressed in the dichotomy of mass production at low cost on one hand and cost-intensive individual/ small-series production on the other, is a topic of uttermost importance for production technology in high-wage countries. Selective Laser Melting (SLM) as a method of tool-free production offers a high technological potential to resolve this dilemma. However, currently available SLM systems are not suitable for high volume production due to their yet too low process efficiency. To increase the efficiency of the process and thus allow series production of individual components, a high-power SLM process is being developed and being implemented in an according SLM-machine. The result is the first laser system in the kW range to be integrated in an SLM-machine. Components manufactured from high alloy steel 1.4404 can be built up with a construction rate of about 21 mm^3/s. Compared to the industrial state of the art, this is an increase of more than 1700%.

Section 3.4 "Modularisation for Die-based Production Technologies"

Marek Behr, Andreas Bührig-Polaczek, Walter Michaeli, Stefanie Elgeti,
Markus Probst, Yann Queudeville and Christian Windeck

Tool-based production technologies are classically suited for high lot sizes. In order to put up with the increasing competitive pressure and allow for individual customer requirements to be taken into account, a diversification of the product range is being considered, thus enabling a more flexible production.

In this project, a modularisation methodology is being presented to design the construction and manufacturing process of the tools more efficient. The methodology enables the user to systematically analyse the tools, identify modules and determine product variants much faster. Following the example of a die casting tool, the methodology is being described, presenting a way to lay out tool modules for the individual design of a variant. For this matter the layout of the flow channel for a plastic profile tool is being supported by an automatically running optimization algorithm.

Section 3.5 "Efficient Order Processing for Individualised Machine Tools Through Modelling of Product Families"

Christian Brecher, Werner Herfs and Andreas Karlberger

For machine tool manufacturers in high-wage countries, the combination of pre-planned machine variants (Configure-to-order) and customization (Engineer-to-Order) holds the potential to take advantage of the economies of scale while maintaining a high degree of differentiation. The implementation of this strategy requires a strict separation between order-neutral and order-specific development. The call for early validation nowadays requires the use of means of description that support both the modeling of product variants and the early cross-disciplinary validation of requirements to be supported. For the elaboration at the level of CAx systems, concepts for the efficient development and configuration of product families are needed. In this chapter, a concept is presented, which proposes the use of distributed configurators that are synchronized on standard feature models. For the cross-disciplinary modeling, a Systems Modeling Language (SysML) is used.

Chapter 4 "Virtual Production Systems"

Section 4.2 "Platform for Integrative Simulation"

Christian Bischof, Sabina Jeschke, Torsten Kuhlen, Thomas Beer, Tobias Meisen, Rudolf Reinhard and Daniel Schilberg

The planning of production processes is generally characterised by unlinked simulations of specific aspects, which are based on standard assumptions. As a consequence, the impact of upstream production processes cannot be taken into account. In order to increase simulation quality, these individual simulations have to be linked and combined to a simulation chain. This chapter comprises a method to link simulation resources, data sets and access rights to the generated integrative simulation platform. Furthermore, the general concept also contains subsequent analysis of simulation results by methods of visualisation. The text also presents an approach to visualise large amounts of data on different temporal and spatial scales.

Section 4.3 "Factory Planning of the Future—Connecting Model Worlds"

Achim Kampker, Günther Schuh, Peter Burggräf, Bastian Franzkoch, Jan Nöcker, Marcus Rauhut and Cathrin Wesch-Potente

Existing approaches of factory planning are based on the perception of a linear planning process, which can be subdivided into deterministic and discrete stages.

Planning activities are performed consecutively. The according software tools are designed on the premise that planning results are transferred into the documentation of the corresponding planning stage. Industrial practice however shows that there is no such standard element of factory planning and that the planning process requires individual adjustment due to specific boundary conditions and unforeseen scenarios before and during the planning process. As part of the Cluster of Excellence a planning method was developed, which allows the individual configuration of the planning object as well as a modular-parallel planning procedure. For this matter, planning content was encapsulated into planning modules to provide defined interfaces, which contain the necessary input information as well as the expected results. The modules themselves comprise corresponding planning methods and tools. This enables modules to be re-combined and repeated according to the specific initial condition and to possible changes in the course of the project.

Section 4.4 "Integrative Process Chain-simulation for Material and Production Technologies"

Georg J. Schmitz, Stefan Benke, Gottfried Laschet, Markus Apel, Jan Rosenbaum, Ulrich Prahl, Patrick Fayek, Sergey Konovalov, Hendrik Quade, Stephan Freyberger, Thomas Henke, Markus Bambach, Tim Arping, Thomas Baranowski, Barbara Heesel, Urs Eppelt, Ulrich Jansen, Eduardo Rossiter, Britta Kuckhoff, Thomas Gries, Markus Linke, Tatyana Kashko, Kirsten Bobzin
and Nazlim Bagcivan

The resolution of the dichotomy of planning orientation requires approaches that either significantly reduce planning effort and/or significantly increase product value. To simultaneously achieve both of these goals, an effective and efficient combination of heterogeneous simulation models of different time and length scales is needed. The concept of an open, modular, standardised and upgradable simulation platform—the Aachen (Aix) Virtual Platform for Materials Processing AixVipMap®—is a first step towards an "integrative computational material engineering" (ICME). The functionality and the advantages of the platform were verified against a number of economical and scientific test cases. First results such as (i) the computation of yield curves of ferrite-austenite steels, (ii) the optimisation of experimental effort in the investigation of yield curve groups, (iii) the computation of fine grain stability with regards to a shorter carbonising process of micro-alloyed steels, (iv) the computation of sphaerolithical structures including molecule orientation and a subsequent forecast of the mechanical properties of plastics, (v) the computation of the effective, anisotropic elastic modulus of metal infiltrated ceramic weaving as well as (vi) the identification of segregation in the solidification and mechanical stresses resulting the heat treatment and processing as a result of the martensite transformation as well as the resulting failure of a stainless steel workpiece are described and clearly highlight the advantages of a platform-oriented planning approach. In conclusion, this chapter features a short evaluation of the future development of the AixVipMap® platform in the dimensions of product and production design and durability forecasting as well

1 Introduction

as its connection to data from ab-initio approaches on an atomic scale and logistical parameters on the factory scale.

Section 4.5 "Integrative Simulation of Machine Tool and Manufacturing Technology"

Christian Brecher, Fritz Klocke, Wolfram Lohse, Gustavo Francisco Cabral, Matthias Rasim, Johannes Triebs, Marcel Fey, Meysam Minoufekr, Stephan Bäumler, Thomas Bergs, Werner Herfs, Lothar Glasmacher and Hagen Wegner

Efficient and reliable machining of complex products in high-wage countries requires a comprehensive support of manufacturing. Existing planning, evaluation and simulation approaches are not sufficient to meet these challenges. One major cause is the disregard of the physical properties of machine tools and machining processes as well as the corresponding interdependencies. Hence, planning processes in the CAM environment are only based on idealised assumptions of the dynamic system response to machine tool and manufacturing processes. In order to determine these effects before running the NC-processing on a real-world machine tool, a Virtual Manufacturing System (VMS) has been developed and implemented. The VMS integrates the control systems and mechanics of the machine tool's structure, the process characteristics as well as the energy demand. It is integrated into the CAD-CAM-NC-chain in a way that the simulation results can be processed with corresponding visualisation and evaluation tools, which have been specially designed for the VMS. In addition to the VMS, this chapter also evaluates methods which contribute to the combined planning of various technologies and which are transferred into the so called CAx-framework. Both the VMS and the CAx-framework allow an efficient and precise planning of the manufacturing process and thus contribute to an increase of productivity, quality and hence profitability of machining processes in industrial production.

Chapter 5 "Hybrid production systems"

Section 5.2 "Methodology for the Development of Integrated Manufacturing Technologies"

Fritz Klocke, Steffen Buchholz, Kai Gerhardt and Andreas Roderburg

Advancing manufacturing processes plays an important part in securing the competiveness of industrial production as it pushes currently existing boundaries of manufacturing technologies and taps new fields of applications. Hybridisation of manufacturing technologies is such a process. The issue therefore is the systematic development of hybrid manufacturing technologies and the early assessment of possible applications. Based on actual case studies, the requirements of a systematic development are defined, implemented in a special development method and subsequently validated in exemplary processes. The developed methodology is used to analyze effect relationships of the boundaries for individual manufacturing processes and supports the development of hybrid technology concepts that dissolve these boundaries. Results show that the integration of knowledge of different disciplines of production engineering is crucial to its success. The methodology points out how cause-effect relationships can be implemented and used to derive new technological solutions.

Section 5.3 "Shortening Process Chains for the Production of Plastic/Metal Hybrids Using Innovative Primary Forming and Joining Processes"

Kirsten Bobzin, Andreas Bührig-Polazcek, Edmund Haberstroh, Walter Michaeli, Reinhart Poprawe, Uwe Reisgen, Dustin Flock, Oliver Grönlund, Matthias Jakob, Pia Kutschmann, Andreas Neuß, Andreas Rösner, Sven Scheik, Markus Schleser and Johannes Wunderle

Growing demands on increasingly complex components can often only be met by the combination of different materials. Due to their different but complementary characteristics, the material combination plastic/metal is of particular interest for this matter. New manufacturing technologies for plastic/metal hybrids have been developed and analysed, culminating in the combination of the primary shaping processes die casting and plastic injection molding in one hybrid process. Different thermal joining processes and efficient pre-treatment processes are being used to join plastic and metal in one component. The developments hold the opportunity to realise selected composites of plastic and metal with high strength while significantly shortening process chains. The new hybrid manufacturing technologies allow the production of a diversity of plastic/metal hybrid components. Products range from compact highly integrated electronic components to large-scale reinforced structural components with outstanding mechanical properties.

Section 5.4 "Improving the Efficiency of Incremental Sheet Metal Forming"

Wolfgang Bleck, Gerhard Hirt, Peter Loosen, Reinhart Poprawe, Uwe Reisgen, Markus Bambach, Georg Bergweiler, Jan Bültmann, Jörg Diettrich, Alexander Göttmann, Ulrich Prahl, Alireza Saeed-Akbari, Lars Stein, Marius Steiners, Jochen Stollenwerk and Babak Taleb Araghi

This chapter presents the work of the project "Development of a machine platform for hybrid incremental sheet forming (ISF)." The project goal is to overcome existing process boundaries of ISF through a combination with other forming and operating principles as well as a broadening of the technology's range of application. Based on state of the art, the process limits (a) strong thinning in areas with high wall angles, (b) comparable low geometric accuracy, (c) long process times and their causes are identified. Furthermore, the potentials of using high-strength sheet metal materials (e.g. titanium alloys) and tailored hybrid blanks in the field of lightweight construction, are shown. The solution approaches—combination of the ISF with stretch forming and heating of the deformation zone with a laser beam—are presented and validated using manufactured parts. The producibility of components in welded aluminium-steel composite panels is shown in simple principle components. Finally, the results and their relevance for industrial applications are discussed.

Section 5.5 "Shortening Process Chains by Process Integration in Machine Tools"

Christian Brecher, Fritz Klocke, Kristian Arntz, Stephan Bäumler, Tobias Breitbach, Dennis Do-Khac, Michael Emonts, Daniel Heinen, Werner Herfs, Jan-Patrick Hermani, Andreas Janssen, Andreas Karlberger and Chris-Jörg Rosen

The integration of hybrid process chains in a machine tool has the potential to reduce the production times by saving changeover times, as well as to increase the attainable accuracy by not chucking the workpiece. This chapter describes a concept and prototypical implementation of a 5-axis machining center with integrated different types of laser machining processes. In two equivalent workrooms processing of parts can simultaneously take place by the machine spindle and the robot. In addition to the training of new laser technologies and the machine tool, a heterogeneous control architecture was developed and tested. On the one hand, it was shown that value-added processing steps can be carried out by robots, on the other hand, the complexity of the operation and process planning is reduced to a minimum. Several editing sites may be better utilized in this way, while more complex process chains can be depicted in a machine tool.

Section 5.6 "Shortening Process Chains for Manufacturing Components with Functional Surfaces via Micro- and Nanostructures"

Kirsten Bobzin, Andreas Bührig-Polaczek, Walter Michaeli, Reinhart Poprawe, Nazlim Bagcivan, Stefan Beckemper, Stephan Eilbracht, Arnold Gillner, Claudia Hartmann, Jens Holtkamp, Todor Ivanov, Fritz Klaiber, Micha Scharf, Maximilian Schöngart and Sebastian Theiß

Products of functional surfaces on the basis of micro-or nano-structures have a large economic potential through new functionality in flow behavior, tribology, feel and optic. Different methods such as lithography, micro-milling and laser ablation are available for the production of these surfaces but are limited by high production costs. To solve this problem, a new process chain is integratively studied, in which the functional surfaces can be prepared directly in the mass production-ready primary shaping process together with the component. The investigated process chain begins with the production of an adapted tool that is micro-structured and wear-resistant in terms of coating. Then an adjusted variothermal injection molding or extrusion process follows for the production of plastic components or adapted precision casting process to produce metal parts with a functional surface. The functionality of the surfaces is demonstrated with a superhydrophobic effect for the plastic components.

Chapter 6 "Self-optimising Production Systems"

Section 6.2 "Integrative Self-optimising Process Chains"

Robert Schmitt, Reinhard Freudenberg, Mario Isermann, Matthis Laass
and Carsten Wagels

Cognitive technologies can enable production systems to adapt self-optimizingly to variable conditions. The primary objective here is to ensure product quality by aligning the considered process chain to the fulfillment of the required product features by designing inter-process control loops. This allows the production system to dynamically permit deviations of other goals (such as cost, quality, time) during the production process or to compensate deviations by specific reactions and increases its flexibility and thus its competitiveness. With Cognitive Tolerance Matching (CTM), a cognitive architecture has been developed which is able to regulate and improve complex production processes. The architecture combines different approaches of cognitive information processing. The key modules are data analysis of data mining techniques, process modeling of artificial neural networks and optimisation of the Soar cognitive software. The CTM-architecture is evaluated as a prototype in the production of sophisticated rear axles of the BMW Group.

Section 6.3 "Integrative High-Resolution Supply Chain Management"

Günther Schuh, Fabian Bauhoff, Tobias Brosze, Sascha Fuchs, Niklas Hering,
Till Potente, Maik Schürmeyer and Simone Runge

The efficient use of the dynamic conditions of manufacturing companies is one of the essential tasks of supply chain management in high-wage countries. The close to real time availability and processing of the relevant planning information plays a key role. It provides the basis for realistic planning and control of production. The main challenge is the complexity of the information diversity and its management as well as the effective integration of human intuition and experience in the control circuit of the supply chain management. High Resolution Supply Chain Management (HRSCM) describes an approach to enable organizational structures and processes based on a high information transparency to adapt in a self-optimising manner to constantly changing conditions by decentralized production control systems in terms of a cascaded control loop model.

Section 6.4 "The Road to Self-optimising Production Technologies"

Ulrich Thombansen, Thomas Auerbach, Jens Schüttler, Marion Beckers,
Guido Buchholz, Urs Eppelt, Yves-Simon Gloy, Peter Fritz, Stephan Kratz,
Juliane Lose, Thomas Molitor, Axel Reßmann, Heiko Schenuit, Konrad Willms,
Thomas Gries, Walter Michaeli, Dirk Petring, Reinhard Poprawe, Uwe Reisgen,
Robert Schmitt, Wolfgang Schulz, Dražen Veselovac and Fritz Klocke

In times of short-living products and faster innovation cycles, new challenges are arising in relation to the competitiveness of manufacturing. The result is the

1   Introduction

necessity to create, beyond the pure increase in processing speed, processes even for small quantities that robust, that a required product quality is obtained. The concept of "model-based self-optimisation for manufacturing systems" points the way to integration of process models in technical systems, that are able to determine its operating point in parameter space itself, because of cognitive components. Hereupon based system solutions will command a multi-dimensional and highly non-linear optimization task. The monitoring strategies, setting assistants and regulations, which are realised in that project, demonstrate already today the potentials on the way to self-optimizing manufacturing technologies.

Section 6.5 "Integrative Product and Process Design for Self-optimising Assembly"

Christian Brecher, Peter Loosen, Rainer Müller, Robert Schmitt, Kamil Fayzullin, Sebastian Haag, Adam Malik, Alberto Pavim, Nicolas Pyschny and Matthias Vette

In the sub-projects "Flexible assembly systems for self-optimising automation" and "Model-based flow control for the self-optimising automated assembly" principles of self-optimisation in the automated assembly of complex and more varied products have been developed to contribute to the resolution of the plan value dilemma. Here, a flexible hardware platform and a robust agent-based control have been developed, which have been combined with a model-based flow control, to solve tasks of the planning act autonomously. As a demonstrator, a miniaturized laser system was used with very high requirements for precision assembly, which was developed in close coordination with the system concept. One result of the integrative development approach is the ability to communicate between product and assembly control, whereby at the run time, the product can be optimised individually regarding to its function and component tolerances can be actively compensated. This chapter presents the concepts and results of this pioneering work on flexible and robust precision assembly of sophisticated products.

Section 6.6 "Self-optimising Assembly Systems Based on Cognitive Technologies"

Marcel Ph. Mayer, Barbara Odenthal, Daniel Ewert, Tobias Kempf, Daniel Behnen, Christian Büscher, Sinem Kuz, Simon Müller, Eckart Hauck, Bernhard Kausch, Daniel Schilberg, Werner Herfs, Christopher M. Schlick, Sabina Jeschke and Christian Brecher

In this paper a novel approach of a so-called cognitive control unit, short CCU, is presented for self-optimizing integrated assembly systems. For the first time, with the result of the research, the design, development and application of cognitive mechanisms in automation technology based on the control of a robotic assembly cell could be demonstrated integratively and analysed scientifically. With cognitive automation of production systems a technology is available, that allows the same or even reduced planning effort to automate varied product lines in an efficient and robust way even in small quantities for each variant, and thus virtually to achieve a customer-oriented mass production. The use of cognitive mechanisms in automation enables in particular companies in high-wage countries to achieve significant competitive advantages and thus to contribute directly to the assurance and development of local production sites.

Section 6.7 "Reconfigurable Assembly Systems for Handling Large Components"

Burkhard Corves, Rainer Müller, Martin Esser, Mathias Hüsing, Markus Janßen, Martin Riedel and Matthias Vette

With regard to increasing cost pressures, large variety and shorter innovation cycles, the demand for a flexible assembly and handling system for the realisation of an individualised production becomes more and more important. Within the Cluster of Excellence "Integrative Production Technology for High-wage Countries" in the project "Reconfigurable and self-optimizing component handling" a manipulation system as a cost-effective approach to adapt to component-dependent tasks is developed, that can be configured as required by the task at hand. The modular concept creates the requirements for a flexible and responsive arrangement of several handling devices, which allows, by a corresponding control, the joint handling of large components by several gripping points. Together with the extended configuration options, new degrees of freedom are developed. With self-optimizing functions, it is possible to use these advanced degrees of freedom particularly economically. It also requires a new way of planning such assembly systems, since existing modules should be reused and reconfigured for different assignments.

Chapter 7 "Integrative Business- and Technology Cases"

Section 7.2 "Process Planning and Monitoring of Blisk Production"

Fritz Klocke, Thomas Auerbach, Lothar Glasmacher, Stephan Kratz, Meysam Minoufekr and Dražen Veselovac

The long-term security of the competitiveness of manufacturing companies in high-wage countries is the primary aim of today's production technology. The necessary condition to achieve this goal is the simultaneous controllability and adaptability of process systems with given dynamically changing demands and boundary conditions. One of the key industries in Germany, which currently is facing these challenges, is the manufacturing industry for engines. This business case focuses on the production of safety critical components such as blisks (blade integrated disks), which are produced by 5-axis milling using roughing and finishing strategies.

Section 7.3 "Integrative Production of Micro-lasers"

Peter Loosen, Max Funck, Alexander Gatej, Valentin Morasch
and Jochen Stollenwerk

The previously manually implemented installation of micro-laser systems for marking applications with an average output of up to 10 W needs automated solutions to address the global trend of miniaturization in high-wage countries more flexible and competitive. To accomplish this task a simultaneous development and adaptation of product design, joining technology and assembly strategy is needed. By renouncing mechanical supports, a concept is developed, which allows, based on SMD technology, a direct joining of optical components of standardised geometries on a planar substrate using resistance soldering. Developed specifically for the automated assembly, the optical design uncouples capabilities and increases the robustness of the system. In addition, the component selection takes place using so called Tolerance matching processes, in which manufacturing tolerances are compensated by specific pairs of lenses. An algorithm based on tolerance analysis determines the order of assembly of the components depending on necessary boundary conditions. The integration of electronics in the laser system is modular and allows communication between product and assembly facility to support the assembly.

Section 7.4 "Integrative Process and Product Development for Hybrid Plastic-metal Structural Components"

Walter Michaeli, Andreas Neuß and Johannes Wunderle

Combinations of different material groups such as particularly plastic and metal in a component hold enormous potential for future applications. Thus, in recent years, hybrid technology has become increasingly important. With this background a new manufacturing process for the manufacture of plastic/metal structural components has been implemented in the Cluster of Excellence, which combines the forming process pressure casting of metals and injection moulding of plastics for hybrid multi-component pressure casting. In cooperation with the industrial partner LANXESS Germany GmbH, Dormagen, and based on the work of the Cluster of Excellence, especially in the fields "hybrid production," "individualised production" and "virtual production," a guide is developed. The guide aims to provide comprehensive information on holistic and integrative implementation of specific production scenarios for the hybrid multi-component pressure casting in future applications in the automotive industry. The basis for creating the guide and to identify future perspectives is a survey of experts from industry and research based on the Delphi method.

Section 7.5 "Integrated Rear Axle Drive Production for Cars"

Robert Schmitt, Mario Isermann, Matthis Laass and Christian Niggemann

Today value-adding chains are typically considered only in sections on the basis of simplified model assumptions. Interactions between the processes, materials, production resources and people as well as the effect of changes to the product are usually not completely considered. To address these deficits is to provide self-optimizing production systems. These systems are the key to both use value-stream-oriented approaches in the production technology, and to enhance the efficiency of planning by reusing already acquired knowledge and its transfer to similar new production cases. These systems copy cognitive behavior or human decision-making capacity by technical architectures. The paper describes the design and use of a self-optimising system using the example of the production of passenger car- rear axle drive. These are characterised by high variety and complex production steps. Thus frequent process adjustments and high planning efforts are required. With cognitive methods the production of rear axle drives can be flexibilised, the planning efforts can be reduced and dynamically adapted to a changing environment. Here the focus is on function-based optimization of the product under more flexible adaptation of individual process steps during production.

Section 7.6 "Integrative Levelling of Production"

Günther Schuh, Arne Bohl, Sascha Fuchs, Till Potente and Stephan Schmitz

The Business & Technology Case "Leveling of the Production" based on the cooperation with a single and small batch production of mechanical and plant engineering. The aim is to design a product- production system, which enables a high degree of variation from the customer's ability in highly standardised and leveled production processes by a loop system. For this matter, based on constituent characteristics of individualized products and the pollution situation in production, the consequences of technology as well as the resources and tools have been configured systematically. The steps of work planning are based so far on experience or are generally not included, so that potential economies of scale in production processes get lost in spite of similar products. The contribution of integrated assessment and configuration logic leads to the result, that technological, process- and product-side potentials for economies of scale in the planning of individualised products are more useful. For this purpose the integrative, High-resolution Supply Chain Management contributes to increasing the current value orientation in the production control at the configuration of the work plan and determination of standard times. By a distributed control system between machine-level data collection and work planning on the one hand, and construction on the other hand, the overall efficiency of production in integrated technology sequence planning is additionally increased.

Section 7.7 "Integrative Production of TiAl-products"

Georg J. Schmitz, Julio Aguilar, Ulrike Hecht, André Schievenbusch,
Robert Guntlin, Günther Schuh, Achim Kampker, Bastian Franzkoch,
Peter Burggraef, Jan Noecker, Cathrin Wesch-Potente and Rick Hilchner

Based on many years of development for production of components from titanium alloys, the use of different—in particular numerical—methods for further optimisation of the manufacturing process developed in the Cluster of Excellence and the properties of these components with a view on their commercial application, the specific objective of this cluster-B & T-case was the application and development of methods of plant design in relation to the planning of a foundry for components from titanium alloys in Aachen. The specific role of this B & T-case is the application of interdisciplinary, integrative methods developed in the Cluster of Excellence for planning an appropriate foundry for TiAl components and ensuring the planning results by simulation as well as by economic evaluation.

Chapter 2
Integrative Production Technology for High-wage Countries

Christian Brecher, Sabina Jeschke, Günther Schuh, Susanne Aghassi, Jens Arnoscht, Fabian Bauhoff, Sascha Fuchs, Claudia Jooß, Oliver Karmann, Stefan Kozielski, Simon Orilski, Anja Richert, Andreas Roderburg, Michael Schiffer, Johannes Schubert, Sebastian Stiller, Stefan Tönissen and Florian Welter

Contents

2.1	Importance of Domestic Production for High-wage Countries	18
2.2	The Polylemma of Production	20
2.3	Target Sectors and Product Segments of the Integrative Production	23
2.4	Research Program of the Integrative Production	25
	2.4.1 Long Term Goal and Scientific Core Result	26
	2.4.2 Individualised Production	27
	2.4.3 Virtual Production Systems	30
	2.4.4 Hybrid Production Systems	31
	2.4.5 Self-optimising Production Systems	33
2.5	Development of a Production Theory	34
	2.5.1 Abstract	36
	2.5.2 Industrial Relevance	36
	2.5.3 Evaluation of Existing Production Theories	40
	2.5.4 Starting Points for the Further Development of the Production Theory	48
2.6	Technology Roadmapping for Production in High-wage Countries	50
	2.6.1 Technology Roadmaps for Integrative Research	51
	2.6.2 Integrative Development of Technology Roadmaps	54
	2.6.3 Individualised Production	55
	2.6.4 Virtual Production Systems	58
	2.6.5 Hybrid Production Systems	60
	2.6.6 Self-optimising Production Systems	62
2.7	Organisation and Management of Integrative Research	64
	2.7.1 Abstract	64
	2.7.2 State of the Art	65
	2.7.3 Cross Sectional Processes and Procedures for Data Generation	67
	2.7.4 Industrial Relevance	73
References		73

C. Brecher (✉)
Werkzeugmaschinenlabor WZL der RWTH Aachen, Steinbachstr. 19,
52074 Aachen, Deutschland
e-mail: c.brecher@wzl.rwth-aachen.de

Manufacturing companies in high-wage countries are increasingly set under pressure in international competition due the lower production costs in low-wage countries. In order to counteract this development, many manufacturing companies respond to this issue by relocating their production facilities. Due to industrial production's high dependence on other sectors such as the services sector, this trend threatens Europe's medium and long term prosperity since outsourcing generally leads to subsequent relocation of services as well as of research and development activities (Lau 2005). With over 40% of German employees assigned to the manufacturing sector, production plays a key role in the national economy of Germany. Consequently, relocation of production poses huge risks for the country's future economic development (Statistisches Bundesamt 2010).

In order to maintain production sites in high-wage countries and to improve their capacity to act, countries with high unit labor costs must not respond passively to those changes. Instead, they need to actively shape competition depending on their needs and beliefs. To support this, RWTH Aachen University concentrates its research in the Cluster of Excellence "Integrative Production Technology for High-Wage Countries", which was granted in 2006 by the German Federal Ministry of Education and Research and the German Research Foundation as part of its Excellence Initiative.

The overall aim of the Cluster of Excellence is the resolution of the essential dichotomies of production for the manufacturing of customized products at mass production rates with minimal planning effort. This primarily involves the development of methods of organisation and of the technology implementation, which are specifally designed to meet the specific requirements of high-wage countries. This approach adresses all production factors, in particular human resources, in order to gain competitive advantages for the production in high-wage countries.

2.1 Importance of Domestic Production for High-wage Countries

According to Western Europe's manufacturing industry, the need for relocating their production is based on the new global competitive pressure. New competitors mainly arise from low-wage countries ("best-cost countries"), which generate their economical advantages primarily by making use of the low wages while improving their technological capabilities at the same time. This development can clearly be identified in the distribution of the worldwide production quantities in the course of the last centuries (cf. Fig. 2.1) (Tseng 2003).

The percentage of the western industrial nations—that is Europe and North America—constantly increased until approximately 1930. Until that time, competitve pressure was mainly generated by these countries. This was also reflected by the high gradient in wealth and economic power. Around 1930 however, as a consequence of the growing global trade and the increasing technological capability, this situation slowly started to change. China, Japan, Russia, Brasil, Mexico and other

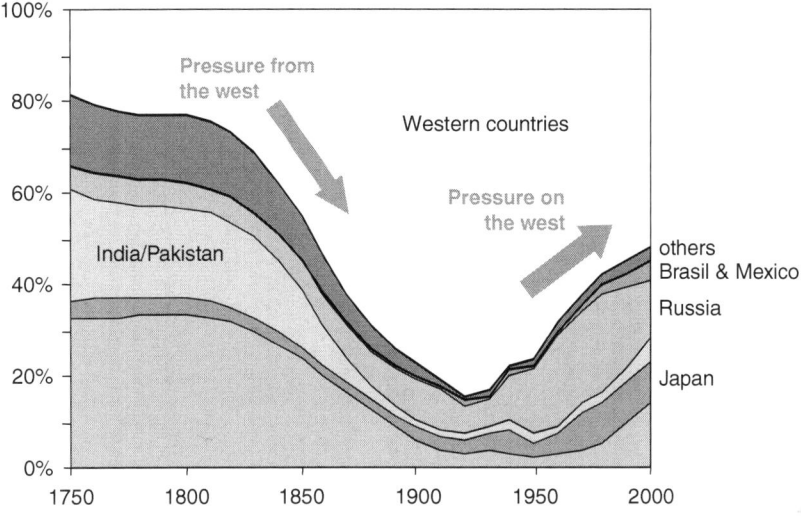

Fig. 2.1 Distribution of worldwide production in the course of time

emerging countries were able to sustainably increase their percentage of wordwide production quantities at the expense of the traditional western economic powers.

Production accounts for the main fraction of German national economy. In the second quarter of 2010, the industry nearly employed 7.7 million people, which is around one fifth of the country's working population. A distinctive feature inherent to the manufacturing industry is its multiplier effect: intra-industry changes have a significant inter-industrial impact, especially on the other industries value creation and on the level of employment. Contrary to this, services have rarely any inter-industrial effects employment-wise. This is rooted in the fact that its value creation chain only comprises a small number of levels and rarely requires any preliminary products in comparison to the manufacturing industry. Beyond that, services generate no specific after-sales revenue. Industrial production can therefore be regarded as the main driver of Germany's value creation and employment.

Nearly all forecasts predict a downswing of employment in Germany's manufacturing industry. Main cause is the relocation of value creational production steps to foreign countries, which is based on the fact that manufacturing and especially labour costs are significantly cheaper in most cases. While at the same time corresponding manufacturing technology and know-how mostly still remains in Germany, it is unlikely that this can be maintained in the long run. In this context it seems only logical that corresponding technology follows the relocation of production. In the long run, Germany therefore runs risk to lose the basis of its international competitiveness. In order to respond to this issue and sustainably secure Germany's level of employment, changes in the manufacturing industry must be recognized and corresponding adjustments must be implemented at an early stage.

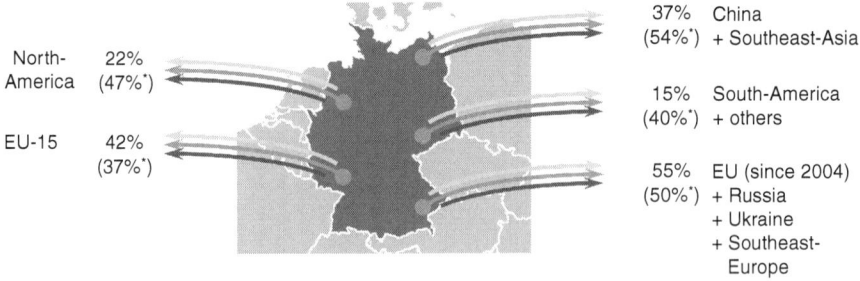

* Proportion of companies undertaking investments in the region (multiple answers possible)
** Proportion of total production-related relocation

Fig. 2.2 Foreign investments of German companies

Companies of the manufacturing industry in high-wage countries are constantly put under increasing pressure by international competition. One reason for this development are the manufacturing costs of low-wage countries, which are significantly cheaper in comparison. As a consequence, margins shrink and sales decrease. In addition, new competitors not only take advantage of their advantage in labour costs, but constantly upgrade their technological capabilites.

According to a survey carried out by the German chamber of industry commerce DIHK, nearly one in four German industrial companies reacts to this threat by relocating production to foreign countries, mostly to the new members of the EU, South America and Asia (Deutsche Industrie- und Handelskammer (DIHK) 2010). Besides cheaper labour costs, the vicinity to local customers and foreign production sites is the main reason. Among relocated R&D activities, design, technical development, testing, software development and even fundamental research are the most prominent (Rose and Treiver 2005). As mentioned before, the relocation of production sites is likely to result in a relocation of services and R&D. For high-wage countries like Germany, securing domestic production therefore is a vital part of securing national wealth (Fig. 2.2).

With the Cluster of Excellence "Integrative Production Technology for High-Wage Countries", RWTH Aachen University responds to this development with concepts of how to increase competitiveness of western high-wage countries by enabling them to implement a more economically efficient production. Based on the inherent advantages of domestic production sites, new approaches in different fields of science are integrated into a comprehensive approach, in order to meet global competition and strengthen innovation. Main part of the concept is the resolution of the so-called "Polylemma of Production".

2.2 The Polylemma of Production

Manufacturing companies have to define their position in the field of tension between production and planning profitability, thereby striving towards their specific ideal operating point. In this process, labour costs are of high importance. Hence, the position

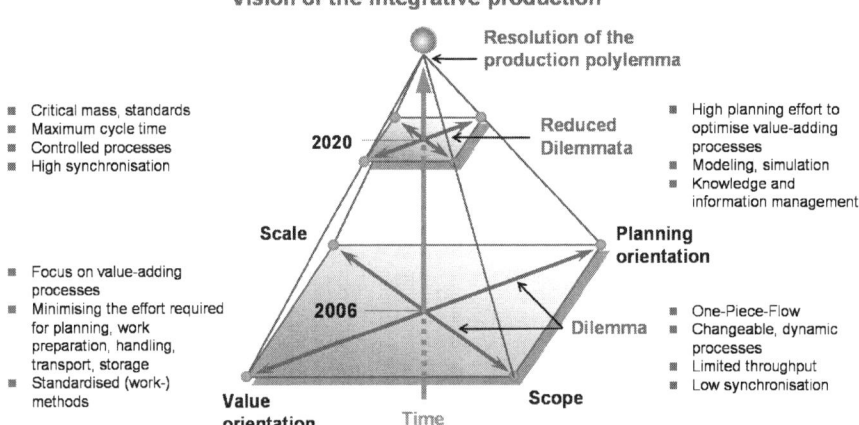

Fig. 2.3 The polylemma of production

in the above mentioned area of conflict is mainly location-dependent. In terms of their production efficiency, companies in low-wage countries generally focus solely on mass production (economies of scale) while companies based in high-wage countries need to balance their production system between mass production (economies of scale) and production of customized products (economies of scope). Regarding the second dimension in the area of conflict, the planning efficiency, manufacturers in high-wage countries continuously make efforts to optimise their processes using sophisticated, hence capital-intensive planning tools and production systems. In contrast to that, companies in low-wage countries generally employ simple and robust value stream oriented process chains. Recent developments, however, show that despite a seemingly perfect corporate positioning in the area of conflict, relocation of production from high-wage to low-wage countries still continues. Thus, the key to strengthen the competitiveness of high-wage countries is the resolution of the contradictions in the fields of production and planning efficiency. These dichotomies, which are specific to the scientific disciplines, form the so-called polylemma of production. It illustrates the complexity of the task to dissolve the fields of tension in the production and planning economy (Fig. 2.3).

In the context of production economy, low unit costs can be achieved by focussing the production system on the economies of scale. Hereby, increases in efficiency necessary for the utilisation of scale effects can be accomplished through process standardisation. This includes standardisation in terms of both the organisation's business processes and its technical and manufacturing processes, which are part of the production system. In mass production this includes interlinked machineries with a high degree of automation. However, the downside of this configuration is that, when put into practice, standardisation often leads to restrictions in terms of production flexibility. Beneficial economies of scale like these are gained by a low adaptability of the production system to changing boundary conditions, like e.g. a modified market behavior. Contrary to that, adaptability is of top priority when configurating a production system for the economies of scope. This involves business processes and technical systems, which are designed in a way that they allow a high degree of freedom for variants of producible goods. However, this comes at the price of additional investment or higher proportion of manual work, leading to higher costs per unit compared to a production system optimised for the economies of scale.

The dimension of planning economy features a similar dichotomy. A high degree of planning orientation leads to an extensive use of models, simulations and optimisation approaches. These support operational processes as well as planning and decision-making processes (production planning and control, design of production processes and machinery, etc.). As a result of this planning oriented approach, high expenditures in personnel are required. Since those kinds of activities do not immediately add value, corresponding approaches are contrary to the concepts of lean management, which aims to maximise the value added by employing efficient planning concepts. In the dichotomy of planning efficiency, this position is represented by a high degree of value orientation. Such an approach waives elaborate planning and/or control processes in production.

The resolution of both the dichotomy between the economies of scale and the economies of scope as well as between planning and value orientation therefore is the key to sustainable preservation of production in high-wage countries and thus needs to be in the focus of production research. The solution hypothesis to this objective is to secure the system's integrativity, i.e. combining approaches of different fields of science into a comprehensive strategy to resolve the polylemma of production.

Towards an integrative production, companies in high-wage countries must regard globalisation as a chance rather than a threat. They must not respond to new market requirements hastily. Past has shown that, despite the discontinuities already taking place, production can still successfully be maintained in high-wage countries. This fact is prominently demonstrated by the high export revenues of Germany's mechanical engineering sector. Furthermore, relocation of production sites to low-wage countries must be seen as part of a global corporate strategy, which distributes manufacturing capacities according to specific local benefits. Operating foreign production facilities and international inter-corporate cooperation are main drivers of growth and revenue in industrial production. However, high-wage countries must aim to enable their companies to retain economically significant parts of the global value-adding chain in their country of origin. This facilitates achieving a high product and process quality as well as securing specific product and process knowledge in terms of protecting intellectual property.

2.3 Target Sectors and Product Segments of the Integrative Production

The intended increase of value-adding production activities in high-wage countries requires an intense focusing of production research on both established and potential fields of action. In addition to the quality of the result itself, the evaluation of research therefore needs to implicate factors of economic relevance. This criterion comprises the evaluation of the industry sector's focus as well as the significance of the industry sector for the national economy and the consideration of relevant requirements from the industrial practice.

The following text focusses on some sectors, which are relevant for Germany's manufacturing industry, in order to explain the approaches of an integrative production. These sectors are:

- Automotive industry: production of automobiles including the upstream value-adding chains of its suppliers
- Aviation industry: development, production, processing and assembly of aircraft, components and
- Mechanical and plant engineering and
- Power engineering: production of components of compressors and turbines.

These sectors alone are responsible for nearly 30% of the net production value of Germany's manufacturing industry and rate among the high-technology sectors of industrial production. In addtion to the focus specific to the sector, the relevant product segments of industrial production in high-wage countries are subject to the cluster's research as well. Hereby, two perspectives are differentiated:

- *Production technology as a product*: Production technology is regarded in terms of machinery, equipment, information technology and corresponding services, which are distributed globally.
- *Production technology as an instrument for manufacturing of products*: In this perspective the main focus is on the production system, which is used for manufacturing products in high-wage countries. Hence, the production technology offers the instrument for the production of goods.

A sustainable strategy for the development of industrial production in high-wage countries is to include both perspectives. The production technology (perspective 1) developed in high-wage countries has to address both markets in high- and low-wage countries. Furthermore, concepts to establish long term competitive advantages for manufacturing companies in high-wage countries must be developed.

Production technology can be classified into premium, medium and low-tech based on technological capacity. Whereas premium production technology requires a high technological capacity, medium- and low-tech production systems only require lower technological capacities. In addition, the relative proportion of premium production technology is lower than those of medium- and low-tech production technology.

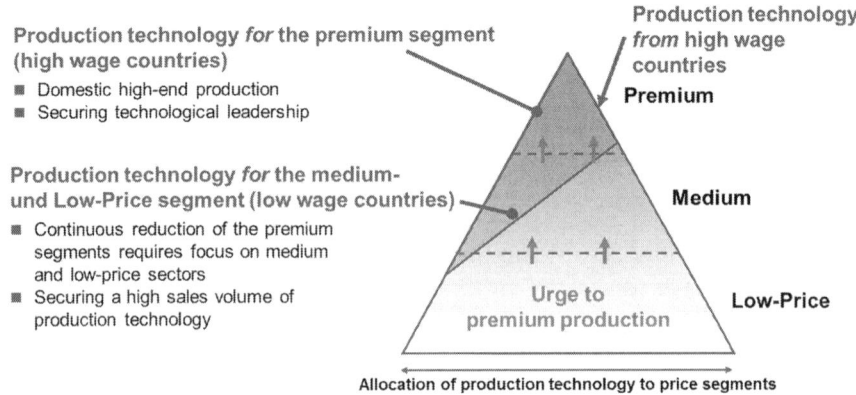

Fig. 2.4 Classification of production systems

An important characteristic of premium production systems is the vertical differentiation, which arises from the difficult controllability of high standard production systems. Thus, only technologically advanced companies have the skills to control these systems. High-wage countries have to continuously develop this claim of technological leadership in order to increase the competitive pressure on low-wage countries. Thus, high-wage countries are to develop production technologies and to protect their technological advance by taking appropriate actions. Furthermore, these premium production technologies have to be implemented into the domestic production (Fig. 2.4).

In the course of their technological advance, low-wage countries such as India and China develop from the low-tech and medium-tech-sector to markets for premium production technology. With increasing pressure of growing competition in small-sized premium markets, high-wage countries must be able to provide production systems for the medium- and to some account also for the low-tech sector in order to maintain a critical amount of value creation.

With this diversification high-wage countries reduce their dependency on the premium sector and simultaneously increase the competitive pressure on low-wage countries, while at the same time benefiting from the big production volume of medium- and low-tech sectors. However, this requires the resolution of the dichotomy between scale and scope, since production goods in the medium and low-tech sector often call for different distribution, organisation and technology in comparison to products of the premium sector, which are generally produced in smaller numbers.

Another reason for high-wage countries to offer both production technology for the premium and medium sector is the dependency of the optimum intensity of technology implementation on the local wage level. This intensity directly connects to the required performance of production facility, which is reflected in dimensions like quality level and required R&D expenditures (Fig. 2.5).

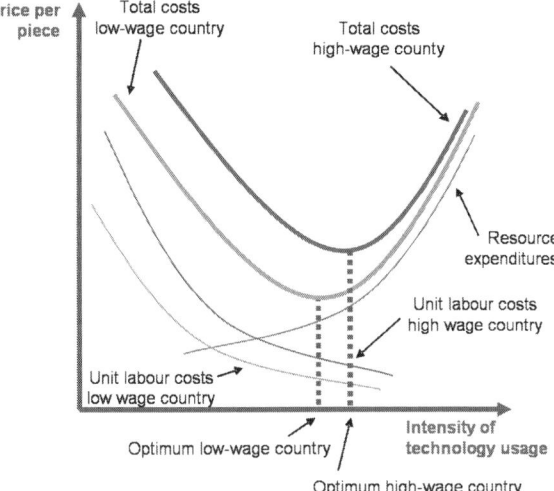

Fig. 2.5 Determination of the ideal intesity of technology implementation

For this purpose, resource costs and unit labor costs are distinguished, which together add up to the costs per unit. Resource costs (for material, energy and investment) correlate with the intensity of technology implementation but without much deviation in a worldwide comparison. Since an increased technology implementation leads to substitution of human work by machinery and an increase in output, it consequently decreases the labor costs as well. In contrast to resource costs, labor costs are subject to a high variation depending on the location of the production site. High-wage countries are characterised by higher labor unit costs than low-wage countries with the same intensity of technology implementation. Hence, the ideal intensity for low-wage countries is lower than for high-wage countries. A fully automated production is economically not suitable for low-wage countries. For these countries, medium-tech production systems offer optimal solutions in terms of costs, whereas high-wage countries can efficiently minimise their costs by positioning in the premium sector.

2.4 Research Program of the Integrative Production

Within the Cluster of Excellence, the aims of production research are defined both in terms of research content and reference. Research topics were to make a significant contribution to an increase in efficiency of processes both directly and indirectly linked to industrial production and were orientated with regards to the four main production factors process, machinery, energy and workforce.

Following the aforementioned, the resolution of the dichotomies in the context of the target system layed out above can only be achieved by interdisciplinarity, which therefore is an integral part of integrative production research. The research within

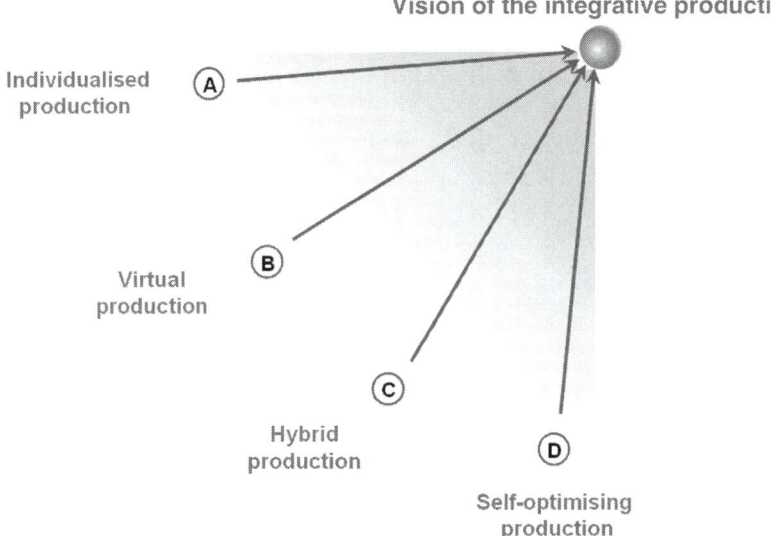

Fig. 2.6 Fields of research within the integrative producton

the Cluster of Excellence focuses on the four key areas of individualisation, virtualisation, hybridisation and self-optimisation of production systems. Institutionalised in the organisational structure of the cluster as so-called "Integrative Cluster Domains" (ICD), these research areas combine the specific competencies of the individual sub-projects. In total, the sub-projects contribute to the resolution of the polylemma of production and to the competitiveness of industrial production in high-wage countries by generating structuring, explaining and adaptive approaches. (cf. Fig. 2.6).

2.4.1 Long Term Goal and Scientific Core Result

As outlined in Sect. 2.2, a separate approach addressing just one of the dimensions of production economy or planning economy is no longer sufficient to achieve a sustainable competitive advantage for industrial production in high-wage countries. In fact, the research questions have to aim for simultaneous resolution of the two dichotomies and thus increase the system capacity of industrial production:

> The Cluster of Excellence "Integrative Production Technology for High-Wage Countries" has the long-term goal to increase the competitiveness of industrial production in high-wage countries such as Germany. Overall solution hypothesis is a higher level of integrativity in production technology.

Industry in high-wage countries has found its competitive advantage for mature markets in the individualisation of products. Despite the efficiency of the individual processes themselves, this development has lead to a decrease of the economies of

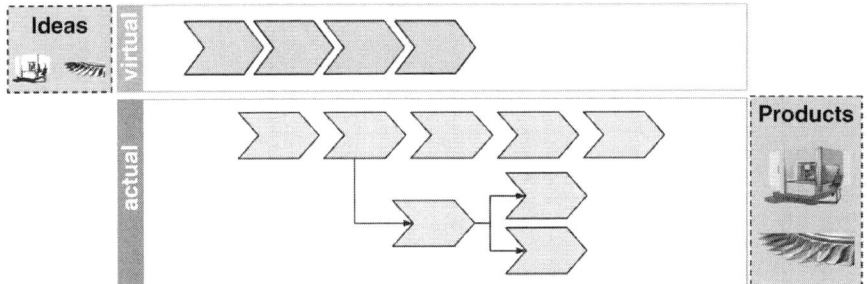

Fig. 2.7 Refernce frame: optimised coordination between virtual and real elements

scale along the value chain. By applying modularity and configuration logic into products and production systems and by implementing both suitable tool-dependend and tool-independend production technologies, the dichotomy of scale and scope can be resolved.

When dealing with a high product variety, the successive design of a production system requires high planning expenditures. A suitable approach is the successive virtualisation of manufacturing processes. Following this, both dichotomies can be resolved by decreasing planning expenditures while increasing the number of alternative solutions or by decreasing expenditures for production engineering while introducing an efficient first-time-right philosophy.

Some products require an approach which integrates multiple technologies such as different manufacturing processes and materials. Another approach is a self-optimising production system, which is considerably more competitive when operated by highy qualified personnel typically found in high-wage countries.

> The scientific core result of the Cluster of Excellence is the development of a combined theory of value creation and dynamic system performance of industrial production. The theory comprises comprehensive models for the description, explanation and design of production systems and is the culmination of the cluster's sub-projects' results.

Based on this theory, the fundamental understanding of industrial practice is deepened, which contributes to the modelling of actual interdependencies in industrial corporations. Eventhough production systems must be regarded as socio-technical systems and thus elude perfect predictability, the theory promotes both the reduction of complexity by quantitatively modelling predictable elements and the control of complexity by transforming non-predictable elements into cybernetic explanatory models (Fig. 2.7).

2.4.2 Individualised Production

The concept of individualised production ("mass customisation") combines the two contradictory concepts of mass and single-part production in order to allow for an economically efficient production of customer-specific products. Thus, the concept of individualised production contributes to the resolution of the classic dilemma between scale and scope.

The individualisation of industrial production can be looked at from two different perspectives. From a customer point-of-view, the question of the suitable degree of individualisation arises depending on the category of the specific product. The suitability is based on a cost-benefit consideration of the prospects of an individualisation on the one hand and the costs involved on the other. From a company's point-of-view, however, it's a question of how much individualisation is required in order to effectively address markets of adequate volume. Furthermore, this point-of-view questions the economic feasibility of the individualisation of product ranges.

In this context the main research questions are

- How much individualisation is required for industrial production in high-wage countries in order to successfully operate in their relevant markets? (Customer point-of-view)
- How can the required individualisation organisationally and technically be implemented in a way that costs remain within a competitive range? (Production point-of-view)

Since the degree of individualisaton required by the customer can hardly be influenced and may vary significantly depending on the specific type of the product, the second research question is especially important for industrial production. This question specifically addresses the need to qualify industrial production in high-wage countries for individualised production of relevant volume ("mass customisation").

From a logistics point-of-view, the one-piece-flow is the ideal production concept for an individualised production. It comprises a production free of buffers, in which workpieces/products of batch size 1 are moved along the process chain without any idle time. Thus, non-value adding processes and resulting costs are minimised, while at the same time the product's variability can theoretically be maximised. Hence, the production concept of one-piece-flow is typical for small batch sizes (scope). Until now, it has therefore not been effectively implemented for larger batch sizes—especially not for batch sizes of economically relevant size.

Based on the qualification of the one-piece-flow, the research questions mentioned above will be further concretised. One of the main issues is related to the ideal design of the production system's elements. It covers the harmonisation of the product structure and the structure of the production system with each other in a way that the one-piece-flow production system is operated at an ideal point. In order to address this issue, research approaches of both explanatory and structuring nature are required. Explanatory approaches have to highlight dependencies and interdependencies between the elements of the production system and of the product's characteristics. The understanding created during this process provides the basis for the structuring approaches, which address the development of a suitable procedure of a harmonised design of product and production systems. This may relate to both manufacturers and the customer, who then is—with respect to certain boundary conditions—enabled to design products according to his or her own needs.

The second important research approach addresses the simultaneous increase of productivity and flexibility of production technology itself (manufacturing technologies and corresponding machinery), i.e. production technologies have to be fitted for individualised production in large quantities.

Rapid manufacturing technologies, which generate products directly based on corresponding CAD data without any tools specific to the product, are a good example for a suitable manufacturing technology. However, generative manufacturing technologies have the substantial disadvantage of a throughput inapplicable for mass production and hence are too expensive. Therefore, the resolution of the dichotomies of scale and scope cannot be achieved with existing technology. However, by increasing its capability in terms of the generative rate the dichotomy can be reduced.

Among generative manufacturing technolgies, "selective laser melting" is the most prominent. It uses a laser beam to melt layers of metal powder, which are applied sequentially, and thus generates the product's geometry with hardly any design limitations. Within the Cluster of Excellence, researchers develop ways to increase this technology's productivity. Besides using laser beam sources with increased beam quality and power output, they focus on the development of laser optics, which work with lasers of different focus diameters. This enables an adaptive manufacturing strategy: if high generating rate is required, a laser beam with high output and a wide diameter is employed and if high accuracy is required, a laser beam of small focus diameter is employed.

Furher savings can be achieved by developing suitable machinery for the individualised mass production, i.e. machine tools, which manufacture a range of products to a certain quality, and the actual tools themselves.

The classic engineering process in machine tool manufacture is predominantly characterised by the sequential design of the machine tool's mechanics, its electronics and its control software. Since most customer-specific orders feature a high proportion of recurring functions, individual machine tools can mostly be manufactured using existing, scalable components. Especially in the design and production of variants and modified constructions, required documentation such as drawings, part lists, circuit diagrams and control programs can automatically be compiled using modular design systems. Within the Cluster of Excellence, such mechatronic-oriented modular design systems for the processing of orders are subject to research.

Despite their high degree of individuality, the tools themselves feature recurring functions as well and can therefore be produced using pre-manufactured components made in small batch sizes. By focussing on the forming parts of the tool, new design engineering can be avoided and cycle times as well as manufacturing costs are reduced. As part of the research program of the Cluster of Excellence, new modular design concepts for casting and extrusion tools are evaluated and an approach to automate the design of flow channels of extrusion tools is undertaken.

First results show that in terms of the development process and corresponding methods, the approach of an individualised production requires a new mindset in the design departments and must not be underestimated. On the other hand, it offers good prospects to significantly reduce design efforts and must therefore be subject to further research. The results of the research field of individualised production are compiled in Chap. 3.

2.4.3 Virtual Production Systems

Since the requirements of product and process quality are usually higher than in low-wage countries due to general framework determined by the market, simulation tools are of high importance for companies in high-wage countries. As such tools are non-value adding in the first instance, the capability of virtual production systems in terms of the planning efficiency is to be continuously increased. This results in the following, sometimes contradictory research goals:

- Increase of the results' quality in terms of conformity of the calculations and the actual process and of accuracy of information or data gained from a technical system, which is relevant to certain decisions.
- Increase of the universality of the technical support within the production process. This addresses integration efforts of different levels in the company and hence different simulation applications with a different level of detail.
- Minimisation of the efforts necessary for setting up and operating simulation systems. This issue deals e.g. with the increase of the level of automisation of the technical support and with the reduction of such application's complexity. This is especially important with the background of training necessary for users.

As a consequence, the approaches in the research field of the virtualisation of production systems are of explanatory and structuring nature. An integrative research is to translate a company's organisational as well as technical sub-aspects using new, networked simulation tools and methods. Thus, it contributes to an efficient support for companies in high-wage countries.

From an organisational point-of-view, the planning of production facilities with the help of computational methods still bears significant potential. Since corresponding software tools are predominantly developed inhouse, they are usually not universally applicable and their upgradability is limited. Until now, the system's interconnection via standardised interfaces mainly focuses on the tools and fails to compensate for the lack of coordination inherent to every planning process. Therefore, a comprehensive approach is developed within the Cluster of Excellence in order to improve control over planning and sequence dependencies as well as coordination and interface difficulties.

However, the difficulty of developing comprehensive simulation approaches does not only apply to planning processes. In fact, in order to reach higher levels of simulation quality, material science and machine tool simulations require a combination of different simulation models and tools as well. The same applies to the individual simulation techniques themselves. In terms of material science, the Cluster of Excellence's research is focused on the microscopic property changes, which a material undergoes along the entire process chain. This information is then used to transform the microsopic into macroscopic properties. Therefore, this can be regarded as a fundamental step towards a comprehensive simulation.

Likewise, this knowledge is necessary in the field of process and machine tools simulation as well. The implementation and ramp-up of new manufacturing processes requires an iterative approach, which is often very time-consuming, especially when it comes to complex parts. Due to the technical boundary conditions of the machine tools and further equipment involved, optimisation of these manufacturing processes is partly affected—sometimes with no satisfying results at all. As a solution to this problem, existing simulation systems are connected to form an integrated simulation chain, which combines the individual computational results. Thereby, the optimisation involves not only effects of the machinery's control system, its dynamic behaviour and the process itself, but also interdependencies between them. By linking the elements and creating a virtual simulation chain, the entire manufacturing process is represented and hence can be comprehensively optimised. Further information on the research of virtual production systems will be given in Chap. 4.

2.4.4 Hybrid Production Systems

While a virtual product development allows great freedom in terms of design, actual development processes are rather restricted. Those boundary conditions are at best hardly possible to exert influence on. Therefore, new approaches must equally push technical boundaries and involve a logistically optimal one-piece-flow, while at the same time increasing flexibility and productivity. Hence, hybridisation of manufacturing processes is a promising approach, which often allows tapping potentials in all the dimensions mentioned above.

For manufacturing engineering in high-wage countries, securing employment poses a challenge to expand its technological leadership. In this context, the concept of hybrid production can contribute to these goals by increasing efficiency and flexibility of manufacturing processes and machine tools as well as their degree of automation. As a consequence, hybrid production increases the complexity of the system and the corresponding processes and thereby makes imitation difficult. On the downside, the resulting lack of controllability poses a challenge for a successfull implementation of hybrid processes and systems into industrial practice.

In the context of a production system, the term "hybridisation" has different definitions. These are

- an integrated combination of normally separated manufacturing processes in order to manufacture suitable products made of different materials (like e.g. the casting process of integrated plastic-metal-products),
- integrated application or combination of different mechanisms of action (like e.g. laser-assisted machining) or

- the processing of different types of manufacturing operations in an integrated machine tool (like e.g. autonomic machine centre for manufacture and repair of forming tools: machining, measuring and built-up welding in a single machine centre with an integrated CAM-NC-measurement chain).

Generally, a combination of manufacturing processes, which are normally applied separately, leads a significant reduction of machining times and hence to a shortening of manufacturing chains. Likewise, restrictions of processing demanding products and materials are lowered as well, which strengthens high-tech locations. Possible applications for highly integrated manufacturing processes are products of the mould and tooling industry and the clock and watch industry, medical implants as well as certain parts of turbine engines.

As part of the Cluster of Excellence, a machining centre for milling operations is combined with a robot and a laser system and hence converted into a multi-technology platform. The machining process of the five-axis milling operation itself is complemented by laser-machining processes. For this, corresponding laser machining heads are developed. A fibre laser system is used for wire built-up welding, hardening and ablation whereas for laser structuring, a short pulsed laser together with a laser scanner is used. Since in a five-axis machining operation both the robot and the machining centre can process the workpiece at the same time, a high-precision synchronisation of both systems' movements is a fundamental requirement for a successful realisation. Among others, a high-frequency connection of the machining centre's and the robot's control systems, which have mostly been autonomous until now, is therefore subject to the research activites. The same applies to the laser system's control system. In total, this contributes to a comprehensive control system. In order to compensate for the resulting complexity in production planning and to tap the full potential of the integrated machining centre, corresponding research also involves the complete integration of all components into a single CAM-system.

Besides the shortening of manufacturing chains, hybridisaton also leads to a "local" realisation of the concept of one-piece flow. As shown before, this has the disadvantage of an increased complexity of the hybrid machine tool and the hybrid process respectively. Since the advantages mentioned above are put into perspective by the fact that they are downsized by the non-value-adding efforts necessary, the economic risks of the hybridisation of production processes apply mostly to the planning efficiency. Therefore, corresponding research must also involve planning-sidely integrated approaches to the design of hybrid processes for practical applications. Given its success, hybridisation contributes to the resolution of the production polylemma of high-wage countries.

However, this example of a successful hybridisation is only an excerpt of the research concerning "hybrid production systems" done in the Cluster of Excellence. Other examples are e.g. tapping potentials in sheet metal forming by combining incremental forming of highly-individualised workpieces and local warming in order to lower process forces and machining time. Further information is given in Chap. 5.

2.4.5 Self-optimising Production Systems

The self-optimisation of a production system is the fourth dimension of the Cluster of Excellence and leads to increased integrativity of industrial production and a more efficient use of the production factors in Germany. Just like business and production processes, entire production systems are also based upon hypotheses, which only lead to a step-by-step examination or to decided technological interaction. Interdependencies between processes, materials, machinery and human resources as well as their effects on the product itself are mostly not entirely understood. Implications of possibles effects on the entire value creation are not predictable since mostly only single elements of the production system are in the focus of optimisation due the complexity of a comprehensive approach. Sometimes, this can even lead to a decreased performance of other elements.

Yet, a resolution of this conflict is possible if the production system is designed in a way that it adjusts its goals depending on its boundary conditions. Whereas in most cases systems are optimised externally by e.g. human beings, an internal optimisation by the production system itself often carries significant potential. Past developments in automation engineering however have shown that even for relatively simple issues, a comprehensive optimisation has not been successfully implemented so far. Therefore, human activities are still of high importance in many cases. In this context, the implementation of self-optimising system features certainly offers substantial potential to reduce the dichotomy of planning economy.

The nature of a self-optimising system can be explained on the basis of a classic control feedback system. A control loop controls a system using parameters, which are provided externally. A control loop is called adaptive if it not only controls the target state of the parameters but simultaneously adjusts them according to any detected change of the system. In this context, self-optimisation focuses on the continuous adjustment of the target system. A self-optimising system can therefore only adapt itself using a set of decisions, which have been made internally. As a result, it defines its targets based on the situation and not only tries to achieve predetermined targets using classic feedback control or adaptive parameters.

In industrial production there are plenty of possible applications for self-optimising production systems. A good example are assembly operations, which have not been automated yet due to the requirements of the concept of one-piece-flow or to ambiguous interdependencies. The aspect last-mentioned plays a key role in the production of laser systems, which is still characterised by a high degree of manual processes and whose quality is therefore mostly determined by the know-how of the employees involved. In the Cluster of Excellence, the evaluation of the potential of a self-optimising assembly is therefore performed exemplary on the production of a small fibre laser system. Here, an assembly-compatible design is of high importance just as the development of new joining technologies and the realisation of the assembly system in a way that it features all the necessary microscopic and macroscopic degrees of freedom.

The realisation of an effective interaction of the different robotic systems for joining, assembly and control operation is particularly challenging in the conceptual design of a flexible assembly system. Furthermore, the disadvantageous ability to plan and predict specific challenges creates a broad field of application for self-optimising assembly operations. This is caused by the specific characteristics of the optical components, which are mostly unknown. The assembly system's ability to dynamically adapt itself to changing conditions serves to the analysis of the contribution of self-optimisation to a possible increase in efficiency and economic effectiveness of an automated assembly.

In the context of self-optimising systems, research often refers to the concept of cognition. Cognitive systems are adaptive system, which closely follows the learning behaviour of the human brain. Hence, the development of an "intelligent" production system has to go along with research on cognition of technical systems. It can be regarded as a part of information processing, in which data is recorded, processed into information and independently applied in a problem-oriented way. Contrary to common systems, cognitive systems must not contain fixed sensor-actuator elements in order to provide the necessary learning aptitude. Therefore, cognitive systems differ significantly from non-adaptive, reactive systems.

The essential approach of the concept is the separation of knowledge, which represents certain objects and objectives of a specific domain, and the competence of processing, which is given by the control itself. Therefore, different modelling approaches and techniques are evaluated in the Cluster of Excellence with the premise of generating parameters for planning assembly processes. In order to aggregate environmental information and to establish an appropriate and coordinated planning process, flexible sensor-actuator modules are used.

Since the assembly planning is performed by the cognitive system, the user is to be provided with situation-relevant information in order to monitor the system and, if necessary, take appropriate action. Furthermore, an autonomic system is not necessarily possible to anticipate. Therefore, the user is to be supported selectively in order to gain understanding of the system's dynamics and to be able to operate the system.

As for now, the development and realisation of cognitive systems is still non-satisfactory and still requires basic research. Yet, it has great potential with regards to the resolution of the production polylemma and therefore is of high relevance for production engineering. The results of the Cluster of Excellence's research activities concerning self-optimising productions systems is comprised in Chap. 6.

2.5 Development of a Production Theory

Günther Schuh, Jens Arnoscht, Fabian Bauhoff, Sascha Fuchs,
Andreas Roderburg, Michael Schiffer, Johannes Schubert,
Sebastian Stiller and Stefan Tönissen

The Cluster of Excellence "Integrative Production Technology for High-Wage Countries" pursues the long-term goal to strengthen the competitiveness of German production technology. The overlying solution hypothesis is to achieve a higher level of integrativity in production technology. The scientific key result is the development of a common theory of value creation and of (dynamic) system behavior in production that includes holistic description, explanation and design models for production systems and joins the results from the research carried out in the Cluster of Excellence (Production Theory).

The theory delivers a more fundamental and deeper understanding of real word scenarios and will allow a better modeling of real world correlations. Even if the "behavior" of production systems can never be fully predicted, due to its sociotechnical composition, the theory should deliver quantifiable relationship formulas for a large part of predictable relationships (Complexity Reduction) and reproduce non-predictable parts through a common cybernetic model (Complexity Control).

While the research areas of national economy and, later, business economy, focus solely on the economic analysis of production, the engineering and manufacturing sciences often focus only single areas of production. Thus the technical disciplines are lacking of a broad technical approach for a holistic understanding of the relationships within a company's production which can be linked to macro and microeconomic research of economic analysis.

There is great added value in such a linkage. It makes it possible to answer higher-order questions on how a production system has to be planned, configured and managed. This question invariably includes economic as well as technical subaspects that cannot be completely separated from one another. Besides costs, still other evaluation factors such as product variation and quality, delivery times and reliability and flexibility are gaining importance for the measurement of a production system's performance (Chen 2008; Suwignjo et al. 2000; Grubbström and Olhager 1997). These attributes are mainly determined by the technologies in use and their underlying technical dependencies, which are largely neglected in economic considerations (Fandel 2005). Accordingly, greatly expanded opportunities for forecasting and configuration can be opened up in practice through an integrated treatment.

Against this background, an integrative view is required to determine the economically optimal operating point of a production system. It is essential to link or complement existing economic science-orientated production theories with approaches and models from manufacturing engineering. At the same time, it is necessary to theoretically describe the influence of various aspects—for example, production volume, product variation, quality, logistical issues, or the adaptability of a production system—on its operating efficiency.

In the following, the industrial relevance of a more fully developed production theory is presented. Subsequently, existing production theories are examined for their potential and weaknesses with regard to the problem described above. Taking these weaknesses into account, approaches are presented that make it possible to quantitatively consider the various influencing variables and their mutual dependencies within a production system. This ought to contribute to a better understanding of these dependencies, forming a foundation to identify the economically optimal operating points of a production system.

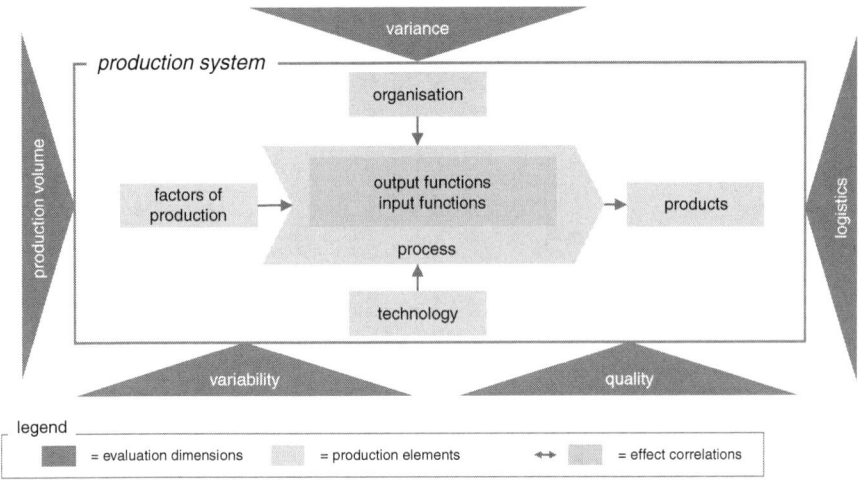

Fig. 2.8 Variables influencing the economic efficiency of a production system

2.5.1 Abstract

As described in the introduction, the main challenge to developing a production theory orientated to production engineering lies in incorporating the (production-) engineering sub-disciplines into a theoretical, descriptive model. In view of the existing production theories, it is necessary to expand the model with a treatment of economic inputs and outputs.

This requires a theoretical consideration of the influence of manipulable variables in various evaluation dimensions (cf. Fig. 2.8) on the economic efficiency of a production system.

Here it is necessary to link the relevant influencing variables and their mutual dependencies into a model, which represents the basis for the determination of the optimal operating points of the production system. In this model, formal sub-models have to be analysed and integrated, assuring that the state of research from various technical disciplines in production engineering, such as manufacturing technology, machine tools, logistics and production planning and control (PPC), are used to quantify the economic effect of the influencing variables.

2.5.2 Industrial Relevance

The necessity for an integrative production theory to model cause and effect in a production system becomes apparent in many producing companies day-to-day. But current theories only incompletely depict the reality and thus have only slight practical relevance (Nyhuis and Wiendahl 2010). Change measures with the goal of

Fig. 2.9 Logistical target parameters. (Wiendahl 2005)

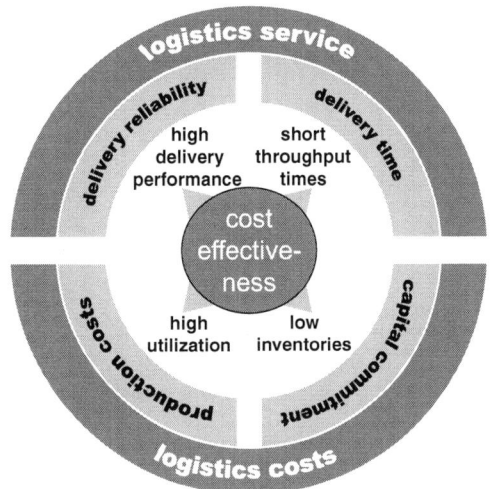

increasing productivity or reducing costs must be evaluated based on their efforts and costs involved and the benefits resulting from them. As a rule, the objective is to primarily maximise company profits. But in practice, not all measures can be broken down into costs, so there are other target parameters that have a qualitative character. Classically, factors such as adherence to deadlines, through-put time, product quality or service level are connected with customer satisfaction. This means that a monetary evaluation of improvements in these variables can only be carried out partially, otherwise it rests on assumptions (cf. Fig. 2.9). Even basic monetary parameters such as machine hour rates display shortcomings. This is because the calculations involve assumptions in the form of past average values, which can change in the course of operation.

The abovementioned can be clarified with the example of a kanban implementation. Besides the inventory target parameter, the introduction of a material requirement planning logic directed to use increases the availability of material, along with the level of service or delivery reliability. At the same time, a reduction in lead time can be achieved. Yet this normally leads to higher inventories. In addition, a kanban component translates into even lower efforts and costs in the materials planning area, since no scheduling intervention is required due to the self-managing character of the kanban. It also means reduced efforts in the area of operations planning and scheduling, as routing plans mostly do not need to be created within the framework of the regular workflow. In addition they are directly available in the kanban itself. Even in this simple example, the difficulty in determining the overall balance becomes clear.

Besides the difficulty in evaluating changes, e.g. from the conversion to a kanban control loop, a great timidity often prevails on change measures that cross departments. There is a reluctance to undertake measures that change the design of a product but provide a better producibility and do not merely imply continued development

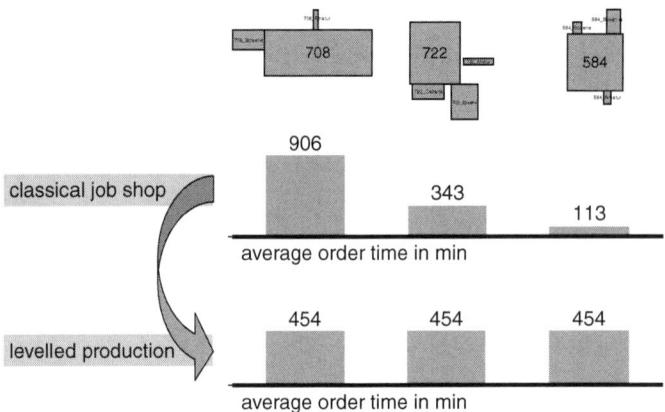

Fig. 2.10 Levelling of a manufacturing process

of technical characteristics. For the most part, product characteristics are seen as unchangeable and not included in the change process.

These kinds of boundaries must be observed in production as well. Once investments are made, they are called into question only rarely, such that the focus is put on logistical processes to optimise the production system. For this reason, the solutions that result from these efforts are clearly sub-optimal. But they would lead to significantly greater improvements if the area under consideration was expanded (e.g. into product manufacturing and technologies or production processes). The reason should be sought in the lack of an opportunity to broadly evaluate the interconnections under consideration.

One reason that companies find it difficult to unerringly identify the best line of attack for production improvements lies in the variety of different adjustment levers and in their complex interdependencies. For example, in the levelling of a manufacturing process (cf. Fig. 2.10), it is not merely job content that can be shifted. One can undertake changes in machine concepts or adjustments in product geometry or construction. This in turn requires other logistical concepts. At first, it is more simple to hold several adjustment levers constant and only carry out changes with particular ones. In this way, effects can be assigned to the appropriate causes.

Not infrequently, the line of attack is based on the assumptions or the conjecture that an improved situation will materialise. The experience of participating employees takes a central role here. On the other hand, one seldom finds a situation where the decision is made based on a comparison of figures for various scenarios. In turn, this is based on the lack of an integrative approach, as described previously. In this way, qualitative target parameters play a major role as well. For example, there can be no generally accepted answer to the question of whether a complete revision is better than a levelled process chain consisting of a number of machines. In most cases, the existing models of a technical nature also describe special cases exclusively. So the claim of general validity is not met here as well (Brinksmeier et al. 2006).

If the necessary calculation models exist, difficulties can likewise arise. The calculation variables are subject to fluctuations over time. For example, if the product mix changes, a static calculation is insufficient for the most part. Simulations have to be carried out. This is a general problem that has been met so far with the use of mean values. Depending on the desired result, this procedure can be sufficiently precise. In many cases, however, such an approach is inadequate.

Apart from the previously cited problems with the identification and evaluation of improvement measures, there is frequently still a flawed understanding of causal interactions in manufacturing. The state of constant problem-solving is already part of employees' daily work life. They no longer call it into question. They are thus part of a system that cannot change from within.

The weakness in existing production theories lies in the high degree of aggregation. This is exhibited in descriptions of production processes (Fandel 2005). The reality can only be painted imprecisely and in highly simplified form. Consequently, the demands for a production theory from the business side rest in the possibility of describing the interactions in its production completely deterministically. In this way, an optimal operating point can be determined under given boundary conditions. One first needs to depict the important elements that have an influence on production in an adequately precise fashion and then provide them with functional interactions. Depending on the manufacturing technology used, various functions must be employed. Grinding processes exhibit a different behaviour than turning processes, for example. Besides pure optimisation in a mathematical sense, improvements should be possible across technological boundaries, in contrast to existing theories (Wibe 2004).

Production theories to date are dominated by a business management point of view. They cannot sufficiently depict the causal interrelationships of real production systems. From the manufacturing field, however, there are already numerous models, e.g. various manufacturing processes, machine tools, logistics or PPC. But it has not been possible to practically link these models to an integrative theory in any over-arching approach so far (Nyhuis and Wiendahl 2010). In achieving sustainably successful manufacturing in international competition, it is a crucial advantage to gain an understanding of the multifaceted causal interactions between the processes and elements within a production system. It is possible to determine theoretically optimal operating points. Improvement measures can be derived on this basis. The interactions and dependencies between the individual—non-value creating as well as value-creating—process steps must be better depicted for this, enabling an effective evaluation of various possible scenarios. Furthermore, a linkage of the individual operation and technology areas would enable the design of individual processes orientated to the economic optimum for the overall production system. This would avoid local optimisations that come at the expense of the performance potential of the overall system. One example to cite is the local time and cost optimisation of individual manufacturing processes within the process chain, taking into consideration an assumed machine-hour rate (Klocke and König 2008). For example, from a logistical standpoint, they do not involve any advantages. Instead, the costs are higher, if they do not result in a bottleneck. Technology decisions are frequently accepted that take restrictions on other areas for granted, proceeding from one point of view. Increased transparency for the implications of individual decisions in other planning areas (e.g. logistics) has significant potential. Restrictions that exist within the technology chain from sequential process steps should be considered early in process chain design and the selection of specific manufacturing technologies through a better depiction of a component's manufacturing history (Klocke et al. 2009). In this way, a local cost optimisation of a forming technique at the start of the process chain can mean a disproportionate increase in the manufacturing and monitoring efforts and costs for a manufacturing process at the end of the process chain (Klocke and König 2006).

It must be possible to link various kinds of models with one another to enable the various areas to be combined into an integrative theory. On one side, there are some physical-analytic models that have great general validity. But they are mostly insufficient for the creation of a complete model of the manufacturing system under consideration. Instead, many decisions today are based on practical knowledge and empirical models derived from it. They often have a sharply limited range of validity. Thus, a priori, the process behaviour or result is often only vaguely predictable. Nonetheless, relevant model variables and their interactions for an overall model are frequently important. The applicability of such technology- and application-specific models must therefore have the ability to be evaluated against a background of transferability.

2.5.3 Evaluation of Existing Production Theories

In economics as well as engineering, manufacturing is a central object of consideration as a transformation of input into output objects. It is initiated by humans in a targeted way and applied to the supply of specific yields (Dyckhoff 2009; Corsten and Gössinger 2009; Fandel 2005; Eversheim and Schuh 1996).

In the course of history, various models have been developed in these two disciplines to describe the transformation process. As a rule, the core of the existing production theories is thus formed by so-called production functions. They characterise the transformation process and describe the interdependence between the production possibilities and the production factor inputs (Steven 1998).

In this section, an overview will be presented on developments and the state-of-the-art in economic and production-engineering manufacturing theories. To this end, a historical outline of important production theories will be given and the developmental direction of these theories with regard to their content will be discussed.

The second section presents the important findings of the theories and their basic modelling building blocks. In the third section, they are critically scrutinised and evaluated with regard to their shortcomings and potential.

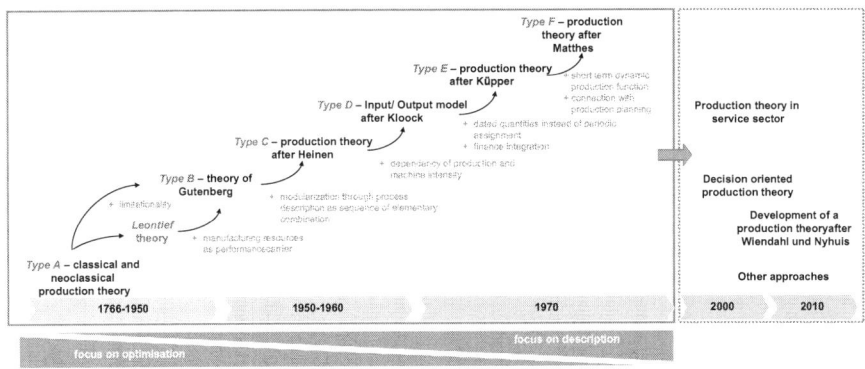

Fig. 2.11 Trend in production theories

2.5.3.1 Development of Production Theories

The first attempts to develop a production theory emerged in the 18th century (cf. Fig. 2.11), a time when business economics was not a separate economic discipline (Steven 1998). Thus, economic questions, in particular, are the focus in its early phases of development: Spurred by agricultural problems, TURGOT develops the law of diminishing returns in the 18th century. It examined the harvest that can be achieved when the labour input is varied and the surface cultivated is held constant (Fandel 2005). In the 19th century, the model is constantly developed further and ultimately emerges in the neoclassical production theory. It transferred economic concepts to describe transformation processes in industrial applications (Schwalbach 2004).

As business economics increasingly establishes itself, the focus of production theory increasingly shifts to technical production processes within companies. To this end, Leontieff resolves the paradigms of classical and neoclassical production theories with a limitational production function. To date, this has always assumed substitutionality, the possible replacement of one production factor with another (Fandel 2005; Leontieff 1951).

In 1951 GUTENBERG develops a new production function, which is the best known to date. Its added value lies in the fact that it takes the performance intensity of the machine into consideration. Input and output are no longer directly determinative but rather the performance of the process is (Gutenberg 1983).

In 1965, HEINEN first expands GUTENBERG'S production function, which makes possible an even more precise and realistic description of production management processes. KÜPPER, KLOOCK and MATTHES make further significant contributions in the following decades. By describing production through input/output analyses, which are underlying these models, the detailed and realistic view of process chains was continuously improved through the consideration of their dynamics, that is the changes in production factors and functions over time, and the integration of still more model building blocks (logistics, financial management, etc.) (Steven 1998; Fandel 2005).

Newer approaches break through the rather continual development of production theories to that point. Based on a cybernetic framework for production economics and management, DYCKHOFF develops a production theory orientated to decision-making (Dyckhoff 2009). Still other approaches follow the modelling of vague production functions or transfer existing know-how into the service sector. In general, a clear trend in the orientation of the production function can be discerned since the initialisation of the first production theories. While the first theories particularly focus on the quantitative description and the optimisation of production, the ability to realistically describe the process and explain interdependencies in production management stands at the centre of the considerations in newer models. Due to the theories' growing complexity, there has increasingly been a decline in their empirical content, factual verifiability and the corresponding degree of validation.

2.5.3.2 Short Description of the Production Theories

In the following segment, there is a brief characterization of the production theories presented in the historical outline above as well as the DYCKHOFF production theory, the first theory orientated toward decision-making.

Classical and Neoclassical Production Theories

Classical and neoclassical production theory is based on production functions in terms of the law of diminishing returns, which are also designated Type A production functions. While the two approaches premise a (limited) substitutionality, they are largely differentiated from one another in the course of their production functions.

The model for the classical law of diminishing returns assumes first a rising and then a falling marginal yield through variation in a given production factor while all the remaining factors are simultaneously held constant. The characteristic S-shaped course (cf. Fig. 2.12) of the production function is formed in this way (Steven 1998; von Stackelberg 1932).

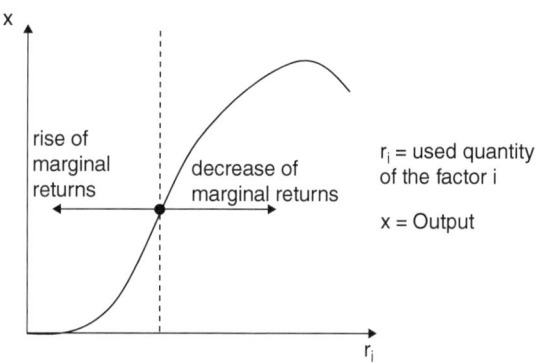

Fig. 2.12 Law of diminishing returns production function. (Schwalbach 2004)

The evaluation of the production factors using prices leads to the classical cost function. It describes the operational minimum, that is the lower price limit at which the company can sell products on the market in the short term, and the operational optimum, the point where the relationship between costs and output quantity is most favourable.

On the other hand, the neoclassical production function is not derived from technical or empirical observations, but is rather postulated as a highly aggregated consideration of production systems (Steven 1998). The characteristics of this production function are identified by constant or decreasing economics of scale, positive, simultaneously decreasing marginal returns and decreasing marginal rates of substitution. Thus, neoclassical production functions are special cases of the classical functions, whose function course is only determined in a section of the curve shape of the classical production functions. One prominent representative of neoclassical production functions is the Cobb-Douglas production function, whose operational existence was able to be repeatedly demonstrated economically in various sectors.

Production Theory Based on Gutenberg

The Gutenberg production function, also called the Type B production function, basically describes a limitational production (Fandel 2005). That means that a quantitative rise in output requires an increase in a number of inputs that are dependent on one another. In this regard, technical equipment in the form of capital goods is assumed to be a constant. On the input side there is a differentiation between consumption factors and usage factors. With the help of so-called consumption functions, individual production units (aggregates) are depicted; they indicate the ratio of aggregate input to output, depending on intensity. Intensity is directly influenced by the technical characteristics of the aggregate (pressure, temperature, speed), whereby an intensity with optimal factor consumption as well as an intensity with optimal costs can be reached (Fandel 2005; Gutenberg 1983).

On the output side, intermediate products, end products and waste products are differentiated. The consideration of waste products makes it possible, for example, to consider disposal costs in the optimisation of production.

The linkage of consumption functions gives rise to a complex overall production function. It can be influenced in three ways (Fandel 2005). For one thing, the operating time of the capital equipment can be curtailed or extended with a timing adjustment. Furthermore, capacity can be influenced with a quantitative adjustment, in which the number of capital goods utilised in parallel per production step is changed. Lastly, the quality-neutral production speed can be changed with an adjustment in intensity.

In consideration of costs for individual factors, an adjustment of production at minimal cost can be found. It is based on the substitutability of the abovementioned adaptations.

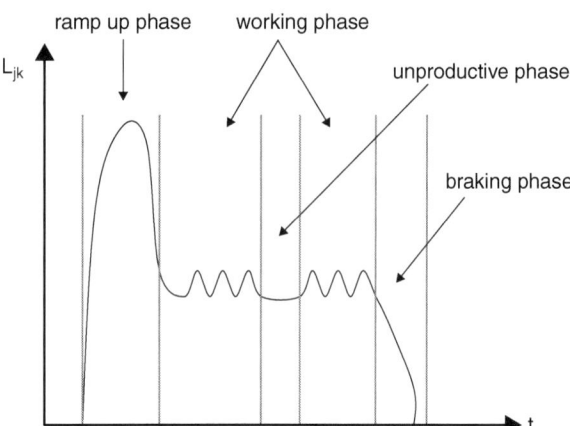

Fig. 2.13 Chronological sequence of the load (L_{jk}) on capital goods. (Steven 1998)

Production Theory Based on Heinen

The Heinen production theory, which is largely based on the so-called Type C production function, is a further development of the Gutenberg production theory. In the process, the differentiation of various Gutenberg production factors for the description of production is expanded with the help of elementary combinations (Fandel 2005). If one first assumes constant factor consumption over time in Type A and Type B functions, Heinen considers various phases—the launch phase, the processing phase, the braking phase and the idle phase—in the course of production (cf. Fig. 2.13).

The production event is broken down into small sub-units, which are modelled and analysed separately. According to the theory, the product is generated through the one-time or multiple execution of one or several elementary combinations in a specified time frame. Repetition functions indicate the frequency of the processing of the elementary combinations. Hence, HEINEN initiates the development of dynamic production theories considering the change in the production event over time (Heinen 1983).

Business Management Input/Output Analysis Based on Kloock

Business management input/output analysis, as well as the Type D production function following from it, considers the interconnection of input and output on various levels of aggregation. Due to the very general formulation of the theory, multi-step production processes with cyclical inter-dependence can be investigated (Steven 1998).

Since the interconnections are represented by means of a transformation function, it is possible to formulate all the production functions developed to date as special cases of the Kloock theory. Due to the variability of the levels of aggregation, the business management input/output analysis has a wide range of applications. In this way, qualitative interactions can be represented and analysed by means of graphs and matrices.

Systems of equations that lead to a transformation function are used to represent quantitative interactions. From the systems of equations, a complete calculation of requirements can be established by means of a matrix operation.

Besides its use in production planning, the transformation function can be expanded with cost variables. In this way, the input/output analysis can be used for the calculation of internal prices and lower limit prices for end products (Kloock 1969).

Production Theory Based on Küpper

Production theory according to KÜPPER encompasses a short-term dynamic production function (Type E). It explicitly considers time as an influencing variable, in addition to production factors, but not changes in production conditions over time.

KÜPPER sets his theory on the foundation of the KLOOCK input/output analysis, which he expands with a consideration of chronological sequences and additional operational production areas (Steven 1998). In the process, the input/output model is provided with period index for each type of goods. This makes possible the assignment of factor consumptions to periods. Depending on the residence time of the factors in the process—residence time of zero, residence time greater than zero, combination—different basic equations apply.

Likewise the theory can take account of important operational circumstances such as the lead time offset, job-lot production, set-up process or stock balance (Küpper 1979).

Production Function Based on Matthes

The MATTHES production function, also called the Type F production function, involves an expansion of the KÜPPER production function. The addition of structural, process-engineering, financial and social secondary conditions provides a refinement in the production process sequence portrayed. An additional refinement and dynamism are effected through the use of dated lots instead of period-related lots (Steven 1998).

This methodological presentation of manufacturing is achieved through a network-supported approach that has the manufacture of an end product as its result. Using the network technique, financial processes can similarly be represented as time-delineated variables such as production orders. The individual processes within manufacturing are represented by means of production functions based on GUTENBERG and HEINEN. In this way, a comparatively detailed representation of reality is possible.

The large number of parameters can cover a variety of different input data and variables. In addition, there are relationships through which interdependencies, such as consumption and payment functions, between variables and data are taken into consideration. Furthermore, restrictions can be incorporated into the model that assigns limits to predefined interrelationships (Matthes 1979).

Thus, depending on the degree of aggregation, individual processes or an entire production system can be examined considering financial aspects as well (Steven 1998).

Production Theory Based on Dyckhoff

The DYCKHOFF decision-orientated production theory is divided into three levels. At the lowest level, the interrelationships of input and output are defined using transformation functions, with-out consideration of other aspects. In the process, multi-step and cyclical production process sequences can be depicted and waste products taken into consideration. This level is also called the technology level (Dyckhoff 2009).

At the mid-level, the production outcomes are analysed by means of cost and earnings ratios from the perspective of the producer. The result is an apportioning of the output into types of goods, types of waste and types of neutral objects. The data on this technology level are analysed in the form of a production function in terms of efficiency aspects. In this way, the term results level arises from the quantity-related results that are identified (Dyckhoff 2009). The third and uppermost level considers the success of production in terms of benefit and loss aspects. The economic success function, whose maximum is sought by means of cost minimization, is established through a comparison of costs and revenue. Here bottleneck factors and opportunity costs can be incorporated. The evaluation of success has led to the term success level (Dyckhoff 2009).

Evaluation of Production Theories and Theory Deficiencies

The basic types of production functions, such as the law of diminishing returns, the COBB-DOUGLAS production function, the Leontieff production function or the GUTENBERG production function, describe the basic economic, input-output relationships in manufacturing processes (Fandel 2005) or the functional interrelationship between quantities of factors used and quantities of production output. These approaches are distinguished by a high degree of aggregation in the representation of the production process. The technical interconnections underlying the economic input-output relationships are largely neglected (Fandel 2005).

GUTENBERG brings out the significance of basic technical conditions in manufacturing in the depiction of productivity relationships and places technically orientated questions more firmly in the foreground (Sonntag and Kistner 2004). In the example of an engine, he takes into consideration the interdependency of fuel consumption and the output from the rotational speed, or shows how tool wear in turning or milling machines varies according to the rate of work (Gutenberg 1983). There have been numerous other endeavours to achieve better consideration of technological characteristics and relationships in manufacturing (Fandel 2005). For example, HEINEN examined a production process broken down into sub-processes, which enabled a clear derivation of technical and economic performance and considered medium-term, variable parameters (e.g. storage of a machine tool), along with long-term constant technological parameters (e.g. span length of a turning machine) (Heinen 1983).

The so-called engineering production functions feature a distinctly stronger technical detailing (Chenery 1949). They aim for the development of analytical transformation functions for the description of individual production processes based on technical-scientific laws (Schultmann 2003). Production factors are thus described with concrete technology parameters (temperature, viscosity, resistivity, etc.). This approach has been mainly used in process industries (e.g. transport of electrical energy, mass transport with flow or pump processes) (Schweyer 1955; Smith 1961). The slight extent of this model and its small importance in practise is associated with the substantial effort for the description of the input-output relationships using analytical functions (Fandel 2005). When necessary, system behaviour must be examined experimentally, with the collection of data at great expense, if there is a lack of analytical description.

For a predetermined quantity of production factors, production functions by definition deliver the maximum quantity of output that can be produced with the underlying technology. This linkage is based on the assumption that the production process is configured in a manner that the greatest possible production efficiency is assured (Wibe 2004). In contrast to the kind of efficiency considerations based on pure quantity relationships, the means of production is efficient within the framework of cost theory if there is no other input-output configuration that generates at least the same product quantity at lower costs (Bogaschewsky and Steinmetz 1999). The configuration or optimisation of production systems based on input-output relationships, however, obscures the view that the optimal configuration of a production system is substantially determined by the technology in use and by concrete technological performance parameters (Wibe 2004). Approaches such as the engineering production functions indeed take into account technological parameters. But, as explained above, the specific models depict only a special problem and therefore are mostly just isolated from one another. Thus, a generalization and generic illustration of the linkages continues to pose a major challenge.

Aside from costs, aspects such as product variation in quality, lead times and delivery reliability or flexibility increasingly stand in the foreground in the measurement of production system performance (Chen 2008; Suwignjo et al. 2000; Grubbström and Olhager 1997). These attributes are substantially determined by the technology being used. In addition, these factors are not, or only insufficiently, considered in the existing production functions.

2.5.4 Starting Points for the Further Development of the Production Theory

Beginning with the previously described weaknesses in existing production theories, some starting points for the further development of production theories are presented in the following from an integrated production-engineering point of view.

In industrial practice, due to complexity, or a lack of descriptive models and/or insufficient avail-ability of data, individual elements of a production system are often viewed in isolation and optimised without considering a classification and analysis of effects in their overall interaction. As a result, while local optimums can indeed be achieved, operating points chosen in this way do not normally correspond to the overall optimum. An integrative optimisation of production systems is a challenging task. The crucial influencing parameters and the mutual functional interactions are frequently unknown. Improvements regarding a specified direction of optimisation often have negative effects on another decisive variable in the production system. Special restrictions often diminish the opportunity for optimising a sub-aspect. Thus, the theoretically achievable overall optimum is frequently not attained when consideration of an aspect is isolated. For example, increased variation in a product characteristic while other influencing parameters are held constant leads to a reduced maximum product volume, decreased delivery performance, increased throughput times and rising manufacturing costs. This is due to the increased secondary processing time and greater tool variation. At the same time, however, the variant fit improves, causing the customer value to rise due to an improved satisfaction of needs.

Against this background, an integrative view is necessary for the evaluation of a production system and for the determination of its economically optimal operating point. In particular, the currently existing economics-orientated production theories must be linked with or complemented by approaches and models from production engineering. With the developed descriptive model, it becomes possible to theoretically describe the influence of various aspects—for example, production volume, product variation, quality, logistical aspects or the mutability of a production system—on the economic efficiency of a production system. To this end, the fundamental elements and dimensions as well as the disciplines involved in a production system have been examined and depicted in the model. In addition, in the individual dimensions, the economically relevant influencing variables and their mutual dependencies have been identified and evaluated (cf. Fig. 2.14). In this way, an interdisciplinary and intradisciplinary descriptive model results. It supports a coordinated adjustment of the individual operating points for the purpose of achieving an overall economic optimum.

Thus the overarching effects of a decision, quantified in the inputs and outputs of the production system, can be depicted with the help of the causal interactions identified among the dimensions being considered. Figure 2.15 presents the effects of an increased variation on the inputs and outputs in the dimensions of logistics, variation and production volume.

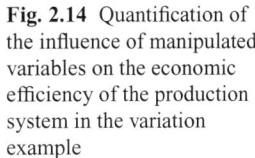

Fig. 2.14 Quantification of the influence of manipulated variables on the economic efficiency of the production system in the variation example

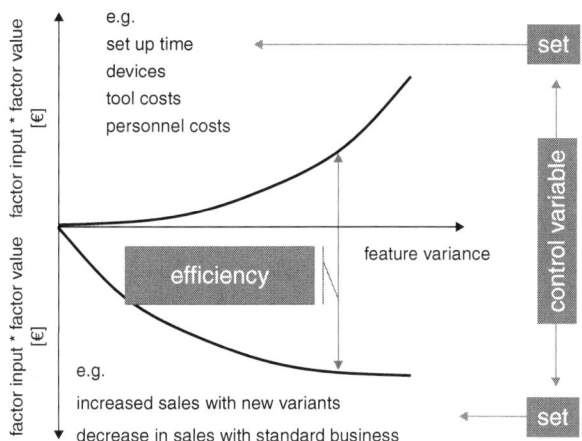

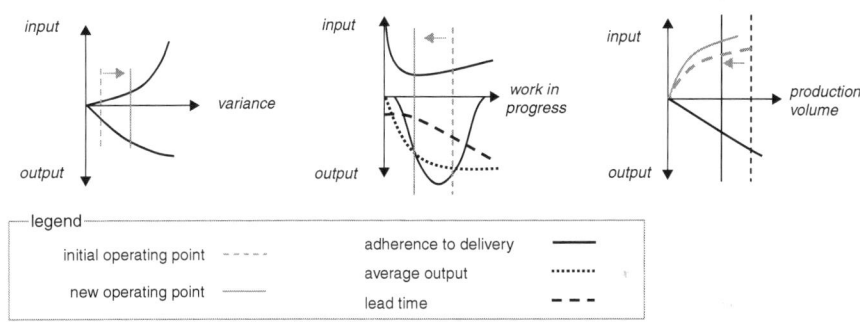

Fig. 2.15 Effects of increased variance

In addition, partly lacking engineering knowledge e.g. in the form of formula-like interactions and technical restrictions can be added to the existing economic production theories with the help of the descriptive model. This will assure that the state of research from specialist fields relating to production-engineering technical areas, such as manufacturing engineering, machine tools, logistics and PPC, will be used to quantify the economic influence of the various dimensions' parameters. The quantifiable influences on the necessary input as well as output and the dependencies between the parameters can be used to determine the optimal operating points of the overall system (Fig. 2.16).

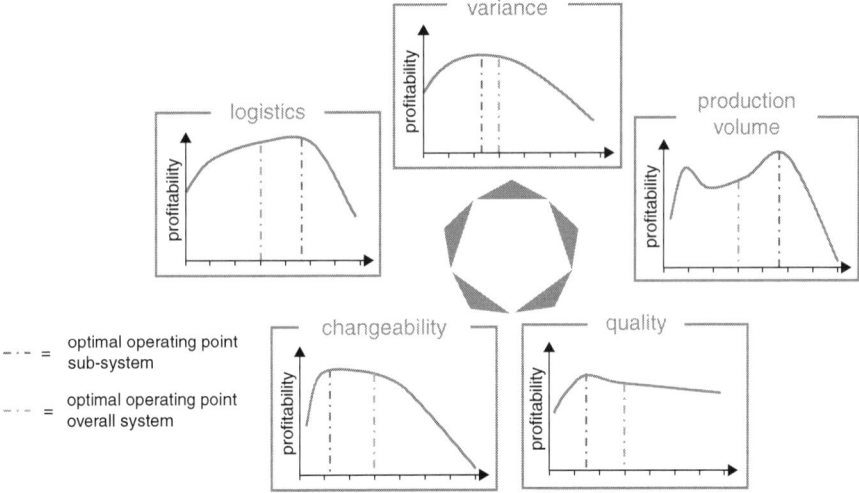

Fig. 2.16 The optimal operating point of the overall production system

2.6 Technology Roadmapping for Production in High-wage Countries

Günther Schuh, Johannes Schubert, Simon Orilski and Susanne Aghassi

As an integrative planning method, technology roadmapping is increasingly being viewed as an important instrument for assisting in projecting future developments and for enabling alignment of uniform objectives and strategies (Walsh 2004). Roadmapping is an instrument which serves to increase transparency and thus to increase the ability to act in complex situations (Orilski et al. 2008). Roadmaps are used in many different areas, such as in politics, science or industry. They permit medium to longterm orientation and support the systematic development of a structured perspective. At the same time they allow relevant technological developments to be recorded and evaluated in detail. A transparent information base of this nature provides a platform for extrapolating, selecting and visualising future development paths (Vinkemeier 2008). Technology roadmaps are used frequently in various technological fields and industries, in particular because extensive internal and external expertise can be used and integrated in the process of creating the roadmap.

The technology roadmaps which have been developed for high-wage countries cover the following four areas of research and application, which are also important for the structure of the "Integrative Production Technology for High Wage Countries" Cluster of Excellence (Schuh et al. 2007; Schuh and Orilski 2007):

- Individualised production,
- Virtual production systems,
- Hybrid production systems,
- Self-optimising production systems.

2.6.1 Technology Roadmaps for Integrative Research

There are very many different types of roadmaps in existence. Depending on the application and the aims of the roadmap, there are different forms of presentation and specific procedures for drawing up the roadmap. Phaal et al. provide an overview of the different types of technology roadmaps with regard to their purpose, graphical form and usage (Phaal et al. 2004). Technology roadmaps are typically used for the following purposes: product planning, planning of increases in capacity, strategic planning, long-term planning, knowledge building planning, programme planning, process planning or technology integration planning (see Phaal et al. 2004). Technology roadmapping assists manufacturing companies in developing a technology strategy, which contains both objectives and fundamental ways of achieving these objectives (Beckermann et al. 2008; Schuh et al. 2011a). Early recognition of new technologies and the analysis of the organisation's own expertise and resources normally form part of the process of drawing up the technology strategy and the roadmap (Beckermann et al. 2008).

A generalised version of the layout of a technology roadmap has been proposed by the European Industrial Research Management Association (ERIMA). The suggested constituent elements of a roadmap are shown in Fig. 2.17. It shows the main elements as being the timeline for ordering the planning objects chronologically, the planning levels (the market, the product and the technology) for structuring the content of

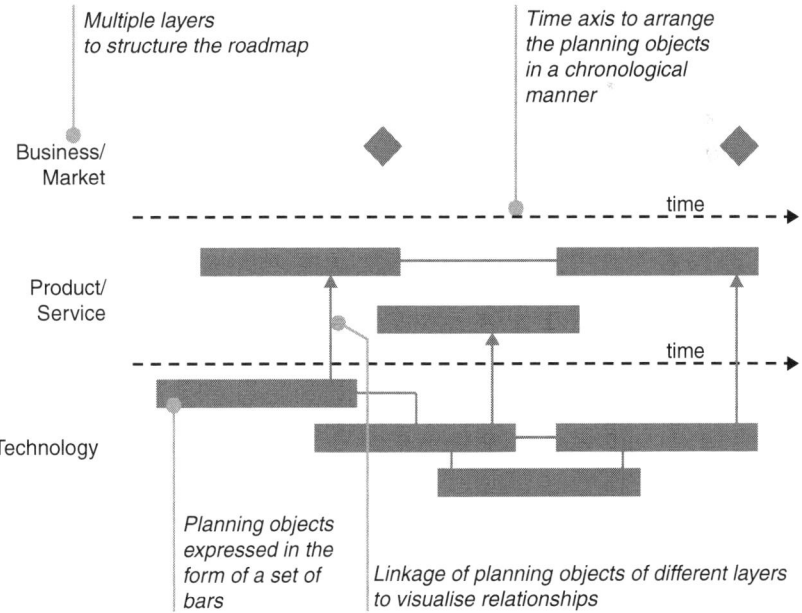

Fig. 2.17 Constituent elements of a technology roadmap. (European Industrial Research Management Association (ERIMA) 1997; Schuh et al. 2011a)

the roadmap, the presentation of the planning objects in horizontal bar format and the linkage between the planning objects, e.g. between the different levels to help visualise the causal relationships (Beckermann et al. 2008).

In the same way that technology roadmaps are used to track many different applications and goals, equally there are many different ways of presenting the roadmaps in graphical form. The use of one or more planning levels, horizontal bar charts, tables, graphs, illustrations, flow diagrams or text-based roadmaps are among the most common formats (Phaal et al. 2004). The form of presentation chosen in each case is dependent on the form and level of detail of the information to be visualised, any specific constraints and the purpose.

The use of technology roadmaps in research, however, calls for new approaches and new ways of presenting roadmaps, in order to do justice to the following specific requirements in the research environment: research roadmaps should give an overview of a new field of research. At the same time information must be included which has been researched and compiled in a variety of different ways. Due to the fact the information being presented is so new, the field has first of all to be structured, definitions established for sub-areas and information from widely differing sources reviewed, processed and standardised. Preliminary information on new technologies presents a particular challenge, as these are in some cases still in the research phase. The lack of certainty over the chronology of the development of the technology, from the basic research phase through to start of production must continue to be taken into account. Frequently, problems and challenges in the development of the technology cannot yet be anticipated in the early stages. Additionally, future application fields for technologies may still not be clear at the beginning of their development. In order to arrive at a coherent overall picture, the information gathered must be aggregated across the whole area of research and presented in a clear and concise format.

In order to meet these requirements, a modified technology radar, as shown schematically in Fig. 2.18, was developed to display technology roadmaps produced for manufacturing companies in high-wage countries. This format is based on the suggestion for a technology radar presented by Rohrbeck et al. as an instrument for early recognition of technologies and for developing a strategy for innovation (Rohrbeck et al. 2006). The technology radar used here gives an overview of relevant technologies in a given field. The technology radar does not have a timeline because, in the early phase of technology development or in the research phase in particular, it is very difficult to forecast accurately the timescales that will apply for the development. Particularly when researching new technologies, the development time required from basic research through to start of production is dependent on many influencing factors. The volume of investment, for example, aid granted for projects or the discovery of synergy effects from other research programmes can all change the development time considerably. The evaluation and display of the maturity of a technology using the radar is one way of eliminating the high level of uncertainty involved in estimating development times. Technologies can be allocated, for example, to "Basic research", "Concept development", "Prototype stage", "Qualification for production" and "Ready for production" levels of maturity. The individual levels of

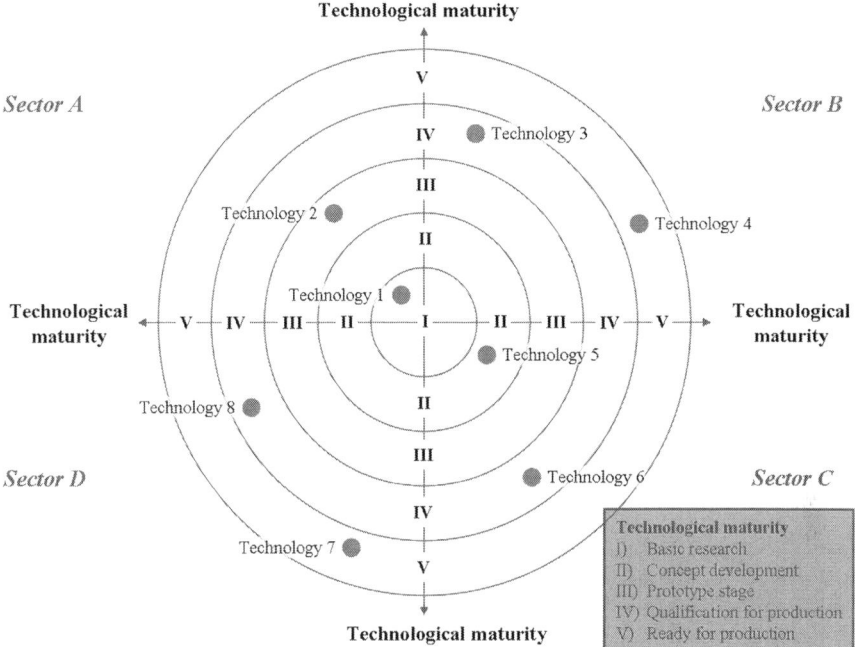

Fig. 2.18 Example of how to structure a technology radar. (following Orilski et al. 2008 and Wellensiek et al. 2011)

development are then displayed in concentric circles, starting with "Basic research" in the centre of the radar, in order to highlight the relevant research topics. The closer a technology is positioned to the outer edge of the circles, the further developed it is and thus the closer it is to series production, making it more attractive to manufacturing companies. Apart from technological maturity, other relevant parameters (such as processing speed) which represent important technology development objectives may be used as the basis for evaluation.

The technology radar is able to aggregate information to a very high degree. Additional information on the individual technologies can be summarised in accompanying profiles (Wellensiek et al. 2011). In addition to the advantages and disadvantages of the technologies, other information such as research requirements as well as (potential) application examples for the technologies can be recorded. In this way, technology radars together with accompanying profiles give a structured overview of the areas of research that have been examined and a summary of the further actions required. Technology radars visualise the path from basic research to industrial implementation. They therefore make it possible to forecast the future of technologies, they display the levels of maturity of different technologies, provide detailed approaches to current technology planning and development and assist strategic planning, both in the area of research and development. In particular, technology roadmaps facilitate communication between science and industry and thus promote the transfer of researched technologies into industrial practice.

The process of creating technology radars consists of the following activities, which equate to the technology intelligence process (Wellensiek et al. 2011):

- Determining the information demand,
- Acquiring the information,
- Evaluating the information,
- Communicating the information.

In order to acquire and use information in a systematic and targeted way, firstly the information required must to be determined. The information demand must be defined as precisely as possible. To do this, in a first step, the fields of research were structured and the criteria for evaluating the technologies defined. Then the sources of information for acquiring the information were ascertained and determined. A basic distinction can be made between explicit and implicit information sources (Wellensiek et al. 2011). For one thing, the creation of technology roadmaps drew on explicit knowledge, for example in the form of studies which had already been completed or other publications. But implicit, unpublished knowledge plays an important role, as the knowledge acquired in the early development stages is still being formulated. In these predominately complex and developing fields of research it is rare to find studies that have already been published: the information base is either incomplete or not transparent, owing to degree of novelty of the research field. For this reason especially the implicit, unpublished knowledge of experts on the areas of technology in question (both in the world of science and in the world of industry) was used. In such cases, discussions with experts on roadmapping provide the opportunity to take account of current trends and developments (Schuh et al. 2009). Additionally a study was conducted specifically for each of the four research topics to obtain targeted answers to questions from research institutes and manufacturing companies. The collated information must then be evaluated. The information gathered in the form of profiles (containing, amongst other information, description of the technology, advantages and disadvantages and principal research requirements) was used to evaluate the level of maturity of the technology. The evaluation carried out on the basis of this data was discussed with experts from the worlds of science and industry and thus further developed on a continual basis. Additionally, the results obtained were made public in the form of publications and presentations at meetings and conferences.

2.6.2 *Integrative Development of Technology Roadmaps*

Technology radars were developed for each of the four areas of research "Individualised production", "Virtual production systems", "Hybrid production systems" and "Self-optimising production systems". These structure the relevant technologies in the field examined and provide information on the current status of development for each technology. Additional background information in the form of standardised profiles was compiled on the technologies being looked at. Alongside this, a study seeking to identify relevant trends, technological challenges and developments and research challenges in each of the four areas was carried out.

Building on the analysis of existing studies, literature and roadmaps in each field of research and application, a workshop was conducted with technology experts for the purpose of developing hypotheses. These hypotheses were used to define the areas to be examined in the study, in particular the challenges, developments and requirements for each area. A questionnaire was developed together with the technology experts from the Cluster of Excellence. A total of 178 companies and research institutes took part in the study. The focus of the study was on the machine-building and automotive industries. Both SMEs as well as large enterprises were covered in the study. On the basis of further interviews and discussions with technology experts, the roadmaps were developed. The key results of the roadmapping process and the study are summarised in the sections that follow.

2.6.3 Individualised Production

Individualised production can be defined as a concept for the design and layout of all the elements of a production system, which permits a high degree of variability in the production programme where production costs are comparable to mass production. Individualised production is targeted at satisfying specific customer requirements (Tseng and Jiao 1996). The implication for manufacturing companies of the general trend towards individualisation in society is a requirement to manufacture a rising number of variants and to offer customisation of products.

At the same time, a high level of product standardisation and the robustness of the production system play an important role in the area of individualised production. Key elements of the production system are thus the product programme and the product architecture, the manufacturing technologies and processes deployed, and the resource structure.

The technology radar for individualised production displayed in Fig. 2.19 covers the three principal areas "modular products", "customised mass production processes" and "rapid manufacturing".

Modularly-designed products serve to reduce the number of variations and the costs associated with these in the manufacturing process. There are various types of modular product: tools in medium quantity (e.g. modular extrusion dies), complex products in medium to high quantity (e.g. modular vehicle structure), equipment of low complexity in high quantity (e.g. white goods), machines of medium complexity in medium quantity (e.g. agricultural machines) and complex industrial machinery in small series (e.g. injection moulding machines). In each instance the degree of modularisation was evaluated for the modular products (Brun and Zorzini 2009) and displayed on the technology radar. At the same time a distinction has been made between an integral design, a physically or functionally modular design, a modular

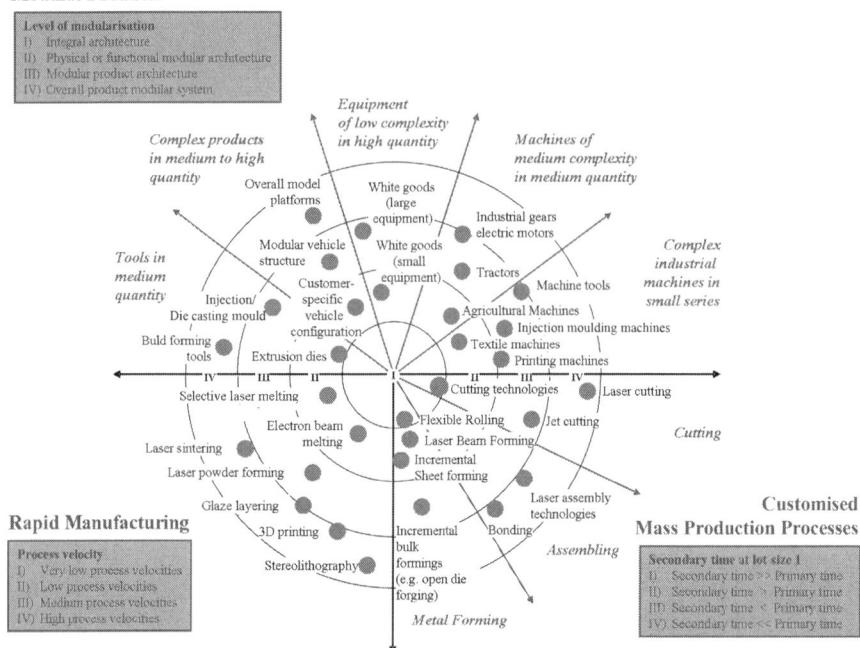

Fig. 2.19 Technology radar for individualised production

product architecture and a modular system applying to all products. The other sectors of the technology radar cover rapid manufacturing technologies and customised mass production processes. Although conventional mass production processes typically are unsuitable for the production of individual products, different technologies in the areas of separation, joining and forming make the flexible manufacturing of customised products possible. The criterion used to evaluate suitability for individualised production was the secondary time in relation to the primary time. Rapid manufacturing technologies allow the direct manufacture of customised products based on 3D CAD data (Eyers and Dotchev 2009). The technologies looked at were positioned in terms of their process velocity on the radar.

2.6.3.1 Challenges and Trends in Individualised Production

The impact of market dynamics on manufacturing companies will increase further in the future. Continuing globalisation and the need to generate competitive advantages are significant drivers for technology-oriented firms and have ramifications for strategic planning, the development of innovative products and the implementation of new production technologies and systems. The study makes it clear that the range of products offered by participating companies is strongly characterised by a high level of product complexity and high tolerance requirements and a high technological demand. Although the variability of the products manufactured is already judged to be good, a further increase in variability is expected. Besides this, new products will be launched on the market at a higher frequency in future, or existing products will be modified more frequently (product variation). On average, about 40% of the companies surveyed modify their products within a year as a result of changing market conditions. This implies a rising demand for adaptable or reconfigurable production systems.

More than 70% of the participants consider the demand for individually-designed products to be high to very high. The majority of firms already offer their customers a wide range of possibilities when it comes to influencing product design. These include specific development contracts, manufacture according to the customer's design specification, a choice between pre-determined variants or the modular configuration of a product using a product configurator. The principal advantage of modularly-designed products as being perceived by the participants is the fact that modular product designs offer the opportunity to create derivatives cost-effectively. Apart from this, modular products make it possible to produce many variants based on only a few individual components; and modular products make it easier to modify existing products. The development of modular products is, however, very time-consuming; and the construction of modular products presents a big challenge. Despite these challenges, the study confirms the trend towards modular product designs against the background of the increasing complexity of products. With regard to the manufacture of those products, the results of the study show that mass production processes for customer-specific production are judged to offer the greatest potential for reducing the manufacturing costs for individualised products. Alongside this, modular tools and rapid manufacturing processes also offer great potential.

2.6.3.2 Relevant Research Topics in the Area of Individualised Production

The results of the study point to several potential scientific approaches and research topics for the individualised production. In the area of modular products, further development of modularisation on different levels is needed. A systematic approach to derive modules and sub-modules, e.g. to achieve physical and functional separation at component level, continues to be regarded as a relevant research task. Apart from this, further standardisation and the definition of interfaces is of significance for the development of overarching modules. On the basis of combination rules, the reconfigurability of products can be increased and simplified through modular systems. Design methods for modular product architecture and concepts for its ongoing maintenance must continue to be developed further and implemented in supporting software systems. In the area of mass customisation, the reduction in non-productive time represents an important task for scientific researchers, for example, through the development of processes that do not require tooling, or by the transition to tools which can be flexibly deployed. Approaches to reduce production time include shortening process chains, integrating process steps, parallel processing or increasing process speed. In order to get to a point where rapid manufacturing technologies are widely deployed there are many barriers to overcome. The principal tasks are to increase set-up rates and process speeds. In addition, the surface qualities have to be further improved, the range of materials extended and the production of large component parts enabled.

2.6.4 Virtual Production Systems

Virtual production systems are deployed in the efficient development of new products. One of the main objectives of virtual production is the reduction of the time and resources used for non-productive planning activities prior to value creation. Typically, product and process development take place at the same time and are decentralised activities. Decisions frequently have to be taken on the basis of incomplete information. It is often the case, for example, that production and process technologies cannot be planned in the early stages of product development because relevant product information is missing or is not certain (Eversheim et al. 2002). Virtual production systems address these problems by supporting parallel product and process development and calculating missing data using simulations and software-assisted tools (Gottschalk et al. 2007). In addition to these benefits, virtual production systems offer the potential to reduce development time and usage of resources, and to achieve better product quality through the support offered by simulations of product and process developments (Potinecke and Slama, 2005).

Virtual production technologies can be divided into five areas: "factory planning"; "product planning"; "technology planning"; "machinery and control simulation"; and "materials and process simulation" (Schuh et al. 2011b).

The technology radar for virtual production is shown in Fig. 2.20. The sub-area "factory planning" covers applications using digital systems to assist the planning of production locations, factory layouts or capacity as well as the optimisation of cycle times or material flow. The "product planning" area summarizes tools for planning, developing and detailed designing of components, sub-assemblies and products (geometry, fulfilment of function, strength, kinematics, heat conducting properties etc.). Systems for technology planning are used to generate and evaluate technology chains (selection and linkage of suitable technologies). The "machinery and control simulation" sub-area comprises the simulation and optimisation of the manufacturing process, in particular taking into account interactions between process and machine behaviour. In the "materials and process simulation" sub-area, methods and tools to simulate the characteristics of materials and components are examined depending on materials and process chains (Schuh et al. 2011b).

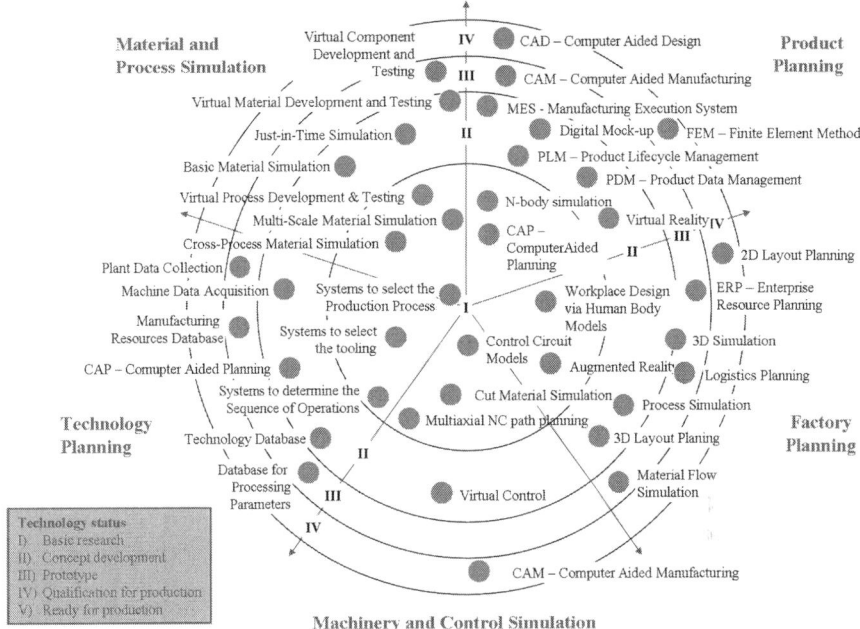

Fig. 2.20 Technology radar for virtual production

2.6.4.1 Challenges and Trends in Virtual Production

Virtual technologies are currently mainly deployed in product and factory planning. Systems for technology planning, in contrast, are rarely used: only about 13% of the companies surveyed use appropriate software solutions in this area. Virtual production tools are used in the first instance by large and medium-sized companies, whereas small businesses with fewer than 50 employees only use virtual production tools in exceptional cases. The main criterion for deciding to use software tools is not the quantity to be produced, but rather the technological requirements. Virtual tools are mainly used to support the planning of high-tech products and production processes. The companies surveyed see the greatest benefits of using virtual systems in better quality and flexibility of the planning processes and a reduction in the overall planning costs, particularly through avoiding planning errors.

In contrast, there are obstacles to the use of virtual production tools in industrial practice: firstly, the lack of knowledge about the virtual production tools available and about their usage represents a barrier to their deployment. Secondly, both the large amount of resources needed to integrate virtual production tools into a company's existing software systems and the adaptation of virtual tools to the specific requirements of the company can be identified as further, relevant obstacles.

The relevance and role of people in virtual production systems was also examined in the study. It can be concluded that human creativity and the ability to make decisions cannot be replaced by software systems. Software tools can help to identify and weigh up possible solutions within a given solution space, but only humans are capable of finding new solutions that are going beyon the boundaries applying at the time. It therefore follows that people will continue to play a decisive role within the context of virtual production. At the same time, integrating and being heedful of human capabilities in software tools in an appropriate way represents a significant challenge.

2.6.4.2 Relevant Research Topics in the Area of Virtual Production

Virtual production systems are used mainly in high-tech applications in order to increase quality and flexibility, and to reduce costs in the planning process. Besides improving individual software tools and methods, the integration of different tools into one platform represents a major field of activity. The study confirms the importance of linking and integrating different tools in order to arrive at a holistic systems approach. By implication, the development of a uniform data basis (including the processing of data to an appropriate level of detail) for all planning tools plays a decisive role. Apart from this, the standardisation of interfaces and the development of open systems architecture are also relevant topics.

Additionally, improvements (in particular of man-computer interactions and visualisation of the results) in the usability of virtual software solutions are important, in order to extend its deployment to small and medium-sized enterprises. A further development task is to increase the user-friendliness of virtual tools. It should be possible to use the software solutions intuitively, i.e. use them without special expert knowledge as is currently the case. By implication, the development of user-friendly concepts and the targeted visualisation of simulation results are important tasks.

2.6.5 Hybrid Production Systems

Hybrid production technologies can be defined as the combination of production technologies based on differing physical principles or the integration of separate production processes into a single, new production process (Feiner et al. 2005). The integration of production technologies leads to a shortening of the value creation chain, as several production steps are replaced by a single hybrid process and planning resources can be focussed on this single process. Hybrid processes additionally allow for new materials to be processed, which could not be processed economically using conventional processes (Klocke et al. 2007).

The technology radar for hybrid production technologies shown in Fig. 2.21 consists of the four sectors "hybrid machines", "combined principles", "combined process steps" and "hybrid products" (Schuh et al. 2009). In each sector representative technologies have been evaluated with regard to their level of maturity.

2 Integrative Production Technology for High-wage Countries

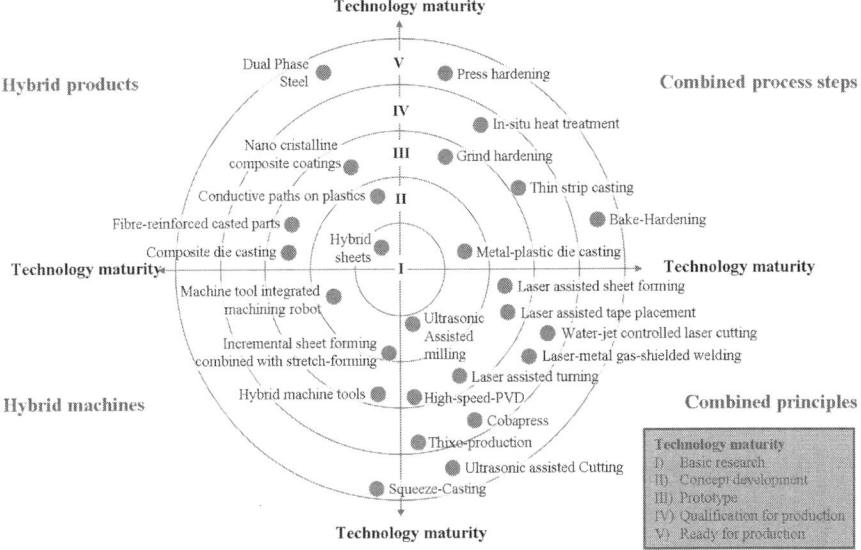

Fig. 2.21 Technology radar for hybrid production

"Hybrid products" cover components or systems with a hybrid design or a hybrid function in which the linkage of specific benefits from the individual components creates a unique benefit at systems level. Hybrid products are often created at materials level, e.g. through hybrid materials based on metal, ceramic, glass or polymers, lightweight materials, organic materials, coatings or bonds. "Hybrid machines" make it possible to carry out very different processes within a single machine and even in a single setting. This enables set-up costs to be reduced significantly. "Combined principles" are geared to combining processing methods with differing physical working mechanisms with the aim of increasing output (e.g. laser-assisted or ultrasonic-assisted processes).

In contrast, "combined process steps" cover manufacturing processes based on the parallel execution of process steps which would normally be carried out sequentially (e.g. integrated plastic injection moulding and metal die-casting to manufacture hybrid plastic and metal components) (Schuh et al. 2009).

2.6.5.1 Challenges and Trends in Hybrid Production

Hybrid production technologies are still rarely used in industry today. Roughly only a quarter of the companies surveyed deploy hybrid production technologies regularly. A major reason for this is the lack of knowledge about hybrid technologies. Although approx. 80% of the participants know of this field of technology, only about 30% have detailed knowledge of hybrid technologies. The greatest potential for use lies in mass production and in high-tech applications. Respondents expect an increase in the use of hybrid production technologies particularly in the area of combined process steps, because of the technological and financial benefits that can be achieved. More than half of the participants in the study see opportunities for deploying hybrid production technologies for their own range of products. More than 40% expect to use hybrid technologies on a regular basis within the next five years due to technological advantages.

Hybrid technologies are usually developed for a specific application. More than 80% of the hybrid production equipment used in industry was modified or developed for a specific application area. The development of hybrid machinery is primarily carried out in the companies' own R&D departments (approx. 36%). This is followed by hybrid machinery developed by machine tool manufacturers (approx. 31%) and research institutes (approx. 15%). Standardised hybrid machine tools are only used in exceptional cases (approx. 16%). The standardisation and qualification of hybrid production machines for a wide range of applications is thus a very promising field of development in order to exploit the potential for deployment in industrial practice and to reduce specific development costs.

2.6.5.2 Relevant Research Topics in the Area of Hybrid Production

In the area of hybrid materials, the integrated modelling and simulation of material behaviour over different scales (at nano, micro and macro level) in general, and the development of linking mechanisms on the different scales in order to create hybrid products, are both important challenges. An improvement of process control for hybrid processes as well as the simulation of hybrid processes must be driven forward. Additionally, it is necessary to look at the design and development of the tools required.

Besides the identification of potential applications for hybrid production technologies, the identification and evaluation of useful technology combinations represents a significant challenge. The technological limits of a single technology's performance can be overcome through such combinations and synergies can be exploited in order to create manufacturing processes which are not possible with single technologies. As a trial-and-error approach to identifying useful technology combinations is neither reliable nor efficient, it is necessary to develop a methodology for identifying and evaluating suitable technology combinations.

2.6.6 Self-optimising Production Systems

In order to achieve greater process flexibility, self-optimising production systems possess an inherent "intelligence" and have the capability to adapt themselves autonomously to changing ambient conditions (Gausemeier et al. 2006). The concept of self-optimisation, both in technological and organisational contexts, is based on the following three steps: continuous recording of the current status of the system;

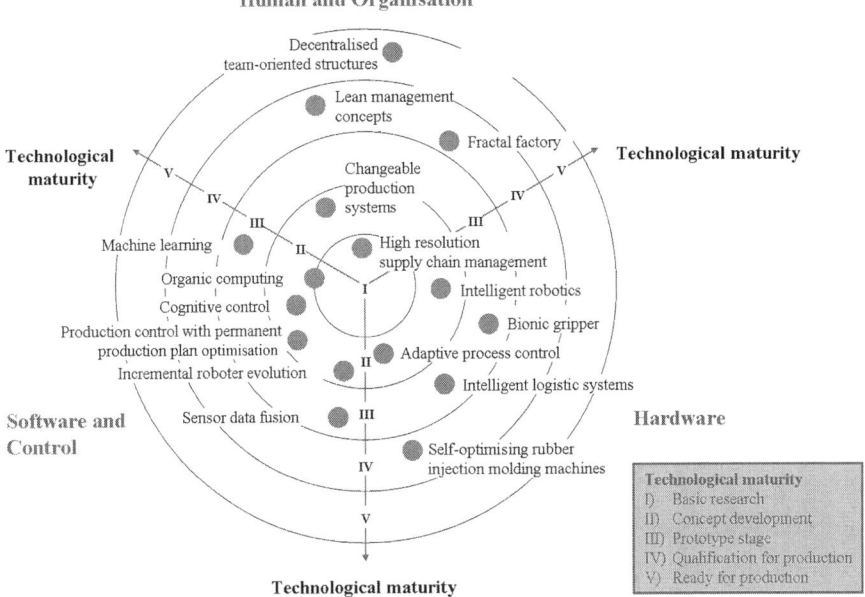

Fig. 2.22 Technology radar for self-optimising production

evaluation of the situation based on previous experiences, models and knowledge stored in the memory and the autonomous and dynamic modifying of the system in response to changing parameters, with the functioning of the system assured at all times (Schuh et al. 2007; Weller 2008; Schuh and Gottschalk 2008). Thanks to the ability to use stored knowledge and to transfer it to new production situations and tasks, planning effort can be reduced and diverted to the manufacturing process.

The area of self-optimising production systems can be subdivided into the areas of "hardware", "software and control" and "human and organisation" (Fig. 2.22). "Hardware", in self-optimising production systems, covers the hardware components deployed, such as intelligent robots and bionic grippers, while "software and control" contains topics related to automation and robust process control. The collection of themes under the header of "human and organisation" covers topics related to production planning and control as well as organisational methodologies and concepts.

2.6.6.1 Challenges and Trends in Self-optimising Production

Although more than half of the respondents know that there are significant benefits to be derived from self-optimising production systems, currently only very few of those taking part in the study have detailed knowledge of self-optimising technologies. The reason for this may be that many of the concerned technologies are still at research and/or development stage and are therefore not being deployed in the industrial companies questioned. The participants in the study see the benefits of self-optimising production systems in the first instance to be the contribution they make to an increase in the robustness of a production facility and to greater adaptability in production. For this reason it is expected that the main application areas for products that are highly demanding as regards technology will be in a dynamic and flexible production environment. Furthermore, self-optimising technologies enable product quality to be enhanced and can lead to an increase in ramp-up speeds. In contrast to these benefits, the immaturity of the majority of self-optimising technologies (still either at research stage or concept development) and the lack of suitable solutions for manufacturing companies are cited as significant obstacles to their deployment in industrial practice.

2.6.6.2 Relevant Research Topics in the Area of Self-optimising Production

Relevant fields of activity can be identified in the integration of the system "man—organisation—technology". At the same time it is necessary to identify and describe the principal connections and interdependencies. Modularly designed and configured production elements with standardised interfaces need to be defined in order to bring about improvement in the planning, recording and control of an integrated system of this nature. Suitable sensors for systems diagnostics, particularly of the man—organisation system, must continue to be developed and combined, in order to record the current system status. In the sub-area "software and control", it is necessary to develop new and robust algorithms to control technical systems (including self-learning systems), so that problems, errors or bottlenecks can be recognised and appropriate decision rules derived. Work must continue to be done on expanding the solution space of self-optimising production systems (e.g. the number of problem-solving options that can be autonomously selected). Additionally there is a requirement to define clearly and develop further the integration of humans into the production process as well as man-machine-interfaces. In addition to these topics, hardware components should be further developed with a view to flexible application.

2.7 Organisation and Management of Integrative Research

Florian Welter, Claudia Jooß, Anja Richert, Sabina Jeschke and Christian Brecher

2.7.1 Abstract

Over the last few decades there has been a tendency to move away from the increasing segregation of basic research, applied research and corporate research and development (R&D) towards a more dynamic and interactive process of generating interdisciplinary knowledge. In order to respond to this tendency, the federal government and the federal states of Germany have initiated the Excellence Initiative. This aims to strengthen and maintain the status of Germany as a science and academic hub. A number of clusters of excellence were set up in 2006 under the Excellence Initiative, including the RWTH Aachen cluster "Integrative Production Technology for High Wage Countries". This brings together academics from a wide variety of disciplines, principally those of production engineering and materials science, in a common research network with the aim of researching integrative production theories.

The highly complex and heterogeneous network in this cluster includes nineteen university institutes and Chairs, seven affiliated institutes, a number of non-university research establishments (such as the Fraunhofer-Gesellschaft) and various industrial and scientific advisors. Its organisational structure with the Cross Sectional Processes thus represents an approach to the integrative networking of the cluster participants to meet the challenge posed by its interdisciplinary nature and its organisational heterogeneity in the context of production engineering and materials sciences. This challenge lies in current academic trends such as a stronger integration of cognitive and information technology elements in production engineering and mechanical engineering. In addition, there are also external requirements placed on interdisciplinary research, such as those set by the funding strategies of the German Research Foundation (DFG), and the German Council of Science and Humanities (Wissenschaftsrat – WR) or the European Union.

The aim of Cross Sectional Processes is the academic interlinking of the integrative cluster domains and their research processes. For this reason, suitable individual measures for networking all cluster participants are being developed and implemented and unified solutions are being worked on for the management of an integrative Cluster of Excellence.

The Cross Sectional Processes pose the question:

> How should networked, intensive academic and highly complex research co-operations within the Cluster of Excellence be organised in order to ensure high-quality output?

This leads to the following objectives:

- Reports on the design elements for supporting the development of networks between highly complex research associations
- The development of an instrument for performance measurement in highly complex research co-operations
- The generation of a transferable application model for designing highly complex research co-operations.

2.7.2 State of the Art

In network theory, a distinction is made between four basic approaches, each of which takes as its theme a different aspect. Compared with social network analysis, neo-institutional network research and networks from a system theory perspective, structural analysis network approaches (Sauer 2005) show the greatest convergence in respect of guaranteeing learning and knowledge processes in highly complex research co-operations. A structural analysis network perspective sees networks, like system theory, as *"social systems that essentially expand and reproduce as a result and medium of interorganisational practices"* (Sydow and van Well 2006). The approach is subject to the acceptance of the duality and recursiveness of structure[1]. Consequently, the dealings of actors relate to existing structures, which are reproduced or transformed recursively by the behaviour of the actors (cf. ibid.). Network development therefore depends crucially on common social practices and network actors who, as *"knowledgeable agents"*[2] (Sauer 2005) adapt them reflexively to the given environmental circumstances. The embedding of the network actors in both the network and the home organisation gives rise to a co-evolutionary development via the transfer of specific networker knowledge from both organisations (Sydow 2003). Furthermore, Sydow makes a distinction between strategic, regional, virtual and project networks (Sydow 2006). Based on the following properties, the network type that the research network most closely resembles is that of a project network:

- time limits and selected partners with a highly fluctuating membership (Sydow 2006)
- a coordinating research unit and individual projects (Hunecke and Sauer 2003)
- alignment of the innovation targets with specific topic areas (Sydow 2006)
- a new combination of implicit and explicit knowledge components holds a high innovation potential (Ahrens 2003).

These qualities lead to the expectation that Cross Sectional Processes conceal both opportunities and challenges for participating organisations and research networks. According to Siebert, the best possible coordination in the network requires the following core conditions: firstly, the *effectiveness condition* of networks requires a higher degree of productivity than other types of coordination such as a market or hierarchy. For research networks, this means that the research output from the cooperation must be greater than that from conventional, non-integrative research. Secondly, the *efficiency condition* requires that all network actors have to be satisfied. In the case of research networks, this means that the utility from the cooperation must be greater than the costs of communication and cooperation borne by each actor (Siebert 2006).

Since this represents a challenge for interdisciplinary networks, the structure and organisation of clusters of excellence and the corresponding Cross Sectional Processes should also consider the following aspects to guarantee a high quality cluster output:

---

[1] The structures are according to Giddens, rules of signification and legitimation and resources of domination (Giddens 1984).

[2] According to Windeler the actors monitor, supervise and control what happens "in that they recognise the potential and create test events, understand these and extend them. By doing this they control their behaviour or that of others by directing activities and securing them." (Windeler 2002).

- The heterogeneity of the cluster participants may lead to structural, cognitive and cultural challenges in respect of their cooperation and coordination, as different disciplines with different scientific methods and concepts come together (e.g. from production engineering, material sciences, information technology or economics).
- This creates both new academic potential and challenges in view of communication within the cluster and cooperation between individual disciplines (Sauer 2005), (Röbbecke et al. 2004).
- One of these challenges, for example, is the process of understanding between interdisciplinary actors. These cultural challenges offer corresponding opportunities and risks for the performance of research networks.
- Project-orientated networks in particular are additionally characterised by a high degree of fluctuation of the actors. The continual coming and going of participants can have both a positive and a negative effect on the coherence of the entire network and may delay the process of network development (Sydow 2006). This is a particular challenge for cross sectional tasks, as it is particularly high within a university research network because of time constraints on doctoral projects.

With reference to current challenges in the interdisciplinary management of academically orientated networks, the Cross Sectional Processes combine two approaches:

On the one hand, Cross Sectional Processes offer special services that aim to link the cluster participants in order to increase cooperation within the cluster and the academic output. Corresponding "Network Services" (Sydow and Zeichhardt 2009) (Buhl and Meier zu Köcker 2009) (Jooß et al. 2010) represent organisational measures such as colloquia for collaborators or digital knowledge management instruments. On the other hand the Cross Sectional Processes are in themselves a branch of academic research within the Cluster of Excellence that is continually examining how to optimise the interlinking of interdisciplinary academic actors. The results of a performance measurement system are used to this effect for instance for the continued strategic development of the whole Cluster of Excellence and of individual "network services".

2.7.3 Cross Sectional Processes and Procedures for Data Generation

In view of the current challenge of the interdisciplinary and integrative network management described above and against the backdrop of the next funding phase of the Cluster of Excellence, there is a need for further research with respect to achieving efficient linking of academic processes and therefore developing a model for the management of Cross Sectional Processes in the context of academic networks. In the example of the "Integrative Production Technology for High Wage Countries" cluster, the Cross Sectional Processes are developing a model that shows specific measures to promote interdisciplinary cooperation and the development of

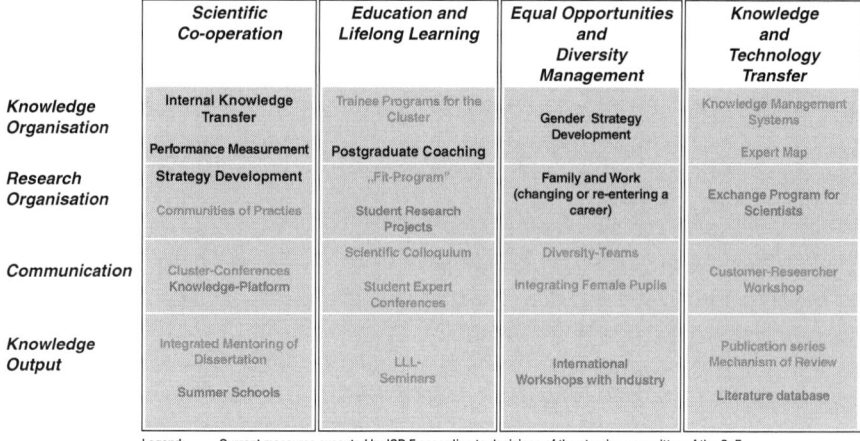

Fig. 2.23 Toolbox of the cross sectional measures

the whole cluster. The aim is to transfer this to comparable networks and clusters.[3] A toolbox comprising cross sectional measures and addressing various levels of the internal cluster academic cooperation (knowledge organisation, research organisation, communication and knowledge output) is being implemented to this effect. The field-tested measures that proceeded from the SENEKA project of the BMBF[4] ("service networks for further education and training processes") have been adapted to meet the requirement of an engineering research cluster. Based on wide project experience with respect to linking heterogeneous actors, the first measures in the toolbox were implemented with the agreement of the steering committee and the management board of the Cluster of Excellence (cf. Fig. 2.23).

For example, since the start of the programme two colloquia for employees have been held each year as part of the area "Internal knowledge transfer". One of their functions is to provide transparency regarding current research and work processes. With regard to "Performance measurement", an annual Balanced-Scorecard-based

---

[3] At this point it should be mentioned that every research project, including the example quoted here of the Cluster of Excellence "Integrative Production Technology for High Wage Countries" at RWTH University in Aachen, is subject to certain project-specific constraints (e.g. spatial and temporal constraints) and that there is no standardised pattern for projects in interdisciplinary research networks. This means that an effective model must adapt to the individual contextual conditions of an academically intensive organisation.

[4] SENEKA was given a project duration of five years (1999–2004) within the framework of the pilot project initiative "Making Use of Internationally Available Knowledge for Educational and Further Educational Innovation Processes" of the German Federal Ministry of Education and Research (BMBF). The participants were 20 small and medium-sized enterprises, 3 large companies, 6 research institutes, and associated partners included more than 38 organisations and enterprises in Germany and the rest of the world (further information is available online at http://www.seneka.rwth-aachen.de/).

evaluation takes place over the full duration of the project. In addition, the area "Strategy development" deals with the organisation of strategy workshops. "Postgraduate coaching" takes place using a cluster-specific advanced education programme with seminars on project management and academic tasks in technology-orientated action systems. Moreover, the "Gender strategy development" area is realized apart from other means by a series of lectures with the topic "I did it my way–career paths from university into industry". If one looks at the cross sectional measures in current use in Fig. 2.23, the focus is primarily set on the segment of scientific cooperation, stressing the overall challenge of cluster coordination.

A methodical approach that contains both quantitative and qualitative elements is needed in order to examine the success of these measures against the background of setting up a management model. For this reason three different methods were used, details of which are given below.

2.7.3.1 Indirect Evaluation

The first method represents the conversion of a cluster-specific Balanced-Scorecard-approach that relates to a model by Kaplan and Norton [1992] and describes a "performance measurement system" that was originally designed as a communication, information and learning system for companies (Kaplan and Norton 1992). The approach was adapted to suit the needs of an academic Cluster of Excellence (for example, the evaluation of knowledge and cooperation processes is at the forefront of the approach) and has been implemented at RWTH Aachen both for the "Integrative Production Technology for High-Wage Countries" cluster and also, in comparable form, for the "Tailor-Made Fuels from Biomass" cluster (Welter et al. 2010). The approach is characterised by the measurement and comparison of performance indicators at various levels such as that of the sub-project leaders or management board and is based on a survey among all cluster participants to measure, among other things, the quality of academic cooperation or the academic output. The approach that is specific to the Cluster of Excellence follows that of Kaplan and Norton's original Balanced Scorecard and constitutes a "performance measurement system". It includes four adapted perspectives that relate directly to the overall vision of the cluster, namely the solution of polylemma of production technology. The perspectives are:

- the internal perspective/research cooperation,
- the learning and development perspective,
- the output or customer perspective,
- and the financial perspective.

The breaking down of the overall vision into four perspectives and sub-objectives enables the measurement of academically orientated performance indicators such as the degree of interdisciplinary academic exchanges or the satisfaction of the employees. An internal cluster evaluation takes place annually in order to measure the achievement of predefined sub-objectives and targets. This annual iteration enables information to be gathered and analysed and, working with the managers of the cluster, recommendations for action to be worked out. This is how the performance of the cluster as a whole is reflected and a controlling of the academic network is made possible.

To date three evaluations with an average of 117 participants have been implemented. Unlike Kaplan and Norton, who do not describe a specific method for the generation and collection of performance indicators, a semi-standardised questionnaire is used. This was developed by the persons in charge of the Cross Sectional Processes and the management board of the cluster. In this manner, performance indicators such as the cooperation quality of the academic network were made measurable. Descriptive statistics were used to analyse and compare the indicators. Figures including the generated arithmetic means and standard deviations of the individual questions are used as a basis of annual comparison in order to derive recommended actions for steering the Cluster of Excellence.

2.7.3.2 Direct Evaluation

Added to the annual Balanced-Scorecard-based evaluation, which represents an indirect form of evaluation, cluster-specific meetings and other events (e.g. colloquia for employees) are evaluated directly with semi-standardised questionnaires by the participants at the end of the event concerned. This method is used to obtain additional feedback on the use of individual cross sectional measures at any particular stage of the network development of the Cluster of Excellence.

2.7.3.3 Guided Expert Interviews

Compared with the two previous approaches of performance measurement and other forms of direct evaluation, guided expert interviews are a purely qualitative method of generating data. For the Cluster of Excellence, 25 interviews were conducted at various levels of the hierarchy, covering members of all sub-projects in the cluster. Due to the fact that professors, leading researchers, members of the management, sub-project supervisors and research assistants were interviewed it was possible to obtain comprehensive information about already implemented and potential future cross sectional measures.

2.7.3.4 Overall Analysis of all Data and Triangulation

With the grounded theory approach, a theoretically based style of research from the Chicago school of symbolic interactionism, it is possible to use, for example, interviews, field observations, documents and statistics to develop a theory ("*grounded theory*") based on the data (Glaser and Strauss 2008; Strauss and Corbin 1996; Strübing 2008). This "grounded theory" method in qualitative social research is widely used for developing organisational management theories and models that reflect a complex social reality. For this reason, it is particularly important to collate the data that has been previously generated by using the three methods described above. During this triangulation process, hypotheses are formulated that describe reports on advancing and restricting factors of network development and aim to secure the high quality of the academic output at any given stage of the network's development (initiation, stabilisation, continuation) (Ahrens et al. 2004).

To obtain further insights into the internal structures of clusters of excellence, an external analysis is also looking at the cross sectional data and processes of other clusters of excellence in Germany.

2.7.3.5 Preliminary Results

The preliminary results of the annual performance measurement, the direct evaluations and the guided expert interviews highlight various aspects that must be taken into account in order to support the high quality of the academic output of the Cluster of Excellence.

Based on the triangulation, the data make obvious that more intensive cross sectional activities at the level of integrative cluster domains[5] are important for a successful cluster development. These should preferably be implemented in the initiation phase of the cluster and can, for instance, take place during regular meetings of the staff in order to promote exchange of the sub-project leaders and research assistants in a domain.

As a reaction to the results of the Balanced-Scorecard-based evaluations carried out in 2008–2010, it was necessary to optimise the linking of various integrative cluster domains. A corresponding form of a network with a stronger content has been implemented by the management of the cluster with so-called 'demonstrators'. These can be regarded as highly networked sub-projects of the cluster that integrate research assistants from different integrative cluster domains and/or individual industrial partners. In doing so a greater level of scientific networking can be achieved on the one hand and a more intensive inclusion of applied research is realized on the other hand.

The results of the Balanced Scorecard highlight the fact that the personal benefit to academics of attending cluster-specific meetings diminishes as the number of persons attending the meeting increases. If this aspect is borne in mind, then regular meetings of smaller groups (e.g. sub-projects and integrative cluster domains) are beneficial to internal cluster cooperation, compared with frequent meetings in the style of a plenary session.

In view of the reports from the guided interviews, the role of the (sub-) project leaders requires a redefinition according to their role as *"knowledgeable agents"* (Sauer 2005). This is the principal group of people who integrate and transfers

---

[5] In the Cluster of Excellence "Integrative Production Technology for High-Wage Countries", interdisciplinary research is processed in six integrative cluster domains (ICDs).

project-specific knowledge, both into the relevant integrative cluster domains and into the overall Cluster of Excellence. Cluster-specific further education offers preparing (sub-) project leaders to fulfil their task in the cluster must therefore be explicitly implemented and communicated in the initiation phase and supported by the individual institutes. Consequently the project leaders must also be offered the capacity to reduce the trade-off between work in the Cluster of Excellence and work within the context of their other obligations to the institute.

Moreover, the data analysis underlines that in spite of a highly rated use of digital tools for the networking of actors, the tools currently implemented (such as the common data platform) are not regularly used. For this reason, possible approaches to a solution include greater integration of the internal cluster data platform into the official cluster home page, and the integration of a semantic network and methods of knowledge retrieval into the internal data platform in order to associate content intelligently with individual academics and projects. A more interactive form of digital networking is also desirable by the cluster participants according to the analysis of data, as this can have a beneficial effect on increasing the intensity and quality of cooperation.

The regular use of the Balanced-Scorecard-approach has proved to be an important cross sectional measure, as it allows the research activities of the Cluster of Excellence to be shown by a comparable set of (particularly) quantitative indicators and data at all stages of the network's development. The performance measurement system supports the reflection and redesign of already implemented activities and in addition enables the correction of medium- and long-term management strategies. As to the process in which long-term strategies are critically examined by internal management, Kaplan and Norton refer to *"double-loop learning"* (Kaplan and Norton 1997).

In view of the production engineering-orientated focus of the cluster, the Balanced Scorecard approach serves to measure the status quo of the cluster performance regularly in order to support, among other things, the high quality of the academic output or the training of excellent academics. Efforts must be made to further adapt the tool to meet the current requirements of a production engineering-orientated cluster by holding regular workshops with management, as this makes the results easier for all groups of actors to use. A stronger bottom-up approach during the definition of cluster-specific indicators for future projects is also advisable in order to improve the intrinsic motivation of all the actors in the Cluster of Excellence to play a greater part in the ongoing Balanced Scorecard-based evaluation.

With regard to the cross sectional measures to be used in a Cluster of Excellence it should be noted that the high staff turnover of the university environment leads to a delay in the development of the characteristic network phases (initiation, stabilisation and continuity). Network management instruments such as plenary sessions, strategy workshops, colloquia for employees and also data and knowledge management systems therefore compensate for the effects of personnel fluctuations at all stages of network development, as these also serve to transfer knowledge within the cluster.

2.7.4 Industrial Relevance

Within the framework of the Cross Sectional Processes a management model for the networking of scientific orientated clusters of excellence was developed with the aim of transferring these management practices to comparable networks and clusters. In the light of the sustained excellence initiatives throughout Germany, this is particularly promising, as there is a great need for research on a scientific concept of processing network tasks. Given the global trend towards enhanced interdisciplinary work in projects, the model for managing Cross Sectional Processes, like the tools that have been examined for networking purposes in the context of the Cluster of Excellence, can be evaluated as industrially relevant. Partners from industry are being included in project networks to an augmenting extent and co-operations within them need to be successfully initiated and made continuous.

References

Ahrens D (2003) Was sind Netzwerke? In: Henning K, Oertel R, Isenhardt I (eds) Wissen – Innovation – Netzwerke. Wege zur Zukunftsfähigkeit. Springer, Berlin, pp 44–55

Ahrens D, Frank S, Franssen M, Riedel M, Schmette M (2004) Phasen der Netzwerkentwicklung und des Netzwerkmanagements. Das Netzwerk-Kompendium – Theorie und Praxis des Netzwerkmanagements. Shaker, Aachen, pp 17–24

Beckermann S et al (2008) Roadmapping: Geschäftserfolg durch zielgerichtete Technologieentwicklung. In: AWK Aachener Werkzeugmaschinen-Kolloquium Wettbewerbsfaktor Produktionstechnik. Apprimus Verlag, Aachen, pp 513–536

Bogaschewsky R, Steinmetz U (1999) Effizienzbetrachtungen in der Theorie der betrieblichen Produktion. Dresdner Beiträge zur Betriebswirtschaftslehre 22

Brinksmeier E et al (2006) Advances in modeling and simulation of grinding processes. Ann CIRP 55(2):667–696

Brun A, Zorzini M (2009) Evaluation of product customization strategies through modularization and postponement. Int J Prod Econ 120:205–220

Buhl C, Meier zu Köcker G (2009) Kategorien von Netzwerkservices. Bundesministerium für Wirtschaft und Technologie Innovative Netzwerkservices. Netzwerk- und Clusterentwicklung durch maßgeschneiderte Dienstleistungen, Berlin. Online verfügbar unter: http://www.kompetenznetze.de/service/bestellservice/medien/publikation_netzwerkservices_internetversion.pdf

Chen CC (2008) An objective-oriented and product-line-based manufacturing performance measurement. Int J Prod Econ 112(1):380–390

Chenery HB (1949) Engineering production functions. Q J Econ 63:507–531

Corsten H, Gössinger R (2009) Produktionswirtschaft. Einführung in das industrielle Produktionsmanagement, 12. Aufl. Oldenbourg, München

Deutsche Industrie- und Handelskammer (ed): Auslandsinvestitionen in der Industrie (2010) Ergebnisse der DIHK-Umfrage bei den Industrie- und Handelskammern; DIHK, Berlin, Brüssel

Dyckhoff H (2009) Produktionstheorie. Grundzüge industrieller Produktionswirtschaft, 5. Aufl. Springer, Berlin

European Industrial Research Management Association (ERIMA) (ed) (1997) Technology roadmapping – delivering business vision, Working Group Report. ERIMA, Paris

Eversheim W, Schuh G (1996) Produktion und Management; Betriebshütte. 7. Aufl. Springer, Berlin

Eversheim W, Schmidt K, Weber P (2002) Virtualität in der Wertschöpfungskette – Durchgängig von der Produktentwicklung bis zur Produktionsplanung. wt Werkstattstechnik Online 92(4):149–153

Eyers DR, Dotchev KD (2009) Rapid manufacturing for mass customisation. In: IPROMS conference proceedings 2009, Cardiff, UK

Fandel G (2005) Produktion I. Produktions- und Kostentheorie, 6. Aufl. Springer, Berlin

Feiner A et al (2005) Werkzeugmaschinen für die Produktion von morgen im Spannungsfeld: flexibel und einfach, schnell und genau. In: Eversheim W et al (eds) Wettbewerbsfaktor Produktionstechnik, Aachener Werkzeugmaschinen Kolloquium 2005. Shaker, Aachen, pp 373–409

Gausemeier J, Frank U, Schmidt A, Steffen D (2006) Towards a design methodology for self-optimizing systems. In: Advances in design, Springer series in advanced manufacturing, Part 1. Springer, London, pp 61–71

Giddens A (1984) The constitution of society. Outline of the theory of structuration. Polity Press, Cambridge

Glaser B, Strauss A (2008) Grounded theory. Strategien qualitativer Forschung. Huber, Bern

Gottschalk S, Kupke D, Lohse W, Vitre M, Wesch C (2007) Virtuelle Produktionssysteme. In: Schuh G, Klocke F, Brecher C, Schmitt R (eds) Excellence in Production. Apprimus Verlag, Aachen, pp 75–88

Grubbström RW, Olhager J (1997) Productivity and flexibility: fundamental relations between two major properties and performance measures of the production system. Int J Prod Econ 52:73–82

Gutenberg E (1983) Grundlagen der Betriebswirtschaftslehre, Bd 1 Die Produktion, 24. Aufl. Springer, Berlin

Heinen E (1983) Betriebswirtschaftliche Kostenlehre, 6. Aufl. Gabler, Wiesbaden

Hunecke H, Sauer J (2003) Innovation in Netzwerken. In: Henning K, Oertel R, Isenhardt I (eds) Wissen – Innovation – Netzwerke. Wege zur Zukunftsfähigkeit. Berlin, Springer, pp 55–60

Jooß C, Welter F, Richert A, Jeschke S, Brecher C (2010) A Management approach for Inter-disciplinary research networks in a knowledge-based society – case study of the Cluster of Excellence "Integrative Production Technology for High-Wage Countries". In: Proceedings of the international conference of education, research and innovation (ICERI) 2010, Madrid. Online access: http://library.iated.org/authors/"Claudia+Jooß"

Kaplan R, Norton D (1992) The balanced scorecard measures that drive performance. Har Bus Rev 001:71–79

Kaplan R, Norton D (1997) Balanced Scorecard – Strategien erfolgreich umsetzen. Schäffer-Poeschel, Stuttgart

Klocke F, König W (2006) Fertigungsverfahren 4, Urformen, 5. Aufl. Springer, Berlin

Klocke F, König W (2008) Fertigungsverfahren 1, Drehen, Fräsen, Bohren, 8. Aufl. Springer, Berlin

Klocke F, Zeppenfeld C, Roderburg A (2007) Hybride Produktionssysteme. In: Schuh G, Klocke F, Brecher C, Schmitt R (eds) Excellence in Production. Apprimus Verlag, Aachen, pp 89–106

Klocke F, Timmer A, Bäcker V, Tönissen S, Frank P, Kauffmann P, Wächter D, Hollstegge D (2009) Manufacturing-related product properties: challenges and changes. 1st international conference on product property prediction, Bochum, pp 667–696

Kloock J (1969) Betriebswirtschaftliche Input- und Output-Modelle. Gabler, Wiesbaden

Küpper HU (1979) Dynamische Produktionsfunktionen der Unternehmung auf der Basis des Input/Output-Ansatzes. Z Betriebswirt 49

Lau A (2005) Going International. Erfolgsfaktoren im Auslandsgeschäft. Zusammenfassung der wichtigsten Ergebnisse. DIHK – Deutscher Industrie- und Handelskammertag, Berlin

Leontieff W (1951) The structure of the American economy, 1919–1939. New York

Matthes W (1979) Dynamische Einzelproduktionsfunktion der Unternehmung (Produktionsfunktion vom Typ F). Betriebswirtschaftliches Arbeitspapier Nr. 2, Seminar für Fertigungswirtschaft der Universität zu Köln

Nyhuis P, Wienhahl HP (2010) Ansatz zu einer Theorie der Produktionstechnik. ZWF 105, Jg 1–2. Carl Hanser, München, pp 15–20

Orilski S, Beckermann S, Schuh G (2008) Technology roadmapping: aligning technology planning. In: IAMOT 2008 proceedings of 17th international conference on management of technology, Dubai, UAE

Phaal R, Farrukh CJP, Probert DR (2004) Technology roadmapping – a planning framework for evolution and revolution. Technol Forecast Soc Change 71:5–26

Potinecke T, Slama A (2005) Virtual engineering optimal nutzen. Ind Manag 21(2):55–58

Röbbecke M, Simon D, Lengwiler M, Kraetsch C (2004) Inter-Disziplinieren. Erfolgsbedingungen von Forschungskooperationen. Ed. Sigma, Berlin

Rohrbeck R, Heuer J, Arnold H (2006) The technology radar – an instrument of technology intelligence and innovation strategy. In: The 3rd IEEE international conference on management of innovation and technology, Singapur, pp 978–983

Rose G, Treiver V (2005) Offshoring of R&D – examination of Germany's attractiveness as a place to conduct research. DIHK, Berlin

Sauer J (2005) Förderung von Innovationen in heterogenen Forschungsnetzwerken und Evaluation am Beispiel des BMBF-Leitprojektes SENEKA. Aachener Reihe Mensch und Technik, Wissenschaftsverlag Mainz

Schuh G, Gottschalk S (2008) Production engineering for self-organizing complex systems. Prod Eng 2(4):431–435

Schuh G, Klappert S, Orilski S (2011a) Technologieplanung. In: Schuh G, Klappert S (eds) Technologiemanagement. Handbuch Produktion und Management 2. Springer, Berlin, pp 171–222

Schuh G, Kreysa J, Orilski S (2009) Roadmap "Hybride Produktion". ZWF – Z Wirtsch Fabrikbetr 104(5):385–391

Schuh G, Orilski S (2007) Roadmapping for competitiveness of high wage countries. In: Proceedings of the XVIII ISPIM Annual Conference,Warschau, Polen

Schuh G, Orilski S, Kreysa J (2007) Chancen für die Produktion in Hochlohnländern. In: Schuh G, Klocke F, Brecher C, Schmitt R (eds) Excellence in Production. Apprimus Verlag, Aachen, pp 31–53

Schuh G, Orilski S, Schubert J (2011b) Roadmapping for the virtual production. In: Proceedings of the international conference on advances in production management systems (APMS 2010). Cernobbio, Italien

Schultmann F (2003) Stoffstrombasiertes Produktionsmanagement. Betriebswirschaftliche Planung und Steuerung industrieller Kreislaufwirtschaftssysteme. Schmidt, Berlin

Schwalbach J (2004) Produktionstheorie. Wahlen, München

Schweyer HE (1955) Process Engineering Economics. McGraw-Hill, New York

Siebert H (2006) Ökonomische Analyse von Unternehmensnetzwerken. In: Sydow J (ed) Management von Netzwerkorganisationen. Beiträge aus der „Managementforschung". Gabler, Wiesbaden, pp 7–27

Smith V (1961) Investment and Production. Cambridge

Sonntag S, Kistner KP (2004) Die Gutenberg-Produktionsfunktion. Eigenschaften und technische Fundierung, 1. Aufl. DUV, Wiesbaden

Stackelberg H (1932) Grundlagen einer reinen Kostentheorie. Springer, Wien

Statistisches B (2010) Bevölkerung und Erwerbstätigkeit. Struktur der sozialversicherungspflichtigen Beschäftigten. Stand: 31. Dezember 2009. Statistisches Bundesamt, Wiesbaden 2010

Statistisches B (2011) Produzierendes Gewerbe. Beschäftigung und Umsatz der Betriebe des Verarbeitenden Gewerbes sowie des Bergbaus und der Gewinnung von Steinen und Erden 2010. Statistisches Bundesamt, Wiesbaden

Steven M (1998) Produktionstheorie. Gabler, Wiesbaden

Strauss A, Corbin J (1996) Grounded theory: Grundlagen qualitativer Forschung. Beltz, Weinheim

Strübing J (2008) Grounded theory: Zur sozialtheoretischen und epistemologischen Fundierung des Verfahrens der empirisch begründeten Theoriebildung. VS Verlag, Wiesbaden

Suwignjo P, Bititci US, Carrie AS (2000) Quantitative models for performance measurement system. Int J Prod Econ 64:231–241

Sydow J (2003) Dynamik von Netzwerkorganisationen: Entwicklung, Evolution, Strukturation. In: Hoffmann WH, Grün O (eds) Die Gestaltung der Organisationsdynamik. Konfiguration und Evolution; Festschrift für Professor Dipl.-Kfm. Dr. Oskar Grün zum 65. Geburtstag. Schäffer-Poeschel, Stuttgart, pp 327–356

Sydow J (2006) Management von Netzwerkorganisationen – Zum Stand der Forschung. In: Sydow J (ed) Management von Netzwerkorganisationen. Beiträge aus der „Managementforschung". Gabler, Wiesbaden, pp 385–469

Sydow J, van Well B (2006) Wissensintensiv durch Netzwerkorganisation – Strukturationstheoretische Analyse eines wissensintensiven Netzwerks. In: Sydow J (ed) Management von Netzwerkorganisationen. Beiträge aus der „Managementforschung". 4., aktualisierte und erw. Aufl. Gabler, Wiesbaden, pp 143–186

Sydow J, Zeichhardt R (2009) Bedeutung von Netzwerkservices für den Erfolg von Netzwerken. Bundesministerium für Wirtschaft und Technologie Innovative Netzwerkservices. Netzwerkund Clusterentwicklung durch maßgeschneiderte Dienstleistungen, Berlin, pp 21–29

Tseng MM, Jiao J (1996) Design for mass customiziaton. Ann CRIP 45(1):153–156

Tseng MM (2003) Industry development perspectives: global distribution of work and market. CIRP 53rd General Assembly, Montreal, Canada

Vinkemeier R (2008) Gesamtkonzept zur langfristigen Steuerung von Innovationen – Balanced Innovation Card im Zusammenspiel mit Roadmaps. In: Möhrle MG, Isenmann R (eds) Technologie-Roadmapping. Springer, Berlin, pp 279–295

Walsh ST (2004) Roadmapping a disruptive technology: a case study. Technol Forecast Soc Change 71:161–185

Wellensiek M, Schuh G, Hacker PA, Saxler J (2011) Technologiefrüherkennung. In: Schuh G, Klappert S (eds) Technologiemanagement. Handbuch Produktion und Management 2. Springer, Berlin, pp 89–169

Weller W (2008) Automatisierungstechnik im Überblick. Was ist, was kann Automatisierungstechnik? Beuth, Berlin

Welter F, Vossen R, Richert A, Isenhardt I (2010) Network management for clusters of excellence: a balanced-scorecard approach as a performance measurement tool. Bus Rev Camb 15(1):171–178

Wibe S (2004) Engineering and economic laws of production. Int J Prod Econ 92:203–206

Wiendahl H-P (2005) Betriebsorganisation für Ingenieure, 5. Aufl. Hanser, München

Windeler A (2002) Unternehmensnetzwerke. Konstitution und Strukturation. Durchgesehener Nachdruck (1. Aufl. 2001). Wiesbaden

Chapter 3
Individualised Production

Günther Schuh, Marek Behr, Christian Brecher, Andreas Bührig-Polaczek, Walter Michaeli, Robert Schmitt, Jens Arnoscht, Arne Bohl, Damien Buchbinder, Jan Bültmann, Andrei Diatlov, Stefanie Elgeti, Werner Herfs, Christian Hinke, Andreas Karlberger, Daniel Kupke, Michael Lenders, Christopher Nußbaum, Markus Probst, Yann Queudeville, Jerome Quick, Henrich Schleifenbaum, Michael Vorspel-Rüter and Christian Windeck

Contents

3.1	Research Programme of Individualised Production	78
3.2	Integrative Assessment and Configuration Logic for Production Systems	80
	3.2.1 Abstract	81
	3.2.2 State of the Art	84
	3.2.3 Motivation and Research Question	86
	3.2.4 Evaluation Model and Configuration Logic for Production Systems	87
	3.2.5 Industrial Relevance	130
	3.2.6 Future Research Topics	133
3.3	Tool-Less Production Technologies for Individualised Products	135
	3.3.1 Abstract	135
	3.3.2 State of the Art	135
	3.3.3 Motivation and Research Question	143
	3.3.4 Results	144
	3.3.5 Industrial Relevance	167
	3.3.6 Future Research Topics	170
3.4	Modularisation for Die-based Production Technologies	170
	3.4.1 Abstract	170
	3.4.2 State of the Art	174
	3.4.3 Motivation and Research Question	177
	3.4.4 Results	178
	3.4.5 Illustration of Industrial Relevance	205
	3.4.6 Future Research Topics	207
3.5	Efficient Order Processing for Individualised Machine Tools Through Modelling of Product Families	209
	3.5.1 Abstract	209
	3.5.2 State of the Art	211
	3.5.3 Motivation and Research Question	216
	3.5.4 Results	217
	3.5.5 Industrial Relevance	228
	3.5.6 Future Research Needs	230
References		231

G. Schuh (✉)
Werkzeugmaschinenlabor WZL der RWTH Aachen, Steinbachstr. 19, 52074 Aachen, Germany
e-mail: g.schuh@wzl.rwth-aachen.de

3.1 Research Programme of Individualised Production

The concept of mass customisation connects the opposed production concepts of mass production and customisation to enable an economical production of customer-specific products (Piller 2006; Pine 1993). The individualised production enables companies to do this and contributes to the solution of the dilemma between the economies of scale and economies of scope by delivering suitable approaches and solution concepts.

The aspect of individualised production has to be considered from two perspectives. From the customer's point of view, the arising question concerns the extent of desired individualisation. An increasing extent of individualisation enables the fulfilling of heterogeneous needs. At the same time individualisation and comprehensibility of the offered product variety by the customer must be ensured (Schuh et al. 2011a; Schwartz et al. 2002). From the producer's point of view, it needs to be decided what level of individualisation is necessary to sustainably serve a market of adequate size. At the same time it is important to control the variety of products and processes in terms of cost (Schuh et al. 2011a, 2011b).

The answer to the aforementioned questions interacts with the reached level of coordination of structure-forming elements in the product production system: For example only with an aligned architecture of products and processes it is possible to meet a high level of individualisation economically. Therefore the main research question regarding the field of "Individualised Production" is:

> How can product and process structures be matched in order to manufacture small quantities in a significantly more cost-efficient manner?

The customers' requests for individualisation may vary enormously among different product categories, which can only be influenced in a limited way. Because of that, this research issue is essential to production research. It addresses the necessity to enable producing companies in high-wage countries to produce at an increasingly higher degree of individualisation in an economically relevant range (Fig. 3.1).

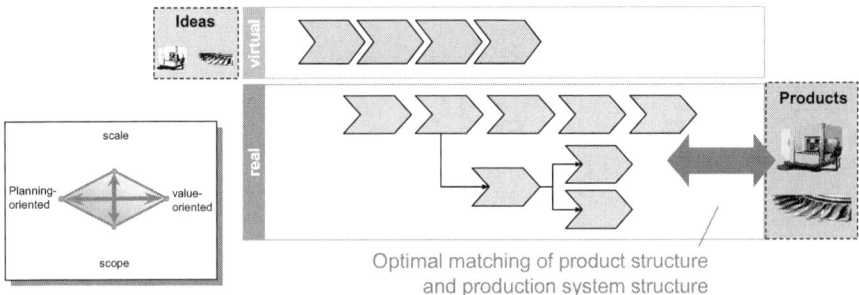

Fig. 3.1 Research question of individualised production

From the logistics point of view, the ideal of individualised production is represented by the concept of one-piece-flow. One-piece-flow describes a buffer-less manufacturing, in which the work pieces or products are transferred from work station to work station without waiting time with a batch size of 1. As a result, non-efficient periods and inventory costs are minimised. At the same time variability of the products can be maximized within the technological and physical boundaries. One-piece-flow is a typical concept for small batch sizes and piece production. For larger batch sizes—particularly for economically relevant production—one-piece-flow concepts can yet only be applied insufficiently.

The objectives of the one-piece-flow specify the research question. The questions refer to the optimal layout of the elements of a production system, i.e. how could the structure of a product and the structure of a production system be matched in a way that a sufficiently optimal operating point can be reached according to the concept of one-piece-flow? For this purpose, explanatory and structural research approaches are necessary. The explanatory research approaches have to point out the relations, dependencies and interactions between the elements of the real production system and the characteristics of the considered product portfolio. The gained knowledge allows to structure research approaches that deal with the development of appropriate strategies for the coordinated design of products and production systems. This can refer to the producer as well as to the customer, who is enabled to design the products in line with his own demands as long as the constraints of feasibility are respected.

The second essential research focus is on the simultaneous increase of productivity and flexibility in the manufacturing technologies, thus concentrating on the technological processes and the corresponding manufacturing machines.

On the one hand, this means that production technology for the individual production must be qualified for producing larger quantities. Classic examples for such production technologies are rapid manufacturing processes that are able to manufacture components directly from CAD data without using any product-specific tools. However, those mostly generative production technologies feature the fundamental disadvantage that their processing speed is too slow for series production and therefore causes high manufacturing costs. By increasing the build-up rates of these technologies, a reduction of the aforementioned dilemma can be achieved.

The manufacturing technology "selective laser melting" is one well known representative of generative production processes. Layers of metal powder are sequentially applied on a base body and then locally melted by a laser, making it possible to produce nearly any component geometry. Therefore, as part of the Cluster of Excellence, extended possibilities of an increase of productivity of this technology are developed. A first increase can be achieved by using a laser beam source with a higher beam quality and a higher laser output power. In addition, beam optics are developed that are able to work with lasers with diverse focus diameters. Thus, a laser with a high output power and a large focus diameter can be used to reach a high build-up rate with low accuracy demands, while simultaneously applying a laser with a smaller focus diameter in order to build up filigree areas of the component.

Besides an increase of performance of generative manufacturing processes a individualisation of mass production is tracked. Manufacturing plants for mass production consist of two complementary elements: the machine tool that allows the production of diverse components within defined limits and the product-specific tool itself.

The classic engineering process in machine tool manufacture is predominantly characterised by the sequential design of the machine tool's mechanics, its electronics and its control software. Since most customer-specific orders feature a high proportion of recurring functions, individual machine tools can mostly be manufactured using existing, parameterisable components. Especially in the design and production of variants and modified constructions, required documentation such as drawings, part lists, circuit diagrams and control programs can automatically be compiled using modular design systems. Within the Cluster of Excellence, such mechatronic-orientated modular design systems for the processing of orders are subject to research.

Despite their high degree of individuality, the tools themselves feature recurring functions as well and can therefore be produced using pre-manufactured components made in small batch sizes. By focussing on the forming parts of the tool, new design engineering can be avoided and cycle times as well as manufacturing costs are reduced. As part of the research programme of the Cluster of Excellence, new modular design concepts for casting and extrusion tools are evaluated and an approach to automate the design of flow channels of extrusion tools is undertaken.

First results show that in terms of the development process and corresponding methods, the approach of an individualised production requires a new mindset in the design departments and must not be underestimated. On the other hand, it offers good prospects to significantly reduce design efforts and must therefore be subject to further research.

3.2 Integrative Assessment and Configuration Logic for Production Systems

Günther Schuh, Robert Schmitt, Jens Arnoscht, Arne Bohl, Daniel Kupke, Michael Lenders, Christopher Nußbaum, Jerome Quick and Michael Vorspel-Rüter

According to PORTER a company can decide between three generic competitive strategies (Porter 1998). These are the strategy of cost leadership, the strategy of differentiation and the strategy of focussing on niches. Due to its minor relevance, the latter one will be excluded from the following discussion. The strategy of cost leadership is based on the achievement of "economies of scale", by generating advantages from the benefits of cost reduction, learning curve effects and automation. In contrast, the strategy of differentiation focuses on "economies of scope" by meeting customer-specific requirements with individualised products. However, this strategy generally leads to an increase in the complexity of products and processes. In the past it was assumed that these two strategies were mutually exclusive, as an increase in the economies of scale basically leads to a reduction in the economies of scope, and vice versa. But in order to survive in a globalised environment companies in high-wage countries increasingly need to offer custom-tailored products at competitive prices. Therefore the target to be aimed at is to provide individualised products at the cost of mass production and resolving the dilemma between economies of scale and economies of scope. Due to the high level of interdependence, an alignment of all the structural elements regarding both the product and the production system is necessary for the resolution of the referred dilemma (Schmitt et al. 2010; Schuh et al. 2007a, 2010a; Stich and Wienholdt 2009).

Focussing on this over-all alignment, the area of observation in the following will be the whole product and production system (abbreviated as production system). This not only includes the resources and processes of a value creation system, but also the products produced and offered on the market as one coherent entity. In order to tackle the challenges mentioned above, it is necessary to enable measuring and assessing the current position of a given production system in the dichotomy between economies of scale and economies of scope. Based on this a synchronised redesign of the entire production system is made possible. For this purpose, a methodology for the integrative evaluation and design of production systems is presented below.

3.2.1 Abstract

As described above, a significant problem for companies producing in high-wage countries is to resolve the dilemma of combining economies of scope and economies of scale. On the one hand individual customer needs are to be satisfied, while on the other hand companies are trying to achieve benefits of scale by achieving a maximum level of standardisation in production processes. To solve this dilemma the structural elements of a production system need to be aligned in relation to their scale-scope orientation and to be empowered through the implementation of suitable principle solutions to resolve the scale-scope dilemma. As a basis for this an evaluation tool is needed, which can capture all determinative, structural elements of a production system and analyse these in relation to their simultaneous scale and scope capability.

By breaking the production system down into an internal and an external perspective, as well as considering the product and production aspects, four relevant areas of tension can be determined. These areas of tension can be described as the *Product Programme,* the *Product Architecture,* the *Production Structure* and the *Supply Chain.* These four areas of tension interact with each other and therefore require an integrated alignment in order to enable a resolution of the scale-scope-dilemma. For this, each of the four areas of tension is broken down into operationalisable main dimensions, which together form an integrated evaluation model (Fig. 3.2). Originating from the superior scale-scope dilemma, the two dimensions of an area of tension show a dichotomous relationship to each other. The structure of the evaluation model will be examined more closely below.

The *Product Programme* represents the range of products offered from an external customer perspective. Against the stated background there are two determinant,

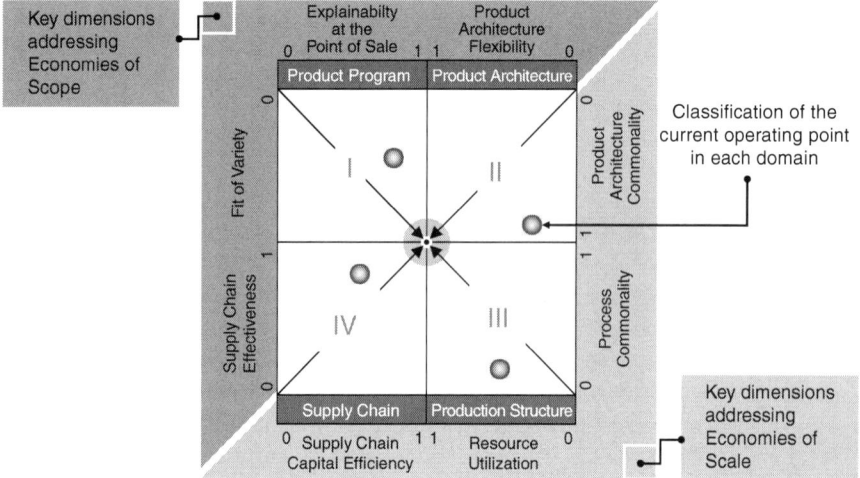

Fig. 3.2 Integrative evaluation model

contradictory objectives which exist for the *Product Programme*: on the one hand the need to develop and produce products meeting individual customer needs and on the other hand the multiplicity of products that this engenders must be easy enough for the sales channel to be explained, so that a customer can identify the correct product variant for himself and understands the added value this gives him. To render both these goals operational the main dimensions of "*Fit of Variety*" and "*Explainability at the Point of Sale*" were defined and quantified by performance indicators.

The second product-related area of tension is described as the *Product Architecture*. This represents the internal structure of the product range in terms of its functional structure and components. This relates to an internal point of view of the product. The main dimensions of this area are the *Flexibility of the Product Architecture* and the *Commonality of the Product Architecture*. By its flexibility the ability of a product architecture is described to allow for a derivation of those product variants demanded by the market. The commonality, on the other hand, relates to the level of sameness or similarity of the product variants in relation to requirements, functions, technologies or components.

The third area of tension, the *Production Structure*, relates to the company-internal resources and the production processes carried out by applying them. The dichotomous main dimensions of the *Production Structure* are *Resource Utilisation* and *Process Commonality*. The dimension of *Resource Utilisation* here includes the efficiency of usage of existing resources. The rating for *Process Commonality*, on the other hand, indicates the level of standardisation of the process flows. There is also a dichotomous relationship between these two main dimensions, as a high level of *Resource Utilisation* generally can only be achieved by dynamic allocation of processing steps. However this collides with the requirement for a high *Process Commonality*.

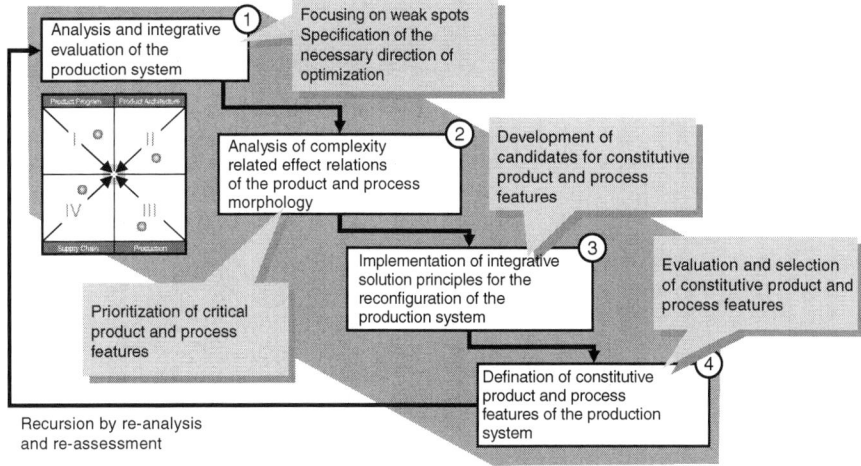

Fig. 3.3 Integrative configuration logic

The last of the four area of tension is described as *Supply Chain* representing the logistical interface to the customers. The main dimensions of this area are the *Supply Chain Capital Efficiency* and *Supply Chain Effectiveness*. The *Capital Efficiency* compares the cost of inventory for the production of the envisaged product range to the expected turnover. The main dimension of *Effectiveness* is determined by the delivery reliability of the Supply Chain.

The four areas of tension and their main dimensions described here combine to form an evaluation model which enables an integrative analysis of a production system. It can be used as a basis to identify bottlenecks relating to complexity and over-capacity. Eliminating these weak points by aligning the structural elements of the production system to each other is supported by an integrative configuration logic (Fig. 3.3) which builds on the evaluation model described above.

Starting from an identification of weak points relating to the resolution of the scale-scope dilemma by applying the evaluation model (1) an analysis is made of the causal relationships caused by complexity, between the four areas of tension (2). This particularly focuses on the product- and production-related variances, as well as their mutual dependencies. As a result of this analysis, for example, it can be derived what effects the characteristics of a product programme-related product feature will have on product architecture, production structure and supply chain. This enables the identification of the triggers of misalignment between the areas of tension in the production system.

In order to mitigate these misalignments, four integrative principle solutions are introduced, each focussing on one of the four quadrants, but positively affecting the production system across multiple areas of tension, and thereby being integrative. *Feature clustering* has been developed as a principle solution specifically for the product programme. This principle solution allows a systematic alignment of the product programme to the needs structure of customer segments. The *Decoupling of the Product Architecture* on the other hand aims at sub-dividing the product into independent functional complexes. In relation to the supply chain the principle solution of *Postponement* is proposed. This aims at shifting the order penetration point to later supply chain stages, and substituting forecast-driven processes for order-driven processes. For a concomitant increase in forecast quality collaborative requirements planning is introduced. In the area of the *Production Structure* the solution principle of *Integrative Balancing* is presented. This approach aims at defining process standards for modular production cells, which enables load balancing between resources.

The results of applying the integrative principle solutions are candidates for so-called *constitutive features* of the production system. These properties define the underlying rules of standardisation for a product-production system. As the final configuration step, these candidates are analysed to determine their contribution to resolving the scale-scope dilemma, and defined as *constitutive features*. The process described in outline here is expanded and described in detail in the following sections.

3.2.2 State of the Art

The dilemma between scale and differentiation benefits has already been the focus of many studies from various technical disciplines. These studies can be basically divided into three main types of approaches: The first type seeks to assess the degree economies of scope or scale achieved (e.g. Schuh 1989; Caesar 1991; Heina 1999; Junge 2004; Cronjäger 2005; Martin and Ishii 1997). The second type of study focuses on the systematic creation of a product or production system, with the intention of increasing economies of scale and scope (e.g. Marti 2007; Kohlhase 1997; Erixon 1998). In addition there are combinations of the two approaches (e.g. Martin 1999; Rathnow 1993; Nyhuis et al. 2008).

A particularly relevant study for the above combination approach is the "Global Variant Production System" (GVP) by Nyhuis et al. (2008). The GVP method aims at designing globally distributed production systems for products with high variety and with scalable production quantities at low costs. This integrative approach considers both the product and the production system equally by the application of four methodical modules: Product Structuring, International Co-operative Relationships, Technological Differentiation, and Production Stages and Logistics Design. Basically each of the four modules can be split into an analysis and a design part. The results of each of the modules is used as an input value or pre-requisite for the other modules of this approach. It is worth stressing that the strength of the GVP approach is the holistic view of product and production. It would, however, be desirable to have an integrative evaluation of the four areas in order to enable a simultaneous and focussed optimisation within the product and production system.

Another methodology which can be classified as a combined approach is the "Structuring and Assessment of Modular Product Platforms" by Kohlhase (1996).

In this product modularisation focused methodology the design process is broken down into three main parts: planning the modular systems, developing the module structure and the development of the modules and modular products. All three elements are supported by inter-connected individual methods. As well as creative individual methods, this methodology includes analysis and evaluation tools. The latter focus particularly on a monetary-economic evaluation of product structures which is supported by an IT-based tool. As the three process elements show high levels of interdependency, Kohlhase proposes an iterative creative process. The evaluation tools here act as the control function for the design of the modules. But the integrative view of all structural elements of the product-production system, which is necessary for the objective being set here, cannot be delivered by this approach. In addition our objective requires a value model, which goes beyond a directly monetary approach.

A further approach which aims at a simultaneous improvement in both economies of scale and economies of scope, is the "Modular Function Deployment" (MFD) by Erixon (1998). This approach supports in particular a systematic modularisation of products. The method consists of five design phases, which build on each other. The first two phases represent common systematic approaches to product development. The core of the method is represented by the "Module Indication Matrix", which evaluates the influence of so-called "Module Drivers" on the components of a product. The "Module Drivers" incorporate the drivers for modularisation of a product and represent criteria for encapsulating product components into modules. The MFD method provides a generic set of twelve "Module Drivers" which support simultaneous improvement in the economies of scale and the economies of scope. The advantage of this method is based on its practicability and facile implementation. But it should be noted that, although a strong emphasis is laid on the product design, the integrative design of product and production system is barely addressed.

A third approach, which particularly addresses an integrative evaluation of the scale-scope dilemma is the "Modularisation Balanced Scorecard" by Junge (2004). This approach adapts the balanced scorecard approach to the evaluation of modular product groups. Using a four-step method the product programme and the product architecture are analysed and evaluated by applying indicators. Junge proposes a generic set of performance indicators for this, which are derived from both development and production system viewpoints. The indicators are consolidated in a Balanced Scorecard, which supports a product design aligned with the modularisation goals. This approach is lacking the necessary integrative aspect, as although it aims at an optimisation of product design, it does not include any optimisation of the corresponding production system.

Similarly the approach by Martin and Ishii (1997) can be classified as an evaluation-based approach. The basis of this approach also is a system of indicators. Like the Design for Variety by Martin (1999) this approach focuses on the product architecture and its optimisation. Further relevant aspects of a product-production system, such as production, are indeed included in the system of indicators, but are not seen as a possible design focus. In addition there is no analysis of interdependencies between product and production, so that this approach has to be qualified as insufficiently integrative regarding the objectives set above.

The approaches outlined do deliver some starting-points for the evaluation and configuration of production systems. However it became clear that, particularly in respect to an integrative evaluation and design of products and production systems, there are still some gaps in methodology.

3.2.3 Motivation and Research Question

The increasing individualisation of products confronts production systems in high-wage countries with a dilemma: On the one hand, they are supposed to maximise the benefits of scale throughout the value chain, on the other hand they must increase benefits to the individual customer. The challenge is to resolve this dilemma.

The economies of scale throughout the value chain are improved by using standards in both products and processes. These relate to the technologies, components, assemblies and resources applied, in both the product and throughout the whole value added chain. On the other hand it is necessary to identify those product features which customers perceive as a differentiation and personalisation and valuate accordingly. For fixed, external variants in the product range offered, therefore, the main challenge is to identify or define those invariable parameters (in both product and process) which can be used to achieve the highest level of standardisation in the product-production system.

The *central research question* therefore is as follows:

How can an optimum level of standardisation be achieved for a product-production system, so as to increase the benefits of reduction in unit-costs (scale) combined with maximum fit to customer needs (scope)?

The so called "optimum level of standardisation" referred to above is dependent to a large extent on the mutual alignment of structural aspects of the production system. For example, only limited benefits of scale can be derived from a highly standardised product programme if the underlying product architecture does not efficiently apply this standardisation. Conversely, achieving differentiation benefits by having a flexible product architecture can be cancelled out by the product being too difficult to explain in sales presentations (see also Schuh et al. 2010a). These interdependencies are shown in the Fig. 3.4 below.

To determine the balance between the structural aspects of a production system which are needed to respond to the main issue of finding the optimum level of standardisation, the following *supplementary questions* need to be resolved:

- *How can a product-production system be described in sufficient detail?*
  - *What are the most important parameters in describing a product-production system?*
  - *What are the cause-and-effect relationships between these parameters?*

3 Individualised Production

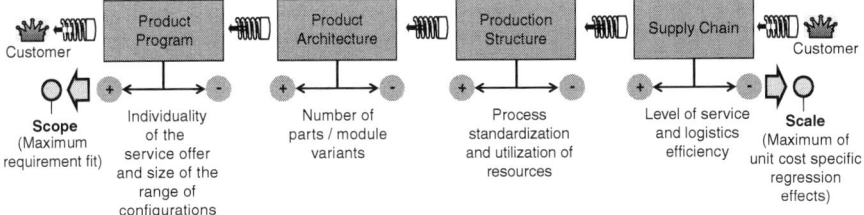

Fig. 3.4 Interaction of structural elements in a production system

- *How can the quality of the configuration of a product-production system be evaluated?*
  - *What are the decisive measures for evaluating the product-production system on the scale-scope dilemma?*
  - *What connections exist between these measures and how do they affect the achievable economies of scale or economies of scope?*
  - *How can an evaluation model be constructed, so that clear recommendations for measures regarding the design/configuration of the product and production system can be derived?*
- *How can a process be designed for the integrative configuration of product and production systems?*
  - *How can the areas of a production system in which optimisation is needed be determined?*
  - *Which process steps are necessary for the integrative configuration of a product-production system?*

These questions are addressed below and answered from the results of research. Initially we will look at the first two groups of questions, focusing the design of an evaluation model for production systems.

3.2.4 Evaluation Model and Configuration Logic for Production Systems

In line with the first two groups of issues, the following section will start by describing an integrative evaluation approach for production systems faced with the dilemma of scale and scope. Based on this evaluation model the next step will present an integrative configuration approach for production systems, which systematically reduces the scale-scope dilemma.

3.2.4.1 Fundamental Structure of the Evaluation Model

The contradictions between economies of scope and economies of scale reside mainly in the fact that product and process complexity arise – a high level of differentiation normally is associated with a loss of benefits of scale. Therefore a key to resolving the dilemma between economies of scale and economies of scope is the ability to manage complexity. In order to create a fundamental organisational framework for the intended evaluation model, the concept of complexity needs to be subdivided into non-overlapping but complete sub-areas.

According to Malik (2000) and Kaiser (1995) complexity can be split into a company-internal dimension and a company-external dimension. External complexity is induced by external conditions such as legal requirements on products and processes, or arises from interaction with the customer. Internal complexity, on the other hand, arises from the company's own structures and processes. In contrast to this sub-division Wiendahl and Scholtissek [1994], differentiate product- and production-related complexity. If both of these structuring levels are combined, four necessary areas of tension can be derived for the evaluation model (Fig. 3.5). These areas of tension are interpreted in sequence below and detailed as elements of the evaluation model.

The creation of the *Product Programme* can be viewed here as an external and product-related area of tension. The Product Programme is the range of products offered on the market as seen from the customer point of view (see first sub-section in 3.2.4.2). An internal view of the product range is provided on the other hand by the *Product Architecture,* as another area of tension in the evaluation model. The *Product Architecture* is to be understood as the internal structure of the product in the form of functions and components (see second sub-section in 3.2.4.2).

From production side, the areas of tension of *Supply Chain* and *Production Structure* can be derived from the framework stated above. The Supply Chain describes the external view of production in the sense of interfaces from the production process to the outside environment (see third and fourth sub-sections in 3.2.4.2). The remaining internal part is understood as the *Production Structure*. This includes the company's

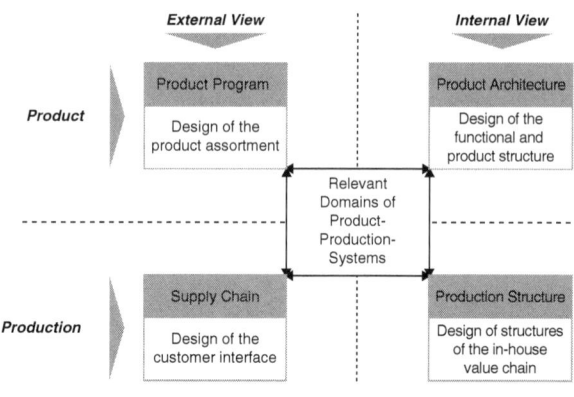

Fig. 3.5 Significant areas of tension in a product-production system

3 Individualised Production

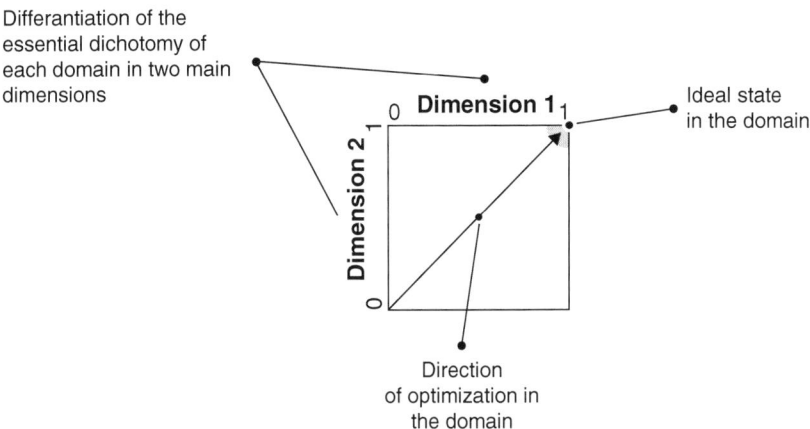

Fig. 3.6 Dichotomous main dimensions and clear direction for optimisation

own value-adding resources, as well as the processes carried out applying them (see third sub-section in 3.2.4.2).

The four areas of tension presented here require a consistent evaluation approach, in order to capture the effects of complexity relating to the *Product Programme, Product Architecture, Production Structure* and *Supply Chain*. To do this the above-mentioned scale-scope-dilemma of a production system is broken down and interpreted specifically for each of these four areas. In each area of tension, the dilemma is represented by two dichotomous main dimensions (Fig. 3.6).

The value range of the main dimensions is standardised to a range between zero and one. This allows to achieve comparability between the results of evaluation—both between individual areas of tension and across companies. The dichotomous main dimensions in the four fields are selected such that if both area-specific main dimensions were maximised simultaneously in all four areas this would mean that the scale-scope dilemma would be resolved for the entire production system. For this reason each of the four areas of the constitutive framework shows a clear direction for optimisation regarding the resolution of the scale-scope dilemma and corresponding to an optimum balance between externally generated product variety and caused internal complexity. The adaptation of this evaluation approach to the framework introduced above is shown in Fig. 3.7.

Using the evaluation model a production system's operating point regarding all four areas of tension can be assessed by quantifying the eight main dimensions. Based on the defined optimum position per field the difference to the theoretical optimum can be determined.

For a more detailed exploration of the evaluation model it is necessary to derive the important dichotomous objectives in each area of tension. On this basis the main dimensions in the four areas will be defined. For each of the four defined areas of tension *Product Programme, Product Architecture, Production Structure* and *Supply Chain*, the two central and dichotomous objectives which are relevant

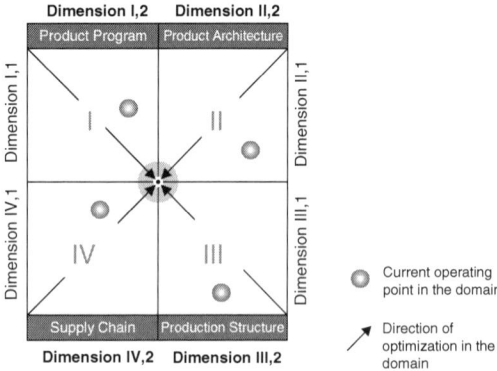

Fig. 3.7 Fundamental structure of the evaluation model

for the resolution of the scale-scope dilemma in connection with the evaluation and management of product and process complexity are presented below.

3.2.4.2 Operationalisation of the four areas of tension

For a more detailed exploration of the evaluation model it is necessary to derive the important dichotomous requirements in the areas of tension, on the basis of which the main dimensions in the four areas are defined. For each of the four defined areas of tension, *Product Programme, Product Architecture, Production Structure* and *Supply Chain*, the two central and dichotomous questions are presented below, as well as relevant indicators are defined and explained.

Product Programme

The *Product Programme* represents the product variants on offer by a company which are visible to customers. Two central issues are of particular importance for the *Product Programme*. The first issue can be framed as follows: "Are the correct product features and variants being offered to ensure maximum coverage of customer requirements?" Alongside this issue the area of tension for the *Product Programme* is defined by the following issue: "Can the offered product variants be explained persuasively to the customer?" Depending on the complexity of the product programme on offer, this is a common problem—the variants offered provide good coverage of customer requirements, but the sales force fails in its ability to explain to the customer the exact product variant needed. On the basis of these two core issues, the following main dimensions are defined for the *Product Programme*.

- Fit of Variety
- Explainability at the Point of Sale

Fit of Variety and *Explainability at the Point of Sale* represent dichotomous values which both equally need to be optimised. An increase in the external product programme complexity creates the basis for satisfying individual customer requirements, by offering product variants which are as customer-specific as possible. But the effect this has is to increase the need for explanation at the point of sale and to make it difficult to ensure sufficient ease of explanation of the product programme. Only a balanced increase in the *Fit of Variety* (increase in the external complexity of the product programme) and of the *Explainability at the Point of Sale* leads to an actual increase in revenues, as in this case the improved variant can still be communicated to the customer.

Fit of Variety

The multiplicity of products on offer should satisfy as many of the customer requirements in the addressed market segment as possible. If necessary adjustments for individual customers can be made to meet specific customer requirements and to reach as many customers as possible. The *Fit of Variety* needs to measure the size of the overlap between the variants requested by customers and those offered by the company within the product programme. To do this it is necessary to evaluate the congruence of the offered and requested product features and options.

The product programme can be visualised as a feature tree, in which the available option combinations supported by the configuration logic can be displayed and thus the product variants within the product programme (Fig. 3.8).

An example from the automobile industry can easily demonstrate how extensively the multiplicity of variants has increased in the last few years. Twenty years ago at Audi the Audi 80 with a 1.3 L engine and 55 hp accounted for 40% of the sales by volume. Currently at Audi there are around 10^{20} different ways to combine the various components of a vehicle, at BMW they are even talking about 10^{32} different vehicle variants (Andres 2006). Given these examples it is clear that it might have been useful to look at the Fit of Variety by starting from the product variants a few years ago,

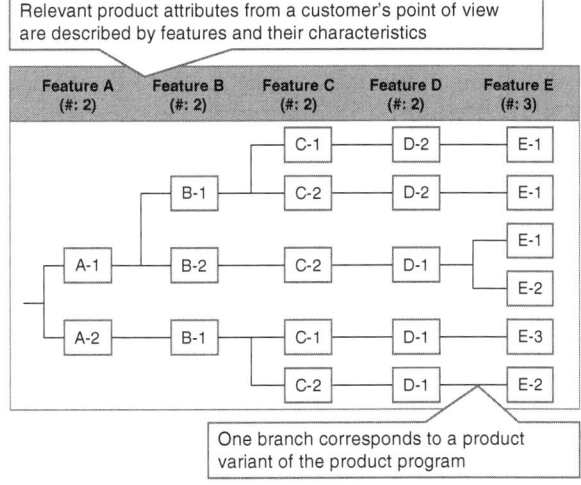

Fig. 3.8 Structuring the product programme in a feature tree

but in light of the common multiplicity of variant options today, an approach of this kind is unlikely to deliver as desired. If the multiple ways of combining features and the theoretically possible product variants this can create are compared with typical sales volumes, it becomes obvious that almost no end-product variant is exactly the same. Therefore, a suitable indicator needs to be defined, by which the variants on offer in the product programme can be sensibly captured and to which the variants actually requested by customers can be compared.

The variant requested by the customers is defined by the feature options of the proposed product programme which are actually sold. For each feature option a defined value is retained for the assumed quantity to be sold at the time of the definition of the feature. For example, an automobile manufacturer will define expected sales levels for various roof covers (grey roof cover, beige roof cover, etc.). This is not only useful for looking at customer acceptance of various finish feature options, but also with a view to procurement and production planning.

For planning features in advance, the configuration logic of the product programme is used as a basis. But only those features, which are relevant for the customer, are taken into account. These are those features and their options, which influence the buying decision. Purely technical features and their options, which are not relevant from a customer point of view, are not taken into account here. To calculate the Fit of Variety the feature options actually sold over a representative period are measured ex post and compared to the options planned for (Fig. 3.9).

The sales volume ratio is a measure of the correction of the product programme planning for the nth option from the product programme on offer. For the evaluation of the *Fit of Variety* all the variations across all feature options must be taken into account (3.1):

$$FV_{PP} = \frac{1}{n} \cdot \sum_{i}^{n} \frac{N_{ist,i}}{N_{plan,i}} \qquad (3.1)$$

FV_{PP} Fit of Variety of the Product Programme
N Number of feature options perceived by customers
$N_{ist,i}$ Actual sales quantity for feature option i
$N_{plan,i}$ Planned sales quantity for feature option i

The value for the Fit of Variety is found by taking the average value of the sales quantity ratios for all—customer-perceivable—feature options (1). For those cases

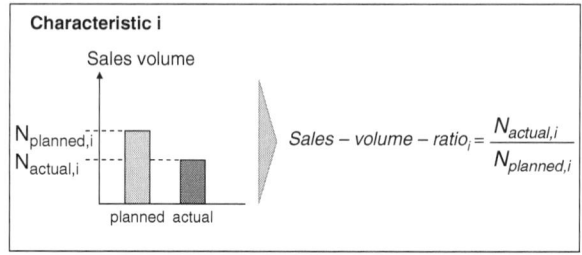

Fig. 3.9 Comparison of plan and actual sales quantities for a feature option

where the actual sales quantity is greater than the planned sales quantity, the sales quantity ratio is set to 1 to simplify, for those cases where the actual sales quantity is smaller than or equal to the planned sales quantity the exact sales quantity ratio is used in the calculation of the *Fit of Variety*. The overshooting of a sales quantity of feature options is therefore not given greater weighting in the calculation of the Fit of Variety than the exact hitting of the sales quantities target, as overall it is not better to have sold a higher quantity of specific feature options than planned. So the optimum value for *Fit of Variety* is achieved if for all feature options the actual and planned sales quantities are the same. The theoretically lowest value which can be achieved for the *Fit of Variety* will occur if no feature options are sold, which is the same case as if not a single product variant was sold. In this case the value for the *Fit of Variety* is zero.

Explainability at the Point of Sale

A major challenge for the sales force consists in being able to inform each potential customer about the best fitting product variant, which meets his/her specific requirements. Two central issues are decisive here. The existing product programme's need of explanation and the ability of the company to explain it need to fit each other. The matching of the need for explanation and the ability to explain are measured by the main dimension of *Explainability at the Point of Sale* which forms the second main dimension in the *Product Programme* area.

The product programme complexity perceived by the customer is mainly determined by the following parameters:

- Number of product features
- Number of product feature options
- Similarity of product feature options
- Number of configuration rules and restrictions in the product programme
- Rate of change of the Product Programme on offer

The number of product features and product feature options which the customer can perceive defines the choices which a customer has in choosing for a particular product variant. Depending on the similarity of the feature options it may become more difficult for the customer to comprehend the differences between the various product variants. In addition the pre-defined configuration logic from the company, with its rules and restrictions on configurations is another aspect which the customer has to understand when choosing a product variant. The rate of change of the product programme on offer also makes the product selection process more difficult for the customer, as over time different features and feature options are included in the product programme or are removed from the product programme.

The customers' expertise, and in particular the ability of the customers in the target market segment of the product programme to understand the product, is a major factor which influences the need for explanation. The knowledge of and ability to articulate the requirements on the part of the customer are a basic pre-requisite for being able to create a fit between the product programme on offer and the customer requirements. Depending on the level of understanding of the product by the customer, the challenge is greater or smaller to translate the customer's requirements into specific product features and options, in order to identify the matching product variant among the product programme. Significant tasks for the sales organisation at the point of sale include understanding the dependencies between customer requirements and the product programme, in order to identify the best-fit product variant as precisely as possible. It must then be possible to explain to the customer the match between requirements and product features as persuasively as possible. The need for explanation thus stands directly opposite the available *Explainability at the Point of Sale*. The ability to explain is primarily determined by the knowledge of the product programme on the part of the sales personnel. Depending on the complexity of the product programme, it may be more or less difficult to ensure solid knowledge of the product programme by the sales channels.

The *Explainability at the Point of Sale* requires the ability to explain and the need for explanation to be in balance. In the ideal case—the need for explanation is fully met by the ability to explain—all the different product variants which could meet each of the specific customer requirements could be explained persuasively to the customer at the point of sale. In order to measure the *Explainability at the Point of Sale* the following formula (3.2) is defined:

$$EF_{PoS} = \frac{\frac{1}{n}\sum_{i}^{n} V_{i,expl}}{V_{abs}} \qquad (3.2)$$

EF_{PoS} Explainability at the Point of Sale
N   Representative number of sales people
V_{expl} Number of product variants, which can be explained by sales person i
V_{abs} Number of all product variants

For the defined indicator the average of the number of products which can be explained by an indicative number of sales persons is given as a ratio to the total number of variants offered. In the ideal case each sales person can explain every product variant convincingly to every customer. This would give the theoretically possible value of one for *Explainability at the Point of Sale*. As the number of features, feature options, their similarity to each other and the rate of change of the product variants on offer increase, it is increasingly likely that the *Explainability at the Point of Sale* will fall.

Product Architecture

The *Product Architecture* provides a model as well as being a fundamental determinant of the product design. This overall model comes from linking two partial models: the functional structure of a product and also its product structure.

The functional structure presents a function-related model of the product. A function is understood as the abstract and generic description of the causal relationship between the inputs and outputs of a product or part of a product (Eversheim 1998; Koller 1998; VDI 1997). If the whole product is considered, the corresponding function is described as the overall function, in other cases we are dealing with a partial function (Pahl et al. 2005). Using functional decomposition the overall function of a product can be broken down into partial functions (Ulrich and Eppinger 2004). Due to their related inputs and outputs the partial functions are linked into a functional structure (Eversheim 1998). The functional structure can therefore be interpreted as a subdivided and networked model of the functions of a product (Hubka 1984).

The product structure is the complementary model to the function structure. The elements of this model are, in contrast to the function structure, the physical components of the product. Depending on the levels of sub-division of the product structure the components can be viewed as assemblies or individual parts of the product. Assemblies here include components with deeper structural levels (Schuh 2005). Rapp and Ungeheuer define the product structure as a model which reflects the composition of the product from parts and their relationships to each other (Rapp 1999; Ungeheuer 1986). Similarly to this definition the product structure includes both the physical interactions between components in the form of interfaces, and the hierarchical structure of a product made up of assemblies and individual parts (Steffen 2007).

As described above, the product architecture presents an overarching model of a product, which brings together the product structure and the function structure by mapping the relationships of the partial functions and components (Göpfert 1998; Ulrich and Eppinger 2004; Ulrich 1995). In relation to the dilemma focussed here, it is particularly the degree of physical and functional decoupling within a product architecture, i.e. the product modularisation, which is a determining characteristic of the product architecture. If a product shows a high degree of decoupling within its architecture, the individual elements or modules can easily be substituted or changed independently (Ulrich and Tung 1991; Ulrich 1995). The basis for this decoupling is a number of rules and regulations at a physical, technological or functional level, which form the common product architecture of the product programme. These rules are referred to below as the *constitutive product features*. Examples for this are standardised module interfaces, or unified product technologies used to implement specific functions. By intelligent definition of a suitable number of constitutive features a high level of standardisation is achieved at the relevant points in the product architecture, while at other points variation parameters are retained to allow for adaptations to market-specific requirements.

In the explanations above, for the sake of simplification, we reduced the scope to the architecture of a single product. In the following the scope must be abstracted to cover the entire product programme. The measure against which the individual product architectures need to be compared will be referred to as commonality. A high level of commonality enables the exploitation of benefits of scale, for example by cross-product use of the same parts (Ulrich 1995). Overall the increase in commonality is achieved by standardising the product architecture. The result is typically a reduction in the flexibility of the product programme and the underlying product architecture, which leads to a drop in the fit to customer requirements and to a loss of differentiation benefits (Neuhausen 2001; Secolec 2005). If instead, the number of variants offered to the customer is increased and with it the flexibility of the product programme, then in general this leads to a fall-off in commonality of the product architecture and to a reduction in benefits of scale (Kaiser 1995).

The *Product Architecture* therefore is the connector between the differentiation benefits linked to the *Product Programme* and the benefits of scale achieved in the *Production Structure* and in the *Supply Chain*. These two trends, and the resolution of the dichotomy between them, represent the core challenge for the product architecture design. In relation to the potential differentiation benefits, the question arises as to whether a product architecture can show sufficient flexibility to be able to satisfy customer requirements adequately. In contrast to this is the question, whether the Product Architecture enables a suitable level of commonality within the product programme to achieve benefits of scale. Starting from these two opposing questions we can derive the two main target values for the *Product Architecture*. These are the *Flexibility* and the *Commonality of the Product Architecture*.

Flexibility of the Product Architecture

Similarly to what was said above, the *Flexibility of the Product Architecture* includes how well this can enable satisfaction of individual customer requirements, without compromising the constitutive features. Relevant for this is how suitably a product architecture and its constitutive features are designed in order to be able to provide the product derivatives and variants which customers are asking for. The indicator for the *Flexibility of the Product Architecture* must therefore measure how many of the relevant product variants desired by the market can be derived without compromise of constitutive features. As Eq. 3.3 shows, this can be represented by the proportion that the product variants derived from product architecture represent in relation to the total number of product variants requested in a particular time frame.

$$F_{PA} = \sum_{i}^{v} \frac{U_{s\,\tan d, i.}}{U_{ges}} \qquad (3.3)$$

F_{PA} Flexibility of the Product Architecture
$U_{s\,\tan d,\,i.}$ Turnover of product variant configuration i, which does not compromise the constitutive features of the product architecture
$U_{ges.}$ Turnover of all product variants

The number of configured product variants here is necessarily smaller or equal to the number of requested variants, so that the theoretical optimum in flexibility would give the result of One. A constraint for calculating the flexibility is the compliance with the constitutive features in the configuration of a product variant from the product architecture. Figure 3.10 shows how the number of configured product variants is derived as the intersection of the configurable and requested product variants. A subset of these again are represented by the product variants derived from the product architecture without compromising the constitutive features.

Fig. 3.10 Delimiting the product variants under consideration

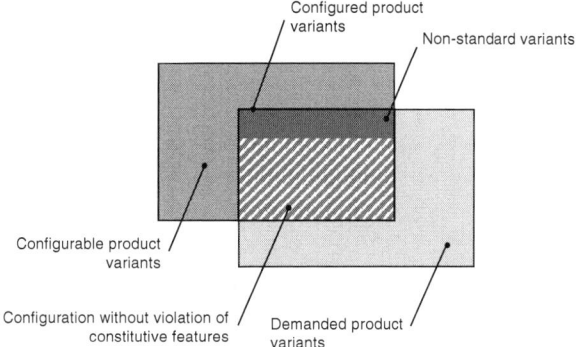

Flexibility can be substantially influenced by two approaches. On the one hand, by abandoning commonality within the product architecture, the constitutive features can be diluted. But in this case, as described above, the potential for benefits of scale is reduced. On the other hand, by clever choice of constitutive features, a suitable degree of freedom can be created for the product architecture to enable satisfying individual customer requirements while still enjoying benefits of scale (see for this also Sect. 1.1.4.4). Figure 3.11 below shows these influences and their effects on the *Flexibility of the Product Architecture*.

A particular challenge in defining suitable levels of freedom within the product architecture is the anticipation of dynamically developing customer requirements. Further development of customer requirements typically leads to a gradual reduction in the *Flexibility of the Product Architecture*, as the overlap between configurable and requested product variants shrinks due to a shift in customer requirements. Similarly to the preceding case, a trend like this is generally compensated for by an infringement of constitutive features and therefore a reduction in the *Commonality of Product Architecture*. To avoid this, from the outset a choice of suitable constitutive features

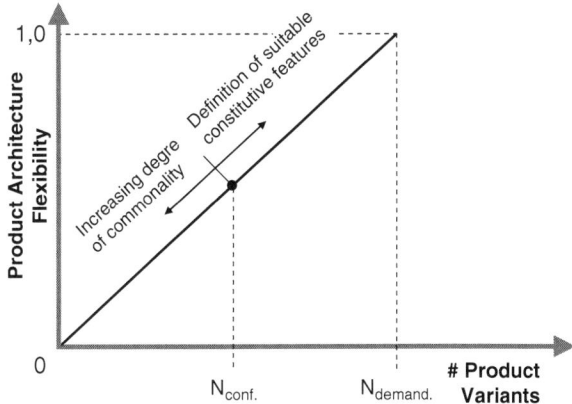

Fig. 3.11 Possible factors which influence the flexibility of the product architecture

Fig. 3.12 The influence of dynamically changing customer requirements. (According to Schuh et al. 2009)

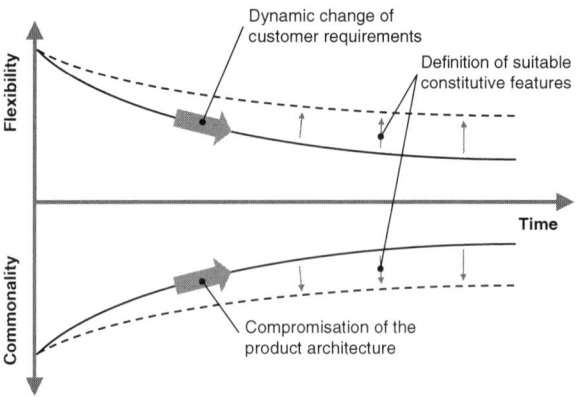

must be made, which specifically enable the level of freedom required for future development of the product in the sense of active release management (Schuh et al. 2009). Figure 3.12 shows the simultaneous reduction in *Flexibility* and *Commonality* as well as the potential influence of the choice of suitable constitutive features.

Commonality of the Product Architecture

The second main dimension in the area under review of the *Product Architecture* is represented by the *Commonality of the Product Architecture* which is achieved via the Product Programme. As already mentioned, by commonality we understand equivalences between products or product variants. Commonality enables, for example by the use of the same parts or standardised assemblies across multiple products, the creation of benefits of scale within the production structure and the supply chain (Meier 2007). The interpretation of the term commonality chosen here is, however, not limited to just the physical level of individual product components. Equivalences can be achieved in the form of cascading levels. The first and lowest of these levels is the commonality of requirements. Here the products are designed against common requirements. This provides the basis for the next layer of commonality, the creation of common functions or functional structures. Common functional structures in turn enable the use of the same technologies to provide the functions. In addition, this enables reaching the highest level of commonality, the use of standardised physical components. As Fig. 3.13 below shows, these four levels of commonality deliver increasing benefits of scale, in the sequence listed above.

As described previously, each individual level of commonality builds on the previous one. The level of physical commonality, which provides the greatest benefit of scale, requires the achievement of the underlying levels first. Despite focussing on the level of physical commonality in our further comments, these levels are therefore included by default.

To enable quantifying the physical commonality in the sense of a level of commonality, the whole range of different components required for the production programme needs to be brought together and evaluated. Martin and Ishii for example offer the

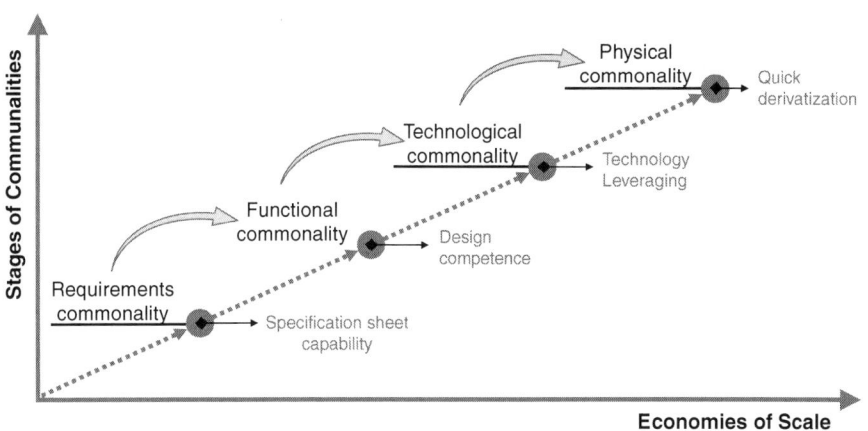

Fig. 3.13 Model of levels of types of commonality

Commonality Index as an indicator for this (Martin and Ishii 1997). But to quantify the *Commonality of the Product Architecture* a new indicator was developed, which includes both the value contribution of a component as well as a precise definition of the re-use frequency of a component. This indicator captures, as Eq. 3.4 shows, the average value-based weighted re-use frequency of the components of the product architecture. The value contribution of a component here is understood as its procurement or production costs in relation to the comparable cost of all components of the entire product architecture under consideration.

$$K_{PA} = \frac{1}{m} \cdot \sum_i^m w_i \cdot \frac{Abs_i}{Abs_{ges}} \tag{3.4}$$

K_{PA} Commonality of the Product Architecture
m Number of all components of the Product Architecture
w_i Value contribution of component i in the Product Architecture
Abs_i Sales quantity of Component 1
Abs_{ges} Sales quantity of all components of the Product Architecture

In relation to the possible value areas within the *Commonality of Product Architecture* there are two extreme cases to be considered: In one extreme case there is no commonality between the product variants within a product programme, such that every variant is entirely assembled from individual parts. In this case the *Commonality* would reach a value of Zero. In the opposite case, the theoretically ideal situation, all product variants are built from the same set of standard components. In this case the *Commonality* would have a value of One.

As Fig. 3.14 shows, the *Commonality of the Product Architecture* can be influenced both by the average, absolute frequency of component re-use, in the sense of using the identical part, and also by the complexity of the product programme, in the sense

Fig. 3.14 Possible influence on the *Commonality of the Product Architecture*

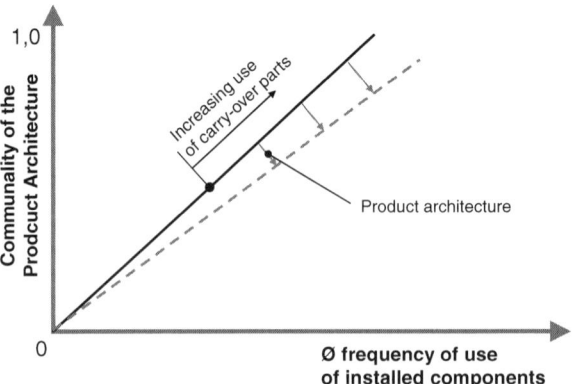

of the multiplicity of the incorporated components. Increased use of identical parts leads directly to a higher level of commonality. An increase in the multiplicity of incorporated components leads to an increase in the points of reference and so to a flatten its gradient.

In the area of the *Product Architecture* the two target values of *Flexibility* and *Commonality* were therefore defined. *Flexibility* here reflects the possible differentiation benefits, while *Commonality* reflects the benefits of scale. Analogously to the dilemma between the benefits of scale and the benefits of differentiation, there is a dichotomous relationship between *Flexibility* and *Commonality* in the product architecture.

Production Structure

By production we mean the human-controlled process of creation of products. It is achieved using labour, technical equipment, materials, energy and services (compare Arentzen 1997; Westkämper and Warnecke 2006).

The main objective in production is to reduce unit costs by achieving benefits of scale. The benefits of scale are cost reductions (on a unit basis) in the production of standard products, which are achieved by increased output quantities without product extension (compare Hungenberg 2006). Benefits of scale are based primarily on:

- The reduction of the proportion of fixed costs (fix cost decrease) per unit,
- Learning curves and specialisation benefits (increased efficiency in the activity),
- The increasing cost-effectiveness of system automation with increased quantities or use of new technologies (compare Lindstädt and Hauser 2004).

Fixed costs are costs which are not related to the produced quantity, such as e.g. machine or equipment costs or building rental costs. Increasing production quantities with a constant set of fixed costs reduces the unit costs (Fig. 3.15).

Learning curves and specialisation benefits complement the fixed cost reduction benefits and further reduce unit costs (Lindstädt and Hauser 2004). Learning curve

3 Individualised Production

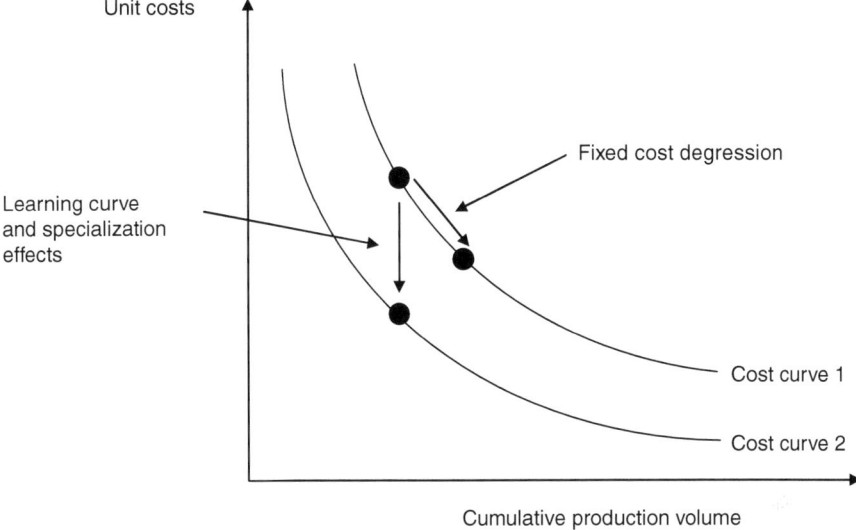

Fig. 3.15 Options to reduce unit costs

benefits occur if employees can carry out a job more efficiently due to doing it repetitively. An example of the learning curve effect is the time taken to initiate a new employee: initially new and time-intensive activities, with repetition, become rapidly executed routine tasks. Specialisation benefits are enabled by breaking down a work process into its separate sub-tasks. Then employees can specialise in different sub-tasks. As a result of specialisation, employee productivity rises and unit costs decrease (Schuh 2007).

The challenge in designing the production structure lies in reducing unit costs despite an increase in the number of the variants. In particular the achievement of benefits from learning curves and specialisation is made difficult by an increase in the number of variants. Redesigning the production structure can lead to a reduction in unit costs despite increases in the number of variants. The design of the production structure can be broken down into three areas of focus:

- *Resource structure:* In the context of production processes, resources are generally machines and resources. Resources are the entities which are available to an organisation to be used in production (Palupski 2002). Resource planning tasks are the definition of the type of necessary production sites, and the definition of the type of spatial structures and the flow system (Schenk and Wirth 2004).
- *Sizing:* As part of sizing, the quantity of necessary resources, the necessary areas, as well as the staff required for operations per organisation or per product line are determined (Arnold et al. 2008).
- *Process structure:* The production process planning is concerned with the definition of the long, medium and short-term production programmes and the assignment of resources in the form of work plans (Domschke and Scholl 2005).

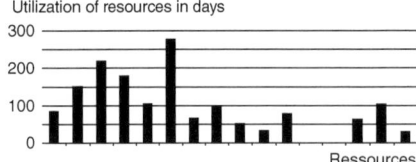 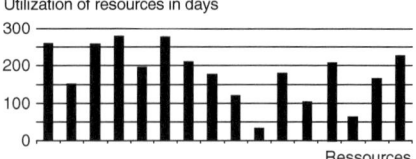

Fig. 3.16 Resource utilisation

The work schedule defines concrete working process sequences within the production levels which cover production from input to final output (Kettner et al. 1984). In addition, the necessary operational equipment as well as the predicted times are defined (Spur and Stöferle 1994). For factory planners, the work schedule is one of the most important planning tools, because it defines the work processes and their sequence, as well their assignment to work locations (Kettner et al. 1984).

In evaluating the production structure there are two issues which are particularly important: in connection with the resource structure and the sizing, the decisive issue is: "What is the level of resource utilisation". In connection with the process structure the question to be answered is: "How much does the process sequence between resources vary?" These two issues can be measured using the indicators "resource utilisation" and "process commonality".

Resource Utilisation

Resource Utilisation shows how effectively the existing resources are being used. Maximum utilisation of individual machines (so-called bottlenecks) often leads to many other resources showing only low levels of utilisation. When calculating resource utilisation, it is not the maximum utilisation of one machine which is relevant but rather the spread of utilisation across all machines. All time when a resource is not in use has a negative impact on the overall resource utilisation. By relieving the bottleneck resources, (e.g. by replanning orders onto other resources by adjusting the work schedules) the quantity produced can be increased, and therefore the fixed costs per product can be reduced (fixed cost reduction). Figure 3.16 shows an example with a low and a high level of resource utilisation.

To calculate the resource utilisation for a given time period, the used production capacity (total order times) and the available production capacity (total of the available resources are multiplied by the length of the time period) are viewed as a ratio (formula 3.5).

$$A_i = \frac{K_i}{T_i} = \frac{\sum t_{p,i}}{t_i \cdot m} \qquad (3.5)$$

A_i Utilisation of all resources in time period i
K_i Capacity used in time period i

3 Individualised Production

T_i Available capacity in time period i
$\Sigma t_{p,i}$ Sum of order times in time period i
t_i Length of time period i
m Number of available resources

Process Commonality

Process commonalities express the degree of similar process sequences. An increase in process commonalities leads to a standardisation of process sequences. Standardised process sequences also lead to an improvement in the learning curve benefits described above. Generally an increase of product variants will have a negative impact on the process commonality, as new variants often require use of different process sequences. Unification of work schedules across different variants can enable an increase in process commonalities. Process commonalities can be calculated by comparing the theoretically possible process sequences, and the process sequences actually applied between the available resources. Formula (3.6) shows the calculation of process commonalities:

$$P = 1 - \frac{r}{R_{max}} = 1 - \frac{r}{n^2} \qquad (3.6)$$

P   Process Commonality
R   Number of process sequences used across resources
R_{max} Theoretically possible number of process sequences across resources
N   Number of available resources

When calculating process commonalities it is not the multi-step process chain as a whole but each of the individual process steps which is considered. In order to capitalise from learning curve effects and specialisation benefits, only two decisions are important: on which resources the process step is carried out, and to which resource the order is then passed on. The decisions which occur after the following step are not relevant for the resource and so not for the employee. By taking into account all resources or process sequences, all possible process chains are covered.

Figure 3.17 shows three examples with different process commonalities. The number of resources is identical in all three examples. The used process sequences

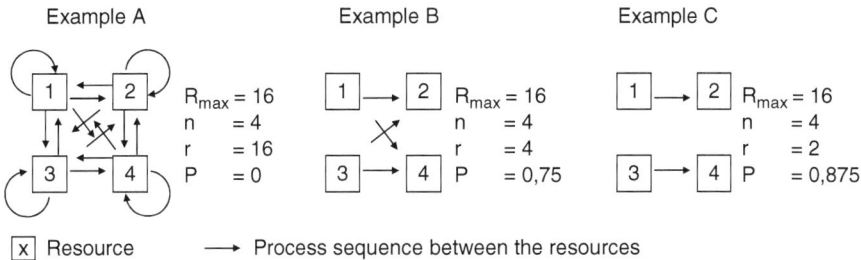

Fig. 3.17 Calculation of process commonalities

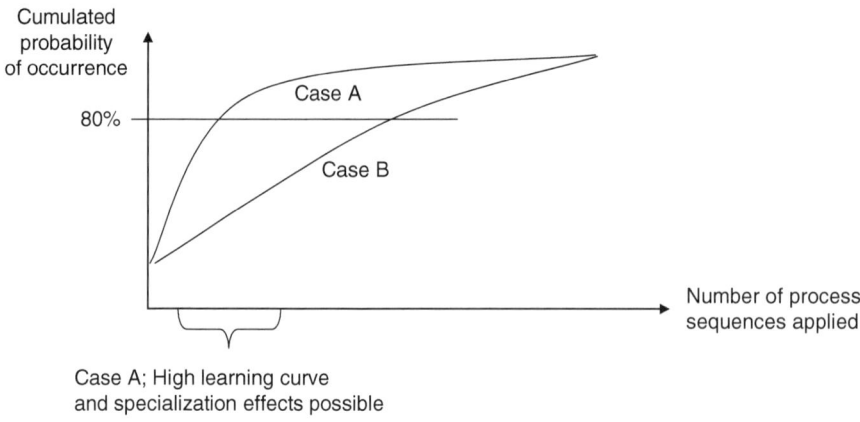

Fig. 3.18 Area of focus for process sequences

vary. Example A shows the minimum possible process commonality. All theoretically possible process sequences occur in actual production. There is no standardisation of process sequences. Example C shows a very high level of process commonality. For each of resources 1 and 2 there is only one fixed follow-on resource. The maximum possible process commonality exists, if work plans with a single process step exist. This case occurs very rarely in practice.

As learning curve benefits depend on the frequency of repetition, only the most frequent process sequences (e.g. 80%) are taken into account when calculating the process commonality. In Fig. 3.18 there are two different applications with the same number of process sequences in use. Assuming the same number of resources the process commonalities between case A and B are identical. In case A some process sequences occur more frequently compared to other process sequences. In case B the process sequences occur more or less evenly. Therefore in Case A greater benefits can be achieved from learning curve and specialisation effects, because the process sequences which occur frequently can be thoroughly standardised.

Resource Utilisation and process commonalities are two dichotomous target dimensions of production design which need to be optimised jointly to achieve maximum benefits of scale. The reduction of bottlenecks by rescheduling orders has a positive effect on resource utilisation, because a larger production quantity can be produced. But rescheduling frequently leads to additional process sequences and therefore reduces process commonalities. On the other hand, standardisation of process sequences generally leads to individual resources being high utilised but many other resources being barely utilised at all. This reduces the degree of resource utilisation and therefore the reduction in unit costs.

Similarly to the previous areas of tension, some constitutive features can be defined for production. These determine, for example, the standards for the process technologies being used, the production processes and the tools and equipment used.

A suitable choice of constitutive production features can, as shown in the third subsection 3.2.4.4, help to resolve the partial dichotomy in the area of tension of the production structure, and so help to resolve the scale-scope dilemma.

Supply Chain

A Supply Chain can be generally described as the logistical network consisting of the company internal production and logistic sites, the sites of external suppliers, the sites of the logistics service providers as well as customer locations, whose components are connected by material, financial and information flows. The main task of a Supply Chain in a wider sense is understood to be the creation of value from the customer's point of view in the form of physical products or services (Christopher 2005; Stadtler 2008; Chopra and Meindl 2010; Schönsleben 2007). The optimisation of the Supply Chain in the direction of higher performance and efficiency is the object of Supply Chain Management. The procedural focus of Supply Chain Management is on the customer-facing processes, given the core task described above (Stadtler 2008; Chopra and Meindl 2010). When evaluating the performance and efficiency of a Supply Chain in general, a distinction is made between external or customer-oriented, and internally-oriented indicators (Supply Chain Council 2010; Sürie and Wagner 2008; Beamon 1998). While externally-oriented Supply Chain indicators mainly address the evaluation and analysis of the performance of the customer-facing Supply Chain processes, the internally-oriented indicators focus either on the Supply Chain costs generated in the Supply Chain, or on the capital tied up in the Supply Chain being studied. The focus when looking at the Supply Chain quadrants therefore lies on the performance and capital efficiency of the customer-facing Supply Chain processes, i.e. particularly the distribution processes. The upstream Supply Chain processes are described and evaluated as part of the *Production Structure* (see third sub-section of 3.2.4.2).

To adequately analyse and optimise the performance and capital efficiency of a Supply Chain in relation to fulfilling customer needs, it is necessary to define adequate evaluation dimensions (Beamon 1998; Sürie and Wagner 2008). The Supply Chain quadrant is therefore built up using the dichotomous evaluation dimensions of *Supply Chain Effectiveness* and *Supply Chain Capital Efficiency*. The dimension of Supply Chain Effectiveness addresses the issue of how well the logistical customer requirements are met for a certain product or order, in relation to delivered quantity, quality and due date. This reflects a customer-oriented, performance-related evaluation dimension, and represents the scope dimension of the production technology polylemma. The dichotomous dimension of scale can also be described by the dimension of the Supply Chain capital efficiency, which describes the relationship between the value of the finished and unfinished products tied in the Supply Chain, and the product or product programme related turnover.

Supply Chain Effectiveness

Various possible indicators are offered in the literature to determine Supply Chain performance (Beamon 1998; Sürie and Wagner 2008; Arnold et al. 2008; Supply Chain Council 2010). In this area the indicators of delivery time, delivery reliability, delivery flexibility and delivery quality are particularly worthy of attention. Delivery time can be defined as the time period between order placing by the customer and the delivery of the goods. Delivery reliability is a measure of the accordance of the delivery conditions agreed with the customer in respect of the delivery due date, the delivery quantity as well as implicitly the delivered quality. Delivery flexibility describes the ability of a Supply Chain to react swiftly to unforeseen customer demands. Delivery quality describes on the one hand the quantity and quality related conformity of the delivery to what was ordered, and on the other hand the condition of the goods delivered to the customer (Pfohl 2010; Arnold et al. 2008). A strong correlation can generally be observed between these indicators (Straube and Pfohl 2008). As delivery reliability implicitly includes the indicators of delivery time and delivery quality as well as, depending on how the time period is measured for capturing the promised delivery dates, it is a valuable and easily comparable indicator both for make-to-stock and make-to-order manufacturers (Straube and Pfohl 2008), and serves as the fundamental indicator to describe the *Supply Chain Effectiveness*. Operationalizing delivery reliability of a Supply Chain can be achieved thanks to the indicator of the service level (Alicke 2005; Tempelmeier 2005). In the expert literature there are generally three different approaches to defining service levels: event-, quantity, and both time- and quantity-oriented approaches. With a quantity-oriented approach the service level is defined ex-ante as the ratio between the expected values for on-time and correct quality delivery of the quantities ordered and the sum of all order quantities with the delivery date as requested by the customer within the same period t over an observation time frame of n periods. For ex-post calculations the specification of the expectation values is done as the arithmetic mean over the observation time frame of n periods. The acceptable value range of the delivery service level (Formula 3.7) lies between "0" or 0% (minimal delivery reliability) and "1" or 100% (maximum delivery reliability). This can be stated mathematically as follows:

$$LSG_{SC} = \frac{\frac{1}{q}\sum_{i=1}^{q} BM_{FQ,i}}{\frac{1}{q}\sum_{t=1}^{q} BM_{ges,i}} \qquad (3.7)$$

LSG_{SC} quantity-oriented Supply Chain delivery service level
Q   Number of periods in the observation time frame to be used for the calculation of the quantity-oriented delivery service level
$BM_{FQ,i}$ ordered quantity delivered on time with correct quality in Period i
BM_i Quantity ordered in Period i

Supply Chain Capital Efficiency

Supply Chain capital efficiency looks upon the working capital which is tied up in the form of products or a product programme and which can be substantially influenced by management decisions regarding the Supply Chain. This represents an evaluation dimension of the Supply Chain which relates to the working capital of a company, and therefore is internal. Working capital can be broken down by the way the funds are used, basically into capital and current assets (Heesen and Gruber 2009). While capital assets are affected by investment decisions relating to the longer-term (e.g. decisions about locations), the results of decisions about logistic planning the of supplier-side, business internal and customer-side Supply Chain processes mainly affect the inventories, which is shown on the balance sheet under the heading of current assets (Sürie and Wagner 2008; Supply Chain Council 2010). From the point of view of financial reporting, (e.g. under IAS 2 of the International Financial Reporting Standards), the inventories tied-up in a Supply Chain are made up of raw materials, supplies and consumables, finished products and goods as well as unfinished goods.

The logistic process approach distinguishes the tied-up inventories more precisely into seasonal buffer stocks to compensate or smooth variations in demand behaviour, safety stocks, work-in-progress inventories, as well as stocks in transit and base stocks driven by batch sizes (Sürie and Wagner 2008). A monetary valuation of the inventories tied up in a Supply Chain can be done based on the derived cost of capital or inventory costs, which are calculated by multiplying the average inventories within a reporting period by weighted average capital costs for the company (Drukarczyk and Schüler 2009). The rate of cost of capital is substantially determined by the financial structure of a company, as well as long.-term expectations of the company and its sector. Depending on the sector, average rates of cost of capital can be seen to vary from 7% (consumer goods, pharmaceuticals) to 10.5% (machine and equipment construction, Software) (Drukarczyk and Schüler 2009).

The decisive drivers for the inventory costs in a company, alongside the cost of capital, are the variants in the product programme, the sales risk or the volatility of customer demands and the structural configuration of the Supply Chain (Mayer et al. 2009). To define a relative, cardinally-scaled, and therefore comparable indicator for the *Supply Chain Capital Efficiency* a suitable reference base needs to be chosen for the inventory capital tied up in the Supply Chain. A suitable reference basis for this is the product or product programme related sales turnover (or sales revenues) within the same reporting period. In line with the requirements stated above for scalability of the dichotomous dimensions of the quadrant, it is also necessary to dimension the scale for measuring the *Supply Chain Capital Efficiency* such that the maximum and minimum values of the Supply Chain capital efficiency is set within the limits of "0" or 0% (minimum *Supply Chain Capital Efficiency*) and "1" or 100% (maximum Supply Chain capital efficiency). The indicator for measuring the *Supply*

Fig. 3.19 Example of classification of three basic German industrial sectors in the Supply Chain quadrant

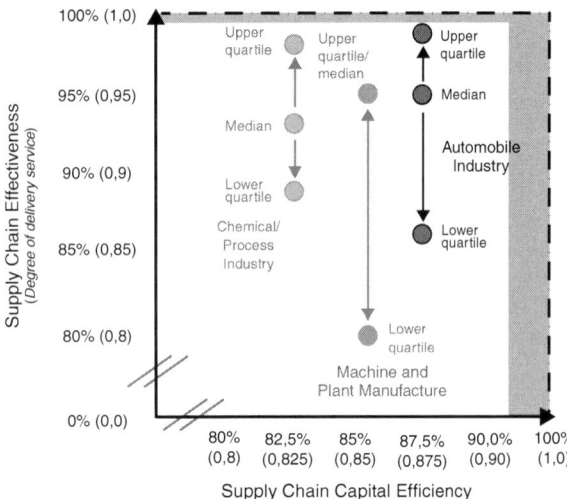

Chain Capital Efficiency can therefore be formulated as follows:

$$\text{SC-KE} = 1 - \frac{\sum_{j=1}^{m}(\text{SC-dVM})j + \text{SC-iVM}}{\sum_{j=1}^{m} Uj} \tag{3.8}$$

M   Number of products or product variants from the product programme under consideration

$(\text{SC-dVM})_j$ inventory capital directly attributable to a product or product variant j (primarily finished goods)

(SC-iVM) inventory capital attributable indirectly to a product or product variant j, but directly to the product programme (e.g. raw materials and unfinished goods)

U_j cumulative turnover per product variant j in the observation period

Based on the dimensions defined and rendered operational for Supply Chain Effectiveness and *Supply Chain Capital Efficiency*, measures to achieve a greater effectiveness in the Supply Chain can be introduced. In a Supply Chain, thanks to its function as a link between production and the customer, both benefits of scope and of scale can be considered.

An example of the classification of three industrial sectors in Germany according to the two dimensions specified for evaluation in the Supply Chain quadrant for the years 2008/2009 is shown in Fig. 3.19[1] (Straube and Pfohl 2008; Mayer et al. 2009).

---

[1] To determine the inventory capital for the calculation of Supply Chain capital efficiency, the following rates for cost of capital were applied (from Drukarcyk and Schüler 2007). Automobile industry: 7.9%; Machinery and installations: 10.5% and Chemicals and processing industry: 8.2%.

3 Individualised Production

Fig. 3.20 Threshold level considerations in the Supply Chain quadrant

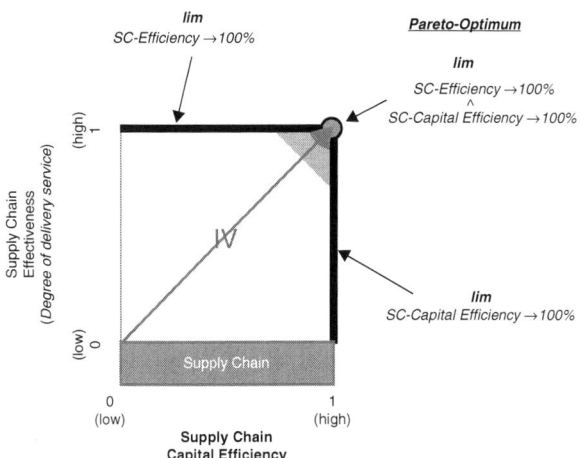

Although a high level of performance already exists in relation to both evaluation dimensions it is possible to see differences in relation to performance and capital efficiencies both between sectors and between companies within the same sector.

The upper limits or extreme values within both evaluation dimensions are both calibrated to "1" or 100% (Fig. 3.20). Even if when configuring a Supply Chain, in theory you are trying to achieve these limit values for both evaluation dimensions (so-called Pareto-optimisation (Wiese 2010)), achieving this goal remains purely hypothetical from an ex-ante perspective.

So although the upper limit (*lim*) of Supply Chain Effectiveness or the service level can indeed be reached in the ex-post evaluation for shorter observation periods and for a limited range of the product programme, this is not possible when taking an ex-ante approach. As a statistical normal distribution must be assumed for the anticipation of customer requirements in relation to the variants for future order quantities, for this case a theoretically infinite inventory level of each product variant would have to be held in the distribution warehouses (for make-to-stock manufacturers) or infinite capacity in production (for make-to-order manufacturers).

Achieving an optimal value in relation to *Supply Chain Capital Efficiency* is purely hypothetical both ex-ante and ex-post. According to the Eq. (3.2) the result of a 100% capital efficiency would be achieved in the following two theoretical cases:

- Throughout the observation period there is no inventory employed for the whole product programme observed. But at the same time turnover or sales revenue is realised on the product programme.
- The cumulative turnover per product in the observation period is infinite.

In both cases the subtractor in the equation becomes zero and the *Supply Chain Capital Efficiency* would hit the upper limit of 100%. But when looking at the Supply Chain processes in a company for the production and distribution of a product the limitation applies that, according to the above breakdown of the inventory capital, for every product which is produced, stored, transported and sold, inventories must exist at the various levels of the Supply Chain. Equally product turnover which is close to infinity must be seen as extremely unrealistic. Given that the optimal values on each evaluation dimensions taken in isolation are not achievable in business practice, the achievement of a Pareto-optimum within a Supply Chain configuration is equally impossible. Nevertheless the goal of the Supply Chain configuration using the following principle solutions is still to get as close as possible to this hypothetical optimum.

Interim Conclusion

The higher-level dichotomy for production systems between economies-of-scope and economies-of-scale, when managing product complexity, is addressed by the definition of four central stress fields and eight main dimensions corresponding to these. For the stress fields, namely *Product Programme, Product Architecture, Production Structure* and *Supply Chain,* equations are defined in the form of ratio indicators to quantify the main dimensions defined. Measurement parameters for capturing both the economies-of-scope and the economies-of-scale were defined. As shown in Fig. 3.21, the main dimensions defined can be assigned unequivocally to one of the higher-level targets (realising economies-of-scope and economies-of-scale) of a production system it.

Using the indicators it is possible to show quantitatively the current position of a production system, starting from the external product complexity of the *Product Programme,* to the internal complexity of the *Product Architecture,* and the internal and external production-related effects in the *Production Structure* or the *Supply*

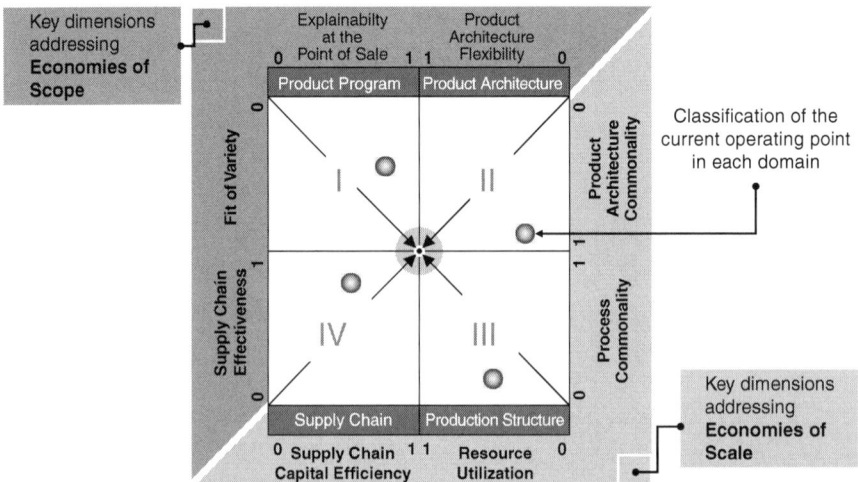

Fig. 3.21 Main dimensions of the evaluation model

3 Individualised Production

Chain. In the following sections, the indicators of the eight main dimensions are combined into an evaluation model in order to provide at an integrative evaluation.

3.2.4.3 "Degree of Efficiency" of the Complexity of a Production System

Using the main dimensions which have been defined and the clear optimisation direction in the stress fields, an evaluation of the effects of production complexity is possible. Further to assess how successfully a company has operated in each stress fields, we need to combine the dimension indicators into a single, aggregated and measurable value. The objective of the following sub-section is therefore to translate the evaluation logic described in the previous sections into a quantitative and integrative evaluation approach for resolving the scale-scope dilemma. At the core of this is a methodology for evaluating product complexity which is analogous to measuring the degree of efficiency of technical systems.

In general, degrees of efficiency are used in a technical context to describe the efficiency of systems (Lucas 2008). The degree of efficiency (Formula 3.9) describes the ratio of outputs to inputs, or in more general terms, the ratio of effort and results.

$$\eta = \frac{\text{Output power}}{\text{Input power}} = \frac{\text{Benefit}}{\text{Cost}} \quad (3.9)$$

The difference which arises between what is input and what is produced are described as losses or, more specifically, as dissipation of effort. The effects of varying the parameters on technical systems can then be understood and discussed across different disciplines by taking reference to the generally understood value of the degree of efficiency, since a result which can be stated as an overall degree of efficiency can be understood without having to have a detailed knowledge of each individual discipline. The physical dependencies which result in the specific degrees of efficiency therefore only need to be understood in detail by the relevant experts. By looking at degrees of efficiency you can see which parameters influence improvements in a technical system. Only with certain knowledge of these relationships is it possible to act effectively to improve the system.

The advantages listed above regarding the evaluation of technical systems using degrees of efficiency will be applied below by using an analogy for the evaluation of product complexity.

The complexity-related degree of efficiency of a company will be determined by specific partial degrees of efficiency in the stress fields identified above which are related to product complexity.

For each of the stress fields, specific partial degrees of efficiency η_j are defined. These partial degrees of efficiency represent a unit of measurement to resolve the dichotomies in the areas of tension. Overall, these partial degrees of efficiency are defined such that a simultaneous improvement in the main dimensions will lead to an improvement in the partial degree of efficiency (Fig. 3.22).

The partial degrees of efficiency achieve a value of one, if both main dimensions reach their maximum, and have a value of zero when both main dimensions reach

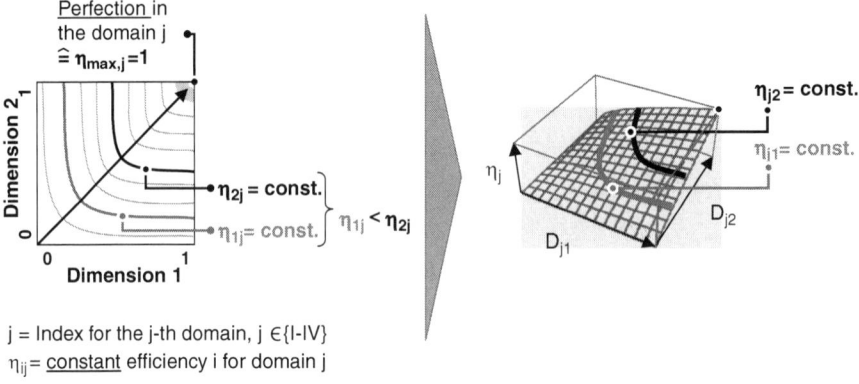

j = Index for the j-th domain, j ∈{I-IV}
η_{ij} = <u>constant</u> efficiency i for domain j

Fig. 3.22 Partial degree of efficiency dependent on two main dimensions

a value of zero. There are lines of constant partial degrees of efficiency (isoquants) for particular value combinations of the main dimensions. This can be interpreted as follows: For a given company there can be various different conditions in the stress fields which, in terms of the economies-of-scope or economies-of-scale which can be achieved, account for the same benefit to the company. The main dimensions have fixed compensatory relationships, i.e. if the value of one main dimension is increased then a specific reduction becomes possible in the value of the dichotomous second main dimension in that same stress field, which, when taken together, net out to the same level of benefit. The degree of efficiency recognises this fairly common occurrence in practice.

The general equation of the Cobb-Douglas-Function (Formula 3.10) (for a detailed description of the Cobb-Douglas production function we refer you to the relevant literature (e.g. Cobb and Douglas 1928)) is defined as follows for the adapted equation of the degree of efficiency:

$$\eta_j = D_{j1}{}^{\alpha_j} \cdot D_{j2}{}^{\beta_j} \qquad (3.10)$$

η_j Degree of efficiency for the area of tension j
j   Index for the j-th area of tension, with j ∈ (I–IV) for the four areas of tension
D_{j1} Indicator value for the first main dimension in the j-th area of tension
D_{j2} Indicator value for the second main dimension in the j-th area of tension
α_j Elasticity for the indicator for the first main dimension in the j-th area of tension
β_j Elasticity for the indicator for the second main dimension in the j-th area of tension

In the equation for the degree of efficiency, the non-constant level parameter typically used in the Cobb-Douglas function is not applied. Apart from this difference the equation is constructed analogously to the Cobb-Douglas function. The elasticities α_j and β_j enable weighting the relevant indicators for the main dimensions in the stress fields.

3 Individualised Production

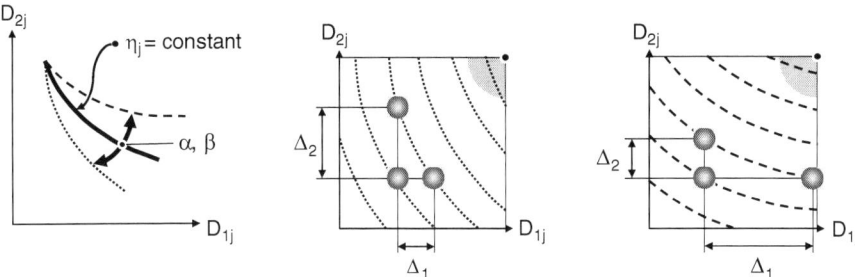

Fig. 3.23 Isoquant patterns varying by elasticity

The weighting of the main dimensions is defined depending on the requirements and strategic direction of the type of company. The influence of the elasticity parameters on the isoquant patterns is shown in Fig. 3.23.

Depending on the elasticity parameters the isoquants may be stretched or bunched in the direction of the relevant principal dimension. An improvement in one of the main dimensions can therefore result in a different degree of efficiency depending on the weighting of the dimensions.

The indicators defined in Sect. 3.2.4.2 are to be used in the general equation for the degree of efficiency, in order to create the equivalent equations for the different areas of tension namely *Product Programme, Product Architecture, Production Structure* and *Supply Chain*.

In Fig. 3.24 the degree of efficiency indicators are shown qualitatively in the form of isoquants for the four areas of tension in the evaluation model (The isoquants are shown as having identical patterns. That would represent a special case for a given production system. To simplify the presentation this case was chosen for the illustration).

With the help of the defined degree of efficiency an absolute measure was defined, by means of which production systems can be rated regarding their achievement of the economies-of-scope or economies-of-scale in the four central stress fields.

The degrees of efficiency provide an important indication in relation to the optimisation measures in the four stress fields. For example, decisive measures taken in relation to the *Product Architecture* can be assessed to evaluate, if they have a significant impact on the efficiency of the *Product Architecture*, as well as on the other degrees of efficiency in the other stress fields. The efficiency rating which can be achieved depends strongly on the principle solutions applied in the product and production systems. For example, a modular construction for a product programme can make a significant contribution to provide the necessary flexibility in the right areas, and at the same time ensure a sufficiently high *Degree of Commonality*, which in turn leads to a correspondingly high degree of efficiency in *Product Architecture*. These principle solutions can be adapted—at least up to a point—and are independent of the specific type of company. Therefore it is important to know to what extent various principle solutions in the stress fields can contribute to a simultaneous

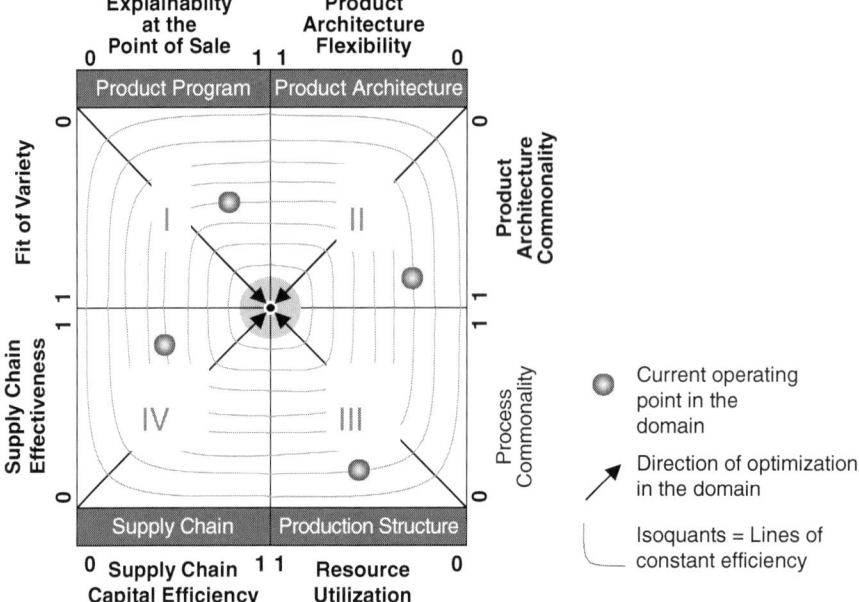

Fig. 3.24 Degrees of efficiency for evaluation of product complexity

improvement in both the dichotomous main dimensions and so also improve the degree of efficiency. A basic challenge here is the evaluation of the principle solutions which have already been applied. Furthermore the evaluation of the impact of alternative principle solutions that were not yet tried out in the company, on the degree of efficiency is rendered possible. Figure 3.25 shows the depicted relationship.

Depending on the current operational level of the production system, the question should be asked whether the existing principle solutions should be maintained, or whether better alternative principle solutions should be applied, in order to improve the degrees of efficiency.

Depending on the currently applied principle solutions, there may be levels of the partial degrees of efficiency which are not achievable (areas Fig. 3.25 shown in grey). That means that, despite repeated attempts to optimise and improve the principle solutions being applied, there will be a particular level of the relevant partial degree of efficiency which simply cannot be exceeded. The Otto motor can be used as an analogy for the evaluation of technical systems by using degrees of efficiency. By adhering to this principle for converting energy, the system has physical limits which apply to the theoretically achievable thermodynamic efficiency. If you assume an asymptotic approximation to the efficiency rate achievable with the current principle solutions, (shown as $\eta_{optimal}$ in Fig. 3.25) then awareness of the position on the efficiency level curve is particularly interesting. By applying alternative principle solutions it becomes possible to shift the achievable partial degree of efficiency to higher levels, (shown as $\eta_{optimal}'$ or case 2). By applying

3 Individualised Production

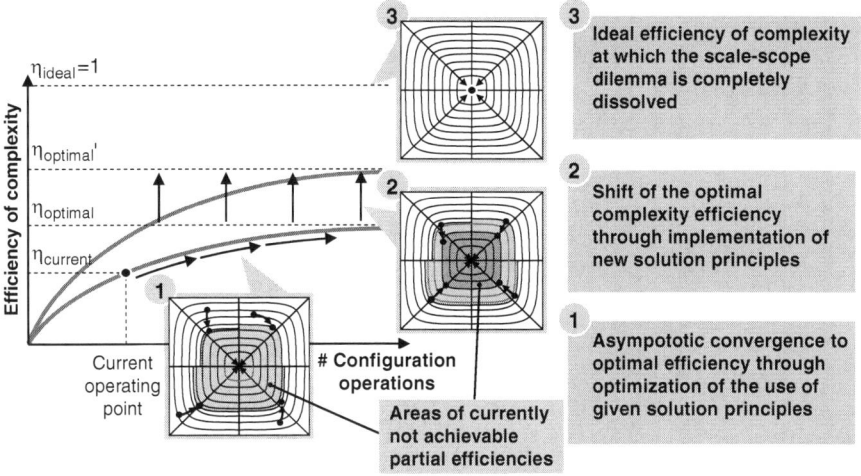

Fig. 3.25 Convergent degrees of efficiency depending on principle solutions applied

"ideal" principle solutions, which can then also be optimised to their maximum level, theoretically an efficiency level with the value $\eta_{ideal} = 1$ can be achieved. This case would represent fully resolving the scale-scope dilemma and is the ideal situation where all indicators of the dichotomous main dimensions simultaneously reach their maximum value of one.

3.2.4.4 Integrative Configuration of Production Systems

Using the model described above for integrative and complexity-related evaluation of production systems it is possible to identify the necessary optimisation approach each time. In the following, an integrative design approach is presented, with the help of which a methodical improvement can be achieved, in line with the defined optimisation approach. This design approach, referred to below as the *integrative configuration of production systems*, includes four effective principle solutions which work across the areas of tension and which are described in detail in the following sections. Moreover the design approach also represents the shared core of these principle solutions. Figure 3.26 below shows the procedural logic for integrative configuration, as well as its interaction with the evaluation model described above. The individual configuration steps, their sequence and their place in the evaluation model are described in detail below.

Analysis and Integrative Evaluation of a Production System

The starting point of the configuration is the overarching analysis of the production system, based on the previously described integrative evaluation model. This analysis enables identification of weak points, as well as of the necessary direction for

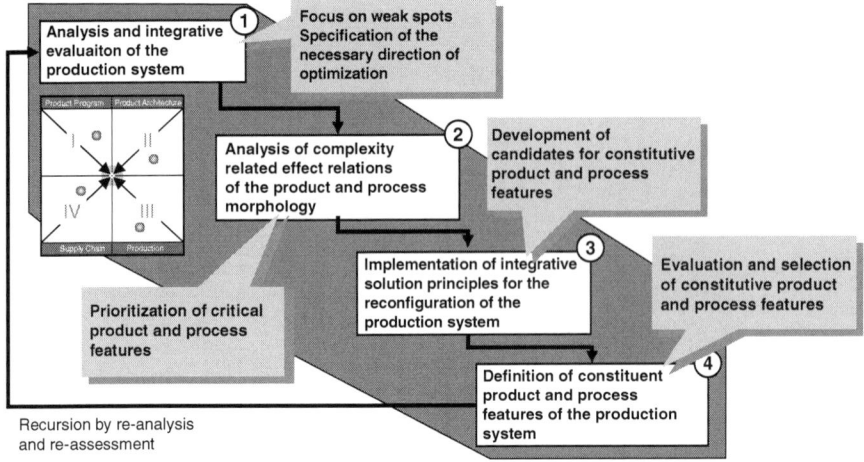

Fig. 3.26 Integrative configuration logic for production systems

the optimisation required to resolve the scale-scope issue. Prescribing a direction for optimisation enables focussing on the configuration steps which follow. The integrative evaluation model serves further as a set of guidelines for the configuration of the production system.

Analysis of Complexity-Driven Causal Relationships in the Product and Process Morphology

The prime task of the *integrative configuration of production systems* is the targeted optimisation of complexity-driven causal relationships between products and production. If products are offered within a product programme, or as variants of a product architecture, then they vary from each other in terms of different options in specific product characteristics. The sum total of these options for a single characteristic are referred to in the following as the *product morphology* and represent the result of the decisions made in the product programme and the product architecture. As a counterpoint to the *product morphology*, by applying a *production morphology*, we can describe the individual process variants in terms of the options of their process characteristics. Figure 3.27 below shows the product and production related morphology for a special rolling bearing programme. These morphologies are sub-divided according to the four areas of tension from the evaluation model.

There are multiple causal relationships between the features and characteristics of the product and process morphology. Thus the individual options within a product feature, such as for example a specific load rating of a rolling bearing, determine other product and production features, such as specific bearing materials, production processes or process types. Starting from the analysis of these causal relationships,

3 Individualised Production

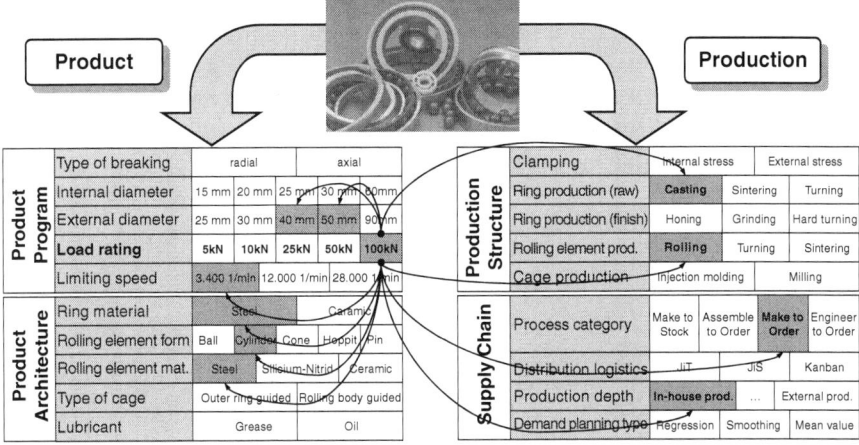

Fig. 3.27 Product and process morphology for a special rolling bearing programme

the variant-related influence of a characteristic can be defined quantitatively by applying formulas (3.11) and (3.12). Here we distinguish between active and passive variant influences. An active variant influence corresponds to a direct cause of a multiplicity of options in other product and production characteristics. Passive variant influence on the other hand reflects a high degree of variation within a feature, which is caused by other features and their options.

$$V_a = \sum_{i=1}^{n}\left(\sum_{j=1}^{m} g_{i,j}\right) \quad (3.11)$$

$$V_p = \sum_{k=1}^{n}\left(\sum_{l=1}^{m} g_{l,k}\right) \quad (3.12)$$

V_a Active variant influence
V_p Passive variant influence
N Number of options within the characteristic being observed
M Total number of characteristics
$g_{i,j}$ Weighting of the causal relationship of option i of the characteristic being observed on characteristic j
$g_{l,k}$ Weighting of the causal relationship of characteristic l on the option k of the characteristic being observed

Starting from formulas (3.8) and (3.9) it can be determined from the causal relationships between the product and process morphology, which variable product and production features are variance-causing or-driven to an exceptional degree. This degree is clarified by weighting individual causal relationships. This weighting reflects the level of the complexity costs generated by the multiplicity. Standardisation of exactly those product and production characteristics which display a high active or passive variant influence, will therefore deliver a disproportionate reduction in the product and process complexity or a resolution of the scale-scope dichotomy. By implementing the integrative principle solutions described below, this kind of standardisation can be pursued. The analysis of product and process morphologies as well as of the complexity-driven causal relationships helps to further focus configuration activities.

Implementation of Integrative Principle Solutions

In this step of the *integrative configuration of a production system,* the integrative principle solutions are applied. A necessary property of these principle solutions is that they each display an integrative effect. In relation to each of the four areas of tension in the evaluation model, the application of the principle solutions has the effect of improvement, or at least of no deterioration of the operational level. The individual principle solutions each have a focus of effect which lies within one of the four areas of tension. The configuration measures linked to each solution approach are therefore assigned specifically to the relevant area of tension. Figure 3.28 shows four integrative principle solutions which are described in detail below, together with their associated areas of tension.

By looking at the current operational level of a production system within the evaluation model, and the necessary optimisation direction, it can be deduced which of the principle solutions should be applied for any given case. The principle solutions can be freely combined.

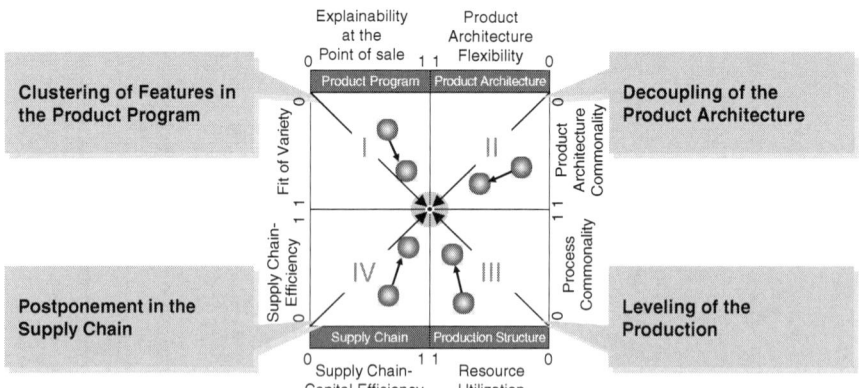

Fig. 3.28 Four principle solutions for integrative configuration

3 Individualised Production

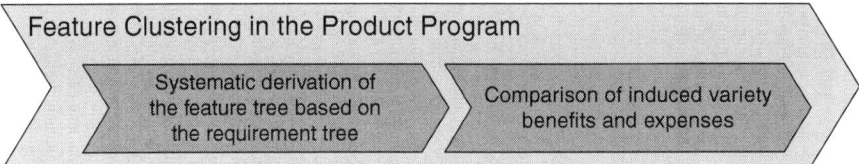

Fig. 3.29 Logical approach to feature clustering

Feature Clustering in the Product Programme

The main aim of feature clustering is to analyse the product features and options systematically in terms of their contribution to providing a good requirements fit to customer needs, and to cluster them according to their contribution.

In Fig. 3.29 the logic of the approach to feature clustering is integrated in the *integrative configuration of a production system*. Feature clustering consists basically of two consecutive steps (Fig. 3.29). These two fundamental steps of feature clustering are described below.

As a first step, the customer requirements need to be analysed and defined according to their main features and options. The various requirement options are structured using a so-called requirements tree (Fig. 3.30). All the different types of requirement options as seen from a customer viewpoint are entered on this structure and visualised. Each branch in the requirements tree therefore represents a particular customer type profile.

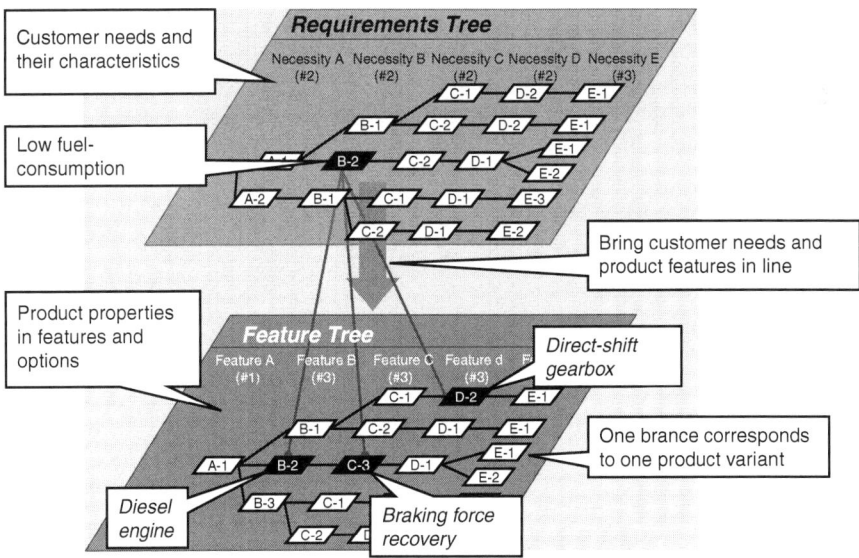

Fig. 3.30 Systematic derivation of the features tree from the requirements tree

Fig. 3.31 Feature clustering in the complexity portfolio

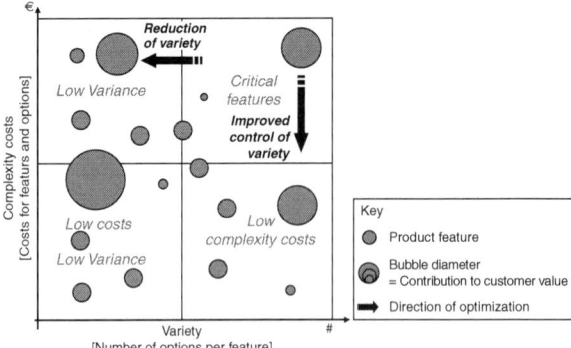

The customer requirements which have been structured in this way need to be projected onto the structure of the product programme. To do this, first the features and options necessary to describe the product programme are collected and transferred to a so-called features tree (see Fig. 3.30 below). By means of a systematic comparison of the requirements tree and the features tree it can be checked whether the multiplicity of options in the product programme is justified or not, in the light of the multiplicity of options of the customer requirements.

In a following step, the various features are analysed by the complexity costs and benefits they entail, as well as the multiplicity of their options (Fig. 3.31).

In the complexity portfolio, features with an imbalance between the benefit delivered, because of the multiplicity of options and the induced cost of complexity are identified. Features where a high multiplicity of options has little impact on delivering customer value, but complexity costs are relatively high, need to be reduced in the number of their options. Features with a balanced cost-benefit ratio in relationship to the multiplicity of their options, need to be handled with special care when defining the product architecture. Particularly for critical features, (high customer benefit with simultaneous high complexity costs), suitable management of the complexity is needed with a skilled definition of the architecture.

Decoupling the Product Architecture

The core task of the principle solution *Decoupling the Product Architecture* is the subdivision of the product architecture into sub-systems which only have slight inter-dependencies between themselves. These mainly independent sub-system in the product architecture, each of which includes a group of product functions and components, will be referred to from now on as modules (Blackenfelt 2001; Göpfert 1998; Koppenhagen 2004; Pahl et al. 2005; Wagner 1999). The breakdown of the product architecture into such modules enables the substitution or the change of a module, without any impact on the other modules. A product which is modularised in this way can be adapted by combining modules to suit individual customers, while still achieving benefits of scale thanks to the standardisation and product-wide re-use of the modules themselves (Koppenhagen 2004; Reichwald and Piller 2006; Schuh

3 Individualised Production

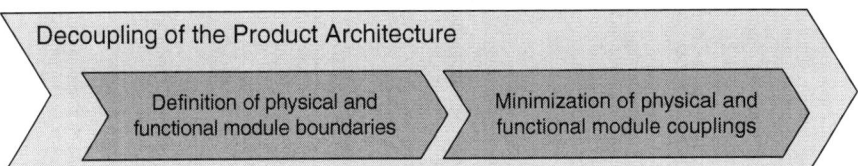

Fig. 3.32 Approach for *Decoupling the Product Architecture*

2005; Ulrich and Tung 1991). The basis and requirements for decoupling are the constitutive product features, which were already introduced above (see second subsection of 3.2.4.2). As Fig. 3.32 below shows, the decoupling can be split into two steps: definition of module boundaries, and minimising module links.

The inter-dependencies, which are to be minimised within the product architecture are referred to as links in the following. These links, which exist between two components or functions of a product, can occur as different forms of link. If you focus on looking at components, then a functional or a physical link can exist between them. In the case of a functional link, both components interact via an exchange of material, energy or information in order to perform a product function (Pahl et al. 2005; Roth 2000; Wiener 1948). In the case of a physical link, on the other hand, the components have an interface to each other with physical distance constraints. Component links can be described mathematically and visualised. One such option is represented by the Design Structure Matrix (DSM). This is a square matrix, whose rows and columns each represent one of the components of the product under review. The entries on the matrix each describe the strength of the link between the row and the column element (Steward 1981).

In a first step towards decoupling, product components are grouped into modules which are as independent as possible. The definition of the module boundaries here follows the component links described using the DSM. Particularly strongly linked components are here bundled together into modules.

In the second and actual decoupling step, there is a targeted shifting of the component links. The focus here is primarily on component links which span different modules. As well indirectly changing component links by adding and removing components, links can also be modified directly (Lindemann et al. 2006). By geometric manipulation the physical links can be affected here. An example of this is the standardisation and over-dimensioning of geometrical module interfaces, so that these allow the swapping out or the modification of a module. Functional links, on the other hand, can be influenced by changes to the functions of the product, their context, or the assignment of functions to product components. The latter leads to a clear assignment of related areas of functionality to modules within the product. The aim of this second step is to minimise or eliminate interdependencies between modules or circular relationships across modules (Lindemann et al. 2006). The result of this step in turn interacts with the defined constitutive product features. Therefore, standardised module interfaces can be defined as a constitutive feature of the product architecture for example.

The decoupling of modules described here influences mainly the area of tension related to product architecture. By adapting and varying single modules, individual customer requirements can be met and so a higher level of flexibility is achieved, while thanks to the product-wide re-use of modules, the commonality level of the product architecture is increased. These two effects have a direct impact on the remaining three areas of tension in the evaluation model described above. The extension of the level of flexibility allows a better fit to customer needs on the product programme side. On the other hand supports the standardisation of modules the principle solutions described below of balancing production and of postponement in the Supply Chain.

An example where targeted decoupling of modules interacts with the definition of the constitutive product features is the MQB platform of the VW Group. The MQB is a cross-brand and cross-product class platform to be used for cars with a transverse engine (Rumpelt 2009). This platform includes several constitutive product features, such as a uniform front length. This means a standardisation of the positions of the engine, gears, axle, air conditioning unit and pedals, with a standard distance between the middle of the front axle and the pedals. This standard is defined to be the same for all products which will be based on the MQB (Löschmann 2009; Schuh et al. 2009). Another example is the uniform interface for rear axles in the form of a standardised axle housing (Löschmann 2009). Each of the constitutive product features mentioned correspond to a decoupling in the product architecture. For example, the standardised rear axle housing enables the rear axle module to be changed without any impact on other modules. Using the standardised interface on a bodywork structure either composite shafts or multi-joint axles can be attached (Löschmann 2009).

Production Balancing

The main aim of reshaping the production structure is to reduce unit costs through benefits of scale. To achieve maximum benefits of scale, we need to increase both the utilisation of resources (fixed cost reduction) and the process commonalities (learning curve and specialisation effects). Under ideal conditions, the sequence of processing steps, i.e. the process chain, is identical for all orders, and order deadlines are constant for all process steps and all product variants (Fig. 3.33).

In this way resources would be fully loaded and use the same process chains between the resources. Each and every new order would lead to an even increase in load for all resources. In reality, process chains and delivery deadlines often vary considerably from order to order.

Different products and product variants are produced using different process chains. Even for a single product variant, different possible process chains may exist within the company, which are selected depending on the availability of machines or employees (Fig. 3.34).

Process chains can be reduced for individual product variants. Standardisation across separate product variants is also possible. But complete unification of all process chains across all products and product variants is often impossible, because products and their components vary significantly in their design. The integrative

3 Individualised Production 123

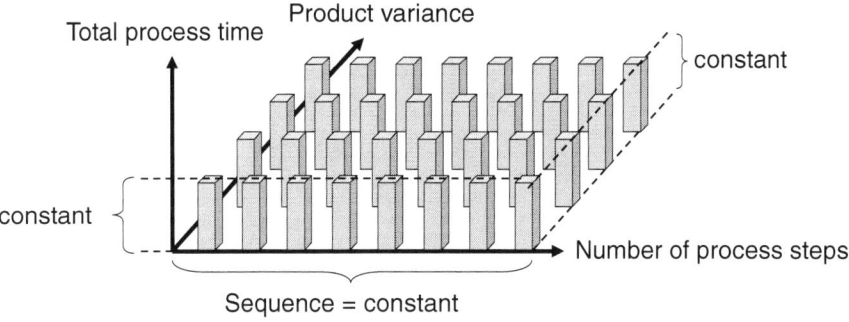

Fig. 3.33 Ideal production conditions

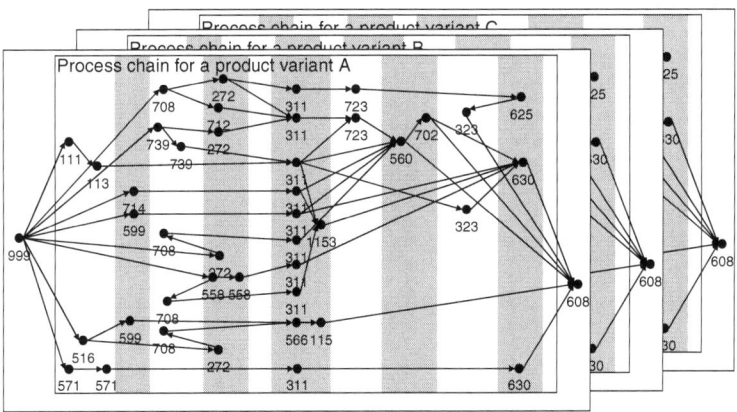

Fig. 3.34 Variability of process chains

definition of constitutive product and process features (see second and third subsections of 3.2.4.2) does help to keep the process chain variants as few in number as possible. By defining the maximum dimensions of components, for example, the number of machines required can be limited.

As described in the third sub-section of 3.2.4.2, the variability of the process chain can be calculated by examining process commonalities. Process commonalties, in this case, only take account of the process sequence between two resources. By looking at all resources or process sequences, all possible process chains are covered.

Process commonalities are uncovered by deriving the (process steps and resources of) orders from the company's ERP system. The Laboratory for Machine Tools and Production Engineering (WZL) at RWTH in Aachen has developed a tool, as part of the work in the excellence cluster, which uses the ERP data to evaluate process commonalities and presents this graphically (Fig. 3.35).

In Fig. 3.35 the existing resources (e.g. processing machines) are each represented as a circle. The lines between the resources are the process sequences which actually

Fig. 3.35 Visualisation of process commonalities

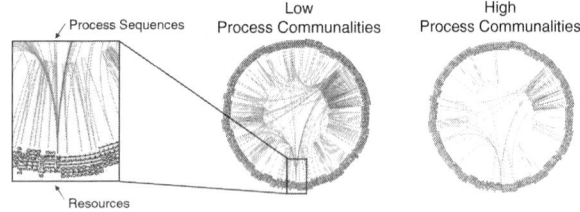

occurred. A large number of lines leaving an individual resource is an indication of a low level of process commonality.

Alongside the different process sequences, there is the additional problem that the resources are, in reality, often loaded to very different levels (Fig. 3.36). This is normally caused by two phenomena:

- On the one hand, the resource utilisation is influenced by production planning. Newer resources have very high hourly machine rates because of their higher acquisition costs. To reduce these hourly machine rates and so to be able to argue for a benefit to using the new resources, the machines are often loaded up to their limits. The other machines are thereby sometimes only planned to be used in a minor way.
- On the other hand, order deadlines vary for individual orders, which can lead to both increased waiting and queuing times for orders waiting for resources, or to downtime for the downstream resources.

The principle solution to increasing resource utilisation and at the same time increasing process commonality, can be broken down into three steps (Fig. 3.37), which are described below in detail.

Segmenting Resources in the Layout: Segmentation has been an accepted concept for structuring production design since the 1980s. A manufacturing segment is a limited, mainly stand-alone, organisational unit in which individual functional elements are

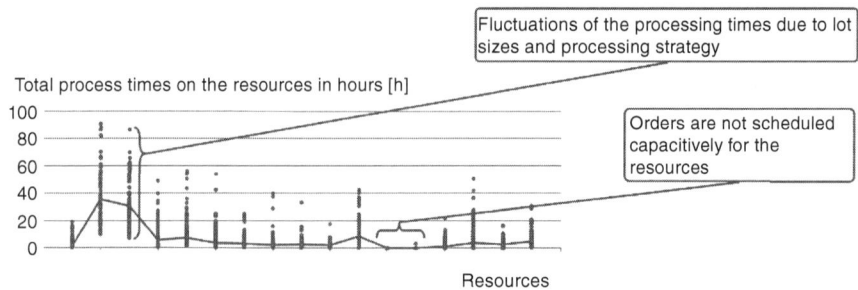

Fig. 3.36 Unfavourable conditions for balancing production

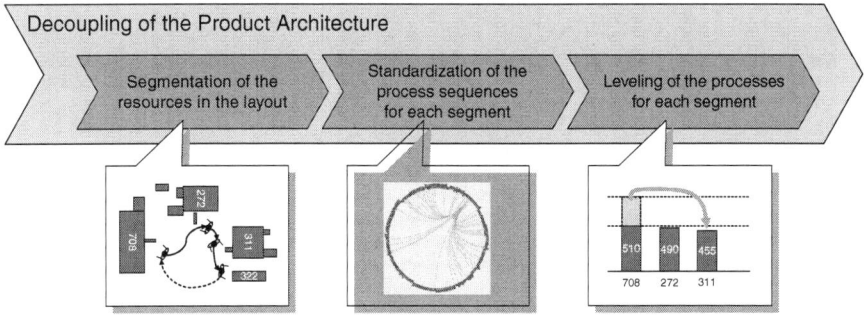

Fig. 3.37 Approach to achieving an overall increase in benefits of scale

combined to produce larger, product-oriented segments (Wildemann 1998). In a segment, orders or products are combined, which have the most similar possible work content or processing sequences. This allows process commonalities to be increased. To simultaneously provide a high load level, there need to be sufficient orders to be processed by each segment. But this in turn leads to not every product having the same work content or processing sequences.

Unification of Process Sequences per Segment: To further increase the process commonality within a segment, the process sequences need to be standardised. This is done by adjusting the existing work plans when work is being prepared. The degree of freedom to adapt the work plans is determined by the product design. This requires close collaboration between the development department and production or work planning. Clear instructions from production mean that products have to be designed in such a way as to reduce the variations in the processing sequences. The result is work plans which have the maximum possible identical process sequences per segment. The resources can subsequently be organised according to the process sequences.

Balancing Sequences per Segment: In addition to the adjustment of process sequences, the utilisation of individual resources needs to be levelled by balancing the processes. This is done by replanning the processing tasks within the segments. For example, milling and drilling can be moved from a universal tool machine to a milling and drilling centre. Again, with this type of adjustment, the degree of freedom available is determined by the product design. Therefore an integrative approach is also needed for this step.

Segments constructed in this way enable further improvements in production control. As is shown in the context of integrative, high resolution Supply Chain Management (Sect. 6.3), segmentation supports the creation of control sections. These control sections in turn allow a reduction in the throughput time by One-Piece-Flow. The application of the described approach is presented in more detail in Sect. 7.6 as part of the application example "Integrative Production Balancing".

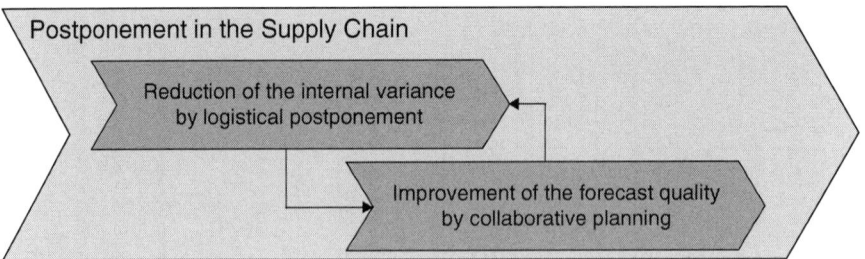

Fig. 3.38 Approach to postponement in the Supply Chain

Postponement in the Supply Chain

The dimensions and values described above for evaluating a Supply Chain form the basis for an objective and comparable evaluation of the customer-oriented Supply Chain in a product-production system. The optimisation of Supply Chains towards the Pareto optimum is task of the Supply Chain configuration, the heart of which includes both of the partial principle solutions of *"Logistic Postponement"* and *"High-resolution Supply Chain Management"*, as shown in Fig. 3.38.

Supply Chain Configuration generally is used to describe the purposeful design of system elements and their substitutes (especially the flows of materials and information) into a logistical network (Chandra and Grabis 2007). Depending on the approach, you can view these system elements as physical Supply Chain structures (e.g. the distribution structure), or as Supply Chain tasks (e.g. sales and procurement planning).

The decisive factors for Supply Chain configuration, along with the existing process configuration, are the product programme and the product architecture. Under the integrative evaluation model for configuring a product-production system, while configuring a Supply Chain, initially the logistically relevant product characteristics and the characteristics' options for the customer-oriented Supply Chain planning process are compared (see Fig. 3.27).

The product characteristics relevant to logistics or Supply Chain are, for example, the product type, the predicted length of the product life cycle, the number of product variants offered to customers, the vertical integration of production, the type of transport structure, the overall level of market uncertainty, the expected pattern of customer demand and the sales channels. While analyzing the process characteristics in the customer-facing Supply Chain, the fundamental Supply Chain process category must be identified. This determines the order penetration point, i.e. the point in the value-adding process from which further value creation, such as the generation or assembly of a customer-specific product variant, is only initiated if a customer order actually exists. Upstream of the order penetration point, allocation of products, materials and capacity is carried out based on a plan or forecast. Based on the usual multiplicity of relationships to different customers, any further process characteristics are mainly to be regarded as customer-specific. The general stability of the business relationship, its duration, as well as the proportion this customer represents of total product-based turnover, along with the technology aspects, are all decisive factors

3   Individualised Production

which need to be taken into account in the context of a "High Resolution Supply Chain Management (HRSCM)" approach.

The analysis of the effectiveness and dependency relationships between the product and process morphologies represents the first step of the integrative configuration logic. With the results of this, the initial configuration alternatives can be identified or can be excluded from further consideration. For example using an "engineer-to-order" process to produce consumer goods with a short life can be ruled out. On the other hand, an A-customer where a long-term, stable and trust-based business relationship, exists can be involved in working towards a higher level of collaborative planning or negotiation, by stepping up the level of exchange of information.

In the same way, the constitutive process features of a Supply Chain can already be defined (see also the third sub-section of 3.2.4.2). These arise on the one hand from the previously defined constitutive product features and on the other hand from the basic requirements according to the customer structure as well.

The second step in the Supply Chain configuration is the analysis of Supply Chain capital efficiency and delivery service level sensitivity to internal and external variations. The internal variation applied here is the breadth of the product programme. Offering a wide range of variants can have a negative impact on both the Supply Chain capital efficiency as well as on the delivery service level (see also Fig. 3.39). For example, containers may need to be provided for each variant in the distribution centres, or in production, otherwise the delivery period to the customer may vary accordingly. The external variance corresponds to fluctuations in customer demand for the relevant products over time. This variation is normally covered by increasing safety stocks and by increasing forecast levels above the actually expected demand and its fluctuations.

Both the principle solutions of "logistic postponement" and especially HRSCM influence the external and internal variations of the product-production system. While the principle of logistic postponement especially reduces the internal variations by shifting the logistical decoupling point in the product programme closer to the customer, the HRSCM principle increases the quality of the plan- or forecast-based Supply Chain processes within a customer-supplier relationship and so reduces

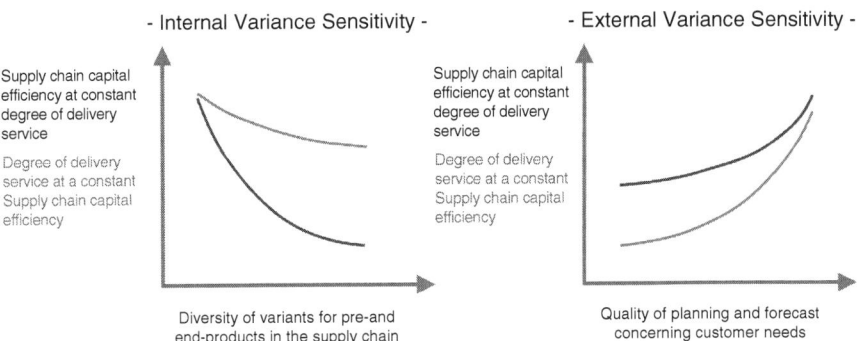

Fig. 3.39 Qualitative impacts of internal and external variations on the evaluation values of the Supply Chain quadrant

external variations. In a broader sense Supply Chain process categories, such as openness of information, can be regarded as constitutive features of the customer-facing Supply Chain.

Logistic Postponement: Logistic Postponement is described as a shift of the customer order decoupling point within the Supply Chain (Vahrenkamp 2007; Schönsleben 2007; Bretzke 2010). The aim of postponement is to move the creation of variants to the latest possible structural step in the Supply Chain (e.g. fitting, distribution stages) in order to reduce market risks, inventory and storage costs, and so to increase concomitantly the cost efficiency of the Supply Chain. By decreasing the overall number of variants of intermediate and final products at all structural stages of the Supply Chain, (see also Fig. 3.40), the work-in-progress and finished goods stock in the fitting and distribution inventories are reduced.

Achieving a logistical postponement has direct effects on the structure of the product architecture. Therefore it is vitally necessary for the product architecture to be built in such a way that proliferation of variants, (for example by selecting a colour or loading the software), only occurs at one of the later structural Supply Chain stages.

High Resolution Supply Chain Management (HRSCM): Coordinating the flows of materials and information within a Supply Chain can be applied in two different ways: by hierarchical or vertical coordination, or by non-hierarchical or horizontal coordination (Dudek 2008). Hierarchical coordination is normally applied within individual planning domains, e.g. companies. The aim of hierarchical planning is

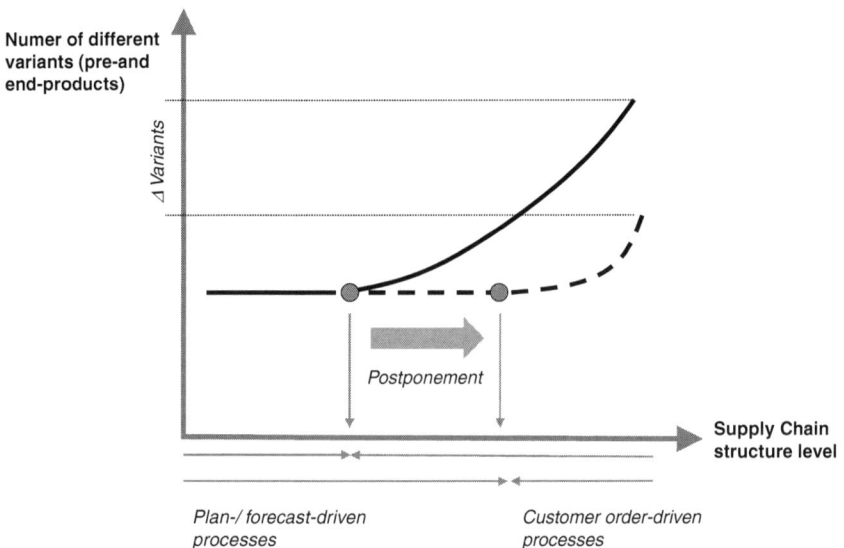

Fig. 3.40 Reducing internal variants by logistic postponement

to use IT support (such as ERP or MRP systems) to generate synchronised plans for lower planning levels. In a distributed Supply Chain the planning quality as well as the quality of the decision-making process can be substantially improved by taking account of additional and frequently amended information about horizontally-networked planning areas (suppliers, customers) (Kilger et al. 2008).

The High Resolution Supply Chain Management approach therefore aims to improve the quality of planning for non-hierarchical coordination of planning domains for sales, procurement and production planning in the various independent planning domains, i.e. in the participating companies of the Supply Chain which is studied, by increasing the transparency of information, in combination with decentralised and self-optimising control loops (Schuh et al. 2011c). The basis of the HRSCM approach lies in the intensive, extensive and frequent exchange of relevant information (inventory, capacity, planning data, delivery periods etc.) between the planning areas, which is needed in order to achieve a better level of accuracy in plans and forecasts for sales, production and procurement planning. The availability of real-time information in the Supply Chain has been made possible by technological advances in the area of data capture (e.g. RFID) and the integration of information technology in planning activities (Fleisch 2008). In order to make this information usable in planning areas for planning and control the other component of HRSCM is the decentralised organisation of planning tasks. A decentralised organisational structure for planning tasks is the determinant pre-condition for ensuring a controllable Supply Chain process.

The advantages of an HRSCM approach for a company are, on the one hand, the improved quality of forecasts and therefore more reliable planning for sales, production and procurement. On the other hand, thanks to the availability of information, the better utilisation of production and logistics capacities and the reduce of safety stocks.

After the application of the outlined principle solutions, an evaluation of the results is carried out as part of a re-evaluation of the subsystem's complexity within the Supply Chain. If there has been a successful reduction of the internal and external variation of the product-production system, then positive effects could be expected on the delivery service level and on the capital efficiency of the Supply Chain (see also Fig. 3.41). The final step in the configuration process is the decision on a possible recursion of the process.

Definition of Constitutive Features of the Production System

As was described above, a core result of the application of integrative principle solutions lies in the identification of potentially suitable constitutive product and process features to help resolving the scale-scope dilemma. In the final step of configuration, therefore, there has to be an analysis and evaluation across all areas of tension, of the proposed constitutive features, in order to be able to make an informed decision on their implementation.

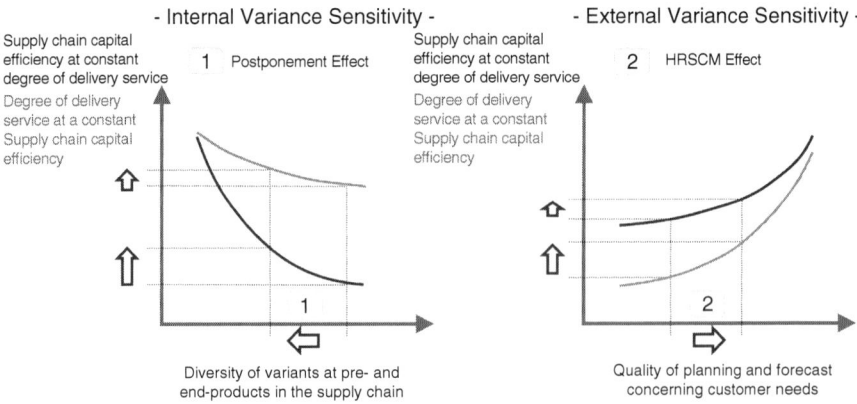

Fig. 3.41 Evaluation of the impact of postponement and HRSCM on the evaluation measures of the Supply Chain

The basis for this is presented in the evaluation matrix in Fig. 3.42. The candidates identified during the application of the integrative principle solutions as constitutive product and process features, are assessed using this matrix for their ability to influence the levels of variation. An entry on the matrix therefore represents the extent to which the definition of each of the constitutive features reduces the variants of another product or product feature. The basis for this evaluation is provided mainly by the previously analysed causal relationships between the features of the product and production morphology. Individual contributions by a potential constitutive feature managing the variants are assessed on a scale from 0 to 9, and added together as a line total. A pre-requisite here is that external, pre-existing variants must be respected. For example, no constitutive product feature can be defined which prevents the creation of necessary product variants. The line total for a constitutive feature enables its evaluation and prioritisation based on its suitability for resolving the scale-scope dilemma.

Supported by the derived priorities, the constitutive features are defined for the product-production system. A check on the implementation and its effects using a feedback loop is achieved by an iterative implementation of the integrative evaluation model.

3.2.5 Industrial Relevance

Different empirical studies demonstrate a relationship between the complexity of the product programme, the product architecture and the productivity of a production system (MacDuffie et al. 1996; Fisher and Ittner 1999). This relationship corresponds to the complexity-related causal relationships described in the product and production morphology. In addition these studies show a degree of ambivalence in

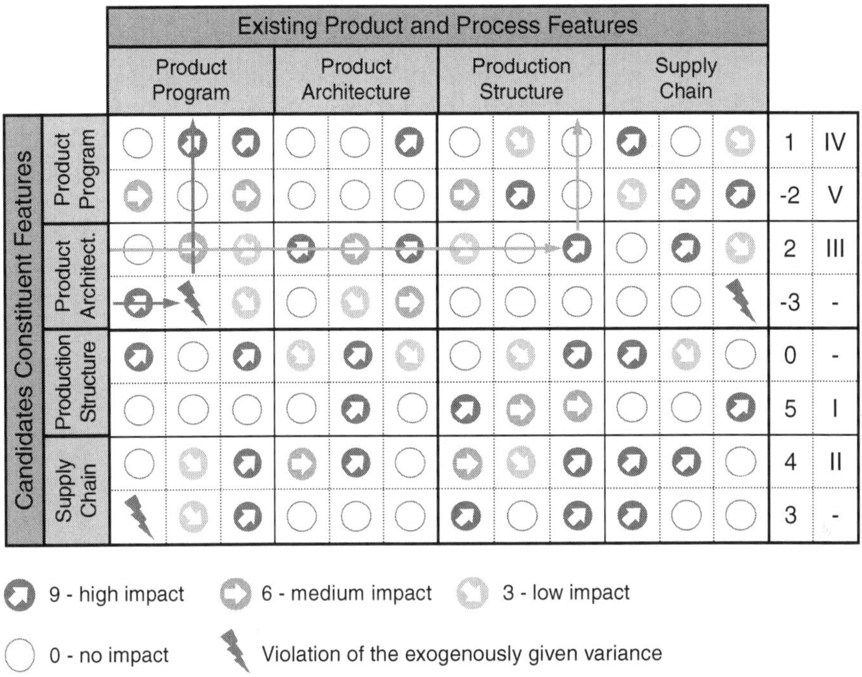

3 Individualised Production

Fig. 3.42 Evaluation and hierarchical organisation of constitutive features

the relationship between product complexity and productivity: several variable features generate disproportionate complexity disadvantages, while other features under certain circumstances only have little or no effect. According to this fact, there is a need to analyze products and production systematically and in an integrated way when configuring the production system and defining its constitutive features.

To confirm industrial relevance an empirical study was run in parallel to the development of the configuration logic (Schuh et al. 2010b, c). This particularly investigated the status quo of complexity management in manufacturing industry. Based on this research, it can be demonstrated that 87% of the companies asked are aware of the need to resolve the scale-scope dilemma. A huge majority (94%) of the companies are working on the development of a modular product approach as a possible measure. From these modular products companies are hoping, on the one hand, gaining differentiation advantages, particularly in relation to improve the reaction time to markets, and also improved product quality. As well as these scope benefits and scale benefits are being targeted in form of cost reductions. The latter are spread across the entire product life cycle, but with the accent being placed on the areas of Supply Chain and production.

The development of these modules is, however, in most cases (70%) based on a conventional approach to the development of individual specific products. An overarching design logic, as presented in the above text, is only being adopted by a few companies. Using success criteria such as the level of commonality achieved, solidity of planning or financial indicators, it can, however, be demonstrated that within the sample surveyed, it is mainly the top performers who are applying this kind of overarching design logic. For example, constitutive product features are being defined in the form of standardised product functions across model ranges. A noticeable concentration of the application of this kind of methodology is found in the automobile sector.

Against the background of these empirical research findings, the widespread need for a methodical, overarching approach can be found. As a second determining factor of industrial relevance, the ability to apply the proposed approach in an industrial environment must be demonstrated. To this end, the integrative evaluation model was applied to a privately owned medium-sized German manufacturing company (for reasons of confidentiality the company analyzed is described here anonymously) (Schuh et al. 2011a). Within this company, alongside actual production, there are product development and sales and logistics activities, so that all four areas of tension could be addressed in full. To simplify the analysis, and to achieve clear results, the observations were focused on a specific Business Unit and product line. However, an observation period of four successive financial years was adopted, in order to be able to see dynamic changes.

Within the application of the evaluation model it was possible to discern, over and above the general determination of indicators which were developed, that the model provided useful conclusions. Figure 3.43 below shows the quantitative results of this analysis.

As can be seen from Fig. 3.43, the results of the analysis support the hypothesis of a dichotomous relationship of the main dimensions in the *Product Architecture* and *Supply Chain* areas of tension. The time-based operational points here lie on the relevant diagonal curves. For reasons of data capture, it was not possible to determine a similar relationship for the product programme and production structure, as the main dimensions of *Explainability at the Point of Sale* and *Resource Utilisation* could only be quantified on the basis of static estimates.

The results obtained from the analysis, despite these limitations, do match the qualitative self-assessment of senior managers responsible for those areas within the company. The product programme which had grown up over time, and its product variants, are tailored to the applications of a small group of long-term customers. Within this group a high *Fit of Variety* and a high level of *Ease of Explanation* is always achieved. Given the increasing pressure for global sales of the products, and the related increase in size of the customer group, the *Fit of Variety* in particular is falling. A similar development in the *Ease of Explanation* is suspected by those responsible, but cannot be proven due to a lack of base data. A product programme which grew up organically, and was adapted to suit specific customer requests, is reflected in unclear architectural standards in the *Product Architecture* area of tension, and creates high *flexibility* with low *commonality*. At this level there is also a dynamic substitution of *commonality* by *flexibility* which can be traced back to the introduction of new variants. In relation to the production structure, given a high *process commonality*, this creates lower *resource utilisation*. Within the Supply Chain area of tension, the

3 Individualised Production

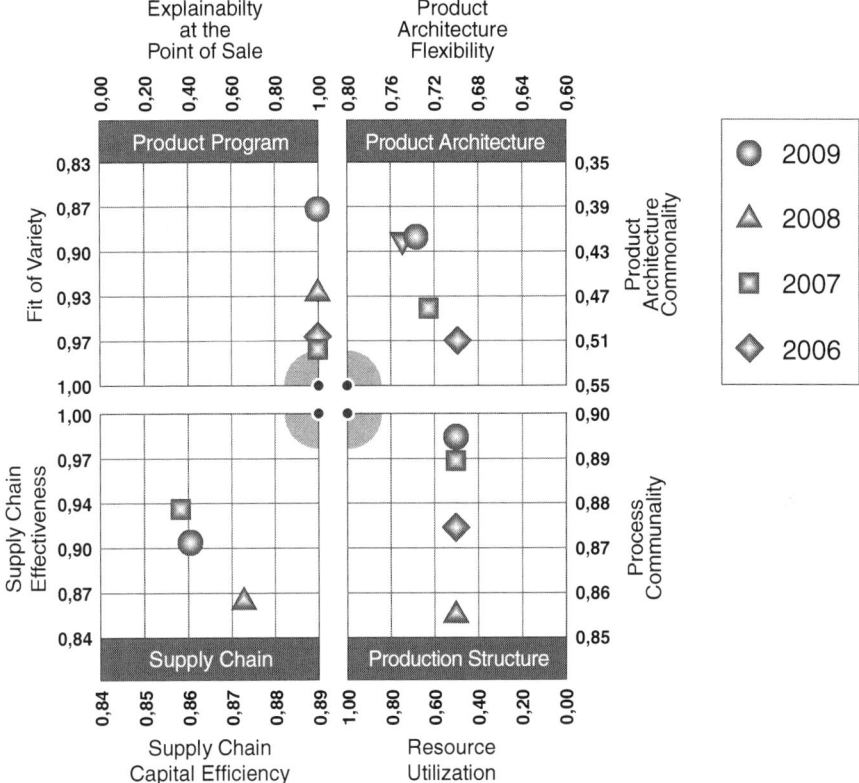

Fig. 3.43 Results of integrative evaluation in a practical example

result of the high process commonality has a high value for each, the Supply Chain capital efficiency and the Supply Chain delivery service levels.

In order to fill in the gaps left by this experimental application, as well as to provide a broader empirical basis, some further examples of application in practice are needed. Validation in this way of the evaluation model and configuration logic took place, for example, as part of the application example "Integrative Production Balancing" (see Sect. 7.6).

3.2.6 Future Research Topics

Starting from the results described above, there are various issues and tasks for future research in the area under focus. As well as further empirical investigation and underpinning of the configuration logic, another medium-term goal which needs to be defined here, is the development of an integrated design methodology for production and product architectures. The configuration logic presented can be used as a framework to organise the methodology to be developed.

A further cornerstone developing an "integrated construction methodology for product-production systems" is a better understanding of the complexity-based causal relationships between product and production morphologies. Based on this understanding, there could then be systematic exploitation and influencing of these causal relationships when designing product-production systems, in the sense of a construction methodology which is adapted to the specific problem. To enable this better understanding there are several research tasks to be undertaken: As a preparatory step to actual understanding, a way needs to be found to describe the causal relationships. It is necessary to develop a formal descriptive model for the causal relationships in the product and production morphology. This descriptive model must allow for the identification of all relevant characteristics of a causal relationship, or of a whole network of causal relationships. In addition, the descriptive model must include a quantitative-mathematical description of the causal relationships. Starting from this quantitative-mathematical description there can then follow a systematic analysis and evaluation of the causal relationship networks of a product-production system. Going beyond this, the formal descriptive model must also provide the basis for the development of a typology of causal relationships, which supports the use of a characteristics-based classification and a complexity-based evaluation of a causal relationship. In order to be able to apply the descriptive model and the derived typology of causal relationships in an industrial setting a method must be defined in parallel for identifying, classifying and describing causal relationships of a product-production system. This gives the person applying the method some process-oriented support for implementing the instruments described above.

A systematic approach to the evaluation of alternative product architectures and value-adding processes is needed as a further module of the described design methodology to enable methodical support for research into designs for product architectures and value-adding processes in relation to their sensitivity to variants. In addition, this evaluation approach will allow the identification of weak points and the necessary optimisation direction to remedy them, in relation to the product architecture and value adding design.

In order to achieve the indicated medium-term goal of an integrative design methodology for product-production systems, the modules described need to be pulled together into an overall methodology. This enables the systematic, iterative and fine-tuned development of product programmes and their architectures, as well as the associated value-adding systems. The interaction of evaluation logic and the mathematical description of causal relationships should enable the fit of product architecture and values add structure to be measured. On the basis of these measurements, there is then an iterative alignment of the product-production system elements to each other.

The results of this future research activity, particularly the approaches to operationalising complexity-based causal relationships, will provide a core element in creating a more general theory of production.

3.3 Tool-Less Production Technologies for Individualised Products

Damien Buchbinder, Jan Bültmann, Andrei Diatlov, Christian Hinke and Henrich Schleifenbaum

3.3.1 Abstract

One of the primary aims of the "Integrative production technology for high wage countries" excellence cluster is to resolve the dilemma of scale and scope, i.e. to eliminate the dichotomy of mass production with low costs on the one hand and cost-intensive production of individual products and small scale production on the other. Selective Laser Melting (SLM), a die-less process, is one of the most likely ways to achieve this aim. Nevertheless, its process and cost efficiency is as yet unsuitable for mass production.

To increase the efficiency of the process and therefore enable mass production of products individually tailored to customer requirements, a new production machine is being designed and implemented at the Fraunhofer ILT. It incorporates a laser system in the kW range, the first time in this type of system. To transform this laser power into an increased build-up rate and thereby increase the overall efficiency of the process, an optical system is being implemented which enables generative production with a significantly increased build-up rate. The initial samples made of high-alloyed 1.4404 steel have thus been manufactured with a build-up rate of approximately 21 mm^3/s. In comparison with the industrial state of the art this represents an increase of more than 1,700%.

However the accuracy and detail of additively-manufactured components lowers as the melt pool size increases, i.e. as the diameter of the beam increases. To address this dilemma, a multi-beam concept is being designed and implemented which enables processing with different laser powers and focal diameters depending on the required component features. In analogy with the conventional roughing and finishing processes, component cores can be generated with high build-up rates (wide focus) while the required detail resolution and surface quality can be guaranteed in the shell area by reducing the rate (narrow focus).

In future therefore at least short runs can be manufactured at significantly lower costs using Selective Laser Melting without loss of the increasing individuality required by the customer. In order to optimise the system further, work needs to be carried out initially on widening the existing process window with the aim of optimising build-up rates and secondly on developing suitable build-up strategies for the skin-core interface.

3.3.2 State of the Art

Advancing globalisation and the associated saturation and segmentation of the markets is leading to an increase in the number of product variants and to the demand for customer-specific individualised products (Kinkel 2005; Schuh 2005). The resulting requirements for highly flexible and individualised production of products are seen as one of the greatest challenges for production technology today (Wiendahl et al. 2000). One possible solution to the economic production of small to medium unit quantities is die-less production processes. Since the fixed costs are far lower than in moulding processes, die-less production processes can enable the cost-effective production of smaller unit quantities. So-called additive production processes play a particular role in this. These processes, also known as Additive Manufacturing, 3D Printing, Solid Freeform Fabrication, Rapid Manufacturing, Layer-based Manufacturing, Digital Direct Fabrication or Direct Digital Manufacturing are distinguished in that complex three-dimensional components are built up in layers directly from a 3D data model (Gibson et al. 2010; Gebhardt 2007; Hopkinson 2006). In terms of metallic materials, processes are basically divided into local material feed and powder bed-based processes.

3.3.2.1 Additive Production Processes with Local Material Feed

In the case of Laser Metal Deposition (LMD) a powdered additive is melted with the laser beam and metallurgical fused with the substrate. Basic process features are the high-precision automated application of layers with thicknesses from 0.1 mm up to several centimetres, metallurgical fusion of the layer with the work piece, minimal heat input into the component, a wide choice of materials and the processing of practically all metallic alloys. The LMD process can be used for various tasks such as wear and corrosion protection, repairing worn or incorrectly machined work pieces, shape deposition welding and the manufacture of 3-D parts. By varying the process control and by the additives, alloying or reinforcement of the edge zone of a component is also possible. The most important application areas are in mechanical engineering, tool- and die making as well as power train and engine building. The principle of the Laser Metal Deposition process is outlined in Fig. 3.44

The company POM (Precision Optical Manufacturing) uses CO_2 lasers with outputs of up to 6 kW for its DMD (Direct Metal Deposition) process with several powder feeders for simultaneous processing of different materials. In this way, it is possible to implement in one component both composite layers in which areas of different materials can be built-up as well as graded layers in which the proportions of different materials can be gradually varied. The material range comprises tool steels, stainless steels, copper alloys and aluminium alloys. POM pursues two strategies in tool making. Firstly, milled preforms made, for example, of copper alloys are coated with wear-resistant layers which are then conventionally finished. Secondly, tools are built up completely from tool steel. Build-up using low melting-point base material such as copper in the required voids enables the manufacture of producing structures up to 90° from the perpendicular such as contour-adapted cooling ducts. The base material is melted out after the DMD process and the corresponding void remains in the component. DMD-manufactured tools made of H13 (1.2344) exhibit a yield strength and a tensile strength of 1,500 MPa and/or 1,820 MPa, 6% elongation and a

Fig. 3.44 Principle of the Laser Metal Deposition process

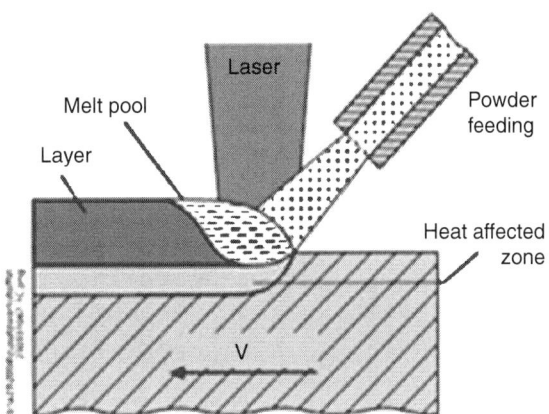

hardness of 46–50 HRC. The minimum beam diameter of the lasers used is 0.6 mm and the typical layer thickness is between 0.13 and 0.38 mm.

A comparable process, Laser Engineered Net Shaping (LENS), is supplied by the company Optomec. The material palette also comprises tool steels, stainless steels, copper, nickel and aluminium alloys. As in the case of POM the manufacture of graded and composite layers is enabled by the simultaneous use of several powder feeds. Unlike POM, Optomec uses fibre coupled Nd:YAG lasers and machining systems with up to five axes which gives greater geometrical freedom. A specific tool-making strategy also enables the manufacture of producing structures such as contour-adapted cooling ducts. In this case the void is filled with the same powder material as the component being produced and is fused onto the surface by a laser beam without further powder feed. This creates a closed component surface on which to build up the part with further powder feed. Tools made of H13 tool steel exhibit a hardness of 48–58 HRC. There is no information about further mechanical properties. The minimum beam diameter of 0.5 mm is comparable to that of POM. The processing of TiAl6V4 using LENS produces components with yield strengths of 833–1,066 Mpa and tensile strength of 900–1,112 MPa at elongations of 0.8–13.1%. A decisive factor for the properties is firstly the post-processing of the samples with stress relief annealing or Hot Isostatic Pressing (HIP) and secondly the alignment of layers into the test geometries (parallel or vertical to the layer build-up).

The company Aeromet uses the LasForm process to manufacture large structural components from various titanium alloys for the aviation industry. They use a 14 kW CO_2 laser to manufacture the blanks which are then machined and heat treated. The components are therefore oversized by between 0.8 and 5 mm. The aim is to avoid costly forging processes, attachment of additional components to forged parts and repair of components. The aircraft industry favours the process due to the better material exploitation of the additive process as compared to conventional material removal processes, which provides considerable commercial benefits in the light of high material prices and so-called buy-to-fly ratios of on average 5:1. Tempered and evacuated LasForm parts made of TiAl6V4 have a yield strength and/or tensile strength of 839 MPa and/or 900 MPa at 12.3% elongation. The maximum part size is $2.5 \times 1 \times 1\,m^3$ and overhangs of up to 60° to the vertical are implementable.

The combination of an additive process with a material-removing process such as milling enables the exploitation of both better material usage in the case of the generation method coupled with the higher quality surfaces and accuracy of size in the case of the material removal method. One such combination process is the CMB-process (Controlled Metal Buildup) which has been developed by Fraunhofer IPT and is used by the company Röders. The generative process step is carried out using Nd:YAG or diode lasers and wire or powder feeders and it is combined with a high-speed milling head. Once a single layer has been built up, laser machining is carried out to achieve high dimensional accuracy and a high-quality surface finish. Tools made of Cronitex RC44 (Similar to X38CrMoV5-1) exhibit hardness values of between 41 and 59 HRC. The build-up of overhangs of more than 20° to the vertical and particularly post-machining in the form of milling is problematic due to the use of the 3-axis machine. The diode laser beam diameter is 1.2 mm, which in combination with a wire of 0.8 mm enables a minimum web thickness of 1 mm.

The benefits of these processes lie in the use of single-component metallic materials, the possibility of simultaneously processing several materials and the production of graded and composite layers. The resulting components exhibit high densities, a fine-grained microstructure and good mechanical properties. The disadvantage at present is the lack of detail resolution due to the focal diameter of the laser beams used (minimum 0.5 mm) and the minimum layer thickness of 0.13 mm. In addition, protruding structures beyond a certain angle from the vertical are impossible or require considerable extra work in the use of supporting material or filling the voids and subsequent sintering.

3.3.2.2 Powder Bed-Based Additive Production Processes

Unlike the process with local material feed, the powder-based principal involves applying a layer of material corresponding to the layer thickness and selective melting by energy beams (Fig. 3.45). The common principle is a repeated process comprising three steps. There are several commercial systems in existence which differ in terms of the materials processed, the energy source and the additional process steps.

Selective Laser Sintering (SLS) is a widespread process which can refer to both indirect and direct SLS. In indirect SLS the process technology is used for the additive manufacture of thermoplastics in order to produce metallic components. This is achieved by coating the individual metal powder grains with a thin layer of plastic. When exposed to the laser beam, only the thermoplastic outer shell is melted. The resulting green compacts are low in density and strength and require subsequent processing to increase their strength. In the case of direct SLS however, special multi component powder systems are used. The individual components are metallic powders wherein a low melting point component is used as a bonding agent. This

Fig. 3.45 Principle of the powder bed-based process on the basis of Selective Laser Melting (SLM)

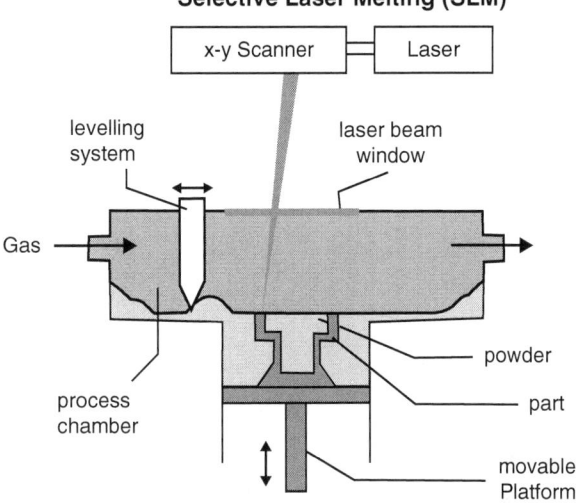

bonding agent is melted by exposure to the laser beam. Similar to liquid-phase sintering, the material is made denser by reordering the fixed high melting point components in the liquid phase. Indirect SLS is used commercially by the company 3D-Systems (formally DTM), and direct SLS is offered by the company EOS. The materials go by the trade names Rapid Steel 2.0, LaserForm St100 (3DSystems) and Direct Metal and/or Direct Steel (EOS).

Along with commercial suppliers, many institutions are involved in research into the modification and refinement of the SLS process, primarily in the area of material development. One approach is to develop new material combinations such as the Laser-Tool Material, an iron-based alloy which has better mechanical properties than EOS's Direct Steel 50 and substantially less susceptibility to cracking (Simchi et al. 2003). The residual porosity has been reduced from 5 to 1% and the hardness increased from 180–320 HV0.3. A further research point is the use of a different high-fusing phase for the fluid phase sintering of composite powders. Stainless steels, heat treated and high-speed steels, tungsten carbide, titanium carbide, ceramic and nickel alloys are just some of the materials used (Over 2003).

Another system for the additive production of metallic components in the commercial area is 3D-Metal-Printing from the company ProMetal, which is based on MIT's 3-D printing process. Unlike SLS, the layer information is transmitted by means of a print head and a fluid binder to a single component metal powder bed (steel, nickel, aluminium). Like indirect SLS, a pre-form is created which is infiltrated with bronze in a furnace process and which then exhibits a composition of 60% steel and 40% bronze. The benefit of this process is that it does not require a laser source. The resulting parts are therefore subject to less delay which increases commercial viability. Its disadvantage is that the components shrink by up to 1.5% in the infiltration process. Characteristic mechanical values are available only for stainless steel parts. When using SS420 steel the parts exhibit a hardness of 26–30 HRC, a tensile strength of 696 MPa, a yield strength of 676 MPa and an elongation of 9%.

All presently known processes have the disadvantage that they use materials or material combinations which have been specially adapted for the process. This means that the manufactured parts have different characteristics to the series production products and their use is therefore limited. The use of series materials without additives is therefore another focus for research and development.

As well as new approaches such as Solid Freeform Powder Moulding or Metal Printing with mask exposure, the developments in the single component area based either on fixed phase sintering or on the attempt, similar to the process with local material feed, to fully melt the material. In the case of fixed phase sintering, the material is heated to just below melting point and sinter necks form between the individual particles. (Kathuria 2000) produces a porous structure which can be compacted for example using HIP to higher densities of up to 99%. The materials used are stainless steels, nickel-based alloys such as Inconel 625, iron-based alloys and ceramic. Wohlert [2000] proposes the combination of SLS and HIP of TiAl6V4. The aim of the investigations is to manufacture parts using SLS with core regions of 80% density and edge regions of >98% density. The compaction is entirely carried out by HIP. This involves the use of a modified SLS system with vacuum technology, a 130 and/or 250 W Nd:YAG laser and infrared heating up to 650 °C (powder bed temperature).

Direct Metal Laser Remelting (DMLR) is based on the principle of fully melting a single component metal powder. The subject of the investigations is stainless steel powder (316L), in which individual layers are generated using a pulsed Nd-YAG laser beam. Adjustment of the process parameters indicates a significant influence on the fusion of the individual scan lines to form unified layers and the exclusion of cracks and delay. Multiple coating and exposure of individual layers enables the build-up of elements with a density of >99%. There are no available results of mechanical tests. There are also a few commercial systems available for complete plasticisation. The company Arcam uses an electron beam instead of a laser in a vacuum chamber in order to build up components from a tool steel powder bed. Tools manufactured using this technology have produced some 60,000 parts in the die-casting process and over 100,000 parts in injection moulding. The company Concept Laser uses a different approach with a machine which can be used both for laser labelling and 3D laser cutting as well as lasercusing. Lasercusing, according to Concept Laser, can be used to build up components with 100% density from single-component powdered standard materials. Nevertheless, the only material that can be used on this machine at the present time is 1.4404 stainless steel.

3.3.2.3 SLM, Selective Laser Melting

The Selective Laser Melting process originated in Rapid Prototyping and or Rapid Manufacturing. SLM can be used to manufacture prototypes and functional elements

3 Individualised Production 141

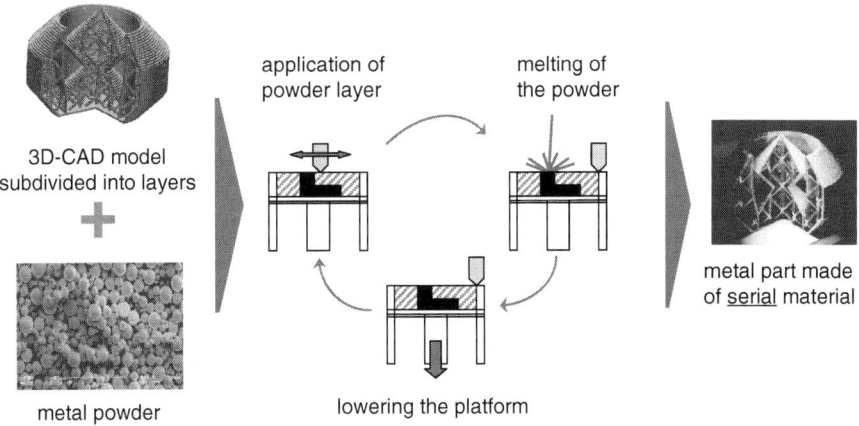

Fig. 3.46 Principle of the Selective Laser Melting process, SLM

from single component series-identical materials (Over 2003; Meiners 1999). The basis is a 3D-CAD volume model which was broken down into layers. The component is then built up in layers from a powdered material according to the CAD data. The spacing between the layers corresponds to the sections in the CAD model. The geometric information of the individual layers is transmitted by laser beam to the powder bed wherein the regions to contain solid material are scanned (Fig. 3.46).

By absorption of the laser beam the powdered material is completely plasticised in a localised area. It then solidifies to form a dense material which is recognisable as a solid material scan line. The mechanism is a melting process wherein the compaction is determined only by the flow and wetting behaviours of the melt (Zhang 2004). The adjacent positioning of such scan lines creates surfaces of solid material. After selective melting of a surface, the component is lowered by the amount of one layer thickness and new powder is applied by a coating unit. The laser process is then recommenced. In the process the newly melted material is metallurgical fused to the preceding layer. This overlaying of layers can be used to create almost any contour.

Figure 3.47 below shows the essential characteristics of additive production processes with local material feed based on the example of Laser Metal Deposition (LMD) and of powder bed-based additive production processes based on the example of Selective Laser Melting (SLM).

3.3.2.4 Process Efficiency in the Case of Selective Laser Melting (SLM)

Research activities in the context of SLM have so far focused primarily on the qualification of new materials and their industrial implementation. As yet there has been no known systematic investigation of the process parameters for increasing the build-up rate in the case of SLM. In order to be able to analyse the build-up rate and therefore the process efficiency in detail, the overall process time is divided into

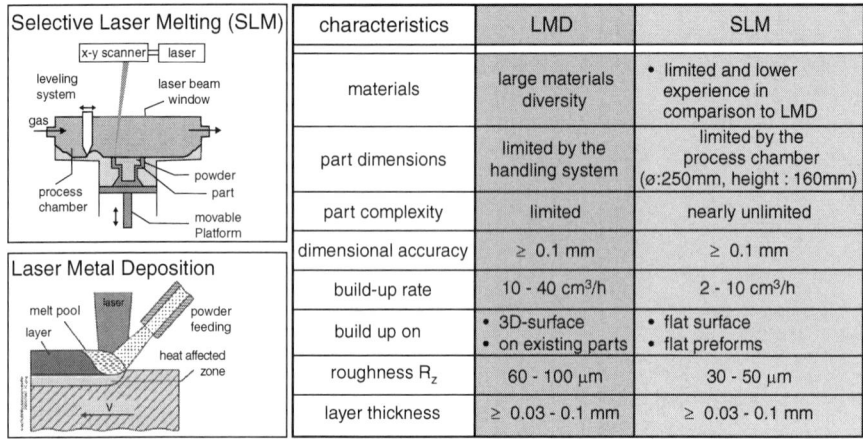

Fig. 3.47 Comparison of Laser Metal Deposition (LMD) and Selective Laser Melting (SLM)

primary and secondary processing time as for conventional production processes. The primary processing time is the time required by a laser beam to melt the powder layer. The secondary processing time is the sum of the times required for the process chamber to be prepared for the primary period. These include for example lowering the component platform and the coating process. The primary processing time is determined by the process parameters of layer thickness (D_s), scanning speed (v_{scan}) and the scan line spacing (Δy_s). The process-related build-up rate is determined according to the following mathematical formula:

$$\dot{V}_{Process} = D_s \cdot v_{scan} \cdot \Delta y_s \tag{3.13}$$

The layer thickness and scanning velocities are limited amongst other factors by the available laser power. The scan line spacing is limited by the diameter of the beam. The process parameters are generally material-dependent and must be adapted accordingly. The material and process parameters summarised in Fig. 3.48 have been investigated as far as the current level of knowledge permits.

Some of the results in the top rows of Fig. 3.48 are based on individual scan line trials while others give no details about the density of the samples produced. The (theoretical) values of the build-up rate (right-hand column) therefore indicate limit values depending on the material used and the system technology under laboratory conditions. The values in the bottom rows of Fig. 3.48 however represent the current industrial state of the art for the material 1.4404. These values have been verified by the Fraunhofer-Institut für Lasertechnik ILT on the "TrumaForm LF" production machine belonging to the company Trumpf and a benchmark of the "Laserinstitut Mittelsachsen" in collaboration with "Fraunhofer-Allianz Prototyping" and "LBC GmbH". On the basis of these values, efforts are being made to increase the build-up rate while preserving the industrial viability of the process and of the components.

Source	Material	Laser source	Max. Laser power* [W]	Beam diameter (Scan line spacing)** [mm]	Scan velocity [mm/s]	Layer thickness [mm]	Theoretic build rate*** [mm³/s]
(Meiners 1999)	stainless steel (X2 CrNiMo17 12 3, X2 CrNi24 12), tool steel (1.2343), nickel	Nd:YAG (cw)	105	0.2 (0.14)	< 200	< 0,1	< 2.8
(Over 2003)	tool steel (X38 CrMoV 5-1), titanium (TiAl6V4)	Nd:YAG (cw)	120	0,2	< 250	< 0,1	< 3.5
(Zhang 2004)	aluminium (AlSi25, AlSi10Mg, etc.)	Nd:YAG (cw)	330	< 0.4	< 250	< 0,1	< 7
(Wagner 2003)	stainless steel (1.4404), hot-work steel (1.2714), ni-base alloy (IN718)	CO_2	200	0,1	50	0.05 - 0.1	0.5
(Petersheim 1997)	stainless steel (X38 CrMoV 5-1 & X40 CrMoV 5-1)	no information	no information	~0.4	no information	< 0.4	——
(Kobyrn 2001)	Titan (TiAl6V4)	no information	no information	< 0.5	no information	0.13 - 0.38	——
(Su 2003)	tool steel	Nd:YAG (pulsed)	550 (150)	0.9 (0.6)	< 10	0.4	< 2.4
(Wang 2002)	WC-Co	Nd:YAG (cw)	60	0.8	30	0.2	2.5
(Kruth 2005)	Cu, Ni,Fe3P	CO_2	60	0.3 (0.2)	< 100	0.2	< 4
(Meiners 2007, NN 2005)	stainless- & tool steel (1.4404 & 1.2343)	Nd:YAG (cw)	250	0.2 (0.15)	160	0,05	1.2

Fig. 3.48 Material and process parameters in Selective Laser Melting (SLM). *In the case of a pulsed laser beam, the details in brackets indicate the average laser power. **If no scan line spacing is specified, this is determined according to the empirical ratio Dys @ 0.7 × beam diameter, ***calculated

The build-up rate achieved in (Meiners 2007; NN 2004) thus serves as a reference for the results which follow.

3.3.3 Motivation and Research Question

An essential production technology issue in high wage countries is the dilemma between scale-orientated mass production and scope-orientated individual production. In high wage countries the trend is towards ever-smaller production units of highly specialised and/or individualised components. Company strategies such as "Mass Customization" or "Open Innovation" intensify this trend because the demand for individualised products at the same costs as mass-produced products is increasing significantly (Reichwald and Piller 2006). From a technological perspective, generative production with its practically unlimited geometrical freedom is one of the process groups which offers the greatest potential towards resolving this dilemma. Particularly Selective Laser Melting (SLM) developed at the Fraunhofer ILT for processing single-component metallic series-production materials permits the rapid manufacture of complex components with series-identical characteristics. The main cost driver of the SLM process is the manufacturing time of the parts which is primarily determined by the volumes required. The build-up rate is limited to the existing system technology and laser power (approximately 200 W) so that additive processes at the current time are primarily used in preproduction and in manufacturing individual parts.

Fig. 3.49 Volume build-up rate in the case of Selective Laser Melting, SLM

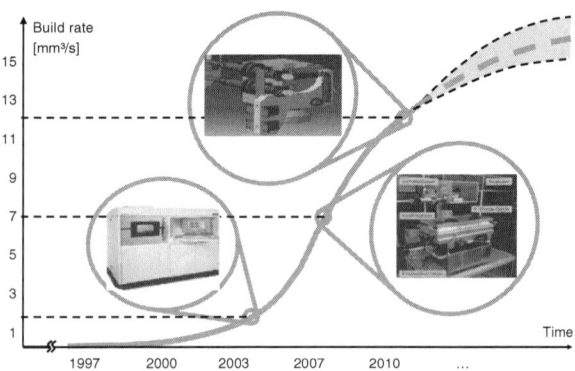

A central research aim is therefore to increase the build-up rate to enable the commercial production of at least short runs.

A key secondary condition in the increase of the build-up rate is the retention and/or improvement of the currently achievable level of detail in order to reduce or avoid additional cutting cycles. It should be possible to significantly increase the build-up rate and thus the entire process efficiency while retaining the currently achievable accuracy (Fig. 3.49).

The material vaporisation rate may be expected to increase by increasing the intensity through increasing the laser power with a constant beam diameter. The energy hereby consumed is no longer available for the actual melting process. Spattering also occurs, which can be expected to multiply process instabilities. Therefore the key research question is whether and to what extent the scaling of the laser power from 200 W to 1 kW is associated with a (linear) scaling of the build-up rate.

3.3.4 Results

3.3.4.1 Model and Diagnosis-Based Process Development

The approach for the model and diagnosis-based process development is illustrated in Fig. 3.50. The aim of increasing the build-up rate by a factor of 10 can only be achieved by optimised process control. For this purpose the key basic physical processes (Fig. 3.50 left) are modelled and their interactions with the essential process parameters in the case of SLM are analysed (Fig. 3.50 right). The corresponding results are illustrated in the following sections.

3.3.4.2 Machine Concept

Unlike the die-related production process, production costs in the case of the additive production process are basically influenced by the component volume but not by the

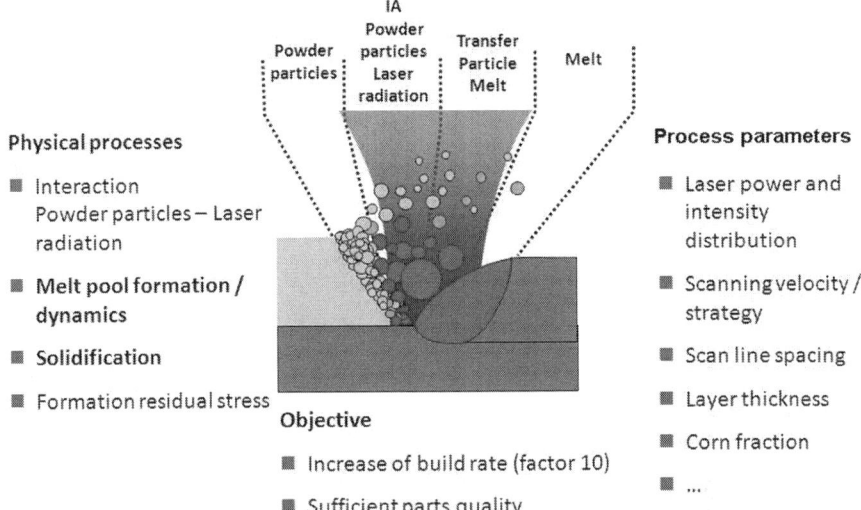

Fig. 3.50 Model and diagnosis-based process development in the case of Selective Laser Melting (SLM)

component complexity. The key cost driver of the additive production process is therefore the volume melted per time unit, the so-called build-up rate.

In investigations into increasing process efficiency, Schleifenbaum et al. show that the potential for increasing the build-up rate with a constant beam diameter is limited (Schleifenbaum et al. 2010). It is also known from (Meiners 1999; Schleifenbaum et al. 2008, 2010) that the detail resolution and the surface quality of the parts produced deteriorate when the beam diameter is increased. In general the components to be produced feature different requirement criteria. Thus the requirement for minimum surface roughness and maximum detail resolution is limited to functional surfaces and external contours of the components. Within the component interior however, such requirements are fewer. In this case it is necessary only to provide a supporting structure which fulfils the function. Thus, in analogy with the conventional roughing and finishing production approach, the so-called skin-core strategy is applied (Fig. 3.51).

To achieve a sufficient detail resolution and surface quality, the skin of the component can be generated with a beam diameter of 0.2 mm to guarantee the required surface quality and the core can be generated with a larger beam diameter to increase the build-up rate. In preliminary tests aimed at increasing build-up rates, beam diameters from 0.8 to 1.2 mm were investigated (Schleifenbaum et al. 2008). In accordance with these results, the approach focuses on increasing the beam diameter to approximately 1 mm. For this approach, a new optical system is required. As well as providing the necessary optical energy for the High Power SLM process, this system is also required to vary the beam diameter automatically in the processing plane at least within a range from 0.2 to 1 mm during the construction process. In

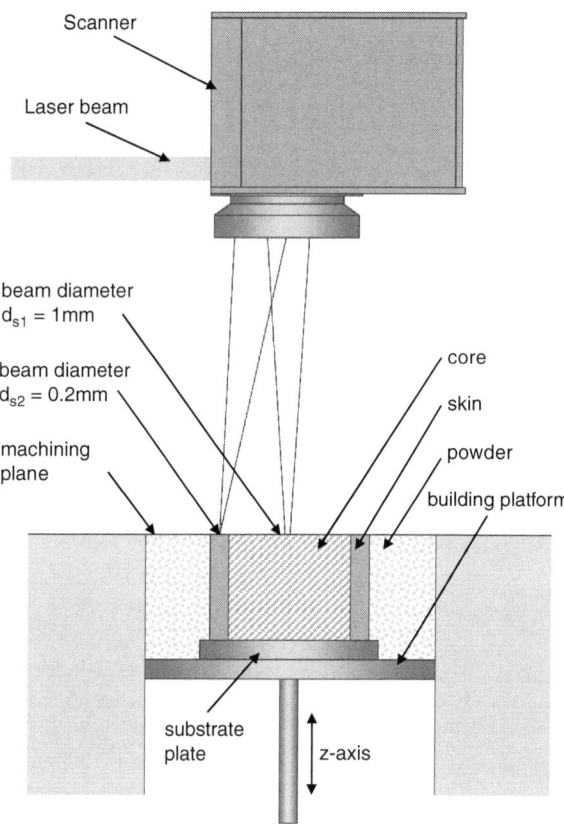

Fig. 3.51 Skin-core production principle

this case as evenly-shaped an intensity distribution as possible ("Top-hat") should be aimed at the processing location (Schleifenbaum et al. 2008).

In order to achieve a top-hat intensity profile at the processing location it is necessary to work within the focus of the laser beam because the intensity distribution of non-fundamental modes fluctuates along the propagation direction. Therefore the design of the optical system must guarantee a sharp image of the fibre end at the processing location. Various optical elements are available which enable the focal diameter to be varied. These include for example collimation optics, flat field lenses and the diameter of the fibre-optic cable (FOC). The formal relationship between the focal diameter and the aforementioned optical elements can be described as follows:

$$d_{Focus} = \frac{w_l \cdot f'}{w'_l \cdot f} \cdot d_{LLK} \qquad (3.14)$$

In this case d_{Focus} is the focal diameter at the processing location, w and w' are the laser beam radii on the focusing and/or collimation lens, f and f' are the focal lengths

Fig. 3.52 Schematic illustration of the automated fibre-optic cable switching

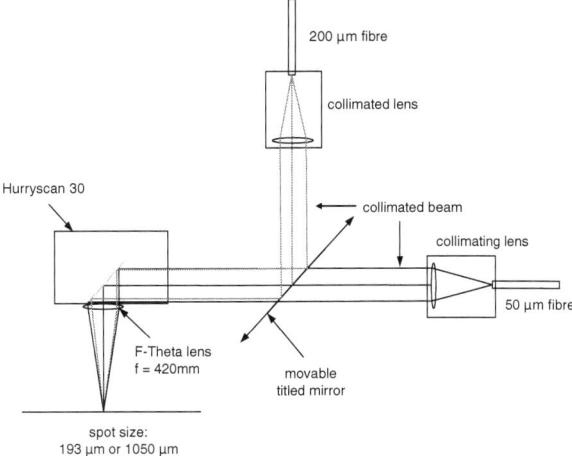

of the collimation and/or focusing optic and d_{LLK} is the core diameter of the fibre-optic cable. The focal diameter in this setup is varied by the automatic switching of two fibre-optic cables (core diameters 50 and 200 μm) (Fig. 3.52). With this system the focal diameter can be varied between 193 and 1,050 μm.

Because fibre-optic cables are sensitive to mechanical stresses such as torsion, crippling etc, the automatic switching is implemented at the work piece end by a mobile 90° beam deflector cube. This can travel along a linear axis in the beam path of the fibre-optic cable, which is arranged concentrically to the scanner input opening in the case of an active fibre-optic cable arranged vertically to it. In this case the beam deflector cube travels in the mirror plane so that the end positions of the mirror path do not represent critical values for the co-axiality of the two laser beams (Fig. 3.53). After entering the scanner the laser beam is deflected by two galvanometer-operated deflector mirrors to the processing location. The collimated laser beam is focused by a flat field lens (F-θ lens). This arrangement guarantees that the overall optical system is sealed off from its environment. This is an essential criterion for series operation of the machine in industry in conjunction with the powdered based material.

To extend the field of application of the machine, the optical system is designed such that with the use of different optical components, for example collimations or F-θ lenses of different focal lengths, different focal diameters can be set at the processing location.

3.3.4.3 Fibre-optics

In house investigations have shown that the scope for increasing the build-up rate through increasing the laser power in connection with a Gaussian intensity profile is limited. Firstly the intensity in the centre of this kind of output density distribution is higher than the average intensity by a factor of 2. This means that more powder grains

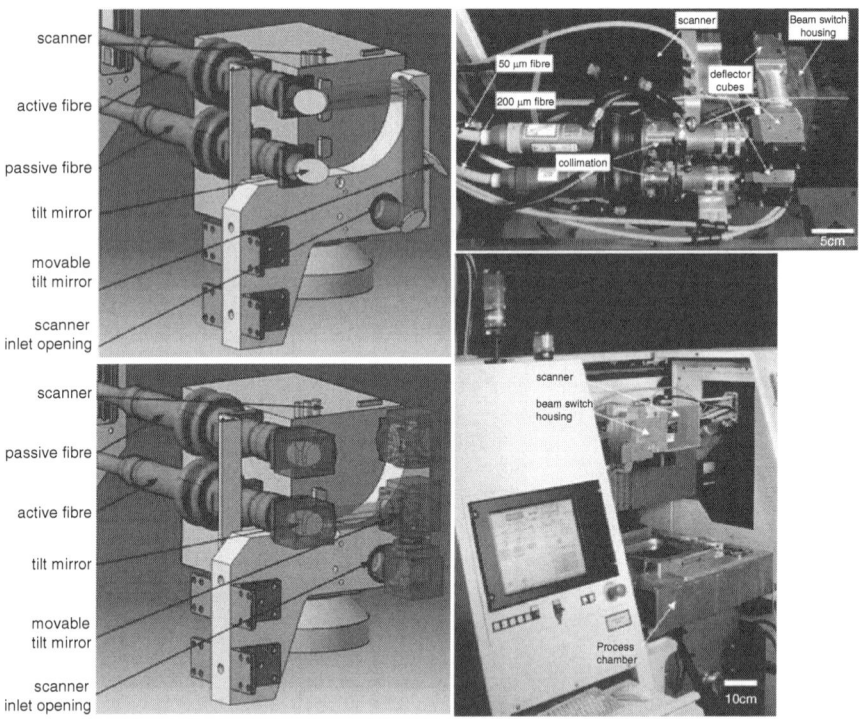

Fig. 3.53 3D model (*left*) and work piece-end illustration (*right*) of the optical system with Beam-Switch

exceed the vaporisation temperature thus increasing spatter and process instability. However the maximum trailing edge steepness of a Gaussian intensity profile is limited such that the powder cannot fully melt in the edge region. This in turn limits the possible spacing between the individual scan vectors, the scan line spacing Δy_s. This is incorporated into the calculation of the build-up rate on a linear basis however and should thus be maximised. The intensity distribution at the processing location thus plays an essential role in the design of the beam source and the optical system.

A top-hat output density distribution is beneficial to the SLM process, particularly for increased build-up rates, because this is how the scan line spacing can be maximised on the basis of the steeply ascending intensity flanks. This kind of beam profile requires the superimposition of many transverse electromagnetic modes over the run-length of a fibre-optic cable. However not all modes can be propagated within a fibre-optic cable (Pedrotti et al. 2002). The maximum number of modes, m_{max}, is reached for the smallest angle of entry; the limit angle of the total reflection. This angle is determined by the varying indices of refraction of the fibre core and the fibre surface and is incorporated into the so-called V parameter (V) of an optical fibre.

This parameter can be used to differentiate between single and multimode fibres. In the case of a multimode fibre, the number of modes (N) in a cylindrical fibre is:

$$N = \frac{4}{\pi^2} \cdot V^2 \qquad (3.15)$$

The probability of achieving an even top-hat intensity distribution increases with the number of superimposed modes in the optical fibre. The theoretical number of modes of the optical system described above is 114 in the case of the 50 µm fibre, and 1,825 in the case of the 200 µm fibre. The theoretical values of the mode number indicate that the intensity profile of the fibres with a small core diameter (50 µm) due to the relatively small number of modes exhibits neither an ideal top-hat shape nor a Gaussian intensity profile ($N=1$). However a top-hat output density distribution can be expected in the case of a FOC with a core diameter of 200 µm.

3.3.4.4 Output Density Distribution and Scan Line Width

In order to estimate the influence of the intensity distribution on the possible scan line width, the interaction of powder and laser beam is analysed in a theoretical model. The thermalised optical energy in a powder grain can be calculated using the following integrals:

$$E = A \cdot \iiint I(r,\varphi) \cdot r \cdot dr\, d\varphi\, dt \qquad (3.16)$$

In this case A describes the absorption coefficients of the powder material, I the intensity distribution, r the radius, ϕ the angle and t the time. Taking into account a rotationally-symmetrical Gaussian output density distribution at any distance Z along the propagation direction of the laser beam, the intensity distribution can be expressed as follows:

$$I(r,z) = I_0 \cdot e^{\frac{2r^2}{(w(z))^2}} \qquad (3.17)$$

In this case I_0 is the maximum intensity at the centre of the beam, w is the beam radius and r is the run variable in the radial direction. This means that powder grains in the outer area of the beam absorb less energy than those near or at the centre of the beam. If the absorbed energy is less than the thermal energy content of the grain, (complete) plasticising does not occur. This in turn can lead to pores and flaws within the components.

On the contrary, an even intensity distribution, a so-called top-hat, can be described with the following equation:

$$I(r,z) \approx I_0(z) \qquad (3.18)$$

i.e. the intensity distribution of the laser beam is no longer dependent on the beam radius (Fig. 3.54) so that with adequate intensity, the powder can plasticised over

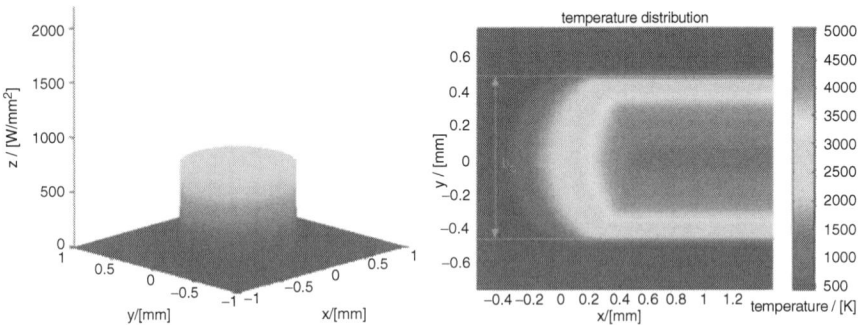

Fig. 3.54 Illustration of an ideal top-hat profile (*left*) and corresponding temperature distribution in the powder bed (*right*)

the entire diameter of the laser beam without increased spatter at the centre. In comparison with the Gaussian beam profile, therefore, the possible scan line spacing and therefore also the possible build-up rate (Eq. 3.13) can be increased. The numerical integration of Eqs. 3.16 and 3.17 allow us to determine mathematically the temperature field of this kind of top-hat profile (Fig. 3.54).

The ideal top-hat profile enables the powder to plasticise over a wider range of the beam cross-section, provided sufficient energy is applied, without generating overheating in the centre (Fig. 3.54). The drop in the absorbed energy towards the edges is smaller in this profile but still distinct. Because of the circular beam cross-section, the time in which the powder grains absorb energy is smaller at the edge of the cross-section than in the centre. The simulated melt pool width of the top-hat profile is 20% wider than that of Gaussian output density distribution.

As well as the fibre itself, the actual intensity distribution depends on other factors such as the optical setup of the laser at the input end of the fibre-optic cable and the beam quality which in turn is determined by the resonator. Since these characteristics cannot be influenced in terms of a plug-and-play solution, the actual beam surface and output intensity distribution must be measured and compared to the theoretical values. In this way it is possible to guarantee that the required characteristics of the laser beam are achieved at the processing location.

The output intensity distribution within the focus is measured with the help of a rotating pinhole which scans the beam cross-section line by line. The pinhole isolates a small part of the radiation in the process. This radiation is deflected via a mirror to an infrared detector which measures the incoming radiation and generates the corresponding measurement signal. The overall measuring head can travel along integrated y- and z-axes so that a scan of the beam surface provides a complete picture of the laser beam parameters. Figure 3.55 left shows the results of the laser beam scan of the 50 μm fibre.

As stated above, the beam profile exhibits neither a distinct top-hat nor Gaussian form. The edges of the profile are steeper than those of a fundamental mode laser while the maximum intensity starting from the centre of the beam is less even than that of a top-hat profile. The focal radius is 97.7 μm and the diffraction dimension is

3 Individualised Production

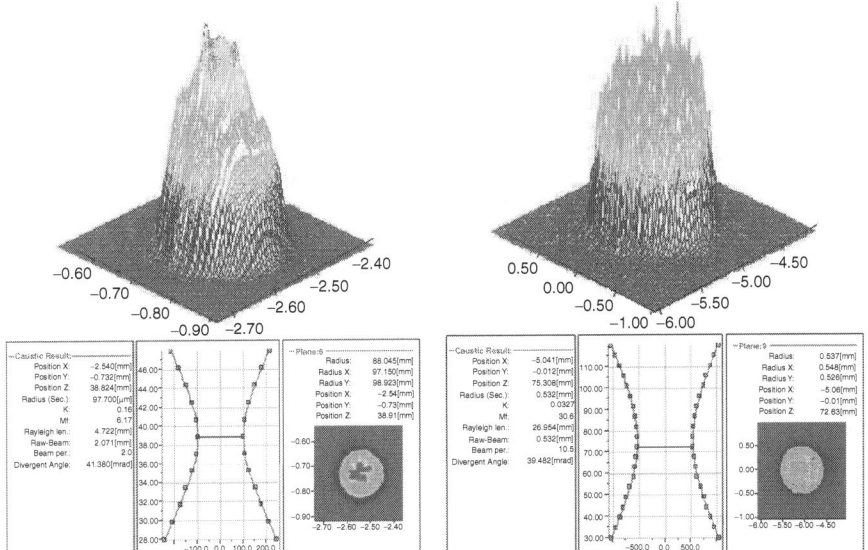

Fig. 3.55 Beam surface of the 50 μm FOC (*left*) and of the 200 μm FOC (*right*)

6.17. On the contrary a distinct top-hat profile is evident in the case of the 200 μm FOC. Due to the higher mode number we can see homogenisation of the intensity profile along the run-length of the FOC (Fig. 3.55 right).

The beam radius is 532 μm in this case while the diffraction dimension is 30.6 in accordance with the larger mode number. The measured values of the focal diameters of both beam path thus correspond to the theoretical values given in 6.1.

In order to guarantee series-identical mechanical characteristics it is necessary for the additively-manufactured components to exhibit a density of >99%. Large flaws can significantly detract from the static as well as the dynamic characteristics. Therefore increasing the build-up rate without retaining a component density of >99% is not practical. In order to determine suitable parameter windows within which the manufacture of series parts is possible, test samples are built up in a first step to determine the density at different build-up rates. To determine the density, cross sections are produced and examined under a light microscope. The resulting images are statistically filtered to suppress noise and are then separated into fused material and flaws, pores etc.

To demonstrate the potential performance of the prototype, initial samples were built up with a beam diameter of approximately 1 mm and a laser power of up to 750 W. Figure 3.56 illustrates the example of a cross-section of one such sample which was built up with a laser power of 600 W.

Although some flaws are visible, the density of the component is >99%. The build-up rate achieved with this set of parameters is approximately 8 mm^3/s. This represents an increase of more than the 650% in comparison with the state-of-the-art.

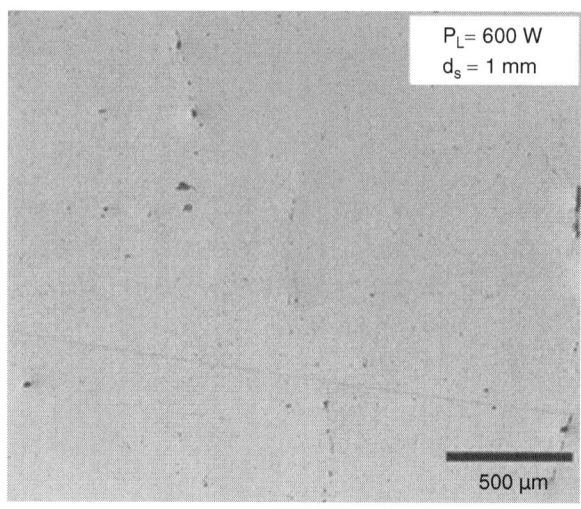

Fig. 3.56 Cross-section of a test sample

$P_L = 600$ W
$d_s = 1$ mm

500 µm

3.3.4.5 Melt Pool Dynamic and Scan Line Stability

As well as process-dependent values such as laser power or laser speed, the material itself has a decisive influence on the surface produced in the process. Every material has different physical characteristics such as surface tension, thermal conductivity, viscosity, melting temperature and solidification interval. Different reactions with the ambient atmosphere also play a key role. In the case of alloys, the alloying elements have influences of varying intensities on these characteristics, some of which are mutually interactive. These can thereby change the melt pool dynamic, i.e. the movements and reactions of the melt. This has a direct effect on the stability of the melt pool and thereby influences the size, form and solidification behaviour of the melt. This in turn has a substantial effect on the shape of the solidified scan lines. Even scan lines lead to an even and therefore less rough surface, i.e. to a higher surface quality. Additionally the solidified scan line should exhibit a flat surface structure so that the next layer can be applied evenly onto it (Kruth et al. 2007). A flat surface is dependent both on the surface tension of the melt as well as on its flow and viscosity.

The surface tension is an important physical characteristic of a material because it permits predictions about the welding potential and surface structure of a solidified component (Sauerland 1993). In fluid mechanics, the surface tension is of great significance in relation to the Marangoni convection (flow within the melt induced by temperature variations). The flow direction depends on gradients of surface tension within a melt pool and therefore significantly influences its geometry (Beyer 1995). Precise knowledge of the influence of surface tension within Selective Laser Melting allows the process to be controlled along with related processes for example fusion welding. In both cases, the material is melted by energy and the quality of the process is determined by the resulting shape of the solidified melt pool amongst other things.

Metal fusion with low surface tension has a better wetting behaviour and therefore better welding potential (Keene et al. 1985). The wetting of the solidified preceding layer by the melt pool also plays a deciding role (Kruth et al. 2007; Schiaffino and Sonin 1997). The wetting of liquid metal on its actual solid form is problematic when the temperature of the solid metal is similar to the temperature of the liquid metal because there is no driving force for wetting (Kruth et al. 2007). Therefore an additional energy input is necessary to melt the preceding scan lines and guarantees a good bond (Das 2003). Nevertheless the temperature can be increased by this additional energy such that material in the laser beam interaction zone is vaporised and so leads to instability of the melt pool and to spattering (Das 2003).

When the surface tension is high and the melt pool reaches a specific length, the scan line retracts into individual beads in order to minimise the limit surface energy (Rombouts et al. 2006; Abe et al. 2000). A continuous solidification front is no longer guaranteed in this case and flaws occur in the component. This increases the porosity and roughness of the component.

The surface tension of a metal melt is temperature-dependent. In the case of pure metals, the surface tension coefficient is negative. i.e. the surface tension is decreased as the temperature of the melt rises. Adding alloying elements however can restore the surface tension coefficient to a positive value. In this case, the surface tension of the metal melt increases as the temperature rises.

The algebraic sign of the surface tension coefficient determines the flow direction within the melt pool. The temperature within the melt pool is unevenly distributed. The temperature in the middle of the melt pool is higher than the temperature in the edge region due to heat conductivity processes. This temperature difference causes a shear force on the surface of the melt pool which results in its flow. This kind of flow is called the Marangoni effect (see Fig. 3.57). In the case of a negative surface tension coefficient, the surface tension is greater at the edge than in the middle of the melt so that the resulting shear stress takes affect from the edge towards the inside. The melt flows over the surface in the direction of the greatest surface tension and therefore to the edge. The flow is deflected at the solid/liquid phase boundary and

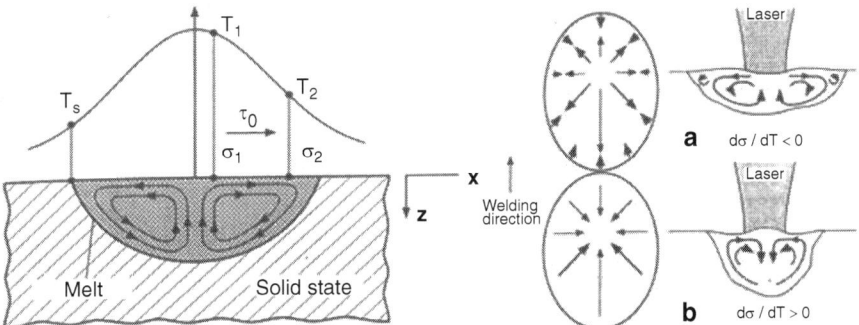

Fig. 3.57 a The Maragoni effect due to the temperature difference within the melt pool, **b** the melt pool geometry depending on the algebraic sign of the tension gradients. (Beyer 1995)

a ripple structure is formed. In the case of a positive surface tension coefficient the liquid metal has a lower surface tension in the middle of the melt than at the edge so that the flow direction of the melt is directed inwards from the edge of the melt pool (Herziger and Lossen 1994). The Marangoni effect directly influences the melt pool geometry. In the case of a positive surface tension coefficient, thus in the case of a flow from the edge towards the inside, deeper and narrower melt pools are formed, while broader and shallower melt pools are formed in the case of a negative surface tension coefficient (Beyer 1995). The Marangoni effect has not yet been observed experimentally in the SLM process. However since it has been detected in processes with similar melt pool sizes, such as in laser beam polishing (Willenborg 2005), its existence in the case of SLM can also be assumed.

Due to the addition of alloying elements the surface tension of Fe alloys is usually decreased. The surface tension at high temperatures mainly depends above all on the amount of surfactants contained in the melt (Keene et al. 1985; Keene 1993; Bergquist 1999).

Oxygen is an active element in liquid iron. If alloyed in a certain quantity the surface tension of the iron is no longer decreased but increased (Czerner 2005). The influence of carbon on the surface tension is minimal in comparison to oxygen and sulphur (Keene 1993). Nevertheless, in the presence of surfactants such as oxygen and sulphur, the activity of the carbon increases. In this case the surface tension falls as the carbon content increases (Schrinner 1993).

As the copper and chrome contents increase, the surface tension of the liquid metal also decreases, nevertheless the effect is minimal in comparison to oxygen. Depending on quantities, vanadium and manganese either increase or decrease the surface tension of the iron melt. The influence on surface tension of an amount of manganese of 0.5% by weight is minimal. Vanadium exhibits a maximum effect at 4% by weight (Czerner 2005).

Viscosity is a measurement of the internal friction of an actual liquid (Sahm et al. 1999) and therefore describes a value for the viscosity of the melt. Viscosity is a temperature-dependent value which falls as the temperature rises (Brillo et al. 2008). In the Selective Laser Melting process the viscosity changes significantly transversely to the length of the melt pool within the laser beam interaction zone. The viscosity of the metal melt depends on the alloying composition. In (Rombouts and Froyen 2005) Rombouts investigations the influence of alloying elements on viscosity with respect to surface roughness are discussed. Bergquist [1999] also provides a short summary of viscosity depending on the alloying content. It is found that viscosity depends on the amount of surfactants such as oxygen and sulphur and is only minimally affected by a small change in the alloying content.

The influence of the solidification range on the melt pool behaviour is discussed in Rombouts et al. [2006] and Rombouts and Froyen [2005]. Materials with a large solidification range have a large temperature difference between the melted and solid aggregate states. This semi-liquid, semi-solid state leads to high viscosity which contains the weld pool kinetics and therefore reduces the balling process. The roughness of Selective Laser Melting components can thus be improved by the addition of alloying elements which increase the solidification range of the alloy.

3 Individualised Production

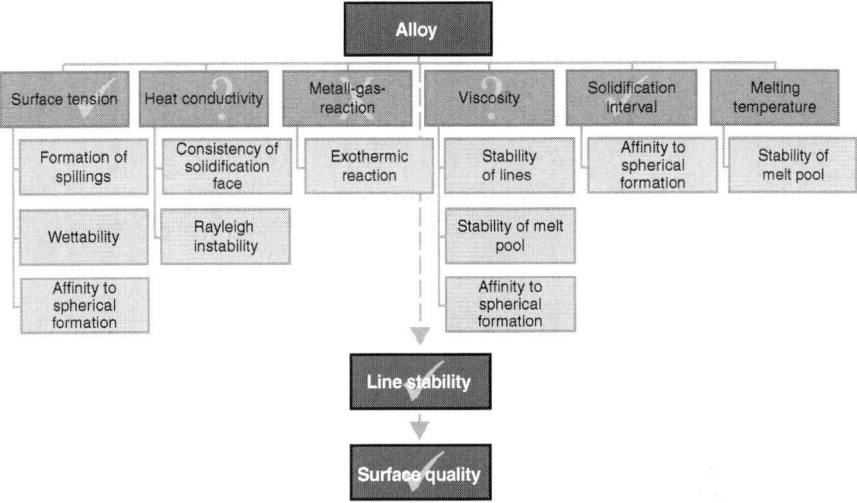

Fig. 3.58 The influences of alloys on surface quality as a result of physical characteristics and reactions of the melt pool

Figure 3.58 gives a summary of the influences of the alloy on the melt pool dynamic and therefore on the scan line stability and the surface quality.

While titanium and silicone are predominantly unfavourable due to their exothermic reactions with oxygen and the resulting unwanted additional heat input, carbon and copper can improve surface quality (Rombouts et al. 2006; Simchi et al. 2003).

The investigations however relate to the addition of elements to pure iron. Technical alloys, e.g. tool steels or chemically resistant austenitic steels are not investigated.

The hot-working steel X38CrMo5-1 (material number 1.2343) was investigated during the project. 0.4% by weight of carbon was added by alloying to one variant and 0.4% by weight of carbon and 2% by weight copper were added by alloying to the other variant. To quantify the influence of the alloying elements on the surface roughness of the samples produced by means of Selective Laser Melting, the side surfaces are investigated with respect to their surface topography and roughness.

As shown in Fig. 3.59 the measurement of the surface tension using the *oscillating drop* method using *electromagnetic levitation* in all three materials reveals a surface tension which rises at the start with the temperature and whose line of best fit exhibits a breakpoint. In the case of the unchanged material X38CrMoV5-1 the surface tension rises further after the break point, but with a lower gradient. The two modified materials exhibit a negative surface tension coefficient after the break point. The increase of the surface tensions of the three materials vary in their inclination up to the break point. The increase of the unmodified X38CrMoV5-1 is the strongest. The addition of carbon reduces the surface tension coefficient. The additionally alloyed copper reduces the value still further. The increases of the two variants of

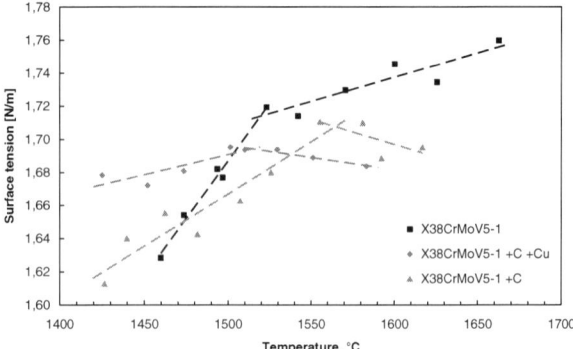

Fig. 3.59 Measurement of surface tension depending on the temperature of the three alloys investigated

X38CrMoV5-1 are similar after the break point wherein here again the copper leads to a flatter curve path.

The viscosity can be measured using the oscillating cup method within a temperature range of between 1,450 and 1,850 N/m. The results are illustrated in Fig. 3.60. For all three materials the viscosity is lowered as the temperature rises. The compensating curve exhibits a similar profile over the temperature range for all three materials. While the compensating curve of the carbon variant is slightly below that of X38CrMoV5-1, copper results in higher values. The best-fit curve of the copper-enriched variant is above that of the carbon-enriched variant and above that of the unchanged material.

The solidus and/or the liquidus temperature can be measured on the melted powder samples using Differential Thermo Analysis (DTA). The results of the investigated materials are shown in Table 3.1. The solidification range are the same for all three materials within the context of the measurement reproducibility. The values for the liquidus and solidus temperatures are at their highest for the unchanged X38CrMoV5-1 and around 50°C lower for the variant with the carbon added by alloying. The values for the copper variant are only around 15°C lower.

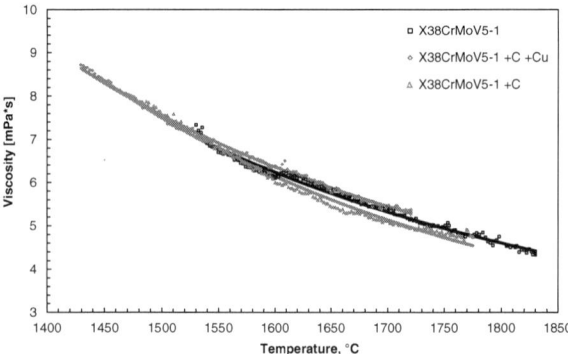

Fig. 3.60 Measurement of viscosity depending on the temperature of the three alloys investigated

3 Individualised Production

Table 3.1 Solidification range of investigated materials after DTA

	Solidus temperature (°C)	Liquidus temperature (°C)	Solidification range (°C)
X38CrMoV5-1	1,493	1,516	23
X38CrMoV5-1 + C + Cu	1,475	1,500	25
X38CrMoV5-1 + C	1,443	1,467	24

To quantify the influence on the component roughness, cube-shaped samples with edges of length 10 mm were manufactured from the different materials using the Selective Laser Melting process with a laser power of 140 W and a scanning velocity of 100 mm/s.

In Fig. 3.61 macroscopic photographs of the lateral surfaces of the manufactured samples and corresponding topographical images taken with a white light confocal microscope are illustrated. The roughness determined from the topographical images indicate a clear improvement of the surface quality of the sample with the carbon added by alloying with a roughness index of Ri = 74 mm as compared to the standard sample with roughness index Ri = 103 mm. The sample with the copper added by alloying exhibits an even lower roughness index of Ri = 44 mm and therefore a higher surface quality.

Both carbon and copper have a positive effect on the surface quality in the proportions added by alloying in this case. All three materials have the same solidification ranges and the viscosity profiles differ very little over the temperature range. The

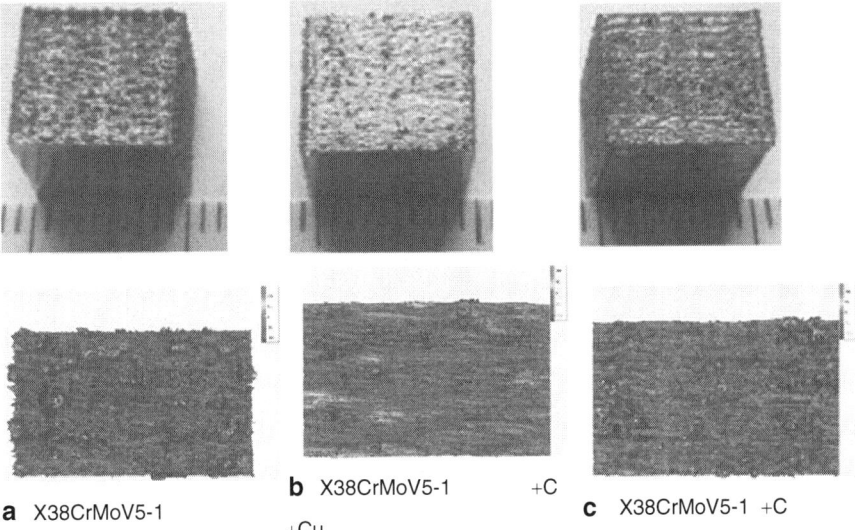

a X38CrMoV5-1   **b** X38CrMoV5-1 +Cu   **c** X38CrMoV5-1 +C

Fig. 3.61 Macroscopic and topographical images of the side surfaces of the U-shaped samples produced in the SLM process. The size of the topographical images are $4 \times 2.5\,\text{mm}^2$

melting temperatures and the surface tensions are distinctly different. In the case of melting temperatures it is found that the best surface quality is not achieved with the material with the highest nor that with the lowest melting point. In fact the best surface quality is achieved with the material with the medium melting point. In the case of surface tension, the variant with copper and carbon exhibits the smallest increase both before and after the break point in the curve profile. A low surface tension coefficient leads to a small shear stress between the cooler edge of the melt pool and its hotter centre and therefore to lower Marangoni convection. This results in a more stable melt pool and an even scan line.

It can be shown that adjusting the alloying system of a steel can successfully improve the melt pool and therefore the scan line stability and surface quality. But only a small number of alloying elements which are used in steel has been investigated adequately. The extent to which the melt pool dynamic of steel with a considerably higher alloy content, e.g. non-rusting or austenitic steels, can be improved by the alloying elements investigated so far. More work is therefore required in the area of SLM- compatible alloying.

3.3.4.6 Process Control and the Skin-core Concept

Laser Power The energy required to melt the powder is coupled in the melt pool and in the powder bed through absorption of laser radiation. The laser power P_L must be determined by taking into account the scanning velocity, the layer thickness and the beam diameter at a pre-specified powder particle grain size such that the powder layer and the upper part of the layer beneath are fully melted in order to guarantee a density of approximately 100%. If the laser power is too high (at a constant beam diameter) a part of the powder vaporises due to the excessive intensity in the interaction zone. This results in a comparatively poor surface quality of the manufactured layer and poor component quality which results in the in the worst case in the termination of the process.

The maximum laser power is specified by the components of the laser system, e.g. laser beam source, optics etc. To guarantee a maximum productivity (in this case: output per time period) the maximum possible scanning velocity is required. Based on the negative correlation of the scanning velocity (v_{scan}) and the energy input per unit length (E_{st}) the laser power should be adjusted according to the increased scanning velocity such that there is sufficient energy available to guarantee full melting of the material (Eq. 3.7). The densities at different laser powers and scanning velocities of a layer thickness of 50 μm are shown in Fig. 3.62 using cross sections of die-casted aluminium alloy AlSi10Mg samples.

The high reflection and thermal conductivity of aluminium compared to steel means that a laser power of at least 150 W at a scanning velocity of 50 mm/s is required to achieve a density of approximately 100%. At a laser power of 250 W the scanning velocity can be increased to 500 mm/s which results in a higher build-up rate.

Fig. 3.62 Sectional view of a sample depending on laser power and scanning velocity

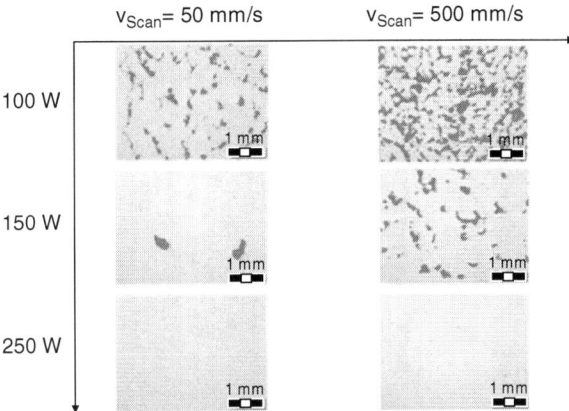

Scanning Velocity The scanning velocity is the speed at which a laser beam moves along the scan vectors of the component. In order to build up the associated layers it is important that the melt pool does not comprise several individual areas but is a coherent self-contained area (Meiners 1999). The energy input per unit length (E_{st}) which is required to plasticise the powder along a scan vector at a constant beam diameter and scan line spacing based on laser power (P_L) and scanning velocity (v_{Scan}) can be expressed with the following formula:

$$E_{St} = \frac{P_L}{v_{Scan}} \left[\frac{W \cdot s}{mm} \right] \quad (3.19)$$

Firstly, the energy input per unit length is decreased as the scanning velocity is increased, which substantially influences the density of the component. Secondly the build-up rate and therefore the productivity of the SLM process depend on the scanning velocity. A satisfactory compromise must be found therefore between the build-up rate and the density. The relationship between the component density and the scanning velocity is shown in Fig. 3.63. As the scanning velocity is increased, the density is decreased beyond a scanning velocity of 100 mm/s from $\rho \approx 100\%$ to a density of <94% at a scanning velocity of 500 mm/s. A density of >98% can be achieved at this laser power (150 W) up to a scanning velocity of 250 mm/s.

Layer Thickness In addition to the parameters of scanning velocity, beam diameter, scan line spacing and laser power, the layer thickness is an important parameter in additive manufacturing. The powder mass is increased as the layer thickness is increased. To create dense components it is also necessary to increase the energy required per specific volume. As a result, either the scanning velocity must be reduced or the laser power must be increased. As mentioned above, the maximum laser power generally depends on the laser beam source and the optical components so that as the layer thicknesses increase, the scanning velocity has to be reduced. However, both the scanning velocity and layer thickness factors affect the build-up rate to the same degree. The layer thickness also affects the surface quality and the possible achievable detail resolution by the so-called "step effect" (Fig. 3.64). When choosing

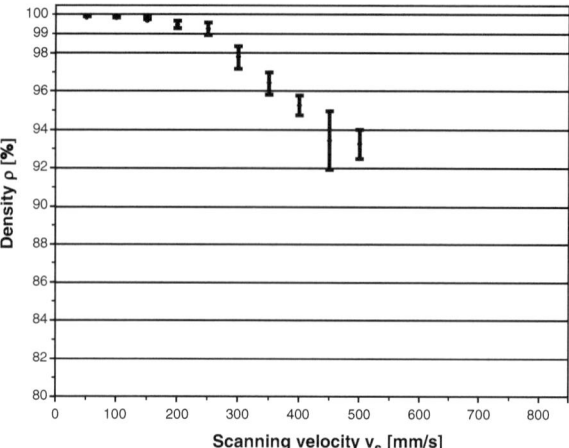

Fig. 3.63 Relationship between density and scanning velocity

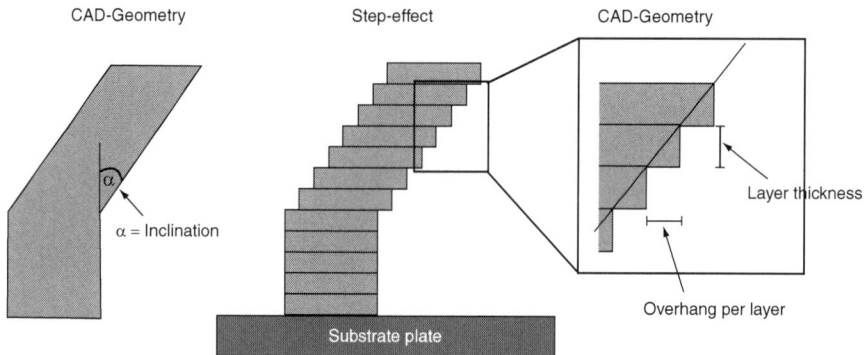

Fig. 3.64 Step effect

the parameters for an optimum layer thickness a compromise must be found between density, surface quality/detail and build-up rate.

In currently available commercial SLM machines layer thickness variations (of metal powder) are usually between 20 and 100 μm.

Scan Line Spacing To manufacture components built up in layers, numerous scan lines and/or scan vectors must be fused one after another. The result of these scan vectors is described as "hatching". As well as fully melting of the powder material, the bonding between the scan vectors and the metallurgical melting compound of the layer beneath must be guaranteed to produce even, dense "hatches". Meiners [1999] shows that there is an empirical relationship between the beam diameter (d_s) and the scan line spacing (Δy_s) (Eq. 3.20):

$$\Delta y_s = 0.7 \cdot d_s (\text{mm}) \tag{3.20}$$

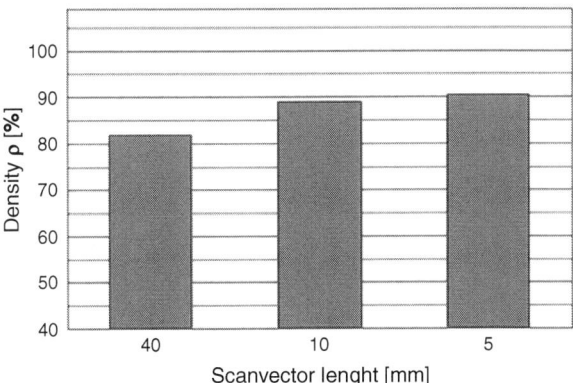

Fig. 3.65 Relationship between density and scan vector length

Scan Strategy The density and surface quality are primarily influenced by the scan strategy. Shorter scan vectors (typically approximately 5 mm) have a positive effect on the density of components due to the local preheating particularly in the deflection points. In Fig. 3.65 is illustrated that an improvement of density of approximately 10% can be achieved by reducing the scan vectors from 40 to 5 mm.

Due to the higher temperature in the interaction zone of the laser and powder bed, shorter scan vectors lead to a continuous melt pool whereas larger scan vectors generate a "diffuse" melt pool due to the energy loss through thermal conductance. The process related efficiency level can be increased with shorter scan vectors.

In general, each layer is built up by the exposure of several scan vectors H (hatches) and one contour vector C. The surface quality is affected by the arrangement of inner scan vectors H and the contour vector C.

In Fig. 3.66 the general effect of the C/H- and H/C-strategies is shown taking into account different scanning velocities. An improved surface quality in terms of the medium roughness R_z can be achieved with the c/s strategy. Based on the partial

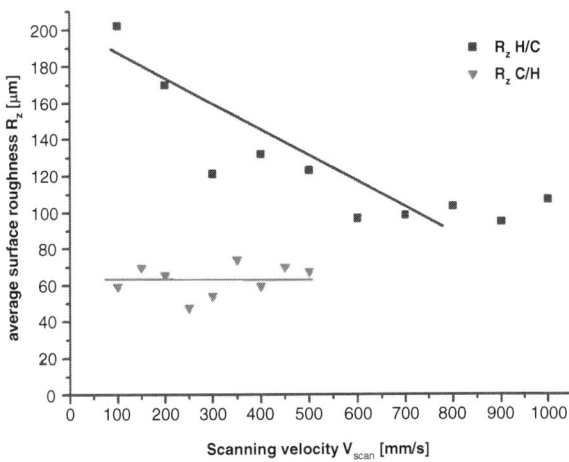

Fig. 3.66 Medium roughness R_z depending on the scanning velocity and scan strategy (C/H, H/C)

reflection of the laser beam on the metallic surface of already melted contour vectors (C/H strategy) numerous small powder particles adhere to the outer surface of the component resulting in increased surface roughness.

The deciding factors in the increase of build-up rate are scan line spacing, scanning velocity and layer thickness. While the imaging optics is limited by the beam diameter (approximately 0.7 times the focal diameter at 1.4404) (Meiners 1999), investigations by Schleifenbaum et al., found that there is little scope for increasing the build-up rate by increasing the laser power and scanning velocity with a constant beam diameter (Schleifenbaum et al. 2008).

In order to solve the problem of applying the "right" amount of energy (in terms of location and time) to the powder layer it is possible to vary the scan vector length. By using decreasing scan vector lengths means that the laser beam changes its direction more times. The result is that the number of deflection points is increased as the scan vector length is decreased. A temperature increase can be observed in the region of the deflection points since the energy input is greater at the deflection point than in the rest of the hatching area because of the longer dwell period. The resulting overheating of the melt pool in these areas causes more powder particles to vaporise which leads to increased spatter and therefore process instabilities. This means that reducing the number of deflection points reduces the vaporization rate which has a positive effect on the process stability, particularly in the case of thinner layers. Figure 3.67 illustrates the density as a function of the layer thickness in the case of different scan vector lengths.

Because of the local increase of the energy input in the case of a scan vector length of 5 mm, the density of approximately 94% of a layer thickness of 100 μm increases to approximately 97% at a layer thickness of 150 μm. As a result of the larger powder mass to be melten as the layer thickness increases, overheating, vaporisation and spatter all decrease so that the process becomes more stable overall. Process instabilities are too large at greater layer thicknesses (particularly at deflection points) to guarantee the manufacture of dense components. The density of the sample isn't

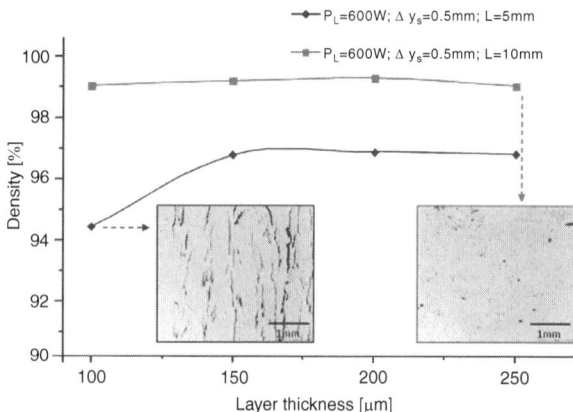

Fig. 3.67 Density versus layer thickness, comparison of the scan vector lengths, beam diameter: 1.05 mm

3 Individualised Production

increased as the layer thickness increases but remains constant at a layer thickness of 250 μm at approx. 97% (Fig. 3.67, continuous line).

While all other process parameters are constant, changing the scan vector length from 5 to 10 mm significantly increases the density (Fig. 3.67, dotted line). At a layer thickness of 100 μm the density difference in the case of 5 and 10 mm scan vector lengths is approximately 5%. The stabilisation of the process by increasing the layer thickness reduces this difference to 2% in a layer thickness of between 150 and 250 μm. Fewer deflection points result in a significant improvement of the process stability and the density of $\geq 99\%$ is achievable irrespective of the layer thicknesses investigated by implementing longer scan factors.

It is evident that the build-up rate can be increased by increasing the laser power and beam diameter with the implementation of longer scan vectors, particularly in layer thicknesses of < 150 μm. The maximum build-up rate achieved in these tests is 20 mm^3/s. If we compare the build-up rate with the state of the art (see state of the art) this corresponds to an increase of more than 1,500%. However, accuracy, detail resolution and surface roughness within this parameter range are inadequate to meet the requirements of near-shape components. To avoid these disadvantages while increasing the build-up rate caused by the greater beam diameter and layer thicknesses, the outer shell of the sample can be manufactured with a smaller beam diameter (0.2 mm) and layer thicknesses of less than 100 μm, e.g. 50 or 30 μm. To implement this kind of skin-core strategy the samples are divided into an inner core and an outer shell (Fig. 3.68). The outer shell is manufactured with a smaller beam diameter (0.2 mm) and a layer thickness of for example 50 μm. On the contrary, the inner core is generated with a beam diameter of 1 mm. The layer thickness of the inner core can only be a multiple of the layer thickness of the outer shell. It follows therefore that if the outer shell layer is 50 μm thick, the inner core must be 100, 150, 200 μm, etc. (Fig. 3.68).

A deciding factor for the metallurgical connection of the core and the shell is the overlap. This interface of the core and the shell is processed twice with different

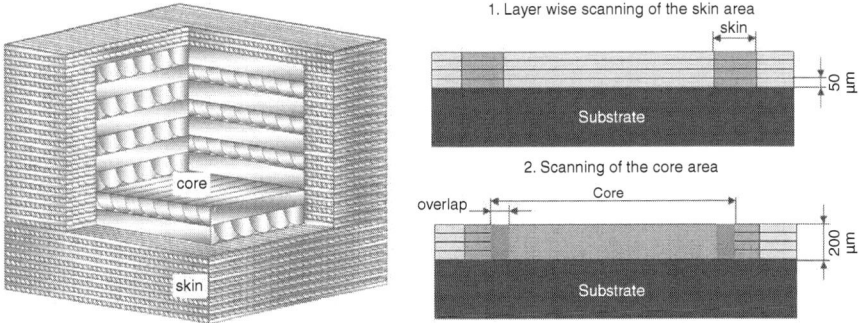

Fig. 3.68 Schematic diagram of the skin-core principle (*left*) and the core and shell manufacturing process (*right*)

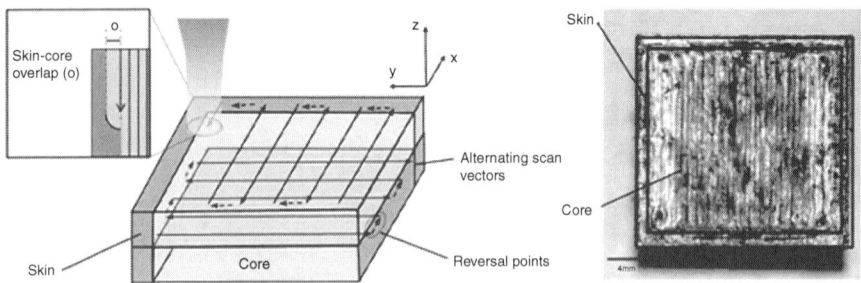

Fig. 3.69 Schematic diagram of the skin-core overlap (*left*) and the core and shell samples, top view (*right*)

process parameters (Fig. 3.69 left). To enable the two regions to join, the overlap must be >0 mm. Figure 3.70, right, shows samples manufactured according to the skin-core principle. The core of the sample was manufactured with a beam diameter of 1 mm and a layer thickness of 200 μm. The shell however was produced with a beam diameter of 0.2 mm and a layer thickness of 50 μm.

Depending on the layer thickness ratio between the skin and the core, the density in the joining zone increases as the overlap increases. The density of the bonding zone in relation to the overlap is illustrated on the basis of layer thickness ratios of 1:2 and 1:4 in Fig. 3.70.

For both layer thickness ratios (1:2 and 1:4) the number and size of flaws (and thus porosity) improves as the overlap increases, In the case of a layer thickness ratio of 1:2 (50 μm skin layer thickness and a 100 μm core thickness) a metallurgical bonding with a density of >99% can be manufactured with an overlap of >0.5 mm. A layer thickness ratio of 1:4 (50 μm skin layer thickness and 200 μm core layer

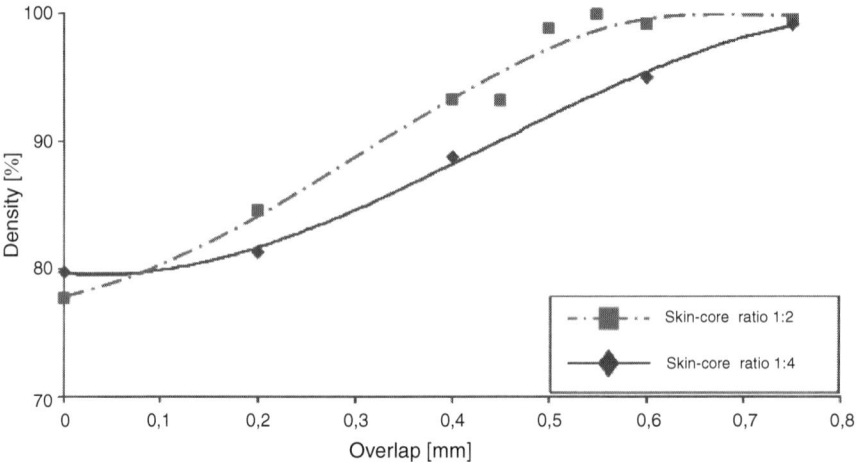

Fig. 3.70 Density of the skin-core transition versus overlap, skin-core thickness ratios of 1:2 and 1:4

3 Individualised Production

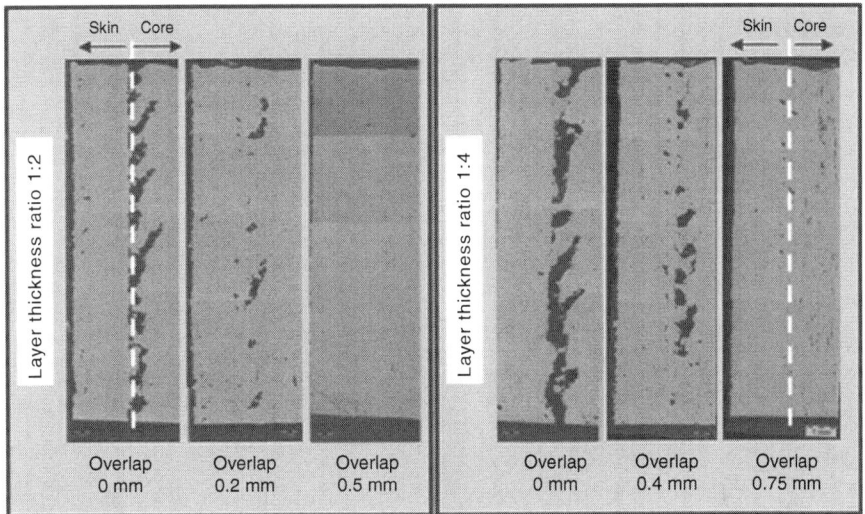

Fig. 3.71 Shell/core sample sections

thickness) requires an overlap of at least 0.75 mm for a dense metallurgical bonding. In Fig. 3.71 cross sections of the skin-core transitional region with an increasing overlap in the case of layer thickness ratios of 1:2 and 1:4 are shown.

The density of the bonding zone of the core and the skin is about approximately 100% at an overlap of 0.5 mm (layer thickness ratio 1:2) and/or 0.75 mm (layer thickness ratio 1:4). Individual flaws cannot be assigned to any of the three areas of skin, core or skin-core interface. The quantity and size of the flaws corresponds to the components manufactured using the conventional SLM process. It is therefore possible to manufacture at least simple cube samples with layer thickness ratios of up to 1:4 (skin: 50 µm, core: 200 µm). To show the potential of the skin-core strategy for the additive production of complex components and parts, the results obtained above are transferred to a real functional component (Fig. 3.71).

The component that is illustrated in Fig. 3.71 is a tool insert for plastic injection moulding. Due to its complex internal cooling channels this component is generally very expensive to manufacture and requires several process steps (e.g. by Laminated Object Manufacturing). The functional structures of the part are basically limited to the outer contour and the location, form and arrangement of the cooling channels due to the cooling near to the outer surface. The inner part of the tool insert however has only a supporting function and absorbs the forces occurring in the plastic injection moulding process which must be deflected to the tool and/or the machine structure. Corresponding to these requirements the component can be divided into skin and core sections (Fig. 3.72).

In order to maximise the build-up rate while retaining the detail resolution and the surface roughness, a layer thickness of 200 µm is used for the core region and 50 µm for the shell region. For the layer thickness ratio of 1:4, an overlap of 0.75 mm is

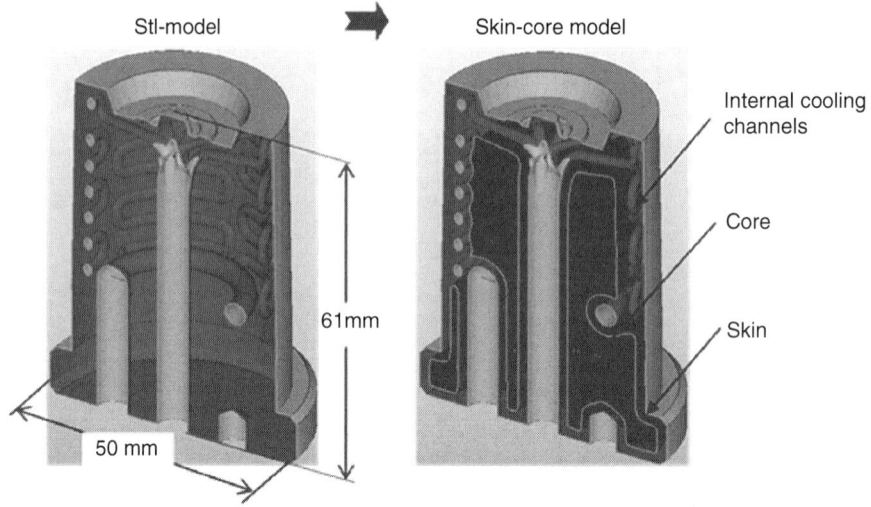

Fig. 3.72 3D model of a tool insert (*left*) and shell-core model (*right*)

predetermined by the prior experimental investigations. The additively manufactured tool insert is illustrated in Fig. 3.73, left. In order to investigate and evaluate the density of the skin and the core as well as the shell-core interface, several vertical and cross sections of the tool insert are taken as shown in Fig. 3.73.

The density of the tool insert exhibits a density of approximately 100% and the number detected flaws is equal to that of conventionally manufactured SLM components. Flaws and pores cannot be assigned to any specific area of the component, i.e. neither skin, core nor interface.

Based on the investigations into additive production according to the skin-core principle it is now possible to manufacture even complex functional parts using layer thickness ratios of up to 1:4. By increasing the laser spot diameter up to 1 mm, corresponding scan line spacing and thicker layers in the core (200 µm) the build-up rate in this region is approximately 21.6 mm^3/s. A beam diameter of 0.2 mm and

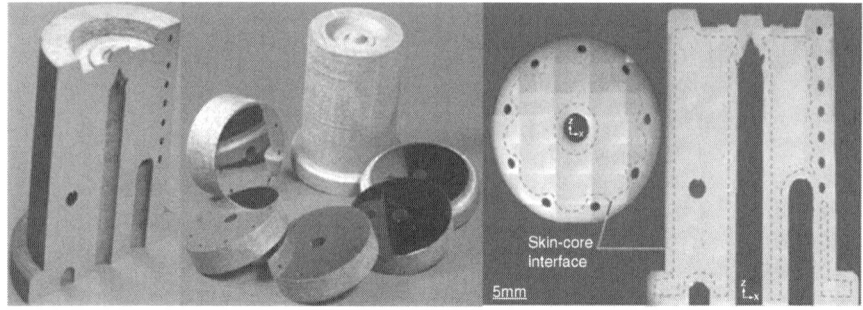

Fig. 3.73 Additively-manufactured skin-core tool insert (*left*), sectional view (*right*)

the shell layer thickness of 50 µm is used in the skin region in order to achieve a high detail resolution and a high surface quality. The build-up rate in this region is therefore approximately 3 mm^3/s. The volume ratios of the skin and core in the tool insert shown above result in an overall build-up rate of 12.7 mm^3/s. This corresponds to an increase of the build-up rate of more than 1,000% compared to the state of the art for the steel material used here.

3.3.5 Industrial Relevance

Selective Laser Melting (SLM) is used to combine the benefits of unlimited geometrical freedom of the additive manufacturing process with the usability characteristics of the standard series material. SLM therefore permits rapid manufacturing of functional prototypes with series-identical characteristics. It is also ideal for the direct production of individual parts and short runs, particularly of parts with completely new and unique functional features. The potential of Selective Laser Melting is now outlined on the basis of a few typical applications already successfully trialled in industry.

3.3.5.1 Tool and Mould Production

The manufacture of a pre-form from powdered material enables the flexible and direct manufacture of tool inserts without using or machining semi-finished products. The insert shown in Fig. 3.74 was produced in 6 h from tool steel 1.2343 using SLM. The final contour of the pre-form is already fully shaped including the 0.8 mm slot thanks to the good detail resolution. The precision and surface quality required in tool and mould engineering is achieved with conventional post-processing such as milling and erosion. Highly polished surfaces can be achieved without problems due

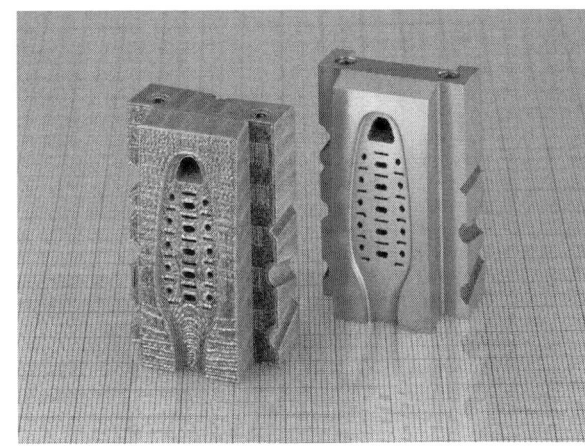

Fig. 3.74 1.2343 mould Insert. *Left*: SLM blank. *Right*: finished mould insert (in collaboration with BRAUN GmbH and TRUMPF Werkzeugmaschinen GmbH & Co. KG)

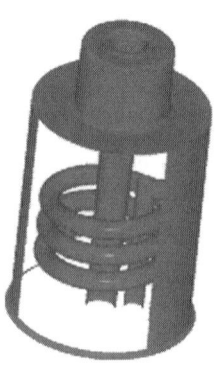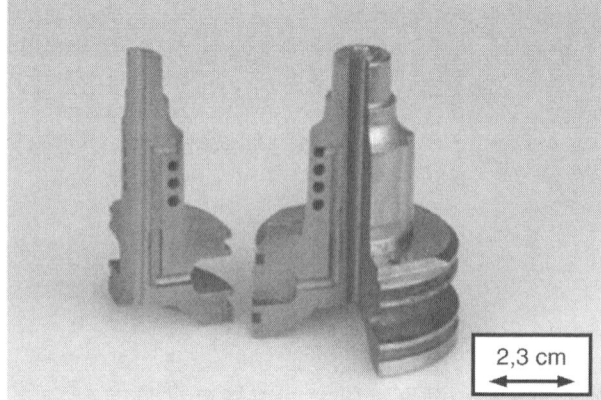

Fig. 3.75 1.2343 mould insert manufactured using SLM with contour-adapted cooling ducts. *Left*: CAD model *Right*: finished mould insert, cut open to demonstrate the cooling ducts, (in collaboration with BRAUN GmbH)

to the high density and homogenous microstructure. The omission of the roughing process can reduce the entire production time by up to 30%.

As well as producing complex external geometries, the layered build-up of SLM enables the manufacture of contour-adapted cooling channels with any complex shape or location within series tools. In this way the quality of the manufactured parts is increased while cycle times are reduced in series production. The insert shown in Fig. 3.75 with contour-adapted cooling channels is made of 1.2343 on a prefabricated, pre-hardened part made of turned tool steel. Once completed, the insert exhibits a hardness throughout of approximately 53 HRC. Post-processing involves turning, eroding and polishing. A thermal image analysis reveals a very even temperature distribution and a significantly higher cooling capacity as compared to the series insert which is made of a copper-based alloy (Ampco 940). Its use in injection moulding can therefore enable a reduction in the cycle time of 20%. The insert exhibits no visible wear after 62,000 plastic parts.

3.3.5.2 Medical Engineering

Patient-specific implants may be required as a result of accidents, tumour resections and hereditary defects (Fig. 3.76). So far these kinds of implants have been primarily manufactured by casting, forming or cutting. This is very costly in time and money. The use of Selective Laser Melting offers significant commercial benefits in this area which can be demonstrated for example on the basis of an individual implant for the facial area made of the biocompatible material Titanium Gd II.

The implant is manufactured in eight hours directly from the CT/ CAD data so there are no shaping tools required. Manufacture involves no loss of material because any powder not used in the component can be reused.

3 Individualised Production

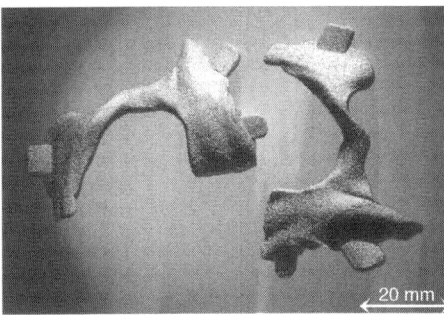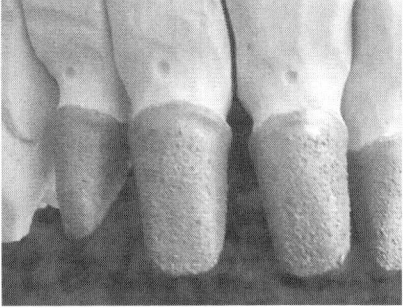

Fig. 3.76 Individual components manufactured using SLM. *Left*: implant for the facial area made of titanium. *Right*: denture made of CoCr-alloy (in collaboration with Peter Brehm Chirurgiemechanik and Bego Medical AG)

Selective Laser Melting is already used for the industrial mass production of individual components. The company Bego Medical AG produces metal dentures (crowns, bridges etc.) from the materials titanium, gold and any CoCr-alloy using direct laser forming. Compared to the casting process normally used in dental engineering, a time saving of approximately 80% is achieved. This application represented the first step for direct laser forming from Rapid Prototyping towards a Rapid Manufacturing series production process.

3.3.5.3 Ultra-Lightweight Construction

The unlimited complexity of components which can be manufactured using direct laser forming opens up new opportunities for ultra-lightweight construction (Fig. 3.77). Along with the use of lightweight materials it is now possible to implement any lightweight construction geometry. Demonstration parts have been made for example out of both TiAl6V4 and AlSi10Mg with an internal grid structure which could not be produced using conventional manufacturing methods.

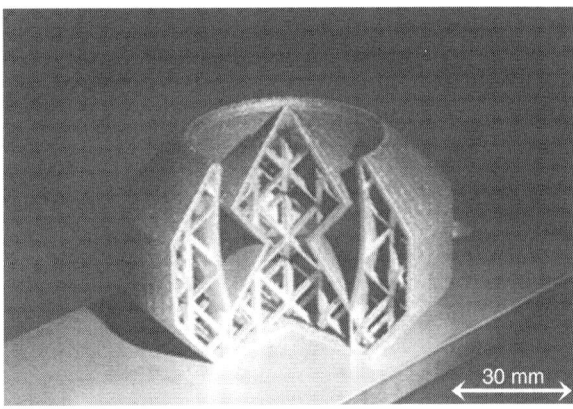

Fig. 3.77 TiAl6V4 demonstration component with internal cavity structure for ultra-lightweight construction

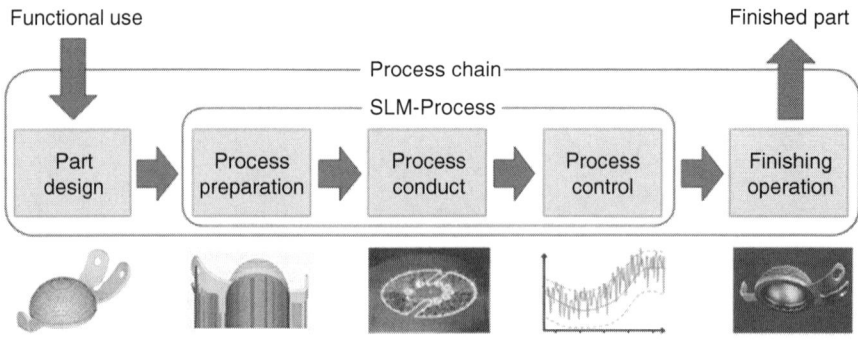

Fig. 3.78 Selective Laser Melting (SLM) process chain

The grid elements are 0.5 mm thick and the manufacturing time is approximately 11 h. There is great application potential particularly for the aerospace industry where weight savings are of great importance. However SLM can also be used to construct and manufacture true-functioning, series-identical components without any technical production restrictions for other areas of mechanical engineering.

3.3.6 Future Research Topics

Along with further increases in the process efficiency of additive production in general and of Selective Laser Melting (SLM) in particular, the unified analysis of the overall process chain is playing an increasingly central role (Fig. 3.78). Precisely in innovative production processes such as SLM, process-inherent benefits can only be fully exploited when these benefits are taken into account at the component design stage. Only by analysing and modelling the complex interactions between the functional requirements, component design and manufacturing process is it possible to utilise the overall technical potential of an innovative manufacturing process.

3.4 Modularisation for Die-based Production Technologies

Marek Behr, Andreas Bührig-Polaczek, Walter Michaeli, Stefanie Elgeti, Markus Probst, Yann Queudeville and Christian Windeck

3.4.1 Abstract

An integral feature of markets in industrialised nations is a high demand for product variants and products specifically tailored to the needs of the customer. For manufacturing companies in high-wage countries whose strategic orientation is generally not aimed at the low-price segment, the manufacture of high-price individual products represents a profitable market. Successfully implementing such a strategy provides immense challenges to those companies whose production technologies are associated with high investment costs. Since selling large quantities of the product is necessary to be profitable, these technologies include typical mass production processes. Therefore, this issue requires manufacturing systems and technologies to create a synthesis between mass production and individual production. This objective may be achieved in two ways. On the one hand, the technologies used for individual production must be developed further so that they are also profitable when used to produce large quantities. On the other hand, mass production technologies must become more flexible in order to be able to take individual customer requirements into account.

The project presented here contributes to the second category. It examines production technologies based on forming dies to produce mass products with high lot sizes. The aim of the project is to simplify the time- and cost-intensive design and production of dies. Additionally, the fast realisation of product variants will be enabled. To achieve this, two approaches are combined. Firstly, a methodology for the modularisation of dies has been developed to identify cost drivers and commonalities in the die portfolio. This methodology provides a systematic process for devising dies of modular design. Secondly, a software package is implemented which enables the computer-assisted design of flow channels in the modules. With the help of this software, a quick and cost-efficient calculation of the optimum shape of new module variants is realised, without relying exclusively on practical knowledge. The modularisation methodology is illustrated using the example of a pressure die casting process and the functionality of the automatic design process is demonstrated using a profile extrusion die. However, it is possible to apply both approaches to other die-based mass production technologies, in particular in combination, as will become apparent in Sect. 3.5.6 regarding future research projects.

Modularisation is an accepted approach to reduce production costs while guaranteeing high product variety. Despite numerous research projects in this field, there are no standard definitions of the terms modularisation or module and there is no systematic approach for implementing modularisation concepts in the production process. In the field of pressure die casting there is usually a one-to-one correlation between the die and the produced part. There are only isolated, company-specific applications with dies of modular design. Improvements in the design process can be traced back to evolutionary developments within the companies. The modularisation methodology proposed here offers a systematic approach to the creation of die modules. In this context, the term module describes a component which contains at least one standardised interface so that it may theoretically be replaced by different components with the same interface. In terms of modularisation methodology there are three different types of modules, depending on the modularisation potential: standard parts, standard modules (industry and company standards) and modules with individual features. Modules with individual features are a crucial success factor in a market differentiation strategy.

In order to produce individual products with mass production dies, specific parts or functions of a module must remain distinct without losing the overall advantages of modularisation. The methodology consists of three phases: initiation, analysis and design phase. The frame of reference is narrowed further with every phase to produce a final specific set of instructions for the production and design of die modules. This takes into account both the technological and operational constraints. In the initiation phase, the constraints of the company's various business segments are first analysed in order to identify the cost drivers along the entire process chain. Classification according to die type and the detailed analysis of the production process assess which field has the greatest potential for modularisation. In the following analysis phase, a product architecture is created for this die type, i.e. the interrelationship between the functions and components is analysed. An evaluation of the dependencies between the components provides precise information, regarding which components are most suitable for modularisation. Such components are then transformed into modules in the design phase. The characteristics of the components are considered on the basis of the available die portfolio and the possibilities of variation examined taking into account the technological and economic constraints. It is then possible to specify the number and design of the discrete variants of the newly created module. Following subsequent assessment of the suitability of the module within the entire system on the basis of the economic viability criteria determined in the initiation phase, the next component undergoes modularisation.

The second pillar of the cost-efficient manufacturing of small batch sizes with mass production processes is the numerical design of die modules (or even an entire die). The example of plastic profile extrusion demonstrates how it is possible to make the die design process more efficient by reducing the dependency on empirical rules and the practical knowledge of the die-manufacturer. In practice, the melt distribution in the die on the extrusion line is analysed and the die post-processed iteratively until the required product quality is guaranteed. These running-in trials represent a considerable time and cost factor which make it difficult to implement individual customer requirements. The use of simulations and optimisation methods in the design process is currently only established in isolated cases and based either on simple 2D models of the flow channel or offers only limited opportunities to vary specific parameters of the flow channel geometry. The geometry optimisation framework introduced here provides a simplified flow channel design which is subsequently optimised to produce a homogeneous melt discharge. An intermediate level is introduced, the shape of which is derived from the profile geometry according to systematic rules and which is responsible for the good pre-distribution of the melt. This takes into account differences in wall thickness and width, the compression ratio and the centre of area, among others. The transition between the circular extruder cross-section on the one side and the exit geometry on the other is initially determined using a linear morphing algorithm. The optimisation framework uses a parameterisation of the flow channel with T-splines, which is compatible with all established CAD programs, in order to allow complex deformations. The practical knowledge of the design engineer is condensed into one function able to measure the quality of the deformations. The framework uses high resolution 3D-simulations of the flow behaviour in order to calculate the optimum flow channel geometry without further user interaction. This may be re-imported into a CAD programme in order to

3 Individualised Production

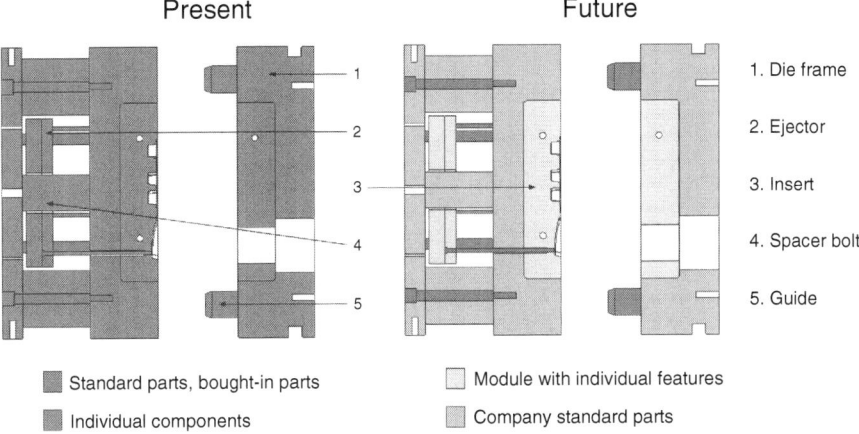

Fig. 3.79 The present and future of pressure die casting dies. (According to Queudeville 2009)

immediately begin with the design. In comparison to the manual design method, the number of running-in trials is reduced to a minimum. Automatic design also reveals non-intuitive, optimum geometries.

Figures 3.79 and 3.80 illustrate the two approaches described above which create flexible mass production technologies for the cost-efficient production of small batch sizes of individual products. Figure 3.79 illustrates how the structure of a typical pressure die casting die may be changed by applying the modularisation methodology in the future. A few standard and bought-in parts are complemented by a number of modules with individual features. Standard interfaces make it possible to quickly exchange the modules. Figure 3.80 illustrates how the optimisation framework is used to automatically design the shape of a flow channel for an L-profile in order to achieve the most homogeneous velocity distribution. The initial design draft can easily be generated. The optimisation algorithm calculates a slight camber along the edge. Due to the already very homogeneous velocity distribution of the optimised geometry at the die exit, only a few iterations are required in the reworking of the die.

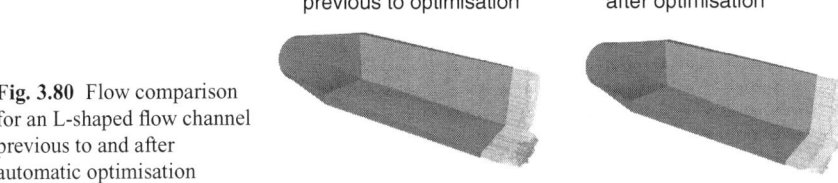

Fig. 3.80 Flow comparison for an L-shaped flow channel previous to and after automatic optimisation

3.4.2 State of the Art

3.4.2.1 Modularisation

The number of publications available on the subject of modularisation has grown constantly over the last ten years (Fixon 2007). Yet despite the numerous papers concerned with the concept of modularisation, there is no standard definition of the terms modularisation or module (Gershenson 2003). The term module is used in this paper to refer to a product component which contains at least one standardised interface so that it may theoretically be replaced by different components with the same interface.

According to Göpfert, the generation of a product architecture is an important aspect of product development. Product architecture may be regarded as a combination of the functional and the physical product description (Göpfert 1998). This systematic approach to the design of a new product has established itself as an effective step towards a modular architecture (Pahl et al. 2007). The attempt to create one-to-one relationships between functions and components can lead to low dependencies between the components of a product (Ulrich 1995). This is a promising way to generate standardised interfaces between the components.

Several studies have investigated the effect of modularisation on matters such as product performance, quality, costs and time (Gershenson 2003). However, in general, it is always important to investigate how practical modularisation is for the technical field in question. The entire product creation process, including strategic product planning, design and production should be examined. Each of these areas offers potential for modularisation. Any modularisation should therefore start in the cost-intensive areas, if possible. The detailed transformation of individual modules requires consideration of the specific manufacturing processes. This project focuses on the formulation of a methodology for the development of modular dies for series production using the examples of pressure die casting and profile extrusion dies.

The following describes the two processes in more detail with regard to the technical state of the art with respect to modularisation.

Pressure Die Casting The pressure die casting process is a highly automated industrial casting process which is used in particular for the mass production of geometrically complex components. Almost 70% of German manufacturing takes place in the automotive industry (Bundesverband der Deutschen Gießerei-Industrie 2007). In pressure die casting, the molten metal is pressed into a die at high speed using a piston under high pressure. The pressure die casting dies used to produce the casting are usually developed and manufactured by specialised medium-sized companies (Bundesverband der Deutschen Gießerei-Industrie 2010). Several attempts have been made to establish a form of modular construction for pressure die casting dies. A modular die with replaceable insert modules has been developed for the production of engine blocks with 2, 3 or 4 cylinders (Hummler-Schaufler et al. 2004). A further example is a pressure die casting die with four core sliders and replaceable pusher heads for the production of various small components made from magnesium (Anonym 2002).

The illustrated examples represent merely some of the isolated applications of successful modularisation. In the majority of cases, there is still a one-to-one correlation between the die and the die casting component. There is at present no universal or cross-company standardisation of individual die components. In fact, improvements made in the design process of companies follow an evolutionary path. They are mostly attributable to the experience and style of the design engineers and the established computer assistance of the company.

Pressure die casting die-manufacturers use the products of standard component manufacturers, particularly those who offer standard parts for injection moulding dies. Guide elements and partially processed and unprocessed semi-finished products may also be used in the production of pressure die casting dies. However, the dimensions are often insufficient for the pressure die casting die. The modular construction of injection moulding dies is much more advanced. This is partly due to the simpler mould filling, the lower moulding temperatures and the lack of release agent. In addition, molten aluminium has a considerably lower viscosity compared to plastics which requires to a great degree of accuracy on sealing faces. The technical constraints of pressure die casting therefore make it more difficult to design modular dies.

In summary it can be said that there is currently no standard modularisation of pressure die casting dies. The majority of pressure die casting die components are designed specifically to order.

Profile Extrusion The profile extrusion process is a continuous process. Plastic melt is delivered through a die using an extruder and thus formed into a profile. After exiting the die, the profile is calibrated and cooled. The process may be applied to both simple and complex profile geometries (from edging strips to skirting boards and window frames). The profile extrusion industry mainly consists of small- and medium-sized enterprises which usually specialise in specific groups of profiles. Still, each die is typically designed and manufactured on an individual basis. In contrast to simple extruded geometries, such as slit-shaped films, sheets or an annular tube, profile geometries are so complex that it is not possible to design those using analytical formulae. In practice, an iterative trial and error process has established itself. This involves producing a real die for moulding a plastic profile and analysing the melt distribution on an extrusion line. If the profile does not meet the necessary quality criteria, the die is reworked and the flow reassessed. Depending on the experience of the die-manufacturer and the complexity of the profile, between five and 15 running-in trials are necessary. This running-in process thus represents a high time and cost factor (Michaeli et al. 2009; Michaeli 2009).This dependency on the personal experience of the die-manufacturer constitutes a considerable amount of uncertainty in the development and design process. For these reasons it is therefore not possible to predict the exact development time and total costs, although they are usually high. For example, one project partner states that, depending on the product mix, approx. 10–15% of the available plant capacity of the elastomer profile manufacturer is permanently required for running-in and sampling inspection processes. With large batch sizes, the relative costs of die design and production are almost negligible.

However, the market is increasingly demanding smaller batch sizes with individual features (Ebli 2010). With small batch sizes, the design of the mould and the resulting costs of production and sampling are important factors for the die-manufacturer and processor. Modularisation of die design can therefore reduce both the costs and the level of dependency on empirical values in the design process. As profile extrusion is characterised by the customisation of the profile, the degree of standardisation is relatively low. Standard parts are primarily used for measuring sensors at the periphery (temperature and pressure measurements). Furthermore, company standards often specify that the die is connected to the extruder via an adapter. Blanks of standardised sizes are likewise usually used for later dies. It is usually more difficult to use standard modules the nearer the components are to the flow channel. The flow channel itself is either incorporated in a long die block or in a die composed of single sections. This layered structure has the advantage of making it easier to post-process the die.

3.4.2.2 Numerical Design

The potential of computer simulations in the design of dies has currently only been utilised to a limited extent. Numerical calculations are usually only consulted in such instances in which a die used in production gives results of insufficient quality and the reason for this needs to be analysed. In this instance, simulations provide information on the changing thermal residual stresses which appear in a pressure die casting die during operation (MAGMA GmbH 2003) or indicate which areas of a flow channel create an unfavourable melt distribution for the extruded profile (Michaeli et al. 2001). However, using numerical methods already in the design phase is not yet widely established. This is usually due to the simplified assumptions of the methods which make application in actual industrial practice problematic. State of the art technology is illustrated below in more detail in the field of plastic extrusion since the framework for the numerical design of dies developed in the project will initially be used for this production process.

The use of numerical methods can be linked to considerable potential savings. For example, COMPUPLAST International Inc. of Zlin in the Czech Republic supplies software for flow simulations in profile dies which allows variations in the height of the flow channel in order to achieve a homogenisation of the flow. Many of the company's customers report that their die development times have been halved by using this software (COMPUPLAST 2010). Knowing that this savings potential exists led to research projects being carried out over many years into the automatic design of profile extrusion dies (Nóbrega et al. 2003; Sienz et al. 2001). In addition to diverse research activities aimed at being able to better model the extrusion process (e.g. by developing new material models), two important fields of work must be identified: first, the development of criteria for quantifying the quality of a flow channel, and second, geometrical representation.

Quantifying quality can be summed up by the term *objective function*. The primary quality criterion is usually the homogeneity of the velocity distribution at the die exit.

In all the publications known to the authors, homogeneity is calculated using a least-squares statement. Differences arise depending on whether the average speed or the flow is used as variables, for example. There are also the options of evaluating the objective function according to nodes or cells, or dividing the die exit into a certain number of sections over which the required size is calculated (Kaul 2004; Lotfi 2005; Smith et al. 1998; Ettinger et al. 2004; Zolfaghari et al. 2009). In addition to the velocity distribution criterion, some work groups include secondary quality criteria in the examination. These could be pressure loss or dwell time, for example (Smith et al. 1998). There is also an example of an objective function which goes one step further in the process chain and takes into account viscoelastic swell behind the die (Debbaut et al. 2006).

There are two fundamentally different ways of approaching geometrical representation. On the one hand, the finite element mesh for the flow solution may also be used for the geometrical representation. In this instance, changes are made to the geometry by shifting individual nodes, i.e. the positions of the nodes represent the design parameters. This approach has the advantage that model generation is considerably easier and the deformation is not subject to any restrictions. However, drawbacks include the fact that filters must be used in order to achieve a smooth surface, the number of design parameters is very high and the optimised geometry is not easily read in a CAD programme (Bletzinger et al. 2010; Le et al. 2011). On the other hand, it is possible to use parameterisation for geometrical representation. The type of parameterisation ranges from very simple approaches such as varying the height and length of the flow channel (Smith et al. 1998; Ettinger et al. 2004; Carneiroa et al. 2001; Zolfaghari et al. 2009) to more elaborate approaches in which free-form surfaces can be represented using splines (Lotfi 2005). The advantage of parameterisation is that it considerably reduces the number of optimisation parameters, thus giving greater flexibility in the selection of the optimisation algorithm. Furthermore, it is easy to transfer it to a CAD system. Parameterisation essentially restricts deformation. This can be viewed as both an advantage and a disadvantage. While parameterisation limits the optimisation range, this is exactly what is required in some situations. The key words here are symmetrical deformation or adherence to manufacturing constraints.

What cannot yet be combined using state of the art technology is the calculation of a complex 3D geometry in conjunction with flexible parameterisation. This is the point at which this research project comes in. The use of NURBS (Non-Uniform Rational B-Splines) for parameterisation allows a high degree of flexibility which may also be applied to 3D geometries in conjunction with a highly parallel flow solver.

3.4.3 Motivation and Research Question

The increasing saturation and diverse segmentation of markets in the industrialised nations have considerably raised demand for product variants and individual products (Kinkel 2005; Schuh 2005). This trend is making all processes increasingly complex and is a great challenge for manufacturing systems and technology (Kinkel 2005). The synthesis of mass production and individual manufacturing is becoming increasingly important in many markets.

The main objective of this project is to bridge the existing gap between individual manufacturing and mass production (Queudeville 2009). This means that the economic batch size for a specific manufacturing process must be reduced. At the same time, companies must be able to offer a very diverse range of products at justifiable costs. This research focuses on processes used in the field of primary shaping. The modularisation of dies is a core strategy in the customisation of products in this field. Studies show that companies, who use modularisation concepts for their products, dies, architecture, etc., are more successful than those companies who cling to their traditional individual design (Eitelwein and Weber 2008).

Achieving this objective requires developing a flexible methodology for the modularisation of dies. The design engineer should be supported by the methodology in a conceptually closed way when developing modular dies. The methodology encompasses modularisation in a broad sense, i.e. it not only includes hardware modularisation but also the modularisation of the design process. This can raise the degree of standardisation in die design.

The modularisation of design and construction is approached in such a manner that the still distinct areas of components can be identified and designed more favourably using simulations and optimisation algorithms. This leads to the following research questions:

How must mass production dies be designed to achieve the economical production of individual products?

- Under which conditions can dies be constructed using a modular design with standardised and individual modules and how can these be identified?
- How can discrete characteristics be defined within reasonable margins for the identified standard modules?
- What form should the design of the standardised interfaces between the modules take?
- How should the individual sections be designed?

3.4.4 Results

3.4.4.1 Development of a Methodology for the Modularisation of Dies for Mass Production Processes

As seen in the previous section, there is usually a one-to-one relationship between the pressure die casting and profile extrusion dies and their products. Some isolated applications for modularisation already exist in both processes. However, there is so far no standardised approach for the modularisation of dies for mass production.

3 Individualised Production

The first stage of developing such a methodology is to identify the general requirements. With respect to the general objectives of the previous section, the methodology must primarily fulfil the following requirements:

- The design engineer should be supported throughout the entire die development process.
- The methodology should be easy to implement.
- It should be flexible and transferable to other production processes.
- No restrictions should be placed on the finished product (cast part or extruded profile).

Bearing in mind the requirement that the end product must remain distinct, it is impossible to define suitable standard modules for all the forming components of the die. This means that certain die components, or specific characteristics or features of a component must remain distinct, just like the resulting end products.

According to Mikkola, it is not practical to modularise every product component (die or end product) as products constructed entirely from standard components may be easily copied by competitors (Mikkola and Gasmann 2001). In order to produce individual products with mass production dies, certain features of a module must remain distinct. These distinct features represent the knowledge and skill of the company and serve as the company's unique selling point. Figure 3.81 illustrates how successful modularisation could change the design of a pressure die casting die in the future.

The image on the left illustrates current technology. The majority of components are individually designed. The die is supplemented with few standard or bought-in parts. The vision is to use predominantly in-house standard modules with individual features which are then supplemented by some standard or bought-in

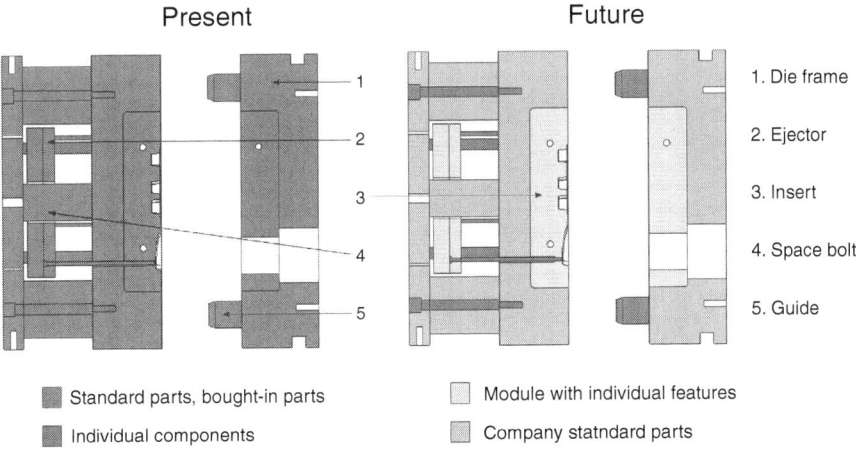

Fig. 3.81 The present and future of pressure die casting dies. (According to Queudeville 2009)

parts. This project defines three categories of module with regard to the potential for modularisation.

- Standard parts or components defined according to official standards and which are already available as standard components.
- Standard modules: standard parts defined according to in-house standards with standardised interfaces. These may be used in the die without post-processing.
- Modules with individual features: components with a low overall modularisation potential but high potential with regard to certain parameters.

Components with very low potential for modularisation remain distinct piece parts although as part of the drive for modularisation there should be an attempt to eliminate these individual components insofar as the product programme in question will allow.

A procedure has been developed based on the methodology developed by Boos (2008) for the modularisation of common dies, which starts with the analysis of the superordinate structure of a company and the existing product programme and ends with the engineered design of modular components. This therefore follows a top-down approach. This newly developed methodology is divided into three phases (Fig. 3.82). These are described in more detail below.

Initiation Phase The aim of this first phase is to quantify the potential for modularisation of several of the company's business segments. This initially requires examination of the constraints of the die manufacturer. This includes, for example, the company's position in the value-added chain, the competitive environment and identification of the strategic success factors. Furthermore, it is important to identify which die components are standard parts or bought-in parts and which are produced internally or externally.

The second stage in the initiation phase is die classification. The manufacturer must categorise its dies according to specific criteria, particularly those with respect to process type. By way of illustration: criteria for pressure die casting dies should include the number of core sliders and cavities within the die. For the extrusion of plastic profiles, the product material or the existence of cores (displacement bodies in the flow channel) could serve as criteria. Armed with this information, the manufacturing company is able to select the die types with the greatest potential for modularisation.

The last stage in the initiation phase is the detailed analysis of the processing of orders. This analysis helps to identify the part of the manufacturing process most suitable for modularisation. Modularisation may be practical in various areas and take on different forms depending on the company profile. This can range from a virtual modularisation of the design (partially automated, parametric CAD system) to hardware modularisation with reusable modules.

Analysis Phase The selected specific category of dies identified in the initiation phase as suitable for modularisation are now analysed with regard to their design. The first stage, the deduction of requests and functions, is based on the general concept of engineering design theory, similar to the VDI 2221 standard (Verein

3 Individualised Production

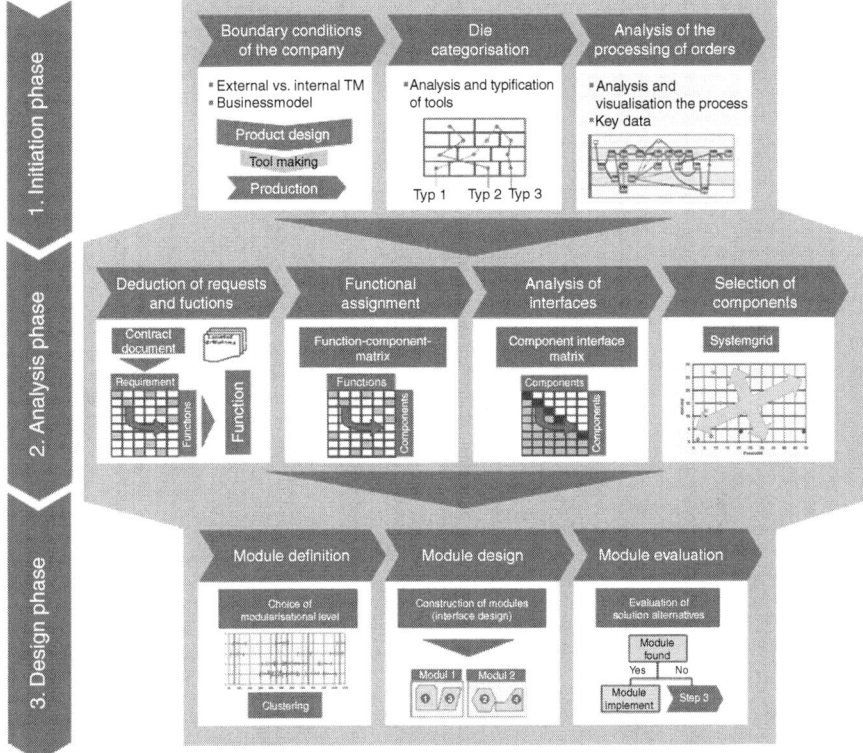

Fig. 3.82 Superordinate structure of the methodology for the modularisation of dies

Deutscher Ingenieure 1993). The central idea is to generate a systematic overview of the dependencies between the various die components. The requirements placed on modular dies are primarily based on future business strategies and individual customer requirements. In addition to the general technical requirements, the strategic economic requirements, such as minimum batch size of the finished product or the focus on a particular product range, must also be taken into account. These requirements are then turned into functions which must be fulfilled by the end product—in this case a pressure die casting die.

In the second stage, functional assignment, the main die functions are combined with the main components to form a product architecture (Göpfert 1998). This product architecture provides a clear overview of the involvement of components in various functions. In general, the complexity of a product increases with the number of functions which have to be fulfilled.

In the next stage, analysis of interfaces, the interfaces are analysed bearing in mind the dependencies between the adjacent components. Both the directions as well as the scale of the dependencies are considered. If the component actively influences another component, it is assigned an active value. If the component is influenced by other components, a passive dependency exists which provides a specific passive

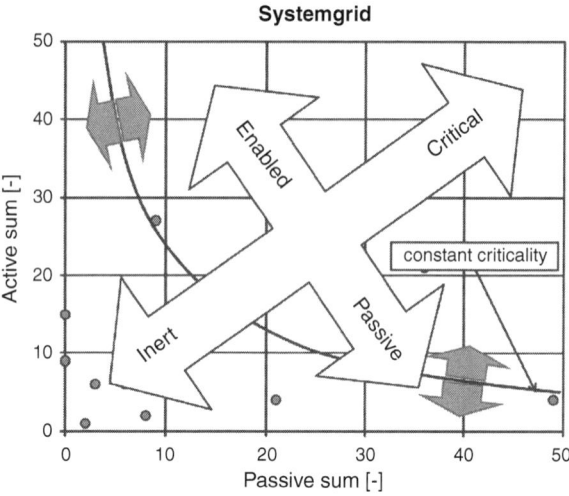

Fig. 3.83 Illustration of the system grid

value. Combining all the dependencies of one component to all other components produces active and passive sums. The active sum indicates the extent to which a component influences all other components. In contrast, the passive sum indicates the extent to which a component is influenced by all other components.

In the last stage of this phase, the selection of components, all the previously defined dependency factors (active and passive sums) are mapped on a system grid (Fig. 3.83). The quantitative illustration of the dependencies between the components provides an estimate of the modularisation potential of all components under examination. The system grid may be divided into four main areas: enabled, passive, active and critical. It generally applies that the lower the dependencies between the components, the greater the possibility of designing them as modules (Ulrich 1995). According to Lindeman, the components with high active sums should be identified early in the development process due to their effect on other components (Lindeman et al. 2009). Components with a high active sum but a low passive sum are ideal components for a modular design. Interfaces defined here are transferred to adjacent components.

An additional factor used to draw conclusions regarding the potential for modularisation is criticality. Criticality is defined as a product of the active and passive sums. Lines with constant criticality may be represented by hyperbola running symmetrical to the bisecting lines of the system grid (Lindeman et al. 2009). Components with similarly high active and passive sums (high criticality) indicate complex interactions with several components. The potential for modularisation is therefore low. Taking into account the hierarchy in the development process, it is possible to modularise components with a high active sum provided these are defined at the beginning of the development process. Ideally, modularisation should begin with active components before working counter-clockwise round the system grid. Having designed the active components, attention should be turned to the inactive components. The inactive components exert only a very weak influence on other components and are only influenced themselves to a minor extent. Low dependencies on and interactions

3 Individualised Production

with other components make it easier to design the interfaces with other components. Passive components should only be designed later in the design phase. These then take on the previously specified interfaces of the active components.

Design Phase The purpose of the design phase is to modularise all the components identified as suitable in the analysis phase. During this phase, the die is examined on a component level. This phase must therefore be performed at least once for each of the components to be modularised. The module category greatly influences the way and means in which the design phase is carried out (cf. Fig. 3.81). On a component which is being redesigned as a module with individual features only those interfaces to other components and a couple of its parameters are standardised. All core features of an in-house standard module are standardised. The individual stages of the design phase are described in more detail below.

The first stage, module definition, involves the systematic comparison of the technical and economic constraints on which the number and design of discrete variants of a module are based. The procedure is structured as follows.

The examined dimensions of the components are collected for all dies of a die type. For example, entering the dimensions of a component into a 3D diagram gives a free spatial distribution as illustrated in Fig. 3.84.

A cell is then positioned at each point which describes the technical constraints of the dimensions of this component. These technical constraints can have very

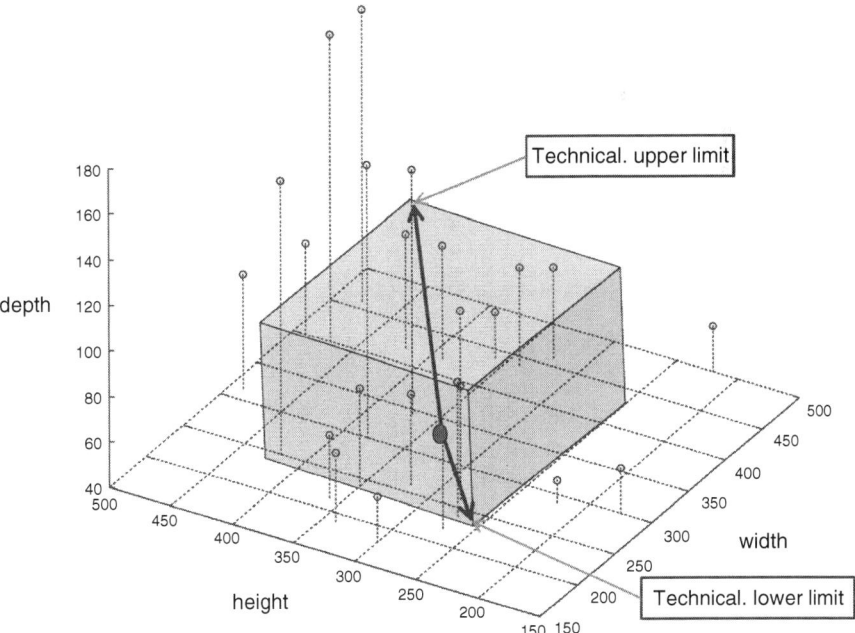

Fig. 3.84 Static analysis of the main dimensions of a component with its upper and lower technological limits

different origins. Some examples include the maximum stress, the installation space or sufficient thermal conductivity of the components.

In addition to the technological boundaries, economic constraints may also be taken into account using the cells. Economic constraints are particularly dependent on development time, cost of materials, available semi-finished products and manufacturing costs. Figure 3.84 illustrates this procedure using the example of the dimensions of the pressure die casting die "insert" component. The black dots represent the existing variants of this component. The cube depicts the upper and lower technological limits for three selected "insert" variants. The upper vertex of the cube represents the technical upper limit and the lower vertex the technical lower limit.

The number and variants of the module are defined so that each of the components available in the die range is covered by at least one module. The number of module variants is determined by means of a cost analysis. This analysis takes into account the actual product range and the company-specific constraints. The number of variants is selected so that the total costs are minimised over a defined number of dies.

The following stage of the module design methodology focuses on the definition of interfaces between interacting modules. In order to ensure good interchangeability between the modules, it is necessary to standardise the interfaces. These interfaces should be simple, robust and able to cope with changes to the associated modules. In addition to interface design, all features which were not covered in the "module definition" stage must be defined. Different aids may be used to improve the design process, from CAE systems to automatic geometry optimisation strategies, depending on the company and process.

In the next stage, module evaluation, the previously developed modules are examined according to their suitability within the entire system. First of all the developed module is checked to see if it fits in the selected module system. At this point all the constraints previously defined in the initiation phase must be fulfilled. If the test is passed, the design phase can continue with the next component in the design process hierarchy. Previously defined interfaces are passed on.

3.4.4.2 Application Example: Modularisation of Pressure Die Casting Dies

The methodology presented will now be illustrated using pressure die casting as an example to provide a better understanding of the individual intermediary stages. The achievable degree of modularisation is highly dependent on the profile and product programme of the company under examination.

This example discusses a fictitious company. However, all assumptions are based on real constraints. During examination of the pressure die casting dies, close attention is paid to the cost and R&D intensive components. Small parts, in particular fasteners such as screws, are disregarded due to facility of inspection.

Initiation Phase for a Fictitious Company The initiation phase defines the modularisation objectives important to the company. A suitable die type is selected. Analysis is also conducted to identify the area in which the company is most able to apply modularisation.

Boundary Conditions of the Company The company in this case study is a medium-sized foundry with its own tool- and die-shop. All of the larger die components are manufactured in house. Plates of specific dimensions are procured from various wrought material manufacturers for this purpose. These are then subsequently machined in-house. Standard parts such as screws, guide bushings and rods are also procured externally. The foundry has 10 casting cells and all pressure die casting machines are of the same type. The foundry is a contract foundry and mainly supplies the automotive supplier market. Due to the available pressure die casting machines, it is only possible to produce small- to medium-sized castings.

Thirty new dies are developed and manufactured per year. The company is expecting to see a financial return on the transition to modular dies within one year. The company is assuming that the outside dimensions and volume of the components to be casted will turn out the same in the coming years. The company hopes that by changing the dies design, set-up times will be reduced from the current four hours to one hour. This should also reduce stock and assembly costs. This is of considerable importance particularly with regard to the economic production of smaller quantities.

Die Categorisation In this stage, the existing die range is divided into types. Each die of the same type should have the same functional structure and therefore the same dependencies between die components. The dies of the same type should have a similar geometry and the same interface for the peripherals, particularly the pressure die casting machine and tempering units. A die type differentiates itself in terms of the number of core sliders or the number of inserts in one half of the die, for example. Three types of die were identified for this company:

- Type A are dies without core sliders,
- Type B are dies with two core sliders,
- Type C are dies with three core sliders.

The classification of the existing dies revealed that 80% of the dies were Type A. The high number and low level of complexity of the Type A dies make them ideal candidates for modularisation. Type A dies are constructed as illustrated in Fig. 3.85. All of the components illustrated below are taken into consideration in the following stages.

Analysis of the Processing of Orders On close examination of order processing within the company, the modularisation factors described below are of interest. As the company is a contract foundry, no constraints may be placed on the components. On successful conclusion of an order, the company is obliged to manufacture the required quantity of components in a specific period of time. The die components are produced in house to discrete dimensions from semi-finished products. The dies belong to the company and are serviced and stored in house. As the foundry and die shop belong to the same company, the modularisation process should attempt to reuse various die modules. This will save inventory costs.

Analysis Phase for a Pressure Die Casting Die Type During the analysis phase of the modularisation concept (Fig. 3.82) a functional analysis is first performed on the

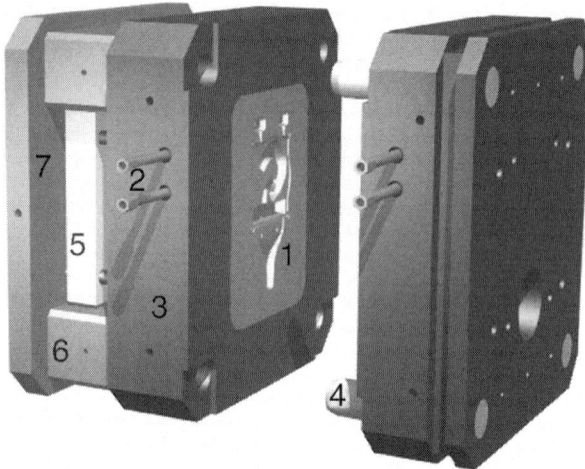

1. Insert
2. Tempering comp.
3. Die frame
4. Frame guides
5. Ejection plate
6. Spacer strip
7. Clamping plate

Fig. 3.85 Type A pressure die casting die

Type A pressure die casting die. The pressure die casting die components are then allocated to individual functions. The knowledge thus acquired about the functions makes it possible to determine the dependencies between the die components. It also enables the modules to be designed according to the functional structure later on.

Deduction of Requests and Functions At the beginning of the analysis phase, the modular die requirements are gathered from the customer requirement specification, for example. These requirements are then transformed into functions which the pressure die casting die must fulfil. Figure 3.86 illustrates the functional structure of a Type A pressure die casting die.

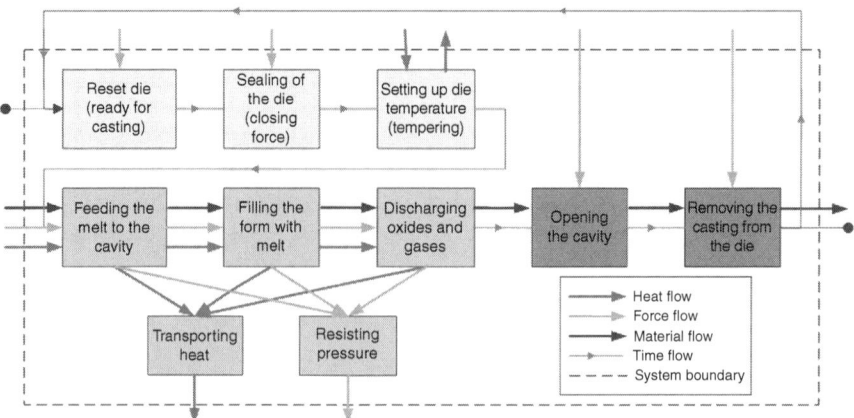

Fig. 3.86 Functional structure of a pressure die casting die

3 Individualised Production

All the main functions, such as material, signal and energy flows, are taken into account. The functional structure provides information on which die components are positioned in a force or heat flow, for example, or which come into contact with molten melt. No signal flow exists for the casting of components in the sense of the functional compliance of the pressure die casting die. The process of deriving the function of a pressure die casting die is the same for each die type due to the solution independent construction and consistent requirements of the functions to be fulfilled during operation. This functional structure can therefore be used as a master function for all pressure die casting dies of a specific type.

Functional Assignment In the functional assignment stage, a relationship should be established between the main functions of a die and the components which perform them. Although castings exist in many shapes and sizes, the main die functions of a specific die type remain the same. The functional allocation of a die type therefore only need to be performed once and then reapplied to all dies of this type.

The following stages analyse the moveable half of the die. The representation of the product architecture considers all the main components (Fig. 3.87). This product architecture already makes it possible to draw some initial conclusions regarding the complexity of the die components (Fig. 3.87) (Göpfert 1998). It is already possible to identify that some components exhibit a high degree of functional integration. Due to the subsequent high level of complexity of these components (e.g. insert) it is already possible to conclude they have a low potential for modularisation.

Functions \ Components	Insert	Tempering comp.	Die frame	Frame guides	Ejector	Ejection plates	Ejection plate guides	Spacer bolt	Spacer strip	Clamping plate
Reset die (ready for casting)				X	X	X	X			
Sealing of the die (closing force)	X		X					X	X	X
Setting up die temperature (Tempering)	X	X								
Feeding the melt to the cavity	X									
Filling the form with melt	X									
Discharging oxides and gases	X									
Transporting heat	X	X	X							
Resisting pressure	X		X		X			X	X	
Open the cavity	X		X						X	X
Removing the casting from the die					X	X	X			

Fig. 3.87 Product architecture of a Type A pressure die casting die

Analysis of Interfaces The next stage defines the interfaces between the die components. Each existing interface is described in detail. To achieve a better understanding of the type of dependencies, the interfaces are divided into different groups. The most important interfaces between the components are:

- fastening interfaces (usually in the form of threaded joints)
- functional interfaces (generally relative movements or guides)
- cast part contour (all areas which come into contact with the melt)
- contact areas (the number indicates how many surfaces are in contact)
- heat flow
- force flow
- pass (non-contact interface in a general clearance hole)

The interfaces between individual components are examined and defined in comparison to all components. The interface analysis derives the dependencies between the cast part and the die components. As the die is usually designed around the cast part, the dependencies usually follow a certain hierarchy. Furthermore, components which share an interface with the cast part are far more dependent on this than components located on the peripheral zone of the die. The qualitative analysis based on the Quality Function Deployment (QFD) valuation method determines the degree of dependency between individual die components (Schmitt and Pfeifer 2010). The ratings are as follows:

- 1 = low dependency
- 3 = medium dependency
- 9 = high dependency.

Dependencies between two components may be uni- or bilateral. If a dependency is unilateral, for example between a casting and an insert, there is a clear hierarchy between the individual components. If the geometry of the cast part changes, the insert must change in response to this. A bilateral dependency, for example that between the ejector plates and the ejector guides, means that there is no clear hierarchy between these components. In addition to the dependencies derived from the interfaces, dependencies exist which have no physical interfaces (cf. Fig. 3.88). For example, the height of the spacer bolt should correspond to the height of the spacer strip.

Selection of Components The system grid provides a good graphical overview for visualising the passive and active sums of the individual components. The four areas of the system grid are illustrated in Fig. 3.83.

The position of a component in the system grid allows conclusions to be drawn regarding the modularisation potential of this component (Fig. 3.89). The components are assigned to the respective module categories. Components with small active and passive sums, such as spacer elements, are easy to modularise because they have little influence on other components and are only slightly influenced by other components. This weak interaction makes it easy to develop the interfaces further. Some components such as the ejectors have a high active sum and a very low passive sum. These components impose their interfaces on other very passive components such as the die frame. Components which are already standard components are positioned

3 Individualised Production

	Cast	Insert	Tempering comp.	Die frame	Frame guides	Ejector	Ejection plates	Ejection plate guides	Spacer bolt	Spacer strip	Clamping plate	Passive sum
Cast												0
Insert	9		3			3						15
Tempering comp.												0
Die frame		9	3		9	3		3	1	1		29
Frame guides												0
Ejector	3	3										6
Ejection plates	3					9		3	3			18
Ejection plate guides						3	3					6
Spacer bolt	1					1				1	1	4
Spacer strip				3		3	1		1		1	9
Clamping plate				1					1	1		3
Active sum	16	12	6	4	9	22	4	6	6	3	2	

Legend: ■ Interface   ■ Dependency without interface   □ Interface without direct dependency

Fig. 3.88 Dependency analysis between the components of a Type A pressure die casting die

on the left and particularly the lower half of the system grid. In contrast, components with a high active and passive sum, such as the insert, are difficult to modularise.

The system grid makes it possible to select die components with a high modularisation potential. As the edges of the areas in the system grid are flexible, however, it is essentially possible for components which have both a high active sum as well as a high passive sum (critical elements, e.g. the insert) to still have a certain potential for modularisation. In order to identify these components, it is necessary to examine the dependency of the individual component features. Although some features of a component may be standardised (e.g. the main dimensions of an insert), the rest of the features remain distinct which leads to modules with individual features such as those in Fig. 3.81.

Design Phase for Pressure Die Casting Die Components The actual modules for the new die type are selected in the design phase. Before the models are designed, the module variants and discrete parameters are determined using a cost calculation. Then

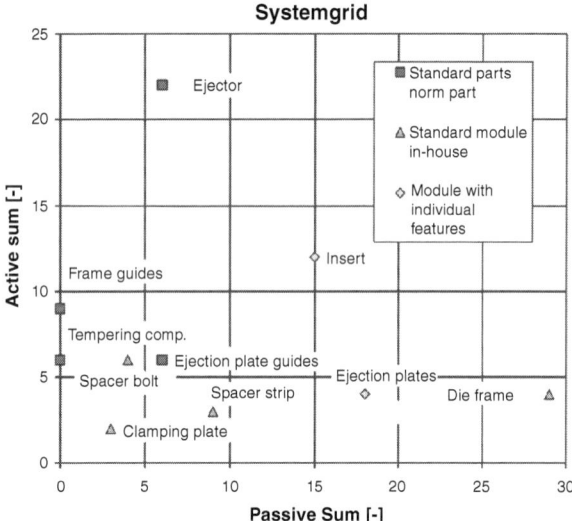

Fig. 3.89 Graphical representation of the modularisation potential of die components

the standards for the interfaces are determined and the remaining module features defined.

Module Definition Figure 3.89 illustrates the direction and scale of the dependencies. Setting up module assemblies allows interdependencies to be bundled together within an assembly. In this instance, reorganisation created two large module groups (Fig. 3.90).

- Module group 1 (MG1): inset, ejectors, ejector plates, ejector plate guides, tempering components and spacer bolts.
- Module group 2 (MG2): die frame, spacer strips, frame guides and clamping plate.

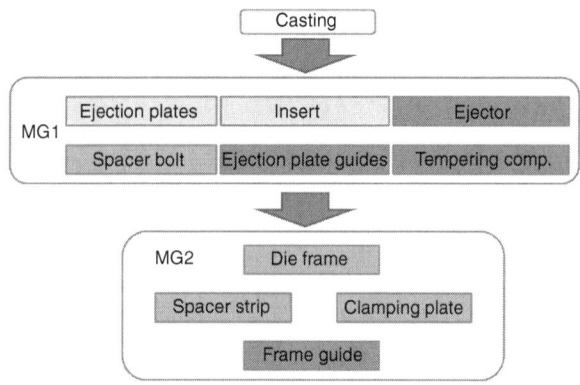

Fig. 3.90 Component structure after setting up module groups

3 Individualised Production

Fig. 3.91 Illustration of the old and new die principles

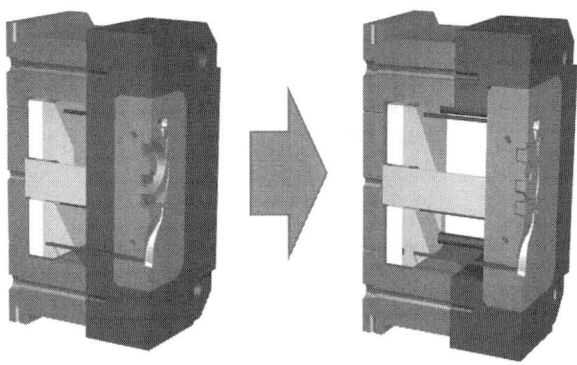

Module group 1 depends directly on the end component (casting). This module group is therefore almost entirely comprised of modules with individual features and standard parts. This means that the modules must be post-processed for a new order. These components therefore do not have any real level of re-usability.

Module group 2 should only consist of standard modules. In order to achieve this, it is necessary to eliminate any weak dependencies between MG1 and MG2. In MG2, the spacer strips and clamping plate have only a very weak relationship to the components in MG1. In order to develop the die frame into a standard module, it is necessary to minimise the relationship between the die frame and MG1. In principle, redesigning the die will reduce or eliminate the dependencies (cf. Fig. 3.91). The ejectors no longer share a direct interface with the die frame. The spacer bolts and the ejector guides are screwed into the insert and not the die frame. It is therefore now possible to design MG1 so that they may be exchanged in the system as an entire unit. This considerably reduces set-up times. In the following cost analysis it is assumed that MG2 will be used up to ten times again in the future.

Before the modules are broken down further and designed, a rough profitability analysis should be used to estimate how many variants or sizes are suitable for a module. In module group 1, the insert has the greatest dependency in relation to the casting. As it also has the highest active sum of the components under development, the insert should be designed first (cf. Fig. 3.89). The dimensions of the insert are currently derived from the dimensions of the casting. These values may be used as the lower technological limits for the insert dimensions. The upper technological limits of the insert depend on the mounting table of the pressure die casting machine and the minimum wall thickness of the die frame. From a technological point of view, the insert dimensions may lie within this range. For a particular dimension of the casting x_{cast}, the dimension of the insert x_I may lie between $(x_{cast\ +\ k})$ and x_{Imax}. Specific FEM calculations may be used to minimise the factor k thus increasing the margin of the technologically feasible dimensions for the insert as per Fig. 3.84. From a technological point of view, the dimensions of an insert may change within these ranges without affecting the function of the die or the geometry of the second cast.

These geometric relationships are defined for all the components under development and are included in the cost considerations.

There are currently n different inserts and die frame dimensions for n end products (castings). When modularising the die type, it is now necessary to examine how many variations or sizes of modules are reasonable. All reasonable module combinations must be considered via a rough profitability analysis. All main components and their geometric dependencies are considered in the calculation. The cavity in a die frame therefore depends on the outside dimensions of the insert. Each component consists of three cost pools (manufacturing costs, material costs and development costs). All costs depend on the number of module variants. Certain factors must be approved by the design engineers:

- the component development time,
- the time-saving made by using modules,
- the level of re-usability of the standard modules.

If the total tooling costs for a given combination are the lowest, these variants should be designed as follows.

Module Design The module shall be designed using the example of an insert. In the previous stage, the discrete outside dimensions were determined for the insert. Now all the other properties have to be defined. The insert can be broken down into seven properties (dimensions, radius, bevel, mounting, ejector holes, cooling channel holes and cavity). It is possible to define standards for the first four properties. Each property has interfaces to other die components. These must be defined for each property and transferred to the components involved.

As a result of further studies, such as deciding on a universal screw size, it is possible to simplify assembly in particular. It is therefore possible to assemble a pressure die casting die using only a few hand tools. The storage of standard parts is also reduced by the larger proportion of identical parts.

The individual features (ejector holes, cooling channel holes and cavity) are only developed for a new request. However, it is possible to complete the module this far using a CAD model so that only these few individual features need be designed. The components may also be available pre-cast in stock so that only the individual features need be post-processed. In this way it is possible to optimise development time on a virtual level and the production process on a real level.

Module Evaluation The last stage involves inspecting the module. The various module variants are combined in the CAD system and all the interfaces are checked to ensure they are aligned with one another. Following a successful inspection, the newly defined modules can be included in the product programme.

3.4.4.3 Results of the Numerical Design of Profile Extrusion Dies

Analogous to 3.4.4.2, the modularisation concept presented has also been applied to the primary shaping process of profile extrusion in order to identify those components in the system which may be standardised or which must remain individual. In the analysis phase, the flow channel of the die has been identified as the area which has potential for standardised design stages and interfaces. In addition automatic flow channel optimisation supports individual design. The die is therefore examined further as the central component below.

Standardised Process for Die Design While standardised components are already being used for extrusion in geometrically simple flow channels (such as annular or slit geometries), this has not yet occurred within profile extrusion. The word standardisation therefore addresses both the use of analytical, transferable formulae for the design process and the use of standard components. Transferability to profile dies is only possible on a limited basis as the profile geometries have a variety of geometric shapes which also require the profile die to have a certain level of customisation. The variety of profiles which can be produced by the extrusion process range from window profiles, shutter profiles, door frames, sealing profiles, guide rails and rail systems to multiwall sheets and edging strips. These flow channel geometries clearly vary in their complexity and the demands they place on the design. In conjunction with the shear thinning melt behaviour of plastics it follows that the rheological design of profile dies is an iterative process. The expertise of a die-manufacturer is evident in the number of stages of empirical post-processing of the die that are required to produce a profile to a previously defined quality standard.

As design is usually only carried out empirically and aligned to the individual product, the standardisation of profile die design must include a methodical, structured design process as well as the identification of potential standard modules or modules retaining individual areas. In order to systematically gather the personal knowledge of the die-manufacturer which is involved in die design and to make it more readily available, the empirical design procedure must be assimilated and structured. The co-operation partner Döllken Kunststoffverarbeitung GmbH, Gladbeck, supports this work with empirical design rules and real profile geometries.

As the flow channel of the profile die must be automatically optimised, the initial geometry of a flow channel need not be designed in full, but only good enough for the optimisation algorithm to find an optimum in relation to the design criteria. A standardised procedure for creating an initial geometry therefore has the potential to considerably shorten the design phase.

In the course of the project so far, the examination of a profile extrusion line stops immediately behind the die. All components located before and after the die and the effects which occur there and likewise have an effect on the profile geometry are therefore disregarded. These effects include volume contraction, drawdown through a puller or the viscoelastic swell of the melt at the die exit. These cause the geometrical differences between the profile required and the exit geometry at the die and are taken into account in practice using appropriate empirical values when determining the outlet geometry. However, as this project focuses solely on die design, these effects shall initially be disregarded.

The design of the channel within the die is entirely free. Apart from the exit geometry of the die and the cross-section geometry at the extruder outlet (which may

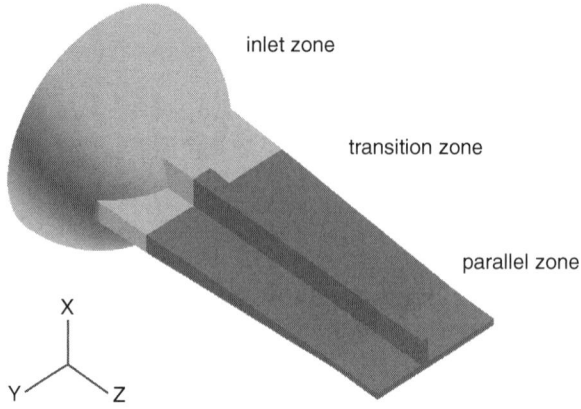

Fig. 3.92 The flow channel of a profile die divided into three sections

be adjusted using an adapter) the geometry of the flow channel may take any form. However, there are a few constraints which must be considered:

- The melt should be accelerated continuously in order to be placed under as few mechanical stresses as possible.
- Dwell time should be kept low. This will keep productivity high and the thermal stress on the plastic low.
- Undercuts should be avoided as the melt can dwell for a long time at these points and under certain circumstances may thermally degrade.
- Just before the melt exit, the melt is guided along a constant flow channel cross-section (parallel zone) in order to homogenise the melt.

Some of these points are material-specific. The length of the constant parallel zone at the die exit should as a rule be e.g. for ABS (acrylonitrile butadiene styrene) 10–20 times the thickness of the profile, while for polyvinyl chloride (PVC) due to the less elastic properties just five times the profile thickness is required.

For the analysis carried out as part of this project, the flow channel has been divided into three sections (Fig. 3.92). It is specified that the inlet section constitutes the first third of the die and is connected to the transition section. An intermediate section constitutes the interface between the inlet and transition sections. It is possible to affect the coarse distribution or pre-distribution of the melt in the inlet section. However, an even finer distribution is determined in the transition section. The final section is the parallel zone as this is where the flow channel cross-section usually remains constant and is not available for flow channel optimisation.

The entire length of the die should be short in order to save die material and also keep the pressure loss which the extruder must overcome to a minimum. However, the die should also not be too short. Making large changes to the cross-section along the flow channel (necessary in a very short flow channel) can damage the melt as a result of high shear stress. Typical die lengths in the investigations performed are around 150 mm with an angle of inclination in the tapered flow channel of up to 25°.

Fig. 3.93 An alternative geometry is created by abstracting the profile using basic geometrical elements

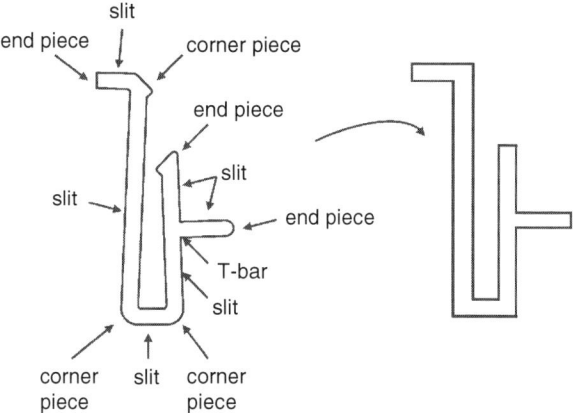

The following procedure is recommended when designing a flow channel for a specific profile: all considerations should be based on the profile cross-section. This is initially abstracted by approximating the cross-section using a combination of basic geometrical elements in order to lead the number of different profile cross-sections back to a common basis. The abstracted alternative geometry is used as the basis for the mathematical formulae to decide upon the geometry of the interface between the inlet section and the transition section. The flow channel is created between the established cross-sections by linear morphing.

During abstraction, the entire profile geometry is assembled using a combination of simple basic geometrical elements. Possible basic geometrical elements include slits, corner pieces, end pieces and T-pieces. An example of an application of the basic geometrical elements is illustrated in Fig. 3.93 using a "Döllken" profile geometry. In angled profiles, the geometrical elements are arranged at 30° increments. The alternative geometry does not form part of the subsequent flow channel, but forms the initial geometry in order to create a free and systematic shape for the interface within the flow channel using basic geometrical elements.

The interface in the flow channel is created by scaling the alternative geometry. Each geometrical element is processed according to certain rules. The most used geometrical element is a slit which is defined by its width being considerably larger than its height. The height and width of the slit are both scaled differently. The slit width in the interface between the inlet and the transition section $w_{interface}$ are determined according to (3.21).

$$w_{interface} = w_{outlet} + \underbrace{b_{outlet} \cdot 0.15 \cdot \frac{l_{die} - l_{inlet}}{l_{die}}}_{\text{Scaling term}} \qquad (3.21)$$

whereby w_{outlet} is the width of the respective slit of the alternative geometry, l_{die} is the length of the die excluding the parallel zone and l_{inlet} is the length from the die inlet to the said interface. The width of a slit in the interface is therefore determined

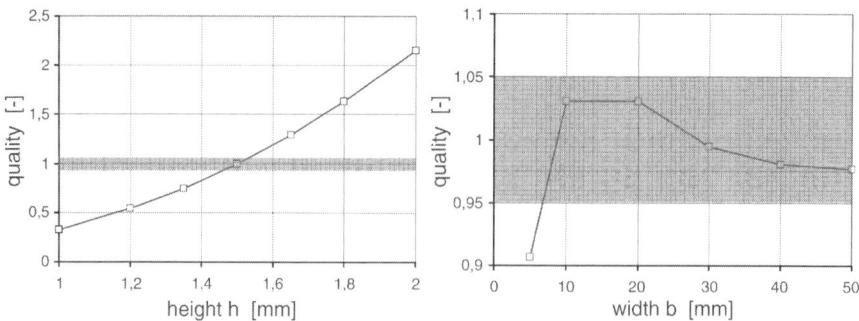

Fig. 3.94 Changing the height of a slit clearly influences the velocity distribution

by the width of a slit geometrical element in the alternative geometry and a scaling term. The smaller the distance between the inlet in the inlet section and the cross-sectional plane, the larger is the width. This formula is recorded in empirical values observed in the industry whereby a scaling factor of approx. 15% is recommended for the width of a slit.

The Polyflow programme of Ansys Germany GmbH, Darmstadt, is used to calculate the velocity distribution at the outlet in flow simulations of a flow channel for moulding an L-shaped profile (the alternative geometry consists of two slits, two end pieces and a corner). The initial geometry consists of two equal slits with a width of 30 mm and a height of 1.5 mm. In each of the variants, three of these values remain constant while merely a width and/or height are modified. As Fig. 3.94 illustrates, a change in the width has only a small effect on the velocity distribution at the outlet. Quality in this respect is defined as the ratio between the average speeds of the two straight pieces and is ideally exactly 1. Changing the height, on the other hand, clearly alters the homogeneity of the velocity distribution so that this parameter may only be modified by approx. 5%. Greater differences in height of the various sections of geometry must therefore be considered during scaling as the exit speeds may become too heterogeneous and may no longer be balanced in the calibration process.

The height of the slit on the interface $h_{interface}$ is determined by (3.21):

$$h_{interface,i} = h_{outlet,i} \cdot 3 \cdot \left(\frac{h_{mean}}{h_{outlet,i}}\right)^3 \tag{3.22}$$

As scaling the height of one slit upwards without taking into consideration the different heights in various slit geometrical elements would lead to a flow channel that would increase the differences in flow along the geometrical elements, the flow channel cross-section must instead be correspondingly larger than the slit areas of the profile with short heights in order to reduce the flow resistance of this path. The heights of the slits in the interface are therefore scaled according to their deviation from the mean value. The various heights of the slit geometrical elements are used to calculate an average height h_{mean} with an upper scaling factor of 300 %. The ratio of the slit height to the mean slit height increases or decreases based on the deviation

3 Individualised Production

from the mean value to the power of three (3.22). The power of three is used as the pressure loss (and correspondingly the flow resistance) also has a cubic relationship to the flow channel height.

This improves the pre-distribution and therefore the initial situation of the required further design (e.g. by means of automatic flow channel optimisation).

An important parameter for accelerating the melt along the flow path is the area ratio between the area in the intermediate section and the exit surface. From experience, a ratio of approximately 3 is used so that the melt accelerates continuously without being damaged by the high shear stresses (3.23).

$$\frac{A_{interface}}{A_{outlet}} \approx 3 \qquad (3.23)$$

The corner piece is another basic geometrical element. The melt flows more quickly in a corner piece than in the adjacent flow channel areas as flow resistance is lower there. A corner piece should therefore be designed so that the inner vertex is positioned slightly towards the outer vertex. This reduces the size of the channel and increases local flow resistance. However, this does not yet need to be considered with regard to the interface as it is sufficient to have a narrowing just before the parallel zone.

Analysis shows that even in T-pieces, the higher speed of the melt in this geometrical element cannot be taken into consideration in the intermediate section. Instead, dents and notches for increasing local flow resistance in the parallel zone are more effective for homogenising the melt speed.

End pieces which are usually connected to slit geometrical elements should be taken into consideration accordingly in the interface. This makes it possible to round the end pieces with large radii. As the flow channel of an end piece contains a relatively high flow resistance, it is recommended that the radius of the end piece until shortly before the start of the parallel zone is taken from the interface or only slightly reduced in order for more material to flow though these areas.

Another design criterion is consideration of the centre of shear. At the die inlet this lies at the centre of the circular strand and indicates the main melt delivery. The centre of area of every cross-section along the die should likewise lie along the axis of the centre of area of the extruder in order to prevent any additional stresses in the melt. In this instance, at least the inlet area, the interface and the exit surface of the die are aligned with it.

The geometry between the interface and the parallel zone is morphed linearly. The geometry between the inlet diameter and the intermediate section may likewise be morphed linearly on one side. Another possibility is to use a spherical or hemispherical ingate into which the linearly morphed flow channel geometry from the intermediate section is lengthened and connected. This has the advantage of being much simpler to produce and that it may also build on the use of standardised components.

Figure 3.95 illustrates the systematic design of a flow channel taking into account the guidelines mentioned above. The velocity distribution at the outlet point is not yet sufficiently homogeneous. This homogenisation takes place in the following optimisation phase.

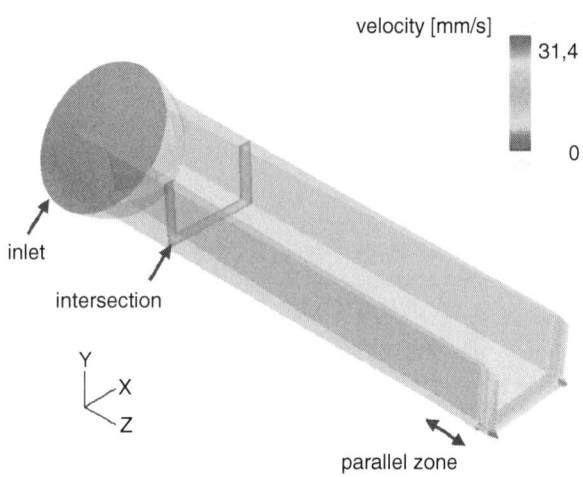

Fig. 3.95 Velocity distribution in a systematically generated U-shaped channel with varying thickness

Optimisation Framework As the initial geometry of a flow channel does not usually provide a homogeneous distribution of plastic melt, the flow channel geometry must be reworked. Instead of performing a number of practical experiments and reworking stages on the real die, an automatic numerical optimisation of the flow channel should be performed during the design stage. The optimisation framework consists of a flow solver, a model of the geometry deformation, an optimisation algorithm and an objective function (also known as the quality function), all of which are described below. The successful application is then presented by way of two examples.

Flow Solver The flow of the plastic melt through a profile extrusion die takes place at a Reynolds number much smaller than 1. This situation is referred to as a creeping flow. In this instance, the Stokes equations can be used to model the flow:

$$-\nabla(\eta(\dot{\gamma})\nabla \mathbf{u}) + \nabla p = \mathbf{0} \tag{3.24}$$

$$\nabla \cdot \mathbf{u} = \mathbf{0} \tag{3.25}$$

Here, η is the viscosity which for plastic melts depends on the local shear rate $\dot{\gamma}$. The variables in this differential equation are the velocity vector $\mathbf{u}$ and the pressure p. Gravitation is generally disregarded. The material models are more complex than the differential equation. Plastic melts are shear thinning and viscoelastic. As the viscoelastic properties mainly affect the behaviour of the melt once it has left the die (viscoelastic swell) only a shear-thinning material model is initially implemented.

There are several mathematical models which describe shear thinning and which are used to approximate the flow curve in the usual shear rate areas of the respective manufacturing process. The power law of Ostwald and de Waele and the Carreau model (Carreau 1968) are well established in the field of plastics processing, and extrusion in particular (Fig. 3.96). In areas of medium and high share rates both

Fig. 3.96 The dependency of viscosity on the share rates according to the Carreau model

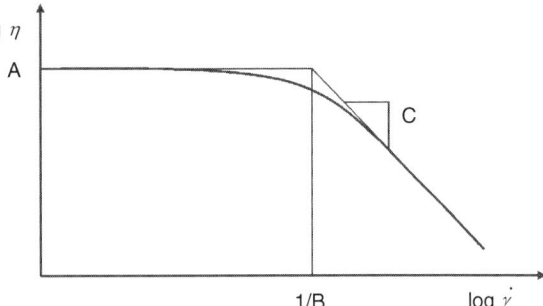

models deliver comparable results. They therefore cover the majority of the shear rate areas relevant to processing. The Carreau model also has the advantage of providing a realistic approximation of the viscosity function even at lower share rates. The Carreau model is therefore used as illustrated in (3.26).

$$\eta(\dot{\gamma}) = \frac{A}{(1 + B \cdot \dot{\gamma})^C} \quad (3.26)$$

In (3.26), the shear rate $\dot{\gamma}$ dependent viscosity η is described by the quotients of the zero shear-rate viscosity A in [Pas], the reciprocal transition velocity B in [s] at the transition between the Newtonian and shear-thinning behaviour and the slope C on the viscosity curve in the shear-thinning zone.

The flow solver used is a proprietary development known as XNS. It is based on P1P1 finite elements, i.e. both velocity and pressure are linearly interpolated. It is recognised that this combination of interpolation functions does not satisfy the Ladyzhenskaya-Babuska-Brezzi (LBB) compatibility condition. This means that the pressure is highly vulnerable to non-physical oscillations if it is not suitably stabilised. XNS uses the Galerkin/Least-Squares stabilisation method for this reason. In this type of stabilisation, the stabilisation term consists of a weighted element-wise least squares form of the original differential equation. Despite the additional stabilisation effort it is worth using interpolation functions of the same polynomial order since they provide the basis for a very efficient parallelisation of the solver. XNS is capable of using all established communication interfaces for distributed computing systems (SHMEM and MPI). The flow solver was tested on a variety of platforms and on the Blue Gene system it demonstrated a scalability on up to 4096 processors (using the MPI parallelisation) (Wylie et al. 2007).

Geometry Deformation The definition of the mapping of the optimisation parameters on the computational domain considerably influences the optimisation results. On large meshes it is impossible to move every node of the finite element mesh with its own parameter. In fact it is worthwhile reducing the number of parameters as far as possible in order to keep the effort involved in optimisation as low as possible. For example, it is possible to make the displacement of each node dependent on a functional relationship. While it is very easy to implement this kind of shift, it considerably restricts the possible types of deformations.

The framework used for this project contains a geometry kernel which provides a much higher level of flexibility. The edge of the computational domain is approximated using non-uniform rational B-splines (NURBS) whereby the optimisation parameter are given by the position and weights of the control points. A mesh moving algorithm may be used to align the position of the inner nodes to the position of the edge nodes without needing to remesh. From a mathematical point of view, this means that the parameter vector (optimisation parameter) α is first mapped on the geometry γ (hyper surface) and then on the mesh $\mathbf{x}$ which discretises the computational domain Ω. The following explains these two individual stages in more detail.

$$\text{Boundary Deformation: } \alpha \rightarrow \gamma \tag{3.27}$$

The shape of the boundary is given by a control point mesh containing m control points. Each control point is associated with a co-ordinate $\mathbf{P}_i$ and a weight w_i. The die geometry is represented using an NURBS curve for 2D calculations and an NURBS surface for 3D calculations.

Since they were first used in the 1970s, NURBS have become an industry standard for exchanging geometrical data processed by computer. For example, they form the basis of all established CAD programs. This is mainly due to their flexibility. They are suitable for precisely representing both analytical shapes such as circles and squares as well as free-form surfaces. Both points are strong arguments for using NURBS in an optimisation context. Firstly, it is possible to export the geometries from CAD drawings and to re-import the optimised geometries into a CAD/CAM system. Secondly, it is possible to profit from the flexibility of the NURBS during the optimisation process.

Each node on the finite element mesh corresponds to a local coordinate τ_k on the spline. These node positions may be clearly specified via these local coordinates using the control point coordinates and weights as per (3.28). If the spline changes shape, the positions of the corresponding points and nodes on the finite element mesh also change in accordance with this displacement.

$$\gamma_k = \frac{\sum_{i=1}^{m} \mathbf{P}_i w_i N_i(\tau_k)}{\sum_{i=1}^{m} w_i N_i(\tau_k)} \tag{3.28}$$

A definition of the basic function N_i can be found in (Piegel and Tiller 1997). Figure 3.97 illustrates an NURBS curve with four control points. The comparably higher weighting coefficient at P_2 means that the curve is pulled in this direction.

In order to link the design vector $\boldsymbol{\alpha}$ to the control point mesh, the following relationship is introduced:

$$\begin{pmatrix} \mathbf{P}_i \\ \beta_i \end{pmatrix} = \mathbf{a}_i + \mathbf{A}_i \boldsymbol{\alpha} \quad \text{with } i = 1..m \tag{3.29}$$

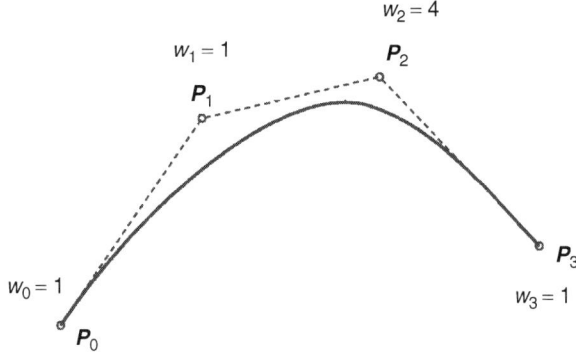

Fig. 3.97 The effect of the increased weighting coefficient w_2 on the shape of the NURBS curve

whereby $\mathbf{a}_i$ is the original position of the control point i. This considerably reduces the optimisation space and is thereby the only way of rendering instances in which deformations with symmetrical or other restrictions are required. In such a case, the shifting of several control points each underlies the same matrix $\mathbf{A}$, whereby n is the number of scalar optimisation parameters α_i and is less than m.

$$\text{EMUM: } \gamma \rightarrow x \tag{3.30}$$

The so-called "Elastic Mesh Update Method" (EMUM) (Johnson and Tezduyar 1994) is used to adjust the computational mesh to the deformation of the edge. In this method, the computational mesh is regarded as an elastic body which reacts to the externally applied deformation. A linear elasticity equation is used to resolve this:

$$\nabla \cdot \sigma_{mesh}(\mathbf{v}) = \mathbf{0} \tag{3.31}$$

whereby:

$$\sigma_{mesh}(\mathbf{v}) = \lambda(tr\varepsilon_{mesh}(\mathbf{v}))\mathbf{I} + 2\mu\varepsilon_{mesh}(\mathbf{v}) \tag{3.32}$$

$$\varepsilon_{mesh}(\mathbf{v}) = \frac{1}{2}(\nabla\mathbf{v} + (\nabla\mathbf{v})^T) \tag{3.33}$$

The structural mechanics require the so-called Lamé parameters, λ and μ, in order to construct the rigidity tensor of isotropic and linear elastic materials. As part of the elastic mesh update, these parameters are used to set the unique rigidity of individual elements, thereby preventing the premature failure of the computational mesh. The mesh update Eq. (3.31) is solved using the conventional Galerkin Finite Element Method.

Optimiser State variables must be exchanged within the optimisation framework between two independently working programs (optimiser and flow solver). The latter is executed in parallel on many processors. One of these processors is used for the optimiser (Nicolai and Behr 2007). The optimiser contains both gradient-free as well as gradient-based optimisation algorithms. The current version of the flow solver does not yet provide the gradient of the objective function. An approach for the analytical calculation of the gradient using the discrete adjoints is currently being developed and integrated into the framework. For a small number of optimisation parameters, it is acceptable to approximate the gradients using finite differences. However, in the application scenarios described later gradient-free methods proved to work well.

The optimal design parameters in the examples were found using BOBYQA (Powell 2009). This algorithm approximates the objective function J using a quadratic function Q, in which the functional value must correspond to function J at a given number of interpolation points. The remaining degrees of freedom are given in two successive iterations by minimising the Frobenius norm of the change in the second derivative:

$$Q_{k+1}(x_j) = J(x_j), \; j = 1, \ldots, m \quad \text{and} \quad \left\| \nabla^2 Q_{k+1} - \nabla^2 Q_k \right\|_F \to \min. \quad (3.34)$$

The quadratic approximation is improved by adjusting the number of interpolation points and adjusting the area of confidence according to the reduction in the objective function on each iteration. The BOBYQA method is very efficient: the number of necessary functional evaluations lies in the order of magnitude of the number of optimisation parameters. It is also possible to take into account simple geometric constraints in the form of a *box constraint*.

Objective Function The framework developed as part of this project should be used to optimise the flow though profile extrusions die. However, the optimum property cannot exist alone. It must include information about an optimisation criterion. The first stages in obtaining an objective function therefore involve analysing the properties of an accurate profile and investigating which factors affect these properties. According to (Szarvasy et al. 2000), an accurate profile is achieved when

- the profile outline corresponds to that anticipated and
- the thickness distribution is as required.

There is a whole range of influencing factors known to affect the quality of the profile but these are only qualitative and immeasurable. According to Michaeli (2009) these include:

- homogeneous velocity distribution at the outlet
- low pressure loss
- uniform acceleration of the plastic melt
- short dwell time
- uniform dwell time
- among others.

However, in practice, the dominant criterion for achieving a high quality profile extrusion has been identified as a uniform velocity distribution. This criterion has been translated into a mathematical formulation. The formulation devised and applied here is based on the segmentation of the die outlet into subsections (Fig. 3.98). These areas do not necessarily have to have the same surface area. Areas of particular importance may be given a greater weighting by having a larger number of subsections. The

Fig. 3.98 Subdivision of an L-profile into sections in order to calculate the objective function

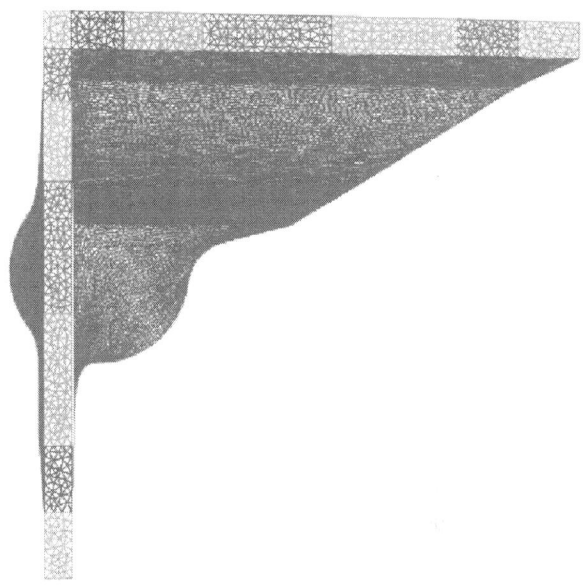

variance of the local mean velocity of these sections is calculated in relation to the global mean velocity over the entire outlet.

$$J = \sum_{section} (\bar{v}_{section} - \bar{v}_{total})^2 \qquad (3.35)$$

This objective function is usually dependent on the flux. This problem may be addressed by scaling if various extrusion velocities need to be considered. Scaling is not required if only one operating point is considered.

Application Examples This section presents two increasingly complex examples of the application of the developed design methods. The first example uses a slit profile extruded from a die with a slit-shaped outlet geometry and which is solely used to check the implementation. A die to manufacture a floor skirting profile is then designed in order to review the method on an industry-oriented profile.

Slit Profile The aim of this test is to check the functionality of the optimisation framework. First, a previously optimised flow channel geometry is deformed in order to start the optimisation algorithm on the resulting geometry. This checks whether the framework is actually capable of working its way back to the original optimum.

The selected slit profile has a width of 100 mm and a height of 1.5 mm. The Eqs. (3.21) and (3.22) produce an interface with a width of 110 mm and a height of 4.5 mm. The overall length of the extrusion die is 100 mm. The interface is positioned at 1/3 of the length. The connection between the inlet, interface and outlet are designed using a linear morphing algorithm.

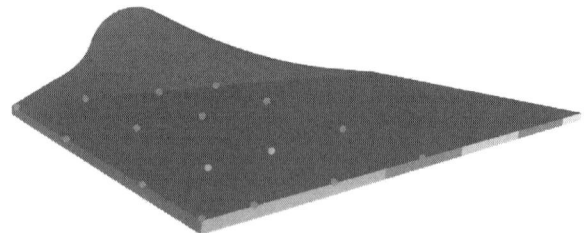

Fig. 3.99 Parameterisation with two moveable control points (*green*) on the slit profile

Looking at just the inner part of the slit channel (i.e. not the effects of the end pieces), an optimum velocity distribution already exists along the mean line. As parameterisation of the end pieces will be a topic of future research, this is not considered in this simple example. This means that the two outer sections are not included in the analysis of the objective function. Figure 3.99 illustrates how the inner area is subdivided into four equidistant sections in order to measure the homogeneity of the velocity distribution.

One half of the upper area of the extrusion die is parameterised. It is represented by an NURBS surface with 16 evenly distributed control points. The selected movement allows the green control points to be shifted in the direction normal to the parameterised slit area. Both control points are controlled by the same optimisation parameter α.

In order to begin optimisation, the initial shape of the finite element mesh is deformed by changing α from 0 to 1. This shifts each of the free control points outwards by 1 mm. Using BOBYQA on a finite element mesh with 2,194,731 elements gives an optimum α of -4.18×10^{-2}. In this instance, the objective function was reduced by 99.05% thus successfully automatically calculating the required channel geometry.

Floor Skirting A more complex extrusion die is illustrated in Fig. 3.100. It is used to produce floor skirting. As the height of the profile remains constant along the entire outlet, the critical points with regard to velocity distribution are to be found on the T-intersection. Here the material is at a greater distance from the wall, i.e. the wall adhesion effect is reduced and velocity increases. At the same time, the velocity within the bridge drops considerably as a result of the reverse effect of the stronger wall adhesion in this area.

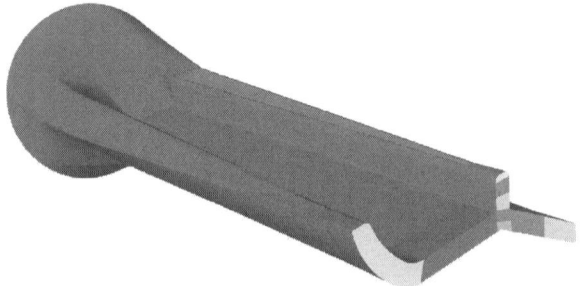

Fig. 3.100 Segmentation of the die outlet in order to determine the homogeneity of the melt distribution

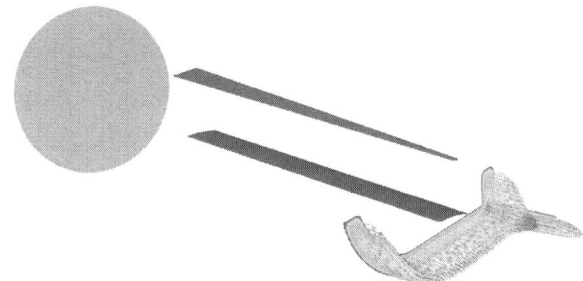

Fig. 3.101 The deformable part of the flow channel (*red* areas) are illustrated using NURBS areas and controlled using two geometrical parameters

The subdivision of the outlet into sections is performed in such a way as to specifically emphasise the critical spot in the middle of the profile. In the areas where only a small change in velocity is expected over long distances, there is a coarser subdivision than in the area near the bridge. In total, there are 9 areas with a surface area of between 4.72 and 46.37 mm^2.

In order to improve velocity distribution, a parameterisation is selected which renders the upper area of the bridge and the surface opposite the bridge deformable. The aim is to reduce flow resistance within the bridge while increasing the wall area opposite the bridge in order to increase flow resistance there. This adjustment of the flow resistances should press the material up into the bridge.

This deformation is based on just two geometrical parameters: one for controlling the lower area, and one for controlling the upper area. All control points are moved normal to the direction of flow (Fig. 3.101).

The optimisation framework is used to calculate the optimum parameter values. The associated deformation of the flow channel geometry was used to reduce the objective function by 35.4%. The velocity distribution and the final shape of the parameterised areas are illustrated in Fig. 3.102.

3.4.5 Illustration of Industrial Relevance

The degree to which costs, time and resources can be saved by a modularisation of mass production dies depends heavily on the constraints. A die-manufacturer with a

Fig. 3.102 Velocity profile and shape of the NURBS area following optimisation

very specific product range can take a modularisation of his dies further than a die-manufacturer with a broad product programme. However, the strategic constraints of these companies are also crucial. In an in-house tool- and die shop, the product in the sense of simultaneous engineering may be influenced so that the geometry variants of the products can be accommodated by the modularisation of dies. If the tool- and die shop is not in-house, it is usually not possible to place requirements on the products. A company must therefore be able to react as flexibly as possible to new orders. Modularisation may likewise contribute to a better estimation of costs and effort required in this respect. In addition, modularisation, at least on a virtual level, can save time and therefore also costs with regard to the development of new dies. It is also possible to optimise the production of die components by using a modular construction. Modules which are produced repeatedly can thus be used more quickly and to a high level of quality.

This methodology helps to take into account all these constraints during modularisation. It helps the design engineer gain an overview of the product programme and understand the interaction and dependencies between the components. While developing the actual modules, the dependencies must be taken into account in order to design standardised interfaces for suitable components.

The basic structure of the methodology presented here can be transferred to many mass production dies. A structured die modularisation process has been developed which may be substantiated by each design department according to their own needs.

In the rheological design of profile extrusion dies, the numerical optimisation of the flow channel geometry provides an enormous savings potential in a variety of ways: profile extrusion is a typical mass production processes. The fixed costs of the die and the design are relatively small if they are allocated across large batch sizes. With small batch sizes, the ratio is worse and the process less economical. In order to benefit from economies of scale while being able to produce individual (and therefore usually small) batch sizes economically, die design must be more systematic and efficient. This is possible with automatic optimisation as a previously optimised design is produced therefore reducing the number of running-in trials. Fewer running-in trials means lower material and energy consumption and therefore a lower utilisation of extruder lines and shorter reworking times. The company therefore avoids using whole extruder lines merely for running-in trials. These lines may then be used for actual production. On the other hand, the additional incidental costs incurred for the software and hardware used for the optimisation calculations are reasonable, particularly for many (individual) dies. Overall, numerical optimisation increases the window of profitability for profile extrusion with regard to small batch sizes.

The manual reworking of the die may not be completely eliminated even after simulation and optimisation. The assumptions made in the simulation with regard to constraints and material models are not capable of entirely mapping the real process, which is subject to many variations. There will therefore always be some variation between the simulation results and actual production. However, simulations used in combination with geometry optimisation models can clearly reduce the number of reworking stages required. The transfer of practical knowledge to the systematic generation of the initial geometry and the realistic transfer of constraints help in this regard. This is yet another advantage for industry. Current practical knowledge exists with the die-manufacturer, bound to individuals. If those employees leave the company, the company loses its knowledge. As basic design knowledge is incorporated into the software, it remains within the company and strengthens its competitive position.

3.4.6 Future Research Topics

The current project has followed two approaches to the customisation of mass production processes. One was modularisation, the other numerical design, whereby both approaches were considered largely independent of one another. In order to illustrate more closely one of the central results of the first phase of research projects, the need for modules with individual features, future projects should involve the linking of the two models and their implementation using generative processes. Modules with individual features are characterised by the fact that only some of their features (e.g. their outside dimensions) are specified using the modularisation concept, while other features, such as the position of tempering channels for example, may be post-processed on an individual basis. At this point in the individual post-processing it is possible to use numerical design in order to make the design process much more efficient. The automatic design approach is complemented by the subsequent production of the designed die components in a generative process, such as Selective Laser Melting (SLM). Generative processes have the necessary flexibility and efficiency in this regard to be able to turn the results of the numerical design into reality. These results often better satisfy other criteria than conventional designs. The shape of the optimised geometries consists of free-formed surfaces which may only be moulded at great expense in one die using conventional manufacturing processes.

The selected example processes, pressure die casting and profile extrusion, should also be considered in future research questions since their dissimilarity (cyclic compared to continuous, metal compared to plastic) guarantees a wide applicability of the methods tested on both processes.

The use of inserts in particular as modules with individual features has been identified in the field of pressure die casting dies. The most development-intensive features of inserts include the sprue and tempering device. Both features are heavily influenced by the product and die-cast component and current technology requires them to be designed manually. Future research projects should enable numerical design. In comparison to the development of a numerical design methodology for extrusion dies carried out in the current project, the situation in pressure die casting illustrates that very sophisticated simulation programs already exist for flow conditions during the filling phase (relevant for the optimisation of the sprue) as well as the temperature balance within the die (relevant for the optimisation of the tempering channels). This means that at this point no additional work is required with regard to the solver integrated in the optimisation framework. In fact, a high level of sensitivity of the optimum solution should be expected in the unsteady pressure die casting process, particularly in the field of sprue optimisation, i.e. even small changes in the shape of the ingate can lead to large changes in the quality of the component. This requires the use of very robust algorithms. A significant part of the research project will deal with the determination and mathematical formulation of suitable quality criteria (objective functions). As numerical geometry optimisation is still relatively new in the field of pressure die casting, there are few previous projects to which we can refer.

Broadly speaking, considering the formulation of new quality criteria, is the continuation of the work in the field of profile extrusion. As part of the first phase of the Cluster of Excellence (CoE), the extrusion die has been regarded as an isolated unit and the melt modelled as a shear thinning fluid. This restriction of the phenomenological and material-specific observations has allowed the focus to be placed on the development of the geometry optimisation framework. The framework allows the direct calculation of an optimum flow channel geometry which provides a homogeneous velocity distribution at the die outlet and therefore a uniform melt outflow.

Building on this work, another effect essential to product quality which should be taken into account in future and is based on the viscoelastic properties of plastics is the swell of the melt once it leaves the die. This requires the extension of the existing geometry optimisation framework so that the swell of the melt is considered as part of the calculation of the optimum die geometry. This means that the flow channel and specifically the die outlet are designed so that the required profile geometry is formed following local swell after the die (reverse problem solving or even: inverse design). This therefore first requires the integration of a viscoelastic material model into the existing software. The Giesekus model is a suitable example and is often used in plastics engineering (Giesekus 1994). In addition to forecasting sufficiently accurate swell sizes, it also displays numerical stability and is an acceptable additional calculation expense. Following the integration of the viscoelastic model, and taking into account the swell effects, the geometry optimisation framework is able to calculate the optimum die geometry, the manual design of which is—if at all possible—currently extremely costly. This represents an important step in the integrated examination of the extrusion line as part of the automatic design of profile extrusion dies. The success of the inverse examination can be evaluated according to comparisons with the registered swell behaviour in real profiles.

The results of the numerical design should be transferred to real components using generative manufacturing processes in both processes. Numerical design techniques often produce free-form surfaces which sometimes do not describe intuitive geometries. The production of these geometries in very small quantities requires a very flexible manufacturing process which may also be applied with little preparation to individual geometries. Generative processes are an example of processes which satisfy these requirements. In order to use these to produce modules with individual features, it is necessary to define suitable constraints for the generative processes. These are not geometrical constraints, which are a result of the geometry optimisation process in any case, but requirements placed on strength or surface quality. The size of the die or insert is designed according to the calculated required local strengths. This enables the use of minimum volumes and therefore a design suitable for production with SLM. Innovative methods must be found, particularly in the field of surface quality, for adjusting the typically relatively raw surface of a component manufactured using SLM, for example, to the required dimension.

The combination of efficient design using numerical models with efficient production using generative processes under the umbrella of modularisation is a promising step on the way to individual production. The concept of automatic individual design is extended by means of a highly individual yet efficient manufacturing process which places ever less importance on the individual part of the module with individual features both in terms of time and cost.

3.5 Efficient Order Processing for Individualised Machine Tools Through Modelling of Product Families

Christian Brecher, Werner Herfs and Andreas Karlberger

3.5.1 Abstract

Manufacturing enterprises have to define their strategic position in the conflicting area of economic planning and production and are searching for the optimum operating point. Also machine tool manufacturers have to find ways – especially regarding the increasing complexity of machines and equipment – to extend their ability of meeting individual customer needs while at the same time exploiting economies of scale in development, production and operation.

A basic classification within the field of Individualised Production is provided by the two manufacturing strategies of build-to-stock (BTS) and build-to-order (BTO). By BTS is meant customer-independent manufacture, while under BTO the production only starts after an order is received. In industry there are generally hybrid versions of both approaches, where an attempt is made to move the order penetration point—i.e. the transition point from order-independent to order-specific manufacture—as far down the assembly process as possible, in order to benefit from scale effects (see also Schuh 2005). Within the BTO strategy there is a further distinction made between the categories *configure-to-order* (CTO) and *engineer-to-order* (ETO) (see also Fogarty et al. 1991). Under the CTO strategy, products with different variants are planned in total in advance. With the support of a configuration system, the customer orders can be passed directly into the ERP or MES system, using mainly automated processes or work flows. In a purely ETO approach, the product is planned from scratch each time. The typical steps including product planning, process planning and production planning are mostly carried out in sequence nowadays.

Machine manufacturers can be roughly categorised by the terms *Special machines* meaning an ETO focus (scope) and *Standard machines* as meaning a CTO focus (scale). As customer requirements cannot be exhaustively predicted, machine

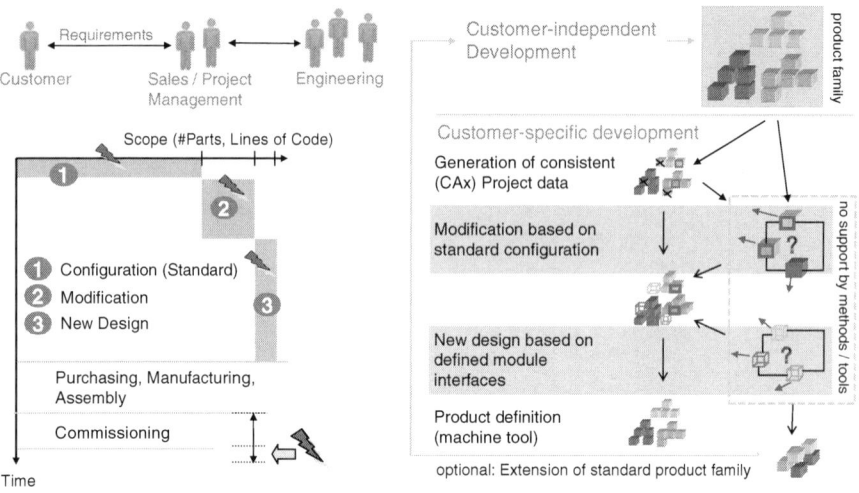

Fig. 3.103 Activities, time requirements and sources of errors. (Brecher et al. 2010a)

producers in high-wage countries practice a combination of the two approaches. Machine tools are therefore defined from the development stage up as product families, and then extended with new options and variants during their life cycle. During order processing, a large part of the requirements can be met from the pre-planned standard versions, while a relatively small proportion requires adjustments or a development from scratch to meet customer-specific needs. During order processing, a number of disruptions occurs today, which can considerably impair the competitiveness of a company.

The main causes of disruption are explained using Fig. 3.103. During order-independent development, the design and development of a product family is done first. In industrial practice, in most cases, the necessary systematic approach to create modular structures, supporting both configuration in the area of standards as well as downstream customised development, is missing. Due to the lack of a methodological underpinning and automation, the risk of initialising the corresponding CAx systems with inconsistent project data is increased. For downstream individualisation, tools are lacking at present for interdisciplinary design, for impact analysis of changes, as well as for early verification and validation. A further source of disruption when combining CTO and ETO currently exists in the time-consuming manual comparison of planning data in ERP or MES systems. In summary, it can be seen that a combination of CTO and ETO creates a considerable increase in complexity in the company's processes. As this complexity is not managed sufficiently well at present, disruptions occur, which finally leads to a very large amount of time for a rather low percentage of individualised components in relation to the entire project scope. Besides the increased development time, it is in particular the large discrepancy between error occurrence and error detection which has a very negative effect on competitiveness, as iterative cycles during commissioning can lead to lengthy delivery time.

In the sub-project "Mechatronic-oriented Modular System" a concept has been developed considering the implementation of product families, the conception of a configuration system and the support for customer specific development using models. For the interdisciplinary design of product families, as well as for the design phase in developing order-specific adaptations, the Systems Modeling Language (SysML) is used. The system model here is to be synchronised using service interfaces with distributed CAx planning systems, in order to be able to uncover characteristics which cross discipline boundaries, and to have early validation of requirements. One of the substantial challenges here is the representation of variants for the modelling of product families. The concept presented by Bachmann for decentralised creation and synchronisation of software artefacts by using higher level feature models, was adopted as a solution approach for interdisciplinary configuration at the level of CAx systems (see Bachmann and Northrop 2009). Here the configurators, which are normally provided as standard in CAx systems, are used to develop product families within the individual expert disciplines. The data structures are passed across via the service interfaces and the variants defined are displayed in the form of uniform feature models. By linking the different feature models specific to each discipline to those of the external variety a synchronisation of the CAx configurations can be achieved. The feature models also provide a human readable presentation of the variants, while at the same time supplying the necessary information about the combination logic to the configuration tools. The models which are created in the configuration process can then be used to control the decentralised configurators to initialise the project data.

3.5.2 State of the Art

In the overall context of module-based development and production of mechatronic systems, different approaches to rationalisation can be identified. The different approaches are each flavoured by the specialist discipline from which they originate, which are now extending their design with approaches from related disciplines or adopting and adapting concepts from other branches of industry. The main impetus comes from the following areas:

- Production planning
- Mechanical construction
- The software industry (including embedded systems) as well as
- System modelling and simulation.

Examples of the adaptation of approaches from related branches of industry can be seen particularly in the areas of production planning and the software industry. As a reaction to the increasing complexity of software systems three significant, and partially contradictory, approaches to increasing the efficiency of software creation have emerged: Model-based development, Agile development, and the software product lines approach. A fundamental trend in the software industry is the increasing level of abstraction in programme development, which can be explained by the fact that the average size of software is undergoing exponential growth in analogy to Moore's law (see Curtis 1992). While the very first programmes were initially written in machine code, there have followed several layers of encapsulation and automatic code generation for the lower levels. The graphical modelling language OMG Unified Modeling Language UML was initially used for software design, specification and documentation purposes, but at the same time it was proposed to design and implement software systems entirely at the modelling level. Parallel to the increase in abstraction layers in the 1980s and 1990s, special procedural models were developed with the aim of making the software creation process manageable, and so to reduce the risk of failed development. Popular examples are the waterfall model, the spiral model as well as the V-model (Herfs 2010). The use of model-based development techniques and procedural models can be seen as a primarily planning-driven approach. With the increasing spread of lean production in production technology, a counter-trend also appeared in the software industry with the Manifesto for Agile Software Development, which aims to put value-adding and communication at the core of business activities (Beck et al. 2011). The concept of executable UML (xUML) described by Mellor has the purpose of creating models which can be executed directly and thereby to resolve the problems relating to a lack of flexibility in model-based software development. As an approach to reduce the dilemma between planning and value-adding activies it is considered that by directly executing the model there is immediate feedback to the developer, so that an agile approach can be achieved, while still working on abstract model level (Mellor and Balcer 2002). The model-based development approaches using xUML were followed up logically by some producers of commercial UML development tools.

With the increasing complexity of mechatronic systems today, the need has grown in the last few years for methodologies and tools to support the design of complex systems. The VDI (Association of German Engineers) Guideline 2206 "Design methodology for mechatronic systems" defines the term mechatronic as "the close synergistic cooperation between the disciplines mechanical engineering, electrical engineering and software engineering in the design and production of industrial products and the design of processes". The guideline defines a procedural model which is characterised by an iterative progress through interdisciplinary design, discipline-specific development, system integration and validation. This cycle is run through several times until the product meets all the requirements (VDI 2004).

A direct application of this approach to machine tool engineering is limited due to the rather general nature of the guideline. The paradigm of Systems Engineering goes beyond the concept of mechatronics. The International Council on Systems Engineering (INCOSE) defines the concept as follows: *"Systems Engineering is an interdisciplinary approach [...]. It focuses on defining customer needs and required functionality early in the development cycle [...] considering the complete problem: Operations, Performance, Test, Manufacturing, Cost & Schedule, Training & Support, Disposal. Systems Engineering integrates all the disciplines and specialty groups into a team effort forming a structured development process [...]. Systems*

Engineering considers both the business and the technical needs of all customers with the goal of providing a quality product that meets the user needs" (INCOSE 2004).

The Systems Modeling Language, specified by the OMG in 2007, is based on UML2 and aims to support interdisciplinary development of any type of system, with the focus being on the early phases of the development process. Using SysML an attempt is made to support the paradigm of Systems Engineering using a mostly formal, graphical modelling language. Procedural models for Systems Engineering are compared in Estefan (see Estefan 2007). Model-based Systems Engineering is based on the recognition that, to have early validation of complex systems, formal models are required which can be interpreted by computers. For this reason, the shift from document-based to model-centric engineering is typical for Systems Engineering (Friedenthal et al. 2009).

In summary, it can be noted that in Systems Engineering the focus is on the new development of complex products. At the core is the capture of customer requirements and their systematic conversion into an interdisciplinary design which is validated on the basis of formal models. But none of the development processes researched by Estefan (see Estefan 2007) addresses the systematic separation of order-independent and order-specific development, which can be regarded as an important characteristic for the production of personalised machine tools. This gap is also currently reflected in the SysML language, which in its base version does not as yet support the creation of models with multiple variants. Thus, built-in mechanisms for the extension of UML, so called stereotypes and profiles, have to be used for this purpose.

In the software industry the paradigm of mass customisation is becoming increasingly common in the form of the software product-line approach. According to Clements and Northrop a software product line (SPL) is a collection of software-intensive systems, which display common characteristics to meet the specific needs of a market segment and which are created from a common range of available core elements in a defined way (Clements 2009). At the heart of this is the systematic reuse of various artifacts such as requirement definitions, system architectures and software components, where there is a consistent distinction made between order-independent (domain engineering) and order-specific development (application development). The so-called production plan describes how in the field of application development the product line artifacts are used to create concrete software products.

An important element for the planning of SPLs are feature models. These describe the problem and solution space which corresponds to the external and internal variety. The first approach using feature modelling was known as Feature Oriented Domain Analysis (FODA). Here the configuration space was described using the feature types "mandatory, optional, alternative" and "mutually exclusive" (Kang 1990). On the basis of the model developed in FODA, the feature modelling approach has continued to be further developed.

Tool-based support for the construction of an SPL is i.e. given by IBM Rational Rhapsody® using corresponding Stereotypes and mechanisms for variant selection. External frameworks are, for example, offered by the software system *Gears* from BigLever Software™ and *pure::variants* from pure-systems GmbH, see (Anon

2006). But these systems are intended to handle the creation and operation of SPLs and do not offer as standard any support for integration with MCAD and ECAD systems.

In the area of mechatronic simulation Matlab/Simulink and the open modelling language Modelica® have become established. Both approaches enable the hierarchical structuring of sub-systems, which can be interchanged flexibly. The disadvantage is the lower level of flexibility compared to the extendable UML based modelling languages regarding target platforms and tool integration. In relation to behavioural modelling, Matlab and Modelica have now established themselves as quasi-standards. SysML modelling tools, which manage the creation of models, generally have interfaces to Matlab/Simulink® in order to be able to test the software against behavioural models. For the integration of SysML and Modelica solutions for specific integration scenarios have been developed (see Paredis et al. 2010).

A further Lifecycle which can be seen as a driver for innovation in the area of engineering is Product Lifecycle Management (PLM). PLM systems are usually derived from the field of mechanical construction and attempt from there to cover a series of further disciplines and development phases. In the area of Systems Engineering the Teamcenter/NX and DELMIA/Catia systems are increasingly including functionality for system validation. Siemens PLM uses its Mechatronics Concept Designer to address mainly the early design phase, while CATIA integrates the Modelica modelling and simulation environment Dymola. The concepts of interdisciplinary configuration at the CAx level required by small and medium-sized enterprises (SME) are currently insufficiently supported by PLM systems, as these systems are mainly intended for use with mass production without order-specific modifications.

Strengthening the links between the various disciplines, and closing the gaps which currently exist for engineering data transfer are being addressed by the Automation Markup Language (AutomationML) data format (see Drath 2010). As AutomationML is mainly focussed on the needs of the automotive industry, specific requirements in the machine tool and equipment construction industries are not sufficiently addressed in the areas of product families and support for early interdisciplinary design.

A further aspect of Systems Engineering which has been developed by the software industry is requirements management. In sectors such as the automotive or aerospace industry high quality standards have been defined for the development of third party control software. Depending on the maturity level, corresponding standards, such as the Capability Maturity Model Integration (CMMI) and ISO 15504 (SPICE) partially demand bi-directional traceability of changes down to the level of the source code. As requirements engineering is already being successfully applied for software development, it can be assumed that the standards will be extended to other sectors such as hardware and mechanical development in future (Müller 2007). Based on the widely varied nature of discipline specific development and primarily document-based requirements management, a direct transfer of the methods used in the software industry to the interdisciplinary development in machine-tool engineering would be rather restrictive. The combination of requirements management and a model-based development process, however, offers the potential to analyse changes on the basis of model information. For this purpose vendors of SysML modelling tools already offer interfaces to commercial requirements management tools.

The full integration of modelling tools in Systems Engineering in today's heterogeneous IT landscape was researched by Peak et al. (2007). The vendors of SysML modelling tools, and also third-party vendors, are today already supplying commercial extensions to connect CAx tools via proprietary interfaces. A generic approach to CAx data integration based on ontologies is e.g. proposed by Brecher et al. (2010b).

In relation to the interests of machine tool and plant engineering, an approach has already been developed under the BMBF (German Federal Ministry of Education and Research) project Föderal for the construction of mechatronic-oriented modular systems, which is now being distributed by the companies Mind8/EPLAN under the name Eplan Engineering Center (2011). This approach mainly addresses automatic creation of documents for multi variant series production. Interdisciplinary design and its early validation are therefore not its prime focus.

In relation to the specific requirements of small and medium-sized enterprises (SMEs), it can be seen that no methodological support for the construction and use of interdisciplinary and flexibly extensible product families exists today.

The area of interdisciplinary engineering shows a high level of variability both in relation to the experts involved, and in relation to the planning systems they use. Therefore different tools and methodologies have grown up within each domain, attempting to provide interface or to cross over to neighbouring domains in order to win market share.

The modelling of product variants is becoming increasingly important, but at present the necessary methodologies and supporting tools are lacking in many sectors. Hierarchically structured modules with many different variants are significantly more difficult to model structurally compared to single products. Vendors of development tools, programming or parameterisation systems mostly offer little or no support for the creation of product families. Therefore this task must be done manually or on the basis of exported data outside of the tools which is time-consuming and less flexible.

The ERP and MES level systems nowadays mainly define work flows, which do support the definition of series products with multiple variants. The dynamic reconfiguration of existing product structures involves a high level of manual effort, and therefore is a potential source of errors. PLM systems on the other hand are mainly designed for the new construction of series products, and so offer little support for a combination of standard components and subsequent customer-specific adjustments including the required changes to the corresponding information within the ERP system.

The use of an interdisciplinary modelling language, such as for example SysML, for modelling product families is promising in principle, as the fraction of software for machine tools is steadily increasing (Wünsch 2008), and the development tools already established in the software industry show a high level of maturity. The possibility of extending SysML by using profiles and adjusting it to suit individual domains requires integration options to existing IT infrastructures and work flows. Currently there are gaps, mainly in the still inadequate support for modelling product families.

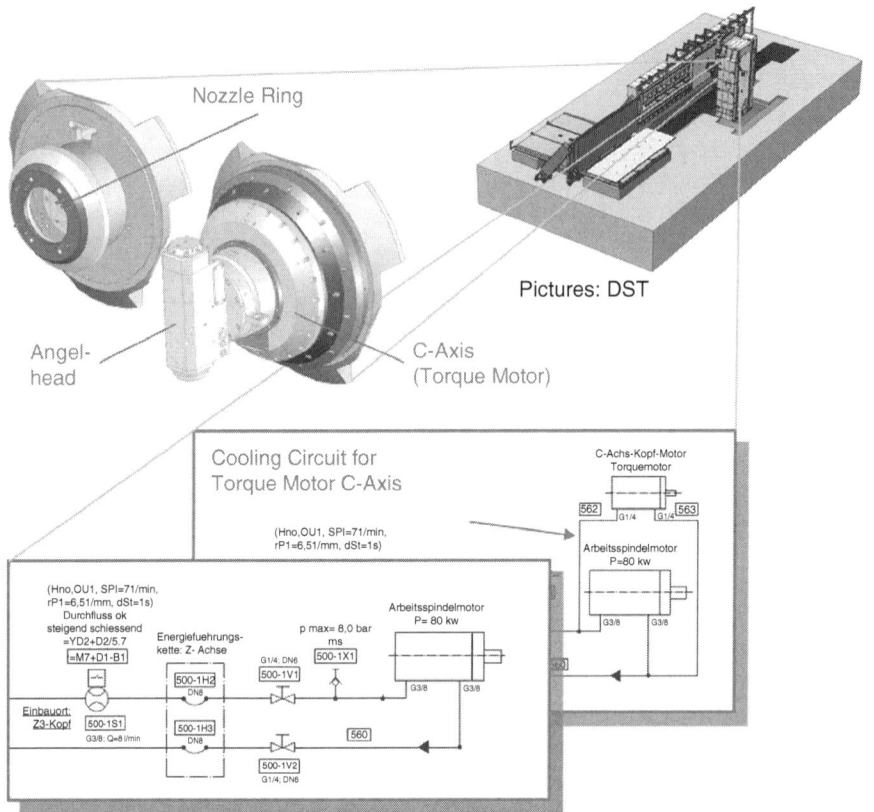

Fig. 3.104 Example of variance within different disciplines

3.5.3 Motivation and Research Question

The rapidly growing function range of machine tools is already leading to a high degree of complexity which provides major challenges in both planning of product families and in customer-specific modifications. Given the progress in information and communication technologies, we can see an increasing ability to network and communicate between planning systems, which can provide a significant contribution to the necessary integrative view on specific sub-goals when designing product families, and in applying customer-specific modifications to complex machinery. To make the increasing welter of information useful for the design of product families and the customer-specific development process, a formal description of the products is necessary—both in terms of their internal interactions and their interactions with the production system of the machine-tool manufacturer. The information to be taken into account here can be traced back to the level of the physical causal relationships and control sequences within the machine. Figure 3.104 shows an excerpt from an application example "Ecospeed F", a milling machine from project partner Dörries Scharmann Technologie GmbH. The machine has a processing head which can be positioned and oriented using parallel kinematics. As an additional option, the machine can be delivered with an angle milling head, which can be automatically mounted and rotated around the C-axis using a torque motor to increase the available swivel range for processing. While the decision on the required configuration from the customer viewpoint is made mainly on the basis of workpiece features, at the technical realisation level, however, one can identify differences spanning all disciplines and a multitude of components. For example, the additional cooling of the motor requires an increase in the overall volume of the water flow, which correlates to a drop in pressure and to the pump performance required. In order to support the design of both product families and customer-specific adjustments, descriptive models are needed, which enable an early analysis of this kind of interaction.

The models must be readable for humans as well as for the computer. Languages which are already in use in Systems Engineering, such as e.g. the Systems Modeling Language (SysML) meet this requirement to a large extent. The application of these approaches fails at present because, among other things, there is inadequate support in relation to the *modelling of product families as well as the integration of model-based approaches under heterogeneous IT architectures* in SMEs. Both aspects are necessary pre-requisites for successful implementation in the machine tool sector. The research question therefore is:

> How can model-based interdisciplinary engineering of product families be effectively implemented in relation to the configuration of customer-specific adjustments and to IT-integration in the machine-tool sector?

3.5.4 Results

The results arising from the research question in Sect. 3.5.3 are presented below. A significant partial result is provided by the approach to model-based development of product families using SysML. In order to make sure that requirements which are relevant in practice were taken into account in the overall design, interviews were held with experts involved in product design at the project partner company Dörries Scharmann Technologie GmbH. Based on these, selected sub-systems of a milling machine with multiple variants were defined jointly as a demonstration scenario and used to validate the concepts.

As a further application example the development processes from the sub-project "Hybrid Machining Center" was considered, where a 5-axis milling machine from Chiron-Werke GmbH was extended by an articulated robot and two laser systems.

The overall design is first described below using two overview illustrations. Based on this, single aspects of the implemented concept are described in more detail.

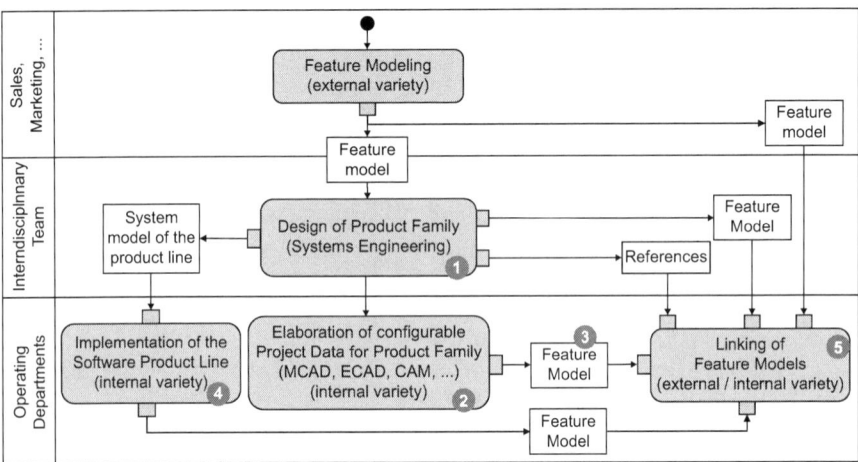

Fig. 3.105 Overall architecture for the development of product families

3.5.4.1 Overall Design for Interdisciplinary Engineering of Product Families

Figure 3.105 shows on the left three areas of responsibility, to which individual activities are assigned horizontally. The definition of essential variants (feature modelling of the external variety) is an elementary step of strategic product planning, and can be either revolutionary or evolutionary in nature when dealing with new developments. The outputs of this step are feature models which are passed on in a standardised format to the design phase of the product family (1), as well as to a link module for feature models (5). The design of the product family is carried out using a system modelling tool. The results of this step are used as a specification (2) or as a template for model-based software development (4) during the elaboration in each of the expert departments. In principle, this can be carried out using the same modelling languages or tools as for the system modelling, but makes use of additional domain-specific extensions and functions for code-generation for specific target platforms (NC, SPS, RC, etc.). Modelling the variability in the control and operating software, and in the system model, is based on the principles of the software product-line approach. In the product family elaboration phase there are discipline-specific tools employed for mechanical construction (MCAD), electrical construction (ECAD), process planning (CAM) etc. These tools increasingly offer mechanisms to define variants, and so enable the development of product families as a configurable project. Using standard interfaces, the data structures are read from the tools and converted into uniform feature models using model transformation. These are passed to the connection module (3) which provides the mapping between internal and external variety (5).

Figure 3.106 shows the steps during configuration and the subsequent individual adjustments for customisation. Configuration and specification normally are assigned

3 Individualised Production

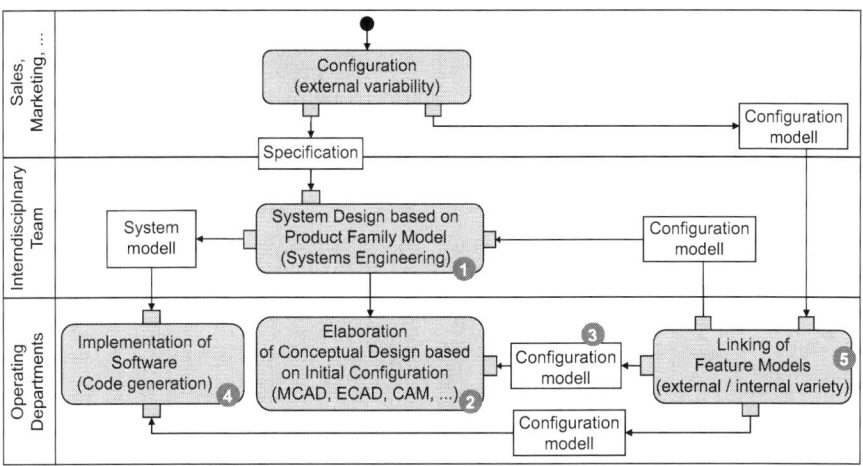

Fig. 3.106 Overall architecture for synchronised configuration and individual adjustments

to the sales department. Standard configurations are frequently extended to include individual requirements. The effects of these changes must be uncovered precisely during the quotation stage, in order to be able to carry out a reliable estimation of the required development effort. By transforming the configuration model from the external to the internal variety, discipline-specific configuration models can be automatically created. These are passed via standardised interfaces to the relevant CAx planning systems (3) and are used there as input for the internal configurators. Therefore pre-configured, consistent project data is available as the basis for customer-specific adjustments, supported by a model-based development process (1/2/4).

Figure 3.107 shows the system architecture of a distributed development environment for the implementation of the necessary communication mechanisms. As middleware, the concept of the enterprise service bus (ESB) is used, which enables flexible connection of distributed applications via service interfaces. For the coordination of the information flows between the endpoints (CAx-systems, systems engineering tool, configuration tool etc.) so-called service engines can be connected to the bus. A significant feature of the ESB concept is increased flexibility by avoiding point-to-point connections between target endpoints. The ESB used here is an implementation of the standard *Java Business Integration* (JBI) and supports the use of different communication mechanisms which are addressed via so-called binding components. Binding components have the job of providing external interfaces for specific communication technologies (e.g. webservices) and of translating these into a standardised XML format for internal communication. Therefore the services of the individual applications communicate solely via a technology-independent abstraction layer which simplifies the networking of applications (Ten-Hove and Walker 2005).

First tests have already been carried out for the architecture shown. The CAx-tools used have been equipped with service interfaces for reading and writing data

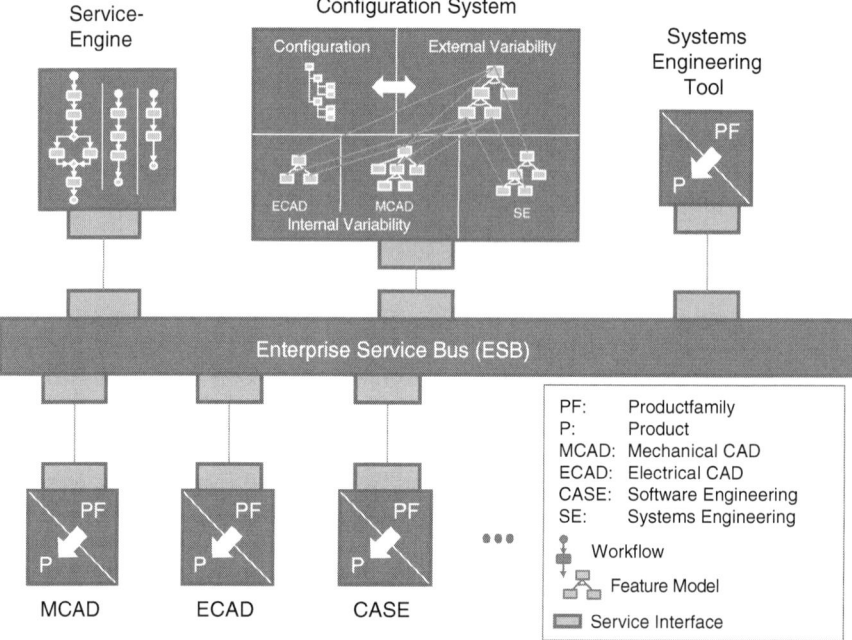

Fig. 3.107 Service-oriented architecture for interdisciplinary development of product lines

structures and component features, which were put onto the ESB using the binding components. To coordinate the services, the Business Process Execution Language (BPEL)—service engine provided with the bus is used.

The transition from product families to a concrete product shown in Fig. 3.107 in the individual disciplines, assumes that the corresponding project structures have already been created with the help of decentralised configurators in a methodical way. In what follows we demonstrate this process using a mechanical construction example.

3.5.4.2 Variant Modelling Using Integrated Configurators (MCAD)

Figure 3.108 shows the transfer station of the WERO tool changing system from Dörries Scharmann Technologie GmbH. In the actual example being studied, the two variants of the external variety *Angle milling head required* and *not required* are to be reviewed (see Fig. 3.104). The angle milling head can be automatically placed on the spindle, but is too large to be handled as a tool in the magazine. Therefore a second transfer station is needed, which is extended for the changing of this head. To increase the number of re-used components, the motion unit is defined as a module, which itself consists of several hierarchically organised sub-modules. Their configuration is specified each time by a parameter on a superior level. The

3 Individualised Production 221

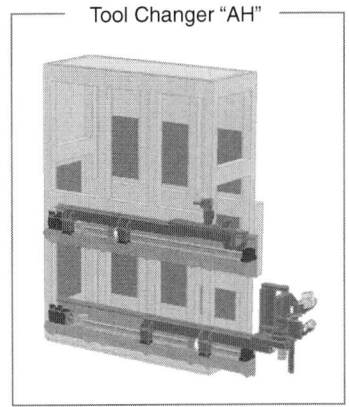
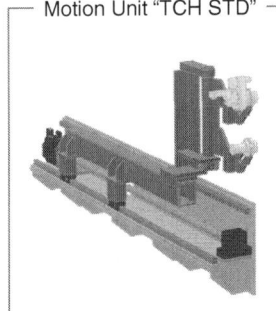
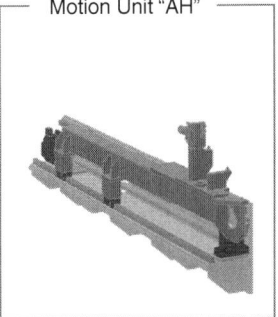
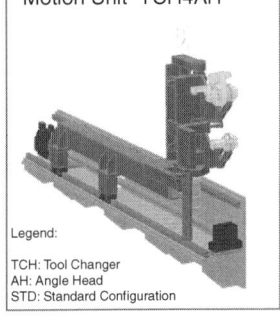

Pictures: DST

Fig. 3.108 Configurable product family of a tool changing system in MCAD

configuration parameters can then be connected to each other, so that two resulting global MCAD configurations at the top level allow switching between the settings. By taking this approach, the lower level gripper module is adjusted accordingly when the top-level transport unit is switched over. This relatively easy-to-understand example shows that the complexity of data structures needed to define the product families grows drastically in relation to single product definitions.

3.5.4.3 Feature Modelling

Figure 3.109 shows the meta-model which can be used to automatically transform a product family definition like that of the tool changing system into a standardised feature model. The root node (*RootAsset*) corresponds to the highest aggregate level, i.e. in this case the tool changing system. A RootAsset has one or more abstract nodes which can be substantiated using elements such as AND (mandatory components), XOR (alternatives) or OPT (optional). Each feature node has further nodes or abstract

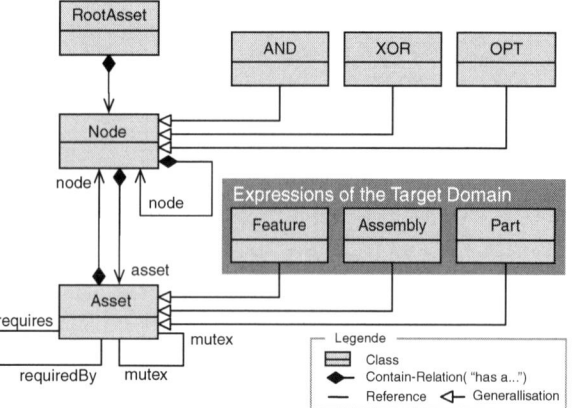

Fig. 3.109 Metamodel for feature modelling of the MCAD data

Assets, which can be substantiated using an element of types *Feature*, *Assembly* and *Part*. Assets can reference each other via the types *requires* or *mutex* (mutually exclusive), which reduces the number of possible combinations.

The terms *Feature*, *Assembly* and *Part* are domain-specific expressions of the MCAD domain. The flexible structure of the metamodel enables adaptation to the terms of other domains. Figure 3.110 shows a feature model that is derived from the MCAD data structures of the product family, and created as an instance of the metamodel shown above.

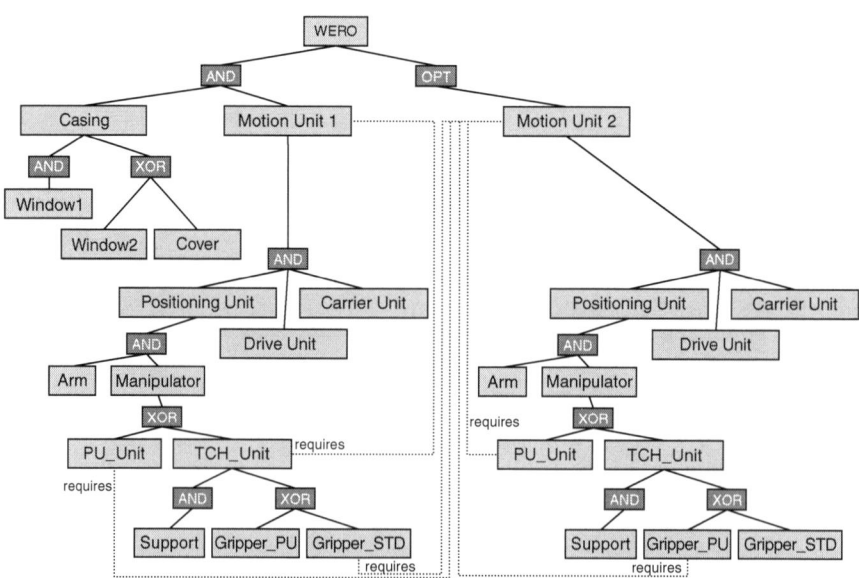

Fig. 3.110 Feature model of the tool changing system (simplified)

3 Individualised Production

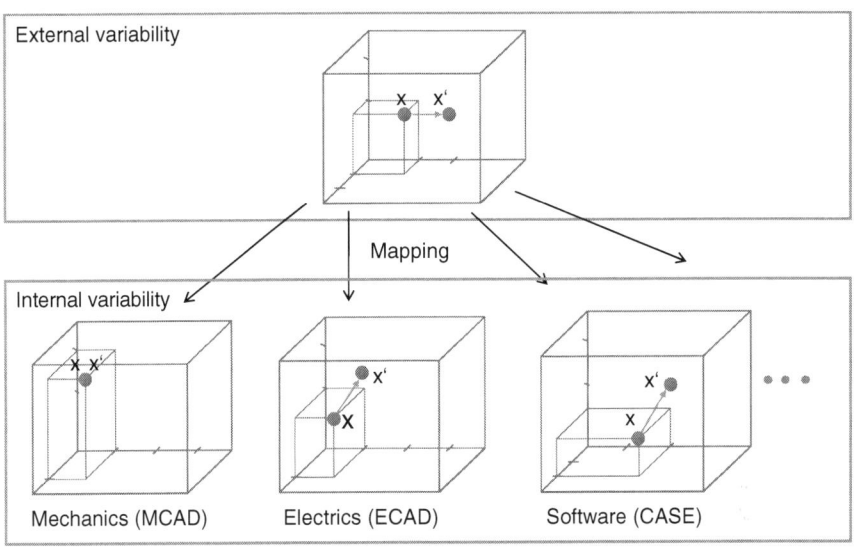

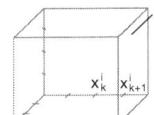

Legendt:
Solution space
x^i_k: Feature i with value k
x : Standard configuration (CTO)
x' : Customised configuration (CTO+ETO)

Fig. 3.111 Connecting external and internal variety

This model describes the solution space of the internal variety, which can be linked to the external variety by a mapping of the feature models.

3.5.4.4 Connecting External and Internal Variety

To clarify the linkage between external and internal variants Fig. 3.111 shows schematically the configuration levels which are described in the literature as problem space and solution space. The dimensions of the configuration spaces are set by the various features whose possible options are represented by different positions on each of the axes. There can also be continuous ranges if, for example, the features in the CAD model include flexible value ranges. The size of the configuration space can vary depending on each discipline. So, for example, it is possible for the differentiation of delivery features to be realised solely in the software and electrical domain, while the mechanics remains unchanged.

Point X represents a specific configuration from among the pre-planned variants; the individualised part of the specification is represented by X'. The challenge during the quotation phase consists of estimating the required changes as precisely as possible, and with the least possible cost. After the subsequent receipt of an order, the efficient implementation of the individualised elements within each specialist department is required. Hereby, the starting point for the individualisation is always the most fitting configuration, i.e. in the case of X' the planning systems are initialised with X to start with.

Both the restrictions within the configuration space, and the mapping rule of the problem space to the solution space can be formulated with the help of specific languages as rule bases. As the system model for the product family brings together requirements, structure, behaviour, as well as references to CAx data, this information can be used to reduce the planning effort for creating the mapping rules to a minimum. The information contained in the feature models may be used further to initialise a configuration tool editor for user interaction (e.g. tree editor with checkboxes). There are two types of application we can distinguish here: On the one hand, during configuration itself a dynamic calculation of the permitted next configuration steps can be made after each user action. In the case of selecting an alternative component, for example, the choice of some other features may then be blocked. The second possible approach is off-line validation, where the consistency check for the configuration can be requested by the user at any time.

3.5.4.5 Modelling Multi-variant Products on System Level

The use of interdisciplinary descriptive models to plan product families is based on the assumption that the existing approaches in the area of system engineering are to be extended to include the modelling of variants. Although there are some tools supporting a software product line approach, in their current forms these are not yet suitable for representing variety in deeply nested structures. The issues with variant modelling using SysML are explained using Fig. 3.112. Based on the concepts originated in Jacobson two main stereotypes are applied to the structural elements: The so-called variation point identifies an abstract component which can be substituted by

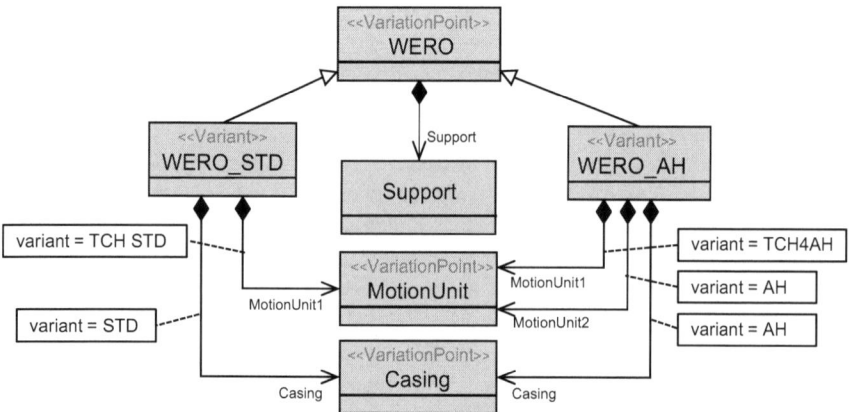

Fig. 3.112 Definition of a configurable modular tool change system (excerpt)

components of the *Variant* stereotype. In our example these are the blocks $WERO_{WFK}$ (with angle milling head) and $WERO_{STD}$ (standard version), which are assigned to the configurations *Angle milling head required* or *not required* (external variant). The lower transport unit is included in both configurations, but is built in two different versions (*WZW STD* or *WZW4WFK*). Common core components which occur unchanged in all configurations, i.e. which do not have any subordinate configurations, can be attached directly to the variation point and are inherited unchanged by all the different variants (example: *Support*)

Like the data structures in the MCAD, parts and components are only defined once and are reused many times by references. At run-time only a limited range of elements, such as e.g. the activated sub-configurations, or the component colour, can be changed in MCAD instances, while, for example, the basic geometry can only be changed at class level. When transferring these concepts to systems modelling using SysML it therefore follows that the information relating to specific, configured subcomponents needs to be defined within the model as an instance specification. This can—as shown in Fig. 3.112—in principle be supported by an extended description of the relationship between components and the variation point.

3.5.4.6 Creation of Component Libraries

The use of modules across product families is practised today by series manufacturers, and therefore also has to be taken into account when modelling product families. Package structures in UML/SysML are the obvious choice to basically organise a model (Friedenthal et al. 2009). In Fig. 3.113 a suitable model breakdown is shown for

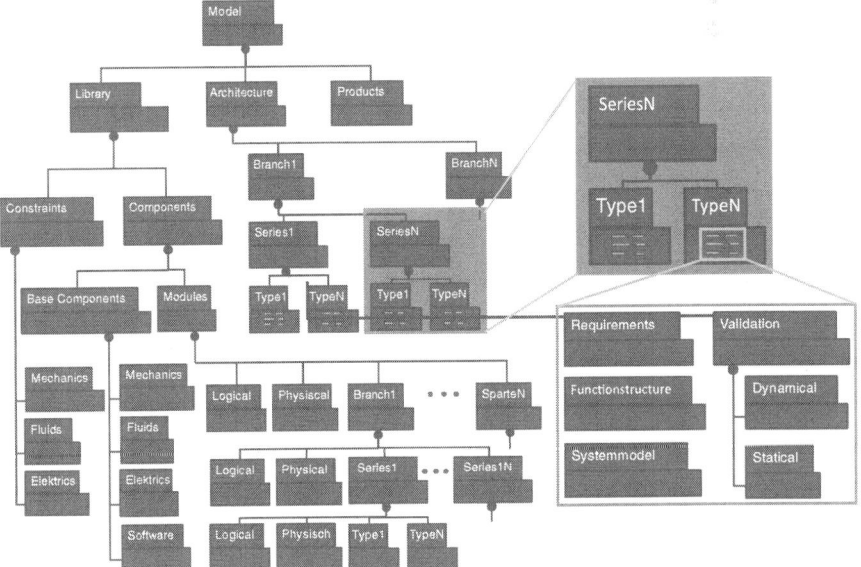

Fig. 3.113 Model structures for enterprise-wide modelling of product families

the creation of mechatronic product families. The planning horizon for this can stretch from an individual product to the entire product range for the company. The basic structure is provided by the packages *Library*, *Architecture* and *Product*. The library is further divided into *constraints* and *components*, which in turn are split into *basic components* and *modules*. Both for basic components and also for constraints there is discipline-specific subdivision, so that the corresponding areas can target the entry of model information themselves. In their breakdown, the *module* and *architecture* packages reflect the product range of the company, in that the packages from the whole enterprise are nested hierarchically, via divisions, product series and product families down to individual products. For each area of use, the system components are assigned to the corresponding level. Read and write permissions can be assigned to the package structures, so that the right to make changes can be linked to areas of organisational responsibility (product line manager, module manager, etc.).

3.5.4.7 Validation of System Properties

The early validation of system properties is one of the most important aspects of model-based development. With the help of SysML parametrics diagrams it is possible to model complex dependencies. The created constraint systems can be evaluated with tools that automatically calculate mathematical constraint expressions—known as solvers—and so enable the validation of requirements and also the calculation of the necessary performance characteristics on different system levels.

In Fig. 3.114 the lower motion unit of the tool changing system is shown, of which the slide is positioned by a ball screw drive (BSD). To select the spindle pitch, which has a significant impact on the acceleration behaviour and the maximum speed, a constraint evaluation was carried out based on a model of the BSD, using the

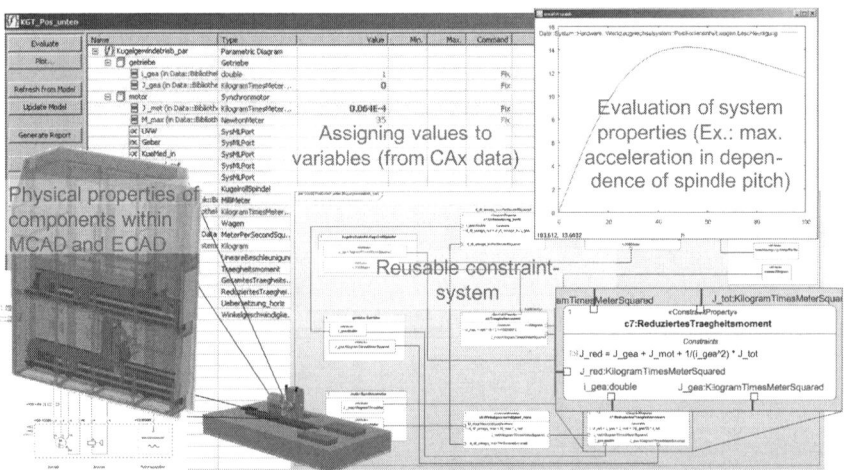

Fig. 3.114 Calculation of overarching system properties using parametric diagrams

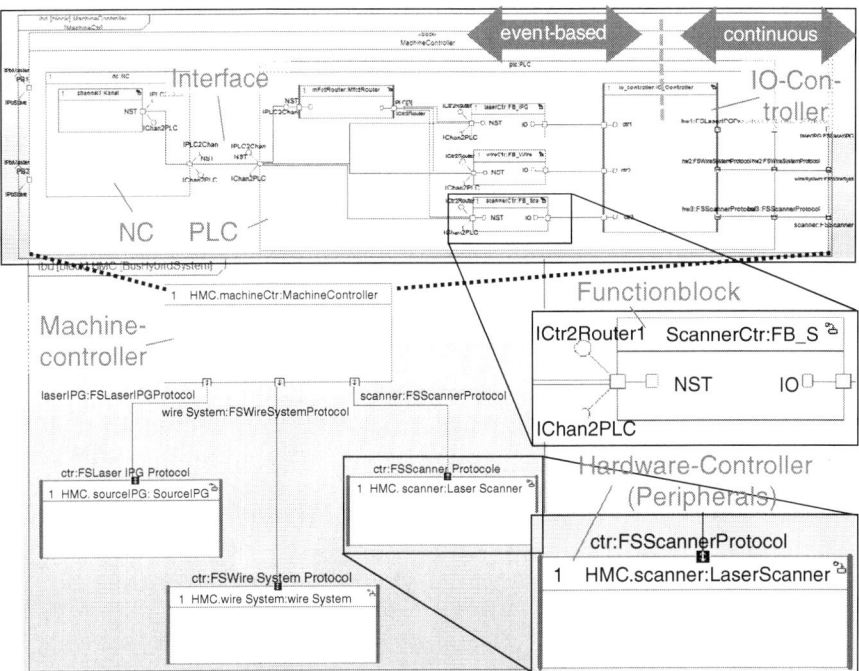

Fig. 3.115 Software modelling using a Hybrid Machining Center as an Example

Parametric Constraint Evaluation (PCE) add-on of IBM Rational Rhapsody®. The concrete values needed for this, such as mass, moment of inertia, current constants, etc. come from the CAx system used to provide the details. They can be fed directly into the model as estimated values even in early stages. This enables a step-by-step enhancement with short iteration cycles for system validation and even in the design phase it allows conclusions to be drawn about the overarching system properties.

The interpretation and validation of components is generally done through static operation points, while validation of software in particular requires a time-based behavioural analysis. In model-based software development the focus is increasingly shifting from modelling individual procedures and instructions to modelling total mechatronic systems. The use of state diagrams to describe the input and output behaviour of complex sub-systems here leads to a clear improvement in the transparency of the software development process. Validation is done iteratively against the behavioural model set up as a first step, for the control devices, actors and sensors. Using the example of the "Hybrid Manufacturing Center" (compare Sect. 5.5), the existing behaviour of the sub-system controls (laser sources, scanners, wire feeder) was modelled in the form of state machines. With the help of a strongly simplified model of the NC-PLC-architecture, to provide an example, parts of the modularly constructed PLC-Software used to control the subsystems was described in SysML as an executable model using IBM Rational Rhapsody® (Fig. 3.115).

One of the main challenges in the use of standard modeling languages for hardware-related software development is the creation of appropriate reference architectures and domain models, which serve as a framework for the engineering process and facilitate the transfer of models to the specific target platform. Against this backdrop it was therefore investigated, how to describe the classical control architecture of a machine tool with model elements provided by SysML. Excerpts of the developed model structure are depicted in Fig. 3.115, representing an "internal block diagram".

The machine controller, visible in the center of the diagram, is connected with the peripheral components of the hybrid system via flow ports. The model currently consists of elements representing the numerical controller (NC), the programmable logic controller (PLC) and the input-/output-controller (IO controller) with respective sub-components. The integration of additional components such as machine operator panels and drive controller is the subject of current research.

An important aspect of modelling is the distinction between event-based and continuous signal processing. The comparison of alternative approaches show that a continuous flow-oriented representation of signals for controlling peripheral components is associated with a lower level of abstraction and therefore more intuitive and easier to apply. The internal communication of the controller, however, relies on event-based signal interchange, which is characterised by a higher degree of flexibility, reusability and a better applicability to target platforms. The IO controller shown in Fig. 3.115 implements the required functionality for the exchange between the event-based section of the PLC and the continuous range of the available peripheral control signals. The distinction between synchronous and asynchronous functions calls from the NC programme to the PLC (through so-called M-functions) is an important design option during the conceptual software design and is featured by the modelled NC-PLC-interface. The behaviour of each component is defined by state machines and activity diagrams and can be transformed into executable code. By means of the presented prototype architecture, behavioural models of complex peripheral control components can be connected and controlled in order to form an overall model of the system under development (compare Fig. 3.116).

The basic procedure hereby starts with modeling the temporal input-output behaviour of the peripheral components, followed by the modeling of the IO controller. The creation of function blocks is based on state diagrams, which can be optionally equipped with interfaces for M-function calls. The modeling of NC cycles is based on activity diagrams which correspond to the primary sequential processing of NC programs. Using animated sequence diagrams the signal flow between the system components can be visualised dynamically during programme execution. Sequence diagrams hence provide an important mean for the conduction of tests and the analysis of complex communication structures.

3.5.5 Industrial Relevance

An increasing efficiency in engineering has many positive effects for the competitiveness of companies in machine and plant manufacture. Studies confirm that those

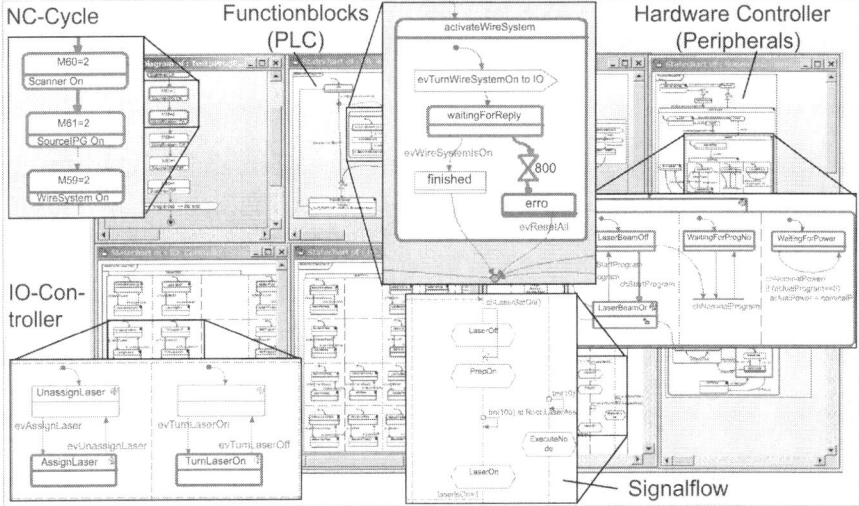

Fig. 3.116 Executable model of controller software

companies which make use of product configurators and who synchronise planning systems across the company are generally more competitive (Boucher and Houlihan 2008). In relation to the area of mechatronics, it can be shown that companies which practice software development using automatic code generation perform better sticking to estimated cost and delivery time, and can thus exploit their competitive advantage (Boucher and Barnett 2008).

When comparing the two strategies of configure-to-order (CTO) and engineer-to-order (ETO) it can be shown that the tools available in the PLM and ERP area can support both strategies mainly in their pure forms. However, particularly machine tool manufacturers can benefit from the combination of both strategies, since as a result of the increased productivity in the standard area, also customer-specific development can profit by scale effects and matured modules. Today, the integration of suitable or newly created modules into existing architectures is fraught with many challenges, as the individual adjustments frequently effect the supposedly standardised sections. Therefore the ratio of variant costs to variant benefits needs in future to be regarded much more from the point of view of changeability towards individual requirements. The decision about the configurations to be activated at the various variation points should be driven more by model based evaluations of objective functions at different system levels. Simple "If… Then" relationships or calculation operations, as currently used in many configuration systems, are too costly here in terms of set-up and maintenance. A clear increase in planning efficiency and in flexibility in relation to future development of product families will therefore come from the application of formal descriptive models across multiple phases for product families. These enable evaluations on the basis of physical descriptive models, which allows for an optimisation of configurations at the level of sub-modules and components in the development process.

Centralised planning of product families and their decentralised development within specialist departments, using configurators which are built into development tools, represents a balanced mix of planning and value oriented approaches to the development of product variants. As CAx planning tools are continuously improving —mainly focusing on the the needs of the corresponding disciplines—it can be assumed that the functionality for the definition of product variants will continue to improve. The approach developed here of automatic extraction of feature models for synchronisation purposes in connection with an overarching feature model has the potential to significantly reduce the effort for handling engineering project structures representing multiple variants by splitting them up into manageable sub-problems.

In the area of system design and software development, using graphical modelling languages which are readable across the disciplines, can provide a clear improvement in integrative collaboration, since models are easier to understand than for example looking at pure source code.

3.5.6 Future Research Needs

In order to get the necessary differentiation while still profiting from scale effects in machine tool engineering, focussed planning of product families is necessary, taking numerous criteria and prerequisites into account. The top level optimisation goal when designing product families is in general the ratio of variant benefit to variant costs (see Schuh 2005). The necessary input values come from various fields such as marketing, procurement, process planning, production planning, etc. For an efficient combination of all this information, there is an increasing need for standardised data formats and domain models that can be automatically evaluated by the different planning systems.

A significant increase in efficiency in the configuration and customer specific design is not solely sufficient to significantly reduce the dilemma of scale and scope in the production of machine tools. Rather, right from the point of planning the product families onwards, flexible architectures need to be developed so that even the modules which are affected indirectly by changes can be easily adapted to meet new requirements using standard components. The goal is to be able to profit from the stable production processes of large scale standard machines, without adversely affecting these by individualised modules.

In order to support the integrative collaboration between people from different expert disciplines—particularly in the earlier phases of development—generally understandable descriptive models are vital. To this end, modelling languages, tools and methodologies for model-based system engineering need to be developed further and adapted to the needs of machine engineering. The use of formal modelling languages can also be seen as an interface between humans and technical systems and therefore represents an important aspect of the integrative approach.

For computer-aided automated validation, formal requirement definitions are necessary. On the one hand in the area of component descriptions (actors, sensors), standards need to be developed taking more account of the physical properties as nowadays listed in product data sheets. On the other hand, descriptive models from the area of process planning should be used for the definition of requirements, to enable greater synchronisation between the product development at the customer and the order specific development at the machine manufacturer.

To reduce the effort required for the creation of configuration rules, it seems promising that formal models which describe the properties of mechatronic components could be successfully employed, particularly in relation to customer-specific modifications. As the abilities of optimisation processes rapidly reach their limits when increasing the number of dimensions, a systematic reduction of the solution space is necessary. For this, it would be helpful to be able propagate the requirements step by step starting from the process description through the machine structure and match requirements at each variation point with properties of existing modules.

Seen over the long-term, the increasing use of models not only has positive effects on the planning and implementation stages, but also on the operating phase at the customer, if the evaluation of mechatronic descriptive models is carried out directly by the machine control system. This can—in the sense of self-optimising production—form the basis for intelligent decision making.

References

Abe F, Osakada K, Kitamura Y, Matsumoto M, Shiomi M (2000) Manufacturing of titanium parts for medical purposes by selective laser melting. Proc Rapid Prototyping 288–293

Alicke K (2005) Planung und Betrieb von Logistiknetzwerken. Springer, Berlin

Andres MS (2006) Die optimale Varianz. Brand Eins 1:65–69

Anonym (2002) Multi-Slide-Druckgießverfahren nun auch für Kleinteile aus Magnesium nutzbar. Gießerei 11:60–62

Arentzen U (1997) Gabler-Wirtschafts-Lexikon. Gabler, Wiesbaden

Arnold D, Isermann H, Kuhn A, Tempelmeier H, Furmans K (2008) Handbuch Logistik, 3rd edn. Springer, Berlin

Bachmann F, Northrop L (2009) Structured variation management in software product lines. In: Proceedings of the 42nd Hawaii International Conference on System Sciences

Beamon BM (1998) Supply chain design and analysis: models and methods. Int J Prod Econ 55:281–294

Beck K et al (2011) Manifesto for agile software development. http://agilemanifesto.org/. Accessed 14 Mar 2011

Bergquist B (1999) New insights into influencing variables of water atomization of iron. Powder Metall 42(4):331–343

Beyer E (1995) Schweißen mit Laser, Grundlagen. ISBN 3-540-52674-9, Springer, Newyork

Blackenfelt M (2001) Managing complexity by product modularisation. Dissertation, Royal Institute of Technology, Stockholm

Bletzinger KU et al (2010) Optimal shapes of mechanically motivated surfaces. Comput Methods Appl Mech Eng 199:324–333

Boos W (2008) Methodik zur Gestaltung und Bewertung von modularen Werkzeugen. Apprimus Wissenschaftsverlag, Aachen

Boucher M, Barnett R (2008) Tailoring products to customer preferences: configuring profits to order. Aberdeen Group, Boston

Boucher M, Houlihan D (2008) System design: new product development for mechatronics. Aberdeen Group, Boston

Brecher C, Karlberger A, Herfs W (2010a) Wettbewerbsvorteile im Werkzeugmaschinenbau. Effiziente Auftragsabwicklung durch Kombination von Produktlinien-Ansatz und modellbasierter Entwicklung. ZWF 11:991–996

Brecher C et al (2010b) Integration und Durchgaengigkeit von Information im Produktionsmittellebenszyklus. ZWF 4:271–276

Bretzke WR (2010) Logistische Netzwerke. Springer, Berlin

Brillo J, Brook R, Egry I, Quested P (2008) Density and viscosity of liquid ternary Al-Cu-Ag alloys. High Temperatures-High Pressures 37:371–381

Büdenbender W (1991) Ganzheitliche Produktionsplanung und -steuerung. Konzepte für Produktionsunternehmen mit kombinierter kundenanonymer und kundenbezogener Auftragsabwicklung. Springer, Berlin

Bundesverband der Deutschen Gießerei-Industrie Statistik Ne-Metallguss (2007) Düsseldorf

Bundesverband der Deutschen Gießerei-Industrie (2010) Düsseldorf

Caesar C (1991) Kostenorientierte Gestaltungsmethodik für variantenreiche Serienprodukte. Variant mode an effects analysis (VMEA), Dissertation, RWTH Aachen

Carneiroa OS et al (2001) Computer aided rheological design of extrusion dies for profiles. J Mater Process Technol 114:75–86

Carreau PJ (1968) Rheological equations from molecular network theories. Ph.D. thesis, University of Wisconsin

Chandra C, Grabis J (2007) Supply chain configuration. Concepts, solutions, and applications. Springer, Berlin

Chopra S, Meindl P (2010) Supply chain management. Pearson, Boston

Christopher M (2005) Logistics and supply chain management. FT Prentice Hall, Harlow

Clements P (2009) Software product lines. Practices and patterns, 7. Print. Addison-Wesley, Boston

Cobb CW, Douglas PH (1928) A theory of production. Am Econ Rev 18(1):139–165

COMPUPLAST (2010) Mitteilung zur Profilextrusion der COMPUPLAST International Inc., Zlin, Tschechien. http://www.compuplast.com/ProfileDieProductDescription.shtml. Accessed 17 Nov 2010

Cronjäger H (2005) Komplexitaetscontrolling am Beispiel Nutzfahrzeugbau. Dissertation, RWTH Aachen

Curtis B (1992) Maintaining the software process. In: Proceedings of the conference on software maintenance. Orlando, Florida, pp 2–8

Czerner S (2005) Schmelzbaddynamik beim Laserstrahl-Wärmeleitungsschweißen von Eisenwerkstoffen, Diss. Univ. Hannover

Das S (2003) Physical aspects of process control in selective laser sintering of metals. Adv Eng Mater 5:701–711

Debbaut B et al (2006) Numerical simulation of the extrusion process and die design for industrial profile, using the multi-mode pom-pom model. Polymer Processing Society 22nd Annual Meeting, Yamagata, Japan

Domschke W, Scholl A (2005) Grundlagen der Betriebswirtschaftslehre: Eine Einfuehrung aus entscheidungsorientierter Sicht, 3rd edn. Springer, Berlin

Drath R (ed) (2010) Datenaustausch in der Anlagenplanung mit AutomationML. Integration von CAEX, PLCopen XML und COLLADA. Springer, Berlin

Drukarczyk J, Schüler A (2009) Unternehmensbewertung, 6th edn. Vahlen, München

Dudek G (2008) Collaborative planning in supply chains. A negotiation-based approach. Springer, Berlin

Ebli M (2010) Greiner ToolTec nutzte die wirtliche Entwicklung für eine erfoglreiche Neuausrichtung. Extrusion 3:42–43

Eitelwein O, Weber J (2008) Unternehmenserfolg durch Modularisierung von Produkten, Prozessen und Supply Chains. Book on Demand GmbH, Norderstedt

Eplan Engineering Center (2011) http://www.eplan.de/produkte/mechatronik/eplan-engineering-center.html. Accessed 14 Mar 2011

Erixon G (1998) Modular function deployment. A method for product modularisation. Dissertation, KTH Stockholm

Estefan JA (2007) Survey of model-based systems engineering (MBSE) methodologies. http://www.omgsysml.org/MBSE_Methodology_Survey_RevA.pdf. Accessed 14 Mar 2011

Ettinger HJ et al (2004) Parameterization and optimization strategies for the automated design of uPVC profile extrusion dies. Struct Multidiscip Optimization 28:180–194

Eversheim W (1998) Organisation in der Produktionstechnik, vol 2: Konstruktion. Springer, Berlin

Fisher ML, Ittner CD (1999) The impact of product variety on automobile assembly operations: empirical evidence and simulation. Manag Sci 45(6):771–786

Fixon SK (2007) Modularity and commonality research: past developments and future opportunities. Concurrent Eng 2:85–107

Fleisch E (2008) High-resolution-management. Konsequenz der 3. IT-Revolution auf die Unternehmensfuehrung. Schaeffer-Poeschel, Stuttgart

Foerst J (1994) Entscheidungsmodell zur unternehmensspezifischen Auswahl von Funktionen des Qualitaetsmanagements. Shaker, Aachen

Fogarty DW, Blackstone JH Jr, Hoffmann TR (1991) Production and inventory management, 2nd edn. South-Western Publishing, Cincinnati

Förster HU (1988) Integration von flexiblen Fertigungszellen in die PPS. Dissertation, RWTH Aachen. Springer, Berlin

Friedenthal S, Moore A, Steiner R (2009) A practical guide to SysML. The systems modeling language. Elsevier, Morgan Kaufmann OMG, Amsterdam [u. a.]

Gebhardt A (2007) Generative Fertigungsverfahren. Hanser, München

Gershenson JK (2003) Product modularity: definitions and benefits. J Eng Design 3:295–313

Gibson I, Rosen DW, Stucker B (2010) Additive manufacturing technologies. Springer, New York

Giesekus H (1994) Phänomenologische Rheologie – Eine Einführung. Springer, Berlin

Göpfert J (1998) Modulare Produktentwicklung. Zur gemeinsamen Gestaltung von Technik und Organisation. DUV, Wiesbaden

Große-Oetringhaus WF (1974) Fertigungstypologie unter dem Gesichtspunkt der Fertigungsablaufplanung, 1st edn. Duncker, Berlin

Gudehus T (2006) Dynamische disposition. Springer, Berlin

Heesen B, Gruber W (2009) Bilanzanalyse und Kennzahlen: Fallorientierte Bilanzoptimierung, 2nd edn. Gabler, Wiesbaden

Heina J (1999) Variantenmanagement. Kosten-Nutzen-Bewertung zur Optimierung der Variantenvielfalt. Dissertation, TU Cottbus

Herfs WJ (2010) Modellbasierte software in the loop simulation von Werkzeugmaschinen, 1st edn. Apprimus Wissenschaftsverlag, Aachen

Herziger G, Lossen P (1994) Werkstoffbearbeitung mit Laserstrahlung, Grundlagen – Systeme – Verfahren. Hanser, München

Hieber R, Schönsleben PH (2002) Supply chain management. A collaborative performance measurement approach. VDF, Zürich

Higginson JK, Bookbinder JH (2005) Distribution centres in supply chain operations. In: Langevin A, Riopel D (eds) Logistics systems: design and optimization. Springer, Berlin

Hopkinson N (2006) Rapid manufacturing: an industrial revolution for the digital age. Wiley, Chichester

Hoppe N, Conzen F (2002) Europaeische Distributionsnetzwerke. Voraussetzungen, Projektablauf, Fallbeispiele. Gabler, Wiesbaden

Hubka V (1984) Theorie technischer Systeme. Grundlagen einer wissenschaftlichen Konstruktionslehre. Springer, Berlin

Hummler-Schaufler B, Schlumpberger C (2004) Produktentwicklung auf hohem Niveau. Druckguss Praxis 1:39–42

Hungenberg H (2006) Strategisches Management in Unternehmen: Ziele – Prozesse – Verfahren. Gabler, Wiesbaden

INCOSE (2004) What is systems engineering? http://www.incose.org/practice/whatissystemseng.aspx. Accessed 14 Mar 2011

Johnson AA, Tezduyar TE (1994) Mesh update strategies in parallel finite element computations of flow problems with moving boundaries and interfaces. Comput Methods Appl Mech Eng 119:73–94

Junge M (2004) Controlling modularer Produktfamilien in der Automobilindustrie. Entwicklung und Anwendung der Modularisierungs-Balanced-Scorecard. Dissertation, Univ. Mainz

Kaiser A (1995) Integriertes Variantenmanagement mit Hilfe der Prozesskostenrechnung. Dissertation, Univ. St. Gallen

Kang KC (1990) Feature-oriented domain analysis (FODA). Feasibility study; technical report CMU/SEI-90-TR-21-ESD-90-TR-222. Software Engineering Inst. Carnegie Mellon University, Pittsburgh

Kathuria YP (2000) Laser assisted metal rapid prototyping in micro-domain. In: 8th international conference on rapid prototyping, Tokyo, pp 160–165

Kaul S (2004) Rechnergestütze Optimierungsstrategien für Profilextrusionswerkzeuge. Dissertation, RWTH Achen

Keene BJ (1993) Review of data for the surface tension of pure metals. Int Mater Rev 38(4):157–192

Keene BJ, Mills KC, Brooks RF (1985) Surface properties of liquid metals and their effects on weldability. Mater Sci Technol 1:568–569

Kettner H et al (1984) Leitfaden der systematischen Fabrikplanung. Hanser, München

Kilger C, Reuter B, Stadtler H (2008) Collaborative planning. In: Stadtler H, Kilger C (eds) Supply chain management and advanced planning. Concepts, models, software, and case studies. Springer, Berlin, pp 263–284

Kinkel S (2005) Anforderungen an die Fertigungstechnik von morgen. Mitteilung aus der Produktionsinnovationserhebung. Fraunhofer ISI, Karlsruhe, p 12

Kobyrn PA, Semiatin SL (2001) Mechanical properties of laser-deposited Ti-6Al-4V solid freeform fabrication symposium. Austin, USA

Kohlhase N (1996) Strukturieren und Beurteilen von Baukastensystemen, Strategien, Methoden, Instrumente. Dissertation, Univ. Darmstadt

Kohlhase N (1997) Strukturieren und beurteilen von Baukastensystemen. Strategien, Methoden, Instrumente. VDI, Duesseldorf

Koller R (1998) Konstruktionslehre fuer den Maschinenbau. Grundlagen zur Neu- und Weiterentwicklung technischer Produkte mit Beispielen, 4th edn. Springer, Berlin

Koppenhagen F (2004) Systematische Ableitung modularer Produktarchitekturen. Dissertation, TU Hamburg-Harburg

Kruth JP et al (2004) Selective laser melting of iron-based powder. J Mater Process Technol 149(1–3):616

Kruth JP et al (2005) Statistical analysis of experimental parameters in selective laser sintering. Adv Eng Mater 7(8):750–755

Kruth JP, Levy G, Klocke F, Childs THC (2007) Consolidation phenomena in laser and powder bed based layered manufacturing. Ann CIRP 56:730–759

Le C, Bruns T, Tortorelli D (2011) A gradient-based, parameter-free approach to shape otimization. Comput Methods Appl Mech Eng 200. Accepted manuscript (in press)

Ley W (1985) Entwicklung von Entscheidungshilfen zur Integration der Fertigungshilfsmitteldisposition in EDV-gestuetzte Produktionsplanungs- und -steuerungssysteme. Dissertation, RWTH Aachen

Lindemann U (2009) Structural Complexity Management. An Approach for the Field of Product Design. Springer, Heidelberg

Lindemann U, Reichwald R, Zäh M (2006) Individualisierte Produkte. Komplexitaet beherrschen in Entwicklung und Produktion, 1st edn. Springer, Berlin

Lindstädt H, Hauser R (2004) Strategische Wirkungsbereiche des Unternehmens. Spielraeume und Integrationsgrenzen erkennen und gestalten. Gabler, Wiesbaden

Löschmann F (2009) Innovative Antriebe und Perspektiven bei Volkswagen Sachsen. Internationaler AMI Kongress, Leipzig

Lotfi A (2005) Optimal shape design for metal forming problems by the finite element method. Proc Appl Math Mech 5:429–430

Lucas K (2008) Thermodynamik. Die Grundgesetze der Energie- und Stoffumwandlungen. Springer, Berlin

MacDuffie JP, Sethuraman K, Fisher ML (1996) Product variety and manufacturing performance: evidence from the international automotive assembly plant study. Manag Sci 42(3):350–369

MAGMA GmbH (2003) Firmenprofil MAGMA Neuigkeiten in der Druckgußsimulation. Druckgusspraxis

Malik F (2000) Systemisches Management, Evolution, Selbstorganisation: Grundprobleme, Funktionsmechanismen und Loesungsansaetze fuer komplexe Systeme, 2nd edn. Haupt, Bern

Marti M (2007) Complexity management. Optimizing product architecture of industrial products. Dissertation, Univ. St. Gallen

Martin MV (1999) Design for variety. A methodology for developing product platform architectures. Dissertation, Stanford Univ.

Martin MV, Ishii K (1997) Design for variety, development of complexity indices and design charts. ASME Design Engineering Technical Conferences, Sacramento

Mayer S, Thiry E, Frank C, Kara G, Menke A (2009) Excellence in logistics. Studie AT Kearney, Chicago

Medion AG (2009) Geschaeftsbericht 2009. Medion, Essen

Meier J (2007) Produktarchitekturtypen globalisierter Unternehmen. Dissertation, RWTH Aachen

Meiners W (1999) Direktes Selektives Lasersintern einkomponentiger metallischer Werkstoffe. Dissertation, RWTH Aachen

Meiners W (2007) Personal communication, Fraunhofer-ILT

Mellor SJ, Balcer MJ (2002) Executable UML. A foundation for model-driven architecture. Addison-Wesley, Boston

Michaeli W (2009) Extrusionswerkzeuge für Kunststoffe und Kautschuk. Hanser, München

Michaeli W et al (2001) Entwicklung vorausberechnen – Prozesssimulation in der Praxis – eine Standortbestimmung. Kunststoffe 7, 91:32–39

Michaeli W et al (2009) Towards shape optimization of extrusion dies using finite elements. J Plast Technol 5:411–427

Mikkola JH, Gasmann O (2001) Managing modularity of produkt architectures: toward an integral theory. IEEE Trans Eng Manag 22:204–218

Müller M (2007) Automotive SPICE in der Praxis. Interpretationshilfe fuer Anwender und Assessoren. 1st edn. dpunkt-Verl., Heidelberg

NN (2011) Variant Management with pure variants. Technical White Paper. http://www.pure-systems.com/fileadmin/downloads/pv-whitepaper-en-04.pdf. Accessed 14 Mar 2011

NN (2004) VDI-Richtlinie 2206 – Entwicklungsmethodik fuer mechatronische Systeme. VDI Fachbereich Produktentwicklung und Mechatronik. Beuth, Berlin

Neuhausen J (2001) Methodik zur Gestaltung modularer Produktionssysteme fuer Unternehmen der Serienproduktion. Dissertation, RWTH Aachen

Nicolai M, Behr M (2007) Portable optimization framework for serial and parallel machines. Second European Conference on Computational Optimization

Nische mit hohem Trendfaktor (2005) In: Werkzeug and Formenbau, März, pp 36–38

Nóbrega JM et al (2003) Flow balancing in extrusion dies for thermoplastic profiles part I–II. Int Polym Process 3:298–312

Nyhuis P, Nickel R, Tullius K et al (2008) Globales Varianten-Produktionssystem. Globalisierung mit System, 1st edn. PZH, Garbsen

Over C (2003) Generative Fertigung von Bauteilen aus Werkzeugstahl X38CrMoV5-1 und Titan TiAL6V4 mit "Selective Laser Melting". Dissertation, RWTH Aachen

Pahl G, Beitz W, Feldhusen J, Grote K (2005) Konstruktionslehre. Grundlagen erfolgreicher Produktentwicklung, 6th edn. Springer, Berlin

Pahl G et al (2007) Pahl/Beitz Konstruktionslehre: Grundlagen erfolgreicher Produktentwicklung. 7th edn. Springer, Heidelberg

Palupski R (2002) Management von Beschaffung, Produktion und Absatz. Leitfaden mit Praxisbeispielen. Gabler, Wiesbaden

Paredis C et al (2010) An overview of the sysML-modelica transformation specification. INCOSE Intl. Symposium, Chicago

Peak RS et al (2007) Simulation-based design using sysML – part 2: celebrating diversity by example. In: INCOSE Intl. Symposium, Sandiego

Pedrotti F, Pedrotti L, Bausch W, Schmidt H (2002) Optik für Ingenieure. Springer, Heidelberg

Petersheim J (1997a) Selektives Laserschweißen – Ein neuartiges Verfahren zur schnellen Fertigung mechanischer Bauteile. Elektrowärme Int 55(B3):B80–B85

Petersheim J (1997b) Technologien zur direkten Herstellung funktioneller metallischer Prototypen. VDI Fachtagung Rapid Prototyping – Neue Wege zur schnellen Fertigung von Modellen und Prototypen

Pfohl H (2010) Logistiksysteme. Springer, Berlin

Pickhardt J (1999) Entwicklung eines morphologischen Merkmalschemas zur Gestaltung der EDV-Unterstuetzung des Produktdatenmanagements im Rahmen der technischen Auftragsabwicklung, 1st edn. Shaker, Aachen

Piegel L, Tiller W (1997) The NURBS book. Springer, New York

Piller FT (2006) Mass Customization, Ein wettbewerbsstrategisches Konzept im Informationszeitalter. 2nd edn. Gabler DUV, Wiesbaden

Pine JB (1993) Mass customization, the new frontier in business competition. Harvard Business School Press, Boston

Porter ME (1998) Competitive strategy: techniques for analyzing industries and competitors. Free Press, New York

Powell MJD (2009) The BOBYQA algoritm for bound constrained optimization withaout derivates. Department of Applied Mathematics and Theoretical Physics, Cambrigde, England: s.n

Queudeville Y, Ivanov T, Nußbaum C, Vroomen U, Bürig-Polaczek A (2009) Decision and design methodologies for lay-out of modular dies for high-pressure-die-cast-processes. Mater Sci Forum 618–619:345–348

Rapp T (1999) Produktstrukturierung. Komplexitaetsmanagement durch modulare Produktstrukturen und -plattformen. Dissertation, Univ. St. Gallen

Rathnow PJ (1993) Integriertes Variantenmanagement. Bestimmung, Realisierung und Sicherung der optimalen Variantenvielfalt, 1st edn. Vandenhoek and Ruprecht, Göttingen

Reichwald R, Piller F (2006) Interaktive Wertschöpfung, 1st edn. Gabler, Wiesbaden

Rombouts M, Froyen L (2005) Roughness after laser melting of iron based powders, virtual modelling and rapid manufacturing – Bartolo, Taylor and Francis Group, London. ISBN 0415390621, pp 329–335

Rombouts M, Kruth J-P, Froyen L, Mercelis P (2006) Fundamentals of selective laser melting of alloyed steel powders. CIRP Ann 55(1):192–197

Roth K (2000) Konstruieren mit Konstruktionskatalogen. Springer, Berlin

Rumpelt T (2009) Lean besser durch die Krise? Automobil-Produktion 12:20–21

Sahm PR, Egry I, Volkmann T (1999) Schmelze, Erstarrung, Grenzflächen, eine Einführung in die Physik und Technologie flüssiger und fester Metalle. Vieweg, Wiesbaden

Sames G, Büdenbender W (1990) Überblick erhalten. Das morphologische Merkmalschema als praktikables Hilfsmittel fuer die technische Auftragsabwicklung. Maschinenmarkt 96(32):54–59

Sauerland S (1993) Messung der Oberflächenspannung an levitierten flüssigen Metalltropfen. Dissertation, RWTH-Aachen

Schenk M, Wirth S (2004) Fabrikplanung und Fabrikbetrieb: Methoden fuer die wandlungsfaehige und vernetzte Fabrik. Springer, Berlin

Schiaffino S, Sonin AA (1997) Formation an stability of liquid and molten beads on a solid surface. J Fluid Mech 343:95–110

Schiegg P (2005) Typologie und Erklaerungsansaetze fuer Strukturen der Planung und Steuerung in Produktionsnetzwerken. Dissertation, RWTH Aachen

Schirmer A (1980) Dynamische Produktionsplanung bei Serienfertigung. Betriebswirtschaftlich-technologische Beitraege zur Theorie und Praxis des Industriebetriebs, vol 6. Gabler, Wiesbaden

Schleifenbaum H, Meiners W, Wissenbach K (2008) Towards rapid manufacturing for series production: an ongoing process report on increasing the build rate of selective laser melting (SLM). International conference on rapid prototyping and rapid tooling and rapid manufacturing, Berlin, Germany

Schleifenbaum H, Meiners W, Wissenbach K, Hinke C (2010) Individualized production by means of high power selective laser melting. CIRP J Manuf Sci Technol 2:161–169

Schmitt R, Pfeifer T (2010) Qualitätsmanagement. Strategien, Methoden, Techniken, 4th edn. Hanser

Schmitt R, Vorspel-Rueter M, Wienholdt H (2010) Handhabung von Komplexitaet in flexiblen Produktionssystemen. Ind Manag 26(1):53–56

Schomburg E (1980) Entwicklung eines betriebspsychologischen Instrumentariums zur systematischen Ermittlung der Anforderungen an EDV-gestuetzte Produktionsplanungs- und -steuerungssysteme im Maschinenbau. Dissertation, RWTH Aachen

Schönsleben P (2007) Integrales Logistikmanagement. Springer, Berlin

Schrinner H (1993) Einfluss verschiedener Legierungselemente auf die Oberflächen- und Grenzflächenspannung von Stahl und Schlacke. Dissertation, RWTH Aachen

Schuh G (1989) Gestaltung und Bewertung von Produktvarianten. Ein Beitrag zur systematischen Planung von Serienprodukten. Dissertation, RWTH Aachen

Schuh G (2005) Produktkomplexitaet managen. Strategien – Methoden – Tools, 2., ueberarb. und erw. Hanser, München

Schuh G (2006) Produktionsplanung und -steuerung. Grundlagen, Gestaltung und Konzepte, 3rd edn. Springer, Berlin

Schuh G (2007) VDMA: Effizient, schnell und erfolgreich. Strategien im Maschinen- und Anlagenbau, 1st edn. VDMA, Frankfurt a. M.

Schuh G, Schöning S, Jung M, Uam JY (2007a) Individualisierte Produktion. ZWF 102(10):630–634

Schuh G, Schöning S, Jung M, Uam JY, Hinke C, Kupke D, Meyer J, Vorspel-Rueter M, Wienholdt H (2007b) Integrierte Produktionstechnik. In: Schuh G, Klocke F, Brecher C, Schmitt R (eds) Excellence in production. Festschrift fuer Univ.-Prof. Dr.-Ing. Dipl.-Wirt. Ing. Dr. techn. h.c. Dr. oec. h.c. Walter Eversheim. Apprimus Wissenschaftsverlag, Aachen, pp 31–53

Schuh G, Klocke F, Brecher C, Schmitt R (2007c) Excellence in production, 1st edn. Apprimus Wissenschaftsverlag, Aachen

Schuh G et al (2008) Lean Innovation. Auf dem Weg zur Systematik. In: Schuh G, Klocke F, Brecher C, Schmitt R (eds) Wettbewerbsfaktor Produktionstechnik. Aachener Perspektiven. Apprimus Wissenschaftsverlag, Aachen, pp 473–512

Schuh G, Lenders M, Arnoscht J (2009) Focussing product innovation and fostering economies of scale based on adaptive product platforms. CIRP Ann 58(1):131–134

Schuh G, Lenders M, Nussbaum C (2010a) Maximaler Wirkungsgrad von Produktkomplexitaet. Kosten und Nutzen integriert Bewerten. ZWF 105(5):473–477

Schuh G, Arnoscht J, Lenders M, Rudolf S (2010b) Effizienter innovieren mit Produktbaukaesten. Werkzeugmaschinenlabor WZL der RWTH Aachen, Aachen

Schuh G, Arnoscht J, Rudolf S (2010c) Integrated development of modular product platforms. PICMET 2010 proceedings, pp 1928–1940

Schuh G, Arnoscht J, Bohl A, Nußbaum C (2011a) Integrative assessment and configuration of production systems. CIRP Ann 60(1):330–334

Schuh G, Arnoscht J, Bohl A, Kupke D, Nußbaum C, Quick J, Vorspel-Rüter M (2011b) Assessment of the scale-scope dilemma in production systems. Prod Eng 5(Special Issue):1–10

Schuh G, Stich V, Brosze T, Fuchs S, Pulz C, Quick J, Schuermeyer M, Bauhoff F (2011c) High resolution supply chain management – optimised processes based on self-optimizing control loops and real time data. Production Engineering, Sonderheft 1

Schwartz B, Ward A, Monterosso J, Lyubomirsky S, White K, Lehman DR (2002) Maximizing versus satisficing, happiness is a matter of choice. J Personal Soc Psychol 83(5):1178–1197

Secolec R (2005) Produktstrukturierung als Instrument zum Variantenmanagement in der methodischen Entwicklung modularer Produktfamilien. Dissertation, ETH Zuerich

Sienz J, Bulman SD, Pittman JTF (2001) Optimisation strategies for extrusion die design. Liège, Belgium, pp 275–278

Simchi A, Petzoldt F, Pohl H (2003) On the development of direct metal laser sintering for rapid tooling. J Mater Process Technol 141:319–328

Smith DE, Tortorelli DA, Tucker III CL (1998) Optimal design for polymer extrusion. Part I: sensitivity analysis for nonlinear steady-state systems. Comput Methods Appl Mech Eng 167:283–302

Spur G, Stöferle T (1994) Handbuch der Fertigungstechnik, vol 6: Fabrikbetrieb. Hanser, München

Stadtler H (2008) Supply chain management. An overview. In: Stadtler H, Kilger C (eds) Supply chain management and advanced planning. Concepts, models, software, and case studies. Springer, Berlin, pp 9–36

Steffen D (2007) Ein Verfahren zur Produktstrukturierung fuer fortgeschrittene mechatronische Systeme. Dissertation, Univ. Paderborn

Steward D (1981) The design structure system. A method for managing the design of complex systems. IEEE Trans Eng Manag 28(3):71

Stich V, Wienholdt H (2009) Flexible configuration logic for a complexity oriented design of production systems. POMS 20th Annual Conference, Orlando, FL

Straube F, Pfohl H (2008) Trends und Strategien in der Logistik. DVV, Hamburg

Su W-N et al (2003) Investigation of fully dense laser sintering of tool steel powder using a pulsed Nd:YAG (neodymium-doped yttrium aluminium garnet) laser. In: Proceedings of the institution of mechanical engineers. Vol. 217 Part C: J Mech Eng Sci, pp 127–138

Supply Chain Council (2010) Supply chain operations reference model version 10.0. The supply chain council, Cypress, TX

Sürie C, Wagner M (2008) Supply chain analysis. In: Stadtler H, Kilger C (eds) Supply chain management and advanced planning. Springer, Berlin, pp 37–64

Szarvasy I et al (2000) Computer aided optimization of profile extrusion dies: definition and assessment of the objective function. Int Polymer Process 15(1):28–39

Tempelmeier H (2005) Bestandsmanagement in supply chains. Books on Demand, Norderstedt

Ten-Hove R, Walker P (2005) Java business integration (JBI) 1.0. Final Release. http://jcp.org/aboutJava/communityprocess/final/jsr208/index.html

Thonemann U, Behrenbeck K, Kuepper J (2005) Ruder übernommen. Handel. Logistik heute 27(12):46–48

Ulrich K (1995) The role of product architecture in the manufacturing firm. Res Policy 24(3):419–440

Ulrich K, Eppinger S (2004) Product design and development, 3rd edn. McGraw-Hill, Boston

Ulrich K, Tung K (1991) Fundamentals of product modularity. Proceedings of the ASME Winter Annual Meeting Symposium on Issues in Design/Manufacturing Integration. Atlanta, GA, pp 73–79

Ungeheuer U (1986) Produkt- und Montagestrukturierung. Methodik zur Planung einer anforderungsgerechten Produkt- und Montagestruktur in Unternehmen der Einzel- und Kleinserienproduktion komplexer Produkte. Dissertation, RWTH Aachen

Vahrenkamp R (2007) Logistik. Management und Strategien, 6th edn. Oldenbourg, München

VDI-Fachbereich Produktentwicklung und Mechatronik (1997) VDI 2222. Konstruktionsmethodik. Methodisches Entwickeln von Loesungsprinzipien. VDI, Düsseldorf

VDI-Richllinie 2206: Entwicklungsmethodik für mechatronische Systeme. Düsseldorf: VDI-Verlag 2004.

Verein Deutscher Ingenieure (1993) Methodik zum Entwickeln und Konstruieren technischer Systeme und Produkte. VDI Richtlinie 2221. VDI, Düsseldorf

Wagner H (1999) Modularisierung von neuronalen Netzen. Dissertation, Univ. Bielefeld

Wagner C (2003) Untersuchungen zum Selektiven Lasersintern von Metallen. Dissertation, RWTH Aachen

Wang XC et al (2002) Direct selective laser sintering of hard metal powders: experimental study and simulation. Int J Adv Manuf Technol 19:351–357

Wannewetsch H (2010) Integrierte Materialwirtschaft und Logistik. Springer, Berlin

Werner H (2008) Supply chain management. Grundlagen, Strategien, Instrumente und Controlling. Gabler, Wiesbaden

Westkämper E, Warnecke HJ (2006) Einführung in die Fertigungstechnik. Teubner, Stuttgart

Wiendahl HP, Scholtissek P (1994), Management and control of complexity in manufacturing. CIRP Ann 43(14):533–540

Wiendahl H-H, Rempp B, Schanz M (2000) Turbulenzen erschweren Planungssicherheit. IO-Manag 69(5):38–43

Wiener N (1948) Cybernetics. Herman, Paris

Wiese H (2010) Mikrooekonomik. Springer, Berlin

Wildemann H (1998) Die modulare Fabrik. Kundennahe Produktion durch Fertigungssegmentierung. TCW, München

Willenborg E (2005) Polieren von Werkzeugstählen mit Laserstrahlung. Dissertation, RWTH-Aachen

Wohlert M (2000) HIP of direct laser sintered metal components. Dissertation, University of Texas

Wünsch G (2008) Methoden fuer die virtuelle Inbetriebnahme automatisierter Produktionssysteme. Forschungsberichte IWB, vol 215. Utz, München

Wylie B et al (2007) Performance analysis and tuning of the XNS CFD solver on BlueGene/L. Proceedings of the 14th EuroPVM/MPI conference. Springer, New York, pp 107–116

Zhang D (2004) Entwicklung des Selective Laser Melting (SLM) für Aluminiumwerkstoffe. Dissertation, RWTH Aachen.

Zolfaghari A et al (2009) An innovative method of die design and evaluation of flow balance for thermoplastics extrusion profiles. Polym Eng Sci 49:49

Chapter 4
Virtual Production Systems

Wolfgang Schulz, Christian Bischof, Kirsten Bobzin, Christian Brecher,
Thomas Gries, Sabina Jeschke, Achim Kampker, Fritz Klocke,
Torsten Kuhlen, Günther Schuh, Markus Apel, Tim Arping, Nazlim Bagcivan,
Markus Bambach, Thomas Baranowski, Stephan Bäumler, Thomas Beer,
Stefan Benke, Thomas Bergs, Peter Burggräf, Gustavo Francisco Cabral,
Urs Eppelt, Patrick Fayek, Marcel Fey, Bastian Franzkoch,
Stephan Freyberger, Lothar Glasmacher, Barbara Heesel, Thomas Henke,
Werner Herfs, Ulrich Jansen, Tatyana Kashko, Sergey Konovalov,
Britta Kuckhoff, Gottfried Laschet, Markus Linke, Wolfram Lohse,
Tobias Meisen, Meysam Minoufekr, Jan Nöcker, Ulrich Prahl,
Hendrik Quade, Matthias Rasim, Marcus Rauhut, Rudolf Reinhard,
Jan Rosenbaum, Eduardo Rossiter, Daniel Schilberg, Georg J. Schmitz,
Johannes Triebs, Hagen Wegner and Cathrin Wesch-Potente

Contents

4.1	Research Program of Virtual Production Systems	242
4.2	Platform for Integrative Simulation	243
	4.2.1 Summary	243
	4.2.2 State of the Art	244
	4.2.3 Motivation and Research Question	251
	4.2.4 Results	252
	4.2.5 Future Research Topics	277
4.3	Factory Planning of the Future—Connecting Model Worlds	280
	4.3.1 Abstract	280
	4.3.2 State of the Art	280
	4.3.3 Motivation and Research Question	286
	4.3.4 Results	287
	4.3.5 Industrial Relevance	314
	4.3.6 Future Research Topics	318
4.4	Integrative Process Chain Simulation for Material and Production Technologies	319
	4.4.1 Abstract	319
	4.4.2 Introduction	320
	4.4.3 State of the Art	322
	4.4.4 Motivation and Research Question	326
	4.4.5 AixViPMaP Platform	327

W. Schulz (✉)
Fraunhofer-Institut für Lasertechnik ILT, Steinbachstr. 15,
52074 Aachen, Germany
e-mail: wolfgang.schulz@ilt.fraunhofer.de

4.4.6	Test Case Line Pipe	333
4.4.7	Test Case Transmission Component	339
4.4.8	Test Case Plastic Component in Automotive Interior	344
4.4.9	Test Case Textile Reinforced Piston Rod	352
4.4.10	Test Case Stainless Steel Casting	360
4.4.11	Industrial Relevance	370
4.4.12	Future Research Topics	372
4.5	Integrative Simulation of Machine Tool and Manufacturing Technology	373
4.5.1	Abstract	373
4.5.2	Planning and Simulation of NC-Controlled Machining Processes	375
4.5.3	Motivation and Research Question	382
4.5.4	Results	383
4.5.5	Industrial Application Options	416
4.5.6	Future Research Topics	418
References		420

4.1 Research Program of Virtual Production Systems

The use of simulation systems is of significant importance for companies in high-wage countries as the requirements of product- and process quality are generally higher than in low-wage countries due to conditions of the market. Since the implementation of simulation tools is not value-adding in the first place, the performance of virtual product development chain must therefore be continuously increased in terms of greater planning efficiency. Research in the field of virtual production systems therefore addresses the following issue:

> How can the benefits of various production processes be consistently described and how can these beneficial effects be comprised into scientific context (such as formulas)?

This requires the following—partially contradictory—objectives to be addressed:

- Improving the quality of the generated simulation results. This, for example, refers to the results' quality as well as to the their conformance to reality or to the quality and the relevant data, which has been generated by a production system (Research on the genesis of active relationship formulas)
- Increasing the standardisation of technical support within the product development chain. This point addresses the integration efforts across multiple corporate levels and thus also across different simulation tools with different levels of detail (Research on structures and references for the quantitative description of process interdependencies)

The research approaches in the field of virtual production systems therefore are primarily of explanatory nature. Integrative research has to transform a high proportion of the interdependencies of a production system into quantitative formulas and thus to contribute to a reduction of complexity by adding to the theory of value creation.

The comprehensive planning of production facility using computational methods is a good example of the research result's potential. Existing software tools are generally developed in-house. As a consequence, they are not universally applicable and have a poor upgradability. Until now, efforts to establish flexible interconnections via standard interfaces have been focused on the level of tools and thus have neglected coordination problems in the actual planning process. In the Cluster of Excellence research focuses on a comprehensive approach to generally increase efficiency in planning processes in order to increase controllability of planning- and sequence-interdependencies as well as coordination and interface-related problems.

Difficulties in the development of standardised simulation application do not solely relate to planning activities. Material science and simulation of production facilities require a combination of different models and simulation systems as well in order to improve corresponding simulation However, one must not forget that this also requires a significant increase in simulation quality within the specific simulation tools and methods as well. In the field of material science, the Cluster of Excellence focuses its research on the microscopic material properties and their dynamic response to manufacturing processes along the whole production chain. In the next step, this dynamic microscopic structure is translated into macroscopic characteristics. This approach represents a fundamental advance towards a comprehensive simulation.

Likewise, this comprehensive approach is essential to the simulation production facilities as well. New processes often require an iterative approach until series production conditions have been established. This is especially true when it comes to complex geometries. Due to potential technical boundary conditions of the machinery, process optimisation may be affected. The Cluster of Excellence addresses this issue with a comprehensive chain of existing simulation systems, which will be linked in a way that simulation results can be transferred across the systems. By doing this, simulation results do not only include effects of the machineries' control system, its dynamic response and of the process itself, but also the interdependencies between them. This virtual chain therefore represents the complete manufacturing process and thus contributes to its optimisation.

4.2 Platform for Integrative Simulation

Christian Bischof, Sabina Jeschke, Torsten Kuhlen, Thomas Beer, Tobias Meisen, Rudolf Reinhard and Daniel Schilberg

4.2.1 Summary

The use of simulation tools in production process planning within industrial environments is already well established. However, this generally involves conducting individual simulations of specific sub-processes using default settings as the boundary parameters. This method does not take into account production history influences on the individual sub-processes. In order to improve planning quality using simulations, the individual simulations need to be linked to form a continuous simulation chain.

The methodology for linking simulations described in this paper allows flexible extension of the overall system, allowing incorporation of a variety of heterogeneous simulation chains. In addition to linking distributed simulation resources on an infrastructural level, Internet-based access is also provided, which enables partners to collaborate in setting up simulation chains. There is a data integration component to ensure the correct syntactic, structural and semantic transformation of data between the individual heterogeneous simulations and to assure that all the required simulation data is integrated into a common database. To enable integrative analysis of the simulation data for a whole process, there is an interactive visualization component, which can be used on a variety of visualization systems, from regular workstation computers through to dedicated virtual reality systems with numerous projection surfaces. Interactive modification of dataset timing within the context of analysis is a key aspect in this respect.

In future, the system will be expanded so that interactive analyses can be conducted on the integrated database and bidirectionally coupled with the visualization component in real time. This will enable intuitive access to the integrated simulation data, even across process boundaries, thus providing optimal support for planning or modification of production processes.

4.2.2 State of the Art

The need to couple and automate simulation rather than manually drive them is obvious: according to a study conducted by Lang in 2003, the process development time using manually linked simulations incurs an overhead of 60%, caused solely by manual operation of the individual simulation components and conversion tools of a simulation chain (Lang 2003). Hence optimising the links between simulation tools offers huge potential to save on planning costs with respect to simulating aspects of production, i.e., the domain of virtual production.

Since numerous studies have highlighted this issue, it begs the question as to why such automated environments are not found in practice. Our opinion is that many approaches fail due to a lack of flexibility and, consequently, a lack of user acceptance. The often relatively inflexible facilities for coupling just a few selected simulation tools may be the reason behind this. Coupling is limited to the facilities offered by the manufacturers, which in turn are generally limited to the existing tools of each individual manufacturer. This severely limits the choice of simulation tools that can be linked and, therefore, the simulation scenarios that can be mapped.

Thus at present there is no universal solution to the problem of simulation coupling. In essence, there are three broad areas that must be taken into account when coupling simulations:

4 Virtual Production Systems

- The technical level – infrastructure, data exchange
- The data level – syntax, structure and semantics of the data formats and models
- The analysis level – analysis and evaluation of the simulation results from linked simulation

On the technical level, simulation tools are special software components that run on corresponding hardware resources. Accordingly, any technical solution for linking simulations must take into account both components, i.e., the software and the hardware. The solution for coupling on a technical level is discussed below.

4.2.2.1 Grid Computing (Technical Level)

Grid computing, which emerged in the mid-1990s, could help meeting the aforementioned challenge on a technical level. Using grid computing, many national and international projects seek to combine computing resources and provide uniform access to them (Berlich et al. 2005). Due to the rapid proliferation and expansion of the Internet, basic networking of resources is not a problem in terms of infrastructure.

However, since the early days of computer technology, the coupling of software components in distributed systems has always posed a challenge. As in all other areas of computer science, ever higher levels of abstraction are being created, while concepts and approaches are becoming more generic and can therefore be used in an ever-widening range of applications. Starting with remote procedure calls (RPC) (Srinivasan 1995; Reynolds and Ginoza 2004), the Common Object Request Broker Architecture (CORBA®) (Henning 2008, OMG 2011), web services and concepts like service-oriented architectures (SOA) (Natis 2003) or ones specific to grid computing such as the open grid services architecture (OGSA) (Foster et al. 2005), there are currently a number of different concepts at various levels of abstraction for creating a distributed software architecture. Hence the whole area of "simulation coupling" presents a very heterogeneous landscape of concepts and different views as to what is meant by "simulation coupling".

For coupling beyond the boundaries of an individual software program, low-level mechanisms such as RPC, CORBA® or MPI (The MPI Forum 1993) can be found for distribution. These concepts allow for coupling that delivers high performance and granularity with respect to data exchange between simulation components. However, at a low level of abstraction in terms of software technology, implementation of standardised communication protocols must be taken into account. So, every software program involved on the technical implementation level will require modification and maintenance. Hence, linking simulation tools at this level is heavily intrusive and should therefore be considered an inflexible form of coupling suitable only for connecting individual components that are readily compatible with each other. Accordingly, such simulation model "couplings" should rather be seen as distributing the modules of a single, overall software system for the purpose of performance enhancement. Using these tools, the actual simulation can be physically run on multiple computers. On the face of it, this kind of construction acts as a closed system.

Such approaches are suitable primarily for environments where the entire code base of all components comes from a single source and can be developed and maintained as an overall system. However, with respect to extensibility and adaptability to changing conditions, such as the incorporation of simulation components not previously considered, this kind of approach is inadequate because it requires regular intervention by the software manufacturer. On the contrary, user intervention is not possible.

The steering and visualisation framework CHEOPS (Schopfer et al. 2004), which is based on CORBA, presents a similar approach to coupling simulations. The CHEOPS framework uses wrappers and adapters to combine existing simulation codes. A variety of applications have also been implemented using the FlowVR toolkit (Allard and Raffin 2006), but again, coupling is only possible with modifications to the source code of each application. With such coupling approaches, compatible codes can be linked, even across various computers, to achieve a high performance level. However, integrating commercial tools into such approaches is at least difficult, but, in fact, in practice it is usually impossible.

SimVis (Kalden et al. 2007) is a simulation and visualization tool developed by the company VEGA IT and used by the European Space Agency (ESA) to implement a "concurrent engineering" methodology. It comprises a component-based simulation system in which not only the simulation components themselves, but also related tools such as mission editors, report generators or spacecraft configuration tools as well as visualization components are integrated. While the shortening of cycle times alone confirms the benefits of the integrated system, another important goal has been achieved: data consistency can be maintained across the numerous simulation steps as well as within the respective individual applications. This was achieved by introducing the XML-based Simulation Model Portability standard (SMP). This standard was specified by the SMP Configuration Board, which consists of several European space agencies and industry partners involved in the development of SimVis. A hybrid type of coupling has been implemented and this allows some direct couplings, e.g., between simulation and visualization components, but mainly it enables a consistent approach at the data level through the use of a global data model.

In addition to linking the software tools, there is also a need to connect the corresponding hardware resources. The field of grid computing is generally concerned with the issue of how distributed computing resources can be made uniformly accessible and, in the ideal scenario, used as a single virtual resource. There are some middleware architectures available in this field, which, on the one hand, provide various tools and software libraries for resource management and, on the other, offer process monitoring as well as security and authentication modules. Popular grid middleware agents include Globus (2011, http://www.globus.org), g-lite (2011, http://glite.cern.ch), UNICORE (2011, http://www.unicore.eu) and Condor (2011, http://www.cs.wisc.edu/condor) (Thain et al. 2008), although the latter should not be seen as grid middleware, but as a component that may be used to build grid middleware. Most of these middleware systems provide mechanisms for connecting other grids that work with other middleware systems. Hence a resource pool managed by Condor can, for instance, act as an individual resource within another grid. Another function of grid middleware is the dynamisation of resources. Hence it is common for data centres to provide resources that allow users to work interactively during the day, for example, and, in a further step, assign those resources to a grid overnight in order to increase its overall performance. Similarly, a company's workstation computers can be incorporated into a grid, which releases those resources for other purposes, such as CPU-intensive simulations, based on either time or system utilisation.

In order to reap the benefits of grid infrastructures, appropriate software components are required. Grid middleware has proven to be the most flexible when dealing with Java applications. From a technical point of view, this seems logical because the Java code is executed by a virtual machine (VM) and can therefore be used regardless of any specific hardware architecture. If an application is not only to be executed "somewhere in the grid", but the application itself is to be optimised for the grid, e.g., by adding monitoring interfaces, etc., there are appropriate software libraries.

With respect to the flexibility of grid middleware, however, things do not look quite so good outside the "Java world". Especially with regard to the aforementioned middleware systems, when this component was selected, native Windows applications could almost exclusively only be integrated with Condor.

Here, the nature of grid architectures becomes apparent: the focus lies on the provision of computing resources and not on the provision of applications or services. Hence an end user cannot use the "grid" directly, but relies on the appropriate methods of access and applications available that are able to communicate with the grid and integrate its resources.

In recent years, the term "cloud computing" has become well established, even in the consumer sector. At first glance, this refers to harnessing grid architectures for end users, i.e., no grid resources per se are provided, rather services and applications are made available, which run, transparently to the users, on distributed system architectures. However, the requirements of cloud computing sometimes vary greatly from those of grid computing: while grid computing is primarily about the distributed processing of CPU-intensive tasks (like a simulation chain), cloud computing focuses on users working interactively with various cloud services running transparently on distributed resources.

In addition to the linking of computing resources, it is essential that the simulations are also connected at the data level. Otherwise, benefits cannot be reaped from such a link-up. The link has been successfully created when data that is actually relevant can be exchanged between the simulations. The process of linking simulations at the data level is described in the following section.

4.2.2.2 Application and Information Integration (Data Level)

In order to map simulation chains, the individual simulation tools must be technically coupled. Furthermore, the information required in order to run the simulation tools, e.g., a finite element model, must be provided to facilitate interoperability of the simulation tools. In this respect, simulation tools use different methods for describing a model. Hence the information may be provided in various formats, for example, or the structure used for mapping the information within a format may vary. Likewise, the underlying semantics of the information, i.e., the meaning of the respective data, may also vary.

In information technology, problems that may result from this heterogeneity are classed as integration problems and in this respect, one differentiates between application integration and information integration (Leser 2007). Application integration is concerned with the development and evaluation of methods and algorithms that allow application functionalities to be integrated along processes (Conrad 2006; Gorton et al. 2003). Examples of this can be found in corporate environments, in particular, and relate to the integration of enterprise resource planning (ERP), for example, or customer relationship management (CRM) as well as production planning and control systems (PPS) along the business process chain (Herden et al. 2006). Information integration is concerned with the evaluation of methods and algorithms that can be used to merge information from different sources (Visser 2004; Halevy et al. 2006). Data warehouses are a popular example of the use of information integration methods. One could also cite metasearch engines, which gather and display information from numerous search engines, as examples of the use of information integration. What application and information integration do have in common is that both can only be successful if the heterogeneity between the pieces of information or applications can be overcome.

A variety of preliminary studies have identified different heterogeneity conflicts (Gagnon and Michael 2007; García-Solaco et al. 1995; Spaccapietra and Parent 1991). These can generally be classified as syntactic, structural or semantic heterogeneity conflicts. Syntactic heterogeneity relates to problems caused by the use of different data formats and models. Differences in the representation of floating point values are an example of this type of heterogeneity. Structural heterogeneity relates to differences in the use of the structure provided by a data format or model. Hence the XML data model, for instance, provides the ability to map information via attributes or in elements so that one and the same piece of information can be represented by different structures although the data model is the same. Semantic heterogeneity conflicts ultimately refer to differences in the meaning and representation of the data. A simple example of semantic heterogeneity is the use of synonyms to describe one and the same piece of information. So, information about the structure may be described by the term "microstructure", for example, in one simulation and by the term "structure" in another.

In the past, a variety of methods and algorithms have been developed to overcome these heterogeneity conflicts. There have been attempts to overcome syntactic heterogeneity through the introduction of standards, for example. In the field of production technology, numerous exchange format standards have been introduced, including the *Initial Graphics Exchange Specification (IGES)*, which was developed in the U.S., the French standard for data exchange and transfer "*Standard d'Exchange et de Transfer (SET)*" and the German neutral file format for the exchange of surface geometry within the automobile industry "*Verband Deutscher Automobilhersteller—Flächen-Schnittstellen (VDAFS)*" (Eversheim and Schuh 2005). Such standards are usually limited to specific disciplines and this inhibits all-embracing, cross-disciplinary integration. The Standard for the Exchange of Product Model Data (STEP), on the contrary, aims to define a standard not limited to specific disciplines (Anderl and Trippener 2000). This kind of standard is characterised by complex specifications. Implementation is therefore usually associated with enormous costs. Also, with this type of standard, modifications are slow to come into consideration and any necessary adjustments are not usually implemented promptly, if at all, because they are too specific for the respective discipline (Conrad 2006). The is due to the use of particular standards already established within the discipline as well as the creation of in-house data formats that offer sufficient accuracy at a low level of complexity and are therefore not dependent on the use of standards.

Approaches pursued to overcome structural and semantic heterogeneity include the application of the methods and algorithms of schema- and ontology-based integration (Rahm and Bernstein 2001; Shvaiko and Euzenat 2005). This research looked in particular at data schemas based on the relational or XML data model (Euzenat and Shvaiko 2007). The aim of schema- and ontology-based integration is to transfer source schema S into a target schema T, whereby data mapped in S should be merged in T. At first a list of correspondences is created, which connects the equivalent components of the schemas with one another and displays them. In the second step, based on these correspondences, transformation rules are applied in order to map the data from S in T. In doing so, correspondences are identified in light of the markings, the structure of the schema and the data mapped in S. In this respect, knowledge bases containing background knowledge about the application domain are often used with a view to improving the integration results (Gagnon and Michael 2007; Giunchiglia et al. 2006). The background knowledge is represented using ontologies, which have become established as an explicit, formal specification of a conceptualisation, i.e., concept formation (Gruber 1993; Studer et al. 1998).

There are a variety of research projects in this field, which have produced different systems for schema- and ontology-based integration. COMA++ (Engmann and Maßmann 2007; Aumueller et al. 2005), developed at Leipzig University in Germany, contains a series of algorithms for identifying schema-based correspondences and for data transformation. COMA++ provides a repository for mapping internal data, which is based on a homogeneous, proprietary format. Background knowledge can be provided in thesauruses and dictionaries in the form of ontologies. COMA++ uses this background knowledge to improve the identification of correspondences between schemas. In addition to COMA++, the FOAM Framework (Ehrig 2007), developed at the University of Karlsruhe in Germany, also contains a variety of algorithms and strategies for schema- and ontology-based integration. Both systems feature the analysis of the structure of the schema in light of the identifiers used. Semantic information not contained in the structure cannot be captured by these systems. Conversely, these systems are unable to identify different dataset semantics in a schema.

Besides syntactic, structural and semantic heterogeneity, integration system architecture is also a key research topic. For application integration, service-based architectures with message-oriented communication have been found to be suitable. The enterprise service bus (ESB) has become well established as an architectural approach to application integration (Chappell 2004; Karastoyanova et al. 2007).

The ESB is a concept that allows heterogeneous applications to communicate in a distributed, service-based computer network. Information integration is further subdivided into materialised and virtual integration. In materialised integration, the data is merged in a data storage device, or, in other words, in a "data warehouse" (Rizzi et al. 2006). Here, the extract, transform and load (ETL) process is the general approach used (Simitsis et al. 2005). There are a number of systems that employ the ETL process for materialised integration, including the Pentaho Data Integrator (PDI) (Lavigne 2006) and the Talend Open Studio (JasperSoft 2007) developed by JasperSoft. In virtual integration, however, the data is not merged until a query is made. The data remains distributed across various data storage devices and is only merged for the purpose of the query. Integration systems based on virtual integration are referred to as federated information systems (Conrad 1997).

Once the simulations are linked at the technical and data level, chained simulations can be run on the underlying grid architecture and simulation results can be produced. However, linked simulations are only of additional benefit if the simulation data generated can also be analysed, allowing new insights to be gained compared with separate analysis of the individual simulations.

4.2.2.3 Visualization (Analysis Level)

There are numerous applications for visualising simulation data. Current simulation tools often comprise in-house solutions tailored to specific simulations. However, these are only suitable for visualising the datasets generated by this particular simulation software; generally, it is not possible to analyse data from multiple tools in a common context. Generic visualization solutions, such as ParaView (The ParaView Guide 2008), AVS (2011), (www.avs.com) or OpenDX (2011), (www.opendx.org), for example, are better suited to this purpose. However, these applications are designed to extract and display visual features, such as isosurfaces, from individual datasets. Furthermore, they lack of means of interaction required to deal with all the simulation results of a simulated process chain in a common context.

With respect to embedding visualizations in distributed simulation environments, some studies have already been undertaken, but these were more concerned with the technical connection of visualization tools in the broadest sense, similar to a video transmission (DeFanti et al. 1996; Karonis et al. 2003). Solutions are available for integrating distributed, real-time simulations and visualizations, as in FlowVR (Allard and Raffin 2006), and for coupling heterogeneous visualization systems (AVS/Express and ViSTA) (Düssel et al. 2007). However, the known approaches involve the possibility of intervention in the tools concerned, or only the coupling of very generic aspects, e.g., the observer and object positions within the visualization, which can be exchanged between various programs with relative ease. However, even for this very simple type of coupling, appropriate interfaces to the applications are required. Such approaches can only be used to a very limited extent for a generic concept of simulation coupling and visualization of data in one context.

COVISE (2011), (www.hlrs.de/organization/av/vis/covise) is a modular, object-oriented software package for visualising scientific data. There are also modules from the field of computational steering (visualization and interaction with simulations whilst they are running) as well as links to interactive post-processing. With COVISE, it is possible to set up simulation tools, steering components and post-processing/visualization components as a distributed system. Again, the focus is on providing software components in order to develop such distributed systems. It is not a solution that allows existing simulation tools to be integrated into such a system without modification.

4.2.3 Motivation and Research Question

Optimisation of production processes in terms of efficiency and quality is a general aim pursued by the industry. Simulation software is already used in many research cases. This saves on both effort and costs compared with real experiments. However, the focus is on individual aspects of processes with associated simulation models, which in most cases are barely linked or not linked at all. In practice, in such simulations, literature values or other approximated default settings are taken as the boundary parameters. This does not take into account any influences between the individual simulated aspects.

Here, there is huge potential for optimising the simulation processes in terms of their accuracy. Besides modelling appropriate simulation processes by linking various simulation models, which would have to be done by experts in each of the simulated processes (see Sect. 4.4), there is also the question of how this link-up can be achieved technically. Both in research and in industrial practice, the human and technical resources involved are usually distributed across various institutes and departments and consequently across different buildings or even different locations entirely. Another issue concerns the data that has to be exchanged between the individual simulations: besides the technical implementation of data exchange, it is particularly important to ensure semantic consistency of the data. Furthermore, the methods and existing algorithms must be identified and refined in order to make the complex data structures in the field of material processing manageable and to facilitate the integration of other simulation tools used in production technology. The discussion and development of ontology-based integration methods in recent years provide a good starting point for the development of the integration solution required. In this respect, the construction of an appropriate ontology is a key challenge. This applies in particular to the study of ontology-based methods that allow data exchange despite dynamic simulation processes. New developments in artificial intelligence may make an important contribution here.

Once the simulation tools and data have been successfully linked, both technically and semantically, access to the created network of simulation resources has to be enabled. For this purpose, a suitable user interface must be devised with facilities for collaborative development of simulations.

Ultimately, to gain new insights from the coupled overall simulation, the analysis of all the simulation data thus generated must also be integrative. In particular, this calls for interactive access to the visualizations of all the datasets of a simulation within a common context.

4.2.4 Results

In this chapter, the results are presented according to the order in which the modules are found within the workflow of the proposed simulation platform: first, a simulation chain is defined and parameterised via a web interface with global access. This chain can then be run automatically (Sect. 4.2.4.1). Conversion of the data between the individual simulation tools and provision of the data for the subsequent visualization is performed by the integration system developed (Sect. 4.2.4.2). Once all the simulation results are available, a manual processing step is executed in preparation for integrative visualization (Sect. 4.2.2.3). 'Integrative' refers to both the visualization itself and to the analysis of the simulation results as facilitated by the visualization.

4.2.4.1 Grid and Web Interface

The usability of a distributed simulation environment largely depends on whether users can implement their simulation projects with it or not. In particular, users cannot be expected to produce configuration files for grid middleware, which may also need to take account of modifications in the system infrastructure. The solution to this problem is the development of a web interface that allows the visual design of simulation graphs (see Fig. 4.1). Using this workflow manager, configuration data for the Condor grid middleware is generated automatically from the graphical representation. First, a "submission file" is created for each individual simulation. This file is used to define which resource the simulation will run on and which parameters should be applied. The variable parameters for each simulation (usually an initial geometry plus a proprietary control file) are entered in separate windows in the workflow manager. Certain additional options can be selected for each application. These are displayed separately on the screen (see Fig. 4.2). To reduce the complexity, only the parameters required for the test cases implemented are actually shown on the screen.

The ability to integrate new simulation tools easily and with low maintenance input into the workflow manager is an important requirement of the system in terms of ensuring its practical usability. This is accomplished in that the web interface is managed on the platform server in an XML structure from which the graphical components are dynamically generated in the web interface during run time. In this structure, the types and names of the input fields are specified in a very simple form (see Fig. 4.2), which means no particular programming skills are required.

4 Virtual Production Systems

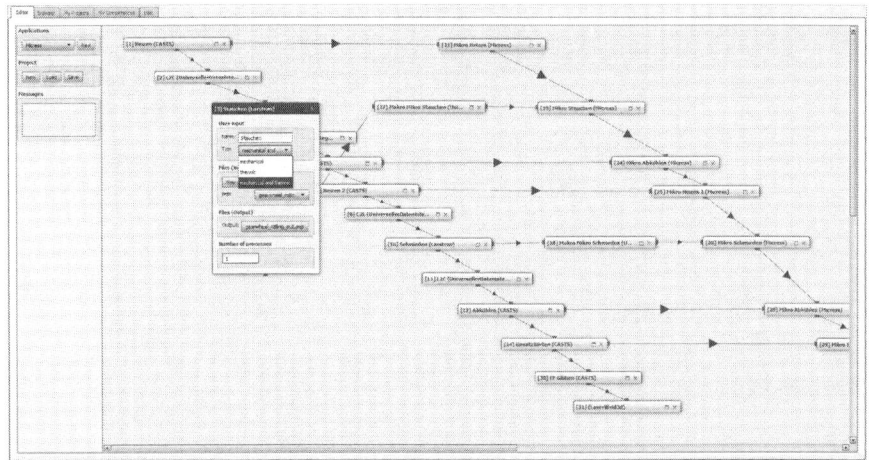

Fig. 4.1 Workflow manager for visual design of simulation graphs

Fig. 4.2 *Left:* XML structure for describing a simulation, *right:* resulting element in the workflow manager

Once the individual simulations have been selected and linked, the resulting simulation graph can be saved. As an individual user is rarely able to configure every individual simulation in a chained simulation, a collaboration component has been integrated into the system: in their profile, users can add information about their competencies with respect to individual simulation tools. This information can be used to support the selection of colleagues for particular simulations and to give them access to those simulations (see Fig. 4.3). Once a simulation graph has been

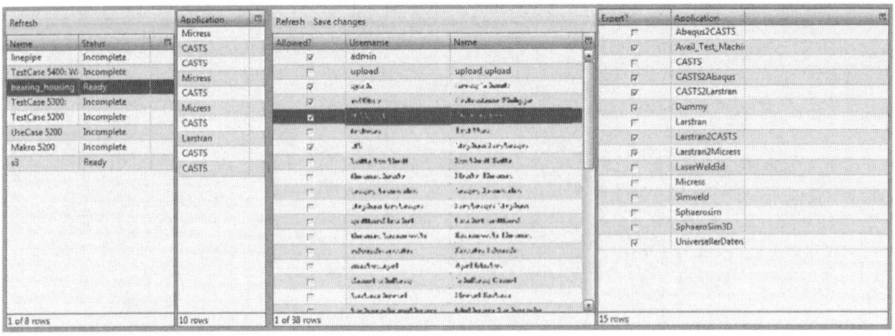

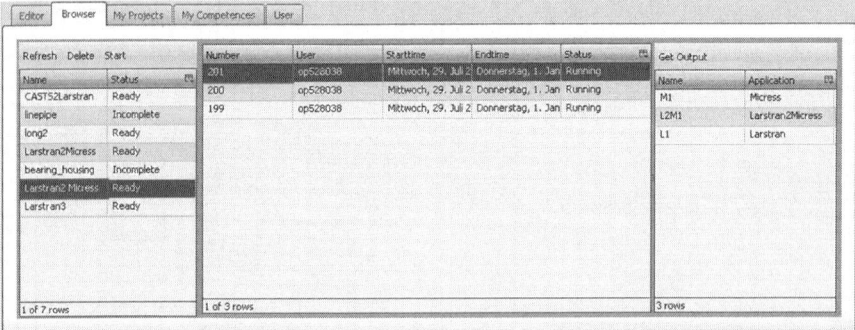

Fig. 4.3 Rights management. From *left to right*: projects, individual simulations used in the selected project, user list, competences of the user selected

Fig. 4.4 Access to simulation chains already created, calculated instances thereof and the associated results

fully configured, it can also be launched from the web interface. Subsequent to a successful run, the results can be downloaded (see Fig. 4.4).

On the technical level, the execution sequence is controlled and the specific resources are selected using Condor tools. The execution sequence for the individual simulations is controlled by the Directed Acyclic Graph Manager (DAGMAN). The DAG files are generated directly from the sequence graphically modelled in the web interface (see Fig. 4.5, left). These files contain references to the submission files that are created for each individual program (see Fig. 4.5, right)—these control when an individual simulation is called up. It is possible to specify a particular resource in its entirety or just certain criteria, such as the operating system or the available disk space. If the latter is the case, the system may use any resource located at the simulation platform that matches these criteria. The submission file is also used to control which data is needed by a resource in order to launch the simulation and which data must be transmitted back once the simulation has been finished. In fact, provided it does not breach licence conditions, a simulation program as such can also be dynamically transferred to any suitable resource for the duration of the simulation.

4 Virtual Production Systems

```
JOB 1_Magmasoft CASTING0  ./Larstran/1/Magmasoft CASTING0_submit
JOB 13_heattreatment0 ./Micress/13/heattreatment0_submit
JOB 19_Magma2vtk ./CASTS/19/Magma2vtk_submit
JOB 20_Magma2vtk ./CASTS/20/Magma2vtk_submit
JOB 9_CUTTING0 ./CASTS/9/CUTTING0_submit
JOB 15_APPLICATION0 Abaqus  ./CASTS/15/APPLICATION0  Abaqus _submit
JOB 11_HEATTREATMENT0 ./CASTS/11/HEATTREATMENT0_submit
JOB 16_application0 ./Micress/16/application0_submit
JOB 14_MACHINING0 ./CASTS/14/MACHINING0_submit
JOB 8_casting0 ./Micress/8/casting0_submit
SCRIPT PRE 13_heattreatment0 /opt/condor_jobs_new/Micress/extract.sh
SCRIPT PRE 16_application0 /opt/condor_jobs_new/Micress/extract.sh
SCRIPT PRE 8_casting0 /opt/condor_jobs_new/Micress/extract.sh
PARENT 1_Magmasoft CASTING0 CHILD 19_Magma2vtk 20_Magma2vtk
PARENT 13_heattreatment0 CHILD 16_application0
PARENT 19_Magma2vtk CHILD 9_CUTTING0
PARENT 20_Magma2vtk CHILD 9_CUTTING0
PARENT 9_CUTTING0 CHILD 11_HEATTREATMENT0
PARENT 15_APPLICATION0 Abaqus  CHILD 16_application0
PARENT 11_HEATTREATMENT0 CHILD 16_application0
PARENT 14_MACHINING0 CHILD 15_APPLICATION0 Abaqus
PARENT 8_casting0 CHILD 13_heattreatment0
```

```
Executable = /opt/condor_jobs_new/Larstran/Larstran.sh
Universe = vanilla
Log = Larstran.log
Output = Larstran.out
Error = Larstran.error
should_transfer_files = YES
when_to_transfer_output = ON_EXIT
InitialDir = bearing_housing/1290091363/Larstran/1
Requirements = (LARSTRAN_CAPABLE)&&(OpSys=="Linux")
Transfer_input_files = bearing_housing/1290091363/
Arguments = agthia
+wwwowner = tb552214
+jobname = 1
+app = Larstran
Queue
```

Fig. 4.5 Listings of a DAG description (*left*) and of a submission file (*right*)

Grid middleware is designed precisely for this type of dynamic resource management. It helps achieving maximum flexibility and availability for each simulation on the overall platform. In practice, however, most simulations are firmly bound to one hardware resource for technical licensing reasons.

The usage of the aforementioned tool from the field of grid computing allows a non-intrusive form of coupling. Hence, applications do not need to be modified in order to be used with regard to the simulation platform presented in this paper. There are just a few prerequisites the software must fulfil in order to be linked. In essence, these can be summed up in two criteria: the software must be accessible using the command line and executable on an operating system for which Condor is available. In this respect, the approach described here is very different to most of the other approaches to simulation coupling in that they tend to require modification of all the software components involved. Since adequate implementation would be difficult, if not impossible, a solution is proposed that renders intervention in the individual simulation tools redundant.

Nevertheless, the system architecture presented does allow such intervention in the simulation tools in order to, for example, decouple hardware and software resources. Like other grid middleware, Condor provides a programming interface that can be used to make in-house applications "grid aware", i.e., the fact that the application can be executed within a grid can be explicitly taken into account in the software. Hence, applications prepared like this can be dynamically (e.g., during a running simulation) transferred to other resources to ensure optimum utilisation of the resources available within the overall system. Another approach for decoupling hardware and software resources is the use of virtual machines. In this case, complete hard disk images from simulation computers can be transferred to, and launched on, any resource that can work with virtual machines. This improves controllability of the run time, even

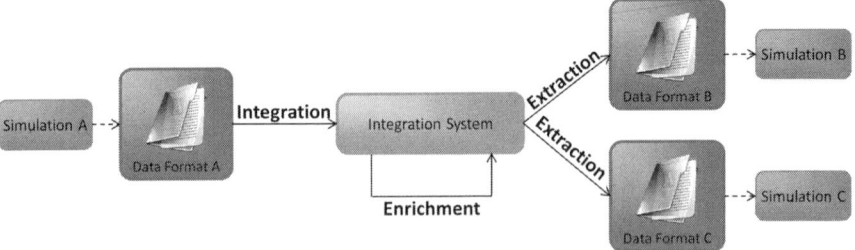

Fig. 4.6 Semantic interoperability between simulations

without intervention in the actual simulation software. However, it also results in a high proportion of data transmission, which is a huge disadvantage, especially in large networks.

Such cases have not yet been realised with the implementation presented herein. However, at this juncture, it is worth pointing out that, if required, there are still lots of technical options for dealing with unforeseen problems in the future.

4.2.4.2 Integration System

The grid solution presented in the previous chapter enables interoperability of the applications on a technical level. The core objective of the integration system is to achieve interoperability of the simulation tools on a semantic level without negating the autonomy of the individual applications. In this respect, autonomy means the degree to which the various applications can be developed and operated independently of one another (Blumauer 2006). The development of standards and an application's need for implementation of such standards, in particular, constitute intervention in the autonomy of the individual applications. Often such intervention is not possible for technical, legal or competition-related reasons. Simulation tools developed within the field of production technology are characterised by different data formats, terminologies or definitions and the use of various models. The integration system developed creates a basis for overcoming the heterogeneity between the simulation tools, as described in Sect. 3.2.2, thus enabling interoperability on a semantic level (see Fig. 4.6).

Integration, preparation and extraction are key functionalities in this respect. Through integration, data provided by a simulation tool in one data format is transferred to the central data store of the integration system. Then, using extraction, this data is converted into the data formats of other simulation tools while the semantics of the data are retained. To this end, there is a data preparation step prior to the actual extraction in which the data provided is transformed using semantic transformations to make it suitable for extraction into a specific data format. Hence, some material processing simulation tools require, for instance, specification of the outer surfaces of a component's geometry. This information, however, cannot be contained in the

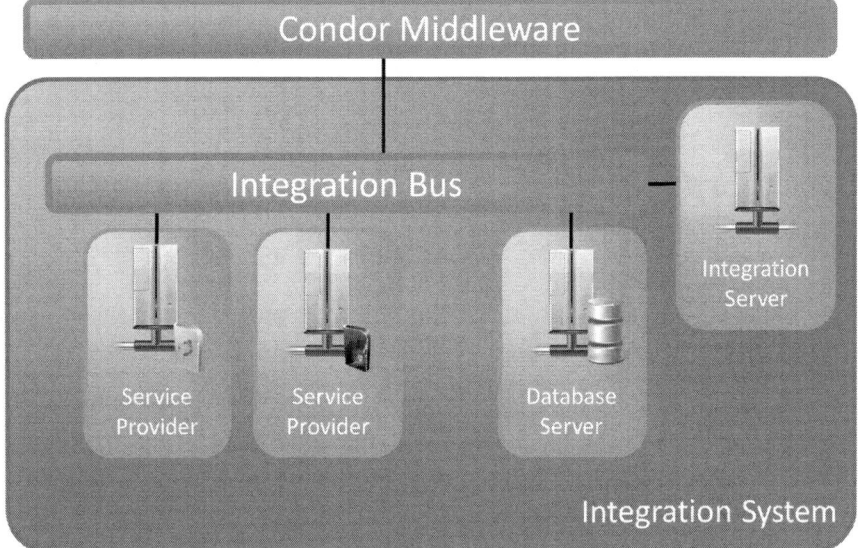

Fig. 4.7 Architecture of the integration system

captured data. Within data preparation, these surfaces are automatically identified and enhanced in the data. Extraction takes this data into account and can therefore deliver a valid geometry description.

The integration system functionalities described above enable various, heterogeneous simulation tools to be coupled to simulation processes as required for implementation of the simulation platform in the sub-project "Virtual process chains in material processing" (see Sect. 4.3).

Architecture The integration system architecture is based on the architectural concept of the enterprise service bus. This concept is service-oriented and therefore guarantees that distribution of functionality across numerous computers and flexible extension of the integration system is possible. Figure 4.7 illustrates the architecture of the integration system.

The integration server coordinates the integration, preparation and extraction processes and provides information about the status of individual conversions. Here, conversion means the sequential execution of integration, preparation and extraction. A conversion is therefore the transfer of data from one data format into another. Conversion requests are provided via the Condor middleware. Message-oriented communication between services and the integration server takes place via the integration bus. A database server is used for materialised integration of the data generated in the simulation process.

The basis for the integration system is the AIIuS framework (Adaptive Information Integration using Services) developed in the Department for Information Management in Engineering at RWTH Aachen University, Germany. The AIIuS provides a

Fig. 4.8 Data conversion process

basic framework for integration systems in which applications with complex data models and formats are integrated along a process allocated during run time. It was built in the Cluster of Excellence in order to facilitate the development and networking of other integration systems in the field of production technology. The framework is based on process-driven information and application integration. This means that each request is handled by means of a process of pre-defined work steps. In this respect, application integration involves tracking the provision of data for the simulation tool used in the next simulation process, while information integration entails the integration of data provided into the central data store of the information system.

The following section presents these processes and work steps as applied for the integration system used within the sub-project "Virtual process chains in material processing". These processes and the associated work steps form an integral part of the AIIuS framework.

Processes for Integration and Extraction As mentioned above, the Condor middleware communicates conversion requests to the integration system developed. A conversion request is characterised by the result data of a simulation tool and the location in the simulation process. The integration system must first integrate the result data into the collected simulation process data and then provide the data for the next simulation in the simulation process. In doing so, it is essential to overcome the heterogeneity of the data formats and models of the simulation tools by applying transformations. The simulation tools used in the sub-project "Virtual process chains in material processing" (see Sect. 4.3) stores their result data in files. The AIIuS framework also supports the provision of data in databases or via application programming interfaces (APIs). Conversion requests are transmitted to the integration system via a Condor adapter implemented for the integration system, which forwards the Condor middleware request to the integration system in message form. Figure 4.8 shows the basic data conversion process.

The integration stage of the process involves the transfer of data from the data model of the data source into the central data model of the integration system, the so-called "canonical data model" (Hohpe 2004). Given the volume of data and the complexity of the data structures, the canonical model used in the integration system is a relational database model. Besides the relational model, the AIIuS framework also supports other canonical data models. Hence the data can be deposited in the XML data model, for example. The integration process is based on the ETL process, whereby at first, the data source is opened to allow the extraction of data and, in a further step, its transformation so that it can be loaded into the canonical data

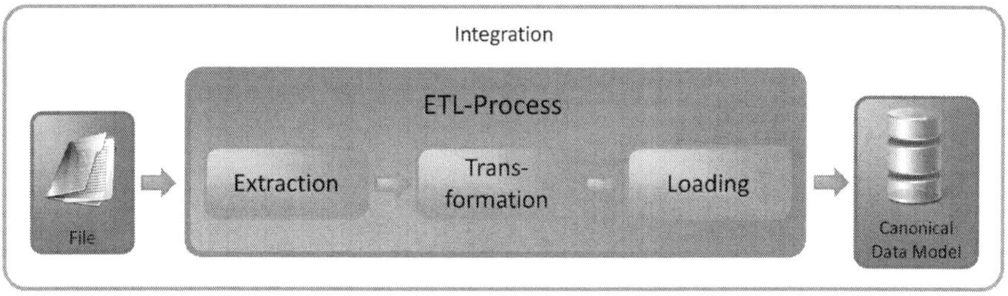

4 Virtual Production Systems

Fig. 4.9 Integration process

model. Data transformation as a process to achieve the necessary syntactic and structural adaptation does not, in this case, produce any changes in the semantics of the data. Semantic transformation is not executed until the data preparation stage. In order to distinguish between the different types of transformation, the transformations designed to overcome syntactic and structural heterogeneity will hereinafter be referred to as data transformations, while the transformations executed in the data preparation stage to overcome semantic heterogeneity will be referred to as semantic transformations. Figure 4.9 shows the integration process.

Once the data has been integrated into the canonical data model, it can be prepared for extraction. Within the AIIuS framework, this is achieved through a combination of methods from semantic information integration and artificial intelligence planning. Data preparation results in transformed data with semantics that meet the requirements of the data format to be extracted. In the sub-project "Virtual process chains in material processing", these requirements concern, for example, the element topology used in the component geometry, the indexing of nodes and elements or the units of state variables such as the temperature field or the strain tensor. Data preparation also comprises the enhancement with new information, e.g., the temperature profile in selected points of the component, or the enhancement with information about the outer surfaces as described earlier. This preparation must be able to take place even if information about the simulation process and the actual data basis cannot be made accessible to the integration system during run time. Hence processes and methods that enable data preparation under these circumstances, too, must be integrated into the integration system. Figure 4.10 illustrates the process underlying this preparation stage.

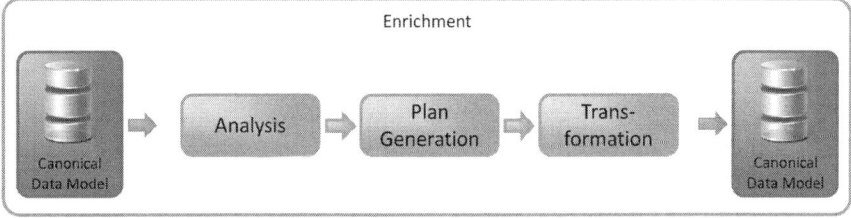

Fig. 4.10 Preparation process

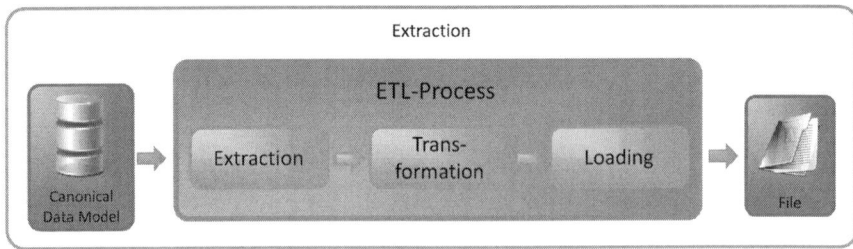

Fig. 4.11 Extraction process

Data preparation uses the structure and methods of schema integration. In this respect, the focus is on overcoming semantic heterogeneity. Generic schema integration processes, such as DELTA (Data Element Tool-based Analysis) (Clifton et al. 1997), DIKE (Database Intensional Knowledge Extractor) (Palopoli et al. 2003) or Cupid (Madhavan et al. 2001), which operate on the basis of the identifiers and the structure of the schema, may well provide an initial approach, but they are not sufficient for identifying the required semantic transformations of the data accumulated in the integration system. Likewise, instance-based processes like iMap (Dhamankar et al. 2004), Automatch (Berlin and Motro 2006) and Clio (Haas et al. 2005), which are designed to identify correspondences using the datasets included, are also inadequate. This is because the actual transformation process does not depend on the schema, but on a specific dataset, for which it is solely applicable. For example, one and the same schema may contain a hexahedron-based and a tetrahedron-based element topology for defining the component geometry. If, however, a simulation tool only supports tetrahedrons as the element topology, then, in the former case, data preparation must be performed and the hexahedrons must be converted into tetrahedrons. In the latter case, however, no preparation is required for the component geometry to be used in the simulation tool. Therefore, the precise transformation process depends on the state of the component geometry, whereby this can be described through attributes such as element topology, material and other global and spatially distributed properties. In a first step, the preparation process analyses the data concerned according to the kind of attributes mentioned above and identifies the requirements placed on the attributes by the target schema. In a second step, a plan is generated, which contains the transformation process for preparing the data. The actual semantic transformation of the data takes place in the third step. The methods and algorithms used for data enhancement will be described after this chapter.

Below, a description of the extraction process is depicted, which ensures that after it has been prepared, the data is transferred into the required data format and model. Similar to the integration process, the extraction process is also based on an ETL process, which is illustrated in Fig. 4.11.

During extraction, a first step consists in the extraction of data from the data store. Its structure and syntax is then adapted via data transformations to match

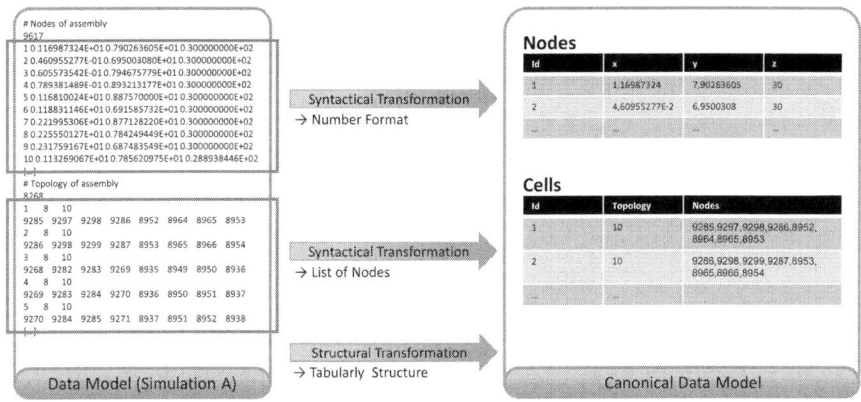

Fig. 4.12 Integration example

the required data format and model. Finally, the data is loaded into the specific physical file. Implementation of the processes requires certain functionalities, which are provided by corresponding services. These services are described briefly below.

Services The individual functionalities required in the processes are provided by services in a distributed system. Hence a process is implemented by linking various services. In this respect, there are different types of services. Integration and extraction services are used to overcome the syntactic and structural heterogeneity, but do not alter the semantics of the data. For integration services, the functionality therefore lies in converting the data model of a simulation tool into the canonical data model of the integration system. By way of example, Fig. 4.12 shows the syntactic and structural transformation involved in the integration of a component geometry defined by nodes and elements.

Transformation services provide the data transformations required for data preparation, whereas the supporting functionality, required in particular at the point of data preparation, is made available by support services. While integration, extraction and transformation services are implemented in the actual integration system, support services are already implemented in the AIIuS framework.

The message-based communication between the services and the integration server takes place via the integration bus. Due to the large data volumes, data is not attached directly to messages, but can be requested via attached connectors instead. By way of example, Fig. 4.13 shows how a conversion process works by linking services.

Service execution is monitored by the integration server, which can advise the status of individual conversion requests through this step. The services communicate with each other via the integration bus using a message protocol defined for the integration system. The specific integration and extraction service in each case depends on the data source and the data destination, while the specific transformation services are the result of the planning service. The services are provided by various service

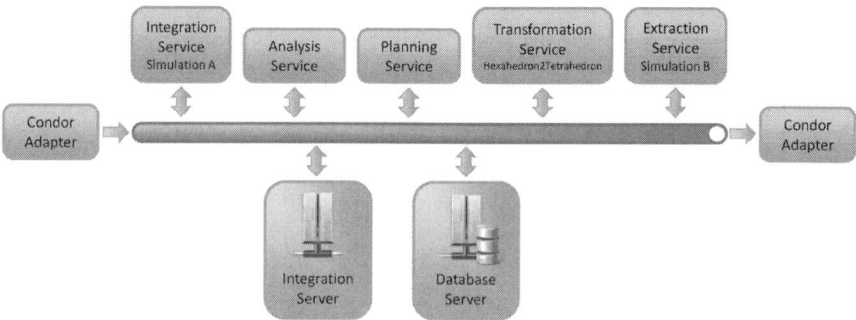

Fig. 4.13 Example of concatenation of services

providers (see Fig. 4.7), whereby a single service may be offered by multiple service providers. This increases the number of conversion requests that can be executed in parallel.

Structures, Methods and Algorithms Selected structures, methods and algorithms used in the AIIuS framework and the integration system based on that framework are described below. The methods and algorithms are based on the processes presented in this paper and implement the functionality required in the services.

The integration and extraction services support the use of established tools for executing real-time ETL and ETL processes. The AIIuS framework already contains implementations for the connection of the Pentaho Data Integrator (PDI), an open-source ETL tool in which most of the integrators and extractors for the aforementioned integration systems were developed. In-house implementations, which require the use of third-party software, for example, are also supported. As described earlier, the aim of the integration and extraction services is to transfer data to or extract data from the canonical data model. The canonical data model of this integration system is a relational data model. This model incorporates a data schema suitable for capturing the data generated during the simulation of the material processing procedure. Figure 4.14 shows an extract from the data schema for capturing FE models.

This representation is not complete and is used here merely to facilitate the following explanations. On this basis, a geometry is composed of nodes, elements and position-dependent properties. Depending on its topology, an element is composed of any number of nodes, whereby one node can be associated with numerous elements. Furthermore, a position-dependent property is assigned to either nodes or elements.

The data depicted in this data schema is analysed through analysis services. To this end, the algorithms developed use the information modelled in ontologies. In general, ontologies are used to model the specialist knowledge of an application domain. The ontology provided in the integration system contains information about concepts, relations and axioms from the domain of material processing procedure simulation. The ontology is modelled using the Web Ontology Language (OWL)

Fig. 4.14 Simplified data schema for FE model data

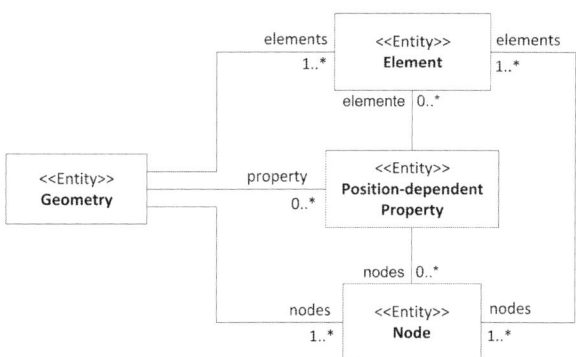

Fig. 4.15 Modelling the concepts geometry and nodes

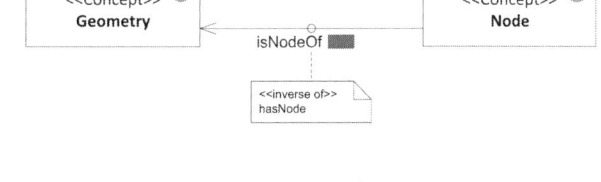

(Stuckenschmidt 2009). In this respect, analysis requires information about the existing data structures. The relationship shown in Fig. 4.14 between a geometry and its nodes is implemented in the domain ontology, which is illustrated in Fig. 4.15. The colour coding corresponds with the colour scheme used in the ontology editor Protégè (Gennari et al. 2002).

In addition to the modelling of data structures of the application domain, data analysis also requires attributes to be defined. An attribute is a property fulfilled by one or more data structures. In the domain ontology, an attribute is formally defined by subsumption, i.e., the conditions that implicitly describe an attribute are defined.

For example, the indexing of nodes or elements is an important attribute of the geometry models used in the simulations studied. Indexing in this respect can be closed or open. With closed indexing, there are no gaps in the numbering of nodes or elements, whereas such gaps can occur with open indexing. Likewise, the start index varies, too. Hence the nature of the indexing in question as well as the respective start index used can be described as an attribute of a geometry. To determine such an attribute, the AIIuS framework allows rules to be defined within an ontology using the Semantic Web Rule Language (SWRL) (Horrocks et al. 2004). One rule used to check whether the indexing of a geometry is closed or not is as follows: num is the number of nodes within the geometry, max_{Id} is the highest and min_{Id} the lowest index used within the geometry. This rule states that a geometry has a closed node index if the number of nodes num is equal to the difference between the maximum index max_{Id} and the minimum index min_{Id} plus one.

$$max_{NodeID}(geo, max_{ID}) \wedge min_{NodeID}(geo, min_{ID}) \wedge num_{NodeID}(geo, num)$$
$$\wedge equal(num, max_{ID} - min_{ID} + 1) \Leftrightarrow ClosedNodeIndex(geo)$$

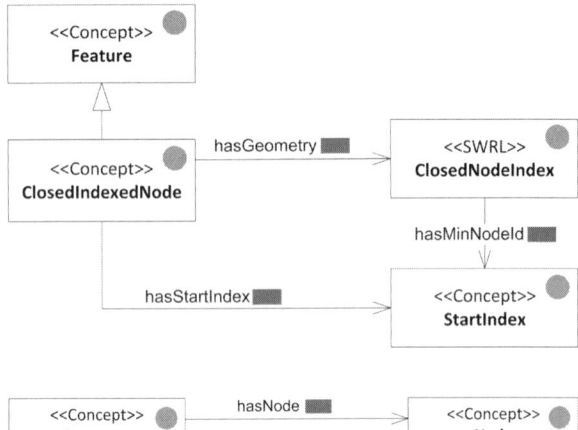

Fig. 4.16 *ClosedIndexedNodes* attribute

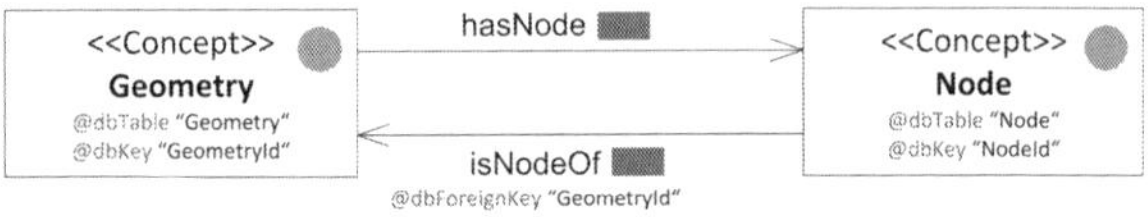

Fig. 4.17 Annotation-based mapping on a data schema

The minimum and maximum node index and the number of nodes are modelled separately within the domain ontology as properties of a geometry. Based on this rule, the required attribute is described via the definition contained in Fig. 4.16.

The ClosedNodeIndex attribute is hereby defined as a subsumption of a geometry, which is fulfilled by the SWRL ClosedNodeIndex rule. The explicit designation of the start index for the attribute adds this to the information provided by the attribute.

Using this approach, it is possible to model attributes that provide answers to questions such as the following: "Which element topologies are used?", "How are the nodes and elements indexed?" or "Are outer surfaces explicitly marked?" To evaluate the attributes, the analysis services implemented in the AIIuS framework use reasoners, which allow conclusions to be reached on the basis of the information modelled in the ontology. However, due to the huge volumes of data, the run time of reasoners such as Pellet (Sirin et al. 2007) or FaCT++ (Tsarkov and Horrocks 2006) is unacceptable for the analysis services. Hence, similar to the KAON2 reasoner (Motik and Sattler 2006), the analysis services developed in the AIIuS framework support the definition of mappings between the data structures modelled in the ontology and the relational data model. Such a virtualised ontology allows the use of index structures and database query optimisation in order that conclusions can be reached quickly, especially where large volumes of data are concerned. Unlike the implementation in KAON2, however, the mapping between the relational data model and the ontology is not defined in a separate configuration file, but via an annotation in the ontology, similar to annotation-based programming. Hence the relationship between geometry and nodes (see Fig. 4.15) is mapped by entering annotations, such as the following, on the data schema as illustrated in Fig. 4.17.

At present, the analysis services implemented in the AIIuS framework support five annotations, which can be used to describe a mapping. In addition, other annotations are available to allow the useage of the aggregation functions of a relational database, such as those employed in the example mentioned above.

The algorithm for the analysis of a given database works as follows:

```
01: For each attribute m in the domain ontology o
02:     Generate request a using the annotations in m
03:     Submit request a to database and obtain result r
04:     Check r for fulfilment
05:     If r fulfils the attribute m
04:         Determine assignment b of m from r
05:         Add m with assignment b into the attribute list l
06: Return attribute list l
```

Listing 4.1 Algorithm for analysing attributes

The result of the algorithm is a list of n-tuples for each fulfilled attribute, whereby the data structure for which the attribute is valid is always in the first place. For example, the tuple

$$NodeProperty(geo1, temperature, °C, double),$$

describes that the geometry geo1 has a spatially distributed property with the name temperature, which is specified in Celsius, and is presented as a double-precision floating point value. Assignment of the attribute is composed of the geometry, temperature, unit and data type.

In addition, the analysis service determines a list of attributes to be fulfilled. To this end, the data format and data model requirements are also modelled in the domain ontology. Figure 4.18 shows how a requirement on the indexing of the nodes and elements of a data format is modelled in the domain ontology.

The data format given in this example is the VTK data format used for visualization (Schroeder and Kitware 2006). This format requires that geometry indexing with nodes and elements begins with zero.

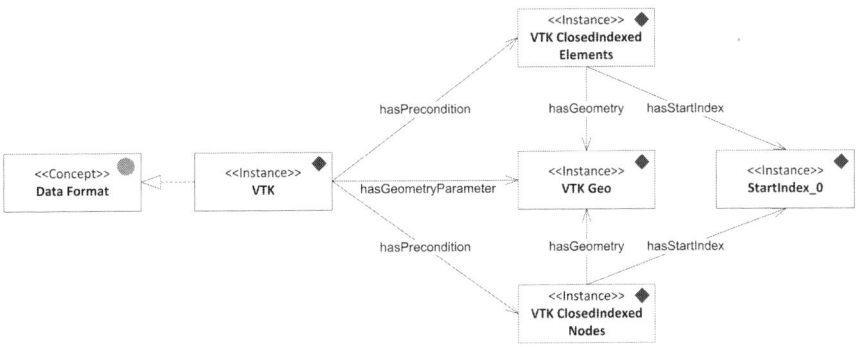

Fig. 4.18 Definition of a prerequisite in the VTK data format

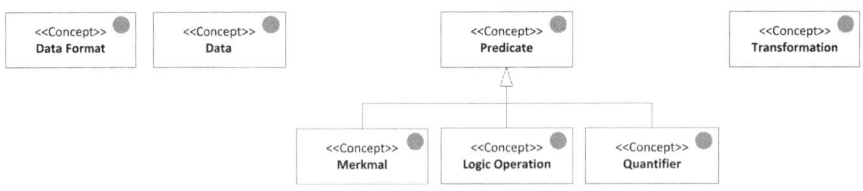

Fig. 4.19 Concepts of the framework ontology (extract)

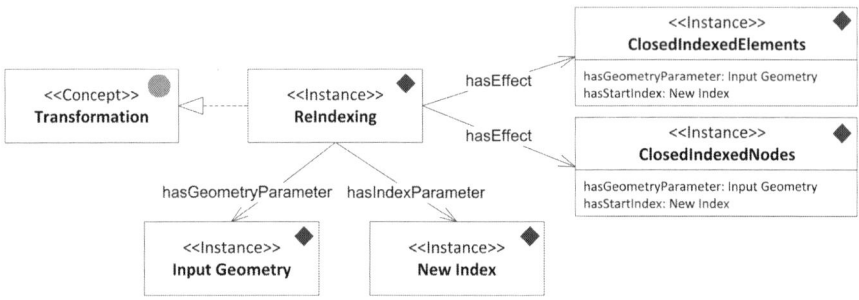

Fig. 4.20 Re-indexing transformation

To provide the functionality described above, the analysis services do not have to be adapted to a specific integration system; they are contained directly in the AIIuS framework. The analysis services are adapted to the respective application domain through provision of the domain ontology. For this to be possible, the domain ontology must import a framework ontology. The framework ontology defines the basic concepts required by an analysis service. The analysis services are thereby able to infer the specific data structures and attributes from the domain ontology. The most important concepts of the framework ontology are listed in Fig. 4.19.

The concepts depicted in the framework ontology, which are needed for the modelling of the data structure, must be subsumable concepts of data. Attributes are modelled as subsumable concepts of attributes. Prerequisites of data formats can be described by using logical links such as And, Or and Not. The concept of transformation is not of any significance for the analysis services; it does gain relevance when not being connected with the planning services.

Planning services are used to determine a transformation process that allows semantic transformation of the data provided. The results of the analysis services, i.e., the actual state and the target state, are used for this purpose. One planning service produces a plan of transformation services that must be executed in order to generate the target state. The planning service also uses the domain ontology in this regard to specify a list of available transformations. Here, in addition to prerequisites, a transformation is also characterised by the effects that are generated through the execution of the transformation. Figure 4.20 shows the modelling of a transformation for re-indexing a geometry.

Fig. 4.21 Extraction of process data into a uniform data format

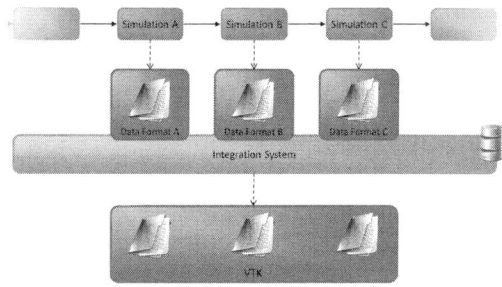

In this diagram, for the sake of clarity, the relationships and their assignments are modelled in the instances as attributes. Re-indexing offers two parameters: the geometry, with changed state, and the start index to be used for re-indexing. The result of re-indexing is that the indexing of the nodes and elements of the input geometry uses the new start index and is therefore closed.

That way, any state change to the data structures of the application domain can be described through transformations. Planning systems from the field of artificial intelligence are used to generate the transformations. The AIIuS framework can connect to any planning systems provided they support the Planning Domain Definition Language (PDDL). PDDL is a standardised language for defining classical planning problems. In this respect, the current and the target state are described as well as actions that can change the state (Fox and Long 2003).

A translator has been implemented in the AIIuS framework to allow the planning problem in question to be expressed in PDDL. Using this translator, the data analysis results and the list of available transformations are transferred into the PDDL. Likewise, the translator interprets the results of the planning algorithms used in the AIIuS framework and creates the transformation plan on this basis. Finally, this plan is executed using the transformation services.

The integration system is therefore able to react to various situations and to transfer the data into the required data format. In particular, this means that all the data generated in the simulation can be provided for visualization in the VTK data format (see Fig. 4.21).

As shown in Fig. 4.21, the integration system is designed in such a way that once process data has been integrated, the system can extract it into any desired data format. Hence all the data delivered in a simulation process can be extracted into a uniform data format. The next section of this paper describes the integrative analysis that can be implemented as a result of the step aforementioned.

4.2.4.3 Integrative Visualization

For the simulation system described, visualization must take into account various aspects, which conventional visualization tools are not addressing so far. Problems

Fig. 4.22 Interactive analysis of simulation data in a fully immersive virtual reality system (CAVE)

such as time synchronisation, for example, do not occur until the various visualizations have to be coordinated. Accordingly, such problems are seldom resolved using current approaches (Fig. 4.22).

The approach presented herein addresses the following problem areas, which are discussed in detail below:

- Uniform pre-processing of datasets
- Decimation methods
- Interactive, simultaneous visualization of multiple datasets
- Metamodel for dynamic management of visualization data during run time
- Methods for interactive modification of the chronological arrangement of datasets
- Scalability of the visualization with respect to the terminals that can be used

Uniform Pre-Processing of Datasets Uniform pre-processing of datasets is facilitated by the transformation methods of data integration into the VTK data format as described in the previous chapter. In the context of the platform, the application concept is designed to allow, on the one hand, individual visualizations to be prepared in a manual post-processing step using the open source tool ParaView and, on the other hand, the geometric data thereby created to be visualised in a common visualization solution. In this step, the volume meshes, which are part of the use cases, can be reduced to one surface mesh that is more suitable for graphical representation. This enables standard visualization methods of the software (in the simplest case, cut surfaces or isosurfaces (see Fig. 4.23) to be included in the data, thereby enabling potentially important information to be included in the common visualization.

Decimation Methods In some of the use cases studied (see 3D microstructure simulations, Sect. 4.3), the surface meshes created in the pre-processing step of the individual simulations are still so heavy that, using current hardware, they cannot be visualised with the refresh rates required for interactive visualization (20 Hz). To achieve these refresh rates, the surface meshes of these datasets still need to be decimated further.

4 Virtual Production Systems

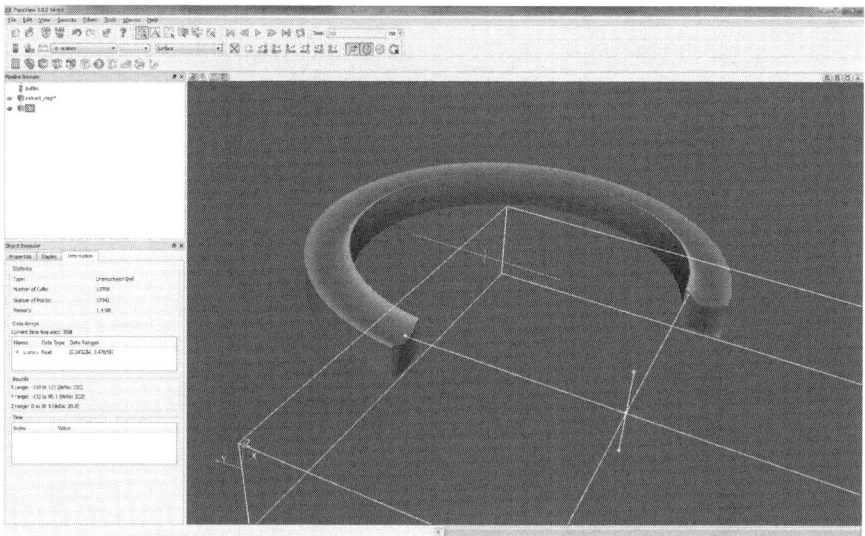

Fig. 4.23 Insertion of cut bodies into a dataset using ParaView

Decimation algorithms themselves represent a major area of research. There are many different approaches to reducing triangle meshes or other types of mesh topologies and datasets (Heok and Daman 2004). The scenario in hand focuses on the reduction of triangle meshes. The quality of a decimation algorithm depends to a large extent on the type of datasets to be processed as well as the intended use of the decimation data. Thus, in order to evaluate a decimation algorithm, the application domains must be factored in. Accordingly, no algorithms can deliver optimal decimation results for all use cases.

With commonly used algorithms, the default mesh is iteratively modified in order to obtain a representation of the dataset that is reduced, yet as accurate as possible in terms of the given metrics. This process continues until a stipulated criterion is met, e.g., the number of triangles remaining (Schroeder et al. 1992).

Another approach is to create a new mesh that can, for example, be generated on the surface of the original mesh. In this case, the hard edges of the starting mesh are usually retained, while remeshing is performed in the regions of low curvature. With respect to the use case in question, algorithms from both categories were assessed for suitability for this particular application.

A key aspect, particularly attributable to the algorithms from the first category mentioned above, is determining and parameterising the metrics for the cost function. In the simplest case, the geometric metrics of the surface mesh itself can be used and these can be inherently derived from the mesh. In this respect, it is mainly the local curvature of the surface that is employed. This approach is problematic when using meshes containing nodal points connected to data values—which are precisely the type of mesh encountered in the field of simulations. Only in a few cases does the geometric surface curvature correlate with the attributes of the data values. To correctly determine the amount of distortion (or "costs") of a decimation step, therefore, all data values for the points concerned must be included in the cost function with an appropriate metric, whereby the decimation problem becomes a high-dimensional problem.

In this area, both research and algorithms are existing that try to take account of the data values through metrics. One well-known researcher in this field is Hoppe (1999). Hoppe's algorithm, a widely used implementation, has been incorporated in the VTK library (Schroeder and Kitware 2006), but with limited success: for one thing, the implementation appears to be unstable if, in addition to geometric attributes, data fields must also be taken into account in the cost function and, for another thing, the necessary weighting of the individual data fields represents a problem for the cost function in that initially the weighting can only be executed for scalar quantities. To take account of the higher-dimensional data such as vectors and tensors, which are typical of simulation information, scalar metrics would need to be defined for this data in advance. The difficulty here is twofold: the implementation of the decimation algorithm in question does not support this process, hence a further pre-processing step would be required to calculate the scalar metrics; and, even after such an extension, there would still be the problem that these metrics, in particular the weightings for the cost function, would highly depend on their individual semantics within the dataset.

In the context of the simulation platform presented herein, the focus lies on developing the most generic methods possible for integrating various datasets into an overall visualization. As of yet, it is not possible for special metrics and weightings that differ for each type of dataset to be taken into account in the overall architecture.

Instead of gradually coarsening a starting mesh, there is another category of algorithms that creates a new mesh from the start in order to approximate the original mesh. One example of this is the Valette approach based on Voronoi cells (Valette and Chassery 2004). In this approach, uniformly distributed surface meshes are generated. Above all, this means that the dataset points, which would be removed by decimation algorithms based on geometric error metrics, can be retained (see Fig. 4.24). Hence decimated surface meshes with a high information content can be created, even without consideration of special metrics.

As part of remeshing, the data values need to be approximated to the new points of the surface mesh. To this end, the data of the points located in the vicinity of a point and its adjacent points in the new mesh is interpolated from the original mesh. Interpolation involves the weighting of the source data according to its distance from the new point (see Fig. 4.25).

To formally describe the quality of the decimated mesh, the individual semantics of each data field should be used with regard to a specific use case, which, in fact, is difficult. However, from the results, as shown in Fig. 4.24, it is intuitively (and therefore without explicit formal description) obvious that the quality is higher—at least for datasets with important attributes in the data values in geometrically flat areas.

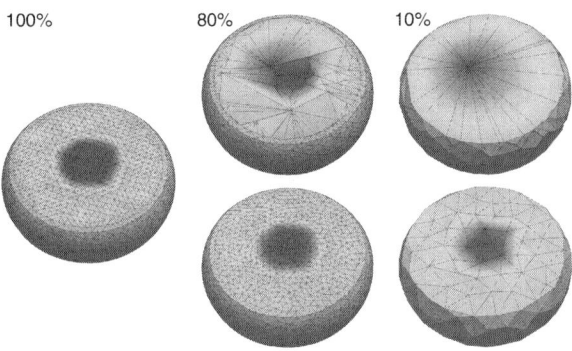

Fig. 4.24 Results of a traditional decimation operation (Schroeder et al. 1992) (*top row*) and remeshing based on Voronoi cells according to Valette and Chassery (2004) (*bottom row*). Original mesh (*links*): 2500 points, decimated to 80% (*centre*) and 10% (*right*)

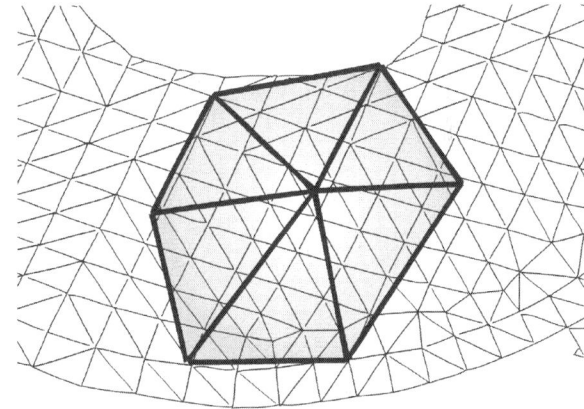

Fig. 4.25 Interpolation of the data values of all points of the source mesh located in the area of the adjoining triangles of the approximating mesh, weighted according to their distance

Another special case regarding decimation occurs in the processing of surface meshes that have been extracted from a regular, three-dimensional grid, e.g., using the well-known Marching Cubes algorithm (Lorensen and Cline 1987). Due to the rectangular block structures generated by this step (see Fig. 4.26 above), an algorithm that is purely based on the geometric composition of the surface would reach its limits. Since all blocks have the same properties locally (i.e., a 90° curvature to the adjacent areas) and, due to their strong curvature, are recognised as important outer edges and are therefore not removed, a strong decimation can only be achieved if this special treatment of outer edges of the mesh is eliminated. This, however, can create holes in the mesh (see Fig. 4.26 below). On the contrary, remeshing can deliver significantly better results because it is not constrained to the vertices of the original block structure (see Fig. 4.26, centre).

The results attainable with remeshing and subsequent data interpolation, which differ significantly from the results of the standard VTK decimator, cannot be achieved without compromise. Run time and memory consumption, for example, are increased between two and five fold depending on the composition of the data. However, depending on the nature of the data, this is acceptable if no usable results with high decimation rates could be achieved with the other methods tested. The

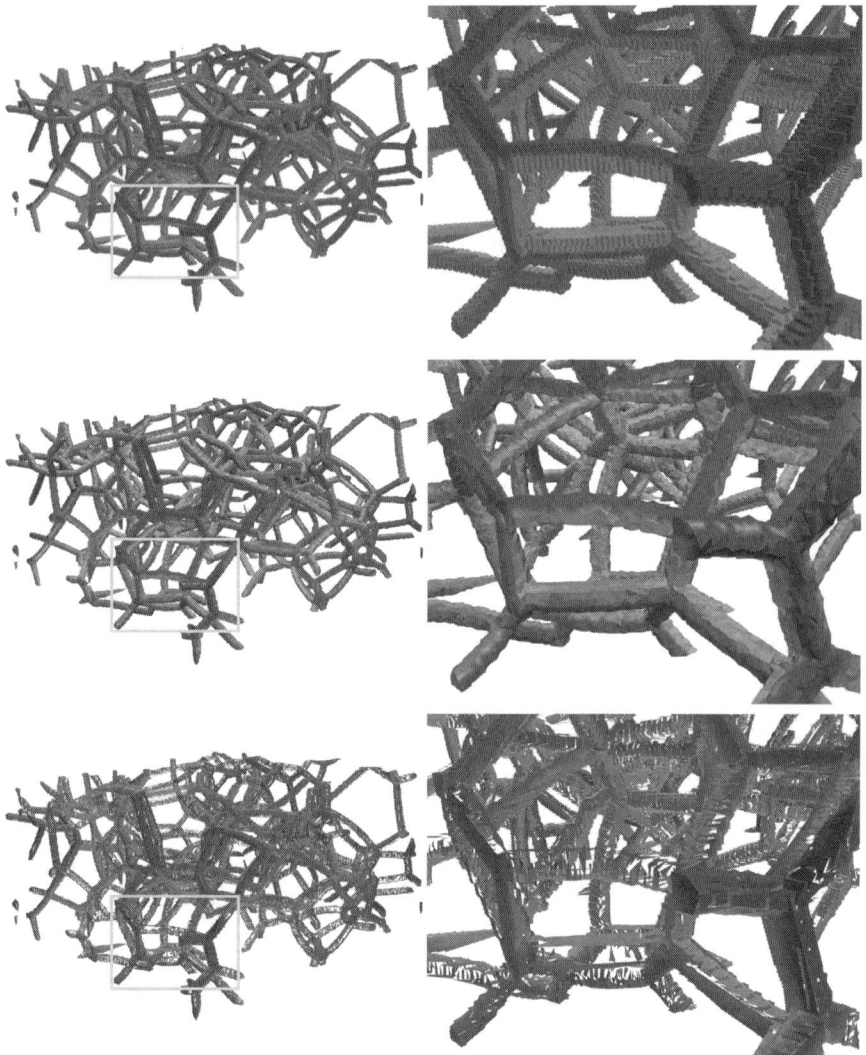

Fig. 4.26 Microstructure grid, 626,000 points (*above*), decimated to 125,000 points (20%) according to Valette and Chassery (2004) (*centre*) and (Schroeder et al. 1992) (*below*)

attempts to integrate data into the VTK implementation of Hoppe's decimation algorithm (vtkQuadricDecimation), for example, failed completely, which confirmed the experiences described in (Knapp 2002).

In the approach presented herein, the underlying solution can decimate with either one of the two algorithms. The choice of algorithm can either be made manually or integrated in the description files for the visualization (see next section). Hence the entire decimation process for all visualization results of a simulated process chain can be automated.

Interactive, Simultaneous Visualization of Multiple Datasets Special software has been developed to visualise the aforementioned geometry data extracted from numerous simulation steps in a common interactive context. This permits the implementation and investigation of methods for interactive manipulation of multiple simulation results in a common visualization context, which is overlooked with traditional visualization tools. In the project in question, the actual visualization algorithms used to extract visual attributes from the raw data are not the main focus. This project focuses more on the manipulation of multiple datasets, both on a technical level in the application context and with respect to the interaction techniques that allow the user to navigate through the datasets and thus analyse them for the first time.

One of the technical challenges is the visualization of lots of large datasets in a common context. As demonstrated earlier, many datasets have to be reduced in order to guarantee a smooth visualization. For data analysis, however, it is preferred that full resolution can always be provided if needed. Hence the study also investigates methods that allow dynamic switching between various resolution levels—or levels of detail (LOD)—of the datasets visualised. Furthermore, other methods that enable the LOD to be selected automatically according to various criteria were also investigated. Depending on the underlying target system, user interaction features, such as the user's line of sight in an immersive VR system, can also be included.

Traditionally, automated LOD selection methods for 3D geometries are used in the visualization of expansive virtual worlds like those found in computer games (Heok and Daman 2004). The selection strategies employed in this context are usually based on a distance metric according to which the LOD is selected. Objects that are far away, and therefore appear small, are visualised in a lower resolution, while objects that are close are displayed with a higher level of detail. These objects are usually relatively static with regard to the level of geometric complexity and the costs associated with visualization. In this kind of system, when and where objects are going to appear is known in advance. Hence, various graphical effects and modelling tricks are used to optimise the geometric data in respect of visualization.

When considering such selection strategies in the context of visualising multiple dynamic simulation datasets, several problems emerge, which the traditional selection strategies do not address: simulation data, for example, is time-variant, so the complexity varies greatly with each time step. Furthermore, during interactive analysis, the chronology of datasets, and therefore the combination in which the various time steps of each dataset have to be represented simultaneously, is constantly changing. By comparison, in a classical virtual world, the objects move continuously, like in the real world, and their geometric complexity changes rapidly. Accordingly, the selection consistency for levels of detail between consecutive instants is always high. For scenarios in which simulation data is visualised, such consistency and predictability only exists as long as the user does not have any interaction with regard to the chronological arrangement of the data.

To date, there are no suitable selection strategies for scenarios such as the ones mentioned above, although in the future, they will be researched using, amongst other things, the visualization solution presented in this paper.

Metamodel for Dynamic Management of Visualization Data During Run Time When dealing with multiple datasets in one context, additional issues arise, which require metainformation about the respective datasets. With a parallel visualization of multiple datasets, spatial and temporal position information must be available in order to provide a useful starting point for interactive analysis. Since this kind of data is usually negligible for an individual partial simulation, it is rarely present in the calculation results. The same applies for the units of measurement. Here, differences are of particular significance that occur when software tools from different manufacturers and/or problem domains are used, which generally apply different systems of units, but from which data must be extracted and put into a common context. In the case in question, the integration component ensures the semantic consensus of the datasets; in practice, however, time and again datasets crop up, which do not originate directly from the simulation platform, but which are to be integrated into a visualization.

For a visualization, in terms of the information required for a correct arrangement of the datasets with respect to one another, whether it be of a temporal or spatial nature, the information for structuring the data as such is more important to start with. Simulation results are usually stored in the form of individual time steps. In this respect, it is essential to know the mapping of time on the time step discretisation in order to enable a chronologically correct visualization as well as the synchronisation of multiple datasets with one another.

Furthermore, the physical storage structure of individual time steps and the distribution of various data fields or sub-objects of a dataset on multiple files are not regulated in VTK. Hence the correct loading of a dataset also requires metainformation about the storage structure. While this requirement can be reasonably easily fulfilled when dealing with individual datasets as the datasets for a specific visualization case are usually stored in the same structure, this step is problematic when dealing with various types of datasets. In an integrative visualization solution, it must be possible for datasets to be loaded without a user having to manually assign individual times to a logical dataset.

Hence, in the case concerned, a metastructure was introduced, which contains the persistent description of a visualization scene in an XML file and which, more importantly, has a run time data structure that allows different modules of the visualization solution (e.g., LOD generation, LOD selection, various interaction modules or the visualization component itself) to have uniform and abstract access to a common context of datasets (see Fig. 4.27). Here, the object-oriented programming paradigm as proposed by Fowler (1998) is used amongst other things. This helps achieving a low degree of dependency between the various modules of the software. Nevertheless, all modules can operate on a common structure. Hence the metainformation model is highly flexible and offers dynamic extensibility both in the form of a run time

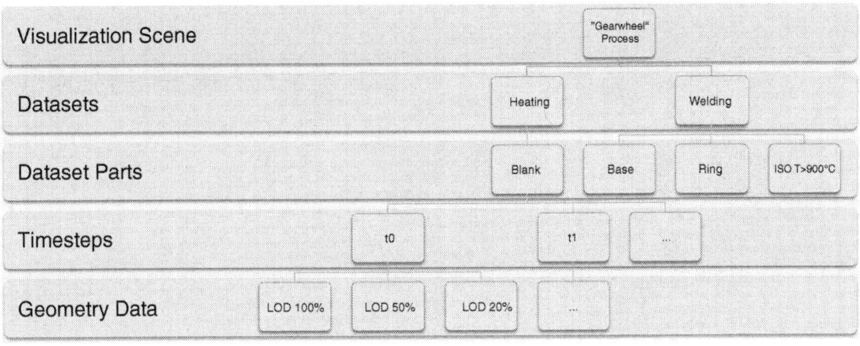

Fig. 4.27 Generic metastructure for describing discrete visualization data

data structure and a persistent scene/dataset description. On this basis, the metainformation model remains adaptable, even in view of rapidly changing requirements, without having to modify the basic structures of the software.

Methods for Interactive Modification of the Chronological Arrangement of Datasets In this project, using Wolter's time model (Wolter and Assenmacher 2008), a generic methodology was created to allow discrete and even heterogeneously distributed time steps to be mapped on a continuous time basis. This makes it possible to control the individual datasets on a continuous time basis, whereby an external view of the datasets is maintained, which is not dependent on considering the actual internal discretisation of time. To achieve chronological arrangement of all datasets, a continuous model for mapping and transforming time information is also included on this external level. In interactive analysis, this continuous model is dynamically modified using graphical control elements so that during data analysis, the user can interactively modify its chronology. Thus, any time steps and datasets can be displayed synchronously in time so that, for example, similar process steps of an overall process can be compared with one another. In addition to control elements being directly embedded in the visualization context, there is also the facility to interactively display and modify the chronology of datasets together with metainformation on a separate surface (see Fig. 4.28).

Scalability of the Visualization with Respect to the Terminals that can be Used A visualization solution for interactive analysis of the results of collaboratively developed simulation chains should be adaptable to various types of output devices and presentation situations. Individual project team members should be able to use integrative visualization at their normal workstations or on small VR systems with a stereoscopic display (see Fig. 4.29, left). Cooperative analysis or presentations in fully immersive high-end virtual reality systems (see Fig. 4.22 and 4.23, right) should also be possible. To this end, specific features in the overall architecture of

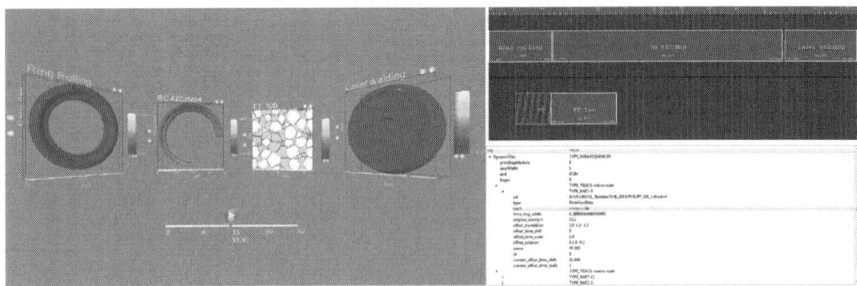

Fig. 4.28 *Left:* Visualization with embedded control elements for local timing (*beneath the individual datasets*) and global timing (*bottom edge of image*); *right:* Interactive 2D view of the chronology and metainformation of datasets

Fig. 4.29 Fully immersive high-end virtually reality system (CAVE) (*left*) Mid-range virtual reality system with head tracking and stereoscopic representation (*right*)

the visualization software must be taken into account since the bigger virtual reality systems run in a cluster on distributed systems. This significantly increases the complexity of the system compared with a conventional desktop application.

To achieve such flexibility with respect to input and output devices, the approach described herein uses the "Virtual Reality for Scientific and Technical Applications" (ViSTA) toolkit (Assenmacher and Kuhlen 2008), which, amongst other things, offers abstraction levels for this type of equipment.

With the research presented, a platform was created that enables individual simulation tools to be linked to a dynamic overall system. In addition to the purely technical link-up, the data levels were also encompassed to enable the semantic transformations of data required between the individual tools. To analyse the data generated, a visualization solution was introduced, which supports the analysts in gaining insights from the linked simulations. Based on the systems developed and the experience gained in this highly interdisciplinary environment, other new research topics will gain in importance in the future. These topics are described below.

4.2.5 Future Research Topics

In addition to coupling heterogeneous simulation tools, the integration system created within the framework of the Cluster of Excellence also enables the data generated in the simulation process to be integrated into a canonical data model. This allows systematic and system-supported evaluation and visualization of the data, taking into account the underlying simulation process. Furthermore, systematic monitoring of the interim results can also be implemented.

The next section will address the challenges and the issues requiring further research that have arisen from the creation of such an analysis component. First, the concept of "virtual production intelligence" will be derived. Then, the aforementioned methods of data analysis and data monitoring as well as the required infrastructure will be presented.

The concept "virtual production intelligence" was selected on the basis of the idea of "business intelligence", which became popular in the early to mid-1990s. "Business intelligence" refers to procedures and processes for systematic analysis (collection, evaluation and representation) of a company's data in electronic form. The goal is to use the insights gained to make better operative or strategic decisions with respect to business objectives. In this context, "intelligence" refers not to intelligence in the sense of a cognitive dimension, but describes the insights that are gained by collecting and processing information. With reference to the research topic of "virtual production intelligence" presented in this chapter, the data is analysed using analytic concepts and IT systems, which evaluate the simulation process data collected with a view to gaining the desired knowledge.

"Virtual production intelligence" aims to collect, analyse and visualise the data produced in a simulation process in order to generate insights that allow for the integrative assessment of the individual and the aggregated simulation results. The analysis is based on expert knowledge as well as on physical and mathematical models. Through integrative visualization, the requirements for "virtual production intelligence" are entirely satisfied.

The collection of the data, i.e., its integration into a uniform data model, has already been accomplished. Likewise, initial solutions for visualising the simulation results have already been implemented. Analysis and visualization of the analysis results in an integrative environment are research projects that, upon successful completion in the future, would enable "virtual production intelligence".

Integrating the results of a simulation process into a uniform data model is the first step towards gaining insights from these databases and being able to extract hidden, valid, useful and action-oriented information. This information encompasses, for example, the quality of the results of a simulation process and, in more specific use cases, the reasons why inconsistencies emerge. To identify such aspects, at present, the analysts use the analysis methods integrated in the simulation tools. The implementation of the integration system, however, opens up the possibility of a unified consideration of all the data since it encompasses both the integrative visualization of the entire simulation process in one visualization component and the study and analysis of the data generated along the entire simulation process. Various exploration processes can be called upon for this purpose. What needs investigating in this respect is the extent to which the information extracted through exploration processes can be evaluated. Furthermore, it should be investigated how this data can be visualised and how information, such as data correlations, can be adequately depicted. To this end, there are various feedback-supported techniques that experts can use via visualization feedback interfaces to evaluate and optimise the analysis results.

The aforementioned data exploration and analysis may be further undertaken as follows: First, the data along the simulation process is integrated into a central data model and schema. Then, the data is analysed at the analysis level by the user by means of visualization. In doing so, the user is supported by interactive data exploration and analysis processes, which can be directly controlled within the visualization environment. Since it is possible to send feedback to the analysis component, the user has direct control over the data exploration and can intervene in the analyses.

From the present perspective, data analysis begets the following questions: Which exploration techniques are needed in the application field of virtual production for analysing the result data of numerical simulations? How can the results of data analysis be used in feedback systems to optimise the analysis results? How can the results of data analysis be represented in an integrated, interactive visualization?

In addition to retrospective analysis by experts, it is equally useful to monitor the data during the simulation process because such process monitoring enables, amongst other things, compliance with a valid parameter selection or other boundary conditions. If a simulation tool were to offer an invalid parameter value, this would lead to termination of the simulation process. Previous results could be analysed by experts using the methods available in data analysis in order to tailor the simulation parameters and thus guarantee that only valid parameters could be selected. Furthermore, the interim results could be evaluated using power functions, which could be checked after the simulation has been run and the simulation results have been integrated. Similarly, process monitoring could also enable the extraction points-of-interest (POIs) based on functions, which then would be highlighted in the visualization.

In data monitoring, process monitoring is integrated into the simulation process as a cross-sectional function. Within process monitoring, the following questions arise: How can power functions, POIs and monitoring criteria be formally specified within process monitoring? How can the information extract from process monitoring be used for process optimisation?

Figure 4.30 shows the components of a system for implementing "virtual production intelligence".

4 Virtual Production Systems

Fig. 4.30 Components of VPI

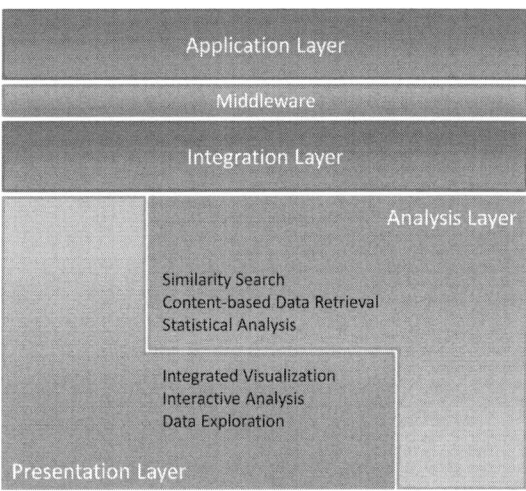

The application level comprises the simulations that are called up along a defined simulation process. These are linked to one another via middleware, which performs data exchange and which is responsible for ensuring data integration and extraction within the simulation process. An integration server is provided for this purpose. Using a service-oriented approach, the integration server provides services for integration and extraction. The database server represents the central data model and serves as the central data store for all data generated in the process. The integration level was already implemented within the framework of the Cluster of Excellence and the application level was already connected to the integration level. Likewise, a visualization component was developed, which serves as the starting point for implementation of the presentation level.

The presentation level is tasked with presenting the simulation results. As an example, dashboards could be provided for statistical control of production processes or for determining key performance indicators. Another possibility for data visualization is a reporting tool, which, amongst other things, would make it possible to evaluate erroneous processes so that the sources of error may be identified. It would even be possible to use tools from the field of artificial intelligence, which could, for example, draw conclusions from temperature gradients and thereby draw attention to any anomalies in the results. This in turn could lead to the revision of simulation process modelling or of individual production processes. These examples provide an initial overview of potential deployment scenarios that could be supplemented with cooperative and interactive features. The user is given interactive control over the analysis. To accomplish this, the results achieved in the first phase of the project will be extended so that the user can modify the analysis parameters interactively from the visualization and the amended analysis results will be fed back into the visualization.

The analysis level and the presentation level—the core of "virtual production intelligence"—are formed on the basis of these components. Implementation of these components would, for the first time ever, enable integrative analysis of heterogeneous simulation processes.

4.3 Factory Planning of the Future—Connecting Model Worlds

Achim Kampker, Günther Schuh, Peter Burggräf, Bastian Franzkoch, Jan Nöcker, Marcus Rauhut and Cathrin Wesch-Potente

4.3.1 Abstract

Existing approaches to factory planning perceive the planning process as a linear and deterministic process that can be divided into discrete phases. Planning tasks are executed consecutively in phases. The software tools used in factory planning are designed on the assumption that the planning results will be transferred into a results document for the respective planning phase. Practical experience, however, shows that there is no standard factory planning project and the planning process is adapted, both before and during planning, to specific boundary conditions and new decision-making situations.

Within the Cluster of Excellence, a planning procedure has been developed that allows individual configuration of the planning project and facilitates a modular, parallel planning process. To this end, the planning content was encapsulated in planning modules equipped with defined interfaces which describe the input information required as well as the results to be achieved. Appropriate planning methods and tools are also encapsulated in the modules. This means modules can be re-combined and run repeatedly according to the specific starting situation for each planning project and in the event of changes over the course of the project. The following chapters present the research conducted by the ICD-B.

4.3.2 State of the Art

Factory planning is a complex planning task for which many models have been described in literature. These models have different emphases, but they all cover four basic topics:

- Motive
- Subject
- Object
- Procedure

It should be noted that factory planning as a scientific discipline features a high proportion of German authors. This sub-chapter will first explore a general view of the object area and then look more closely at the motives behind factory planning.

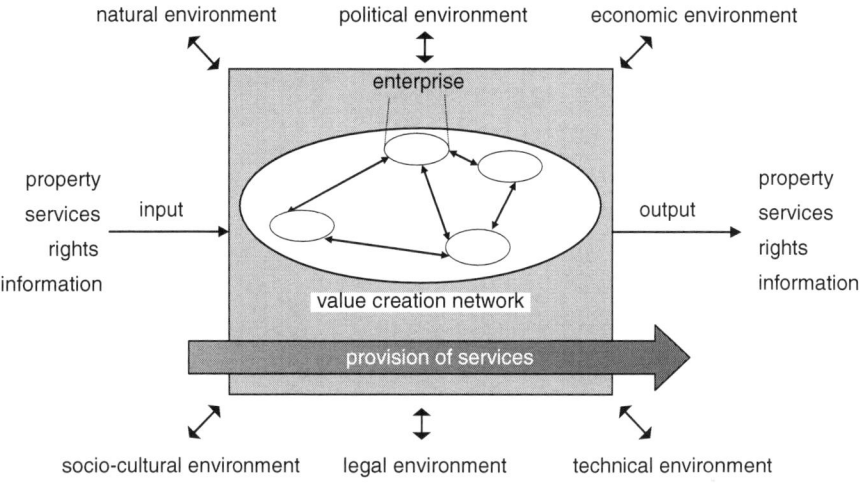

Fig. 4.31 Value creation of a business. (According to Dyckhoff 2006, p. 5)

The planning depth in factory planning ultimately comprises various levels of consideration, ranging from positioning a factory within the value-adding network to designing a production line through to local configuration of a workstation.

Since the documented research focuses on the development of a process model for factory planning, the project nature of factory planning will be discussed in another sub-chapter.

4.3.2.1 The Factory as a Place of Value Creation in a Continuous Transformation Process

The modern industrial production of a business is a complex system of interrelationships with a broad and deep range of products, spatially distributed sites and a variety of production types used in the mix (Dyckhoff 2006, p. 4). Accordingly, planning an industrial production facility is also complex. Numerous drivers of complexity, presented in this chapter, are responsible for this.

A factory is a facility for manufacturing a product by transforming the production factors of land, labour, capital, energy and information. Here, by contrast with agriculture or forestry, the work-sharing and production engineering aspect is paramount (from Bergholz p. 19: Felix, Unternehmens- und Fabrikplanung, 1998, p. 32) (Fig. 4.31).

Transformation is the qualitative and quantitative modification of objects. Transformation initiated by people and purposefully directed and systematically executed with a view to achieving an increased benefit (added value) is called production. In this case, the purpose of the transformation is not necessarily the immediate satisfaction of one's own needs. An economic entity that is dedicated mainly to added value is called a business. In addition to trading, production is therefore the core function

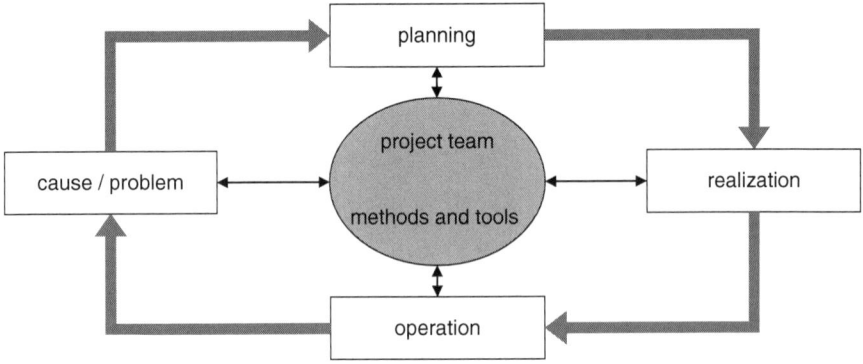

Fig. 4.32 Factory planning as a permanent planning task. (According to Pawellek 2008, p. 17)

of any business, or any company (profit-oriented businesses). The modern industrial production of a business is a complex system of interrelationships with a broad and deep range of products, spatially distributed sites and a variety of production types used in the mix (Dyckhoff 2006, p. 3).

From a methodological point of view, the factory planning process represents a transformation process, i.e., input data is transformed, e.g., through the application of rules, calculation formulas, variant decisions, depending on the problem, into interim or final results (projects, studies) and therefore into output data (Grundig 2006, p. 18). In this respect, factory planning can be seen as a continuous planning task (Fig. 4.32).

This begins with a motive or problem, born of factory operation, which is followed by the planning and implementation. A constantly recurring motive is the limited life cycle of the factory. When this motive for planning arises, it is delimited and the factory planning project is formulated. The divisional and corporate goals have to be addressed in order to develop a common desire for action in contrast to the actual situation. Planning is required if the structure of the problem is complex and initially unknown, or if the time needed to resolve the problem is a medium or long period. Alternatively, planning can be waived and the problem solved by "improvisation" or "snap decisions". In implementation, the planning result is achieved. Thereafter, the results achieved are continuously compared with the planning results. This is followed by continuous operation, which can be maintained without renewed planning unless a new motive arises (Pawellek 2008, p. 17 ff.). The cycle of continuous planning is complete once the actual state of the factory satisfies the time-variable requirements placed on it.

4.3.2.2 Motives for Factory Planning

Planning generally falls into three basic categories: new planning of a factory (greenfield planning), re-planning and/or extension of existing facilities (brownfield planning), or dismantling of factories (see Fig. 4.33).

4 Virtual Production Systems

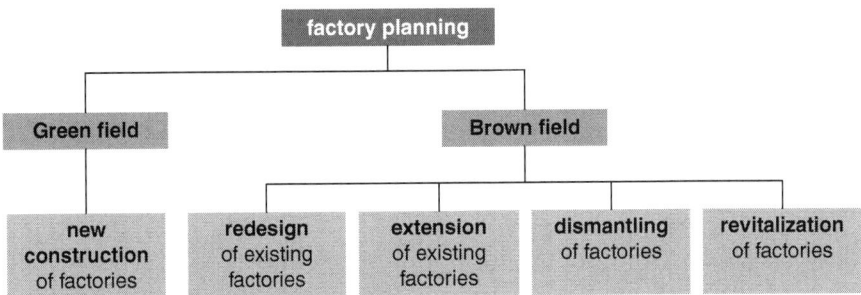

Fig. 4.33 Basic planning categories in factory planning

With new planning, the factory is built on a "greenfield site" with the maximum level of freedom. Such a high level of freedom allows the optimum factory design to be achieved. However, with this increased level of freedom comes increased complexity and planning effort because with new planning, additional tasks arise such as choosing a site. When redesigning a factory, the existing facilities and processes are modernised and adapted to market-driven changes in the production range. Replanning is a particular challenge because of the need to allow for existing constraints such as the current infrastructure, e.g., foundations and supply and waste lines, as well as crane equipment. Extending the existing factory is always an option when the available capacity is no longer sufficient. As with new planning, extension may also involve choosing a site for the new building that will provide the additional capacity. Dismantling is the opposite of extension. If there is excess capacity (e.g., due to adaptation of the value-adding structure, outsourcing or a decline in sales), it can be eliminated through restructuring and redimensioning. Revitalisation of factories basically refers to renovation measures intended to allow a new use or re-use of abandoned industrial plants (see Grundig 2006, p. 14).

4.3.2.3 Factory Planning from the Value-Adding Network Through to the Workstation

Factory planning includes planning the added value distribution in a production network, planning the actual factory with its various segments and planning individual lines and workstations. These planning levels are hierarchically linked. Hence, the interaction across individual levels also has to be planned (Fig. 4.34).

Whether a product should be manufactured internally in the factory or acquired externally via the value-adding network (make-or-buy decision) is determined at the value-adding network level. Ascertaining capacity shortfall is closely associated with this. If an enquiry cannot be met due to a lack of capacity, the options are to provide new capacity or to pass the order on to a supplier. Capacity can only be expanded through the new or re-planning of factories (Stepping 2007, p. 79 ff.).

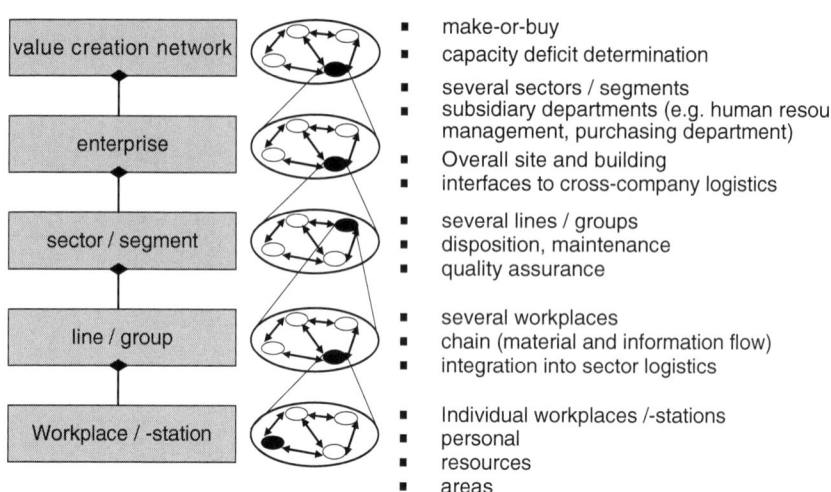

Fig. 4.34 Hierarchy levels in factory planning. (According to Bergholz 2005, p. 129)

The factory level comprises the general building plan or factory layout as well as the internal and external material flow. Besides the actual production, the linking of indirect departments (production management, production planning, material requirements planning, logistics departments) is also determined at this level (Wiendahl 2001, p. 189). A typical issue is here the degree of area utilisation (proportion of constructed area to plot area), for example, and the direction of factory extension.

At the following divisional or segment levels, various lines and groups are structured and indirect activities, such as quality assurance or maintenance, are taken into account during layout planning.

At the line or group level, individual workstations for operating stations or groups are lumped together and spatially arranged based on their position in the handling process. In doing so, particular consideration must be given to the material, personnel and informational flow.

The lowest level is the individual workstation with its equipment and personnel resources. Here, the layout of the machine work area, for example, is determined, taking into account the operating, supply and maintenance areas (see Bergholz 2005, p. 129 ff.).

4.3.2.4 Project Nature of Factory Planning

Generally, a project is an undertaking characterised by a unique set of goals, its duration, finance, personnel and other constraints as well as by project-specific organisation (DIN 69901) (Fig. 4.35).

To ensure the project is executed properly, on schedule and within budget, projects are organised, planned, controlled and coordinated through project management

4 Virtual Production Systems

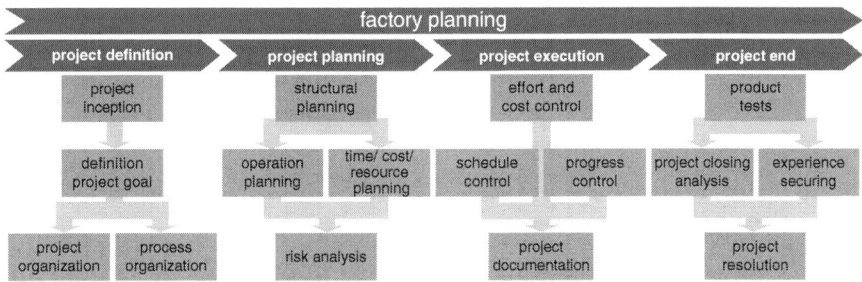

Fig. 4.35 Project management within the factory planning domain

(Schuh et al. 2005, p. 22). The tasks of project management can be divided into four main components. In the project definition, the project goals are clarified and project tasks are set. The second component is project planning. In this stage of the process, a project structure plan is developed to organize execution of the project in terms of time, taking into consideration the account dates and deadlines (for example for acquiring a new machine). The project execution phase is when the project is actually implemented. During this phase, differences between the actual and the target state are highlighted and compliance with the agreed deadlines is checked. Upon completion of the project, the project team reflects on the entire project. They focus on the strengths and weaknesses of the planning as well as the findings that may be of significance for other tasks in the future.

Preventative project management ensures proper execution, on schedule and within budget, of all sub-projects in all the phases of factory planning and identifies potential deviations at an early stage. Project management considers the needs of all departments participating in the project. The more precise the planning of the project execution is, the fewer the problems in the later stages of the project. However, project planning should not be seen as a one-off procedure that is complete as soon as all the planned objectives have been laid out. Rather, the level of detail of the planning can only be sufficiently accurate for the initial development stages, especially with long-term and large-scale development projects. As the project progresses, in-depth planning is usually required for the subsequent development stages, i.e., the project planning procedure is influenced by the project history and is therefore path-dependent. Also, with unforeseen modifications due to requirements expressed at a later stage, planned objectives may need to be adjusted or new objectives may even need to be planned.

Since the planner expects a different planning situation and a different remit for each project, factory planning is marked by the fact that there is no standardised solution procedure to rely on (Aggteleky 1987, p. 49 ff.). Since new types of problems are frequently encountered, factory planning is characterised by a high proportion of innovative knowledge and often relies on heuristic methods to find the (optimum) solution by comparing and assessing planning alternatives (Schady 2008, p. 5, 16).

4.3.2.5 Critical Reflection and Conclusion

In summary, it is fair to say that factory planning is a complex, socio-technical problem that extends far beyond merely organising resources and processes. Factory planning is complex in the sense that the planning objects are dynamic and constantly adapting and there is a high degree of interconnection amongst the individual components of the factory. Factory planning is a complex socio-technical problem because both the factory and factory planning are systems involving people from different departments with different skills and perspectives.

4.3.3 Motivation and Research Question

Approaches to factory planning in existence prior to 2009 perceive the planning process as a linear and deterministic process that can be divided into discrete phases. Thus the relevant approaches described in literature can be incorporated in one model with five stages from preparation to execution. In the consecutive approach, the content of a phase is processed completely before the next phase begins. In this connection, the software tools used are not designed primarily for fast and flexible data exchange because it is assumed that the results of each planning module will be transferred into a results document for the respective planning phase. Practical experience, however, shows that there is no standard factory planning project and the planning process is adapted, both before and during planning, to specific boundary conditions and new decision-making situations.

The production engineering dilemma between plan and value orientation addresses this discrepancy between theory and practice. The existing approaches are designed to support a plan-oriented procedure, while practice has to contend with decentralised planning, frequent changes and adaptations as well as a high degree of self-organisation. Due to the turbulence in planning projects, iterations must be taken into account explicitly in a planning procedure and supported with appropriate software tools.

> In the project "Networked Models of the Digital Factory", research activity was essentially devoted to the question of how efficiency and effectiveness losses in planning processes can be reduced, along with the necessary software support, by implementing planning processes that can be individually configured.

In this regard, modularisation and networking of the planning procedure are the basic prerequisites for networking the software. The region of interest here covers not only traditional factory planning, but also the planning modules and software tools for product and technology planning. In this connection, the flexible link concept is based on the principle of modularisation together with content encapsulation and exchange via defined interfaces. The network model describes the interfaces of the planning modules and the information exchange (input/output relationships). The content of planning tasks is encapsulated in the planning modules and methods and tools are assigned. This serves to achieve standardisation and professionalisation in the modules. Through the defined dependencies between the modules and the use of initial values/corridors, which are iteratively limited over the course of the project, it is possible to begin using the planning modules even if not all the information is available yet or if data and restrictions are unclear.

In accordance with the planning motive and the company-specific prerequisites, the appropriate modules, methods and tools are selected and dimensioned while the planning sequence is individually adapted to the boundary conditions and restrictions. Over the course of the process, further reconfigurations may be required due to internal and external changes. Synchronisation of the planning tasks and the project team during the planning process present further challenges; concepts for meeting these challenges are developed within the framework of the project.

Development of the individually configurable planning process as a holistic approach to reduce efficiency and effectiveness losses in planning processes is complemented by the following research sub-questions:

- How is the company-specific and situation-specific planning process configured based on the planning modules and how is it adapted over the course of the project?
- How can individual configuration of the planning process be evaluated?
- How can the efficiency of planning processes be increased with reduced centralised control and greater self-management within the teams?
- How can the efficiency losses associated with decentralised, parallel planning be reduced through the team structure and synchronisation mechanisms?

Hence, the holistic approach moves out of the deterministic corner of "planning" and sits in-between the conflicting poles of plan orientation and value orientation. It does so by means of flexible value orientation of planning through:

- ascertainment of the planning scope and level of detail required depending on the planning tasks
- coordination of decentralised planning and goal synchronisation
- efficient and effective planning and project organisation

The dilemma between scale and scope is addressed by individualisation designed to meet specific needs:

- individual configuration of the planning project based on standardised modules
- situation-specific adjustment of the sequence

4.3.4 Results

4.3.4.1 Goal System of Factory Planning

A goal system combines the various individual objectives of the business activities into one homogenous goal image. In factory planning, the goal system incorporates four different aspects.

First, the strategic positioning of the entire company is relevant. This determines the strategic orientation of the production system. The goals of the production system are often conflicting. Hence, the production system must be positioned within the dichotomy of the goals. This positioning of the production system is the second aspect incorporated into the goal system of factory planning. Third, the operational goals of the planning project, factory start-up and factory operation are incorporated into the goal system. Finally, with a holistic approach, goals arising from company culture must be taken into account in factory planning too.

Strategic Orientation for Factory Design Factory planning must take into account the strategic orientation of the company since this represents the superordinate direction. The decisions made by factory planning have long-term effects on the development of the business and must therefore be seen as part of corporate planning (see Fig. 4.36).

The superordinate goal of the strategic orientation of the company is to achieve an enduring competitive advantage (Knight 1997). The following approaches have become well established within strategic management as bases for achieving an enduring competitive advantage:

- market-oriented approaches
- resource-oriented approaches
- competence-oriented approaches

Market Orientation Market-based strategic orientation was mainly influenced by the work of Porter in the 1980s and at that time, it was the dominant strategic concept (Zahn et al. 2000). Porter defined strategy as the market position that a company occupies or would like to occupy, together with all those company activities that are

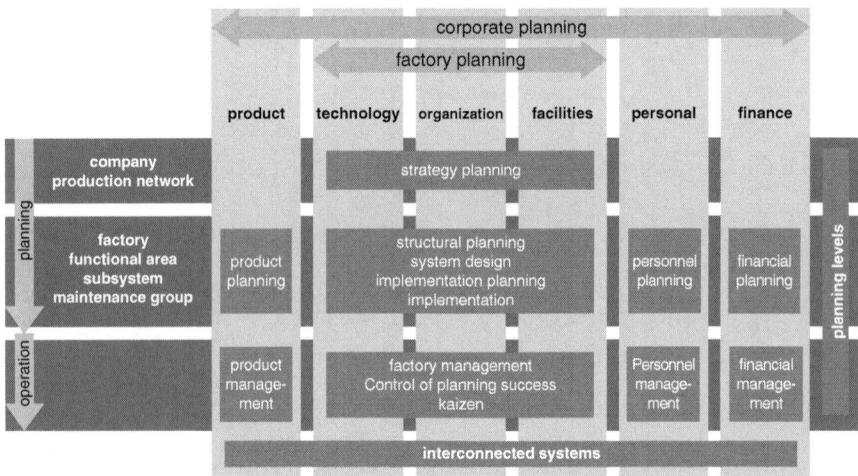

Fig. 4.36 Position of factory planning within corporate planning. (According to Pawellek 2008)

4 Virtual Production Systems

designed to achieve occupancy of a particular market position (Porter 2009). The prevailing view is that a strategy has to be derived from the existing market structure. Implementation of market-based strategy orientation is typical for companies that have a very strong customer and sales orientation and are characterised by highly innovative products (differentiation strategy) or are subject to hard-fought price wars (cost leadership strategy).

In this case, the factory is planned based on the market and is therefore responding to market requirements. Porter describes the five forces that influence the company from the external environment. These forces therefore also influence factory planning (see Fig. 4.37).

In view of the *rival competitors*, the design of the production system must enable competitive production. Under identical general conditions, only the increased efficiency of a process creates competitiveness in relation to another process. The central concern in designing a production system must therefore be to eliminate the seven types of waste (waste due to overproduction, unnecessary transportation, inventory, motion, defects, over-processing and waiting) in order to achieve competitive target costs.

Analysis of the *supplier structure* is an essential input variable in factory planning. For example, if the supplier delivers just in time (JIT), then the material can be supplied directly to the workstation. If JIT delivery is not possible, then storage areas must be provided. If the supplier distinguishes itself through reliability, this allows a lean production system design. Otherwise, appropriate counter measures, such as buffers, must be factored in during production system design. Also, it may be necessary to change supplier, in which case only a sufficiently flexible production system can react appropriately.

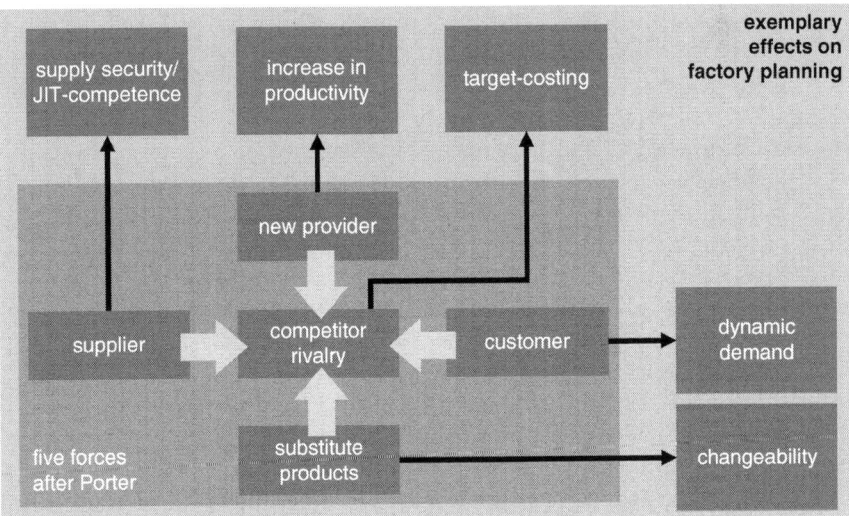

Fig. 4.37 Porter's five forces and the effects on factory planning

If a *new provider* enters the market, this could result in falling market shares. Factory planning can take preventative measures in this respect, for example implementing restricted access to the factory to avoid technology being transferred to competitors and to make entry more difficult for new providers. Should a new provider enter the market (nevertheless), the factory has to be able to react appropriately. In this case, both increasing productivity, which enables prices to be lowered, and upgrading the product through design modifications may be suitable counter measures. Hence, flexibility must be considered in factory planning in order to allow subsequent changes in the production process or to the product.

Detailed analysis of the *customer requirement* structure must also be incorporated in factory planning. In particular, this concerns the following variables: product quality, demand dynamics, the need for variance, lead times and market size. This analysis incorporates both historical data and forecasts for the future so that trends can be identified and taken into account.

Should a competitor launch a *substitute product*, this often produces a sudden change in demand. The capacity gained in the factory must either be rectified or used for other products. The latter, however, requires resource flexibility if other products are also to be produced with the existing resources. If the company introduces the substitute product itself and the new product replaces the old one entirely, a ramp-up strategy will need to be defined (see section on ramp-up management).

Resource Orientation In the *resource-oriented approach,* competitive advantages are not based on the structure of the market, but on the individual resources available to a company in each case (Wernerfeld 1984). Resources cannot be arbitrarily allocated to each company and used by that company. Machinery and equipment are often immobile and investment in them irreversible. Heterogeneity of resources, so to speak, leads to varying performance efficiency (Collins and Montgommery 1995). These differences in efficiency are responsible for varying profits and therefore for varying competitive positions (see Fig. 4.38).

Heterogeneity of resources is particularly evident under application of a broader understanding of resources, which takes into account intangible resources as well as physical resources. An example of intangible resources would be the competencies available to the company. Competencies are developed through long-term learning processes and are often deeply rooted in the company (Sanchez and Heene 1996). These competencies cannot really be transferred to other factories, or at least this would require a suitable transfer and adaptation period. Since physical resources have a long lifespan and take some time to build up, the resource-oriented strategy approach should be considered for long-term decisions in particular and, consequently, when planning factories of a considerable size (Kampker et al. 2009).

Resource orientation in factory planning focuses on the value of resources for the production company, which form the central basis for long-term, value-oriented development of the production system. Resources of high value require a high level of attention during long-term development, while those of low value can be left to short-term and dynamic development (Kampker et al. 2009). This way, efforts are directed towards resource development and competencies towards problems that

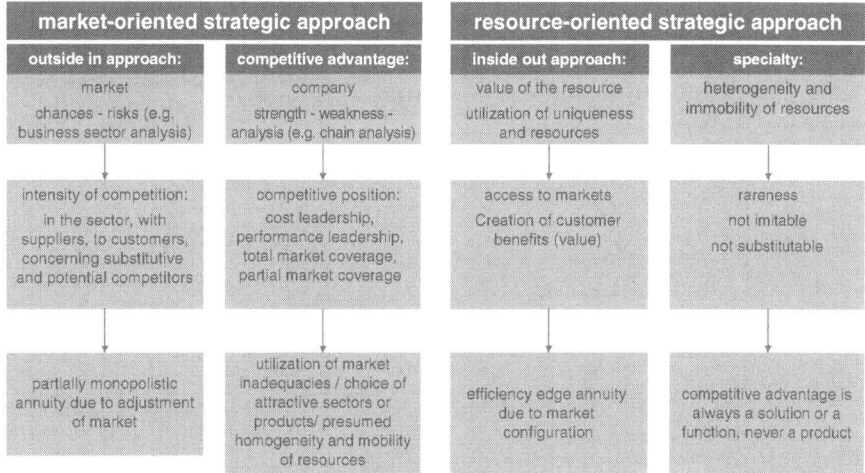

Fig. 4.38 Strategy approaches in factory planning. (See Pepels 2006, p. 463)

represent decisive factors for success. This can also help identify which resources require a stable environment and which can be used flexibly. Resource-oriented value development of a production system is therefore becoming a central theme in continuous factory planning.

Competence Orientation The *competence-oriented approach* combines the resource-oriented and market-oriented strategy approaches (Hamel 1994). A competitive advantage results from the exclusive ability to meet specific customer needs using the company's own resources. With this approach, competencies represent the link between the resources and the products of the company (Prahalad and Hamel 1990). By focussing on a few competitive core competencies, managers can be organised into specific groups. Core competencies are distinguished from ordinary skills through the following criteria: efficiency, effectiveness, non-substitutability and non-imitability (Fig. 4.39).

Defining the core competencies of the company is an important prerequisite for factory planning. Both the product strategy and the resource strategy can be derived by defining the core competencies. This way, the strategic decisions regarding product and production development are synchronised.

Production-based competencies result from the company-specific combination of various resources and require long-term, targeted development (Koruna 1999). Factory planning can influence this development in many ways. For a start, the planning motive can be used to improve the core competencies. Production processes can be optimised, the machinery and equipment technology can be improved and product features can be adapted thanks to the enhanced competencies. Also, factory planning influences the continuous improvement process, which further develops the competencies after construction and alteration of the factory. The ability to improve

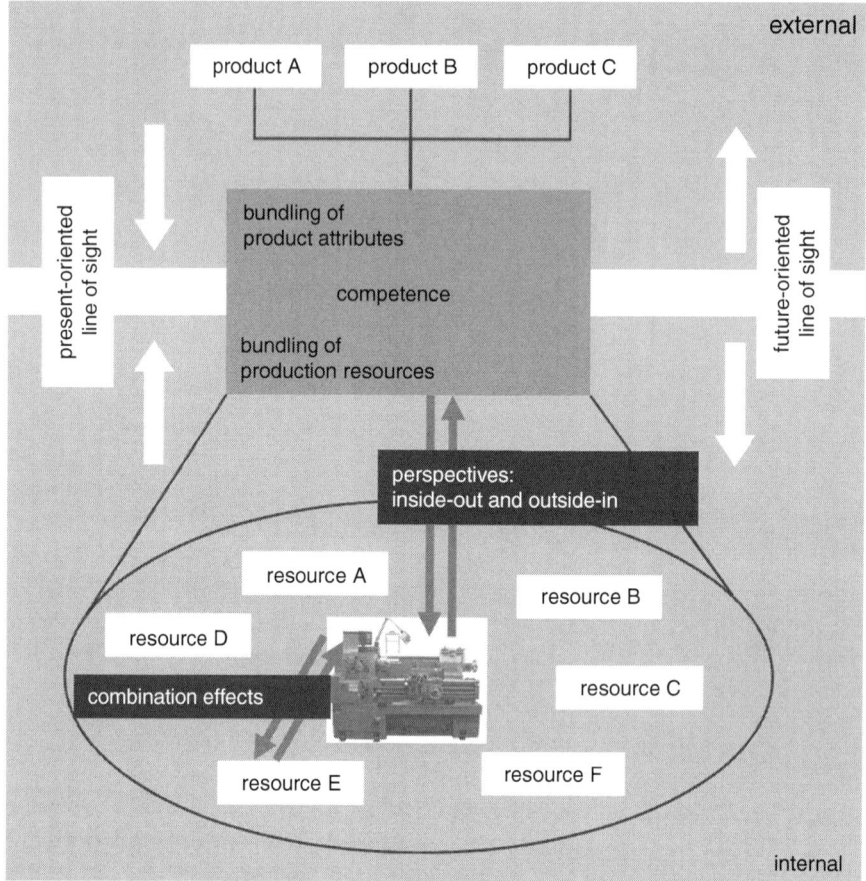

Fig. 4.39 Competencies as the interface between market-orientation and resource-orientation

the competencies, in particular, must be preserved and not hindered by restrictive structures.

Sustainability Orientation Since the Brundtland Report in 1987 and the United Nations Conference in Rio de Janeiro in 1992, sustainability has become increasingly important (Bullinger et al. 2003, p. 814). The concept of sustainability goes back to the Brundtland Report, according to which development is sustainable if it meets the needs of the present without compromising the ability of future generations to meet their needs. The inter-generational equity this calls for is seen as the normative premise of society, which requires no further ethical justification.

According to prevailing opinion, sustainability has three equal dimensions: economy, environment and society. The economic dimension focuses on the development of a viable basis for long-term economic prosperity. The environmental dimension

Fig. 4.40 The three dimensions of sustainable development

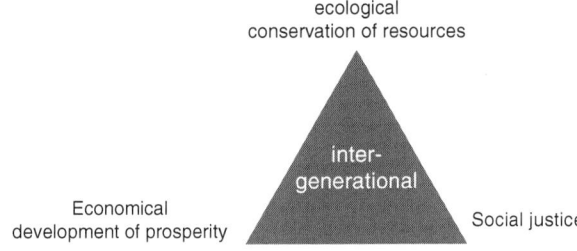

aims to preserve the natural environment for future generations. And the social dimension is geared towards sustainable social development that offers a life worth living for all members of society (Fresner et al. 2009, p. 10) (Fig. 4.40).

This understanding, also known as the "Three Pillars Model", forms the basis for sustainable corporate strategy and equally sustainable business models. The crucial underlying hypothesis here is that economic success is not possible in the long term without equal consideration of the environmental and social consequences of entrepreneurial activities. A sustainability-oriented corporate strategy is therefore developed in accordance with economic, environmental and social criteria and generates long-term competitive advantages on this basis (Grunwald and Kopfmüller 2006; Grunwald et al. 2001; Hauff 1987).

Dichotomies of Factory Planning and Positioning Generally speaking, complex systems, including factory planning projects, require a trade-off between conflicting goals (Schuh and Gottschalk 2008). The main typical goals of factory planning include minimum processing times, maximum capacity utilisation, minimal delays and minimal work-in-progress (Krauth and Noche 2004) and (Wiendahl et al. 2005). This often entails dichotomous goals. A dichotomy is generally understood to involve mutually exclusive elements of consideration. For example, maximum capacity utilisation and minimum inventories are usually mutually exclusive because, in the event of supply irregularities, low stock levels lead to immediate production shutdown and therefore high capacity utilisation usually requires significant buffer stocks in order to remain impervious to supply fluctuations (Fig. 4.41).

Positioning is required during goal setting, i.e., a decision must be made about the goal profile within the mutually exclusive goals (Pawellek 2008). Goal criteria are chosen by the people involved in the decision process and the selection is based on the company's own corporate strategy. The people involved in this regard must include the commissioner of the project and various specialist planners, e.g., those responsible for buildings or infrastructure (Zink 2009).

Efficiency Versus Flexibility *Efficiency* in this case means cost optimisation and is generally a prerequisite for a successful company in the market role of cost leader. In an efficient factory, the production processes are run on the threshold of what is technologically feasible whilst maintaining quality and safety. This leads to optimal utilisation of resources. In this case, the resources (e.g., operating supplies) are

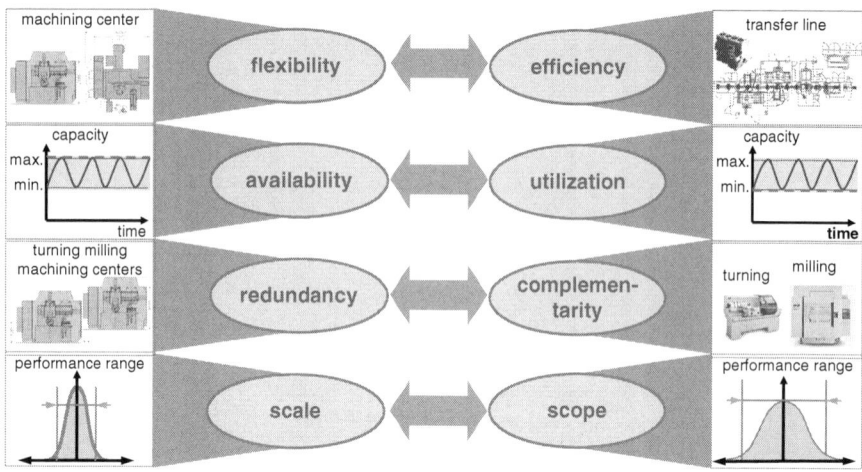

Fig. 4.41 Goal dichotomies in factory planning

matched precisely to the production requirements. Production at the optimal operating point requires the most stable environment possible. This is usually achieved by producing lots of identical goods over a long period so that the fixed costs can be spread over the highest number of parts sold (Burckhardt 2001).

Flexibility is, generally speaking, the ability to adapt to changes in the boundary conditions in the respective environment. Examples of altered boundary conditions include increasing market pressure, changes in environmental regulations and enhanced technologies. There are countless definitions of the term flexibility depending on the field of application (Sethi and Sethi 1990). In this respect, there are three dimensions of flexibility: range of possible specifications of a production system, mobility in the sense of effort required to modify the production system and uniformity as the variance in performance capability depending on the specification of the production system within the range (Upton 1994).

Given the long service life of factories, flexibility has to be taken into account in order to be able to respond to changes with the greatest possible mobility. From a flexibility point of view, the performance capability of a factory depends on how these changes can be absorbed into the factory environment with minimal impact on performance (Pawellek 2008). In a production system, flexibility is a measure of the range of states that can be adopted. The basis for a flexible production system is a modular factory structure that facilitates continuous adaptation. Expandability, integration capacity and the ability to learn are also basic characteristics required to achieve flexibility (Lindemann et al. 2006). In order to use the existing flexibility of a production system, the shortest possible planning times and sliding or rolling factory planning are required (Grundig 2006).

Maximum efficiency and flexible production systems are often in contradiction with one another. Flexibility requires, for example, equipment with additional functions that are rarely used. These additional functions come at extra cost, which affects efficiency. Hence, to define appropriate and consistent guidelines for the planning project, positioning between these extreme positions is required.

Capacity Utilisation Versus Availability The *capacity utilisation* of a production system is a ratio used to express the work time of a production resource actually used in relation to the total available work time of that resource. Maximisation of capacity utilisation leads to maximisation of fixed cost degression and therefore to a cost reduction per production unit. From an economic point of view, a company should aim for the maximum possible utilisation of the capacity of its machinery and equipment. This can be achieved through technical and organisational measures, for a start, designed to reduce downtime, e.g., for routine maintenance or machine set-up. Furthermore, building up an order backlog for a machine is also a prerequisite for high capacity utilisation of the machine. In order to build up an order backlog, the available production capacity should not be greater than the order intake. Utilisation is therefore influenced, in the medium and long term, by the installed production capacity and in the short term by order release. If a high utilisation rate is desired, only a low level of production capacity is held in reserve and early order release is preferable. The disadvantage is that, due to the limited capacity, it may be impossible to meet rising demand and therefore the opportunity to generate additional turnover is lost.

High *availability* provides the ability to accept further orders on a flexible basis and therefore to generate additional turnover. High availability arises when average demand is lower than the installed production capacity. By reducing utilisation and lowering stock levels, shorter processing times can be achieved, creating greater flexibility (Schmigalla 1995). This is an important prerequisite for processing lucrative urgent orders and satisfying peaks in demand and also it is the reason why short processing and lead times as well as lower stock levels are increasingly gaining in significance. In general, fluctuations in demand are becoming more dynamic so responsiveness represents an important competitive advantage (Fig. 4.42).

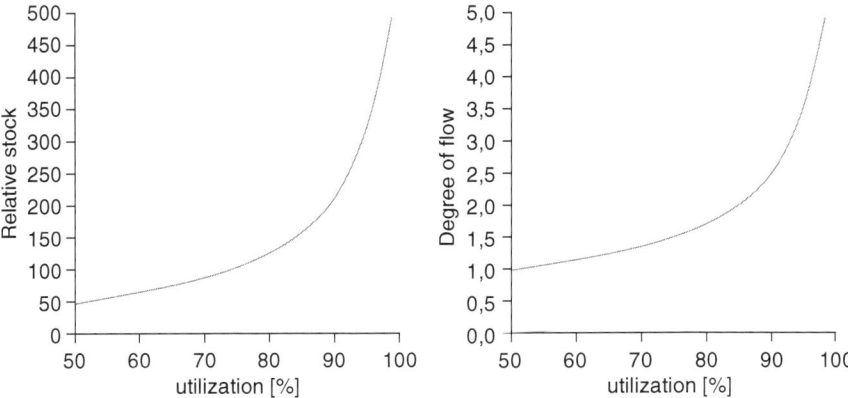

Fig. 4.42 Reducing system capacity utilisation enables an overproportional reduction of stocks and processing time. (Nyhuis and Wiendahl 2002)

Although capacity utilisation is often pushed into the background as a tradition goal in the dynamic market situation, it must be considered a key target parameter in factory planning. Whilst it is certain that investment will be required to create the excess capacity, it is not certain that an increase in turnover will be achieved and, in general, it is uneconomical to install production facilities with significant excess capacity (Grundig 2008). Under these general conditions, positioning is required regardless of whether, in view of the anticipated capacity requirement, the maximum available capacity is designed for maximum utilisation of resources or for maximum availability.

Scale Versus Scope The strategy of price leadership is based on manufacturing large quantities of similar products. The economies of scale generated means that such standard products can be produced very economically. The economies of scale are effective when a small increase in the input factors produces an overproportional increase in the output factors (decreasing marginal costs) (Panzar and Willig 1977). This is due to the

- exploitation of learning effects: with the repeated execution of a task, the performance level of the employees involved in the process increases. Necessary coordination processes become more efficient and the processes more synchronous.
- increase in the level of mechanisation: with the growing piece number, it is possible to consider manufacturing processes and machinery and equipment concepts that offer increased mechanisation and are suited to the production of large quantities.
- exploitation of cost degression effects: costs unrelated to quantity, e.g., investments and set-up costs, can be apportioned to a larger product quantity and the unit costs can be reduced.

The strategy of *differentiation* does not depend on economies of scale to the same extent. Instead, economies of scope are exploited. Pooling and shared use of input factors results in multi-product production cost benefits. This is the case if operating two production lines within one company is less expensive than operating the lines separately (Teece 1980; Panzar and Willig 1981). In this respect, the economies of scope come from two sources (Gorman 1985):

- the exploitation of local economies of scope through cost complementarity, whereby increasing the production quantity of one product reduces the production costs of other products.
- the reduction of pro rata fixed costs per product.

To exploit the economies of scope, various products need to be produced with the tendency to increase product variance. The piece numbers of a differentiated product programme, however, are considerably lower, thus exploitation of economies of scale is reduced and becomes less significant (Goldhar and Jelinek 1983). Economies of scale and economies of scope therefore represent two opposing positions, which are initially incompatible and therefore describe a strategic area of tension.

Redundancy Versus Complementarity *Redundancy* refers to the existence of overlaps, i.e., there are functionally similar resources which are used in trouble-free operation, but not in redundant mode. Redundancy is created by the availability of similar resources. Redundancy of resources ensures production continues in the event of failure because similar equipment means production can be transferred in the event of failure or overload. This results in a decentralised structure, which has a positive influence on adaptability. The provision of redundant resources is associated with additional costs and it is impossible to predict for sure whether a fault will occur and the redundant resources will be used for this purpose (Specht 2007).

Complementarity refers to the existence of initially contradictory, yet complementary properties. This leads to mutual strengthening of the complementary partners in the sense of Aristotle's quote: the whole is greater than the sum of its parts. An example of complementarity in production is the combination of various single-purpose machines. This involves the use of machines that are each optimised for one processing task. Another example of complementary relationships is the cooperation of companies. Here, complementarity is an important factor for success (Kampker 2008) because the partners can share their strengths and eliminate weaknesses. The disadvantage of a highly complementary system is its sensitivity to the failure of one single element. Should, for example, a function of one single machine fail, the other machines could not take over that function. Either stocks are built up (the unaffected machines continue to produce) or the entire system ceases to produce.

Complementary means of production by definition have various capabilities. Redundancy requires means of production with identical capabilities. The dichotomy and the need for positioning are obvious. In addition to the profitability assessment, positioning also requires a risk assessment. While redundant systems are more resilient to faults and failures and therefore represent a lower risk of failure, they cannot offer the economic advantages of complementary systems.

4.3.4.2 Goals for the Planning Project

If a company decides to build a factory, this is the result of a long process involving lots of people as well as obstacles. This process has to be well planned and controlled; after all, a factory with a service life of over 30 years is an important foundation for the economic success of a manufacturing company (Pawellek 2008; Kettner et al. 1984). The factory and the associated factory planning undergo various phases that pose different problems and requirements. In this respect, the life cycle of a factory is divided into three different sections, each of which, importantly, has a different goal system.

Goal System of Planning Factory planning is generally seen as a means to an end. On closer examination, however, it becomes clear that factory planning can contribute to adding value (in the positive case) or wasting value (in the negative case). The goal of the planning project is therefore not just to implement strategic objectives

Fig. 4.43 The return on planning as a KPI of factory planning

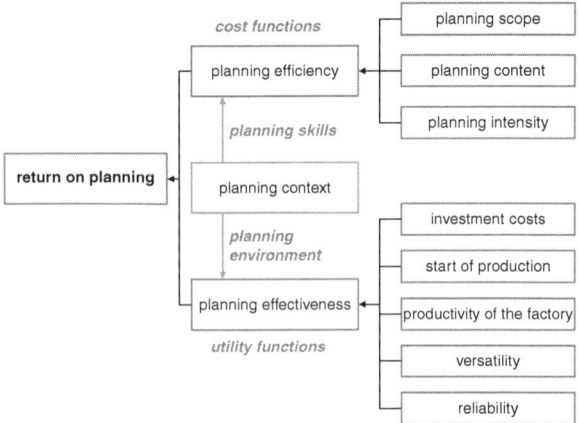

planned within the framework of corporate development, but also to create measurable added value in order to secure the long-term competitiveness of a company (Bea et al. 2008; Pawellek 2008). In this regard, the added value achievable depends on two interdependent variables: planning efficiency (cost function of planning) and planning effectiveness (utility function of planning) (Schäffer et al. 2001). Both put into correlation with one another result in the *return on planning,* which reflects the efficiency of the planning project (see Fig. 4.43).

Planning efficiency is the KPI for the input factors of planning activities. It is a measure of the economic viability of the decision-making and decision-implementation function (Schäffer et al. 2001). In relation to the efficiency of planning, typical questions asked include: "Is the planning to complex and too expensive?" and "Will resources be wasted during planning?" Planning efficiency comprises the following factors: planning scope, planning intensity and planning content. The *planning scope* reflects the size of the planning object (in the dimensions of planning width, planning depth and size of contemplated solution space). The unit of measure for the planning scope depends on the *planning contents*, which are determined by the choice of planning modules. *Planning intensity* is the measure for synchronicity of the planning activities and, in addition to the planning effort, reflects the communication and coordination effort. The costs for a given scope, content and intensity depend on the planning competence, which incorporates the respective experience, the available tools and capabilities, and the respective cost rates.

The value of the factory planning project is defined by the level of goal attainment (Corsten et al. 2008). The basic target parameters of factory planning include:

- Investment costs: i.e., the costs for the investments in the fixed assets and the current assets. These include, amongst other things, costs for the building envelope, the machinery and equipment, the necessary stocks of material and other one-off costs incurred in setting up the factory.

- Production start: The factory can generate revenue from the moment production starts. The sooner revenue is generated, the faster the investment costs can be repaid through the positive cash flow. Furthermore, additional gains are achieved if the market environment enables additional revenue to be generated by launching innovative products early. Planning intensity affects the flow of the planning.
- Factory productivity: the key parameter for assessing planning results is the productivity of the factory once in operation. As a fundamental parameter, productivity defines the ratio of output quantity to input quantity. The structures defined by FP for process and resource design significantly influence productivity.
- Adaptability: Furthermore, to ensure a high degree of adaptability, during factory planning all factory objects must be designed so as to be adaptable to turbulence due to internal and external influences.
- Reliability: In addition to the afore-mentioned target parameters, the quality of the planning is also expressed through the reliability with which the planning results can actually be achieved in reality.

Payment flows can be assigned to the target parameters to allow the monetary effects to be depicted. In this respect, it is important that cause and effect are clearly assigned in order to ensure that only the effects caused by factory planning are taken into consideration. Furthermore, the relevant planning context must be also taken into account.

Goal System During Ramp-Up Production ramp-up generally refers to the time when a prototype is transferred to actual series production (Schuh et al. 2005). The challenges of factory *ramp-up* differ from those of factory planning. Moreover, the processes are not yet well-practised and the production processes are not yet running as they will in later operation, hence different goals emerge in respect of factory operation.

The return on invested capital is an essential factor for the success of a company in the long term. Short and efficient production ramp-up phases are crucial here for reducing the payback period for capital invested in the ramp-up of a new product (Scholz-Reiter and Krohne 2006). The primary goal of ramp-up management is therefore to ensure that the time between product development and market release is as short as possible (see Fig. 4.44).

In addition to the deadline goals, there are cost and quality goals, forming the classic "goal triangle" (Peters and Hofstetter 2008). In the goal system for the ramp-up, these general goals are converted into three goal classes: effectiveness goal, deadline goal and efficiency goal.

The effectiveness goal of ramp-up is to achieve a certain output level with the production of a new product. The goal includes achieving the highest possible quality for the product and the process. The three goal contents for the effectiveness goal are therefore:

- Enable a high production capacity
- Provide large quantities of good parts
- Enable a high quality rate

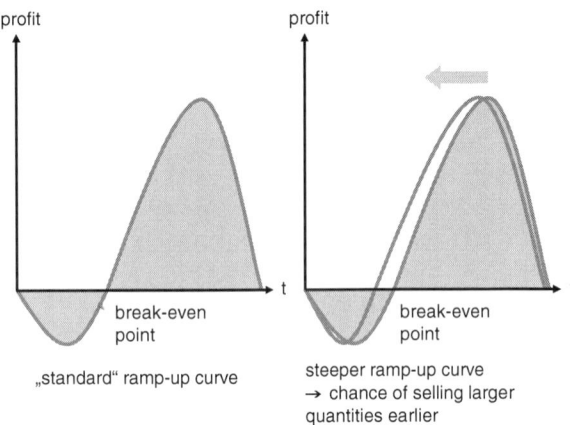

Fig. 4.44 Steep ramp-up curves as the main goal of ramp-up management

The deadline goal involves keeping control of the ramp-up schedule. Project completion plays a decisive role in this respect. Project completion should not deviate from the planned deadline because any delays to production start-up may damage the company's image or result in contractual penalties. The goal content for the deadline goal is therefore to enable early completion.

Due to the significant burden on companies during ramp-up, the deadline and effectiveness goals should be achieved as efficiently as possible. Hence, the goal content of the efficiency goal is low ramp-up costs.

This goal system is also reflected in studies. A survey of automotive suppliers conducted in 2005 revealed that the main target parameters of the ramp-up strategy behind ramp-up management are adhering to deadlines (25%) and achieving the designated quality goals (21%). The survey showed that the following are also important: adhering to the cost and budget goals (17%), keeping control of the product and/or process complexity (16%) and a short overall ramp-up time (11%). Factors of less significance include high flexibility with respect to piece numbers and variants (6%) and a high utilisation rate (4%) for the resources used (Schuh et al. 2005).

Besides developing the goal system, standardising the ramp-up goals across company functions is also important. In particular, during ramp-up, the often local goal of the decision-maker conflicts with the global ramp-up goals (Gottschalk and Hoeschen 2008). A typical problem is that, right up until shortly before market launch, product development may still need to make changes to the product in order to achieve a more flexible response to altered customer requirements. This need, however, does not coincide with the wishes of the production staff who like to be sure about the product to be manufactured at the earliest possible point so they can plan and set up the processes accordingly. Both the obvious and the hidden goal conflicts must be made transparent to the staff. Furthermore, it is necessary to create incentives that support the attainment of the global ramp-up goals. In addition, the ramp-up system must have escalation mechanisms to resolve goal conflicts by appointing higher authorities.

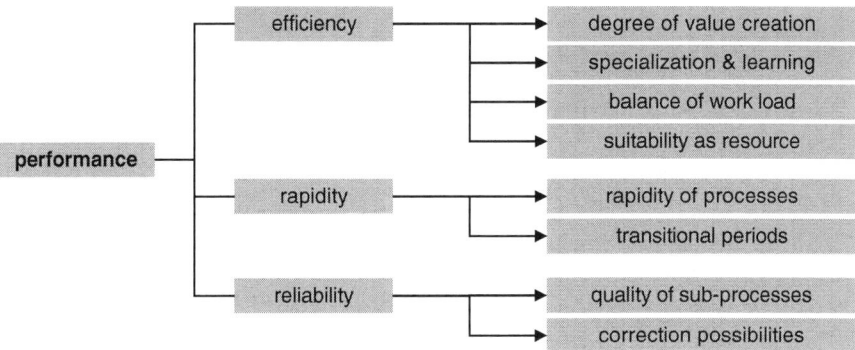

Fig. 4.45 Goal system for factory operation

Goal System for the Plant The most important goal for the production plant is to generate profits (Westkämper 2006). Profits are generated when the output of the production system is greater than the input. To characterise a production system in respect of the potential to generate profits, the output is compared to the input. This gives the transformation variable productivity, or profitability, when the output and input variables are measured in monetary terms.

Often, due to the complex causal relationships of the production system, the pursuit of profit, productivity and profitability cannot be directly influenced by decisions in the production department. Hence, to reduce the complexity, based on these superordinate goals, subordinate goals are formed until the effects of operative decisions demonstrate a transparent goal reference. In this respect, efficiency, speed and reliability are on the first level (Fig. 4.45).

- *Efficiency:* The primary goal of production is to minimise the costs incurred. Only then can a company achieve the pricing ability to sell products on the market and generate profits. In this regard, costs are incurred not only in the direct value-adding areas, but also in the indirect areas. Both areas attain operative goals through the efficiency variable.
  - The *degree of value creation* is the ratio of value-adding to non-value-adding work time. Value-adding activities lead to planned changes to the product. The aim is to maximise the proportion of value-adding activities (Lotter 2006).
  - *Specialisation and learning* are the result of dividing the work content. The more specialised a person's activity is, the shorter the repetition cycle and the faster the learning effects ensue. The learning effects produce a degressive increase in productivity across the repetitions of the work process (Ullrich 1995).
  - *Workload balancing* denotes the capacitive synchronisation of all subprocesses so that they all take the same amount of time and none has to wait for another. This reduces the waiting time of employees and maximises equipment utilisation (Kratzsch 2000).

- *Suitability of resources*: Resource suitability, i.e., the degree to which a resource is suitable for a task, also has an effect on efficiency. An overlap causes unnecessary costs while a shortfall results in the work task not being accomplished at all or not as desired (Kern 1962).

- *Speed:* Customer satisfaction is influenced not only by the product features, but also by the lead time and availability of the product. While product development bears the main responsibility for developing the product features, production is responsible for order lead times. The aim is to process orders as quickly as possible and reduce the lead time.
  - *Speed of the processes:* According to REFA, the cycle time of the production process is divided into primary processing time, secondary processing time, additional time and stoppage (N.N. 1997). The aim is to reduce all of these time slices through technological or organisational measures in order to reduce the cycle time.
  - *Transition periods:* Another influence on the speed within a chain consisting of several subprocesses is how much time a product needs between the process steps. The ratio of value-adding time to idle time is usually in the lower, single-digit percent region. Idle and waiting time depend on the stocks in the production department (length of the queue) and the order of execution (Lödding 2008).

- *Reliability* in production means the predictability of fulfilment of a given delivery promise. It is comprised of the quality of the subprocesses (process capability) and the possibility of correction. The aim is to increase reliability in order to reduce the quality costs while maintaining the same product quality.
  - The *quality of the subprocesses* can be described as the spread of results around an expected value. Based on this, statistical means can be used to describe to what extent the processing quality or the processing duration of a process fluctuates. The aim of increasing the predictability of the results is to minimise the spread of results of production processes and of indirect processes (e.g., in logistics) (Pfeifer and Masing 2007).
  - *Correction possibilities* are required if fault occurrence cannot be excluded. In the simplest case, detection of faulty products is followed by a reject declaration and disposal. If rectification or fault removal is more cost-effective than disposal and re-production, then appropriate measures must be taken to this end. The aim is to correct faults in a reliable, cost-effective and sustainable manner with minimal delay (Pfeifer and Masing 2007).

4.3.4.3 Considering Corporate Culture

Corporate culture describes, on (the) one hand, the capabilities and the knowledge of a company and, on the other hand, staff motivation, i.e., their attitude towards the company, the product, the management, colleagues and the task as well as their own perception of and goals for all processes (Bleicher 1999). Employee attitude is hugely important for the survival of the company, especially from the point of view of *sustainability*. Competitive production calls for creative, motivated staff who identify with their work. This kind of employee attitude results from the company-employee relationship being developed over many years. A suitable means for describing and developing such a relationship is the *stakeholder value* concept.

Here, *stakeholders* are all the internal and external groups of people who are, or will be, directly or indirectly affected, currently or in the future, by the entrepreneurial activities. Stakeholders would therefore be customers, for example, or employees and suppliers, but also the state and the public. The heterogeneity of the stakeholders is clearly evident. According to the stakeholder value concept, the management has to consider not only the interests of the shareholders, but also all the stakeholders because without their support, the company would not be able to survive (Freeman 1984). The aim here is to satisfy the needs of all stakeholders.

Being responsible for the direct working environment of the staff, factory planning has to allocate their needs a central role. The *needs of the staff* are therefore a key input value for factory planning. At the same time, the factory also has an impact on the attitude of the employees because the company expresses to the workers its appreciation of its workforce through the design of its work environment. Hence, factory planning and corporate culture influence each other. On (the) one hand, the structure of the factory is culture-forming and, on the other, the circumstances of the existing corporate culture must be taken into account in factory planning.

A suitable means of taking stakeholder value into account during factory planning is the definition and subsequent pursuit of social goals. *Social goals* determine which general practices a company will pursue in the future with respect to its employees, society and the natural environment. Social goals express the company's social responsibility and ultimately depend on the ethical and moral values of individuals. Social goals can be set by law, but the majority are to be determined within the company itself (Rosenstiel 1993). They can be divided into *employee-related social goals* and *society and environment-related social goals*. For factory planning, this means operational implementation of the social goal guidelines from strategic management as well as incorporation of its own goals as appropriate.

In particular, this means that a modern factory has to be a *living space* as well as a production facility. As a *social and communication space*, it must also facilitate social interaction between the employees. Furthermore, the factory is also an expression of *corporate identity* (Fig. 4.46).

The Factory as a Living Space If a living space is defined as life in an environment in which an individual feels comfortable and at home, then it is imperative that the people who will subsequently work in the factory are involved, to a reasonable extent, in planning the factory in the form of *participative factory planning* (Menzel 2000).

As the meaning of work changes from merely a source of livelihood to part of individual fulfilment, the factory must also change. This makes it necessary to overcome the strict separation of workspace and living space that has been ingrained since the beginnings of industry and to create a factory that is also a living space.

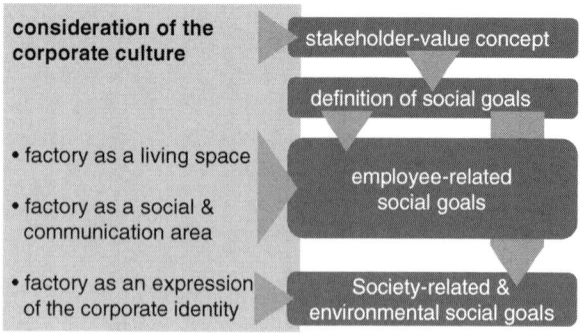

Fig. 4.46 Corporate culture within the context of factory planning

Social goals that serve to make the factory a living space too are usually *employee-related social goals*. Employee-related social goals are based on the individual preservation and development objectives of the employees that have to be met (Klages 1990). It is therefore inherently impossible to define universal rules in this system.

In principle, any measure that makes the employees' time in the factory more pleasant can contribute towards a "life-friendly" factory design. Numerous laws, regulations and decrees prescribe a *health-friendly* design, for example with regard to noise levels and air quality. Naturally, they also take into account ergonomics in workstation design. In view of demographic developments, one obvious recommendation would be a disabled-friendly design that allows for possible limitations in people's physical abilities (Reinhart et al. 2010).

Furthermore, there are some approaches that go beyond basic health protection and actively promote health. *Health promotion* includes all activities designed to strengthen and support all the healthy structures in employees. Health promotion aims to support the mental and emotional welfare of employees and enable effective self-management, leading to good health and well-being.

The factory as a Social and Communication Space A factory is a social and communication space if it facilitates interpersonal contact between the employees who work there. While the only purpose of working used to be to earn a living, employees today demand some quality of life in their profession. This includes the possibility of *interpersonal communication*. Social goals that serve to make the factory a social and communication space too are therefore usually also *employee-related social goals* and hence are also based on the individual preservation and development objectives of the employees.

Since building up social relationships between employees is a long-term group process, good planning is required in this respect too. Shift schedules are one aspect of this and should ensure that existing groups are not subject to unplanned separation. Workstations that do not totally isolate the worker can also help. Against this background, providing social spaces is a necessity and, furthermore, is a legal requirement when there are more than ten employees or if required for health and safety reasons.

Fig. 4.47 Effect of factory planning on corporate image

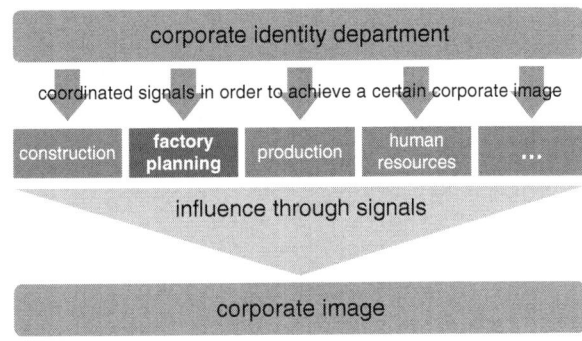

It is also important for the factory to support the company's communication structure. Often, easy and barrier-free communication between employees and their superiors is desired and this can be facilitated through proximity between direct and indirect functional areas.

The Factory as an Expression of Corporate Identity Corporate identity is a holistic and integrative way of thinking that links all the individual measures together conceptually, coordinates them strategically within the context of the whole and is geared towards identity creation, sustainability and image formation (Regenthal 2009). Corporate identity is the self-image of a company. *Corporate image*, on the other hand, is the image that outsiders have of the company (Regenthal 2009).

The primary goal of corporate identity is therefore to convey within the company a culture involving a set of practices and standardised guidelines according to which the employees should act and make decisions. The secondary goal is to reconcile the corporate image with the corporate identity. Hence, *all* the company's external communications must correspond with the desired identity.

Naturally, this also applies to factory planning. The factory is required to convey the corporate culture both internally and externally. It must reflect the corporate identity, as it is intended to be perceived, just as immaculately as a brochure or showroom (Fig. 4.47).

The most superficial role of the production facilities in terms of public perception of the company is reflected in particular by the increased importance of the aesthetics of an industrial building, glass factories and the interest in plant tours or television documentaries.

Since the factory has a major impact on the day-to-day working lives of the employees, it must express the corporate culture internally too. Continuity and credibility, in particular, are important for anchoring the corporate culture. Social goals that serve to make the factory an expression of corporate identity too are usually *society and environment-related social goals* that express the external social responsibility of the company. Fundamental conditions indispensable for life should be amongst the top corporate goals (Hahn 2009). This applies in particular to demands for

- protection and improvement of the environment
- innovations, but manageable technologies
- preservation of the social market economy as a competitive economy
- safeguarding the free and democratic constitution

For the purpose of achieving the company's main goals, this means voluntary limitation to a socially acceptable goal system if necessary.

4.3.4.4 The Aachen Factory Planning Model

The Aachen factory planning procedure is a modular approach to planning that can be adjusted according to company-specific constraints and circumstances. The key components are the planning framework and configuration logic, the project structure, the planning procedure and integrated project and ramp-up management.

Planning Framework and Project Configuration Similar to modularisation of the "factory" planning object, modularisation of the planning tasks allows standardisation of the contents within a module (processes, tools, etc). The dependencies between the planning modules are described via defined interfaces (input information and results). This framework forms the basis that makes it possible to individually configure the planning process from the outset and to adapt it to altered boundary conditions whilst the project is ongoing.

Planning Modules Planning modules have the function of encapsulating the content of planning tasks into subtasks and reducing the interdependencies, and therefore the complexity, of planning. This allows decoupling, parallelisation and decentralised planning of subtasks via defined modules. Hence the planning modules each relate to a specific planning task for a specific object area, e.g., layout planning. Each module is assigned appropriate procedures (block layout planning, detailed planning), methods (affinity matrix, Sankey diagram), tools (factory planning table, CAD) and evaluation criteria. These procedures, methods and tools to some extent form alternatives that differ in terms of the level of detail, planning effort and quality of results. Appropriate procedures, methods and tools can be selected according to the desired result, capacity and budget, e.g., block layout planning in 2D using Powerpoint or workstation design 3D using suitable visualisation and detailing. Planning modules can be run several times in one project with varying levels of detail (Fig. 4.48).

The planning module structure is divided into the encapsulated/inner area and the interfaces. The inner area contains the alternative procedures, methods, tools and evaluation criteria. Via input and output information, the interfaces describe the dependencies and interactions with other planning modules. This structure allows the planning approach to be standardised with regard to the modules that need adapting for use in the specific context. This means that the planning modules in the project are instantiated (brought to life). Hence, the workstation design module can be set up several times for various workstations in the assembly and production areas and can also be processed by various teams. The actual planning project can then be

Fig. 4.48 Planning module with input and output information

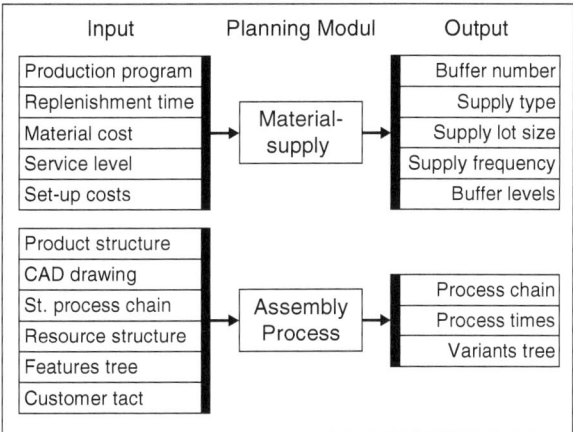

individually compiled from these modules. This way, module-specific experience can be gained, which will improve module processing and achieve learning effects.

Dependencies and Causal Relationships The dependencies and causal relationships between the planning modules are generically described by the interfaces. Based on the input information of a planning module, which is determined by the information requirements (e.g., sales volume, batch size, etc.) of the methods and tools used, the preceding modules and the necessary boundary conditions and decisions can be identified.

The planning results of a module form the initial information available for other planning modules or even for re-running the original module. Hence, input information is obtained from various preceding modules and results are made accessible to various succeeding modules. Just like the planning modules, the generic interfaces must also be adapted for use in the specific project and detailed in order to ensure communication between the modules (Fig. 4.49).

These prerequisites and dependencies determine the module processing sequence within the planning process. At various points between the planning modules there are circular arguments (chicken and egg problem). This means that two modules are each waiting for the input information from the other in order to generate their own results. These circular arguments can be resolved using initial values/assumptions, which are set at the outset and then confirmed or adapted in a subsequent iteration. It is useful to have initial values that describe a planning corridor or set limit values that must not be exceeded or undershot, for example a piece number corridor that has to be taken into consideration in the planning scenario. Provided the values are within these limits, work can then commence without the input information affecting other planning modules. If these limits are breached, or if narrower limits are required in order to preclude planning alternatives, new values must be set at the interfaces. For example, more precise piece number scenarios may be needed in order to decide

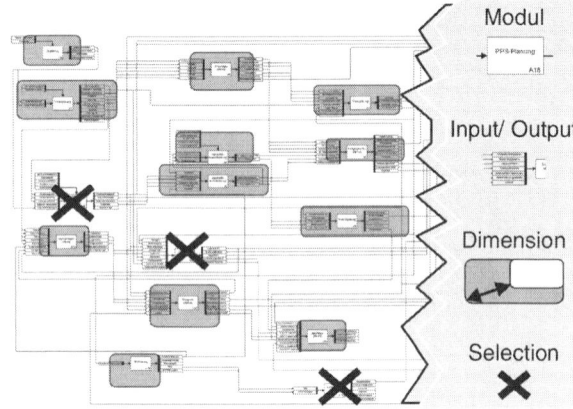

Fig. 4.49 Selection and dimensioning of planning modules for configuration of the planning process

whether resource capacities are sufficient or need to be increased. This requires coordination, especially in cases of step-fixed, rather than linear, dependencies.

Knowing about the dependencies and interactions between the planning modules makes it possible to decide which planning modules can be parallelised to what extent, and how closely the coordination and collaboration between the planners must be synchronised.

Configuration Logic The basis for project configuration is provided, on the one hand, by the planning tasks (planning object and restrictions) and, on the other, the framework, comprising planning domains and modules. In the first step, the planning domains and planning modules needed for the project are selected according to the planning tasks. Hence, when reorganising an assembly area, location planning is not required. Based on the causal relationships, other planning modules that generate input information can be identified and must therefore be integrated as preceding modules if the relevant data is not yet available.

The "standard modules" must be instantiated according to the planning task. In this respect, instantiated means that the modules are also defined and detailed according to the specific project. Similar to building the subsystems of the factory, the modules are instantiated according to subtasks. The planning modules are sized taking into account the available planning budget and the existing resources, i.e., the level of planning detail, and the appropriate methods and tools are selected.

To avoid circular arguments, initial values are set and these are iteratively adjusted during the project. Furthermore, during planning it is possible to identify missing or inaccurate input information, which can only be generated by the integration of new planning modules. In such situations, the procedure has to be reconfigured (Fig. 4.50).

Adaptive Project Organisation and Structure Experts from all sections of the "old" production system are involved in the planning, design and ramp-up of a "new" production system (entire factory or sections thereof). In addition, external

Fig. 4.50 Timing of the modules in the planning project

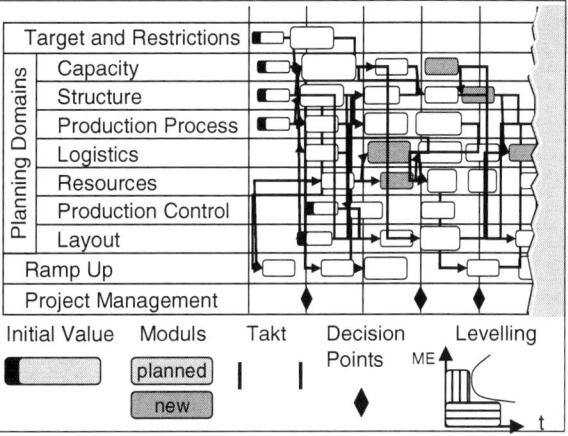

consultants and service providers are also often called in to help with the planning and implementation. As systems theory would understand it, the factory planning process is therefore the creation of a new system from an old system. The composition, structure and size of the planning team, as well as the situation-specific adaptation and suitable communication structures, have a decisive influence on the planning project.

Planning Participants and Stakeholders Depending on the size of the object area of the planning task, up to 100 different professionals may be involved in the planning over the course of the project. The planning team is comprised of experts from the concerned company departments, external consultants and service providers as well as representatives of other stakeholders and they all bring their own perspectives, motivation and goals to the project. Hence, the individual planners pursue very different goals during design and interpretation of the planning object or individual subsystems and these need to be harmonised with the superordinate project goals (Fig. 4.51).

A planner involved in the project may take on several functions/roles and therefore may collaborate on several subprojects. Whilst a planner may be responsible for managing one subproject, in another she/he may take on the role of assistant, thus simplifying the exchange of information between subprojects.

Moreover, the majority of the planning participants are only involved in the project for a limited period to fulfil specific planning tasks. In particular, staff from operational departments, such as production and assembly, can often only be made available to participate in the planning to a limited extent. Involving operational staff with expert knowledge at an early stage makes it possible to consider and, if applicable, preclude alternative solution concepts, thereby avoiding planning errors.

When putting together the project team, including staff assigned to work on the future production system not only helps with error avoidance, but also facilitates acceptance of the planning results.

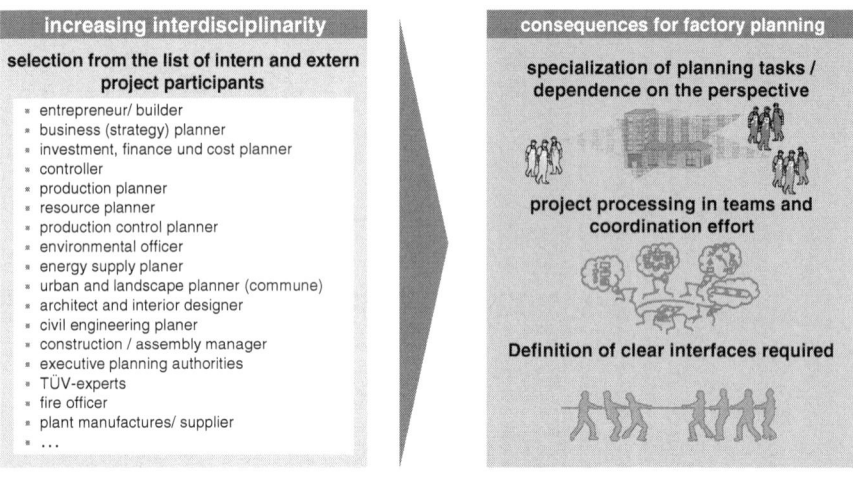

Fig. 4.51 Interdisciplinary planning participants in factory planning projects

Project Structure and Composition of the Team To ensure the project is executed on schedule and within budget, the project must be fully structured into partial assignments. This partial assignment formation is referred to as the project structure and comprises all the assignments necessary to implement the planning project in the form of a structured order tree. The project structure is derived from the modularisation of the "factory" planning object and the existing arrangement of the factory planning process. Configurable planning modules allow for the changeable and iterative nature of real factory planning processes (Fig. 4.52).

As with the planning assignments, the planning team can also be divided into subteams. For the most part, the initial team is comprised of a small group of closely networked employees who work directly with one another. They come from the staff department/ unit for factory planning/plant structure planning, if the company has one. However, with smaller companies, in particular, that do not have such a department, the planning team is made up of employees assigned with important roles in the future production system who are sometimes released from their usual day-to-day duties to assist with the planning. The planners are taken out of their previous organisational structures to form a new unit with other (usually heterarchical) structures. The initial phase is characterised by fast, uncomplicated communication in small adhocratic teams (Fig. 4.53).

Synchronisation of the Factory Planning Process In the factory planning process, coordinating lots of employees from various disciplines leads to an increased planning effort. For efficient and effective synchronisation, introducing takt time into the planning process is an alternative to traditional upfront project management. Takt time is oriented towards the organisational principles of cybernetic management, because the objective of better synchronisation with reduced planning effort can only be met through better self-organisation of the system under consideration.

4 Virtual Production Systems

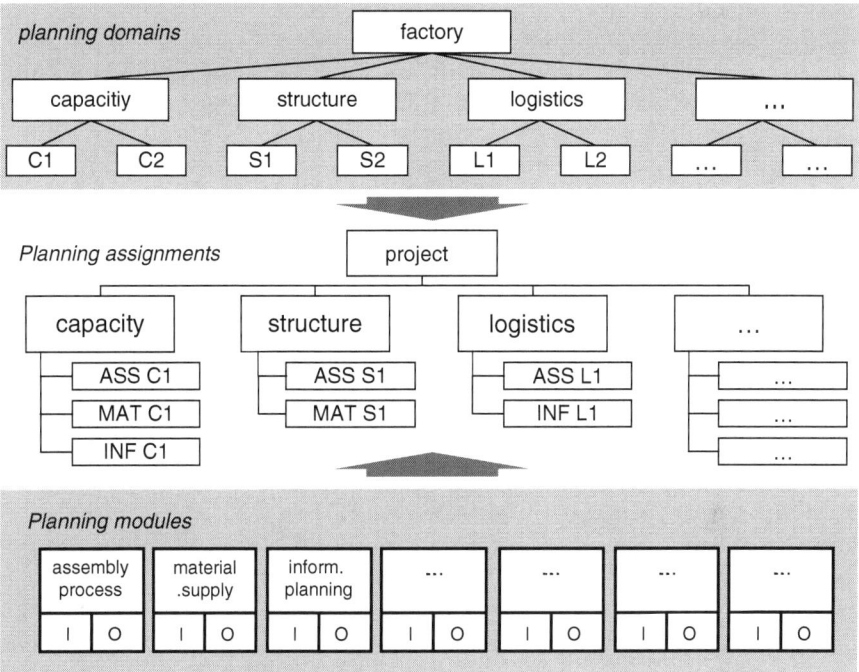

Fig. 4.52 Planning factory planning orders

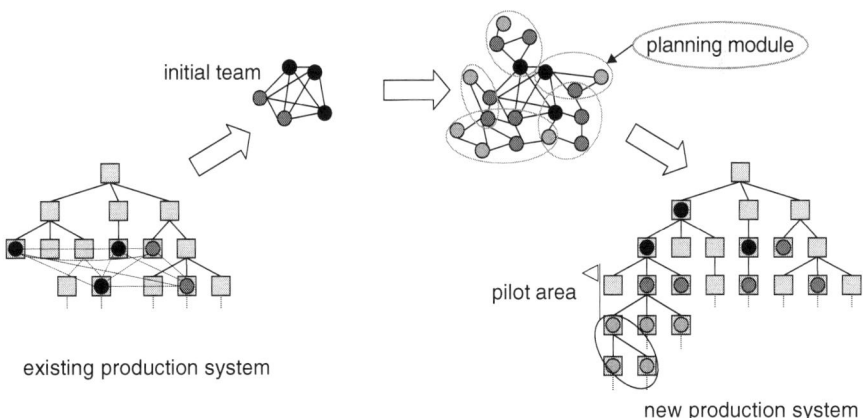

Fig. 4.53 Planning of a new production system out of an old one

The self-organisation approach for complex systems is therefore a central research topic of cybernetic management (Malik 2004).

Cycle-Based Triggering of the Planning Process The introduction of takt time to factory planning processes is derived from the Beer's Viable System Model (Beer 1980). The diagram below shows the process for cycle-based triggering of factory planning processes. The arrows represent interfaces between the individual activities of the process. The task of assignment planning is to transfer the product and process objects to be implemented into *orders*, which can then be handed over to the teams (1). In order planning, the project manager prioritises the individual orders required for implementing a process object, which allows the processing sequence to be derived. The list of prioritised orders is called the *order list*. The project team then plans the next cycle and the orders to be processed (2). In this respect, the work required to process the orders on the order list is assessed based on the experience of the experts on the project team and translated into *tasks* which the experts execute by the end of the cycle. The result of the cycle planning is reported back to the project manager via the information system (3), but in this regard, the only feedback given is which orders can be completed in the next cycle. The project manager identifies the delta between order planning and task planning (4) and then has the opportunity to influence the project team (2) (Fig. 4.54).

Upon completion of the task planning, the project team begins processing the assignments (a). In doing so, the team members take control of the work progress for one another (b) and report completed tasks back to each other (c). Should there be any extraordinary obstacles that prevent tasks from being executed, task re-planning is initiated (d). Several tasks of one or more project team members together constitute an order. Controlling the progress or assignment processing takes place in daily team meetings (5). The progress achieved and the work packages remaining up to the end

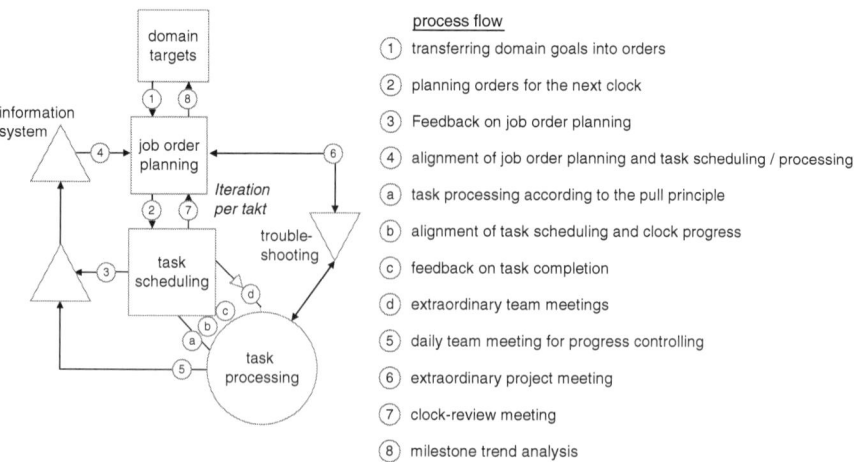

Fig. 4.54 Process for cycle-based triggering of factory planning processes

of the cycle are communicated to the project manager via the information system. Should, at any point, problems arise that cannot be resolved by the project leader or a member of the team, an extraordinary project meeting is called (6) to find a solution to the problem. The extraordinary project meeting is also used to address tasks of an exceptional project nature which, due to their uniqueness and the unpredictability of the method of solution, would interfere with the process nature of takt-time. At the end of a takt, a *takt review* is conducted (7). In this review, the organisational problems of the last cycle are recorded in an *issue list* in order to allow any changes required for the next cycle to be derived. In addition, the process objects are evaluated based on a standardised *result report*. For each order, the result report reflects the result of a quality assurance method which was applied to the relevant process object. At the end of each cycle, the *milestone trend analysis* is updated and, based on this analysis, the steering committee evaluates the progress of the development project so they can intervene, if necessary, in terms of regulating critical deviations.

Prioritising Planning Orders Defining and prioritising orders is the project manager's prime opportunity to control the organisation of the factory planning process. A complete assignment, therefore, consists of the following four elements:

- Designation of the element to be planned
- Criteria of acceptance for the results to be achieved
- Indication of the priority of the assignment
- Designation of the team responsible

The project leader defines individual development assignments based on the structure of the factory planning task. In addition to stipulating the results to be generated, a factory planning assignments also indicates the priority with which this order should be processed and assigns responsibility in this respect. These stipulations and indications are designed to clarify the planning order for the team and to shore up the discussion about the orders during takt planning. While the main structuring methods for elements one and two of a development order have already been explained in the previous section, the next section will explain the procedure for prioritisation and for designation of the team responsible.

Prioritising planning assignments is executed in two steps. First, the project manager assesses the relative weighting of the development orders. Then, the project teams organise the assignments into a design structure matrix in order to take account of the main dependencies. These are in turn fed back into the relative weighting system and thus the orders are put into a provisional sequence. The sequence produced is provisional because the order list may be revised at any and every cycle during the project.

Wiegers' relative weighting method is an appropriate method for prioritising development orders (Wiegers 2005). This is an 8-step method for producing a prioritised order list:

- First, all assignments are composed in a list.
- The project manager estimates the relative value of each order on a scale of 1–9, whereby 1 indicates the lowest value and 9 the highest.

- The project manager estimates, on a scale of 1–9, the relative negative impact that could be incurred if the assignment were not executed. If, for example, it were not possible to produce a particular result, this could have a severe negative impact on the progress of the entire project.
- The sum of the relative value and the relative negative impact reveals the overall value contributed. The value contribution can also be given as a percentage.
- The project manager estimates the effort required to process the assignments in man-days [MD] and calculates the relative effort as a percentage.
- The project team estimates the risk associated with each assignment on a scale of 1–9, then calculates the relative proportion as a percentage.
- Next, the priority number P is calculated using the formula below:

$$P = \frac{\text{Value proposition}[\%]}{\text{Costs}[\%] + \text{Risk}[\%]} \qquad (4.1)$$

- Finally, the assignments are sorted by priority number in descending order (see Table 4.1).

When planning a technical facility like a factory, in addition to the value contribution, the effort and the risk, technical dependencies also determine the planning sequence. For this reason, following prioritisation, the design structure matrix (DSM) is used to structure these sequence dependencies (Eppinger et al. 1994). In optimising the sequence using the DSM, however, it is important to check whether the sequence produced through the prioritisation can be implemented despite technological obstacles. The sorted DSM reveals which development orders are independent of each other and which must be executed consecutively because of interdependencies. Along with the priority number, the prioritisation of the development order is also produced and this must be indicated in the description of the development assignments.

The unique time allowance that a takt specifies for assignments processing can tempt teams to vary the degree to which the work results are achieved and to avoid consistent validation of the results. However, to prevent erroneous or incomplete results of one cycle and planning module being transferred to the next cycle or planning module, all results generated must be subjected to a quality inspection for which a suitable validation method is used. The results can be validated through simulations, calculations and reviews, but in particular they can be validated by transferring them to related planning modules. If individual validations are particularly time-consuming, they can be scheduled into the next takt as additional tasks for other project teams.

4.3.5 Industrial Relevance

The methodology developed has already been applied in numerous projects in the Laboratory for Machine Tools and Production Engineering (WZL) and is presented here with reference to a case study involving re-planning in machine tool engineering.

4 Virtual Production Systems

Table 4.1 Relative weighting of factory planning orders. (According to Wiegers 2005)

Planning order	Relative advantage	Relative penalty	Value added	Value added (%)	Effort (MT)	Effort (%)	Relative risk	Risk (%)	Priority number
Layout hall	5	3	8	32	8	15	1	6	1,52
Layout cell 1	6	2	8	32	20	38	4	25	0,52
Layout cell 2	1	1	2	8	3	6	2	13	0,42
Resource requirement cell 1	3	1	4	16	13	25	3	19	0,36
Resource requirement cell 2	2	1	3	12	8	15	6	38	0,23
Total			**25**	**100**	**52**	**100**	**16**	**100**	**3,04**

4.3.5.1 About the Company

The company belongs to a dynamically expanding group in a world-leading position in the machine tool engineering industry. At the site in question, the company produces components such as spindles and rotary tables for machine tools. Following strategic realignment of the added value distribution, the range of parts at the site in question is being extended with additional manufacturing and assembly parts from other brands and sites. This results in different capacity requirements which must be met by improving the existing manufacturing structures. The initial structure is characterised by the division of production according to the workshop principle without the material flow being reflected in the current layout. In the case study in question, segmentation of production according to product type was examined in terms of feasibility.

4.3.5.2 Application of the Methodology Developed

At the beginning of the application, the creation of project-specific module maps and the associated identification of all planning modules relevant to the project were dependent on the planning tasks. In the first step of the network-based task planning, the core modules Layout Planning (LAY), Implementation Planning (UMS) and Investment Planning and Cost Planning (INV) were selected first of all. Their importance was derived directly from the project order (Fig. 4.55).

Based on the networking of the selected core modules, in the second planning step, other planning modules were automatically highlighted in the module map and

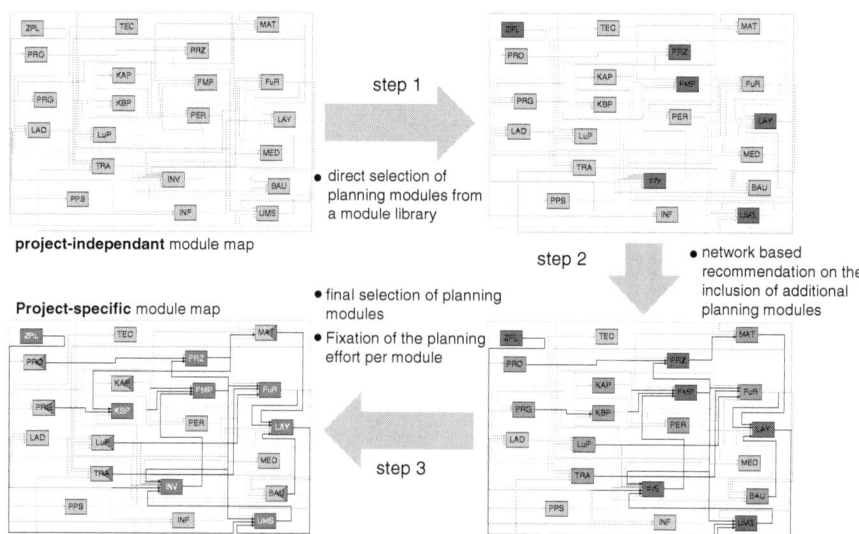

Fig. 4.55 Process to select a project specific module map

4 Virtual Production Systems 317

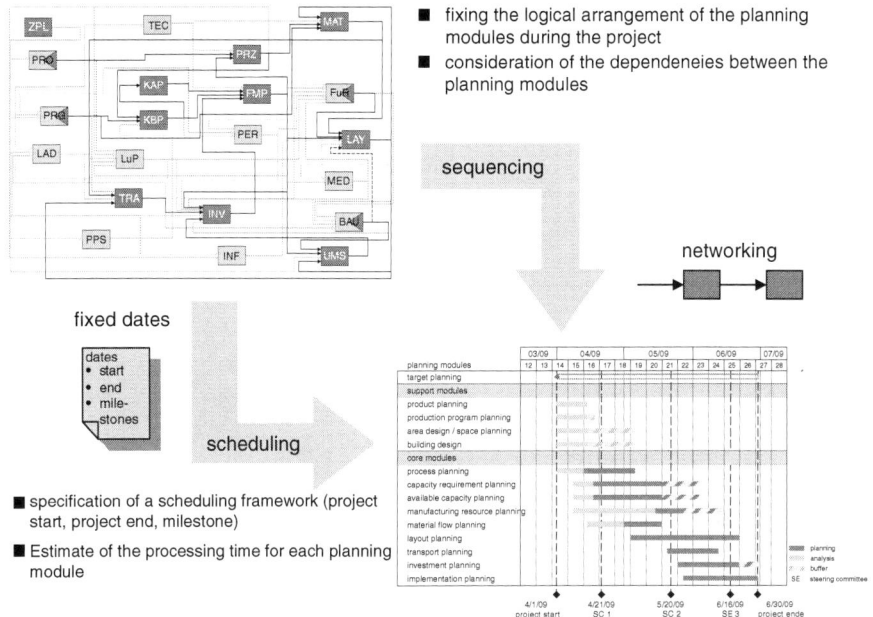

Fig. 4.56 Creation of a project-specific schedule

thus suggested for consideration in the project. Via the input network of the Layout Planning module, the planning modules Material Flow Planning (MAT), Manufacturing Resource Planning (FMP), Area and Space Planning (FuR) and Information Planning (INF) were highlighted accordingly in the module map. To allow for restrictions during layout planning, the modules Building Design (BAU) and Media Planning (MED) were also indicated.

In the next step (step 3), taking account of the planning context and the detailed project task, the processing depths of the previously identified modules were determined. Based on the project-specific networking of the planning modules, the dependencies of the planning modules were analysed, as was the logical arrangement of the modules within the course of the project as necessitated by these dependencies. Any potential time overlap of the planning modules was established in each module based on the data already available at the start of the project and the foreseeable progression of the planning (Fig. 4.56).

In the course of the project, it was revealed that extending the range of parts with additional manufacturing and assembly parts from other sites required a new planning of the capacity requirements and corresponding adaptation of the available capacity. The project-specific module map made it possible to obtain an overview of the project structure and the project-specific networking of the planning modules. As the project progressed, the module map also served as a communication interface during execution of the planning activities. Hence, it was possible to effectively

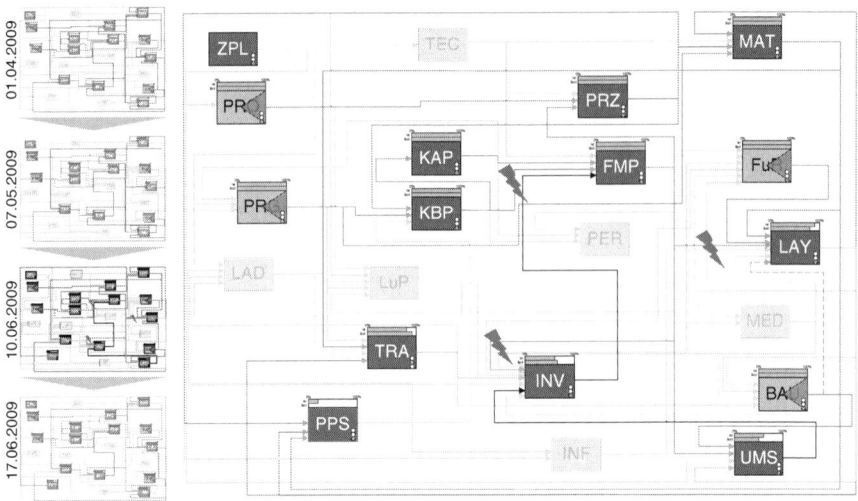

Fig. 4.57 Planning status query

combat the ever-recurring dynamics and turbulence involved in factory planning projects (Fig. 4.57).

The project was completed on schedule despite the schedule being tight. The necessary flexibility was maintained as the project progressed without, however, losing sight of the goals of profitability and efficiency.

4.3.5.3 Conclusion for Industrial Application

It has been demonstrated that explicit consideration of the relevant aspects in the regulatory framework (decision-oriented, appropriate for the complexity, value-adding and participative) is a suitable model for successfully planning factories in a dynamic and turbulent environment. In particular, with this approach, it is possible to satisfy the need to shorten the planning time, which is highly relevant for factory planning in high-wage countries. Hence, this helps contribute towards the competitiveness of high-wage countries.

4.3.6 Future Research Topics

Building on the existing work on modularisation of the planning procedure and the supporting planning tools in ICD-B1.1 as well as the development of the simulation platform in ICD-B1.2, a planning and control cockpit for production systems can be built in the next step. An appropriate cockpit provides the ability to support both the planners during design and simulation of the production system and the decision-makers during operation of the production system with respect to dealing with faults and minor reorganisations. This ensures all planning and decision levels, from the manager to the specialists through to the employees working at the machine, are provided with the necessary information at an appropriate aggregation level.

For this purpose, the planning and simulation models developed in the initial funding period, as well as other commercially obtainable models, must be linked via the simulation platform developed and existing gaps, in terms of the production system levels not addressed, must be closed. Furthermore, in addition to the planning data and simulation results, existing data from production, e.g., historical transaction data depicting the production range over recent years, can be used to supplement missing data.

Of central importance is the development of logics that make it possible to aggregate or detail data in a useful manner across the levels, i.e., from the process and material level through to the factory level, in order to provide the right granularity of information. This user-friendly presentation and adaptation of results and information can then be made available or visualised via the cockpit. Visualisation will make it possible to test various scenarios against one another and to try out and evaluate alternatives. The underlying database and simulation platform must therefore be able to guide the planners and decision-makers with intuitive models and representations without overwhelming them with the potential density and availability of information.

4.4 Integrative Process Chain Simulation for Material and Production Technologies

Georg J. Schmitz, Stefan Benke, Gottfried Laschet, Markus Apel, Jan Rosenbaum, Ulrich Prahl, Patrick Fayek, Sergey Konovalov, Hendrik Quade, Stephan Freyberger, Thomas Henke, Markus Bambach, Tim Arping, Thomas Baranowski, Barbara Heesel, Urs Eppelt, Ulrich Jansen, Eduardo Rossiter, Britta Kuckhoff, Thomas Gries, Markus Linke, Tatyana Kashko, Kirsten Bobzin and Nazlim Bagcivan

4.4.1 Abstract

An increasing complexity of products and their manufacturing processes combined with the demand for tight tolerances and outstanding functional properties is leading to an increasing use of integrative simulation approaches in the development of materials, of manufacture, production and design processes. The properties of materials and components are determined not only by their chemical composition but particularly by their microstructure. The evolution of this microstructure starts with the solidification of the parts from the homogeneous, isotropic stress-free melt and continues throughout the entire chain of manufacturing processes (e.g., shaping, heat treatment, cutting/joining, coating etc.). After exposure to operating conditions and a possible repair process, microstructure evolution eventually ends with failure and subsequent recycling of the component. The microstructure defines the properties of the material relevant for its application. The microstructure has also a particular influence on the manufacturing processes themselves. An integrative approach, allowing a description of the entire component history both at the macroscopic scale of the component as well as at the scale of the microstructure enables the development of new materials being tailored to special applications as well as identifying and optimizing the process chains for their manufacture. The sheer number of parameters involved in the history of a component makes this kind of integrative analysis only possible by means of computer simulations. Even these require linking of a heterogeneous variety of simulation models for different process steps at different time-and length scales. The basis towards such an "Integrative Computational Materials Engineering" (ICME) is now available in the form of an open, modular, standardised, expandable simulation platform, the Aachen (*Aix*) *Vi*rtual *P*latform for *Ma*terials *P*rocessing concept ("AixViPMaP"). The functions and the benefits of such a platform are investigated on individual test cases of economic and scientific relevance. These test cases entail questions about the influence of process chains on the mechanical material properties (e.g., tensile strength, flow curves, isotropic and/or anisotropic), on optimised process conditions in terms of preceding "upstream" processes (e.g., optimised temperature/time profiles in heat treatments), on the stability of the microstructure (e.g., fine-grain stability) and on the influence of the process chain on the geometry/distortions of components during their manufacture and when subject to operational load. The materials of the components analysed in the test cases comprise different types of steel, plastics and a textile-reinforced composite material.

4.4.2 *Introduction*

In many application fields the capabilities and functions of a product are largely determined by the materials used. Therefore the functionality of the material is becoming increasingly important. Examples include semiconductor materials, shape-memory-effect materials, high-temperature materials, construction materials, lightweight construction materials or biocompatible materials. Knowing how to achieve required functionalities thus plays an important role in product design. This is reflected e.g., in the 7th framework programme of the European Union whose topic are "Knowledge-based functional materials" (European Commission 2007).

In the context of the production of goods, success in the market is determined not only by the basic functionalities of the material but also by a very broad spectrum of non-functional requirements. These non-functional requirements relate to aspects such as e.g., raw material and manufacturing costs, shipping and warehousing costs, costs for infrastructure, product size and weight, product quality, product safety and liability, life-cycle or recycling capability and their corresponding costs. Approaches to solutions which simultaneously address a variety of these non-functional requirements are reflected particularly in current megatrends such as "ever lighter".

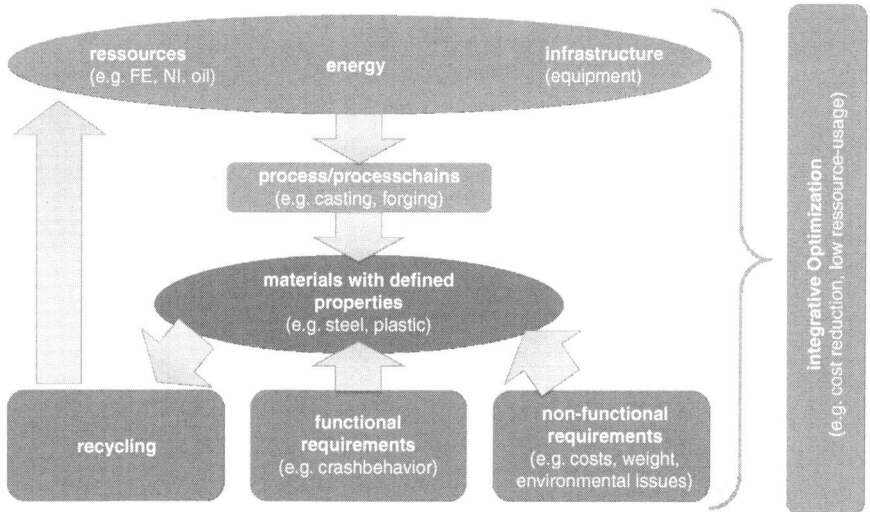

Fig. 4.58 The material in the area of conflict between functional and non-functional requirements and available resources. (Reprinted from Schmitz and Prahl (2011) with kind permission of Wiley VCH)

A specific feature of these non-functional requirements is that they are subject to the ever-changing boundary conditions resulting from a global market, e.g., fluctuating raw material prices or limited availability of raw materials/plant capacities. The producer—and eventually the economy—that can respond fastest to varying constraints and adapt its production rapidly while at the same time maintaining guaranteed functional and non-functional requirements of its product, will have advantages in competition in an evolutionary sense of "survival of the fittest" and will therefore be successful in the market for the particular product (Fig. 4.58).

This requires *identification* of the required basic functionality and of all relevant non-functional requirements and *changes* to these non-functional requirements profile based on the *observation* and *evaluation* of changes of global constraints. The rapid *implementation* of a modified, non-functional requirement profile requires a new design and/or a re-definition of the product and/or its production process. Particularly in the metalworking industry, the high plant investment costs imply that such a redefinition in general is performed by a change in the choice of the material resp. the alloy system or by variations in the process parameters. The rapid availability of a new parameter set which reliably leads to the required product properties and matches the existing plant infrastructure thus provides immediate competitive advantages in the case of modified alloying concepts. In this context, it is notable that more than 70% of the 2,500 steel grades currently available on the market all have been developed in the last two decades (Weddige 2002). Theoretical analyses of the possible alloy configurations (Hougardy 1999) suggests furthermore that by now only a very small part of the full potential has been tapped. Increasingly under investigation and development thus are targeted alloy compositions (Howe 1998) and their optimising in terms of further workability (Reynoldson 1996).

The requirement for a functional material design is therefore a planning-oriented approach. The core of this approach is the provision of effective and efficient methods for developing and processing a suitable material by means of improved models and simulations. When trying to solve the dilemma between the planning efforts spent to develop a production process and the value of the respective, final product it is necessary to trade off costs and benefits of such approaches. In this context the following questions have to be answered:

- What does a requirement-driven material design process look like from a product perspective?
- Which models are required to describe materials, their manufacturing processes and their properties and how can missing model components be developed?
- Is it possible and meaningful to combine a number of models in terms of an integrated, simulation-based engineering approach?
- Is this kind of engineering approach economically viable?

These problems require a systematic horizontal and vertical integration of existing material science models and simulation approaches. In this case, horizontal integration entails the chain of different manufacturing processes while the vertical integration relates to the physical dimensions at the "micro" (the length scale of typical structural features in the material) and "macro" (the length scale of typical component sizes) up to production plant levels.

4.4.3 State of the Art

Material science and engineering can be structured according to various aspects into different levels and functional fields:

- *Length scales:* from "quantum dots" through molecule complexes up to components at the scale of meters
- *Material classes:* polymers, ceramics, metals and alloys, biological materials, composite materials (laminates, fibre and textile composites)
- *Product cycle:* manufacture, e.g., primary shaping, forming, heat treatment, joining, coating, further processing, e.g., cutting, assembly, application, repair and recycling
- *Geometry:* thin layers, thick layers, solid materials, semi-finished products and mouldings
- *Application and functions:* electrical and magnetic functions, e.g., conductors, insulators, semi-conductors, superconductors, magnets etc., mechanical functions, e.g., construction, light-weight structures etc., medical-biological functions, e.g., biocompatibility, biological degradability etc., thermal and thermo-mechanical functions, e.g., high-temperature resistance, ductility at low temperature etc.,

chemical functions, e.g., high-temperature or wet-corrosion resistance etc., optical functions, e.g., design, optical applications etc., fluid-dynamic functions, e.g., capillaries, foams etc. and all conceivable combinations of these individual functionalities in different applications as products, tools, sensors/actuators etc.

Besides the macroscopic shape, its surface and the material from which a component is made, especially the microstructure and defects within this microstructure in particular determine the properties of the material and the component characteristics and functionality. The microstructure is already largely determined during the primary shaping phase; the initial forming from a structureless liquid. Primary shaping processes e.g., are solidification of a component from a melt or the crystallisation of a crystal from a solution. Curing of duroplastic polymers or sintering, i.e., the "baking" of powder into a predetermined form, are also considered as primary shaping processes. Primary shaping is followed by further processes such as forming, heat treatments, coating, joining with other materials or machining which in general all lead to further changes of the microstructure and the material characteristics, respectively.

During the solidification of a component from a melt, physical processes at all length scales play an essential role. Before the beginning of solidification, clusters with different numbers of atoms repeatedly form and disappear within the melt. The average size of these clusters increases as the melt cools down. Slightly below the melting temperature, individual clusters then form solid nuclei which continue to grow. These nuclei are just a few nanometres in size. Their number and position within the melt determines the final number and distribution of grains (crystallites in the microstructure) in the solid component. Numerous nuclei lead to many tiny grains and in an extreme case to a nano-crystalline microstructure. A single nucleus in contrast leads to the growth of a single crystal.

Real components and materials in general are polycrystalline. A well-known example are solar cells which are made of polycrystalline silicon, whose granular microstructure is easy to detect even without a microscope. The efficiency of such solar cells largely depends on the transport of electrical charges generated by sunlight to the contacts. Since these charges have to cross the borders between individual grains the number of grains, their mutual orientation and contaminations at the grain boundaries at an atomic level are crucial.

Technical materials contain a variety of alloying elements. They reveal a variety of grains as in the case of polycrystalline silicon and moreover they reveal different phases also determining the microstructure and the properties of the component at different length scales. Finely-dispersed precipitates in a matrix phase substantially improve the strength of a material at high temperatures, the friction- or wear-characteristics of a component, the superconducting properties of superconductors and many other material properties being important for applications.

A complex technical class of materials are steels which are by far the most important construction material, in terms of both quantity and financial trading volume. Its main characteristics such as strength, forming capability, corrosion resistance or even magnetism, electrical conductivity and much more are all determined by its polycrystalline microstructure and especially by the defects at the structural level (dislocations, splits, flaws, secondary phases, grain boundaries) and constrained by other effects e.g., due to segregations.

A deep understanding of the formation of structures and patterns at different length scales in the evolution of microstructures in technical components is of utmost importance for the improvement of existing materials and their production processes as well as for the development of innovative materials.

Only a few decades ago, the relationships between the chemical composition of a material, its manufacture and its properties were still such complex that their description was almost inaccessible to scientific understanding. However a scientific description of the complicated and sometimes still complex relationships between process, microstructure and properties nowadays becomes increasingly possible. This is due to steady improvements of experimental and analytical methods right down to atomic length scales and to the continuous development of ever more powerful computers and more complex algorithms in dedicated software for rapid and reliable numerical processing of extremely large systems of differential equations (some with parallelized algorithms). At least a qualitative prediction or quantitative empirical descriptions are now possible for many cases.

Computer simulations of macroscopic processes in material processing have achieved great significance in the last two decades and have found more and more fields of application, see also (Zacharia et al. 1995; Radaj 1999). Finite Element Methods (FEM) and Computational Fluid Dynamics (CFD) have reached a high degree of maturity and corresponding software packages are frequently used to simulate and/or to optimise individual processes or in some applications even partial process chains during production.

Simulations in the area of materials research in the last decade have been aiming increasingly towards integrative process simulation (Gottstein 2007) from CAD data to the finished product, wherein particularly the reduction of development times and lower development costs have been and still are in the focus of interest. Integration takes place across several scales and several process steps as well as across different methodological modelling approaches.

Theoretical developments such as the phase field theory e.g. (Kobayashi 1993) and multi-phase-field models (Steinbach et al. 1996) or mathematical homogenisation e.g. (Terada and Kikuchi 2001) in combination with thermodynamic models like the CALPHAD method (Calphad 2010) and corresponding databases like Thermo-Calc (Thermocalc 2010), JMatPro (JMatPro 2010), FactSage (Factsage 2010) or Pandat (2010) have reached a status allowing first engineering applications for simulating the evolution of microstructures at a microscopic scale (Micress 2010). Such models provide the key to the description of microstructures and effective properties of technical materials and products. Industrial requirements ask for an integrated virtual chain from the manufacture to modelling of product lifecycles e.g. (Goldak and Akhlaghi 2005; Pavlyk and Dilthey 2004).

The data and parameters required for respective models can be obtained from even more fundamental models e.g., density functional theory, atomistic modelling and molecular dynamics simulations, the CALPHAD approach and other methods.

BeamSim	Simulation tool for laser- and electron-beam welding (Dilthey et al. 2003, Gumenyuk 2004)
CAOT	(Computer Aided Optimization Tool) automatic optimisation of forming processes (Posielek 2004)
CASTS	Computer Aided Solidification Technologies Flow and mould-filling simulation of single- and multi-phase flows, temperature, macro-segregation, solidification, heat transfers, mechanical stresses, VOF, free surfaces, UDF. (Jakumeit et al. 2003). In 2009 CASTS was linked to commercial package STAR CD to form STAR-Cast (STAR-cast 2009).
Elthe	Rapid FDM simulation with interface to StrucSim (Karhausen 1992),
HoMat	Simulation tool for mathematical homogenisation (Laschet 2004)
LARSTRAN	Commercial FE package: LARSTRAN for forming processes (Larstran 1994),
LaserWeld3D	Simulates the dynamic 3-dimensional heat flow during laser beam welding (Michel et al. 2004) plus free phase boundaries, melt flow and propagation of the laser beam
MESES	Multiphase simulation of solidification and flow; simulation of nucleation and grain growth under the influence of multiphase flow (Pustal et al. 2003)
MICRESS	MICRrostructure Evolution Simulation Software MICRESS Multi-phase field software-package for the description of diffusion-controlled phase transformations (microstructure evolution) in technical alloys in 2D and 3D. Coupling with thermodynamic and kinetic databases via Thermocalc's TQ-interface e.g. (Warnken 2003) (Eiken et al. 2006, micress 2010). Special module for the description of elastic stresses, simulation of recrystallisation.
MicroPhase	Multiphase micro-segregation model for solidification processes for one-dimensional problems. Coupling with MagmaSoft (Magmasoft 2010). Principally used so far in the field of cast iron.
APSFlux	Simulation of the particle flight path, of the particle melting and temperature distribution in the particle during thermal spraying processes (Hurevich et al. 2002)
PEP	3d pre-/postprocessor for Larstran/Strucsim (Franzke 1998),
SoliCon	Split Solid Model for the simulation of porosity formation and macro-segregation in cast parts (Ehlen et al. 2003).
SphäroSim	Calculation of crystallinity, nodule diameter and molecule orientation in polymers, identification of process parameters (local, temperatures, velocities) in the injection moulding process e.g. (Michaeli & Bussmann 2006). Coupling e.g. with Sigmasoft.
StrucSim	Microstructure simulation with calculation of dynamic and static recrystallisation (Wolske & Kopp 2000), (Karhausen 1992),
SimZTU	Modelling phase transformations in steel (Lorenz 2004)
SimWeld	Simulation of welding processes (Dilthey et al. 2005a, Dilthey et al. 2005b)
T-Pack	FEM-simulation of texture, dislocation density and anisotropy (Neumann et al. 2003)

Fig. 4.59 Overview of different proprietary simulation software packages at the RWTH Aachen University. (Reprinted from Schmitz and Prahl (2011) with kind permission of Wiley VCH)

In principle, an unconditioned, physically-based, integrated and thus predictive simulation of component properties can thus be realized starting from an atomistic level. However it seems impossible to describe all effects—even if only approximately—quantitatively or to get values for all necessary parameters as computing power and algorithms will not be sufficient to do so for still quite a time.

Materials modelling based on an integrative simulation approach has already been intensively investigated within the context of the special research area of "Integrative Material Simulation" (SFB 370, 1993–2005 at RWTH Aachen University) (Gottstein 2007). The respective conceptual approaches have been transferred to the European steel and aluminium industry within the context of "through-process-modelling" programs for commercial materials (Hirsch 2006). Founding of the "Aachen competence centre for process chain simulation SimPRO" in 2000 and the setup of the transfer area 63, "Practice-relevant modelling tools", underline the necessity for a transfer of such approaches to industrial applications. The strategic significance of scale-bridging and process chain simulations in the form of Integrated Computational Materials Engineering "ICME" has been emphatically confirmed recently in a study of the American National Research Council (2008) while in Europe various groups in Sheffield (2010) Stockholm (hero-m 2010), Zürich (ETH-Zürich 2008), London (Li et al. 2007) and Bochum (ICAMS 2010) are conducting scientific research into ICME.

At the RWTH Aachen University, several institutes maintain a wealth of commercial and proprietary software packages in the area of material sciences. Several of their proprietary models and simulation programs have been developed over a period of many years and are still being further developed, Table 4.1. These simulation tools, the respective expertise in model-development and the work in SFB 370 have led to the definition of a standard for a simulation platform which will resolve the basic problem of linking heterogeneous material simulation software and models (Fig. 4.59).

4.4.4 Motivation and Research Question

Starting from data elaborated by quantum-mechanical methods for individual atoms or for ensembles of similar or different atoms, the overall vision of present materials research is to achieve an integrated computational materials engineering covering the following aspects:

Quantitative, predictive simulation of the properties of three-dimensional materials, semi-finished components and products as a function of

- material composition
- manufacturing process along the manufacturing chain
- stress history under operational load (where applicable)

Property-optimised manufacture of semi-finished products and components with desired properties. Possible optimisation targets include:

- Raw material costs, e.g., avoiding costly alloying elements while retaining desired properties
- Energy costs, e.g., lower temperatures for process steps or substitution of energy-intense process steps
- Identification of potential process scenarios. Selection of suitable and optimum process scenarios depending on the availability of system capabilities and capacities
- Process times, e.g., reduction of heat treatment times/costs or fewer process steps

- Extended operating times, e.g., achievement of optimum properties after a specific operating period instead of continuous degradation from the time of entry into service
- Process reliability, e.g., "first time right" in the case of complex or large parts or for parts requiring destructive testing, improved quality assurance, defined process variation in the case of fluctuating input values e.g., varying raw material quality; defined tool design
- Product variability, e.g., avoiding/decreasing ramp-up periods
- Recyclability
- Optimised materials properties

Lifecycle models and lifecycle calculation for parts with special consideration of

- long term fatigue
- lifetime prediction
- creep and failure
- self-healing mechanisms (examples: zinc coatings on steel, aluminium oxide on aluminium)
- extreme conditions at transitions (e.g.,: brittle-ductile transition in metallic materials)

Incorporation of further phenomena like electrochemistry, corrosion, radioactivity and others.

Inverse modelling: possible material combinations and process scenarios might be identified by inverse models on the basis of e.g., the required part properties, non-functioning requirements profile and required load scenarios.

Present integrative descriptions and modelling of process steps influencing the microstructure and/or the properties of a component along of the production chain essentially draw back on continuum models. Starting from the homogeneous melt via heat treatments, forming processes, recrystallisation, joining technologies (welding, soldering, diffusion welding) to thermal coating technologies and machining they by now focus on metals, composites and polymers exclusively.

4.4.5 AixViPMaP Platform

4.4.5.1 Standardisation

The exchange of simulation data between individual software packages, both proprietary and commercial, plays a key role for the integrated modelling of the process chain during manufacture of a complex component. Each process step influences the state of the material and accordingly its properties. To model process chains, each process step has to be described by an individual modelling tool being specialised for the particular process and having its own individual format for storing process data and results. A comprehensive description of a material and a respective component, however, requires an effective exchange of the entire component history between different modelling tools and therefore the development of a uniform, standardised data format allowing an easy exchange of process data and results.

The basis for the development of a suitable standard has been identified in the open source data format VTK (Schroeder et al. 2006). The advantage of using an open standard is the free availability of software for the visualisation of the data as well as software tools for the further processing and conversion of the data. Besides a straightforward exchange of results and model data within the simulation chain, it is necessary to create a further standard for the exchange of material laws between the different model levels of micro- and macro-modelling. For this purpose a standardised user programming interface (API, Application Programming Interface) based on Fortran subroutines has been developed permitting the exchange of newly developed material laws between different software packages. All these standards satisfy the requirements of future GRID computing.

The virtual process chains in the processing of materials and the manufacture of components span several abstraction levels and length scales from the microscopic scale via the mesoscale to the macroscopic calculation of parts including necessary parts of the production plant. Documentation of the product and material history is available at the end of this virtual process chain and can be used as a basis for simulation of the product lifecycle.

A number of different, highly specialised simulation programs are used for the simulation of the individual process steps at the micro- and macro-levels. These calculate e.g., distortions of a component at the macro-level but also the evolution of state variables within the macroscopic part or within a representative mesoscopic volume element (RVE). State variables can be temperature distribution, order parameters for describing the microstructure, stress and strain distributions, concentration fields and many more. Coupling between the individual scales using suitable algorithms permits the transfer of effective material properties of the RVE to the macroscopic simulation programs, while coupling of the macro-level to the micro-level is implemented via boundary conditions.

Since the simulation programs differ both in terms of the methods used (FEM, FDM, CVM) as well as in the required type of numerical grid, implementation of a simulation platform requires a uniform data format definition for the exchange of model data along the process chain as well as for coupling the scales. The goal is a standardised file format for exchange of the geometry description and results along the process chain while both retaining and/or changing the scales.

Therefore the Visualization Tool Kit (VTK) file format is used to describe the meshed model geometry and the stress/deformation temperature history in the discrete grid points. VTK is open source software distributed for professional applications by Kitware Inc. It is written in C++ and provides a series of defined functions for file handling, for manipulation of the meshes and results data and numerous graphical display functions in the form of a software library. The software is developed by a large community and numerous functionalities are thus freely available. There is also a range of readily available commercial and free software applications.

Along with netCDF and HDF the VTK file format is one of most frequently used file formats for numerical results.

Since the VTK file format, however, only contains information about geometry and the results itself, it was necessary to extend this file format for the use within a comprehensive simulation platform. The general VTK data format lacked information about the upstream simulation steps, about material data, unit systems and time steps in the case of transient data. In addition, the descriptions for the results were standardised to enable an automated coupling of individual simulations. While the general VTK file format is comprehensively defined on the VTK website (Schroeder et al. 2006) its extension being elaborated during the development of the AixViPMaP-platform and the present standard data format have been documented separately (AixViPMaP 2010).

4.4.5.2 Further Development of Individual Models

Microstructure simulation models particularly required an extension in order to close gaps in the model chain. By now it is already possible to describe diffusion-controlled phase transitions such as solidification, solid state phase transformations and recrystallisation for technical materials with a number of alloying elements. Elastic stresses and their influence on the microstructure can also be taken into account in this context (Micress 2010). From these microstructures effective data have to be extracted. This data reduction or extraction is based on the methods of mathematical homogenisation e.g. (Bakhvalov and Panasenko 1989) or by virtual tests. Mathematical homogenisation allows the calculation of effective properties of a multiphase polycrystalline material based on the knowledge of the (i) properties of the individual phases and (ii) their three-dimensional topological arrangement. Traditional homogenisation models for periodic structures have been developed for linear material properties also for typical non-periodic structures of real materials (Laschet et al. 2008a). Virtual tests numerically mimic experimental material tests and have been carried out as virtual tensile tests or virtual dilatometer tests (Apel et al. 2009a).

Material properties being used in process simulations by now are usually macroscopically averaged, homogeneous and isotropic. Their dependence on process parameters such as e.g., temperature or chemical composition is either neglected or only approximated. In future, respective values will be replaced by local, effective and even anisotropic material properties whose evolution in the upstream process steps is reflected in the simulated microstructure.

Extension of Existing Microstructure-simulation Models The primary objective of the extension of present microstructure-simulation models is to close gaps in the model chain. It is already possible to describe diffusion-controlled phase transformations e.g., solidification processes and solid state transformations as well as recrystallisation processes for technical alloys (Micress 2010). Elastic stresses and their influence on the origin of phases and also the influence of evolving phases on the stress state itself can be taken into account. A methodology for describing the material history and/or the microstructural evolution thus is available in principle for process steps such as casting/solidification, heat treatment and joining processes. However, there is still a substantial gap in relation to continuum descriptions of hot and cold-forming processes. These processes require the extension of present microscopic models to enable a description of the viscoplastic behaviour relevant to deformation and strain-hardening processes and their reverse-coupling with the evolution of the microstructure. First viscoplastic models have been developed and verified in this context (Benke 2008). Future developments aim at coupling the phase field method, which is in general applied to diffusion-driven phase transformations, to a CP-FEM approach ("crystal plasticity") (Roters et al. 2010).

The results of such micro-models offer a maximum depth of information in the form of spatially and temporally dissolved microstructures. Obtaining relevant information for simulation at larger scales requires a data reduction and the determination of effective material properties.

Extraction of Relevant Material Properties from Simulated Microstructures In the framework of "micro-/macro-coupling/effective properties" an interface is created between the macroscopic scale of the processes and microscopic scale of the materials. The calculation of the local microstructure in this case is based on macroscopic boundary conditions which are mapped by the process simulation.

Two special methods—virtual tests and mathematical homogenisation—are used to determine homogenised or "effective" material properties (including anisotropic properties as required) from simulated or experimentally determined microstructures (porosity, phase transformation, etc.). Respective properties may enter macroscopic simulations as spatially-resolved material characteristics.

Virtual tests are based on a direct numerical replication of experimental materials tests. The experiments are reproduced directly on the basis of the simulated microstructure by adequately adjusting the boundary conditions. Mathematical homogenisation uses asymptotic analysis for periodic structures to perform calculations bridging several scales (Terada and Kikuchi 2001; Bakhvalov and Panasenko 1989). It is similar to the micro-mechanical methods of the mid 70s (Bensoussan et al. 1978; Sanchez-Palencia 1980) and is based on the assumption of statistical homogeneity of a material, i.e., the existence of a statistically representative volume element (RVE), whose dimensions are substantially smaller than those of the macro-component. For materials exhibiting an obvious periodicity, the geometric unit cell is selected as the RVE. Special methods are required for materials with non-periodic, random microstructures being in general present in technical materials e.g. (Laschet et al. 2008a). Along with approaches of so-called stochastic homogenisation, one of the current research topics is the adjustment of the typical spectral analysis for signal analysis to define the dimensions of characteristic structure periodicity (Bobzin et al. 2009). Methods of non-linear homogenisation have been developed for calculating the parameters of non-linear material behaviour like plasticity. The development of concepts for a self-consistent coupling with the micro level, i.e., to calculate the microscopic status based on the macro-load, completes the description of the interaction between the micro- and macro-levels. Particularly for solidification processes, e.g., an iterative method has been developed for self-consistent coupling of microscopic temperature/time profiles and the release of latent heat at the microscopic scale (Böttger et al. 2009).

Expansion of Process Simulation Models Toward Local Material Data Many properties being relevant for a product or a component are macroscopic in nature e.g., geometric shape, distortion and hardness of an entire component. Macroscopic simulations based on the available geometry of the component and continuum models are used to predict such properties. These models in general are based on material properties which are frequently assumed as being constant, homogenous and isotropic across the component. The complexity and computing power required for present micro-models by now prevent the simulation of entire components with a resolution at the scale of the microstructure.

The macroscopic properties of products or individual components however are frequently determined by the local properties of the materials involved, whose evolution along the process chain can only be simulated at a microscopic level. The current macro-simulations have been extended in certain aspects and now may draw on material values which have evolved in accordance with the upstream process steps in micro-models and which therefore generally possess properties such as anisotropy or inhomogeneity. These values were previously specified only macroscopically and indirectly (e.g., tabular values) and their dependence on specific process parameters (e.g., temperature) by now has been neglected or only approximated. Values determined in micro-models are converted by mathematical homogenisation or by virtual tests into respective input-values for the macro models and provide substantially more information in comparison to the data used previously. Precise knowledge of the material data resulting from the upstream process steps and depending on the process parameters of the macro-model like temperature, pressure, stresses and strains then allows for simulations in the frame of a consistently-closed process chain. Based on such a precise knowledge of the initial and boundary conditions and of the local description of the material parameters the quality of the simulation results and thus eventually also the quality of the simulated components and products can be significantly improved.

An integrative description of the material properties at each point of the product lifecycle with a maximum depth of information in microstructures requires:

- Networking of existing academic and commercial simulation applications in the material sciences
- Expansion of microstructure simulation models to close the gaps in the model chain
- Expansion of simulation applications available at the process level to take into account local material properties
- Data reduction by extracting relevant values from simulated microstructures
- Verification of integrated simulation chains by examples of economic and scientific interest

The challenges and requirements for optimising virtual process chains are (i) the best-possible use of existing and validated models with the possibility of model-refinement or reduction within the context of an expandable modular concept, (ii) the closure of gaps in existing model chains for a comprehensive mapping of the entire product history, (iii) speed-up of individual models by adapting information and selecting models adequate to the actual problem or by efficient programming and particularly by (iv) accelerating the flow of information between the different models by means of standardised data formats.

The overall requirement is thus to develop a comprehensive, standardised, modular, scale-bridging and expandable platform for the efficient simulation of processes and process chains in the manufacture and processing of materials and components.

The current status of the development of an open, web-based and modular platform for multi-scale simulation of processes and process chains involved in manufacture, processing and applications of materials and components, the AixViPMaP® (Aachen (Aix)—Virtual Platform for Materials Processing) (Schmitz and Prahl 2009) is described in the following. The processes already implemented on the platform include casting, injection-moulding, hot-forming, heat treatment, cold-forming, joining, surface treatment and the prediction of distortion resulting from welding, heat treatment or during operations. This approach is unique in its form and can—on a modular basis—be complemented with new process models and simulation tools. This platform will increasingly enable an effective simulation, development, optimisation and control of process chains. This platform particularly contributes to solving the "planning versus value" dilemma by providing a substantially increased planning effectiveness, efficiency, reliability and quality.

Various scientific and economic test-case scenarios have been specified to test the platform functions:

Line pipe:
Material properties, toughness/strength, flow curves at different locations of the component

Gear component:
Optimisation of process times in relation to case hardening, fine grain stability, distortion

Plastic component in automotive interior:
Properties based on varying local inner for properties

Stainless steel castings for bearing applications:
Manufacturing-induced distortion arising during use

Textile-reinforced piston rod:
Anisotropic component properties of light metal matrix components reinforced by a braided structure

Partners from industry—both simulation users and software providers—were involved for all test cases. Individual results for these test cases are illustrated in the following chapters.

4.4.6 Test Case Line Pipe

The product "pipeline" has been chosen as an example for the integrative simulation of a process chain because of its immense commercial importance. A competitive energy supply is a key factor for productivity, growth, secured working places and effective environmental protection. Oil and gas are the blood in the arteries of the world economy and those arteries are the pipelines. There are currently three million kilometres of oil pipelines ensuring the energy supply of the industrial nations. Around 80% of crude oil used in German crude oil refineries is transported through long-distance crude oil pipelines.

The use of ultramodern simulation tools and newly-developed material models is intended to improve the properties of the product "pipeline". The focus in this case is on toughness and strength properties. Increased strength properties enable thinner pipe walls, which can save on raw materials and reduce transport and installation costs. Improved toughness properties increase the reliability of the pipeline in operation and enable the use of pipelines in extreme environmental conditions.

The integral analysis of the process chain permits the optimisation of the required properties of the pipeline steel. The integrative process and material simulation also enables the most rapid responses to changes in the market, enabling production to be changed or adjusted accordingly. Production and alloying concepts for example can be adjusted to price developments of individual alloying elements in the shortest possible time. Recently, for example, the increase in the price of nickel from 15 to 65 k€ per ton led to a great pressure for replacing this element in stainless steels or steels designed for low temperatures. The method developed in the context of the simulation platform enables a rapid response to fluctuations in prices for specific raw materials, even fully avoiding their use, as well as optimising the relative prices of raw materials when using an available production infrastructure. The resulting cost savings represent a strong competitive advantage in the global market.

The strength and toughness of line pipes are essentially determined by the microstructure of the steels used. The microstructure in turn is determined by the manufacturing process and the chemical composition. One example was the establishment and implementation of an integrated process simulation of the manufacture of line pipe for X65-quality steel. The simulation was applied at the process-, machine- and microstructure levels. The processes being considered are hot rolling, comprising heating, hot forming and cooling, subsequent mechanical processing, U-forming and O-forming and submerged-arc welding. Relevant mechanical properties like strength and toughness were extrapolated from the simulation results for the material.

Heat treatment processes generally comprise heating, dwelling at defined temperatures for a certain time under atmospheric conditions, and subsequent cooling. These processes are frequently combined with hot forming (covered by the Larstran simulation code) to form a thermo-mechanical process. The basic heat treatment steps are simulated with the help of the CASTS program (Benke and Laschet 2008) at the macro-scale, which can take into account both solidification and thermo-mechanical

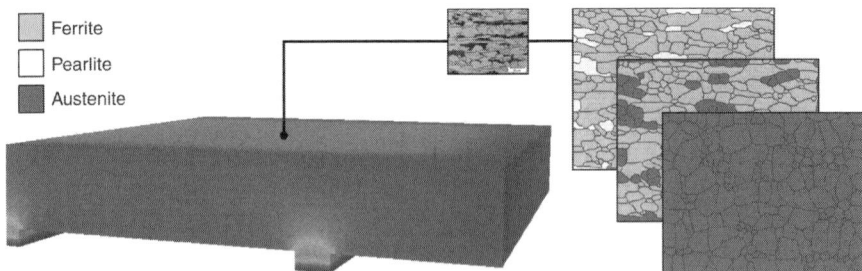

Fig. 4.60 Principle of coupled macro- and micro-simulation of the heat treatment. The guide rails for the slab and the geometry-dependant heat conduction lead to an inhomogeneous temperature distribution within the slab and affect the evolution of the microstructure at the characteristic points. Macrosegregations, which can occur in the continuous casting process, are a further possible source of inhomogeneity. Both types of phenomena affect the microstructure evolution and can be mapped by this approach. (Reprinted from Schmitz and Prahl (2011) with kind permission of Wiley VCH)

processes. This program can also be used to describe phase transformations (Benke et al. 2008) and grain growth phenomena based on the Leblond model (Leblond et al. 1984). For selected characteristic or critical integration points of the simulation, the temperature-time profiles were extracted from the macro-level simulation to serve as boundary conditions for the microstructure simulations at the selected locations. The microstructure evolution is simulated using the MICRESS software, which is based on the phase field method being coupled with thermodynamic databases (Micress 2010). The area in which the microstructure simulations are carried out can be considered as a local representative volume element (RVE). Effective mechanical properties of the area under consideration are thus determined on the basis of this RVE (Fig. 4.60).

This approach enables a preliminary prediction of the microstructure of the material, for instance during or after heating or cooling processes or welding processes. This is of particular interest if the mechanical properties of the material are to be described based on its microstructure. The phase field method also enables to track the microstructure evolution during thermal processes starting from a real microstructure (Fig. 4.61).

The U- and O-forming is simulated using the finite element program ABAQUS (Abaqus 2009), which requires the geometry of the rolled sheet metal and its local mechanical properties as input values. The geometry and several other parameters are taken from the upstream simulation of the cooling process. The effective material properties, however, particularly the elasto-plastic material behaviour, are determined on the basis of the local microstructures. A new type of multi-scale approach is used for this purpose.

The geometry and mesh of the slab, its local temperature- and phase-distributions, concentration etc. resulting from preceding thermal processes are transferred in a standardized data format (Benke et al. 2010; Schroeder et al. 2006) as initial condition for the downstream simulation of the hot rolling process.

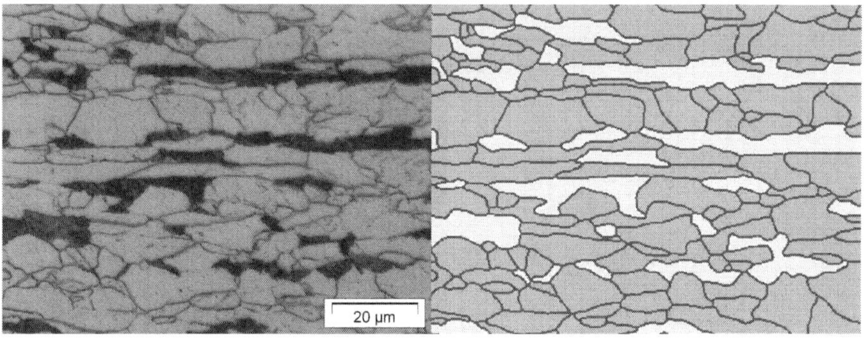

4 Virtual Production Systems

Fig. 4.61 Photograph of an actual microstructure (*left, dark*: pearlite, *grey*: ferrite) and digitised microstructure as input in MICRESS (*right, dark*: ferrite, *pale*: pearlite). The evolution of the microstructure starting from the homogenous melt can be replaced by the incorporation of experimental microstructures at a later time. This experimental information then serves as the initial condition for the simulation chain. (Reprinted from Schmitz and Prahl (2011) with kind permission of Wiley VCH)

The hot rolling process is simulated with the help of the finite element software LARSTRAN/-Shape (LARSTRAN 1994), which is coupled with StrucSim (Karhausen and Kopp 1992). This software combination is able to describe both the macroscopic deformation as well as microstructural phenomena such as the evolution of dislocation density, static and dynamic recrystallisation phenomena and grain growth processes. The hot rolling of a line pipe steel normally takes 10–20 passes. During the rolling process it may also be necessary for the sheet to be rotated and rolled in the lateral direction. The software is able to model all processes such as rotation of the product at a macroscopic level and is particularly designed with its integrated microstructure models to encompass the entire hot rolling process. Besides predicting the rolling forces, the temperature and strain fields in the rolling gap etc., it is also possible to obtain predictions about the grain size and the fractions of recrystallised microstructure.

The results of the hot-rolling simulation are transferred in VTK format to the downstream simulation of the cooling phase, which is modelled with CASTS. In this case it is also possible to simulate special features such as high cooling rates or quenching processes.

Machining for the preparation of welding grooves by now is taken into account by a simple removal of grid elements from the geometry. The stresses and strains are then redistributed. This simulation step is also implemented using CASTS.

Multi-scale simulation enables, for example, the description of the deformation and flow behaviour of a multiphase steel material based on knowledge of the properties of the individual phases and their topological arrangement in the microstructure. In this approach three scales are introduced in the case of X65 steel: the micro-scale, at which a ferrite/cementite bilamella, represent the pearlite phase; the mesoscale, at which a representative volume element (RVE) of a ferritic matrix with embedded

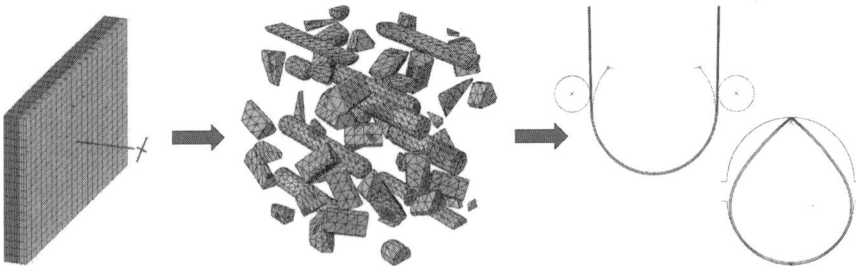

Fig. 4.62 Illustration of the three scales of the multi-scale simulation, *left*: bilamella (lamella width l = 330 μm) comprising cementite (*green*) and ferrite (*red*), *centre*: RVE with embedded pearlite microstructure zones in a ferritic matrix (not shown) and *right*: macroscopic sheet metal during U- and O-forming. (Reprinted from Schmitz and Prahl (2011) with kind permission of Wiley VCH)

pearlitic phases is considered and the macro-scale at which the rolled sheet metal and the forming tools are analysed, see figure (Fig. 4.62).

The effective thermoelastic properties of a pearlite bilamella comprising ferrite and cementite are determined using the HoMat mathematical homogenisation software (Laschet and Apel 2010). Since cementite reveals a high degree of stiffness, this phase is initially considered as a purely elastic phase. The elastic properties of the cubic ferrite phase and the orthorhombic cementite phase are incorporated into the calculation. The latter are extrapolated from ab-initio calculations according to Nikolussi (2008). As the crystallographic orientation relation between ferrite and cementite in the pearlite, the Isaichev relationship, as experimentally determined by Zhang et al. for hypoeutectoid steels (Zhang and Kelly 1997), is assumed for the X65 hypoeutectoid steel.

The elastoplastic properties of the ferrite matrix are described based on the flow curve model of Rodriguez and Gutierrez (Rodriguez and Guitierrez 2003). RVEs with a random phase distribution of the pearlite have been generated on the basis of metallographic sections. In this case the volume fraction pearlite $V_f = 14\%$, determined based on the metallographical analysis, was kept constant. This fraction was then combined from prismatic inclusions (7.2%) and extended spherocylinders (6.8%), Fig. 4.63. In order to incorporate the influence of the pearlite volume on the effective flow properties and locally varying pearlite volume fractions, two further RVEs were produced with $V_f = 10\%$ and $V_f = 7\%$. Uniaxial tests were carried out for these RVEs to obtain the effective non-linear stress-strain curves of the ferritic-pearlitic microstructure depending on the load direction (Laschet et al. 2010; Quade et al. 2010).

Initial results show substantially higher stresses in the rolling direction (X) than in the transverse directions. Due to the random distribution of the prismatic pearlite grains a similar deformation behaviour for the Y and Z directions is observed, which leads to an isotropic material behaviour in the transverse directions. Figure 4.64 (left) shows an experimental flow curve in comparison to the numerical flow curves. There are obvious deviations, which however were also observed by Gutierrez and Altuna

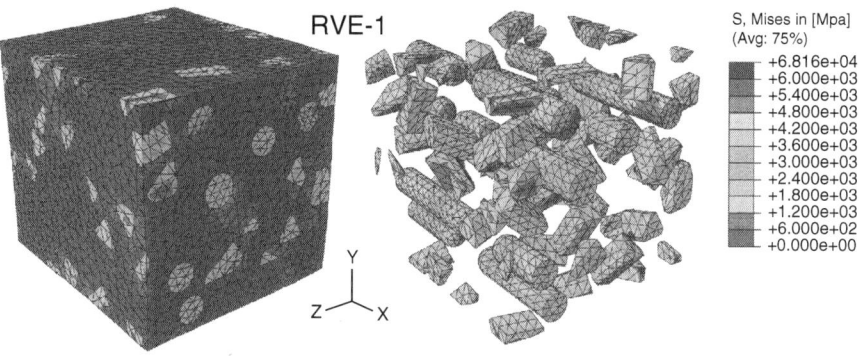

Fig. 4.63 Uniaxial virtual tests of the reference RVE with a pearlite volume of $V_f = 14\%$ ($\varepsilon_{xx} = 0.07$). Von Mises-stresses in the entire RVE (*left*) and in the pearlite inclusions (*right*). (Reprinted from Schmitz and Prahl (2011) with kind permission of Wiley VCH)

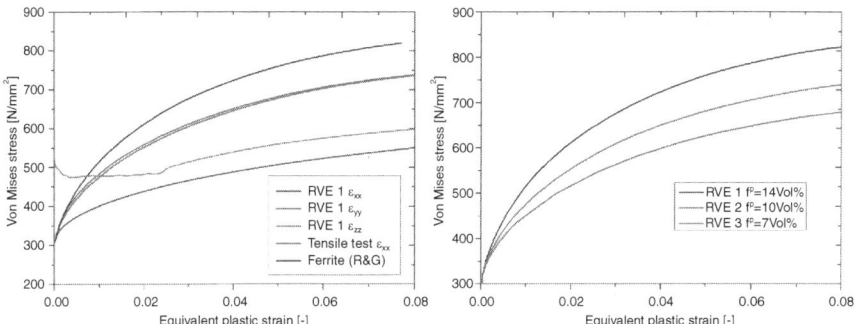

Fig. 4.64 Flow curves of the RVE 1 with $V_f = 14\%$ pearlite in the X-, Y- and Z-directions in comparison with the experimentally determined flow curve and the assumed flow curve for ferrite (*left*) and flow curves in the X-direction of the RVE with a varying pearlite volume according to Rodriguez and Gutierrez (2003). (Reprinted from Schmitz and Prahl (2011) with kind permission of Wiley VCH)

(2008) for different steel qualities. The experimentally determined flow curve exhibits a pronounced yield strength, which is caused by Lüders bands and/or dynamic strain-aging based on interstitially dissolved atoms. This effect cannot be captured using the Rodriguez and Gutierrez model and due to the severe aging tendency of body-centred cubic-phase steels it leads to a marked discrepancy between models and experiments in the region of minor elongations.

Increased hardening is observed in the region of larger elongations, in contrast to the experimental findings. Possible reasons for these discrepancies are (i) the idealised forms and orientations of the pearlite regions, (ii) the assumption of an elastic pearlite, (iii) the absence of residual stresses, possibly caused by upstream process steps and (iv) the deduction of the effective Hooke matrix for cementite at room temperature by extrapolation from ab-initio calculations at $T = 0$ K. The

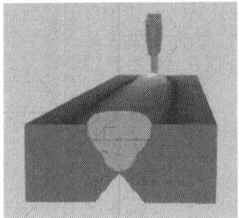 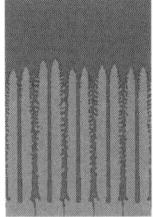 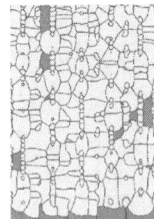 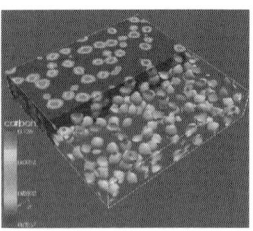

Fig. 4.65 Weld geometry, as calculated in SimWeld (*left*). The temperature curve of the simulation is input as a boundary condition for microstructure calculations to describe the solidification of the ferrite (*centre left*), the subsequent austenite formation (*centre right*) and phase transformation phenomena, recrystallisation and grain growth in the heat-affected zone (*right*). (Reprinted from Schmitz and Prahl (2011) with kind permission of Wiley VCH)

prediction will be improved in future by assuming an elastoplastic material behaviour for pearlite, a more precise mapping of the pearlite morphology and its orientations and by consideration of existing residual stresses.

The flow curves calculated for RVE 2 ($V_f = 10\%$) and RVE 3 ($V_f = 7\%$) exhibit a hardening being reduced by 10 and 17.5% with respect to RVE 1. The flow curves of these RVEs represent locally different material properties based on locally varying pearlite fractions. They are used in the forming simulation and assigned to different areas of the sheet to map the varying pearlite volumes throughout the sheet metal and their influence on the macroscopic stress-strain distribution.

The welding process is simulated at the macro level with SimWeld (Bleck et al. 2010) and SYSWELD. SYSWELD (Rieger et al. 2010) is used to calculate the residual stresses and the distortion of the line pipe. The heat input, the heat distribution, the weld pool geometry, the size of the heat-influenced zones and the weld geometry are calculated in SimWeld. The geometry and other data of the areas to be joined are transferred from the upstream simulation of the O-forming process. Both the thermal and the electromagnetic properties of the substrate, of the flux and the wire are taken into account in the simulation. Respective values are imported from the SimWeld material database (Fig. 4.65).

The welding process is considered as a stationary process in SimWeld. This means that the power input characterized by voltage and current, is calculated and distributed into different parts. The first component is required to melt the weld flux and the wire. The second part is dissipated into heat and is required to simulate the arc. The geometrical form of the exposed molten surface, its changes and the shape of the weld pool are calculated in SimWeld based on the magnetic effects of the arc, the exposed surface, the melting of the substrate, the wire and the weld flux.

The results of the SimWeld simulation of the welding process are used as an input for the thermomechanical simulation using SYSWELD. The welding wire geometry and all the necessary information are transferred with the help of a SW2SW interface (Reisgen et al. 2009) between SimWeld and SYSWELD, which enables the geometry to be integrated into the mesh of the O-formed line pipe. The residual stresses of the material before the welding process are imported from the upstream

O-forming simulation and are also integrated into the model. By this method it will be possible in the future to calculate any residual stresses present after welding and any eventual distortion of the workpiece on the basis of residual stresses existing before the welding process.

The weld is usually the weakest part of a component and is frequently subject to material failure due to degradation of the mechanical properties in the welding zone. The degradation of the properties is caused by the significant change in the microstructure during the process. The microstructure evolution is modelled using the MICRESS phase field program (Apel and Böttger 2009). The temperature-time profiles for characteristic integration points within the melt pool and the heat-affected zone are calculated with SimWeld and integrated into the simulation at the micro-level.

This approach allows the mapping of different phenomena occurring during welding. The O-forming and the preparation of the welding channels are cold-forming processes, which lead to a locally increased dislocation density. This means that these areas of the microstructure are prone to recrystallisation during the welding process significantly influencing the final microstructure. The expected grain size in the molten area can be analysed with solidification simulations. Solid state transformations and growth phenomena in the solid phase are analysed in the heat-affected zone. At present, this approach can thus be used to estimate the grain size in both the melt pool and in the heat-affected zone. Since this has a substantial influence on the toughness of a material, at least qualitative predictions can be made about the degradation of properties caused by the welding process.

4.4.7 Test Case Transmission Component

The gear wheel is another principal component ideally suited for comprehensive analysis by integrated simulation due to its enormous significance in mechanical engineering for the German economy. Gear wheels are found in almost any modern machine. Gear wheels transmit forces and set speeds and they come in all orders of magnitude from minute cogs in a wristwatch to the giant gears in an excavator for daylight mining. The gear wheel material, which itself is influenced by the entire process chain during its manufacture, determines the mechanical properties and therefore the efficiency of the gear wheel. Significant experimental efforts are necessary to investigate the complex process chain from casting the semi-finished product through different heat treatment cycles, both thermo-mechanical as well as thermo-chemical, to joining the gear wheel with other components. In contrast, integrative process simulation enables new alloying concepts or changes to process parameters to be described quickly with minimum testing and at low cost. More efficient gear wheels can minimise friction loss and significantly increase the lifecycle of machines. Greater efficiency reduces maintenance efforts, saves resources and protects the environment.

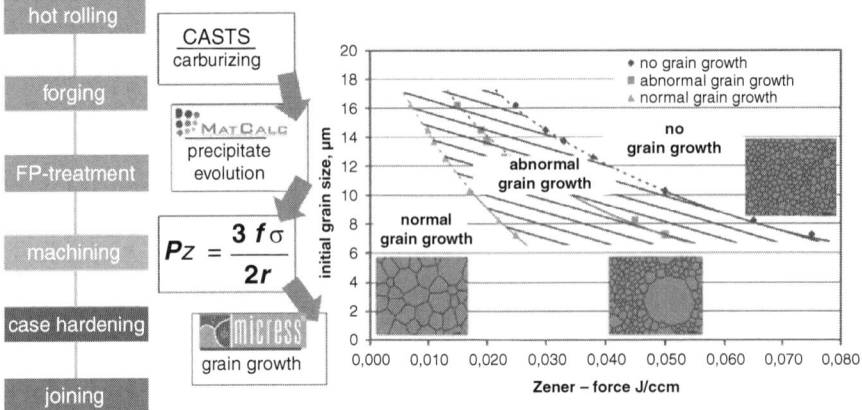

Fig. 4.66 Particle management and process window identification for microalloyed case-hardened steel for high temperature carburisation. (Reprinted from Schmitz and Prahl (2011) with kind permission of Wiley VCH)

The objective of gear wheel manufacture in general is to obtain a very fine-grained microstructure in order to achieve high strength and good toughness characteristics (Grosch et al. 1981; Pacheco and Krauss 1990) so that the material can withstand high cyclic loads. Most important for the formation of such a fine microstructure is the precipitation status of microalloying elements during the individual process steps (Hippenstiel et al. 2002; Klenke and Kohlman 2005). A description of this precipitation status both qualitatively as well as quantitatively for each stage of the process is necessary.

One option for a more efficient manufacture of gear wheels is the reduction of the heat treatment time by increasing the temperature during case hardening. This temperature increase is not only constrained by the furnace technology but also by the stability of the fine grains in the material which have a significant influence on the lifecycle of case-hardened parts. Fine-grain stability can be improved with new niobium-based microalloying concepts.

Respective microalloyed case-hardened steels as a material-based solution to meet the demand for carburisation processes at increased temperatures have been in development for several years and are now used in some industrial applications. Microalloying concepts with very small additions of aluminium, titanium, niobium and nitrogen use precipitates at the nm-scale to reduce the mobility of the grain boundaries even at high temperatures and therefore guarantee the stability of fine-grains. A particular challenge, however, is the control of the long process chain upstream the carburisation process in which the size distribution of the microalloying elements and e.g., carbonitride particles are repeatedly altered by dissolution, nucleation and growth. These concepts thus highly depend on the manufacturing sequence and the selected process parameters meaning that besides adding microalloying elements, the entire process chain must be optimised (Fig. 4.66).

Fig. 4.67 Calculations for an alloying concept with a reduced aluminium content to ensure fine-grain stability. (Reprinted from Schmitz and Prahl (2011) with kind permission of Wiley VCH)

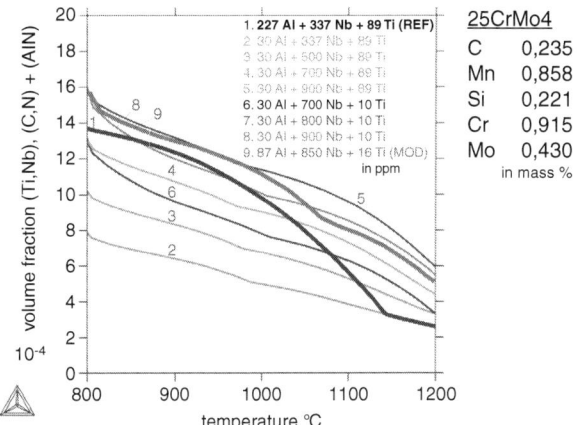

Traditionally, aluminium is added to these alloys because its finely distributed nitride precipitates ensure fine-grain stability for normal carburisation temperatures. However aluminium oxides potentially occurring in this traditional approach can cause premature failure of the component. An alloying concept guaranteeing the required fine-grained stability without aluminium thus reduces the probability of the occurrence of undesired aluminium oxides (Fig. 4.67).

There is sufficient knowledge for a respective alloy design based on thermodynamic simulations. For the test case, a defined amount of niobium was added to the 25MoCr4 case-hardened steel, a typical microalloyed steel for high-temperature carburisation, and investigated in detail. The process parameters for simulation and the experimental validation were adjusted according to typical parameters relevant to industry. The industry-relevant process chain comprises the following basic steps: multiple hot-forming processes (hot rolling followed by drop forging), annealing to a ferritic/pearlitic microstructure (FP), machining, high-temperature carburisation and downstream joining to other components.

Along with the simulation of the process chain itself, a comprehensive laboratory process chain comprising melting/solidification, hot forming and heat treatment as well as case-hardening of materials and components in a carburising furnace is of fundamental significance for the present work. A simple geometry with well-defined boundary conditions is beneficial for validating the simulations and for minimising the corresponding experimental work. Simulations and experiments were thus carried out on a simplified cylindrical shape. This cylinder is first heated to typical hot-rolling temperatures, then formed at this temperature and subsequently cooled. In the next step, the deformed cylinder is indented in the centre with a small forging die. The component is then annealed according to a typical FP annealing cycle. In the final phase the component is carburised and then welded to a ring of the same material by a laser welding process.

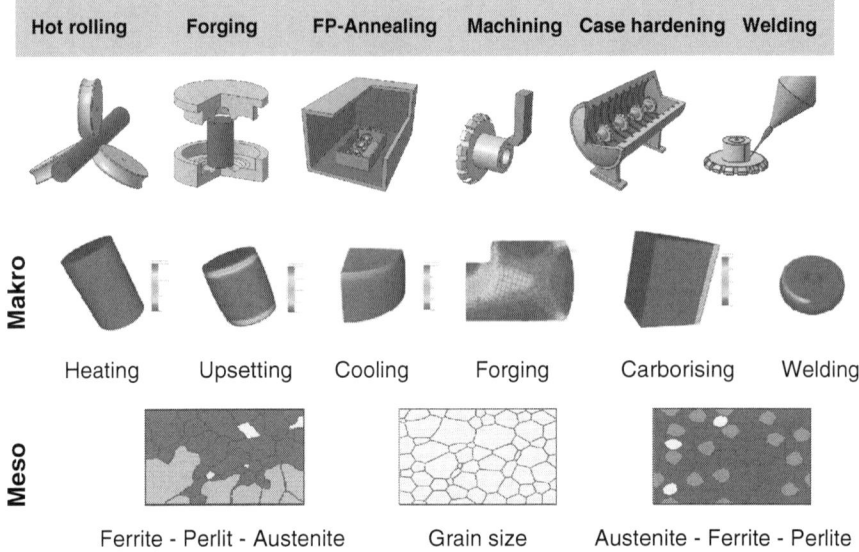

Fig. 4.68 Typical process chain for the manufacture of transmission components and corresponding simulation steps. (Reprinted from Schmitz and Prahl (2011) with kind permission of Wiley VCH)

From a material science perspective, the focus of the experiments and numerical investigations is on the prediction of the fine-grain stability of the parts being carburised at high temperatures. The precipitation profile along the entire process chain is of utmost importance for this purpose. Both the size and the size distribution of the precipitates are the key to adjust the fine-grain stability during case hardening. In particular, the initial hot-forming process steps must be analysed because substantial dissolution and precipitation of the microalloying elements takes place in these process steps (Fig. 4.68).

The first hot-forming processes are simulated after determining the initial status and calculating the respective initial conditions (e.g., temperature distribution within the component). At the macro-level the hot rolling and hot-forming simulations each comprise three steps: heating, deformation, cooling. Two different simulation tools are needed to calculate all effects. The thermomechanical simulation of heating and cooling, including any phase transformations, is performed using CASTS (Benke and Laschet 2008). The forming simulation is carried out using LARSTRAN (Karhausen and Kopp 1992). The calculation results can be transferred between these two programs using the standardised VTK data format and transferred to the meso- or micro-level as boundary conditions. At the mesoscale, the microstructure evolution and a phase transformation for a defined precipitation status (corresponding to an effective Zener-pinning-strength) is modelled using MICRESS (Rudnizki et al. 2010a). The precipitation profile itself is simulated at the submicrometer scale with the aid of MatCalc (Kozeschnik et al. 2004) and an effective Zener-pinning strength is derived from the distribution of the precipitates.

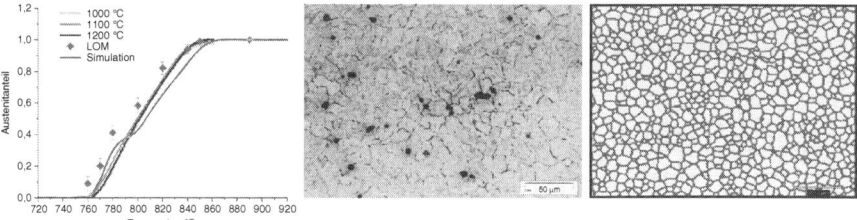

Fig. 4.69 Comparison of simulation results with experimental results. Dilatometer curves and simulation results for the austenite formation from a ferritic/pearlitic microstructure (*left*) and comparison of the microstructures for the austenite grain growth (*right*). (Reprinted from Schmitz and Prahl (2011) with kind permission of Wiley VCH)

The heat distribution, the diffusion of alloying elements, particularly of carbon, and the phase transformation during the heat treatments (FP-annealing and high-temperature carburisation) can be calculated with CAST using empirical approximations. The temperature profile is used by MICRESS to calculate the phase transformation and the microstructure evolution. MatCalc calculates the precipitation profile during heat treatment at the micro-level. This is then transferred to the mesoscale as an effective Zener strength. This Zener strength influences the grain growth (Apel et al. 2009) and phenomena such as abnormal grain growth can be described in this way (Rudnizki et al. 2010b).

Combining of two components into a joint mesh is the first requirement for modelling laser-beam welding. For this purpose the mutual positions of the VTK component data of the gear wheel and of the ring to be joined are defined and transferred to a single VTK file. The laser welding process is simulated with the help of LaserWeld3D (Jansen 2009) which calculates the geometry of the welding capillary in the component and the heat input into the component. The results of this calculation are then fed into the simulation at the meso- and micro-scales as boundary conditions for simulations of the microstructure and the evolution of the precipitate distribution (Fig. 4.69).

To finalise the integrated simulation chain from "steel rod to application", the above process steps are followed by the simulation of machining processes such as milling or grinding which are by now tackled in a simplified manner with the cutting method i.e., by removing volumes numerically and redistributing the stresses and strains. In real cutting and polishing processes, besides the stress redistribution resulting from the removal of the volume, microstructural modifications of the outer layer are to be expected as a result of the thermal load. Corresponding simulations are the subject of current developments. For this purpose programs already being available and existing simulation tools e.g., a CAx framework will be added to the AixViPMaP. Machining simulations will then be performed aiming at describing the influence of the production methods on the properties based on the processes preceding and following the case hardening step.

Coupling of the simulation chain for the manufacture and the eventual use of a gear wheel at the component scale will be completed in future by a transfer of the simulation data accumulated during manufacture to the simulation of the component in its operational environment. The calculation of the contact load will e.g., be based on the ZaKo software package e.g. (Hemmelmann 2007). Work is currently underway to link this package, which calculates the loads involved in tooth contact and for different gear wheel variants, to the VTK platform standard format as an interface with the ZaKo software.

In this way it will be possible to complement the tool and process development with application data and to predict the loads on real components when being in use as a function of their manufacturing history. This approach will provide an initial scientific basis for lifecycle predictions.

4.4.8 Test Case Plastic Component in Automotive Interior

The properties of plastic components made of semi-crystalline thermoplastics are predominantly influenced by the inner properties of the plastic such as molecule orientation or crystallisation microstructure. These in turn depend on the conditions of the manufacturing process. Thus the material properties of the component such as stiffness, strength, optical properties or media resistance can be influenced by varying the process parameters. Despite their relevance, it has by now not been possible to take these properties into account in the design of plastic component simulations. So far, approaches have been limited to describing the occurrence and extent of inner properties with isolated software packages (Raab and Godora 2005). Despite the great interest of many groups of researchers (Janeschitz-Kriegl 2010; Beek et al. 2006; Fulchiron and Cascone 2008; Housmans 2008) in prediction of inner properties and the ever-increasing demand from industry for integrative simulation solutions, there is currently no way of coupling the prediction of microstructures with injection moulding simulations and/or structural analyses.

In the context of this test case, this integrative coupling of the above mentioned simulations is investigated using the example of the injection moulding process as applied to the actual component of a "top box" from the automotive interior. The mechanical component properties are analysed here as they are typical macroscopic target values. In addition the knowledge of the internal microstructure of a plastic part can be used for the more accurate description of numerous other phenomena. The simulation chain (Michaeli et al. 2009a; Michaeli and Baranowski 2010a; Baranowski and Brinkmann 2007) is divided into four fundamental calculation steps which are embedded within the grid structure developed in the context of the Cluster of Excellence (CoE) and which are addressed in detail in the following sections (Fig. 4.70).

A macroscopic simulation of the injection moulding process is carried out initially using the Sigmasoft software of the company Sigma Engineering GmbH, Aachen. The process values obtained, e.g., temperatures, are then transferred to the newly

4 Virtual Production Systems

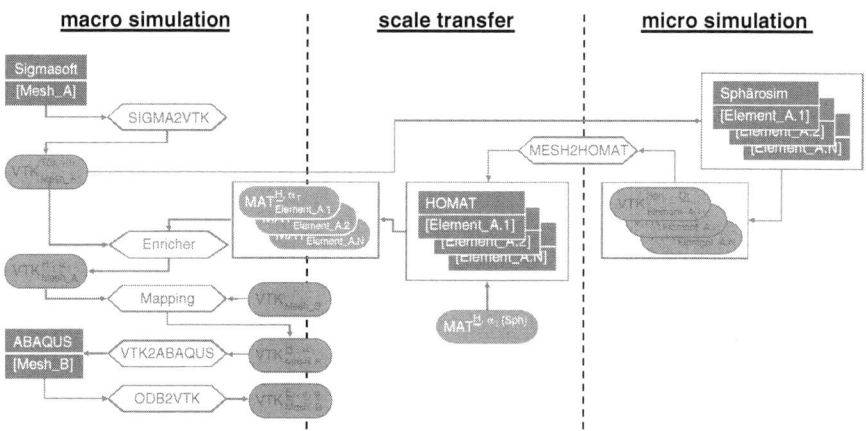

Fig. 4.70 Summary of the dataflow in the "top box" simulation chain. (Reprinted from Schmitz and Prahl (2011) with kind permission of Wiley VCH)

developed simulation software which calculates the microstructure evolving at the micro-level. The mechanical material properties based on this microstructure are determined with the help of refined homogenisation approaches (HoMat). The locally determined mechanical properties are then linked using a user-defined non-linear material model into a macroscopic structural simulation which is carried out using the Abaqus software (Abaqus 2009). Besides as taking into account spherulitic microstructures, structural simulations also account for molecule orientations generated during the manufacturing process. Initial results are illustrated below.

Microstructural Simulation Microstructural simulation is the first step towards an integrative inter-scale simulation of thermoplastic materials. The "SphaeroSim" software has recently been developed for the three-dimensional calculation of microstructure growth. This software is linked to the Sigmasoft process simulation software and is able to transfer the necessary information (temperatures, pressure, flow speeds etc) for the calculations in the standard platform format (VTK) (Michaeli and Baranowski 2010b). The description of the crystallisation, models for describing the nuclei formation (Housmans 2008) and the crystal growth (Hoffman et al. 1976) have been implemented in the software and numerically linked with a cellular automaton for a polypropylene. The calculation results are presented for an element with an edge length of 1 mm, being located near the tool wall (Fig. 4.71).

The calculation results are validated and the models modified using demonstration components on the basis of experimental investigations (Michaeli and Baranowski 2010c). The simulation results agree closely with real microstructures in injection moulded parts for the material PP 505 P, (Sabic Europe, Sittard, Netherlands) (Michaeli and Baranowski 2010d; Michaeli et al. 2010a) (Fig. 4.72).

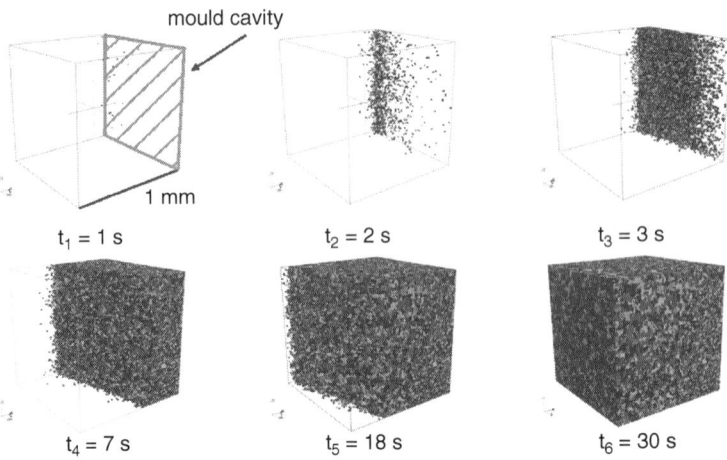

Fig. 4.71 Calculation results of the microstructure growth for an element. (Reprinted from Schmitz and Prahl (2011) with kind permission of Wiley VCH)

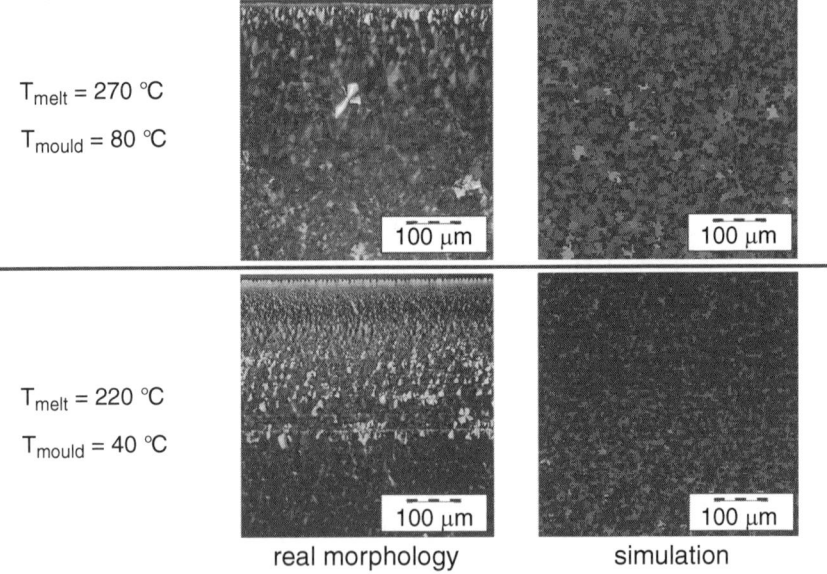

Fig. 4.72 Comparison of real and simulated microstructures (PP). (Reprinted from Schmitz and Prahl (2011) with kind permission of Wiley VCH)

In addition an algorithm for the calculation of the degree of crystallisation is implemented in SphaeroSim. The calculations (both for the microstructure and for the degree of crystallisation) can be carried out for nearly arbitrary geometries and are illustrated in the figure for a plate with 80,000 elements (Fig. 4.73).

Fig. 4.73 Calculated degree of crystallisation in a demonstration plate component. (Reprinted from Schmitz and Prahl (2011) with kind permission of Wiley VCH)

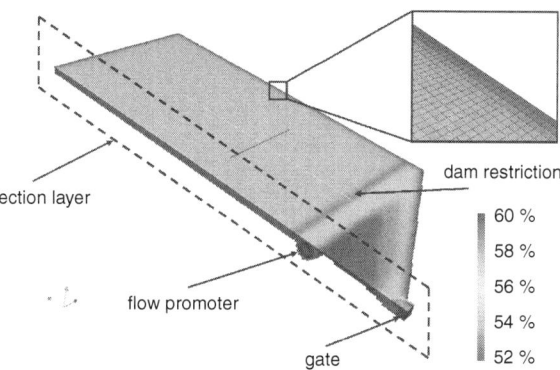

Calculation models (Pantani et al. 2004) are integrated into and modified in SphaeroSim to predict molecule orientations. Here again, the results of the injection moulding simulation (pressure, temperature and flow and/or sheer speeds) are used. In future the calculation results will be validated in detail with various measurement methods (Michaeli et al. 2009b) on the manufactured components (Fig. 4.74).

Finally the microstructure and molecule orientation are simulated for the typical "top box" component.

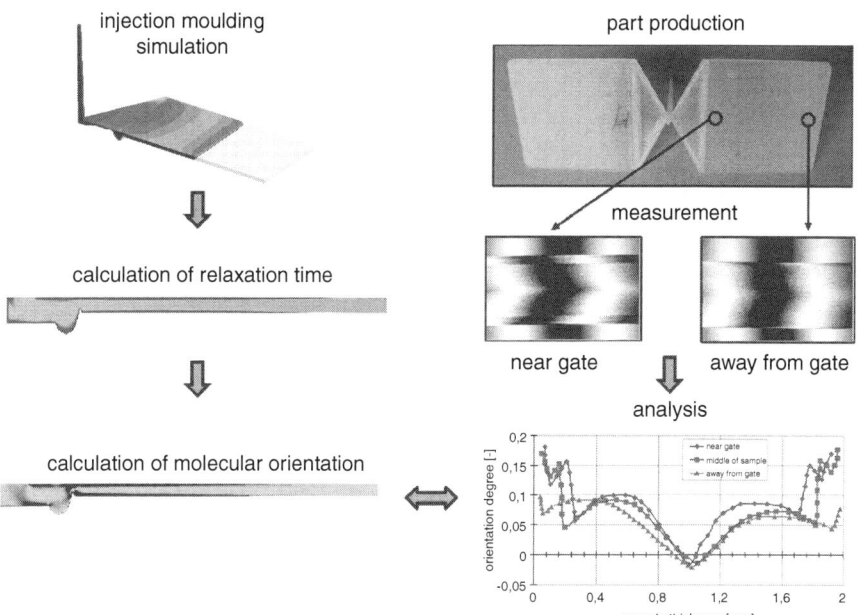

Fig. 4.74 Procedure for validating molecule orientation measurements

Homogenisation Methods for determining effective material properties can be broadly divided into two classes: virtual material testing and mathematical homogenisation methods (Sanchez-Palencia 1980; Bensoussan et al. 1978). Virtual material testing is a simple homogenisation process in which an experimental material test is replicated virtually. A complete description however of, e.g., the mechanical properties of the material, requires the definition of six different load cases. On the other hand the mathematical homogenisation method calculates the effective properties within a single simulation operation.

Both methods are based on the definition of a representative volume element (RVE). Normally the definition of this RVE also involves the variation of the volume section under analysis, particularly for materials with a random microstructure (Kanit et al. 2003; Bobzin et al. 2007). In the case of semi-crystalline thermoplastics it is impossible to differentiate visually between the amorphous and crystalline phase components of the microstructure. However since the individual spherulitic structures can be separated from one another, the spherulite itself could be used to determine effective properties using a homogenisation approach. The literature claims an empirical dependency of the modulus of elasticity on the average diameter of the spherulite (Hoffmann 2003) but this dependency must be determined by experiments for each specific thermoplastic material. To this end, the methods of inverse homogenisation have been developed and implemented in order to extrapolate the properties of the individual phases from experimentally measured effective values and the experimental spherulite distribution. However since it has been so far impossible to clearly identify the 3-D microstructures of those samples investigated, the empirically determined dependency of the modulus of elasticity on the spherulite diameter was used initially in the homogenisation calculations.

To reduce the calculations involved, the spherulites were divided into classes depending on their diameter. Therefore each spherulite class is analysed as an individual phase rather than each spherulite. Reference calculations were carried out for a SphaeroSim-microstructure (see Fig. 4.75) to validate this approach. This microstructure comprises 207 spherulites. The modulus of elasticity of the individual spherulites was calculated based on the empirically-determined dependency on diameter (Hoffmann 2003). Since the spherulite distribution is homogenous, the homogenisation results showed no marked orthotropy. Assuming that each spherulite corresponds to a single phase, an effective modulus of elasticity of $E_{eff} = 4.49$ GPa was determined. On the other hand, an effective modulus of elasticity of 4.747 GPa was determined in the case of three defined spherulite classes. The difference between the two effective values is 5%, and shows that allocation to three spherulite classes is a good approximation. The virtual material test leads to an $E_{eff} = 4.428$ GPa and is therefore only 1.38% below the homogenisation result.

Another homogenisation approach developed in the context of this project is based on the geometrical form of a spherulite. In the model it s assumed that amorphous and crystalline phases will exhibit quasi-radial distribution within a spherulite in the so-called bilamella (Fig. 4.76). This model allows us to analyse a bilamella (Fig. 4.76c) as the RVE in the context of a two-stage homogenisation approach. The anisotropic

Fig. 4.75 Simulated 3D spherulite structure. (Reprinted from Schmitz and Prahl (2011) with kind permission of Wiley VCH)

Hooke matrix of the crystalline phase was taken from the dynamic molecule calculations of Tashiro and Kobayashi (1996). The description of the amorphous phase is based on the work of Bédoui et al. (2006).

The homogenisation of the bilamella provides the anisotropic effective Hooke matrix in the local axis system. These values are then entered into the homogenisation of an entire spherulite. We hereby encounter a radial distribution of the bilamellae in the spherulite (see Figure 4.76b). Initially a 2D-model was developed (Laschet et al. 2008b) and used to predict the effective anisotropic behaviour in the central region of a PP plate. The 3D model with a random distribution of spherulite centres provides better results than the 2D model. In the next step, 3D unit cells, calculated

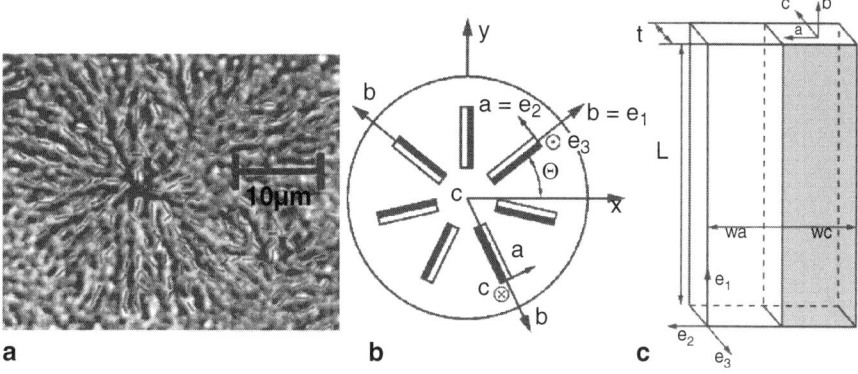

Fig. 4.76 Model of a spherulite: photograph of microstructure (**a**), schematic representation in a 2D-spherulite model (**b**) and a bilamella (**c**). (Reprinted from Schmitz and Prahl (2011) with kind permission of Wiley VCH)

by SphaeroSim are analysed using this two-stage homogenisation process and the radial 3D spherulite model.

Macrostructure Simulation The simulative prediction of the microstructure within the component and the determination of effective material characteristics using the above mentioned homogenisation methodology can be used to include the influences of internal e.g., local degrees of crystallisation and molecule orientations into the mechanical structural analysis (Michaeli et al. 2010a, 2009c). In the first stage the micro/macro couplings based on the prediction of the degree of crystallisation and based on the prediction of molecule orientations are analysed separately from each other. For this purpose two different material models were developed for the macroscopic simulation, which describes the mechanical material behaviour of the thermoplastic material depending on the degree of crystallisation or on the molecule orientation. These material models are explained in the following two sections.

Material Model for the Description of the Influence of Crystallisation The development of the material model to take into account the influence of crystallisation requires consideration of two fundamental effects. Each element must be assigned local material parameters corresponding to the assigned degree of crystallisation from the microstructure simulation. Also homogenisation for the material model calibration of effective material characteristics can take place only in the elastic deformation zone. Secondly however it must be noted that the deformation behaviour of reinforced thermoplastic materials is non-linear even in the area of minor elongations. These requirements were implemented in a multi-linear elastic material model. In this case the local component stiffness is formulated depending on the existing elongation level (Glißmann 2000). It is thus possible to describe the non-linear stress/strain profile of thermoplastic materials despite the use of purely linear elastic material characteristics (Michaeli et al. 2010b) (Fig. 4.77).

The effective material properties transferred from the homogenisation are imported for each element. Mechanical material characteristics are therefore assigned locally by taking into account the calculated degree of crystallisation. The decrease of material stiffness with the increasing deformation of the material is taken into account by the formulation of the material stiffness depending on the prevailing elongation status. The number of different elongation levels and therefore the varying component stiffness can be specified by user inputs into the input file in order to control the description accuracy of the of the non-linearity. The decrease in material stiffness with increasing elongation can be calibrated on the basis of material tests.

Material Model for the Description of the Influence of Molecule Orientations Along with the crystallisation status, molecule orientations also influence the mechanical material behaviour. Molecule orientations can be rendered visible using polarisation microscopy (see Fig. 4.78) and they cause anisotropic material behaviour in the material. This can be measured e.g., using tensile tests on different samples. Process parameters e.g., melt temperature and mould temperature also influence the molecule orientation and thereby the mechanical material behaviour (see Fig. 4.78).

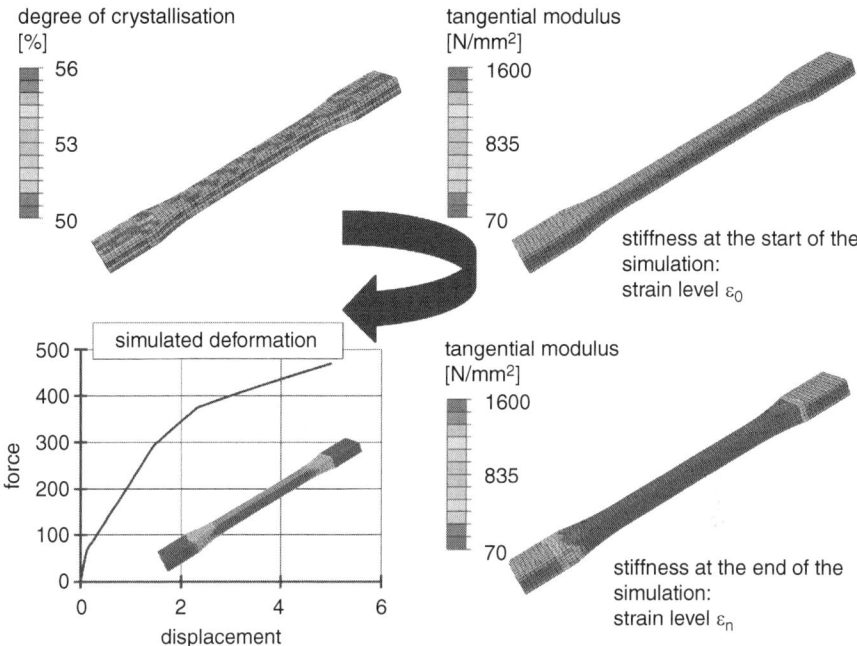

Fig. 4.77 Degree-of-crystallisation-based, multi-linear elastic macro-simulation of thermoplastic materials (c). (Reprinted from Schmitz and Prahl (2011) with kind permission of Wiley VCH)

An orthotropic material model is implemented in the context of the project to calculate the direction-dependent material behaviour in the macro structural simulation. Orthotropic material behaviour is examined on the basis that the material entails three orthogonally arranged preferential directions. In this case, with regard to the primary axes, normal stresses only affect elongations and tangential stresses only affect shearing. Nine material characteristics (E_1, E_2, E_3, G_{12}, G_{13}, G_{23}, ν_{12}, ν_{13}, ν_{23}) dependent upon the orientation status are required for calibration of the orthotropic models (Altenbach and Altenbach 1994).

Unlike the material model based on the degree of crystallisation, homogenisation of a multiphase material as a coupling between the two simulations is not carried out to take into account the molecule orientation in the macrostructural simulation because molecule orientations cannot be described as material phases. The calibration of the orthotropic model is therefore entirely based on performing material tests. This calibration is currently the focus of research activities.

Summary and Outlook for Test Case Plastic Component in Automotive Interior The calculation both of spherulite microstructures as well as molecule orientation is possible on the basis of the results of an injection moulding process

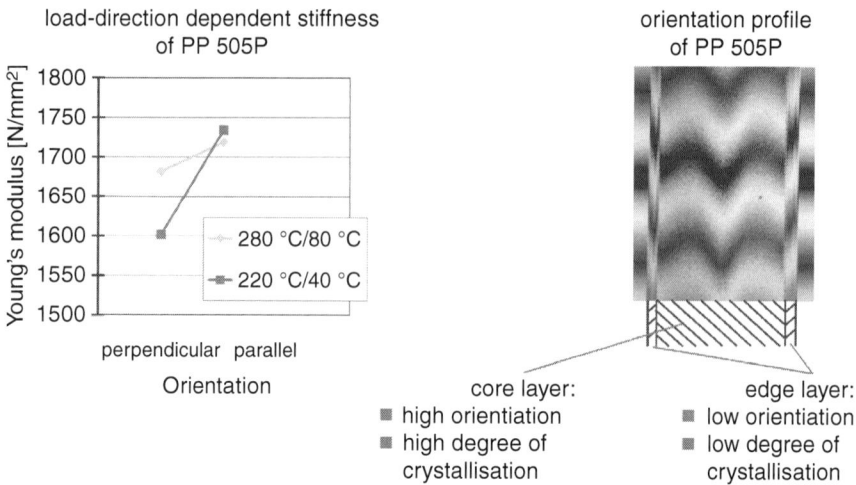

Fig. 4.78 Molecule orientation-based anisotropy of thermoplastic materials by the example of polypropylene (PP). (Reprinted from Schmitz and Prahl (2011) with kind permission of Wiley VCH)

simulation. The additional information obtained about the material status of thermoplastics can be transferred to the microscale in simulations using methods such as homogenisation and used, for example, to predict mechanical material behaviour. Work is currently underway on the automated implementation of the necessary calculation and simulation steps in order to be able to fully simulate even complex synthetic structures such as the "top box" with the help of a web-based infrastructure (Cerfontaine et al. 2008).

In future it is intended to develop a method on the basis of the integrated simulation which will be able to predict the shrinkage and distortion behaviour of plastic parts with a significantly higher accuracy. Knowledge of the crystallisation behaviour can be used in this case to predict more precisely the varying shrinkage potential in the injection moulded component. This should replace the current tool manufacture process for injection moulded parts which is iterative and cost intensive because prediction of shrinkage and distortion effects is not yet sufficiently accurate.

4.4.9 Test Case Textile Reinforced Piston Rod

Modern manufacturing processes, comprising many levels and interactions, are highly complex. The focus of this test case is the development of modern models and simulation tools in the innovative area of textile-reinforced light metals. Textile reinforced metal matrix composites are a new class of materials opening up new applications and more efficient construction methods as compared to conventional non-reinforced metallic materials. Textile reinforced metals are particularly required in lightweight construction in which a well-defined fracture behaviour, high compressive strengths and high strengths at high operating temperatures are required. Textile reinforced plastics cannot or only partly provide such characteristics. The reinforcement components for metallic alloys are 3-dimensional textile structures made of temperature-resistant, high-modulus fibres such as ceramic fibres. The possible application areas of this material group are highly-dynamic components within car engines, high-speed machinery and gas turbines.

The near-net-shape manufacture of fabrics of ceramic fibres allows precise definition of the fibre positions within the component and low-wear machining. During the manufacture of this kind of components, a braid is initially infiltrated with molten aluminium. The molten aluminium then solidifies around the braid. Solidification kinetics is largely determined by the prevailing reinforcement structure. Interactions between the manufacturing process, the textile reinforcement architecture, the matrix infiltration and subsequent solidification and eventually the mechanical component characteristics are to be expected. Investigating these interactions in general involves costly trial and error processes which ideally should be avoided. This requires linking of various software tools for an integrated simulation of a composite fibre material with a metal matrix. Planning costs for new components can be significantly reduced when being able to predict the influences of various production parameters on the component properties in future. This testcase scenario is aimed at predicting the mechanical properties (stiffness) of a textile re-enforced component as a function of the microstructure which is described over several scales starting from the individual roving via the braid and eventually the scale of the component (Fig. 4.79).

The process chain for textile-reinforced components is quite long. The chain starts with the simulation of the braiding process and modelling of the textile geometry. A further step involves analysing the infiltration of the metal and its subsequent solidification process. Finally, the effective component properties per unit cell are determined on two levels (roving and braid). Due to the anisotropic properties and the complex production processes of textile-reinforced metals, material and process simulations are essential to manage this class of materials.

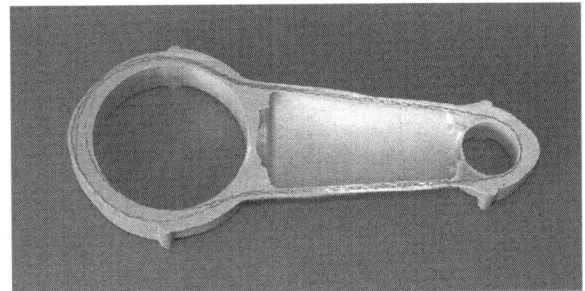

Fig. 4.79 Textile-reinforced aluminium piston rod. (Reprinted from Schmitz and Prahl (2011) with kind permission of Wiley VCH)

Simulation of the Braiding Process The simulation of the braiding process has been developed based on the PAM-Crash finite elements software (ESI Group). PAM-Crash is usually used for the dynamic analysis of the crash behaviour of structures. It provides interfaces for creating and exporting grids, changing model parameters, calculating the crash process and analysing the results.

A PAM-Crash-compatible set up is used to simulate the braiding process. The spool pattern is modelled using CABRUN; a software which was developed to generate a series of movements on a rotary braiding machine in 3-D (Stüve and Gries 2009). A rotational movement was mapped in order to use this software as the basis for the simulation of the braiding process. In a second step the orientation of the coils with respect to the direction of the braiding point was taken into account; the orientation of the braids fibres now corresponding to a radial braiding machine with 144 coil carriers. The model can also be adapted to various other braiding machines. The software output e.g., are the coil carrier paths. These can be imported into PAM-Crash using a track editor. A track function is thereby assigned to each fibre. The user in PAM-Crash defines materials, components, boundary conditions, 3D-forces and PAM control in accordance with the actual braiding set-up. This enables a description of complex braiding set-ups in which different machine parts such as braiding rings and a braid core including its travel path are taken into account in the simulation. All fibres and machine parts can be assigned with different material models.

The braiding process is simulated with PAM-Crash by calculating all collisions between the core and the fibres as well as the mutual collisions of the fibres along their paths. Braid fibres are modelled in PAM-Crash in three sectors: The lower part of the braid fibre corresponds to the coils, the central part of the fibres is used to simulate the braid and the upper part of the fibre is located at the extraction point. This categorisation enables different material properties to be assigned to each part of the fibre. The upper and lower parts of the fibres comprise a "long" 1D element whereas the central section of the fibre comprises an interlinked chain of short 1D elements which are connected at their ends via node points to which the force is coupled. The length of these elements ranges from 1 to 5 mm and is set by the user. A simulation may include 0°-fibre orientations and can take different fibre materials in the braid into account (Fig. 4.80).

The material model used can be interpreted as a general Kelvin mechanism. The properties of the spring and of the damping element can be set as linear or defined as non-linear functions wherein this type of elements can be allotted forces but no torque (Fig. 4.81).

The fibre tension is taken into account in the coil model. A coil behaves like a spring which applies a constant force to the fibre resulting in a constant tension. The parameters being varied for the central portion of the fibre include mass (m), density (ρ), linear stiffness (k) and the viscous damping coefficient (c).

The boundary conditions describe the mechanical degrees of freedom of all components in the simulation. To create the boundary conditions and determine the forces for the newly occurring bodies, the components are assumed as "rigid bodies". A

Fig. 4.80 Node- and rod elements at different time points t_x. (Reprinted from Schmitz and Prahl (2011) with kind permission of Wiley VCH)

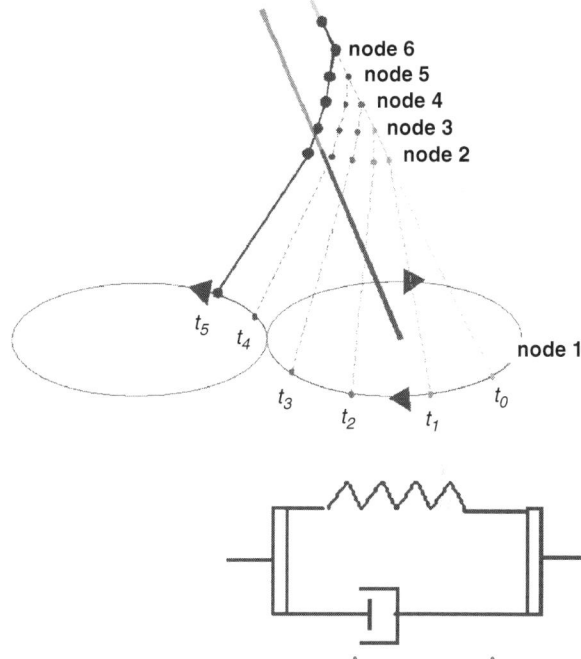

Fig. 4.81 Kelvin model. (Reprinted from Schmitz and Prahl (2011) with kind permission of Wiley VCH)

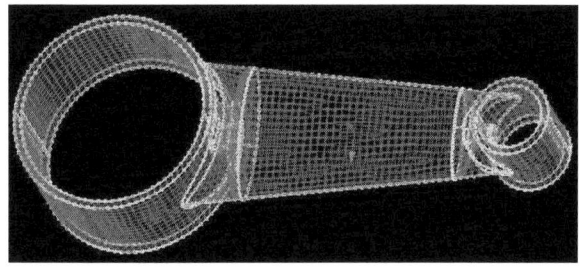

$$F(\delta, \dot{\delta}) = k^*\delta + c^*\dot{\delta}$$

"rigid body" comprises a set of nodes. In PAM-Crash for instance the relevant centre of mass, the local coordinates, the total mass and the components of the inertia tension are calculated in the global frame of reference (Fig. 4.82).

Friction is described in the simulation from the perspective of the contact plane. The friction between fibres and core are treated separately in relation to the mutual friction of the fibres amongst one another. A master/slave contact is selected for the core/fibre friction. While the master is compressed, the slave applies force to the master. This differentiation is made to reduce the computing time for the contact search algorithms. In this case, the core is the master and all fibre node points are the slaves. The overall friction force is divided into three components:

Fig. 4.82 Braided-core piston rod. (Reprinted from Schmitz and Prahl (2011) with kind permission of Wiley VCH)

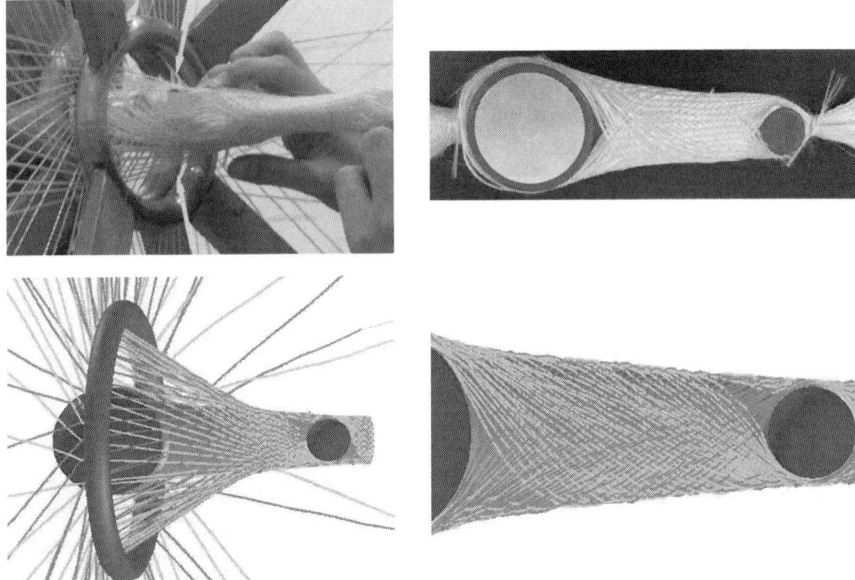

Fig. 4.83 Comparison of an actual braiding process with the simulation. (Reprinted from Schmitz and Prahl (2011) with kind permission of Wiley VCH)

- Damping-dependent component
- Constant friction coefficient (μ)-dependent component
- Contact thickness depending on contact force

The contact thickness corresponds to a distance from the master surface which is used in the search algorithm to determine expected impacting slave-node points. If an impacting node point is discovered, a force is applied to it in the normal direction with respect to segment surface such that the node point does not penetrate the "rigid body". Reality and simulation can be calibrated with the aid of a factor (Fig. 4.83).

Simulation of the Infiltration As a part of this chain, the melt flow through a unit cell of a braid structure is simulated in FLUENT to estimate the pressure drop across the braid structure. The influence of several braid layers on the pressure drop and on the loss coefficients is analysed. In this case the following applies: The thicker the braid the smaller the roving interval and the more layers of braid, the greater will be the pressure loss. Accordingly, the necessary pressure for infiltration is greater. These considerations identify the pressure loss throughout the braid structure as characteristic values for the braid. The pressure loss is therefore a relevant target value for the simulation because the braid structure influences the infiltration quality.

The braid analysed in this case represents an obstacle for the flow. It contributes to friction and/or dissipation losses in the flow and therefore corresponds to an "entrapment" in the flow field. This results, as described above, in a loss of pressure. The braid structure can be compared for example with a filter or a sieve through

4 Virtual Production Systems

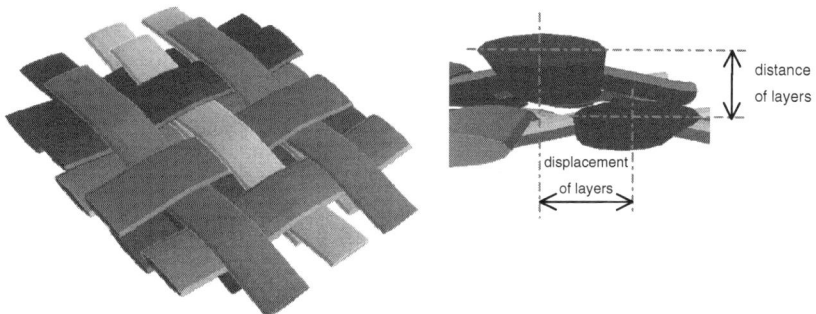

Fig. 4.84 Geometrical arrangement of the braid layers. (Reprinted from Schmitz and Prahl (2011) with kind permission of Wiley VCH)

which the melt passes. There are empirical approaches for estimating the pressure loss through filters and sieves, e.g., according to (Bohl and Elmendorf 2005):

$$\Delta p_{Sieb} = \zeta_{Sieb} \cdot \frac{\rho}{2} \cdot v^2 \tag{4.2}$$

where v is the average flow velocity in a pipe cross-section and Δp_{Sieve} generally depends on the sieve geometry, on the Reynolds number and on the dynamic pressure of the flow. There are tabulated values for ζ_{Sieve} (Bohl and Elmendorf 2005).

The objective of the simulation was to estimate the pressure loss throughout the braid structure under the conditions of a stationary, incompressible and laminar flow. The pressure loss depends on the free volume of textile. The free volume of braid in turn is dependent on (Fig. 4.84):

- The distance and orientation of the rovings in a single layer
- the number of layers of braid lying on top of each other
- the orientation and/or offset of the layers with respect to each other

The distance and orientation of the layers depend on the braiding process and the machine settings. In this case it was assumed that these parameters are fixed and only a defined braid was used. In practice, several layers are generally used for a textile preform. These layers are frequently randomly displaced with respect to each other. This fact is hard to take into account in the simulation.

To describe a multilayer braid, unit cells for each layer are combined in TGrid. Defined rotation and translation of the geometries to be combined is possible. Therefore several layers of braid (lying one on top of the other) can be modelled with or without displacement with respect to each other. Pressure losses and loss coefficients thus were identified for a single-layer and an 8-layer structure. For the 8-layer structure, a relationship between infiltration velocity and pressure loss could be established in the form $\Delta p = f(v)$.

The influence of several layers on the pressure loss is thereby simulated at different flow speeds while the influence of the offsetting is simulated at a constant speed. The basic findings are:

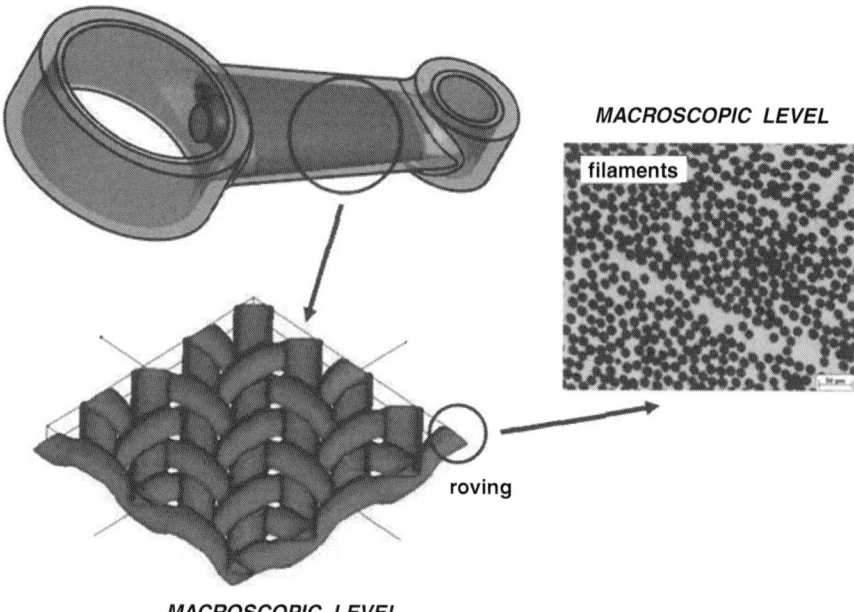

Fig. 4.85 Levels of structural analysis in the homogenisation of the piston rod material. (Reprinted from Schmitz and Prahl (2011) with kind permission of Wiley VCH)

- the more layers are present, the greater is the pressure loss
- the greater the offset between the layers, i. e. the greater the packing density of the braid, the greater is the pressure loss

Homogenisation of the Properties of the Composite Material The homogenisation method being developed to simulate the piston rod material takes into account firstly changes of the roving properties caused by infiltration of a filament structure and an entire roving and also the grain size influence when describing the matrix within and between the rovings. A two-stage homogenisation scheme was developed to describe these phenomena. The influence of the matrix infiltration in the roving is observed at the microscopic level. In this case the bulk modulus of individual rovings comprising several filaments is calculated. The effective modulus of the whole piston rod was calculated at the macroscopic level. In this macroscopic calculation the result of the homogenisation at the microscopic level was used as the modulus of the reinforcing phase instead of the value of aluminium oxide (Fig. 4.85).

Asymptotic homogenisation was used at both levels (Sanchez-Palencia 1980; Bensoussan et al. 1978). Three-dimensional spectral analysis was used to define the dimensions of the adequate representative volume element (RVE) in this method. The so-called windowing method (Torquato 2002) was also used to define the RVE geometry. This method defines the RVE as a section (window) of the original microstructure with corresponding RVE-dimensions. Such an approach can precisely

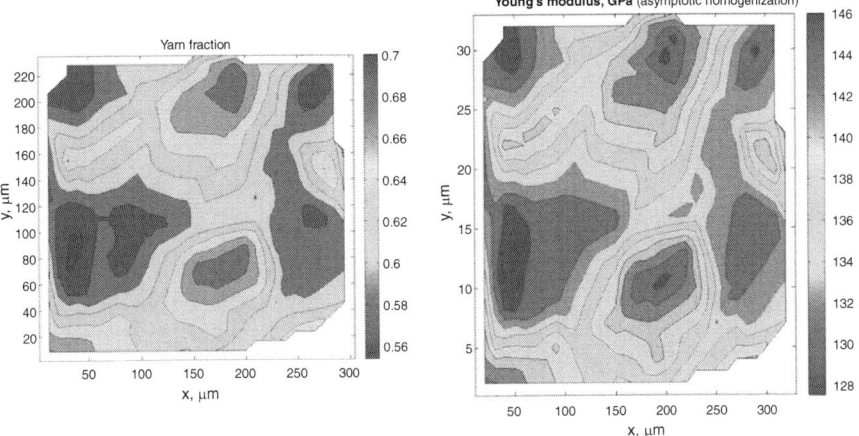

Fig. 4.86 Distribution of the filament component and effective bulk modulus of elasticity on the roving cross-section. (Reprinted from Schmitz and Prahl (2011) with kind permission of Wiley VCH)

replicate the phase distribution in the microstructure (e.g., fibre distribution in a roving) with the disadvantage however that the resulting effective properties depend on the actual position of the section. This case results in a property distribution as the key factor and the quality of the RVE definition is assessed on the basis of the mean value and the standard deviation of this distribution. In order to investigate the distribution of fibre component- and bulk modulus values, 50 RVEs were randomly sectioned out of the original microstructure (Fig. 4.86).

The mean value of the distribution of the filament fraction is *0.624* and the standard deviation is *0.0433* or approx. *6%* of the mean value. The mean value of the distribution of the effective bulk modulus, calculated using asymptotic homogenisation, is *136.9 GPa* and the standard deviation is 5.51 GPa or approx. *4%* of the mean value. This minor variation of the distribution gives 136.9 GPa as bulk modulus of the infiltrated braid material for further use in the calculation at the macro-level.

The results of the virtual material test at the macro level are illustrated in the following table. It is clear that the infiltration of the ceramic filaments with the softer matrix reduces the stiffness of the overall material (Fig. 4.87).

$E_{textile}$	E_x	E_y	E_z
136.9 GPa	98.77 GPa	99.00 GPa	91.50 GPa
227 GPa	129.52 GPa	129.59 GPa	103.15 GPa

Fig. 4.87 Effective modulus of elasticity of the composite material. (Reprinted from Schmitz and Prahl (2011) with kind permission of Wiley VCH)

	Virt. Material test	DIGIMAT
E_x	129.52 GPa	197 GPa
E_y	129.59 GPa	197 GPa
E_z	103.15 GPa	170 GPa

Fig. 4.88 Effective modulus of elasticity of the composite material: comparison between virtual material test and mathematical homogenisation (DIGIMAT). (Reprinted from Schmitz and Prahl (2011) with kind permission of Wiley VCH)

For reasons of comparison calculations at the macro-level were further carried out using the DIGIMAT software. Since it is not possible to use the actual braid as the RVE geometry in this software, the braid was broken down into individual rovings. These were assumed as cylinders. A comparison of the results of the virtual material test with DIGIMAT is depicted in the following table (Fig. 4.88).

Abstract The simulation of the process chain for textile-reinforced components is comparatively long. In accordance with the actual real process chain, it starts with mapping of the braiding process. The individual braid fibres are divided into nodes and rod elements. The braiding process is simulated taking into account friction values, machine parts and the braid core. The braid angle as the output variable is incorporated into the unit cell. The unit cell also takes into account the properties of the roving used. The unit cell also provides the basis for (i) estimating the pressure loss in the infiltration simulation and for (ii) the representative volume element for the two-phase homogenisation. The properties of the infiltrated roving and of the infiltrated braid in the second stage are calculated with the two-phase homogenisation. In this project therefore the foundations have been laid for the simulation of the process chain of the textile-reinforced piston rods

4.4.10 Test Case Stainless Steel Casting

Increasing complexity of products and their manufacturing process combined with the demand for tight tolerances and excellent mechanical properties are leading to increased use of integrative production and material simulation during development, manufacture and design of products. Particularly highly-precise machined castings; representing the focus of this test case, require tolerances within a few micrometres and other strict requirements in terms of material properties. A most precise prediction of the material properties is therefore of utmost importance. Such a prediction requires information about the evolution of the microstructure and its properties throughout the entire production- and lifecycle starting from the homogeneous, isotropic and stress-free melt up to failure under operational load.

Machine parts made of highly alloyed austenitic stainless steel, e.g., bearing housings, in general are produced by sand casting. The basic properties of the microstructure, such as grain size and the micro- and macrosegregation patterns are determined during solidification. Residual stresses caused by locally varying cooling conditions and thermal shrinkage are induced during solidification and the component geometry, originally determined by the cavity in the sand mould, changes. After solidification and subsequent cooling of the casting down to room temperature, feeding and gating systems are removed. This releases the elastic energy of the eigenstresses and the cast part accordingly deforms. In a further process step, the component is heat treated at 1,150 °C and then quenched in water. This homogenisation heat treatment dissolves undesired precipitates and residual stresses from the casting process are eliminated by creep and inelastic processes. The component then is quickly cooled by quenching in water which freezes the homogenised microstructure. Local differences in the cooling rate cause an inhomogeneous, thermally-induced volume contraction. The component is then deformed by new inherent stresses. These locally exceed the elastic limit of the material thus causing major plastic deformations, predominantly in the surface layer. In the final stage of the production process the part is machined to remove the casting surplus. This causes the residual stresses to be redistributed and leads to distortions of the part due to the release of the elastic energy evoked by the removal of material. The removal of surface material, being primarily in a state of compression during machining, reduces the internal tensile stresses. The majority of the newly exposed surface is still under pressure but to a smaller extent.

Determination of the behaviour of the component during its lifecycle cannot be separated from the composition of the alloy, microstructural properties and the presence and size of microstructural defects. External factors such as residual stresses originating from the production sequence and the cyclic operational load create a stress condition which interacts with the microstructure and determines the lifecycle.

The scope of this test case is the integrated simulation of the manufacturing process and the application of a bearing housing made from a highly-alloyed austenitic stainless steel-GX5CrNiMoNb19-11-2 using the AixViPMaP-simulation platform.

After a brief summary of the process and the corresponding simulation tools, the results of the simulation of the consecutive steps of casting, heat treatment, machining and application are illustrated in the following sections (Fig. 4.89).

Fig. 4.89 Bearing housing made of austenitic stainless steel GX5CrNiMoNb 19-11-2. (Reprinted from Schmitz and Prahl (2011) with kind permission of Wiley VCH)

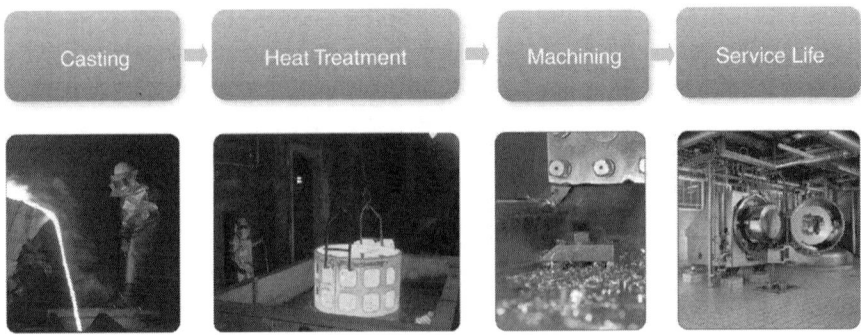

Fig. 4.90 Manufacturing process of high-precision casting. (Reprinted from Schmitz and Prahl (2011) with kind permission of Wiley VCH)

The Manufacturing Process The manufacturing process for the bearing housing comprises three steps before the component is delivered to the customer, Fig. 4.90. The microstructural properties such as grain size and segregation evolve during solidification. Undesired precipitates such as the sigma phase are dissolved and residual stresses originating from the casting process are eliminated in a 4-h heat treatment at 1,150 °C. Quenching in water following solution-annealing induces new residual stresses. Removal of material during machining redistributes these stresses. Cyclic stresses during operation of the component predominantly occur in the bearing seat.

The lifecycle properties of the bearing housing are a function of the history of all process steps. Simulation of the lifecycle therefore requires information about the component's entire production process. While properties such as residual stress can be described sufficiently at the macroscopic level, microsegregation phenomena lead to local variations in the stability of individual phases and must therefore be analysed at a small length scale.

Different simulation tools were used in the context of the AixViPMaP® platform for the simulation of the manufacture and use of the bearing housing, Fig. 4.91. The information exchange between the individual codes in the frame of the platform is characterised by arrows. Automatic coupling of the micro- to the macro-level is the subject of ongoing work.

Simulation of the Different Process Steps The first step in the production process chain is casting the steel into a sand mould. The simulation of the process chain therefore starts with the simulation of the mould filling. Predictions of the local chemical composition in the casting and predictions of microstructural properties such as grain size and the delta-ferrite component are of primary interest. The information obtained here defines the initial and boundary conditions for the microstructure simulation. The mechanical properties are already largely determined in the casting process. While they can still be improved during heat treatment, they cannot be fundamentally changed. Therefore already the casting process influences the lifecycle of the

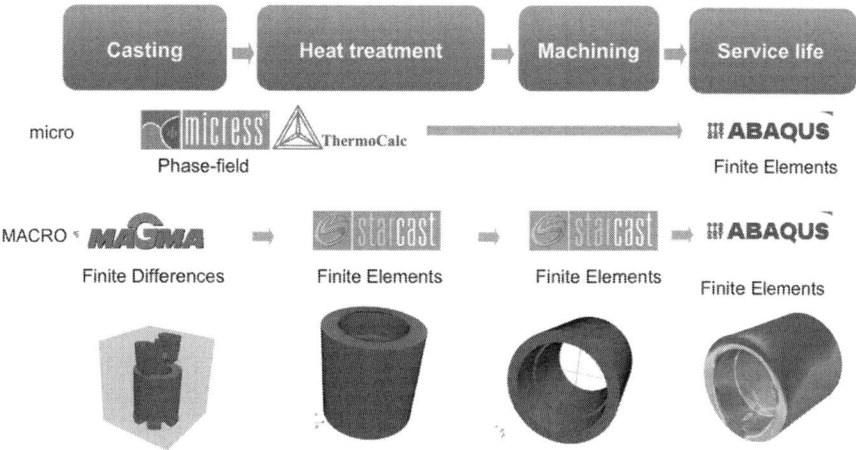

Fig. 4.91 Illustration of the software tools used to map the production process at the macro- and micro-levels. (Reprinted from Schmitz and Prahl (2011) with kind permission of Wiley VCH)

final component. The solidification-induced residual stresses are of minor interest as these are completely relaxed during the heat treatment.

As there are no exactly matching data readily available, the calculation of the macrosegregation is based on available data of the steel GX5CrNiMo18-12, which also is austenitic but does not include niobium as carbide-forming element.

The results of the macrosegregation calculations are illustrated with the model implemented in the casting simulation (Schneider et al. 1995) in the Fig. 4.92. Random checks of the composition with the help of EDX on production castings reveal qualitative agreement with the simulated macrosegregation profiles of silicon, chrome and manganese. The predictions however are incorrect for nickel and molybdenum. This is probably due to the fact, that niobium is not taken into account in the material

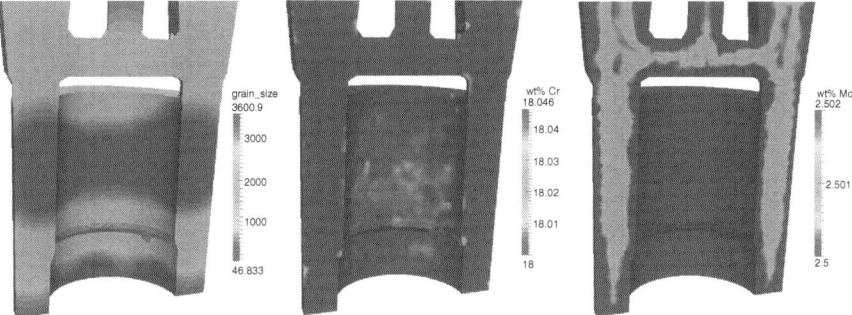

Fig. 4.92 Grainsize (*left*) and macrosegregations of Cr (*middle*) and Mo (*right*). (Reprinted from Schmitz and Prahl (2011) with kind permission of Wiley VCH)

data. The diffusion of nickel and molybdenum in the liquid might be different in presence of niobium.

The average cooling rate during solidification has the greatest influence on the grain size. As expected, the rate of temperature change is greatest in the vicinity of the chill with a rate of 23.5 K/s. At a short distance from the cooling pad, it drops by two orders of magnitude and decreases within the casting down to 0.038 K/s. The grain size and delta-ferrite content are calculated on the basis of this information.

Three main results from the casting simulation are transferred to the subsequent simulation: the grain size, the fraction of delta-ferrite and the local cooling conditions all determine the local mechanical properties in the downstream processes and during the lifecycle. The casting simulation is also used to provide typical temperature/time curves for simulation of the microstructure evolution during solidification.

Micro-segregation of chromium in the austenite grains leads to local variations in stability at room temperature. Depending on the local composition, martensite can be formed if the mechanical load exceeds certain limits. This induces elasto-plastic strains based on the volume increase.

The microstructural properties depend greatly on the local cooling rate. For selected areas these local cooling rates are therefore transferred from the macro- to the microscopic simulations.

The solidification process in the investigated component at the scale of the microstructure was calculated with the MICRESS software package using the phase field method. This approach permits not only simulation of the solidification kinetics but also the redistribution of alloying elements and therefore the evolution of microsegregations. The 2D simulation involved a calculation domain of $250 \times 400\,\mu m^2$. The alloying elements were limited to C, Cr, Mn, Mo, Nb und Ni. The solidification process is peritectic. Delta-ferrite initially solidifies from the melt and austenite evolves at a later stage.

The simulation was carried out at 1,090 °C. At this temperature, 0.2% of the melt is still present due to the high C- and Nb concentrations. Based on the composition of the saturated melt, the carbide fraction can be calculated as 0.018% NbC in these areas using ThermoCalc (Fig. 4.93, 4.94, 4.95).

Simulation of the Heat Treatment In the next step of the production process the casting is heated to a temperature of 1,150 °C and annealed for four hours to dissolve unwanted precipitates originating from the solidification process and to produce a homogeneous, austenitic microstructure with a specific delta-ferrite content. During this process the residual stresses originating from the solidification process are released by time-dependent creep processes. Only minor grain growth takes place within the microstructure and most dislocations disappear. Potentially induced distortions are negligible due to the compact dimensions and wall thickness. The mechanical material behaviour for the simulation of the heat treatment and of the machining is

4 Virtual Production Systems

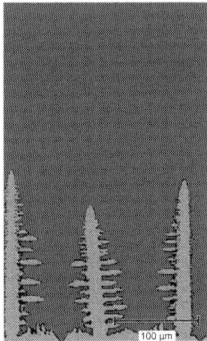 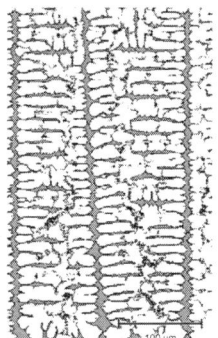

Fig. 4.93 Simulation of microstructure at 1,430 °C (*left*) and 1,090 °C (*right*) at a cooling rate of 6 °C/s; liquid/melt—dark grey, ferrite—pale grey, austenite—white. (Reprinted from Schmitz and Prahl (2011) with kind permission of Wiley VCH)

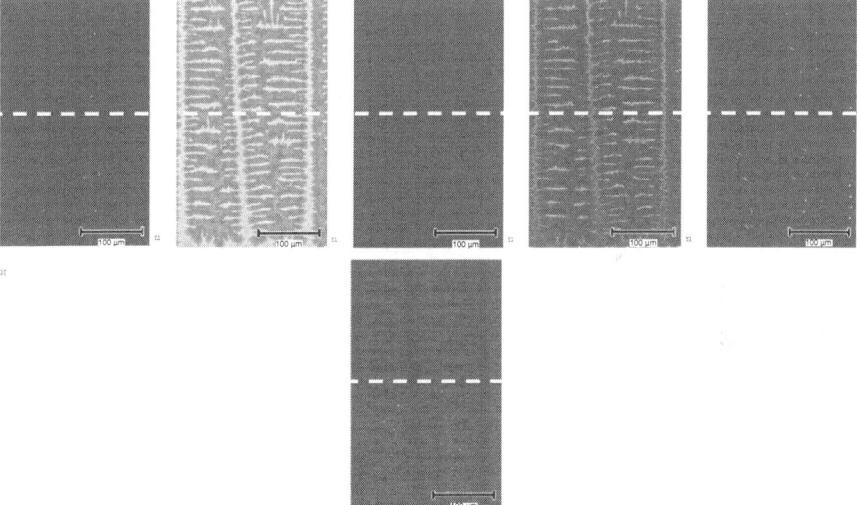

Fig. 4.94 Concentration distributions of C, Cr, Mn, Mo, Nb and Ni (from *left to right*). (Reprinted from Schmitz and Prahl (2011) with kind permission of Wiley VCH)

modelled based on the grain size determined from the casting simulation using the Hall-Petch relation.

The heat treated bearing housing is rapidly quenched by immersion into water. The heat loss mechanism during the transport from the treatment furnace to the water bath is assumed to be based on diffuse radiation. The heat loss from the immersed surface is modelled with an effective heat transfer coefficient. This coefficient takes into account the heat conduction through steam at high temperatures, a heat transition

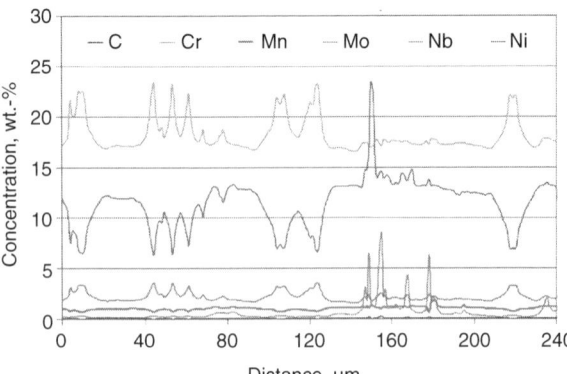

Fig. 4.95 Concentration of C, Cr, Nb, Mn, Mo and Ni along the x-axis through the centre of the simulated domain. (Reprinted from Schmitz and Prahl (2011) with kind permission of Wiley VCH)

based on film boiling between 400 °C and 200 °C and a convective boundary condition on the surface of the component below 100 °C (Sierra 2008) (Fig. 4.96).

Distortions and persistent residual stresses evolve during the quenching process. The evolution of these stresses can be divided into three phases. In the first phase, the outer layer of the component cools, shrinks and develops stresses. The stresses locally exceed the elasticity of the material thus leading to local plastic deformations.

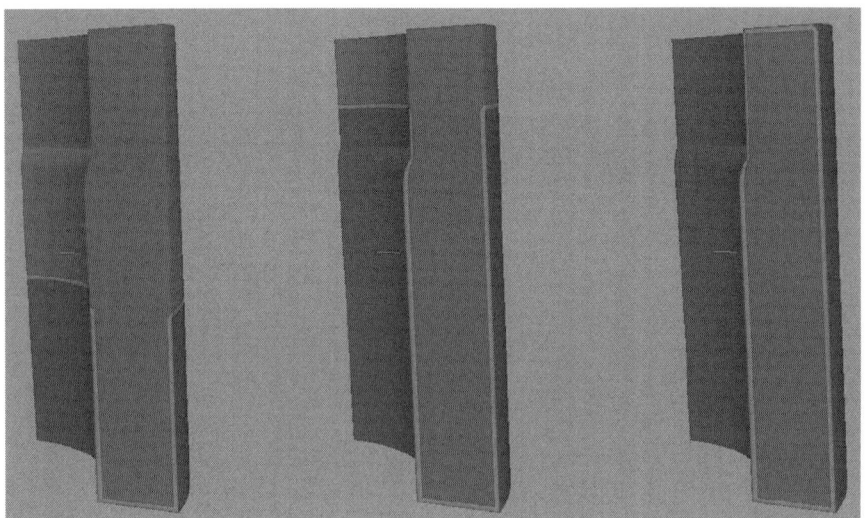

Fig. 4.96 Immersion of a quarter of the cast part into the water bath. The planes of symmetry and the surfaces in contact with air are marked red, immersed surfaces are shown in blue. (Reprinted from Schmitz and Prahl (2011) with kind permission of Wiley VCH)

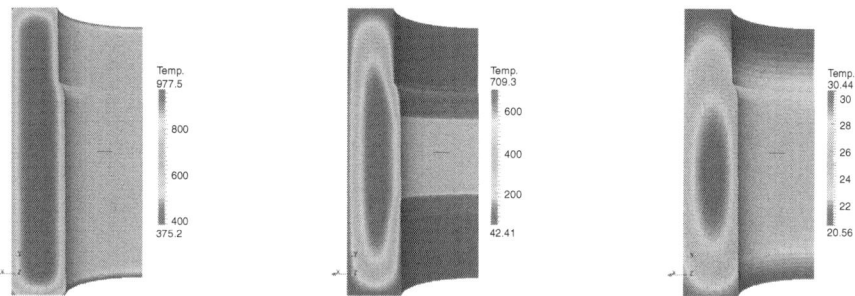

Fig. 4.97 Temperature profile of the component during the quenching process [°C]. Complete coupling is implemented by the exchange of all thermo-mechanical results between the simulation of the heat treatment/quenching and the machining process. (Reprinted from Schmitz and Prahl (2011) with kind permission of Wiley VCH)

The stresses and the inelastic deformations further increase until the temperature difference between the centre and the outer areas of the component reaches a maximum. In the second stage of quenching, the internal area cools and the temperature difference between the centre and the outer surface becomes smaller. This reduces the residual stresses to a minimum level. In the third phase the centre of the component cools further and shrinks. This compresses the outer areas which were subject to inelastic expansion in the first phase and simultaneously subjects the internal region to tensions. The stresses are thereby reversed in comparison to the first phase of cooling. At the end of the quenching process, the residual stresses reach their maximum level again. The predicted residual stresses are now in equilibrium, which comprises tensions in the surface region and, only a small distance away, the stress inside the casting changes to a compressive stress. The largest internal compression stresses exist at the edges of the cross-section and the greatest tensions occur in the central region of the component due to the temperature profile and the corresponding plastic elongations during the quenching process (Fig. 4.97).

Simulation of Machining Simulation of the machining is carried out by step-by-step removal of finite elements, similar to the actual machining process, starting at the outer surface of the casting. A new static, thermo-mechanical equilibrium of the remaining volume is calculated at each time interval. The influence of the tool on the component is neglected. This assumption seems reasonable because the region in which the tool influences the tensions is very small in comparison to the overall component size.

The machining is modelled by removing surface material which is primarily under compression. The residual stresses are reduced by 10–20 % in this processing step. The majority of the newly exposed surface is still under compression but to a smaller extent (Fig. 4.98).

Microsimulation of the Transformation In order to simulate microstructural influences on the phase transformation and thus the component distortion, investigations

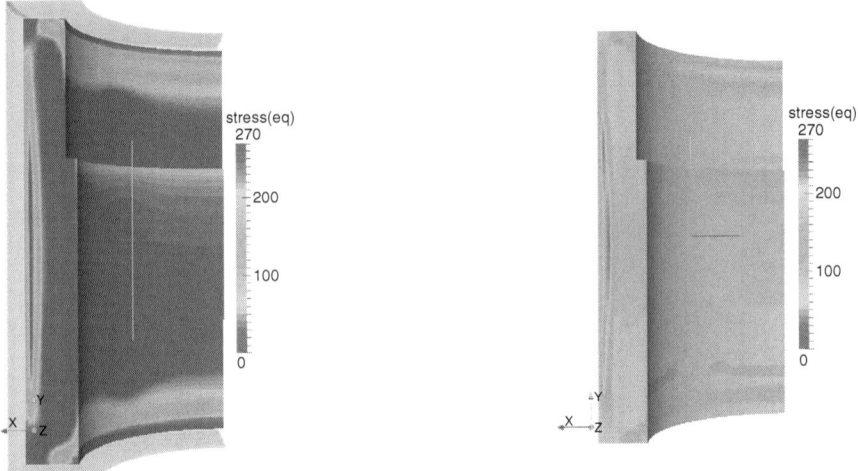

Fig. 4.98 Comparative stresses (MPa) in the component before (*left*) and after (*right*) machining. (Reprinted from Schmitz and Prahl (2011) with kind permission of Wiley VCH)

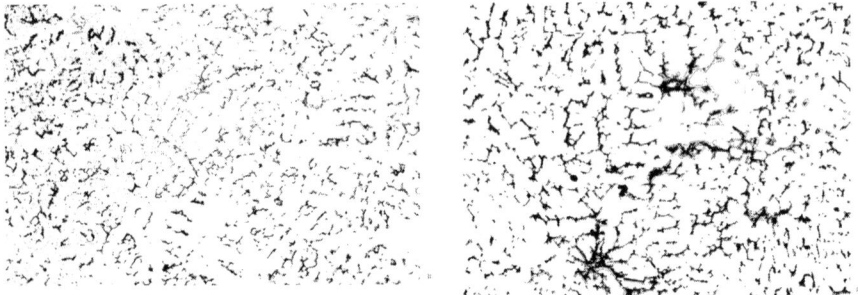

Fig. 4.99 Microstructural regions of the samples CGC (*left*) and CGI (*right*). (Reprinted from Schmitz and Prahl (2011) with kind permission of Wiley VCH)

were carried out on real microstructures of the casting. The corresponding regions, due to varying cooling rates, exhibit different microstructures (CGI: coarse-grain zone, internal (component centre), CGC: coarse-grain zone centre (below the runner)). The simulation was executed with the help of the commercial finite elements program Abaqus®. The austenite/martensite transformation was incorporated via a subroutine (UMAT–user defined material) in Abaqus® (Fig. 4.99).

The material model used takes into account a distortion-induced austenite/martensite transformation which assumes that the parent phase undergoes an initial plastic deformation before martensite forms. A dislocation density-based flow curve approach was used to describe the elasto-plastic material behaviour of the delta-ferrite.

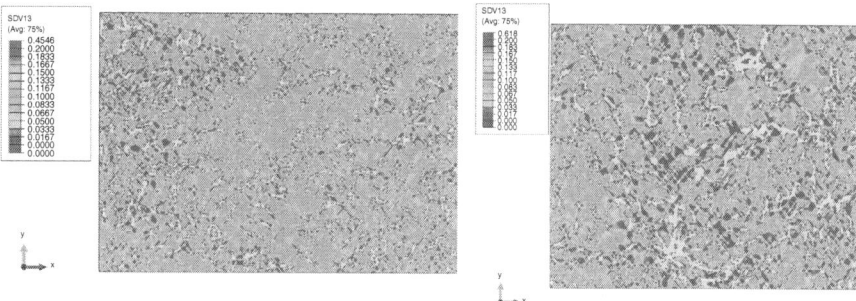

Fig. 4.100 Martensite volume fraction in the microstructure CGC (on the *left*) and GKI (*right*) after 2.5% deformation in the x-direction (*red*: martensite, *green*: austenite, *white*: delta-ferrite). (Reprinted from Schmitz and Prahl (2011) with kind permission of Wiley VCH)

Experimental investigations confirmed similar deformation behaviour. The simulations reveal preferential partial transformation dependent on the local morphology. In addition, transformation is preferentially observed in the vicinity of delta-ferrite regions (Fig. 4.100).

This effect is particularly evident in the GKI microstructure. The influence of the morphology is also evident from the analysis of the stresses within the microstructure. The heterogeneity of the stresses is further increased by the partial phase transformation (Fig. 4.101).

Macrosimulation of the Operating Load The macroscopic simulation of the operating load was carried out using finite element program, Abaqus. The homogenised flow curve from the microstructure simulation was used as the material description (Fig. 4.102).

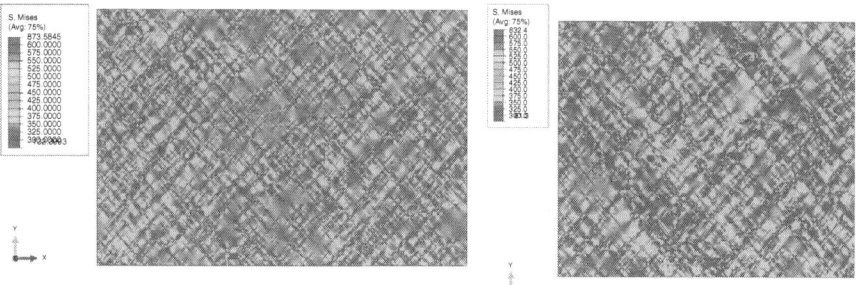

Fig. 4.101 Equivalent stress according to von Mises in the CGC microstructure (*left*) and GKI (*right*) after 2.5% deformation in the x-direction. (Reprinted from Schmitz and Prahl (2011) with kind permission of Wiley VCH)

Fig. 4.102 Locally increased loads within the part due to Hertzian compression of the bearing in operation. (Reprinted from Schmitz and Prahl (2011) with kind permission of Wiley VCH)

Results and Discussion Based on the example of a bearing housing made of the austenitic stainless steel GX5CrNiMoNb19 11-2, the integrative multi-scale process modelling method of the open Aachen (Aix) Virtual Platform for Materials Processing (AixViPMaP®) was used to determine the microstructural parameters and the distribution of residual stresses after the manufacturing process and to predict the evolution of residual stresses and distortion throughout the process chain and the use of the component until failure. The results show the development of residual stresses during production and their influence on the mechanically-induced martensitic transformation under operating loads. The experimental verification of this thermo-mechanical modelling shows that the simulated component reproduces the experimental results very well in qualitative terms. A deviation of approx. 40% between the quantitative results and the measured values however still exists. This is due to numerical reasons and inaccuracy of the material data for the simulation of the heat treatment and quenching processes.

The localised martensitic transformation also means that the maximum stress values within the component are extremely localised. The phase transformation therefore evolves inhomogeneously in the subsequent load conditions. From a microstructural perspective, a morphological influence on the phase transformation could be shown. The combination of residual stresses, operating load and local microsegregation patterns originating from solidification lead to a martensitic solid state phase transformation. Associated volume expansion causes deformations of the bearing housing during the operating period and thus the failure of the machine.

4.4.11 Industrial Relevance

The primary target group for the application of the approaches to the solutions described above and simulation tools being interlinked along the process chains in the context of the AixViPMaP are companies and research institutes involved in metal and plastic production. The metal and plastics processing industry employs at least 25% of the employees in the manufacturing sector in Germany with a sales volume of at least 50 billion € (Bundesbank 2010). This does not include automotive manufacturers and railway/airplane industry and their suppliers which generate a substantial further sales volume in the fields of metalworking and castings.

These companies can benefit directly from the optimisation of existing process lines for existing products/materials, e.g., through optimized process times, lower process temperatures, increased cycle times, less cycling material, increased process reliability and robustness, improved product quality, failure prediction and/or identification of causes for faults.

An advantage to be achieved in the medium term is the use of existing process lines for new products/materials. Here, simulations open up the possibility of predicting parameters for achieving better material properties or for describing optimal properties for specified, non-functional requirements (e.g., raw material prices, recycling, weight, safety, lifecycle, costs, ecology, installation space) corresponding to the megatrends e.g., "lightweight" or "small" without introducing restrictions in functionality.

In the long term, integrated simulation of materials and production processes permits a flexible, *rapid reaction to the ever changing boundary conditions* in the area of raw materials/energy and other resources and to the demands for new products with new properties which often are accompanied by the development of new materials. All material developers are at the beginning of the value chain and improved material properties have an immediate effect with a large lever on production and logistics.

Analytical solutions are not applicable in the light of the complexity of structure evolution processes and physically-based simulations therefore seem to be the only way to gain a deeper insight and to allow for a future integration of further phenomena into the description. The quality and predictive capability of the models being developed can be evaluated in different ways. Models with less predictive capability are based on a limited verification of only partial aspects or are validated on the basis of data entering into the model. Models with a pronounced predictive capability may lead to unexpected results which can then be experimentally tested. They can describe a number of phenomena or mechanisms within a single model or even enable the prediction in areas revealing transitions between different mechanisms. Of particular significance is the combination and adjustment of a variety of models to describe complex, technological materials and their manufacturing processes.

Future potential applications of the simulation platform include e.g., the electronics industry in the field of lifecycle prediction and/or the reliability of solder joints. As integration densities increase, eventually resulting in several thousands of solder joints, the reliability of each individual connection is intimately linked to the productivity for the overall system: a single bad joint causes an entire system to be rejected. The failure of electronic systems, i.e., lack of reliability, occurs substantially at the scale of the microstructure of the solder joint. Precise understanding and control of the microstructural evolution is thus fundamental for reliable solder joints. Description of the microstructure evolution in such highly-alloyed systems is only possible on the basis of high-performance numerical simulations. These form the only path towards an in-depth understanding of the soldering process and thus to the identification of new, promising process routes for reliable solder joints, wherein even microstructural changes in subsequent operation, e.g., due to electro-migration effects, may be significant.

4.4.12 Future Research Topics

The long-term objective of AixViPMaP® is to establish a sustainable, open, modular, standardised simulation platform which enables the efficient design and/or redesign of materials, components and products. Both commercial software providers as well as academic developers may offer their simulation tools on this platform. Particularly for small and medium-sized companies who are unable or unwilling to maintain their own simulation expertise, this kind of platform offers the possibility of "Calculation on demand" for their own production line modelling. The platform also offers an excellent potential for academic and industrial training, particularly as an integrative means of analysing the production process.

Besides further developing the existing platform functions, their application to other material classes, to new products and to further processes or to extended process chains, future research tasks are the expansion of the platform and its particularly in the following directions (Fig. 4.103):

- predicting the lifecycle of individual components under static and dynamic load conditions, their reliability, crack-initiation and crack-evolution, corrosion,
- ageing under operating conditions, repair/refurbishment and ultimately recycling

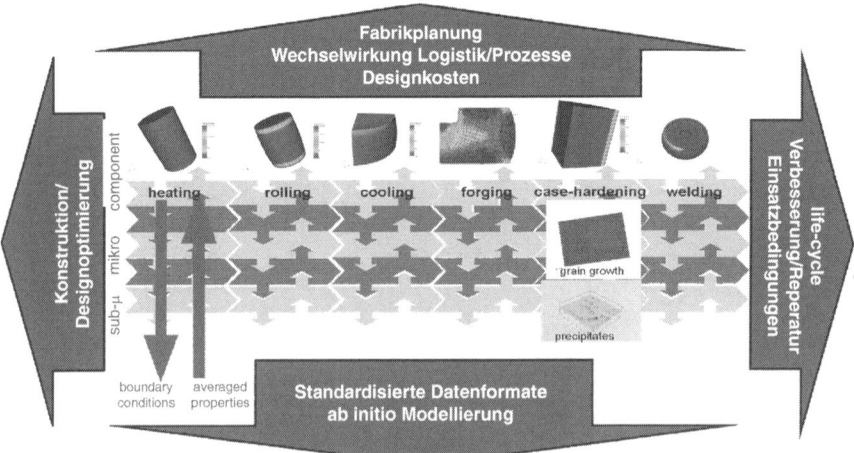

Fig. 4.103 Future developments. (Reprinted from Schmitz and Prahl (2011) with kind permission of Wiley VCH)

- the development and coupling of simulation methods for component design with optimised geometries for specific cases
- the definition, development and standardisation of methods and data formats for generating data, e.g., interfacial energies and properties of pure phases from ab-initio models particularly for complex, technical alloying systems
- the expansion towards factory planning models with the coupling of process-relevant values and production logistic-relevant values (e.g., cycle times), exploitation of logistic variables to optimise process times (e.g., storage at high temperatures, mutual influence of hot components on each other) and many more.

These ambitious aims can only be achieved on the basis of an open, modular, expandable simulation platform in which a variety of disciplines, companies and institutions are able contribute their respective expertise and mutually benefit from the results of a variety of combined software tools.

4.5 Integrative Simulation of Machine Tool and Manufacturing Technology

Christian Brecher, Fritz Klocke, Wolfram Lohse, Gustavo Francisco Cabral[1], Matthias Rasim, Johannes Triebs, Meysam Minoufekr, Stephan Bäumler, Thomas Bergs, Werner Herfs, Lothar Glasmacher and Hagen Wegner

4.5.1 Abstract

Machining of complex products in today's production environment is subject to various planning uncertainties, which require elaborate iterative optimisation, thus leading to nonproductive allocation of manufacturing resources. In this way, the availability of machine tools is reduced and highly qualified personnel resources become tied up.

To counter this, simulation systems may be applied during CAM planning that provide a better understanding of the process. However, available solutions do not consider the physical characteristics arising from the behaviour of the mechanical machine structure and machining process. Furthermore, any influence on the tool path that is caused by NC controllers is not considered. While disregarding this aspect is often justifiable for simple manufacturing tasks, complex machining processes result in significant differences between the virtual and the real process. Conclusions drawn on this basis are therefore subject to high levels of uncertainty.

---

[1] Scholarship from the Brazilian Institution CAPES.

Simulation tools are available to describe physical effects; however, they are hardly used during the CAM planning stages. They usually represent isolated applications that reflect the characteristics and behaviour of a single subsystem, such as the machine structure or the machining process. These systems exhibit significant differences in the computing times, reliability and scope of the analysed phenomena and accuracy, so that some adjustments for use in the CAM environment are required. The development of coupled simulations that can virtually map the interactions between the subsystems of a machine tool are given priority in the research environment; however, they are mostly not developed with the goal of CAM integration.

Therefore, a software platform is being implemented in the Cluster of Excellence sub-project "Virtual Manufacturing Systems" that meets the requirements of the machining planning process for milling and grinding technologies. The Virtual Manufacturing System (VMS) links models to consider the NC control system, drive control circuits, mechanical machine structure and process forces, thereby enabling the virtual prediction of both individual effects and interactions. In addition, a close connection is sought at the planning level through the integration of coupled simulations in a CAM system, so that simulative results may contribute to optimised process planning without breaks in information flow.

In VMS, control simulations form the starting point that permit a precise reflection of the NC behaviour. During the milling process, multi-body models are added as mapping of the mechanical structure of a five-axis machining centre. They are connected with calculation algorithms based on engagement calculations and empirical correlations for determining process forces. The latter approach is also applied during the grinding process. Here, due to the different process characteristics physical models with lower levels of detail are used for the virtual mapping of the machine structure. The simulation options are completed by models to identify the energy requirements of processing operations with which related aspects of energy efficiency may be included in operations planning.

In order to consider more easily the integration of a new functionality from the design and simulation environment, advanced software architecture approaches were analysed in the sub-project "Virtual Manufacturing Systems" that can be used to create information technology frameworks. From these studies, the Fraunhofer CAx Framework for advanced technology and path planning has evolved that provides basic functions and connection points for embedding new CAM functionality.

Based on the VMS and the CAx-framework, companies from high-wage countries can switch to manufacturing processes more economically, optimise and implement those processes. By employing qualified personnel that can apply these software tools, the resulting flexibility and quality may lead to significant competitive advantages. Moreover, machine and plant builders, a successful branch in high-wage countries such as Germany, can benefit economically from the use of VMS and improve its highly complex products, i.e., production machinery.

4.5.2 Planning and Simulation of NC-Controlled Machining Processes

In the manufacturing industry of a high-wage country, products must meet high standards with respect to their characteristics without requiring lengthy break-in and optimising processes for their production equipment. Crucial to the process design and associated production costs are the specified tolerances that must be maintained reliably during each manufacturing step to achieve the component characteristics. How wide the tolerance bands are depends primarily on the intended functionality of the final product. This applies especially to expensive and complex parts or small quantities. Here, a largely zero-defect production start is desirable. Therefore, a safe, reproducible production of the component characteristics is an essential goal in the planning and design of manufacturing processes.

The following observations relate to the machining technologies of milling and grinding. In both cases, machine-understandable processing descriptions (NC programs) of the workpiece machining process must be produced, and their subsequent implementation must lead to the desired products. During the conventional process design, physical properties are only being considered in rudimentary form, although individual sections already exist for simulation tools. Following this separation, conventional NC process chains with emphasis on computer-aided manufacturing (CAM) and their organisational classification will be highlighted first, before currently available simulation systems in areas of process-related manufacturing are being considered. Finally, an overview of existing options to accommodate the energy demand in machining operations is given.

4.5.2.1 Conventional CAM Planning

Figure 4.104 is an abstract image of the production where the level of detail decreases from bottom to top while the considered scope increases. At the lowest level, the product can be found with its manufacturing history and the resulting properties. These will be affected by the manufacturing process during which technological settings interact with the workpiece via the tools. Above this is the production machine, whose potential and characteristics also affect the result of the process. By not just considering an isolated machine, but a manufacturing cell or line, the system level is reached on which Manufacturing Execution Systems (MES) controllers coordinate manufacturing processes.

On the topmost level, there are production systems in which operational processes such as order processing are the focal point. Planning within the CAM system takes place in particular on the levels of the manufacturing process and the production machine, whereby the workpiece and its characteristics are intensely involved. In addition when using manufacturing resources, an exchange of information with the resource management systems in the MES area takes place.

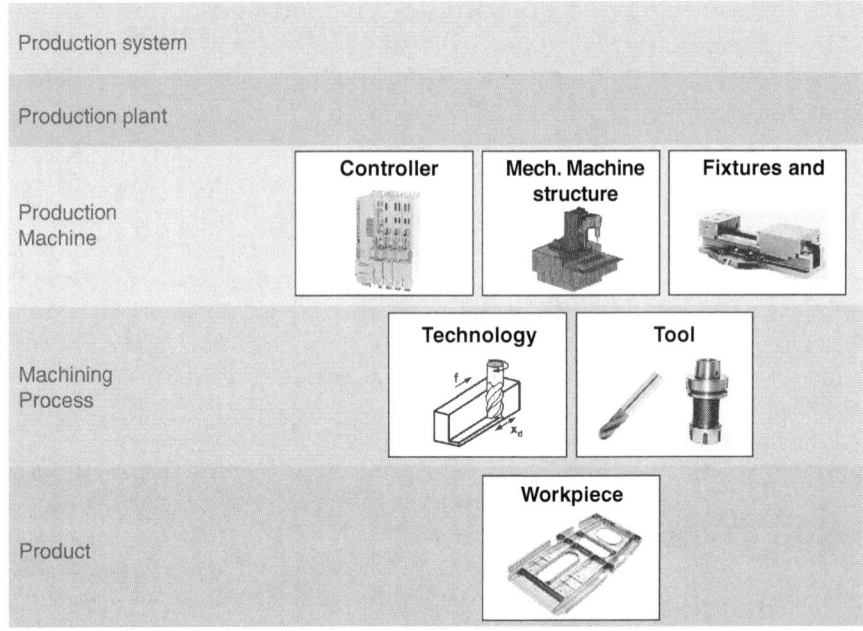

Fig. 4.104 Level model for "Virtual Production Systems"

The CAM planning usually begins with the import of CAD geometries that reflect the workpiece to be produced and possibly the unfinished part. By using these geometries, setups may be modelled; for example, as CAD assemblies with the workpiece and the clamping devices. In order to plan future production, machining operations must be defined. In this case, it must be known which kinematic properties are associated with the machine in use, so that the calculated movements revert only to available machine axes. After the selection of appropriate tools, the technology parameters such as feed, supply, and rotational speed can be adjusted. On the one hand, this data relies on catalogue values; on the other hand, it is based on properly filed information or employee-related knowledge and experience. The simulation-based verification of the chosen technology settings and the consideration for dynamic machine properties are not supported at all or only supported to some extent.

After completion of the path planning, the machining operations are simulated separately with or without presentation of the material removal whereby multiple stages are common. Generally, the tool and the material are being first considered without the machine. In order to perform collision tests, often the integration of a kinematic machine representation is possible. Here, the axes are moved based on location data from the CAM system; a post-processor run for the generation of the NC program is therefore not necessary. A higher of level of detail can be reached by executing the generated NC program on control emulations, which consider the behaviour of real control systems with vendor-specific aspects and have a higher accuracy in position calculation than the generic CAM simulation algorithms. Furthermore, virtual NC systems are increasingly integrated to calculate the position (see Sect. 4.5.4.1). Physical aspects and the exploitation of data generated by control system emulations and simulations going beyond the calculation of position values are not being considered.

While simple machining processes can be dealt with adequately in this manner, more complex parts require further planning, assessment, and analysis options. Particularly in the case of simultaneous five-axis machining processes such as being employed in the tool and die or the aircraft industry, cost-intensive parts are being produced in small series or as individual pieces. For the optimisation of the process on the real machine, this leads to high assignment and material costs. Moreover, modern manufacturing systems have options to use multiple technologies, such as in the field of laser technology. At present, a cross-CAM-planning is only possible by using multiple software tools developed in part specifically for a single application such as surface-layer welding. In this area, software tools that are based on a framework can reduce the integration effort. For example, Brandes (2008) introduced methods and tools for positioning technological interfaces between manufacturing processes. Komoto et al. (2007) and Hara et al. (2009) address the modelling approach to integrate function, production, and service activities in a software tool.

4.5.2.2 Simulation Approaches and Applications for Machining Processes

The use of models and simulations gains momentum in industrial applications and is becoming increasingly important. At various points, there are first solutions to describe machine tools in simulations. Virtual NC controls represent just one example (Baudisch 2003; Volkwein 2006; Brecher et al. 2007a; Brecher and Herfs 2007b). Limitations emerge, however, when considering other control components. On the one hand, programmable logical controllers (PLC) must be mentioned here, which have not yet reached the status of development of the NC area when considering virtual mapping. On the other hand, manufacturer-specific drive modules with direct integration into the virtual control network are missing.

In addition to technical control simulations, diverse applications exist in the field of structural dynamics that focus on the structural mechanics during the virtualisation of the product equipment. Here, static, dynamic, and to some extent, thermal properties are taken into account. During machine design CAD systems, and in particular finite element simulations (FE simulations), are used. This allows investigating the structural behaviour of individual components up to the examination of the entire machine. This application is hardly ever used in the operations planning. Further analysis can be extended to multi-body simulation (MBS). The bodies that serve as mapping of structural components can be developed as ideal rigid, or constructed as flexible and capable of vibration. In addition, coupling points are modelled as fixed joints or spring-loaded damper elements. Together with models of actuator and sensor systems (e.g., drive equations), they allow the identification of the current position of the bodies to one another, taking into account the deformation of the coupling points and bodies (Shabana 2005). Moreover, it is possible to study electro-mechanical

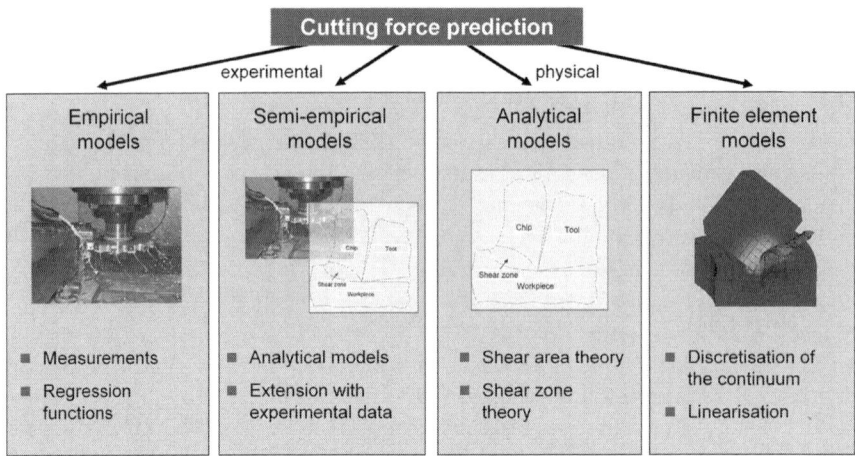

Fig. 4.105 Model forms of the cutting force. (Modified per Clausen 2005)

interactions by integrating systems from the domain of Computer Aided Control Engineering (CACE). Here, actuator models and structural mechanics of machine tools in a scientific environment can be investigated (Brecher and Witt 2006; Brecher et al. 2005).

To model machining operations, process simulations with the aim of predicting different process characteristics are used. Currently empirical, analytical, and numerical approaches are applied when modelling (van Luttervelt et al. 1998; Clausen 2005; Brinksmeier et al. 2006). Here, based on predictable characteristics, one must come to a compromise between the computation time and the availability of relevant input data (Fig. 4.105).

The empirical cutting research goes back to the work of Taylor (Stegen 2007; Taylor 1907). He identified mathematical expressions to describe the defined cutting process, such as the Taylor Wear Formula that connects tool life with the tool's cutting speed. During the simulation of machining processes with a geometrically undefined cutting edge, empirical or semi-empirical models are important. This is due to the irregular shape of the grains and the effect caused on the grinding tool by the grain's stochastic formation (Brinksmeier et al. 2006).

The analytical simulation of machining processes was established by Merchant (1945) whose shear plane theory depicts the process with physical and analytical correlations. This type of modelling uses known practices from the fields of physics, mechanics, and materials science. The chip formation theories often agree only moderately with observable real chip formation mechanisms, leading to some inaccurate results (van Luttervelt et al. 1998).

The simulation of machining processes using the finite element method (FEM) has been increasingly the subject of several scientific papers (van Luttervelt 2001). Due to the discretisation in the FE mesh, the accuracy of this method depends on the net density and thus, the computing time of the modelling depth. In addition to the simultaneous consideration of plasto-mechanical and thermal processes during the machining process, micro-geometric effects can be incorporated into the simulation results by mapping complex-shaped tools (Altan et al. 2004; Denkena et al. 2003; Denkena and Jivishov 2005; Fleischer et al. 2004). During the simulation of machining processes with a geometrically undefined cutting edge—as opposed to a geometrically defined machining process—due to their complex topography, cutting tools are usually shown only macroscopically neglecting the individual grains and grain edges. For a detailed geometric analysis procedure, the simulation is also used for individual grain contacts.

Because of the high temperatures during the grinding process, the simulation of thermal aspects has become common use (Brinksmeier et al. 2006). Basically, a large number of input variables are required for the numerical simulation of the machining process. Besides the tool and workpiece model, this includes the definition of contact and boundary conditions (e.g., friction, heat convection, and heat radiation), the process kinematics, the elasticity modulus, and a detailed description of the plastic material behaviour. The accuracy of the prediction depends to a large extent on the quality of these inputs. The beneficial use of FE simulation within the industry usually fails due to the lack of adequate availability of input parameters for the machining simulation and long computing times. This leads to the fact that only small regions of an overall process can be considered.

Experimental-empirical mechanistic modelling is a combination of experimental and analytical modelling processes. According to Altintas and Lee (1996), mechanistic modelling describes the entire system, comprising tool, workpiece, machine components, and relevant equations that represent physical relations on the basis of the dynamic behaviour of the process (Zabel 2010). However, these equations can often be complex and require data to be gained from experimental studies such as force coefficients and material behaviour. The disadvantage of this modelling method is that the determined characteristics are valid only for certain areas and parameters. With newly added cutting parameters, the variables must be determined by repeated experiments. The advantage of this mechanistic approach is that models based on this approach can be implemented with less computational times (van Luttervelt et al. 1998). The existing demand for empirical information is covered through experimental trials or on the basis of previously developed technology databases.

The relationships in grinding technology are different from those of milling. Thus, grinding tools are characterised by a large number of cutting edges. The outcome of this is a different initial situation for modelling and simulating process forces that result from chip removal with multiple cutting edges on undefined, randomly distributed grinding grains on the grinding wheel.

There is a variety of approaches available for the simulation of grinding processes, such as fundamental analytical, kinematic, finite elements or physical-empirical models, artificial neural networks, or rule-based models. Brinksmeier et al. (2006) provides an overview of the availability. Furthermore, it can be distinguished between microscopic and macroscopic approaches. During microscopic approaches, all components involved in the process are virtually mapped in detail. Due to the high complexity of these models, they are often used only for the simulation of grain removal. Macroscopic approaches describe the grinding process without considering the details of the grinding wheel topography (Aurich et al. 2008). The implementation of detailed 3D models of the grinding wheel with a representation of the individual grains and grain edges is costly and are therefore usually realised in a simplified form. There are approaches that represent only some individual grains in detail, but describe the resulting grinding wheel topography only by parameters (Pinto et al. 2008). Other models simplify the presentation of the grains by using spheres to approximate the grain (Koshy et al. 1997; Chen and Rowe 1996). So far, only Hegeman (2000) has developed a three-dimensional topographical model; however, it has not been validated by real grinding processes.

Tönshoff et al. (1992) have summarised a series of grinding force models. Their commonality is that they take into account the influence of the workpiece and grinding wheel through the use of empirical elements. Because of the complexity of the process and the large number of individual cutting edges, whose shape and orientation are undefined, and due to grinding tools with different specifications and shapes, it has not been possible to use a general model for determining the forces during a grinding process. Depending on the application, existing models must be adjusted based on empirical data, or new force models must be derived (Weinert et al. 2007).

So far, coupled simulations can be found primarily in the research environment. The approaches presented by Albersmann (1999), Stautner (2006), and Surmann (2006) describe methods that allow to visualise milled surfaces prior to the actual process and compare it with the CAD model of the finished part. The mathematical models that describe manufacturing processes in the area of high speed cutting (Stautner 2006) are used in a simulation system that supports the planning of multi-axis milling. Besides geometrical considerations, Surmann (2006) also includes physical effects during the milling process in his simulation model. Coupling mechanisms between drive control loops, machine structure, and process are analysed for example in the Priority Program 1180 (Denkena et al. 2006; Brecher and Esser 2008; Aurich et al. 2008), funded by the DFG. The research is focused on process machine interactions that are also addressed by many other works (e.g., Witt 2007). An overview is given by Altintas et al. (2005) and Brecher et al. (2009a) who present different modelling methods and concepts for co-simulations. In these approaches, influences of the NC control system and the consideration of complex geometries are generally neglected or simplified. A comprehensive coupled simulation was created by Rehling (2009) who considered controllers, machine structure and process forces; however, the solution is not closely integrated into CAM systems.

4.5.2.3 Scope of Energy Demand Simulation

The simulation of the energy demand presents itself as a new research field. In this area, several studies show that the rising energy costs for production machinery and machine tools gain more and more in importance (Neugebauer et al. 2008).

In the past, many studies on cutting forces occurring in machining processes have been carried out, analysing the casual connections of the chip removal and developing models to map the relationships during machining. Recent studies link the findings that are related to the cutting forces occurring during machining with the energy consumption of machine tools. Thiede and Herrmann (2010) study the impact of cutting forces and feed rate on the cutting energy by using model calculations and the total energy consumption of machine tools. Klocke et al. (2010a) developed a model to determine the average cutting power as a function of process parameters such as the removal rate and cutting speed. Klocke et al. (2010b) discuss the effects of individual parameters based on the average cutting model and measurements.

The assessment of energy consumption of machine tools by means of simulative approaches is discussed in various publications. An important contribution is presented by Dahmus and Gutowski (2004). Here, alongside a system description, the authors illustrate various studies on energy consumption at the component level. In addition Gutowski et al. (2006) compiled an exergy system to reflect the different energy flows of machining processes.

Dietmair and Verl (2010a), Götze et al. (2010a, b) and Thiede and Herrmann (2010) focus on the modelling and simulation of energy flows. Thiede and Herrmann (2010) present a model for the energy consumption of the entire machine in different operating conditions such as start-up, standby, or production mode. The work by Götze et al. (2010b) aims to analyse energy flows in machine tools and to predict and to optimise the efficiency of different machine configurations. Based on a dynamic model or mechatronic simulation, an energy balance model for machine tools is developed. Focus of the illustrated work are spindles and axis drives. Neither Götze et al. (2010b) nor Thiede and Herrmann (2010) compare modelling and test results. In addition to the mathematical description of the process, performance and power consumption, Götze et al. (2010a) discuss base load and part load conditions. Dietmair and Verl (2010a, b) focus on the use-dependent energy consumption of machine tools for turning and milling operations and define nine profiles of the various conditions. Additionally, the electrical power consumption of machine tools is divided into four parts. These are as follows: basic consumption, power consumption necessary for any kind of movement, electric process output and heat storage operations of a machine tool that influence the power consumption (e.g., for cooling). In addition, work to optimise the power consumption by changing the tool path is described. Here, the optimum path is determined by means of a feature-based concept, which divides the machining process into a sequence of elementary characteristics. Haas et al. (2009) presents a calculation based on a mathematical description of the kinematic behaviour of auxiliary drives of machine tools. This calculation is used to determine the collective movement of auxiliary drives, and based on that, it compares the energy exchange for various collectives. Further studies on the relationship between the tool path and energy consumption are described in Dietmair and Verl (2010b). The aim of this study is twofold: firstly, to estimate the energy consumption before the actual production; and secondly, to expand the CAM system with functions that optimise the tool path planning in terms of energy consumption.

4.5.3 Motivation and Research Question

It is obvious that a uniform and CAM-integrated simulation chain in which control engineering, machinery mechanics, and process characteristics are recorded does not exist currently. Only partial implementations are available that focus on specific sections and do not meet the required interoperability. Deviations and uncertainties in the produced and measured functional features of a manufactured workpiece must be accepted despite the use of these systems.

Therefore, until a satisfactory process reliability can be assured, complex machining processes often require lengthy and iterative optimisation cycles in order to reduce the undesirable deviations from the component's planned quality. A major reason for this is the insufficient detection of reciprocal relationships that act at different levels between the manufacturing process and machine tool. Hence, a planning system must be in place to take these interactions into account in order to compensate for expected deviations in advance. The long term goal for planning a reproducible production of component characteristics is a holistic approach of the simulation along the entire chain, starting from the control system to the process and integrated into an overall CAM system. Such a solution requires universal methods and models that provide a reliable mapping of machines, processes, workpieces, clamping devices, controls, and tools in a common framework. Besides the attention to detail, these models must meet other requirements as well. Here, especially short computation times and less modelling effort should be named, representing the essential conditions for an economic use of any simulation. Section 4.5.2 shows that these requirements are not covered in full by any solution currently available; a virtual manufacturing system in the above-mentioned sense does not exist at the moment.

In particular, integrated approaches are suitable to increase prediction quality and the realism of simulation results. The key research question thus concerns the system's architecture and the system's integration:

> How can methods and models that map machining processes, machine tools and their interactions be connected consistently in a CAM-integrated simulation considering NC controllers, mechanical behaviour of machines and process characteristics?

For many of the components involved, there are simulation solutions available (see also Sect. 4.5.2.2) that represent only a part of the overall system. However, they clearly show differences in the temporal and spatial resolution, computation time, and simulation goal. Models that should be coupled, and that are necessary for a comprehensive consideration, are often incompatible and therefore elude a direct link. The forecast of production effects, resulting from the interaction of numerical control, machine and process thus requires the conceptualisation of a complete virtual manufacturing system that meets the above-mentioned requirements.

Unfortunately, the preparatory work by the planning engineers or technicians receives little support during the identification and elimination of critical production effects. Here, the necessary momentum for the integration of physical simulations into the context of CAM is lacking.

In addition, shortcomings in the mapping of machining processes exist. To date, available models for the simulation-based CAM planning are either computationally intensive or not accurate enough. The former refers specifically to finite element models in which the separation of materials is determined based on elements of failure. But even here there are restrictions in terms of accuracy; because in many cases only certain parts of the processes such as single chip formation is considered or process influences, e.g., coolant effects, are totally neglected. The second group includes mainly analytical approaches; however, they generally fail due to the geometric complexity of simultaneous five-axis grinding and milling processes. Therefore, there is a need for qualifying process models that satisfy the prediction of cutting forces depending on the machining situation. The aim of the research presented here is not based on the exhaustive recording of all process characteristics, but rather in the construction of process models that represent an appropriate compromise between the process planning of computing time and accuracy.

During the planning and optimisation of NC programs, in addition to determining component properties, certain process parameters should also be aspired. Here, manufacturing time and the consideration of energy demand fall into this category whereby the rising energy costs are becoming more and more relevant. A consideration of this process characteristic during the planning phase requires adequate models. On one hand, these can be created with the use of economic effort; and on the other hand, they can provide a target-oriented image of the energy needs of a component manufacturing process.

In addition to the simulation-related challenges, shortcomings also exist at the CAM level. The integrative cross-technology CAM planning process is mostly limited to a few areas. For example, that implies that loss of data and an overall expensive planning process must be considered due to system change-related breaks in data continuity. Here, concepts are lacking to simplify the integration of new planning algorithms.

These shortcomings were addressed in the Cluster of Excellence sub-project "Virtual Manufacturing Systems". Section 4.5.4 shows the results achieved so far. Initially, the structure of the CAM integrated Virtual Manufacturing System (VMS) is explained. The description of the individual models, the application scenarios, and the results of coupled simulations for the mapping of milling and grinding operations are contained in Sect. 4.5.4.3 and 4.5.4.4. In addition to the simulation activities of the VMS, Sect. 4.5.4.5 introduces a software framework for virtual manufacturing systems, the CAx Framework for advanced technology and tool path planning. Based on the knowledge gained, the applicability of the integrative research approach is clarified in Sect. 4.5.5, and for which Sect. 4.5.6 provides a preview of future topics.

4.5.4 Results

4.5.4.1 Design of a CAM Integrated Virtual Manufacturing System

The Virtual Manufacturing System (VMS), researched and developed by the Cluster of Excellence, is an interlinked simulation for the machining technologies of milling and grinding operations. It addresses the existing weaknesses in the design of CAM machining processes mentioned in Sect. 4.5.2. The software tool supports engineers when attempting to form the first predictions about the expected quality of a workpiece, such as dimensional accuracy and surface finish, cycle times, and energy requirements. In order to verify newly developed or modified machine designs, this software is suitable for machine and tool manufacturers.

The VMS consists of various individual simulations that are connected to one another (Brecher et al. 2009b). The selection of the considered systems is closely related to the effects on the process result. The characteristics of the NC machine control system clearly affect the tool path while transformations, multi-axis interpolation, pre-control algorithms, etc., clearly influence the axis' movements based on the interpretation of the NC program. Drive modules with control circuits provide the interface between control systems and structural mechanics and are therefore taken into account in VMS. The structure-mechanical properties are another significant source of influence on the overall production result. Last but not least, the cutting process also plays a key role in the interaction of systems that often cannot be differentiated clearly from each other. The items listed above are combined to form the VMS and integrated into a commercial CAM system (NX, Siemens PLM), resulting in the overall system shown in Fig. 4.106.

Combining different systems to an integrative simulation tool allows a more accurate prediction of process outputs since effects that are invisible in isolated systems can thus be identified. Here, for example, tool deflection, inertia-caused deviations from the planned path of the tool, and the overlapping of these effects must be mentioned. Even with a subset of coupled simulation systems, benefits are obtainable. During the planning of the tool path, the analysis of nominal values, for which only

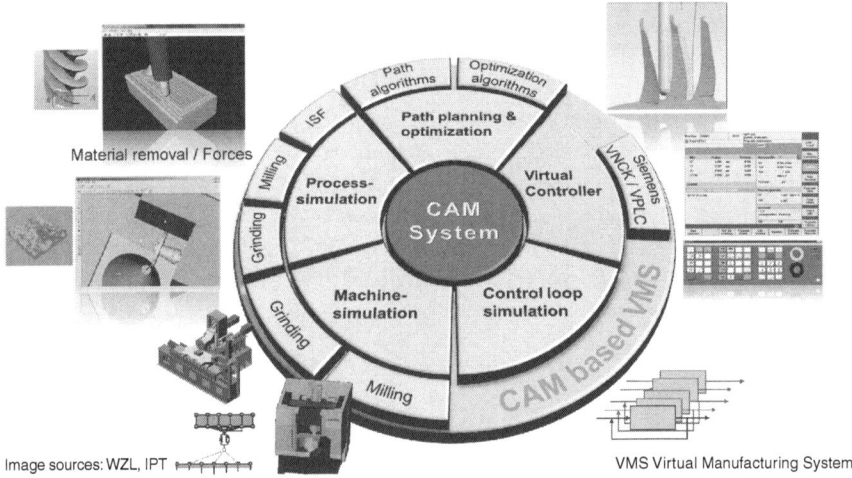

Fig. 4.106 Structure of a CAM integrated Virtual Manufacturing System (VMS). (Brecher et al. 2009b)

the physical boundary conditions of the machine's parameters are important, provides already useful additional information with regard to expected deviations. The combination of control system and machine model (without consideration of the process force) provides insight into the expected dynamic behaviour of a selected machine, which may lead to changes in the NC program. Another example of a meaningful partial link is the feedback of calculated process forces into the CAM planning system, whereby these forces are thus available for technological assessment of the expected force progression. Since all listed systems interact with one another, only a completed network can provide the maximum information feasible with VMS.

To interlink the systems, mechanisms for data exchange and synchronisation are required. From an information point of view, a linking capability of the simulation systems that can be achieved with little effort may be defined as a standardised file format that contains all essential information about the overall process. Required data from previous steps are imported into each system, and in the course of the simulation or after the simulation is finished, newly calculated data will be added. This method can be combined with batch processing in which all associated applications pass through in sequence. Here, the user is not required to start and initialise the various systems or subsequently to export data manually. This approach is particularly suitable for preliminary analysis of virtual processes since results can be obtained without extensive information-technological effort. The development of a comprehensive file format is also suitable in order to achieve the first results on the way to a functionally interlinked solution.

For a direct link via interfaces, a key element is required that establishes the connections between the subsystems. The resulting data and information flow can be organised in a forward facing system chain and a subsequent simulation loop (Fig. 4.107). The forward step sequence begins at the planning level within a CAM

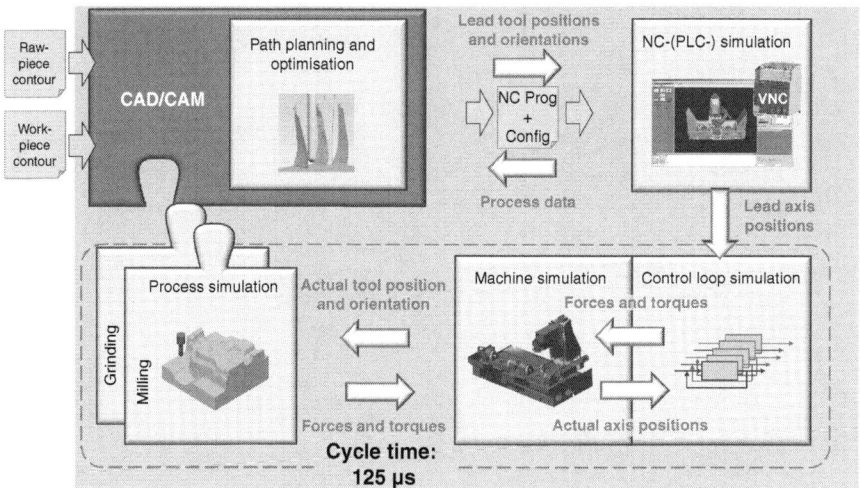

Fig. 4.107 Information and data flow in VMS. (Brecher et al. 2009b)

system; here, the required machining operations are defined including the corresponding tool paths. In addition to the geometry of the raw and finished part, input information consists of which tools are being used, which technological parameters are applied, e.g., cutting data and feed rates, characteristics of the machine, etc. The result of this step is a machining program that is transmitted to the virtual NC system. During the processing the system generates the target values for the machine's axes, thus establishing the position and orientation of the tool. This data is transmitted to the inner simulation loop.

In this loop, the interactions between drive controllers, structural mechanics and the cutting process are simulated; by setting the target values needed in the control simulation, the influence of the NC control is also taken into account. In order to consider the resonance behaviour of the machine and control circuits, the simulation loop is run through in time with the electric current controller. Upon receipt of new input values from the control system, the control circuit models provide a breakdown of the tracking error by calculating the driving torque for each machine axis and pass this data on to the mechanical machine model. This results in the change of the positions of the machine axes and with that, leads to a change in the virtual engagement situation between the tool and workpiece. The forces determined on this basis are induced into the machine model at the Tool Centre Point (TCP) and produce mechanical deformations of the structural components and the coupling points. The modified actual position, which is transmitted via a measurement system model considering noise and dead time effects, closes the loop in a final step.

The systems used in the simulation loop are applications employing complex calculations from the domain of the finite element method (FEM) and geometric information processing. Lengthy processes, which require running times of a few minutes to several days, cannot be simulated completely with this configuration. Therefore, a multi-level simulation is provided (Fig. 4.108). First, a simulation is

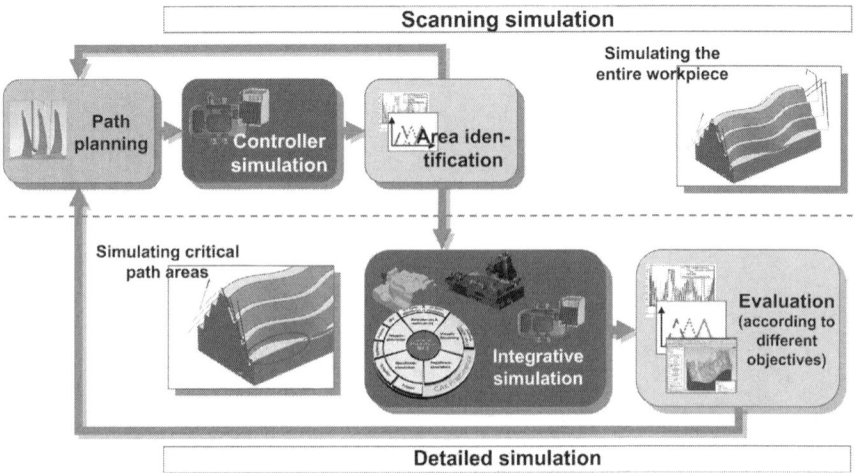

Fig. 4.108 Multi-level simulation. (Lohse 2010)

carried out with a reduced level of detail, for which the isolated control simulation is suitable. In this case, when compared to the full VMS, the calculation time is significantly reduced. Subsequently, the generated nominal values can be tested for conspicuous signal patterns in order to derive potentially critical areas. If already during this analysis shortcomings and associated causes can be detected, a return to the tool path planning occurs, in which the NC program is optimised by adjustments of control parameters or the replanning of machining operations. If potentially critical areas are recognised, the full VMS is used to simulate these areas only by using the complex machine structure and process force models. Finally, the results of the calculation are available for detailed analysis of cause–effect relationship.

To couple these systems, the software tool Matlab/Simulink is employed into which several modules can be integrated. This approach allows the application of proprietary interfaces that can be used with the multi-body system and a modular development of the control loop and process force models. Moreover, interfaces are available, upon which the integration into the commercial CAM system is based.

Only by the integration into the CAM system, it is possible to process and explore the required simulation data adequately. On the one hand, the simulation is thus directly embedded into the planning process, so that the use of the VMS does not imply a break in the informational and organisational flow. On the other hand, the data are available in a planning tool that allows further analysis.

The basis for such an analysis is the path-synchronous visualization of data assigned to the processing time. Hence, recorded data is imported into the CAM system and pre-processed according to a user-defined display configuration. Subsequently, the process data is displayed along the tool path (Fig. 4.109) in the three-dimensional

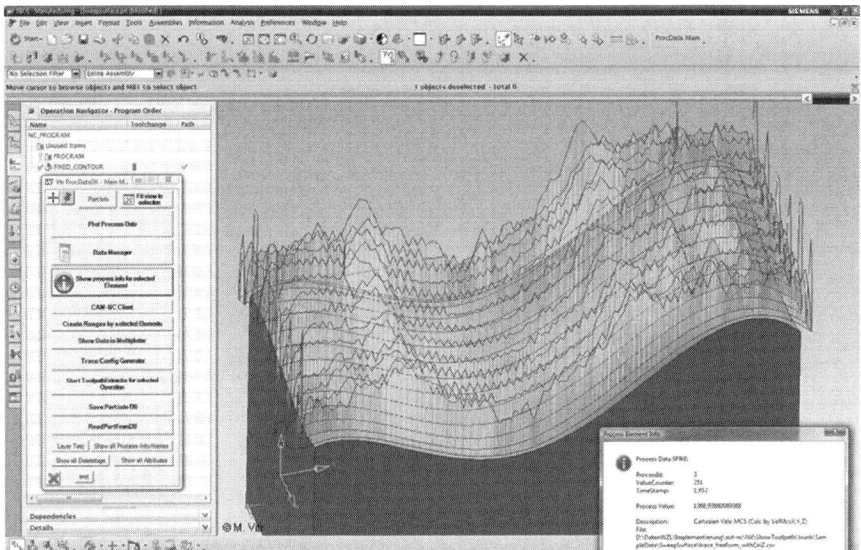

Fig. 4.109 Tool path synchronous process data display in a CAM system. (Vitr 2008)

graphics context of the CAM system. By using different highlighting configurations, planning results can be interpreted more intuitively. In this way, in addition to simulation results, real process data can be analysed, which are available from the close connection of machine control and CAM systems described by Brecher et al. (2007a, b). Figure 4.109 exemplarily shows a workpiece with a double-spline freeform surface. To assess movement dynamics better, the Cartesian velocity of the TCP is determined by the help of virtual process signals traced with the simulation described in Sect. 4.5.4.2. Subsequently, the velocity is visualised synchronously to the tool path.

In areas where the movement is almost linear (in the vicinity of vertices of the tool path) the Cartesian velocity increases noticeably. In turn, at path segments with high curvature (particularly near the sides of the workpiece), the NC controller diminishes the velocity distinctly. The manual evaluation using this CAM visualisation enables users to draw first conclusions concerning possible surface errors or dimensional inaccuracies that may occur during production.

However, visualising data is not sufficient for supporting users, and a manual examination of the available comprehensive data base entails a great effort. For a better use of this data base, a semi-automatic analysis of real and simulated data are developed that places evaluation functions in the hands of the user. These functions partially coincide with the identification of potentially critical areas for a multi-level simulation (see Fig. 4.108), so that the signal pattern analysis can also be employed for supporting the planning process. This approach also includes the aggregation of data in form of criteria, with which signals can be evaluated with regard to different quality parameters such as dimensional accuracy and surface quality. From the combination of these criteria conclusions about the achieved NC planning quality can be drawn on different levels of aggregation (see Sect. 4.5.4.3). In the medium term enhancements to a user-supported optimisation are being considered. Here, conclusions are drawn from the criteria concerning effective control parameters in the area of tool path planning and control system settings (Brecher and Lohse 2011).

4.5.4.2 Virtual Control Systems

To capture the interactions of control system and machine, an accurate picture of the tool path behaviour is required, including grinding mechanisms, exact stop settings, and NC block compression (Volkwein 2006; Brecher et al. 2009b). CAM internal position calculations and control emulations reflect this behaviour only with deviations. On the one hand they are based on different calculation algorithms and on the other hand they are incapable of processing the multiple configuration parameters of today's NC control systems without losses.

Therefore, in spite of the associated increased computational intensity VMS still employs a control system simulation. This component is a software tool with built-in virtual NC (e.g., Siemens VNCK). Since the latter uses the same source code as a real NC system, real and virtual controller are largely consistent in their behaviour.

Restrictions can be seen only in communication; however, for the intended application its role is negligible. During the execution of an NC program nominal values for the axes in the machine and the workpiece coordinate system are generated, which include not only positions but also velocities and accelerations. Moreover, positions and orientations for the TCP are computed. This information is recorded by the controller simulation and saved in a data structure. Depending on the design of the coupling, transmission across communication mechanisms or files is possible. In the latter case, after the data acquisition an export to a text-based format (comma separated values) follows.

The integration of a virtual PLC for realistic processing of control commands, such as those for tool changes or safety functions, will be waived in the VMS. Primary processing times are the main consideration with regard to the above-mentioned machining processes. Thus, neglecting machine periphery-related secondary process time or employing measured average values for this purpose does not significantly affect the accuracy. This is associated with significant performance gains, since the PLC simulation and the coupling behaviour of a mechatronic model (Brecher et al. 2007a) are omitted.

4.5.4.3 Coupled Simulation and CAM Analysis of Milling Processes

Machine and Process Force Simulation The virtual NC controller generates target values for the machine's axes that must be processed in the simulation loop. The key elements in this loop are the control loop models, the model of the machine structure and the process force calculation (see Fig. 4.107). The former are represented by transmitting blocks in the Matlab/Simulink system, in which the subsequently described physical simulations of the machine and process will be embedded as a closed loop controlled system.

The mapping of the structural mechanics is based on the multi-body simulation (MBS), which is suitable to represent the machine's kinematics. The fundamental principle of MBS is based on rigid bodies that are attached and move relative to one another via joints or flexible connectors, such as spring-damper elements.

When formulating the equation of motion for a body with six degrees of freedom, the MBS program will define six so-called generalised coordinates for each body. Information about kinematic constraint conditions are recorded through non-linear algebraic equations, which describe the dependencies between the generalised coordinates of various bodies. Given constraint movements of the bodies caused by external forces are defined in the drive system equations, which depend on the generalised coordinates and the time. Using the external forces and moments that are summarised into a generalised force, a mixed system of equations of algebraic and differential equations is the result (Shabana 2005). To solve the equation system, most available programs use the iterative Newton-Raphson method (Negrut et al. 2004). When using the multi-body simulation, only the kinematics between the bodies can be examined, without considering the deformation of the body itself.

The deformation of complex structures can be calculated using the finite element method (FEM). Here, the body is divided into a finite number of elements. The deformation behaviour of an element under the effect of external forces is described by the direct stiffness method, which consists of approximation functions. By superimposing the stiffness matrices at their nodal points between the elements, this leads to the overall stiffness matrix. When investigating the dynamic behaviour of a component, besides its mass and stiffness, its damping effect must also be taken into account. In contrast to the multi-body simulation, FEM may be used to calculate the structural behaviour of individual components; however, their displacement due to kinematic constraints cannot be computed.

Flexible multi-body simulation is a combination of MBS and FEM, thus expanding the applicability of both methods. In order to consider the elastic properties besides the kinematic constraints that exist between components, the rigid bodies are replaced by flexible bodies in the MBS. The flexibility characteristics of the structure components are characterised by modal stiffness, mass and damping matrices that are generated with the Craig-Bampton method from an FE model. This modal reduction minimises computing time, but also limits the force transmission point on the pre-defined force nodes. During a simulation, forces can be introduced only at these nodes. Therefore, it is only possible to use the flexible MBS to examine the structural mechanical behaviour of the machine kinematics at a single position and not during extended movements.

The modelling of the hybrid five-axis machining centre described in Sect. 5.5 is based on the flexible path simulation developed by Hoffmann (2008). The modelling of a linear guide is constructed as follows. On each slide, each guide carriage is defined by a force node. Along the side of the machine bed, the guide tracks are described by several force nodes in regular intervals. The force nodes of a guide path act alternately as so-called measurement and force application nodes. A computation routine controls the forces that act between the force nodes of the slide and the force nodes of the guide tracks. The guide tracks are stressed only at the points where there is a guide carriage. The inducted forces reflect the stiffness and damping characteristics of the guide. During the simulation, the relative displacement and relative speed against the direction of travel between the measurement nodes and the force nodes on the guide carriages are measured. The result is multiplied by a coefficient of stiffness or a viscous damping coefficient. In order to achieve an even load distribution, the stiffness and damping forces are spread not only over one node, but depending on their spacing, the forces are spread over three nodes of the guide track. Figure 4.110 shows the schematic structure of the guidance modelling routine.

In this way guide ways can be represented by movable spring-damper elements. There are also similar possibilities to simulate ball screw drives, linear direct drives and rack and pinion drives.

Not only the drive torques specified by the control loop act on the MBS model, but also process forces, which are calculated on the basis of mechanistic approaches (see Sect. 4.5.2.2). The process forces and thus the process dynamics determined by the simulation are related to the geometrical engagement conditions, resulting from the

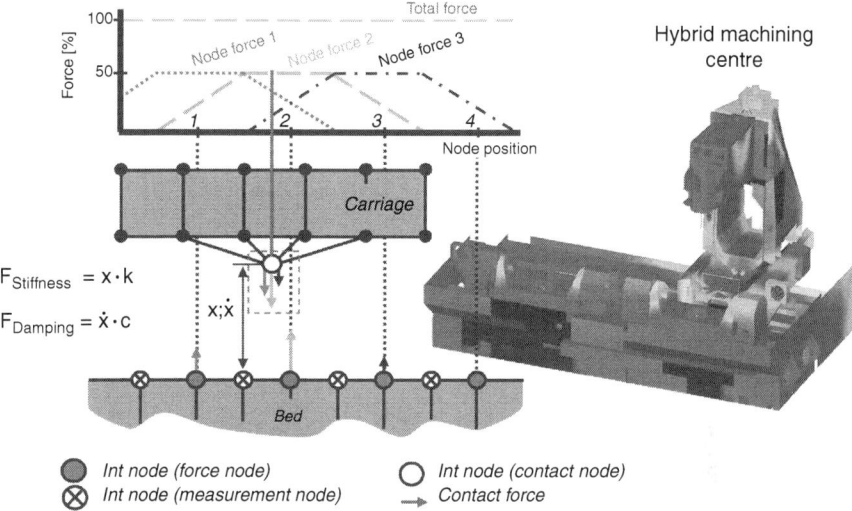

Fig. 4.110 Modelling of the guide tracks with flexible, movable multi-body models

contact between cutter and workpiece. Meinecke (2009) tries to apply geometrical engagement parameters and with this, he attempts to determine the originating forces empirically through force models by means of force coefficients (Klocke and König 2008).

For the engagement situation, a distinction is made between macro and micro contact conditions. Macro conditions are material engagement parameters that arise due to the contact between the tool's bounding geometry with the workpiece (Fig. 4.111, left). Microscopic conditions consider the edges of the cutting tool at the time of contact between cutter and workpiece (Fig. 4.111, centre). According to a classification by Zabel (2010), macro-simulation is a geometric modelling approach, while micro-simulation is based on analytical models.

Basing on macro conditions, the more complex micro-conditions can be approximated. During the study of process forces and process dynamics, these micro

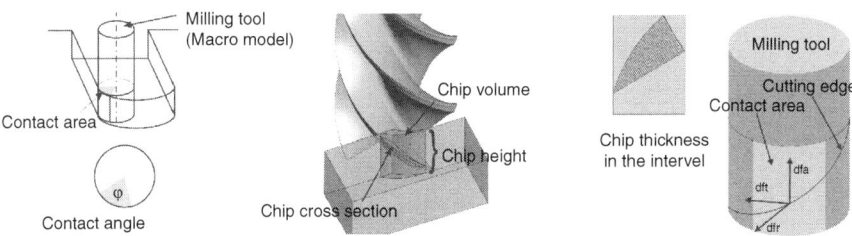

Fig. 4.111 Macro and micro contact conditions

conditions play an important role (Fig. 4.111, right). The macroscopic contact conditions (macro conditions) form the basis for the determination of microscopic parameters (Meinecke et al. 2008).

During macro simulation so-called numeric material removal models are used. Based on these, the geometrical engagement parameters are calculated. Here, analytical methods for modelling process variables are combined with numerical simulation techniques and discrete models.

The procedure of how to derive the process forces is summarised in Fig. 4.112. The approximation of macroscopic engagement conditions is based on raw data obtained from geometric material removal models. The necessary engagement parameters for the force simulation can thus be derived from the approximation of the contact conditions between the workpiece and the macro model of the tool. As shown in Fig. 4.112, the calculation process to determine the engagement conditions and the forces is based on the material removal simulation. Latter takes the workpiece, the tool, and the tool path as inputs (Fig. 4.112, left). Macro and micro simulations then generate intermediate results that are the basis for determining process forces (Fig. 4.112, centre).

The objective of micro simulation is to analyse the important chip thickness h_{sp} for the local force of a cutting element and to evaluate the relevant chip cross section A_{sp} for the process dynamics. Based on the different geometric-kinematic boundary conditions, this analysis may vary, depending on type of milling process used.

During the micro simulation, the tool is discretised by disk elements along the direction of the tool's axis. During three-axis end milling the engagement conditions for each tool disk are calculated as proposed by Martellotti (1941, 1945). By omitting the cycloidal shape of the cutting trajectory, Martellotti provides a mathematical function to describe the thickness of the material removed during the machining

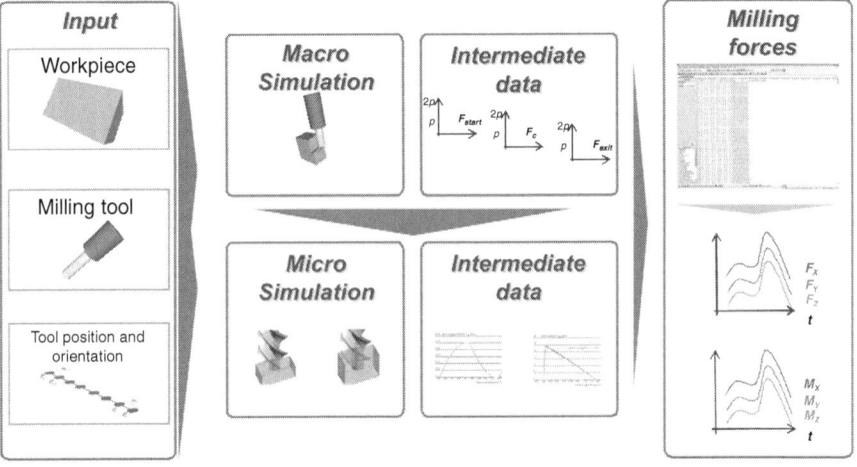

Fig. 4.112 Procedure for the extraction of geometric parameters

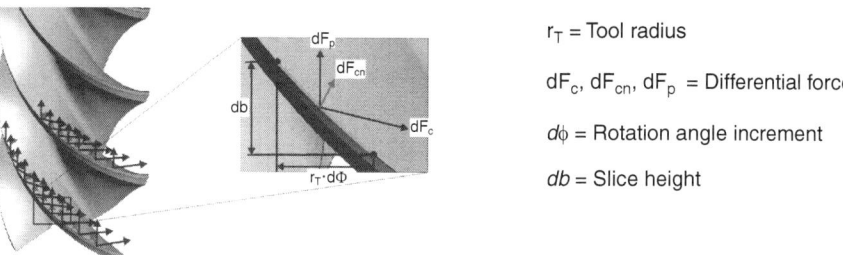

Fig. 4.113 Shear force calculation by breaking down differential forces

process:

$$h_{sp,Martellotti}(\varphi) = r_T + f_z\sin\varphi - \sqrt{r_T^2 - (fz \cos \varphi)^2} \quad (4.3)$$

r_T = Tool radius
f_z = Feed
φ = Angle of rotation

By composing cutting elements, which are calculated for each disc, the entire geometry of the cut chips can be determined. Based on this approach, mechanistic force models can be used, whereby the shear force F can be decomposed into three components, namely, the cutting force F_c, normal cutting force F_{cn}, and passive force F_p. The force component reverts to the Kienzle model (König 1982). When using this approach, experimental tests must be conducted in order to determine the force coefficients (k_i and m_i):

$$F_i = k_{i1,1}bh^{1-m_i} \quad \text{whereby } i = c, cn, p \quad (4.4)$$

The sum of the discrete shear force differences can be applied to determine the cumulative shear force acting upon the tool. For each tool layer, the afore-mentioned shear force differences are engaged with the tool and must be considered along the entire length of the tool, taking into account the contribution along each cutting edge (Fig. 4.113).

Since the force coefficients vary as the tool is cutting into material, the force model according to Kienzle is only pertinent within the context of a small process window. Figure 4.114 shows the experimental and simulated forces of a single cutting tooth determined according to Kienzle's model. These characteristics widely concur, hence a high quality of the model can be assumed for the area under consideration.

Integrating simulated forces into the coupled simulation necessitates the transformation of the force components into the coordinate system of the workpiece. This is done by applying a rotation matrix that is derived from a given tool orientation. The rotation matrix is of great importance when considering five-axis milling processes, since the orientation of the tool varies constantly along the path of the tool.

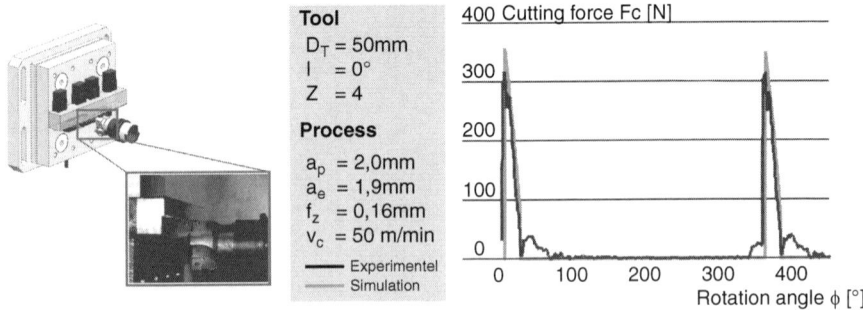

Fig. 4.114 Simulated and experimental cutting force of a single cutting tooth during end milling

Process simulation of five-axis spherical milling cutters is considerably more complex than the simulation of three-axis milling processes. Altintas (2000), Altintas and Lee (1996, 1998) provided considerable contributions in this area. This work was enhanced by Ozturk and Budak (2007) who also consider tools with varying pitch and processes, such as five-axis rough-machining. This approach is well suited for coupled simulations due to its minimum computation time. The Virtual Manufacturing System uses a similar approach, where the geometry of the chip is defined by analytical methods. Here, as in three-axis modelling, the translation and rotation of the tool is omitted. Five-axis machining with spherical head tools is mainly applied during finishing processes. The cutting speed during these processes is much greater than the feet rate; therefore, any resulting errors due to this approximation can be neglected.

Figure 4.115 includes tool geometry according to Ozturk and Budak (2007) and Urban (2009). The angular position of the cutter is defined by the contact angle ϕ. The portion of the cutter edge engaged with the material is described by the setting angle κ and the radial direction, indicated by the vector $\vec{r}$. The tool's radius at TCP R_0 and the disc radius r_n are illustrated in Fig. 4.115b. The tool's coordinate system FCN and the workpiece's coordinate system TCS (xyz) are shown in Fig. 4.115c;

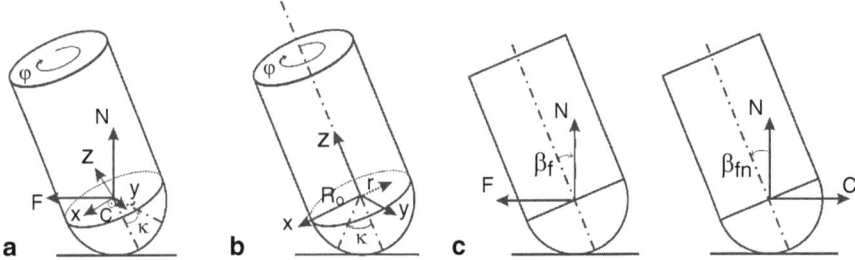

Fig. 4.115 a Coordinate systems, **b** spherical head geometry, **c** definitions of the tool position. (Ozturk and Budak 2007)

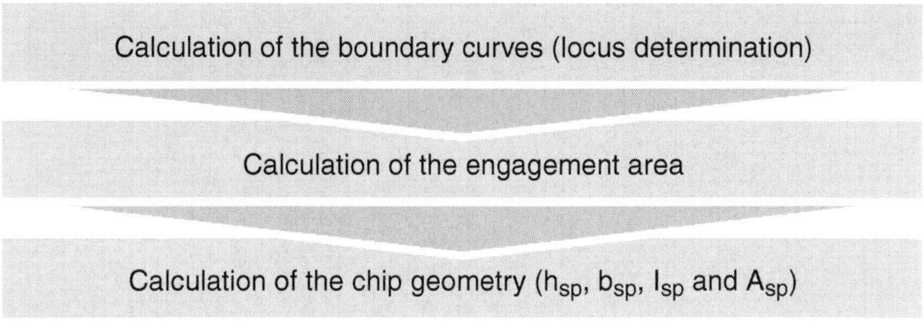

Calculation of the boundary curves (locus determination)

Calculation of the engagement area

Calculation of the chip geometry (h_{sp}, b_{sp}, l_{sp} and A_{sp})

Fig. 4.116 Steps for calculating the chip geometry

where F represents the feed axis, C the bi-normal (or cross axis) and N the axis normal to the surface.

The result of the simulation is the geometry of the chip, comprising chip thickness h_{sp}, chip width b_{sp}, chip length l_{sp} and the chip's cross section A_{sp}. These parameters are entered into the force calculation and analysis of the five-axis spherical milling process. The micro simulation is conducted in three steps (Fig. 4.116).

First, the geometric location of the chip is determined. For this purpose, the uncut chip is divided into three different sections. For each section, the limiting curve that defines the area of engagement is determined (Table 4.2). The label C+ indicates that the workpiece is located on the positive side of the C axis (Fig. 4.117).

For any given workpiece located on the negative side of the C-axis (case C-), the boundary conditions can be calculated similarly.

Table 4.2 Contact conditions of workpieces located on the positive side of the C axis (C+)

Case	Section	Condition	
		F	C
C+		$F \geq 0$ and $F \geq \sqrt{r_n^2 - C^2} - fz$	$C \geq r_n - a_{en}$ and $C \geq \sqrt{r_n^2 - (F+fz)^2}$
		$-\frac{fz}{2} \leq F \leq 0$	$C \geq \sqrt{r_n^2 - (F+fz)^2}$, $C \geq \sqrt{R_0^2 - N^2} - a_{en}$ and $C \geq r_n - a_{en}$
		$F \geq -\frac{fz}{2}$	$C \geq r_n - a_{en}$, $C \geq \sqrt{R_0^2 - N^2} - a_{en}$ and $\frac{F^2}{2a_{en}} - \frac{a_{en}}{2} \leq C \leq -\frac{a_{en}}{2} + \frac{F^2}{2a_{en}} + \frac{fz}{a_{en}}F + \frac{fz^2}{2a_{en}}$
		$-\frac{fz}{2} \leq F \leq \sqrt{r_n^2 - C^2} - fz$	$r_n - a_{en} \leq C \leq \sqrt{r_{an}^2 - (F+fz)^2}$ and $C \geq \sqrt{R_0^2 - N^2} - a_{en}$
		$F \geq -\frac{fz}{2}$	$-\frac{a_{en}}{2} + \frac{F^2}{2a_{en}} + \frac{fz}{a_{en}}F + \frac{fz^2}{2a_{en}} \leq C \leq r_n - a_{en}$

Fig. 4.117 Engagement variables for ball cutters

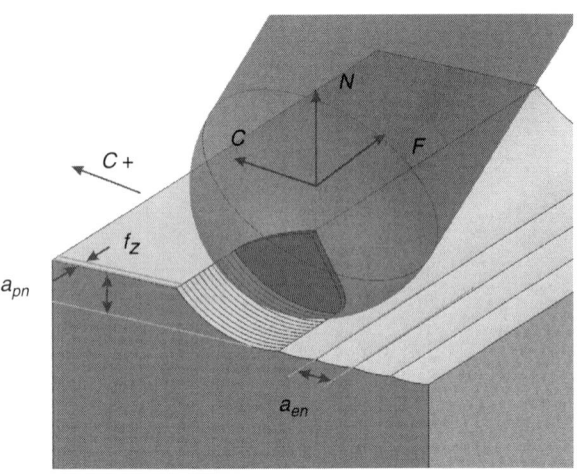

Subsequently, the area of the workpiece will be determined in which the engagement occurs. To do this, the tool is set based on a given orientation. It is of utmost importance that the rotation occurs around the centre point of the sphere, preventing a change of the form and position of the chip. Depending on the simulation's increments, the chip is being discretised. Based on previously computed limiting curves, it must be determined whether each discrete point has engaged the material. Subsequently, the chip's geometry is calculated (Fig. 4.118).

The thickness of the chip h_{sp} represents a significant dimension for the local cutting force and thus, is a parameter suitable to explain the phenomena of various wear and chip building mechanisms. This also forms the basis for numerous attempts to model the process force (Strempel 1963; Piekenbrink 1956). The chip's thickness at one point of the cutting edge is the calculated scalar product between the standardised normal surface $\vec{n}$ and the feed vector $\vec{f}$ (Ozturk und Budak 2007):

$$h_{sp} = \vec{f} \cdot \vec{n} \tag{4.5}$$

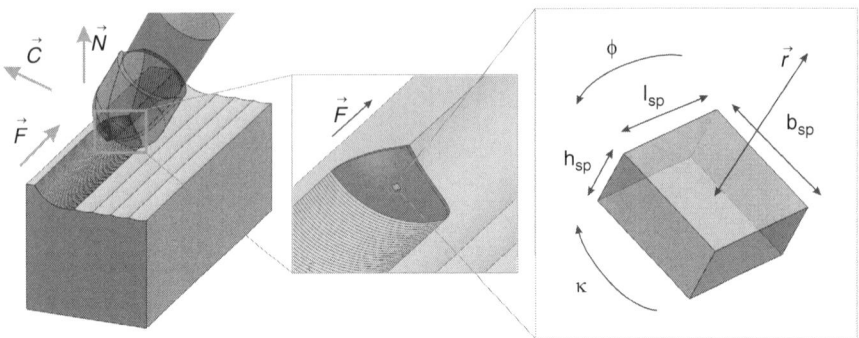

Fig. 4.118 Definition of the chip geometry during five-axis spherical milling

The feed vector in the coordinate system FCN is defined as:

$$\vec{f} = \begin{pmatrix} 0 \\ 0 \\ f_z \end{pmatrix} \quad (4.6)$$

The chip's width of a single cutting element depends on the increment of the tool cutting edge angle, the λ and the height of the disc:

$$b_{sp} = \frac{db}{\cos \lambda} \cdot d\kappa \quad (4.7)$$

The length of the current chip is defined in the direction of the rotation angle φ:

$$l_{sp} = r_\kappa \cdot \varphi_c \quad (4.8)$$

r_κ = local tool radius (depending on κ)
φ_c = wrap-around angle of tool disc

While the chip thickness provides evidence of the forces acting upon the local cutting elements, it is the chip cross-section A_{sp} that makes computation of the resulting overall load of the tool possible. From it, an assessment of the process dynamics can be derived (Meinecke 2009). A_{sp} is defined as "cross-sectional area of a chip to be removed, and measured perpendicular to the cutting direction" (DIN 6580 1985). The entire chip cross section is therefore calculated by the summation of individual cutting elements:

$$A_{sp} = \sum_{j=1}^{z} \sum_{i=1}^{k} h_{sp,ji} \cdot b_{sp,ji} \quad (4.9)$$

z = number of cutting edges
k = number of discs

The implemented method allows determining of the total chip geometry. Figure 4.119 shows the exemplary calculation results for various engagement positions.

Force calculations are based on approaches by Altintas et al. (1991), Budak et al. (1996) and Armarego and Deshpande (1989). Here, in contrast to the approach taken by Kienzle, the differential force consists of the chip deformation part and contributions of the friction and squeezing processes.

$$d\vec{F} = \vec{K}_{cutting} h_{sp} + \vec{K}_{edge} \quad (4.10)$$

In this approach, the coefficients are determined from a database for orthogonal cutting tests (Budak et al. 1996). Based on the empirical relationships, the simulated process forces are the result of the coefficients and the determined chip geometry.

The described algorithms for determining the macro-geometric parameters, the micro-geometry analysis and force calculation are converted into actual components

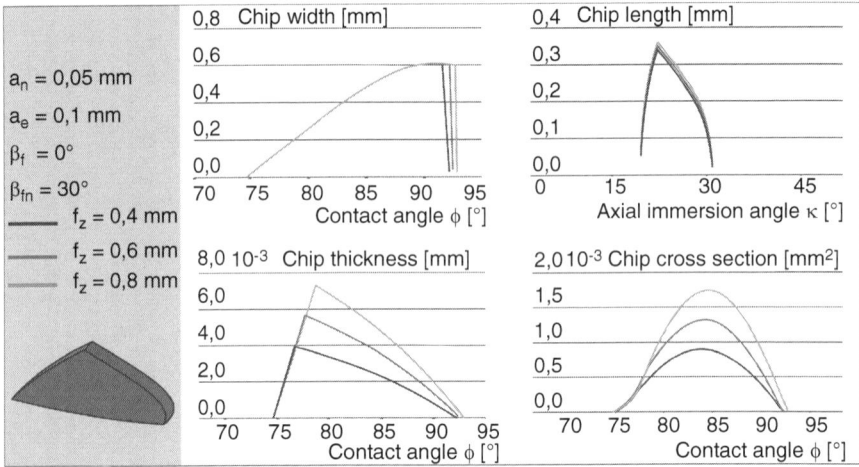

Fig. 4.119 Chip geometry when varying the feed per tooth

of the coupled simulation. With this, all components are described that make up the Virtual Manufacturing System (see Fig. 4.106). The integration into a common framework is largely completed, so that results from partial couplings are available. This particularly applies to the physical simulation of five-axis machine movements that result from the calculation of target values from the virtual NC. In this situation, the control simulation, the control circuit models and the structural-mechanical machine model are integrated.

The use of VMS requires a specific production scenario in which the machining operations can be mapped. As production equipment, the hybrid machining centre (see Sect. 5.4) is chosen. The virtual workpiece to be produced is shown in Fig. 4.120.

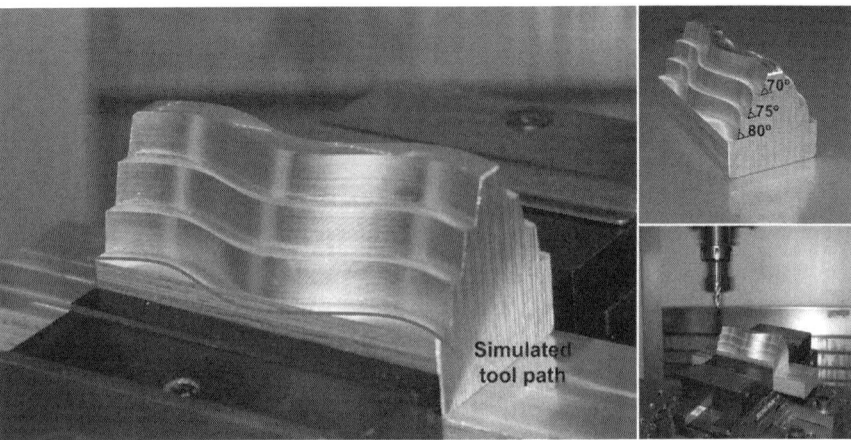

Fig. 4.120 Simulated workpiece. (Lohse 2010)

For the latter, five-axis milling operations were planned for all three flank levels in the CAM system and output by an appropriate post-processor in form of an NC program. To interpret the latter, the virtual controller had to be started beforehand, using the original NC archive of the machine tool, i.e., using a real set of parameters to configure the relevant machine information on the side of the control system.

After the virtual run of the NC program, the recorded axis values were transmitted to the coupled simulation loop (see Sect. 4.5.4.1) in which, using a cycle time of 125 μs and omitting the process forces, the peripheral milling operation on the third level (Fig. 4.120) was virtually reproduced. The simulated process time was 32 s, of which Fig. 4.120 displays only the first part of the lowest flank, that is, the path to the middle vertex of the workpiece. The calculated curves are plausible and demonstrate the effectiveness of the VMS (Fig. 4.121).

Further findings about the machine behaviour can be read from the speed and acceleration curves. Again, a comparison between target and actual values is applicable. In this context, lead values are internal control variables which are not processed further in the control circuit models. They are used by the NC controller for the calculation of reference points, between which interpolation is applied to generate commanded position. With that, the speed and acceleration target values display the expected dynamic behaviour of the controller, so that different actual values can be considered as unplanned process deviations. As an example, the target and actual speeds for a translational axis of the drive axis (Y-axis) and the rotational axis of the rotary Table (C axis) are displayed in Fig. 4.122.

In both cases, the lag of the actual speeds compared to the set values is evident For the Y-axis the controller commands high lead dynamics, which the machine cannot follow, due to dynamic constraints. Later, the difference diminishes; however, the lead value requirements are not met here either. The C axis displays a similar behaviour. It too cannot follow the dynamics required by the control system, in particular within the time frame of 16.5 and 18.0 s

The consequences on the production of the workpiece, arising from a combination of accelerations, velocities and positions are case-specific and must be assessed for each case individually. Relevant investigations are part of ongoing research activities. The same applies to the integration process of the force calculation, which will be completed in the near future.

CAM-Analysis to Determine the NC Planning Quality Evaluation methods for CAM systems, which were introduced in Sect. 4.5.4.1, were also applied to the test work piece as an example (Brecher and Lohse 2011; Lohse 2010). Since the research interest lies in in the identification of the correlation between surface defects and signal patterns, a pre-optimisation of the associated NC programs was waived. The workpiece produced for this investigation (see Fig. 4.120) as well as the recorded position, velocity and acceleration values served as a basis for subsequent evaluation.

From a visual examination of the workpiece's surface comparable results emerged for all three cutting levels (Fig. 4.120, right), allowing to omit geometrical influences exerted by flank angles. The remainder of the result discussion is therefore limited

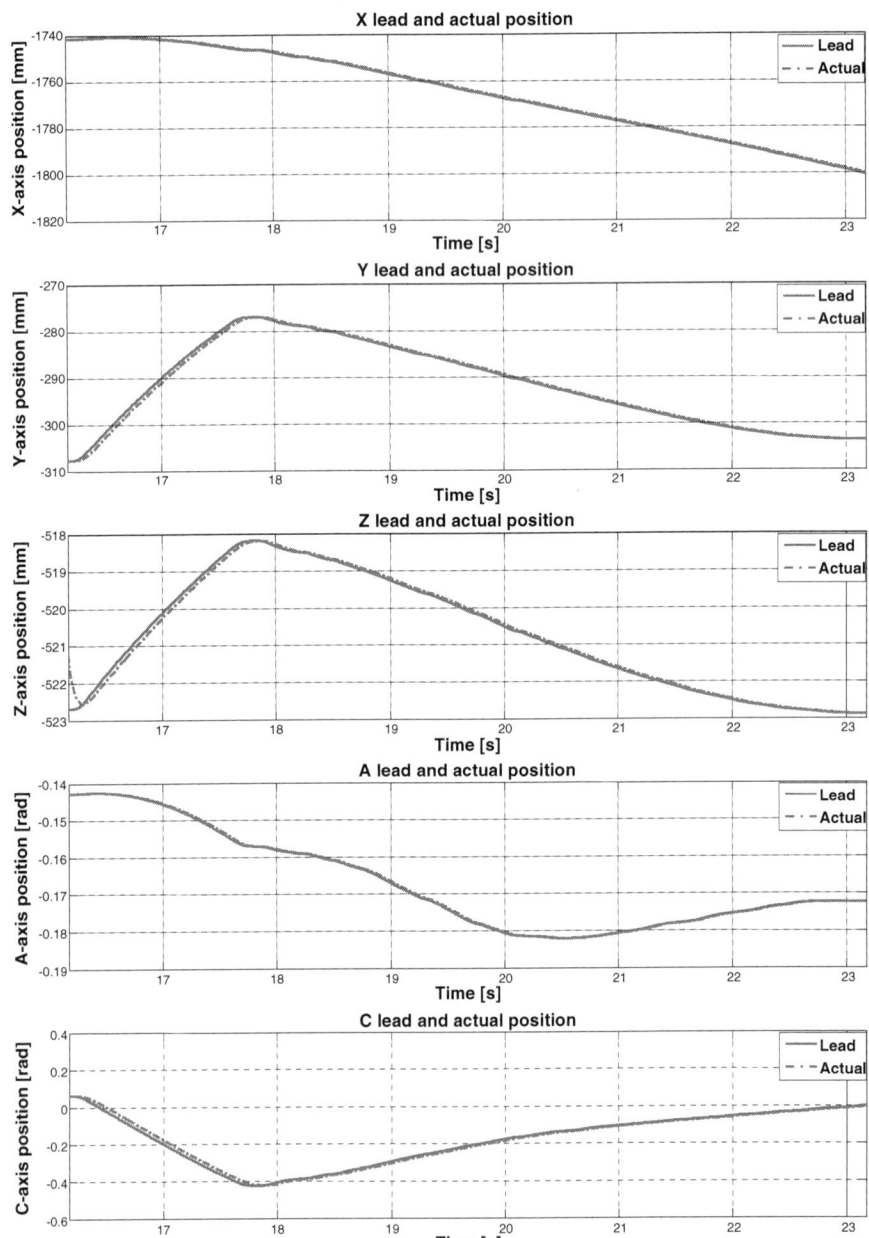

Fig. 4.121 Lead and actual values of the machine axes

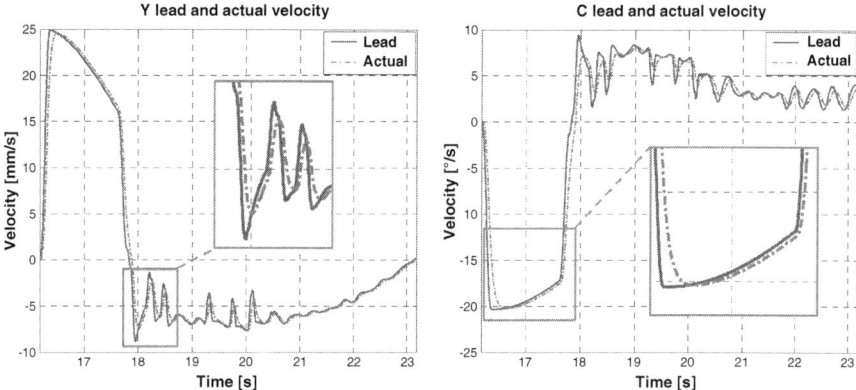

Fig. 4.122 Target and actual speeds of the Y and C axis

to the lowest level, for which the coupled simulation with the VMS was performed (Fig. 4.121 and 4.122).

In total, eleven surface defects occurred, of which especially the outer two, approximately 5 mm from the respective faces away, were distinct. A short-term deceleration or a stay of the end mill cutter and a strong jerk-prone start represent possible causes for these marks. This is why the speed of the Tool Centre Point (TCP) is suitable for the first semi-automated process data analysis. While the measured course along the tool path exhibits a constant level equal to the height of the programmed path feed, the TCP speed drops in the immediate vicinity of the marked surface defects to almost 0 mm/min (Fig. 4.123). The cause for this decline is due to the machine's movement. While the tool path is continuous, the machine path takes a sharp turn at

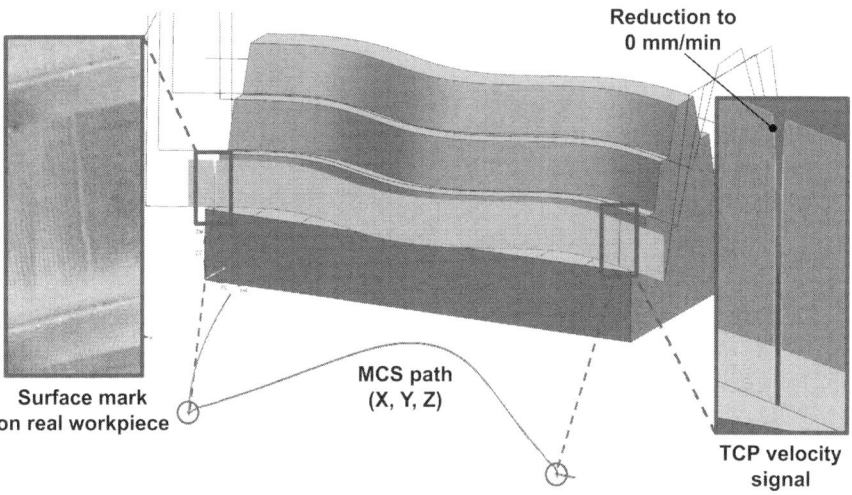

Fig. 4.123 TCP path speed with machine axis path. (Brecher and Lohse 2011)

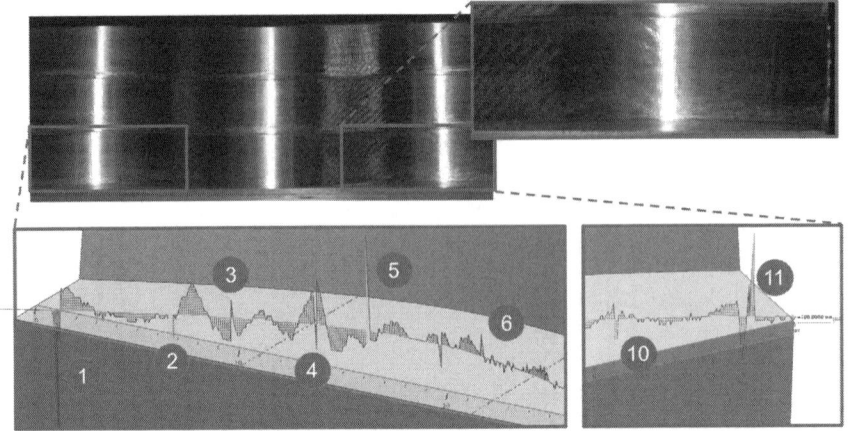

Fig. 4.124 Acceleration of the A axis with correlated surface marks. (Lohse 2010)

this point, because the Y and Z axis reached a reversal point. When starting up after passing the zero point a jerk occurs, which leads to the described surface defects. Consequently, in addition to the TCP speed, the speed of the axis must be taken into consideration when evaluating quality of the NC program (Brecher and Lohse 2011).

Besides these two accented surface marks, there are others which are mainly due to unsteady dynamics courses. Most suitable for the analysis are the acceleration signals of the axes (Fig. 4.124). Since the consideration of absolute limits constitutes a lack of transferability and a neglect of the movement history, the pursuit of an evaluation approach on this basis is not worthwhile. Instead, the difference between the signal and the output signal smoothed by a filter algorithm, gives an adequate evaluation parameter. If a signal difference that was generated at a certain time, exceeds a threshold to be defined, an unstable course and with that surface errors may occur.

In order to determine the threshold value, absolute limits may be applied (if applicable depending on the target machine tool or the production task). Moreover, relative thresholds, basing on statistical parameters describing a specific tool path section, represent a possible choice also. For the presented test piece, the latter approach proved to be purposeful, making it possible to identify eight additional surface marks (Brecher and Lohse 2011).

With increasing simulation quality an analysis based on virtual process data is planned for the long term in order to make statements already during the planning process about the NC program quality and with the goal of a first optimisation. However, building a broader criteria base for the identification of surface marks and similar production errors is required before it can be transferred to the CAM-integrated VMS.

Energy Requirement Simulation During industrial usage, the energy consumption of machine tools is generally unknown. This leads to the conclusion that energy-efficient components which are available on the market are often not employed due to their higher procurement costs. Several investigations highlight for few machines the distribution of energy consumption. However, neither are systematic statements on energy consumption for various machining and machine tools available, nor is the determination of energy consumption prior to actual machining possible at this point. A simulation of the machining-related energy consumption of machine tools and their components prior to actual processing contributes to an increase in energy efficiency of machine tools.

The simulation can be applied in different areas along the value-added chain. Thus, the simulation can be used in the customer-specific selection of machines, such as comparing different auxiliary units in terms of their operating costs and at a given range of work pieces. Another application is to use the energy simulation for different machines during their utilisation phase, e.g., if the range of work pieces is known, optimisation potentials for various over-designed auxiliary units can be revealed, without implementation of elaborate and expensive measures. To achieve this, additional knowledge about the process is needed by the user. On the one hand the information obtained by means of simulation can be used to dimension future machine tools more effectively. On the other hand, this information can be used as a basis for decisions regarding retrofit considerations; e.g., the simulation of various scenarios supports the user to answer the question whether it is worth to exchange a secondary component in the present range of work pieces or any future work piece spectrum, and after which time a more expensive but more energy-efficient secondary component pays off due to reduced operating costs.

The simulation is built component-based and includes models of the spindle, drives and auxiliary equipment like the hydraulic, coolant lubrication, and cooling system. Further partial simulations, such as handling systems, may also be considered. The simulation is described and applied in the following example of a machining centre. The quality of the simulation is then analysed by means of a test work piece.

The input variables for the simulation of the energy consumption of the components are machine data, technical data for the spindle, for all drives, motors and pumps and the NC program. In addition, information about the process is required. This information is required on the one hand to determine the power consumption of the spindle and drives. On the other hand, they form the basis for determining the cooling power (Großmann and Jungnickel 2008). The outputs of the simulation are the current real power and energy consumption for each individual component listed above and for the entire machine.

Basically, two different approaches are used to determine the energy consumption of the spindle and drives on the one side and the machine periphery on the other side. The determination of the energy consumption of the spindle and drives is based on a database collected by measurement of energy consumption for elementary process movements. The NC program of the manufactured component is assessed using an analysis unit. The energy consumption for each actual movement is calculated with the assistance of measured values stored in the database (Fig. 4.125). The results can be used to optimize the energy consumption of the spindle and drives by adopting the NC program.

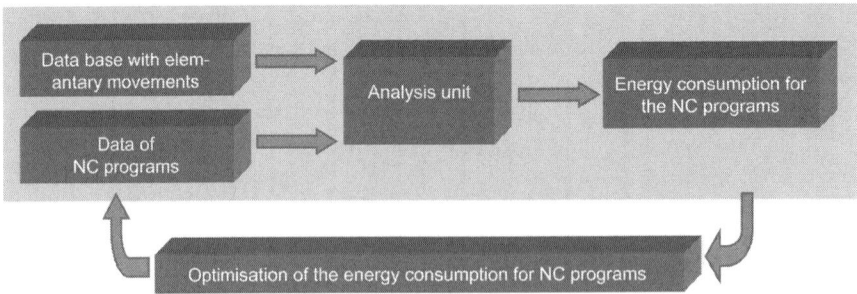

Fig. 4.125 Scheme for determining the energy consumption of the spindle and drives

The integration of the coupled simulation that is described in Sect. 4.5.4.3 for determining the energy consumption of the spindle and drives is too complex. In spite of the long computing times, such procedure results only in an insignificantly higher accuracy in the determination of energy consumption. In addition, studies show that the energy consumption of the spindle and drives for machining at a constant RPM and constant speed for many operations with a low removal rate is negligible.

To determine the energy consumption of the secondary components, hydraulic, coolant lubrication, and cooling system information from the NC program is required. By means of an automated analysis of the NC program, the running times of the individual units are calculated. Furthermore, in the coolant lubrication system the running times of individual pumps can be determined. A constant output value per secondary component is sufficient in certain cases (Dietmair and Verl 2010a), for example in conventional hydraulic systems with constantly operating pumps, simple coolant lubrication units or cooling systems. For more complex units with multiple operating states or more electrical consumers that can be turned on and off independently, a constant performance value, however, is not sufficiently accurate (Fig. 4.126).

For example, larger cooling lubricant systems use several pumps that can operate independently, and are used to provide coolant to the machine. These can be selected and de-selected via machine instructions. Additionally, pumps are used for the supply and regeneration of coolant lubricants. These pumps run either constantly or are activated or deactivated at predefined time intervals or via level sensors.

Figure 4.126 shows a comparison between measured and simulated power consumption for spindle, drives, coolant lubrication system, cooling and hydraulic system of a machining centre based on the air cut of a test work piece. Compared to the measured active power curves, a first good estimate is obtained.

In addition, Fig. 4.127 compares the average effective power consumption of the test work piece between the real measurements and simulation for each component. It is clear that, particularly in the area of coolant lubrication units and the spindle and drives, there is still need for improvement of the models in order to realize a higher simulation accuracy for these components. However, the influence of the effective power consumption of the spindle and in particular of the drives on the total

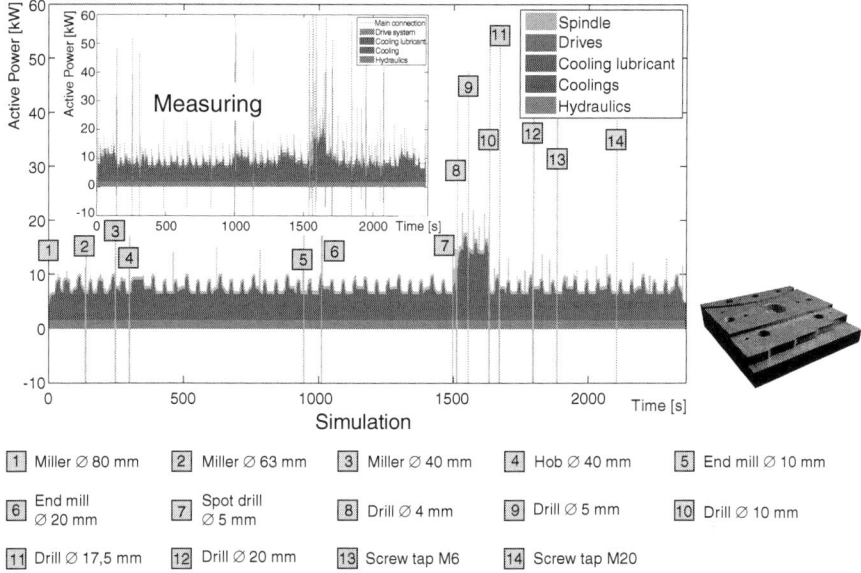

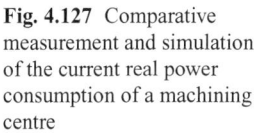

Fig. 4.126 Comparative measurement and simulation of the current real power consumption of a machining centre

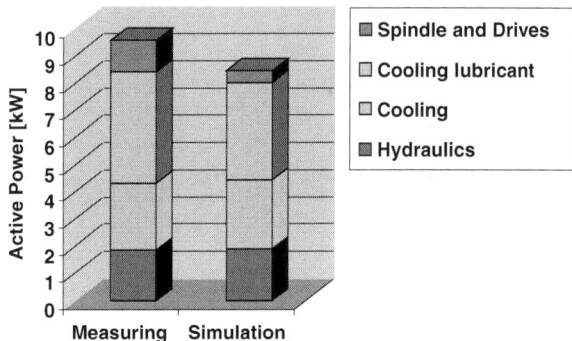

Fig. 4.127 Comparative measurement and simulation of the current real power consumption of a machining centre

active power consumption is rather low. For the hydraulic and cooling system the deviation of the simulated values of the real measurements is less than five percent, and therefore acceptable to derive further statements.

4.5.4.4 Coupled Simulation of Grinding Processes

In precision machining, which includes grinding, the dynamic behaviour is of great importance, since even small process vibrations influence the process adversely. In conjunction with the dynamic elastic behaviour of the machine/grinding

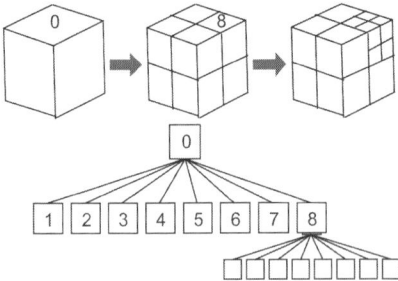 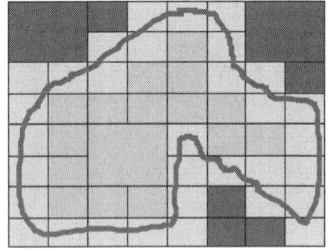

Fig. 4.128 Depiction of an octree model

wheel/workpiece system, dynamic instabilities may occur, for example, which are reflected as a grinding error in the form of waviness on the workpiece and may lead to the rejection of the component.

In order to describe and simulate grinding processes, a variety of approaches exists (see Sect. 4.5.2.2), of which the approach of the kinematic-geometric model was implemented in the project "Virtual manufacturing systems". It is based on the geometric penetration of the workpiece and the grinding wheel. To calculate this penetration, the grinding wheel and the workpiece can be described by models. First, an image has been realized through octree models. Octree models are an extension of simple cell models in which the three-dimensional space is approximated by identical three-dimensional volume elements (cubes). The accuracy of this model depends on the degree of discretization. This can be based on the expansion of a body in the three spatial axes set (i, j, k), whereby the determined memory requirement grows cubically with the same extent in all three axes. Therefore, the cell models are a memory-intensive modelling type. The octrees building upon this are among the spatially partitioned models. Here, the cell space is divided into eight smaller sub-cubes (Fig. 4.128, left). The principle of the spatial division, as it is applied to an octree, is shown in Fig. 4.128, right. This is demonstrated in the form of a two-dimensional example. Only the cubes (blue), which are cut from the surface of the volume are divided until the desired accuracy of the model is achieved. The advantage is that homogeneous regions are described by only one sub-cube. From a certain depth of the tree respectively a corresponding accuracy in connection with the object size, the memory requirement in this type of model is very large.

Due to the large memory requirements and excessive computation times for the penetration calculation, in a further step modelling the workpiece and the grinding wheel by Dexel models was implemented. These offer the advantage of a relatively high accuracy with low computational effort. The discretization of solids through Dexel models is based on a grid from which identically oriented rays are projected onto the solid. By identifying and storing the intersections of the rays with the surface of the solids, it can be approximated by the individual beam sections (Fig. 4.129).

Fig. 4.129 Single Dexel Ray Casting procedure by Weihan and Ming (2009)

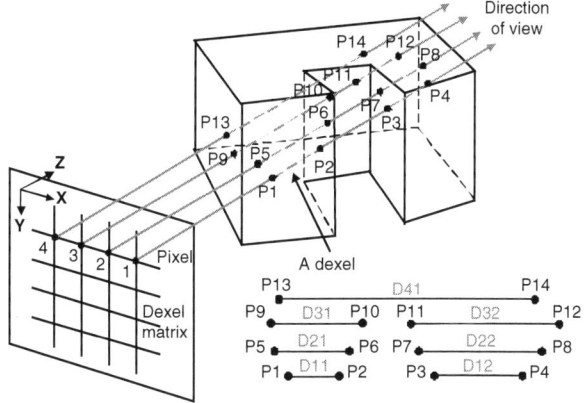

The weakness of this method lies in areas in which the solid surface is nearly parallel to the ray orientation, as significant depiction inaccuracies can occur here (Fig. 4.130).

Therefore, rather than applying simple Dexel models, multi Dexel models are used. They are based on three mutually orthogonal grids, which increase the accuracy of the models significantly.

After modelling the workpiece and the grinding tool, the penetration calculation of the solid is applied, using simple Boolean operations, with which the individual Dexel of grinding wheels and workpiece model are compared. Here, a distinction of eight cases is made that examines two Dexel for partial or complete coverage, cutting or non-contact (Fig. 4.131). Depending on the case, a new segment is created (case 5), an existing segment is changed (case 3 and 6), a segment is deleted (case 4) or no change

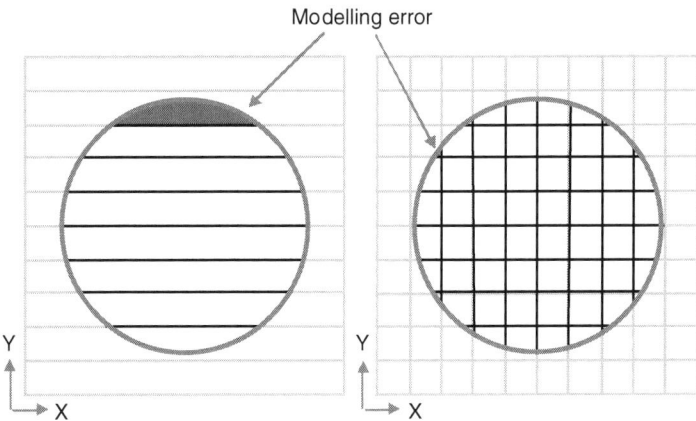

Fig. 4.130 Model error caused by a single and dual Dexel model

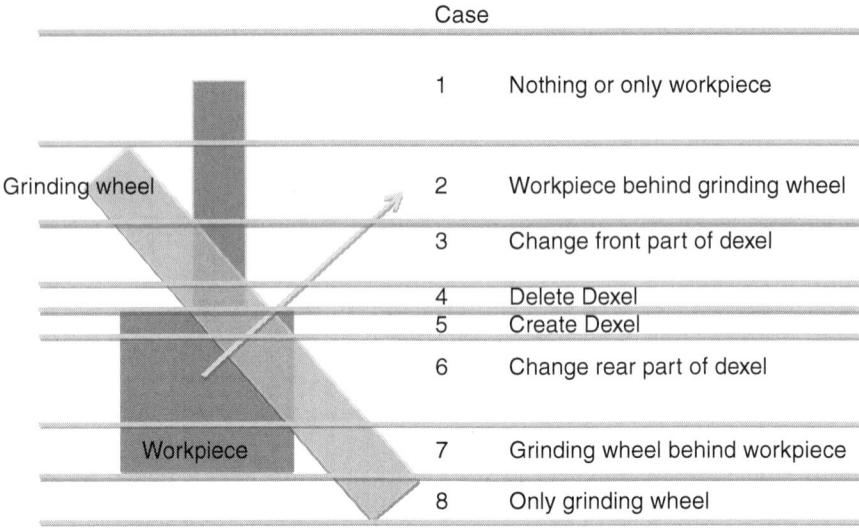

Fig. 4.131 Classification of the cutting operation based on different cases

is induced (case 2 and 7). A major advantage of Boolean operations is its simplicity, since they represent only a comparison of the operations of one-dimensional values.

The computational complexity of this algorithm for all Dexel in a single Dexel field is in the order of O (k × m × n), where k represents the average number of segments that lie on the ray of a grid point, and m and n determine the resolution of the Dexel grid. Due to its low memory requirements, the Dexel model allows for a high-resolution model. As the number of Dexel increases, the number of Boolean operations to be carried out also increases. To perform the intersection efficiently, it must therefore be limited to those areas where a model change is to be expected. This requires the use of special accelerating structures, as they are used in the displaying of graphics. These are passed through during the penetration in order to test which areas of the tool come in contact with the workpiece. The acceleration structure can be applied to both the workpiece and the tool.

In the converted acceleration structure Dexel are combined into groups which are described by an envelope. Since the Dexel of the workpiece is aligned along the line of sight, the envelope is implemented as Axis Aligned Bounding Box (AABB) (Fig. 4.132). During the collision phase, individual groups of Dexel can now be considered for extensive testing using Dexel groups. To test the collision even more efficiently, the defined groups can now be combined to form new groups. Thus, a Dexel hierarchy tree is created.

On the side of the tool an AABB cannot be used; instead an Oriented Bounding Box (OBB) is used (Gottschalk et al 1996). This is justified by the fact that the tool is not aligned along the spatial axis in regard to the workpiece. By using the OBB for the tool and AABB quad trees for the workpiece, it can be assured that only in

Fig. 4.132 Acceleration of the penetration calculation by accelerating structures

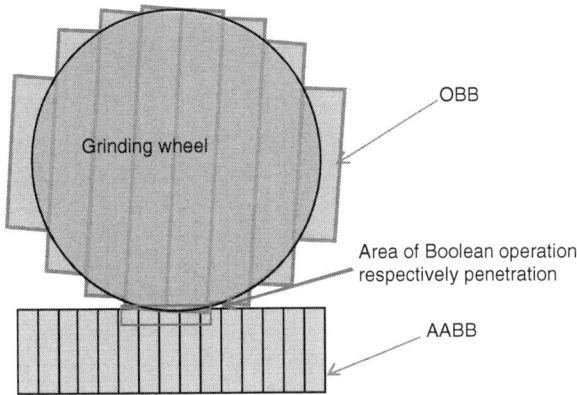

areas that are engaged, the Boolean operations are applied to the Dexel and thus unnecessary operations are avoided (Fig. 4.132).

In order to calculate the material removal that a tool creates on a workpiece, the relative movement of the tool in respect to the workpiece is described by a suitable discretization. For this purpose, the envelope, which results from a movement of the tool, is approximated by a few individual small movements over a certain time. The distance travelled caused by a movement between two time intervals must be less than the width "w" of a Dexel field. Within each time interval the movement of the tool is then calculated by the penetration of the grinding wheel and workpiece.

Next to the most general description of the kinematics, the challenge in developing the material removal calculation lies in the calculation of process parameters such as chip surface and material removal rate Q_W. Here, the required computation time must also be considered. These process parameters are the prerequisite for coupling of the material removal models with a force model. The calculation of the process forces can take place either via the chip cross-section or by means of the material removal rate. Chip cross-section can only be determined indirectly from the Dexel models. Based on the Dexel points, a triangle mesh can be generated and thus the surface can be determined. Due to multiple intersections gaps in triangle mesh may occur, which allows merely an inadequate statement about the chip cross section. Therefore, a computation of the process forces via the material removal rate is preferred and was implemented in the empirical force model. According to Werner (1971), the normal grinding force can be determined as follows.

$$F_n = \int_0^{l_g} k \cdot A_{cu}(l) \cdot N_{kin}(l) \cdot b_{s,eff} dl \quad (4.11)$$

It is the result of the integrated product of local chip cross-section A_{cu} and the number of kinematic cutting edges N_{kin} along the contact length l_g. The calculation of the chip cross A_{cu} is based on formula 4.10, using the material removal rate Q_w and the cutting speed v_c.

$$A_{cu} = \frac{Q_W}{v_c} \qquad (4.12)$$

The constant cutting speed during the process is available to the simulation as an input variable; thus the force can be calculated based on the material removal rate and the specified empiric parameters.

The implemented simulation of the material removal process and the force model assume ideal process conditions. Grinding wheel wear, coolant lubricant, temperature effects, and the dynamic behaviour of the objects are not considered. In principle, during the process simulation one can distinguish between macroscopic and microscopic approaches, which take into account the detailed topography of the grinding wheel with single grains. During the implemented simulation, the penetration is calculated on the macroscopic level. The grinding wheel and workpiece can be described by the envelope. Thus, the complexity of the simulation is relatively low and high computational speeds can be achieved. Currently, no information on the generated surfaces is being provided. For the shape of the workpiece and grinding wheel multiple input formats, such as B-Rep (Boundary Representation) or STEP files are supported. They will be automatically converted to Dexel models.

To increase the accuracy of the simulation process, the process model is coupled with a control simulation and a machine model. The entire simulation circuit is shown in Fig. 4.133.

After the control simulation generated the axis movements, they are discretized. For each discrete time interval, a simulation step is performed, in which first the penetration of the workpiece and the grinding wheel and workpiece and grinding wheel model are calculated. The resulting material removal rate is then passed on to the force model as input. Based on the material removal rate the process forces can be calculated for the empirical forces model. After that the machine model generates machine deformation caused by the machining forces, which is equivalent to the

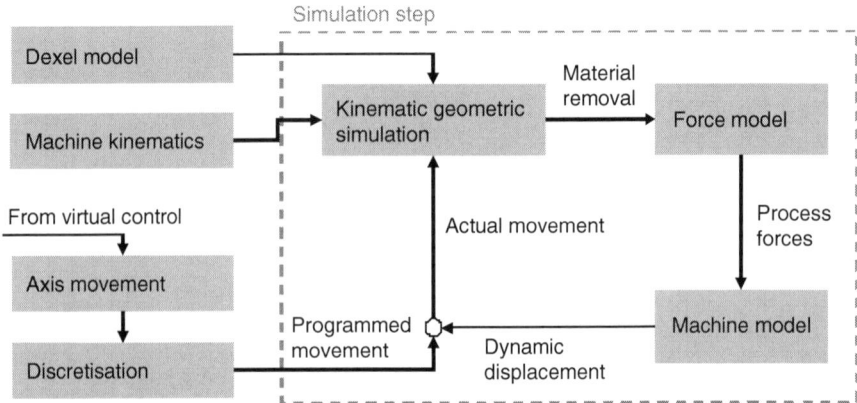

Fig. 4.133 Structure of the coupled grinding simulation

displacement of the workpiece and grinding wheel. Caused by the overlapping of the machine's displacement by the target movement, which is specified by the control simulation, the result is the actual movement for the next simulation step.

There are different approaches for modelling grinding machines. Both pure FEM and multi-body simulation models (MBS models), combined models from FEM and MBS part models, as well as systems of single-mass oscillators are used to map the structural and mechanical behaviour. These differ mainly in the simulation's accuracy, which is associated with an increase of the computational complexity.

Based on previous investigations in SPP 1180 "Prediction and manipulation of the interactions of structures and processes", which is supported by the DFG, it has been established that single-mass models are sufficient for the simulation of grinding with regard to the prediction quality for the time being, and were therefore selected because of the low computation time. The converted machine model is based on the flexibility frequency characteristics measured on the machine that describe the relationship between force and machine deformation or relative displacement between tool and workpiece. In order to utilize the measured flexibility frequency characteristic $G(j\omega)$ during the simulation, a mathematical preparation is required, which can be used to describe the flexibility of the machine's behaviour. As it is known from the modal analysis, the analytical data must first be described and then transferred by means of the curve fitting method within the time frame. Thus, the frequency of single-mass oscillators approximates the flexibility frequency characteristic $G(j\omega)$, which is equivalent to decoupling. These decoupled one-mass oscillators can be transposed into a differential equation of the third degree, resulting in the displacement of the machine that can be calculated due to the force. For each mode of the machine's oscillation, the transfer function can be established in a general format (Alldieck 1994).

$$G(z) = \frac{X(z)}{F(z)} = \frac{b_0 z + b_1 z^{-1} + b_2 z^{-2} + b_3 z^{-3}}{a_0 z + a_1 z^{-1} + a_2 z^{-2} + a_3 z^{-3}} \tag{4.13}$$

The current machine displacement $x(t)$ of each vibration mode results in:

$$x(t) = \frac{1}{a_0} \cdot (b_0 F(t_0) + b_1 F(t-T) + b_2 F(t-2T) + b_3 F(t-3T)$$
$$- a_1 x F(t-T) - a_2 x F(t-2T) - a_3 x F(t-3T) \tag{4.14}$$

The total displacement $x_{ges}(t)$ can then be determined from the individual displacements:

$$X_{ges}(t) = \sum_{i=1}^{m} X_i(t) \tag{4.15}$$

Due to the minimum effort required to implement the simulation, these machine models can be easily implemented for new grinding machines. The design of the overall model to five-axis process does not provide any kinematic constraints for the simulation. Accordingly, all grinding processes can be mapped. By way of example, Fig. 4.136 represents the results of the simulation of a grinding process for machining of an end mill cutter. The material removal rate and the process forces show course that was to be expected.

To realise this coupled simulation, on the one hand it was necessary to integrate knowledge about the grinding process into the process model. On the other hand it was necessary to combine process knowledge in order to couple process and machine model with the knowledge of the dynamic behaviour of machine structures. Furthermore, in order to couple the process and machine simulation with the control simulation, a requirements profile for the data structure had to be defined, as well as the necessary interfaces between sub-models. Accordingly, the integrative research involving the three areas of process, machine dynamics, and control are a key factor in the implementation process of coupled simulations.

4.5.4.5 Cross Process Technology Planning

The virtual manufacturing system relates primarily to the integration of coupled simulation tools within the CAM environment. In addition, and in particular, advanced planning methods and process efficiency can increase productivity in high-wage countries. For this purpose, the software framework "CAx framework" was developed (Minoufekr 2010), which represents a fundamental basis for CAx solutions in the five-axis milling of turbo machinery components and the development of CAM strategies for 5-axis milling in the tool and forming industry (Glasmacher 2011). In addition, experience in cross-technology planning was collected, which are discussed below.

On the one hand, the CAx-framework includes internal framework features that simplify the integration of any CAM modules. On the other hand, the framework also points to interaction and control mechanism and common data models in order to achieve cooperation and coordination between modules.

The framework ensures that the sub-systems can use as many common components as possible. In order to establish a solid basis for the developments, architectural samples from software engineering were analysed and appropriate concepts were transferred to each application. The key objective is to avoid redundant development through the creation of the framework. During the development of extensions, the supplementary effort to plan and implement the framework leads to time and quality benefits already after a few uses. This way, due to the higher quality, development costs especially in the field of software testing, can be saved (Lichter and Ludewig 2007). In order to achieve a high degree of reuse, a new application to be developed does not only take over individual parts of the framework, but also its architecture and control flow. For this purpose connection points (hot spots) are defined, which are connected to new applications. The enhancements are integrated in these areas and are activated by the framework in due time in order to invoke application-specific functions.

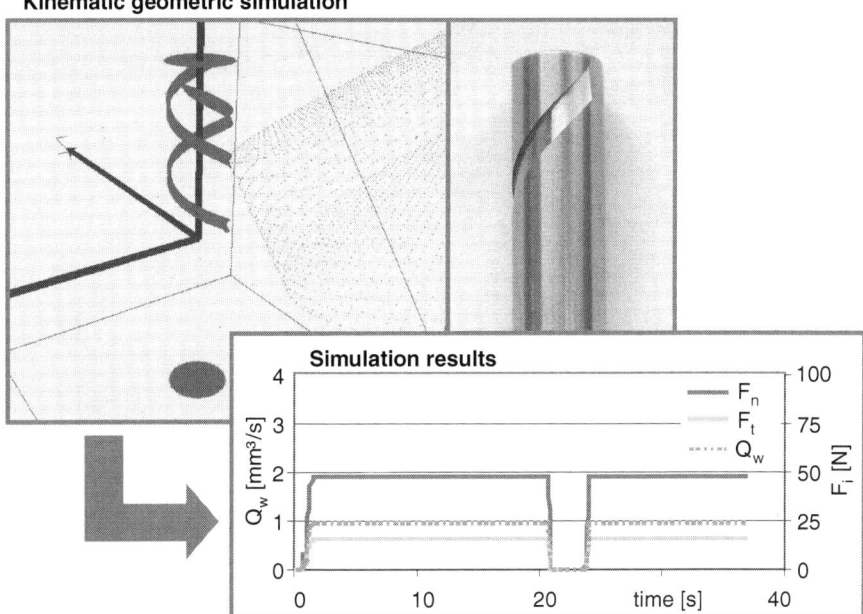

Fig. 4.134 Simulation results of the grinding process of an end mill cutter

In Fig. 4.134 the appropriate use-case diagram is shown, illustrating the crude functionality of the CAx framework. A specialized CAM module, which is based on the CAx framework, has, in the abstract sense, a function for the "input of process parameters". Furthermore, a machining process in several processing steps ("Create operation") is shown, whereby for each processing operation the corresponding tool path is generated ("Calculate tool path"). The tool path generation is characterized by process-specific parameters that are distinctly differently, depending on the application.

Based on the applications and the request to the target system, a draft of the CAx-framework was prepared, in which both the basic architecture and the control flow of the specific instances are specified (Fig. 4.135). The structure is based on the Eclipse platform whose concepts for the requirements of the CAx-framework were adapted. Modifications of Eclipse-based functionality were in particular necessary for integrating new modules. Thus, CAM extensions obtain explicitly the possibility to define even more hot spots by using the "Extension points". Figure 4.136 (below) displays the basic components of the CAx framework. "Extension points" contains everything needed to define extensions. The component "Algorithm" defines mathematical functions—such as the transformation of geometries. "Basis elements" implements only technical features, i.e., it contains elements that may facilitate the development itself (lists, trees, etc.). There is also the subsystem "CAx Components"; it groups the previously mentioned components that are necessary in the

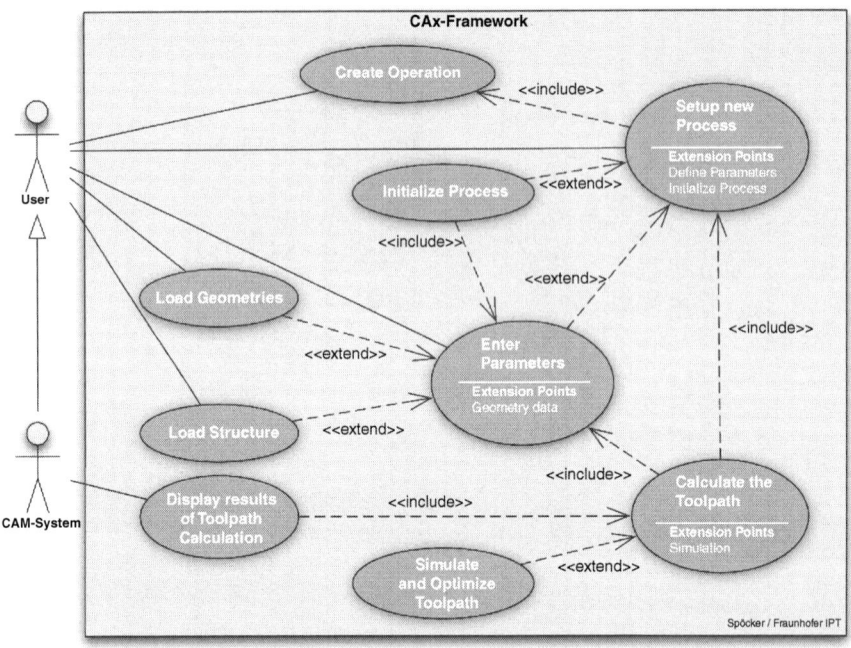

Fig. 4.135 Use-Case diagram of the CAx Framework with the most important Use-Cases

Fig. 4.136 Basic architecture and control flow of the specific instances

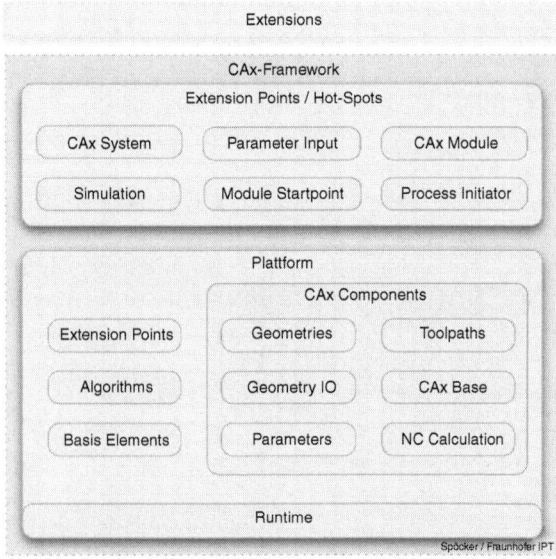

CAx technology. "Geometry IO" allows the serialization of geometries for saving and loading files. Several different file formats are supported. Within "parameters" different models are implemented to retain parameters that are necessary for the processes. These include simple containers for strings, floating point numbers and complex data. "NC Calculation" provides algorithms for computation of the tool path.

In the component "CAx Basics" abstract classes for operations, processes and other models in the context of CAx technology are available. These individual parts form the platform to develop new CAx modules that are based on it. All expansions can be included via the initial extension points "CAx module", "Simulation" "Module start point" and "Process initiator". This way, a distinct flexibility can be achieved, since both the CAx system and the user interface for parameter input can be easily replaced.

Finally, the task of the "runtime" component is explained. It forms the basis for the entire framework by merging with the information about extensions and takes care of initialising the entire system. Thus, in case a test environment is used, this component must initialise the system as a standalone application and respond to any command line. In addition, the "runtime" component occupies a central point of communication between the various subsystems (Fig. 4.137).

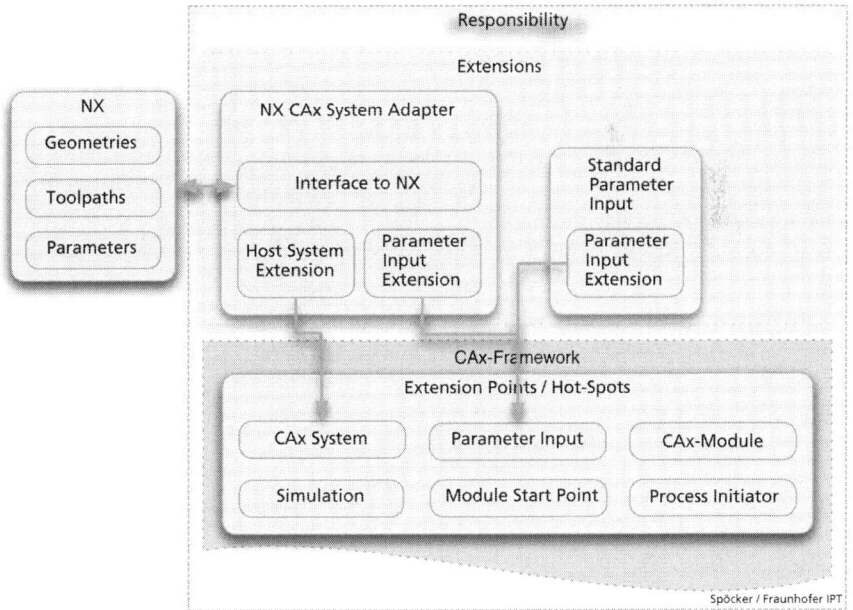

Fig. 4.137 Use of the provided interface (API) of NX

4.5.5 Industrial Application Options

The virtual manufacturing system allows the coupled simulation of manufacturing processes for milling and grinding, during which interactions between control systems, machine structure and process forces are being considered. Several user scenarios are derived for the VMS, in which different industrial companies from high-wage countries may be involved. These include manufacturing companies, for which in particular the production planning through the use of the VMS may be beneficial. In addition, a transfer to the product development of these enterprises is feasible. Machine and control system manufacturers are another group of potential users of the Virtual Manufacturing System, as they can improve their product, i.e., machines or machine components, based on this software tool. Service companies in the NC process optimisation may work with the VMS to improve their efficiency. Applications outlined here briefly are discussed below in detail.

4.5.5.1 Virtual Manufacturing Systems in Production

The first group addressed by the VMS includes users of machine tools. The production of complex parts is usually associated with significant costs that can be attributed in particular to expensive materials and tools, investment-intensive means of production, and lengthy process qualifications. Examples for such parts are flow-related such as bladed discs ("Blisks"), whose production can take several days. Since blisks be used under extreme conditions in turbines, high temperature and heavy-duty materials are required, which generally are associated with poor machinability. This characteristic will manifest itself in increased tool wear and thus a considerable increase in tool cost. Material costs caused by scrap should be excluded at this point, since workpieces outside of the tolerance field must be reworked. The rework, however, entails with additional costs related to means of production, namely measuring machine, milling centre, and tools, and the highly qualified personnel. The use of CAM integrated VMS can help to reduce costs, whereby the manufacturing process is first been run virtually. If iteration loops can be partially avoided by the use of VMS, run-in costs of individual workpieces can be lowered. This applies especially to small series, where an individualised production gains a lot of meaning, and can result in a significant competitive advantage.

The reduced run-in costs face increased planning costs, which are caused by the virtual process. These costs depend on the regularity with which the VMS is used. Based on the software's complexity, an economic application requires the routinely use within the context of available functions and the proper interpretation of the results. Other constraints may arise from the process effects that are neglected by the VMS. In addition, simulations show accuracy limits that need to be assessed properly to ensure the obtained results are transferred effectively from the virtual to real production. Given a further development of the VMS approach towards industrial software applications, an increased contribution to economic efficiency in the production of high-wage countries can be assumed, despite higher CAM planning costs and the simulation-related uncertainties.

The scenario can also be applied to tool and die making, where process boundary conditions are similar. Here, the large number of variants plays an important role, since the companies in this industry are often suppliers to many customers. Since the objective usually consists in a high degree of machine utilisation, set-up and run-in times must be kept low in spite of the frequent process changes. Thus, the demand for an efficient conversion of NC machining processes, whose efficiency can be increased through the use of VMS, raises constantly. The activities shifted from real to virtual machine, that is to say, from the production hall into the CAM process planning, leads to the aspired higher degree of utilisation. Again, however, the expansion of planning capacities has to be considered, so that profitability must be assessed case by case.

In addition to the targeted quality, the inclusion of other variables for the optimisation of the production is feasible. For example, the objective may be reduction of the main process time. Companies that pursue this goal usually cannot be found in the area of single production, since the optimisation effort is not worthwhile due to the low quantity. On the other hand, however, the unit costs for setting up and optimising processes on real production means may be considered as insignificant because they are distributed over a large amount of workpieces. In both cases, a cost-effective application of VMS is only useful in exceptional cases. In contrast, using the VMS for small and medium series, the time savings generated during the planning stages may contribute to increased competitiveness. For example, a virtual testing of elevated cutting values, to some extent also in the vicinity of performance limits are possible, without risk of damaging the target machine. Thus, knowledge gained from the virtual manufacturing processes can be used systematically. Compared to the real process data acquisition, this results in a significantly greater data availability since the need of existing sensors is eliminated. These are countered by model simplifications or incorrect or inaccurate simulation output data that do not allow a purely virtual process design.

Another area of growing significance is the assessment of energy requirements. For instance, when using the VMS, a process with low energy demand can be designed, where quality, dimensional accuracy, and production time represent boundary conditions. In addition, piece-specific costs during the provision of energy can be simulatively estimated, enabling more transparent allocation formula for internal accounting.

4.5.5.2 Virtual Manufacturing Systems During Machine Development

In addition to users, machine manufacturers are also potential users of the CAM integrated virtual manufacturing system.

Companies that invest in the means of production area too often want to link the acceptance of a machine with a process scenario that must be met. Through the use of virtual machine tools in the field of control engineering, the functioning of processes involving virtual programmable logic control systems and mechatronic, models of behaviour can be proven; however, a statement about the process capability is not possible. This indicates a potential application field for the VMS. Due to the realistic simulation of manufacturing processes, machine manufacturers can discuss about the expected machine characteristics with their customers in the development phase and thus accelerate the subsequent acceptance. But, since the precise mapping of the machine structure currently requires measurement calibrations, the possibilities in this area are still limited.

The VMS also provides advantages in the actual machine development. Thus, from the interactions between control, machine, and process requirements can emerge that require action by the mechanics, the fluid or the electrical design and software development. Variations can be virtually tested in a realistic context. During the early stages, a product evaluation is often acceptable even with a simplified representation of the interactions, so the CAM-integrated virtual run-in can be used for the distribution of valuable inputs for the machine development, despite the restrictions mentioned above.

4.5.5.3 Virtual Manufacturing Systems During Product Development

In the long run, the use of VMS for production-oriented design of products is feasible. Based on this, favourable components features can be identified for the processing on a machine and included in the design guidelines. This application option requires a conceptual expansion of the VMS, since it requires a syntactic and semantic link to the design areas of the machine user.

4.5.6 Future Research Topics

The virtual manufacturing system is used for comprehensively modelling NC manufacturing processes. With the approach of integrative simulation, various systems are linked, so that interaction effects are visible. Even if the basic project objectives have been reached, not all parameters involved have been captured. Here, linking points for further research must be considered. Furthermore, from the industrial fields of application, requirements for a further integration could also be derived.

When considering the process, the structure-mechanical behaviour of a machine tool can be captured almost realistically with the flexible, movable multi-body simulation. In the VMS, the forces from the machining process are returned to the machine as well. On examination of the power flow is noticeable, however, that the transmission elements between the machine structure and process in the mechanical model are considered as ideal rigid elements. This simplification must be assessed in accordance with the application. Thus, this modelling format provides for massive tools and forms a sufficiently accurate representation of the workpiece. In other cases, for example in blisk production, workpiece have thin walls at the end of machining and are prone to deformations and vibrations. Modelling this behaviour contributes to a more accurate virtual mapping of the manufacturing process, particularly in terms of stability.

A similar result is provided when observing tools. These too, are ideal rigid in the current VMS. Short mill cutters, supported by holders can be represented this way with a minimum loss of information. However, production of workpieces with features difficult to access often leads to the application of longer tools that exert a negative influence on the stability and bending behaviour. Even for conventional three-axis milling of dies, tools may cantilever strongly because they cannot be tilted for improving accessibility. Neglecting the bending line is only acceptable for small process forces, such as forces during finishing. In examinations regarding process stability, a distortion of the result is possible. The influence determination of flexible tools and parts on the process thus requires further research.

In the area of machine modelling, constraints arise in the use of VMS due to the dependence of resilience frequency responses. This dependence often goes back to the hitherto not reliably predictability of the damping influence. Along with the work on integrative simulation, studies in this research environment on various basic research projects are being conducted (Großmann et al. 2010). Merging these results with those of the multi-body modelling is particularly meaningful for the use of VMS by machinery manufacturers who lack so far an existing MBS model during the development phase. In addition, the measurement adjustment leads to a significant modelling effort, which clearly reduces the efficiency of the simulative approaches followed here. Reliable information based on models about the damping behaviour is therefore necessary for a faster virtual representation of a machine. Although the damping effect (not because of the often unknown boundary conditions in the coupling points) has not been fully captured at present, the correlation of the achieved progress in modelling the VMS present itself as a relevant extension option.

As part of the grinding process simulation, the modelling of abrasive products contains future potential. Because of the many statistically arranged grinding grains, whose shapes are irregular and unknown, the detailed description of the grinding wheel topography is a complex task that still requires a high degree of computational effort. Therefore, in the implemented work the wheel modelling takes place on a macroscopic level by envelops, that neglect individual abrasive grains or neglect even cutting edges. The feasibility of detailed modelling of the grinding wheel's topography will be investigated in the future under a separately DFG-funded research project entitled "Mathematical modelling of the grinding wheel structure".

Further research concerns the process force models. Although, in the investigated cases, the developed models reflect the application process behaviour adequately, however, the application is subject to certain conditions. Thus, under consideration of a thin chip thicknesses, for which the applied empirical correlations loose their validity, a point at which the used models can be expanded, is feasible. In addition, the processing of the results from other simulation systems based on lower scales, represents a useful direction for future research. Here, the platform researched by the Cluster of Excellence for virtual process chain AixViPMaP (see Sect. 4.4) must be mentioned. The latter provide workpieces with spatially resolved mechanical properties, which influence the machinability. The interpretation of these values in a common simulation framework has significant potential for a realistic description of the process. The feedback of information from the VMS to the AixViPMaP can also extend the analysis capabilities. For example, the return of temperature that occurs during the milling or grinding process allows a virtual reproduction of induced structural changes and the resulting internal stresses in the virtual workpiece. The coupling of the two platforms, thus leads to a continuous mapping of process chains with preliminary shaping, forming, and machining steps that may emerge from a complete production history of the virtual workpiece.

The use of virtual manufacturing system is affected also by the low predictability of the resulting output quality confidence intervals are still missing, which provide information on the reliability of the simulation. This approach requires an integrative analysis of the coupled systems, since an isolated analysis is not significant due to the neglect of interactions. In upcoming research, errors of individual models and their impact on the overall composite obtained simulation results must be evaluated. In addition, determining the required level of detail with respect to different applications is desirable in order to keep the modelling effort to a minimum and thus increase the efficiency of the VMS approach.

References

Abaqus (2009) Dassault Systèmes Simulia Corp., Providence, RI, USA; Standard user manual, version 6.9, Hibbit, Karlsson and Sorensen

Abdul-Ghafour S, Ghodous P, Shariat B, Perna E (2007) A common design-features ontology for product data se-mantics interoperability. In: Proceedings of the IEEE/WIC/ACM international conference on web intelligence, 2–5 Nov 2007, Silican Vally, USA, pp 443–446, 2–5 Nov 2007

Aggteleky B (1987) Grundlagen – Zielplanung – Vorarbeiten, unternehmerische und systemtechnische Aspekte – Marketing und Fabrikplanung. München

Albersmann F (1999) Simulationsgestützte Prozessoptimierung für die HSC-Fräsbearbeitung. Dissertation Universität Dortmund, Vulkan, Essen

AixViPMaP (2010) The Aachen Virtual Platform for Materials Processing: www.aixvipmap.de. Accessed 20 July 2011

Aleksovski Z, Kate WT, Harmelen FV (2006) Exploiting the structure of background knowledge used in ontology matching. In: Proceedings of the international Semantic web conference, Athen, GA, USA, 5–9 Nov 2006

Allard J, Raffin B (2006) Distributed physical based simulations for large VR applications. In: Proceedings of the IEEE Virtual Reality Conference, Alexandria, USA 25–29 March 2006

Alldieck J (1994) Simulation des dynamischen Schleifprozeßverhaltens, Dissertation RWTH Aachen, Shaker, Aachen

Altan T, Yen Y C, Rech J, Hamdi H (2004) Influence of cutting edge radius of coated tool on chip formation in orthogonal cutting of alloy steel. In: Proceedings of the 7th CIRP Conference on Modeling of Machining, Frankreich May 2004

Altenbach J, Altenbach H (1994) Einführung in die Kontinuumsmechanik. Teubner Studienbücher Mechanik, Stuttgart

Altintas Y (2000) Manufacturing automation: metal cutting mechanics, machine tool vibrations, and CNC Design, 1st edn. Cambridge University Press, Cambridge

Altintas Y, Lee P (1996) A general mechanics and dynamics model for helical end mills. Annals CIRP 45(1):59–64

Altintas Y, Lee P (1998) Mechanics and dynamics of ball end milling. Trans ASME 120:684–692

Altintas Y, Brecher C et al (2005) Virtual machine tool. Ann CIRP 54(2):657–674

Altintas Y, Spence A, Tlusty J (1991) End milling force algorithms for CAD systems. CIRP Ann Manuf Technol 40(1):31–34

Anderl R, Trippener D (2000) STEP: standard for the exchange of product model data eine Einfuı̂hrung in die Ent-wicklung, Implementierung und industrielle Nutzung der Normenrei-he ISO 10303 (STEP). Teubner, Stuttgart

Apel M, Böttger B (2009) Phase-field simulation of the microstructure formation in the weld pool and in the heat affected zone during welding of steel. Presented at the 2nd symposium on phase-field modelling in materials science, Aachen, Sept 2009

Apel M, Benke S, Steinbach I (2009a) Virtual dilatometer curves and effective young's modulus of a 3D multiphase structure calculated by the Phase-Field Method. Comp Mater Sci 45(3):589

Apel M, Böttger B, Rudnizki P, Schaffnit F (2009) Steinbach: grain growth simulations including particle pinning using the multiphase-field concept. ISIJ 49(7):1024–1029

Armarego EJA, Deshpande NP (1989) Computerized predictive cutting models for forces in end-milling including eccentricity effects. CIRP Ann Manuf Technol 38(1):45–49

Assenmacher I, Kuhlen T (2008) The ViSTA virtual reality toolkit. Software engineering and architectures for real-time interactive systems (SEARIS). In: Proceedings of the IEEE VR conference, Reno, Nevada, USA, pp 23–26, 8–12 March 2008

Aumueller D, Do H, Massmann S, Rahm E (2005) Schema and ontology matching with COMA++. In: Proceedings of the 2005 ACM SIGMOD international conference on Man-agement of data, Baltimore, USA, pp 906–908, 13–17 June 2005

Aurich J C, Biermann D, Blum H, Brecher C et al (2008) Modelling and simulation of process – machine interaction in grinding. Prod Eng 3(1):111–120

AVS Advanced Visual Systems. http://www.avs.com. Accessed 22 Feb 2011

Bakhvalov N, Panasenko G (1989) Homogenisation: averaging processes in periodic media. Kluwer, Dordrecht

Baranowski T, Brinkmann M (2007) Durchgehende Simulationskette für den Spritzgießprozess. Sonderheft Integrative Produktion 1:28

Baudisch T (2003) Simulationsumgebung zur Auslegung der Bewegungsdynamik des mechatronischen Systems Werkzeugmaschine. Dissertation TU München, Utz, München

Bea FX, Scheurer S, Hesselmann S (2008) Projektmanagement. Lucius and Lucius, Stuttgart

Bédoui F, Diani J, Régnier G, Seiler W (2006) Micromechanical modelling of isotropic elastic behaviour of semicrystalline polymers. Acta Mater 54:1513–1523

Beek MHE, Peters GWM, Meijer HEH (2006) Influence of shear flow on the specific volume and the crystalline morphology of isotactic polypropylene. Macromolecules 39:1805–1814

Beer S (1980) Brain of the firm: the managerial cybernetics of organization. Wiley, New York

Benke S (2008) A multi-phase-field model including inelastic deformation for solid state transformations. Proceedings in Applied Mathematics and Mechanics. PAMM 8(1):10407

Benke S, Laschet G (2008) On the interplay between the solid deformation and fluid flow during solidification of a metallic alloy. Comput Mater Sci 43(1):92

Benke S, Rudnizki J, Suwanpinij P, Prahl U (2008) Modeling hot rolling: a study on the microstructural changes during the austenite to ferrite phase transformation in dual phase steels. Presented at 8th World Congress on Computational Mechanics (WCCM8), Venice, Italy, 30 June–5 July, 2008

Benke S et al (2010) Definition of a standardized data format for the exchange of simulation data. www.aixvipmap.de. Accessed 2010

Bensoussan A, Lions JL, Pappanicolaou G (1978) Asymptotic analysis for periodic structures. North-Holland, New York

Berlich R, Kunze M, Schwarz K (2005) Grid computing in Europe: from research to deployment. In: Buyya R, Coddington P, Wendelborn A (eds.) Australasian workshop on grid computing and e-research, vol 44 of CRPIT. Newcastle Australia ACS, Australia, pp 21–27

Berlin J, Motro A (2006) Database schema matching using machine learning with feature selection. In: Pidduck A, Ozsu M, Mylopoulos J, Woo C (eds) Advanced information systems engineering, vol 2348. Springer, Berlin, pp 452–466

Bleck W, Reisgen U, Mokrov O, Rossiter E, Rieger T (2010) Methodology for thermomechanical simulation and validation of mechanical weld-seam properties. Adv Eng Mater 12(3):147–152

Bleicher K (1999) Unternehmenskultur und strategische Unternehmnsführung. In: Hahn D, Taylor B (eds) Strategische Unternehmungsplanung – strategische Unternehmungsführung. Physica, Heidelberg, pp 223–265

Blumauer A (2006) Semantic Web. Springer, New York

Bobzin K, Bagcivan N, Parkot D, Kashko T (2007) Microstructure dependency of the material properties: simulation approaches and calculation methods. Steel Res 78:804–811

Bobzin K, Bagcivan N, Parkot D, Kashko T, Laschet G, Scheele J (2009) Influence of the definition of the representative volume element on effective thermoelastic properties of thermal barrier coatings with random microstructure. Special Issue on recent advantages in modelling and numerical simulations. J Thermal Spray Technol 18(5–6):988–995

Bohl W, Elmendorf W (2005) Technische Strömungslehre. 13th edn. Vogel, Würzburg

Böttger B, Apel M, Eiken J, Schaffnit P, Schmitz GJ, Steinbach I (2009) Phase-field simulation of equiaxed solidification: a homoenthalpic approach to the micro-macro problem. In: Cockcroft SL (eds) Proceedings from the XII international conference on modelling of casting, welding and advanced solidification processes MCWASP, Vancouver, pp 119–127, 7–14 June 2009

Brandes A (2008) Positionierung technologischer Schnittstellen. Berichte aus dem IFW, vol 6

Brecher C, Witt S (2006) Gekoppelte Simulation von Antriebsregelung und Strukturdynamik von Werkzeugmaschinen. In: Brecher C (eds) Simulationstechnik in der Produktion. Fortschrittsberichte VDI Nr. 658. VDI, Düsseldorf

Brecher C, Herfs W (2007) Potenziale der Software-in-the-Loop-Simulation: Erweiterte Diagnose- und Optimierungsmöglichkeiten durch eine vollständige Simulation von Fertigungssystemen. In: Adam W, Pritschow G, Uhlmann E und Weck M (eds) Zuverlässigkeit und Diagnose in der Produktion. Fortschrittsberichte VDI, Reihe 2. VDI, Düsseldorf

Brecher C, Esser M (2008) The consideration of dynamic cutting forces in the stability simulation of HPC-milling processes. In: Proceedings of the 1st international conference on process machine interactions, Hannover, 3–4 Sept 2008

Brecher C, Lohse W (2011) A CAM-integrated virtual manufacturing system for complex milling processes. In: Proceedings of the 44th CIRP conference on manufacturing systems, Madison, 1–3 June 2011

Brecher C et al (2005) Integrierte Simulation von Maschine, Werkstück und Prozess. In: Brecher C, Klocke F, Schmitt R, Schuh G (eds) Wettbewerbsfaktor Produktionstechnik: Aachener Perspektiven – Aachener Werkzeugmaschinenkolloquium 2005. Shaker, Aachen

Brecher C, Lohse W, Herfs W, Eppler C (2007a) Anwendungsorientierte Mechatroniksimulation mit realen und virtuellen Steuerungen. In: Brecher C, Denkena B (eds) Ramp-Up/2 – Anlaufoptimierung durch Einsatz virtueller Fertigungssysteme. VDI, Frankfurt a. M

Brecher C, Vitr M, Voss M (2007b) Vision of a smart machining system – integrating CAM, CNC and machining simulation. In: Proceedings of the 4th international conference on digital enterprise technology, Bath, UK, 19–21 Sept 2007

Brecher C, Esser M, Witt S (2009a) Interaction of manufacturing process and machine tool. CIRP Ann Manuf Technol 58(2):588–607

Brecher C, Lohse W, Vitr M (2009b) CAx Framework for planning five-axis milling processes. In: Proceedings of the 6th CIRP-sponsored international conference on digital enterprise technology, Hong Kong, 14–16 Dec 2009

Brecher C, Herfs W, Bauer S et al (2010) Ressourceneffiziente Maschinensysteme. In: Eversheim W et al (eds) Ressourceneffiziente Produktionstechnik – Ein Aachener Modell: Festschrift für Univ.-Prof. Dr.-Ing. Dr.-Ing. E.h. Dr. h.c. Fritz Klocke. Apprimus, Aachen

Brinksmeier E, Aurich JC et al (2006) Advances in modeling and simulation of grinding processes. Ann CIRP 55(2):667–696

Budak E, Altintas Y, Armarego EJA (1996) Prediction of milling force coefficients from orthogonal cutting data. Trans ASME 118:216–22

Bullinger H-J, Warnecke H-J, Westkämper E (2003) Neue Organisationsformen im Unternehmen: ein Handbuch für das moderne Management. Springer, Berlin

Bundesbank (2010) http://www.bundesbank.de/statistik/. Accessed Sept 2010

Burckhardt W (2001) Das grosse Handbuch Produktion, moderne industrie. Landsberg, Lech

Calphad (2010) http://www.calphad.org/. Accessed Aug 2010

Cerfontaine P, Beer T, Kuhlen T, Bischof C (2008) Towards a flexible and distributed simulation platform computational science and its applications – ICCSA. Springer lecture notes in computer science LNCS 5072 Part1

Chappell D (2004) Enterprise service bus: theory in practice, 1st edn. O'Reilly Media

Chen X, Rowe WB (1996) Analysis and simulation of the grinding process. Part I: generation of the grinding wheel surface. Int J Mach Tools Man 36:8871–882

Clausen M (2005) Zerspankraftprognose und -simulation für Dreh- und Fräsprozesse. Dissertation Universität Hannover, PZH, Garbsen

Clifton C, Housman E, Rosenthal A (1997) Experience with a Combined approach to attribute-matching across heteroge-neous databases. In: Proceedings of the IFIP working conference on data semantics, Leysin, Schweiz, pp 428–453, 7–10 Oct 1997

Collins DJ, Montgommery CA (1995) Competing on ressources: strategy for the 1990s. Har Bus Rev (July/Aug)

Condor – High Throughput Computing (HTC) (2011). http://www.cs.wisc.edu/condor. Accessed 22 Feb 2011

Conrad S (1997), Föderierte Datenbanksysteme: Konzepte der Datenintegration, 1st edn. Springer, Berlin

Conrad S (2006) Enterprise application integration: Grundlagen, Konzepte, Entwurfsmuster, Praxisbeispiele, 1st edn. Elsevier, München

Corsten H, Corsten H, Gössinger R (2008) Projektmanagement: Einführung; [mit Aufgaben und Lösungen], 2., vollst. überarb. und wesentlich erw. Aufl. Oldenbourg, München

COVISE (2011). http://www.hlrs.de/organization/av/vis/covise. Accessed 22 Feb 2011

Dahmus J, Gutowski T(2004) An environmental analysis of machining. Proceedings of IMECE 2004, ASME International Mechanical Engineering Congress and RD&D Expo, Anaheim, California, USA, 14–19 Nov 2004

DeFanti TA, Foster I, Papka ME, Stevens R, Kuhfuss T (1996) Overview of the I-WAY: Wide-area visual supercomputing. Int J Supercomputer Appl High Perf Comput 10(2/3):123–131

Denkena B, Jivishov V (2005) Größeneinflüsse auf die Spanbildung, Zerspankräfte und Eigenspannungen beim Drehen – Experiment und Simulation. In: Vollertsen F (ed) Strahltechnik, vol 27, Prozessskalierung. BIAS, Bremen

Denkena B et al (2003) Deviation analysis of FEM ased cutting simulation. In: The 6th international ESAFORM conference on metal forming, Salerno, Italy, pp 587–590, Apr 2003

Denkena B, Tracht K, Deichmüller M (2006) Wechselwirkungen zwischen Struktur und Prozess beim Werkzeugschleifen. wt Werkstattstech 96(11–12):814–819

Dhamankar R, Lee Y, Doan A, Halevy A, Domingos P (2004) iMAP: discovering complex semantic matches between data-base schemas. In: Proceedings of the 2004 ACM SIGMOD international conference on management of Data, Paris, Frankreich, 13–18 June 2004

Dietmair A, Verl A (2010a) Energieeffizienter betrieb von Produktionsanlagen. In: Tagungsband Energieeffiziente Produkt- und Prozessinnovationen in der Produktionstechnik, 1. Internationales Kolloquium des Spitzentechnologieclusters eniPROD, Chemnitz, 24–25 June 2010

Dietmair A, Verl A (2010b) Energy consumption assessment and optimisation in the design and use phase of machine tools. In: Tagungsband der 17. CIRP Konferenz für Life Cycle Engineering, Hefei, China, 19–21 May 2010

Dilthey U, Goumeniouk A, Pavlyk V (2003) Beamsim – simulation software for laser and electron beam welding. In: Paton BE, Kovalenko VS, Paton EO (eds) Proceeding of the international confererence on laser technologies in welding and materials processing, Katsiveli, Crimea, Ukraine. E.O. Paton Electric Welding Institute, NASU, Kiev, pp 108–113, 19–23 May 2003

Dilthey U, Pavlyk V, Mokrov O, Dikshev I (2005a) Software package simweld for simulation of gas-metal-arc-welding processes of steels and aluminium alloys. In: Cerjak H., Bhadeshia HKDH, Kozeschnik E (eds) Mathematical modelling of weld phenomena 7. Graz Technical University, Austria, pp 1057–1079

Dilthey U, Mokrov O, Pavlyk V (2005b) Modelling of consumable electrode gas-shielded multi-pass welding of carbon steel with preheating. Paton Weld J 4:2–6

DIN 6580 (1985) Bewegung und Geometrie des Zerspanvorgangs

Düssel T, Zilken H, Frings W, Eickermann T, Gerndt A, Wolter M, Kuhlen T (2007) Distributed collaborative data analysis with heterogeneous visualization systems. In: Favre JM, Santos LP, Reiners D (eds) Eurographics symposium on parallel graphics and visualization. Eurographics Association, Lugano, pp 21–28

Dyckhoff H (2006) Produktionstheorie: Grundzüge industrieller Produktionswirtschaft; mit 20 Tabellen. Springer, Berlin

Ehlen G, Ludwig A, Sahm PR, Bührig-Polaczek A (2003) Split-solid-model to simulate the formation of shrinkage cavities and macrosegregations in steel casting. In: Stefanescu DM (ed) Modeling of casting, welding and advanced solidification processes. TMS, Warrendale, pp 285–294

Ehrig M (2007) Ontology alignment, 1st edn. Springer, New York

Eiken J, Böttger B, Steinbach I (2006) Multiphase-field approach for multicomponent alloys with extrapolation scheme for numerical application. Phys Rev E 73:066122

Engmann D, Maßmann S (2007) Instance matching with COMA++. In: BTW Workshops-Model Management und Metadaten-Verwaltung, Düsseldorf, pp 28–37, 6th March 2007

Eppinger SD, Whitney DE, Smith RP, Gebala DA (1994) A model-based method for organizing tasks in product development. Res Eng Design Theory Appl Concur Eng, 6(1):1–13

ETH-Zürich (2008) Three years activity on "Through-Process-Modelling of Al-alloys" started in spring 2008 at the institute for virtual production at ETH-Zurich

European Commission (2007) Ausschreibungen der Europäischen Kommission im 7. Rahmenprogramm. www.cordis.lu

Euzenat J, Shvaiko P (2007) Ontology matching, 1st edn. Springer, Berlin

Eversheim W, Schuh G (2005) Integrierte Produkt- und Prozessgestaltung. Springer, Berlin

Factsage (2010) www.factsage.com. Accessed 2010

Fleischer J et al (2004) 2D Tool wear estimation using FEM. In: Proceedings of the 7th CIRP Workshop on Modeling of Machining, Frankreich, Mai 2004

Foster I, Kishimoto H, Savva A et al (2005) The open grid services architecture, Version 1.0. Technical report, Global Grid Forum (GGF)

Fowler M (1998) Analysis patterns: reusable object models. Addison-Wesley, Menlo park (ISBN 0201895420)

Fox M, Long D (2003) PDDL2.1: an extension to PDDL for expressing temporal planning domains. J Artif Intell Res, 20

Franzke M (1998) Zielgrößenadaptierte Netzdiagnose und -generierung zur Anwendung der Finite Element Methode in der Umformtechnik. Umformtechnische Schriften Band 86

Freeman RE (1984) Strategic management: a stakeholder approach, [Nachdr.] Aufl. Pitman, Boston

Fresner J, Bürki T, Sittel H (2009) Ressourceneffizienz in der Produktion, Symposium Publishing

Fulchiron R, Cascone A (2008) Squeeze flow induced crystallization of isotactic polypropylene. In: Proceedings of the 26th annual meeting of the polymer processing society (PPS), Salerno, Italien, 15–19 June 2008

Gagnon M, Michael (2007) Ontology-based integration of data sources. In: Proceedings of the 10th international conference on information fusion, Quebec, Kanada, pp 1–8

García-Solaco M, Saltor F, Castellanos M (1995) A structure based schema integration methodology. In: Proceedings of the 11th international conference on data engineering, Taipei, Taiwan, pp 505–512, 6–10 March 1995

Gennari JH et al (2002) The evolution of Protégé: an environment for knowledge-based systems development. Int J Human Comput Stud 58:89–123

Giunchiglia F, Shvaiko P, Yatskevich M (2006) Discovering missing background knowledge in ontology matching. In: Proceedings of the 17th European conference on artificial intelligence August ECAI, Riva del Garda, Italien, pp 382–386, 29 Sept 2006

Glasmacher L (2011) Integrative CAx process chains for manufacturing and repair of turbomachinery parts. In: ICTM – international conference on turbomachinery manufacturing, Aachen, Germany, 23–24 Feb 2011

Glißmann M (2000) Beanspruchungsgerechte Charakterisierung des mechanischen Werkstoffverhaltens thermoplastischer Kunststoffe in der Finite-Elemente-Methode (FEM). RWTH Aachen University, Dissertation. ISBN: 3-86130-481-3

g-lite Lightweight Middleware for Grid Computing. http://glite.cern.ch. Accessed 22 Feb 2011

Globus Toolkit. http://www.globus.org. Accessed 22 Feb 2011

Goldak J, Akhlaghi M (2005) Computational welding mechanics. Springer, New York, p 321

Goldhar JD, Jelinek M (1983) Plan for economies of scope. Harv Bus Rev 61(6):141–148

Gorman IE (1985) Conditions for economies of scope in the presence of fixed costs. Rand J Econ 16(3):431–436

Gorton I, Thurman D, Thomson J (2003) Next generation application integration: challenges and new approaches. In: COMPSAC '03: proceedings of the 27th annual international conference on computer software and applications, Dallas, USA, pp 576, 3–6 Nov 2003

Gottschalk S, Hoeschen A (2008) Produktionsmanagement im Anlauf. In: Schuh, G, Stölzle, W und Straube, F (eds) Anlaufmanagement in der Automobilindustrie erfolgreich umsetzen. Springer, Berlin, pp 177–185

Gottschalk S, Lin MC, Manocha D (1996) OBBTree: a hierarchical structure for rapid interference detection. In: Proceedings of the 23rd annual conference on Computer graphics and interactive techniques, New Orleans, USA, 4–9 Aug 1996

Gottstein G (ed) (2007) Intergral materials modelling: towards physics based through-process models. Wiley, Weinheim

Götze U, Helmberg C, Rünger G et al (2010a) Integrating energy flows in modelling manufacturing processes and process chains of powertrain components. In: Tagungsband Energieeffiziente Produkt- und Prozessinnovationen in der Produktionstechnik, 1. Internationales Kolloquium des Spitzentechnologieclusters eniPROD, Chemnitz, 24.–25.Juni 2010

Götze U, Koriath HJ, Kolesnikov A et al (2010b) Energetische Bilanzierung und Bewertung von Werkzeugmaschinen. In: Tagungsband Energieeffiziente Produkt- und Prozessinnovationen in der Produktionstechnik, 1. Internationales Kolloquium des Spitzentechnologieclusters eniPROD

Grosch J, Liedtke D, Kallhardt K, Tacke D, Hoffmann R, Luiten CH, Eysell FW (1981) Gasaufkohlen bei Temperaturen oberhalb 950 °C in konventionellen Öfen und in Vakuumöfen, HTM Härterei-Techn. Mitt 36(5):262

Großmann K, Jungnickel G (2008) Thermische Modellierung von Prozesseinflüssen an spanenden Werkzeugmaschinen. Buchreihe Lehre Forschung Praxis, TU Dresden

Großmann K, Rudolph H, Brecher C et al (2010) Dämpfungseffekte in Werkzeugmaschinen. ZWF 2010(7–8):676–680

Gruber TR (1993) A translation approach to portable ontology specifications. Knowl Acquis 5(2):199–220

Grundig C-G (2006) Fabrikplanung. Carl Hanser, München

Grundig C-G (2008) Fabrikplanung. Carl Hanser, München

Grunwald A, Kopfmüller J (2006) Nachhaltigkeit. Campus, Frankfurt a. M.

Grunwald A, Coenen R, Nitsch J, Sydow A, Wiedemann P (2001) Forschungswerkstatt Nachhaltigkeit. Edition Sigma, Berlin

Guitierrez I, Altuna MA (2008) Work-hardening of ferrite and microstructure-based modeling of its mechanical behaviour under tension. Acta Mater 56:468

Gumenyuk A (2004) Modellbildung und Prozesssimulation des Laserstrahlschweißens von Leichtbauwerkstoffen. In: Dilthey U (ed) Aachener Berichte Fügetechnik Bd 4/2004, Shaker, D 82 (Diss. RWTH Aachen), Aachen

Gutowski T, Dahmus E, Thiriez A (2006) Electrical energy requirements for manufacturing processes. In: Proceedings of the 13th CIRP international conference on life cycle engineering. Leuven, Belgien, May/June 2006

Haas LM, Hernàndez MA, Popa L, Roth M, Ho H (2005) Clio grows up: from research prototype to industrial tool. In: Proceedings of the 2005 ACM SIGMOD international conference on management of data, Baltimore, USA, pp 805–810, 13–17 June 2005

Haas R, Löckmann D, Pfau D (2009) Energieeffizienz auch bei Stückzahl 1. Vergleich der Energieumsätze anhand der Bewegungskollektive von Nebenantrieben in Werkzeugmaschinen. wt Werkstattstech 99(3):186–191

Hahn D (2009) Entwicklungstendenzen bei der Ausgestaltung von Planungssystemen. In: Hahn D, Taylor B (eds) Strategische Unternehmungsplanung – Strategische Unternehmungsführung. Springer, Berlin

Halevy A, Rajaraman A, Ordille J (2006) Data integration: the teenage years. In: VLDB '06: proceedings of the 32nd international conference on very large data bases, Seoul, Korea, pp 9–16, 12–15 Sept 2006

Hamel G (1994) The concept of core competence. In: Hamel G, Heene A (eds) Competence-based competition. Wiley, West Sussex, pp 11–33

Hara T et al (2009) Service CAD system to integrate product and human activity for total value. CIRP J Man Sci Tech

Hauff V (1987) Unsere gemeinsame Zukunft. Eggenkamp, Greven

Hegeman JB (2000) Fundamentals of grinding: surface conditions of ground materials. Dissertation Universität von Groningen

Hemmelmann J (2007) Simulation des lastfreien und belasteten Zahneingriffs zur Analyse der Drehübertragung von Zahnradgetrieben. Dissertation RWTH Aachen

Henning M (2008) The rise and fall of CORBA. Commun ACM 51(8):52–57. doi:10.1145/1378704.1378718

Heok TK, Daman D (2004) A review on level of detail. IEEE Computer Graphics, Imaging and Visualization '04, 70–75. doi:10.1109/CGIV.2004.1323963

Herden S, Gomez JM, Rautenstrauch C, Zwanziger A (2006) Software-Architekturen für das E-Business: Enterprise-Application-Integration mit verteilten Systemen. Springer, Berlin

Hero M (2010) Hierarchical engineering of industrial materials (hero-m) activity started in 2007 at KTH Stockholm. http://www.hero-m.mse.kth.se. Accessed Nov 2010

Hippenstiel F, Bleck W, Clausen B, Hoffmann F, Kohlmann R (2002) Innovative Einsatzstähle als maßgeschneiderte Werkstofflösung zur Hochtemperaturaufkohlung von Getriebekomponenten HTM Härterei-Techn. Mitt 27(4):290–298

Hirsch J (2006) Through process modelling. Mater Sci Forum 519–521: 15–24

Hoffman JD, Davis GT, Lauritzen JI Jr (1976) The rate of crystallization of linear polymers with chain folding. In: Hannay NB (ed) Treatise on solid state chemistry, vol 3. Plenum, New York, pp 497–614

Hoffmann F (2008) Optimierung der dynamischen Bahngenauigkeit von Werkzeugmaschinen mit der Mehrkörpersimulation. Dissertation RWTH Aachen, Apprimus, Aachen

Hoffmann S (2003) Berechnung von Kristallisationsvorgängen in Kunststoffformteilen. RWTH Aachen, Dissertation, 2003. ISBN 3-89653-999-X

Hoppe H (1999) New quadric metric for simplifying meshes with appearance attributes. IEEE Visualization '99, 59–66

Hohpe G (2004) Enterprise integration patterns: designing, building, and deploying messaging solutions. Addison-Wesley, Boston
Horrocks I, Patel-Schneider P, Boley H, Tabet, Grosof B, Dean M (2004) SWRL: a semantic web rule language com-bining OWL and RuleML. http://www.w3.org/Submission/SWRL/
Hougardy HP (1999) Zukünftige Stahlentwicklung. Stahl Eisen 119(3):85–90
Housmans JW (2008) Flow induced crystallization of isotactic polypropylenes. Technische Universität Eindhoven, Dissertation, 2008. ISBN 978-90-386-1466-3
Howe AA (1998) Alloy design: from composition to a through process model. Steel World 4(1):46–51
Hurevich V, Gusarov A, Smurov I (2002) Simulation of coating profile under plasma spraying conditions. In: Proceedings of the international thermal spray conference, Essen, pp 318–323, 3–6 March 2002
Immpetus Group Sheffield (2010) A new framework for hybrid through-process modelling, process simulation and optimisation in the metals industry. http://www.immpetus.group.shef.ac.uk/. Accessed Nov 2010
Interdisciplinary Centre for Advanced Materials Simulation (ICAMS) Ruhr University Bochum. http://www.icams.de. Accessed Nov 2010
Jakumeit J, Laqua R, Ivas T, Scheele J, Braun M, Mukhopadhyay A, Pelzer M (2003) Multidisciplinary coupled simulations of investment casting processes using CASTS-FLUENT. Materials Science and Technology, Chicago, Illinois, Modeling, control and optimization in nonferrous and ferrous industry, pp 333–342
Janeschitz-Kriegl H (2010) Crystallization modalities in polymer melt processing. Springer, Wien
Jansen U (2009) Simulation des Schweißens kleiner Bauteile. Diplomarbeit, Lehrstuhl: Nichtlineare Dynamik der Laserfertigungsverfahren, RWTH Aachen
JMatPro (2010) http://www.thermotech.co.uk/jmatpro.html
JasperSoft (2007) An Introduction to the Jaspersoft Business Intelligence Suite, whitepaper
Kalden O, Fritzen P, Kranz S (2007) SimVis A concurrent engineering tool for rapid simulation development. In: Proceedings of 3rd international conference on recent advances in space technologies, Istanbul, Turkey, pp 417–422, 9–11 June 2011
Kampker A (2008) Standortplanung I – Planung des Wertschöpfungsumfangs, Aachen. http://www.wzl.rwth-aachen.de/de/7bfd32120f8ba69bc1256f330029938b/fp_ss09_v2.pdf
Kampker A, Schuh G, Franzkoch B, Burggräf P (2009) Factory planning based on a resource strategy. Asian Int J Sci Technol 2(2):91–98
Kanit T, Forest S, Galliet I, Mounoury V, Jeulin D (2003) Determination of the size of the representative volume element for random composites: statistical and numerical approach. Int J Solid Struct 40:3647–3679
Karastoyanova D, Wetzstein B, van Lessen T, Wutke D, Nitzsche J, Leymann F (2007) Semantic service bus: architecture and implementation of a next generation middle-ware. In: Proceedings of the IEEE 23rd international conference on data engineering workshop, Istanbul, Turkey, pp 347–354, 15–20 Apr 2007
Karhausen K (1992) Integrierte Prozess- und Gefügesimulation bei der Warmumformung. Dissertation, Institut für Bildsame Formgebung der RWTH Aachen
Karhausen K, Kopp R (1992) Model for integrated process and microstructure simulation in hot forming. Steel Res 63:6
Karonis N, Papka M, Binns J et al (2003) High-resolution remote rendering of large datasets in a collaborative environment. Future Generation of Computer Systems (FGCS)
Kern W (1962) Die Messung industrieller Fertigungskapazitäten und ihre Ausnutzung: Grundlagen und Verfahren, Westdt., Köln
Kettner H, Schmidt J, Greim H-R (1984) Leitfaden der systematischen Fabrikplanung. Carl Hanser, München
Klages H (1990) Wertorientierungen im Wandel. Campus, Frankfurt am Main

Klenke K, Kohlman R (2005) Einsatzstähle in ihrer Feinkornbeständigkeit, heute und morgen. HTM Z. Werkst. Wärmebehand. Fertigung 60(5):260

Klocke F, König W (2008) Fertigungsverfahren: Drehen, Fräsen, Bohren, vol 1. Springer, Berlin

Klocke F, Schlosser R, Tönissen S (2010a) Prozesseffizienz durch Parameterwahl – Evaluierung des Fräsprozesses. wt Werkstattstech 10(05):346–349

Klocke F, Lung D, Schlosser R (2010b) Energy and resource consumption of cutting processes – how process parameter variations can optimise the total process efficiency. In: Tagungsband zur 17. CIRP Konferenz für Life Cycle Engineering, 19–21 May 2010, Hefei, China

Knapp M (2002) Mesh Decimation using VTK, Technical Report, Vienna University of Technology

Knight JA (1997) Value-based management: developing a systematic approach to creating shareholder value. McGraw-Hill, New York

Kobayashi R (1993) Modeling and numerical simulations of dendritic crystal growth. Physica D 63(3–4):410–423

Komoto H et al (2007) Integration of a service CAD and a life cycle simulator. CIRP Ann Manuf Technol 57(1):9–12

König W (1982) Spezifische Schnittkraftwerte für die Zerspanung metallischer Werkstoffe. Verlag Stahleisen MBH, Düsseldorf

Koruna S (1999) Kernkompetenzen-Dynamik. Orell Füssli, Zürich

Koshy P, Jain VK, Lal GK (1997) Stochastic simulation approach to modelling diamond wheel topography. Int J Mach Tools and Manufact 37(6):751–761

Kozeschnik E, Svoboda J, Fischer FD, Fratzl P (2004) Modelling of kinetics in multi-component multi-phase systems with spherical precipitates: II: numerical solution and application. Mater Sci Eng A 385(1–2):157

Kratzsch S (2000) Prozess- und Arbeitsorganisation in Fließmontagesystemen. Vulkan, Essen

Krauth J, Noche B (2004) Simulation in Produktionsplanung und -Steuerung, whitepaper. http://www.sim-serv.com/pdf/whitepapers/whitepaper_23.pdf

Lang U (2003) Process Chain Integration for Simulation and Data Processing. Space grid workshop, Frascati (Italy)

LARSTRAN (1994) User's Manual (Revision F). LASSO Ingenieurgesellschaft, Leinfelden Echterdingen

Laschet G (2004) Homogenization of the fluid flow and heat transfer in transpiration cooled multilayer plates. J Comput Appl Math 168(1–2):277–288

Laschet G, Apel M (2010) Thermo-elastic homogenization of 3-D steel microstructure simulated by phase-field method. Steel Res Int 81(8):637

Laschet G, Baranowski T, Kashko T, Brinkmann M (2008a) Effective anisotropic properties of semi-crystalline polypropylene via a two-level homogenization scheme, 2nd GAMM Seminar on Multiscale Material Modeling, Stuttgart, 11–12th July 2008

Laschet G, Kashko T, Angel St, Scheele J, Nickel R, Bleck W and Bobzin K. (2008b) Microstructure based model for permeability predictions of open-cell metallic foams via homogenization. Mater Sci Eng A 472:214–226

Laschet G, Quade H, Henke T, Dickert H-H, Bambach M (2010) Comparison of elasto-plastic multi-scale analyses of the U-forming process of a steel line-pipe tube. In: Proceedings of IV European Conference on computational mechanics, Paris, France, May 2010

Lavigne C (2006) Advanced ETL with Pentaho Data, Whitepaper, Breadboard BI

Leblond JB et al (1984) Acta Metall 32:137

Leser U (2007) Informationsintegration: Architekturen und Methoden zur Integration verteilter und heterogener Da-tenquellen, 1st edn. Dpunkt, Heidelberg

Li P, Maijer DM, Lindley TC, Lee PD (2007) A through process model of the impact of in-service loading, residual stress, and microstructure on the final fatigue life of an A356 automotive wheel. Mater Sci Eng A 460–461:20–30

Lichter H, Ludewig J (2007) Software Engineering – Grundlagen, Menschen, Prozesse, Techniken. dpunkt, Heidelberg

Lindemann U, Reichwald R, Zäh MF (2006) Individualisierte Produkte. Springer, Berlin

Lödding H (2008) Verfahren der Fertigungssteuerung. Springer, Berlin
Lohse W (2010) Virtuelle Fertigungssysteme in der CAM-Planung. In: Tagungsunterlagen zum Seminar Potenziale und Trends im Bereich der CAD/CAM/NC-Verfahrenskette, 17–18 Nov, Aachen
Lorensen WE, Cline HE (1987) Marching cubes: a high resolution 3D surface construction algorithm. Comput Graph 21(4):163–169
Lorenz U (2004) Anwendung von Werkstoffmodellen auf die Phasenumwandlung und die Austenitkonditionierung von Stählen. Dissertation, Institut für Eisenhüttenkunde der RWTH Aachen, Aachen
Lotter B (2006) Montage in der industriellen Produktion: ein Handbuch für die Praxis; mit 16 Tabellen. Springer, Berlin
Madhavan J, Bernstein P, Rahm E (2001) Generic schema matching with cupid. In: Proceedings of the 27th international conference on very large databases, Rom, Italien, pp 49–58, Sept 2001
Malik F (2004) Systemisches Management, Evolution, Selbstorganisation Grundprobleme, Funktionsmechanismen und Lösungsansätze für komplexe Systeme, 4., unveränderte Aufl. Haupt, Bern
Martellotti ME (1941) An analysis of the milling process. Trans ASME 63:677–700
Martellotti ME (1945) An analysis of the milling process, part II-down milling. Trans ASME 1945:233–251
Meinecke M (2009) Prozessauslegung zum fünfachsigen zirkularen Schruppfräsen von Titanlegierungen. Dissertation RWTH Aachen University. Apprimus, Aachen
Meinecke M et al. (2008) Model based optimization of trochoidal roughing of titanium. In: Proceedings of the 11th CIRP Conference on Modeling of Machining Operations, Gaithersburg, USA, 16–17 Sept 2008
Menzel W (2000) Partizipative Fabrikplanung: Grundlagen und Anwendung, Als Ms. gedr. Aufl. VDI, Düsseldorf
Merchant M E (1945) Mechanics of the metal cutting process I, Orthogonal cutting and a type 11 chip. J Appl Physics 16(5):267–276
Michaeli W, Bussmann M (2006) Developments in the field of microstructure simulation semicrystalline thermoplastics. J Polymer Eng 26(2–4):275–305
Michaeli W, Baranowski T (2010a) Skalenübergreifende Simulation der Kristallisation in Spritzgussbauteilen. Der Stahlformbauer 5(25):30–36
Michaeli W, Baranowski T (2010b) Skalenübergreifende Simulation der Kristallisation in Spritzgussbauteilen. Tagungsumdruck 25. Internationales Kunststofftechnisches Kolloquium, Aachen
Michaeli W, Baranowski T (2010c) Simulation of the microstructure formation in injection molded semi-crystalline thermoplastic parts. J Polymer Eng 1(30):29–43
Michaeli W, Baranowski T (2010d) Three-dimensional simulation of crystallization effects in injection molded semi-crystalline thermoplastic parts. Proceedings of the 68th Annual Technical Conference (ANTEC) of the Society of Plastics Engineers, Orlando, 2010
Michaeli W, Baranowski T, Dorscheid W, Henseler J (2009a) Integrative materials modeling: towards improved characterisation of molecular orientation in semi-crystalline thermoplastics. In: Proceedings of the 25th Annual Meeting of the Polymer Processing Society, Goa, India
Michaeli W, Bobzin K, Heesel B, Baranowski T, Kashko T, Parkot D, Bagcivan N (2009b) Skalenübergreifende Simulation teilkristalliner Thermoplaste. Werkstoffe in der Fertigung 5:33–34
Michaeli W, Henseler J, Heesel B (2009c) Structural analysis of semi-crystalline thermoplastic parts considering inner properties. In: Proceedings of the 25th annual meeting of the polymer processing society, Goa, India
Michaeli W, Bobzin K, Arping T, Bagcivan N, Baranowski T, Heesel B, Kashko T (2010a) Integrative materials modeling of semi-crystalline thermoplastic parts in 3D. In: Proceedings of the 26th annual meeting of the polymer processing society (PPS), 4–8 July 2010, Banff, Kanada

Michaeli W, Brinkmann M, Heesel B (2010b) Structural analysis of semi-crystalline thermoplastics with regard to the locally disturbed material behaviour due to varying inner properties. In: Proceedings of the 68th annual meeting of the annual technical conference (ANTEC) of the society of plastics engineers, Orlando, USA

Michel J, Pfeiffer S, Schulz W, Niessen M, Kostrykin V (2004) Approximate model for laser welding. In: Radons G, Neugebauer R (eds) Nonlinear dynamics of production systems. Wiley, New York, pp 427–441

Micress (2010) www.micress.de

Minoufekr M (2010) CAx-Prozessketten zur Fertigung und Reparatur von Turbomaschinenkomponenten. In: Tagungsunterlagen zum Seminar Potenziale und Trends im Bereich der CAD/CAM/NC-Verfahrenskette, 17–18 Nov 2010, Aachen

Motik B, Sattler U (2006) A comparison of reasoning techniques for querying large de-scription logic aboxes. In: Hermann M, Voronkov A (eds) Logic for programming, artificial intelligence, and reasoning, vol 4246. Springer, Berlin, pp 227–241

N.N. (1997) Datenermittlung. Hanser, München

National Research Council (2008) Integrated computational materials engineering: a transformational discipline for improved competitiveness and national security. National Academic Press, Washington. ISBN: 0–309-12000-4

Natis Y (2003) Service-oriented architecture scenario. Gartner Research Note AB-19–6751

Negrut D, Rampalli R, Sajdak T (2004) On the implementation of the alpha-method in MSC.ADAMS. In: MSC.Software Knowlede Base KB12299

Neugebauer R, Westkämper E, Klocke F et al (2008) Abschlussbericht Untersuchung zur Energieeffizienz in der Produktion. Fraunhofer-Gesellschaft zur Förderung der angewandten Forschung e.V., München

Neumann L, Aretz H, Kopp R, Goerdeler M, Crumbach M, Gottstein G (2003) Integrative finite element simulation of the rolling of Al alloys with coupled dislocation density and texture models. Z Metall 94(5):593–598

Nikolussi M (2008) Extreme elastic anisotropy of cementite, Fe3C: first-principles calculations and experimental evidence. Scripta Mater 59:814–817

Nyhuis P, Wiendahl H-P (2002) Logistische Kennlinien. Springer, Berlin

OMG (Object Management Group) (2011), Common Object Request Broker Architecture. http://www.corba.org. Accessed 22 Feb 2011

OpenDX Paths to Visualization. http://www.opendx.org.. Accessed 22 Feb 2011

Oppenheim BW (2004) Lean product development flow. Syst Eng 7(4):352–376

Ozturk E, Budak E (2007) Modeling of 5-axis milling processes. Mach Sci Technol 11(3):287–311

Pacheco JL, Krauss G (1990) Gefüge und Biegewechselfestigkeit einsatzgehärteter Stähle. HTM Härterei-Techn Mitt 45(2):77

Palopoli L, Terracina G, Ursino D (2003) DIKE: a system supporting the semi-automatic construction of cooperative information systems from heterogeneous databases. Soft Practice Exp 33:847–884

Pandat (2010) http://www.computherm.com/pandat.html

Pantani R, Sorrentino A, Speranza V, Titomanlio G (2004) Molecular orientation in injection molding: experiments and analysis. Rheol Acta 43:109–118

Panzar JC, Willig RD (1977) Economies of scale in multi-output production. Q J Econ 91(3):481–493

Panzar JC, Willig RD (1981) Economies of scope. Am Econ Rev 71(2):268–272

Pavlyk V, Dilthey U (2004) Simulation of weld solidification microstructure and its coupling to the macroscopic heat and fluid flow modelling. Modell Simul Mater Sci Eng 12:33–45

Pawellek G (2008) Ganzheitliche Fabrikplanung. Springer, Berlin

Pepels W (2006) Produktmanagement: Produktinnovation, Markenpolitik, Programmplanung, Prozessorganisation. Wissenschaftsverlag, Oldenbourg

Peters N, Hofstetter J S (2008) Konzepte und Erfolgsfaktoren für Anlaufstrategien in Netzwerken der Automobilindustrie. In: Schuh G, Stölzle W, Straube F (eds) Anlaufmanagement in der Automobilindustrie erfolgreich umsetzen. Springer, Berlin, pp 9–31

Pfeifer T, Masing W (2007) Handbuch Qualitätsmanagement, 5. vollst. neu bearb. Aufl. Hanser, München

Piekenbrink R (1956) Kräfte und Eingriffsverhältnisse an Stirn- und Walzenfräsen. Dissertation RWTH Aachen

Pinto FW, Vargas GE, Wegener K (2008) Simulation for optimizing grain pattern on engineered grinding tools. CIRP Ann Manuf Technol 57:353–356

Porter M E (2009) Wettbewerbsstrategie: Methoden zur Analyse von Branchen und Konkurrenten. Campus, Frankfurt am Main

Posielek S (2004) Einsatz kombinatorischer Optimierungsmethoden bei automatischer Optimierung von Umformprozessen. Umformtechnische Schriften, Bd 125

Pustal B, Böttger B, Ludwig A, Sahm PR, Bührig-Polaczek A (2003) Simulation of macroscopic solidification with an incorporated one-dimensional microsegregation model coupled to thermodynamic software. Met Mat Trans B 34B:411–419

Prahalad CK, Hamel G (1990) The core competence of the corporation. Harv Bus Rev (Mai/Juni):79–91

Quade H, Fayek P, Laschet G, Henke T (2010) Multi-scale simulation of the U- and O-forming of a line-pipe tube. In: Presented at the first conference on multiphysics simulation, Bonn, Germany, Int J Multiphysics, 22–23 June 2010

Raabe D, Godora A (2005) Mesoscale simulation of the kinetcs and topology of spherulite growth during crystallization of isotactic polypropylene (iPP) by using a cellular automaton. Modell Simul Mater Sci Eng 13:733–751

Radaj D (1999) Schweißprozesssimulation. Grundlagen und Anwendungen, vol 141 of Fachbuchreihe Schweißtechnik. DVS, Düsseldorf

Rahm E, Bernstein PA (2001) A survey of approaches to automatic schema matching. VLDB J 10(4):334–350

Regenthal G (2009) Ganzheitliche Corporate Identity. Gabler, Wiesbaden

Rehling S (2009) Technologische Erweiterung der Simulation von NC-Fertigungsprozessen. Disseration Leibniz Universität Hannover, PZH, Garbsen

Reinhart G, Egbers J, Schilp J, Rimpau C (2010) Demographiegerechte und doch wirtschaftliche Montageplanung. wt Werkstattstech (1/2):9–14

Reisgen U, Schleser M, Mokrov O, Ahmed E, Schmidt A, Rossiter E (2009) Integrative Berechnung von Verzug und Eigenspannung auf Basis realer Schweißparameter. In: Presented at SYSWELD Forum Weimar, 2009

Reynolds J, Ginoza S (2004) Internet official protocol standards, internet official protocol standard RFC 3700. http://tools.ietf.org/rfc/rfc3700.txt. Accessed 22 Feb 2011

Reynoldson RW (1996) The evolution of heat treatment equipment and processes for metal processing. Mater Forum 20:71–92

Rieger T, Gazdag S, Prahl U, Mokrov O, Rossiter E, Reisgen U (2010) Simulation of welding and distortion in ship building. Adv Eng Mater 12(3):153–157

Rizzi S, Abelló A, Lechtenbörger J, Trujillo J (2006) Research in data warehouse modeling and design: dead or alive? In: Proceedings of the 9th ACM international workshop on Data warehousing and OLAP, Arlington, USA, pp 3–10, 5–11 Nov 2006

Rodriguez R, Gutierrez I (2003) A unified formulation to predict the tensile curves of steels with different microstructures. Mater Sci Forum 426–432:4525

Rosenstiel LV (1993) Der Einfluß des Wertewandels auf die Unternehmenskultur. In: Lattmann C (ed) Die Unternehmenskultur. Physica-Verlag, Heidelberg

Roters F, Eisenlohr P, Bieler TR, Raabe D (2010) Crystal plasticity finite element methods CP-FEM. Wiley, Weinheim. ISBN 978-3527 32447-7

Rudnizki J, Zeislmair B, Prahl U, Bleck W (2010a) Prediction of abnormal grain growth during high temperature treatment. Comput Mater Sci 49(2):209

Rudnizki J, Zeislmair B, Prahl U, Bleck W (2010b) Thermodynamical simulation of carbon profiles and precipitation evolution during high temperature case hardening. Steel Res Int 81(6):472

Sanchez R, Heene A (1996) A systems view of the firm in competence-based competition. In: Sanchez R, Heene A, Thomas H (eds) Dynamics of competence-based competition. Elsevier, Oxford, pp 39–62

Sanchez-Palencia E (1980) Non-homogeneous media and vibration theory. Springer, Berlin

Schady R (2008) Methode und Anwendungen einer wissensorientierten Fabrikmodellierung. Dissertation, Otto-von-Guericke Universität Magdeburg

Schäffer U, Weber J, Willauer B (2001) Zur Optimierung von Intensität und Neuplanungsanteil der operativen Planung. Controlling 13:283–288

Schmigalla H (1995) Fabrikplanung. Carl Hanser, München

Schmitz GJ, Prahl U (2009) Toward a virtual platform for materials processing. JOM 61(5):26

Schmitz GJ, Prahl U (2011) Integrative computational materials engineering-concept and applications of a modular simulation platform. Wiley. ISBN 978-3-527-33081-2 (2011, in preparation)

Schneider MC et al (1995) Formation of macrosegregation by multicomponent thermosolutal convection during the solidification of steel. Met Mat Trans A 26A:2373–2388

Scholz-Reiter B, Krohne F (2006) Prozessänderungen – Engpassorientierte Realisierung von Anlaufzielgrößen. Ind Manage 22(6)

Schopfer G, Yang A, v Wedel L, Marquardt W (2004) CHEOPS: a tool-integration platform for chemical process modelling and simulation. Int J Soft Tools Technol Trans 6(3):186–202

Schroeder W, Kitware Inc (2006) The visualization toolkit: an object-oriented approach to 3D graphics, 4th edn. Kitware, Clif-ton Park

Schroeder W, Martin K, Lorensen B (2006) The visualization toolkit – an object-oriented approach to 3Dgraphics, 4th edn. Kitware, Inc., www.vtk.org

Schroeder WJ, Zarge JA, Lorensen WE (1992) Decimation of triangle meshes. SIGGRAPH 92:65–70. doi:10.1145/133994.134010

Schuh G, Gottschalk S (2008) Production engineering for self-organizing complex systems. Prod Eng – Res Dev 2(4):431–435

Schuh G, Kampker A, Franzkoch B (2005) Anlaufmanagement. wt Werkstattstech 95(5):405–409

Sethi AK, Sethi SP (1990) Flexibility in manufacturing: a survey. Int J Flex Manuf Syst 2:289–328

Shabana AA (2005) Dynamics of multibody systems, 3rd edn. Cambridge University Press, Cambridge

Shvaiko P, Euzenat J (2005) A survey of schema-based matching approaches. In: Spaccapietra S (ed) Journal on data semantics IV, vol 3730. Springer, Berlin, pp 146–171

Sierra N (2008) Int J Mech Sci 50

Simitsis A, Vassiliadis P, Sellis T (2005) Optimizing ETL processes in data warehouses. ICDE 2005. In: Proceedings of the 21st international conference on data engineering, Tokio, Japan, pp 564–575, 5–8 Apr 2005

Sirin E, Parsia B, Grau BC, Kalyanpur A, Katz Y (2007) Pellet: A practical OWL-DL reasoner. Web Sem Sci Ser Agents World Wide Web 5(2):51–53

Spaccapietra S, Parent C (1991) Conflicts and correspondence assertions in interoperable databases. SIGMOD Record 20(4):49–54

Specht D (2007) Strategische Bedeutung der Produktion. Gabler, Wiesbaden

Srinivasan S (1995) RPC: Remote procedure call protocol specification version 2, internet official protocol standard RFC 1831. http://tools.ietf.org/rfc/rfc1831.txt. Accessed 22 Feb 2011

STAR-Cast (2011) Gieß-und Erstarrunsgsimulationssoftware. http://www.starcast.org. Accessed 18 July 2011

Stautner M (2006) Simulation und Optimierung der mehrachsigen Fräsbearbeitung. Dissertation Universität Dortmund, Vulkan, Essen

Stegen A (2007) Ein systemtechnischer Beitrag zur Zerspanungstheorie. Dissertation RWTH Aachen, VDI, Düsseldorf

Steinbach I, Pezzolla F, Nestler B, Seeßelberg M, Prieler R, Schmitz GJ, Rezende JLL (1996) A phase field concept for multiphase systems. Physica D 94:135–147

Stepping (2007) Fabrikplanung im Umfeld von Wertschöpfungsnetzwerken und ganzheitlichen Produktionssystemen. Dissertation, TU Karlsruhe

Strempel H (1963) Ein Beitrag zur Darstellung der Schnitt- und Drangkräfte beim Walzfräsen und Drehen. Dissertation, TU Dresden

Stuckenschmidt H (2009) Ontologien: Konzepte, Technologien und Anwendungen, 1st edn. Springer, Berlin

Studer R, Benjamins VR, Fensel D (1998) Knowledge engineering: principles and methods. Data Knowl Eng 25(1):161–197

Stüve J, Gries T (2009) Advances in the simulation of the overbraiding process using FEM. In: Erath MA (ed) Composites – innovative materials for smarter solutions: SEICO 09; SAMPE Europe 30th international jubilee conference and forum, Paris; Proceedings 2009 – Riehen: SAMPE Europe Conferences, 2009, pp 618–625, 23–25 March 2009

Surmann T (2006) Geometrisch-physikalische Simulation der Prozessdynamik für das fünfachsige Fräsen von Freiformflächen. Dissertation Universität Dortmund, Vulkan, Essen

Tashiro K, Kobayashi M (1996) Molecular theoretical study of the intimate relationships between structure and mechanical properties of polymer crystals. Polymer 37:1775–1786

Taylor FW (1907) On the art of cutting metals. Trans ASME 28:31–279

Teece DJ (1980) Economies of scope and the scope of the enterprise. J Econ Behav Org 1(3):223–247

Terada K, Kikuchi N (2001) A class of general algorithms for multi-scale analyses of heterogeneous media. Comput Methods Appl Mech Eng 190(40–41):5427–5464

Thain D, Tannenbaum T, Livny M (2008) Distributed computing in practice: the condor experience. Concur Comput Pract Exp 17(2–4):323–356

The MPI Forum (1993) MPI: a message-passing interface. In: Proceedings of supercomputing '94, IEEE Computer Society Press, pp 878–883

The ParaView Guide (2008) ISBN 1-930934-21-4

Thermocalc (2010) www.thermocalc.com

Thiede S, Herrmann C (2010) Simulation-based energy flow evaluation for sustainable manufacturing systems. In: Tagungsband zur 17. CIRP Konferenz für Life Cycle Engineering, Hefei, China, pp 99–104, 19–21 May 2010

Tönshoff HK, Peters J, Inasaki I et al (1992) Modelling and simulation of grinding processes. Ann CIRP 41(2):677–688

Torquato S (2002) Random heterogeneous materials: microstructure and macroscopic properties. Springer, New York

Tsarkov D, Horrocks I (2006) FaCT+ +description logic reasoner: system description. In: Automated Reasoning vol 4130, Springer, Berlin, pp 292–297

Ullrich G (1995) Wirtschaftliches Anlernen in der Serienmontage: ein Beitrag zur Lernkurventheorie, Als Ms. gedr. Aufl. Shaker, Aachen

UNICORE Uniform Interface to Computing Resources (2011). http://www.unicore.eu. Accessed 22 Feb 2011

Upton DM (1994) The management of manufacturing flexibility. California Manag Rev 36(2):79

Urban B (2009) Kinematische und mechanische Wirkungen des Kugelkopffräsens. Dissertation Leibniz Universität Hannover, PZH, Garbsen

Valette S, Chassery JM (2004) Approximated centroidal voronoi diagrams for uniform polygonal mesh coarsening. Comput Graph Forum. doi:10.1111/j.1467–8659.2004.00769.x

van Luttervelt CA (2001) Towards predictable performance of metal cutting operations – 50 years of efforts by CIRP STC

van Luttervelt CA, Childs THC, Jawahir IS, Klocke F, Venuvinod PK (1998) Present situation and future trends in modelling of machining operations. Ann CIRP 4(2):587–624

Visser U (2004) Intelligent information integration for the Semantic Web. Springer, Berlin

Vitr M (2008) CAM-NC-Kopplung für einen durchgängigen, bidirektionalen Informationsfluss zwischen Planung und Fertigung. In: Tagungsunterlagen zum Seminar Potenziale und Trends der CAD/CAM/NC-Verfahrenskette, Aachen, 19–20 Nov 2008

Volkwein G (2006) Konzept zur effizienten Bereitstellung von Steuerungsfunktionalität für die NC-Simulation. Dissertation TU München, Utz, München

W3C (2002) Web services. http://www.w3.org/2002/ws. Accessed 22 Feb 2011

Warnken N et al (2003) Multiphase-field model for multicomponent alloys coupled to thermo-dynamic databases. In: Stefanescu DM (ed) Modeling of casting, welding and advanced solidification processes X. TMS, Warrendale, pp 21–29

Weddige HJ (2002) Stahl im Wettbewerb der Werkstoffe. Dissertation, TU BA Freiberg

Weihan Z, Ming C (2009) Surface reconstruction using dexel data from three sets of orthogonal rays. J Comput Inf Sci Eng 9(1):011008

Weinert K, Blum H, Jansen T, Rademacher A (2007) Simulation based optimization of the NC-shape grinding process with toroid grinding wheels. Prod Eng 1:245–252

Werner G (1971) Kinematik und Mechanik des Schleifprozesses. Dissertation RWTH Aachen

Wernerfeld B (1984) A resouce-based view of a firm. Strat Manage J 5

Westkämper E (2006) Einführung in die Organisation der Produktion. Springer, Berlin

Wiegers KE (2005) Software requirements, Dt. Ausgabe der 2nd edn. Microsoft Press, Unterschleißheim

Wiendahl H-P (2001) New methods for improving flexibity of capacities in productionnetworks. Prod Eng 8(1):93–98

Wiendahl H-P, Nofen D Klußmann J H, Breitenbach F (2005) Planung modularer Fabriken, Carl Hanser Verlag, München

Witt S (2007) Integrierte Simulation von Maschine, Werkstück und spanendem Fertigungsprozess. Dissertation RWTH Aachen, Shaker, Aachen

Wolske M, Kopp R (2000) Microstructure simulation of Ni based alloys. In: Proc IV Conf. Int. de Forjamento, Porto Alegre, pp 42–53

Wolter M, Assenmacher I et al (2008) A time model for time-varying visualization. Comput Graphics Forum 28:1561–1571

Zabel A (2010) Prozesssimulation in der Zerspanung – Modellierung von Dreh und Fräsprozessen. Habilitationsschrift TU Dortmund, Vulkan, Essen

Zacharia T, Vitek JM, Goldak JA, DebRoy TA, Rappaz M, Bhadeshia HKDH (1995) Modeling of fundamental phenomena in welds. Model Simul Mater Sci Eng 3:265–288

Zahn E, Foschiani S, Tilebein M (2000) Wissen und Strategiekompetenz als Basis für die Wettbewerbsfähigkeit von Unternehmen. In: Hammann P, Freiling J (eds) Die Ressourcen- und Kompetenzperspektive des strategischen Managements. Gabler, Wiesbaden

Zhang M-X, Kelly PM (1997) Accurate orientation relationships between ferrite and cementite in pearlite. Scripta Mater 37(12):2009

Zink K J (2009) Personal- und Organisationsentwicklung bei der Internationalisierung von industriellen Dienstleistungen. Physica, Heidelberg

Chapter 5
Hybrid Production Systems

Gerhard Hirt, Wolfgang Bleck, Kirsten Bobzin, Christian Brecher,
Andreas Bührig-Polazcek, Edmund Haberstroh, Fritz Klocke, Peter Loosen,
Walter Michaeli, Reinhart Poprawe, Uwe Reisgen, Kristian Arntz,
Nazlim Bagcivan, Markus Bambach, Stephan Bäumler,
Stefan Beckemper, Georg Bergweiler, Tobias Breitbach, Steffen Buchholz,
Jan Bültmann, Jörg Diettrich, Dennis Do-Khac, Stephan Eilbracht,
Michael Emonts, Dustin Flock, Kai Gerhardt, Arnold Gillner,
Alexander Göttmann, Oliver Grönlund, Claudia Hartmann, Daniel Heinen,
Werner Herfs, Jan-Patrick Hermani, Jens Holtkamp, Todor Ivanov,
Matthias Jakob, Andreas Janssen, Andreas Karlberger, Fritz Klaiber,
Pia Kutschmann, Andreas Neuß, Ulrich Prahl, Andreas Roderburg,
Chris-Jörg Rosen, Andreas Rösner, Alireza Saeed-Akbari, Micha Scharf,
Sven Scheik, Markus Schleser, Maximilian Schöngart, Lars Stein,
Marius Steiners, Jochen Stollenwerk, Babak Taleb Araghi, Sebastian Theiß
and Johannes Wunderle

Contents

5.1	Research Programme of Hybrid Production Systems	436
5.2	Methodology for the Development of Integrated Manufacturing Technologies	438
	5.2.1 Abstract	438
	5.2.2 State of the Art	439
	5.2.3 Motivation and Objectives	445
	5.2.4 Results of the Methodology Development	447
	5.2.5 Future Research Topics	478
5.3	Shortening Process Chains for the Production of Plastic/Metal Hybrids Using Innovative Primary Forming and Joining Processes	480
	5.3.1 Abstract	480
	5.3.2 State of the Art	481
	5.3.3 Motivation and Research Question	489
	5.3.4 New Approaches for the Manufacture of Plastic and Metal Hybrids	492
	5.3.5 Industrial Relevance	546
	5.3.6 Future Research Topics	549
5.4	Improving the Efficiency of Incremental Sheet Metal Forming	551
	5.4.1 Abstract	551

G. Hirt (✉)
Institut für Bildsame Formgebung der RWTH Aachen, Intzestr. 10,
52056 Aachen, Deutschland
e-mail: hirt@ibf.rwth-aachen.de

	5.4.2	State of the Art	552
	5.4.3	Motivation, Research Question and Approaches to a Solution	558
	5.4.4	Results	560
	5.4.5	Industrial Relevance	587
	5.4.6	Future Research Topics	590
5.5		Shortening Process Chains by Process Integration into Machine Tools	591
	5.5.1	Abstract	591
	5.5.2	State of the Art	592
	5.5.3	Motivation and Research Question	596
	5.5.4	Results	600
	5.5.5	Industrial Relevance	624
	5.5.6	Future Research Topics	626
5.6		Shortening Process Chains for Manufacturing Components with Functional Surfaces Via Micro- and Nanostructures	628
	5.6.1	Abstract	629
	5.6.2	State of the Art	631
	5.6.3	Motivation and Research Question	638
	5.6.4	Results	640
	5.6.5	Industrial Relevance	679
	5.6.6	Future Research Topics	682
References			685

5.1 Research Programme of Hybrid Production Systems

While virtual product development allows great freedom in terms of design, actual development processes are rather restricted. Those boundary conditions are at best hardly possible to exert influence on. Therefore, future research has to focus both on the realisation of the concept of one-piece-flow while simultaneously increasing flexibility and productivity and on the technological advancement. Hence, hybridisation of manufacturing processes is a promising approach, which often allows tapping potentials in all the aforementioned dimensions.

In the context of a production system, the term "hybridisation" has different definitions. These are

- an integrated combination of normally separated manufacturing processes in order to manufacture suitable products made of different materials (like e.g. the casting process of integrated plastic-metal-products),
- integrated application or combination of different mechanisms of action (like e.g. laser-assisted machining) or
- performing different types of manufacturing operations in an integrated machine tool (like e.g. an autonomic machine centre for manufacture and repair of forming tools: machining, measuring and built-up welding in a single machine centre with an integrated CAM-NC-measurement chain).

Therefore, the research in the field of hybrid production systems focuses on the following question:

> How can the benefits of integrated production technologies be described and how can the integration of production technologies be configured in a generalisable way?

5 Hybrid Production Systems

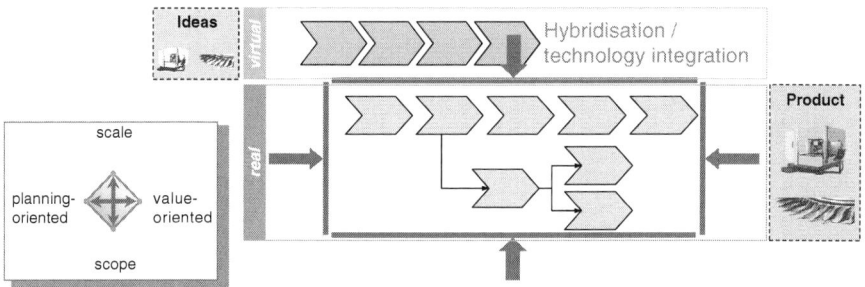

Fig. 5.1 Research questions of hybrid production systems

The main result of the research area "Hybrid Production Systems" and its contribution to the theory of production focuses on the evolution of construction and design methods for products and production systems in terms of the integration of technologies. In the past, such integration efforts have often led to significant extensions of development periods. As a consequence, restrictions of machining sophisticated geometries can only slowly be compensated for. Main fields of application for these integrated manufacturing processes are products of the mould and tooling industry as well as medical implants and certain parts of turbine engines (Fig. 5.1).

As part of the Cluster of Excellence, a machining centre for milling operations is combined with a robot and a laser system and hence transformed into a multi-technology platform. The machining process of the five-axis milling operation itself is complemented by laser-machining processes. For this purpose, corresponding laser machining heads are developed. A fibre laser system is used for wire-based deposition welding, hardening and ablation whereas for laser structuring, a short pulsed laser together with a laser scanner is used. Since in a five-axis machining operation both the robot and the machining centre can process the workpiece at the same time, a high-precision synchronisation of both systems' movements is a fundamental requirement for a successful realisation. Among others, a high-bandwidth connection of the machining centre's and the robot's control systems, which have mostly been autonomous until now, is therefore subject to the research activities. The same applies to the laser system's control system. In total, this contributes to a comprehensive control system. In order to compensate for the resulting complexity in production planning and to tap the full potential of the integrated machining centre, corresponding research also involves the complete integration of all components into a single CAM-system.

However, this example of a successful hybridisation is only an excerpt of the research concerning "hybrid production systems" done in the Cluster of Excellence. Other examples are e.g. tapping potentials in sheet metal forming by combining incremental forming of highly-individualised workpieces and local warming in order to lower process forces and machining time.

5.2 Methodology for the Development of Integrated Manufacturing Technologies

Fritz Klocke, Steffen Buchholz, Kai Gerhardt and Andreas Roderburg

5.2.1 Abstract

In the wake of today's globalised markets, a large number of manufacturing firms in countries with high labour costs find themselves in a situation that is characterised by the increased competitive pressure. As a result, markets are intensified and competition is increased. With this in mind, innovations in production processes and further developments of manufacturing technologies in particular, are an important factor in the competitiveness of manufacturing firms. Innovative processes may open access to new product features or functionalities, and in many cases, their aim is to achieve higher process stability or shorter production times. An important criterion of an innovative new process may be considered overcoming technological boundaries by a radical advancement. For a wide variety of manufacturing technologies such as the radical advancement through existing optimisation methods is often not possible, in most cases technological boundaries can only be approached, but not completely dissolved. Fundamental developments of manufacturing technologies can be implemented, among other things, through a combination of discriminative technologies or technological know-how as the examples of hybrid technologies show. Hybrid manufacturing processes are examples of technology developments exhibiting radical improvement of performance parameters. They provide a way to overcome previous manufacturing technology boundaries and present the opportunity for the development of options for a process chain design. Using hybrid technology solutions, the field of application of conventional methods can be expanded so that new materials and geometries can be adapted. One example is the laser-assisted turning of ceramic components. Only with the use of the laser, the turning of ceramics was made possible (Klocke et al. 2009).

The selective development of hybrid manufacturing process requires the integration of expertise of different individual technologies. It is important to have a comprehensive understanding of production-technological relationships within the context of the research presented here, and to have an understanding of the transparency of cause-effect relationships of manufacturing technologies.

When observing development processes of hybrid technology that are known to this date, it can be noticed, however, that an approach for a targeted and systematic combination and transfer of technology solutions from different areas cannot be seen. Therefore, the aim is to develop a methodology that facilitates a specific combination of different manufacturing technologies or mechanisms of action to hybrid technology solutions. An innovative process-oriented development strategy can also make a significant contribution to the development of technological skills, which are usually more difficult to imitate when compared to product innovations. Thus, this strategy helps to obtain a sustainable technological advantage in international competition, which bears an inherent increased risk of attempted imitation.

In the present study methodology tools are developed that make the targeted detection of cross-technology solutions for manufacturing processes possible. Technological expertise, which is reflected in different types of technology models, is the main analysis object of the methodology. For the acquisition and processing of data received from different technology models, various models of manufacturing technologies in regard to their scope, their ambiguity, and their compatibility will be analysed. An explanatory model for the targeted capture of cause-effect relationships of manufacturing technologies will be presented. Known methods for general problem solving and the generation of ideas are used to derive a solution. These will then be adapted to adjust the requirements of production-oriented tasks. For this purpose, the concept of a solution catalogue for the development of hybrid manufacturing technologies based on the engineering design theory will be presented. This catalogue is built on the principles of a morphological box and contains active principles of known process solutions. It offers the user who is in the search for hybrid integrated technology solutions a wide range of support. Modelling of production-technological knowledge on the cause-effect relationships and the utilisation of the morphological box is finally illustrated in form of an application example.

5.2.2 State of the Art

5.2.2.1 Hybrid Manufacturing Technologies

In the past, the integration of manufacturing knowledge from different disciplines has led to much advancement of manufacturing technologies. Examples of such advanced developments are hybrid manufacturing technologies. Here, the combination of individual technologies into hybrid manufacturing technology solutions presents a special approach to the development of manufacturing technologies and their potential applications. The enhancement of technological capabilities and the ability to surmount boundaries of individual technologies by using a combination of processes and/or their active mechanisms of action are promising. Due to the multitude of combinations available, the potential for expanding production-technological options and sustainable production of improved and new product features are very high (BDI 2005).

The consideration and development of hybrid manufacturing processes are associated with the definition of the parent concept of "hybrid technology". In production technology, depending on the particular design focus of the system that is to be hybridised, the term "hybrid" is defined in different ways. To date, there is no clear definition of the term hybrid manufacturing technologies.

According to Lauwers et al. (2010) and Schuh et al. (2009) hybrid technology can be classified as:

- Hybrid processes as a combination of different modes of action, in which different forms of energy (or energy generated in a different way) are injected into an active zone (e.g., laser-assisted turning) at the same time or in a single process step.
- Hybrid processes as a combination of process steps where process steps that are usually carried out sequentially are implemented in parallel operations, e.g., grind hardening.
- Hybrid machines as an integration of various production processes in a single machine, where individual process steps are performed sequentially, or different methods are carried out in parallel at different locations of the component. Such machines are sometimes referred to as multi-technology platforms.
- Hybrid products, that are characterised as such, based on their hybrid structures or hybrid functions, regardless of their production, e.g., metal-plastic composite components.

An unambiguous classification of individual products, processes, and machinery to a group of hybrid technologies as defined above is not possible in many cases since the boundary of hybrid to non-hybrid technology cannot be defined precisely. This is particularly true in the case of hybrid products, as the definitions provided in relevant literature leave a lot of room for their interpretation. A hybrid machine can be the functional element for a hybrid process; however, it may also include operations or processes that are executed in sequence. The characterisation of hybrid processes, however, is more advanced. This characterisation is approximated in relevant literature by a variety of definition attempts, and it is currently being discussed in a special study group of the International Academy for Production Engineering (CIRP). Thus, today there are several possible definitions for hybrid production processes, some of which are reduced to specific technology groups (e.g., separation processes) or a generally accepted definition applicable to all categories of manufacturing processes:

- Hybrid processing combines two or more processes of material separation. These hybrid processes increase the benefits and reduce the disadvantages that occur in the various technologies (Rajurkar et al. 1999).
- According to Kozak, in general, there are two types of hybrid processes. Firstly, there are those in which participating single processes are directly involved in the material separation or processing. Secondly, there are those in which only one of the single procedures of material separation and processing participates, while the other process executes only a supporting function by influencing the machining conditions of the main process in a positive way (Kozak and Rajurkar 2000).
- In hybrid manufacturing processes, different forms of energy (or energy generated in a different way), are used simultaneously in the same active zone. Hybrid processes are also defined as a combination of effects during a process that can be achieved conventionally in separate but successive processes (Klocke et al. 2010b).

Hybrid processes are characterised by a combination of methods of which individual process mechanisms of action, such as machining or chip formation mechanisms of action, are significantly affected. Through a combination of mechanisms of action, it is intended to achieve significant positive effects during hybrid manufacturing processes or process parameters such as a reduction of forces, decrease of the wear and tear of tools, or improving the machining ability of very hard materials (Lauwers et al. 2010).

The above-mentioned definitions specify various features for the characterisation of hybrid manufacturing technologies. However, these various characteristics of hybrid processes do not cancel one another out. Therefore, when combining the various definitions given, the result is an array of definitions, which can be divided into different classes.

Therefore, hybrid manufacturing processes are processes whereby combining different process mechanisms of action—that otherwise run separately-ensures that the benefits of the different processes are combined; at the same time, this interconnection will reduce the disadvantages of a single process or may prevent such disadvantages altogether. In these cases, however, the main function of the hybrid process can be also met by the respective individual processes. An example of such a synergy effect through hybridisation is the combination of a laser welding process with a conventional MIG welding process. Both MIG and laser welding processes are used as independent methods for joining materials. The hybrid welding process has advantages over the individual processes, whereby the synergy effect of this method combines the advantages of each individual technology. In this example of the hybrid welding process, the hybrid process is more stable than the laser welding method and faster when compared to MIG welding process (The Federation of German Industries—BDI 2005). Within this class, those hybrid manufacturing processes are set apart in which the processing capabilities and efficiency of a single process are broadened by additional mechanisms of action. In this case, additional technologies are introduced by processes that are of supportive character. One example is the ultrasound-assisted grinding process. Here, in specific applications, new processing qualities for the grinding process can be facilitated by heterodyning ultrasonic vibration. In the example of laser-assisted turning of ceramic materials, the application of the turning process for machining the workpiece becomes only possible by the heating effect of the laser. This removes the material-based limitation of the turning process or extends the processing potential of the ceramic material. In summary, it should be noted that in all cases hybrid procedures change the mechanism of single processes, while the output limits of one or more single processes are shifted. According to Bachmann et al. (1999), by using hybridisation one can distinguish between two main strategies (see Fig. 5.2) when expanding technological boundaries. Firstly, when combining technologies for hybrid methods, it is possible to influence the material properties of the processed material by using an additional form of energy such as heat. Thus, the workability of the material is improved or by changing fundamentally the mechanisms of action, machining of the material becomes possible. Secondly, a strategy of hybridisation can be used whereby the mechanisms of action of the process are directly changed without adversely affecting the material of the workpiece. Here, a supplementary energy is added so that a different form or a changed composition of the process mechanisms of action becomes effective.

Strategy 1

Change in the material machining characteristics in the working zone by additional process energy

Strategy 2

Change of physical active principles by adding principles / energy

Hybrid process

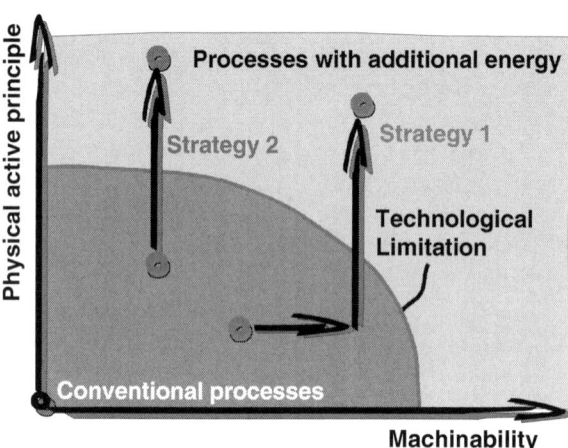

Fig. 5.2 Widening of process boundaries through hybrid processes. (Bachmann et al. 1999)

Based on this understanding of a hybrid process, and according to the classification by Lauwers et al. (2010), among other things, the following hybrid procedures are categorised in these groups:

- Vibration-assisted manufacturing methods (e.g., ultrasonic-assisted grinding, ultrasonic-assisted turning, ultrasonic-assisted drilling, and ultrasonic-assisted EDM)
- Laser-assisted manufacturing methods (e.g., laser-assisted turning, laser-assisted ECM, laser-assisted incremental sheet metal forming, laser-assisted shearing, and laser-assisted pressure forming)
- Production processes combined with EDM/ECM (e.g., EDM grinding, ECM grinding, ECM-supported wire EDM)
- Medium-assisted manufacturing processes (e.g., high-pressure coolant supply during grinding, high-pressure coolant supply for turning)

In addition to the hybrid methods mentioned above, further methods are known, such as the combination of incremental forming and stretch forming, which cannot be classified in the groups indicated earlier. Therefore, the classification can be extended to other groups or individual combinations of hybrid processes. The mechanisms of action of hybrid manufacturing processes referred to in the literature have been studied considering various aspects, and they have been described in great detail. Therefore, depending on the method used, the model and knowledge base differ greatly in their level of detail. In many cases, the potentials and implications of hybrid methods, as well as the individual processes suitable for a suitable combination of hybrid procedures, have not been systematically recorded. To ensure hybrid technology can be developed further, a structured approach for the integral compilation of mechanisms of action of different methods is necessary; unfortunately, such a procedure does not exist at this time. However, this is necessary in order to be able to recognise and evaluate targeted application-specific hybridisation options to increase process capabilities of conventional processes. In order to do that, a holistic and systematic adaptation of technological know-how of current mechanisms of action of production technologies for the purpose of an integrated technology development must be implemented.

5.2.2.2 Design Methodologies

With the goal of a conceptual design of innovative hybrid manufacturing technologies, the development of manufacturing processes generates new solutions by combining technical components, their functions, and the resulting effects. With recent advancements of manufacturing technologies, such as in the development of hybrid manufacturing processes, quite often, new solutions were found only intuitively or by accident. So far, no methodological support is known for a purposeful synergy of different production technologies to find new technological solutions (Klocke et al. 2008a). If, during the ongoing development of manufacturing technologies specific integrated solutions should be gained, a comprehensive overview of all mechanisms of action available or technologies that are currently in development is imperative. Even if initially some processes do not appear to be appropriate for a specific application, it may be possible to attain the desired capability through the use of known mechanisms of action of other processes.

The desired form of a development process is referred to both in the English and in the German language as a "Design process". For the systematic implementation of a design process relevant literature, and particularly in the science of engineering design, numerous teaching methodologies and methodological tools are available. In design processes, a distinction is made between general and product-specific design processes (Koller and Kastrup 1994). For the development of hybrid manufacturing technologies, the general descriptions of design processes are relevant since these general procedures described can be transferred to the particular requirements.

Ideally, at the beginning of each design process the designer is faced with a problem for which there are causes and for which solutions can be found during the design process (Hubka 1976). In addition, it must be considered that any technical solution serves one or more purposes, whereby the implementation of such solution is subject to various requirements and restrictions. Design processes consist of synthesis, analysis, and selection activities (Koller 1994). From the time the problem is identified until the final solution is implemented, the different phases of the design or development process are divided in chronological order as follows:

- Clarifying the task, designing, drafting, and formulating (Hubka 1976; Hubka and Eder 1992)
- Clarifying the task, function synthesis, principle synthesis, qualitative designs, and quantitative designs (Koller 1994)
- Clarifying and refining the task, determining functions and their structures, searching for solution principles and their structures, organising into feasible modules,

designing relevant modules, designing the entire product, developing execution methods and proper utilisation (VDI 1993)
- Problem analysis, problem formulation, system synthesis, system analysis, evaluation, decision (Pahl et al. 2005, 2007)

During real development processes, this results in many iterations. Nonetheless, the course of finding a solution can still be described as a progression from the abstract search for solutions to a concrete solution (Lindemann 2007).

The common goal of the design methodologies is to comprehend available knowledge of development tasks or problems. By doing so, one must apply known solutions, understand, and find one or more appropriate solutions for the given problem (Tomiyama et al. 2009). According to Matchett and Briggs, design is "a process of discovering a way that brings into line all conflicting factors and relationships in a multidimensional situation" (Matchett and Briggs 1966). It therefore follows that one must consider the starting point of a design process as problems in which conflicting objectives of two or more criteria should cancel each other out. Generally, design methodologies are composed of a process model as a logical sequence of development stages. Based on models of the development's objective, specific development steps are compiled. The differences between the methodologies are presented in their region of interest, the information provided by the models used, and the specific combinational logic of individual sub-steps, such as iterations.

Methodologies used to develop technical systems can be divided by type of design object (product, process chain, and manufacturing processes), and each of the desired type of system development (continuous or radical). In *continuous system development*, adjusting parameters within the existing system boundaries are changed, thus changing the result variables of the system towards its optimum. The optimum determines a limit of performance that cannot be exceeded. During *non-continuous* or *radical* development, the performance of a system leaps. Here, a basic change in the system at the level of the system's structure take place, e.g., by adding elements (technologies) or relations (mechanisms of action). By doing this, the performance limit of the system can be lifted or shifted (Klocke et al. 2009).

Radical system developments are usually more expensive than continuous ones; however, they do create conditions that initiate innovations. During the development process of hybrid manufacturing technologies, sudden changes of the process capabilities of individual manufacturing processes are being pursued. When using well-known design methodologies, in which radical developments to improve a system are aimed for, the focus rests mainly on products used as design objects. For example, many TRIZ methods (Theory of Inventive Problem Solving) are used in the inventive and evolutionary development of technical products (Altshuller 1996, 1999, 2002; Klein 2002; Orloff 2002). TRIZ applications for the development of manufacturing processes are relatively rare and are not yet used for the systematic development of manufacturing processes. A possible reason for this is insufficient methodological support to cope with the highly complex interconnection of many manufacturing processes (Klocke et al. 2008a; Roderburg et al. 2009). In recent studies, such as research conducted by Jenke or Tillmann, process chains are viewed as design objects performing their task on a system level higher than the manufacturing process to which they are allocated (Jenke 2007; Tillmann 2009). The effective relationships between input and output variables of the manufacturing process are depicted as "black box", thus they are not considered for the hybrid method's relevant mechanisms of action of individual manufacturing technologies. Besides the methods shown in this example, there are very many other development and design methodologies, in particular, at the product level. An overview of relevant design methods is presented by Lindemann (2007) and Tomiyama et al. (2009).

5.2.2.3 Interim Results

After analysing existing methodologies for the development of technical systems and their application in the field of production technology, a need has been derived to research a design method for a radical development of manufacturing technologies (e.g., by hybridisation). The need for research results from the following deficits among the requirements for the development of hybrid manufacturing processes:

- To this date, detailed knowledge about individual production technologies available in companies and research facilities is not being sufficiently used to support a systematic and efficient development of manufacturing technologies.
- Manufacturing processes for existing methodologies are often viewed as the description of a phenomenon. The potentials available by using targeted integration processes and the profound understanding of different experts cannot be predicted with sufficient accuracy.
- At this time, existing solution knowledge from different areas of manufacturing technologies is not represented on the necessary level of abstraction or provided systematically as analogies for the transfer of solution principles.
- Unfortunately, the necessary knowledge base is not available in an appropriately prepared format.

As a result, an analytical approach for the targeted development of manufacturing technologies does not take place. Due to undiscovered potentials and long test phases, the result is reduced efficiency in the development of inventive manufacturing methods. However, the design methods whose focuses are used mainly for developing technical products form a promising basis for a methodology for the development of hybrid manufacturing technologies.

5.2.3 Motivation and Objectives

In the past, innovations at the level of manufacturing processes have led to sustainable competitive advantages. An innovative process-oriented development strategy can also make a significant contribution to the development of technological skills, which are usually more difficult to imitate when compared to product innovations.

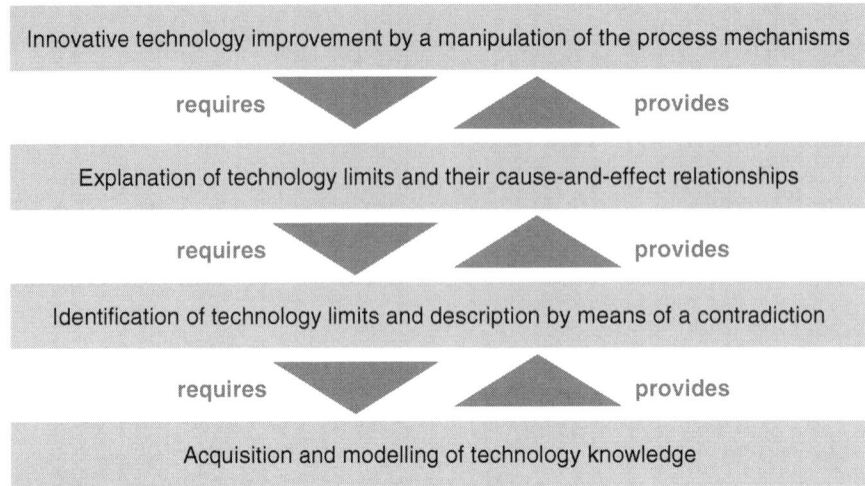

Fig. 5.3 Target system of a methodology for developing hybrid manufacturing technologies

Thus, this strategy helps to obtain a sustainable technological advantage in international competition. The potential of innovative development of manufacturing technologies by integrating various technologies and technological know-how has been described above. Furthermore, the shortcomings in the scope of methodology for the development of technological systems were demonstrated. This results in the paramount objective *to develop a methodology to efficiently analyse the existing boundaries in manufacturing technologies and to develop and evaluate new solutions to overcome the limits of technology.*

As a result, development time and cost savings as well as technological leaps can be realised, which leads to competitive advantages. The benefit of methodology for industry and research lies in the fundamental and efficient development of manufacturing technologies through integration of various technology areas. It also includes using extensive and in-depth technology expertise and the purposeful creation of an innovation edge by overcoming technology limits. Moreover, implementing new features and enhancing customer value is an important factor for improving competitiveness (Klocke et al. 2009).

In the following description of the methodology for developing hybrid manufacturing technologies, the focus is on radical technological developments. The sub-goals for the methodology development shown in Fig. 5.3 were derived from the overall research objective. The sub-goals are building on top of one another and in their entirety lead to the structure of the developed methodology. This leads to the research questions that must be answered. In order to enable the identification of process boundaries and their dissolution by integrating different technology components into a hybrid technology, interactions and dependencies among the mechanisms of action of individual technologies must be better portrayed. It so follows, that the effective assessment of a variety of production engineering solution scenarios is possible.

The design framework of the methodology is created according to the existing technology expertise and is based on the models that describe the production technologies, their effective relationships, capabilities, and limitations. Thus, mainly models of parts, tools, auxiliary materials, and mechanisms of action of the processes are being considered. Machine-specific parameters are not recorded. In the design and application of methods, above all, man and his distinctive knowledge of cause-effect relationships must be seen as the essential factor. Therefore, the implementation of the methodology has to be adaptable to the expertise of the individual or a group of people. Methodology cannot replace the expertise of an individual; rather, through targeted information processing and knowledge in form of technical analogy solutions can be provided. Now, a system can be set to use the existing knowledge of the user as best as possible and thereby develop effective integrative hybrid solutions. By using a guided approach and providing supporting and valuable information about physical relationships and solution strategies, the user as an important part of the development process can focus on the problem at hand and spend time to use his/her individual skills. These skills, which cannot be replaced by methodology, are the creativity, the reflection, and the ability to transfer problem-solving processes and the understanding of solutions into new solutions. Here, the use of man's individual experiences are particularly valuable and should be assisted by the methodology.

5.2.4 *Results of the Methodology Development*

The methodology research for the development of hybrid manufacturing technologies is based on an approach proposed by Ulrich. It is a research methodology used as a design framework for the methodology development carried out here. In the following it is therefore necessary to determine those requirements that specify the methodology based on the empirically-defined needs and challenges encountered in an industrial context for the development of hybrid manufacturing technology. To do this, case studies about individual hybrid manufacturing technologies are used to identify general relationships and properties of hybrid technologies that represent the relevant framework and design methodology that should be employed. Furthermore, by interviewing experts from industry, potentials and challenges for the development and the practical use of hybrid manufacturing technologies are identified. These findings are used to derive the requirements placed on methodology development. From the empirically derived requirements profile of the methodology, a procedure model is initially developed as a coarse structure of the methodology. This model defines a decision logic for the application of the methodology and the targeted linking of individual methods. Building on this model, the individual steps of the methodology are explained in detail. When generating the detailed design of the methodology, methods used in the area of design and graph theory are drawn upon and adapted to the new application context. Methods are being developed that are used for capturing and processing information on known manufacturing relationships and technical solution principles. Known technology solutions are a useful information base for the application of the methodology. They are also used as analogies for the development of hybrid technologies. Use cases will show and validate the application of the methodology in an industrial context. For this purpose, practical examples are used from a variety of different technical fields of manufacturing processes in order to promote integrated applicability of the methodology.

5.2.4.1 Empirical Analysis for the Development of Hybrid Manufacturing Technologies and Their Requirements for a Development Methodology

In order to derive requirements for a methodology to develop hybrid processes, case studies of several known hybrid process technologies have been carried out. During hybridisation relevant mechanisms of action of a single process that contribute to the improved performance of a single process were analysed, and generally accepted solution principles that derive from hybrid processes were evaluated. These analyses were carried out, among other things, based on a survey pertaining to hybrid technologies and conducted by CIRP. An overview of the hybrid processes identified in this survey are shown in Fig. 5.4. Furthermore, additional hybrid processes from different process areas have been identified during these studies, and their cause-effect relationships have been analysed. The process combinations shown here are only representative for a selection of well-known hybrid technologies and must be viewed in the context of other hybrid technologies examples.

The requirements for a methodology for the development of hybrid processes have been determined to be generally applicable principles in an empirical-inductive process deriving from the properties of individual hybrid processes. Therefore, when compared to a single process, developments of hybrid processes in which the boundaries of individual technologies will be lifted, are distinguished in most cases by the following characteristics:

- Sudden improvement of the process result parameter and/or performance characteristic

Process matrix of hybrid processes		Primary Process Technology					
		Laser	EDM/ECM	Cutting	Grinding	Forming	Others
Secondary Process	Vibration		7	8	5		
	Laser	(3)	2	6		6	
	EDM/ECM		4	2	3		2
	Media-assisted			10	5		
	Cutting				1	1	
	Grinding						
	Forming					(1)	

Fig. 5.4 Identified hybrid processes as a result of a survey. (Lauwers et al. 2010)

5 Hybrid Production Systems

- Resolution of a contradiction between at least two result parameters and/or performance characteristics of a process
- Resolution of a conflict of objectives by a single non-hybrid process
- Enhancement of the application range of a single process to a new material
- Enhancement of the technology through technical system elements (tools, supplies, techniques for generating energy fields)
- Significant change or addition of mechanisms of action
- Active influence of given boundary conditions, assumed constant in an individual process, such as material parameters
- Active influence of process parameters that are the result of given conditions in definitive variables, such as forces or temperatures

When developing a methodology, these features of hybrid manufacturing processes intended for the determination of a design framework are relevant. They are the essential elements of consideration for the methodology and are used as target boundaries for individual steps during the development of hybrid manufacturing processes.

Besides the object-oriented approach and the derivation of requirements for the methodology development, an empirical-analytic study to consider the needs of potential users of the methodology took place in cooperation with project partners from the German Cluster of Excellence. For this purpose, a survey was conducted in cooperation with project partners of the Cluster of Excellence and industry representatives. The results presented in the following text support a needs-based objective of the methodology. This points to partial results of the methodology that are required in praxis to be prerequisites for a systematic development process for hybrid manufacturing technologies.

Figure 5.5 shows extensive research challenges during the development of hybrid manufacturing technologies from the perspective of the respondents. Based on this survey, the most important criterion for the widespread use of hybrid technology is the achievement of economic efficiency. Efficiency can be attained through a comparison

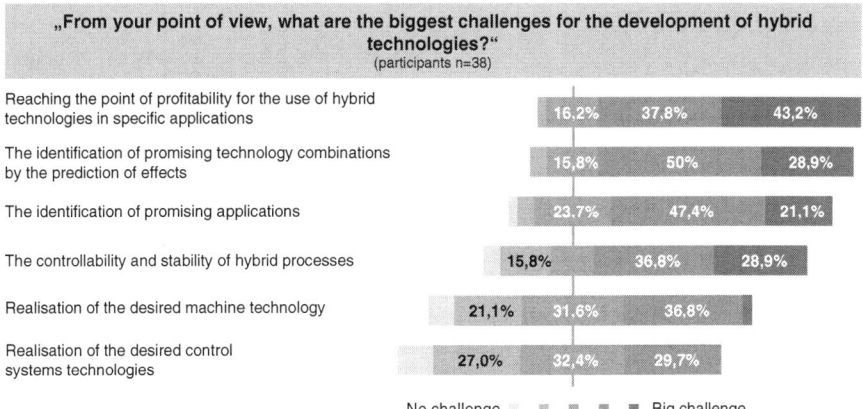

Fig. 5.5 Research challenges during the development of hybrid technologies

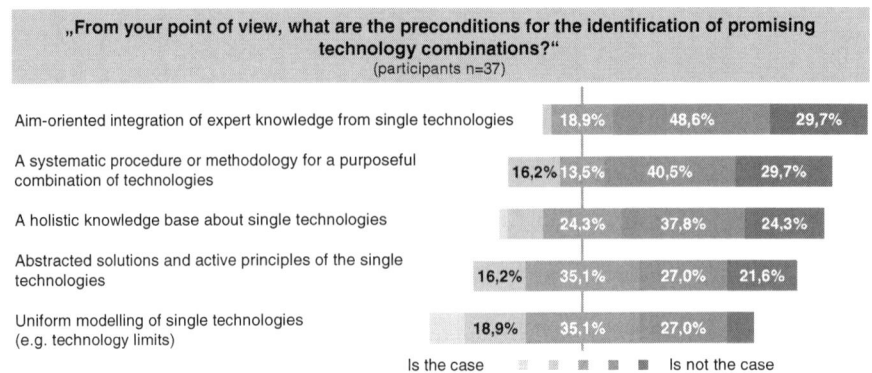

Fig. 5.6 Requirements for the development of hybrid manufacturing technologies

of operating costs and the benefits of new product features. In spite of potentially higher operating costs, the economic use of hybrid technology can be achieved by reducing other costs, such as quality assurance costs or error costs through a more stable process. From the perspective of the respondents, the main challenges are the predictability of these processes and the identification of applications considering a higher quality of planning for the use of new hybrid technologies including the economics assessment and the identification of useful combinations of technology. When implementing machines and control systems, a careful selection is necessary and a clear focus must be maintained.

Figure 5.6 shows the most important requirements for the identification of useful hybrid manufacturing technologies from the perspective of the responding industry partners. This clearly shows the need for systematic support during the development of hybrid manufacturing technologies. In addition, conditions listed here indicate the necessary intermediate results within the methodology that needs to be developed. When combining various technologies, a comprehensive knowledge base pertaining to individual technologies must be generated and expertise must be integrated systematically. This clearly emphasises the high regard for the expert knowledge of the potential user. In order to develop the technology, the knowledge of engineering experts from various fields must be integrated and is therefore an elementary effort to develop this new methodology. The expertise of the user is irreplaceable since it cannot be completely captured but can be only integrated systematically in specific application by the user. On the other hand, it also shows the value of an abstracted solution and working principles of different technologies that are currently known. This can clearly expand the user's options and offer new solution concepts. Unified modelling of individual technologies, however, is seen as less important.

From the empirical research on the needs of potential users of a methodology for developing hybrid manufacturing technologies, the following user-specific requirements for the methodology apply:

- During the design phase of a technology innovation process, the methodology is used by a technology developer or technology planner.

- The user is an expert or group of experts with specific engineering skills. In order to counteract the effect of a "psychological inertia vector", the user will be provided with generally accepted principles or analogy solutions, and a systematic approach will be made available.
- The requirements for the technological and methodological knowledge of the user to apply the methodology should be kept as straightforward as possible. The methodology should include a logic component that allows the automated implementation of the basic methodology to cope with the diversity of the information received at a later expansion stage.

Besides the user-specific requirements, there are object-specific prerequisites, such as the consideration of manufacturing technologies that must be developed with the respective methodology, e.g.:

- A key objective for the development of the methodology is to derive a new specific requirement/technological problem solution from the currently available technologies ("technology pool").
- These can be new (hybrid) technologies as well as available in hybrid or non-hybrid technology objects of the methodology.
- The cause-effect relationships of these manufacturing procedures are to be considered as an essential design object of the methodology.

5.2.4.2 Process Model of the Methodology

The process model for the methodology on how to develop hybrid production technologies are based on the process models of different design methodologies. While the definition of individual steps of this construction method is different, as a whole, they offer a guideline in the form of basic phases for a general development process. The phases of development processes have been described in Sect. 5.2.2. Here, it was shown that design processes in general, and especially for the development of hybrid manufacturing technologies, start with a single problem and therefore can be described as problem-solving processes. A problem-solving process can be generalised and grouped into three main steps (Ehrlenspiel 2009; Lindemann 2007):

- Clarification of goal and problem (analysis)
- Generation of alternative solutions (synthesis)
- Singling out of the solution (evaluation)

From the requirements identified above and based on a higher-level process model of the methodology, a general problem-solving process was derived. As format for the methodology, a process model for the application of individual methods for the further development of manufacturing technologies has been developed. This model distinguishes between a descriptive and an explanatory level of the target and problem analysis. Thus, the importance of a profound analysis of the task for the subsequent finding of solutions is emphasized. The methodology consists of individual methods and a system to link these methods. The methods are defined according to Jenke.

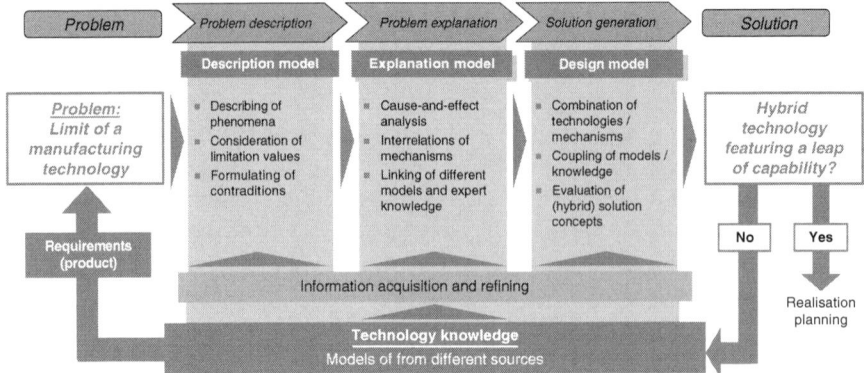

Fig. 5.7 Structure of a methodology for the development of hybrid manufacturing technologies

Here, the methods are properly planned and goal-oriented, using resources (e.g., models) in order to arrive at a result from a single input (Jenke 2007). The system for linking these various methods describes a systematic, rule-based breakdown and procedure for the use of each method. It must be possible to interpret and apply these models and their link to the problems in different manufacturing technologies (Klocke et al. 2009).

The research hypothesis, on which the design for the development of the model is based, states that a systematic methodology can be developed to overcome the limits of production technologies based on models of technology and the general approach of a problem-solving process. Furthermore, it is assumed that with an improved targeted integration of engineering knowledge of different methods, the probability of a fundamental development of manufacturing technologies can be enhanced. As an example of such developments, hybrid manufacturing processes (e.g., laser-assisted machining) are considered. Different models of problem-solving processes can be found in pertinent literature. Based on this, a general approach for solving problems was derived and used for this research to classify individual methods. The stages shown in Fig. 5.7, which are based on the problem-solving process, form a superordinate process model for linking the individual methods.

The starting point of the design model is a technology-specific problem that arises when aligning technology capabilities and market-based technology requirements for a particular product. At this point, the problem defines the technology boundary that must be considered thereafter. In the first stage of the design model, a delimitation of the system under consideration and a description of the problem takes place. In order to identify and describe the technology boundaries, descriptive models are used. During the system analysis and problem statement, the methodology reverts to explanatory models. During this stage, the cause-effect relationships and the effects of the mechanisms are analysed. Based on an analysis of the technological boundaries, the next stage technology solutions are developed and selected at the level of solution models. All stages of the design model and its individual methods are based on the provision of appropriately processed technological knowledge. This is represented by different types of models. The three stages of the design model presented here and the compilation and adaptation of process knowledge are explained in more detail in the following text.

5.2.4.3 Adaptation of Technology Knowledge as the Basis of the Methodology

The methodology research for the development of hybrid manufacturing technology is based on the hypothesis that existing technology knowledge bears a great potential for developing new technology solutions for manufacturing processes; however, this has not been used efficiently at this time. Hence, the question ensues how technological manufacturing knowledge can be efficiently captured and processed for the purpose of the systematic development of hybrid production methods. Within the context of this methodology, comprehension of knowledge implies that sources of knowledge and important data is recognised and obtained. Adapting in this context means that the essential information and the correlation of such data are reduced to the relevant characteristics of the manufacturing processes and are represented in a form that makes the knowledge for the development of other technologies available as well.

According to Probst, *knowledge* is the cognition and skills in their entirety that individuals use to solve problems. This includes both theoretical knowledge and practical everyday rules and instructions. Knowledge is based on data and information, in contrast to this, however, knowledge is usually associated with a person. It is conceived by individuals and represents their expectations about cause-effect relationships (Probst et al. 1997).

Technology is the science of the application of a technique in which raw materials and products are transformed (Borsdorf 2007). As part of this research, techniques are represented by manufacturing techniques. Therefore, the technologies referred to here are specifically identified as manufacturing technologies.

The above-mentioned definitions of *knowledge* and *technology* can be used as a single nomenclature. This follows, that *technological knowledge* is therefore a set of knowledge about the cause-effect relationships in the application of manufacturing techniques and the ability to apply that knowledge to solve production tasks and problems.

The differentiation between a manufacturing task and a manufacturing problem is essential to the methodology. A manufacturing task presents itself when the requirements for a manufacturing technique by known measures (change in process parameters, change of tool specifications, etc.) can be fulfilled with existing techniques. A production problem exists when requirements cannot be implemented with existing techniques. A manufacturing problem and the available technology knowledge together describe the starting point for a methodology that must be applied in order to develop a production technology. Hence, technological knowledge, which is imperative to describe, explain, and solve a production problem, requires a focal research object by developing hybrid technologies.

Knowledge about manufacturing technologies is provided in models as an image of a real system. With this, a model provides general statements about elements, structure, and behaviour of a considered part of reality. It is characterised by deliberately neglecting certain attributes in order to emphasise the essential model properties that are necessary for the modelling purpose (Stachowiak 1973). Validity and quality depend largely on the model's own level of detail (Tönshoff et al. 1992). In many cases, models of manufacturing processes exist as a "black box" image of the process. Here, processes will be represented in a closed system through the interdependencies of input and output variables in order to draw conclusions as to the internal structure of the system. However, most of the time, the internal structure of a manufacturing process features complex cause-effect relationships. Because of that, such models are often unsuitable for the prediction of system behaviour in order to meet changing parameter ranges or changing the boundary conditions. To analyse the system of a manufacturing process, Brinksmeier proposes a differentiation of the black box model in smaller model units. Based on this proposal, subdividing the system into its cause-effect relationships in order to increase the validity of a model is made possible. The divisions are the outcome of a cause-oriented consideration of the work result. For instance, during the manufacturing process of grinding, one can distinguish between mechanically and thermally induced effects and their impact on the material properties of a workpiece's boundary zone (Nachmani 2008). Brinksmeier breaks down the process model even further along the cause-effect chains by dividing the process model into a concatenation of transfer, cause, and effect models. This is no longer sufficient to consider only input and output variables of the process. In addition, process parameters, such as forces, temperatures, or converted energies must be considered (Brinksmeier 1991).

Figure 5.8 shows the principle of dividing a process model in inter-related sub-models. The two strategies for partitioning sub-models shown here can be combined. This leads to a network of sub-models depicting the cause-effect relationships of the entire system. With this procedure, an arbitrarily complex cause-effect model can be created. However, it only serves a purpose when a sufficient level of detail is

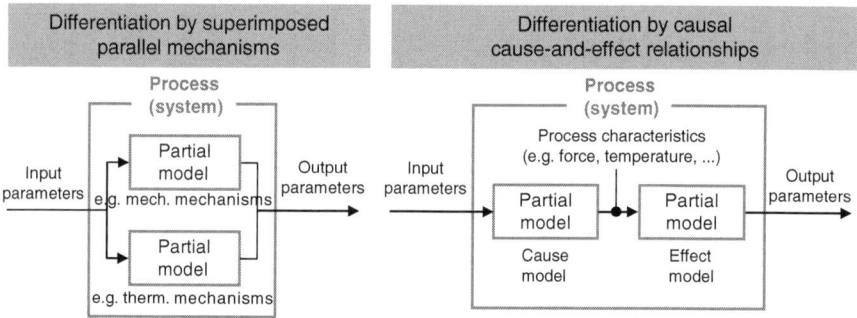

Fig. 5.8 Breakdown of a black box process model by differentiation of sub-models according to Brinksmeier (1991)

Fig. 5.9 Classification of the knowledge base for the methodology of modelling manufacturing processes

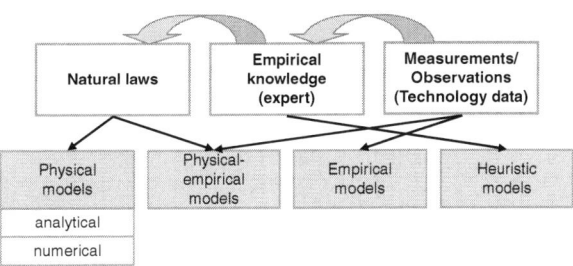

achieved for any given problem. If this is the case, partial models will be displayed and highlighted that describe the essence of the problem.

In order to attempt an analytical derivation of new technology solutions as described here, a pivotal research question must be asked: Which technological knowledge is needed for the various methods, in which forms and in what source can this knowledge be found, and how can the relevant information be made available? In order for the Cluster of Excellence to conduct its research, a frame of reference has been defined for the provision and processing of technology knowledge. The technological knowledge, which is considered the basis for a development process, has been classified for the methodology as shown in Fig. 5.9. Accordingly, the main sources for technology knowledge of individual manufacturing processes are subdivided into three groups: documented data, expertise, and universally recognised laws of nature. The laws of nature are formulated by man and are based on observations.

In order to represent the phenomena of the manufacturing process, physical models derive from the laws of nature, or a law of nature presents itself as such a physical model. Process models can be derived and formulated from data (e.g., stored in technology databases) of manufacturing processes that are already implemented or with the help of human knowledge accumulated on the basis of documented and undocumented observations. Process models of manufacturing processes are divided into heuristic, empirical, physical, and analytical models (Brinksmeier et al. 2006). This classification of process models made by Brinksmeier in relation to a grinding process is, at this point, accepted for other manufacturing processes as well.

There are physical models that are formed by a mathematical explanation of cause-effect relationships based on physical laws. The physical modelling approach distinguishes between analytical and numerical modelling. The analytical model is characterised by an exact mathematical description of real facts. If constant iterative problem solving solutions are only available partially, then the modelling is based on the numerical approach (Tönshoff et al. 1992). The physical models have an extended universal validity; however, these models are usually not present in a sufficient quantity to create a complete system of partial models of a single manufacturing process. Instead, today many process designs are built on the basis of expertise or experimental studies and empirical models derived from these. These models often have a very limited scope of application when considering a particular operating condition and cannot be transferred to other machining conditions. Heuristic models are usually formed by subjective assessments of cause-effect relationships. They are characterised by a high degree of uncertainty or ambiguity of the information. Therefore, due to the empirical and heuristic models, the behaviour of a new process is a priori, and its prediction is often blurred; however, relevant model parameters and their relationships from such models are still important for a complete model.

To form a differentiated process model under consideration of the cause-effect relationships, there are different model types as partial models available. In order to enable the integration of different modes of mechanisms for the explanation and design of a (hybrid) manufacturing process, the different types of models must be interlinked.

Known correlations between cause and effect in manufacturing processes are used for the methodology and represented in form of cause-effect chains, as described in the following chapters. These cause-effect chains resulting from the analysis of known process solutions are used in the methodology as a knowledge base for an integrated solution for finding problems in other processes. The extent of validity of the cause-effect relationships is limited, which is why the recorded and processed relationships can be only considered as analogy solutions, whose validity and applicability must be verified during actual applications. At the same time, while explaining a process' boundary that must be overcome during hybridisation, the cause-effect relationships are also modelled. Thus, by applying the methodology for the solution of a problem, its knowledge base can be extended continuously by the generated models.

5.2.4.4 Description of the Technological Boundaries

Here, the boundary of a manufacturing technology is defined as the margin of a technological capability profile, pointing towards higher levels of performance characteristics. Assuming a single performance parameter for the evaluation of a technology's ability limit, then, this parameter is the result of a multi-criteria, multi-dimensional analysis of different targets. An absolute indication of each technology's efficiencies as well as narrowing down technology-specific levels of performance parameters is to a great extent only application-specific. Therefore, it can be said that a general description of boundaries of individual technologies fails due to the complex interdependencies of the process variables to one another and the dependence of the boundary conditions. It follows that an absolute narrowing down is possible only in the area of an operating condition/working condition and under the same boundary conditions, which are usually not completely known.

This being said and with the objective in mind, in order to describe a manufacturing process, the research question must be asked: How should such a boundary condition be specified? The attempt to find a solution is carried out by starting from a manufacturing process close to the capacity limit and to execute a correlation analysis of individual result parameters and transmitting the trends of this consideration to an area of surrounding operating conditions. This is permissible under the assumption that in a work area surrounding the considered operating condition, the correlation trends of individual target parameters remain identical. The assumption can be justified in that the causes for a capacity limit in a range surrounding the operating condition have the same physical relationships. Thus, the action to change these relationships and dependencies in order to exhibit a similar change in the system performance of different operating points leads to a contiguous operating range. This implies, even if a process limit cannot be determined implicitly, it is possible to resolve this limit by adapting the same actions and starting from an assumed working point of a larger workspace, the same measures taken as for a single operating point. Thus, in view of the desired radical improvement of a production technology, it is possible to generate new technology solutions without any knowledge of the absolute performance characteristics. This may be done by providing a (diffused) description of the process limits by a correlating analysis of individual result parameters.

The key challenge during analytical modelling of technologies lies in the development of models that describe the technological capabilities in a manner that reaches at least partially across different processes. The scientific approach is based on the theory that a process-independent and uniform modelling of different manufacturing methods can be reached through the effect of individual processes caused on a single component. These effects are often described in relevant literature as manufacturing-related product characteristics (Klocke et al. 2008b).

A detailed analysis and description of tasks for which a manufacturing process is to be used are the prerequisites for a suitable selection and development of the manufacturing process. By comparing production task and technology capability, the qualifying production concept is finally obtained. This necessitates a description of the production task in the form of requirement characteristics. Not only do the requirements for the production system include economic and organizational data such as order data or time values but also in particular geometry and material properties (Fig. 5.10).

When planning a manufacturing process, especially a finishing process, the aim is to produce a component's feature in a repetitively reliable manner. In order to optimise the process to achieve this goal, as much information as possible about the manufactured workpiece and its component's features must be available. This includes, among other things, the geometry of the workpiece, which must always be taken in correlation with the permitted fit, form, and functional tolerances. In addition, and depending on the requirements, information describing the properties of the material, the component surface, and its boundary layer is necessary. Standard surface parameter values are used to describe the component surface. These parameters must describe the major properties of the component that are applicable during most applications. Furthermore, it must also be possible to include function-specific generated features in the description of the specific requirements. Material properties

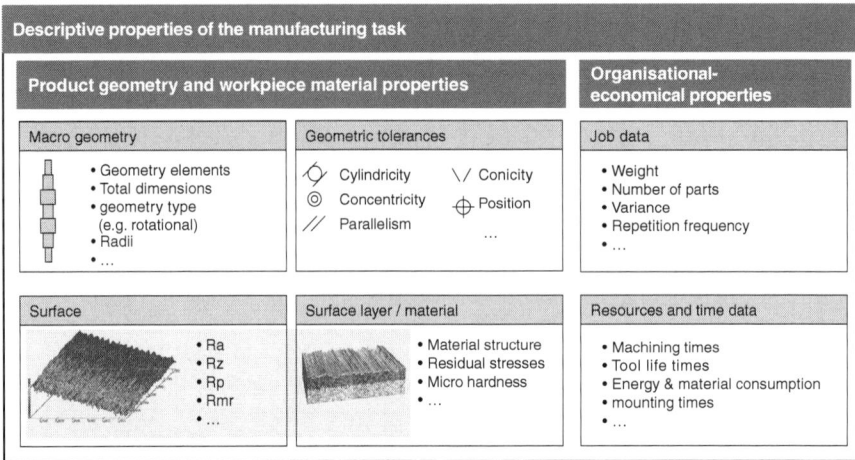

Fig. 5.10 Examples of requirement parameters for the description of a production task

and boundary layer characteristics represent additional important descriptive features of the component. These features consist, among other things, of information pertaining to hardness, residual stress, and microstructure.

In order to describe a technological boundary, a selection of relevant process and component characteristics suffices. At this point, it is assumed that based on a reduced model (e.g., component section) the considered limit of a process is represented fully and the resulting problem can be described. Simultaneously, the component's region of interest is reduced, and thus the system boundaries are defined. In this case, the machining process is viewed as a black box. Thus, the mechanisms are only implicitly considered by the effects of the process on the component. A consistent description of the cause-effect relationships is not possible by such a phenomenological examination.

Based on an assumed process operating point, the boundaries of technology can be described qualitatively. The boundaries can be expressed by contradictions, and thus converted into a problem description. A boundary can either be described by a resultant with an upper or lower limit or shown by an optimisation problem with two outcome variables that indicate opposing tendencies (while one variable improves, the other deteriorates). For further considerations, limits of a process are being considered based on a specific operating point, which subsequently reduces the complexity.

As a starting point for the development of a hybrid technology, trade-offs between different requirements on the manufacturing process are being considered. Especially in complex multi-criteria requirements of a manufacturing process, it is usually inevitable that trade-offs must be considered. For instance, a much higher wear on tools occurs during machining processes with increased feed rates and a resulting reduction in machining time. In most cases, the resulting compromises are the outcome of a cost-benefit analysis of the existing trade-off, which represents at

5 Hybrid Production Systems

			OD	1	2	3	4	5	6	7	8	9	10	11	12	13	14	15	16
Macro geometry	Geometric tolerances	1	+		Z								0	0	Z				Z
	Max. total dimensions	2	+			Z									Z		Z	Z	
	Min. Edge radius	3	−												Z				
	Cylindricity	4	+																Z
Micro geometry	Surface Roughness	5	−								Z	Z	Z	Z	Z		Z	0	
	Material contact area	6	+																
	Spin	7	−												Z				
Surface layer properties	Material structure modification	8	−										0		Z	Z		Z	
	Micro hardness modification	9	−											0	Z	Z		Z	
	Residual tensile stresses	10	−												Z	Z		Z	
	Residual compressive stresses	11	+												Z	Z		Z	
organisational-economical properties	Machining times	12	−													0		Z	
	Tool life	13	+																
	Energy consumption	14	−															0	
	Lubricant consumption	15	−																
	Mounting times	16	−																

0: Varying relationship
Z: Aim conflict
OD: Direction of optimisation

Fig. 5.11 Example of a trade-off matrix for the identification of conflicting process requirements in external cylindrical grinding operation

the considered technical level the technological boundary of a manufacturing process. During an attempt of hybridisation in which the technological boundaries are expected to expand radically, the existing trade-offs are called into question. If in the above example, both a short machining time and minimum tool wear are required; finding a solution provides the special challenge since two contradictory requirements must be solved simultaneously since a compromise cannot be considered. In most cases, it is hard to have a full grasp of all requirements and identify conflicts of interest. Therefore, it is helpful to understand the requirements systematically. In this case, a matrix is useful in which all requirements are compared with one another. Here, the dependencies are entered into the fields of the matrix. A weighting of the interaction is useful to distinguish strong from weak. In relevant literature, this type of matrices often referred to as a consistency matrix or trade-off matrix (Lindemann 2007) (Fig. 5.11).

Based on the trade-off matrix for a particular application of a manufacturing process, relevant contradictory goal criterion characteristics can be identified. In addition, correlation analyses can be used to combine and simplify target parameters and reduce the problem, which is characterised by contradictory result variables to only essential components of the result. For example, during a grinding process a problem can be reduced to the contradiction between short machining time and minimum surface roughness. By decreasing the region of interest to fewer contradictory target parameters, the complexity of the overall system is reduced during the process of the methodology and allows focusing on major reasons that limit the process. Therefore, an analysis of the minimised problem will be carried out first, with the objective to obtain as many solutions as possible. However, it must always be considered that some attempts to solve the minimised problem may also have unintentional effects on neglected result criteria, which may lead to the exclusion of these solutions. Therefore, the selection of all target parameters listed here will be used again in a later process to assess potential measures to solve the minimised problem.

5.2.4.5 Problem Analysis to Identify Hybridisation Potentials

Each manufacturing process uses specific mechanisms to transpose a workpiece from a state before machining into a state after machining. The transformation of the workpiece describes the effect of the manufacturing process. In order to explain the effect of the manufacturing process in detail, in-depth knowledge of the interactions is required. Thus, a previously purely phenomenological problem described above will be traced to the level of mechanisms of action and can be explained by previously hidden variables as well as contradictions in the cause-effect chains. To achieve this, the black box representation of the manufacturing technology must be replaced by an explanatory model, which reflects technical know-how about the effective relationships of technology. With this, the expertise of process characteristics and mechanisms of action are added to the knowledge of input and output values of the manufacturing process. From this a cause-effect diagram is drawn. Figure 5.12 shows a simplified diagram in a case in which the cause-effect relationships exist only in one direction of action. Often the manufacturing process also indicates relationships within a single level (e.g., force and temperature on level of the process characteristics). These interactions can also be shown in the model and are analysed below.

The cause-effect diagram is developed on the basis of at least one result parameter that should be improved. Only those result parameters are considered that describe

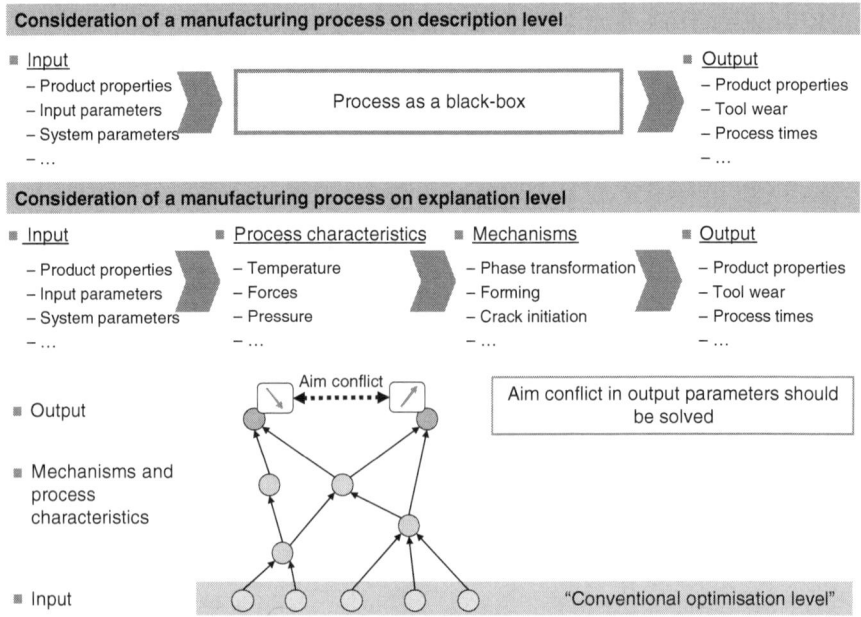

Fig. 5.12 Modelling principle for the depiction of cause-effect relationships of production methods. (Klocke et al. 2009)

the boundary of the manufacturing technology. At the level of input values, major components of the system that contain input variables for the system are identified. Finally, mechanisms of action and process parameters are identified and based on the result variables, interactions are entered. The differentiation into result variable, mechanism of action, process parameter, and input variable offers a guideline; however, compliance with this guideline is not mandatory. Theoretically, the chain of effects between outcome and process input variable can contain any number of variables. However, with the understanding of the process or with the level of detail, the number of variables and the length of the chain of effect increase. When considering a manufacturing system, the cause-effect diagram represents the individual concept of cause-effect relationships. Through the illustrations of a manufacturing process created by various people, different cause-effect diagrams can be created. Ideally, they can complement each other and can be summarised in a single diagram. However, it is also conceivable that a different understanding of processes produces very dissimilar cause-effect diagrams. In this case, contradictory understanding of process ideas by different people is revealed.

The first step in establishing the cause-effect diagram is used to identify relevant variables for the process and the effective relationships between these variables. The effective relationship between two variables can be either positive or negative and can display a strong or weak influence. The strength of the relationship can be described by weighting this relationship. However, the cause-effect relationships are not weighted in a first step, since the necessary information is often not available or very vague and can be determined only with additional effort. If the trends of the relationships are clear, it is possible to identify contradictions and explain the trade-offs. To do this, cause-effect relationships are identified the first time by adding "+" or "−" for each positive or negative effect. This is not considered for further use. However, it serves a better understanding of the cause-effect diagram when viewed by the user, especially if several people are involved in the creation and interpretation of the chart.

In addition to the graphical representation of the relationships of the manufacturing process, mathematical expressions are used. This is valid for different manufacturing processes and provides the basis for linking different physical models. Using mathematical formula of laws of physics can be used to represent very different phenomena from different scientific fields that may be compared or even linked. A mathematical formulation of the cause-effect relationships as a basic common language is important for the proposed combination of phenomena and mechanisms of hybrid processes that are to be used in different technologies of manufacturing technology. In addition, the mathematical description is a prerequisite for an algorithm that can be implemented with the assistance of a computer, and thus saves time and avoids errors in the use of development methodology. With the use of design theory, the mathematical formulations of relationships and dependencies of a system's elements are known by representing them in a matrix. They have been long established because of the following advantages of a matrix application (Köhler et al. 2008; Ulrich and Eppinger 2004):

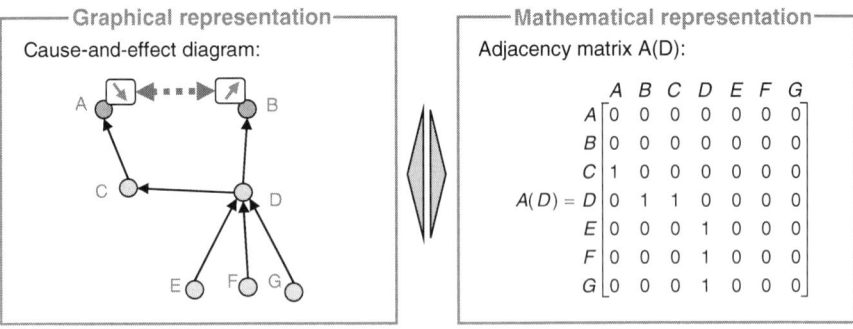

Fig. 5.13 Illustration of cause-effect relationships

- Efficient and quick system analysis
- Compact and clear presentation of comprehensive system contexts
- Simple identification of possible side effects during the search for solutions
- Resulting in high-quality design solutions
- Potential expandability, and thus integration of new system connections

For the first step of the mathematical description of the cause-effect relationship through the Attainability matrix $A(D)$ no weighting of the individual effects are made (see Fig. 5.13). In many cases, those variables by which a result can be improved without affecting a different result negatively can be identified efficiently without weighting the relationships. Such values provide a high hybridisation potential and therefore should be used first for a possible direct influence of an additional technology (hybridisation). Such a parameter is referred to as a potential hybridisation and interconnection point (HIP).

Figure 5.14 depicts a generalised example of a cause-effect diagram. It shows the principle how such potential hybridisation and interconnection points can be determined by a definite mathematical description of the cause-effect diagram and its cause-effect relationships. The following method for the mathematical definition of the dependencies between the variables of the cause-effect diagram is based on the formulation of rules of the graph theory (e.g., Diestel 2006; Harary et al. 1965; Harary 1969). For this purpose, a cause-effect diagram is viewed as a directed graph, also called a digraph. The nodes of the graph represent the parameter or variables of the system under consideration, which are represented by cause-effect relationships and related to one another by edges (arrows). To determine the variables that have a direct or indirect influence on a different variable, the attainability matrix is $R(D)$ will be calculated as follows:

$$R(D) = I + A(D) + A(D)^2 + \cdots + A(D)^{p-1} \quad (5.1)$$

I : Identity matrix
A(D) : Attainability matrix
p : Number of nodes in the digraph

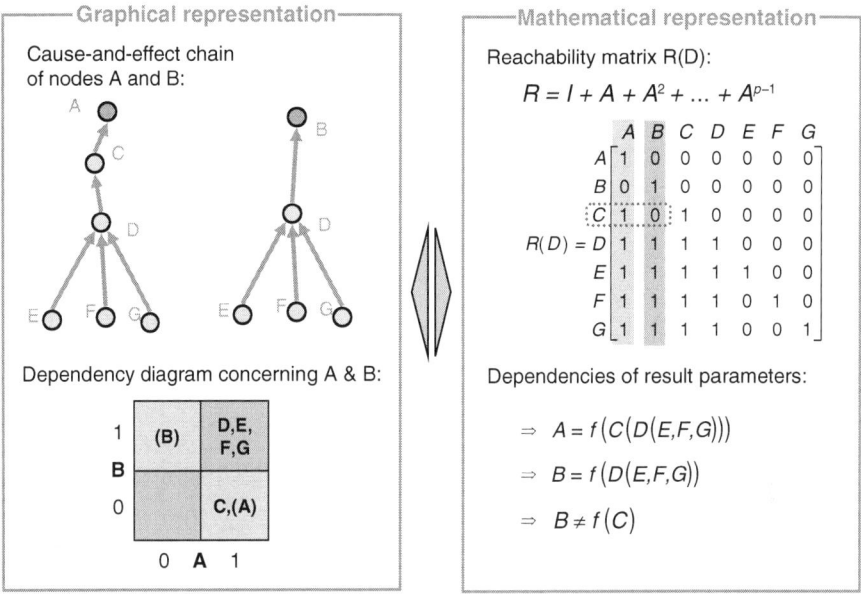

Fig. 5.14 Analysis of the influences on the contrary result variables

A definite HIP exists in the considered case. This means, one parameter was identified that influences only one of the contrary result variables in the cause-effect model. If, as shown here, at least one variable is identified as a definite HIP, weighing of many of the cause-effect relationships could be omitted and thus, the effort could be reduced. In this case, only the implications of changing the HIP will be considered further, and the affected relations of a section of the cause-effect diagram can be weighed as described below. To describe the generally accepted case, the whole cause-effect diagram will still be analysed to determine the weighting.

If no definite HIP can be found, it is no longer sufficient to neglect the weighting of the relations as done so in the first step of the cause-effect modelling. Based on the unweighted cause-effect diagram, the relationships that must be determined are weighed as follows. The necessary knowledge about the cause-effect relationships of a manufacturing technology required for weighting is included in different models and sources of knowledge (see Sect. 5.2.4.3). Depending on the level of detail, different types of mathematically formulated models (physical and empirical models) are analysed or assessed by experts (heuristic). The latter allow vague estimates of correlations between variables that are empirically based and not documented. Dealing with the vagueness generated by the heuristic models leads to a higher complexity of the proposed methodology. To clarify the next steps, one must first assume precise contexts within the methodology. The approaches to extend the methodology for dealing with vague information are based on the procedures considered in the following text. They result from a combination of the possible case-by-case analyses or by inserting the upper and lower limits of vague numbers.

For the weighting, the various cause-effect relationships are considered. Based on the Attainability matrix A, a weighed Attainability matrix A_G can be created, whereby the respective weighting replaces the logical "1" of the entries a_{ij}. In order to be able to describe the relationships by weighting in form of scalar values, the following conditions must be considered. In case of non-linear relationships, the assumed operating point of the process is determined. If the ratio of two variables is non-linear and time-variant, the operating point of the system will be considered at a specified time. According to Bjørke under these conditions the characteristics of a relationship between physical parameters in a technical system are described by the slope of the functional graph at the operating point (Bjørke 1995).

This allows the determination of the relationship between two variables in a clear form by linearisation of the functional graph at the operating point. Mathematically, this linearisation for one or more (influencing) variables can be described by a Taylor series expansion of the first order. By linearisation of mathematical functions that contain several unknowns in the operating point, a formulaic separation of variables occurs in a first-order polynomial. This separation of the influencing variables is reflected in the expression of the total differential of the mathematical function of the observed relationship:

$$df = \sum_{i=1}^{n} \frac{\partial f}{\partial x_i} \cdot dx_i \qquad (5.2)$$

In order to compare all influences of X_i on the variable Y, the weights are determined in normalised form:

$$W_{X_i}^Y = \frac{X_{i,OP}}{Y_{OP}} \left. \frac{\partial Y}{\partial X_i} \right|_{OP} \qquad (5.3)$$

$W_{X_i}^Y$: normalised weighting of the influence of variable X on variable Y in the considered operating point.
$X_{i,OP}$: value of variable X_i in considered operating point
Y_{OP} : value of variable Y in considered operating point

The normalised weighting $W_{X_i}^Y$ indicates at what relative amount $\Delta Y/Y$ ceteris paribus changes the function value y, if an influencing variable is changed by the relative amount $\Delta X/X_i$. Dependencies of variable X_i that potentially were not considered at this point will be considered in a subsequent concatenation of the various cause-effect relationships. By standardising the weights, they become dimensionless. One advantage of this approach is that different types of models can be linked. According to this, after weighting various cause-effect relationships they can be added together when they converge in a node as shown in the cause-effect diagram. Within a cause-effect chain, the influence of variables on the resulting variable is determined by multiplying the weights along the chain and "across several stations". The principle of the normalised weighting factors is used also in economics and is referred to as the elasticity between two variables (e.g.: Feess 2004). Here, weighting is called as

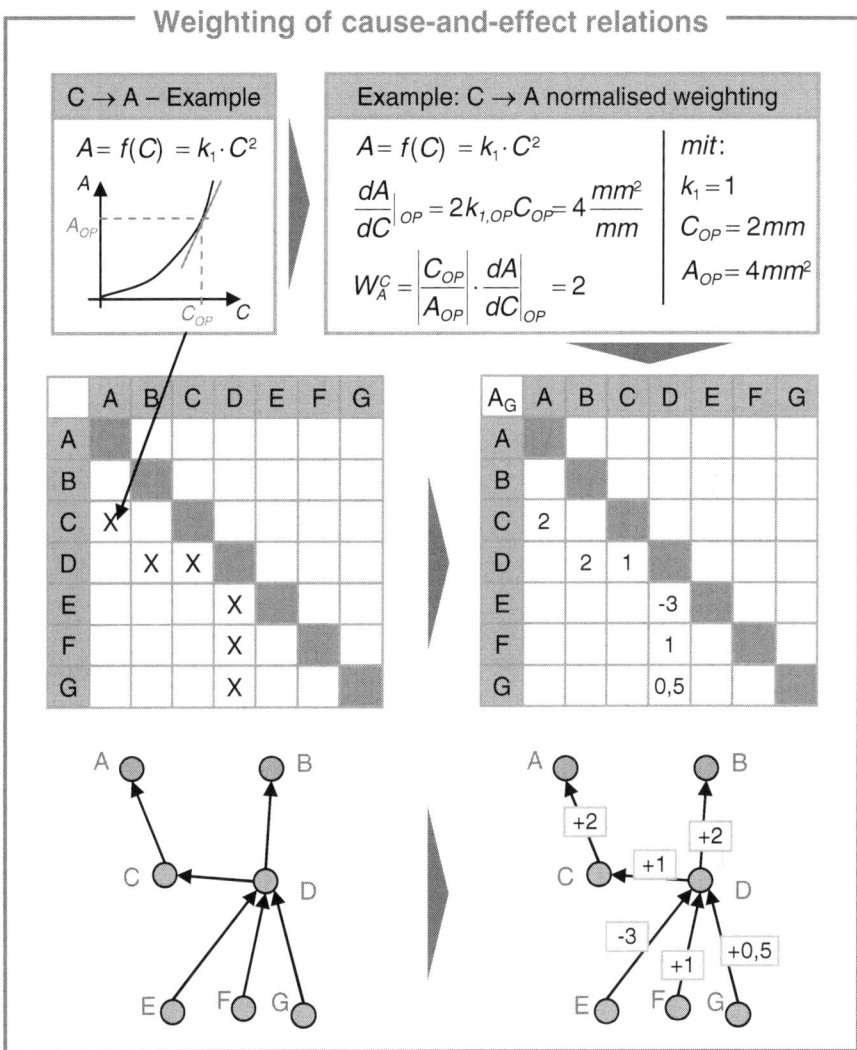

Fig. 5.15 Weighing of the cause-effect relationships after linearisation at the operating point of the process

such and not elasticity, as it serves the purposes of describing the characteristics of a weighted graph.

Figure 5.15 shows the procedure of the edge weighting. Here, the individual variables will be first related to one another in form of a mathematical formula. Subsequently, the normalised weights of the various parameters of the respective influencing variable will be determined using partial derivatives. For clarification, the derived weights can be inserted on the edges in the cause-effect diagram.

After deriving all influences and dependencies between the variables of the cause-effect model, an evaluation is carried out with respect to design options for resolving the conflict of goals under consideration. In order to evaluate influence matrices, it is not sufficient to distinguish variables according to their active and passive sums. Although this leads to a general classification of the parameters according to active variables, critical values, and passive output variables, a purposeful statement about the usability of a control variable for improving a system status cannot be made. Thus, in search of design variables, parameters that have both a large passive and active sum are frequently characterised as unsuitable. Often, however, it is precisely these variables which, in view of hybridisation as HIP, demonstrate a high level of potential. Therefore, a more discriminating view of these variables must be considered. Instead of differentiating variables only according to their active and passive sum, it is rather more important to identify those values which have a defined influence on one of the output variables to be improved, and which, by undergoing change, also eliminate a conflict of goals in the output variable.

In order to evaluate all factors within a cause-effect model (i.e., both the immediate effects of a node to adjacent nodes but also the indirect influences on a non-adjacent node), the influence matrix E is calculated from the weighted Attainability matrix A_G based on the algorithm for calculating the Attainability matrix (Fig. 5.16).

How all considered parameters influence the result variable in the cause-effect model can be read off the influence matrix. Here, element e_{ij} indicates the normalised influence of variable i on variable j. The influence of all parameters on the contrary result variables is illustrated in a diagram. In the example, the objective was to increase A without simultaneously increasing the result variable B. The quadrants, which are bisected in the diagram, mark the areas of those variables that do not contribute to the resolution of the problem. Depending on the direction of improvement of the contrary result variables, the bisecting line may run through the 1st and 3rd quadrant or through the 2nd and 4th quadrant of the coordinate system. The classical variables of the manufacturing processes considered in this methodology are generally located in the vicinity of this line. One must look for values that can be used as HIP and that lie in those quadrants through which the line does not run, or locate variables that have no influence, or when compared with contrary variables, have a negligible impact on one of the two result variables.

Not all problems can be depicted by a clearly defined cause-effect chain as a stationary view of the manufacturing process at a particular time. For example, there are problems whose cause lies in its positive feedback, whereby self-reinforcement can lead to a steady deterioration of a target parameter. One example of such effects is signs of wear on tools, or the formation of surface waviness caused by regenerative chatter. In the case of a problem caused by positive feedback effects, which can be represented by a loop in the cause-effect diagram, a discrete-time observation of the variables can be made. The condition parameters between two points in time are considered to be constant, and a change occurs abruptly at the later of the two points in time. Here, another time increment Δt is introduced as an additional system variable and can be considered an influence parameter within the cause-effect diagram. A further general view of this particular case is omitted at this point.

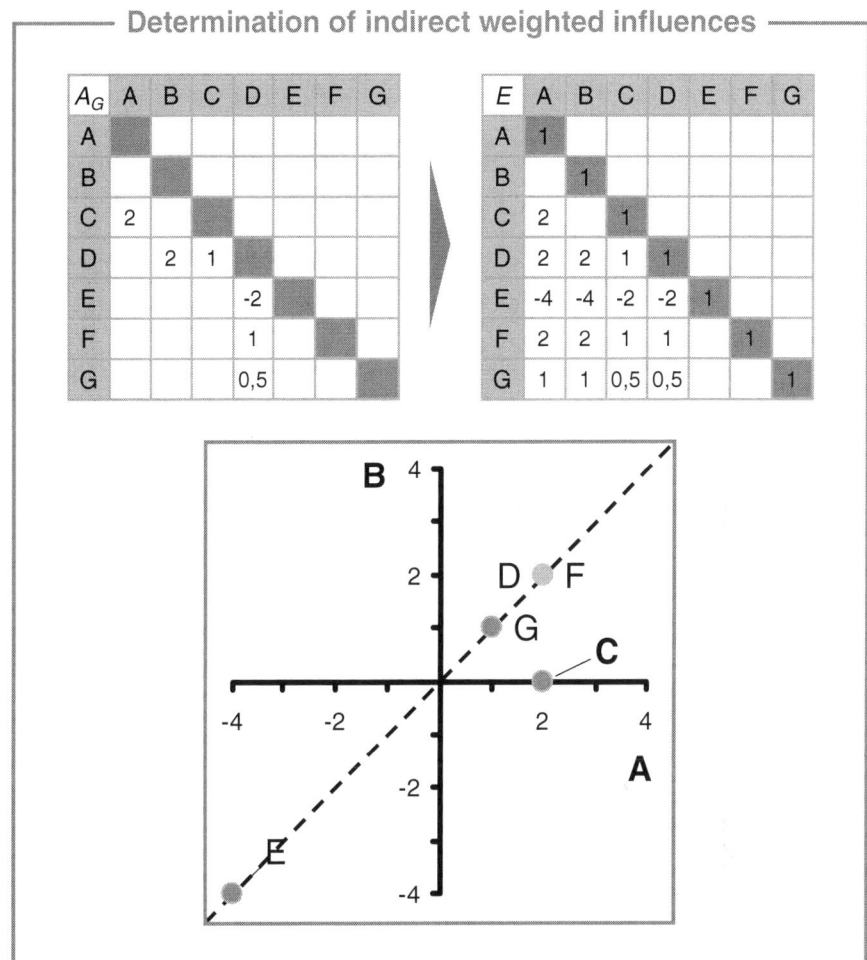

Fig. 5.16 Determining the weighted direct and indirect influences on the result variables

5.2.4.6 Design Model for the Derivation and Evaluation of Hybrid Technology Solutions

In terms of integrative (hybrid) solutions, many developers or development departments in a company lack sufficient comprehensive process knowledge that includes a variety of technologies. In addition, the necessary tools to utilise the best available expertise is also lacking. For this reason, today's interpretation of process chains as well as the evaluation and selection of manufacturing procedures is usually based on the technology planner's individual expertise and knowledge of the process, which is relatively limited (Borsdorf 2007). It is also understood that in search of solutions,

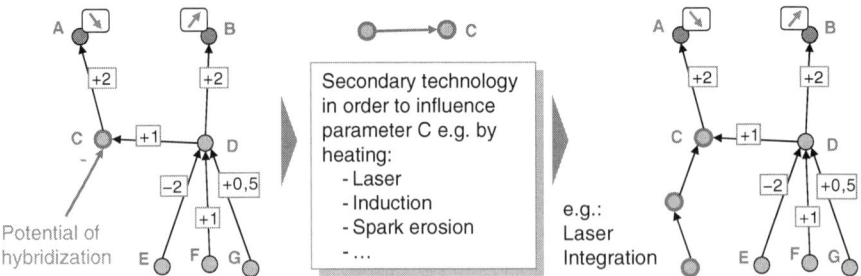

Fig. 5.17 Principle of integration of alternative technologies in the cause-effect diagram of a hybrid technology concept

man tends to prefer to pursue such approaches within realm of his/her individual skill levels. As a result, alternative solutions will be quickly rejected. By such selection, he/she minimises the risk of uncontrollable solutions or "false solutions". This tendency of man to revert always to the same solutions is described as "Inertia vector" (Orloff 2002). Contrary to this, it is a fact that innovative solutions are mostly created by the combination of different scientific disciplines (or different technology concepts) (Klein 2002).

If, based on the analysis of the cause-effect model, potential variables could be identified as HIP to positively influence the result variables, and thus recognised the expanding the technological boundaries; they can be used in the next step as new "targets" to improve fundamentally the system of the manufacturing process. Here, the possibility of influencing the process exists by connecting an additional energy source or technology that directly impacts a mechanism of action or adds a mechanism of action (Fig. 5.17).

In the case of hybrid manufacturing processes, such as laser-assisted machining of hard brittle materials, the principle displayed in this cause-effect model is implemented in praxis. By using laser to heat the workpiece material, additional energy is introduced into the process and positively influences the mechanism of action of the chip formation.

To allow individuals with limited technological knowledge to find potential technology solutions from other special fields, these solutions must be made available in a goal-oriented manner. For this purpose, much of the existing technological knowledge is systematically reduced to a collection of abstract technology descriptions. This abstract description of technology is a result of mapping each individual production technologies by means of the above-described cause and effect modelling. This set of mechanisms of action is an abstract solution, characterized by a high degree of generality and thus applicable to different technology disciplines. Within this collection of solutions, a targeted search for certain mechanisms of action can be conducted, e.g., where these mechanisms cause a desired effect on the influencing variables in the potential hybridisation and interconnection point (HIP). As part for the filing of such a collection of solutions, the concept of the design or solution catalogues has experienced a broad application in the field of engineering design

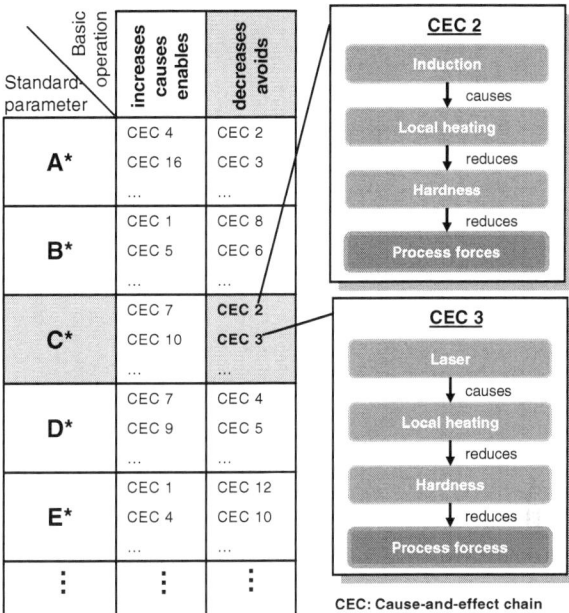

Fig. 5.18 Concept of a design catalogue for hybrid production technologies

theory. Here, the established solution catalogues for recording technical solutions are based on the principles of a morphological box (Koller 1994; Koller and Kastrup 1994). However, since there are a very large number of possible solutions, the solution catalogue does not claim to be exhaustive. Besides the content, there are formal differences in design catalogues. The formal design distinguishes between one-dimensional, two-dimensional, and three-dimensional sub-structures. The most common is the two-dimensional sub-structure. Here, the sub-structure consists of a matrix that contains the input as rows and output as columns. By using the catalogues, principle solutions can be applied for new problems and converted into specific solutions.

Assuming the improvement potentials of conventional optimisation measures are almost exhausted, the manufacturing process may be improved dramatically by adding or changing mechanisms of action. In terms of hybrid production technology, energy and/or material of an external source are integrated directly affecting the hybridisation and interconnection point, that is represented in the above-mentioned example by the variable C. To provide solution principles from other manufacturing technologies and their implementation as analogies, the concept of the morphological box is used based on methodologies from the engineering design theory. Figure 5.18 shows the concept of a design catalogue for the compilation and provision of cause-effect chains as potential solution principles for a hybrid production technology and a specific effect on the previously identified hybridisation and interconnection point.

Based on the technology-specific or non-specifically formulated parameter of the HIP, which is to be affected directly and specifically, one or more of the suitable default parameters of the morphological box are identified (in the example shown, standard parameter C*, which is represented here by way of example as a process of force). In light of the objective to influence the selected default parameters – in this example, to reduce the parameter – other production-technological application solutions are displayed in form of cause-effect chains, where the size of C* is effectively reduced. The general form of the solution presentation looks like this:

- Naming the causal technology or energy source of an effect
- Provision of a known (hybrid) analogous process in which the proposed effect has been realised
- Explanation of the effect by a cause-effect chain

Thus, in order to reduce process forces, an analogy solution is provided that can be formulated accordingly as "Heating through the use of a laser beam during laser-assisted turning with the following cause-effect chain: Laser causes local heating; this reduces material hardness, which in turn reduces process forces".

Due to the combination of cause-effect chains provided in the solution catalogue, the result of the solution search process generates many answers for the current problem. In addition, inspired by the analogies provided, the user will generate more solutions as a result of the creative process. During the first step, the cause-effect chains provided any additional solution concepts, introduced by the user by applying the analogy of applications, that are desired in large quantity. Even, and particularly, those solution concepts that appear to be unworkable at the beginning, may lead to new feasible technology solutions by modifying these concepts. Therefore, during several stages of a selection process, the next step is to reduce a large number of solution concepts efficiently to a smaller number of promising solutions. The principle of this selection process is represented by the model of the idea funnel by Eversheim (2003). A successive selection of ideas based on this model was already widely used in industry and research (Gausemeier and Berger 2004). The objective here is to select from a large number of ideas those ideas that promise to be particularly successful from a technical and economic point of view.

The first step to reduce the number of solutions consists of a plausibility check. The reduction of the solutions is done in the first step of the plausibility check by examining the concepts of the necessary conditions. A necessary prerequisite is a condition without which a desirable effect of the hybrid process does not occur. Colloquially, a necessary condition is also called "knock-out" requirement. By verifying the necessary conditions, a rapid reduction of the potential solutions is achieved with little effort.

In the methods considered here are special cases of a development process for (hybrid) production technologies in which the integration of mechanisms of action should lead to an improved process. This assumes that during the necessary temporal and spatial resolution the effect of an additional process is superimposed onto the mechanisms of the process that needs to be improved. Here, a technology added to a process that should be hybridised shall take effect at specific locations and during a particular period of time. Furthermore, the suitability of the potential solution concepts for the considered material must be checked. Selection criteria and corresponding questions on the verification of the relevant criteria are therefore:

- Material: Is the proposed additional technology applicable as a single technology or as a hybrid technology for the material that needs to be machined?
- Spatial and temporal expansion of an effect: Is superimposing and thus the desired influence of mechanisms of action/cause-effect chains feasible within the temporal and spatial aspects when using the proposed technology?

To verify the above criteria, one can refer to technology data sheets about effect carrier technology, in which relevant information is provided. The material suitability data and the spatial and temporal resolution potentials of a technology's effect are based on application examples of hybrid and non-hybrid production processes. However, in special cases, the impact of a technology on a problem must always be judged separately by the user.

By checking the criteria mentioned above, the functionality of a solution concept cannot be verified; only the solution space can be reduced by the exclusion criteria. One advantage of the plausibility check is that it can be done with minimal effort. The disadvantage is that less obvious reasons for exclusion within their framework are not detectable. Therefore, the following steps to evaluate the hybrid solutions is aimed at identifying and estimating non-obvious effects that can refute the desired impact of solutions.

During this second evaluation stage, those solution concepts that have been rated by the verification of the exclusion criteria as positive or neutral will be subject to detailed tests in terms of their added value as well as the prognosis of the effects' reliability. Alongside the desired impact of technology integration, the unwanted effects are also considered since in addition to the desired effect for the resolution of conflict, a technical solution to unwanted interactions may also lead to undesirable effects. Therefore, the reliability of a new hybrid process must be assessed under aspects of potentially undesirable effects in order to press ahead with the development of the solution. Only after the risk of adverse effects was determined, an evaluation of the benefit/risk ratio and a decision to implement the solutions can take place.

The benefit and the risk of adverse effects of a solution concept are considered as the reference concept in relation to the conventional method, which was initially considered as the description of the problem. Basically, the solution concepts must meet the following requirements:

- The solution can meet the desired functionality in the assumed specific case.
- Under changing boundary conditions and in a defined work area, the process can meet the required functionality.

These two requirements form the structure of a further assessment. While in the case of the first requirement influences and conditions are assumed to be at a stationary state, these requirements are considered to be variable at the second assumption in order to estimate the robustness of the process and thus the validity of the generated solution.

5.2.4.7 Example: Enhancing the Shearing Process

The following example analyses the laser-assisted shearing of high-strength sheet metal materials, based on the cause-effect relationships and used as an element of the morphological box. In sheet metal processing, conventional shear cutting (blanking) is the most commonly used method for trimming, punching, and cutting of sheet metal parts. The primary application areas are the automotive, medical, and home appliances industries. It is characterised, among other things, by its very high output rate. However, in contrast to fine blanking, conventional shear cutting is always subject to a break phase at the generated cutting surface.

The resulting variables for assessing the quality characteristics of the components are shown in Fig. 5.19. For roll-overs and burrs the height and width are measured; for smooth sheared zone, the height; and for the break, the height and depth are measured. The ratio of smooth sheared height to sheet metal thickness is referred to as smooth sheared portion; the angle between rupture surface and sheet metal plane is called the break angle. In addition, the roughness of the smooth sheared surface and the deflection of the component are determined (Klocke and König 2006; Lange 1990; Schmidt 2006; Spur and Stöferle 1985; VDI 1994).

The undesirable rupture zone is caused by exceeding the material-dependent shear strength. Thus, besides the smooth sheared zone, the separation surface of the workpiece always exhibits a rupture zone, which is generally undesirable and therefore often requires post-machining. By reducing the shearing gap, the smooth sheared portion can be increased; however, the tool will also wear faster. It is therefore necessary to find always a compromise between the quality of the cutting surface and the wear of the tool (Emonts 2010).

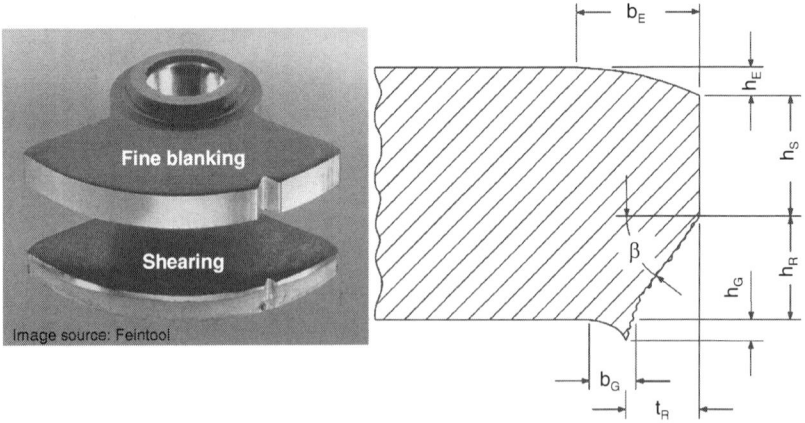

h_E Roll-over height b_E Roll-over width h_S Smooth sheared zone height h_R Rupture zone height
t_R Crack depth h_G Burr height b_G Burr width β Break angle

Fig. 5.19 Quality characteristics of the cutting surface of shear-cut workpieces. (VDI 1994)

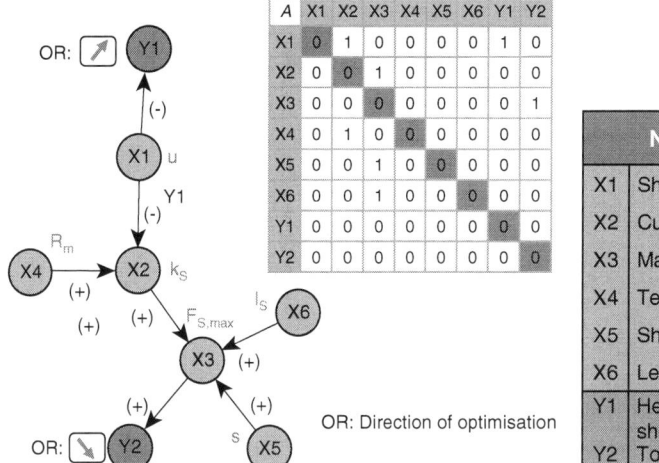

Fig. 5.20 Cause-effect relationships for the contrary variables of smooth sheared zone and tool wear during the shearing process

From these findings it therefore follows that the next step is to examine the contradiction between an increase of the smooth sheared surface and a reduction of tool wear. Starting with the contrary result variables and based on the knowledge of cause-effect relationships, the cause-effect diagram (see Fig. 5.20) and the resultant attainability matrix A was derived. In order to illustrate the contradiction of assumed tendencies of various cause-effect relationships, prefixes are shown in the illustration. In addition to the variables considered here, other system elements and characteristics influence the result variables. However, it is expected that significant correlations of the problem and the resulting solutions can be described independently and validated later for a more comprehensive system approach. In order to simplify the matter, the effects of lubrication, the surface quality of the tool and the edge rounding of the tool, among other things, were omitted.

The following relationships are the results of a system analysis of the described problem. The results are based on documented models or were derived from estimates of people surveyed. They are the basis for both the design shown in Fig. 5.20 of the cause-effect model, as well as the steps of the following methodology.

In the model of the shear cutting process adopted here, it is assumed that the tool wear depends mainly on the maximum cutting force $F_{S,max}$, which occurs during the process. First, the tool wear increases progressively over the cutting power and subsequently continues in nearly a straight line (Klocke and König 2006). By using the following equation, the maximum cutting force $F_{S,max}$ can be determined with sufficient accuracy (Klocke and König 2006):

$$F_{S,\max} = s \times l_s \times k_s \qquad (5.4)$$

s : sheet metal thickness
l_S : length of cutting line
k_S : cutting resistance

It therefore follows that the cutting force depends largely on the thickness s, the cutting length l_s and the cutting resistance from k_s. The influences of these variables can therefore be assumed to be approximately linear and correspond to a normalised weighting of $G_i^{X3} = 1$.

According to Oehler et al., in the case of closed lines for the shearing gap width u the following relationship applies (Oehler and Kaiser 1993):

$$u = c \times s \times \sqrt{\tau_B} \tag{5.5}$$

s : sheet metal thickness
τ_B : shear resistance
c : factor as a function of the targeted cutting surface quality

Factor c is an empirically derived value. In order to achieve a good cutting surface quality, it is assumed that factor $c = 0.005$ and $c = 0.035$ for lower cutting forces. The factor is not a determinant and is therefore not shown in the cause-effect diagram (Klocke and König 2006). For the shear strength τ_B the following approximation is assumed:

$$\tau_B \approx 0.8 \times R_m \tag{5.6}$$

R_m : tensile strength

In order to generate the cause-effect diagram based on the given formula for the width of the shearing gap u, it may be concluded erroneously that it has a direct physical relation to the thickness s of the sheet metal plate and the shear resistance τ_B. To ensure a physical cause-effect relationship, it must be checked whether the shearing gap width u can be changed independently of the parameters s and τ_B. This being the case, the said formula does not represent a physical cause-effect relationship with the shearing gap width u. Instead, it provides a recommendation for the determination of the shearing gap width u as a control variable of the process. This allows an operating point for the shearing gap can be determined; however, there will be no edges between s and u or τ_B and u indicated in the cause-effect diagram.

To create the cause-effect diagram, the following information was recorded:

- It is assumed that with increasing tensile strength R_m of the material, the cutting resistance k_S increases linearly ($k_S \approx 0.6 \ldots 0.9\ R_m$).
- Another important factor influencing the cutting resistance is the shearing gap width u. The cutting resistance k_S decreases gradually with increasing shearing gap width u.
- An influence of the shearing gap width u on the smooth sheared portion was identified and estimated to be gradually decreasing.
- The influence of the maximum cutting force $F_{S,max}$ on the tool wear was assumed to be progressively increasing.

From the relationships mentioned above, with the aid of the cause-effect diagram and the weighted cause-effect relationships, the influences on the result variables of the height of smooth sheared zone (Y1) and tool wear (Y2) shown in Fig. 5.21 are

5 Hybrid Production Systems

A_G	X1	X2	X3	X4	X5	X6	Y1	Y2
X1	0	-1; -0,5	0	0	0	0	-1; -0,5	0
X2	0	0	1	0	0	0	0	0
X3	0	0	0	0	0	0	0	1; 2
X4	0	1	0	0	0	0	0	0
X5	0	0	1	0	0	0	0	0
X6	0	0	1	0	0	0	0	0
Y1	0	0	0	0	0	0	0	0
Y2	0	0	0	0	0	0	0	0

Fig. 5.21 Analysis of the influences on the contrary result variables

determined. It is the aim of the hybridisation of this process to reduce the wear of tools without a reduction of the smooth sheared portion or to increase the smooth sheared portion without increasing the tool wear.

For the variables X1 to X6 impact areas have been identified that result from vague estimates of the weights of individual cause-effect relationships. The estimates are expressed for individual weights by a extreme value analysis shown in Fig. 5.21. As influencing parameters for the resolution of the contradiction between the result variables Y1 and Y2, the variables X1 to X5 were identified. The sheet metal thickness and the average cutting line length may be excluded as variable sizes since they are assumed to be fixed requirements for the process. Therefore, the cutting force $F_{S,max}$, the cutting resistance k_s, and the tensile strength R_m can be considered as possible influencing variables for the resolution of the contradiction.

A reduction in the tensile strength R_m is therefore a potential target to either increase the smooth sheared portion or to reduce tool wear. Figure 5.22 shows the

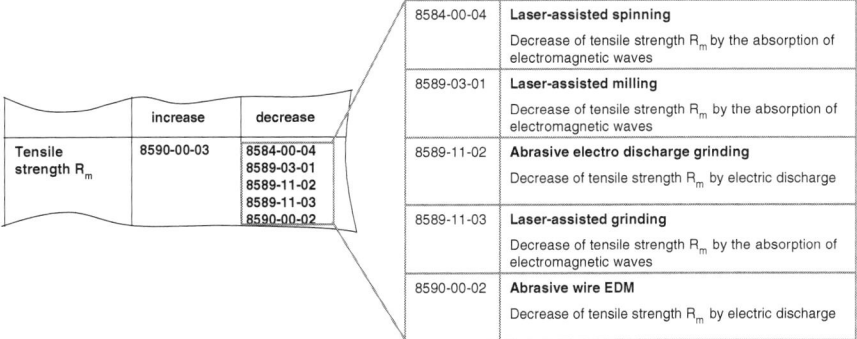

Fig. 5.22 Excerpt from the morphological box on developing hybrid manufacturing processes

corresponding excerpt for a reduction in the tensile strength of the developed morphological box. The morphological box offers known solutions from a wide range of applications of (hybrid) production technologies in which the tensile strength was reduced. These serve as analogies for solutions of the problem under consideration, thus providing the user with concrete examples of solutions to aid the conceptual design of a concrete solution. The assessment of the transferability of solutions needs to be done in relation to the particular problem and, therefore, cannot be integrated into the morphological box as information. It can be seen that a reduction in tensile strength R_m already has led to an increase in performance in other hybrid manufacturing technologies. When using hybrid electrical discharge grinding (Koshy et al. 1995) and abrasive wire-cut EDM (Menzies and Koshy 2008), the abrasive energy input of the grinding process is connected with the electrical discharge energy input of the EDM process and thus achieves an improvement in the removal rates and a reduction in surface damage. Alternatively, during laser-assisted spinning (Mäntyjärvi et al. 2007), laser-assisted milling, and laser-assisted grinding (Bausch and Groll 2003), an additional heat energy using a laser beam was included in the process. As a result of the external heat input, a significant increase in the ductility of the material occurs. The activation of migration of dislocation induced by the thermal energy also leads to an increased deformability. At the same time the tensile strength R_m is reduced by the above-mentioned effects. Thus, in sheet metal forming, a significant reduction in the required process forces is achieved (Geiger and Merklein 2007; Groche et al. 2003). As a further advantage the reduced spring-back behaviour caused by the addition of heat must also be mentioned (Neugebauer et al. 2009; Yanagimoto and Oyamada 2005).

During the pre-selection, it was found that an input of electrical discharge heat for the shearing process is not effective, since this process is also associated with material removal. This is not desired during the shearing process. However, the integration of a laser as a heat source could be considered a purposeful solution for hybridisation of the process. Heating of the sheet metal material during the hybrid process can be achieved by various physical principles. The heating principle must be able to allow rapid heating to meet the cycle times and integration into the process chain. Potential heating principles are induction, conduction, convection, thermal radiation (e.g., infra-red radiation), resistance heating, and laser heating (Groche et al. 2003). Ultrasonic waves offer another heating alternative. Studies of the use of ultrasound in stamping processes showed that the integration of powerful sonotrodes is critical (Peng et al. 2004a,b). There are two heating strategies. On the one hand, heating of the workpiece outside the tool and on the other hand, heating from within the tool. Since not all heating principles are equally suitable for a global or local heating, this characteristic represents another distinctive feature. Moreover, the temperature control is used as an additional criterion. Figure 5.23 shows a list to assess the different heating principles (Groche et al. 2003; Neugebauer et al. 2006).

By using inductive heating, a rapid temperature increase in certain areas of sheet metal can be achieved. The heated area is determined by the inductor's geometry. However, the efficiency is low and the concentration of the specific heat in the desired plate regions is difficult to achieve (Groche et al. 2003; Peng et al. 2004a). Due to the

Fig. 5.23 Evaluation of physical principles of warming according to Groche et al. (2003)

	Conductive	Convective	Inductive	Therm. radiation (IR)	Flame heating	Laser heating	Resistance heating
+ appropriate ± possible − inappropriate							
Local heating	+	−	+	±	±	+	±
Global heating	+	+	±	+	+	±	+
Temperature control	+	±	±	+	±	+	±

low electrical resistance of the material, resistance heating of metal sheets requires a large amount of current, which is reflected in the technical effort needed. With this concept of heating, and due to the high degree of efficiency, a quick and energy-efficient heating can be achieved (Geyer 2008; Groche et al. 2003; Mori et al. 2007). A high rate of heat that flows at short heating times can also be achieved by using conductive heating. Conductive heat is highly energy-efficient and can be integrated into most tooling concepts. The heating area is defined here via the heated contact element and is therefore limited (Groche et al. 2003; Holtkamp and Gillner 2008). By using convective heating in an oven, for example, and based on fixed cycle times, processes can be automated easily. It is a comparatively inexpensive method for heating outside of the tool. However, local heating does not take place but instead a continuous increase in temperature over the entire sheet material (Groche et al. 2003; Livatyali and Terziakin 2005).

The heating principle must be able to allow rapid heating to meet the cycle times and integration into the process chain. When heating by laser, radiation can be controlled geometrically accurately. As a function of laser power and beam parameters, and in dependency of the absorption coefficient, a high heat input can be provided. By using optical elements, the laser beam can be guided along the workpiece. By adjusting the beam guides and optics, the smallest areas can be heated up. With the use of laser scanners, it is possible to produce a homogeneous temperature field in defined areas of the workpiece within a short period (Groche et al. 2003; Holtkamp and Gillner 2008; Keskitalo et al. 2007; Kratky 2009; Peng et al. 2004b).

In initial experiments Emonts showed that by applying laser-induced softening of the sheet metal material in the shear zone, the cutting surface qualities can be improved and the existing process forces can be significantly reduced (Emonts 2010), see Fig. 5.24.

The additional heat energy does not only influence the properties of the sheet metal material during the forming process. After exceeding a material-specific temperature, a significant influence on the mechanical properties and structure of the manufactured components does also exist. As a function of the heating temperature and the cooling characteristic, there is a tempering effect and a risk of loss of the high strength of the starting material in the manufactured components (Keskitalo

Fig. 5.24 Experimental studies on laser-assisted shearing. (Emonts 2010)

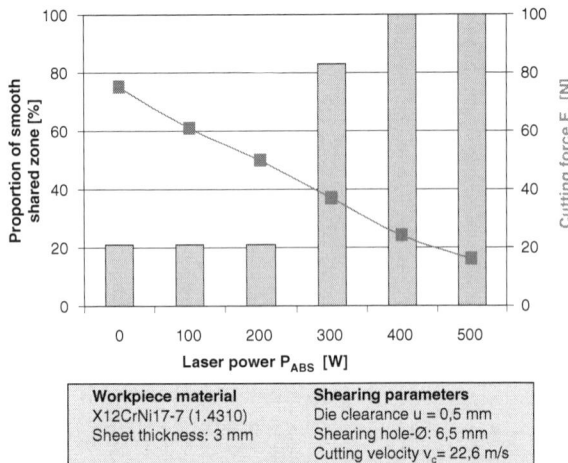

et al. 2007; Weisheit et al. 2005a,b). The manufacture of parts with high surface quality is also more difficult with increasing temperature during forming. The increased formation of mill scale causes the deterioration of the surfaces and leads to poorer tribological conditions (Erhardt et al. 1999; Livatyali and Terziakin 2005; Mori et al. 2008; Neugebauer et al. 2006). Furthermore, during the metal forming processes of heated sheet metal materials, the quality characteristics of the cutting surface are adversely affected by oscillation and warping (Behrens and Olle 2008). These possibly unintended effects of the hybrid process must be studied further in experimental trials.

5.2.5 Future Research Topics

The development and application of the evolved methodology has shown that success is based on existing knowledge and documented models of production technologies, and developments can be systematically supported. Similarly, it has been shown that, for the development of hybrid process solutions for both cross-technology modelling and model links, as well as a deep understanding of the various relationships of cause and effect in each specific manufacturing processes are necessary. From the completed work on methodology development, different requirements for future research topics have evolved. These can be divided into

- the development of technological understanding through intense studies and modelling of mechanisms of hybrid and non-hybrid manufacturing technologies
- the expansion of the application area of the developed methodology and the integration of the work completed with other methods.

5.2.5.1 Expansion of Technological Understanding and Modelling of Mechanisms of Action

While the methodology is applied in specific technology examples, existing knowledge gaps are being uncovered during the analysis of cause-effect relationships. The missing or vague knowledge of system interactions within the system results in a prediction about process behaviour that in many cases cannot occur at all or can only be inaccurate. The added value of the method increases with its scope and the predictability of effects and the cause-effect relationships of different technologies. This results in immediate challenges for the expansion of the underlying knowledge base of the methodology and associated with that, the widespread use of this methodology. The expansion of the stored knowledge base can be done in two fundamentally different strategies. Firstly, this means, that fundamental models must be researched and formulated based on cause-effect relationships of individual technologies in specific applications. Physical-technological dependencies of variables of the process must be examined for new application areas (e.g., by material development), not yet sufficiently explored parameter ranges and boundary conditions (e.g., different temperature ranges). Secondly, the objective should be to examine the transferability of application-specific determined models and to expand the scope of the models. The expansion of models of various production processes onto areas of modified boundary conditions is also relevant in terms of mapping the hybrid processes. Thus, during analogy research activities valuable information about the material's behaviour under high temperatures can be gained allowing a better predictability of laser-supported forming or machining processes. For these purposes test methods must be developed or used that allow efficient data acquisition for the development of manufacturing processes or permit the development of new application areas for individual production technologies.

5.2.5.2 Expansion of the Application Area and Integration of the Methodology

The objective of the methodology developed here was an integration of different mechanisms of action and the development of manufacturing technologies in order to generate new (hybrid) production technologies. The methods developed and modelling approaches considered for this purpose can be a helpful tool for other purposes, e.g., when analysing the underlying cause-effect relationships of manufacturing processes. While here the focus was the consideration of individual process steps in the value chain, the link between cause-effect relationships of individual manufacturing steps for the mapping of a technology chain is conceivable. Thus, the production history of a component can be mapped in order to identify specific interactions between the processes. One application area is the design and evaluation of process chains, while taking into account the interactions between sequentially related manufacturing process steps. This can effectively complement existing methods for the design and evaluation of alternative process chains and can be used to optimise the entire production chain.

When considering new or substantially enhanced technologies, it can be noticed that their industrial application is not or only slowly implemented in many cases. Hybrid production processes are a recent example of this. The main reason is that basic, significant changes, such as the use of new manufacturing technologies, cannot be performed due to lack of information. The technological impact on component properties and their opportunities and/or risks often cannot be predicted adequately. The technological interactions between processes of the technology chain and the impact on the component's properties must therefore be mapped better and evaluated. The valuation of not yet established manufacturing technologies within a company and its interaction with other process chains must therefore be compared to the established practice in the process chain. Sufficiently accurate estimations of risks and potentials for a change within the current production technology must be provided. Therefore, a methodology must be developed that, on the one hand, allows the cause-effect diagrams of the individual sequential processes to link with one another and, on the other hand, maps the influence of the entire process chain onto the component's characteristics (production history). In combination with material models, the lack of information can be removed this way, and the industrial use of new technologies can be promoted.

So far, the evaluation of hybrid technology solutions has been mainly considered in technological and to a lesser extent in economic terms. In order to evaluate the cost, however, the considering aspects must be extended to the hybrid machine concepts and their costs. The possible applications of such machines are strongly correlated with the product-side requirements (lot size, variance, etc.). Depending on this requirement, boundary areas unfold where the use of hybrid technologies under economical terms and in comparison to conventional machining is beneficial. This boundary criteria must be identified, described, and their dependencies to one another must be evaluated. So far, little is known about this boundary criteria and the research should be a subject of future work on hybrid manufacturing technologies.

5.3 Shortening Process Chains for the Production of Plastic/Metal Hybrids Using Innovative Primary Forming and Joining Processes

Kirsten Bobzin, Andreas Bührig-Polazcek, Edmund Haberstroh, Walter Michaeli, Reinhart Poprawe, Uwe Reisgen, Dustin Flock, Oliver Grönlund, Matthias Jakob, Pia Kutschmann, Andreas Neuß, Andreas Rösner, Sven Scheik, Markus Schleser and Johannes Wunderle

5.3.1 Abstract

Growing requirements for increasingly complex components can often only be met by combining different materials. The pairing of plastics and metals is of particular interest due to their different, but complementary characteristics. Technological advances in plastic and metal processing have also opened up development perspectives for new hybrid production processes in recent years. Even though various processes can be used to produce plastic/metal composite parts today, disadvantages of the methods currently available include the number of processing steps required and limitations in terms of productivity and attainable component complexity.

To overcome these disadvantages the development of an innovative combination of plastic injection moulding and metal die casting in the form of a hybrid primary forming process for the manufacturing of plastic/metal composite components in a single mould and on a single machine is required. This new process chain should help to increase productivity by reducing the number of manufacturing steps. In addition, plastic/metal composite parts should be produced using thermal joining processes. As an alternative to current assembly processes (e.g. adhesive bonding) this process also offers great potential for shortening the process chain. In addition, this process can be characterised by a flexible adaption to different joining geometries.

The main focus of development particularly concentrates on a decisive contribution to the understanding of the process- and material-specific influencing variables on the quality of hybrid components produced by primary forming or joining. In this case the bond strength and the bond mechanisms between metal and plastic are at the centre. In addition the applicability of the hybrid production technologies should be presented regarding the manufacturing of industry-oriented components on the basis of demonstrators.

These developments make it possible to achieve high bond strengths in selected combinations of plastic and metal and to simultaneously shorten the process chains significantly. Using these new hybrid production technologies it is possible to produce plastic/metal hybrid components of highest diversity. This results in opening up new application fields for plastic/metal parts and to consolidating existing ones. The products range from compact, highly-integrated electronic components to high-volume reinforced structural components with outstanding mechanical properties.

5.3.2 State of the Art

Metal and plastic are traditionally competitors in many industrial applications. These days, however, the growing demands on the functionality and complexity of components cannot often be met by just one material. Therefore hybrid technologies have been gaining in importance in recent years. Hybrid technologies combine the benefits of both plastics and metal materials in a single part.

Hybrid material combinations such as plastic and metal composites already play an important role in industry. Particularly in automotive manufacture, joining different types of materials presents an enormous challenge. New legal regulations e.g. reduction of CO_2 emissions, are forcing the manufacturer to a further reduction of consumption and exhaust gases. This can primarily be achieved by reducing the weight of vehicles. In this context the trend towards lightweight construction requires reconsideration of the materials selection. A mix of materials is increasingly employed in industry where the single materials are precisely used where their properties provide the maximum possible technical and commercial benefits. The use of different materials which can be adapted to local stresses should open up new ways for further optimisation. However, until now, plastics and metals have often found themselves in competing positions. While plastics are particularly characterised by their low weight, low costs and almost unlimited shaping potential, metals withstand significantly higher mechanical stresses due to their mechanical properties. The hybridisation of components however combines the material-specific advantages of different materials. This leads to components which are simultaneously light and stiff. Furthermore the integration of different functions into one part can improve the production process by saving assembly stages and shortening the production chain. Such composite materials are primarily of interest for applications in the automotive and air travel industry.

5.3.2.1 Processes for Combining Plastic and Metal

To date, various processes have been developed to manufacture plastic/metal components. We can basically divide the manufacture of plastic and metal components into post-mould assembly processes and in-mould assembly processes. Traditionally both plastic and metal materials are combined to a hybrid part by joining in a multi-stage post-mould assembly process. In the joining process the individual raw materials are first transformed into preforms/semi-finished products in separate production steps. In a subsequent assembly or joining process, these are joined for example by clamping or screwing. Besides these mechanical joining processes, particularly adhesive bonding processes are currently used to create a hybrid connection between plastic and metal. Adhesive bonding is understood as being the manufacture of an adhesive bond between two joining partners by means of adhesion (interface bond strength) and cohesion (internal strength). Due to the property of the adhesive to develop adhesive forces with practically any material adhesive bonding technologies have a high potential to even join materials which cannot be joined with competing methods such as welding. Additional characteristic benefits are high vibration damping and the ability to join parts with minimal thermal stresses. This aspect makes adhesive bonding particularly suitable for joining plastics. In adhesive bonding technology however, extensive pretreatment is both the state of the art and essential to achieve sustainably effective bond strengths. Furthermore the disadvantages of adhesive bonding also include the application and the curing of the adhesive. The latter frequently represents the critical path in the supply chain (Fig. 5.25).

Various joining processes are being investigated and (further) developed in the scope of the Cluster of Excellence of "Integrative Production Technology for High Wage Countries" to join plastic and metals directly via so-called thermal joining processes (Fig. 5.26). This approach should significantly shorten the process chain in comparison to adhesive bonding (see also Fig. 5.25).

5 Hybrid Production Systems

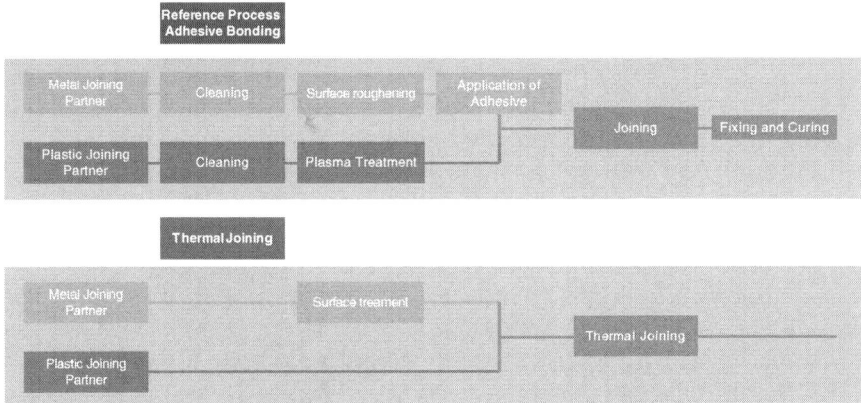

Fig. 5.25 Comparison of the process chain between adhesive bonding and thermal joining

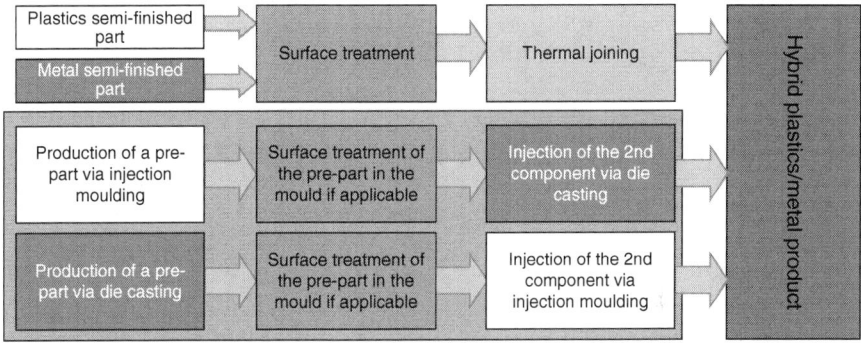

Fig. 5.26 Process chain for the manufacture of plastic and metal hybrids

In this case processes such as laser joining, inductive joining, ultrasonic joining and thermal conductivity joining are particularly analysed in order to join plastic and metal parts produced for example by injection-moulding and die casting. By means of these joining processes, it should be possible to achieve a secure connection without the use of adhesive or positive locking via macroscopic undercutting by physisorption and interlocks at the microscopic level. In addition, adhesion at the atomic level can be achieved by the mechanism of physisorption, for instance by dispersion- and dipolar coupling forces. Moreover, the adhesion can be achieved by the rough surface of the metal, which exhibits microscopically small undercuts. These undercuts are filled by the molten plastic and lock the plastic to the metal after solidification. The effect of an upstream modification of the surface of the substrate, e.g. by laser structuring, flame spraying or plasma treatment is also investigated in terms of their improvement of the bond strength.

Beside the post-mould assembly processes described above, a bond can be created between plastic and metal using in-mould assembly processes, i.e. directly in an (injection) mould. Plastic and metal parts can be manufactured by overmoulding of metallic inserts made upstream by, for example, stamping, or by back-injection moulding of metallic foils. The process of overmoulding can be further divided into the insert, outsert, and hybrid techniques. The insert technique involves inserting a metal part into a mould and injecting around it with plastics material while the outsert technique enables the injection moulding of plastic functional elements being injection moulded onto parts made of different materials, particularly metal, in a single operation. In the case of insert components such as plugs, wires, electrical contacts or housings with integrated threaded bushings the plastic acts as the carrier while the metal insert represents the functional element. In outsert parts the metal is the carrier and the plastic components perform the function, particularly in the form of functional and connecting elements. The outsert technique is thus to a certain extent the opposite of the insert technique. On the basis of the outsert method, so-called composite casting moulds can enable inserts to be cast in-place in the metal die casting process. For instance the magnesium and aluminium composite crankshaft case of BMW AG and the VarioStruct® steel sheet and aluminium hybrid component of Imperia GmbH are both manufactured using composite casting moulds (Kallien 2009; Bührig-Polaczek et al. 2010).

Another possibility for the manufacture of plastic and metal parts is hybrid technique. The so-called "hybrid technique" is the combination of a preformed profiled metal part that is placed in a injection mould to be overmoulded with the thermoplastic material. In this case, both the metal structure and the injection moulded plastic contribute directly to improving the mechanical properties (e.g. stiffness) for example in the form of ribs. This enables the manufacture of relating to the weight stronger and yet highly functional, cost-effective composite parts. The distinction between the hybrid technique and the outsert technique described above can be performed by the particular of the plastic in the composite component: In the case of an outsert part the injected plastic increases the functionality of the overall part by connecting and joining elements. The supporting function of the part is performed by the structure, (usually metallic) inserted upstream. In the case of the hybrid technique, both the metal structure and the injection moulded plastic contribute directly to improving the mechanical properties. Typical applications for the hybrid technique including the automotive industry, for example in the form of front ends, instrument panel reinforcements, engine housings, door/roof modules or roof carriers. Current developments such as combining the internal high-pressure forming of metallic structures with injection moulding proposed by Daimler AG is one hybrid technique trend following the implementation of integrative production cells (VDI (Society of German Engineers) 2010b).

A central quality feature of plastic/metal components is the achievable bond strength of each part. The key question is with which material connection mechanisms (non-positive locking, positive locking or adhesive/cohesive bond) can the highest possible project and application-specific bond strength be reproducibly achieved? These days for instance, bonds are predominantly achieved by macroscopic undercuts for example openings, grooves and overmouldings. In practice and in current developments, bonding agents are also used to create a surface bond between plastic and metal.

Another application field of hybrid components made of plastic and metal is in the electrical/electronics industry. In this case the metallic components are integrated into the plastic component because of their electrical conductivity. Beside the insert technique described above, for example back-injection or back-injection compression-moulding of foils with selective conductive areas can increase the functionality of plastic parts. The miniaturisation of electronic components is increasing the demands on geometrical complexity. Based on the constant demand for ever more rationalised production various processes has been developed for the manufacturing of so-called Moulded Interconnect Devices (MID) featuring respectively greater geometrical complexity. The ability to produce three-dimensionally injection moulded circuit carriers with integrated conductor track structures enables electronic functions to be integrated into a small area. Basically the production of MIDs involves the application of conductive areas in downstream process steps to the circuit carrier which is injection moulded in the first step. Processes for manufacturing electrically conductive coatings and tracks include hot embossing, metallising, laser structuring or the application of conductive paint. Quality features in the electrical/electronics industry include the current-carrying potential, the potential complexity of the strip conductor structures on the plastic carrier, the contact capability of electrical connecting elements and the potential bond strength between the plastic and the metal.

The process chains illustrated for the production of hybrid components have disadvantages in both structural component and electrical engineering applications in terms of the number of production steps required and productivity limitations. There is further potential for improvement in terms of correspondingly cost-intensive process steps and geometry-limiting process steps for instance stamping, bending, embossing, positioning, assembling etc. In view of the fact that, it is generally worth to develop process chains which are shorter compared to the state-of-the-art for the manufacturing of plastic/metal components by linking technical production expertise in plastics processing (especially primary forming of plastic) with technical production expertise in metal processing (especially primary forming of metal) and joining technique expertise (see also Fig. 5.26).

The newly developed process of hybrid multi-component injection moulding for the manufacture of plastic carriers with integrated metallic conductor tracks should reduce the number of assembly steps in production and enable the direct production of electronic components made of plastic and metal in a single mould and on a single machine. In the case of the new hybrid process, a plastic carrier is produced in an initial step by conventional injection moulding on a modified injection moulding machine. With the addition of a metal pressure die casting unit, it is possible to inject a metallic conductor track onto the plastic carrier in a second step on the same machine. A further approach for the manufacture of hybrid structural components, the hybrid multi-component pressure die casting, is based on a combination of metal pressure die casting and plastic injection moulding. In this process, the metallic component is initially manufactured on a modified pressure die casting machine and the plastic is injected in the subsequent step. The basis is a metal pressure die casting machine which is extended by a plastic injection-moulding unit.

5.3.2.2 Adhesion Mechanisms Between Plastic and Metal

Joining plastic and metal combines material-specific benefits of different materials in a single component. The properties of this join are largely determined by a good and durable adhesion within the interface. In order to be able to describe the adhesion phenomena occurring at the product boundary, the surface conditions of the joining partners and the effects exerted by each must be analysed in more detail.

Essential to the adhesion of boundary surfaces is that the atoms and molecules on the surface of an element are present in a more energy-rich condition than in the bulk material. If the two elements are brought into the closest possible contact, this leads to intermolecular interaction resulting in adhesion. The resulting interaction on the surface is separated into several atomic and molecular layers and depends on the atomic and molecular structure of the solid which is different for each material system. All adhesive interaction processes therefore always involve analysis of a surface and/or boundary surface layer.

With reference to a material bond, it can be assumed, according to (Bischof 1993), that the generation and make up of the boundary surface layer which comprises molecular and supramolecular structures is influenced substantially by the following factors:

- the chemical composition and structure, the geometrical and morphological form and the energetic status of the joining part surfaces,
- the technological conditions during the manufacture of the bond (temperature, time, pressure, medium), which can for example lead to internal stresses in areas close to the boundary layer.

Various adhesion models have been developed on this basis for the resultant complex adhesion mechanisms at the phase borders (Fig. 5.27). The mechanical adhesion model states that the plastic penetrates the pores and unevenness of the metallic joining partner with the result that a positively-engaging connection is produced during the cooling process (Bischof 1993). Specific adhesion models include the interpretation of the physical boundary layer processes caused by bonding forces such as dipolar coupling forces, dispersion forces and hydrogen bridge connections. Chemical bonds result from the functional groups of polymers which can form strong bonds with the metallic partner. The wetting and compatibility of the joining partners

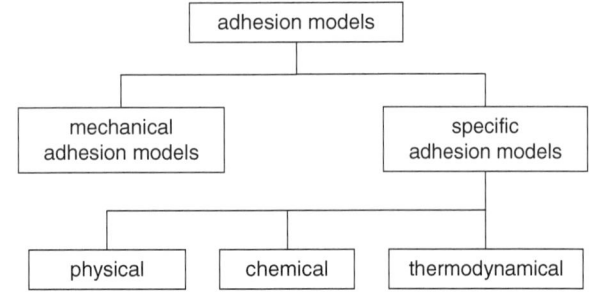

Fig. 5.27 Adhesion models. (Bischof 1993)

can be described by surface and boundary layer energies which are summarised under the thermodynamic models. In the search for approaches to clarifying the adhesion mechanisms of a material bond it should be noted that a complex interaction of adhesion mechanisms takes place within a join. The adhesion models, as they are described in brief below, are based on idealised preconditions which are not present in an actual bond. However, they can provide indicators for the adhesion mechanisms of any bond.

The polarisation theory is based on the proposal that only the existence of polar functional groups in the materials to be joined at a distance of less than 5×10^{-10} m results in a force which can be described as adhesion. Hydrogen bridge connections and dipolar interactions at the border layer of the bond reinforce this adhesion force (Weiss 2004). In relation to connections between steel and plastic, this means for example that the oxide and hydroxide groups always present on the surface of the steel can lead to dipolar interactions. The functional groups present in the plastic can form hydrogen bridge connections with the oxide hydrates and induce dipoles at the active centres of the surface. Covalent bonds cannot be discounted but are rare (Bischof 1993). In addition the electrostatic theory, described by Derjaguin can be assigned to the polarisation theory. This theory describes a constant shifting of the charge carrier within the boundary layer. The resultant potential difference causes a charge displacement and thus an electrical double layer which leads to the formation of electrostatic attractive forces between the surfaces (Wertheimer et al. 1999).

The diffusion theory describes the mutual diffusion processes which occur at the phase boundary between the metal and polymer. In this process, metal atoms defuse in the polymer network or components of the polymer molecules become embedded in the metal grid. Gibbs free energy is the driving force for the diffusion process of a metal atom which is embedded in a polymer matrix. This is less than the enthalpy for a metal atom adsorbed on the polymer surface. This effect is heavily temperature-dependent (Weiss 2004). The strength of the adhesion depends on the number of defused molecules, on the type of bond between the individual atoms and/or molecules and on the molecular involvement.

The chemical adhesion theory states that ionic- or covalent-type chemical bonds at the phase boundary of the material bond cause the cohesion. The metal molecule interactions play a particular role in the formation of the metal and plastic bond. In general, the same internal forces which occur in homogenous solids (primary and secondary valence bonds) are also responsible for the adhesion in the boundary layer (Grundke et al. 1995; Sauer 1999). Table 5.1 shows the possible bond types which can be present in a polymer and metal boundary layer. Analysis of these chemical bonds in the polymer and metal boundary layer however is not immediately possible.

Table 5.1 Possible bond types in a polymer-metal transition. (Grundke et al. 1995; Sauer 1999)

Primary valence bonds	Secondary valence bonds
Homopolar (covalent)	Hydrogen bridge bond
	Dipole-dipole interaction
	Dipole-induced dipole interaction
	Dispersion forces

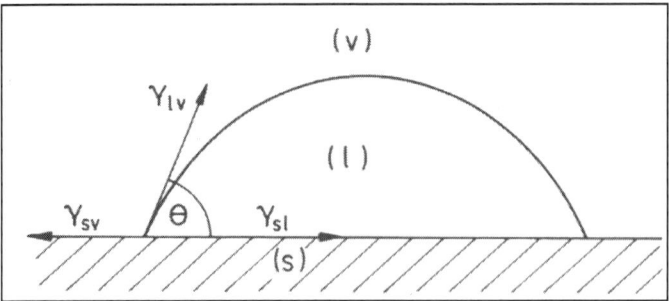

Fig. 5.28 Definition of the forces according to Young. (Sauer 1999; Schultz and Nardin 1999]

The wetting process is the starting point for the thermodynamic analysis of adhesion (also known as the adsorption theory). This theory was developed by Zismann, Fowkes and Good. It states that the better the wetting, the better the adhesion. Correspondingly the adhesion forces comprising polar (dipolar coupling forces) and disperse (van der Waals forces) portions are also higher. The characteristics for determining the wettability are the surface energies of the two materials in contact with one another. The surface energy is defined as the work required to increase a surface by one surface unit. The surface energy is determined experimentally using the contact angle measurement and application of the Young equation (Fig. 5.28). The theoretical background of the contact angle measurements, the calculation of the surface energies with differentiation into the polar and disperse portions can be taken from (Fowkes 1964; Owens and Wendt 1969; Zisman 1964). Ultimately it is possible to divide the forces involved in the adhesion into non-polar London-dispersion interactions and polar, non-dispersive interactions (e.g. hydrogen bridges) in which the total surface energy γ is divided into a dispersive γ_d and a polar γ_p component:

$$\gamma = \gamma_d + \gamma_p \qquad (5.7)$$

Equation according to YOUNG:

$$\gamma_{sv} = \gamma_{sl} + \gamma_{lv}\cos\Theta \qquad (5.8)$$

γ_{sv} : Surface tension of the solid in equilibrium with the saturated vapour phase of the fluid
γ_{sl} : boundary layer tension between fluid and solid
γ_{lv} : surface tension of the fluid in equilibrium with its saturated vapour phase
Θ : contact angle

Metallic surfaces count as energy rich, chemically reactive surfaces. Plastics and other organic solids generally possess energy-poor surfaces. The metal and plastic phase boundary is thus characterised by an energy rich metal surface in connection with a comparatively energy-poor plastic surface. Quantitative conclusions with respect to the adhesion energy can be drawn with knowledge of the respective surface energies (Sauer 1999).

A further adhesion model is the Weak Boundary Layer. In this case it is assumed that the bond between metal and plastic will always result in a weak boundary layer

Table 5.2 Surface pretreatment processes and effects. (Bischof 1993)

Processes	Type of pretreatment	Effect
Cleaning, degreasing	Washing with water and/or solvents	No decisive structural change of the surface
Mechanical pretreatment	Grinding, sanding, sand blasting	Geometrical change of the surface → roughness Removal of contamination layers Activation (tribochemistry) of the surface
Chemical pretreatment	Etching, eroding with acids or alkaline substances (including gases), thermal through flaming or silicoating	Change of the chemical structure of the surface (oxidation, phosphating, concentration of external elements on the surface)
Physical, preferentially electr. and sandblasting pretreatment	Corona, LD plasma, UV electrons, laser and radioactive beams	Change of the chemical and physical surface condition
Combined physical and chemical pretreatment	Corona, ozonation, priming, benzophenone/UV beams	Ditto
Coating	Bonding agent, metal Coating (e.g. galvanising)	Ditto

which results in substantially less adhesion. In the separation of the plastic and metal bonds, failure is almost always a cohesive failure in the boundary layer. Adhesive failure at the phase boundary between the Weak Boundary Layer and the metal or polymer occurs only in exceptional cases.

The Weak Boundary Layer prevents direct contact between metal and polymer. This weak boundary layer is primarily influenced by:

- impurities or low-molecular substances which migrate to the surface of the polymer
- air inclusions between the two phases
- reaction products, e.g. formation of oxide layers on the metal surface (Weiss 2004)

After analysing the adhesion models one needs to question the extent to which the knowledge obtained can be used to influence the adhesion between the plastic and metal. Since the surface conditions of the joining partners are key adhesiveness criteria, a targeted substrate pretreatment can achieve a decisive adhesive improvement. Table 5.2 provides a comprehensive overview of possible surface pretreatment processes and their effects.

5.3.3 Motivation and Research Question

The new approaches to manufacture plastic/metal hybrids can be used to devise contributions to an integrative production technology. With respect to the challenges faced by producing companies and companies from high-wage countries in particular, the integrative approach has the potential to increase the competitiveness of these companies in the long-term.

Traditionally, three different competitive strategies have been used to ensure the success of companies in their specific market environments. These are the strategies of price leadership, differentiation and niche concentration. If the company is aiming the niche market, monopolistic tendencies mean minimum competition for that company. This situation offers a company the possibility of long-term commercial success. The huge variety of companies and of the products they manufacture however mean that only a few companies can manufacture niche products. A far greater number of companies must position themselves in the zone of compromise between the strategy of price leadership, in which the primary positive influence on the purchasing decision of a customer is the low price, and the strategy of differentiation wherein the focus is on maximising user benefits in comparison with competing products. The advantage of price leadership strategy is based on the manufacture of a large quantity of identical products. Due to the economies of scale such standard products can be made very economically. The economies of scope are in constant competition with the economies of scale. The unit quantities of differentiated products are way smaller. The increased number of variants leads to an increase in dynamics and thus complexity within the production system. Differentiated products thus cause higher production costs.

At many levels, globalisation is forcing companies from high-wage countries to adopt the strategy of differentiation to maintain their competitive position. In the long-term, focusing purely on the strategy of differentiation as a means for companies from high-wage countries to remain competitive does not ensure success because companies from low-wage countries can also penetrate these market segments. Therefore companies from high-wage countries must also actively address the price-driven markets. This ultimately results in companies from high-wage countries implementing both economies of scale and economies of scope simultaneously within their production systems. The optimum goal for a company is therefore to simultaneously pursue a strategy of price leadership and a strategy of differentiation. Accordingly, for example a product differentiated by its use benefit should also be attractively priced.

Up until now, high-wage countries have been distinguished from low-wage countries by their advanced expertise in production technology. This advanced expertise in the individual disciplines within production technology and in the material sciences however is steadily decreasing due to the increasing level of education in the low-wage countries. In the long-term, low-wage countries will develop the discipline-specific expertise which will put them on the level of the high-wage countries. To maintain a competitive advantage in the long term therefore, the high-wage countries now need to start linking their expertise across traditional boundaries. We should therefore analyse the challenges of production technology with a greater focus on integrative solutions. Only this integrative thinking will form the basis of implementing the two current strategies of price leadership and differentiation in a single approach.

5 Hybrid Production Systems   491

Using the example of the manufacture of metal/plastic hybrid components, high-performance process chains should illustrate how the integrative approach can contribute to resolving the conflict between economies of scale and economies of scope. Integrative thinking is particularly characterised in this chapter by the link between production technology expertise in plastic processing with production technology expertise in metal processing (especially primary forming of metal). The result is integrative production technology for the manufacture of plastic/metal hybrids. The central question at this point is this: *"How can we manufacture products which are adapted to customer requirements with minimum outlay using the developed integrative production technology?"* In this case particularly the development of new, high-performance processes should contribute directly to resolving the conflict between scale and scope.

An analysis of the existing process chain for the manufacture of plastic/metal hybrids both with respect to manufacture of mechanically stressed structural parts as well as the production of electronic components firstly shows that both are characterised by high unit quantities. This results for example from the combined use of stamping/bending machinery and injection moulding machinery. Flexible adjustment of production to different variants for the purposes of diversification of the product portfolio is almost impossible for this reason. Production is also frequently characterised by intensive planning because the sequence of different production steps requires precise cycle control and may need to be synchronised with buffer stations if these are present.

This is the point at which the new approaches for manufacturing metal/plastic hybrid parts become relevant. They offer new perspectives for both mechanically-stressed structural parts as well as for the manufacture of electronic components within the area conflict outlined above. The potential improvements illustrated in Figs. 5.29 and 5.30 should be emphasised in comparison to existing process chains.

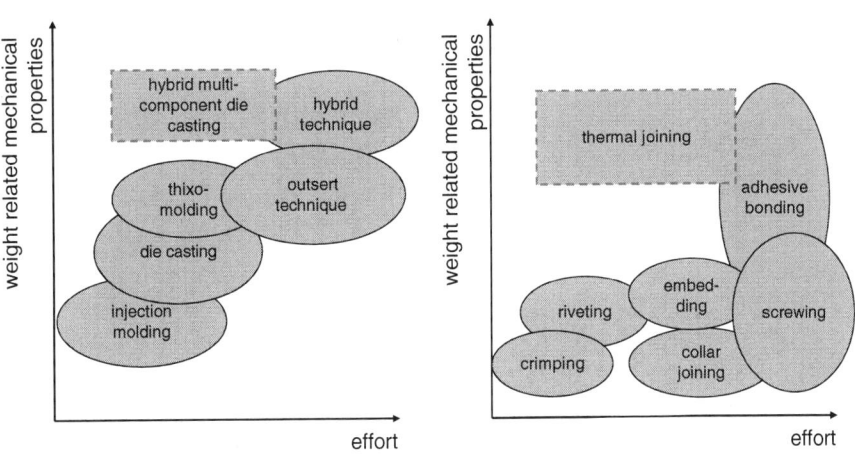

Fig. 5.29 Industrial manufacture of structural parts made of plastic and metal for lightweight construction applications

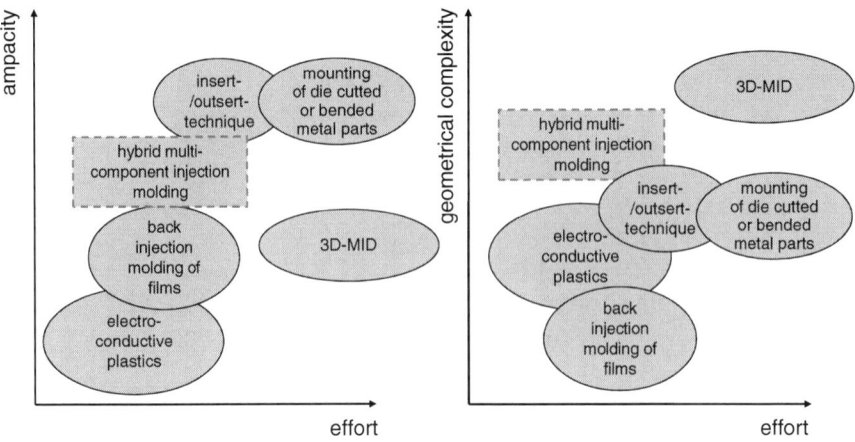

Fig. 5.30 Industrial manufacture of plastic and metal hybrids for electrical and electronic applications

5.3.4 New Approaches for the Manufacture of Plastic and Metal Hybrids

Figure 5.26 presents possible new approaches to the manufacture of plastic/metal hybrids. The individual technologies are now presented in more detail. In this context, we look at both the system-related implementation of the new processes as well as the relevant research results. After a description of the materials used, the findings with respect to surface treatment, thermal joining and hybrid multi-component technology are illustrated in detail. The common feature of the various processes is the creation of a hybrid bond of plastic and metal. Finally therefore comparisons will be drawn between the individual processes on the basis of achievable bond strengths. The chapter concludes with a summary of findings on adhesion mechanisms between metal and plastic.

5.3.4.1 Material Selection

A suitable selection of materials is made to limit the scope of the experiment. On the one hand the experiment should be wide-ranging while on the other, the use of standard materials should ensure practical relevance. In principle, technical thermoplastics with high industrial relevance for applications in lightweight structures, joining technology and electrical/electronic applications are to be used as plastic components for all the process chains analysed. Polyamides (PA 66), polybutylenterephthalate (PBT) and polycarbonate (PC) are used therefore (see also Table 5.3). The plastics used differ in terms of mechanical characteristics due to their differences in chemical composition and their morphology.

5 Hybrid Production Systems

Table 5.3 Materials used

Metals	Plastics
Stainless steel (1.4301)	Polycarbonate (PC)
Aluminium (3.3547)	Polyamide 66 (PA66)
Galvanised steel (1.0226)	Polyamide 66 with 30% glass fibre (PA66GF30)
Pressure die casting alloy (AlZn10Si8Mg, AlMg5Si2Mn)	Polybuteneterephthalate (PBT)
MCP 200 (low melting tin/zinc-alloy)	Polybuteneterephthalate with 30% glass fibre (PBTGF30)

Different materials are considered as metallic joining partners depending on the respective boundary conditions of the individual processes. The metallic joining partners used in the thermal joining area are stainless steel, a wrought aluminium alloy (AlMg4.5Mn0.7) and a galvanised steel with a 7–12 µm zinc coating (see also Table 5.3). Along with steel, aluminium is the most frequently-used metal in automotive engineering and is used in lightweight structural applications due to its low density. In this case the addition of alloying elements is used to achieve the corresponding properties profile. Material properties such as strength, hardness, workability and castability can thereby be adjusted according to the application. An oxide layer (Al_2O_3) with a thickness of 0.001 µm also forms spontaneously on the surface of aluminium by reaction with the oxygen in air and/or water vapour. This oxide layer protects the aluminium against environmental influences. Steel can also be galvanised to protect it against environmental influences and is therefore of great significance for automotive engineering.

The metal alloys used for hybrid multi-component pressure die casting are selected firstly with respect to their workability in a combined process and secondly with the focus on applications in automotive engineering. From these perspectives, two aluminium pressure die casting alloys have been selected (a so-called natural strengthen alloy (AlMg5Si2Mn) and a self-hardening alloy (AlZn10Si8Mg) (see also Table 5.3) which from a technical perspective do not require heat treatment to achieve their maximum mechanical properties. The heat treatment of a hybrid metal/plastic component would run the risk of irreversibly damaging the plastic.

Also, for the hybrid multi-components injection moulding, the selected materials need to be suitable for combined processing. The thermoplastic carrier produced in a first step should not be thermally damaged or inadmissibility mechanically stressed by the downstream overmoulding with metal alloy. Therefore, tin- and zinc-based low melting metal alloys are used to achieve compatibility with the processing temperature range of thermoplastics (see also Table 5.3). These alloys which are manufactured by MCP HEK Tooling GmbH, Lübeck, have mainly been used in lost core injection moulding and in rapid prototyping (MCP-HEK Tooling 2002; Gebhardt 2007). The eutectic composition of the alloy partners results in a melting point of around 200 °C.

The low melting metal alloys have electrical conductivities within the range of those metals that are frequently used for electrical applications. This is illustrated for example in Fig. 5.31 for the alloy MCP 200. The conductivity σ is measured in the unit Siemens per m. This is the reciprocal value of the resistance R as a quotient

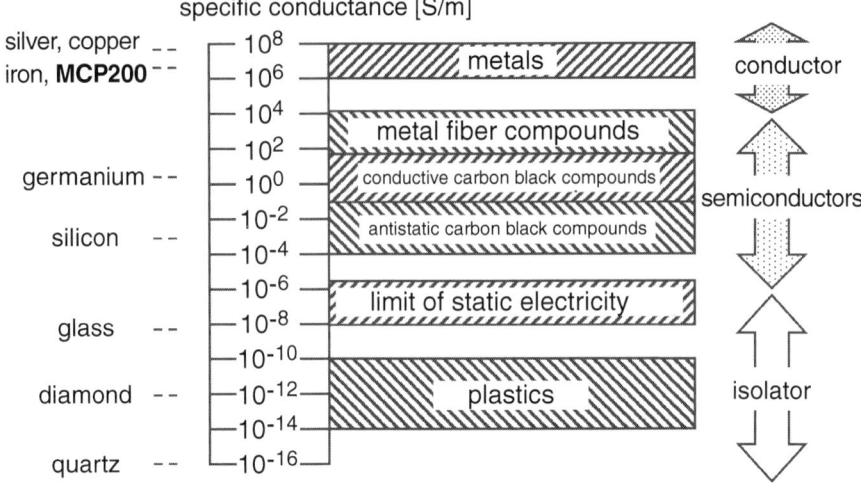

Fig. 5.31 Comparison of the electrical conductivity of different materials

of the applied voltage and current with respect to the lead cross-section A and the conductor length l:

$$\sigma \left[\frac{S}{m}\right] = \frac{1}{U} \times \frac{1}{A} \left[\frac{1}{\Omega \times m}\right] \tag{5.9}$$

Unlike plastics, which are made electrically conductive in the electrical/electronics industry using fillers such as graphite, carbon black or metal filings, the low melting metal alloys exhibit higher conductivities of 2–5 decimal powers.

The physical properties of the low melting metal alloys differ significantly from those of technical thermoplastics. This is illustrated for alloy MCP 200 in Table 5.4 in comparison with a standard polyamide 6 (PA 6).

The low melting-point of metal alloys is characterised by high ductility. The attainable strengths however are comparatively low. At a breaking strain of 32.5% the strength of the alloy MCP 200 is 54.7 N/mm². The mechanical properties are therefore very similar to those of the plastics shown, e.g. for the standard PA 6. However, steels used in mould making, for example 1.2343, frequently exhibit strengths of 1800 N/mm² at a breaking strain of 4%. Consequently the metal alloy selected here

Table 5.4 Physical and mechanical properties of the alloy MCP 200 compared with PA 6

	MCP 200	PA 6
Viscosity (Pa*s)	0,01–0,03	30–120
Heat conductivity (W/(m*K))	61	0.29
Heat capacity (kJ/(kg*K))	0.24	1.95
Thermal conductivity (mm²/s)	35	0.13
Tensile strength (N/mm²)	54.7	80
Breaking strain (%)	32.6	20

is not suitable as a load-bearing component in structural parts. It is useful however in relation to the forming of stress peaks which undergo no major changes in terms of mechanical properties at the boundaries between plastic and metal alloy, as it is the case for instance with insert-/outsert or hybrid components.

5.3.4.2 Surface Pretreatment Methods

The analysis of adhesion mechanisms in plastic and metal composites has shown that the surface conditions of the joining partners are decisive for the achievement of good adhesion. The effect on the bond formation of different surface pretreatment techniques is investigated in order to identify positive influences on the formation of the bond between plastic and metal. Promising surface pretreatments (see also Table 5.2) are selected to improve the adhesion of the plastic and metal joins. Mechanical processes are used, e.g. sandblasting processes, which alter the geometric surface. Laser structuring of metal or plastic substrates both increases the effective joining surfaces and creates undercuts which permit reproducible interlocking of the joining partners. Plasma treatment particularly for the plastic substrate is one approach for improving adhesion by chemical and physical modification of the surface state. Another pretreatment method is the coating of both plastic and metal substrates. For example, applying an intermediate layer to the metal which has chemical affinity to the plastic can be used to create a bond in the joining process between the intermediate layer and the plastic by means of diffusion and intermolecular forces. Thermal spraying is a suitable coating technology for both plastics and metals.

Sandblasting Sandblasting involves abrasive particles propelled by means of compressed air through a nozzle onto a surface. The impact of the particles mechanically abrades the surface (Lugscheider 2002). As well as having a cleaning effect, sandblasting increases the actual surface of the joining partner. A larger surface firstly increases the number of surface-dependent atomic and molecular interactions which has a favourable effect on adhesion (Gleich 2004). Secondly the polymer can mechanically clamp onto the joining surface depending on the final surface roughness of the respective topography.

The process is ideally used for the preparation of metal surfaces for downstream coating processes such as painting, adhesive priming, plastic coating or thermal spraying. Sandblasting is a typical method for the preparation of joining surfaces for adhesive bonding.

Characterisation of the Surface Topographies of the Metal Surfaces To show the effect of the surface topographies on the bond strengths of plastic and metal hybrids, the bonding surfaces, particularly of the metallic joining partner, are mechanically prepared by sandblasting. The sandblasting is carried out using a silicon carbide abrasive at a pressure of 7 bar and from a distance of 20 mm for 4–5 s. To quantify the effect of the pretreatment, a comparison is made with as smooth a sample as possible. For this purpose some of the metal surfaces are lapped.

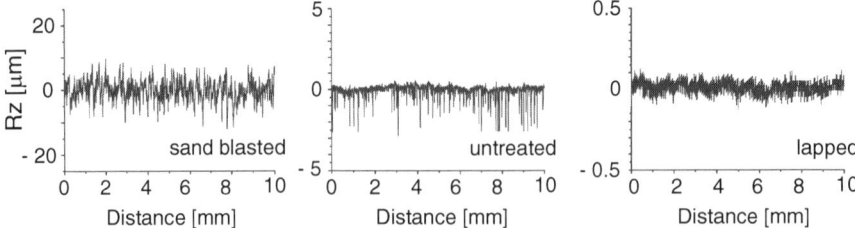

Fig. 5.32 Roughness profiles of variously pretreated steel surfaces

The surface is characterised with a stylus instrument (T2000, Hommel-Etamic GmbH, Villingen-Schwenningen), with which the surface structure is graphically illustrated and roughness characteristics (mean roughness R_a, average roughness R_z and maximum roughness R_{max}) can be determined. Figure 5.32 for example shows the roughness profiles of variously pretreated steel surfaces. The surface of the sandblasted sample exhibits an even roughness with height differences of up to 25 μm. The untreated steel surface however exhibits an uneven roughness profile with roughness of up to 5 μm. The roughness peaks in Table 5.5 are evened out by a lapping process shows the determined roughness values (R_a, R_z, R_{max}) for the steel (1.4301) and aluminium samples (3.3547) after various surface pretreatments. The comparison of results between the aluminium and steel samples is interesting in that the sandblasted aluminium exhibits greater roughness. The pretreatment method was carried out identically on both metals but aluminium is softer than steel so more material is removed in this case.

Laser Micro-Structuring Applying micro-structures with undercuts to the metal surface is an efficient method of creating anchoring points for the mechanical interlocking of plastic and metal. The application of these structures in this shape is only possible by laser beam. The use of this wear-free, flexible and efficient technology can guarantee the required cycle times necessary for an efficient production process. The minimal energy requirement and the highly dynamic flexible beam movement is particularly effective for pretreatment of defined local areas on sensitive components. The beam guidance is realized by galvanometric scan systems which exhibit low inertia due to the small volumes moved and can therefore produce any contour and can reach high surface speeds of up to 15 m/s. This innovative approach which

Table 5.5 Results of the roughness measurement on aluminium and steel samples after various surface pretreatments

Material	Pretreatment	R_a (μm)	R_z (μm)	R_{max} (μm)
Steel (1.4301)	Sandblastet	2.61	17.01	24.47
	Untreated	0.43	2.38	4.17
	Lapped	0.02	0.18	0.31
Aluminium (3.3547)	Sandblastet	4.36	27.03	32.12
	Untreated	0.27	2.80	5.24
	Lapped	0.05	0.41	0.67

Fig. 5.33 a Principle of laser structuring of the metal surface, **b** light microscopic cross section of the structure of material: 1.4301; **c** Cross section PA66

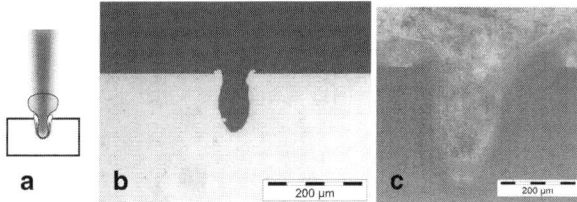

combines sublimation and melting speeds up processing compared to conventional laser structuring methods by factor 50–100. The vaporisation pressure generated by the high intensity in the centre of the structure forces the surrounding melt out of the structure towards the surface (Fig. 5.33 left). This process is repeated until the melt no longer reaches the surface but solidifies in the throat of the structure. This results in the undercut required for the interlock (Fig. 5.33 centre).

To enable the interlocking of low melting metal alloy in plastic in the case of combined injection moulding, the semi-finished plastic products are provided with an undercut structure. Practically all plastics exhibit low optical penetration depth (~10 μm). Therefore a pulsed CO_2 laser with an emission wavelength of 10.6 μm is used. Plastic cannot be vaporised without thermal damage like metals so a different process control is required. The plastic is locally heated and plasticised. The melt must be partially forced out with compressed air. To ensure proximity with the interaction zone it is introduced with a modified cutting head. This enables the production of geometrically similar structures (Fig. 5.33 right).

Results of the Laser Micro-structuring Geometrical reference values are defined to predict the effects of varying the laser parameters and to enable an appraisal of the process-suitability of the laser beam sources designed to carry out the surface structuring. The parameters are the width and the depth of the microstructures. For vertically structured samples, the reference values are determined by metallographic investigations. In the course of the investigations, the influence of the parameters of output power, feed rate and traversing frequency on the reference values are investigated for stainless steel 1.4301. For this purpose a pulsed disc laser is used with a wavelength of 1,030 nm, a maximum output power of 65 W at 45 A and a maximum repetition rate of 30 kHz.

Varying the output power (Fig. 5.34) at a constant feed rate of 600 mm/s and a pulse frequency of 30 kHz shows, as expected, that the structure depth increases as the output power increases. While the structure depth at an amperage of 41 A increases from 20 μm (number of passes (N = 1) to 76 μm (N = 3) and at 43 A from 23 μm (N = 1) to 83 μm (N = 3), maximum structure depths are achieved at 45 A and N = 3 (91 μm)). In this case a significant increase in the structure depth is discernible both when increasing the laser output power as well as when increasing the number of passes.

The laser output power has no influence on the maximum achievable structure width. This can only be changed by increasing the focal diameter. In general the structure width decreases as the number of passes increases (Fig. 5.35). This is

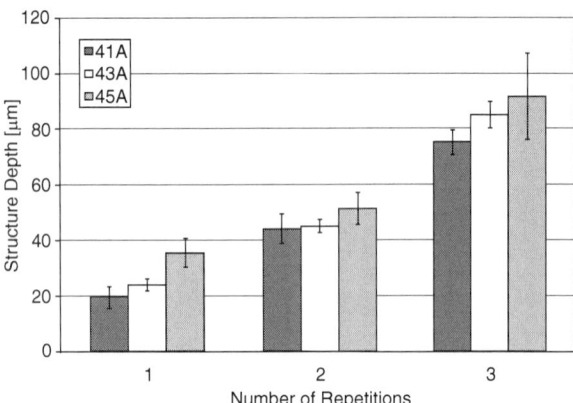

Fig. 5.34 Structure depth with variation of the average power (30 kHz, 600 mm/s, 1.4301)

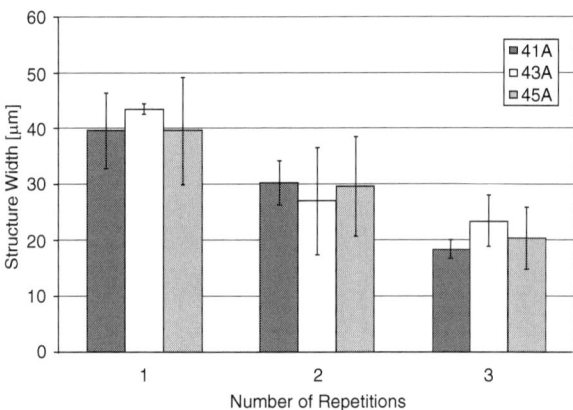

Fig. 5.35 Structure width with variation of the laser output power and number of passes (30 kHz, 600 mm/s, 1.4301)

explained by the solidification of the melt which is forced out from the structure wall. The structure width at the throat of the structure becomes narrower with each pass. This behaviour implies a movement of the melt from the base of the structure to the throat of the structure in accordance with the expected process trend. Further passes cause the structure opening to close so that the plastic cannot flow into the structure making the structure is unsuitable. To produce an undercut requires selection of the process point at which the melt at the throat of the structure leads to a narrowing of the cross-section but before the structure has closed up again.

The cross sections of the structures are analysed to obtain a prediction about the size of the undercut. There is a clear change in the cross-section with each pass from a V-shaped cross-section to a U-shaped cross-section and finally to a structure with an undercut (Fig. 5.36). If the output power is too low, insufficient material is plasticised and transported from the base of the structure to the edge so that no undercut is created (Fig. 5.37).

The use of a continuously emitting beam source enables even more effective material processing. In a series of experiments an undercut was achieved after three

5 Hybrid Production Systems

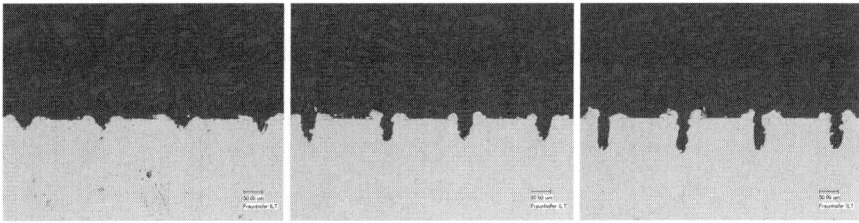

Fig. 5.36 Cross-section of the microstructure for 1, 2 and 3 passes at 45 A, 1.4301

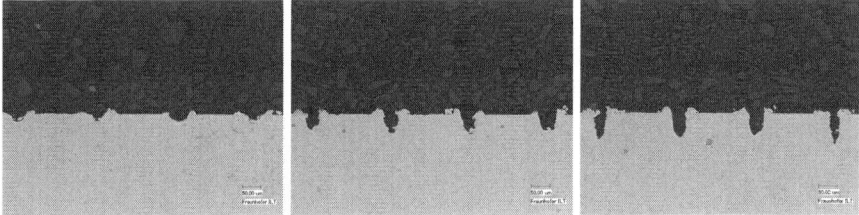

Fig. 5.37 Cross-section of the microstructure for 1, 2 and 3 passes at 41 A, 1.4301

passes at a feed rate of 10 m/s (Fig. 5.38). It is thus possible to structure the area for the test components ($5 \times 20\,\text{mm}^2$) with a structure distance of 0.4 mm within 0.075 s. This gives a theoretical surface rate of $12\,\text{cm}^2/\text{s}$.

Structuring in the Dual-Beam Process In the two-beam process, structuring is carried out with a correspondingly-designed optic (Fig. 5.39) to produce a significantly greater undercut. This consists of a beam splitter which divides the laser beam at an incidence angle of 45° with respect to its wavelength, polarisation and intensity into two almost identical parts. One part of the beam is reflected, the other is transmitted. The two parts of the beam are contradirectionally focused, respectively via 90°-deflection mirrors, at a 45° angle onto the surface in the processing plane. The focussing lenses have a focal length of $f = 80$ mm. This gives a theoretical focus radius w of 8.6 μm.

The changed conditions in the ablated zone give different results to single laser ablation. A minimum of two passes are required to create undercuts. The angle between

Fig. 5.38 Cross section of a structure with a cw laser (10 m/s, 3 passes, 1.4301)

Fig. 5.39 Application of laser structuring in the dual-beam process

the laser beam and the interaction zone as well as the resulting projection means leads to insufficient intensity during the first pass. Ablation is therefore minimal. In the second pass the newly produced groove changes the angle of contact to practically 90° with correspondingly increased intensity. The high intensity with the generation of the vapour pressure enables more effective ablation by driving out the melt. With multiple passes, there is a linear increase of the structure depth (Fig. 5.40).

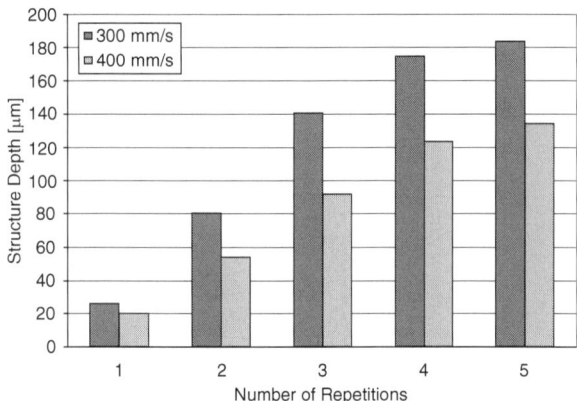

Fig. 5.40 Influence of the number of the passes on the structure depth depending on the feed rate (1.4301)

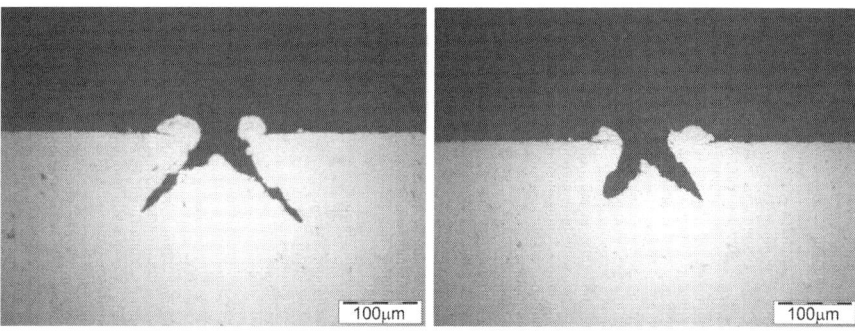

Fig. 5.41 Cross-section of the undercut structure for different feed rates (350 and 300 mm/s, 1.4301)

After five passes however there is no notable enlargement of the undercut depth. This is due to the beam caustic, wherein the intensity at the base of the structure decreases below the vaporisation threshold. This causes the ablation rate to fall as the ablation depth increases. Enlargement of the undercut-depth by increasing the number of passes is limited. The greater the feed rate the smaller the structuring depth achieved per pass. In summary one can conclude that double sided undercut structuring is possible with suitable parameters and therefore provides good conditions for interlocking of the plastic (Fig. 5.41).

Structuring of Plastics The melt behaviour of plastics is not comparable with metal because its viscosity is significantly higher. In particular plastics do not cool and solidify rapidly, as metals do. Therefore important optimisation variables are the laser output power and the pressure of the compressed air to the nozzle. These also affect in principle the feed rate, the energy applied to a specific surface area and thus the melt volume created. Due to the very low thermal conductivity compared to metals, the influence of the feed rate on the structure geometry however is comparatively small. The variation of the laser output power on the other hand influences the quantity of melt produced and therefore also the structure depth. Furthermore the pattern of the structure is influenced by the pressure of the compressed air at the nozzle because this particularly influences the quantity of melt displaced. As a result of the investigations the following parameters are used when structuring the plastic samples: feed $v = 40$ mm/s, laser output power $P = 20$ W and nozzle pressure $p = 1.2$ bar.

Thermal Spraying Thermal spraying is a flexible coating method for the application of practically all technically relevant materials onto the most diverse substrate materials. The thermal spraying process principle is described in DIN EN 657. A wire-, powder- or rod-shaped spray material is added either inside or outside a spray gun, heated to the plastic or melted state and then propelled onto the prepared component surface (Fig. 5.42). The component surface remains unmolten in this process (DIN EN 657 2005). The spray coating is built up by the impact of individual sprayed particles which spread and flatten on the substrate surface and then solidify. The

Fig. 5.42 Schematic diagram of flame spraying as an example of a thermal spraying process

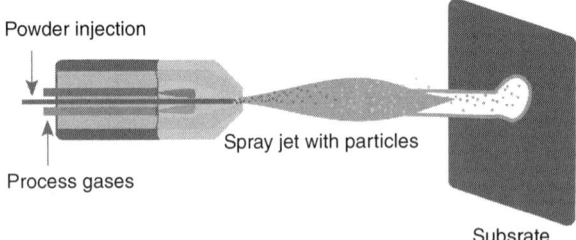

connection between the substrate and the coating material is primarily based on a mechanical clamping which requires a prepared, roughened surface (Lugscheider 2002). Thermal spraying allows layer thicknesses of a few µm up to several mm.

The thermal and kinetic energy required to heat up and accelerate the spray material can be provided by a combustion (flame), a plasma or an electric arc. Amongst the most widespread processes are flame spraying, high velocity oxygen flame (HVOF) spraying, air plasma spraying (APS) and wire arc spraying. Plastic flame spraying is typically used for processing low melting spray material such as plastics. The wire arc spraying process has been proven to be suitable for coating plastics.

In the context of surface pretreatment, thermal spraying is used to heat up plastic powder and apply it with high kinetic energy to metal components. A layer is developed which achieves very good adhesion with the component thanks to a clamping effect and which can be connected to another plastic component using a conventional plastic joining process. The plastic layer therefore operates as an adhesion agent. The thermal spraying method is also used to coat plastic substrates with a low melting metal material which is designed to function as solder in order to join metal to metal.

The coating of the joining surface with thermal spray processes is used as the adhesive agent both in thermal joining as well as in hybrid multi-component injection moulding. For the former processes, plastic layers are also applied to metal substrates using the plastic flame spraying process to prevent the locally applied heat from flowing out of the joining zone. In this case ultrasonic and laser transmission joining are used as downstream joining processes.

The target values of the layer formation for the adjustment of the flame spraying process are a homogenous structure with few pores and an even surface structure. Guaranteed effective particle plasticising during the coating process is required to achieve this. An oxyacetylene flame can only heat and pre-melt the plastic particles. The plasticising and merging of the spray material in plastic flame spraying only takes place on the substrate material. For this reason the substrate material should be preheated. The layer quality is appraised on the basis of surface photographs and cross sections. Since there are only a few plastics available commercially in powder form for thermal spraying (Winkler et al. 2003), a semi-crystalline polyamide PA11 (GTV Plast 35.900.1, GTV GmbH, Luckenbach) with a particle size of $-125+45$ µm and a melting point of 184–186 °C is used as the spray material.

Figure 5.43 shows the results of varying the substrate preheating on the layer structure. It can be assumed that as the layer surface heats up during the pre-heating

5 Hybrid Production Systems

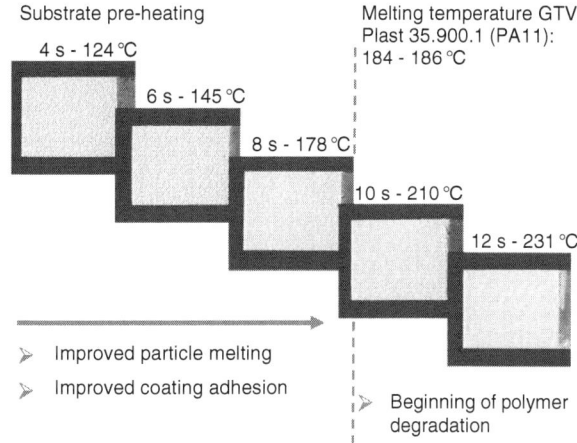

Fig. 5.43 Surface structure of PA11 flame-sprayed layers produced with different substrate preheating parameters (spray parameters: acetylene 0.5 bar, oxygen 2.5 bar, compressed air 2 bar, spraying distance 200 mm)

process it exhibits less roughness which indicates that the merging of the PA11 particles is improved at higher temperatures. At 210 °C however the surface starts to discolour and turn yellow. As the preheating temperature is above the melting temperature of the PA11 powder, this can indicate the start of the decomposition of the plastic. Fourier-Transformation-Infrared-Spectroscopy (FT-IR) of the initial powder of one optimum layer and one layer which has been sprayed on too hot has shown that the powder has oxidised in the spraying process. The increase of the carbonyl groups (CO) if the spraying parameters are too hot is significantly greater than in the case of the cooler substrate preheating processes up to 178 °C.

The typical layer structure of a plastic flame-sprayed coating with previous substrate preheating is illustrated in Fig. 5.44 in cross-section. There is good layer adhesion on the sandblasted substrate which was preheated to 178 °C before the coating. The plasticised plastic particles have penetrated into the unevenness and undercuts of the metallic substrate surface, solidified there and mechanically clamped onto the boundary surface. The homogeneity in the layer structure indicates good merging of the plastic particles wherein the pores are caused by the vaporising of the

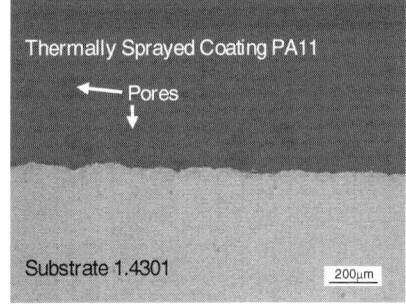

Fig. 5.44 Cross-section photograph of a PA11-flame-sprayed layer on steel 1.4301 (spray parameters: acetylene 0.5 bar, oxygen 2.5 bar, compressed air 2 bar, spraying distance 200 mm, substrate preheating 178 °C)

Fig. 5.45 Arc sprayed zinc coating on PBT (spray parameters: voltage 21 V, current 150–180 A, compressed air 3.5 bar, spray distance 150 mm, surface velocity 800 mm/s, offset 15 mm)

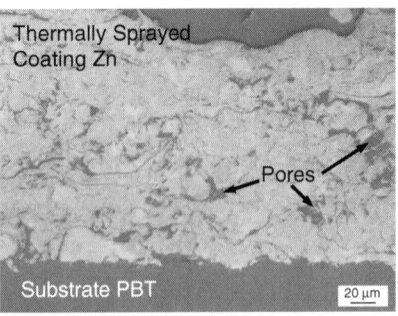

moisture within the plastic powder. The porosity of the layer is 7.5%. With preheating to 145 and 210 °C the microstructure of the layer in cross-section looks similar to Fig. 5.44. The porosity in these cases is 5.4 and/or 5.1%.

Metal layers are applied to plastic samples for hybrid multi-component injection moulding. In a second step they are designed to bond with a low melting tin/zinc-alloy. Zinc is chosen as the coating material to enable diffusion to take place between the coating and the melt. Layers are applied using the arc spraying process. The coating is produced with regard to a good layer adhesion and a high porosity to achieve infiltration of the tin/zinc-melt. Overheating of the plastic substrate must be prevented during the coating process.

Figure 5.45 shows the layer structure of a porous zinc coating on PBT. The porosity continues as far as the substrate to enable infiltration of the metal melt. In addition, a good layer connection should be detected as the precondition for a secure plastic and metal joining mechanism.

Atmospheric Plasma In the atmospheric pressure plasma process a plasma is generated by a high-frequency alternating field between two electrodes and is forced onto the substrate using process gas; oxygen in most cases, via the nozzle (Fig. 5.46) (Höper 2009). As well as a fine cleaning effect which is largely based on the contact of the electrons with a high kinetic energy and the resulting expulsion of contaminants, this process can also be used to chemically change and/or activate the substrate surface of plastic. The energy of the electrons and molecules released during the ionisation on contact with the surface exceeds the bonding energies of the C–C– and C–H bonds within the plastic. Free macroradicals react either during the surface pretreatment with the corresponding process gas or after the surface pretreatment with the atmospheric oxygen and form polar functional groups. This raises the surface energy and therefore the wetting ability of the substrate being treated. Additional reactions can be degradation reactions on macromolecules or wetting reactions (Gleich 2004).

Surface Energies of Joining Partners The surface energies of joining partners are determined according to the method of Owens and Wendt (1969) in which the surface energy is divided into polar and disperse portions. The contact angle measurement on the surfaces (metal: polished, plastic: cleaned) takes place at room temperature

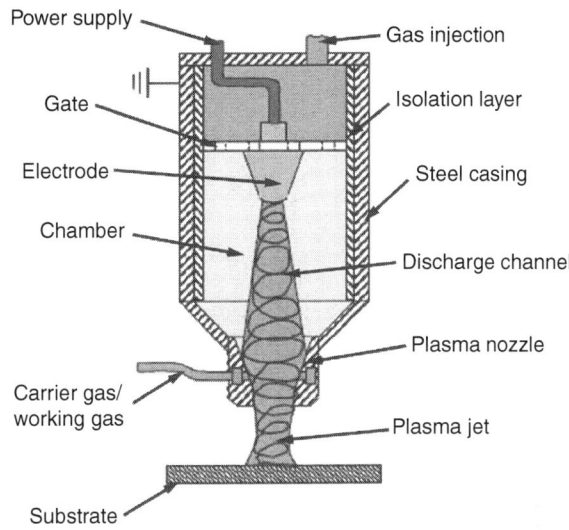

Fig. 5.46 Schematic diagram of the atmospheric plasma process. (Source: IFAM Bremen)

based on four test fluids (distilled water, glycerol, ethylene glycol, diode methane) whose polar and dispersed portions of the surface energy are known.

Influence of Plasma Pretreatment on Plastic Samples Plasma pretreatment of plastics is state of the art in adhesive bonding technology and is essential to achieving good bond strengths in the long-term. Besides cleanliness, the effect of the good adhesion is largely based on the formation of polar functional groups on the plastic surface. Energetic interactions cause these functional groups to attempt to align within the interior of the plastic. This is the basis of the temporary effect of plasma pretreatment of plastics (Hartwig and Albinsky 1999). Plasticising of the plastic facilitates this effort so that a positive effect of plasma pretreatment of the plastic is not essential. The presence of a solid element (e.g. metallic joining partner) increases the probability that the functional groups of the plastic will form physical bonding forces with the metal substrate instead of aligning themselves within the plastic bulk material.

The surface energies of the plastic samples in an untreated state and after atmospheric-pressure plasma treatment (PL) are shown in Fig. 5.47. The surface energies achieve values which can indicate good potential for adhesion. However the dispersed portion of the surface energies prevails in all untreated plastic samples. This is typical for polymers and reduces the wetting capability of the plastic surfaces. PBT and PBT GF30 exhibit in this case the greatest dispersed portion and the lowest surface energies.

Pretreatment with an atmospheric-pressure plasma leads to an increase of the overall surface energy. In this case the polar portion increases partly at the cost of the dispersed portion. In general a better wetting capability is achieved in this way. The increase results from the formation of polar functional groups on the surface. On plasma-treated PC the contact angle measurements with the test fluids resulted in a complete spread such that the surface energy could not be calculated.

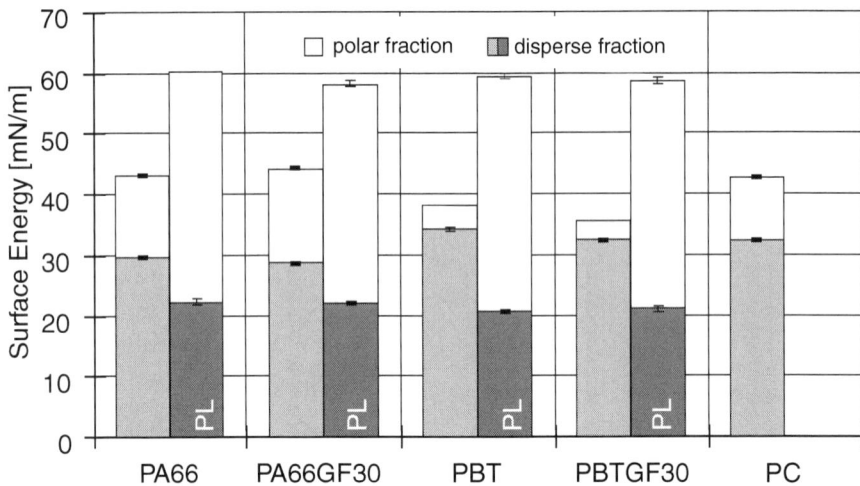

Fig. 5.47 Comparison of surface energies of the plastic surfaces before and after plasma pretreatment (PL)

As well as the surface energy, the chemical composition of the surface also changes. As expected, XPS analysis of the surface of plastic substrates which have been previously treated with atmospheric-pressure plasma particularly showed an increase of the oxygen portion and a reduction in the carbon portion (Fig. 5.48). This results in a greater reactivity of the surface for example for the formation of secondary valence bonds such as hydrogen bridges.

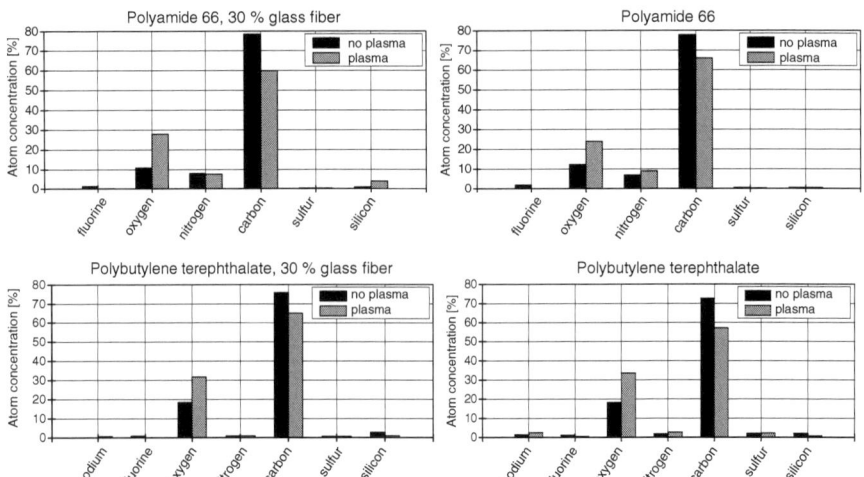

Fig. 5.48 XPS investigations for the influence of a surface treatment with atmospheric pressure plasma

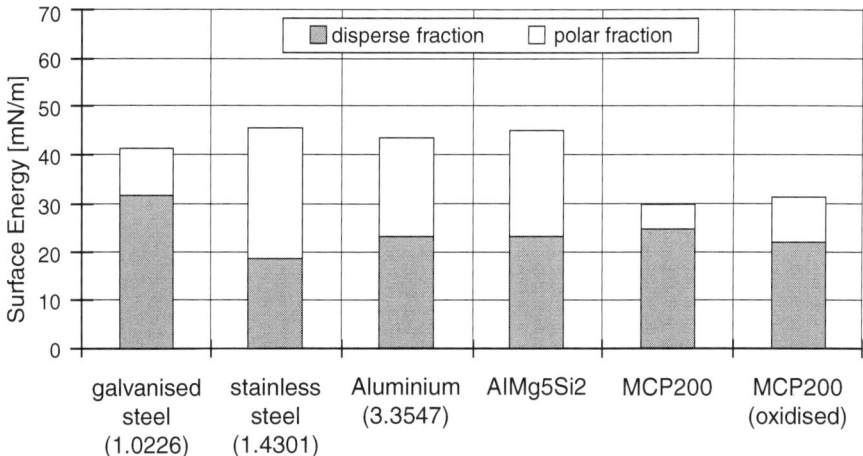

Fig. 5.49 Comparison of the surface energies of the metal materials

Surface Energies of the Metallic Joining Partners The surface energies of the metallic joining partners are graphically illustrated in Fig. 5.49. The surface energies lie in the region of those of plastics PA66, PA66GF30 and PC. In the case of the galvanised steel sample there is a strikingly high dispersed portion of the surface energy, which means poor wetting of the surface by a plastic. Good wetting is a necessary criterion for good adhesion but it is not an indicator of good adhesion.

5.3.4.3 Thermal Joining Processes

Thermal joining processes exhibit benefits with respect to mechanical joining processes and adhesive bonding. The planar force transfer and the great freedom of part geometry are an advantage as compared to mechanical joining methods and the shorter cycle times are an advantage as compared with the adhesive bonding process. Thermal joining processes are valued because of the local, selective and rapid application of energy. As with all series-joining processes, process controllability is essential.

Development of new methods for the thermal joining of metals and plastics requires the constant analysis of different types of energy application to the bonding surface. With regard to temperatures, it must be considered that plastics do not have sharply delineated melting points like metals. They have broader temperature ranges from solid to liquid state. The thermal expansion coefficients for metal and plastic are also generally very different. This can result in stresses within the component during cooling process.

Commercially available joining methods are tested for their suitability for manufacturing plastic and metal bonds against this background in a continuous benchmarking process. Criteria include locally selective and rapid energy application

Fig. 5.50 Schematic diagram of thermal conductivity joining trial

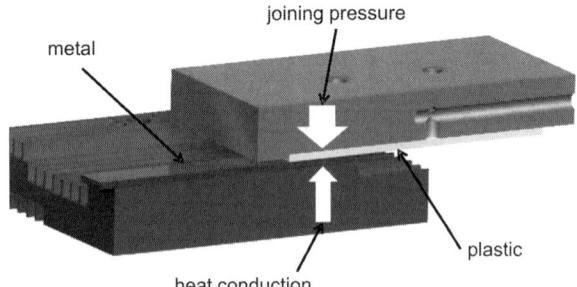

to the joining zone, process control and particularly suitability for mass production. The thermal joining processes of thermal conductivity joining, induction joining, laser transmission joining and ultrasonic joining for the manufacture of plastic and metal hybrid components are presented in detail next.

Principle of Thermal Conductivity Joining The term thermal conductivity joining describes the process which enables the joining of metals with plastics wherein the mechanism of thermal conductivity acts as the thermal transfer process. Heating cartridges are used to apply the heat. These heat the metallic joining partner until the surface temperature corresponds to the required contact temperature. Once this is achieved, a pneumatic lifting action brings the heated metal joining partner under pressure into contact with the plastic. Thermal conductance causes the plastic to plasticise and flow under the prevailing joining pressure into the surface structure of the metallic component. Figure 5.50 shows a schematic illustration of the thermal conductivity joining trial. Due to its reproducible thermal limit conditions, thermal conductivity joining is a process which is used as a reference process for determining adhesion mechanisms in thermal joining.

Results of Thermal Conductivity Joining Thermal conductivity joining is well-suited to identifying underlying influences. The focused temperature control within the process and the arrangement of the heating elements make it possible to adjust the surface temperature of the metal samples homogenously to the desired joining temperature. Therefore the influences of temperature, surface topography and material combinations can also be targeted.

Temperature Influence The limitations of the joining temperatures are firstly, the temperature at which the plastic starts to melt and secondly the temperature at which the plastic degrades. It is advisable to determine a process window between these two limits for any material combination which involves joining plastic and metal. Figure 5.51 shows the above effect based on the example of material combination PA66 GF30 with untreated sandblasted steel. There emerges an ideal temperature for each material combination at which the maximum tensile shear strength is achieved. This is 270 °C in the case of the material combination in the analysis.

5 Hybrid Production Systems

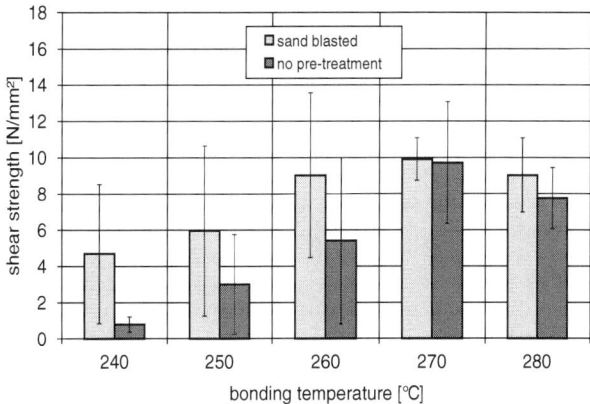

Fig. 5.51 Dependence of the attainable breaking forces in the tensile shear test on temperature for sandblasted and untreated surfaces

Surface Topography The quality of the surface has a substantial influence on the adhesion conditions. Figure 5.51 shows the attainable breaking forces for PA66 GF30. An untreated steel substrate and a sandblasted steel surface are also compared. It is noticeable that the breaking forces are generally higher in the case of tensile shear strength samples with a sandblasted metal surface. This is because plastic is able to create a higher number of bonds due to the expansion of the active surface. The plastic is also able to clamp into the structure of the metal surface. To examine the effect of surface topography in more detail, the surfaces are smoothed by lapping and roughened by sandblasting or structured by laser structuring. The results of the tensile shear strength tests are compared in Fig. 5.52. The attained breaking strengths for PC, PA66 and PA66 GF30 joined to steel with different surface topographies are compared. The samples were joined with a 20 mm overlap at T = 260 °C, the laser structuring was 5 mm wide so the laser-structured samples are only comparable to a limited degree. However it is clear that effective bonding qualities can be achieved with targeted laser structuring on a comparatively small surface. The dispersions of mechanical results in the case of PA66 and PA66 GF30 can be reduced on the basis

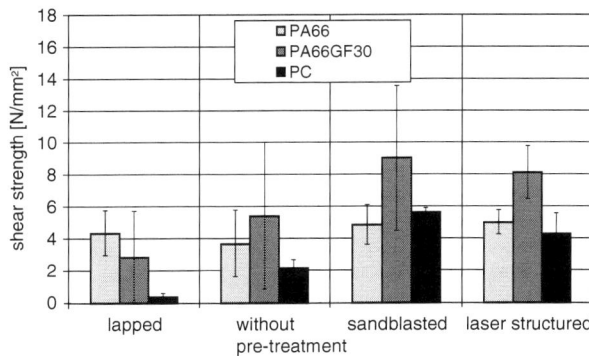

Fig. 5.52 Dependence of the achievable tensile shear strength on the variously pre-treated surfaces (T = 260 °C)

of the connection defined by the undercut. In the case of polycarbonate, the optimum joining temperature is above 280 °C. The polymer is of too high a viscosity at 260 °C to penetrate the laser structured areas. The result is less tensile shear strength.

One reason for the greater adhesion qualities is the rougher structure of the metal surface (see also Fig. 5.32). From a lapped surface to the sandblasted sample the roughness increases by a factor of 20. Confocal microscope photographs confirm a significant enlargement of the surface. However, for process-related reasons these investigation methods are not effective in demonstrating any undercuts.

Material Influence PA66 is basically effectively joined with untreated steel. The achievable breaking strengths lie in the region of the basic strength of the PA66 and the samples fail cohesively. Filling the polymer with glass fibres on an untreated steel surface leads to a slight increase of the tensile shear strength. This increase results from the increase in the basic strength of the plastic resulting from the fillers. Failure therefore occurs at a higher load. The strength of the connection however does not lie in the region of the basic strength of the PA66 GF30.

PBT is not suitable for creating strongly adhesive connections on untreated and thus relatively smooth surfaces. The same behaviour is displayed in the case of glass fibre-filled PBT GF30 in combination with untreated steel. The early adhesive failure indicates a poor connection on this smooth surface.

Pretreatment in the form of sandblasting substantially improves the adhesion conditions. Connections with PA66 display maximum loading capacities which reach the basic strength of the polymer. The breaking pattern is purely cohesive. In this case a precise estimate of the bearing strength of the connection is possible only to a limited extent. Good adhesion can be achieved with PBT on the rough surfaces. The strengths approach the range of the basic strength of the polymer and the breaking pattern remains purely adhesive. The maximum achievable tensile shear strengths in the material combination of PBT GF30 and sandblasted steel, while not in the region of the basic strength of the polymer, nevertheless a high-quality connection can be assumed because there is a mixture of adhesive and cohesive failure in this case. Failure occurs both in the joining zone as well as in the plastic. The same applies for PA66 GF30. PC achieves even higher bonding strengths on sandblasted steel than on untreated surfaces. These values however do not extend to the basic strength of the polymer so there is no obvious potential in this case. Join failures are purely adhesive.

The mechanical tests reveal substantial differences in the tensile shear strengths. These can be traced firstly to the polymers and secondly to the pretreatments. There are hardly any explanations for the adhesion phenomena which occurred. However there are differences with respect to the material in the molecular structure and in the thermal performance of the polymer. With PA66, whatever the substrate, higher strength can be achieved than with PBT which is partly due to the change of the specific volume during the cooling process (Fig. 5.53).

Since PBT has a greater degree of crystallinity there is a greater change of the specific volume in the case of PBT than in the case of PA66 such that the material contracts more during the cooling process. This volume contraction is accompanied

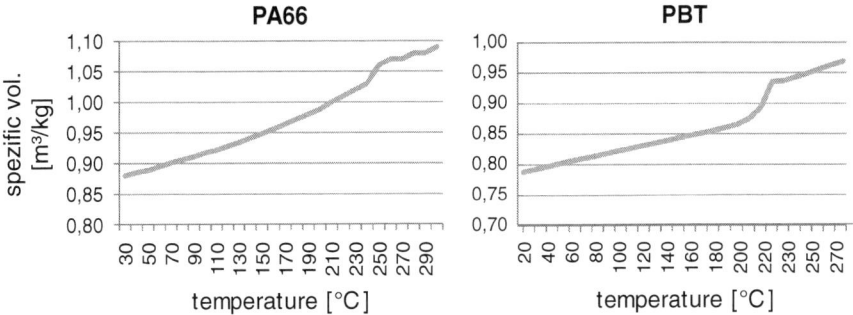

Fig. 5.53 Change in the specific volume in the case of PA66 and PBT depending on the temperature

by lengthwise shrinkage such that the PBT retracts out of the surface structures of the metal. Samples with an initial adhesion of the PBT fail, with no external exertion of force, due to the above-mentioned shrinkage behaviour. Furthermore by the nature of the material PBT shrinks more than PA66 in the 24 h after processing. This also affects the payload of the joins.

High thermal stress degrades polymers in their molecular weight. This thermo-oxidative degradation is associated with degradation of the mechanical properties and can be verified by Gel-Permeation Chromatography (GPC) Analysis. The analysis of the plastic bonded to the metal has shown that molecular weight loss occurs near the surface of the PA66. In this case however the shift of the molecular weight distribution to lower molecular weights favours adhesion this case. Shorter chains have increased movement and can form micro-clamps more effectively. The increase in the proportion of molecular weights in the range 1,000–15,000 g/mol supports the assumption that an adhesive boundary layer forms in the plastic. This effect is not observed in the case of PBT.

A heterogeneous picture also emerges with the pretreatments. Coating steel with a zinc layer does not improve adhesion conditions. Rather, it affects the bonds between the polymer and the metal surface negatively. The result is less strength. Lapping the untreated metal surfaces also considerably reduces the maximum tensile shear forces. Sandblasting has been found to be a suitable means of efficiently and reproducibly improving strengths in the joining zones. In fact a cohesive failure pattern emerges in that the samples fail in the polymer material. The actual surface after a sandblasting process is increased around fourfold and therefore offers the polymer more joining surface.

Initial investigations with laser-structured metal surfaces confirm the assumption that this method leads to an improvement of the bond. In tensile tests cohesive failure occurs in the glass fibre-filled PA66 GF30 polymer. The penetration of the molten thermoplastic into the laser-generated surface structure are the drivers of this phenomenon. Based on the findings from the thermal conductivity joining process recommendations for process control within other thermal joining processes can be derived.

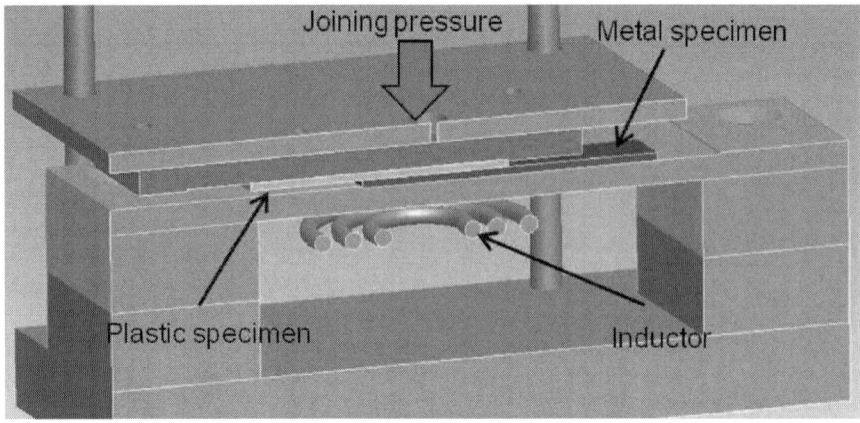

Fig. 5.54 Schematic diagram of induction joining of plastic and metal

Principle of Induction Joining Induction joining is pursued in order to transfer the basic findings of the thermal conductivity joining to a more industrially-relevant process. The principle of induction heating is used in various industrial applications. The advantage of fast, zero-contact energy application qualifies this as a productive process for producing joints between plastic and metal. The joining partners can be brought into contact either before or after induction heating of the metal components. A constant pressure is applied during the joining process (Fig. 5.54) to minimise heat loss and guarantee better wetting of the metal surface. A high frequency alternating current voltage generates an alternating magnetic field around the inductor. This magnetic field induces a turbulent flow within the metallic joining partner. The heating is dependent upon the electrical resistance of the metallic joining partner. The plastic joining partner plasticises as a result of thermal conductance and wets the metal surface. The joining pressure is maintained throughout the cooling phase to prevent cooling effects from loosening the join.

Results of Induction Joining The thermal joining of metal and plastic hybrids by induction heating permits high process speeds. The energy can be applied from the metal side or from the plastic side. The use of robot technology allows great flexibility of the joining geometry. The induction coil can be adjusted according to the joining surfaces. Since current and thereby heat can be induced by the induction coil in many materials as well as in the actual semifinished products to be joined, particularly the technical design of the joining point and the use of handling equipment such as clamps is very complex. Furthermore, process control is a decisive factor in producing permanent joins. Thermodynamic conditions in the joining surface (temperature and pressure profile) are fundamental influencing factors in this case.

Heat Control Metal surfaces can be heated very quickly using the induction technique. With corresponding contact pressure and the associated increase in the thermal transfer coefficient, the plastic can be brought rapidly to its processing state. The "skin

5 Hybrid Production Systems

Fig. 5.55 Bond strength depending on the temperature

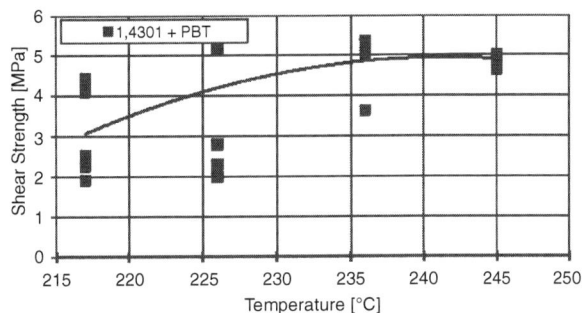

effect" which occurs during the induction technique enables short cooling times, particularly in the case of thick walled metal samples. At the same time, different temperature profiles with different heating and cooling rates can be pre-programmed into the induction equipment controller for example to counteract over-rapid cooling of the plastic. The essential process parameters for influencing the temperature profile are induction capacity, induction duration and distance from the metal surface.

When the plasticisation phase is reached, the plastic can wet the metal surface. At higher temperatures the plastics viscosity is lower. This firstly improves the wetting and secondly it offers the chance of mechanical clamping by the penetration of the molten plastic into the indentations of the rough surface of the metal component.

The dependence of the temperature on the achievable bonding strengths in the tensile shear test is illustrated in Fig. 5.55 based on the example of the PBT/stainless steel (1.4301) material pair. The temperatures were measured and averaged by thermocouples at five points directly within the joining zone. The maximum bond strength is achieved when the temperature is rising since the viscosity of the plastic is reduced and it can penetrate in the surface roughness. Once the maximum level has been exceeded, the reduction of the supporting cross-section due to an increase in the amount of melt forced out. Furthermore the degradation of the plastic at extreme temperatures reduces the bond strength.

Pressure Profile Pressure control is not initially essential for creating physical adhesion forces between molten plastic and a metal surface. However pressure can be used to force the plastic into cavities in rough materials or specifically roughened surfaces and thereby to wet surfaces which would not be wetted purely by the energetic characteristics of materials in contact with one another. In addition, the pressure in the cooling phase can also make a decided contribution to increasing the bond strength between the plastic and metal. As the plastic cools, the process of shrinkage can cause it to escape from the cavities in the metal. In the limit case, the shrinkage forces exceed the existing adhesive forces at the metal surface which leads to immediate failure of the sample during the cooling process. Holding pressure is also essential to guarantee permanent bond strengths.

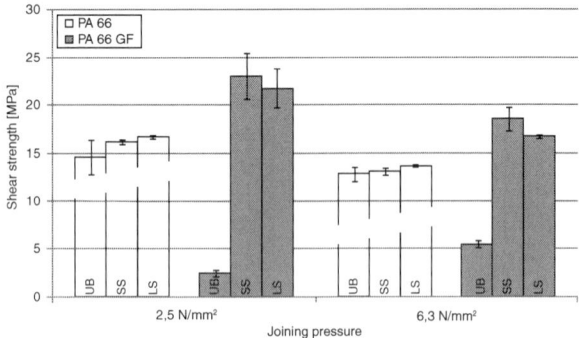

Fig. 5.56 Bond strength depending on pretreatment and pressure

If the pressure is too great however the resulting molten plastic can be forced out of the joining space which will lead to inadequate wetting and a reduction of the supporting cross-section.

Figure 5.56 shows the influence of pressure on the bond strength based on the example of tensile shear test specimen with 5 mm overlaps. The metal and plastic substrates are 20 mm wide and 2 mm thick. A glass fibre-reinforced polyamide (PA66 GF) was used along with an un-reinforced polyamide (PA66). Stainless steel (1.4301) was used on the metal side. Three different metal surface structures were investigated: untreated (UB), laser structured (LS) and sandblasted (SS) surfaces.

Firstly the high transferable forces, even with small overlaps, are evident in the un-reinforced as well as in the reinforced polyamide. While adhesive forces formed in the un-reinforced polyamide as well as in the untreated metal surfaces can enable cohesive failure in the base material of the plastic, there is clearly a need for pretreatment of the surface in the case of the glass fibre-reinforced plastics to achieve adequate bond strengths.

Furthermore, in the case of the un-reinforced polyamides, there is a clear influence of the increased melt flow as the applied pressure is increased. There is cohesive failure of the plastic in the base material at a surface pressure of 2.5 N/mm^2 and at a surface pressure of 6.3 N/mm^2. The bond strengths fall due to the reduction of the supportive cross-section. There is adhesive failure of the tensile shear test specimen with a reinforced polyamide in the metal/plastic boundary layer irrespectively of the pressure. The evident reduction in the bonding strength therefore is not due to a narrowing of the cross-section.

Analysis of the breaking surface of the plastic using a scanning electron microscope (SEM) shows a significant increase in the fibre content in the boundary layer at increased pressure (Fig. 5.57). Spray-coated glass fibre-reinforced components have a low-fibre edge layer. At increased pressure and associated greater melt flow, the proportion of fibres in the boundary layer is greater. These glass fibres cannot form physical bonding forces with the metal which are comparable with the plastic matrix. This explains the loss of bond strength in the case of the PA66 GF30.

Low Pressure   High Pressure

Fig. 5.57 Glass fibre content in the breaking surface in the case of high and low contact pressure

Plasma Pretreatment of Thermally Joined Metal/Plastic Hybrids The Influence of surface pretreatment of the plastic is illustrated using amtospheric pressure plasma which is a state of the art pre-treatment for adhesive bonding. The bond strengths of thermally joined and plasma pre-treated tensile shear testsamples with 20 mm overlaps are illustrated in Fig. 5.58 in comparison with samples which have not been physically pre-treated. Glass fibre-reinforced PBT and PA66 have been joined to stainless steel (1.4301).

The glass fibre reinforced PBT-samples fail significantly below the strength of the base material. Nevertheless the breaking pattern is partly cohesive. The PBT samples which are not plasma-treated differ neither in strength, breaking pattern, nor distribution from the plasma treated samples.

There is a threefold increase in strength with the plasma pretreatment of glass fibre-reinforced polyamides on untreated stainless steel substrate. It is impossible to tell exactly whether the significant increase is based on cleaning effects, embedding of strongly electro-negative atoms or strong dipole forces based on the resulting functional groups. The breaking pattern shows partially a failure in the plastic close

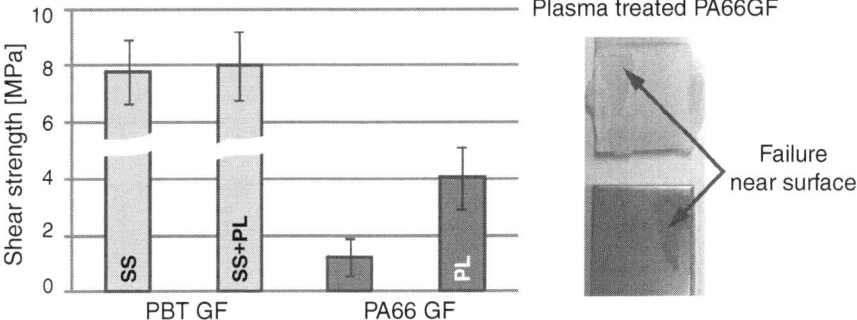

Fig. 5.58 Thermal joining (induction joining) of plasma treated samples

to the boundary layer. Local damage to the plastic due to overtreatment with plasma radiation is improbable due to the considerable increase in adhesion. More likely is the formation of a Weak Boundary Layer wherein the molecules and functional groups located at the surface create physical bonds with the surface of the metal. The cohesive strength of the plastic, which is based on physical bonds amongst other things, is therefore reduced in the boundary layer area and thus forms the weakest point of the join.

Principle of Laser Transmission Joining Laser transmission joining is relevant to industrial practice and is therefore investigated along with induction joining. It is also worth applying the knowledge of the joining mechanisms gained from heat conduction joining to the temperature and process control of this process.

In the laser transmission joining of plastic and metal the laser beam passes through the laser-transparent thermoplastic join partner (plastic) facing the beam source. Electromagnetic energy converts practically all the transmitted radiation into heat in the layer of the metal joining partner close to the surface in the region of the joining zone. The joining zone is thereby heated. The thermoplastic join partner in contact with the metal surface is plasticised by heat conduction. The contact of the two join partners causes wetting of the metal surface under the influence of the externally applied joining pressure.

Due to the focused application of energy, the plastic is plasticised only locally. This thereby largely avoids thermal and mechanical damage of the components. The irradiation strategy for joining plastic and metal, as with laser transmission welding of plastics, can be divided into three irradiation strategies. These are the contour process, the simultaneous process and the quasi simultaneous process. Contour joining (Fig. 5.59a) is the simplest and most flexible irradiation strategy. The processing optic is moved relatively along the joining seam. This is carried out either by moving the laser beam or by moving the part.

In general, the process is characterised in that every point of the welding contour interacts just once with the laser beam. The size of the interaction depends on the beam diameter in focus and the feed rate.

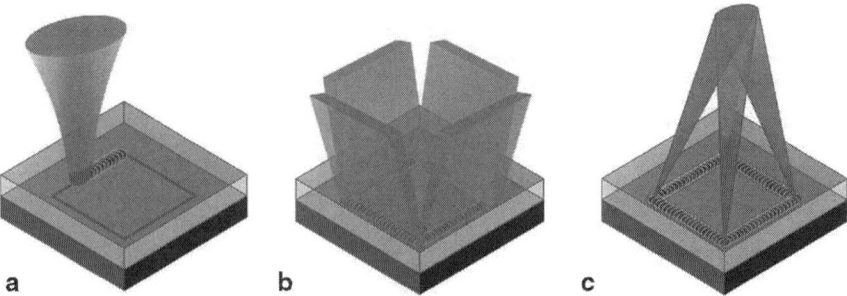

Fig. 5.59 Irradiation strategies with laser beam joining: **a** contour joining, **b** simultaneous joining and **c** quasi-simultaneous joining

As the beam travels along the entire seam contour, fusing times depend on the length of the seam and the selected feed rate. However this means that typically only a small laser output power is required.

In simultaneous joining (Fig. 5.59b) the entire seam contour is irradiated simultaneously. The process time is therefore very short. The longer interaction time as compared to contour joining leads to greater seam strength. Simultaneous irradiation requires high initial laser output powers with low intensity. Due to the short process times < 1 s the simultaneous process is ideal for large quantities. Quasi-simultaneous joining (Fig. 5.59c) combines contour welding with simultaneous joining. The laser beam is generally moved several times across the seam geometry by means of a galvanometric scanning system at a high feed rate so that it is simultaneously plasticised. Flexible beam control also guarantees a high flexibility in terms of the seam contour. The high feed rates mean that a high laser output power with equally high intensity is required.

Simultaneous joining and quasi-simultaneous joining enable the bridging of large gaps and joining with a melt-off path because of the plasticising of the entire seam contour.

Results of Laser Transmission Joining

Influence of the laser output power To ensure comparability with the other joining processes, the design must permit simultaneous irradiation of the entire joining surface (20 × 20 mm²). This is made possible by the adjustment of a focusing optic. A high laser output power (4 kW) with a short processing time is required to heat only the surface of the metallic joining partner after the laser beam passes through the plastic. The laser power is therefore emitted as a dual pulse with a pulse pause of 50 ms. The pulse length is varied to determine its influence on the consequently changing surface temperature. PBT and sandblasted stainless steel are selected to achieve comparability with other joining processes. The results of the tensile shear strength dependent upon the radiation duration are shown in Fig. 5.60.

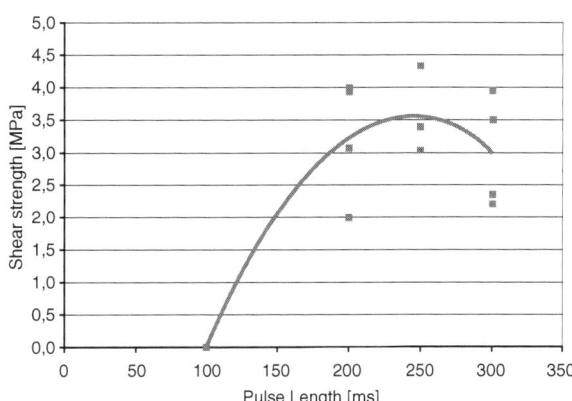

Fig. 5.60 Influence of the pulse length on the tensile shear strength of the combination of PBT and stainless steel (1.4301)

Fig. 5.61 Joining zone in the case of simultaneous joining (plan view)

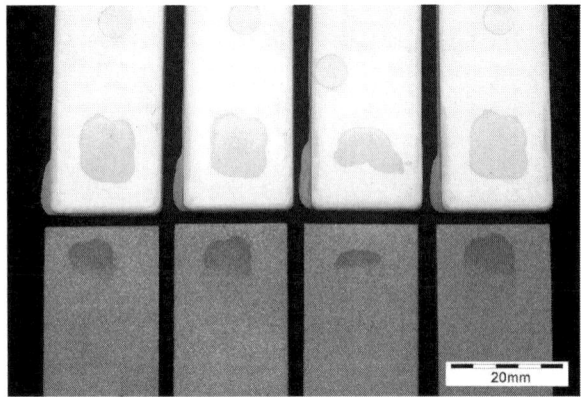

It is found that joining occurs provided there is sufficient surface temperature but it can be improved by increasing the temperature. If the temperature is too high, the plastic degrades and the strength of the join is lower. The analysis of the joining zone shows good adhesion of the plastic only in the centre of the sample over an area of approximately $10 \times 10\,\text{mm}^2$ (Fig. 5.61). The tensile shear strength is nevertheless related to the overlap surface to allow a comparability. The difference is firstly caused by a lower joining surface temperature through thermal conductance in the outer areas of the joining zone. A further effect is scattering of the radiation at the crystalline areas of the PBT such that the laser energy is significantly greater after passing through the highly scattering PBT in the joining zone towards the centre of the sample. Higher temperatures therefore result here.

Influence of Laser Structuring Laser transmission joining tests are used to compare the different laser structures. The same joining parameters (laser output power, feed rate, irradiated area, join pressure) are used for each sample to guarantee comparability. In addition, just one material combination is investigated, namely stainless steel 1.4301 and polycarbonate (PC). The transparent PC does not scatter the laser radiation and is therefore better suited for comparison tests than the semi-crystalline materials PBT and PA66.

The results show that at least two passes are required in the linear structuring process to achieve a reliable bond. This confirms the assumption that bonding is not possible without interlocking. An additional pass narrows the cross section at the throat of the structure and reduces the area which has a decisive effect on the tensile shear strength. This reduces the tensile shear strength but increases tensile strength so that the optimum structure must be determined on the basis of the local stress profile.

The variation of the distance of the linear structures permits the determination of a relationship between structure density and tensile shear strength. The results are shown in Fig. 5.62. An increase is measured thereafter with the reduction of the structure spacing. If only interlocking is responsible for the bond, the tensile

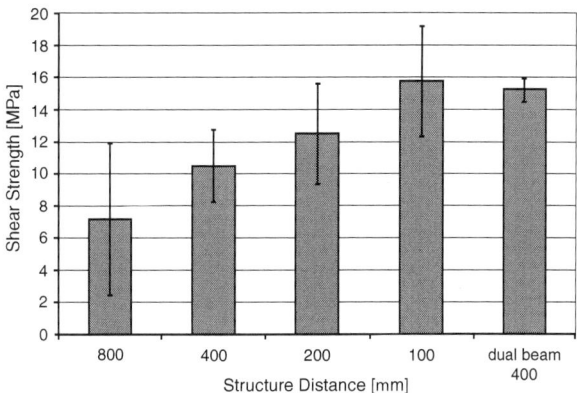

Fig. 5.62 Influence of the structure spacing on the tensile shear strength (PC + 1.4301, overlap length 4 mm)

shear strength would be doubled if the structure spacing was halved. Since this is not the case, adhesion must also increase the bond strength. With these two results it is possible to define the structure spacing and thus the surface pretreatment outlay depending on the stress case. In the comparison with the structures created with the dual beam optic it is clear that these achieve a better tensile shear strength despite a structure spacing of 400 μm than vertically structured samples. Otherwise in the case of vertical structuring the same strengths can only be achieved with four times the structuring effort. It should be noted however that the cross-section and therefore the supporting surface of the dual beam structure is also greater. The significantly lower standard deviation of the tensile shear strength indicates reproducible interlocking. However, the increased process control requirements for structuring using the dual beam optic have to be taken into account. Considerably greater requirements are placed on the accuracy of the focal/intersection points and processing with a flexible galvano scanner is not possible. In the light of increasing miniaturisation, the smaller the area to join, the greater the local strengths through clamping must be so that in this case the dual beam undercutting process is of interest.

Principle of Ultrasonic Joining Along with laser and induction, ultrasound can also be used to apply energy to local component areas for heating. High-frequency mechanical vibrations are used in ultrasonic joining. Based on the hysteresis loss resulting from the damping coefficient, dissipation of mechanical vibration energy causes plasticising of the plastic in the joining zone.

An ultrasonic welding machine comprises basically six essential components. The ultrasound generator, the converter, the booster, the sonotrode, the anvil and the welding press which generates a joining force. Since they are highly efficient, piezo-electric converters are used today to convert alternating electrical voltages produced by the generator into mechanical vibrations. The longitudinal vibrations transmitted by the sonotrode lie within a frequency range of 20–70 kHz and have amplitudes in the region of 10–80 μm. A joining pressure applied by the sonotrode causes the energy director to melt. The molten liquid plastic flows under joining pressure out

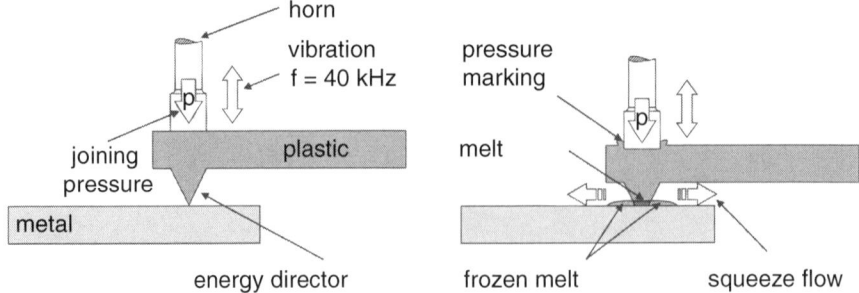

Fig. 5.63 Schematic diagram of test with ultrasonic joining

of the joining zone over the metal surface. Figure 5.63 illustrates one example of this process.

Results of Ultrasonic Joining In the case of ultrasonic joining it was found that no adhesive bonding of metal and plastic hybrid components is possible without pre-treatment of the metal substrate because the ultrasonic energy alone cannot generate enough heat in the joining zone. Preheating the metal substrate to surface temperatures of around 180 °C, close to the melting temperature of the plastic, can produce an adhesive bond between the plastic and metal. This approach however requires long heating times.

An alternative therefore is to coat the surface of the metal substrate with a plastic layer. Plastic-to-plastic welding is carried out in the downstream joining process so that the short joining times of 0.1 s to 1 s, which are typical for the ultrasonic welding of plastics are maintained (Ehrenstein 2004). Figure 5.64a shows the results of the tensile shear tests of ultrasonic welded joining composites of PA11-coated steel samples (1.4301) with the plastic PA66. The layers produced with an optimum

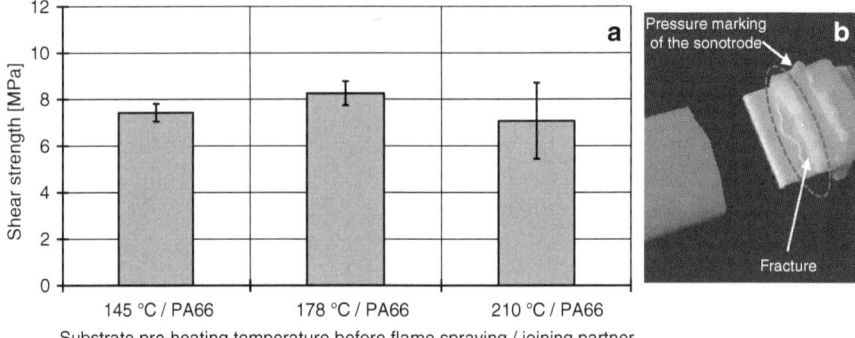

Fig. 5.64 a Result of the tensile shear tests from joins between joining partners of PA11-coated steel samples with PA66, **b** Typical breaking pattern of an ultrasonically welded PA11-coated steel and PA66 combination after the tensile shear test

substrate preheating of 178 °C result in the best bond strengths. The reason for the high standard deviations in the case of preheating to 210 °C may be due to the degradation of the plastic coating. 70% of the bonds failed cohesively (Fig. 5.64b). There is no evidence of failure between the layer and the substrate.

The results of the tests of ultrasonic joining of plastic and metal hybrid parts with a plastic flame-coated interim layer of PA11 thereby show that high bond strengths can be achieved with short process times. The metal samples must be provided with respectively appropriate plastic coatings for use of this process with other thermoplastics.

5.3.4.4 Hybrid Multi-component Pressure Die Casting (M-HPDC)

The development of the hybrid multi-component pressure die casting process for manufacturing metal and plastic parts is based on the further development of the traditional metal pressure die casting process. For better understanding the state of the art and the way the process works, a brief insight will be given into the pressure die casting process at this point.

The pressure die casting process is the commercially most productive casting-based production process for the primary shaping of aluminium or magnesium and is particularly suited to the manufacture of large series components such as engine housings, cylinder heads and all types of housings. It is increasingly used in the production of hard-to-cast thin-walled structural components, e.g. space frame components.

Pressure die casting permits the manufacture of true-to-size castings with an outstanding surface quality ($R_z < 25$ µm). Aluminium, magnesium and zinc are the most common casting materials in this process. Due to the rapid solidification, particularly in the case of thin-walled castings, the castings possess finely developed microstructures whose mechanical properties can be expected to be higher in principle than comparable parts made by using other casting processes. The mould filling is however very turbulent so that oxides and gas inclusions in the type of porosity reduce the theoretically achievable mechanical properties.

With the hot and cold chamber pressure die casting process there are two process variants which differ in terms of connecting the shot sleeve. The hot chamber pressure die casting process is not suitable for the casting of aluminium because molten aluminium is highly aggressive compared to iron materials. However, both magnesium and aluminium can be processed in the cold chamber pressure die casting process. These are the two most important materials in terms of their usability according to technical standards. Moreover, with a view to the new hybrid process only the cold chamber pressure die casting process will be analysed below (Fig. 5.65).

In the cold chamber pressure die casting process, the shot sleeve is located outside the melting bath. The shot sleeve is loaded mechanically from a melting furnace which is separated from the die casting cell. Within pressure die casting, the molten alloy is poured under pressure into a permanent mould. The occurring metal pressures

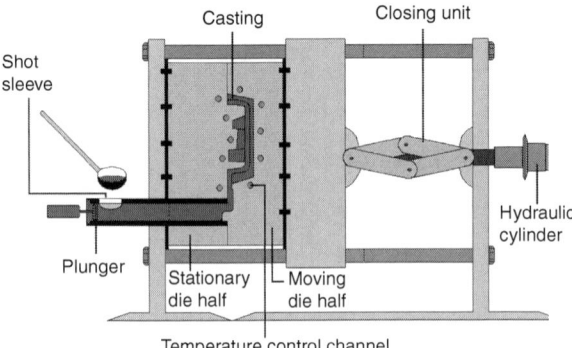

Fig. 5.65 Basic design of a cold chamber die casting machine

can reach values of between 10 and 200 MPa and the pouring rate can reach values of between 30 and 150 m/s. Mould filling times are therefore counted in milliseconds.

Comparing the process times of pressure die casting and injection moulding, the pressure die casting cycle is many times quicker with comparable part volumes and geometries. This is particularly due to the low heating capacity and higher thermal conductivity of the metal. Due to the high melt- respectively casting temperatures, these are approximately at 680–740 °C for aluminium alloys, and the very low viscosity of the molten metal results in a significantly reduced lifetime of the dies as compared to injection moulds. Also, in the case of pressure die casting, at this point there are no known applications comparable with the multi-component injection moulding process. Unlike die casting tools, which are made in two parts (one moving and one fixed mould-half) and which ideally have core pullers and cam finger core pullers, special mould systems have been developed for injection moulding e.g. family moulds, stack moulds, turning stack moulds and tandem moulds (Brunhuber 1991). One of the first machines for hybrid-multi-component pressure die casting was presented at the 11th International Foundry Trade Fair in 2007 by the company Urpe GmbH, Remscheid. This first hybrid machine is a hot-chamber pressure die casting machine with an integrated injection unit for the positive-locking connection of zinc- or magnesium diecastings with plastic. This application demonstrates the great potential of combining primary forming processes for the manufacture of plastic/metal hybrids.

Development of the Hybrid Mechanical Engineering The development of the process technology for the combined process is concentrated on the expansion of a horizontal cold-chamber pressure die casting machine. The concept is of a stand-alone injection moulding unit which is adapted to the pressure die casting machine. The design is limited to a simple control linkage and a modular mechanical coupling. Plastic-add-on units are state of the art in multi-component injection moulding.

For the combined process a stand-alone add-on unit with its own controller is positioned in the L position vertically to the pressure die casting machine at the height of the mould parting level (Fig. 5.66).

5 Hybrid Production Systems

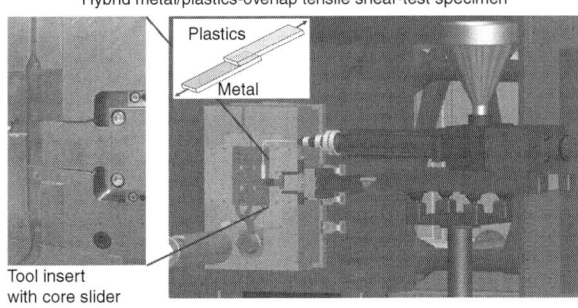

Fig. 5.66 Tool and machine technology for the manufacture of a metal/plastic overlap tensile shear test specimen

The controllers of the two machines are connected together via an interface. In this case controller signals from the casting machine are used and forwarded to the injection moulding machine. For example the signal switching from the shot (2nd phase) to the holding pressure phase (3rd phase) is used by the pressure die casting controller to start the pre-programmed cycle for injection of the plastic through the injection unit. The overall cycle of the new multi-component pressure die casting process looks like this:

- The opened mould is sprayed with a release agent after removal of the hybrid component.
- The mould is closed and the slider is inserted.
- Molten metal is fed into the shot sleeve.
- The plunger retains the melt in the shot sleeve (1st phase).
- The tool cavity is filled by initiating the shot (2nd phase).
- The cycle switches to the holding pressure phase (3rd phase) and the start signal for the add-on unit is initiated.
- The slider is pulled by the pressure die casting machine which frees the cavity for the plastic. Simultaneous operation of the programmed cycle of the add-on injection unit.
- Injection cycle: the nozzle driven up and attached to the die, the plastic is injected, than dosing and retracting of the nozzle.
- The die is opened once there is a sufficient dimensional stability of the hybrid-component.
- The ejector is actuated and the hybrid component is removed.

Development of Hybrid Die Technique The aim of the new multi-component pressure die casting process is to run both processes, casting of the metal and injection of the plastic, in a single die. In this case according to the casting of metal a cavity should be opened, as far as possible when the die is closed in order to inject the plastic. The implementation of a form closure enables a reproducible bond of both materials. However this is extremely difficult to achieve with the current die casting tool technology. For the development and validation of the new hybrid multi-component pressure die casting process the focus, based on reports from the literature (Zhao 2001) and the analyses from the "Thermal joining" project, should be on an adhesive bond between metal and plastic. This should be primarily achieved by temperature control and by working "from first heat". The results demonstrate that the heat from the die casting component casted in the first step can be used to achieve an excellent bond strength.

To investigate the effect of temperature of the aluminium casting on the adhesive bond strength depending on the adjustable process parameters, the tests are carried out initially with a simple geometry. The overlap tensile shear test specimen can be transferred into the die due to its geometrically simple shape with a plate slider. In context of the research work the test specimen operates as a reference specimen against which comparisons can be made within the overall project. Both metal and plastic components have a length of 100 mm and a width of 20 mm. The test specimen can be manufactured with a wall thickness of 2 or 6 mm using corresponding tool inserts. The overlap length is 40 mm.

The overlap tensile shear test specimen is positioned in the modular die of the multi-component pressure die casting process such that the plastic can be injected from the side via the mould parting level as shown in Fig. 5.67. There is also sufficient clearance to implement three heating cycles for each half of the mould. Therefore the component can be respectively heated in the plastic, overlap and metal areas with the heat transfer media oil or water.

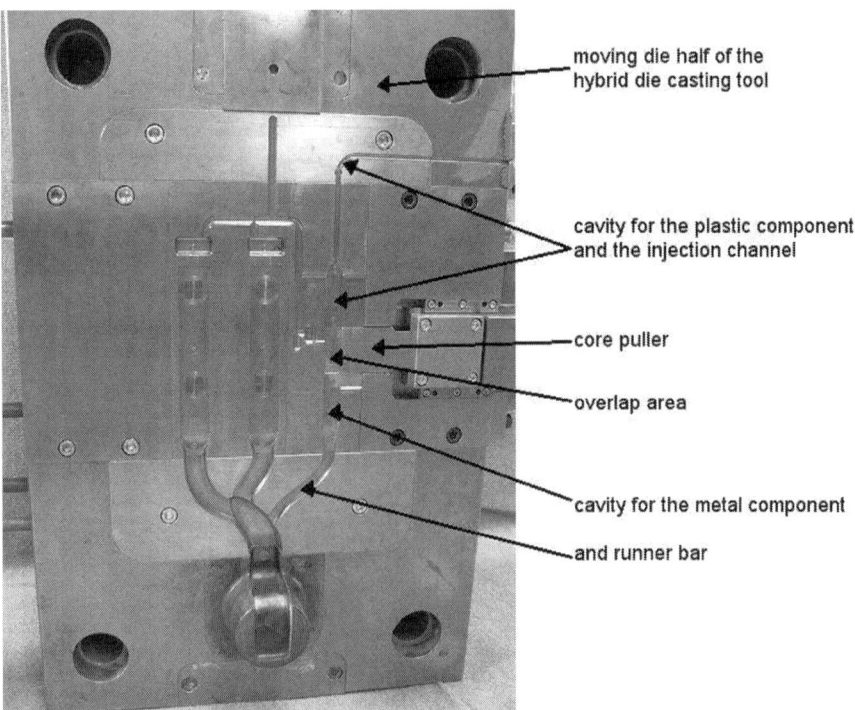

Fig. 5.67 Tool for the manufacture of a plastic/metal tensile shear test specimen

Results of Hybrid Multi-Component Pressure Die Casting At the start of the investigation of the new hybrid multi-component pressure die casting process, the process parameters of the individual processes of die casting and injection moulding are selected such that both melts are cast and/or injected with a minimum loss of heat into the mould cavity. This means for the adjustment of the casting profile in pressure die casting, that the casting temperature of the melt is set to the maximum level according to the manufacturer's information and the velocity of the plunger for the first phase is set to a maximum speed based on the casting simulation data. The parameters of the pressure die casting machine are monitored during the first tests and the tool temperature is reduced sufficiently to produce a high-quality aluminium part with no casting defects. The casting parameters can be kept constant for all tests. This means that in the tests the heat balance of the die and the injection parameters plus the melt temperature of the plastic are changed to enable a process window to be defined for the joining of the overlap tensile shear test specimen. The use of a release agent, currently imperative in the pressure die casting process, is also reduced to a minimum and the overlap area of the hybrid tensile shear test specimen is not sprayed directly with release agent by the spraying system. Nevertheless a wetting and contamination of the cavity with release agent in the area of the bonding surface can be caused indirectly during the casting process by the vapour and the inflowing metal.

The challenge in the evaluation of the process window lies in selecting a suitable die temperature. Temperature adjustment tests are carried out initially to test the cooling profile of the die for the overlapping tensile shear test specimen at different temperature settings and sample thicknesses. For the temperature adjustment tests it has been possible to incorporate a thermocouple into the slider in the area of the bonding surface in order to record the die temperature profile in the overlap area. The slider has to be actuated in all other tests which means that it is no longer possible to use the thermocouple in the slider.

The cooling characteristics of the thermocouple from the overlapping area for the 2 mm and 6 mm thick aluminium samples are illustrated in Fig. 5.68. By observing the cooling process within the temperature range of over 200 °C, which is required for the bond as a result from the findings about thermal joining, it can be realised that the temperature of the 2 mm thin sample is below this value after just two seconds. Due to its geometry the 6 mm thick sample has stored more energy in terms of heat than the 2 mm thin sample. More time is required to dissipate this energy as is shown in the temperature profiles. For the hybrid multi-component pressure die casting process the cooling time of two seconds is too short compared to the five seconds required by the 6 mm sample because the pressure die casting controller cannot actuate the slider for at least two seconds after the shot (2nd phase). The operator cannot change this setting. On this basis, the tests are continued with the 6 mm thick sample.

The manufacturer's recommended die temperatures which were used at the start of the analysis proved to be inadequate to create a bonding between metal and plastic. To further slow down the cooling of the casting it would require an increase of the die temperatures at least in the area where the metal is poured. The temperature settings for the mould had to be increased close to the melting point of the respective plastic

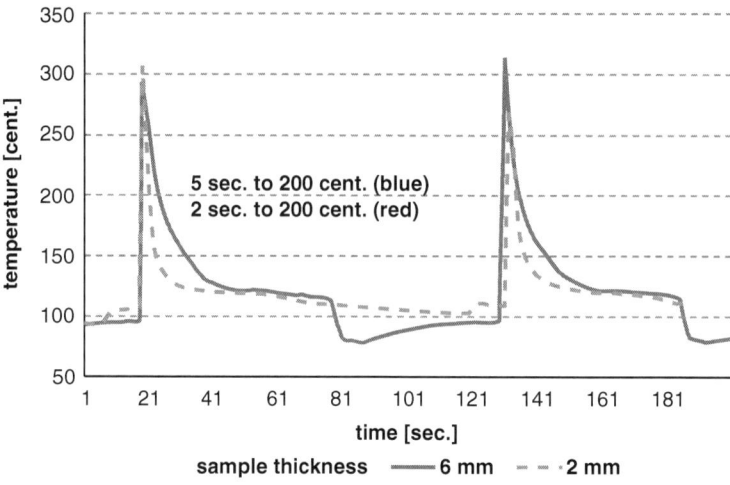

Fig. 5.68 Temperature profile in the overlap area of a 2 and a 6 mm thick aluminium sample (oil temperature 240 °C, water temperature 120/90 °C (fixed/moving mould halves))

to achieve a bond. The resulting slower cooling of the temperature after the injection moulding means that the plastic samples possess reduced dimensional-stability at the opening of the tool so that not every sample removed for analysis is suitable due to a deformation. The controller of the pressure die casting machine allows only a maximum 90-s waiting time before the die must be opened. A decrease of the die temperature can only be done at a minimum interval. Otherwise the plastic will not adhere to surface of the aluminium component. The result of a very slow decrease in temperature is a cycle time which is many times longer than the pressure die casting or injection moulding process. The results for the die temperature settings and the resulting tensile shear strength are shown below based on a material combination of aluminium (AlZn10Si8Mg) and polyamide (PA6 GF30).

Table 5.6 and Fig. 5.69 show the temperature settings with which the series of tests with polyamide 6 GF30 was started. The fixed die half is heated to 110 °C in the plastic area (KB), to 220 °C and/or 240 °C in the metal area (MB) and to 210 °C in the overlap area (ÜB). The moving die half is heated to 110 °C in the plastic and overlap areas and to 240 °C in the metal area. The temperatures for the heating zones of the plasticisation cylinder of the injection moulding add-on unit are taken for example from the recommendations of the polymer manufacturer. They are adjusted to the hybrid process in a preliminary series of rheological tests. The basic settings of the mould temperature control which are shown below result initially in a slight adhesion of the plastic to the aluminium.

Table 5.6 Injection unit temperature settings for polyamide 6 GF 30

Nozzle heating zones	T1	T2	T3	T4	T5
Temperature (°C)	280	285	270	250	220

Fig. 5.69 Mould temperature control at the beginning of the series of tests with plastic PA6 GF30, KB: Plastic area, MB: Metal area, ÜB: Overlap area

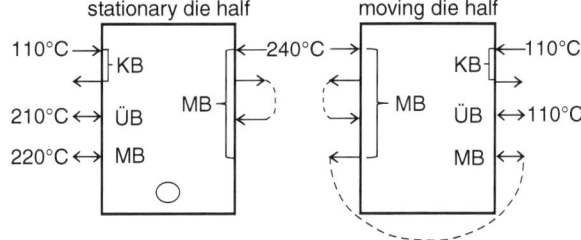

However, these settings are not sufficient to measure tensile shear strength. To obtain a stable adhesion over a large surface area the temperatures are increased gradually in the overlap and plastic area of the die. The corresponding increase in strength of the adhesive bond can be observed in Fig. 5.70.

The temperature in the overlap area was therefore set to 230 °C and the temperature in the plastic area left at 110 °C. While this temperature change permits bonding, the strength achieved is very low. This low strength is caused by the insufficient bonding surface. The plastic is bonded to the metal only in the lower part of the overlap area. In order to extend the bonding surface upwards, the temperature in the plastic area (moving and fixed die halves) was set to 140 °C. This resulted in a marked increase of the bond strength. Increasing the plastic melt temperature by changing the temperature setting in the first and second heating zones of the plasticisation cylinder (HZ 1, 2) results also in an increase in strength. A further increase of the temperature in the plastic area (moving and fixed die halves) to 160 °C shows no clear benefit as compared to 140 °C. Subsequent increase however to 180 °C is

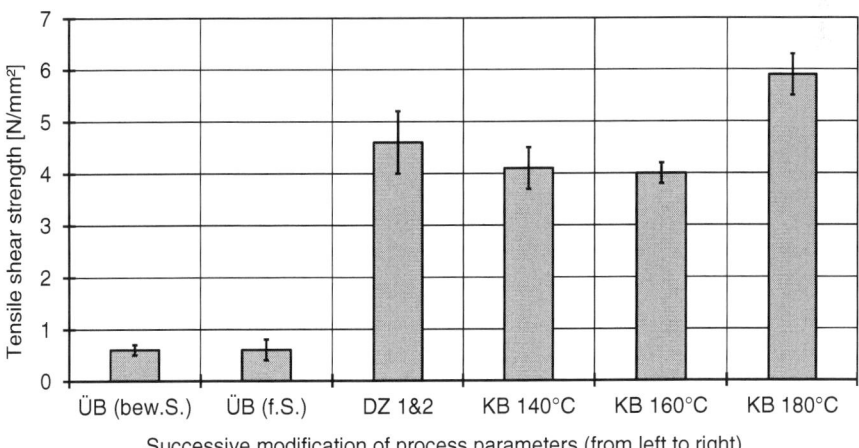

Fig. 5.70 Development of tensile shear strength with successive alteration of the temperature settings in the case of PA6 GF30 in the different areas of the die (see also Fig. 5.69); KB: Plastic area (both sides), ÜB: Overlap area (moving half/fixed half), HZ 1 & 2: Heating zones 1 and 2 at the nozzle of the injection unit

Fig. 5.71 Effect of increasing the temperature in the plastic and overlap areas in the case of PA6 GF30, KB: plastic area, ÜB: overlap area

shown to be very positive for the bond strength. The average tensile shear strength achieved in this case is 5.9 N/mm^2. The maximum measured tensile shear strength is 6.2 N/mm^2. The effect of increasing the temperature in the plastic and overlap area on the increase in the adhesion surface is shown in Fig. 5.71.

The results of this work show that a single-step process is possible for the manufacture of a metal/plastic hybrid component. In this case the heat of the metal component is used to obtain the required bond of the plastic to the metal ("from first heat"). The usage of different die temperatures and the resulting optimisation of the bond surface achieve an adhesive bond and an increase of the tensile shear strength. Nevertheless the hybrid component cool down very slowly at very high die temperatures, which further intensifies the problem of the limited maximum waiting time of the die (90 s). One solution can be to use dynamic temperature control on the die to minimise the negative factors mentioned above. This kind of temperature control in the overlap area can be useful if the heat needs to be dissipated very rapidly after the injection moulding process. The use of dynamic mould temperature control has positive effects on the cycle time as the temperature to remove the hybrid component is reached more quickly.

5 Hybrid Production Systems

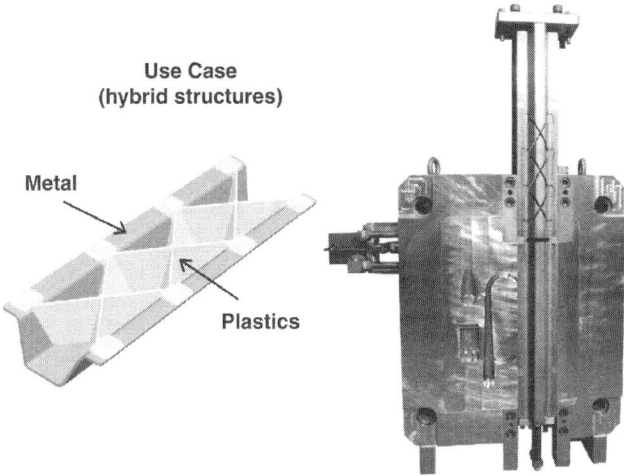

Fig. 5.72 3D-drawing of the demonstrator component and the moving die halves with the vertical slider

We now show the applicability of the process for the manufacture of practically relevant structural components on the basis of a demonstrator with a more complex geometry. Accordingly, a structural component based on the design of the so-called "Erlangener bearing" was selected. The external profile is pressure die casted aluminium and the internal ribs are plastic injection moulded. The component is implemented using an enclosed vertical slider in a closed die. The slider is actuated by two hydraulic core pullers, one of which is used to lock and unlock the vertical slider.

Figure 5.72 shows the hybrid bearing as a 3D-CAD drawing with the respective moving die halves

5.3.4.5 Hybrid Multi-Component Injection Moulding

As well as developing hybrid multi-component die casting for the production of mechanically stressed structural components made of metal and plastic, another focus of development is on the combination of plastic injection moulding and metal die casting to create a hybrid multi-component injection moulding process. This focuses particularly on the production of electrically conductive plastic/metal components on a single machine and in a single mould. In accordance with established technologies for the multi-component injection moulding of plastics, a plastic carrier is produced in the first step. After transferring the thermoplastic preform (e.g. using turntable or transfer technologies) the preform is overmoulded with a metallic conductor track in a second step (Fig. 5.73). So that the plastic is not thermally or mechanically overstressed in the overmoulding process, low melting metal alloys (ideally tin or zinc-based) are used. These have a melting point of approx. 200 °C and correspond in principle with plastic temperature ranges.

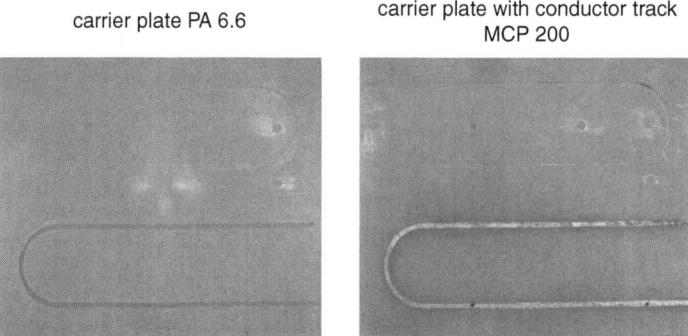

Fig. 5.73 New approach for the manufacture of electrically conductive plasic/metal parts

The hybrid multi-component injection moulding process is based on an injection moulding machine (Ferromatik K-Tec 200 2F-S, Ferromatik Milacron Maschinenbau GmbH, Malterdingen) to which a unit for the processing of the low melting metal alloys is added. While there have been add-on units for converting standard machines into multi-component injection moulding machines in the field of plastic injection moulding since the early 1990s, there are no such modular solutions for metal pressure die casting (Brunhuber 1991; Johannaber and Michaeli 2004). Therefore an existing plastification and injection unit from the plastic processing sector has been modified and optimised for processing the low melting metal alloys.

Development of a Unit for Processing Low Melting Metal Alloys The unit for processing the low melting metal alloy was modified and optimised with two particular perspectives in mind. Firstly, the unit has to enable reproducible injection of the low viscosity metal alloys and secondly small shot weights should be reproducibly metered because of the small shot volumes involved in the manufacture of conductor track structures.

These two development perspectives have resulted in an add-on unit which unifies elements from the pressure die casting, particularly from the hot chamber process, and from the area of micro injection moulding (Fig. 5.74).

The unit has separate metering and injection plungers. The piggy-back configuration of the metering plungers ensures reproducible metering of small shot weights. The metal alloy which is fed in from a hopper is melted by heat conductance in the metering plunger area and is transferred shot by shot into the injection plunger. The injection plunger which has been optimised for pressure die casting ensures reproducible injection of the metal alloy (Fig. 5.75). For this purpose it has two wedge-type rings made of a metal alloy with high wear resistance and good friction properties which can be clamped to the cylinder wall in the heated and mounted state by tightening a nut. The sealing mechanism (Fig. 5.76) between the injection and the metering plungers ensures that there is no material reflux in the direction of the dosing plunger during the injection phase. The unit is mounted on the side of the injection mould.

5 Hybrid Production Systems								531

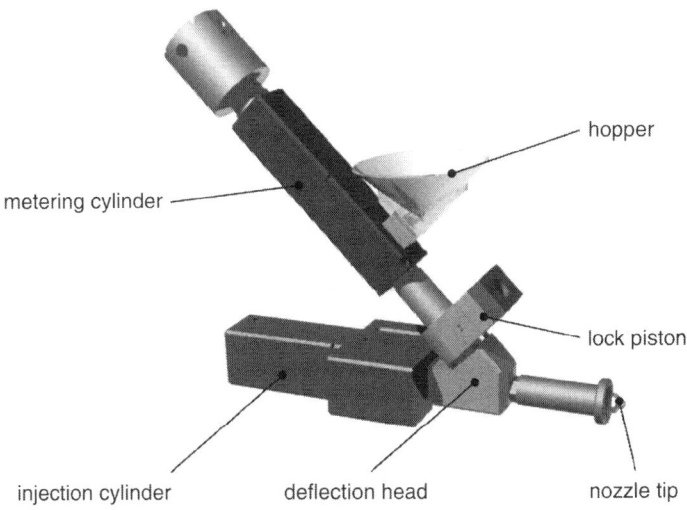

Fig. 5.74 Add-on unit for processing low melting-point metal alloy

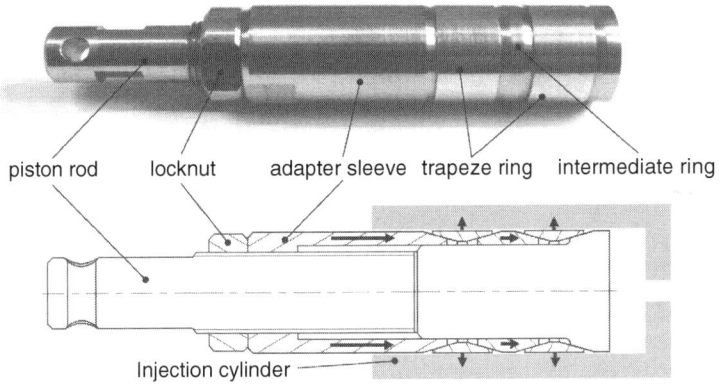

Fig. 5.75 Injection plunger in the unit for processing the low melting metal alloy

Pressure and path sensors are integrated into the unit to monitor the manufacturing process. The deflection head contains a pressure sensor (model 6152A from Kistler Holding AG, Winterthur, Switzerland). Both plungers are connected to their own path sensor to record the movements of the injection and metering plungers (model SL 720, Control Products Inc., New Jersey, USA). The recorded sensor signals (Fig. 5.77) are used to indicate how the unit processes low melting metal alloys.

Molten metal is dosed upstream from the injection plunger at the start of the cycle. The metering plunger is completely withdrawn at this point. The lock piston between the injection and metering units is closed. The injection process begins after around 8 s and is initiated by the injection moulding machine. The melting

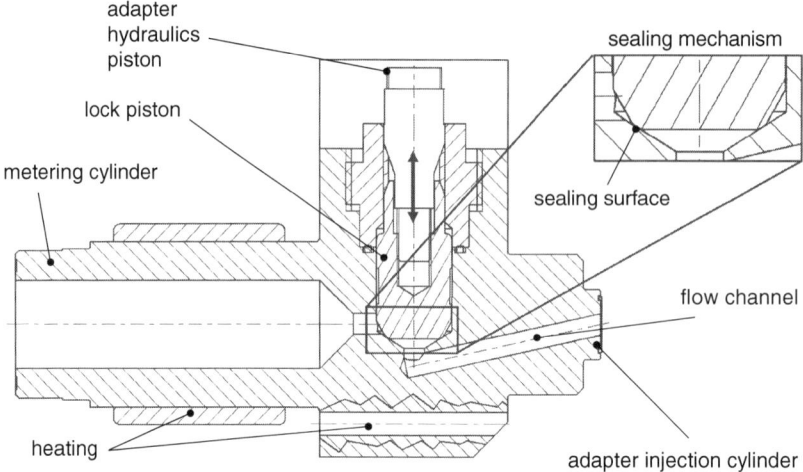

Fig. 5.76 Locking mechanism between the metering and injection units

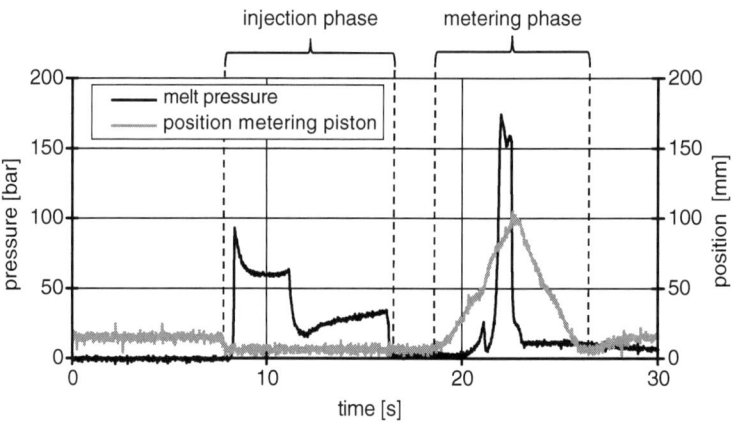

Fig. 5.77 Pressure- and path signals recorded in add-on unit for processing low melting metal alloys

pressure recorded in the deflection head increases to a peak pressure of 90 bar as a result of the incompressibility of the melt. Due to the high flow speeds and the low shot volume, the injection process is completed after less than one second and the two-stage follow-up pressure phase begins. After the end of the injection phase, the lock piston opens and the metering plunger moves forward and feeds the material into the injection unit. This increases the pressure in the deflection head and the injection plunger is forced back until the preset shot volume is reached. The lock piston then closes again and the metering plunger moves into the rear position. Then next injection cycle starts.

5 Hybrid Production Systems

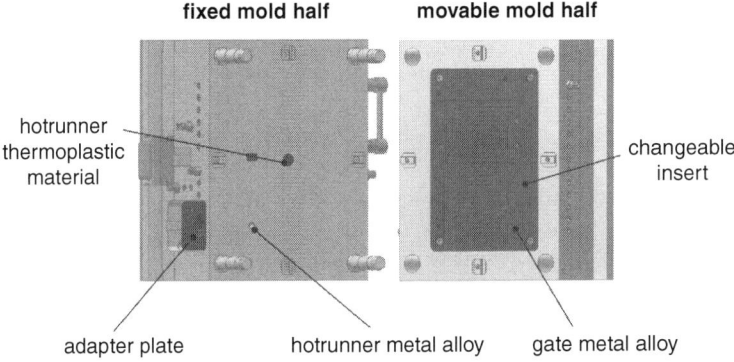

Fig. 5.78 Mould technology for carrying out investigations into hybrid multi-component injection moulding

Pilot Mould for the Hybrid Multi-Component Process As the mould technology is based on similar structural and functional groups for both pressure die casting and injection moulding, a pilot mould to carry out initial practical tests in accordance with accepted construction rules is designed based on a mould from the area of injection moulding (Fig. 5.78).

The pilot mould, as usual in the overmoulding technique, has two separate runner systems for the thermoplastic and the metallic components. The plastic component is injected in this case via a hot runner. The metal alloy runner can be heated in the region of the inserted nozzle and then continues as direct gating. The compact design of the unit for processing the low melting metal alloys means that this can be mounted directly onto the mould with an adapter plate. The pilot mould has interchangeable inserts so that different research questions can be addressed on the basis of the corresponding component geometries (Fig. 5.79). One focus is the investigation of the processing of low melting metal alloys in the one-component process. As the focus of the process chain is on the integration of fine electrical conductor tracks into an injection moulding, the achievable flow path/wall thickness ratio is an important

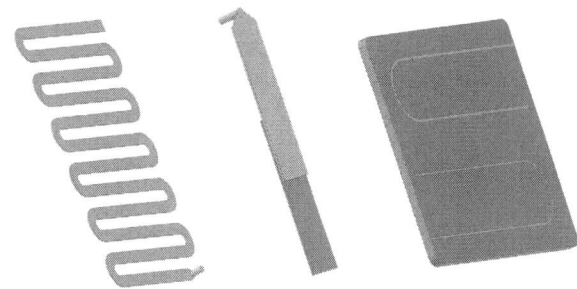

Fig. 5.79 Sample geometries for the investigation of hybrid multi-component injection moulding

target parameter for the investigations. This is determined with a flow meander with a flow cross-section 10 mm wide and 1 mm high. It is used to identify process parameters with a significant influence on this target parameter. It is particularly worth analysing the influence of mould temperature in this case because of the considerably greater thermal conductivity and the associated faster solidification of the metal alloy as compared to plastics. By using a process-defined heat control the longest possible conductor tracks with the smallest possible cross sections should be achieved in further investigations.

A second focus of the investigations is on the adhesion between the plastic and metal alloy in order to be able to generate a bond between the plastic carrier and the conductor track in subsequent applications. These investigations are carried out with the illustrated tensile shear test specimen. The specimen, as with the specimens for the hybrid multi-component die casting, is 160 mm long and 20 mm wide and has an overlap length of 40 mm.

A third research focus is set on the analysis of the implementation of the new hybrid multi-component process for the manufacture of plastic parts with integrated electrically conductive tracks. As well as optimising the process parameters with reference to a maximum flow path/wall thickness ratio in the conductor track area, questions about achievable conductivities and ampacities are of interest. The analyses are carried out based on the circuit board illustrated in Fig. 5.79 with length 105 mm, width 125 mm and thickness 4 mm. Interchangeable inserts allow variation of the conductor track cross sections in the board of between 0.5×0.5 and $1.5 \times 1.5 \text{ mm}^2$. The maximum length of the conductor track is 220 mm.

Results of Hybrid Multi-component Injection Moulding The aim of the hybrid multi-component injection moulding process is to manufacture plastic parts with metallic conductor tracks for applications in electronics and electrical engineering. The results of the investigations are presented below. The results of the analysis of the achievable flow path lengths in the processing of the alloy in the single component process are transferred in this case to the manufacture of plastic parts with integrated metallic conductor tracks and verified. The electrical properties of the components (specific electrical conductivity and amacity) are also monitored. To influence the bond strength between plastic and low melting metal alloy, the plastic component of the tensile shear test specimen undergoes different surface treatments before being overmoulded with metal alloy.

Analysis of the Flow Capacity of the Metal Alloy Investigations are carried out on the one-component process to obtain initial experiences with the processing properties of the low melting metal alloy (tin/zinc-alloy MCP 200, melting point 200 °C). A special focus is set on the influence on the achievable flow path length of the process parameters of injection speed, mould temperature and material temperature. The moulding geometry used is the aforementioned flow meander geometry (see Fig. 5.79).

Figure 5.80 shows the influence of the process parameters of mould temperature and injection pressure on the achievable flow path length for MCP 200. In this case

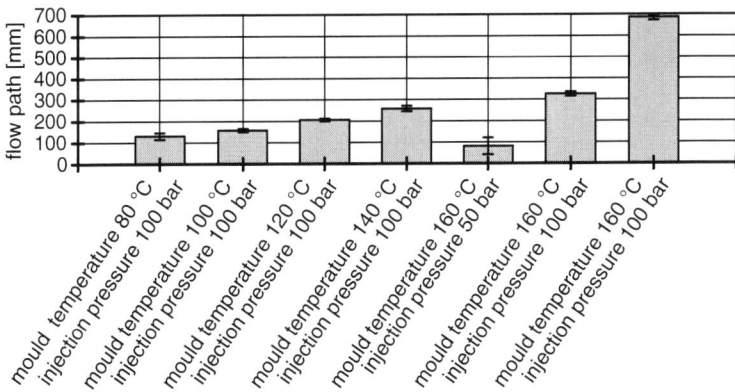

Fig. 5.80 Influence of mould temperature and injection pressure on the achievable flow path length/geometry: flow meander, material: MCP 200

the mould temperature is varied at a constant injection pressure of 100 bar. The achievable flow path length for a mould temperature of 160 °C and for injection pressures of 50 and for 100 bar are also illustrated.

There is an almost linear increase of the achievable flow path length by increasing the mould temperature. Significantly higher flow paths are achieved at a mould temperature of 180 °C, i.e. at a temperature near the melting temperature of the observed metal alloy MCP 200. It is not practical to increase the mould temperature further at this point because the components exhibit inadequate form stability in this case.

Along with mould temperature the injection speed, which is influenced in the analyses by the prevailing injection pressure, also has a major influence on the achievable flow path length. Thus significantly longer flow paths ensue if the available injection pressure is 100 bar rather than 50 bar. Due to the lower injection speed at an injection pressure of 50 bar the flow front stops far earlier due to the rapid solidification of the metal alloy.

In the analyses carried out, no evidence was found that material temperature influences the achievable flow path length. For this reason this value is not shown in the diagrams. This aspect has a positive effect on the investigations of the hybrid multi-component process described below because the melting temperatures of the metal alloy need not to be increased significantly above the melting point for better processing results. Therefore plastic carriers manufactured in the first step will not be excessively thermally stressed.

Plastic Carrier with Integrated Conductor Tracks The findings from the processing of the low melting metal alloy in the one-component process have been verified in further investigations based on the manufacture of plastic carriers with integrated conductor tracks.

Fig. 5.81 Achievable flow path lengths of the conductor tracks depending on the conductor track cross-section

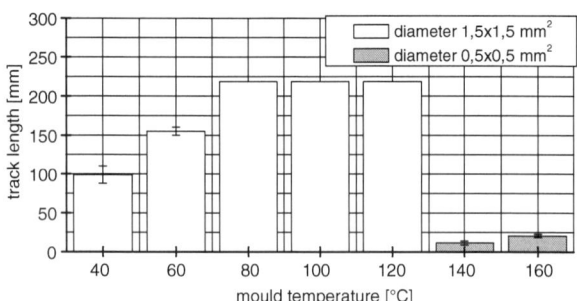

For this purpose the pilot mould described above is used in the first step to manufacture a plastic carrier (see Fig. 5.78). PA66 is the material used in this case. In a second step, the plastic carrier is transferred to the second cavity in accordance with the transfer method. This step is carried out by a machine operator. In the third step, the metal alloy (MCP 200) is injected in an overmoulding process in which the conductor tracks are applied to the plastic carrier.

In the manufacture of plastic carriers with integrated conductor tracks an important question concerns the achievable minimum geometric dimensions of the overmoulded conductor tracks. On the basis of the processing of the metal alloy in the one-component process it is worth gaining experience in the combined process in which a metallic conductor track is overmoulded onto a prefabricated insulating plastic carrier. The insulating effect of the plastic carrier has a positive effect on the achievable conductor track length. For conductor tracks with a cross section of 1.5×1.5 mm^2 the maximum achievable flow path length can be already achieved at comparatively low mould temperatures of 80 °C (Fig. 5.81). When mould temperatures are below 80 °C, however, conductor tracks with a cross section of 1.5×1.5 mm^2 are not completely filled. In comparison, even significantly higher mould temperatures lead only to comparatively small flow path lengths of less than 25 mm in the case of a conductor track with a cross-section of 0.5×0.5 mm^2. Higher mould temperatures could not be achieved in this case because the plastic mouldings will have lost considerable form stability.

Along with the important target parameter of the achievable dimensions of the conductor track, the achievable conductivity within the component is a further important criterion. The conductivity values were determined for alloy MCP 200 on a conductor track length of 65 mm based on the test components with a conductor track cross-section of 1.5×1.5 mm^2. The conductivity values obtained in this case indicate no dependence on the preset process parameters. This is illustrated in Fig. 5.82 on the basis of the process parameter of mould temperature. The achieved conductivity values, at over 8.0×10^6 S/m lie within the range of metals frequently used for electronic applications.

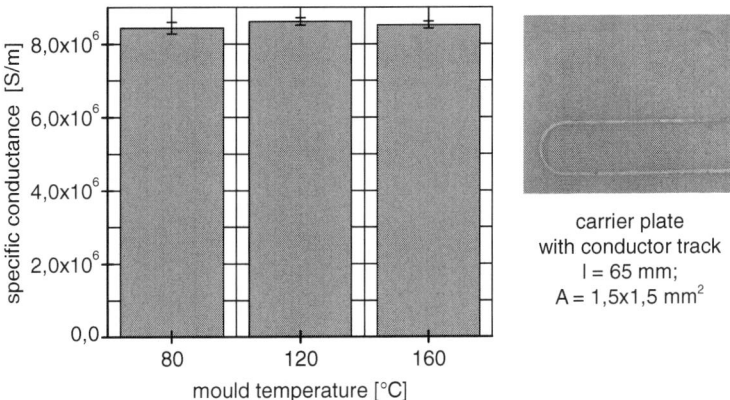

Fig. 5.82 Specific electrical conductivity of the conductor tracks integrated into the plastic carrier plotted against the mould temperature

Similar values for the specific electrical conductivity values ensue for alloy MCP 200 when the conductor track cross-section varies (Fig. 5.83). For the conductor track cross sections 1.5×1.5, 1.5×1 and 1×1.5 mm^2 analysed, specific electrical conductivity values above 8.0×10^6 S/m also occur. Overall the variable conductivity is practically unaffected by the process conditions and the geometric boundary conditions under which the component is manufactured. The conductivity values are sufficiently high for metals in electrical applications. These are furthermore very homogeneous and constant alongside the flow path.

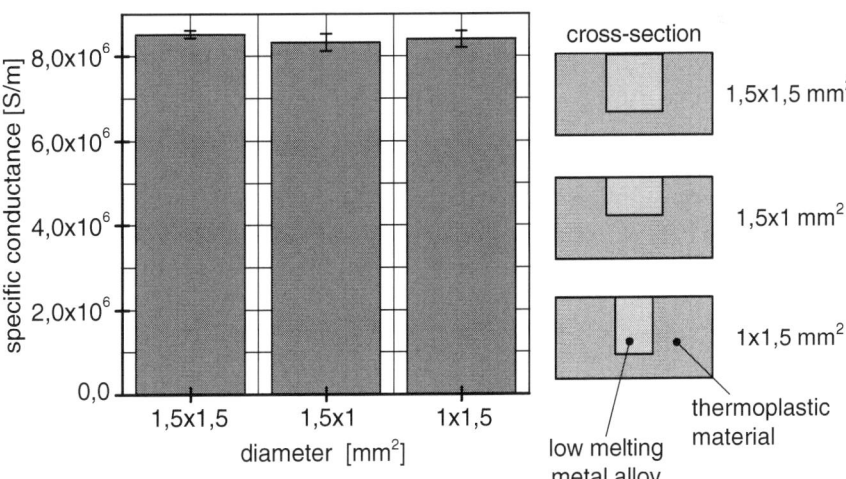

Fig. 5.83 Specific electrical conductivity as a function of the conductor track cross-section with a mould wall temperature of 160 °C

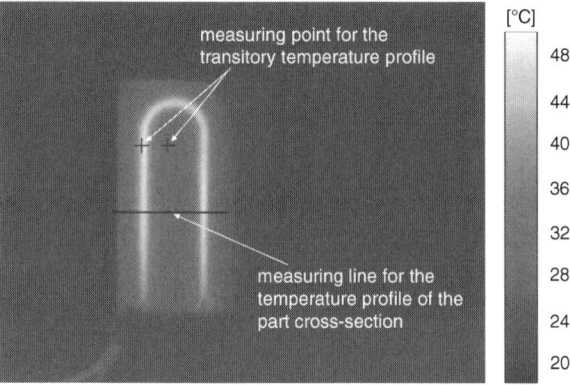

Fig. 5.84 Thermographic photograph of a conductor track structure on a plastic carrier subjected to current

Along with the specific electrical conductivity, the achievable ampacity is a further important quality criteria. Based on the primary manufacturing process the hybrid multi-component injection moulding process is suitable for the manufacture of comparatively fine conductor track structures and for the manufacture of conductor tracks with relatively large cross sections as encountered in the area of power electronics. The maximum electrical load of the conductor track can be evaluated on the basis of the quality feature of ampacity. The ampacity is a measure of the amperage a component can withstand without overheating. Since the melting point of the low melting metal alloy and the temperature range in which the plastic softens are at a similar level, it is not possible to predict which of the two components will fail first. Various conductor track cross-sections are analysed to verify the amapcity of the plastic carrier with integrated conductor track manufactured in the pilot mould. In addition, the contact surface between the conductor track and the plastic carrier is varied.

In order to measure the temperature using infrared thermography the components are coated with graphite in an upstream step which results in a less differentiated emissivity. The conductor track is then subjected to an amperage of 10 A. The resulting temperature is measured with an infrared camera from the company FLIR Systems Inc., Wilsonville, USA. Once the temperature has stopped rising and has stabilised, a measurement curve is described on the thermography image (Fig. 5.84) which permits comparison of the temperature profiles throughout the component.

Figure 5.85 compares the different conductor track cross-sections and the resulting temperature profiles along the measurement curve once the temperature has stabilised. All the temperature distributions exhibit a characteristic profile with two maximum points which are located directly above the conductor track. It is also significant that the conductor track with a greater contact surface with the plastic is less intensely heated than the conductor track with the same cross-section but less contact surface with the plastic. This effect is due to the better heat transfer between the plastic and the metal alloy as compared to the metal alloy and the air as well as the better thermal conductivity of the plastic. An increase of the cross-section by 50% leads to a temperature reduction, which can be explained by the reduced electrical

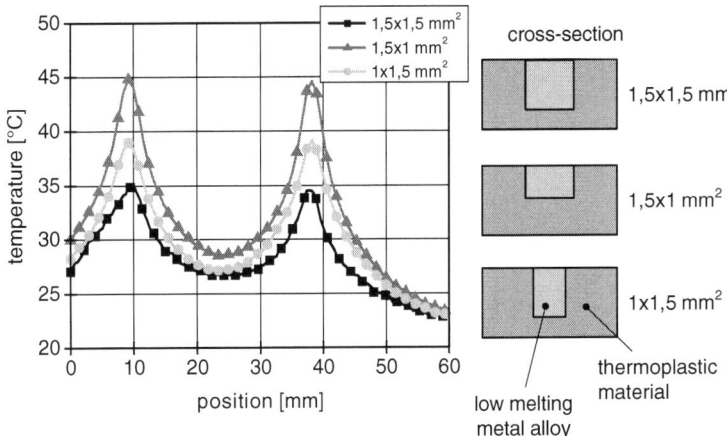

Fig. 5.85 Temperature along the measuring curve at a current load of 10 A for different conductor track cross-sections

resistance of the conductor track. In all temperature profiles it is clear that the area of the conductor track near the edge of the board is more intensively heated than the inner area. This behaviour is caused by the plastic board losing heat from its surface, similar to a cooling rib. Therefore the outer conductor track is connected to a shorter cooling rib and thus loses less heat which leads to more intense heating. Overall the maximum temperatures which stabilise at a current load of 10 A vary within a range which is non-critical both for plastic and for the metal alloy.

A coupled simulation of ohmic heating and the influences of thermal conductance is implemented for the outlined application to enable a prediction of the ampacity in the component design phase. A model of the plastic carrier with integrated conductor tracks is produced with the Comsol Multiphysics simulation software from the company Comsol AB, Stockholm, Sweden. In this case, the defined volume is divided into finite elements. The electrical boundary conditions are now defined in the first module, which calculates the flow of current and the resulting ohmic heating. Amongst other things the surfaces at which the current enters and the component with its electrical material properties are defined in the next step. In the second module the ohmic heating given off by the first module is coupled with the emerging thermal conductance, radiation and convection. The thermal boundary values for the simulation derive from a comparison with the practical investigations of the ampacity with a load of 10 A. Figure 5.86 shows the simulated and the measured temperature profiles over the measuring line shown in Fig. 5.87. In this case the model and samples were loaded with currents of 5, 7.5 and 10 A. The results exhibit realistic values for all currents. The effect described above of the more intensive heating of the conductor track at the edge of the board is also replicated in the simulation model. The coupled simulation of thereby offers the chance to detect unpredictable temperature increases in components of any complexity at the design phase and counteract these if necessary.

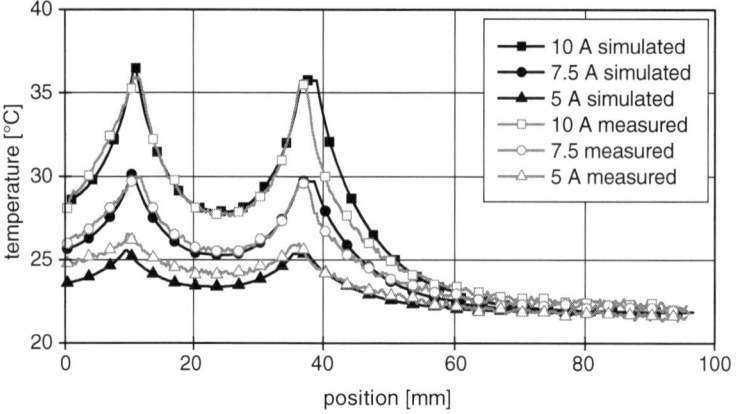

Fig. 5.86 Comparison of temperature profiles in thermography and simulation with a conductor track cross-section of 1.5×1.5 mm^2

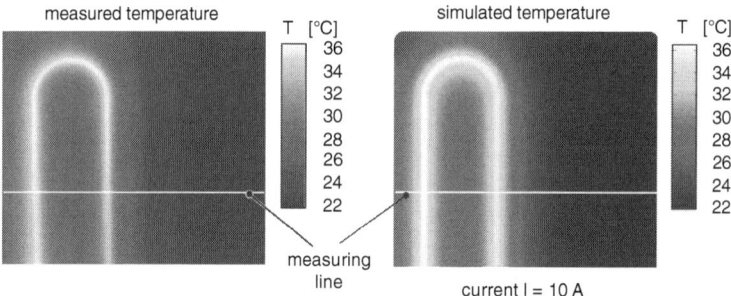

Fig. 5.87 Comparison of values for thermography and simulation along the measurement curve – cross section 1.5×1.5 mm^2

Bond Strength Between Plastic and Metal in Hybrid Multi-Component Injection Moulding Even though in most cases plastic and metal hybrids are not subjected to maximum loads in electrical and electronic industry applications, the minimum bond strength must nevertheless exceed the level required to ensure no loss of components. The state of the art is that a bond between plastic and metal must be implemented by interlock means, i.e. by undercuts integrated into the component.

To investigate the bond strength between plastic and metal, tensile shear test specimen are manufactured with the new hybrid process. In the second step, the low melting metal alloy MCP 200 is injected onto the plastic preform manufactured in the first step. Since ahead a good bond between plastic and metal is not expected based on the thermodynamic boundary values, different surface pretreatments are carried out on the plastic preforms. These pretreated plastic parts are then overmoulded with a low melting point metal alloy. The results of surface treatments using laser microstructuring, thermal spraying and plasma treatment are presented next.

In the first series of tests, a linear microstructure is applied to the surface of the preform with a laser. This microstructure has an undercut so that the low melting metal alloy can form an interlock with the preform. Different orientations of the microstructure relative to the direction of material removal form the basis for different expectations of results in terms of bond strength. Thus the orientations of the microstructure are varied at 0° and 45° with respect to the direction of material removal. In a further series of tests, preforms are coated with zinc in the thermal spraying process and subsequently overmoulded with the metal alloy in the hybrid multi-component injection moulding process. In the thermal spraying process, zinc particles are fired at high speed in the molten state onto the surface of the plastic thus building up a rough, porous layer of zinc of a chosen thickness which adheres to the plastic. The objective of this treatment is to use the microscopic undercuts present in the rough surface as anchoring points to create an interlock engaging connection between the overmoulded metal alloy and the zinc layer. It is also possible that an adhesive bond will form between the zinc and the zinc and tin alloy. In principle, the thermally sprayed zinc layer performs the role of an adhesive agent. Another alternative is to treat the preforms with atmospheric pressure plasma. For this purpose a plasma nozzle is guided over the surface of the plastic preform at a defined distance and at an even feed rate. This treatment step can also positively influence the bond strength. One explanation for this is that the plasma treatment increases the polarity of the formerly low-polar plastic which helps to create an adhesive bond.

Figure 5.88 shows the measured shear strength of selected tensile shear test specimen. The achievable bond strengths in the case of the different surface treatment methods of laser microstructuring, thermal spraying and atmospheric pressure plasma treatment and an untreated sample as a control are compared for different plastics. The maximum breaking strength is measured in the case of the laser structured sample made of glass fibre-reinforced polybutene terephthalate (PBT). On the polyamide substrate the variation of the orientation angle of the applied linear microstructure has a clear influence on the achievable tensile shear strength. If the microstructure is orientated at 0° and is thereby vertical with respect to the direction of material removal, a greater a tensile shear strength can be achieved than with an orientation angle of 45° with respect to the direction of material removal. The glass fibre-reinforced PBT samples coated with zinc in the thermal spraying process exhibit a comparably high tensile shear strength. These good results are due to the interlock mechanisms. However, samples of the same plastic pretreated with plasma exhibit rather lower bond strengths which is due to the adhesive nature of this connection. The samples with unreinforced PBT exhibit even lower strengths which can be explained by the increased shrinkage potential in comparison to the reinforced plastic.

A connection with the metal alloy can be produced with the amorphous polycarbonate (PC) even without pretreatment of the preform. Amorphous materials, whose morphology develops during the solidification, exhibit lower shrinkage values than semi-crystalline materials such as PBT or PA66. Atmospheric pressure plasma treatment can improve the shear strength with simultaneously less dispersion of the results.

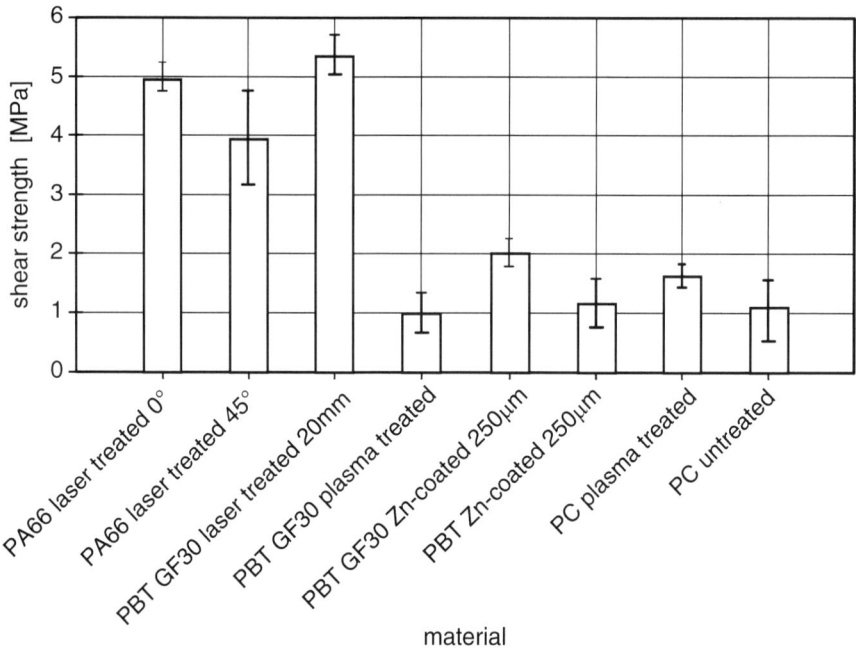

Fig. 5.88 Bond strength of selected tensile shear test specimen

5.3.4.6 Benchmark of Bond Strengths

In summary, a comparison of the bond strengths between metal and plastic achievable with the different processes of thermal joining, hybrid multi-component die casting and hybrid multi-component injection moulding should be regarded as an important common topic of the various production processes. Adhesive bonding as a reference joining process is compared with the joining results of the newly developed process chains. Tensile shear test specimen made of different material combinations with a 20 mm overlap are used for comparison. PBT is the plastic used in this case. Due to its molecular structure and the greater shrinkage potential as compared to PA66, this material presents particular challenges in terms of creating a bond with metal. Figure 5.89 shows the results of adhesive bonding for the material combinations of stainless steel/PBT and stainless steel/PBTGF30 compared with the results of the newly developed processes. A two component, cold-curing epoxy resin adhesive is used as the adhesive (DP410). The manufacturer recommends the adhesive both for metal adhesion and particularly for polar and non-polar plastics. In tensile shear tests with cold-rolled steel, according to the data sheet, tensile shear strengths of 20 MPa should be possible while tensile shear strengths of 10 MPa should be achieved for glass fibre reinforced plastics. Implementation in this case should always take into

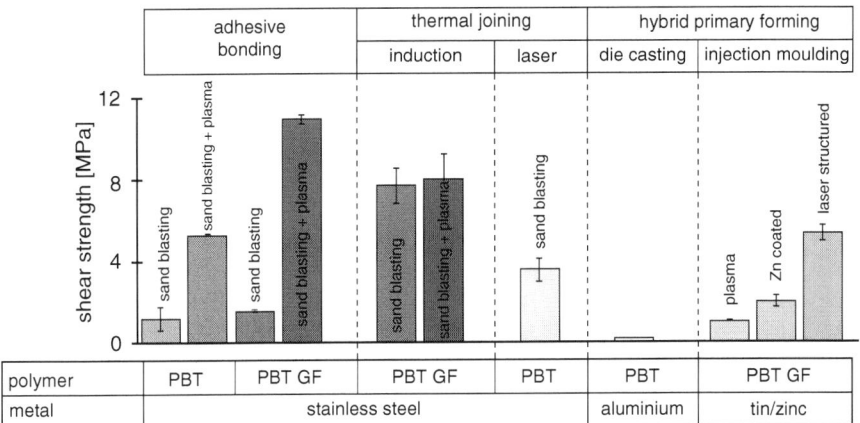

Fig. 5.89 Benchmark of bond strengths

account technical joining boundary conditions such as underlying curing temperatures or the type of fixture. Residual stresses and uneven adhesive curing can reduce the achievable tensile shear strengths as well.

When using PBT it is shown that surface pretreatment is necessary to achieve sustainable bond strengths. This applies both for the adhesive bonding of metals and plastics as well as for hybrid multi-component technology and thermal joining. With optimum pretreatment technologies, bond strengths are achieved in the series of investigations both with adhesive bonding as well as with thermal joining (induction joining) which exceed the basic strength of the plastic in the case of the selected sample geometry. It should be mentioned at this point in the case of the purely mechanical surface pretreatment method of sandblasting, that adhesive forces in the adhesive bonding process are insufficient either with PBT or with PBT GF to achieve high tensile shear strengths. The samples suffer adhesive failure of the plastic/adhesive boundary layer. The inadequate adhesion strengths in the case of hybrid multi-component die casting are process-related because surface pretreatment cannot be carried out within the process chain. The adhesive forces in this case are not strong enough to counteract the strong initial shrinkage of the PBT. Adequate strength can be achieved, depending on the pretreatment, with hybrid multi-component injection moulding on which far fewer demands on bond strength are made since the main area of use is in electronic devices rather than for structural components.

Despite the pretreatment step in the case of the hybrid technologies analysed, the process chain can be shorter than for adhesive bonding. Figure 5.25 at the start of the chapter illustrates the shortening of the process chain based on the example of thermal joining compared to adhesive bonding. Along with cleaning, particularly the pretreatment of the plastic sample can be omitted. Also in the case of thermal joining, the need for curing and/or fixing times can be disregarded because the small quantities of energy introduced for thermal joining dissipate within a short time.

However the possibility of designing and chemically formulating adhesives specifically for the joining task with simultaneous adhesive-appropriate pretreatment, for instance, plasma pretreatment of the plastic, enables strengths which cannot yet be achieved with the newly developed processes without new intelligent pretreatment processes. The possibility of formulating and/or adapting adhesives for specific joining tasks explains the strength deficit of thermal joining in the case of standard semi-finished products. Modification of the plastic sample in accordance with the adhesive bonding technology could close the gap to the adhesive bonding technology with respect to the achievable bond strength.

A further benefit could be the ageing resistance of thermally-joined metal and plastic hybrids. While in the case of adhesive bonding there are two boundary surfaces as well as the adhesive itself exposed to corrosion, there is only one boundary layer in the case of thermal joining.

5.3.4.7 Characterisation of the Influencing Variables on the Formation of Adhesive Forces

The results clearly show the different influencing factors on the adhesion between metal and plastic. Particularly the mechanical, chemical and physical properties of the joining partners were identified as the standard influencing factors, along with the technical process values of temperature and pressure.

The results of the thermal joining processes show the influence of the joining temperature on the adhesion strength. Besides reducing the viscosity and the resulting potential for the melt to wet a greater surface, in this case also the shortening of the polymer chains close to the boundary layer, as verified by GPC investigations, and the associated greater mobility of the polymer chains should be taken into account. By this means, the molecule structures of the plastic are encouraged to get closer to the surface of the metal. This increases the potential for forming physical bonding forces because these only become effective in the range of a few nanometres.

An optimum pressure profile during the process is used firstly to increase adhesion and secondly as a compensating method. Supporting pressure enables the molten polymer to press into the surface structures of the metal during the joining process and can be adjusted after the joining process in the form of a follow-up pressure to counteract the material-specific shrinkage of the polymer.

The analysis of the pretreatment methods have enabled the determination of the material-specific influential variables on the adhesive strength. These can be divided into mechanical, physical and chemical influencing variables. The aim of a join is to achieve a failure within the material featuring a lower basic strength. Failure within the boundary layer has to be avoided from a calculation perspective. Therefore the utilisation of the cohesive strength of the plastic is to be regarded as standard for a good joint. The high elasticity module gradient and the resulting stress peaks between the plastic and metal which, unlike adhesive bonding, result from the absence of the interim layer, is a major reason for the reduction of the maximum achievable bond strengths in the case of failure within the basic material. A further relevant mechanical property of the substrate surfaces is the surface topography. As the thermal conductivity joining and induction joining tests show, changing the surface topography has a distinct influence on the bond strength. In this case we differentiate between positive locking and the effect of increasing the surface area. While the influence of producing undercuts and clamps by means of laser structuring has been demonstrated, increasing the surface area potentially generates a greater number of intermolecular bonding forces. Although both influences have been verified, it is not yet technically possible to take measurements to make a quantitative prediction of individual influences.

The results of the investigations also indicate the considerable influence of the physical properties of the materials. In the comparison of the joining results of the samples produced using unfilled PBT materials and those filled with short glass fibres it was found that the polymer shrinkage after the manufacture of the hybrid component has a considerable influence on the long-term development of adhesive forces. While in the case of the unreinforced polymer, the adhesive forces immediately after the manufacturing process are insufficiently strong to counteract the shrinkage forces, a glass fibre content of 30% reduces the shrinkage potential sufficiently to achieve long-term bond strengths.

As compared to polymers with a different chemical composition, not only the physical properties but also the molecular structure of the plastic must be taken into account. A different molecular structure particularly leads to a different surface energy resulting amongst other things in a different distribution of the types of intermolecular forces which develop between plastic and metal.

The differences of the chemical properties of the bulk materials are described based on the example of (PA66) and polybutylene terephthalate (PBT).

While, as shown, the overall surface energy of the two plastics is similar, the proportion of polar surface energy is considerably greater in the case of polyamide. The proportion of polar surface energy is due to permanent dipole-dipole reciprocities, which are able to form greater adhesive forces than temporarily fluctuating dipole-dipole reciprocity based on the disperse distribution of the electrons within a molecule. The cause of this lies within the molecular structure of the polyamide. The nitrogen atom in the molecular structure has a very high electro-negativity. This means that the electron in a covalent bond with the hydrogen atom is attracted more strongly by the nitrogen atom. The atomic core of the hydrogen atom is thereby highly insulated and forms a strong positive permanent dipole. Approaching an oxidised metal element can cause strong hydrogen bridge bonds between the metal element and the polymer. PBT has no nitrogen atom in its molecular structure so that the majority of the adhesion forces between PBT and the metal element consist of the relatively weaker dispersion forces. The strong hydrogen bridge connections in the case of polyamide can better compensate for the ensuing shrinkage forces.

It is hard to isolate the different values to quantify their influence on the adhesion. Until now there are no suitable test methods or innovative analysis options for the quantification and separation of physical and chemical interactions in the boundary layer.

5.3.5 Industrial Relevance

The industrial relevance of the newly developed technologies for the manufacture of plastic and metal hybrids is confirmed on the basis of two typical industrial cooperations, firstly for surface pretreatment and secondly for hybrid multi-component injection moulding.

5.3.5.1 Manufacture of an Electronic Demonstrator Within a Fully Automated Production Cell

Using the newly developed hybrid multi-component injection moulding process, plastic and metal hybrids can be produced within a relatively short process chain. To demonstrate the performance capability of the new process we have collaborated with industry partners to develop, for example, a compact production cell which enables an industrially-relevant fully automated production of plastic and metal hybrid components for electronic applications. Heated sports goggles with a plastic lens which is heated by an integrated metallic conductor track to prevent fogging are an ideal compact demonstrator to illustrate the benefits of the new process (Fig. 5.90). It has been possible to inject a conductor track with a three-dimensional path and a varying cross-section onto the lens of the goggles. Also possible within the process is the direct contacting of metal inserts with the conductor track. Numerous elements of the demonstrator therefore reflect electronic components in use in industry.

An important step in the implementation of the hybrid process is selection of the material both for the metallic conductor track as well as for the plastic lens and frame of the goggles. The processing of the metal component may not thermally or mechanically damage the plastic carrying the conductor track structure (in this case the lens of the goggles). Low melting metal alloys, as used within the previous investigations, fulfil these boundary conditions. There are also clearly defined requirements for the plastic components. Since these are sports goggles, the lens must be sufficiently flexible and impact-resistant to ensure safety while exhibiting outstanding optical properties. Microcrystalline polyamides have become established on the raw

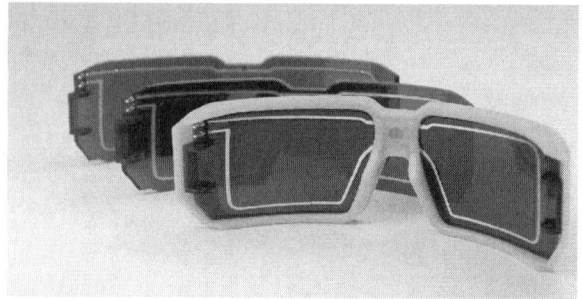

Fig. 5.90 "Heated sports goggles" electronic demonstrator which is build up by a plastic lens, a conductor track with a three-dimensional path, which is contacted inline with metallic contact pins, and a plastic frame component

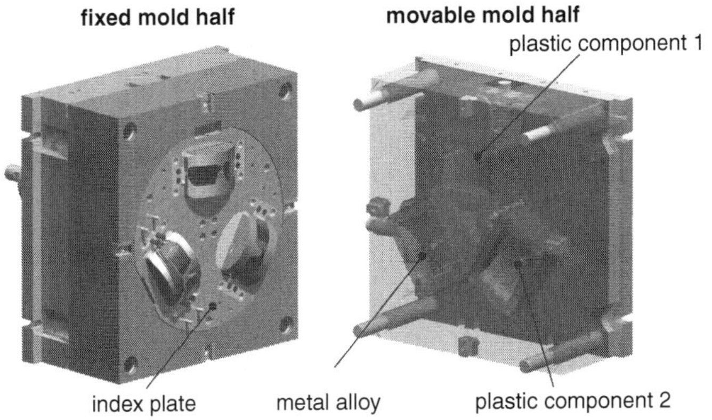

5 Hybrid Production Systems

Fig. 5.91 Three-station index plate mould for the production of the heated sports goggles

materials market as optical materials. They offer high UV-resistance and good mechanical properties and they are resistant to stress tension cracking. A polyamide is also used as the second component for the manufacture of the goggles frame. The types selected for the application, CX 7323 (lens) and CX 9704 (frame) from the raw material manufacturer Evonik Industries AG, Essen, exhibit good adhesion compatibility. Since polyamides are a hygroscopic material, they must be dried before processing to prevent moisture streaking and bubbling in the finished part. The KKT 55 drying and feed system from the company Werner Koch Maschinentechnik GmbH, Ispringen, is dedicated for processing optical-quality polyamides. The equipment works on the sorption principle and ensures that the granulated plastic reaches the injection moulding machine at a consistently high dried quality.

The findings obtained from the pilot mould and from a conceptual mould design study showed index plate technology to be the most suitable mould technology. A three-station arrangement allows primary forming of two plastics and a low melting metal alloy in a single mould on a single machine (Fig. 5.91). Both the two plastic melts as well as the molten metal are fed to the respective cavities by pneumatically closing hot runner shut-off nozzles. Transfer between the individual stations is realised by a servo-electric driven index plate. The cavity inserts in the second station are heated with water via separate circuits. As the preliminary investigations on the flow-ability of the low melting metal alloy show, directed heat control of the mould can extend the flow path. For this reason the second station (for the moulding of the conductor track) is equipped with a highly dynamic variothermal heat controller from the company gwk Gesellschaft Wärme Kältetechnik mbH, Kierspe. There is an additional challenge in placing the contact pins for the metallic heating circuit (conductor track) within the mould as required for the overmoulding and contacting step. A six-axis robot from the company KUKA Roboter GmbH, Augsburg, is used for this purpose. In combination with a special robot hand from the company ASS Maschinenbau GmbH, Overath, it guarantees reproducible positioning of the contact pins and safe removal of the finished goggles. Different quality assurance approaches are pursued to evaluate the quality of the goggles and the conductor track in detail. Firstly pressure- and temperature sensors are used for online monitoring of the processes in the injection mould and in the injection unit for the low melting metal alloy. In the metal processing unit the melt pressure is measured with a sensor (Kistler Holding AG, Winterthur, Switzerland) installed distant from the flow channel within the nozzle which determines a value for the melt pressure from the deformation of the metal part induced by the melt pressure. The manufacture of the plastic components, particularly in the lens area, is monitored by combined pressure and temperature sensors. A further quality criteria is the resistance of the electrical conductor track which is checked and recorded for each part inline within the production cell. The profile of these measured target parameters over several cycles permits a direct conclusion on the process constancy of the hybrid multi-component injection moulding process and on the electrical part properties.

The production cell permits the fully-automated moulding of the demonstrator "heated sports goggles" in a cycle time of 60 s. In this case the conductor track can be moulded very precisely. Also possible within the process is direct contact of metal inserts with the conductor track. A specially developed variothermally heated hot runner nozzle with an integrated pneumatic needle valve makes it possible for the first-time to mould a low melting metal alloy reproducibly and without additional gating. The production cell for the manufacture of the electronic demonstrator using the hybrid multi-component injection moulding process was presented to the trade public at the world's biggest plastics trade show; the "K 2010" in Duesseldorf (Michaeli et al. 2010d; IKV 2010).

5.3.5.2 Manufacture of a Paper Feed Roller by Means of Laser Microstructuring

Along with hybrid multi-component injection moulding, industrial relevance was also demonstrated for the investigated laser structuring as an upstream surface pretreatment method based on an actual example. This was implemented incorporation with the company Hunold + Knoop Kunststofftechnik GmbH, Geseke. Rotating cylinders are used to separate banknotes in cash machines. Piles of bank notes can thus be reliably separated and safely dispensed from the machines. The separation process subjects the rollers to strong forces. For this reason and also to meet the strict requirements for concentricity the main body of the roller is made of steel. To ensure maximum possible adhesion between the banknotes and the roller for the separation process, a thermoplastic is injection moulded around these sections of the roller. The separation process exerts a high momentum onto the metal and plastic connection and so far a primer system has been used to achieve the required adhesive strength. The primer system is applied in an upstream process. The result, along with the high costs of the primer, is a complex and time-consuming application process.

Circumferential laser microstructuring can eliminate the need for the primer. The laser process both statistically roughens the surface as well as producing a defined structure with undercut-geometry in which the plastic can interlock. The structur-

5 Hybrid Production Systems

Fig. 5.92 Roller for banknote separation

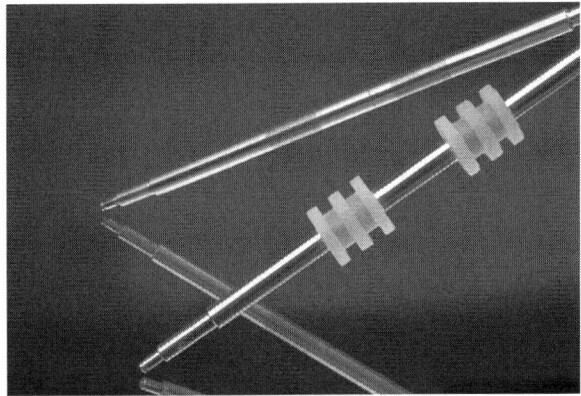

ing is an automated process. After the pretreatment, the roller can be safely injection moulded. The shrinkage of the plastic during cooling additionally improves the adhesion of the plastic to the microstructure and the specified torque transfer requirement is thus achieved. Figure 5.92 shows a laser structured roller and one which has already been extrusion coated.

The use of an economical laser source reduces the investment for the pretreatment process but the required cycle times of the injection moulding machine are still achieved. With corresponding calculation of the costs for the primer and the application the investment pays for itself after just a short time even for small unit quantities.

5.3.6 Future Research Topics

Based on the knowledge obtained so far in the area of hybrid manufacturing processes and surface pretreatment methods, numerous areas of development and research have emerged on the path towards qualifying the new technologies for industrial applications. A central aspect of these further developments lies in the area of mechanical and systems engineering. The implementation has so far clarified important questions for the two hybrid multi-component technology processes analysed. Initial prototypes have been implemented which have some direct links to industrial applications. Nevertheless, consistent further development measures are required to bring the technology to the stage of industrial maturity. In this case it is particularly worth analysing mechanical and mould engineering as one unit in terms of an integrated modular machine platform embedded in a high performance production cell and adapting it for optimum process performance. Here highly different system elements should be addressed such as gating technique, dosage and feed of material and tempering technologies. These should be tailored to the process under analysis along with other comprehensive investigations for process understanding (e.g. the influence and targeted exploitation of temperature control with respect to the resulting adhesion). The development of suitable quality control concepts and the demonstration of machine

Fig. 5.93 Surface modification according to CMT process

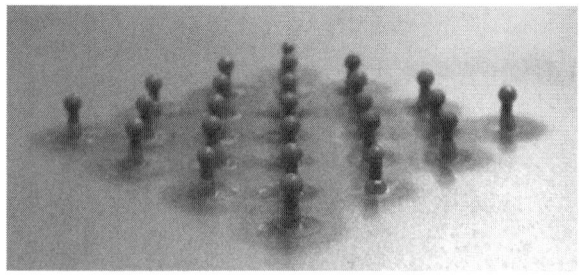

capability of the hybrid machine platform play an important role. Increasing value is also being placed on virtual analysis of the processes in the form of simulations.

In the area of thermal joining processes, it is worth increasing the attractiveness for industrial implementation in terms of modular machine technology for the processes laser transmission joining and induction joining which are especially relevant for industrial applications. In this case it is particularly worth generating solutions for modularising the technology in order to make an adaption to different geometries possible. In addition, new joining and pretreatment technologies with potential for the manufacture of plastic/metal hybrids will be analysed. The investigations of the influence of a surface pretreatment have shown that particularly changing the surface topography of the metal has great potential to improve the adhesion properties. So far however no load-optimised structures have been produced. This would make sense due to the high e-modulus gradients on the metal/plastic boundary layer. Two new surface structuring process technologies should be qualified in the future for thermal joining: pin structures can be manufactured by skilful control of the wire feed with the help of the Cold Metal Transfer welding process (CMT). These pin structures increase the effective surface of the metal structure, thus providing good conditions for mechanical clamping based on the characteristic pin structure. The flexible application layout lends itself to load-adaptable surface pretreatment in order to decrease stress peaks at the edges of the joining zone for example (Fig. 5.93).

Structuring of pins using electron beams is also under investigation. Pin structures with a high aspect ratio can be produced by re-melting material (Fig. 5.94). Pins of 20 µm up to several mm can be produced to bridge the gap between micro and macro hybrid connections. Using a suitable approach, this process can also be used to create undercuts and barbed structures.

The second important development focus which is being pursued along with mechanical and systems technology for the manufacture of plastic/metal hybrids dealing with the product. Central in this case is further refinement, validation and generation of new findings for the creation of adhesion between plastic and metal. These findings should be quantified and made accessible for targeted and reproducible use. With this in mind, the findings should be used particularly for the design of plastic/metal hybrids. In this respect it would be worth developing an integrated qualification methodology. In detail this includes classification of the joining quality and type in the boundary layer with reference to process parameters such as temperature, pressure

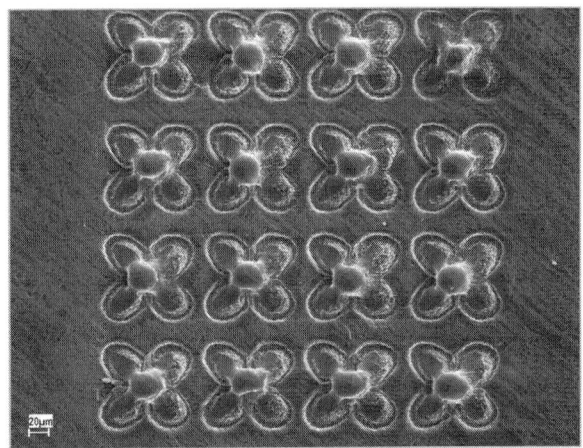

Fig. 5.94 Surface modification through electron beam structuring

and component geometry, the analysis of failure mechanisms of hybrid bonds with respect to positive locking, adhesive or cohesive locking and non-positive locking as well as quantifying the types of flaws in the boundary layer. Particularly predictions of the chemical, physical and electrical long-term stability of the parts are of key significance for the industrial use of hybrid products. In this case extensive investigations, for example of the influence of temperature variation or different media on the resulting mechanical or electrical component properties in accordance with known industrially relevant guidelines, for example VDA 621-415 and VW-P1200 play a decisive role. Not least, the development, adaptation and qualification of non-destructive measurement methods are also an important field for providing the possibilities of 100% quality testing of plastic/metal products available for production lines.

Systematic commitment to these research issues can advance hybrid production technologies and the hybrid products produced with them to the series testing stage.

5.4 Improving the Efficiency of Incremental Sheet Metal Forming

Wolfgang Bleck, Gerhard Hirt, Peter Loosen, Reinhart Poprawe, Uwe Reisgen, Markus Bambach, Georg Bergweiler, Jan Bültmann, Jörg Diettrich, Alexander Göttmann, Ulrich Prahl, Alireza Saeed-Akbari, Lars Stein, Marius Steiners, Jochen Stollenwerk and Babak Taleb Araghi

5.4.1 Abstract

The principle of incremental sheet forming (ISF) with CNC machine tools is based on the idea of replacing one half of the tool set used in sheet metal forming processes (such as deep drawing) with a CNC-controlled round-tipped tool. This tool follows the contour of the desired part and brings the sheet progressively into the desired form by local plastic deformation. The spatial movement of the forming tool is calculated from the CAD model of the desired product. In spite of almost 20 years of development, ISF has not qualified for widespread industrial use even though the process combines high flexibility and short development times with low tooling costs. It is therefore ideally suited for small and medium batch production.

The causes of its limited application in industry can be found in the following process limitations:

- The geometric accuracy of parts produced with ISF does not always meet industrial requirements.
- The sheet metal often experiences an unwanted inhomogeneous thinning during forming.
- The incremental process is slow and requires cycle times lasting from minutes to hours.
- Potentially interesting markets, such as the aerospace market, that require formed sheet parts in quantities of less than 1,000 per annum, require the forming of materials that are difficult to form at room temperature, such as titanium alloys.
- ISF also requires long computing times for numerical process simulation, so that in many cases numerical process design and simulation as commonly used in the sheet metal forming industry are not available for ISF.

This is where the hybrid process combinations described below come in. It will be shown that by combining ISF with stretch forming it is possible to improve geometric accuracy and material distribution and achieve higher productivity. At the same time, this shortens the computing time required for numerical process simulation.

Laser-assisted ISF has been developed for materials that cannot be cold formed, such as titanium alloys that are often used in the aerospace industry. Furthermore, the processing of steel-aluminium composite sheets that can be used as connecting pieces in modern multi-material vehicle structures is studied.

With these developments it is shown that hybridisation can considerably extend the range of applications for ISF.

Hybrid incremental sheet forming demands suitable forming machines to be developed. In the case of the process combination of ISF and stretch forming, the new sheet metal forming machine combines a CNC machine tool with integrated stretch forming modules. Laser-assisted ISF is made possible by the integration of a specially designed laser optic in the basic machine tool. By combining this with further development of ISF, a hybrid sheet metal processing centre has been developed that enables the integrative manufacture of sheet metal parts.

5.4.2 State of the Art

5.4.2.1 Incremental Sheet Metal Forming

Incremental sheet forming with a CNC forming tool (ISF) is an innovative approach to the cost-effective production of prototypes and small batches of complex

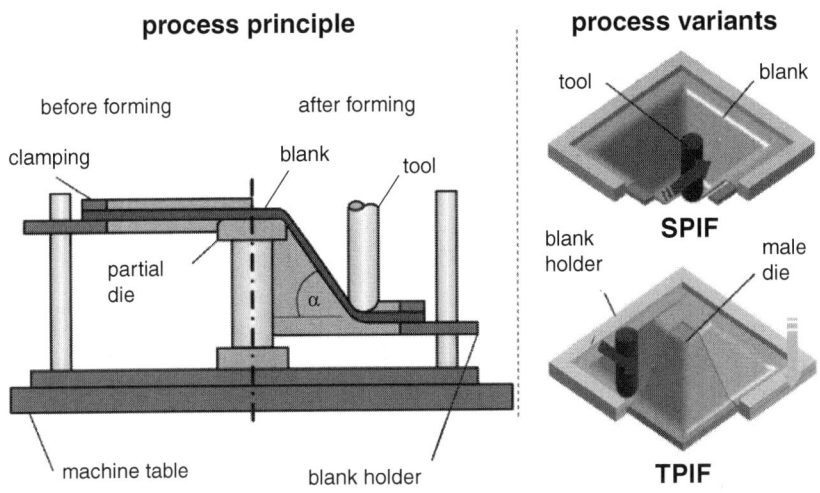

Fig. 5.95 Principle and variants of the process

three-dimensional parts from sheet metal. Figure 5.95 illustrates the principle of the process. In this process, a flat sheet is first clamped in a blank holder. Depending on the process variant, a die is placed under the sheet to support the geometry of the part to be produced at critical points, either fully or in the case of simple parts only partly (see right-hand side of Fig. 5.95). A pin-shaped forming tool starts to form the sheet into the desired form, following a sequence of contour lines. After one plane has been processed, vertical feed starts. If a supporting die is being used, the holder moves downwards simultaneously with the tool pitch.

The process principle described enables flexible production of sheet metal parts with a lower tool requirement than conventional sheet metal forming processes such as deep drawing. Tooling and setup costs can be saved as there is no need for complex segmented tools. Thus, compared with conventional sheet metal forming processes, ISF enables shortened lead times and low investment costs.

In addition to the process variant that uses a supporting tool, known as "two-point incremental forming" (TPIF), it is possible to manufacture parts completely without a die. This variant is known as "single-point incremental forming" (SPIF). In this case, support is provided by a blank holder that has to be matched to the geometry of the part. A comprehensive overview of the state of the art in ISF, on which the current work is based, can be found in the review paper by Jeswiet et al. (2005b). Since conventional tools are largely dispensed, ISF offers a high degree of flexibility when the geometry varies (Tuomi and Lamminen 2004). Typical applications of ISF are prototyping (Jeswiet und Hagan 2001; Jeswiet et al. 2005b), small batch production for sectors such as automotive and aviation industries (Amino et al. 2002) medical technology (e.g. implants) (Duflou et al. 2005) and architecture (Petek et al. 2009). It is not economical to use conventional forming processes for small batch production of particularly large parts because of the need for big machines and expensive tools.

Since forming forces do not increase with geometrical size when using ISF, the size of the part that can be made using this process is mainly determined by the working area of the machine. However, it is absolutely necessary to find solutions for the following process limitations before ISF can be used in industrial applications:

- Extreme thinning of the sheet in areas with steep wall angles
- Limited geometrical accuracy
- Long process duration

In addition, it must be mentioned that FEM process design, as currently used in the industry, is not an economic approach for ISF because of the long computing time required. Nevertheless, there are potential markets for ISF, such as aerospace, where less than 1,000 parts are required each year using a high proportion of alloys that cannot be cold formed, such as titanium.

5.4.2.2 Thinning

The substantial thinning of the sheet in areas where the part shows steep slopes severely limits the industrial use of ISF. A characteristic of parts produced with ISF is an uneven thickness distribution in the sheet after forming. Unlike deep drawing, there is no material flow from the flange areas during forming in ISF. Therefore, only the thickness of the sheet compensates for the increase in surface area (Jeswiet et al. 2005a). In a similar way to spinning, the "sine law" can be defined using volume constancy (Junk 2003):

$$t_1 = t_0 \times \sin(90° - \alpha) \qquad (5.10)$$

t_0 and t_1 describe the initial and the current sheet thickness; α is the wall angle according to Fig. 5.96.

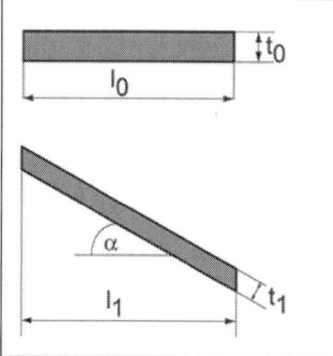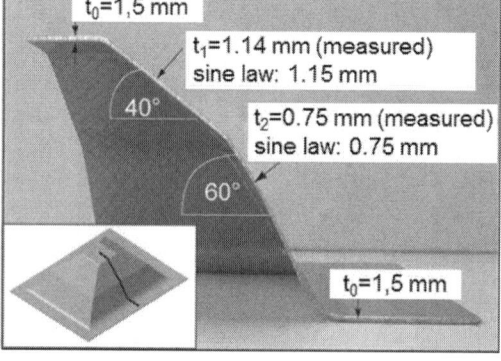

Fig. 5.96 Geometric relationship between the sine law (*left*) and its validation on a demonstration part (*right*). (Junk 2003)

 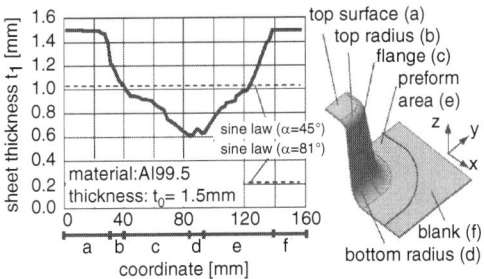

Fig. 5.97 Square cup produced by multi-stage ISF (*left*). Measurement of wall thickness along a radial section (*right*). (Hirt et al. 2004)

To prevent excessive thinning of the sheet, multi-stage forming strategies for manufacturing parts with wall angles greater than 60–70° were developed at the Institute of Metal Forming (IBF) (Hirt et al. 2004). In multi-stage forming, a preliminary shape, several intermediate shapes and the final shape are formed using ISF. Figure 5.97 shows an example of a square cup produced using a multi-stage forming strategy. This strategy required 15 stages and 7 h machining time. The long process time required means that in its present form this approach cannot be transferred to more complex and larger parts.

5.4.2.3 Geometric Accuracy

The geometric deviation from the CAD model, as in all sheet forming processes, is dependent on the part geometry, the part size, the material used, the thickness of the sheet and the rigidity of the part. In the past, various approaches for improving geometric accuracy in ISF have been developed and studied. Micari et al. (2007) have compared a few forming strategies. They identified "overbending" of the tool path as a promising approach to improve the accuracy. The part, e.g. as shown on the left of Fig. 5.98, is produced with a tool path that is "overbent" against the expected form errors (Junk et al. 2003) in order to compensate for the spring-back compared with the target geometry. More recent work at the Institute of Metal Forming (IBF) on the theme of "overbending" shows that the overall improvement in accuracy can sometimes be accompanied by the formation of waves (see Fig. 5.98, right). In Junk (2003) it was shown that if a full die was used, the deviations in geometry could be greatly reduced. Based on this work, optimization of TPIF is the main focus of the work at the IBF.

5.4.2.4 Production Times

The long processing time for ISF is due to the incremental nature of the forming process. The time it takes to manufacture a part is determined by the length of the

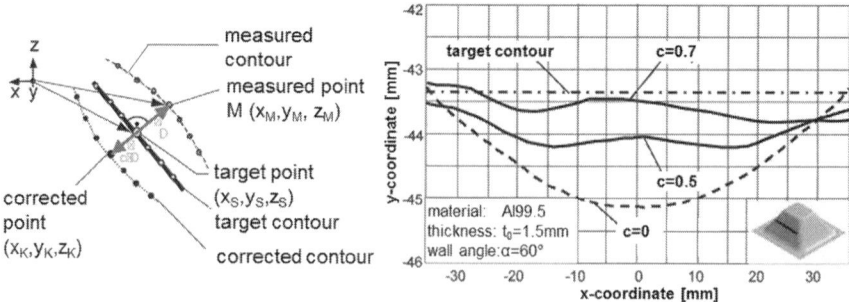

Fig. 5.98 "Overbending" of the tool path to improve geometric accuracy. (Junk et al. 2003)

tool path and the mean feed rate of the forming tool. The process can last from several minutes to several hours. The maximum possible feed rate is determined by the machine specifications and process effects such as friction and tool wear. Therefore, the speed cannot be specified at will. In Bambach (2008) it was shown that the process time required to manufacture a truncated cone of height h, bottom radius $r_B = 2h$ and top radius $r_O = h$ increases quadratically with h.

Instead of increasing the feed rate, the process time can be reduced by developing new machine and tool concepts. Therefore, Kwiatkowski et al. (2010) studied various tool concepts and developed a structure with a number of forming tools acting simultaneously to reduce process time. Increasing the number of forming tools used synchronously reduces the forming time according to the number of forming heads. Using the results of a concept analysis and considering the technical and economic aspects, a prototype tool with two forming heads was built (Fig. 5.99).

Although it is true that the concept realised by Kwiatkowski et al. (2010) shows a possible way of halving forming time, the structure of the tooling concept as realised limits the geometric range to rotationally symmetrical parts.

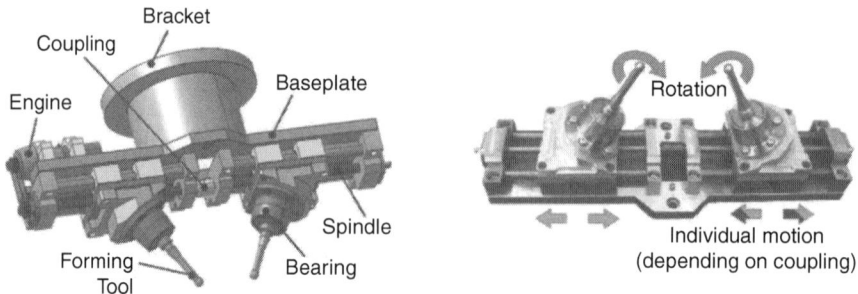

Fig. 5.99 CAD model (*left*) and prototype (*right*) of the new "TwinTool" tooling concept. (Kwiatkowski et al. 2010)

5.4.2.5 Numerical Simulation

In industrial sheet forming practice the Finite Element Method (FEM) is used for process planning. While researching ISF, the possible use of FEM as a tool for process planning and determining part properties was studied. However, FE simulation of the ISF process is very time-consuming because of the kinematic forming and the extremely long tool paths, and it is not currently practicable. In (Bambach 2008) the process time needed to simulate forming the truncated cone mentioned in the previous section with a dynamic explicit solver was determined. The calculation itself took almost 10 days for a tool path with a tool pitch of 0.5 mm, and more than 20 days for a tool pitch of 0.2 mm. Thus process planning is faster using experimental methods ("trial and error").

The long computation times motivated the development of various approaches to shortening the computing time for FE simulations of ISF. One possible approach is based on an area analysis in which the meshed sheet is divided into two areas. The first area is the forming zone, which is treated as non-linear elastic-plastic. The second area, which is outside the forming zone, is defined for FE calculation purposes as linear-elastic. As reported in (Hadoush and van den Boogaard 2008), the computation time for a 12 mm deep truncated pyramid was reduced from 12 to 6 h. Although this halved the computation time, computer-assisted process planning is not practical because the computing times for large parts are still far too long.

5.4.2.6 Materials for ISF

The same materials can be processed by ISF as those by conventional sheet forming processes (deep drawing, stretch forming). In theory, even larger parts can be produced with machines of comparable size because of the localised forming. The limits are mainly determined by the sheet thickness. Materials such as most titanium and magnesium alloys cannot be formed at room temperature due to their hexagonal structure. A possible improvement in formability can be achieved by heating the forming zone, as has been shown in (Duflou et al. 2007) and (Biermann et al. 2009). Furthermore, suitable process control can improve accuracy and reduce forming forces (Duflou et al. 2007).

Another interesting development is the processing of hybrid materials, e.g. sandwich materials or hybrid combinations of different metals, which are mainly used in lightweight construction and aircraft manufacture. The first forming trials on sandwich sheets made from aluminium outer layers and a thermoplastic core (Hylite sheet made by CORUS) are described in (Jackson 2008). The result confirms that it is generally possible to produce parts from such composite materials. A methodical study of the interactions between forming process and hybrid material has not yet been carried out.

5.4.3 Motivation, Research Question and Approaches to a Solution

The state of the art shows that ISF is suitable for prototyping and small batch production. The area of application for producing parts from sheet metal using ISF would be sectors such as the automotive industry, aerospace, medicine and architecture. Nevertheless, the classical ISF process is not capable of widespread industrial use because of the limitations specific to the process. Thus, the implementation of ISF is lagging a long way behind its potential. This has been the motivation for a large number of research projects aiming to develop process strategies for increasing and extending the productivity of ISF. The approaches to overcome the process limits, which have already been cited in the state of the art, have the following disadvantages:

- The approaches to avoid extensive thinning and the uneven distribution of sheet thickness by multi-stage forming cannot be generally applied to all geometries. In addition, they prolong processing times.
- The proposed solutions to increase geometrical accuracy by overbending have side-effects such as waves.
- Process planning using FEM is not possible due to the long computational time.
- Process optimization by "trial and error" and the long process time reduces the cost-effectiveness of the process.
- The production of parts from high-strength materials and materials that are difficult to form at room temperature requires the development of hot forming processes.

The suitability of hybrid process combinations for overcoming the process limits is being studied as part of the Cluster of Excellence.

> The aim is to overcome the drawbacks of the conventional ISF and enable new applications by combining processes, physical principles and materials.

The present article examines and develops the process combinations of stretch forming and ISF (SF + ISF) and local heating via laser-assisted ISF (Laser + ISF). The research on hybrid processes is based on the development of a suitable machine platform that permits the processes to be implemented and executed. Another interesting aspect is the processing of hybrid combinations of materials. One example is the study of the formability of aluminium-steel composite sheets (Tailored-Hybrid-Blanks) using ISF.

5.4.3.1 Combined Stretch Forming and Incremental Sheet Metal Forming

Due to the combination of stretch forming and ISF (SF + ISF), a pre-form can be produced by stretch-forming. Only the areas of the sheet that cannot be formed by stretch-forming are produced using the more time-consuming ISF. Thus, the flexibility of ISF is used in order to form elements such as pockets, beads or corrugations.

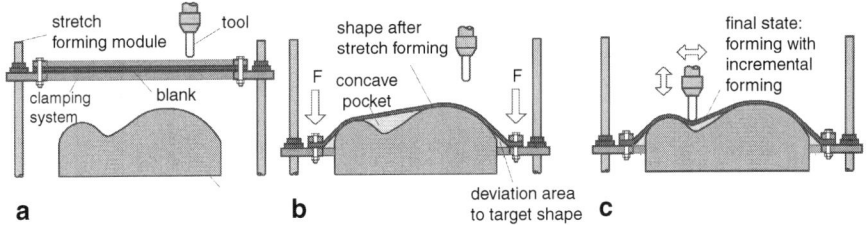

Fig. 5.100 Principle of process combination "SF + ISF". **a** Machine set-up. **b** Stretch forming. **c** Incremental sheet forming. (Taleb Araghi et al. 2009)

Figure 5.100 is a schematic representation of the "SF + ISF" process combination. After stretch forming, the areas of the sheet that are in contact with the die are already in their final form (Fig. 5.100b). The remaining areas of the sheet must be worked to fit the shape of the part by ISF. This reduces the proportion of incremental forming and shortens the processing time.

As was explained in Sect. 5.4.2, the production time in ISF depends on the size of the part and for example, the length of the tool path. In the case of stretch-forming the process duration only depends on the height of the part, or the height of the die. Consequently, preliminary stretch-forming greatly reduces the duration of the forming process. In particular, large and slightly curved parts with a large surface area can be produced more quickly by the process combination. Another anticipated advantage of the "SF + ISF" process combination, compared with ISF on its own, is that the changed material flow in stretch forming results in a more homogeneous sheet thickness distribution compared to ISF. It is expected that the use of preliminary stretch-forming will make it possible, and simpler, to form steep areas on parts. Furthermore, the superimposed tensile stresses from stretch-forming should improve dimensional accuracy and reduce residual stresses. The fact that stretch forming operations can be simulated within an acceptable time is an advantage for virtual process planning.

5.4.3.2 Laser-Assisted Incremental Sheet Metal Forming

Laser-assisted ISF (Laser + ISF) is being developed with the aim of forming high-strength materials flexibly and cost-effectively. Titanium and magnesium alloys are of particular interest in this respect as they are difficult to form at room temperature. Additionally, titanium alloys are often processed using expensive tool-based processes such as superplastic forming. The use of local laser heating is expected to improve the formability of metals that cannot, or can only with difficulty be formed at room temperature. The process is shown schematically in Fig. 5.101. The forming area is heated by means of a laser beam in front of the tool.

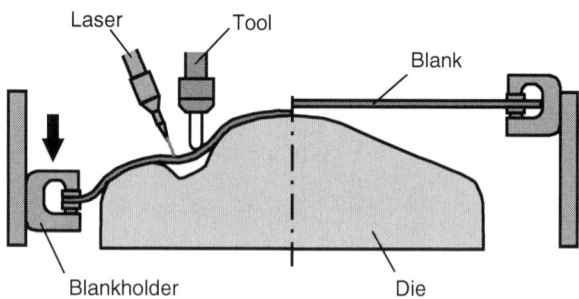

Fig. 5.101 Principle of laser-assisted ISF

5.4.3.3 Forming of Tailored Hybrid Blanks

The use of parts made from hybrid materials (THBs: Tailored Hybrid Blanks) opens up new possibilities in automotive lightweight construction. A new concept for producing THBs has been developed at the Welding and Joining Institute (ISF). This enables an integral joint between aluminium and steel. The processing of aluminium-steel composites by ISF is expected to enable the production of small numbers of multi-material sheet metal parts. The combination of the two areas is expected to show the way towards manufacturing customised individual products, using innovative joining techniques and production technologies. As a basis for this, the forming characteristics of THBs are being studied and optimised.

5.4.3.4 Development of Machines

Research on the process combination is based on the technological implementation of a machine platform and a virtual process planning chain to depict the process. The sheet metal processing centre that has been developed and a CAD/CAM/CAE process chain are presented in the following. Using experimental studies, the hypotheses that were derived were verified and confirmed by experiments.

5.4.4 Results

5.4.4.1 Development of the Machine Platform

A new hybrid sheet metal forming machine was built at the Institute of Metal Forming at RWTH-Aachen University in cooperation with the manufacturer EIMA Maschinenbau GmbH. The machine is based on a standard machining centre. The machine was adapted for the high forming forces. Additionally, four stretch forming modules were implemented.

Four possible approaches were considered when designing the stretch forming elements (see Fig. 5.102).

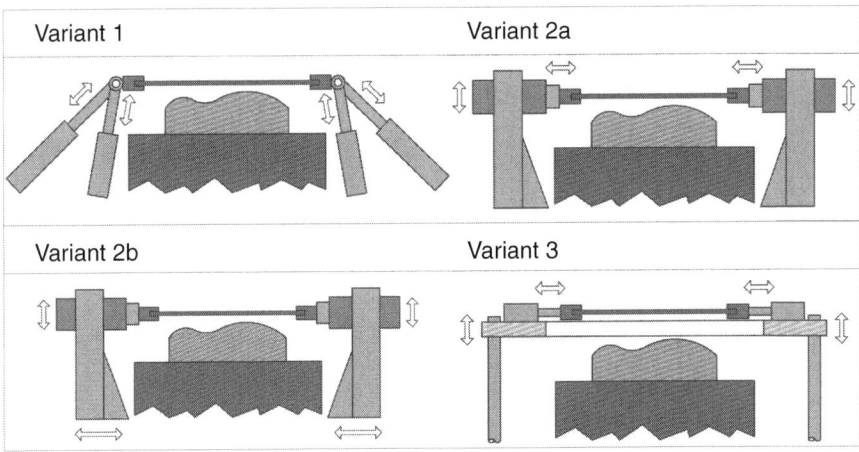

Fig. 5.102 Four possible variants of the stretch forming units. (Taleb Araghi 2011)

Variant 1 is based on two pivoting hydraulic cylinders connected to the machine base. The blank is clamped by means of clamping jaws. In variants 2 and 3 the forming forces are introduced into the sheet by stretch forming modules that move on vertical and horizontal oriented ball screws. The horizontal movement can be achieved via movable clamping jaws, as in variant 2, or via the movement of the entire module, as in variant 3. In variant 4 the horizontal mobile clamping jaws are positioned in a closed frame. The entire frame moves vertically.

A comparison of the advantages and disadvantages of the concepts is shown in Table 5.7. It is obvious that variant 3 offers the best approach. The structure of this design combines flexibility and a high potential forming force.

The basic structure that was built is a portal milling machine tool with 5 machining axes (Type "GAMMA") made by EIMA Maschinenbau GmbH. The table of the "Gamma" machine was adapted so that the stretch forming modules could be incorporated. Two possible module arrangements are considered when designing the

Table 5.7 Comparison of the feasibility of various designs for the stretch forming modules. (Taleb Araghi 2011)

	Weighting (%)	Variant 1	Variant 2	Variant 3	Variant 4
Flexibility	10	2	3	5	1
High force	25	3	4	4	4
High rigidity	15	4	3	3	4
Tangential stretch forming	10	1	5	5	5
Operability	5	2	4	5	4
Setup cost	5	4	3	3	2
Suitability for automation	5	4	4	4	5
Suitability for integration	25	1	4	4	2
Σ	100	2.4	3.8	4.05	3.25

0: not possible; 1: very poor; 2: poor; 3: average; 4: good; 5: very good

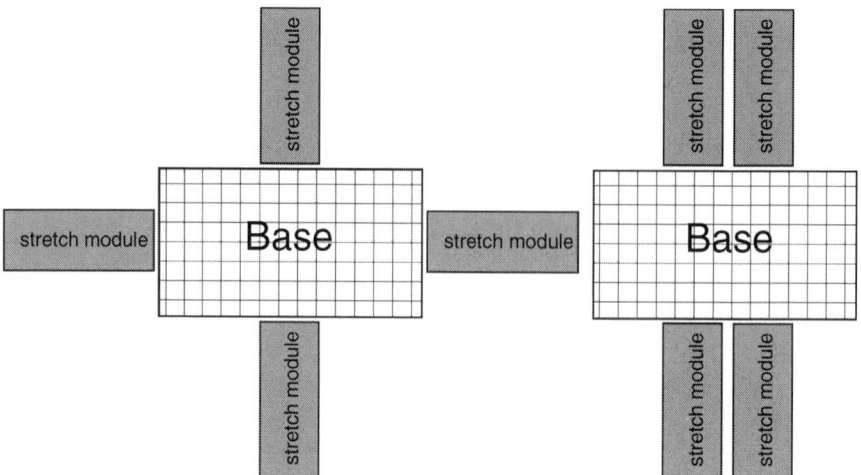

Fig. 5.103 Possible positions for the stretch forming modules. *Left*: star configuration, *right*: two-sided arrangement. (Taleb Araghi 2011)

machine tool table. In one variant, the modules can be arranged in a cross formation to enable stretch forming on four sides. Another possible arrangement is to position the modules for stretch forming on two sides, with two modules positioned next to each other on each side (see Fig. 5.103). The machine, including the control system, is designed in a way that makes both arrangements possible.

Figure 5.104 shows the CAD model for a stretch forming module. The clamping bar moves vertically (z-axis) via a ball screw. The forming forces are absorbed by vertical slides. The module unit is mounted on two horizontal slides which allow the

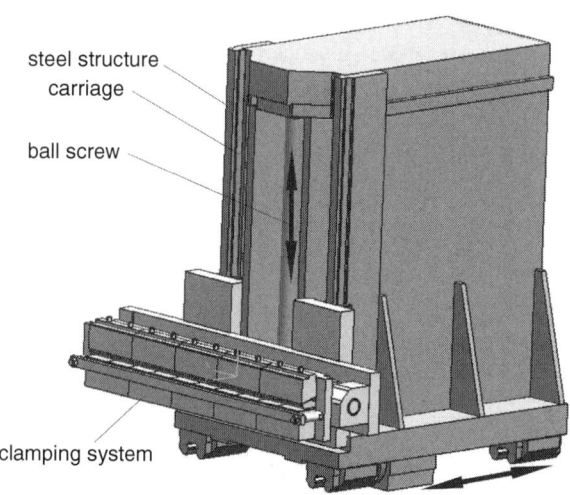

Fig. 5.104 CAD design for the stretch forming units. (Taleb Araghi 2011)

Fig. 5.105 Sheet metal processing centre at the Institute of Metal Forming. (Taleb Araghi 2011)

module to move horizontally by means of sliding carriages. The machine table acts as an abutment that is anchored in a machine foundation.

Figure 5.105 shows the hybrid sheet processing machine at the Institute of Metal Forming at RWTH Aachen University. Design and construction were completed in April 2009. Its special feature is that the same machine can be used to both mill the forming tools needed (e.g. dies) and to carry out the forming operations. Conversion from milling to ISF operation takes very little time. It is done by fixing an ISF toolholder to the machine's milling head.

The working space for milling operations is $2,000 \times 2,000 \times 900 \, mm^3$. For sheet forming, a working space of $2,000 \times 1,700 \times 700 \, mm^3$ is available. The four stretch forming modules each provide a horizontal force of 250 kN and a vertical force of 150 kN. The modules are integrated in the machine control system by additional NC channels. Thus the present machine provides 5 NC axes for milling and ISF and 8 additional NC axes for the stretch forming modules.

In order to implement local heating for laser-assisted ISF, a new laser optic was designed and integrated into the machine in cooperation with the Fraunhofer Institute for Laser Technology (ILT) and the Chair for Technology of Optical Systems (TOS). The laser selected is a type "LDF 10000" diode laser from the company Laserline. The maximum available output of 10 kW (radiation power) is sufficient to heat materials up to temperatures above 1,000 °C. The main advantage of a fibre-coupled diode laser is that the beam can be guided via an optical fibre. Thus, the energy required for heating can be directed flexibly right to the forming area. The movements of the forming tool can be compensated by the optical fibre and a feed device.

Since the optical system cannot be rotated around the tool, it must be designed so that the laser beam does. In order to be able to rotate the beam along a circular path, it must be offset parallelly to the optical axis by two mirrors tilted at 45° with respect to the beam. Rotation of the mirrors causes the laser beam to move in a circle. The amount by which the mirrors offset the beam must exceed the radius of the tool in order to prevent interaction with the latter (Diettrich 2010).

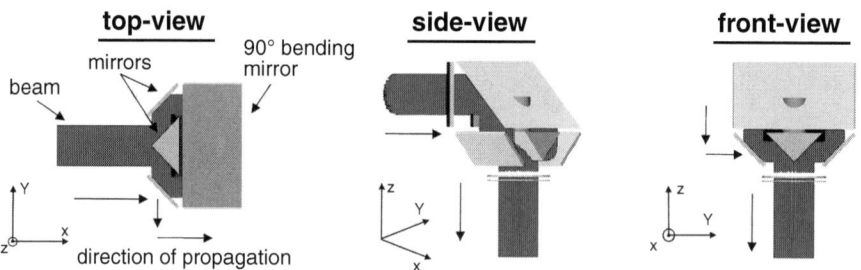

Fig. 5.106 Division of the beam and reconstitution after the deflection by mirrors and mirrored prisms. (Diettrich 2010; Göttmann et al. 2011)

The diverted rotating beam has to be guided around the tool and diverted towards the processing area. Figure 5.106 shows schematically how the beam is split and brought together again. To split it, an arrangement comprising a mirror-prism and two mirrors at the sides is used. The beam is bent 90° by a deflecting mirror and then brought back together by a second identical arrangement of mirrors. This creates an opening in the beam path, into which the tool can be centred without being hit by the laser beam (Diettrich 2010).

Figure 5.107 is a schematic representation of the configuration of the optical system. The combination of the rotating deflection and the mirror components described in Fig. 5.106 splits the laser beam and then recombines it so that it can rotate along a concentric path around the tool.

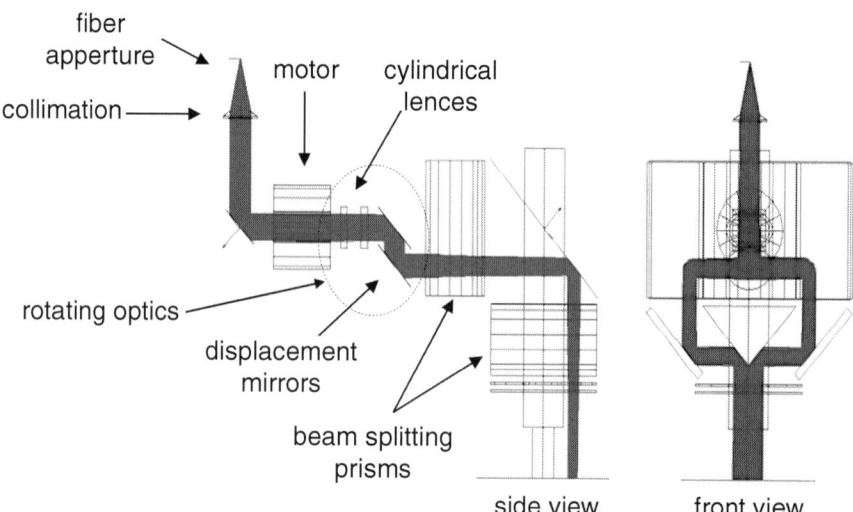

Fig. 5.107 Design of the optical system with a coaxially rotating laser beam. (Diettrich 2010; Göttmann et al. 2011)

5 Hybrid Production Systems

Fig. 5.108 Hybrid forming machine with built-in optical system. (Göttmann et al. 2011)

The shape and position of the laser spot can be influenced by selecting different lenses and varying the distances between the mirror components. In the simple version, a circular laser spot with a diameter of 35 mm is projected onto the surface of the part at a distance 45 mm from the tool axis. To test the optical system and carry out initial trials of laser-assisted ISF, cylindrical lenses are used, creating an elliptical laser spot 15 mm × 45 mm at a distance of 45 mm from the tool centre distance (Diettrich 2010).

The optical system described is fixed to the forming head of the hybrid machine (see Fig. 5.108). The laser source used is outside the machine, so the laser beam is guided to the optic system via a fibre. The optical system moves together with the processing head during the forming process. The laser spot is positioned by a motor that is built into the optical system.

5.4.4.2 CAX Development

In addition to a new machine platform, cost-effective use of the "SF + ISF" and "Laser + ISF" hybrid processes requires the development of a suitable CAX environment consisting of CAD (*C*omputer *A*ided *D*esign), CAM (*C*omputer *A*ided *M*anufacturing) and CAE (*C*omputer *A*ided *E*ngineering). The commercially available software Unigraphics NX5® was selected as the development platform. Further modules were added to the standard CAM module. The development work was done in close collaboration with iCASOD GmbH.

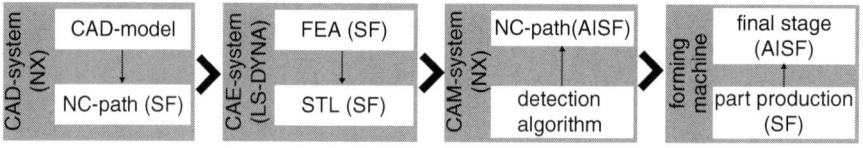

Fig. 5.109 CAX process chain for the SF + ISF process combination. (Taleb Araghi et al. 2009)

In addition to programming the modules to execute stretch forming and ISF operations, calculation of the laser spot position from the forming path was required for the laser ISF. Method engineering for stretch forming was made possible with the aid of an interface between NX5® and the CAE program.

In order to achieve substantial time savings, from the combination of stretch forming and ISF compared with ISF alone, the time spent on ISF after stretch forming must be kept to a minimum. However, after the stretching operation the zones that already conform to the final geometry are unknown a priori. In addition, the stretch formed sheet is subject to elastic spring-back that only appears after stretch forming. To achieve the shortest forming time per part, it is necessary to determine the zones that are to be formed by ISF after stretch forming and only generate the ISF paths for those zones. Furthermore, the form of the sheet after stretch forming must be known in order to avoid collisions between the forming tool and the sheet during the subsequent ISF.

In the case of the stretch forming of complex geometries the analysis is unable to forecast the shape of the stretch formed part. Therefore, it is essential to start by using FEM to determine the shape of the part after stretch forming, and then to identify the zones of the sheet that are to be processed with ISF.

iCASOD GmbH, working with the IBF in the "SIBUFORM" project of the BMBF, has developed a new CAX chain for this purpose. It is used to plan the path for ISF after stretch forming. Figure 5.109 is a schematic representation of the new CAX process chain, which is made up of four individual steps.

First of all, the geometry of the part to be produced is prepared in the NX5® CAD system. For the purpose of the "SF + ISF" process combination, the additional areas needed are added to the shape of the part. After the geometry has been prepared, the NC path for stretch forming is generated in the CAD system. Next, the stretch forming process is simulated in the LS-DYNA® CAE system. If necessary, the stretch forming path can also be optimised at this stage. The sheet thickness and strain distributions, potential wrinkling and maximisation of the proportion of the final geometry to be achieved by stretch forming can be taken into account for this purpose. After the FE calculation of the stretch forming operation, the simulated part shape is imported into the NX5® CAM system as an STL mesh. An algorithm for determining deviations between the simulated geometry and the target geometry, that has been implemented in NX5®, defines the zones that need to be formed using ISF. Lastly the ISF path is planned in NX5®. After the planning process is complete, all NC data (stretch forming and ISF) is transmitted to the forming machine. Afterwards, the part can be produced.

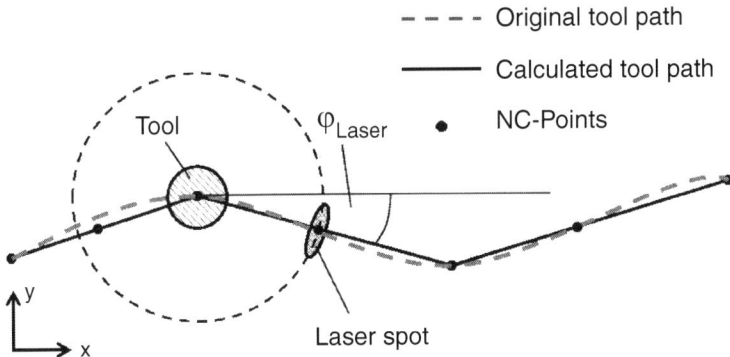

Fig. 5.110 Schematic representation of path calculation for the "Laser + ISF" process combination. (Göttmann et al. 2011)

Virtual planning of laser-assisted ISF has to be possible in a similar way to that for the SF + ISF combination. The process design requires a suitable CAM environment for calculating the position of the laser spot at every instant during forming. In addition, the timing for switching the laser on and off must be reproducibly definable. Lastly, it must be visually possible to check the computed forming process.

Working with iCASOD GmbH, a standalone postprocessor was programmed to calculate the forming paths. Figure 5.110 is a schematic representation of the NC data calculation. First of all, a forming path analogous to conventional ISF is programmed (dashed line). In a second computation, the calculated path, which follows the geometry of the part and can be curved as desired, is divided into partial concrete linear paths (continuous line). The coordinates of the end points of the partial paths represent movement commands (NC points) in the subsequent NC code. For each of the points, a value is calculated for the X, Y and Z axes of the machine and also a laser angle (ϕ_{Laser}) for the motor rotating the optical components. The machine control system regulates the movement of all the axes according to the current position so that all the axes reach the programmed point at the same time.

5.4.4.3 Studies on the Process Combination of Stretch Forming and ISF

As already described in Sect. 5.4.2, the reduction in sheet thickness, or thinning, that occurs with ISF is mainly dependent on the wall angle of the geometry to be produced. Large wall angles thus represent a substantial limit to the process. In Sect. 5.4.3 the combination of stretch forming and ISF is suggested as a possible approach to solving the problem of optimal material flow and thus reducing the extensive thinning and making the sheet thickness distribution more homogeneous. Practical examples have demonstrated that this process combination achieves a lesser degree of thinning than conventional ISF.

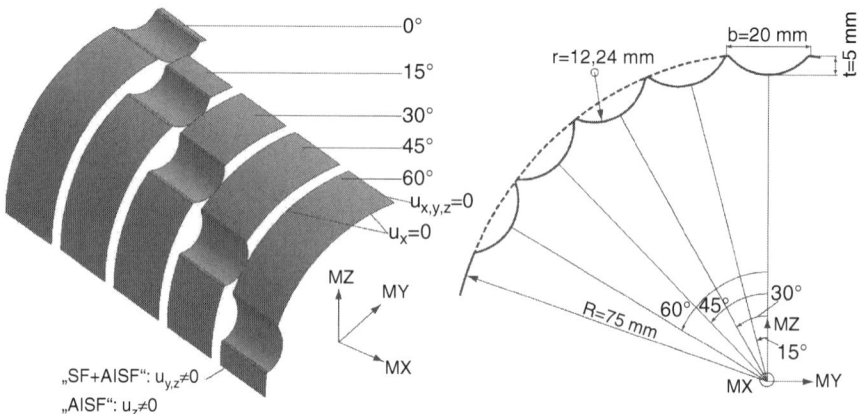

Fig. 5.111 3D geometry inclusive (*left*) and side view (*right*) of the half-shell with varying pocket position. (Taleb Araghi 2011)

In order to study the effect of the combined process on sheet thinning and material flow systematically, an FEM model of a 50 mm wide strip from a half-shell made from 1 mm deep-drawing steel DC04 with a radius of R = 75 was considered. The half-shell has a 5 mm deep pocket of radius r = 12.24 mm. The position of the pocket was varied between 0° and 60° in steps of 15° as shown in Fig. 5.111.

FE modelling was done using shell elements and FE calculation was carried out with the LS-DYNA® explicit solver. The type of mesh used was a quadratic mesh with a 0.75 mm edge. Manufacture of the half-shells was simulated using both ISF on its own and the SF + ISF combination. The Unigraphics NX5® CAM system was used to plan the path for ISF as a vertical section strategy in which the part is machined downwards from the highest point with a constant vertical tool pitch of 0.4 mm. In the case of ISF on its own, the forming was planned and simulated as a single-stage strategy for the 0°, 15° und 30° half shells. For the pocket positions at 45° and 60° two-stage forming was used because of the steep angle that the wall makes inside the pocket. For the SF + ISF process combination, tangential stretch forming was used to form the half shell structure (see Fig. 5.112, left). After stretch forming, the pocket was shaped with ISF using the same path strategy that was used for the variant with ISF on its own.

The mean linear thickness strain ε_{mean} and the sheet thinning t_{mean} were introduced to evaluate and assess the thinning (see Fig. 5.113):

$$\varepsilon_{mean} = \frac{l_1}{l_2} - 1 + \varepsilon_0 \qquad (5.11)$$

ε_0 describes the initial linear thickness strain from a possible prior stretch forming or ISF. The developed pocket length l_1 is constant, in this instance 22.36 mm.

$$t_{mean} = \varepsilon_{mean} \times t_0 \qquad (5.12)$$

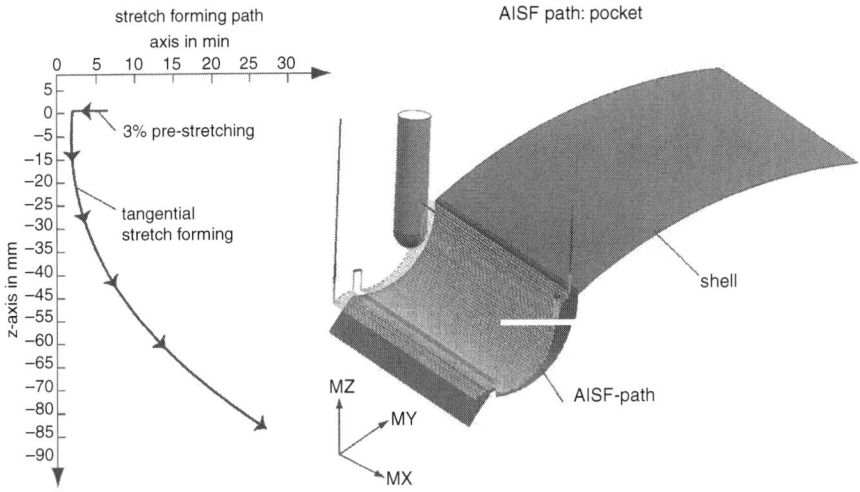

Fig. 5.112 The stretch forming path and ISF path of the pocket. (Taleb Araghi 2011)

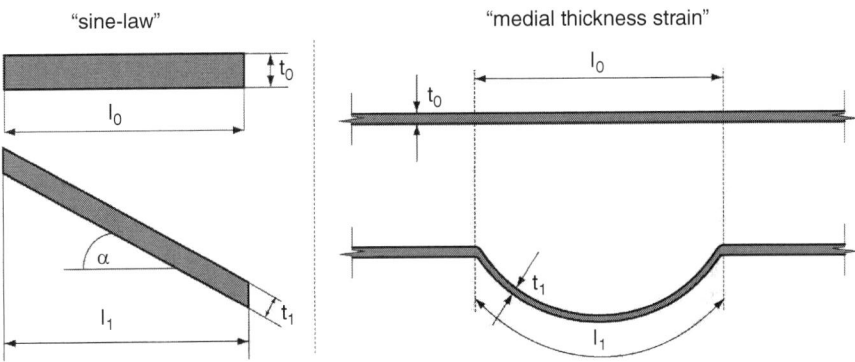

Fig. 5.113 Geometric representation of the "mean strain" and mean sheet thinning. (Taleb Araghi 2011)

Figure 5.114 is a summarized representation of mean thinning according to Eq. (5.12) for the "pure ISF" variant and the combined "SF + ISF" process plotted against the five pocket positions on the half shell. The graph shows that the trend curve for ISF on its own produces less thinning below about 19° than the trend curve for the combined process. At angles greater than 19°, the thinning curve for ISF rises and passes above the "SF + ISF" curve. Furthermore, the trend curve for ISF on its own shows that thinning is greatly influenced by the position of the pocket. In contrast, it is clear that the trend curve for the combined process slopes much more gently and thus shows less sensitivity to the position of the pocket on the half shell.

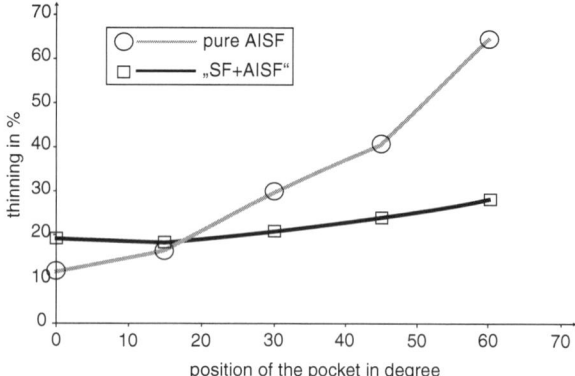

Fig. 5.114 Effect of the pocket position on the half shell on thinning. (Taleb Araghi 2011)

To sum up the result of the systematic study on sheet thinning, it can be concluded that the combination of the stretch forming and ISF processes can limit the thinning of parts with steep bodies.

As part of the experimental trials with "SF + ISF", the sheet thinning and geometric accuracy were studied for a demonstrator. The same geometry was produced with pure ISF for reference and the results compared. The geometry of the component was first modelled in CAD and the corresponding die was milled in the hybrid machine (see Fig. 5.115).

Process planning assisted by FEM was used first in order to optimise the stretch forming operation. This was done to minimise the sheet thinning while at the same time maximising the proportion of stretch forming. Tangential stretch forming, in which the stretch forming modules move both vertically and horizontally, gave the best result. Figure 5.116 shows the result of the stretch forming simulation and the movement profile of the stretch forming modules that are opposite each other. The stretch forming covered approximately 53% of the final geometry. It was followed by incremental forming of the upper pocket and the outer areas of the sheet that could not be produced by stretch forming.

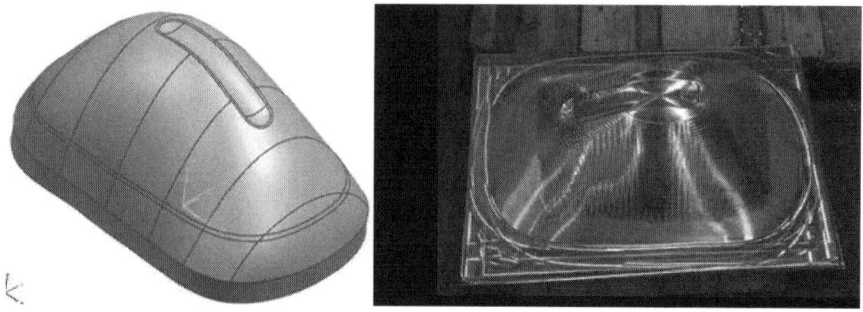

Fig. 5.115 CAD model of the geometry under study (*left*), Stamp after the milling operation (*right*). (Taleb Araghi et al. 2010)

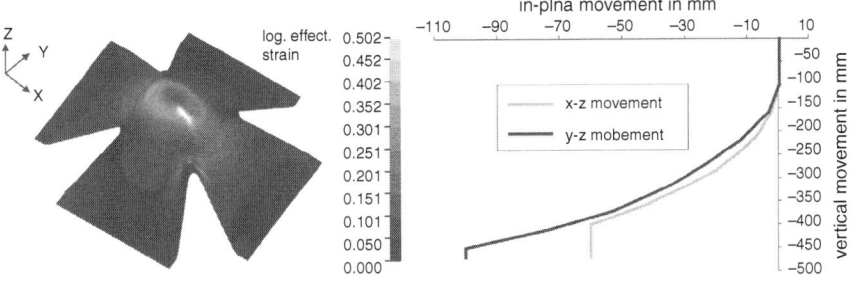

Fig. 5.116 Sheet geometry calculated with LS-Dyna after stretch forming (*left*), movement profile of the stretch forming modules (*right*). (Taleb Araghi et al. 2010; Taleb Araghi 2011)

Figure 5.117 shows the parts made from deep-drawing steel DC04 (sheet thickness 1.0 mm) using ISF on its own (left) and using the "SF + ISF" process combination (right). The combined process achieved a reduction in manufacturing time of approx. 33% compared with ISF on its own. The manufacturing time of the combined process was about 40 min. Using ISF on its own, the process took 60 min.

In addition to the systematic FEM study on sheet thinning, the following experimental results were proposed, confirming that the "SF + ISF" process combination can reduce thinning in critical zones compared to conventional ISF. To determine the distributions of strain and sheet thickness, the sheets used were marked with a regular pattern of points before being formed. After forming, the surface of the part was measured using an optical system to determine the strain distribution. The reduction in sheet thickness is calculated from the strain in the sheet surface, the assumption that plastic deformation takes place at constant volume and from the assumption that only negligible shear phenomena occur in the thickness of the sheet.

The following figures show the results of the strain and sheet thickness distributions along the cross-section of the part. The main changes in shape with ISF on its own (see Fig. 5.118) follow the sine law from Eq. (5.10) as the wall angle increases.

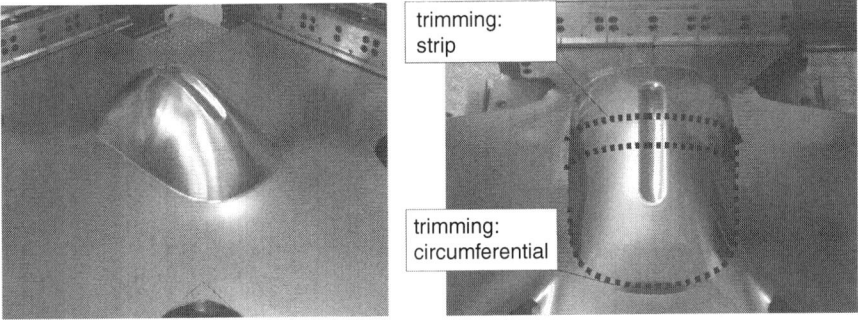

Fig. 5.117 Reference part made with ISF on its own (*left*), and part after forming with the "SF + ISF" combination (*right*). (Taleb Araghi et al. 2010; Taleb Araghi 2011)

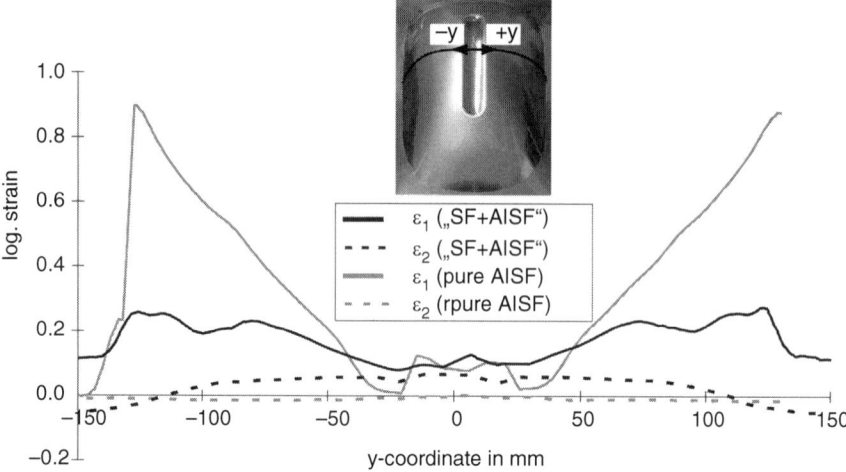

Fig. 5.118 Changes in shape along a cut. (Taleb Araghi 2011)

In this case values reached close to 0.9. There are no subsidiary changes in shape with conventional ISF. Thus, seen in macroscopic terms, the process is subject to a plane strain. Compensation of the principal positive strains takes place exclusively via the negative strains in the direction of the thickness. The main shape changes from the combined "SF + ISF" process have a quasi-homogeneous plot with a maximum value of approx. 0.25. Because stretch forming takes place on four sides, additional positive shape changes can be seen in the central area of the part. In the zone outside the component geometry there are subsidiary negative shape changes. The reason for this is the cutout at each corner of the sheet to ensure that only tensile forces are transmitted. Necking has occurred here as it would in a tensile specimen.

Figure 5.119 shows the evaluation of the reduction in sheet thickness along the cross-section (broken line). The plot for the thickness reduction is similar to that for the main shape changes in Fig. 5.119. Sheet thinning with ISF on its own reaches a maximum of approx. 60% whereas with SF + ISF the maximum sheet thinning is only approx. 25%. The curve for ISF shows a constant increase in wall thinning as the wall angle increases. In contrast, the curve for the combined process shows less wall thinning and a smoother plot. The wall thinning using the combined process is only greater than that with ISF on its own in the pocket area on the top of the part. This is due to the superimposition of the strains from both the stretch forming and the ISF.

To test the hypothesis relating to geometric accuracy, the components were cut away from the supporting surfaces (see Fig. 5.117, peripheral cut). The parts were then digitised using the gom GmbH ATOS system and compared with the desired geometry. In addition, the residual stresses induced in the component by the forming process were studied by cutting out a strip (see Fig. 5.117, strip cuts), and determining the spring-back after cutting.

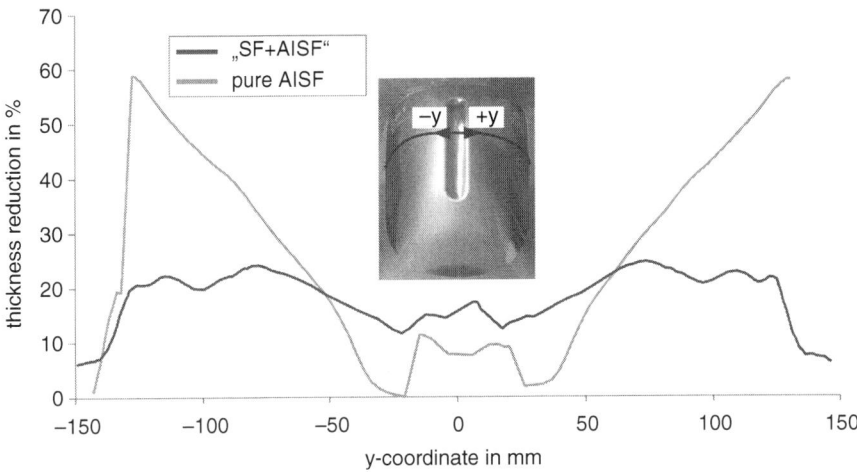

Fig. 5.119 Reduction in sheet thickness along a cut. (Taleb Araghi et al. 2010; Taleb Araghi 2011)

Figure 5.120 shows the result of the geometric comparison with the CAD model after cutting the component out all around. The ISF component shows buckling on the long side areas with a maximum value of approx. 8 mm. Form errors with ISF often occur along large areas that are not very curved. They can be reduced with a carefully chosen forming strategy but not eliminated. Contrasted with the ISF component, the component produced with SF + ISF had discrepancies of approx. 0.5 mm in the top area. In the lower part, which had been incrementally finish formed after stretch forming, there were deviations of up to 3.0 mm. This result shows that

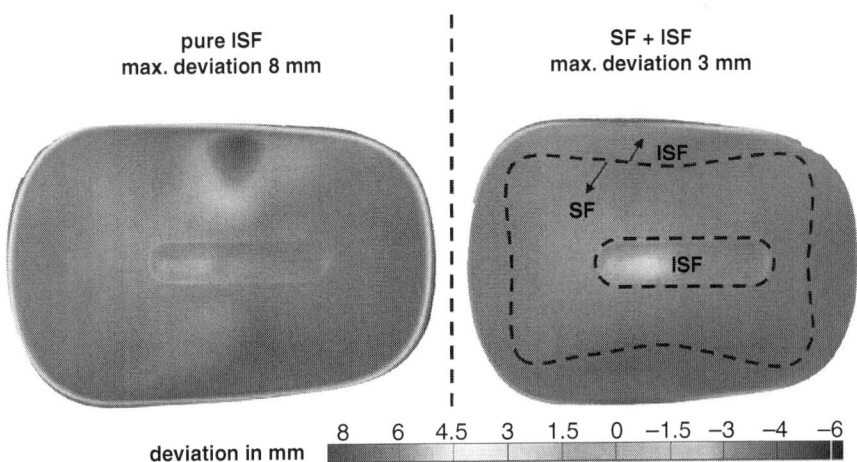

Fig. 5.120 Geometric comparison with target geometry after cutting out, ISF component (*left*), SF + ISF (*right*). (Taleb Araghi 2011)

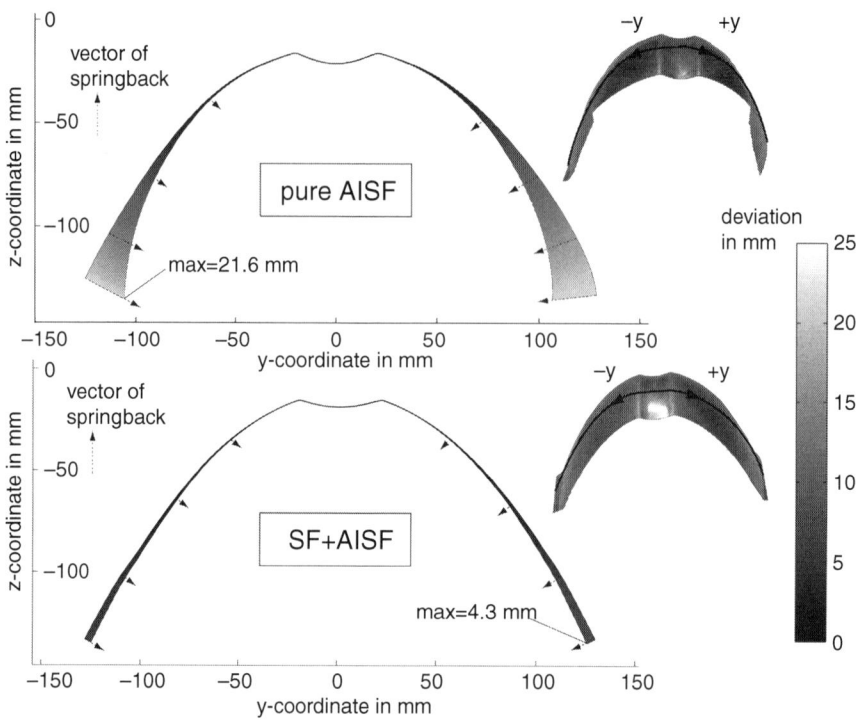

Fig. 5.121 Evaluation of deviations. ISF part (*left*), SF + ISF part (*right*). (Taleb Araghi 2011)

for the selected geometry, superimposing the tensile stresses from stretch forming suppresses buckling.

The spring-back of the strip cut out from the formed part is shown in Fig. 5.121. It can be clearly seen, that the values for the geometry produced with the combined forming process are lower, with a maximum spring-back of 4 mm, than for the part produced exclusively with ISF. In the case of conventional ISF, spring-back of up to 23 mm is directed inwards.

The results illustrated show that the production of the demonstrator with the "SF + ISF" process combination induces lower residual stresses than with ISF on its own.

5.4.4.4 Study of the Combination of Laser Heating and ISF

The approach developed from the state of the art on combining local heating by laser with incremental forming was studied in parallel with the development of a heating unit in simplified tests. The research work was centred on extending the formability limits of materials with limited formability at room temperature.

5 Hybrid Production Systems

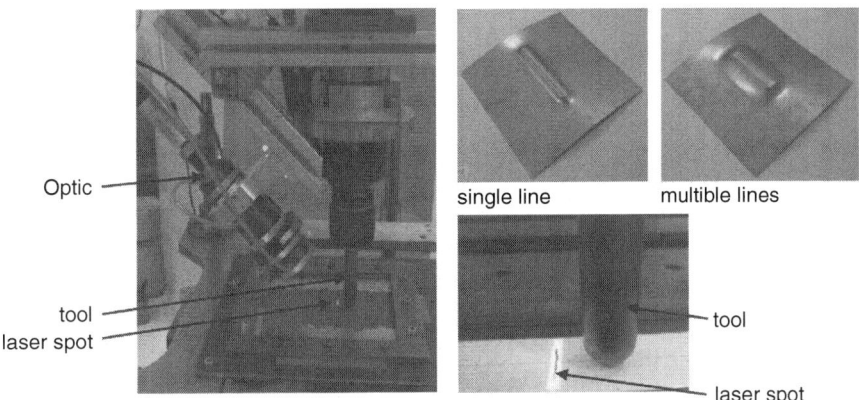

Fig. 5.122 Test rig on the model machine (*left*) for laser-assisted ISF with a linear laser spot preceding the tool (*bottom right*) and the test parts that were produced (*top right*). (Biermann et al. 2009)

The left side of Fig. 5.122 shows the test rig that was used. The area to be formed was heated with a laser ahead of the tool. A laser lens was fastened to the tool head for this purpose. The laser beam was projected onto the sheet at an angle of 45° from a distance of 20 mm in front of the tool (Fig. 5.122, bottom right). An LDF 2000 diode laser from Laserline was used as the laser source.

In this case it was not possible to rotate the laser spot about the tool and for this reason only simple linear geometries could be produced (Fig. 5.122 top right). The structure shown nevertheless allowed the forming behaviour of materials in a laser-assisted incremental forming process to be studied. The tests were carried out on sheets of pure titanium (Grade 2), a titanium alloy (TiAl6V4) and a magnesium alloy (AZ31B) with a sheet thickness of 1 mm.

The formed specimens shown in Fig. 5.122 were produced by repeated plunging of the tool and forming of individual straight-line paths. Between the individual paths, the vertical tool feed was increased by 0.35 mm. The laser was switched on at the plunge point and off when the tool was lifted from the sheet. The output power of the laser beam was set so that different temperatures were reached during forming. The temperature was measured with thermocouples on the back of the sheet. The maximum achievable depth was defined by the vertical feed of the tool at the time of crack formation.

Figure 5.123 is an extract from the results obtained and shows the maximum forming depths that could be achieved by varying the temperature and the geometry of the laser spot compared with cold forming. The studies have shown that local heating has a beneficial effect on the formability of titanium and magnesium.

It is very clear that the formability of magnesium material improves greatly at 250 °C. This is explained by the temperature-dependent properties of magnesium's crystalline lattice structure. The hexagonal structure that is present at room temperature offers too few active glide systems and this leads to limited forming

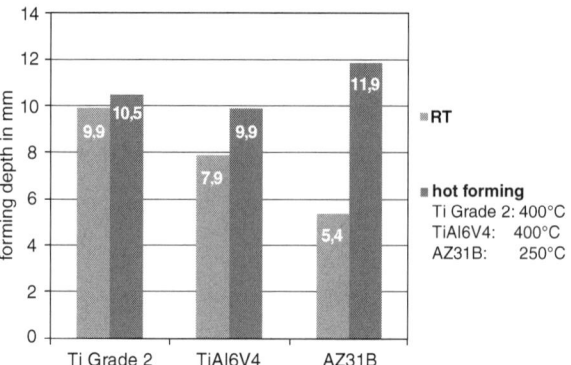

Fig. 5.123 Comparison of the depth reached when forming a single line in pure titanium (Ti Grade 2), TiAl6V4 and AZ31B at room temperature and at elevated temperatures. (Taleb Araghi et al. 2010)

characteristics at the macroscopic level. It is known that at temperatures above 225 °C additional glide systems are activated, leading to increased formability.

Additional metallographic studies show that recrystallisation processes can take place during hot forming. These results were examined to support the evaluation of the forming behaviour of magnesium alloy AZ31, and to develop a detailed description of the thermal and mechanical interactions.

In that regards, the effects of initial grain size and forming speed on the degree of forming during laser-assisted incremental sheet forming process performed on sheets of the magnesium material AZ31B are discussed below. The aim was to investigate the mechanisms occurring during the combined incremental forming and laser heating, and to integrate them in a mechanism map so that the areas of increased formability that were important for process control could be identified. This diagram was intended to be used later to set the specific process parameters.

The first step was to form the as-received sheets with four different initial grain size regimes (mean grain size values of 10, 16, 18 and 20 µm) using three different tool speeds (2, 4 und 6 m/min) at a temperature of 250 °C. The results of this testing setup were used to determine the recrystallization behaviour and the highest formability that could be achieved. Other studies were focused on different temperature profiles during the forming process.

The investigated metal sheets were incrementally formed along a line to a stepped pocket by six steps. The first of these was 2.1 mm deep. Each additional step was 2.1 mm deeper so that the total depth of the sixth step was 12.6 mm. At an infeed of 0.35 mm, six forming steps were required for each level. Figure 5.124 is a schematic representation of this stepwise forming process.

Before the process, a uniform quadratic raster was applied to each plate for the purpose of evaluating the strain achieved in the forming process. After forming, the deformation was recorded with an optical measuring system and then evaluated. The applied strain at each forming stage was determined in this manner. Figure 5.125 shows an example.

In addition, to determine the deformation response, metallography, microhardness and X-ray texture analysis were used, and finally the activated deformation

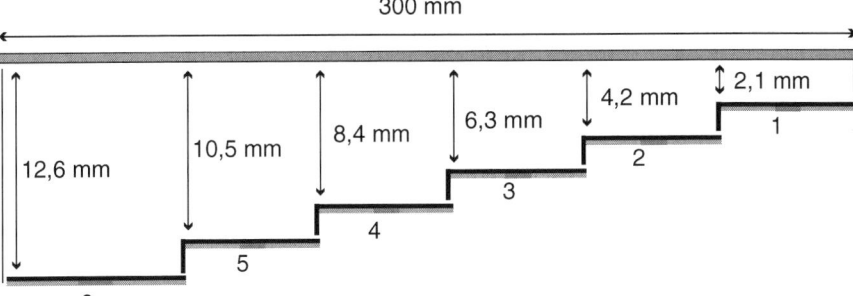

Fig. 5.124 Schematic representation of stepwise forming, including the zones marked red where specimens were removed to examine the structure

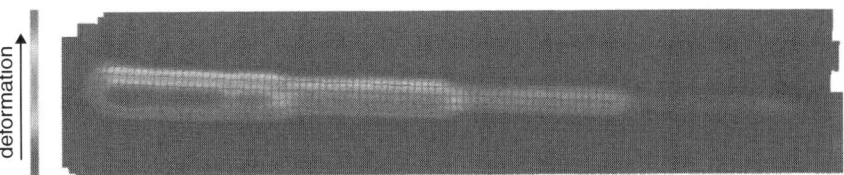

Fig. 5.125 The main strain in the last four stages of a sheet formed using the laser-assisted ISF process shown by different colours

mechanisms, hardness, and the reduction in sheet thickness were examined. The specimens were cut out of the centre of each step for this purpose following an optical evaluation of the grid pattern.

The activation of different slip systems at high temperatures, together with dynamic recrystallization (DRX) and grain boundary sliding, are the major mechanisms that could be used to achieve a greater formability for the magnesium alloy AZ31B (Gottstein 2004; Al-Samman 2008). The dynamic recrystallization—among all—requires a certain critical dislocation density (as the stored energy) and an appropriate temperature to occur. In the current experimental setup, the high temperature and the required dislocation density were provided by the laser heating and the incremental sheet forming, respectively, while the cooling system behind the deformation tip tried to avoid the dynamic recovery after each forming step, to facilitate the dynamic recrystallizations.

The mechanism map drawn up with the experimental data (Fig. 5.126) shows an overlap of the areas with increased formability (>40% thickness reduction) and dynamic recrystallization (40–45% DRX). It also shows that the highest strains were achieved at the medium tool speed with the smallest initial grain size.

The examination of X-ray texture showed that the basal slip systems were more active than the prismatic slip systems (Fig. 5.127a). In addition, the metallographic images (Fig. 5.127b) showed the recrystallised areas which string together like shear bands (Ion et al. 1982). As a result, the active basal slip, together with the mentioned

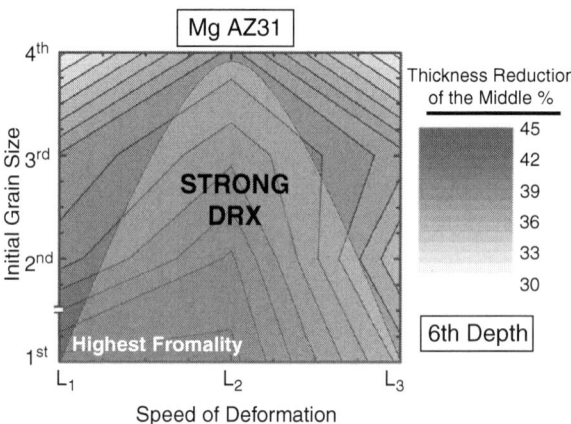

Fig. 5.126 Effect of initial grain size and tool speed on the reductions in thickness achieved in the test. Also shown are the areas of strong recrystallization. (Taleb Araghi et al. 2011)

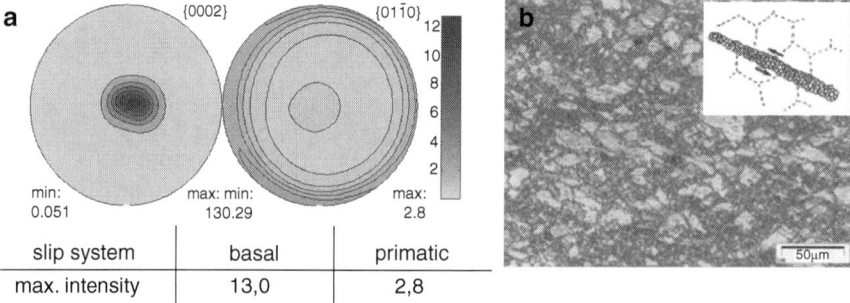

Fig. 5.127 a Analysis of the more active slip systems by X-ray analysis of the sixth formed step, b Metallographic image of the microstructure of the magnesium alloy AZ31 in the sixth formed step with a schematic representation of the slip bands arising as a result of recrystallization. (Taleb Araghi et al. 2011)

shear bands, increased the formability of the AZ31B magnesium alloy during the laser-assisted ISF process at forming temperature of 250 °C.

Different adjustments concerning the temperature-time-deformation schedules can result in different microstructural changes. Figure 5.128 shows the metallographic images of the specimens taken from the sixth formed step after forming at 200 and 250 °C, once with and once without cooling immediately after forming. Part (a) shows the microstructure of the sixth formed step after forming at 200 °C. The serrated grain boundaries are caused by the dislocations which evolved and moved during the forming process, and consequently stored in the grain boundaries. The serrated boundaries are the fingerprints of a deformation state which is just before the onset of dynamic recrystallisation. As seen in Fig. 5.128b, under the condition of high-temperature forming (with enhanced cooling behind the deformation tip), the serrated boundaries are observed while the recrystallization has not yet been

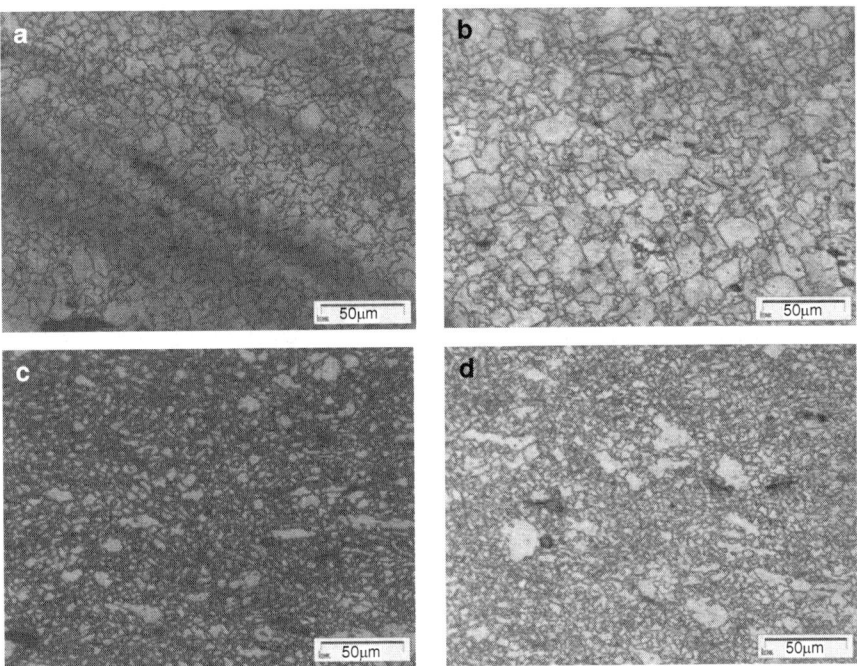

Fig. 5.128 Metallographic images of the structure of the sixth formed step of the sheets (**a**) after forming at 200 °C with subsequent cooling, (**b**) after forming at 200 °C without subsequent cooling, (**c**) after forming at 250 °C with subsequent cooling and (**d**) after forming at 250 °C without subsequent cooling. (Taleb Araghi et al. 2011)

triggered at 200 °C. A Recrystallized microstructure can only be found at a forming temperature of 250 °C, as seen in both Fig. 5.128c, 5.128d.

The microstructure of the sixth step of the deformation given in Fig. 5.129 is due to the combined low-temperature and high-temperature (250 °C) deformations. The parts (a) and (b) of the figure show a large number of deformation twins. If the

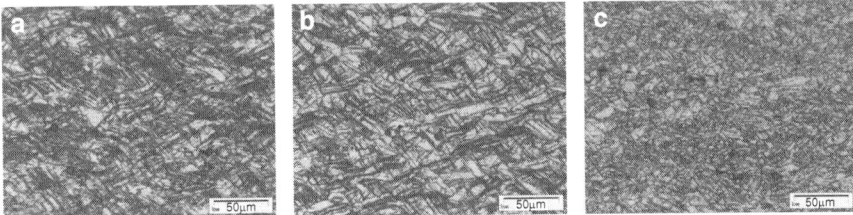

Fig. 5.129 Metallographic images of the structure of the sixth formed step of the sheets (**a**) after two stages of cold forming followed by annealing at 200 °C and forming at 250 °C, (**b**) after two stages of cold forming followed by annealing at 250 °C and forming at 250 °C and (**c**) after four stages of cold forming followed by annealing at 200 °C and forming at 250 °C. (Taleb Araghi et al. 2011)

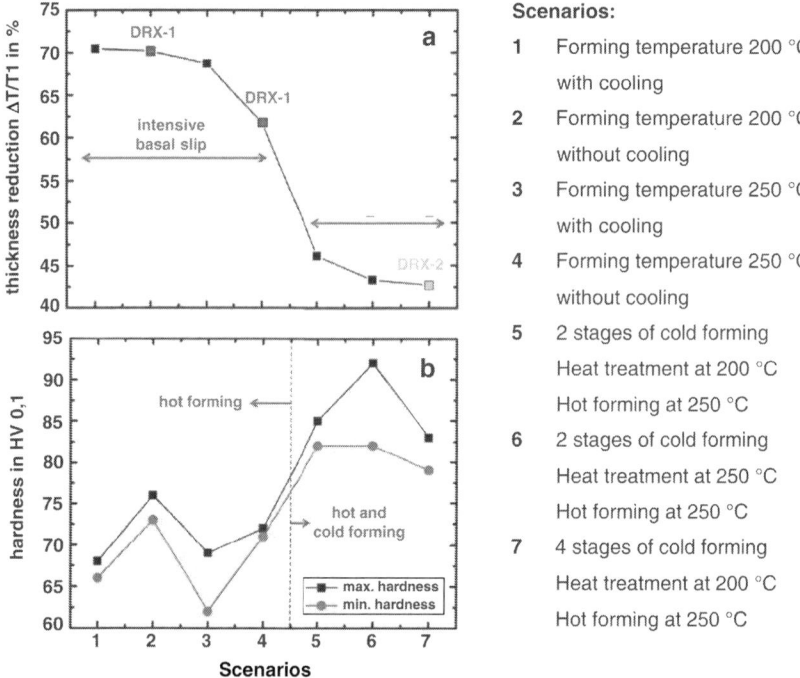

Fig. 5.130 Results of the measurements of (**a**) thickness reduction as a measure of formability and (**b**) the effect of the parameters set for the tests on the microhardness of the resulting microstructure

number of cold forming steps is increased, then the recrystallization occurs in the twinned zones, as can be seen in part (c).

Dynamic recrystallization is considered as one of the main reasons for the increased formability in the laser-assisted incremental sheet forming of magnesium. There are two different morphologies for the recrystallized microstructure, depending on whether forming is performed exclusively at high temperatures or the first forming operations are without heating.

Figure 5.130 shows the results of the sheet thickness reduction measurements as a measure of the effect of the process parameters on formability and on the hardness achieved in the final microstructure.

The cooling operation applied immediately behind the tool in order to bring the material temperature rapidly down from the forming temperature impedes recovery from the high dislocation density that arises during forming. In this way dislocations can initiate an earlier dynamic recrystallization (DRX-1) at 250 °C or support basal slip at 200 °C. The substantial activation of basal slip systems at 200 °C produces a similar formability as for forming at 250 °C plus subsequent cooling. The subsequent cooling results in a finer microstructure but greater hardness in the final state. The process can be carried out at 200 °C for high formability without increased hardness.

The tests at 250 °C with and without cooling after forming both led to a finely recrystallized microstructure with an equal hardness. In the specimens that were not cooled, however, the onset of recrystallization (DRX-1) was much later, not until shortly before failure of the material. Formability is thus much lower than that of the cooled variants. If the aim is to achieve a hard structure without greatly increased formability there is no need for in-process cooling.

The initial cold forming leads to deformation twinning in the microstructure. This causes a marked increase in hardness, but with greatly reduced formability because there is no dynamic recrystallization (DRX-1) in the first forming steps. The dynamic recrystallization (DRX-2) that begins later, during the hot forming stages, mainly affects the zones of intense twinning and does not increase formability any further but reduces the hardness of the microstructure.

To sum up, it can be concluded that the variations in the forming temperature and speed, plus a modified and controlled cooling operation in the forming zones can have a considerable influence on the properties of the parts both during and after the process. These studies lead to a better understanding of the laser-assisted ISF and opportunities for designing the tailored process routes. Finally, the highest degrees of formability were achieved in sheets with an initial grain size of 10 µm, formed using a tool speed of 4 m/min at a temperature of 200 °C without additional cooling. If greater hardness is required, the process can be carried out with the same initial grain size and tool speed at the higher temperature of 250 °C with subsequent cooling.

After the optical system for the sheet processing centre was completed, the results of the forming operations on titanium materials at elevated temperatures that were initially investigated in a small-scale set-up were continued. The aim of the study was to verify the improvement in formability in an actual hot ISF process. In addition, it became possible to test the functionality of the developed machine concept. The tests were performed on the rig shown in Fig. 5.132. A complex geometry was designed for the forming process. It met the requirements of the process control aspects by taking into account the design limitations of the optical system and those of the machine. Figure 5.131 illustrates the applied geometry which was a cone with a wall angle of 60°. The contour of the cone involves radi of different curvatures so that the

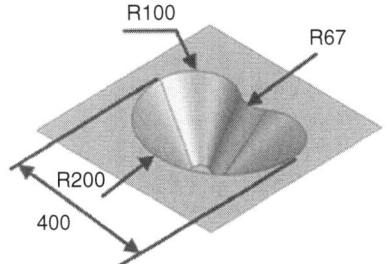
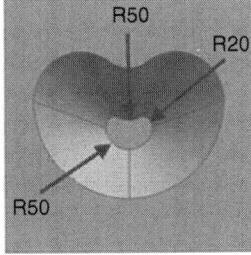

Fig. 5.131 CAD model of a cone designed with two different curves for the manufacture of demonstrator parts from titanium grade 2 and TiAl6V4 in the laser-assisted ISF process without a support tool. (Göttmann et al. 2011)

Fig. 5.132 Laser-assisted ISF. (Göttmann et al. 2011)

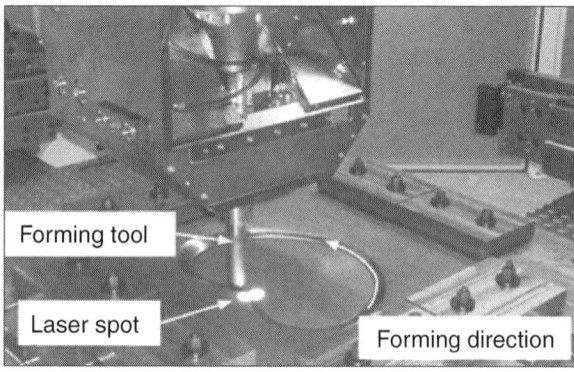

laser spot had to move in both directions about the tool spindle. The total height was fixed at 110 mm in order to avoid a collision between the optical system and the part.

The tools were manufactured from hot-working tool steel 1.2379 and given a PVD coating to reduce friction. Tools with a 20 or 30 mm spherical head were used. The selected forming strategy was vertical section machining with constant feed of 0.35 mm between the forming steps. The programmed tool speed was 4 m/min. The laser output power required to heat the forming zones to at least 400 °C was determined in preliminary trials. During forming a constant initial output power of 1,700 W was set. Figure 5.132 shows a photograph of the forming process. The laser spot, positioned on the tool path ahead of the tool, can be seen clearly.

Tests have shown that the machine design developed is suitable for producing complex components in a combined process. The positioning of the laser spot, the control of the times for switching the laser on and off, and the function of the optical system were thus verified.

The evaluation of the achieved forming depths showed that the formability of TiAl6V4, in particular, could be considerably increased. Figure 5.133 shows the cones formed from TiAl6V4 using the laser-assisted and the conventional ISF processes. The evaluation of the achieved depths showed an increase of approximately 600% (hot: 98 mm, cold: 16 mm).

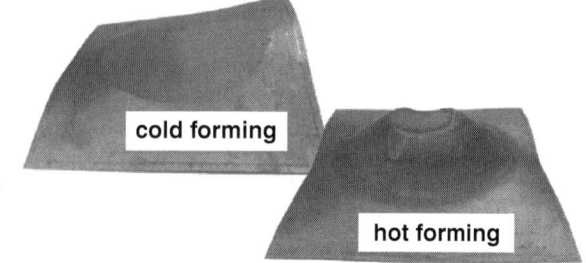

Fig. 5.133 Comparison of the cones produced from TiAl6V4 at room temperature and in a laser-assisted process

5.4.4.5 Production of Parts from Tailored Hybrid Blanks

As already mentioned in Sect. 5.4.3, the use of parts made from hybrid materials (THBs: Tailored Hybrid Blanks) opens up new possibilities in lightweight vehicle body construction. A bonding concept for producing THBs, which enables an integral joint between aluminium and steel to be created, has been developed at the Welding and Joining Institute (ISF). The processing of aluminium-steel sheet by ISF is expected to enable the production of small numbers of multi-material sheet components. The combination of the two areas is expected to indicate possible means of manufacturing customised individual products, using innovative joining techniques and production technology. As a basis for this, the formability of THBs is being studied and optimised.

The challenges presented by the fused/integral joining of steel to aluminium lie in the mastery of metallurgical processes on the one hand and in the great differences in the physical properties of the two base materials on the other. As they are not soluble in the solid state, brittle intermetallic phases and superlattices form when the melt pool solidifies and these have to be controlled, or at least limited to the extent that they are not critical.

Another problem is that the aluminium melt often does not wet the steel sheet very well. This results in poor weld geometries that are not suitable for every application. Improved wetting can be achieved by a more energy-rich process, associated with a more strongly heated melt pool, but a greater heat input also increases the seam thickness of the intermetallic phase.

The joining concept that was developed is based on a combination of a digitally controlled low-energy arc process with low-melting-point filler metals, surface coatings on the steel sheet and a special clamping device. Figure 5.134 illustrates the principle of the process and the welding equipment that was used. Steel-aluminium composite butt joints can be produced without fully melting the steel base material if low-melting-point filler metals based on an aluminium-silicon alloy (AlSi5) are used. This creates a hybrid joint that has the character of a brazed joint on the steel side and a welded joint on the aluminium side (see Fig. 5.135). When making the

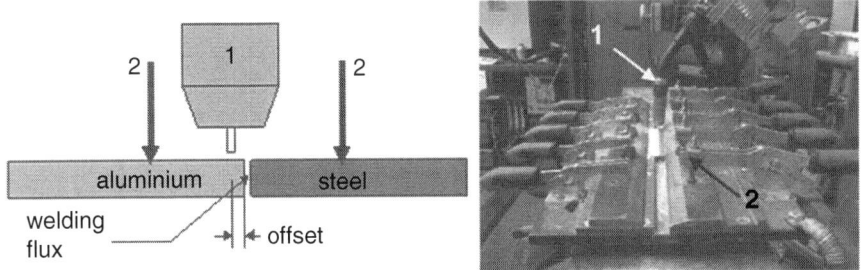

Fig. 5.134 *Left*: Schematic representation of the arc process for producing integral hybrid steel-aluminium joints; *Right*: Welding system used at the Welding and Joining Institute (ISF). (*1*) Torch, (*2*) Clamp

Fig. 5.135 Macrosection of the braze welded joint

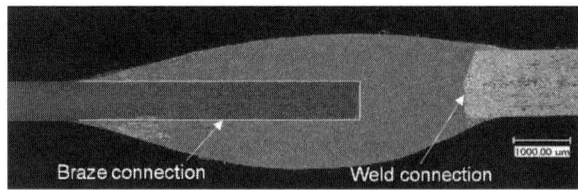

joint all of the arc is positioned over the aluminium sheet so that the steel sheet is only heated by thermal conduction.

The quality of the combination joints made using the concept described is good and surface flaws such as insufficient wetting only occur occasionally. The strengths that can be obtained are within the strength range of the aluminium base material.

In order to produce tailored hybrid blanks from the combination of aluminium and steel, the joints made not only have to possess high static strengths, they also have to be sufficiently formable. Since steel and aluminium behave differently when formed, their different behaviours have to be aligned by adapting the thickness of the two materials (thicker aluminium, thinner steel), so that neither of the partners in the joint deforms faster than the other. This would lead to premature failure of the joint during the forming process. Tensile tests with determination of local elongation clearly showed the effect of sheet thickness on forming behaviour. The coloured areas on the left side of Fig. 5.136 show the strain distribution of a tensile specimen of 1 mm thick DC05 and 1.3 mm AA6016T4.

It can be clearly seen that the strain is limited to the aluminium base material. The weld seam and the steel material are virtually unchanged before the aluminium material fractures. The strengths are higher than those obtained in a pure aluminium welded joint (≈ 205 MPa). The right-hand side of Fig. 5.136 shows the stress-strain

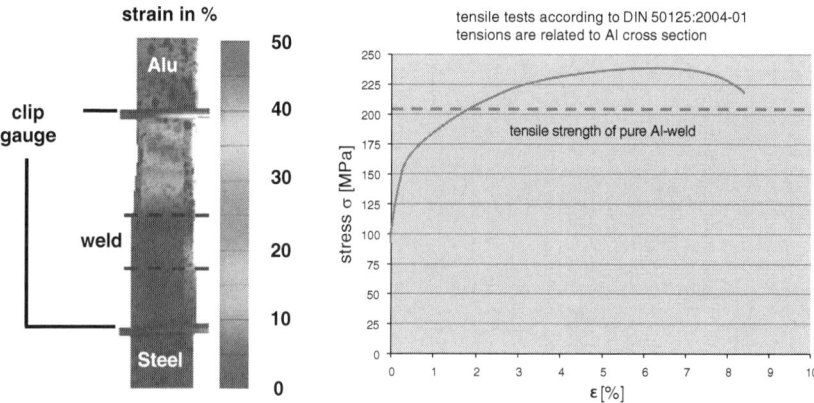

Fig. 5.136 *Left*: Strain distribution of a tensile specimen from DC05 and AA6016T with 1 mm sheet thickness for DC05 and 1.3 mm for AA6016T. *Right*: Stress-strain diagram for the steel-aluminium joint with respect to the initial sheet thickness of the aluminium base material

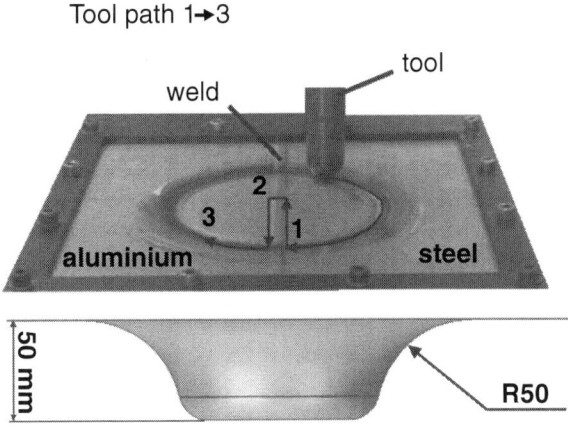

Fig. 5.137 Test rig for producing a rotationally symmetrical radial cone from THB and the traverse path of the tool (*top*). CAD geometry of the radial cone (*bottom*)

diagram. The stresses relate to the initial cross-section of the aluminium base material. The comparatively higher strength values are due to the transverse support provided by the steel. The steel sheet within the joint prevents the tensile specimen from deforming in the transverse direction by inhibiting shrinkage in this area. This leads to increased strength and prevents necking within the joint zone. The area of fracture necking thus moves to the weaker base material.

Cupping tests showed that the steel-aluminium composite joints can be formed to a technically useful extent. A reproducible drawing ratio of 1.9 was obtained for the combination of 1.2 mm thick AA6016 and 0.8 mm DC05. However, the additional height of the seam proved a problem as the clamp could not always level it out. Similarly, the additional height of the seam must be taken into account in incremental forming, otherwise the consequence would be mechanical damage plus additional dynamic bending stress at the location of the joint. Additional stress on the joint can be minimised by "jumping over" it, i.e. lifting the tool in front of the seam and lowering it again behind it. Figure 5.137 shows where the tool "jumps over" the weld.

The first ISF tests on THBs have shown that the weld seam has a significant influence on the forming depth that can be reached. A radial cone with an increasing wall angle was defined as the test geometry (see Fig. 5.137 below). Forming took place without a counter-tool and continued until the part failed. This procedure enabled a statement to be made about the formability of THBs from a simple test rig.

Figure 5.138 summarizes the test results from forming rotationally symmetrical radial cones. First it shows the forming depths obtainable when forming the two base metals steel (DC05) and aluminium (AA6016). These tests show that the maximum achievable forming depth of aluminium is established as roughly 28 mm. Compared with this, a depth of only 18 mm could be obtained in the tests on THBs. The parts in these tests failed at the weld seam. The reason for this is very small irregularities in the joint, which lead to premature failure of the joint and limited reproducibility of the forming results. Closer analysis of the break shows that the cracks are virtually

Fig. 5.138 Comparison of the forming depths of THBs and the base materials in ISF tests (*left*). Imperfections on the surface of the weld seams leading to failure of parts during forming (*right*)

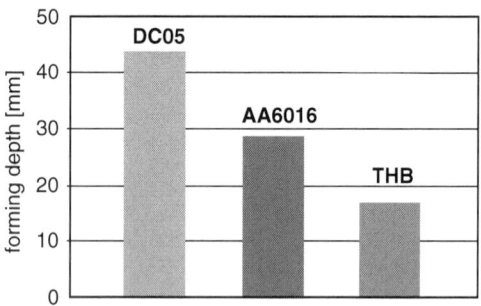

Fig. 5.139 Pyramid made with ISF. Aluminium-steel composite made from 0.8 mm thick DC05 and 1.3 mm thick AA6016

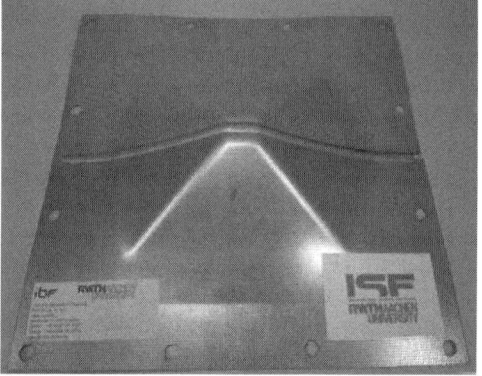

all caused by wetting faults or irregularities in the surface of the seam. These wetting faults and imperfections occur preferably at the front edge of the steel sheet. From there the crack propagates along the front edge until, as forming progresses, the test sheet fails entirely.

The component shown in Fig. 5.139 demonstrates that it is possible to manufacture parts from THBs in an incremental forming process. The forming strategy illustrated was used to manufacture a pyramid 50 mm deep with a side angle of 30°.

Because the weld seam is subject to cyclic loading during subsequent incremental sheet forming, the mechanical properties of the combination play an important part. Vibration tests were therefore carried out on the seam to characterise its vibration stress properties. Two different material combinations were considered (see Table 5.8).

The specimens used were flat tensile specimens to DIN EN 6072 specimen type 3 (see Fig. 5.140) that had not been straightened before the tests. A mechanical 60 kN pulser made by Schenck was used at room temperature with sinusoidal force-controlled loading and a R value (ratio of minimum stress to maximum stress) of 0.1. The test frequency was 25–30 Hz. The load was applied across the weld seam.

Table 5.8 Material combinations used for the dynamic tests

	Combination 1	DC05 (0.8 mm)–AA6016 (1.3 mm)
	Combination 2	DC05 (1.0 mm)–AA6016 (1.3 mm)

Fig. 5.140 Specimen geometry for vibration test to DIN EN 6072, Prototype 3

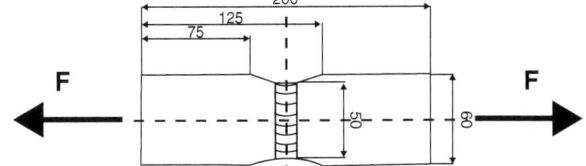

Figure 5.141 shows the results of the dynamic tensile tests. The vibration resistances were divided into two groups for the two combinations. The first group consists of large numbers of specimens that withstood the fatigue test without rupture (stress cycle amplitudes of 60–80 MPa) and were not included in the calculation. The second group comprises specimens that failed prematurely and shift the Wöhler line towards lower stresses. The vibration stability that can be achieved depends on the position in the seam at which the specimen was taken. Specimens taken from the start of the seam were very much more porous along the front of the steel sheet, as the sheets had not yet been sufficiently preheated by the process. Specimens from this zone fail prematurely because the pores weaken the cross-section. As the distance from the start of the seam increases the pore concentration reduces as a result of the advancing heat. This increases service life and the failure occurs in the heat-affected zone of the aluminium or in the base metal area, depending on the thickness of the steel sheet (failure occurs in the steel where this is 0.8 mm and in the aluminium where it is 1.0 mm). Optimising the design of the seam brought about a considerable reduction in the number of pores along the front edge.

5.4.5 Industrial Relevance

Both the combination of stretch forming and ISF and the combination of local heating and ISF have the aim of flexibly producing individual sheet metal parts while at the

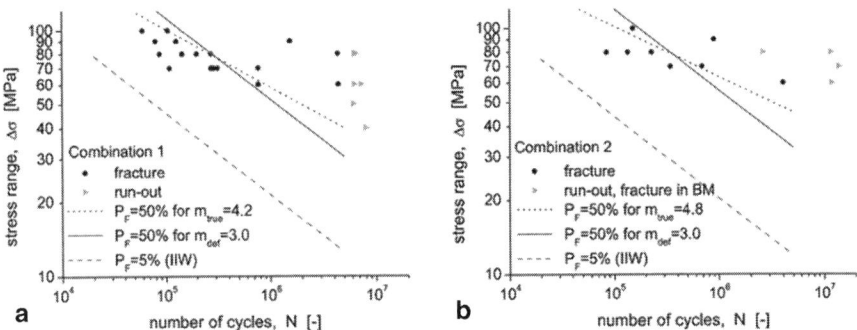

Fig. 5.141 Wöhler curves produced in vibration tests on aluminium-steel composites. (Reisgen et al. 2010)

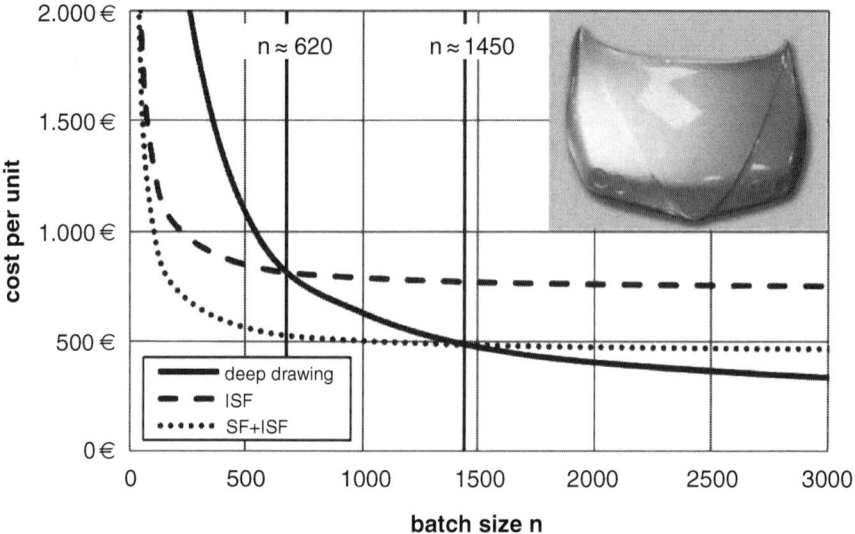

Fig. 5.142 Extended cost calculation from (Ames 2008) for the manufacture of a engine bonnet (*top right*) with the following processes: deep drawing, conventional ISF with full support and the combined process using stretch forming and ISF (SF + ISF)

same time improving productivity and the quality of the products. Applications for the processes developed include the manufacture of sheet metal formed parts for the automotive industry and aerospace.

Traditional sheet metal forming processes such as deep drawing are only cost-effective in large-scale production because of their tool-based nature. In contrast, ISF is suitable for prototyping and small batch production. It is expected that unit costs will be considerably reduced by combining ISF and stretch forming as the time taken to process each part will be shorter.

Figure 5.142 shows the effect of batch size on the anticipated manufacturing costs for a car bonnet. The diagram compares the unit costs using conventional ISF and deep drawing. The calculation is based on the work done by Ames (2008) with the addition of the calculated costs for the use of the "SF + ISF" process combination. The reference geometry considered was a car bonnet of dimensions $1800 \times 2000 \times 150\,\text{mm}^3$. The unit costs K_{Part} were calculated using Eq. (5.13) with K_{Form} as the tooling costs, K_{Sheet} as the material costs, K_{Personal} as the personnel costs, K_{Machine} as the plant costs and batch size n (Ames 2008).

$$K_{\text{Part}} = \frac{K}{n} = \frac{K_{\text{Form}} + K_{\text{Sheet}} + K_{\text{Personal}} + K_{\text{Machine}}}{n} \quad (5.13)$$

The considerable differences in the calculation are due to the very high tooling and machine costs for deep drawing processes. Compared with conventional ISF with full support, the use of a deep drawing process, which for a part of this size is particularly affected by the high fixed costs of the deep drawing machines and tools, only makes economic sense for a batch size of roughly 620 or more. The reason for this is the long processing time for ISF that has been described above. However, the calculation shows that with the shorter process time of the combined "SF + ISF" process, production costs can be markedly reduced compared with those for ISF on its own. The combination of ISF and stretch forming can greatly reduce cycle time and thus close the gap between prototyping and small batch production. In the example shown, 1450 parts can be produced cost-effectively. Furthermore, the combined "SF + ISF" process improves geometric accuracy because of the superimposed tensile stresses from the stretch forming and reduces local thinning because of improved material flow. The increased productivity and improved component properties open up new potential for the use of ISF in industrial practice.

The manufacture of sheet metal formed parts for the aerospace industry requires processes that permit small numbers of very large sheet metal components to be produced economically. Incremental sheet metal forming is an eminently suitable process for producing small batches of such components because it is not tied to expensive tooling. Because of the low numbers of aircraft components, the higher manufacturing costs of these products compared with means of land transport can be tolerated (Peters and Leyens 2002). The potential for lightweight construction can be particularly exploited by using low-density high-strength materials. Consequently high-strength corrosion-resistant and heat-resistant steel and titanium alloys are increasingly used in aircraft and turbine construction.

Sheet metal formed parts from titanium and titanium alloys, particularly TiAl6V4, are mainly manufactured at present by superplastic forming. By making full use of the superplastic material property, components can be made with a high degree of deformation and very good geometric accuracy. A serious disadvantage of this process is the very long forming time and this is reflected in process costs. At the same time the high process temperatures mean that expensive heat-resistant tools have to be used and this leads to additional expense when the component geometry changes. The "Laser + ISF" approach shown demonstrates the suitability of local heating methods used with ISF for forming high-strength materials. From the point of view of the aviation industry, the following requirements can be met by the use of hybrid ISF:

- Cost-effective manufacture of large-format lightweight components in small quantities.
- Forming of high-strength materials that can only be cold formed with difficulty.

In recent years it has been possible to reduce the amount of material used in the automotive industry by using high-strength materials. However, the steady customer demand for safety and comfort in vehicles has led to an increase in vehicle weight. This had to be counteracted by increasing the engine power, but the resulting emissions run counter to the current international efforts to reduce fuel consumption. The production of components made from hybrid sheet metal opens the way to new principles of lightweight construction for automotive bodies. However, these hybrids are not currently sufficiently formable for widespread use. The work described here is an initial approach to the production of body parts from steel-aluminium composites.

5.4.6 Future Research Topics

Further research into new processes is needed, building on the results obtained. The aim of the work is to increase the maturity of the process. At the end of development, reliable implementation of the processes in industrial practice should be possible. A basic precondition for this is mastery of the processes and the definition of process limits so that parts can be designed with confidence.

The work done to date has provided the basic proof of the hypotheses set out for the "SF + ISF" combination. Future studies should now aim to qualify the technology for industrial use. Thus the following questions have priority in future:

- Rapid, simple and reliable process planning with the CAX tools that have been developed,
- Forecast of the effect of process control, the material used and the geometry on the expected geometric accuracy,
- Achievable improvements to forming limits in the combined process,
- Reliable forecast of the influence on manufacturing costs of component geometry, process control and the number of parts to be produced.

The successful integration of a laser heating unit into the sheet processing centre offers a basis for further development of the "Laser + ISF" combination. Deep research into the forming process requires more far-reaching work such as:

- the development of suitable solutions for temperature control during the process,
- the development of tooling concepts for incremental hot forming,
- research into the temperature-dependent forming behaviour of materials such as titanium alloys and high-strength steels.
- and the facilitation of simulation-supported process design.

The degrees of forming that can be achieved when producing parts from THBs are currently inadequate. The main thrust of future work will therefore be to optimise formability in hybrid composites. Here the focus can be extended or shifted to joining techniques such as the friction stir welding of aluminium and steel.

The integration of other manufacturing principles can increase the functionality of the machine used or the degree to which it is capable of integration. By increasing the functionality and maintaining the flexibility of the incremental sheet metal forming process, two opposing manufacturing principles can be realised:

- Complete manufacture on one machine, starting with the CAD draft and including forming, heat treatment and joining until the finished part is obtained by trimming
- Customisation of components by including flexible processes in cycles that are suitable for series production.
- Cost-efficiency must always be borne in mind when integrating new functions.

5.5 Shortening Process Chains by Process Integration into Machine Tools

Christian Brecher, Fritz Klocke, Kristian Arntz, Stephan Bäumler,
Tobias Breitbach, Dennis Do-Khac, Michael Emonts, Daniel Heinen, Werner Herfs,
Jan-Patrick Hermani, Andreas Janssen, Andreas Karlberger and Chris-Jörg Rosen

5.5.1 Abstract

The solution of the scale/scope dilemma requires an answer to the question: How can products which are tailored to meet customer requests be produced at costs similar to standard product lines? The approach presented below offers new possibilities to reduce unit costs for machine tool shops and die-makers. In particular, we are looking at high value components which are currently produced in expensive, sequentially organised technical production chains. An alternative to this time-intensive manner of working is the integration of different production technologies in one common machining platform. This allows the production of highly complex products in an integrated, and therefore significantly shorter, process chain and so enables production without multiple reclamping operations between processes, which achieves higher quality and precision in less time.

This contribution presents the development of a hybrid machining centre, which goes beyond just integrating additional production processes as far as the possibility of running all processing steps in parallel. For this, both a robot and laser technologies were integrated into a milling centre with two separate working areas. Along with the new potential uses of hybrid production machines this creates changed requirements on the sub-systems, as well as for their planning, linking and application. In particular, we looked at and implemented new types of process technology fundamentals, and their integration into the control and planning environment. A particular effort was made to encapsulate as much complexity as possible in order to improve usability.

Over and above the technical challenges of commissioning, it was necessary to ensure that an economic added value can be gained from the hybrid quality of the machine tool. For this purpose, the option for the parallel processing of all production processes in both working areas was created consistently and universally valid. To achieve this, the heterogeneous control architecture was built up for twin channels, so that processes running in one working area do not prevent a different 5-axis milling process from running in the other work space. It is not only when planning production sequences that the multiplicity of production processes presents a challenge. At the same time we found many, so far unresearched, interaction mechanisms.

The prototypes described were used to successfully investigate programming and synchronisation options for mixed control technologies, combined with the process knowledge gained. Equally we were able to assess integrated process planning for milling and laser material processing as an example.

Metrological tests were carried out which enabled an initial insight into the different interaction mechanisms which arise from the mutual impact of mechanical and thermal influences, as well as from the linking of different types of controls.

The information gained so far forms the basis for a future set of design and development guidelines, which could make the development of hybrid machine designs significantly more efficient by using a modularised approach.

5.5.2 State of the Art

The development of hybrid machine tools including integrated laser system technology has, in the past, focused on the sequential combination of laser processing for hardening and deposition welding, and machining on a lathe, or the sequential combination of laser deposition welding and machining in milling machines. For example, in the completed project OptoRep (project code: 02PD2490), funded by the German Federal Ministry for Education and Research (BMBF), a hybrid 3-axis milling machine was developed which, in one clamping, determines the component topography using optical measurement technology and generates NC data in a CAM system using a target/actual geometry comparison. The OptoRep machine carries out repairs using near-net-shape 5-axis powder deposition welding. The base carrier of the repair cell consists of a twin spindle portal milling machine which is equipped with an optical measurement system module, a 2-axis laser beam coating head module and end-to-end CAD-/CAM connections (Bichmann et al. 2005) (Fig. 5.143).

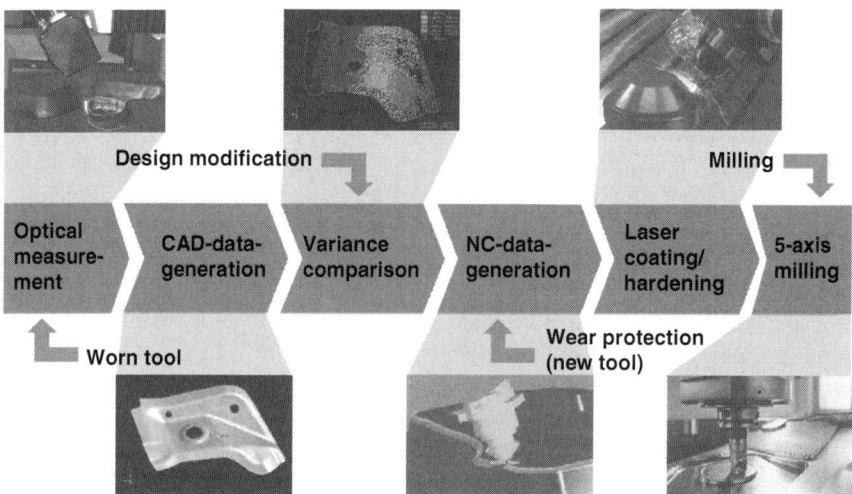

Fig. 5.143 Process chain for the BMBF OptoRep project

In comparison, in the KombiMasch project funded by the BMBF (project code: 02PW2145), a hybrid machine tool with modular design was developed in which rotationally symmetrical components can be completely processed in a single clamping by combining different production technologies. Both the conventional turning and milling work and the laser materials processing (laser hardening, coating and alloying) were combined in one machine in order to reduce throughput times. To do this, modular laser processing heads can be changed in one milling spindle for laser coating. Using wire-based laser deposition welding, wear and corrosion protection coatings can be applied to the surface of the component. The chemical composition of these coatings can be adjusted by means of the filler material. Using laser hardening, areas which are subject to large mechanical loads can be hardened martensitically to minimise abrasive wear during the component usage. With energy inputs which are locally restricted and carefully metered, component distortion due to thermal processes can be reduced to a minimum (Brecher et al. 2008) (Fig. 5.144).

As part of the EU research project CURARE (SME 222317), completed in 2010, a machining centre for 5-axis laser surface machining was developed with an integrated laser machining head. The gantry-equipment Alzmetall GX 1000/5-T-LOB was developed specifically for the laser treatment of component surfaces. And in addition a high-power performance diode laser system and laser machining heads are included. Along with the laser hardening and fusing, both wire and powder-based

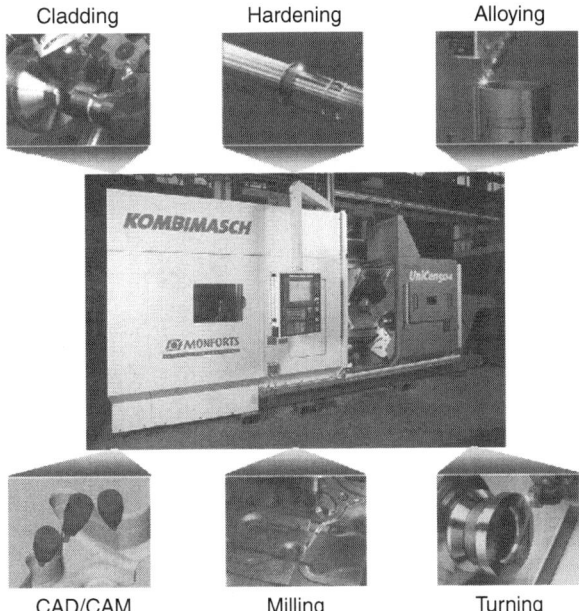

Fig. 5.144 KombiMasch multi-functional machining centre

laser surface processes can be carried out using the equipment. SINUMERIK 840 D Solution Line from Siemens was used for the control architecture, so that the comprehensive adjustment of all relevant machine and process parameters is possible from one central point. This system is only designed for laser surface treatment of components. No cutting operations are included. For this reason and as only one workspace is available, parallel processing is not possible (Klocke et al. 2010a).

The use of robots in machine tools is currently generally restricted to handling operations for automatic loading, repositioning and removing parts. Aside from machine tool applications, robots are used in separate work stations for many different applications, such as welding or painting. A chain of production lines results, where a centralised server calls individual robot programmes whenever the robot programme is not being triggered by the preceding and following workstations. The robot programme therefore runs simultaneously with, but decoupled from, the later production stages, whose downstream workstations are waiting for the robot's programme to complete (Pires 2005).

A further major application for industrial robots can be seen in the area of appliance-supported assembly. Current developments here are aiming at a collaboration between humans and robots and are investigating relevant safety aspects (Morioka and Sakakibara 2010) as well as socio-technological aspects of introducing man-robot-cooperation (Arai et al. 2010).

The majority of digital controls for machine tools (NC) and robots (RC) come from different suppliers. The relevant requirements, which the control programmes must fulfil, as well as the specific programming approaches, mean that in the expert literature there is still a strict divide between the two (Weck and Brecher 2006). For this reason, designs and strategies are needed for direct integration of robots in the workspace of a machine tool to enable cooperation between different automation components (Brecher and Karlberger 2010).

For robot programming, online programming is widespread and is required for many tasks, which is why there are attempts to increase the productivity of robot programmers in this area (Schröter 2008). The offline approaches available on the market have in the meantime been complemented by manufacturer-independent CAM-extensions (Unicam 2011). Programming of machine tools for complex geometries is mainly done with CAM support. All the existing programming methods currently require a homogeneous control architecture.

Current attempts to use robots for laser processes are based on homogeneous control architectures, which are normally developed by one manufacturer (Munzert 2010). Collaboration between robots and production machines across heterogeneous control architectures is managed quite simply by calling the individual robot programmes individually (Brecher et al. 2005), for example by using sensor signals projected differently for each application in practice, or PLC signals. The programme is then running solely at the robot control, while the machine control system waits for a ready signal from the robot controller.

In almost every production process there are individual systems interacting with each other, which leads to compromises in productivity or in the process capability of the production process. For example, for cutting processes, interaction between the process and the machine structure can lead to stability problems during machining (Brecher et al. 2009b). Both when using a simulation approach to the process-machine interaction, and during the design and development and construction phase of a machine tool, simulation tools can be used to provide a mechanical structural analysis of the machine structure, to depict the kinematic qualities and to integrate control loops in the structural simulation (Altintas et al. 2005). If the virtual machine or the measurements are linked to the real machine using process models, then process stability can already be described properly today—as an interaction between the machine and the cutting production process (Brecher et al. 2009c; Esser 2010). This approach was also applied to multi-spindle machining of the same type of process, to optimise it by simulation (Brecher et al. 2010). Also thermal-mechanical interactions in machine tools, which are extremely relevant in the context of the posited hybrid machine concept with integrated laser system technology, have already been under research for many years (Bryan 1968). The current state of the art distinguishes between internal influences (resulting from the processing on the machine), and external ones (resulting from the environment), on the machine's thermoelastic behaviour (Bryan 1990). Based on the temperatures measured, it is already possible today to apply compensation algorithms to correct for thermoelastic distortions (Weck et al. 1995). In another approach, displacements can be corrected directly at the tool, by using revolution speed and performance data from the tool spindle with a machine model (Brecher and Hirsch 2004; Brecher and Wissmann 2009).

- In machine developments carried out so far, with the aim of integrating laser system technology into machine tools for materials processing, no simultaneous main time processing of the components has been achieved. Simultaneous mechanical and thermal processing, such that both processes are available in parallel, without mutual detriment, has not yet been attained. This results in a lower level of tool utilisation, including peripheral system components such as the laser source. This necessarily involves losses from an economic point of view. It still needs to be assessed how, and in what timeframe, the increased capital investment for a platform of this kind might be recovered in operational production.
- In the area of control technology for multi-functional machining centres, approaches to modular, real-time-enabled control coupling, which would allow NC-interpolated control of movements across control boundaries are lacking, particularly with respect to integrating handling systems or robots.
- In relation to laser material processing technology the laser hardening and laser deposition welding processes so far have not been combined in a single hybrid laser processing head. The current state of the art provides separate functional heads which need to be changed for different processes.
- On the facility side no design has yet been put forward in industry and research projects for machine tool integration of a laser system for laser deburring.
- On the machining side, in research projects on laser deposition welding, the laser radiation so far has always traversed the component with a fixed lens and so with a

fixed diameter focal spot. A variable setting for the focal spot size or the shape of the beam, to adapt to the size of the weld bead, is not possible. Equally, oscillating the laser beam perpendicular to the direction of movement, which gives a better welding result when carrying out joint welding, cannot currently be achieved and needs basic process technology research.
- At present, there is no fundamental knowledge about interaction mechanisms which would occur for the combination of several processes in one piece of equipment. In particular, the mechanical-mechanical and the thermal-mechanical interactions when integrating laser technology into machine tools are unknown. Both mechanical influences from the movement of the spindle or axes, and thermal influences from laser machining can negatively affect the machining results. Equally, there has so far not been any research into the long-term, thermal stability of hybrid laser machining heads.
- Research carried out so far in the area of machine dynamics has mainly focused on the process-machine interaction created at the individual engagement point of a tool. In the area of multi-spindle machines, the first approaches have been addressing at full modelling of the milling work, in order to achieve process optimisation in practice. So far there have not been any known or scientific attempts to understand and minimise the mutual interaction of processes of different types.
- The use of robots in machine tools is so far mainly limited to handling tasks. Robot-based cutting is subject to strict limits in relation to the range of materials, the precision accuracy which can be achieved and the quality of the surface (Kief and Roschiwal 2007). Materials processing using laser technology is suitable for use by robots, since it is a force-free process, but it has not yet been completely defined which laser processes can usefully be shifted from the spindle to robots as part of parallel robot usage during the overall processing time. In addition, there are no known overall designs showing how an overarching architecture for control of both robots and machines could provide a single, usable programming approach.

5.5.3 Motivation and Research Question

Within the scale-scope dilemma, particularly for the production of customer-specific products in small and medium sized batches, there is considerable potential for reduction of unit costs (Fig. 5.145). The increase in unit costs for short runs is mainly due to the increased percentage share of changes to set-up and planning activities. Integration of several of the production technologies used in modern tool and die production into a single platform, with concurrent encapsulation of the complexity, could prove a decisive competitive advantage for high-wage countries.

The combination of a 5-axis milling machine with laser production processes offers various advantages. First, process sequence chains can be shortened thanks to the integration of processes and rapid changeover of production processes using a single clamping. This saves time when changing and adjusting the set-up, and

Fig. 5.145 Potential for reduction of unit costs depending on batch size

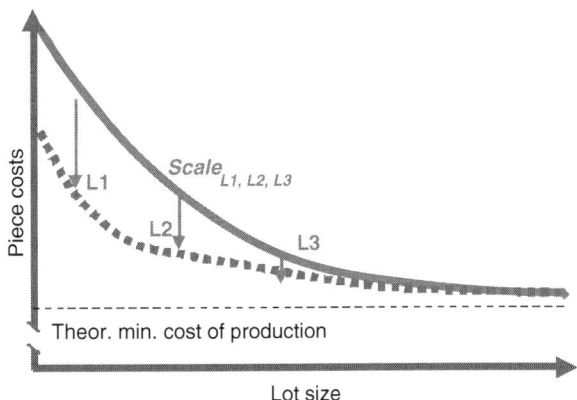

logistical transfers of parts are reduced to a minimum (Fig. 5.146). On the other hand, higher component quality is achieved by the flexible application of all production processes using one single component clamping.

So, for example, in tool and die production, where component areas are subject to heavy loads, a partial material substitution can be carried out, by using an initial machining, followed by a laser deposition welding process and then final machining finishing work. This means that if a more economic base material is used, the area of a component which is subject to heavy loads can be economically replaced with a higher quality material. Another application area which increases total processing tool life in die production is the partial hardening of dies for heavy duty areas, and subsequent finishing work on the tempered surface. Pre-machined components prepared by cutting can also be finalised using follow-on laser structuring and they

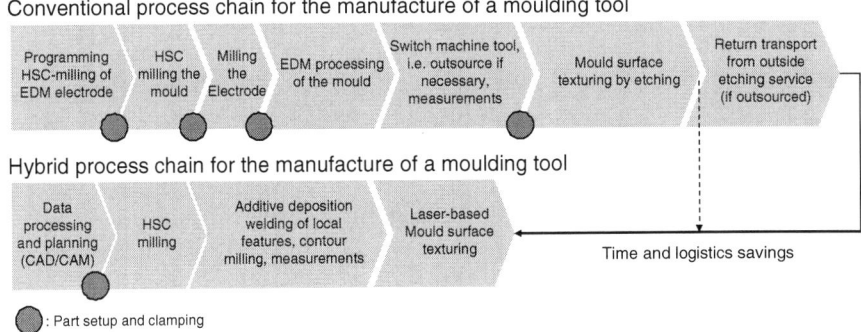

- Integrating functionality enables quick alternation of manufacturing processes in single-clamping setups
- Temporal and spatial proximity of manufacturing processes induce new mechanical and thermal interactions while eliminating workpiece re-clamping and workpiece transfers

Fig. 5.146 Shorter process chains by using hybrid machining centres

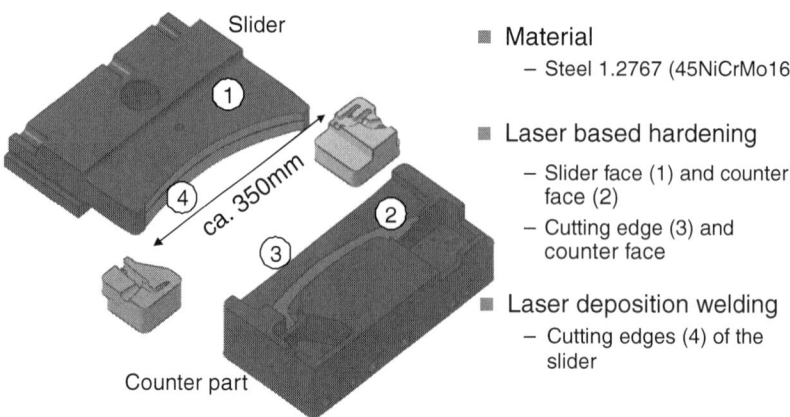

Fig. 5.147 Typical component from tool and die production

may then be refined to meet customer requirements. Figure 5.147 shows a typical injection moulding tool, which was produced using the described production process.

A disadvantage of the existing hybrid machining centres which combine processes that involve cutting with laser production processes, is found in the economic efficiency of the integrated production chain. The combined production technologies are always applied sequentially. While one technology is productively participating in value creation, all the other integrated technologies cannot be used. As the significantly higher investment cost of these hybrid machines can clearly increase the cost of each hour of machine usage, they can not always be used economically. A further disadvantage, which relates to the laser systems employed so far, is that these can only be used for a single laser production process—deposition welding, hardening or structuring.

> The goal of this project is to develop a hybrid machining center, which allows parallel processing of mechanical and laser-based manufacturing processes, taking into account interactions and safety issues.

The new approach used in this project meets the systematic weak points of hybrid machine tools, firstly by extending the machine to have two working areas, which reduces costs by having access to all the available technologies. This means that, at any time, two different technologies can be applied in parallel within the working areas. So the productivity and therefore the cost-effectiveness of the hybrid production machines can be significantly increased (Fig. 5.148).

Secondly, machining of components using all the integrated technologies becomes possible through a single clamping. Thirdly, by using a new type of hybrid laser machining tool which employs a laser scanner for the deposition of both welding and hardening can be combined in one laser tool and for the first time, applied using robots. Application areas for this kind of technology portfolio range from small and medium batch production of individualised injection moulding tools made from

5 Hybrid Production Systems

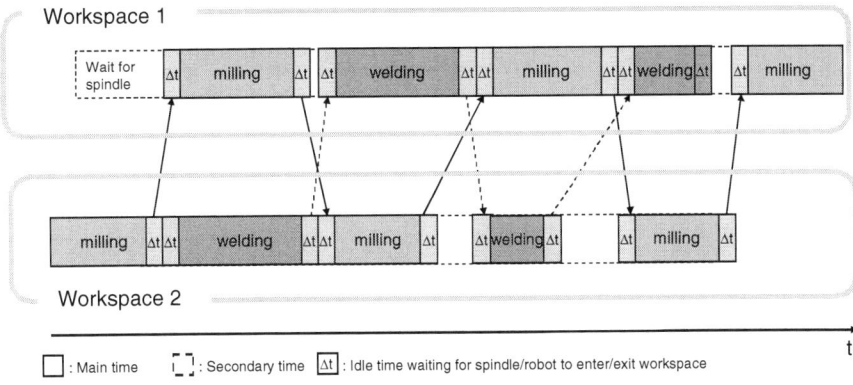

Fig. 5.148 Principle of parallel hybrid machining in the new machine approach

hard-wearing steels, to short runs of high-quality medical implant components made from high strength metal alloys.

Challenges in the approach to hybridisation used here lie in the interaction between the individual manufacturing systems, the operational process reliability of the hybrid laser machining head, the integration of heterogeneous control systems, as well as the planning tools to optimise process management and machine loadings.

In addition to the interactions discussed in the current state of the art within individual manufacturing systems, in the case of parallel application of multiple manufacturing processes within a single system structure, there also exist additional interactions between these processes. For example, there is currently insufficient knowledge of the structural dynamic influence of machining processes on a laser production process. To date, there have been no reasons to run laser processes which require defined precision levels simultaneously with other programmes in neighbouring work spaces. A further interaction to be looked at is the thermal influence between the separate systems. Both the laser radiation which penetrates the component and the reflected laser radiation can have a thermal impact on parts of the machine structure, such that the resulting geometric distortions in the machine structure can lead to a loss of accuracy for subsequent processes. Equally, the influence of the machine structure on the laser system—e.g. vibrations caused by cutting, or positioning failures, can lead to a negative impact on the process capability of the laser process. The interactions of the laser scanner-based deposition welding and hardening within a hybrid laser machining tool are currently unknown. In addition, the approach of using a wire-based laser deposition welding process using an oscillating laser beam has so far been neither implemented nor researched.

An initial implementation of a hybrid laser machining tool needs first to establish its process capability and long-term stability, in order to ensure its effective usability within a complex system.

Because of the different control approaches used for individual components, the implementation of a hybrid process chain is not possible without having an integration approach for mixed control architectures. This would allow all the sub-systems of the hybrid system to be programmable in the same way. With the different control and adjustment approaches which exist, this leads to some more open questions. And in addition the parallel machining in several work spaces places new demands on process and procedural planning and scheduling.

5.5.4 Results

5.5.4.1 Structure of a Hybrid Machining Centre

The multi-technology platform developed for the combined machining of highly refined, complex components in a single work piece clamping is shown in Fig. 5.149. The machine design is based on a conventional milling machine with a traversing column design (travel X: 2,000 mm, Y: 630 mm, Z: 550 mm). The 25 kW main spindle is arranged vertically and reaches a speed of up to 20,000 rpm. A Sinumerik 840D is used for the NC control. Both rotary swivelling tables, separated by the integrated laser magazine, can be combined with the three axes of the column to provide two five-axis machining work spaces.

As indicated in the state of the art description, current designs deliver relatively high machine hourly rates due to the sequential machining of the main spindle and the laser machining unit. For the first ever implementation of parallel working, car-

Fig. 5.149 Structure of a hybrid machining centre

5 Hybrid Production Systems

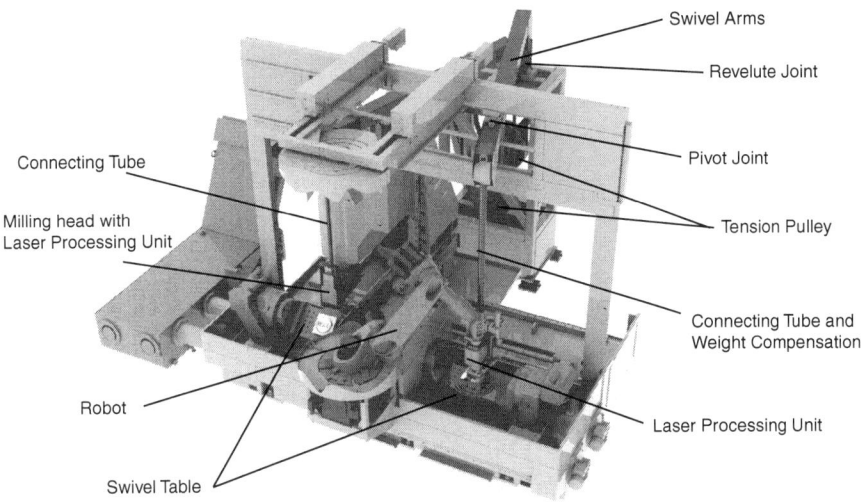

Fig. 5.150 Mechanical arrangement of a hybrid machining centre

rying out machining and laser material processing was focused while designing the hybrid machining centre a jointed-arm robot was integrated in both working areas. Figure 5.150 shows the mechanical structure of the arrangement of the basic machine, the robot and the laser processing units.

In order to be able to construct the entire machine symmetrically and to ensure equal accessibility to both working spaces, the robot is placed in the centre, in front of the two rotary swivelling tables. This arrangement simplifies process planning and control. Possible design alternatives placing the robot sideways in front of the machine were not taken any further, in the interests of attempting to remain compact and/or minimising the complexity of the overall equipment. The integration of the robot into the machine housing for the working space was possible with the help of the platform and of a pivoted, cylinder-shaped robot casing.

For reasons of access and symmetry, the laser processing units are positioned in the middle of the machine, above the counter-bearing for the rotary swivelling tables. To protect the lenses within the magazine, both sides were provided with roller shutters which, when machining with cooling lubricant is necessary, can be closed.

The integration of the laser processing units creates new requirements for the peripheral areas of the machine. Firstly, the laser processing units need to be supplied with laser power, cooling, process gas, compressed air and signals at every point within the working space. Secondly, the robot must not be subject to any unnecessary disturbances when moving the laser processing unit. Disturbance variables include, in particular, the weight of the laser processing unit, and the drag created by the need to move its supply lines with it. Thirdly, the optical fibres must be kept taut and free

Fig. 5.151 Media supply of the laser processing units

of torsion. Dynamic bending radii under 150 mm, or torsion values above 45 ° would cause damage to the optic fibres.

The media supply ducts are therefore placed in a rotatable rack behind the machine, and they enter the working space from above. The active pre-positioning of the media supplies, using electric motors and belt drives, above the rotary swivelling tables, or above the laser magazine minimises the constraint forces, and so the errors, from mechanical drag during processing. Figure 5.151 shows the arrangement of the media supplies.

The tension pulleys in the utility supply rack are positioned to move vertically, so that the utility pipes are kept under slight tension simply due to their weight. The weight of the hybrid laser machining unit, which is too heavy for the robot, is reduced on the one hand by using a compensating weight in the front area, on the other hand by the pre-tensioning of the utility supplies. The whole system is set up in such a way that the robot carries more or less half the weight of the laser machining unit.

When swapping machining units in and out from the spindle or from the quick-change interface on the robot, the bundles of pipes remain firmly attached to the laser machining units. This situation means that there is a permanent disturbance element represented by the bundles of pipes, which in turn means that the robot and the milling machine can pick up only one of the machining units, respectively.

Table 5.9 Assignment of laser procedure to resources of the machining centre. (Brecher and Karlberger 2010)

Laser process	Physical principle	Emission	Required precision (µm)	Laser head guidance
Laser deposition welding	Melting	Continuous	<500	Robot
Laser hardening	Hardening/annealing	Continuous	~1,000	Robot
Laser deburring	Sublimation	Pulsed	~10	Machine
Laser structuring	Sublimation	Pulsed	~10	Machine

5 Hybrid Production Systems

Table 5.9 organises the integrated laser production processes, using the necessary guidance accuracy to the robot or to the spindle.

5.5.4.2 Hybrid Laser Machining Unit for Laser Deposition Welding and Laser Hardening

On the laser system side, a scanner-based laser machining system was developed, which supports high-frequency oscillation of a laser beam for deposition welding using metal wire as filler material. This makes it possible to carry out the fundamental assessment of this technological approach for industrial applications, which for example occur in machine tool and die production. In contrast to existing wire-based laser welding machining heads (Nowotny et al. 2008), the integrated scanner system opens up a wide range of different laser production processes. The following objectives are pursued:

- Extension of the processing range of conventional machine tools by the availability of additional laser processes
- Functional and processing integration for combining several separate processes in multi-kW laser applications
- Identification and investigation of interacting physical interactions, in relation to component quality, process management and productivity
- Research on the influence of laser processes on the components in the system, and development of strategies for measuring and compensating for disturbances
- Derivation of system and process requirement profiles for modular laser machining systems and hybrid technology platforms (Fig. 5.152)

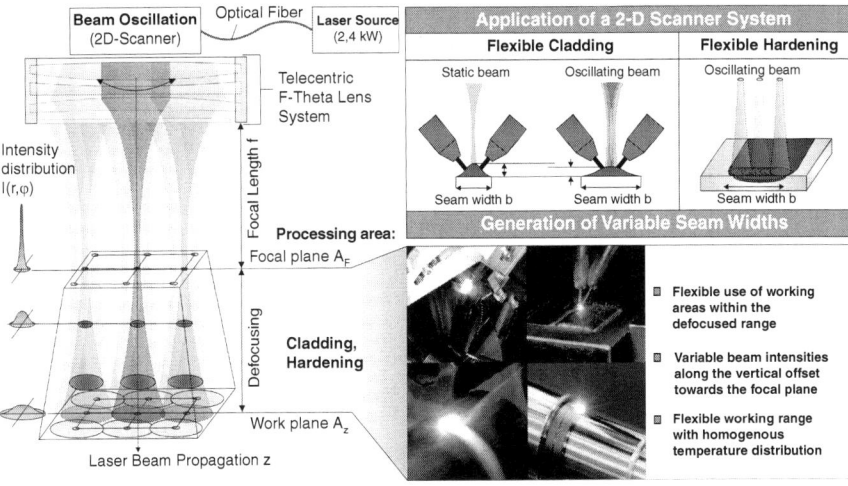

Fig. 5.152 Technological approach for scanner-based laser deposition welding, laser hardening and ablation in a hybrid laser machining unit

The complete robot-based processing system uses a 2D laser scanner, with a fibre-connected solid-state laser for output of up to 2,400 W. This is linked to a wire feeder system with hot wire functionality, as well as integrated shield gas feed and cooling of the wire feed nozzle. The telecentric, flat field lens which is part of the scan system enables flexible processing of a flat scan field. The varying intensity distribution is exploited specifically here to create defined processing areas for different laser production processes. The laser beam in the resulting working plane is defocused, so that the laser spot generated on the work piece surface becomes larger. The laser deposition welding procedure using added wire and laser hardening are both carried out in this defocused working area of the laser beam. In non-stop laser operation (cw mode), where there is a power density of <1 kW/mm^2, this leads primarily to heat being supplied to the component, which in the case of hardening leads to phase transformations in the marginal zone (martensitic hardening up to approx. 1.5 mm depth penetration). Depending on the material being processed, its surface characteristics and the laser beam parameters, a hardness of up to 65 HRC can be obtained. Due to the high intensity of the laser radiation and the steep temperature gradients in the marginal area, hardening is achieved through self-quenching as a result of heat conduction without any external cooling. This means it is particularly suited for partial hardening of surfaces. The thermal energy induced is minimal, and usually there is no need for subsequent post-processing by grinding or hard-milling processes. Via its own wire feeder system, which is attached to the periphery of the scanner system, the process area can be supplied with wire material for the production of variable sized structures. The wire feeder system also includes a material pre-heating option, which has a stabilising effect on the coating process. The use of the 2D scanner system has for the first time made the creation of variable width tracks, and therefore flexible processing paths in the process possible, without the need for a change of lens or of the entire laser processing system. The switching between the deposition welding process with additional wire and the hardening process requires the activation/deactivation of the wire feeder and the insertion/removal of the wire feeder nozzle into/out of the processing area. For this requirement, a hinged kinematic device was developed for the wire feed nozzles, which enables an automated process change-over, and which minimises the resulting disturbance contour in relation to tracking when carrying out laser hardening or laser coating. The complete system also uses a cross-jet to protect the lens from particles which may be emitted during the deposition welding process. Various working fibres can be attached to the scanner, which has integrated collimation and a focussing unit, via a standardised interface. The integrated telecentric flat field lens for beam correction and focussing enables the flexible processing of a flat scan field, with orthogonal beam angles of incidence on the work piece surface being processed. The intensity distribution, which varies within the area of beam propagation, is specifically exploited here to create defined processing areas for the different laser processing steps in which flexible processing conditions are possible. Figure 5.153 shows the arrangement of the construction of the hybrid laser processing unit based on a 2D scanner system.

When hardening, the focal point is scanned at several 100 HZ over an adjustable width of up to 20 mm and is simultaneously guided over the component by the robot.

Hybrid Laser Processing Unit

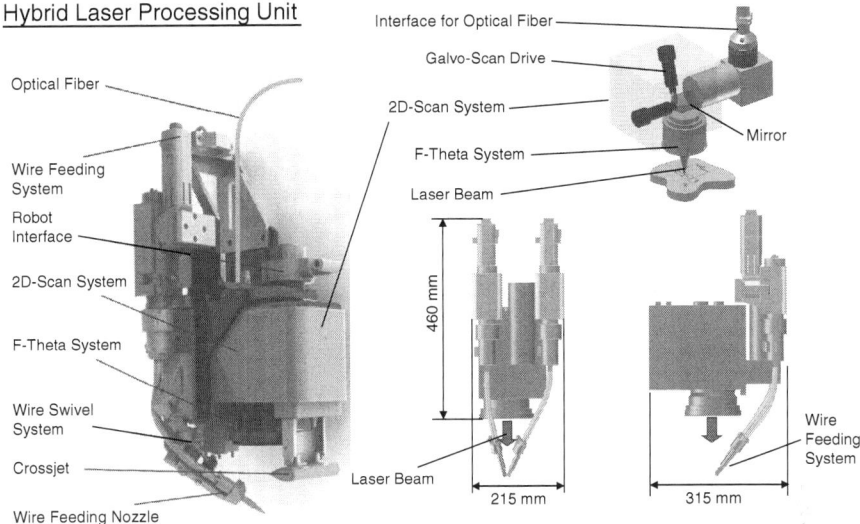

Fig. 5.153 Hybrid laser machining unit for laser deposition welding and laser hardening

This creates the resulting working track, where the width of the track can be adjusted by using the scan amplitude during the relative movement between the robot and the work piece and therefore a flexible hardening shape can be produced. By scanning the focal point across the work piece, a close to homogeneous temperature distribution can be ensured, which has a decisive influence on the process result when hardening at the edges of the work piece.

For wire-based deposition welding on the other hand, the oscillating laser focal spot is used to specifically influence the temperature distribution at the surface of melt pool which is being created. This allows the surface tension to be varied locally, and thereby the width of the resulting coating track can be set to be within defined limits. The variation in the oscillation amplitude makes it possible to change the width of the track by varying the relative movement between the work piece and the dynamic focal spot. This allows flexible shapes to be produced. During the coating process, it can also be adjusted to the current kinematic boundary conditions, such as change of direction and contour variations in the application of the material. During the laser deposition welding process, the targeted control of the temperature profile within the welding zone can be used to selectively influence the surface tension in the molten filler metal, and thereby the desired form of the weld seam geometry. The resulting shape of the weld bead can therefore be varied within certain limits.

In order to use and manage the laser process reliably for technological applications, as well as the development of a suitable laser scanner system, focused research was needed in particular to enable stable process control which could be adapted to the relevant production task each time.

5.5.4.3 Results of Laser Deposition Welding

The aim of the laser deposition welding process is to coat the surfaces to obtain functional properties, taking into account requirements such as, improved durability or corrosion resistance of the component's surfaces. During the deposition welding process, the laser radiation heats both the surface of the component and the added wire material to above the melting point. The surface of the base material is only melted superficially to the dilution mixing between the base and the filler metal. This avoids unwanted creation of alloys between the two materials, which would lead to undesirable mechanical properties in the melt zone. This minimal dilution is one of the principal advantages of laser deposition welding compared to competing processes. Further advantages are the reduced distortion of the component, thanks to the small amount of heat input and the ability to achieve higher surface and coating quality.

Borrowing from tool and die production, an X38CrMoV5-1 (hot working steel) was used for the basic process technology research on the base material, with the similar type of wire filler material RC 44 from CRONITEX. As Fig. 5.154 makes clear, the laser focal spot oscillates during the deposition welding process orthogonally to the direction of movement of the laser—in this case a sinousoidal motion within the prescribed circular geometric element. In the tests both circular and elliptical scan geometries were used.

With regard to the required characteristics of the weld seam, the energy input into the component can be adjusted to suit each application. For this purpose the strategy of how the laser focal spot is moved within the prescribed scan geometry can be changed. For example, unidirectional, meandering pattern or circular strategies are available. By varying the described parameters, the movement of the melt is also affected. By using the various oscillation strategies of the laser focal spot in this way, different temperature gradients can be created which lead to differences in the density of the melt (Kuchling 1996). The described influence of the scanner system on the laser focal spot therefore allows a bigger bandwidth of variations, and requires less effort than influencing the melt pool using locally created electro-magnetic fields (Dolles 2007). All the parameters listed above have a significant influence on the geometry and quality of the weld seams obtained.

At the start of the process technology research five operating points relating to different technological applications were initially defined and validated (Fig. 5.155).

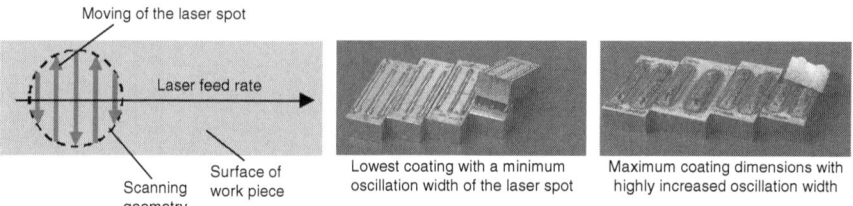

Fig. 5.154 Principle of scanner movement

Fig. 5.155 Defined operating points for specific technological applications

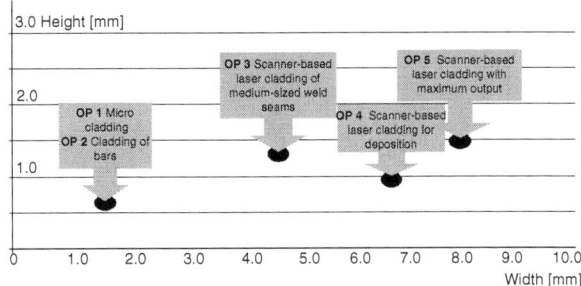

Applications which were addressed are e.g. the coating of components with large surface areas, but also the machining of filigree geometries. These examples help to show the enormous potential of this process in terms of flexibility.

The metallographic cross-sections of each of the operating points and possible applications are shown in Fig. 5.156.

Starting from the validated operating points, the dependency between laser output power and laser movement was determined. The purpose of the investigation is to guarantee a steady weld seam geometry and quality at reduced laser movement speeds. Reduced laser movement speeds particularly occur due to the deceleration of the NC when machining radii and edges. Here it is necessary to be able to adjust the laser output power online. Figure 5.157 shows the correlation between the laser output power and the laser movement speed with all other parameters remaining constant, starting from a defined operating point 3, in order to guarantee a consistently high quality and true geometry. In order to transfer this automated adjustment online, the functional dependency between laser output power and laser movement speed is implemented in the NC. In the NC, the relevant geometry of the weld seam for each operating point is defined as a parameter for the CAM system.

Starting with a maximum laser power output of 2,000 W, a laser feed speed of 333 mm/minute and a wire feed of 2,914 mm/minute the resulting feed ratio is 8.75. This ratio remains constant within the framework of throughout the study. In addition, the start value of the energy input per unit length is 360 J/mm. The energy input per

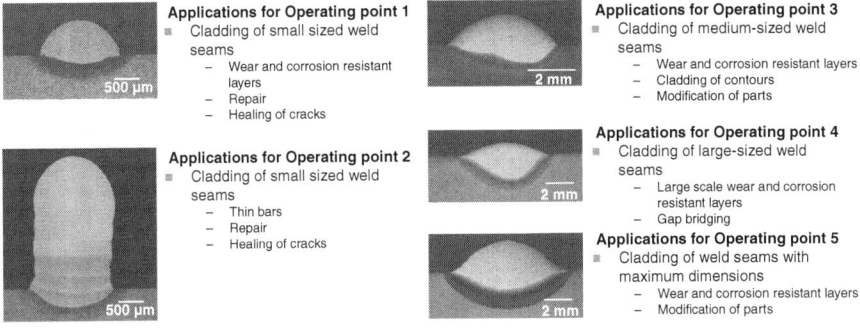

Fig. 5.156 Metallographic cross-sections at the five operating points

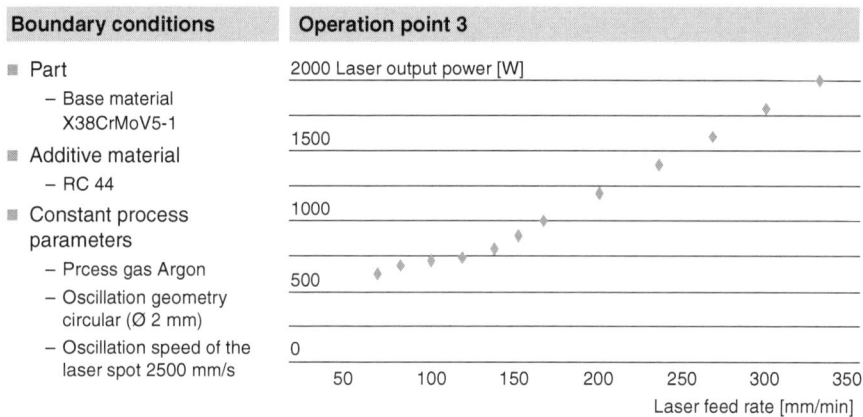

Fig. 5.157 Correlation between laser movement speed and laser output power for operating point 3

unit length is only kept constant for as long as the prescribed weld seam geometry can be achieved by a linear adjustment to the laser power. Between 2,000 and 800 W an unchanged weld seam geometry is obtained, reflecting the linear adjustment of the laser power. Below a laser power of 800 W, the laser power needs to be adjusted non-linearly. This means that the value of the energy input per unit length must be increased in order to achieve the required weld seam geometry and quality. This is clear from the reduced inclination of the derived curve (Fig. 5.157) below 800 W. The input energy per unit length rises further, as the laser feed speed falls. The maximum set input energy per unit length for this operating point shows a value of 500 J/mm. The resulting lowest possible laser feed speed is 67 mm/min with a laser power of 600 W. Any further reduction in the laser feed speed leads to insufficient weld seam geometries, as well as to a reduction in quality of the weld seams created. Taking into account the boundary conditions, therefore, the lower limit for laser feed for this operating point has been reached.

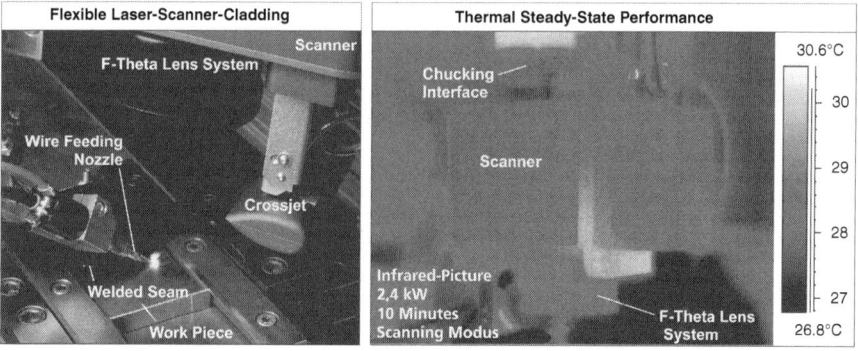

Fig. 5.158 Picture of the laser deposition welding process and proof of the long-term thermal stability of the new type of hybrid laser machining tool for laser deposition welding and laser hardening

This fundamental research warrented, on the one hand a clearly improved coating quality, (small heat-affected area, minimal mixing, minimisation of faults such as pores or other inclusions), on the other hand the insights gained also integrated directly into the control and planning of the processes.

During the process investigations which were carried out, it was also possible to successfully demonstrate the thermal long-term stability of the hybrid laser processing tool both for laser deposition welding and for laser hardening (Fig. 5.158).

5.5.4.4 Laser Deburring

In addition to the hybrid laser machining unit, a short-pulse fibre laser with lower average power was used for deburring of defined edges in combination with an active 3D beam deflector system. Peak pulse power levels of around $1\ GW/mm^2$ enables the removal of material by sublimation. Given the required positioning and path precision, the process control cannot be carried out by the robot. For this reason, the structuring laser system is swapped as a module onto the spindle tool support in the machine tool via the positive taper lock interface.

Deburring processes using laser radiation can be divided into two basic principles (Fig. 5.159). When deburring using laser ablation, the burr is steamed off using pulsed laser radiation and so can be cut off or completely removed. When deburring using laser melting the material in the area of the edge is melted. When it resolidifies a new edge geometry is created. Both process variants have already been researched and applied to special applications (Schmidt-Sandte 2002; Fritzsche et al. 2002). The choice of process is made based on the following criteria:

- the size, position and shape of the burr
- the accuracy of the deburring process
- the resulting edge microstructure
- whether the process can be automated

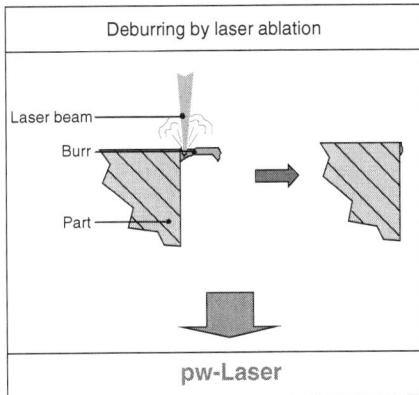
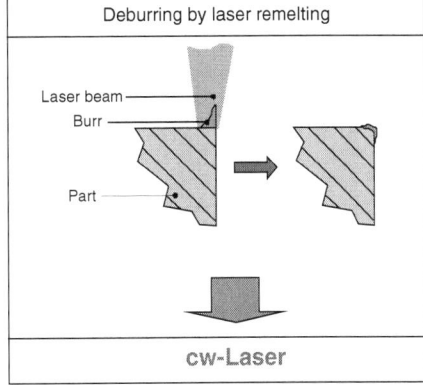

Fig. 5.159 Process for laser deburring

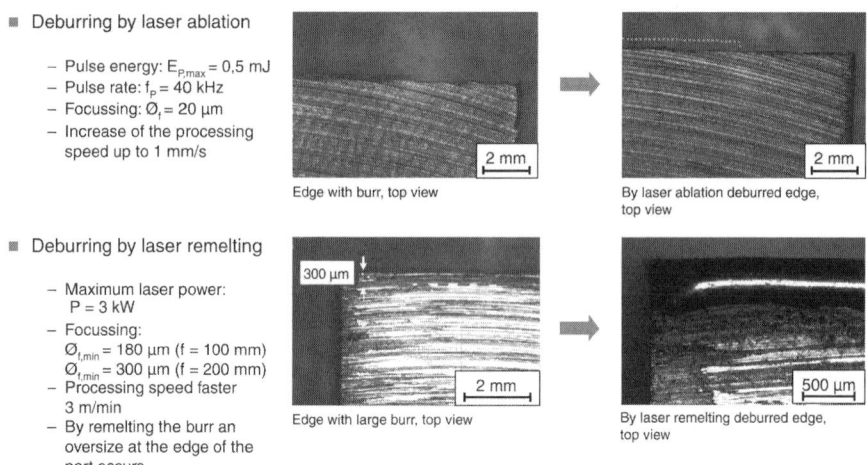

Fig. 5.160 Microscope photos of deburring tests

The aim of the work in this project is an initial integration of laser deburring into a machine tool. Both processes for laser deburring were implemented in hybrid machining centres.

Preliminary investigations showed that both deburring procedures could be realised with the system components. As the laser beam in both processes is fed through the galvanometer scanner-axes, very high levels of accuracy in the range of a few μm could be achieved. This makes it possible to process even the smallest components with minimal burring. The implementation of laser deburring in the hybrid machining centre also enables the fully automated machining of components with complex shapes. Figure 5.160 shows the results of some test examples for both processes. The process speed which can be achieved is determined by:

- the form of the burr, its size and direction
- the material
- the conditions placed on the resulting edge shapes and microstructure
- the geometry of the component

For the selected burrs, it appeared that the achievable process speed for laser melting was substantially higher than for deburring using laser ablation.

5.5.4.5 Control Technology

Even the preliminary tests prior to commissioning the laser systems with an attached PLC showed that the spatial and functional integration of the many sub-systems lead to a clear increase in complexity, compared to conventional machines. The integration approach chosen, for example, to link robot control to machine control, will have a major impact on the performance of the overall system. If there were full

5 Hybrid Production Systems

Table 5.10 Comparison of strategies for robot integration. (Brecher and Karlberger 2010)

Strategy/requirement	Master-Slave (NC→RC)	Master-Slave (RC→NC)	Master-Master (RC←→NC)
Programming in work piece coordinates	Yes	Yes	No
Access to actual track speed	Yes	Yes	Limited
Low synchronisation error	Limited	Limited	Limited
Smooth, yerk free motion	Yes	Yes	Limited
Defined orientation of laser beam in relation to the machine	Yes	No	Yes
Extensibility for cooperative work on *one* part	Yes	No	Yes
Realization by standard component interfaces	Limited	Limited	No

integration of robot kinematics into the NC controls then the robot controls could be dropped, which would mean that the need for communication between two different control systems would be avoided altogether. This would, however, result in high cost, as in-depth modifications would be needed to robot kinematics in measurement systems and drives. In addition, we would have to do without all the adjustment and feed-forward algorithms which are available in robot control systems. In contrast, by linking the NC controls and robot controls heterogeneously, no hardware modifications would be needed, the robot specific control loops would ensure optimum robot precision and handling processes could benefit from the specific functional scope in the robot controls. The disadvantage which must be accepted in this approach, is that the implementation of closed control loops is problematic, as communication between the heterogeneous control systems inevitably leads to additional deterministic dead time.

To try and counter these disadvantages, which arise from basic principles, a control architecture was developed which takes account of the specific features of hybrid machines (Brecher and Karlberger 2010). Here the properties of various synchronisation approaches were assessed, and the optimum implementation option selected for the desired implementation. These are compared briefly in Table 5.10.

As the hybrid machining centre has two machining stations, initially two NC-channels are needed for working simultaneously in parallel on sub-programmes. One channel then executes each sub-programme, which describes all the processing steps required to complete it. The channel here takes on the role of a "work piece agent" whose structural logic is defined by the NC programme. Individual technologies are "requested" from the NC programme, to carry out which either the robot or the machine spindle is selected as the functional operator. The PLC here assumes all the coordination functions. If a resource is requested by a machining station while it is blocked on the other side, there is a waiting period until the resource is released. With a balanced share of milling and laser operations, a high load factor of the machine can be achieved. This type of decentralised process control leads to a lower planning effort being required of the user.

During machining, the robot is controlled by the NC as master. This has the advantage that both the milling and the laser processes can be created and carried out as NC sub-programmes in the same way. The robot acts here in a figurative sense as

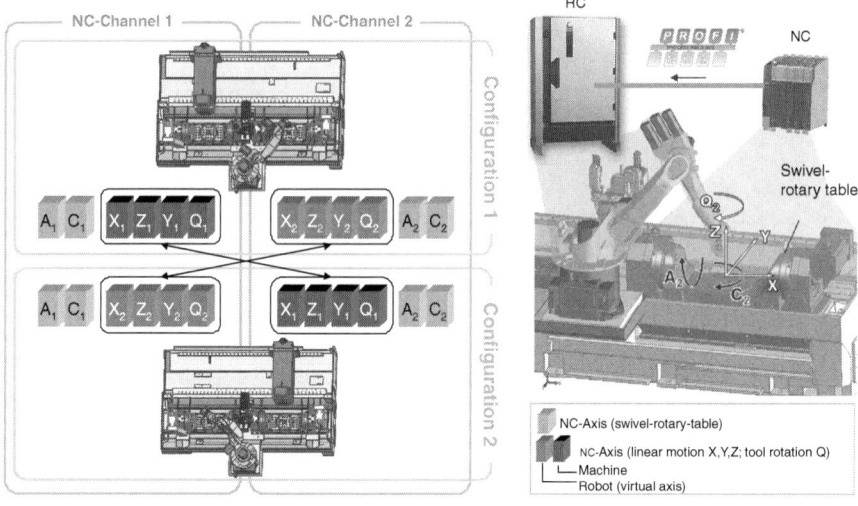

Fig. 5.161 Design for robot-machine-cooperation. (Brecher and Karlberger 2010)

a service supplier, acting on the instructions from the NC online. The Cartesian set points are calculated by the NC using virtual axes and passed to the robot online. This makes it possible to use the 5 or 6 axis transformation available in the NC, which makes simplified programming of work piece coordinates possible. Figure 5.161 shows the design for robot-machine-cooperation. Within the NC an axis system consisting of virtual and real axes is created where the set points of the virtual axes are transferred online via the fieldbus to the robot control (RC) system. When the working areas are changed there is an automatic re-assignment of the axes to the relevant channel.

A substantial advantage of programming using work piece coordinates is the possibility, using the NC internal transformation calculation, to be able to call up the current path speed. As variations from the programmed path speed—e.g. when traversing tight curves—is one of the principal disturbances for laser deposition welding, being able to adjust the set points of *laser output power* and *wire feed speed* depending on the path speed is desirable. To this end, so-called synchronous-actions in the Siemens 840D are used, which allow the current path speed to be read and can be used for the calculation of set points for the laser systems. The information, generated as part of the preliminary technical process tests, about the functional link between the wire advance or laser output power and the path speed, is defined to the NC in the form of parameter records, and then activated by the sub-programme (Fig. 5.162).

In order to achieve a high level of process stability for laser deposition welding, precise synchronisation in the timing of the various process variables—laser output power, wire feed and relative movement—is needed. While the signal transmission time and the scatter involved in the fieldbus communication are negligible, within the

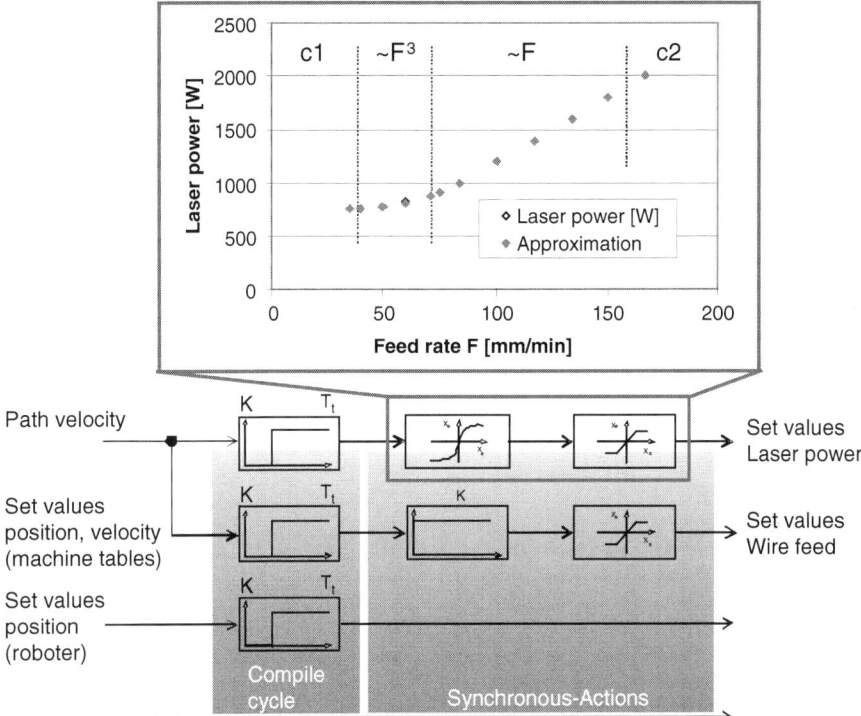

Fig. 5.162 Delay compensation and speed-related set point adjustment

individual sub-system controls delays arise due to the processing of the set points. For this reason, when using a mixture of control architectures, mechanisms to compensate for the different signal run times are required. As a solution, a flexible buffering function for signals was developed using an NC-core extension (Compile-Cycle). This permitted exact synchronisation of the movements of the rotary table and the robot, so that the process requirements from the control system in relation to the precision of the path can be met. This design, implemented to handle dead time compensation, is shown as an example in Fig. 5.163.

For the implementation, tight coupling of the digital machine control system, robot controls system, wire feeder and laser control systems is necessary. Figure 5.164 shows the bus topology developed for the hybrid machining centre. Those components which depend functionally on the robot's movements are assigned to the subordinated fieldbus for robot control (RC) as slaves.

This makes it possible to decouple the entire hybrid system from the machine control system, which simplifies the commissioning of the equipment and the separate operation of the machine tool. The information backbone for hybrid machining is provided in Fig. 5.164 by the *Profibus 2*, which is running isochronically with the interpolation cycle of the NC. Along with the positional set points for the robot

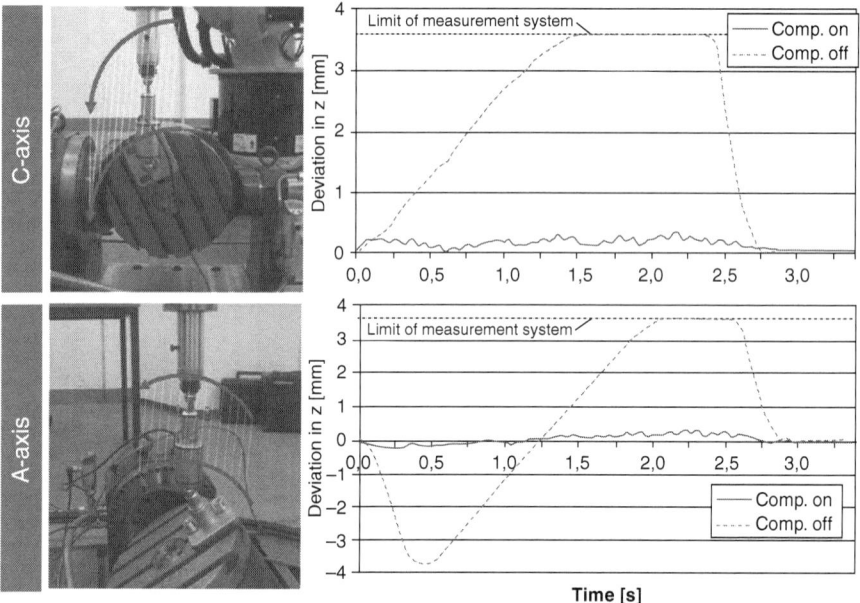

Fig. 5.163 Compensation achieved for dead time between robot and rotary swivelling table. (Brecher and Karlberger 2010)

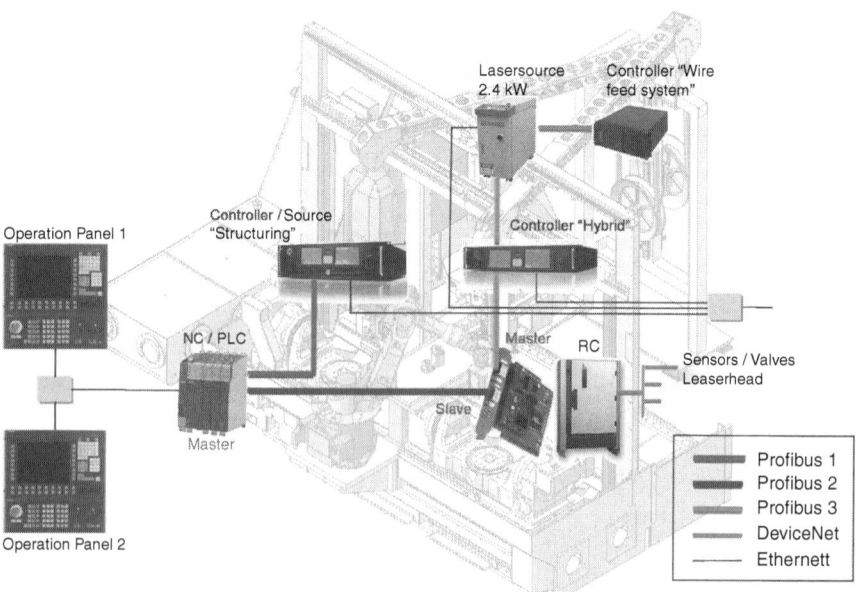

Fig. 5.164 Bus topology for hybrid machining centre (extract)

5 Hybrid Production Systems

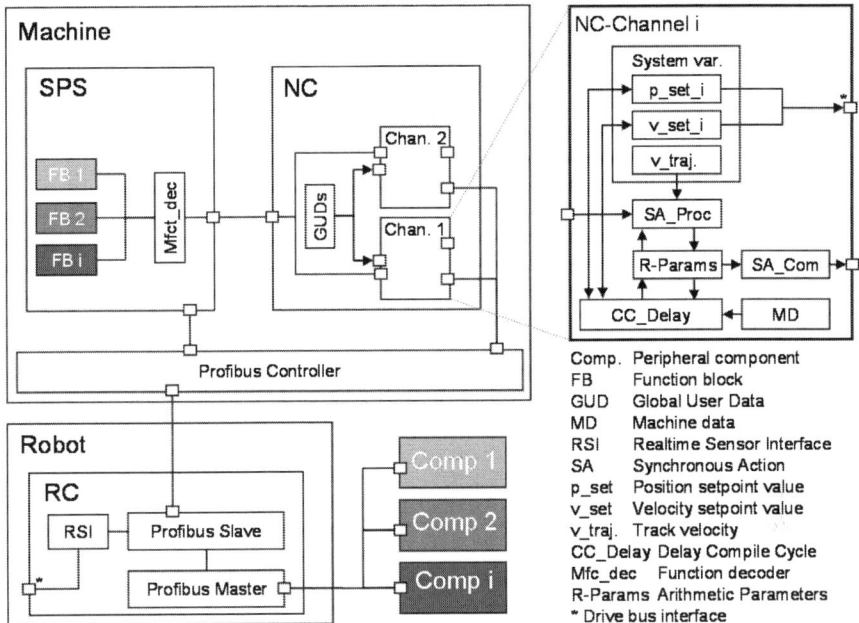

Fig. 5.165 Control Architecture for Laser Processing (simplified)

there are also the path speed dependent set points, such as wire feed and laser output power, which are sent via the RC to the subordinated slaves.

Details of the architecture developed for controlling the hybrid laser system are shown in Fig. 5.165. The architecture is characterised by the allocation of individual functions to different system components which communicate via standardised interfaces. Communication between NC and PLC is realised using synchronous and asynchronous M-functions. Each component of the hybrid system (robot, laser source, wire controller and scanner) is controlled by an individual function block of the PLC.

In order to handle mutual influences among system components – particularly during system faults – relevant information about system states is exchanged among the function blocks. The interfaces of the laser system components provide both, continuous and discrete processing modes for input and output signals. The discrete signals for controlling the system state are assigned to the PLC. Continuous supply of process data in contrast is provided by the NC, calculating set points – such as laser power – through Synchronous Actions and writing them directly to the isochronous Profibus. For open-loop process control, necessary process knowledge is stored within the NC using global user data (GUDs) as model parameters.

During laser processing, M-functions for initialising the velocity dependent process control parameters according to Fig. 5.162 are called first. The R-parameters serve either as switches to enable and disable delay compensation functions, or as an interface to a ring buffer which can be parameterised by machine data. In contrast to using R-parameter values, the buffering of axis set points requires deep interventions in the data processing chain of the NC. Interfaces available in the scope of NC kernel development only provide means for reading access, whereas writing axis set values is not directly supported. Therefore, the kernel interface for applying compensation values is used. By adding the negated current speed and position values, the output to the drive system is effectively suppressed. Adding buffered set values at the same time, however, allows enforcing these values after a defined temporal delay. The amount of this delay can be flexibly changed by the aid of machine parameters. Figure 5.165 shows how this concept works. Dead time delay causes compression, i.e. decompression of the double-ball-bar of several millimetres. The described deterministic time delay in outputting setpoints to the rotational-swiveling table by 88 ms effectively eliminates this effect entirely (refer to Brecher und Karlberger 2010).

Also in relation to the process planning, the integration of laser processes and freely programmable (via the NC controls) robot kinematics requires new solutions which go beyond the functional range of conventional systems. Normally robots are integrated into machines in order to carry out handling tasks for component parts and finished machined work pieces, which are programmed once when they are commissioned, and then are run as unchanging processes.

The fact that the robot is freely programmable led to certain requirements being placed on the programme generation, since programs had to be modified to match the characteristics of the different control systems. The robot kinematics were tied into the mixed control architecture by means of virtual axes, which enabled programming basically analogous to the NC-axes. But the programmes generated for the robot kinematics must adhere to the dynamic boundary parameters of the robot kinematics in line with the mechanically acceptable accelerations when running the programs. This avoids uncontrolled vibration in the robot kinematics, while the dynamic parameters for the more rigid real machine axes are unaffected and can be set higher.

With regard to the scope of the programs generated for both robots and machines, the processing of irregular shapes showed up variations in program scope. If the required precision in the CAM system for the depiction of the planned tool movements in the NC program is defined as high, then many program lines are created to describe short distances of movement. This fact is generally known in the case of programming, for example, milling work. If the precision is set as high for the robot programs as is generally required for milling, then unnecessarily large control programs result. These define the path of the tool with an absolute precision of lower than 20 µm, which robot kinematics generally cannot actually achieve. The increased size of the program may then require critical computing resources when running the program on the NC.

With the increase in capabilities of the processing centre, the requirements on the programmer planning the production steps naturally also grow. These increased demands need to be compensated by support for the programmer. Basically, programming techniques for classical machining processes needed to remain unchanged

5 Hybrid Production Systems   617

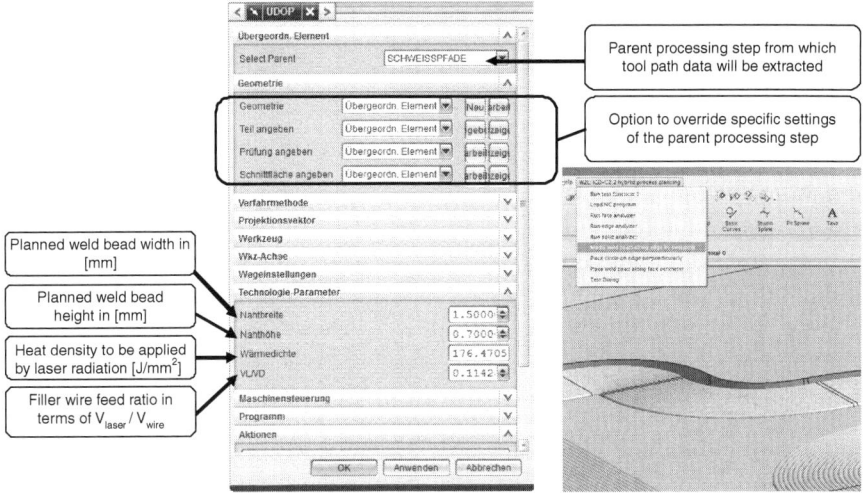

Fig. 5.166 Input screen for process data for a laser deposition welding process

despite the new options, in order to be able to continue to use the processing centre for machining-only production, and to continue to exploit the expertise of the machine operators. With this measure, the opportunities for machine deployment are increased without necessarily increasing the complexity of its operation. For the additional planning of laser processes the software should also give the user as much support as possible. The planning effort for the programmer should be reduced, with a high level of visibility and transparency provided at the same time. In order to be able to plan individual laser processing steps as well, a planning structure which was split into modules by application seemed appropriate. Therefore the laser hardening and the laser deposition welding were created as stand-alone planning modules alongside each other, which offer the user input options which are relevant to the selected application (Fig. 5.166). The user of the CAM system is not just offered a general module for laser process planning, but input screens which are tailored to the application, which also make sensible suggestions for suitable process parameters.

The extensibility of the standard market offerings for NC controls allowed integrative control and activation of the relevant elements which were needed for the laser processes (Brecher et al. 2009a). For the hybrid machining centre, the integrative development approach allowed use of modularity and effective encapsulation of the control architecture within the CAM based process planning, to provide a high degree of transparency. For example, the laser source in the *laser welding* process can be fed with set points for laser output power directly from the NC programme. This allows the additional process parameters to be combined with the planned tool path in a single NC programme. As this programme can include all the necessary commands, there is no need to create a separate programme to control the additional units which are not controlled directly by the NC controls. This significantly reduces

Table 5.11 Typical command sequences at the start of a weld seam

N0210 M54=2	Request wire feed readiness
N0220 M37=1	Request laser source readiness
N0230 M46=2	Start program in scanner controller
N0240 G4F0.2	Wait for 0.2 s
N0250 M38=1	Request laser radiation clearance
N0260 M42=1	Activate wire feed

programme management and maintenance efforts. In the hybrid machining centre—as already mentioned—all the additional units were connected to the NC controls via a fieldbus connection. For process planning and programming it is absolutely transparent to the user of the CAM system, if, and how, the laser source is linked to the hybrid machining centre as an independent additional unit.

For the activation of the controls involved in the laser processes, new NC functions, as well as so-called control or M-functions were created, as well as synchronous actions when needed. As a result these are only comprehensible using a machine specific command table, and by their very nature are not normally known to operators, as can be seen clearly from the command sequence in Table 5.11.

Especially when processing irregular shapes, NC programs may comprise many thousands of lines, which can no longer sensibly be corrected manually. As this processing is predominant in tool and die manufacture, the program processes for new laser processes should follow the existing approach. Therefore the approach taken in respect of the legibility for the machine operator is nothing new, and therefore does not represent any additional limitation. The laser processes should also not be editable by hand in detail, but the processing plan must be presented in a form which is comprehensible to the operator. As with the planning of tool paths, it is useful for the operator if the planning result is presented graphically. This means that at any time he gets an overview of the plan, and if necessary, can react with a change to the planning parameters, see Fig. 5.167. This resulted in a requirement for a visualisation of the parameters entered for additional process variables, intended for the expert

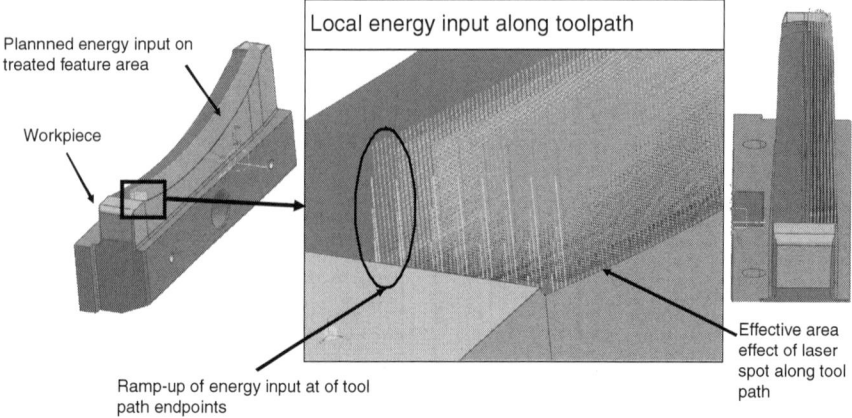

Fig. 5.167 Visualisation of planned energy inputs using an NX extension

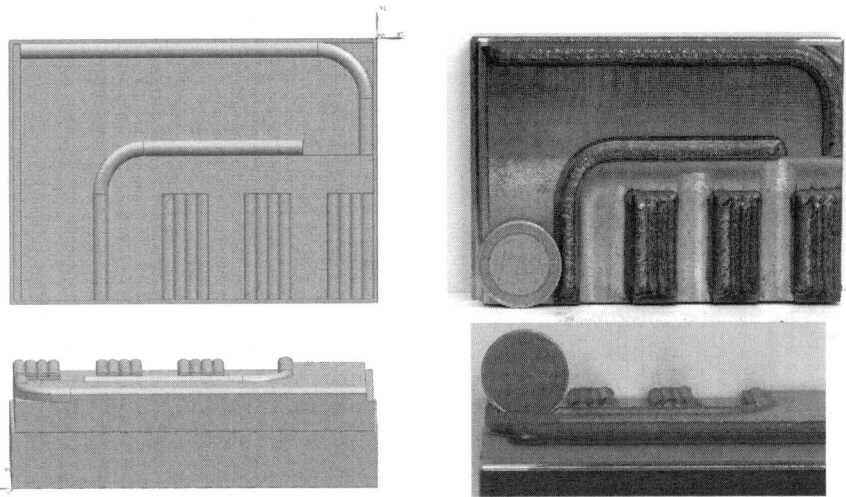

Fig. 5.168 Design for visualisation of planned weld seams in a qualitative comparison to the actual weld test

user. By logically closing the chain of processing steps in the CAM system, the operator is able, even without knowing anything about the specific NC functions, to plan the processing steps in a controlled way.

Another required extension was identified, since unlike machining processes, laser deposition welding needed to be planned as a generative process. To be able to check the correctness of the intermediate steps of the process chain, the CAM planner must be able to estimate the quantity of material generated. To this end, visualisation extensions were developed, which use the basic assumptions about the process to carry out a basic visualisation of the material volumes on the work piece, as shown for example in Fig. 5.168. This offers the planner an overview, whether for example the application of material can be expected to be free of gaps, or if the curves in the planned weld seams will be sufficiently large. When discussing a simulation module to present the laser production process more precisely, the integrative working method in the Cluster of Excellence was able to create synergies with the Virtual Manufacturing System by sharing expertise. In a preliminary test on the hybrid machining centre, it could be confirmed using a deposition welded shape, that the visualisation provided was initially suitable for the qualitatively correct display of the generated volume (Fig. 5.168).

The basic validation of the welding process using the hybrid machining centre provides a reliable process procedure for fixed weld seams. In preliminary tests for the generative use of deposition welding, it was tested by samples whether the process stability theoretically permits application of multiple layers (Fig. 5.169). For the generation of local cutting edges, or features with small dimensions, it is possible to build up these features on a basic work piece and so avoid large-scale rough-working.

Fig. 5.169 Tests using generative deposition welding for one and several superimposed process steps

The successful prototype commissioning of the entire piece of equipment requires a safety design, with particular attention being paid to the laser systems used and the robot, which covers the described technical system requirements, as well as the control and planning approach adopted. In order to meet the machine safety requirements with the prototype, in the logical linking of robot controls system, laser source controls system and machine control system a dedicated safety PLC is needed alongside the machine PLC. The control programme running on this safety PLC must in particular ensure that the emission of laser radiation is always coordinated with the status of the laser protection system. The high performance laser systems which are used (2.4 kW continuous wave or pulsed structuring mode), when combined with the degree of freedom of movement of the robot, represent a significant potential risk for operational staff and the components of the equipment itself. In particular, along with the constant scattering of the beam, there is the risk that the robot could affect the surrounding casing with the laser machining unit. In the event of a directly focussed laser beam radiation on the housing, the laser can cut through the sheet metal surrounding within a few seconds. The housing must meet the requirements of test class T1 (standard IEC/EN 60825-4), in a "Worst Case Exposure Scenario", it must resist for 8 h—a full working shift. Even a T2 resistance time of 100 s is barely achievable, given the laser output power involved, solely by the use of passive measures, in any kind of economic or practical manner.

The current state of the art in this technology are double-walled sections, with the inner space being monitored by sensors. If the first wall is penetrated by a laser beam, then the highly sensitive LaserSpy sensor detects the incoming radiation, opens the safety circuit and switches the beam source off using an emergency cut out circuit, before the beam melts the outer protective wall. Figure 5.170 shows the structure and principle of laser protection walls and Fig. 5.171 shows as an example of a logical link of the laser protection in the safety PLC.

As the sensors can only monitor flat areas, the surfaces of the laser protection casing cannot include any curves or invisible angles. In order, on the one hand, to keep the laser protective casing as small as possible, and on the other hand to avoid enclosing sources of heat unnecessarily, all components which are not required to be in the interior are positioned outside the casing, for example the chip conveyor and control cabinet. As both the laser source and the optical fibres are monitored for any laser radiation leaks of laser light, the laser source itself could also be placed outside of the actual laser protective zone.

5 Hybrid Production Systems

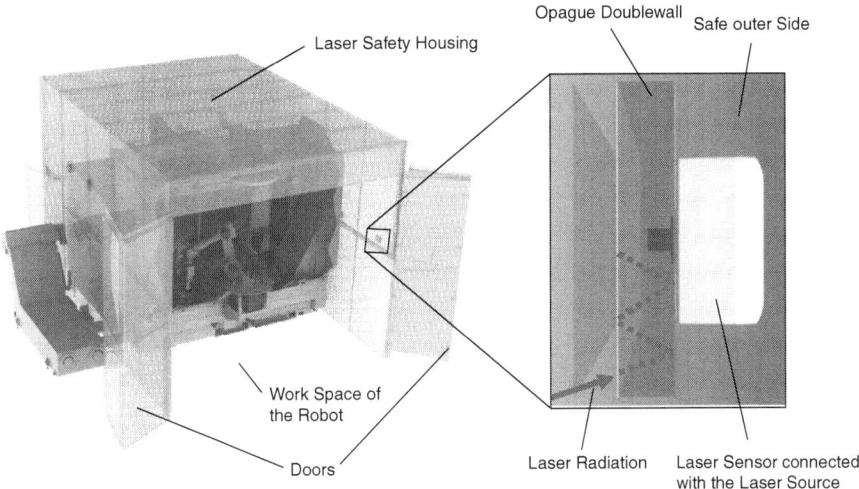

Fig. 5.170 Laser protection booth

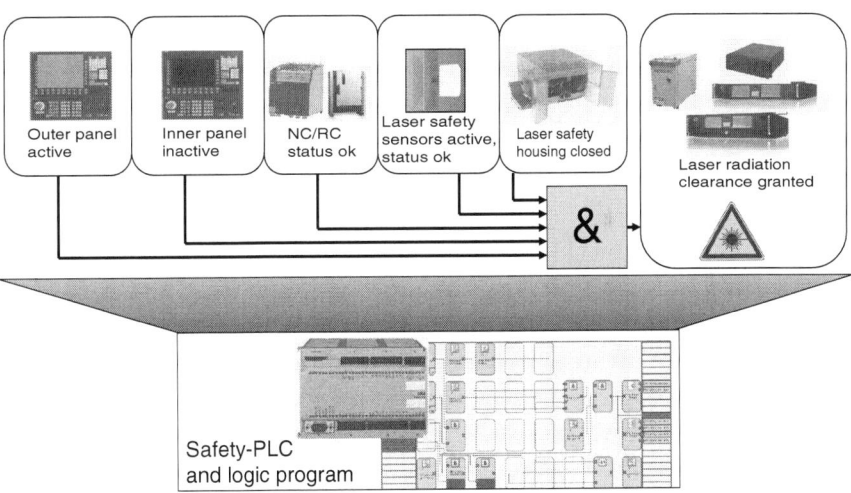

Fig. 5.171 Logical query for granting laser radiation clearance in the safety PLC

5.5.4.6 System-Side Interactions

During operations, unlike in conventional machining centres with only one active process, there is a possibility of mutual interactions. Disturbances to processes or deviations from paths of the Tool Centre Point can arise not only from causes within the process or self-induced interactions or environmental causes, but also from the second production process. In the hybrid machining centre both mechanical and thermal processes can take place either simultaneously or sequentially. Therefore several possible mechanisms for externally induced influences exist. The milling

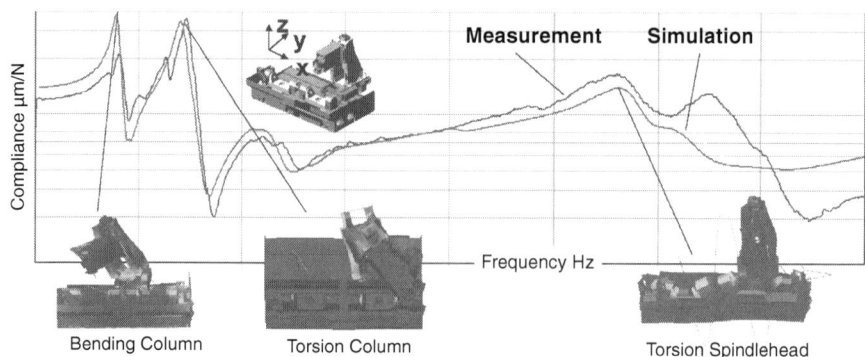

Fig. 5.172 Simulation of resilience frequency response

spindle, for example, can cause vibrations through the entire mechanical structure, and so cause deviations from the set point of the robot. At the present stage of the project, this transfer behaviour has been identified phenomenologically, and during the remaining time of the project it will be investigated in detail, both using metrology and simulation. Equally, the heat arising from the welding or hardening processes can have negative effects on the precision of the equipment.

Figure 5.172 shows the comparison between the simulated and measured frequency responses of the machine, for example in direction X, which already provided some important pointers in the design phase as to how to minimise mechanically weak points in the structure.

Using simulations, the robot mounting was identified as a potential dynamic weak point (see also Sect. 4.5). Figure 5.173 shows the results of the simulated construction of the robot's mounting in the development phase. By designing the structure of the mounting particularly rigid it was possible to reduce the dynamic peaks to a minimum compared to the conventional mounting. In the Z-axis the rigidity compared to the metrological structure could even be increased with the help of substantial struts.

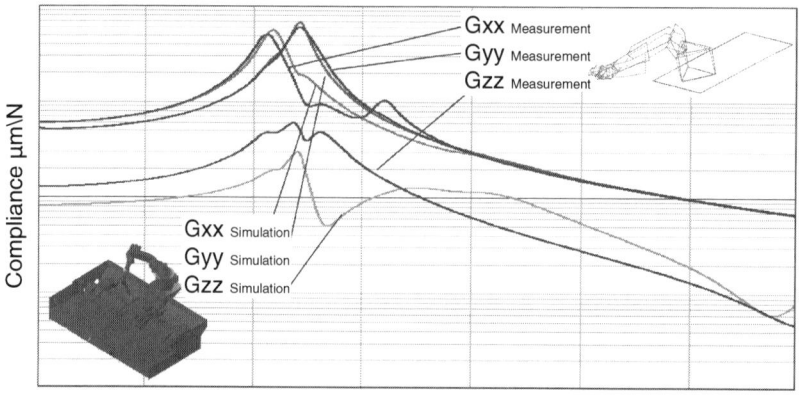

Fig. 5.173 Results of simulated optimisation of the robot mounting

5 Hybrid Production Systems

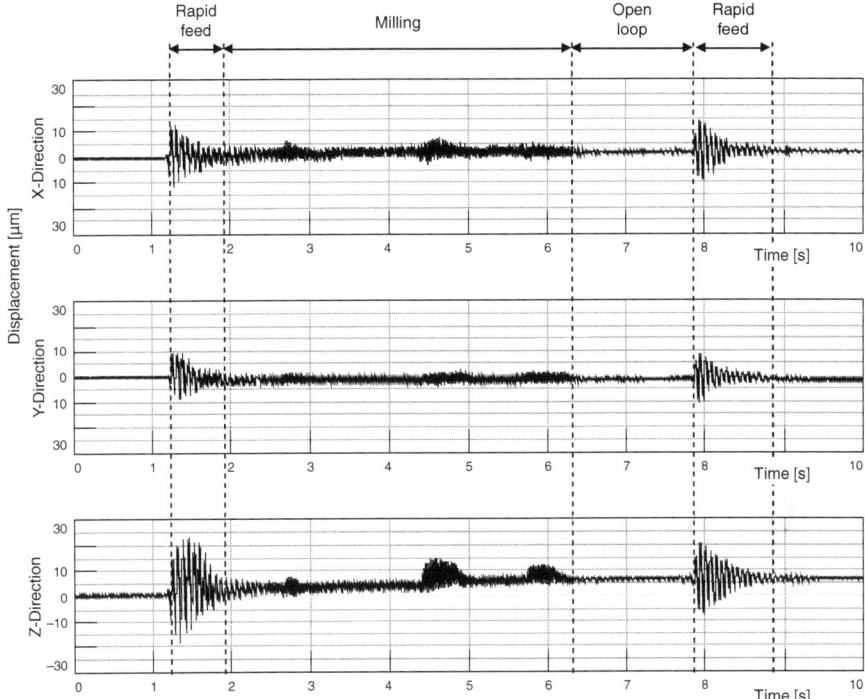

Fig. 5.174 Deviation of the robot in the left working space while milling is carried out in the right working space

Figure 5.174 shows as an example the externally induced vibration of the robot in the left working space when milling is carried out in the right working space. The robot was standing centrally over the left rotary swivelling table while the spindle (revs: 2,500 1/min, movement 2,250 mm/min) was cutting aluminium using a five-cutter milling head (diameter 63 mm) in a rough cut process 50 mm wide and 6 mm deep.

The average vibration amplitudes were around 8–10 μm. The large deflections at the start and the end of machining are caused by the acceleration and slowing down of the travelling column in rapid traverse mode.

The mechanical influence on the hybrid laser system at the robot interface, shown as an example, is caused by the running of the milling spindle. In further tests during the remaining term of the project, the mechanical and thermal interactions of the machine will be investigated based on this. In addition, the resulting impact on the laser machining systems will be assessed, with the emphasis on process capabilities and fatigue limits of the components. In particular, dynamic restrictions on achieving individual precision points of accuracy will be identified. On the aspect of durability, we will look to see how far the mechanical vibrations linked to the laser machining system impact the performance and service life of the galvanometer operated mirrors in the laser machining system.

5.5.5 *Industrial Relevance*

Job retention in high-wage countries presents the challenge to production technology of extending technological leadership. Hybrid production technologies can provide a decisive contribution here, by making production processes more efficient and production machinery more flexible, while simultaneously increasing the level of automation. The downside is the complexity of systems and processes, which on the one hand makes the copying of hybrid technologies more difficult. On the other hand, both hybrid processes and hybrid systems are more difficult to manage as a result of their complexity, which in many cases is a barrier to them becoming established industrially. The basic research work in this project aims to increase the capability of hybrid production machines and therefore make them long-term relevant for industry.

Hybrid production machines are defined by the fact that they carry out mechanical, thermal or maybe chemical processes and procedures within one machine. This, for example, with a combination of conventional machining work (turning, milling, drilling) and laser material processing (laser hardening, laser deposition welding, laser structuring), enables the complete machining in one clamping of components. Thanks to this extension of the functionality of machine tools, manufacturing flexibility and depth is increased, particularly for the production of complex components, e.g. for tool and die production. Attempts in the past to integrate laser system technology into machine tools have been based on the conversion of all processing steps (cutting and laser material processing) into one work space. The resulting limitations from this sequential processing are the relatively high machine hourly costs.

As part of this project, for the first time, a milling machine for 5-axis-machining was combined with laser hardening, laser deposition welding and laser structuring enabling parallel processing of laser processes and machining. This means that the throughput times for work piece completion are significantly reduced.

5.5.5.1 Shortening the Processing Chain and Parallel Processes

The new approach of complete system integration and of splitting the machine into two equivalent, simultaneously usable working spaces, for the first time enables totally parallel processing of two components. This parallel processing increases the productivity of the equipment compared to existing hybrid machine designs with sequential processing sequences.

5.5.5.2 New Type of Laser Machining Heads

Hybrid laser machining heads, which combine laser deposition welding and laser hardening in one system, offer the user greater flexibility compared to conventional single processes. The processes can be activated without a new set-up and so without any down time. In addition, the scanner-based laser deposition welding opens up new application possibilities through the option of a definable track width in many areas. Both flat and wide, or narrow and high, structures can be generated by adjusting the amplitude of the oscillating laser beam. Further research work in the area of

5 Hybrid Production Systems

laser machining heads can seek to optimise the relation between size and weight, to achieve better accuracies, as well as a longer service life for the positioning unit.

5.5.5.3 Testing of New Types of Laser Production Processes

Scanner-based laser deposition welding and also laser hardening show significant advantages compared to the processes currently available on the market for wear protection, such as plasma and thermal sprays or TIG welding. In particular, the significantly improved flexibility and efficiency, but also the improvement in energy input into the component, and the resulting low distortion, as well as the simplicity of automating laser processes offer potential customers a technological and economical alternative way of producing their components with better resistance to wear not only in the field of tool and die manufacturing, the steel and automobile industries.

5.5.5.4 Hybridised Control Architectures and Planning Approaches

From the point of view of the machine manufacturer, the advantages of hybrid technologies are counterbalanced by a considerable development risk, as reference models for suitable system architectures do not yet exist. The specific demands for parallel processing, simple operation and programming, process control, as well as cooperation between robot, machine and laser system components require the development of fundamental design guidelines for designing control hardware and software and the process planning system. Using the solutions that have been developed and implemented, some fundamental insights can be deduced, generalised and transferred to similar machine concepts.

For users, too, the risk of not achieving cost-efficient process planning exists when considering hybrid process chains. For the planning of individual laser processes in other projects in the past, special extensions have been created for CAM systems (Bichmann et al. 2005; Brecher et al. 2008; Freyer 2007; Pursche 2002), which underline the relevance of technologically competent CAM support for the machine programmer. But, in each of these examples, it was assumed that only one NC-control is needed for the production process, and thus that a homogeneous control architecture exists. By including scanner-based beam guidance systems as well as robot and machine controls, boundary conditions and procedures can be derived from the approach pursued here as to how an integrated control design influences machine programming and planning support.

5.5.5.5 Fundamental Research for Mutual Interactions

The technological advantages of multi-technology platforms, with several processes running simultaneously, can only be exploited economically if the processes do not interfere with each other or lead to unacceptable reductions in quality. The investigation of the mutual interactions of several industrially relevant system components is the basis for defining design guidelines for future hybrid multi-technology installations. In particular the risk of potentially unusable developments can be minimised.

5.5.6 Future Research Topics

As a possible example of a hybrid multi-technology platform, the machining centre was researched through integrative cooperation between project partners from research and from industry in the first phase of the Cluster of Excellence, using scientific methods.

As the sub-systems of the hybrid machining centre taken individually already display complex behaviour, the design and the control of integrated systems is one of the main challenges in hybrid machining. For this, many different requirements, such as, for example, process stability, economic returns, usability and safety, extensibility etc. need to be taken into account.

Based on decades of experience in the field of machining processes, companies who build machine tools have focussed almost exclusively on the known technologies, and have built up an immensely high level of expertise in these areas. By introducing additional processes there are now many new requirements to be taken into account, which to a large extent are still unknown as of now.

To reduce the development risk for the machine manufacturers of introducing multi-technology platforms, design guidelines are necessary which set boundaries to the area for development, where solutions need to be found and which enable a targeted and efficient implementation of new technologies. The basis of this approach is first and foremost the criteria of process stability, economic return and usability. In all these points, the multi-technology platform developed provides the necessary basis from which to abstract results which can be generalised.

To simplify the engineering of multi-technology platforms, the integration capabilities of incompatible sub-systems need to be substantially improved. For this, communication models are needed which provide the prerequisites for standardised profiles (e.g. for robot-machine cooperation). Above and beyond that, standardised component descriptions need to be defined and available for the planning phase, so that the development path can be shaped in an overall efficient and—given the heterogeneity of the sub-systems and their suppliers—forward-looking manner.

Alongside development, the testing and approval of sub-systems in test benches is a central and unavoidable part of machine development in future, although this does raise the question, what is the optimal scope and level of detail required. The question continues to be raised, how can the tests can be planned in a targeted way and the results gained be systematically transferred into models, to be able to make use of these both for control and for the process planning system.

It remains to be seen which tests and test sequences on the designed test rigs can best cover, as fully as possible, the requirements which will arise in later total integrations.

In relation to the sub-systems used, the integrated laser processing throws up some far-reaching and detailed questions. Following the successful demonstration of the feasibility and the technical processing potential of hybrid laser processing heads, the question arises how the power of the available scanner and beam guidance systems can be safely increased up to 10 kW for high-performance applications. The necessary compensation for the thermal interactions between the production process and the laser system technology implemented and the optical components of the laser beam guidance system is absolutely necessary for the further improvement of laser applications. The technology of scanner-based laser deposition welding also raises questions in relation to the use of different filler and base materials. The combination of filler and base materials is a decisive factor for the targeted improvement of component surfaces' resistance to wear. Intensive individual tests are needed here, to be able to apply protective layers on high-wear surface areas, using a laser scanner, with a targeted approach to each application. Wide-ranging research according material combinations and the influence that laser spot oscillation can have promises to improve the ability to forecast, for example, how the resistance to specific stresses on a component feature can be increased in a targeted way. In the interactions with the machine as a whole, it currently remains to be seen how the machine and component behaviour can be qualified or quantified in relation to their interactions in various processes, with thermal effects in particular, and the resulting manufacturing tolerances, to be given sufficient attention.

The results in the area of combining the controls of heterogeneous system architectures show that, for example, controlling the movement of a stand-alone robot system using virtual NC-axes is already sufficiently precise with today's existing communication infrastructures. In addition, it can be demonstrated that, with the help of a systematically calculated compensation, precise movement synchronisation of robot and machine or rotary swivelling table as part of a 5-axis transformation is possible. In the long-term this enables, apart from the parallel and interactive processing of multiple NC-programmes on one machine, which is important for laser machining, running processing sequences in parallel in a shared work space. For this, further research is required as to how an efficient discovery of processing strategies, of possible areas of conflict and collision check geometry, as well as programming different systems via planning and simulation systems can be achieved both conceptually and—also long-term—in an economically viable way. As the scenarios listed go significantly further than the common usage in production today, of robots in parallel for handling or processing tasks outside the working area of the machine, there is also a valid question as to which other implementation scenarios could be envisaged by targeted integration of robots into machine tools. From the start, the use of robots for machining should be precluded because of insufficient accuracy and stiffness.

The necessarily higher demands on staff also increase the costs of running extended machine concepts. In relation to the possible user base of such hybrid machining systems, it is also necessary to follow up on the questions relating to suitable man-machine cooperation, in terms of process planning using a CAM system, and for the operation of the machine.

The additional process variables to take into account in multi-technology platforms can today only be integrated in existing control and planning systems with considerable effort. A scientific approach, which has not yet been able to become an industrially practiced standard, was developed as part of STEP-NC and is defined in ISO standard 10303-238. This standard already provides extensions for different processes, but mainly it covers milling, turning and electrical discharge machining technologies. For the inclusion of further processes the question then arises, whether an information and data model could be planned which allows the generic modelling of new processes. In addition, we need to follow up on the question as to how the trend to further hybridisation might continue, without encouraging the development of a plethora of niche developments in the CAM area. During operations, the user needs to be kept informed about the current state of the system status, with context-sensitive and user-appropriate information. Modelling techniques from engineering can make a substantial contribution here in reaching a rapid target-oriented specification of the necessary new functionality.

Research questions whose answers need to be found on the basis of an integrated multi-technology platform relate primarily to the different forms of interaction, which only take on significance within the context of an entire system. From the mechanical point of view, these are both dynamic structural interactions which arise due to layout and operation of the individual components, and interactions which arise from the targeted input of energy into the work piece to achieve thermal short- or long-term effects. By bringing together process production steps in a single machine, the full processing can be carried out from a single clamping. This requires methods and procedures for integrated measurement of components in the working area to be developed and taken into account in the programming systems.

The use of the hybrid machining centre is currently hampered by the low predictability of the quality of results. Based on the high level of functional and physical integration in multi-technology platforms, there are many possible interactions between the mechatronic systems. To achieve a well-founded scientific analysis of these interactions, system modelling methods are needed—taking into account the consumables, energy and material flows between the boundaries of the sub-systems.

The evaluation of economic viability is closely linked to the theoretical system description of the machine as a mechatronic system. In order to have focused further development of hybrid machine concepts, it is therefore useful to have integration of economic viability models with models of process boundaries and machine capabilities. For considerations of economic viability in particular, and to have a comparative view of real existing process chains, the complete implementation is needed of the industrial applications already defined. All the sub-areas of an integrative approach described form an important pre-requisite for the derivation of many new topologies for future multi-technology platforms.

5.6 Shortening Process Chains for Manufacturing Components with Functional Surfaces Via Micro- and Nanostructures

Nazlim Bagcivan, Stefan Beckemper, Kirsten Bobzin, Andreas Bührig-Polaczek, Stephan Eilbracht, Arnold Gillner, Claudia Hartmann, Jens Holtkamp, Todor Ivanov, Fritz Klaiber, Walter Michaeli, Reinhart Poprawe, Micha Scharf, Maximilian Schöngart and Sebastian Theiß

5.6.1 Abstract

It is already possible to use micro- and nanostructured surfaces to give components functional surfaces. Structures with a non-adhesive effect can be used for self-cleaning building façades, for example, which reduces cleaning costs and helps protect the environment by using less detergent. The function of medical products can be extended by means of hydrophobic structures (dosing cups), structures which reduce friction (catheters) or structures which support cell attachment, thus opening up new product ranges. Structured surfaces can also have a positive effect on flow behaviour, saving aviation fuel, for example, or they can be equipped with optical structures for e.g. holograms (product or security markings). Anti-reflective surfaces are another possible area of application. Such functional surfaces are usually manufactured in multi-stage processes in which the functional structure is applied retrospectively. Single-stage processes are usually restricted to smooth surfaces and flat geometries, such as CDs and DVDs.

New production chains are being developed in order to avoid the limitations imposed by current production technology. These combine complex and precise micro- and nanostructuring with replicating methods for mass production processes. The functional microstructured and nanostructured 3D-surfaces are produced directly and quickly together with the macro part by means of wear-resistant structured tools together with an adapted primary shaping process. The examined process chain therefore comprises the manufacturing of the micro- and nanostructures in the moulds, the wear-resistant, anti-adhesive coating of the moulds and the replication by adapted injection moulding, extrusion and precision investment casting processes. Particular consideration is given to the reciprocal effects of the individual process steps and their repercussions.

Two different approaches have been taken with regard to the production of the wear-resistant, structured tools: for structure sizes < 5 µm, the tool is initially coated and then structured, while for structure sizes > 5 µm, the tool is initially structured and then coated. Both approaches require coatings which have a non-adhesive effect on the polymer or wax melts used. Coating systems must be suitable in terms of wear resistance, anti-adhesion and laser structurability respectively coatability of the laser-structured tools. For laser structuring this involves the examination of fast and precise micro- and nanostructuring processes. In order to guarantee accurate contours the ablation process needs to be melt-free. Also strategies for large-scale, seamless structuring must be developed in order to apply structures to the three-dimensional moulding tools without any visible transition.

Manufacturing plastics components with microstructured surfaces in a single primary shaping process requires a variothermal temperature control of the mould. To this temperature control of the moulds heavy demands are placed to ensure both the combined plastics primary processing and forming at melt temperature, and the calibration and cooling of the shaped or formed plastic melt. Two variothermal heating concepts, an inductive and a laser-based heating system, are established and tested for both injection moulding and extrusion embossing. The replication processes used to manufacture plastics components are adapted to the moulding of functional structures.

The reproducible replication of metal components with functional surfaces by means of precision investment casting requires the qualification of the individual moulding steps and specification of their accuracy. Various materials are tested and evaluated for all moulding steps beginning with wax models up to the actual cast.

Within the here presented work coatings in the $(Cr_xAl_{1-x})N$ system, which are deposited using Physical Vapour Deposition (PVD), are also optimised and qualified. These coatings reduce the total surface energy in comparison to the uncoated tool insert. The adapted surface energy reduces the adhesion of the plastics melt. It is both shown, one the one hand an adjustment of the structuring processes allows the precise laser structuring of these coatings and on the other hand the complete coating of previously laser-structured tool inserts with structure sizes greater than 10 µm (width and depth). Ps-multi pulses are used to improve the laser microstructuring of the tool inserts both in terms of the ablation speed and the surface quality. Suitable strategies are used to apply these structures seamlessly over a large area.

Two variothermal heating systems are designed, constructed and evaluated for both investigated primary shaping processes. Heating rates of up to 60 K/s are achieved in the injection moulding process with inductive heating. The newly developed laser-based system consists of a 2.7 kW diode laser, a mould with an integrated collimator lens and a transparent quartz glass mould insert. This allows the direct heating of the cavity and achieves heating rates of up to 300 K/s.

Both variothermal heating processes are simulated for the temperature control of the embossing roll in the extrusion embossing process. The results are transformed into a master curve which can be used to predict and set all the heating parameters. The embossing used for the microstructures in the plastic melt is also simulated in order to determine the causes of any inaccuracies in the moulding process resulting from elastic deformation, entrapped air or premature solidification of the melt.

Moulding experiments are performed for injection moulding and extrusion embossing. The production of superhydrophobic surfaces requires the structure on the mould insert to be a combined micro- and nanostructure. The combination of a suitable ductile material and variothermal process control with the ability to reach high temperature gradients makes it possible to produce very filigree and highly functional surfaces as shown in Fig. 5.175. The functionality in this example is not produced by means of a highly precise moulding, but rather a systematic deformation of the structure during the demoulding process.

In precision casting processes structures with a minimum structure size of 10 µm (width and height) can be moulded. A silicone or polyurethane die is used to produce a wax model which can be demoulded without damaging the microstructure. The wax model is then embedded in a gypsum bonded investment material. This moulding step is the main source of errors with regard to moulding accuracy due to the porous surface of gypsum bond. The degree of porosity depends mainly on the formation of the gypsum crystal mesh and limits the modelling potential of the microstructure in the precision casting process. After melting the wax out of the mould and the subsequent firing process, the aluminium melt is cast in the available cavity. By

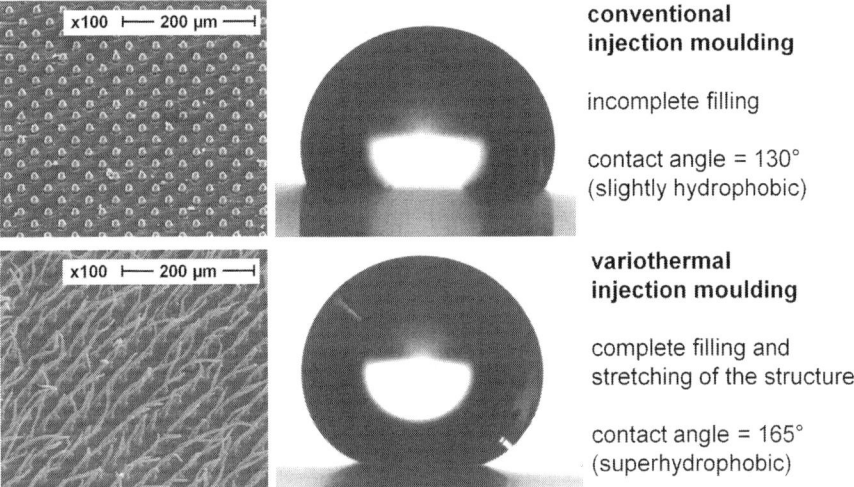

Fig. 5.175 Comparison of the moulding accuracy of a microstructured mould insert with and without variothermal heating. Due to variothermal heating it is possible to produce superhydrophobic surfaces

adjusting the parameters casting temperature, moulding temperature and casting pressure, it is possible to fill the microstructures on the surface of the mould.

The integrated precision investment casting chain, from the laser-structured master pattern to the finished cast, is demonstrated for an air restrictor. This flow-related component from the racing sport is given a shark skin-like structure on its inner surface in order to reduce air resistance.

5.6.2 State of the Art

Microstructured surfaces have not only since the discovery of the lotus effect provided enormous technical and economic potential. The prospective applications as well as the different structures are versatile. Examples of this microstructures have been copied from the botany or wildlife. Examples are the lotus effect or the shark skin effect. The lotus effect results in a very hydrophobic surface (Barthlott and Neinhuis 1997). Such structured surfaces show very good self-cleaning properties when they come in contact with water, as the dirt particles are easily removed. The shark skin structure improves the frictional resistance of bodies in fluids (Tian et al. 2007). It can be used to reduce fuel consumption of aircraft and vehicles. In general microstructured surfaces provide functional properties of the component that exceed the respective material properties of the component.

Processes for producing micro- and nano-scaled surface structures can be subdivided into three types (Bruzzone et al. 2008), deposition, ablation and forming/primary shaping processes. In the deposition processes, e.g. CVD (chemical vapour deposition), PVD (physical vapour deposition) or UV-cured varnishing, structures are deposited on a surface. Ablation processes include e.g. etching, lithography, ultra-precision machining, micro-erosion, grinding, laser ablation and electron beam processes. By local ablation of material the microstructure of interest is generated on the surface. In the primary shaping processes, an existing surface structure is directly transferred from the structured mould to the moulded component. In the forming processes, on the other hand, an existing surface is moulded locally to provide the required structures. Examples of primary shaping and forming processes include molecular migration, casting, rolling, extrusion embossing, injection compression moulding and injection moulding. The processes mentioned above can alternatively be subdivided into direct structuring and moulding technologies.

Plastics components with functional surfaces are often manufactured in several process steps. Once the component has been formed, the surface is functionalised by coating, etching processes or plasma treatment. Another alternative for producing e.g. hydrophobic surfaces is to use additives or material combinations (Klaiber 2010). These process chains, characterised by many individual steps, are time consuming, cost intensive and need a lot of resources. This inhibits the economic production of large quantities. It also makes it difficult to recycle the product as separating particles and substrate materials is an elaborate process (Barberoglou et al. 2009; Callies et al. 2005; CH 268258 1950; DE 102 10 668 A1 2003; DE 19715 906 A1 1997; JP 11171 592 1999; US PS 5599 489 1997). For mass production of components with micro- or nanostructured surfaces moulding processes are best suited. Here due to short process times large quantities can be produced (Bläsi et al. 2002; Michel et al. 1993; Rupprecht et al. 1997; Wagenknecht et al. 2006; Weber and Ehrfeld 1998). Moulding micro- and nanostructures in this way requires the adaptation of the relevant primary shaping process, e.g. injection moulding, extrusion embossing and investment casting. It also requires micro- or nanostructured moulds which provide a high level of stability and yet low adhesion to the material to be moulded. This reduces the demoulding forces and increases the durability of the moulds.

Micro- and nanostructures on tool surfaces can be produced by different means. In this regard, laser structuring exhibits great advantages over other processes as a result of its flexibility and the ability to machine without any additional tools. There are three possible ways of producing structures on tool surfaces by laser structuring: machining with a moving laser beam, e.g. scanning technologies, mask machining and holographic lithography (Duarte et al. 2008). Holographic lithography is an effective means of producing large areas with periodic structures of high resolution, i.e. with very small structure sizes, especially in the range of few µm to less than 1 µm (Campbell et al. 2000). Conventional techniques use two or more coplanar beams to produce one-dimensional linear structures. Several symmetrically aligned beams are used to produce two-dimensional structures in the interference plane. A setup with fixed optical path allows the production of a single structure shape and size. The optical path must be changed in order to obtain different structures. Recent approaches attempt to produce different textures on the workpiece using phase-manipulated multi-beam lithography (Klein-Wiele et al. 2003; Kondo et al. 2001; Yang et al. 2008). The intensity distribution can also be altered by changing the polarisation. Parameter-controlled holographic lithography is becoming a very common and suitable tool for surface structuring (Escuti and Crawford 2004; Su et al. 2003; Vita et al. 2007).

Scanning processes with highly dynamic deflection systems are used for larger and particularly non-periodic structures. One reason for the increasing use of laser structuring is the high surface quality, comparable to that of eroded surfaces (Gillner et al. 2008a). This and the generally wear-free machining make laser structuring a versatile and economically relevant machining technology. Due to the latest development of laser beam sources in the field of high performance ultra-short pulse lasers, including the use of multi pulses instead of single pulses, it is possible to produce fast precise, melt-free surfaces with roughness values in the range of $Ra < 0.5$ μm (Gillner et al. 2008a, b; Hartmann and Gillner 2007).

Surface coatings have proven to be very efficient in protecting laser structures and reducing demoulding forces generated by the moulding process. In recent years, new coating systems have been developed that meet the requirements for processes like injection moulding. Such coating systems can be applied by means of e.g. vapour deposition, whereby for less complex geometries the PVD process is preferred to the CVD process. As it is possible to use process temperatures below 500 °C the heat treatment of the tools is in most cases not affected (Gornik 2005). The PVD coating has little influence on the form and dimensional accuracy of the tool surface. Changes in dimensions are less than 5 μm and the achievable roughness value is $Ra < 0.5$ μm (Menges et al. 1999). Therefore, anti-adhesive coatings play a major role. Particularly in the area around the plasticising unit and the mould inserts, the surfaces which come into contact with polymer have considerable influence on the product quality. The production of optical plastics components, such as glasses, lenses or automotive glazing is subject to high quality requirements. If melt sticks to the conveyor screw when the plastics granules are melted, its dwell time in the plasticising unit is increased which causes thermal degradation of the material. If this material is picked up by the melt flow, it creates component defects such as flow marks or nibs. Very adhesive material also makes it difficult to clean the conveyor screw (Bobzin et al. 2007). Anti-adhesive coatings can provide a solution to such problems. As illustrated by (Bienk and Mikkelsen 1997; Bobzin et al. 2007; Cunha et al. 2002; Heinze 1998), many different coating systems have already been deposited on components of the plasticising unit for optical products and tested in the field. By using the developed TiN, TiAlN, TiAlON, CrN, CrAlN and CrAlON coating systems on substrate materials typical to this field of application, it was possible to reduce the number of defective parts. Furthermore, the cleaning of the plasticising unit components is considerably simplified (Bobzin et al. 2007).

Especially, surfaces of the mould inserts have to be as resistant as possible to wear and adhesion of the plastics melt. Specifically adapted surface coatings are essential, for example those based on titanium nitride (TiN), titanium aluminium nitride (TiAlN), chromium nitride (CrN), chromium carbide (CrC) and carbon based coatings. One application is the production of reproducible, high quality textured surfaces for special polish effects on design parts in the automotive industry while ensuring longer tool life (Mumme 2002; Neto et al. 2009). Tough, corrosion-resistant, smooth and friction reducing coatings on the mould surfaces protect polished and microstructured surfaces. In addition they allow the smooth flow of the melt, which has a positive effect on the quality of the moulded parts as well. The low tendency of the moulded part to stick to or to shrink on to mould and the lower adherence of residue ease demoulding compared to uncoated moulds (Bobzin et al. 2009a, Van Stappen et al. 2001). Coating the surface also affects the solidification of the melt and therefore improves the filling of the mould and the moulding accuracy. This can be used in order to produce dull surfaces on plastic components, and therefore avoid the need to paint it later on (Mumme 2004). PVD coatings in low temperature processes increase the range of possible applications of sensitive tool steels like 1.2767 or 1.2083 or aluminium- and copper-forging alloys in mould construction. The processing of various plastics is made considerably easier because less cleaning and maintenance is required. The lubricant-free operation of the moulds allows for a problem-free finishing of the plastics parts. The variety of materials which can be used and the low coating temperatures of 160–500 °C also make it possible to use PVD coatings on extrusion dies. Titanium nitride (TiN), titanium carbon nitride (TiCN), titanium aluminium nitride (TiAlN), titanium aluminium nitride with carbon top layer (TiAlN/C) and chromium nitride (CrN) coating systems are predominantly used (Seibel 2005).

In addition to the replication of microstructures on a large area, three-dimensional micro- and nanostructures also pose particular challenges for the demoulding process. Apart from the complex determination of the optimum process parameters, one means of resolving these demoulding issues is optimising the draft angle, although this will alter the geometry of the microstructure. It is likewise possible to use certain additives which are cost-intensive and can have negative effects e.g. on the mechanical properties of the components. A further solution is provided by applying an anti-adhesive coating to the cavity surface suitable for the particular plastics and particular mould material. It is therefore possible, as discussed above, to not only improve the moulding and demouldability but also increase the tool life of the sensitive microstructured surfaces up to twenty-fold (Menges et al. 1999).

Replication processes play the most important role in producing microstructured plastics components since they are easy to integrate into the process sequences of mass production forming processes. Additionally, processing time and costs of direct structuring as a production step is usually disproportionately large compared to the other manufacturing steps. The most important criterion for classifying moulding processes for transferring micro- or nano-scaled surface structures to plastics components is whether the transfer of the structures takes place on a continuous or discontinuous basis. The discontinuous processes include injection moulding, injection compression moulding and hot embossing. The continuous processes include roller embossing, extrusion embossing and belt embossing. Discontinuous processes are suitable for manufacturing microstructured parts with confined dimensions but also partially for three-dimensional surfaces. Continuous processes, on the other hand, enable economic manufacturing of large-scale microstructured surfaces on films and sheets. The microstructuring tool can be substantial smaller compared to the structured surface. In between these processes lies the belt embossing processes used for nano-imprinting, whereby instead of a circular belt, a flexible elastometric stamp is used which is pressed onto the component to be embossed by means of several horizontally moving rollers (Seo et al. 2007; Youn et al. 2008).

5 Hybrid Production Systems

Common to all these processes is that the microstructures are applied at temperatures around or even above the softening or melting range of the plastic as the perfect moulding of the microstructured surfaces is otherwise impossible. However, there is a significant difference in whether the shaping of the structures is performed in first or second heat parallel to the primary shaping of the macroscopic component (single-stage process) or in a subsequent second forming step (two-stage process). Roller and belt embossing can be used as a two-stage process in which a semi-finished product is first produced, calibrated and cooled. Thereafter it receives its structured surface following a second heating. However, it can also take place as a single-stage process in which the calibration and the embossing are performed by the same embossing belt or pair of rollers. The latter process option is referred to as extrusion embossing and investigated in more detail. The benefits of the single-stage process include a shorter process chain and therefore saving of plant components. It also reduces the thermal load placed on the plastics by only melting it once. Additionally a better moulding of the microstructured surface and a reduction in the mechanical load on the surface of the microstructured embossing roll due to the lower viscosity and rigidity of the plastics surface to be formed are achieved.

It is also possible to produce structures within a micrometer range by embossing. Plastics sheets or flat components are mainly machined for this purpose. In the so-called hot embossing, plastics, which has already been given its macroscopic geometry, is heated to a temperature above the glass transition temperature T_g. A stamp with the negative microstructure of the desired pattern is used to emboss the component. Embossing forces are between 700 N (Youn et al. 2008) and 1000 kN (Herzinger 2004). Youn et al. (2008) introduce a system which uses a pressure roller to perform the embossing, while (Herzinger 2004) presents hot embossing with two stamps. The disadvantages of these processes are discontinuous operation and restriction to a limited embossing surface. In newer types of machines (Herzinger 2004) up to $3\,m^2$ which can be embossed at once. Kimerling et al. (2006) use a method that uses a variothermal process control known from injection moulding. The surface of the stamp is first heated using infrared radiation. Subsequently the embossing process is performed. The stamp is heated from the inside to a temperature below the softening temperature so that the plastics cools again on coming into contact with the forming surface of the embossing stamp. Mäkelä et al. (2007), Yeo et al. 2009) and (Hennes 2000) all present a continuous embossing process. A plastics film is embossed between two structured rollers. In (Menz et al. 2006), the film is pre-heated by means of an infrared emitter before being formed in the embossing gap. Typical machining speeds are in the range of 10 m/min.

In injection moulding, the high mould temperatures can often only be achieved by dynamic heating due to process related technical reasons. Integrated ceramic heating elements permit heating rates on the surface of the mould of up to 25 K/s (Bürkle and Burr 2007). The cavity is thereby indirectly heated by thermal conduction. Thus, the heating element requires elaborate thermal insulation (Bürkle and Burr 2007). Thermocouples placed a few millimetre away from or behind the cavity are used to measure the temperature inside the cavity. It is not possible to directly regulate a target temperature on the cavity surface as the system features a dead time due to thermal conduction. During the cooling phase, the heat must also overcome the area of the auxiliary heating in the mould plate in order to be conducted into the mould (Ridder et al. 2009). This has a negative effect on the maximum cooling capacity. The best results in terms of short cycle times with high heating and cooling rates is therefore achieved by using external auxiliary heating. For this reason a heating source is positioned in an opened mould in front of the cavity which specifically heats the cavity surface. With this method high heating rates can be achieved. However, it necessitates a precise and highly repeatable positioning of the external heating element in front of the cavity in order to achieve the same thermal conditions on a cyclical basis. With external heating systems, the temperatures balance more quickly in comparison to internal systems as no thermal barriers are built in the tool (Engel 2007; Giessauf et al. 2008; Hinzpeter 2008, 2009).

In contrast to the direct production of surface structures in the primary shaping process, two-stage processes are currently often used for plastics components in which the component or semi-finished product is first produced and then covered with a sub-millimetre thick, UV-curing lacquer coating into which the structure is then embossed (Nezuka et al. 2008; Vogler et al. 2007). The rapid curing of the lacquer coating at room temperature prevents a second heating of the plastics component. However, the disadvantages related to the two-layer structure of the microstructured part and the limited choice of materials must be taken into account.

In the here examined extrusion embossing process, a structured cooling roller (Bärsch 2001) is used to emboss a microstructure into the melt film which is then calibrated (Bagcivan et al. 2010; Michaeli et al. 2009, 2010, 2010b; Michaeli and Scharf 2009). The smooth cooling roller is replaced by a microstructured embossing roller. The surface of the embossing roller has a negative copy of the topography to be formed. This embossing roller is tempered like a conventional cooling roller and differs only in the structure on the roll shell. The melt exits the tool as a film and enters the gap between the embossing roller and the counter pressure roller. There the structure of the embossing roller is pressed into the still molten polymer. The melt remains on the rotating embossing roller and is cooled until solidification. After solidification the foil can be removed with stable microstructures on the surface. The pressure exerted by the rollers in the embossing gap aid the shaping of the microstructures by pressing the plastics into the microstructures of the embossing roller (Bärsch 2001). The patent specification (Coyle 2007) uses a dual heating of the embossing gap. The microstructured embossing roller is heated to a temperature above the glass temperature T_g and the counter pressure roller on the opposite side is tempered to a temperature below T_g. The non-structured side therefore cools down more quickly and gives the film its geometrical stability. The film side which is in contact to the hotter embossing roller on the other hand, is able to mould the structures well. Another method (Coyle 2007) uses a carrier film which is guided into the embossing gap along with the melt. In order to avoid the problem of high temperature gradients over the film thickness or the use of a carrier film, various scientific papers (Brinksmeier and Michaeli 2005; Fink 1997) discuss the variothermal temperature control of the embossing roller. The aim of the variothermal heating of primary shaping and forming tools in the embossing of microstructured surfaces is to raise the temperature of the tool surface above the melting range of the plastics shortly before coming into contact with the plastic melt. After completion of the moulding of the microstructures the temperature is reduced considerably to a lower demoulding temperature. In discontinuous processes such as hot embossing, injection moulding or injection embossing, such variothermal tempering processes are already used in industry. Variothermal heating is not only used for the moulding of microstructured surfaces or the manufacturing of microscopic components, but also for improving the surface quality, surface polish or avoiding optical imperfections as a result of weld lines (Engel 2007; Giessauf et al. 2008; Hinzpeter 2008, 2009; Liu et al. 2008; Mapleston 2008; Ridder et al. 2009). All currently employed variothermal heated tools possess a classic fluid tempering system which controls the base temperature respectively the cooling of the tool. A temporary local or overall increase in the tool surface temperature is created by means of a second heating system which is either integrated in the tool or positioned externally.

Industrial production of microstructures on plastics films in one process step is nearly not implemented. This is to a great extent due to the difficult production of the negative pattern on the embossing roller (Brinksmeier and Michaeli 2005). On the other hand, the production of microstructures on compact components in the injection moulding process has already been implemented in industrial manufacturing (Giessauf et al. 2008).

In addition to producing microstructured plastics components the structuring of metal surfaces is also of great interest. Investment casting in its simplest form was used 2500 years ago and is therefore one of the oldest production process. In modern casting technology for the replication of microstructures, a wax model which acts as a positive model is created by injecting wax into a laser-structured master pattern. In the following process step, this wax model is embedded under vacuum in a lost mould made of gypsum bonded (negative mould). When the mould has gained the maximum green strength by drying, the wax is removed in a furnace and the mould is fired in order to gain the final strength. When the firing process is complete, the metal is cast. After solidification and cooling of the metal, the mould is removed. Gypsum bonded moulds can be dissolved in water.

All research conducted regarding the moulding accuracy of micro-components and microstructured surfaces are done for the investment casting process which is capable of a high precision of components. Different publications show the possibility of casting micro-components in aluminium/zinc alloys (Baumeister et al. 2008; Yang et al. 2010), high-grade steel alloys (Charmeux et al. 2007) and gold/silver alloys (Wöllmer 2000) with geometric characteristics between 700 µm and 50 µm at low tolerances and a high surface quality (Ra between 0.5 and 1 µm). Schmitz and Grohn proved for the first time the mouldability of microstructured surfaces in superalloys on a dummy turbine blade with a microstructure of 50 µm hight and an aspect ratio of 1 (Schmitz et al. 2004, 2007).

Other attempts have been made to cast alloys with a low melting point like bismuth-tin and bismuth-lead alloys. Here, the usual precision casting process chain is replaced by casting directly microstructured plastic or quartz moulds. Due to the high surface quality of the moulds it is possible to reproduce narrow channels with a width of 4 µm and a depth of 200 nm (Baumeister et al. 2008) or riblets at intervals of 400 nm (Cannon and King 2009).

5.6.3 Motivation and Research Question

As illustrated in the previous chapter, conventional methods for manufacturing components with functional surfaces based on micro- and nanostructures exhibit considerable disadvantages. For this reason, process chains for manufacturing these structures together with the component in a single primary shaping process have to be investigated.

The whole process chain is examined, beginning with the production of the tools for the primary shaping process to the functional component itself. Since the individual steps in the process chain influence each other, the process chain must be examined in all. The examined process chains are extrusion embossing and injection moulding for plastics components and investment casting for metal components. The resulting research question is therefore:

> How can process chains for manufacturing metal or plastics components with functional surfaces be shortened by means of structured and coated tools?

In order to address this research question, the process chains are divided into four different areas: the production of wear resistant, anti-adhesive, micro- or nanostructured tools, injection moulding, extrusion embossing and precision casting. The area of tool production is again divided into two machining sequences according to the desired structure size. For structure sizes greater than approx. 5 µm, the tools are first structured and afterwards coated. For smaller structure sizes, the tool is first coated and then the micro- or nanostructure is generated into the coating.

The objective of laser microstructuring is to produce microstructures of between 5 and 200 µm respectively < 1 µm seamless and large-scale on three-dimensional surfaces. The structures of larger sizes are structured directly into the tool steel using scanning technologies while the structures of smaller sizes are structured in the wear-resistant, anti-adhesive PVD coatings. As the structures are molded afterwards, particular attention must be paid to avoid undercuts. In order to enable a spectrum of functional effects as broad as possible, an aspect ratio of one or more is investigated.

The structures which are generated by laser structuring increase the surface area of the tool. This increases the effect of adhesion of the moulding material to the tool, which can affect the demoulding of the component negatively. The small structures are also exposed to considerable wear from abrasion and adhesion. In order to address these challenges, it is necessary to develop hard material coatings which reduce not only the adhesion of the moulding mass but also the wear of the moulding tool. Therefore, various coating systems made by PVD technology are deposited on the tool steel 1.2083 which is commonly used in the plastics industry. The coatings are examined regarding their adhesion towards various plastics. Additionally, the coating process is optimized with regard to the coatability of microstructures in the order of magnitude of 10 μm. The production of a coated, microstructured tool with a structure size less than 5 μm also requires an adaption of the coating structure and coating morphology due to the laser structuring process.

The examination of the primary shaping processes for plastics components with functional surfaces includes both injection moulding and extrusion embossing (Bagcivan et al. 2010; Michaeli and Scharf 2009; Michaeli et al. 2009, 2010b, e). The moulding of structured surfaces with thermoplastic materials using conventional process technologies is possible but restricted. In addition to reducing the mechanical load which is applied on the forming tool and affects the durability of the microstructured surface, the temperature control is of crucial importance to the moulding process. A variothermal process considerably increases the moulding quality without substantially lengthening the process time. However, the variables in the individual process and the interdependencies between the individual processes increase the complexity of the whole process chain.

The here examined variothermal extrusion embossing is a continuous process used for the economical production of films with microstructured surfaces in a single process step (Bagcivan et al. 2010; Michaeli and Scharf 2009; Michaeli et al. 2009, 2010, 2010b). In parallel to the development of the required temperature control systems, the extrusion embossing process for moulding of microstructures is analysed and optimised (Michaeli et al. 2009, 2010, 2010b). On the one hand, the theoretical understanding of the extrusion embossing process is achieved by analytical and numerical observation (Michaeli et al. 2009). On the other hand, the challenges of the real process need to be explored, understood and overcome. Particularly important are the interactions between the extremely complex temperature control and the geometry of the microstructures which have to be moulded with respect to the parameters of the extrusion embossing process. Only with knowledge of the extrusion embossing process it is possible to achieve a good, accurate moulding of high quality. The numerical and experimental approach should therefore combine knowledge and experience regarding the production of microstructured film surfaces in the extrusion embossing process.

Due to increasing the temperature of the moulding or embossing roller to the range of the melting temperature directly before moulding good results can be achieved in both injection moulding and extrusion embossing of micro- and nanostructures. By a variothermal process high temperatures before moulding with sufficiently low temperatures for demoulding are realised while keeping cycle times short. High power transfer rates are provided by heating processes based on electromagnetic induction or laser radiation.

Variothermal mould heating processes are therefore developed using induction and laser radiation. Variothermal injection moulding tests are performed using these system technologies in reproducible test conditions (Bekesi et al. 2010; Michaeli et al. 2008, 2010a, c; Michaeli and Klaiber 2009).

For the generation of microstructured surfaces on metal components in a replicative manufacturing process the microstructure is moulded from a laser-structured master pattern. This eliminates the labour-intensive process step of the subsequent microstructuring each component separately. Within this work investment casting is examined which is a precision casting process and therefore suitable for replicating microstructured surfaces.

Basically it is possible to cast microstructured surfaces of a high mould quality. No profound process understanding of the entire precision casting chain with regard to the production of structured surfaces has been available. The process- and material-related variables are thus examined with regard to the quality of the microstructured surface of the aluminium casting. The modelling accuracy of the microstructures is analysed for all moulding steps from the laser-structured master pattern to the aluminium cast. The usability of the process with regard to the manufacturing of real components is presented by means of a demonstrator. For demonstration of the process chain an air restrictor for a formula student racing car is moulded with a technical shark skin on the inner surface.

5.6.4 Results

5.6.4.1 Manufacturing of Structured, Wear-Resistant Tools

There are two ways to manufacture structured, wear-resistant tools: either the tool is first structured and afterwards coated or the tool is first coated and afterwards structured. Basically, both ways of production are feasible, the choice of the order of the coating and structuring processes depends on the structure size and the coating thickness of the wear-resistant coating. The here presented experiments use coating thicknesses between 1 and 5 μm. Therefore the process orders are allocated as follows (see Fig. 5.176):

- first structuring, then coating for structure sizes > 5 μm
- first coating, then structuring for structure sizes ≤ 5 μm

The structuring and coating process have to be adapted for each case.

Laser structuring The investigations in the field of laser structuring are divided into several areas. On the one hand, the structures have to be produced on the moulds. As

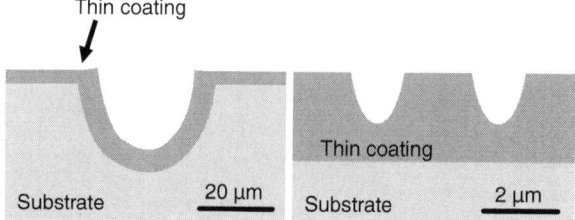

Fig. 5.176 Two machining orders for the production of microstructured, wear-resistant tools. *Left*: for structure sizes > 5 μm, the tool is first structured, then coated. *Right*: for structure sizes < 5 μm, the tool is first coated then structured

Fig. 5.177 It is possible to generated pulse bursts with the used ps-laser. The pulses within a burst have an interpulse separation of $\Delta t = 20$ ns or $\Delta t = n \times 20$ ns

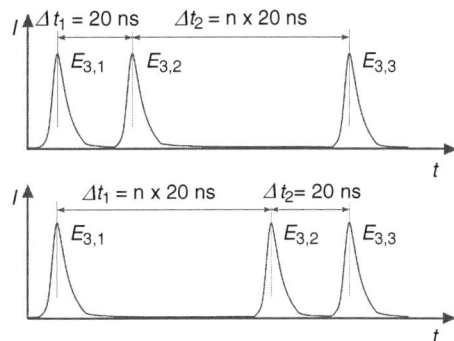

illustrated in Fig. 5.176, this means either direct structuring of the metal mould or the previously applied wear-resistant coating is structured. Therefore suitable processes for a fast and precise micro- and nanostructuring of the used materials are examined. On the other hand, these structures have to be applied on the tools on large areas. Therefore a large-scale, seamless structuring strategy is required.

For laser structuring two different approaches are used depending on the structure size. Structures > 5 μm are machined using scanning technologies, structures < 5 μm are produced by holographic lithography. For the laser structuring on the one hand two frequency-tripled ps-Nd:YAG MOPA Lasers (Master Oscillator Power Amplifier; Rapid resp. Hyperrapid, LumeraLaser GmbH) are used. The lasers run at a wavelength of $\lambda = 355$ nm and a pulse duration of $\tau = 12$ ps. Due to the special design of the pockels cells these lasers can generate pulse bursts. At a given repetition rate two pulse packages can be generated with a separation on n-times 20 ns. Each pulse package itself can consist of several pulses with a puls duration of $\Delta t = 20$ ns. These two packages together form a pulse burst. The energy of the pulse burst E_{burst} is the sum of the energy of all pulses within the burst E_{pulse}, see Fig. 5.177. Due to the use of pulse bursts the ablation rate can be increased by 90% compared to single pulse ablation (Gillner et al. 2008a, b; Hartmann and Gillner 2007).

On the other hand a frequency-tripled ns-laser (Q301, JDS Uniphase Corporation) is used at a wavelength $\lambda = 355$ nm and a pulse duration of $\tau = 35$ ns. The specifications of the laser sources are listed in Table 5.12.

When nanostructuring with holographic lithography, the interference pattern is directly used for structuring. The geometry and size of the structure are defined

Table 5.12 Technical data of the laser sources used for structuring

Laser	Rapid	Hyperrapid	Q301
Manufacturer	LumeraLaser GmbH	LumeraLaser GmbH	JDS Uniphase Corporation
Wavelength (nm)	355	355	355
Pulse duration	12 ps	12 ps	35 nm @ 10 kHz
Repetition rate	1–500 kHz	400–1,000 kHz	1–100 kHz
Max. average power	1 W @ 500 kHz	20 W @ 1000 kHz	10 W @ 10 kHz

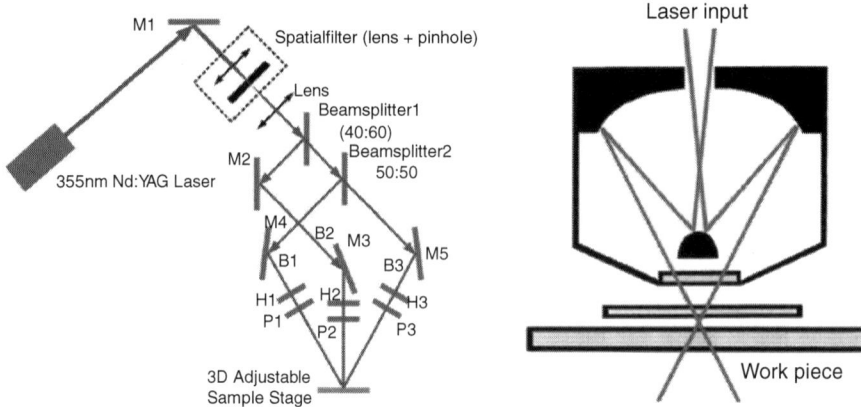

Fig. 5.178 Two different set ups for nanostructuring by holographic lithography. *Left*: adjustable set up with beam splitters for nanostructuring with ns-laser. *Right*: schwarzschild optics for nanostructuring with ps-laser

by the interference pattern. Two different set ups are used for nanostructuring. For structuring with the ns-laser a complete adjustable set up with beam splitters is used, as illustrated in Fig. 5.178, left. As the ps-laser has a very small coherence length, the ns-laser is used for the basic investigations. For the structuring with the ps-laser a setup with a reflective objective, a so-called schwarzschild optics, is used, see Fig. 5.178, right.

The schwarzschild optics consists of a smaller convex and a larger concave mirror. It is therefore free from chromatic aberration. Due to adjustment of the mirrors spherical aberration, coma and astigmatism can be corrected. The precise design of this optics makes it possible to use lasers with very small coherence lengths for holographic interference. In combination with a phase mask, which divides the beam into several partial beams, this optics generates an interference set up. The period of the interference pattern is defined by the numerical aperture of the lens, i.e. by the angle of the partial beams towards each other.

During microstructuring, the laser beam is fast moved by a galvanometer scanner (IntelliSCAN-DE, Scanlab AG) and focused onto the surface of the sample by an F-Theta lens. The ablated structures are linear and dimple structures. Both processes for reproducible manufacturing of micro and nanostructures on the tools and strategies for the seamless structuring of large areas are examined. The structures are analysed by white light interference microscopy (WIM), scanning electron microscopy (SEM) and atomic force microscopy (AFM).

Previous tests have shown that by using ps-double pulses instead of ps-single pulses the ablation rate can be can increased. These analyses have been extended to include triple pulses. One interpulse separation within the triple is fixed to 20 ns due to the set up of the laser, the other interpulse separation can be set to n-times 20 ns. By using ps-triple pulses instead of ps-single pulses the ablation depth per layer is almost doubled using the same total energy E_{burst}, see Fig. 5.179 left (Gillner

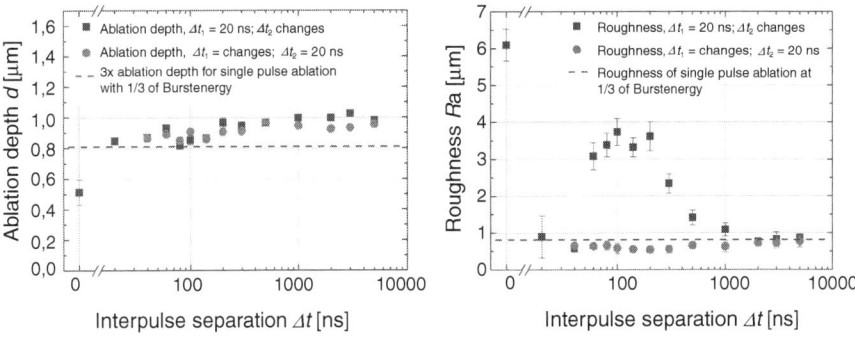

Fig. 5.179 *Left* ablation depth and *right* roughness of the ablated surface for the ablation with triple pulse with canged interpulse separations. The ablation depth is nearly doubled in comparison to single pulse ablation with the same total energy; a surface roughness of $Ra = 0.6$ μm is achievable. Interpulse separation of zero is ablation with single pulses of the same burst energy. (Gillner et al. 2008a, b; Hartmann and Gillner 2007)

et al. 2008a, b; Hartmann and Gillner 2007). Changing the value of the interpulse separation within the bursts Δt_1 and Δt_2 has not much influence on the ablation rate. Comparing the ablation depth using ps-triple pulses to the ablation depth using ps-single pulses with a third of the burst energy, i.e. the single pulse amounts to the same energy as every pulse of the burst, due to triple-pulse the ablation rate is increades by 25%. Furthermore machining with triple pulses is three times faster than machining with single pulses.

The value of the interpulse separation within the burst is important for the roughness of the ablated surface. For $\Delta t_1 = 20$ ns the roughness of the surface depends strongly on Δt_2. For $\Delta t_2 = 20$ ns the roughness of the surface is almost entirely independent of the impulse separation Δt_1, roughness values of $Ra = 0.5$ μm have been achieved, see Fig. 5.179, right, and Fig. 5.180, left (Gillner et al. 2008a, b; Hartmann and Gillner 2007). In comparison to machining with triple pulses, machining with a single pulses of the same burst energy produces a higher roughness on the surface, see

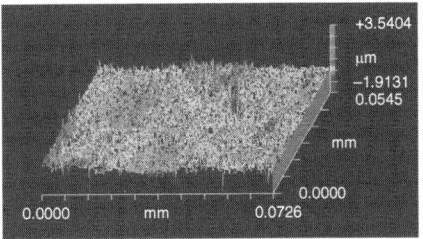 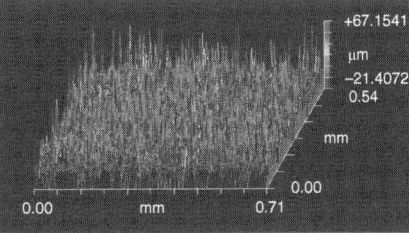

Fig. 5.180 WIM images of laser-structured surfaces with the same burst energy $E_{burst} = 16$ μJ. *Left*: machining with ps-triple pulses produces a smooth surface without any melt residue. *Right*: machining with ps-single pulses generates a rough surface. (Gillner et al. 2008a, b)

Fig. 5.181 Injection mould with dimple structure for moulding of parts with superhydrophobic surfaces. The micorstructure is superposed by a nanostructure (ripple structure)

Fig. 5.180 right. The use of ps-triple pulses can produce very good surface qualities comparable to the quality of surfaces produced by EDM (Gillner et al. 2008a, b).

When microstructuring with ps-lasers a ripple structure is generated on the machined surface. This structure is generated by interference of the laser radiation with surface waves and generates a structure with a periodicity in the range of the incident laser wavelength. This ripple structure is superposed generated on all structures which are generated on the tools. Figure 5.181 illustrates an injection mould used to produce superhydrophobic structures, see Sect. 5.6.4.3.

For nano structuring with holographic lithography, experiments are performed with double- and triple-beam interference, set up see Fig. 5.178, left. Additionally the effect of the polarisation direction of the partial beams is analysed. The intensity distribution of the interference pattern is simulated and compared to the results of structuring experiments. Direct nanostructuring of metals with a ns-laser creates too much melt on the surface and the nanostructures cannot be detected. Therefore photoresists or plastic films (in this case polyimide films) are used as tests samples for the experiments with the ns-laser. For the direct structuring of metals a set up with a ps-laser is required, see Fig. 5.178, right.

All interfering partial beams are directed under the same angle towards the normal of the surface on the sample. The direction of the polarisation of the partial beams is specified regarding the plan which is defined by the partial beam itself and the normal of the surface. The direction of the polarization of the partial beams has a strong impact on the intensity distribution of the interference pattern, see Fig. 5.182 (Huang et al. 2010).

Figure 5.183 illustrates the machining of polyimide films which are produced using the same settings for the polarization direction as simulated in Fig. 5.182. The simulated intensity distributions and the structuring results correspond well (Huang et al. 2010).

By changing the polarisation direction it is possible to adjust the intensity distribution of the interference pattern. It is possible to create holes, round, oval and rectangular burled structures in a hexagonal arrangement as well as uni-dimensional linear structures (Beckemper et al. 2010; Huang et al. 2010).

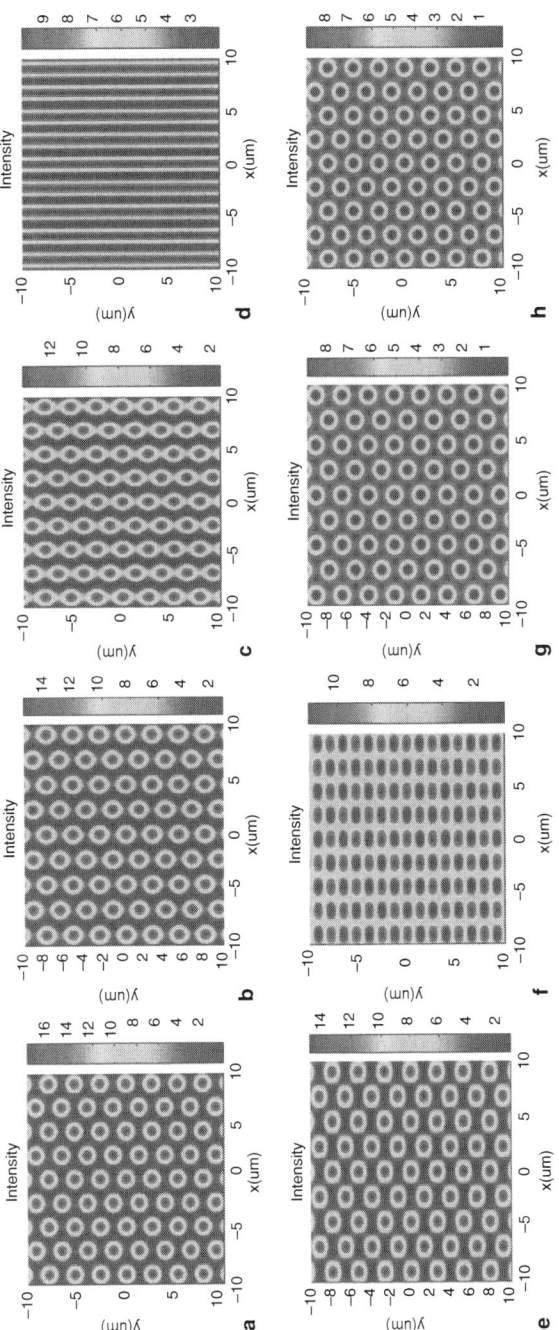

Fig. 5.182 Simulation of the intensity distribution of triple-beam interference with different polarization directions of the partial beams. Direction of polarisation of the partial beams: **a** (−30°, 90°, 210°); **b** (−30°, 135°, 210°); **c** (−30°, 150°, 210°); **d** (−30°, 180°, 210°); **e** (−30°, 120°, 270°); **f** (−30°, 135°, 300°); **g** (−30°, 150°, 330°); **h** (−30°, 210°, 450°)

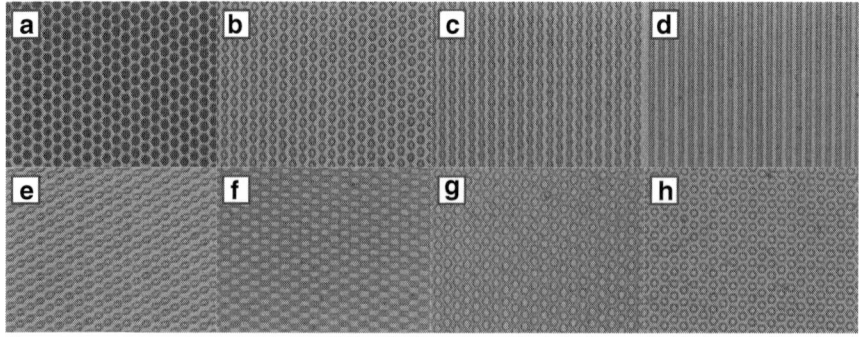

Fig. 5.183 Structuring of polyimide film with triple-beam interference with different polarization directions of the partial beams. Direction of polarisation of the partial beams: **a** ($-30°$, $90°$, $210°$); **b** ($-30°$, $135°$, $210°$); **c** ($-30°$, $150°$, $210°$); **d** ($-30°$, $180°$, $210°$); **e** ($-30°$, $120°$, $270°$); **f** ($-30°$, $135°$, $300°$); **g** ($-30°$, $150°$, $330°$); **h** ($-30°$, $210°$, $450°$)

The seamless structuring of large areas requires an adapted structuring strategy. Either an axis system is used to move the laser beam relative to the workpiece, or the laser beam is moved by a scanner. A movement of the laser beam towards the workpiece by an axes system is very precise, but also slow. The maximum speed is limited by the mechanical axes. When machining with scanners, the laser beam can be moved much faster, up to a few meters per second. However, this is limited to the scan field of the optics. The size of the scan field depends on the used optics. For the small structure sizes which are investigated here optics with small focus lengths are needed. These optics typically have scan files with sizes of a few millimetres to a few centimetres. The structuring of larger fields is performed positioning several scan fields next to one another. When only positioning the scan fields next to each other without any strategy the edges of the scan fields can easily be detected visually. This is because slight variations occur in the structuring over a scan field, which have different optical effects, see Fig. 5.184, left. These edges can be eliminated by a suitable strategy, see Fig. 5.184, right. In the dimple structuring illustrated in Fig. 5.184, right, the dimples at the edge of a scan field are randomly machined in an adjacent scan field. This makes the hard transition between two scan fields, which otherwise always happens at the same position in the grid next to each other, spread over an area and the edges of the scan fields are no longer visible. This strategy has to be adapted according to the pattern of the micro structure.

Coating Development As part of this project, coatings for plastics processing are developed using Physical Vapour Deposition (PVD). This technology uses an energy input to vaporise/atomise a solid raw material in a low pressure process. The addition of a reactive gas, e.g. nitrogen (N_2), causes a reaction and thus produces a coating system on the tool, also referred to as substrate. Two process variants are used which are differentiated by their different energy inputs. One option is the use of Arc Ion Plating (AIP) (Bobzin et al. 2009b). An electric arc is used to melt raw material, also

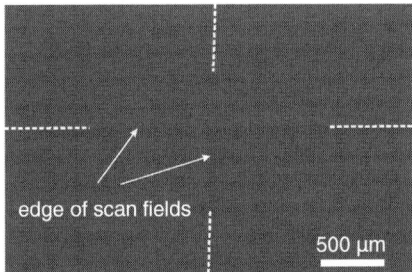 

Fig. 5.184 Adjacent scan fields for the machining of a dimple patterns. *Left*: scan fields are structured directly adjacent to one another without a suitable strategy—the edges of the scan fields are visible. *Right*: scan fields placed adjacent to one another with a suitable strategy. In an overlapping area, the machined dimples are randomly machined in one of the adjacent scan fields whereby the edges of the scan field get blurred

referred to as target material, which is distributed towards the substrate in a cluster-like and atomic manner. The other option is to use the Magnetron Sputter Ion Plating (MSIP) process. Ions which are generated by a glow discharge, sputter individual atoms of the target material which build a coating on the substrate. Based on the state of the art, both extensively used surface coatings TiN and CrN as well as (Cr, Al)N-based coatings are analysed which are very effective at increasing the mould quality and reducing the number of rejects (Bobzin et al. 2007). Fundamental analyses are performed to initially evaluate the coatings. The coating systems are analysed in terms of their chemical composition, their hardness and the Young's modulus. Contact angle measurements are used to determine the surface energy of both the coatings and the examined plastics at room temperature. It is possible to determine the atomic adhesion of various materials using the surface energy. However, due to the experimental set up, it is only possible to perform tests at room temperature and not at the processing temperature of the plastics. The exact experimental approach is presented in detail in (Lugscheider and Bobzin 2001, 2003).

The coatings are examined in a Zeiss DSM 982 Gemini Scanning Electron Microscope (SEM) with regard to their morphology and the coating thickness. As part of these SEM tests, EDX analyses are performed to determine the chemical compositions of the ternary and quaternary coating systems. A nanoindenter XP, MTS Nano Instruments, is used to determine the hardness and the Young's modulus. The penetration depth in this respect does not exceed 1/10 of the coating thickness. The measurements are evaluated according to the method of Oliver and Pharr (1992). A Poisson's ratio of $v = 0.25$ is assumed. The roughness values are determined using a Hommel Etamic T2000 tactile measurement system. Figure 5.185 illustrates the analysis results. The $Cr_{0.50}Al_{0.50}N$ coating, which is deposited by an AIP process and has shown good results in the preliminary tests for plastics processing (Bobzin et al. 2007), exhibits the highest hardness but also the highest roughness value. This indicates a good protection of the moulds against abrasive wear, but can also influence the mould quality of the intended micro- and nanostructures negatively. For this reason, two additional variants of the (Cr, Al)N system with different content

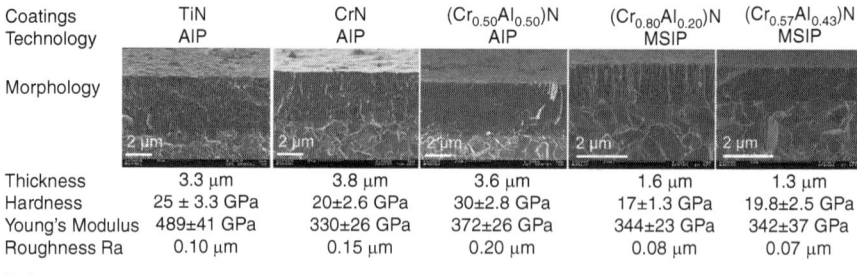

Coatings	TiN	CrN	$(Cr_{0.50}Al_{0.50})N$	$(Cr_{0.80}Al_{0.20})N$	$(Cr_{0.57}Al_{0.43})N$
Technology	AIP	AIP	AIP	MSIP	MSIP
Morphology					
Thickness	3.3 µm	3.8 µm	3.6 µm	1.6 µm	1.3 µm
Hardness	25 ± 3.3 GPa	20±2.6 GPa	30±2.8 GPa	17±1.3 GPa	19.8±2.5 GPa
Young's Modulus	489±41 GPa	330±26 GPa	372±26 GPa	344±23 GPa	342±37 GPa
Roughness Ra	0.10 µm	0.15 µm	0.20 µm	0.08 µm	0.07 µm
Reference:	1.2083, uncoated, HRC 51, Ra = 0.03 µm				

Fig. 5.185 Characterisation of the deposited coating systems with regard to chemical composition, morphology, thickness, hardness, Young's modulus and mean roughness value Ra

of aluminium are compared, which are deposited using MSIP technology. These coatings exhibit a lower level of hardness but also a much lower value of the mean roughness, which is decisive for the coating of structured surfaces. The variants are examined representatively for the (Cr, Al)N-MSIP coating system. It is possible to set the morphology and the mechanical properties of the coating by altering the content of aluminium.

For determination of the surface energies using contact angle measurements, the tests performed in the presented work use the OWENS, WENDT, RABEL and KAELBLE method for high energy surfaces and the WU method for low energy surfaces. Both methods are often used in practice and are based on the Fowkes division of the surface energy into a **d**isperse and a **p**olar portion:

$$\sigma = \sigma^d + \sigma^p \qquad (5.14)$$

The interaction forces in the phase boundary are thereby divided into the dispersion forces, which occur in all atoms and molecules, and the induced and/or permanently polar forces which only affect specific molecules and are attributable to the different electronegativity of the atoms. Generally, a strong interaction and an associated low interfacial tension only occur when similar forces exist. This means that at a phase boundary between a liquid and a solid, interactions only occur between the two polar or the two disperse portions of the surface energies. Firstly the surface energy of all surfaces is determined. In addition to the plastic variants, the uncoated tool steel 1.2083 (X42Cr13) and various coatings are tested. The results are illustrated in Fig. 5.186. The performed tests show that the selected PVD coating systems reduce the total surface energies, and particularly their polar portion in comparison to the uncoated metal surface. This leads to the conclusion that the adhesion forces decrease which should improve demoulding. This is particularly clear in the MSIP variants. It has to be noted that the surface energy is not only defined by the chemical composition but also the crystallographic alignment of the phase contents. In this respect the MSIP method seems to provide better results for this application. The effect of the roughness value can be neglected in these tests since all samples are prepared uniformly for the tests.

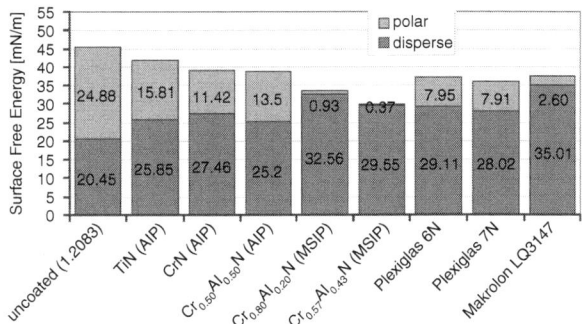

Fig. 5.186 Surface energies of the coating systems and the plastics used in injection moulding at room temperature. The coatings can be used to reduce the polar portion of the surface energy in particular

The adhesion energies of Plexiglas 6N, Plexiglas 7N and Makrolon LQ3147 on the coatings are determined according to the calculated total surface energies and their division into disperse and polar portions. The surface energy values calculated at room temperature are used for the plastics melt, which causes a systematic error. However, for the individual plastics, comparable qualitative statements can be made regarding their interactions with various surfaces. Furthermore, the adhesion energy is a measure of the adhesion of two materials. The adhesion energy or adhesion work W_A indicates the work per area required for the reversible separation of two immiscible phases at the phase boundary. According to Durpè (1869) the following applies to the adhesion energy W_A:

$$W_A = \sigma_l + \sigma_s - \sigma_{sl} \tag{5.15}$$

Whereby σ_l is the surface tension of the liquid, σ_s the surface energy of the solid and σ_{sl} the interfacial tension between the liquid and the solid. The combination with the YOUNG equation (Young 1805)

$$\sigma_{sv} = \sigma_{sl} + \sigma_{lv} \times \cos\theta \tag{5.16}$$

gives a correlation between wetting and adhesion energy and is referred to as YOUNG-DURPÈ equation:

$$W_A = \sigma_l + \sigma_l \times \cos\theta = \sigma_l \times (1 + \cos\theta) \tag{5.17}$$

As shown in Fig. 5.187, the adhesion energies of Plexiglas 6N and Plexiglas 7N regarding the examined MSIP variants lie significantly below the other coatings. This leads us to the conclusion that these coating systems are very well suited to improving demoulding.

Among all the coating systems analysed, the CrAlN-coatings manufactured using the MSIP process excel both in terms of their low adhesion properties and their low roughness values. The low roughness values of the coatings lead to the conclusion that microstructures of the tool surface are not negatively affected and the build-up of mould deposit in the injection moulding process is counteracted. Therefore in the next sections this coating system will be further investigated.

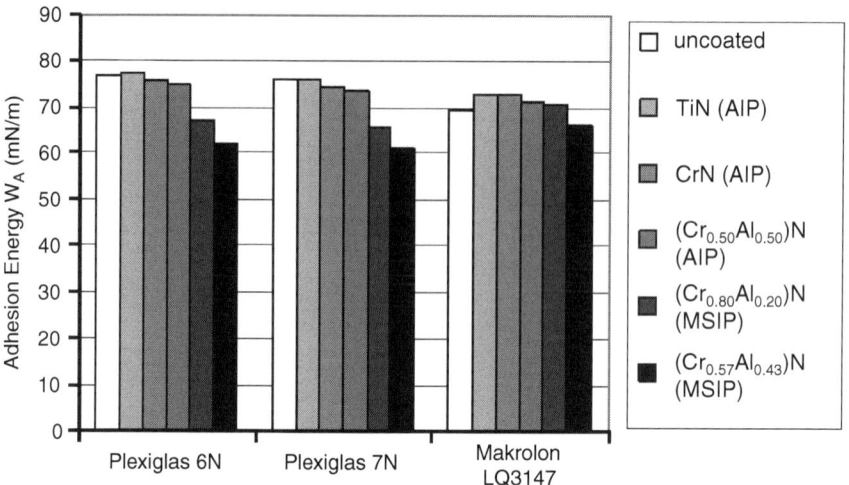

Fig. 5.187 Adhesion energies for Plexiglas 6N, Plexiglas 7N and Makrolon LQ3147 on the examined surfaces. The adhesion is particularly reduced by the MSIP variants

Laser Structuring of Coated Tools When laser structuring the already coated tool, it is possible to apply either very small microstructures using scanning technologies, or structure the coating using holographic interference. When structuring with scanning technologies, it is necessary to ensure that the structure depth is not larger than the coating thickness otherwise the function of the coating is compromised. The two structuring types are considered separately.

For microstructuring purposes $(Cr_{0.79}Al_{0.21})N$ and $(Cr_{0.66}Al_{0.32}Si_{0.02})N$ coating systems which are deposited using pulsed MSIP technology are used. The former corresponds to the system described above, but has been deposited with a greater coating thickness for the structuring test. Silicon (Si) is added to the second system in order to improve the laser structuring. It is known that by doping with silicon the morphology of the coating is affected. The impact on the structuring is investigated. Due to the improved structuring results on metals when using pulse bursts instead of single pulses, which are achieved using an infrared ps-laser, comparable structuring experiments are conducted on the coatings with UV ps-pulse bursts. The parameters burst energy (E_{burst}), number of pulses in the burst (n), number of passes (N) and scan speed (v) for the structures are varied in such a way that always the same cumulative energy is applied per structure. Figure 5.188 illustrates a cross section of a structure machined with double pulses processed with the identical parameters in the two coatings.

Both coatings exhibit a different ablation depth, which is probably due to the different absorption of the two coating materials. The structuring is very precise with ablation depths of a few 100 nm per pass. Also the coating is not further damaged during structuring (Bobzin et al. 2011). In the cross section fracture, a superposed nanostructure can be detected on the surface of the grooved structure, which is

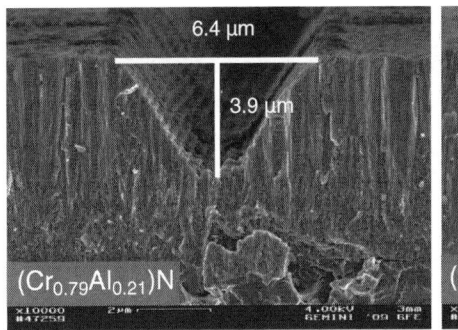 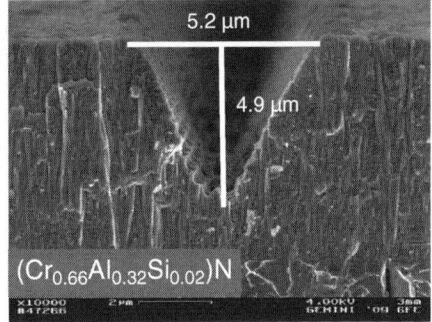

Fig. 5.188 SEM images of the cross section fracture of the coating system after laser structuring with the same parameters: $E_{burst} = 0.4\ \mu J$; $n = 2$; $N = 10$; $v = 100\ mm/s$; *left*: $(Cr_{0.79}Al_{0.21})N$; *right*: $(Cr_{0.66}Al_{0.32}Si_{0.02})N$

explained later. The ablation width and depth are analysed in order to examine the behaviour of the ablation as a function of the pulse bursts. Their behaviour is compared in relation to the number of pulses per burst for both constant burst energy and constant pulse energy of the individual pulses in the burst. The structure width decreases and the structure depth increases with increasing number of pulses per burst and constant burst energy. The total amount of ablated material increases with increasing number of pulses per burst, see Fig. 5.189 (Bobzin et al. 2011).

The structure width increases and the structure depth decreases when increasing the number of pulses per burst but keeping the burst energy constant. The total amount of ablated material still increases with increasing number of pulses per burst, see Fig. 5.190 (Bobzin et al. 2011).

For the structuring of PVD coatings by holographic interference, both structuring with ns-laser and ps-laser is investigated. When structuring the above qualified PVD layers with the ns-laser too many cracks are induced in the PVD coating,

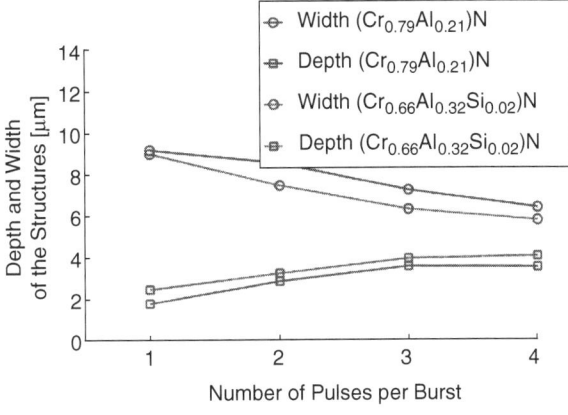

Fig. 5.189 Width and depth of the ablated grooves according to the number of pulses in the burst at constant burst energy. The total amount of energy per structure is kept constant by variation of the scan speed. $E_{burst} = 0.8\ \mu J$; $N = 10$; $v = 200\ mm/s$; $E_{pulse} = 0.8/0.4/0.27/0.2\ \mu J$ at $n = 1/2/3/4$

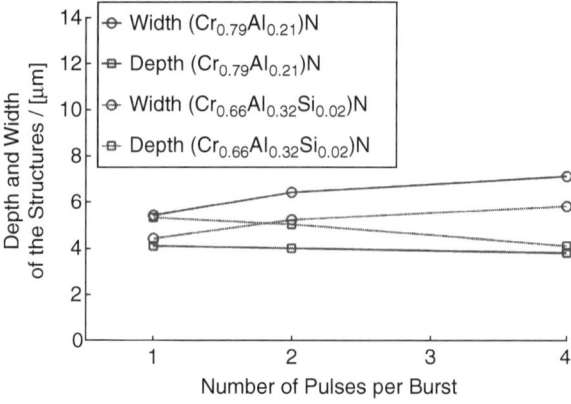

Fig. 5.190 Width and depth of the ablated grooves according to the pulses per burst at constant burst energy. The total amount of energy introduced per structure is kept constant by altering the scan speed. $E_{pulse} = 0.2\,\mu J$; $N = 10$; $E_{burst} = 0.2/0.4/0.8\,\mu J$ at $n = 1/2/4$

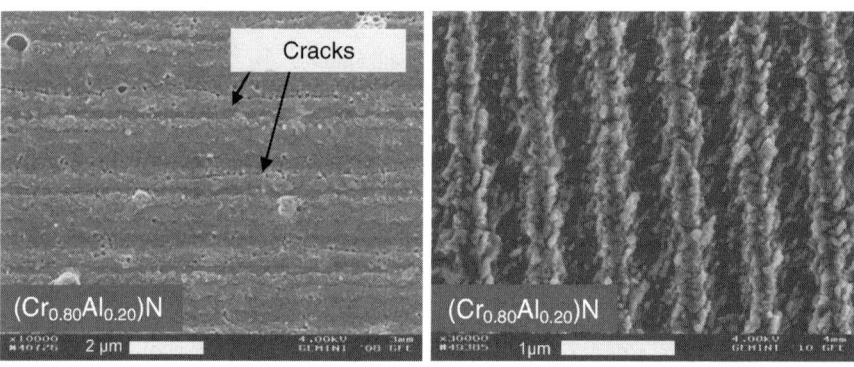

Fig. 5.191 Nanostructuring of a $(Cr_{0.80}Al_{0.20})N$ MSIP coating. *Right*: using a ns-laser generates cracks in the coating. *Left*: using a ps-laser at pulse energy $3.2\,\mu J$, 8000 pulses per spot. A crack-free structure is possible

see Fig. 5.191, left. When structuring with ps-laser, a crack-free structuring can be achieved which is not completely covered with ripples, see Fig. 5.191, right. The parameters to be chosen are highly dependent on the material to be structured. For nanostructuring of PVD coatings first good results have been achieved. Further research has to be conducted in this research area.

Coating a Laser-Structured Tool and Moulding by Injection Moulding The alternative to manufacturing a coated, structured tool is the retrospective coating of a structured tool. As explained at the beginning, this option is restricted to structure sizes $>5\,\mu m$. In this context, it is important to include the coating in the design of the structure. It is essential to know the structurable aspect ratios and the distribution of the coating thickness, which can be achieved with industrial coating units. Figure 5.192 illustrates an example of a tapered 30 μm wide structure which is coated

5 Hybrid Production Systems

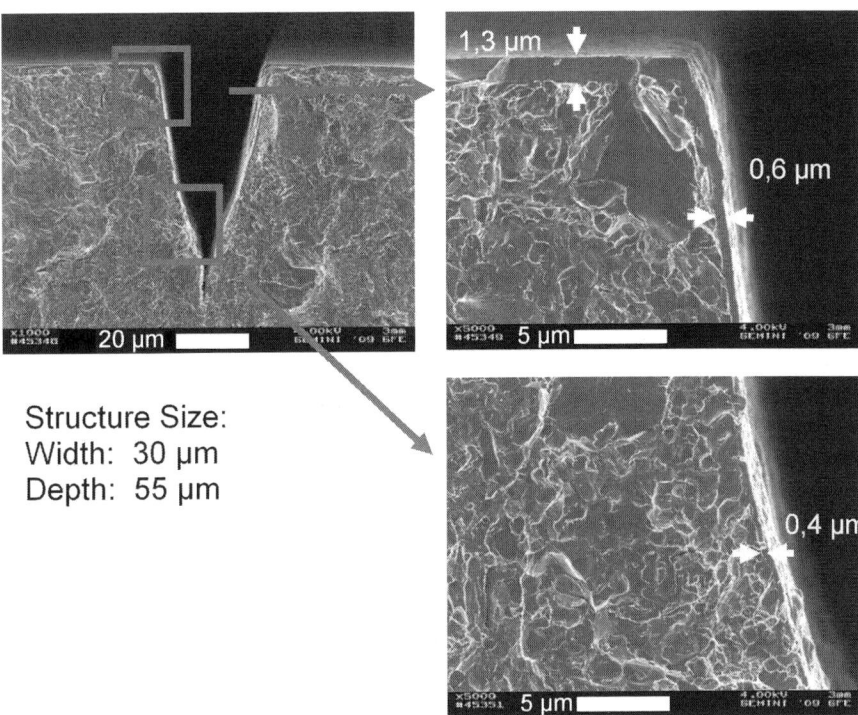

Fig. 5.192 SEM cross section fracture image of a $(Cr_{0.80}Al_{0.20})N$ coated 30 µm structure at different magnifications. The coating is on the complete depth of the structure

with a 1.3 µm thick $(Cr_{0.80}Al_{0.20})N$ coating. The structure is coated to the base despite the difficulties for coating due to the complicated tapered geometry.

Furthermore, the coatings which are deposited by means of pulsed MSIP technology and provide the best results in the basic analyses discussed above are observed. Here two examples are shown where the coatings $(Cr_{0.80}Al_{0.20})N$ and $(Cr_{0.57}Al_{0.43})N$ are applied to a 40 µm wide linear structure (aspect ratio ≈ 1) and a 10 µm wide linear structure (aspect ratio ≈ 2). Figure 5.193 illustrates these structures by means of confocal laser microscopy using an uncoated injection moulding insert.

Figure 5.194 illustrates the SEM image of the surfaces of the uncoated and coated mould inserts. It can be seen that the wall and bottom of the structure are rough due to laser structuring. These structures are attributed to three different structuring phenomena. The following structures can be detected: a linear nanostructure with a structure size of a few hundred nanometres, a linear microstructure and a hole structure (see Fig. 5.195).

The nanostructures are attributed to laser induced periodic surface structures (LIPSS) which can be attributed to the interference of the laser radiation with surface waves (Guosheng et al. 1982; Sipe et al. 1983). The linear microstructures and hole

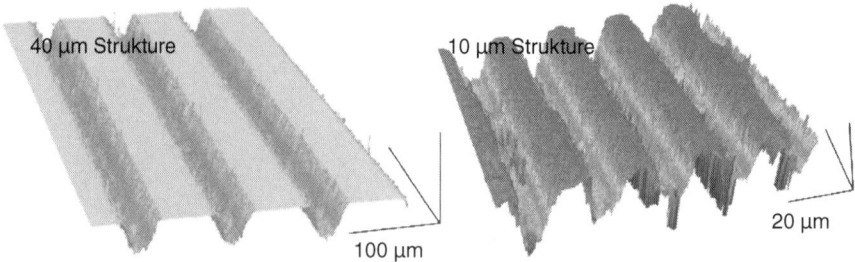

Fig. 5.193 3D image taken by a Keyence VK9710 confocal laser microscope: *left*, 40 μm structure and *right*, 10 μm structure of an uncoated injection moulding insert prior to coating

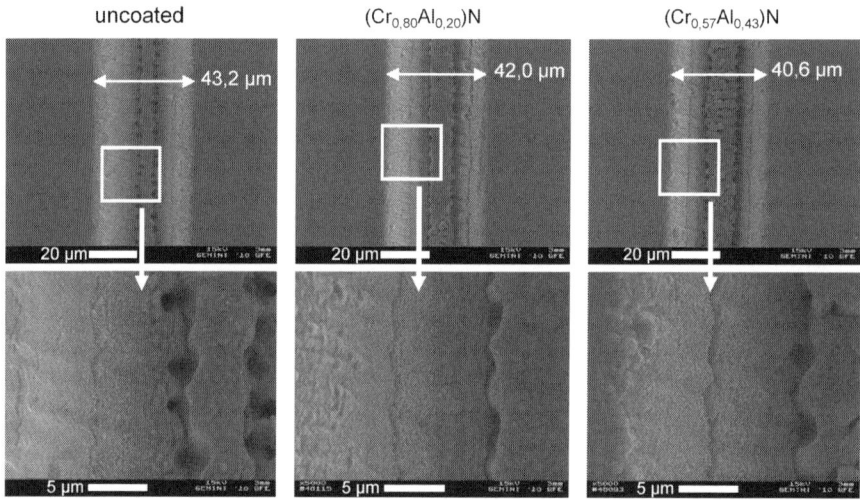

Fig. 5.194 SEM images of the surfaces of the uncoated 40 μm structure (*left*) and the 40 μm structures coated with $(Cr_{0.80}Al_{0.20})N$ (*centre*) and $(Cr_{0.57}Al_{0.43})N$ (*right*). The coating levels structuring errors of up to approximately 2 μm

structures are caused by periodic substructures and the so-called cone-like protrusions (CLP) (Eifel et al. 2010). There is no precise description of the development of these structures, but has to be analysed in order to prevent the occurring of such structures. The roughness leads to slightly different widths of the structures. The coatings are able to completely level the LIPSS and periodic substructures. However, it is only possible level the hole structures slightly. It can be concluded that in addition to its actual function, the coating is also able to cover surface imperfections on the structured tools. The roughness on the corners and walls of the structures, which are particularly noticeable on the uncoated inserts, can cause problems during demoulding in the subsequent injection moulding process. With a good moulding, the plastics sticks in these structures and can therefore cause stretching or tearing of the structures of the plastic component, see Sects. 5.6.4.2 and 5.6.4.3. During the

Fig. 5.195 SEM image of the parasite structures which occur during the structuring of a 40 μm structure

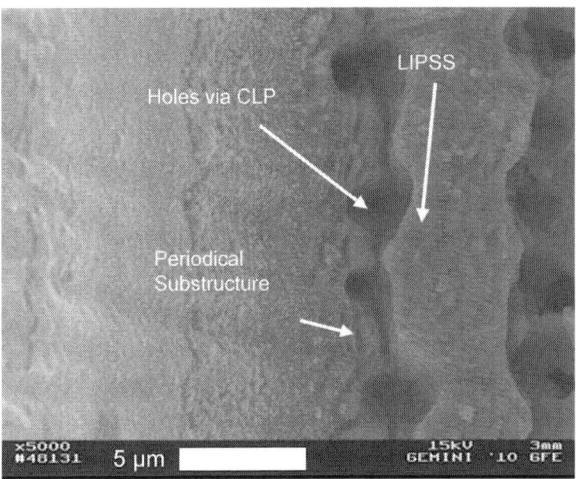

cooling of the moulding mass, volume contraction of the relatively large structures can simplify demoulding compared to smaller structures.

The results of coating the 10 μm structures are illustrated in Fig. 5.196. The aforementioned structuring effects can also be seen here. As discussed above, the volume contraction effect of the plastics is proportionally smaller with these structure sizes, which can lead to problems during demoulding. The probability of stretching or tearing the structure is therefore bigger compared to the 40 μm structure.

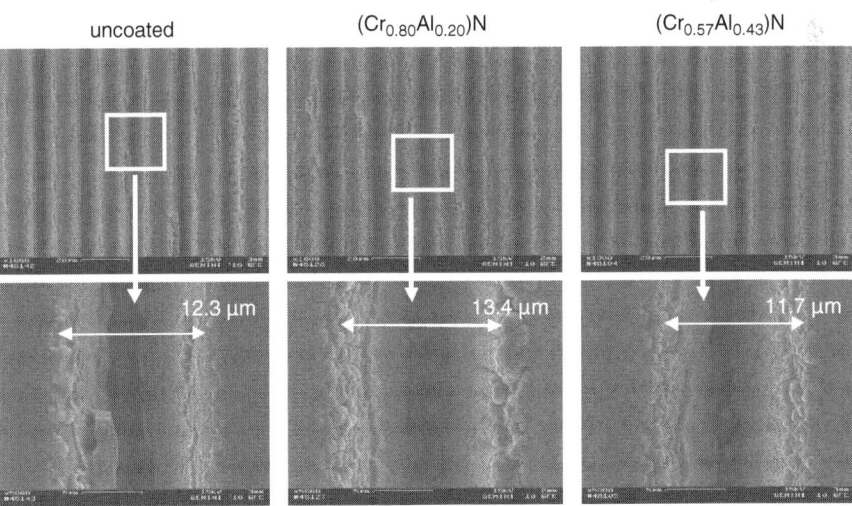

Fig. 5.196 SEM image of the surfaces of the uncoated 10 μm structure (*left*) and the 10 μm structures coated with $(Cr_{0.80}Al_{0.20})N$ (*centre*) and $(Cr_{0.57}Al_{0.43})N$ (*right*)

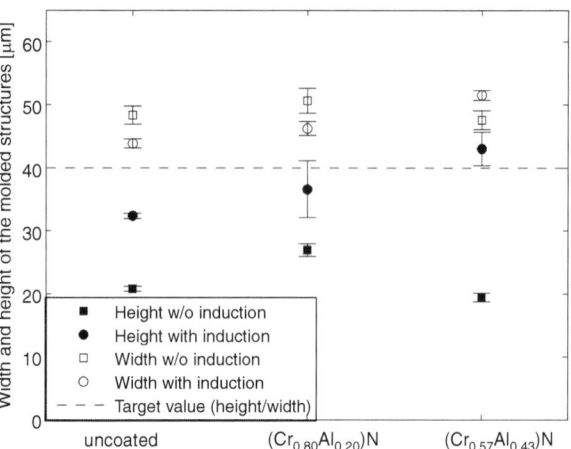

Fig. 5.197 Width and height of the structures on the moulded polypropylene plastic components for 40 μm micorstructures. Moulding the tools coated with $(Cr_{0.80}Al_{0.20})N$ provides good moulding results, even without variothermal tempering

All three mould inserts are tested in an injection moulding process and the moulded structures evaluated according to their depth and width. A standard polypropylene (PP, Sabic® PP513MNK40) is used as moulding mass. Two different processes are analysed. The first is the conventional injection moulding with constant mould temperature control, the second is a moulding process with variothermal heating of the tool which is discussed in detail below. The results of coating the 40 μm structures are illustrated in Fig. 5.197. The dashed line represents the target value of the 40 μm structure (aspect ratio ≈ 1). The $(Cr_{0.80}Al_{0.20})N$ coating in particular provides good results with regard to the intended structure sizes. Furthermore it can be seen that the moulding accuracy is significantly improved by the use of variothermal heating with an inductor. It is not possible to produce tolerable structure heights with a conventional moulding process.

When moulding the 10 μm structures (aspect ratio ≈ 2), it can be seen that also here, the combination of variothermal tool heating and $(Cr_{0.80}Al_{0.20})N$ coating has a positive effect on the moulding accuracy (see Fig. 5.198). With the $(Cr_{0.57}Al_{0.43})N$ coating, stretching appears during the demoulding process due to sticking of the plastics. This is attributed to the smaller structure width, which can be seen in Fig. 5.195.

5.6.4.2 Extrusion Process for Manufacturing Functional Surfaces

Replication processes used to produce microstructured plastics films have particularly high requirements for the heating of the tools. On the one hand, the replication process consisting of the combined primary processing and forming process of the polymer takes place at or above melt temperature. On the other hand the calibration and cooling of the shaped or formed plastic melt takes place piror to haul-off. The replication of the negative structure of the microstructured tool surface involves only

Fig. 5.198 Width and height of the structures on the moulded polypropylene plastic components for 10 μm microstructures. Moulding the tools coated with $(Cr_{0.80}Al_{0.20})N$ provides the best moulding results

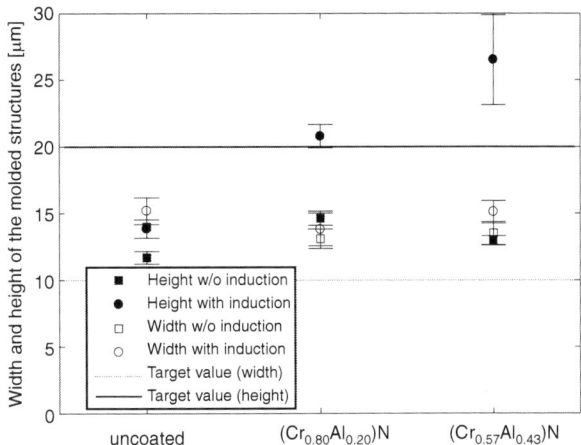

a micrometre thick layer of the surface of the plastics film. In order to guarantee good replication accuracies, the thermal balancing in this surface layer must be targeted.

In the extrusion embossing process (Fig. 5.199) a plastics film is moulded, laid on the embossing roller, embossed with a microstructure and then removed as a geometrically stable film (Bagcivan et al. 2010). To achieve a high quality replication, the contact pressure between the counter pressure roller and the embossing roller must be as high as possible in order to press the molten plastic into the microstructures. Furthermore, the plastics melt must have a high flowability. If the flowability is

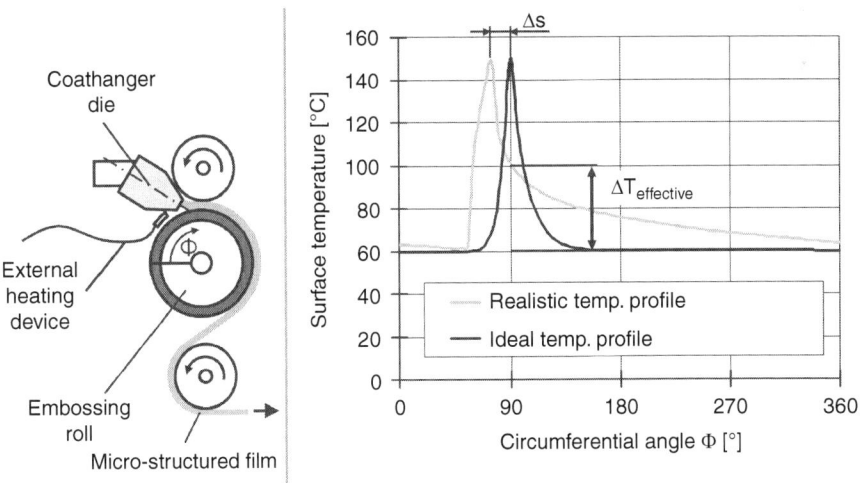

Fig. 5.199 Basic set-up of the variothermal extrusion embossing process (*left*) and schematic temperature profile along the circumference of the embossing roller during variothermal heating (*right*)

insufficient, the microstructures are only rudimentarily formed. The melt must be sufficiently low viscous in order to be able to flow into the microstructures in the embossing zone. The viscosity of the melt is adjusted by selecting a certain type of plastics, or by using additives. Flowability is also improved by means of dynamic roller temperature control. In order to increase the flowability of the plastics melt on the roller and therefore improve the replication accuracy of the microstructures, the surface of the embossing roller is heated to a temperature above the glass temperature T_g or the crystalline melting temperature T_k before the melt reaches it by using an additional external heating system. Therefore, an additional external heating device is integrated to heat the roller on the whole width. This heating system is used to generate a temperature profile along the circumference of the embossing roller surface (Fig. 5.199). Thus, the surface temperature in the area of the heating zone of the external device and the embossing zone is remarkably high. This modification of the extrusion embossing process is referred to below as variothermal extrusion embossing.

Both, experimental and numerical tests have shown that during extrusion embossing, the initial melt temperature has only a negligible effect on the time required until the temperature of the plastics melt surface reaches the temperature of the embossing roller surface (Bagcivan et al. 2010; Michaeli et al. 2005, 2009, 2010b, e; Michaeli and Scharf 2009). In order to improve the replication quality, the embossing roller surface temperature should therefore be as close to as possible or even above the melting range of the polymer. At the same time, however, the plastics melt must be cooled to geometrical stability by the embossing roller once the microstructures are replicated in order to guarantee a damage-free demoulding of the plastics film and the surface structures. It is possible to achieve both by applying variothermal heating.

In order to be able to implement variothermal heating for embossing rollers, the heat output per area of the locally implemented heating must be big enough to generate an accumulation of heat in the surface layer of the embossing roller. The accumulation of heat is characterised by the heat input clearly exceeding the heat output by convection and thermal conduction in the embossing roller. This requires a heat output of 2–6 kW/cm^2 (Benkowsky 1990). Such high outputs can be achieved by induction heating or laser radiation, for example. Usually, an external heating of the embossing roller surface directly in the embossing zone is not possible due to space limitations. In order to reduce the neck-in (the melt recovery behind the coathanger die exit due to the velocity profile and melt properties) of the melt film, the distance between the exit slit of the coathanger die that shapes the melt film and the embossing zone should be kept to a minimum. The size of the heating zone of the external heating system is based on the feasible output power of the heater and the circumferential speed of the roller respectively the heat flow which has to be provided to the roller surface. In this respect, the penetration depth of the temperature depending on the type of energy input plays a crucial role. The lower the penetration depth, the less energy needs to be provided to the roller surface in order to temporarily achieve the necessary surface temperature. Alternatively, as the penetration depth of the temperature decreases, the local maximum surface temperature increases at the same heat output.

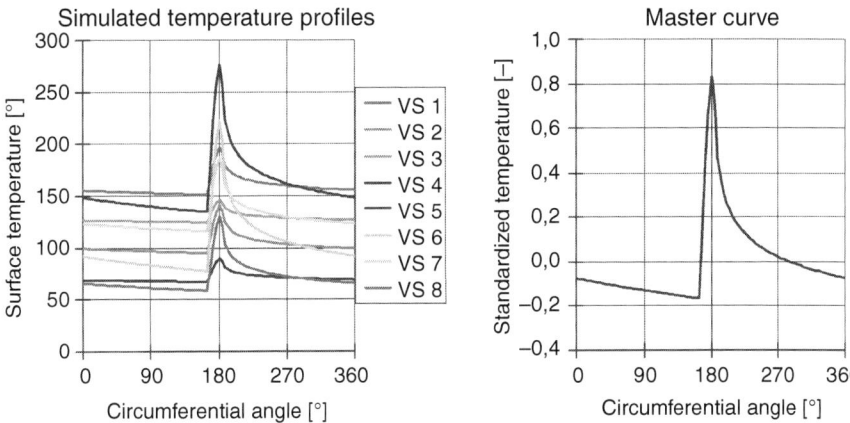

Fig. 5.200 *Left*: variation of the process parameters heat output of the external auxiliary heating per roller width, cooling medium temperature and haul-off speed, each on two levels. *Right*: derived, standardised master curve of these simulated test points (VS 1–VS 8)

Due to the distinct effect of the produced temperature profile, the effect is predicted by simulations and experimentally verified by variation of process parameters and measurement of the temperature profile. Tangentially placed thermocouples are used for measuring the temperature profile. These damage the roller surface and are therefore not suitable for taking measurements on the microstructured roller surface. However, they produce significantly more accurate measurement values compared to infrared measurement as here an accurate measurement of the temperatures on the curved and highly reflective surface of the roller is not possible.

The analysis of the numerically predicted temperature profile indicates that independent of the three selected process parameters heat output of the external heating device (characteristic heat output per roller width), basic temperature of the embossing roller (cooling medium temperature) and haul-off speed, all temperature profiles can be derived from a master curve (see Fig. 5.200). The temperature amplitude and mean temperature can be used to shift the temperature gradients.

The temperature is therefore standardised according to (5.18) and examined via the circumferential angle.

$$\theta(\Phi) = \frac{T(\Phi) - \overline{T}_{mean}}{T_{max} - T_{min}} = \frac{T(\Phi) - (T_{water} + \Delta \overline{T}_{mean})}{\Delta T_{amplitude}} \quad (5.18)$$

With given equations for determining the maximum temperature amplitude $\Delta T_{amplitude}$ and the change in mean temperature $\Delta \overline{T}_{mean}$ according to the process parameters cooling medium temperature, heat output per roller width and haul-off speed, it is easy to define the entire course of the temperature profile along the circumference of the roller (5.19) by transposing (5.18).

$$T(\Phi) = T_{water} + \Delta \overline{T}_{mean} + \theta(\Phi) \times \Delta T_{amplitude} \quad (5.19)$$

The maximum temperature amplitude $\Delta T_{amplitude}$ and the mean temperature $\Delta \overline{T}_{mean}$ are defined according to (5.20) and (5.21). The necessary parameters are defined by determining the master curve from the experiments and are valid in the observed process window with good accuracy. They describe the effects and interactions of the process parameters heat output per roller width, cooling medium temperature and haul-off speed.

$$\Delta T_{amplitude} = A_T \cdot T_{water} + A_{P'} \cdot P'_{heating} + A_v \cdot \frac{1}{\sqrt{v_{haul\text{-}off}}} + A_{P'v} \cdot \frac{P'_{heating}}{\sqrt{v_{haul\text{-}off}}} \quad (5.20)$$

- Influence of basic temperature of the roll
- Influence of output power of the external heating device
- Influence of the haul-off speed
- Interaction between haul-off speed and output power

$$\Delta \overline{T}_{mean} = B_T \cdot T_{water} + B_{P'} \cdot P'_{heating} \quad (5.21)$$

- Influence of basic temperature of the roll
- Influence of output power of the external heating device

The coefficients A_T, $A_{P'}$, A_V, $A_{P'V}$ as well as B_T and $B_{P'}$ are used to calculate the temperature profile for any variation of process parameters. The mean error of 0.08 °C, which is derived from comparison with the simulations, is very low. The maximum error is 1.23 °C. Due to the distinct behaviour of the temperature profile it can be assumed that the given equations are valid in a wider area.

One geometrical parameter of the variothermal temperature control which has so far been ignored is the length of the heating zone l_{heat} in the circumferential direction of the roller. In the heating zone itself (which varies in length depending on the specific heating system) and in a small area behind the heating zone the observed temperature profiles exhibit considerable differences. However, in the range of circumferential angles of $\Phi > 190°$ (see Fig. 5.201), the temperature profiles are almost identical. For practical applications, the length of the heating zone must therefore only be taken into consideration when it is possible to position the heating zone or the external heating device directly in front of the embossing gap.

The temperature gradient for changing the heating zone length l_{heat} can be normalized according to (5.18) into a standardised excess temperature profile $\theta(\Phi)$. As the effect of the length of the heating zone on the temperature profile effectively only affects the heating zone itself and a small adjacent circumferential angle area of $\Delta \Phi \approx 10°$, it is possible to transform the various master curves into a global master curve. This master curve only depends on the geometry and material parameters of the roller, which are not examined in greater detail here, and the heat transfer coefficients of the internal and external surfaces of the roll shell.

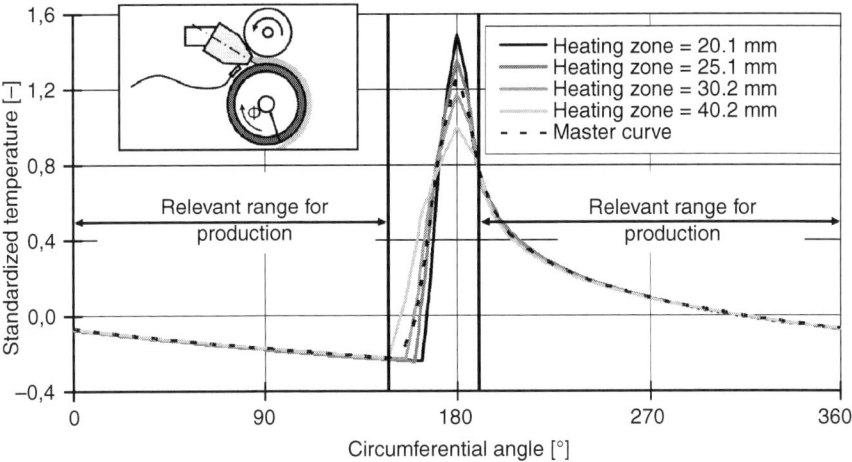

Fig. 5.201 Gradient of the standardised excess temperature θ over the circumferential angle for various heating zone lengths l_{heat}; end of all heating zones $\Phi = 180°$; good conformity at $\Phi < 148°$ and $\Phi > 190°$ ("valid area").

In the experimental verification of the master curve, a very good level of practicability was observed in both the inductive and the laser-based variothermal heating (see Fig. 5.202). As the absolute maximum temperature at the end of the heating zone cannot be measured in experiments, the standardisation is performed using the maximum value $T_{max} = T(\Phi = 190°)$. The change in the mean temperature $\Delta \overline{T}_{mean}$ is represented by the mean value of the temperature in the circumferential angle area of $0° \leq \Phi \leq 148°$ and $190° \leq \Phi \leq 360°$ (i.e. with the exception of the heating zone and a small adjacent area of $\Delta \Phi = 10°$).

As can be seen in Fig. 5.202, the experimentally determined standardised temperature gradients do not match the calculated master curve exactly. One source of error for the experimental determination of the master curve is that only 10 values are measured along the circumference between $110° \leq \Phi \leq 226°$ (inductive heating) respectively $130° \leq \Phi \leq 226°$ (laser-based heating). The temperature profiles

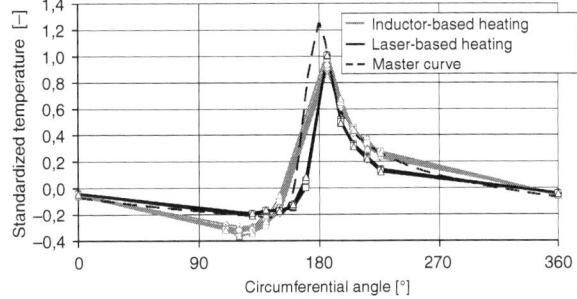

Fig. 5.202 Gradients of the standardised excess temperature θ over the circumferential angle for a steel roller with inductive and laser-based variothermal heating (measured tangentially); end of the heating zone at $\Phi = 180°$

Table 5.13 Coefficients for determining the maximum temperature amplitude and mean temperature

A_T	$A_{P'}$	A_V	$A_{P'V}$	B_T	$B_{P'}$
$[-]$	$\left[K \times \sqrt{\frac{m}{\min}}\right]$	$\left[K \times \sqrt{\frac{m}{\min}}\right]$	$\left[K \times \frac{mm}{W} \times \sqrt{\frac{m}{\min}}\right]$	$[-]$	$\left[K \times \frac{mm}{W}\right]$
−0.015425	0.044424	2.1281	3.9219	−0.040077	3.0960

between these measurement points are interpolated linearly. The change of the mean temperature $\Delta \overline{T}_{mean}$, which is included in the standardisation, is particularly strongly influenced by the two measurement values at the edge of the measurement range.

The values of the coefficients A_T, $A_{P'}$, A_V, $A_{P'V}$ and B_T and $B_{P'}$ used in the master curve in Fig. 5.202 are provided as examples in Table 5.13.

In addition to generate a suitable temperature profile and high replication accuracies during the embossing of the microstructures, haul-off of the film from the embossing roller is also a decisive factor with regard to the quality of the microstructures. The film is usually removed from the embossing roller at the separation point in a tangential direction. However, as the microstructures are positioned on the film in a radial direction, a shear stress occurs in the area of the base of the structures during haul-off. Axial stress has also to be expected in the microstructures. This occurs when the microstructures remain stuck in the structures on the roller during demoulding (Bärsch 2001). The replicated microstructures can be sheared of from the film as a result of the high stresses. The shear forces can be reduced by adjusting the structure size; a large aspect ratio increases the shear forces and reduces the replication quality. In this regard, it should be noted that the aspect ratio is often crucial for the functionality of the microstructures and can therefore only be adjusted with reservation (usually only within a very small range). Axial stress is reduced by smooth walls of the structures. It is also possible to improve the quality of the microstructures in the area around the separation point by selecting suitable materials. The selected plastics should react to the stress with an elastic deformation if possible. From a processing point of view, the temperature at the separation point should be adjusted so that the plastic is suitably elastic. Setting a suitable temperature profile for the embossing roller represents an important opportunity to improve the accuracy of the replication process.

By using inductive or laser-based variothermal roller heating the quality of the produced microstructures in extrusion embossing can clearly be changed. Increasing the heat output of both systems increases the overall replication quality. The laser-based heating has the advantage that by adjusting the beam path (multiple reflections at the counter pressure roller) a bigger difference of temperature along the circumference of the roller is achieved while reducing the increase of the mean surface temperature. This leads to a better overall quality of the microstructures, in particular edges and surfaces which are positioned perpendicular to the direction of extrusion. However, the laser beam not only heats the embossing and counter pressure roller, but also heats the melt film to a certain extent. This can lead to an increase of the melt bulge in front of the embossing zone and therefore to changed process conditions. Furthermore, embossing defects occur which are traced back to gas inclusions in the melt film.

Microstructure at the surface of the embossing roll

Replicated microstructure on the polymer film

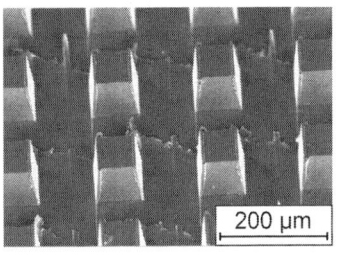
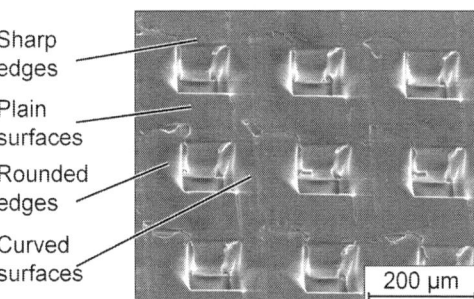

Sharp edges
Plain surfaces
Rounded edges
Curved surfaces

Direction of extrusion

Fig. 5.203 The surface of a microstructured embossing roller (*left*) and embossed plastic film with moulding defects perpendicular to the direction of extrusion (*right*)

Such inclusions lead to surface imperfections and influence the otherwise improved replication of the microstructures negatively. This issue can be avoided by selecting a different type of plastic or varying the individual auxiliary materials or aggregates.

The main factors affecting the replication accuracy are the melt viscosity at the flat film die exit, the temperature of the embossing roller and the pressure used for embossing in the gap between the rollers. As viscosity decreases, the melt flows more easily into the microstructures of the embossing roller. This effect can be achieved in terms of an additive or a higher melt temperature. When increasing the embossing roller temperature, the melt solidifies more slowly. Hence, the polymer can penetrate the structure on the roller more easily. However, before haul-off of the plastic film from the embossing roller, the plastic must be geometrically stable. Otherwise, the just produced structure will be destroyed again when being removed from the embossing roller. In addition to the thermal processes, the mechanical processes which take place in the embossing zone are of great significance to the replication of the surface structures of the embossing roller into the plastic melt. In this area, the viscoelastic melt undergoes primary shaping or embossing depending on the temperature control. Experimental analyses show that the moulding of structural details not only depends on the exact geometry of the details, but also their alignment relative to the direction of extrusion. As Fig. 5.203 illustrates, the edges and surfaces which are aligned parallel to the direction of extrusion are moulded with greater contour precision than those which are aligned tangential to the direction of extrusion, although the cubic structure on the embossing roller surface is identical in both directions.

There are several explanations for the observed replication variations. It can be attributed to friction between melt and embossing roller, which causes local stress peaks, or local differences in the flow behaviour of the plastic melt during the embossing process. Also capillary effects between the melt and the microstructured

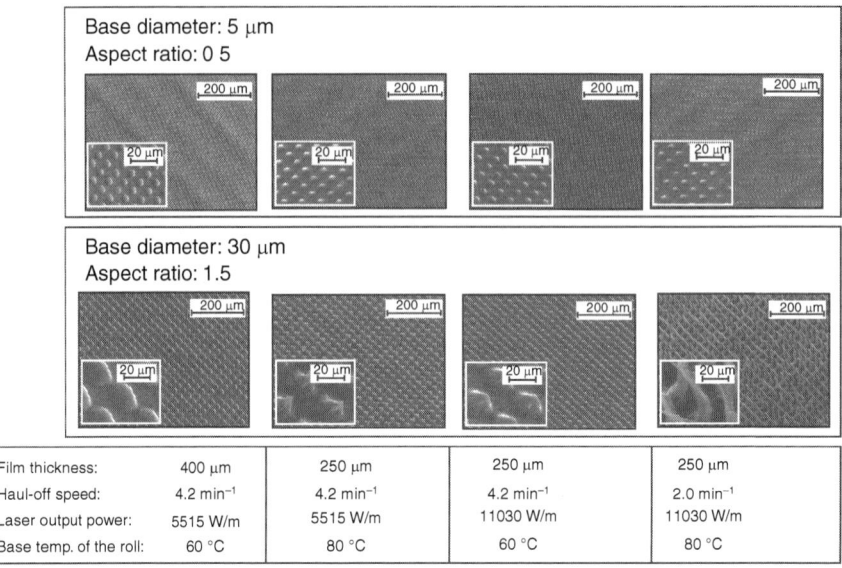

Fig. 5.204 SEM images of spherical surface structures; the changes of the moulded geometries due to variation of the characteristic process parameters are shown for two exemplary geometries (*top*: diameter 5 μm, aspect ratio 0.5; *bottom*: diameter 30 μm, aspect ratio 1.5)

embossing roller which act as driving forces for local flows, and air inclusions between the melt and the embossing roller can be a reason. Common to all these effects is the fact that they cannot be directly observed in the real embossing process. For a better insight viscoelastic simulations of the embossing process and in particular an analysis of the internal stresses in the melt or plastic film are necessary.

In order to examine hybrid, function-integrated microstructures on plastic films, a superhydrophobic, cone-like surface based on the lotus effect without any additional coating is studied. The microstructures are measured using SEM and WIM images. Figure 5.204 presents examples of SEM images of two geometries of the microstructures with different basic diameters D and aspect ratios a. It is obvious that varying the process parameters clearly affects the produced structures. The obtained geometries range from low replication quality, to cone-like structures and further to considerable stretching and build-up of filaments (hair-like structures).

At some samples specific anomalies can be observed while studying the replication of the microstructures. A remarkable difference between the produced microstructure geometry and the intended geometry is observed which requires further explanation. If the melt film is not yet properly solidified or the plastic sticks too tightly in the rough structures on the roller when the foil is removed from the embossing roller, the microstructures can be stretched. The result are hair-like structures with a pronounced melt filament on the tip, as illustrated in Fig. 5.205 left. If, on the other hand, the removal forces and rigidity of the cooled microstructures are too high, fissures in the base of the microstructures or even a complete tear off of the microstructures

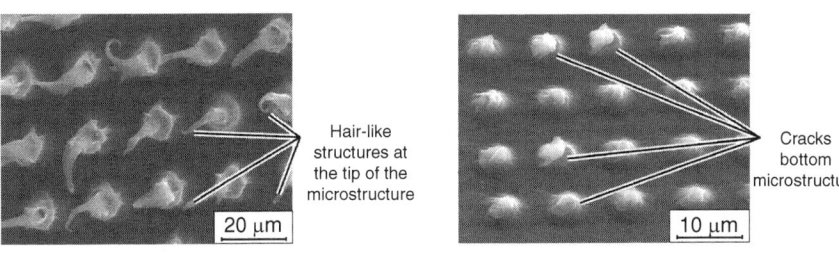

Fig. 5.205 *Left*: embossed microstructures with melt filaments. The plastic has not solidified fully during removal or is anchored in the roughness of the roller surface. *Right*: embossed microstructures with fissures in the basic structure since the removal forces and rigidity of the cooled microstructure are too high

and therefore clogging of the embossing roller surface (see Fig. 5.205 right) can be observed.

Overall, the size of the melt bulge has practically no influence on the product quality due to the dominant effect of the haul-off speed. While lower haul-off speeds cause practically no surface imperfections, they are obvious at greater haul-off speeds.

5.6.4.3 Manufacturing of Functional Surfaces by Injection Moulding

High heating rates can be realised using systems based on the principle of electromagnetic induction. For this reason, as part of the here presented experiments, a system is developed for the external inductive heating of the mould surface. Due to the large temperature gradients, the external inductive heating requires an active and fast temperature control in order to prevent an overheating of the mould surface. The necessary temperature measurements can be conducted in a non-contact manner using a pyrometer.

The moulding tests are performed using two types of injection moulding. The moulding results using the conventional injection moulding method are compared to the results using the variothermal injection moulding. The injection moulding process begins with the closing of the mould. The injection unit moves forward. The hot polymer melt is injected into the cavity. When the mould is filled, the process is switched to holding pressure which is maintained until the sealing point is reached and no more material can be squeezed into the cavity. In the holding pressure phase, the volume contraction of the polymer is being compensated. A residual cooling time follows until the component is geometrically stable and can be removed from the mould. During the residual cooling time, new material is plasticised for the next shot and the injection unit is moved back. Once the mould is open, the finished product is ejected using the ejector. As part of the analyses of the variothermal injection moulding, the course of the process following the demoulding of the moulded part is extended by the following process steps. For variothermal injection moulding with inductive heating, the external auxiliary induction heating is moved into the open mould by a robot and positioned in front of the cavity. The microstructured cavity

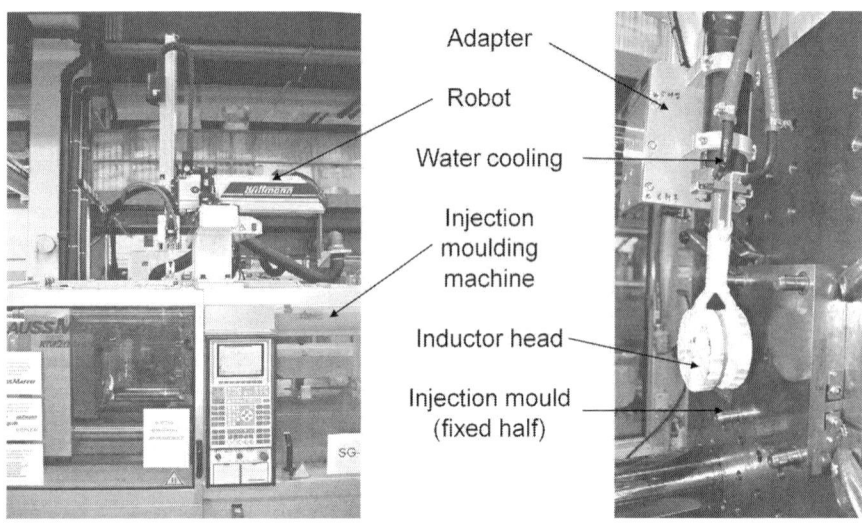

Fig. 5.206 System for the external inductive heating of the surface of the injection mould

is then partially heated with heating rates of 60 K/s and small penetration depths. Following this heating, the auxiliary induction heating is then removed from the mould. The injection moulding machine closes the mould and the cycle is continued as described above. For laser-based variothermal injection moulding, the surface is heated using an integrated laser optics once the mould has been closed. This heating system does not require any additional movements. Once the heating phase is completed, the injection cycle continues with the injection of the plastics melt. All other process parameters remain constant in the conventional and variothermal processes. The injection moulding tests for analysing the effects of the variothermal process on the functionality of the moulded part surface are performed using a fully hydraulic injection moulding machine (KraussMaffei CX 160-1000, KraussMaffei Technologies GmbH). The basic tempering of the mould is achieved by means of water tempering units (P160, Wittmann GmbH).

A system involving an injection moulding machine, a robot and an inductor (Fig. 5.206) is designed, operated and tested for the external inductive heating of the surface of an injection mould. A W721 (Wittmann GmbH) handling system is used. This system is conventionally used as a removal system for finished moulded parts, but is ideal for positioning the inductor head in front of the cavity. The robot system is mounted on the injection moulding machine and can be moved to any geometrical point within the mounting plates of the injection moulding machine by means of a 3-axes continuous path contouring control. The robot is equipped with highly dynamic servo motors and features a solid construction. It guides the inductor head firmly and vibration-free. The robot position tolerance is 0.1 mm. Therefore it is possible to reproduce minimum distances between the cavity and the inductor head.

5 Hybrid Production Systems

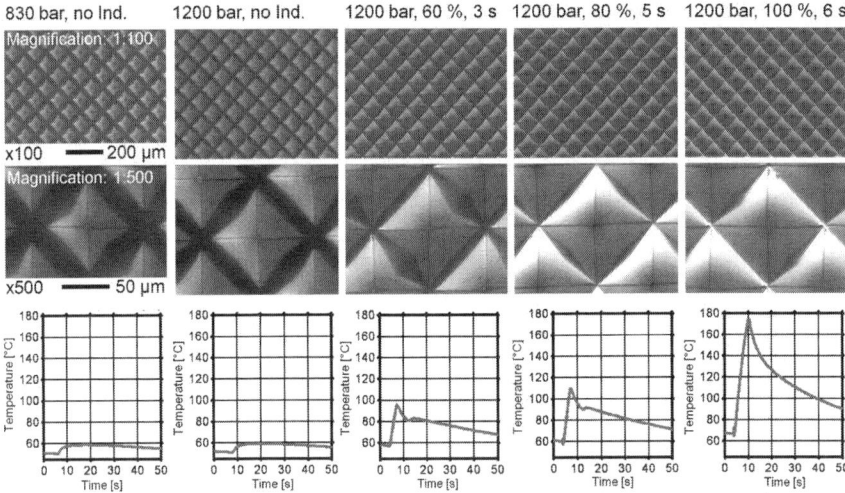

Fig. 5.207 The effect of the parameters holding pressure, inductor heat output and heating time on the moulding quality of the microstructures. The increase in mould temperature into the range of the melt temperature of the plastics by a large inductor heat output in particular, enables the precise moulding of the surface structures

Benkowsky (1990) cites the medium frequency spectrum (10–25 Hz) as an optimum for heating the area near the surface, as needed for the heating of the surface of an injection mould. For this reason, a MFG 15 by Eldec Schwenk Inductions GmbH with a nominal output power of 15 kW is used to heat the cavity with high heating rates. This is a medium frequency induction system which guarantees an optimum frequency range for the application discussed and therefore an optimum penetration depth of a few tenths of a millimetre into the mould plate.

The inductor head is designed specifically for this application. A plate inductor with a diameter of 80 mm is used, which is positioned approx. 7 mm in front of the cavity during testing. The heat output of the unit is changed by regulating the inductor output. High inductor power is initially used for heating up until the target temperature is reached then the output power is reduced.

The variothermal process increases considerably the moulding accuracy of the microstructures even without using the positive effects of coatings (Fig. 5.207). For the first time, a system is set-up and used to produce injection moulded parts in a fully automated variothermal process with an external temperature-regulated induction unit under entirely reproducible conditions.

The technology developed here for the external inductive heating is characterised by a previously unachieved heating rate of up to 60 K/s, whereby a circular area with a diameter of 80 mm is heated very homogeneously. The output power of the inductor in this respect is regulated according to the temperature measured on the surface. The system is successfully used in series of tests in a fully-automated variothermal injection moulding process.

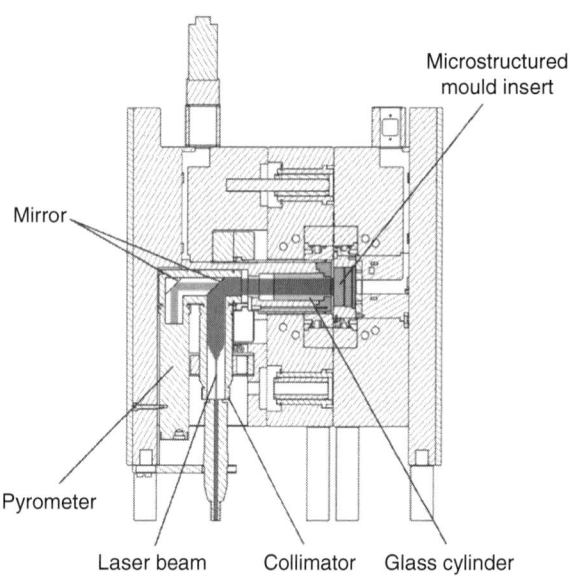

Fig. 5.208 Injection mould with integrated laser optics for variothermal laser heating. The laser beam passes through the cavity and is absorbed by the microstructured mould insert

As the induction heating is an external system, no modifications need to be made to the injection mould. However, there are limitations in terms of the heatable mould insert materials as it is only possible to heat ferromagnetic materials inductively. Injection mould inserts of brass or aluminium, which are used in pilot and short production runs in particular, cannot be used. The coupled input depends heavily on the distance between the inductor and the surface to be heated. The heating of 3D surfaces therefore requires an inductor with a geometry that very accurately matches that of the surface. The inductor design is also restricted by being produced with copper profiles. These reasons make the inductive heating process particularly suitable for simple part geometries and large quantities.

A process has therefore been developed for heating curved surfaces and mould inserts of non-ferromagnetic materials, whereby a laser is used for the first time to locally heat the surface of the injection mould. A system consisting of a diode laser unit, fibre optic cable, collimator lens, pyrometer and laser control is designed and put into operation. The optics is integrated directly into an injection mould designed specifically for this purpose (Fig. 5.208). A transparent mould insert of quartz glass is used to guide the laser beam from the optics exit directly onto the mould surface where it is absorbed, thus heating the mould surface.

The heating behaviour of the laser on the surface of an injection mould is examined by thermography. The possibility of using a collimated laser beam with a diameter of 23.5 mm to heat a selected surface is demonstrated. A maximum temperature gradient of 300 K/s is achieved. A pyrometer-based power control system is used to adjust specific target temperatures. Therefore—analogous to inductive heating— the full output power of the system is used at the beginning of the heating process

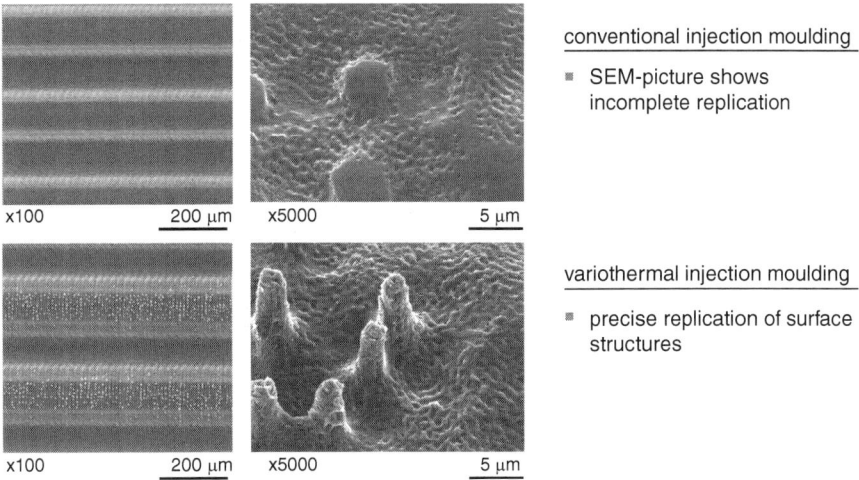

Fig. 5.209 Moulded microstructures. *Top*: conventional injection moulding. *Bottom*: variothermal injection moulding with laser heating. Laser-based heating of the mould surface can be used to transfer fine surface structures to the plastics

and is afterwards reduced in a controlled manner in order to prevent a temperature overshoot which would damage the mould inserts.

Reflection measurements on structured and unstructured mould inserts exhibit large differences when comparing the different materials. The degree of reflection on the polished mould inserts are 99% for copper, 90% for aluminium and 68% for steel at a laser wavelength of 940 mm. The unstructured samples reflect more laser radiation than the structured samples. This is due to the radiation being reflected repeatedly in the structures and each time a part of the radiation being absorbed. This raises the coefficient of absorption.

With regard to laser-based heating, no limitations are placed on the heatable materials. It is also possible to dynamically heat both planar and curved surfaces with laser radiation. Integrating the laser optics into the injection mould introduces no additional process steps and the increase in cycle time is less in comparison to inductive heating.

Diode laser, laser control and injection moulding machine are connected and programmed via digital and analogue interfaces so that the heating phase is integrated seamlessly and fully automatically into the injection moulding process. The system is successfully tested in both manually and fully automated manufacturing operations. The first moulding experiments indicate that due to laser heating mould surface structures of few microns are possible to be moulded with high contour precision (Fig. 5.209). As illustrated in Sect. 5.6.4.1, coating the mould surface can further increase the moulding quality.

Holding pressure, injection speed and residual cooling time are defined as influencing factors in the experiments since holding pressure and residual cooling time in

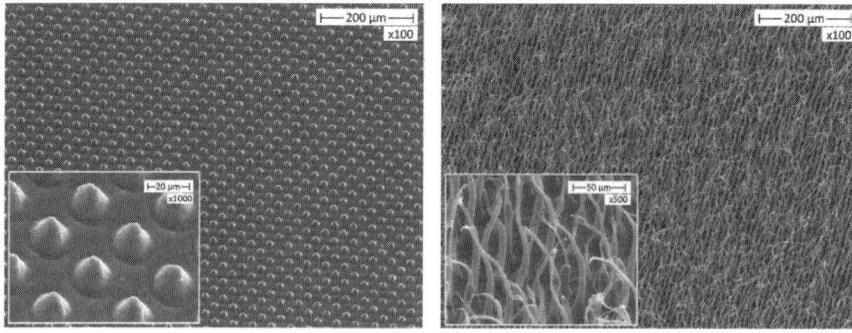

Fig. 5.210 Moulded microstructures. *Left*: conventional injection moulding. *Right*: variothermal injection moulding. In the conventional process, the structures are not formed completely. In the variothermal process, the structures are moulded completely and are stretched during demoulding

particular exert a considerable influence on the quality of the moulded part. The holding pressure profile and the injection speed of the basic setting are therefore scaled up- and downwards. The tests are conducted in a fully-automated cycle whereby each moulded part is manufactured under same conditions and same cycle time.

The moulded parts are assessed qualitatively using the Scanning Electron Microscope (SEM). Figure 5.209 illustrates examples of the moulding results using SEM images. The results of the microscopic analysis demonstrate that the mould quality is reproducible and depends significantly on the variothermal process.

As illustrated in Fig. 5.210, left, the plastic melt solidifies prematurely in the conventional process. The melt cools fast on initial contact with the relatively cold mould surface and the viscosity of the material increases. The higher viscosity causes an insufficient effect of holding pressure within the moulded part. This in turn causes the incomplete moulding of the given microstructures. The result of the variothermal process is illustrated on the right hand side of Fig. 5.210. It is obvious that a completely different surface topography has been formed.

The short-term local increase of the cavity temperature allows the cooling of the melt to be delayed. The corresponding effect of the holding pressure allows the combined microstructure and nanostructure to be moulded completely. A high degree of moulding quality of the rough structures results in a dramatic increase in the demoulding forces. Using brittle polymer materials, e.g. PMMA, the moulded part breaks off and therefore clogs the cavity. The polypropylene material used in this experiment (Sabic® PP513MNK40) is ductile. This behavior causes the individual microstructures to stretch. The demoulding and stretching process is illustrated in Fig. 5.211.

Stretching the individual microstructures leads to a unique filigree surface which has previously not been achieved by injection moulding. The moulding result is also very reproducible. The microstructure and superposed nanostructure in the mould insert in combination with a suitably ductile material and variothermal process control with the ability to achieve high temperature gradients for both heating and cooling, makes it possible to produce filigree and highly functional surfaces. The functionality

5 Hybrid Production Systems 671

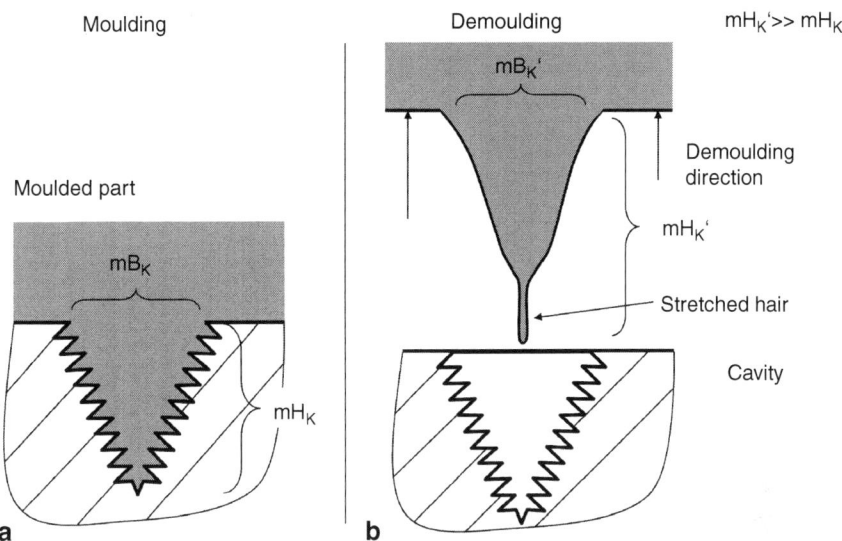

Fig. 5.211 Schematic of **a** the moulding and **b** the demoulding of the microstructures in the variothermal process. The fine structures are completely filled by the variothermal process. With ductile plastics, the microstructures are stretched during demoulding

in this example is not only generated by means of a highly precise moulding process, but also a systematic deformation of the structures during the demoulding process.

The functionality of the moulded parts is evaluated quantitatively by measuring the contact angle on wetting with ultra-pure water (Fig. 5.212). A contact angle of 120° develops on an unstructured, smooth polypropylene surface. Structured moulded

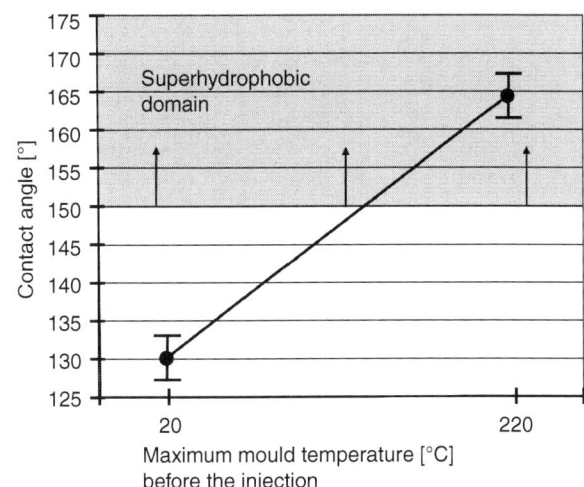

Fig. 5.212 The effect of the variothermal process on the contact angle when wetting with ultra-pure water. It is only possible to produce surfaces with contact angles above 165° by raising the moulding temperature

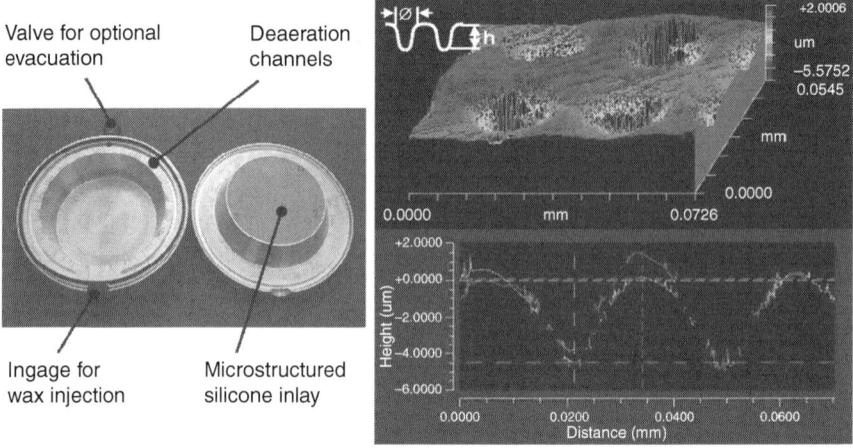

Fig. 5.213 Wax model (*left*) and white light interferometer image of a hole structure: ⌀ = 22 μm, h = 5 μm (*right*)

parts manufactured in a conventional injection moulding process feature a mean contact angle of 130°, while the parts moulded in the variothermal process feature contact angles of more than 160°, see Fig. 5.212.

5.6.4.4 Investment Casting Process for Manufacturing Functional Surfaces

The first process step of the investment casting process involves the production of a microstructured wax model. The molten wax is injected into a laser-processed, microstructured wax die. In precision casting, the standard modelling wax is used to completely reproduce microstructured surfaces with structure characteristics at a magnitude of down to 1 μm (depending on the aspect ratio), Fig. 5.213. The crucial factor for the reproduction capability is the surface tension of the wax which can be influenced by the process parameters injection temperature and injection pressure and can be set for an optimum moulding. The wax die characteristics also influence the mould quality. Metal dies with microstructured inserts made of silicone or polyurethane have proven themselves in this respect.

On the one hand, the metal dies are used in wax presses with high mould locking forces. On the other hand, the low thermal conductivity of the insert aids the penetration of the wax into the microstructured surface before it solidifies. Furthermore, the elastical insert ensures that the microstructured surface can be carefully demoulded. However, it is important that the flexible insert has a certain Shore-hardness as otherwise a risk is given that the microstructured surfaces are deformed by the pressure load of the inflowing wax.

As the results of the plastic injection moulding of microstructured surfaces show (see Sects. 5.6.4.2 and 5.6.4.3), it can be assumed that the wax can also be moulded directly in the microstructured metallic wax die. In this regard, both the tempering and demouldability play a decisive role in the mould quality. The limiting factor for the production of wax models is therefore not the transfer of the microstructured surface to the wax, but the wax properties which are significant for the creation of the macro component. These include, e.g. the viscosity (filling the wax die), the thermal contraction (contraction and shrink marks during solidification) and the thermal expansion (fissures in the gypsum due to de-waxing).

The precision casting mould can only be produced from gypsum bonded investment materials as in dental casting. Other mould materials used in casting technology for lost moulds are either too coarse grained (sand moulds) or too hard (ceramic mould shells, phosphate bonded investment materials). Therefore they cannot be removed from microstructured aluminium cast parts without damaging the microstructured surfaces. Other requirements placed on the mould material include sufficient mechanical properties to withstand any loads which occur during the subsequent process steps. The mould material is stressed during the de-waxing process by the expansion of the wax and by its own expansion and by contraction during the subsequent firing and cooling processes. Furthermore, the mould material must also be able to withstand the stresses, which occur during the later casting of the metal, which can cause mould erosion or penetrations. The comparison of various commercially available gypsum bonded investment materials, characterised by different chemical compositions and grain size distributions, exhibits minimal differences in the moulding accuracy of the microstructured surfaces with structure sizes between 100 and 10 µm (structure height at an aspect ratio of 1). On the other hand, the mechanical properties of the tested investment materials differ significantly. As illustrated in Fig. 5.214, the mould surface consists of gypsum crystals of different size and orientation. In some places, fire resistant elements contained in the investment material, such as quartz or cristobalite, appear at the surface. It is also noticeable that the investment materials show different surface roughnesses. While some investment materials build up a very dense network of gypsum crystals (gypsum modification:

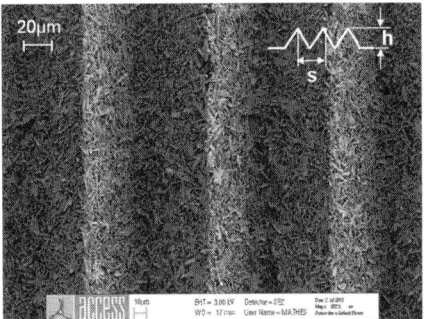 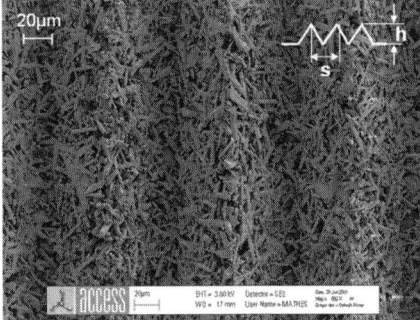

Fig. 5.214 Surface quality of a riblet structure. *Left*: investment material Mo28, Hoben Int. Ltd. and Hinrivest G, *right*: Ernst Hinrichs GmbH. Microstructure geometry: hight h = 50 µm, distance s = 100 µm

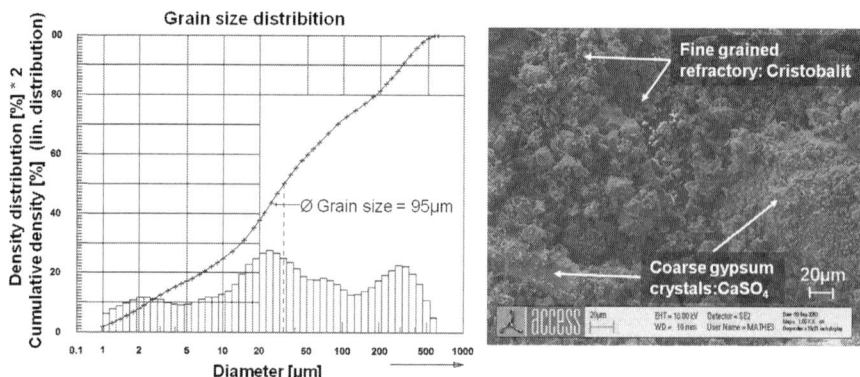

Fig. 5.215 *Left*, distribution of grain sizes and *right*, SEM image of the investment material basic material Mo28, Hoben Int. Ltd.

α-hydrate) and therefore exhibit a lower surface porosity, the surface of other materials is rough or even fissured, see Fig. 5.214. The fineness of the gypsum crystal mesh therefore limits the achievable reproduction accuracy of the microstructured surface.

Chemical analyses show that the main components of all investment materials are quartz or cristobalite fire-resistant components and gypsum ($CaSO_4$) with a ratio of 70% fire-resistant components to 30% gypsum which equates to an optimum composition (Kim et al. 2008). The average grain size of the various investment materials is between 16 and 95 μm, see Fig. 5.215. EDX analyses show that fire-resistant particles occur in a range of < 1–70 μm while the gypsum crystals are much bigger with sizes up to 500 μm and more, see Fig. 5.215. However, as the gypsum dissolves when mixed with water, the coarse gypsum crystals of the original powder have no negative influence on the modelling accuracy of the microstructured surface.

The determination of the compressive strengths and surface hardnesses in both the green state and the fired state show that investment materials with a very dense and interwoven gypsum crystal mesh have considerably greater mechanical properties, Fig. 5.216. The compressive strengths of the different materials in green state lie between 1 and 10 MPa. In fired state, the compressive strength is considerably lower since the gypsum loses the intercrystalline water and therefore its cohesion. They exhibit values of between 0.2 and 2 MPa. Despite the loss of strength, firing the moulds is unavoidable as the intercrystalline water would interact with the molten aluminium. The results of the hardness testing verify the compressive strength values and are between 3 and 20 HV (0.1). When used for casting moulds, investment materials with high strength (absorption of stress during firing and cooling and pressure load during casting) and high hardness (sufficient resistance towards abrasion and mould erosion) are preferred.

The results presented below confirm that gypsum bonded investment materials with a fine gypsum crystal mesh and good mechanical properties are a suitable mould material for casting microstructured surfaces.

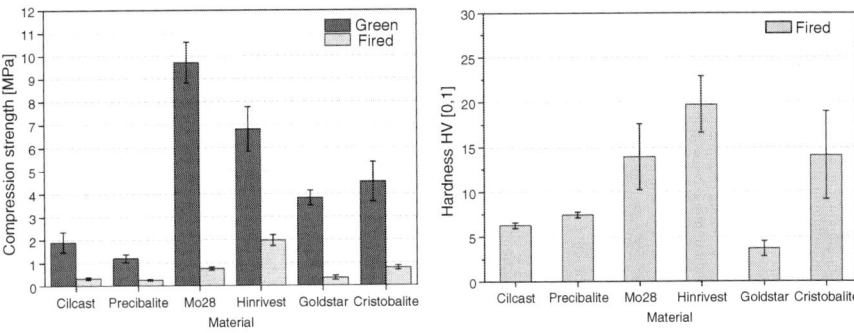

Fig. 5.216 *Left*, comparison of compressive strength and *right*, hardness of various investment materials exhibit different behaviour. Materials with high mechanical properties are preferred in order to withstand the stress during the casting process

The critical material property for the moulding accuracy of microstructures is the surface tension of the metal. This depends on both the temperature and pressure and is influenced by the process parameters casting temperature, moulding temperature and casting pressure, see Fig. 5.217. It is also possible to add alloying elements to the casting alloy to reduce the surface tension. Three standard casting alloys (AlSi7Mg, AlSi12, AlMg5) with good mechanical and good casting properties along with pure aluminium (Al99.99) and pure aluminium with a supplement of 1% bismuth, which considerably decreases the surface tension, are used, see Fig. 5.218. The materials are analysed according to the theoretical filling capability. These results are compared with the moulding accuracy of the microstructured surface.

The filling capability indicates how good a metal can mould contours. This is measured by the so-called bolt test whereby the molten metal is cast in the gap between two contacting, cylindrical steel bolts. The further the metal flows into the

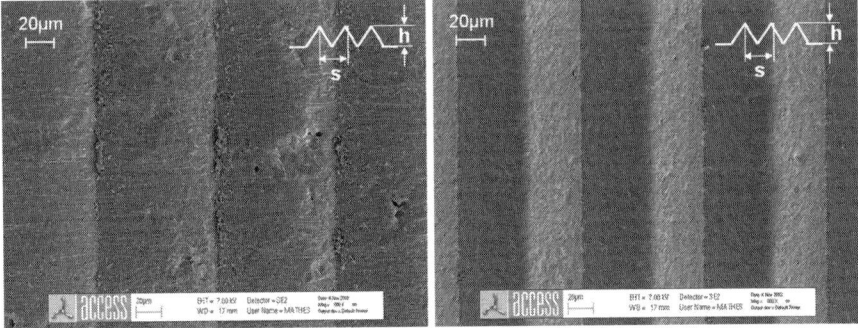

Fig. 5.217 SEM image of the microstructured surface of an AlSi12 alloy cast at a casting temperature of 740 °C and different moulding temperatures (*left*: room temperature; *right*: 420 °C) by gravity casting in an evacuated mould, Microstructure geometry: hight $h = 50$ µm, distance $s = 100$ µm

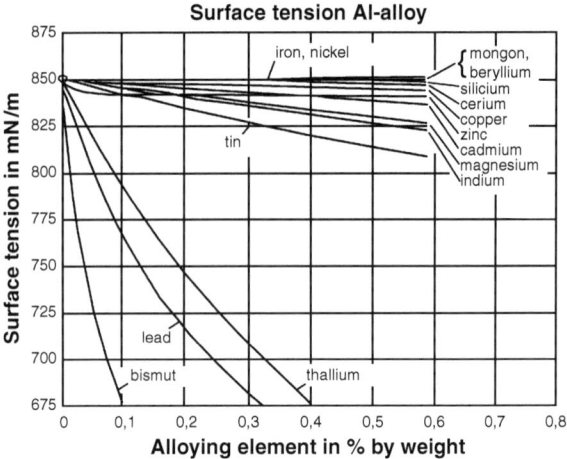

Fig. 5.218 Influence of alloying elements on the surface tension of aluminium alloys. (Lang 1973)

gap, the greater the filling capability. As illustrated in Fig. 5.219, the aluminium alloy with added bismuth exhibits the greatest filling capability.

In contrast to the results of the filling capability, all used alloys as well as pure aluminium mould the microstructurs on the surface of the gypsum bonded mould reproducible over a large area in the magnitude of 100μm to 10 μm (structure height at an aspect ratio of 1), see Fig. 5.220. An influence of the microstructured surface on the formation of the cast structure cannot be identified with regard to the riblet structures examined. Figure 5.221 illustrates an AlMg5 sample etched with Barker's solution. The grains of the structure on the surface are spread out across several microstructure features and do not deviate significantly from the grain sizes within the sample. The large grain size occurs due to the long solidification time (pre-heating temperature of the mould: 420 °C, low thermal conductivity of the gypsum bonded investment material). No change in the frequency of secondary precipitations of the

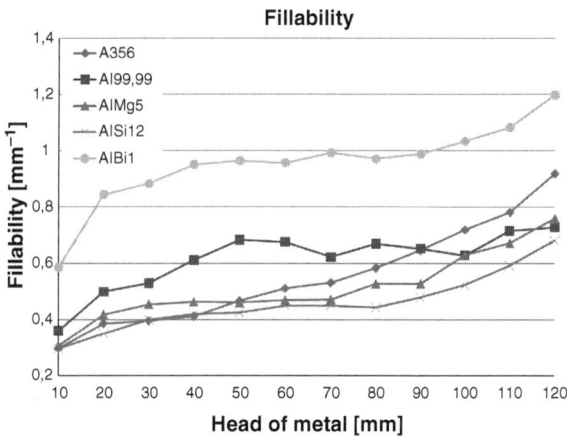

Fig. 5.219 The filling capability and therefore the ability to mould contours is considerably greater for the Al-alloy with added bismuth than for other alloys and pure aluminium since bismuth reduces the surface tension most

5 Hybrid Production Systems 677

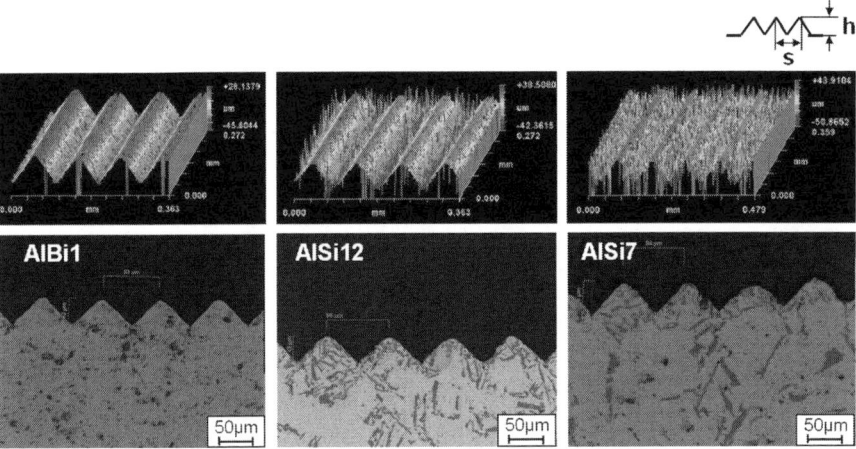

Fig. 5.220 Image of the microstructured surface of various casting alloys: white light interferometer (WIM) image (*top*) and metallographic cross-section (*bottom*), Microstructure geometry: height h = 50 μm, distance s = 50 μm, (noise in the WIM images is due to the different reflection in the casting surface)

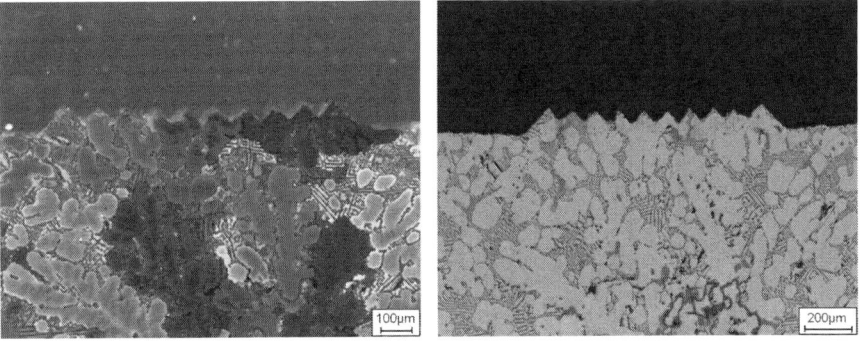

Fig. 5.221 *Left*: metallographic cross-section of an AlMg5 sample etched with Barker's solution. Observing with polarised light illustrates the growth of the structure grains over several microstructure features. *Right*: metallographic cross-section of the same AlMg5 sample. As can be seen, the Al-Mg eutectic is evenly distributed

alloys AlMg and AlSi (which form the eutectic) in the microstructured surface has yet been detected, Fig. 5.221.

With regard to the moulding accuracy of aluminium alloys, it is assumed that mould materials with high contour precision are used to mould microstructured surfaces even smaller than those described.

An air restrictor for a Formula Student racing car is cast as a demonstration cast part, Fig. 5.222. The Formula Student racing series is a design competition for students. The regulations dictate every vehicle to have an air restrictor. This restricts the effective cross-section through which air is drawn in the engine. Technical microstructured surfaces are used here to increase the air flow through the air restrictor

Fig. 5.222 Demonstrator casting: air restrictor with technical shark skin (Formula Student car of the Season 2010)

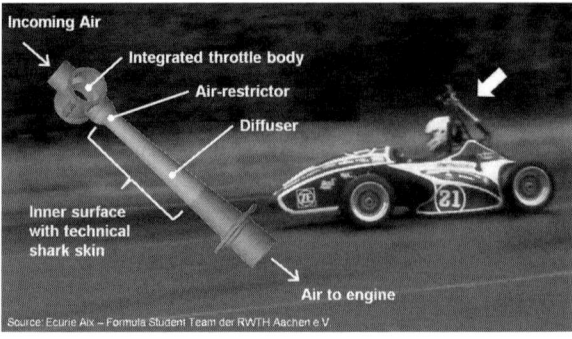

by applying a technical shark skin to the inner surface of the casting. Dut to the greater air flow the engine output power is increased which provides an advantage in the dynamic race assessment of the competition.

The design of the microfeature geometry requires knowledge of the wall shear stresses occurring on the smooth inner surface of the air restrictor. These are simulated using the flow software StarCD. Using simplified assumptions (Ivanov et al. 2010) the riblet intervals of the technical shark skin are calculated according to the equations of Bechert et al. (1997). A microstructured surface consisting of riblets with a riblet distance of 18 μm and a riblet height of 9 μm is determined.

Transferring a microstructured surface onto the inner surface of a component goes along with challenges and requires new tool concepts. There are basically two alternatives to achieve this in investment casting. On the one hand, the microstructured surface can be moulded onto the wax model using an structured original model. On the other hand, a microstructured lost core can be used to mould the micorstructured surface onto the casting. When moulding the microstructured surface onto the inner surface of the wax model, it is necessary to demould the original model after solidification of the wax without damaging the microstructured surface. In the case of the air restrictor, this is only possible with complicated tools containing several cores. Therefore, the alternative microstructuring of a lost core made of gypsum-bonded investment material which dissolves in water after casting is described below.

The microstructured gypsum core is moulded in a core box which is equipped with a flexible silicone inlay. Once the gypsum has dried in the core box, the silicone inlay is removed together with the gypsum core from the core box. The flexible silicone inlay is then peeled from the core so that the microstructured surface is not damaged as it features undercuts with regards to the parting plane, Fig. 5.223. The core is then inserted into the wax die and coated with wax. The advantage here is that the wax need not mould the microstructured surface since the wax is melted out again and the structure remains in the gypsum core.

The component is cast as described above in an aluminium standard casting alloy in a counter pressure casting process whereby the cavity is evacuated. As can be seen from Fig. 5.223 the technical shark skin is moulded in the air restrictor. In a subsequent step, the process is optimised for the reproducible, high quality production of the microstructures over the entire casting area. The air restrictor will then be

5 Hybrid Production Systems

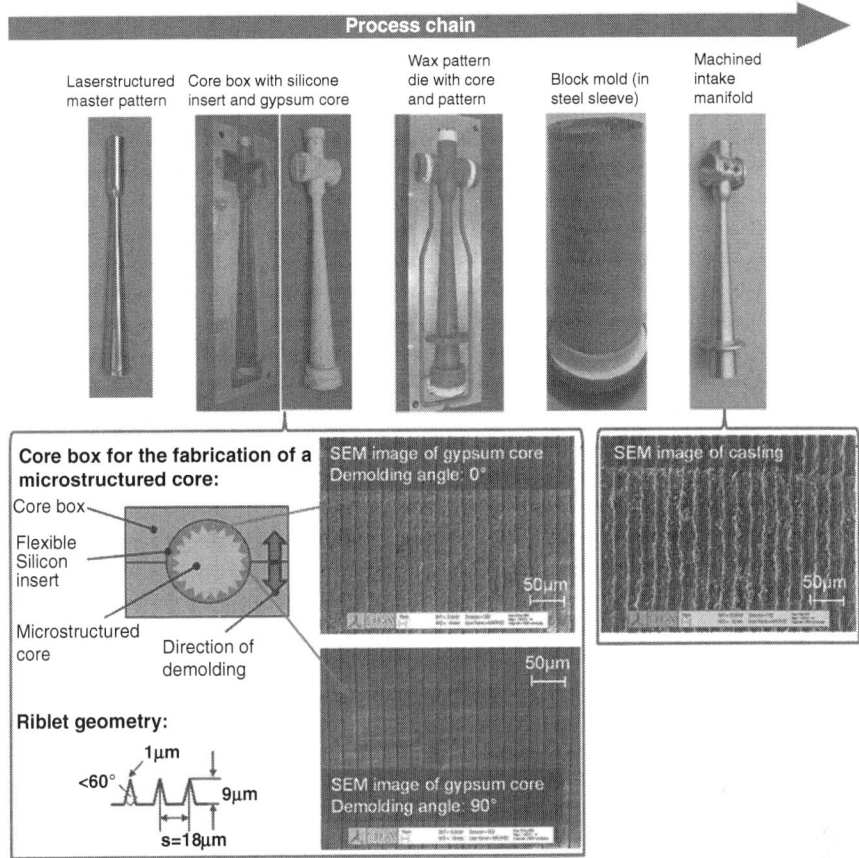

Fig. 5.223 Process chain for manufacturing the air restrictor with technical shark skin on the inner side. A flexible silicone insert is used in the core box to demold the microstructure of the gypsum core without causing any damage

assembled and tested on an engine test bench for the intended increase in the output of the motor.

5.6.5 Industrial Relevance

Microstructured components with functional surfaces are already being used in some initial applications in industry. They are usually superior to conventional products by their additional functionality. The field of application is limited however at present to single use products or optical applications where the surface is totally enclosed by a transparent coating and therefore protected. For long-term applications so far the proof of long-term functionality and/or the production of a wear resistant surface texture is missing.

In order to be able to use these technologies economically, both the primary forming process and the tool technology must be available and adjusted to each other. Only after this adjustment of the tool and the moulding process regarding each other, components with functional surfaces can be manufactured reproducible and economical.

Production of the tools requires that both the structuring and the coating are adapted to one another and to the structure size. Both is possible: coating of laser-structured moulds and laser structuring of coatings. In this respect, laser structuring is a fast and economical alternative to the established processes—depending on the required structure it may even be the only possible economic alternative, e.g. for the production of superposed structures.

A suitable coating enables both the processing of new polymer materials and the opportunity to lower costs, which is achieved by reducing the required cleaning as well as by reducing the wear. Thus it is at the same time possible to achieve an increase in value by new and innovative products and lowering of additional costs. However, it must be noted that not every coating is suitable for every material. A suitable coating must be selected in cooperation between coaters, machine manufacturers and the users of the production technology.

Microstructured plastics films allow a wide range of applications as such films can be applied to conventional components and thereby provide them with functional properties. Some surface structures which can be produced on foil surfaces are presented in Fig. 5.224. The modular combination of component and film provides a fast and economically interesting implementation of the analysed technology.

The simulation of the extrusion embossing process with regard to the generated temperature distribution allows the operating point to be set accurately and quickly. The effects of variation of the process parameters can be estimated in this way and time-consuming "trial and error" methods when running a process can be saved. In this way it is also possible to modify specifically the produced surface structures. A targeted partial filling is therefore also possible with the given tool geometry and structure of the negative mould on the roller in order to consciously limit aspect ratios. It is also possible to achieve an intend stretching which can cause "filament-like" structures which are not given by the mould. The specific combination of tool,

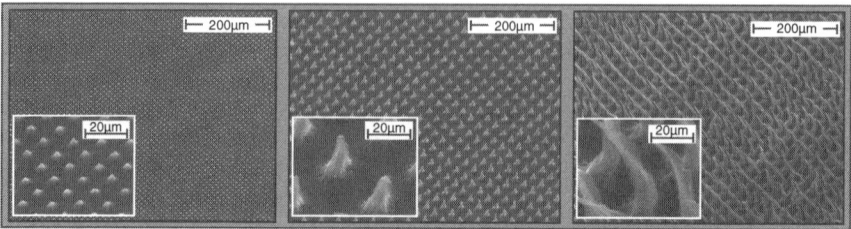

Fig. 5.224 Examples of film surfaces manufactured by variothermal extrusion embossing

Fig. 5.225 Hydrophobic effect on an injection moulded component

material and the process therefore allows influencing the produced microstructures in many areas. A simulated approach can also reduce both the risk and the research effort involved in running a new process.

The developed tools and the associated plant technologies enable for the first time to mould microstructured surfaces with functional effects directly within the injection moulding process. The temporary increase of cavity temperature makes it possible to achieve a high moulding quality of the microstructures. Conversely, this leads to an increase in the surface functionality. The cycle time is not considerably longer than with the conventional injection moulding process as the injection mould is heated up only on the mould surface whereby the cooling time remains almost unchanged.

The systems presented for variothermal tool temperature control run fully automated and are high reproducible in cyclic mode. The integration of both the primary shaping process and functionalisation of the surface into the injection moulding process allows saving the following process steps for functionalising the surfaces, e.g. by etching, varnish or coating. The results of the analysis show that superhydrophobic surfaces can be produced directly as part of the injection moulding process, see Fig. 5.225.

Components with the effect presented above can be designed almost arbitrarily. Neither the structure size nor the surface shape onto which the structure is applied are limited. In this way, it is possible to produce, e.g. completely drainable reservoirs or self-cleaning components by injection moulding. Other effects such as reduced flow resistance or optical effects are also feasible.

In the investment casting process, components with a unit weight of a few grams to 200 kg can be produced in quantities of 20 to more than 100,000 units. The use of lost models and one-piece lost moulds gives the investment casting process a considerable amount of freedom in design. This allows technical solutions which are either impossible or uneconomical to produce with other processes, e.g. freeform surfaces (compressor wheels) or defined inner contours, serrations, etc. The good accuracy of dimensions and high surface quality also reduce the number of the finishing processes of the casting parts considerably. This advantage is particularly significant with regard to materials that are hard to machine. In investment casting, it is possible to process nearly all metals and alloys. The most commonly used investment casting materials are steels, Al-, Ti-alloys, superalloys based on Ni and Co and Cu-based alloys. Therefore, with investment casting a very large range of materials and components can be produced with regard to metal components with microstructured surfaces.

Possible applications include the optimisation of fluid dynamics where the components are already produced by casting. In this respect, flow channels or fluid dynamic parts, for example in engines (suction parts and exhaust gas system, conduits in the injection system, cooling and fuel channels) or even components such as pump housings, turbocharger wheels, propellers, ships screws or compactor and turbine blades offer a great potential for an enhanced efficient use. Applications are only reasonable for constant flow conditions as the microstrures have to be designed for specific flow conditions.

The previously described demonstrator "air restrictor" is based on the flow-related optimisation of a peripheral engine part. The producibility of a microstructured surface on the inner surface of a casting part is shown. Furthermore, the technological challenges are demonstrated which have to be overcome in the field of tooling technology and the necessary moulding accuracy.

In addition to the flow-related applications, other potential applications are in the field of adhesion and tribology. In engine technology it is possible to use microstructured surfaces to improve lubrication between different tribological pairings such as piston and cylinder or crankshaft and camshaft bearings. Also premium lifestyle products, such as casted bathroom fittings or cooking utensils can profit of anti-adhesive surfaces. Castings exposed to extreme weather conditions, e.g. car rims or underbodies of vehicles, provide an interesting application for microstructured surfaces particularly with regard to corrosion protection. Another large application field is offered by industrial chemical plants, power stations or pipelines whose complex pipe, distribution and pump systems are difficult to access and clean. This field of application also includes the food industry with high purity specifications in their technical facilities.

5.6.6 Future Research Topics

The results presented here demonstrate the feasibility of a continuous process chain suitable for mass production of components with functional surfaces. The first components with superhydrophobic surfaces are already being produced. For a wide use of components with functional surface the process chains have to be developed further to understand the reciprocal influences of product, functionality, producibility, reproducibility and process stability. With this process knowledge it is possible to define the structures on the moulds which are needed to generate a given micro- or nanostructure. Furthermore, the surface effects in nature and technology must be better understood to design the structures with regard to the shape, size and tolerances. In order to be able to produce an optimum effect using functional structures, the design of the structures must be integrated in the process chain that the structures are not only designed according to their optimum effect, but also with regard to their producibility within the corresponding process chain.

In addition to the continuing development of the whole process chain further research is required within the individual processes. For tool production, this particularly includes the fast, seamless and large-scale structuring of free-form surfaces. This initially requires a fast scanning technology in order to be able to use the new, even more powerful lasers effectively for machining. High repetition rates (several 10 MHz) and large output powers (some 100 W to kW average output power) of current laser call for future scanning systems with scan speeds in the range of several 100 m/s. A measuring technology must directly be integrated in the system which is able to accurately measure the free-form surface. As a positioning error of even a few micrometers is visible in structured areas, there are considerable demands regarding the accuracy of the measuring technology. The measured data is required for both adjusting the focus and precise alignment of the consecutive scan fields. Furthermore, neither strategy nor software solution is available for the seamless structuring of large free-form surfaces by aligning consecutive scan fields without any visible optical intersection.

In the field of nanostructuring, the currently developed processes must be developed to a homogeneous, large-scale structuring. This includes both a homogeneous structuring within the individual machining spots as well as a strategy for fast machining large areas. Analogous to microstructuring, this must be transferred to system technology.

As part of the coating development, research must be undertaken on highly ionised plasma coating processes as these will enable a better coating of complex geometries and small structures. High Power Pulse Magnetron Sputtering (HPPMS) technology provides an excellent coating uniformity and a low mean roughness value. In this context the coatable structure sizes and aspect ratios as well as the practicability of the HPPMS technology in mass scale for the coating of extrusion rollers need to be evaluated.

This also raises the questions, how well the coating of micro- and nanostructured surfaces can protect against wear in series production and how much the demoulding and replication accuracy of these structures are increased compared to uncoated metal surfaces. A detailed examination of the exact processes during cooling and demoulding of the plastics component from the injection mould has to be done. This includes volume shrinkage, crystallisation, residual stress, orientation and relaxation as well as the dependency of the specific surface energy on the temperature.

In the field of plastics injection moulding, a process has been developed which enables the production of microstructured functional surfaces in a process for mass production by a shortened process chain. A microstructured mould insert with a superposed nanostructure and replication in an injection moulding process with variothermal process control is essential in this respect. The developed and implemented system for the dynamic heating of injection moulds using diode laser radiation is however currently limited to the heating of planar surfaces. To investigate the moulding of three-dimensional components with microstructured surfaces the heating process needs to be developed further. The use of a high-speed laser scanner basically offers the option of heating almost any surface geometry with a focussed, fast-moving laser beam. The system is also suitable for heating large areas of the cavity or several cavities quasi-simultaneously. A mould heating process using a laser scanner must therefore be implemented and validated for injection moulding.

Besides the precise moulding of microstructures by a suitable heating during injection moulding also the subsequent ejection of the component has a major influence on the formation of the structures. The demoulding of three-dimensional, microstructured surfaces must therefore be analysed with regard to sample geometries.

Microstructures manufactured using ultra short pulse lasers have a superposed finestructure which is also filled with plastics melt during the replication in a variothermal injection moulding process. This anchors the plastic part in the microstructures of the mould insert. However, the polymer must be ductile in order to safely demould the component. Brittle plastics break during demoulding of the structures and therefore clog or destroy the mould insert. In order to overcome this process limitation and to be able to mould brittle plastics such as PMMA and PC the finestructures must be avoided. A suitable combination of material properties and process parameters will then be determined to precisely mould and safely demould these structures.

An important issue for the economically relevant use of microstructured plastics surfaces is the durability of the microstructures. This includes both the long-term behaviour during usage, for example when wetting with a liquid, as well as the resistance of the structures to external mechanical effects. Currently no process is available which effectively increases the durability of microstructured plastic surfaces. In future, suitable attempts have to be made to define new areas of application.

For the extrusion embossing processes an integrated manufacturing system for hybrid, functionalised components needs to be developed as well as the determination of a suitable geometric design of the surface structure and the used materials. The interaction between the manufacturing process and the structural geometry is of particular importance in this respect. The moulding of even finer structures with increasing aspect ratio is technically very challenging. In future, the potential of the extrusion embossing process must be investigated.

By implementing solutions to these issues it is possible to manufacture plastic films with different functional surfaces in the field of optics, haptics, tribology and hydrophobia. These foils can be applied on products to create new functionalities and therefore equip them with new functionalities or optimise them for their purpose. The reliable handling of the necessary manufacturing and tooling technologies enables the flexible production of geometrically complex structures of very different aspect ratios, geometries, tolerances, layouts and interfaces, with respect to the application and material. The production of co-extruded films with microstructured surface will become more significant in the future. Apart from the complexity, it is possible to increase the mould quality by combining different materials. A low-viscosity embossed layer combined with a high-viscosity coating layer e.g. can be a future research topics.

However, the challenge for all potential applications is the understanding of the physical and chemical effects in nature for transferring into technical applications. For the future it is therefore absolutely necessary to define design rules to determine the microstructure geometry for various applications.

In this regard, it is necessary to develop quality requirements for the casting process. This raises the issue of e.g. the permissible deviation of the manufactured microstructured surface from the ideal microstructure. The experimental determination of such quality criteria is best done in the investment casting process as it is a cost effective way of producing prototypes. When the manufacturing of microstructured surfaces on cast parts has been proofed, the technology has to be transferred to processes for mass production.

The focus for casting rests upon processes using permanent moulds of tool steel, as they promise a very high mould quality. Two examples are die casting and gravity die casting. Existing knowledge in tool design for plastic manufacturing (e.g. in the systems engineering for variothermal tempering and the demoulding of microstructures from the tool) can be used for this purpose. However, there is a considerable need for research in the casting of molten metals in steel tools as significantly higher process temperatures are required for processing metals (aluminium, magnesium and zinc) in permanent mould casting than for processing plastics. In combination with the low-temperature a high temperature gradient which leads to a greater cooling rate is formed. In this instance it is necessary to analyse the thermal efficiency of the tool and to understand its influence on the modelling of the microstructured surfaces in the casting. As a result of the high temperatures and the chemical affinity of the casting metal to the tool steel, the wear of cast tools is an enormous cost factor and has to be examined. In order to reduce the adhesion of the cast metal to the tool, a release agent is used which can clog the microstructures on the surface. The effect of the release agent on the moulding accuracy of the microstructures is therefore an important focus of future research. An alternative to release agents are anti-adhesive coatings which are also capable of minimizing tool wear specifically oft the microstructures.

The simulation of the process chain is a very useful aid for developing suitable casting processes. It is primarily used to accurately predict the process parameters required for an optimal moulding. To achieve this, functions for criteria have to be defined and implemented in a standard simulation software, such as Magmasoft or Fluent. These criteria for example describe the moulding of the microstructure during filling.

References

Al-Samman T (2008) Magnesium – the role of crystallographic texture, deformation conditions, and alloying elements on formability. Cuvillier, Göttingen

Altintas Y, Brecher C, Weck M, Witt S (2005) Virtual machine tool. CIRP Ann Manuf Technol Keynotes 54(2):115–138

Altshuller G (1996) And suddenly the inventor appeared: TRIZ, the theory of inventive problem solving. Technical Innovation Center, Worchester

Altshuller G (1999) The innovation algorithm: TRIZ, systematic innovation and technical creativity. Technical Innovation Center, Worchester

Altshuller G (2002) 40 Principles: TRIZ keys to technical innovation. Technical Innovation Center, Worchester

Ames J (2008) Systematische Untersuchung der Beeinflussung des Werkstoffflusses bei der Inkrementellen Blechumformung mit CNC-Werkzeugmaschinen. Umformtechni-sche Schriften, vol 140. Shaker, Aachen

Amino H, Lu Y, Ozawa S, Fukuda K, Maki TL (2002) Dieless NC forming of automotive service parts. In: Advanced technology of plasticity. Proceedings of the the 7th ITCP, Yokohama

Arai T, Kato R, Fujita M (2010) Assessment of operator stress induced by robot collaboration in assembly. CIRP Ann Manuf Technol 59:5–8

Bachmann F et al. (1999) Hybride Prozesse – Neue Wege zu anspruchsvollen Produkten. In: Eversheim W et al. (ed) Aachener Werkzeugmaschinen Kolloquium – Wettbewerbsfaktor Produktionstechnik. Aachen, pp 243–278

Bagcivan N, Bobzin K, Eilbracht S, Gillner A, Hartmann C, Klaiber F, Michaeli W, Scharf M, Theiß S (2010) Lotus-Effekt für Massenprodukte. Plastverarbeiter 66(9):104–106

Bambach M (2008) Process strategies and modelling approaches for asymmetric in-cremental sheet forming. Umformtechnische Schriften, vol 139. Shaker, Aachen

Barberoglou M, Zorba V, Stratakis E, Spanakis E, Tzanetakis P, Anastasiadis SH, Fotakis C (2009) Bio-inspired water repellent surfaces produced by ultrafast laser structuring of silicon. Appl Surf Sci 255(10):5425–5429

Bärsch N (2001) Untersuchung von Verfahren zur Erzeugung von mikrostrukturierten Oberflächen auf extrudierten Folien und Platten, Institut für Kunststoffverarbeitung, RWTH Aachen, unveröffentlichte Diplomarbeit, Betreuer: P. Blömer, F. van Lück

Barthlott W, Neinhuis C (1997) Purity of the sacred lotus, or escape from contamination in biological surfaces. Planta 202(1):1–8

Baumeister G, Brando O, Rögner J (2008) Microcasting of Al bronze: influence of casting parameters on the microstructure and the mechanical properties. Microsyst Technol, Springer 14(9–11):1647–1655

Bausch S, Groll K (2003) Perspektiven für die laserunterstützte Zerspanung. Wt Werkstatttechnik online Jahrgang 93(H.6):457–461

BDI (2005) Intelligenter produzieren – 32 Thesen zur Forschung für die Zukunft der industriellen Produktion. Hrsg. Bundesverband der deutschen Industrie e. V (BDI), Fraunhofer-Gesellschaft zur Förderung der angewandten Forschung e. V. (FhG), Verband deutscher Maschinen- und Anlagenbau e. V. (VDMA)

Bechert DW, Bruse M, Hage W, van der Hoeven JGT, Hoppe G (1997) Experiments on drag-reducing surfaces and their optimization with an adjustable geometry. J Fluid Mech 338:59–87

Beckemper S, Huang J, Gillner A, Wang K (2010) Generation of periodic micro- and nano-structures by parametercontrolled three-beam laser interference technique. Proceedings of the 11th international symposium on laser precision microfabrication, ON-021

Behrens B-A, Olle P (2008) Berücksichtigung der Gefügeumwandlung in der numerischen Simulation des Presshärtens. In: Tagungsband zur Internationalen Konferenz „Neuere Entwicklungen in der Blechumformung". Fellbach, 3–4 Juli 2008. Frankfurt: Mat Info, 2008, pp 263–281

Bekesi J, Kaakkunen J, Michaeli W, Klaiber F, Schöngart M, Ihlemann J, Simon P (2010) Fast fabrication of superhydrophobic surfaces on polypropylene by replication of short-pulse laser strucutred moulds. Appl Phys A 99(4):691–695

Benkowsky G (1990) Induktionserwärmung, 5th edn. Verlag Technik GmbH, Berlin

Bichmann S, Emonts M, Glasmacher L, Kordt M, Groll K (2005) Automatisierte Reparaturzelle OptoRep. wt Werkstattstechnik online 11/12;1–8

Bienk EJ, Mikkelsen NJ (1997) Application of advanced surface treatment technologies in the modern plastics moulding industry. Wear 207:6–9

Biermann T, Göttmann A, Zettler J, Bambach M, Weisheit A, Hirt G, Poprawe R (2009) Hybrid laser assisted incremental sheet forming: Improving formability of Ti- and Mg-based alloys. In: Conference proceedings of international conference on lasers in manufacturing (LiM2009), München

Bischof C (1993) ND-Plasmatechnik im Umfeld der Haftungsproblematik bei Metall-Polymer-Verbunden. Materialwiss Werkstofftech 24:33–41

Bjørke Ø(1995) Manufacturing systems theory – a geometric approach to connection. Tapir, Trondheim

Bläsi B, Aufderheide K, Abbot S (2002) Entspiegeln mit Mottenaugenstrukturen. Spritzgießen funktionaler mikrostrukturierter Oberflächen. Kunststoffe 92(5):50–53

Bobzin K, Nickel R, Bagcivan N, Manz FD (2007) PVD-coatings in injection molding machines for processing optical polymers. Plasma Process Polym 4:144–149

Bobzin K, Michaeli W, Bagcivan N, Immich P, Klaiber F, Theiß S (2009a) Chromium based PVD coatings for injection moulding tools. In: Fischer A, Bobzin K (eds) Friction, wear and wear protection. Wiley-VCH, Weinheim, pp 737–743

Bobzin K, Bagcivan N, Immich P, Theiß S (2009b) Arc ion plating process monitoring by optical emission spectroscopy exemplified for chromium containing coatings. Plasma Process Polym 6(S1):357–361

Bobzin K, Bagcivan N, Ewering M, Gillner A, Beckemper S, Hartmann C, Theiß S (2011) Nano structured physical vapor deposited coatings by means of picosecond laser radiation. J Nanosci Nanomater. doi:10.1166/jnn.2011.3468 (in press)

Borsdorf R (2007) Methodischer Ansatz zur Integration von Technologiewissen in den Produktentwicklungsprozess. Dissertation RWTH Aachen. In: Berichte aus der Produktionstechnik, vol 17. Shaker, Aachen

Brecher C, Hirsch P (2004) Compensation of thermo-elastic machine tool deformation based on control internal data. CIRP Ann Manuf Technol 53(1):299–304

Brecher C, Wissmann A (2009) Optimierung des thermischen Verhaltens von Fräsmaschinen. Z Wirtsch Fabr 6:437–441

Brecher C, Karlberger A (2010) Steuerungskonzept für ein hybrides Bearbeitungszentrum. In: Hybride Technologien in der Produktion, Fortschritt-Berichte VDI Reihe 2, Fertigungstechnik. VDI, Düsseldorf

Brecher C, Schröter B, Almeida C (2005) Development and programming of portable robot systems for material handling tasks. In: Digital Proceedings of the 3rd CIRP International conference on reconfigurable manufacturing, NSF Engineering Research Center for Reconfigurable Manufacturing Systems, Ann Arbor, USA

Brecher C, Klocke F, Wenzel C, Emonts M, Frank J (2008) KombiMasch – hybride Werkzeugmaschine zur Verkürzung der Prozesskette. wt Werkstattstechnik online 7(8):525–532

Brecher C, Breitbach T, Do-Khac D, Herfs W, Karlberger A, Klein W, Heinen D, Rosen C-J (2009a) Prozessketten hochveredelter Produkte verkürzen. ZWF 9:739–744

Brecher C, Esser M, Witt S (2009b) Interaction of manufacturing process and machine tool. CIRP Ann Manuf Technol Keynotes 58(2):588–607

Brecher C, Esser M, Witt S (2009c) Simulation of the process stability of hpc milling operations. Machin Sci Technol 13(1):20–35

Brecher C, Klein W, Trofimov Y (2010) Modellierung der doppelspindligen Fräsbearbeitung. ZWF 7(8):643–648

Brinksmeier E (1991) Prozeß- und Werkstückqualität in der Feinbearbeitung. Habilitationsschrift, Universität Hannover, VDI, Reihe 2: Fertigungstechnik, Nr. 234

Brinksmeier E, Michaeli W (2005) Mikrostrukturierung von Prägewalzen und Kunststofffolien, Labor für Mikrozerspanung, Universität Bremen; Institut für Kunststoffverarbeitung, RWTH Aachen, Abschlussbericht zum AiF-Vorhaben Nr. 105 ZN

Brinksmeier E, Aurich JC, Govekar E, Heinzel C, Hoffmeister H-W, Klocke F, Peters J, Rentsch R, Stephenson D-J, Uhlmann E, Weinert K, Wittmann M (2006) Advances in modeling and simulation of grinding processes. Ann CIRP 55(2):667–696

Brunhuber E (1991) Praxis der Druckgussfertigung. Schiele and Schön, Berlin

Bruzzone AAG, Costa HL, Lonardo PM, Lucca DA (2008) Advances in engineered surfaces for functional performance. CIRP Ann Manuf Technol 57(2):750–769

Bryan J (1968) International status of thermal error research. CIRP Ann Manuf Technol 16:203–215

Bryan J (1990) International status of thermal error research. CIRP Ann Manuf Technol 39(2):645–656

Bührig-Polaczek A et al. (2010) Sheet metal components reinforced by light metal cast structures, Session Transport Paper 03, Aluminium 2010 Conference, 09/2010, pp 1–11

Bürkle E, Burr A (2007) In drei Sekunden von 100 auf 140 Grad. Kunststoffe 97(10):210–214

Callies M, Chen Y, Marty F, Pépin A, Quéré D (2005) Microfabricated textured surfaces for super-hydrophobicity investigations. Microelectron Eng 78–79:100–105

Campbell M, Sharp DN, Harrison MT, Denning RG, Turberfield AJ (2000) Fabrication of photonic crystals for the visible spectrum by holographic lithography. Nature 404:53–56

Cannon AH, King WP (2009) Casting metal microstructures from a flexible and reusable mold. J Micromech Microeng 19:095016

Cavallucci D, Weil R (2001) Integrating Altshuller's development laws for technical systems into the design process. Ann CIRP 50(1):115–120

CH 268258 (1950) Revetement hydrofuge. Patentschrift, Schweizer Patentorganisation, 16.08.1950

Charmeux JF, Minev R, Dimov S, Brousseau E, Minev E, Harrysson U (2007) Benchmarking of three processes for producing castings incorporating micro/mesoscale features with a high aspect ratio. J Eng Manuf Part B 221(4):577–588

Coyle DJ (2007) Process and apparatus for embossing a film surface. Patent application publication, US 2007/0126145 A1 06 July 2007

Cunha L, Andritschky M, Pischow K, Wang Z, Zarychta A, Miranda AS, Cunha AM (2002) Performance of chromium nitride and titanium nitride coatings during plastic injection moulding. Surf Coat Technol 153:160–165

DE 19715 906 A1 (1997) Wasserabweisende Beschichtungszusammensetzung und Beschichtungsfilme sowie beschichtete Gegenstände, für die diese verwendet werden. Patentschrift, Deutsches Patent- und Markenamt, 06 Nov 1997

DE 102 10 668 A1 (2003) Vorrichtung, hergestellt durch Spritzgussverfahren, zur Aufbewahrung von Flüssigkeiten und Verfahren zur Herstellung dieser Vorrichtung. Patentschrift, Deutsches Patent- und Markenamt, 25 Sept 2003

Deutsches Insitut für Normung e. V. DIN EN 657 (2005) Thermisches Spritzen – Begriffe, Einteilung

Diestel R (2006) Graph theory, 3rd edn. Springer, Heidelberg

Diettrich J (2010) Koaxiale Strahlführungs- und -formungssysteme für die hybride Lasermaterialbearbeitung. Dissertation, im Prozess der Veröffentlichung

Dolles M (2007) Laserauftragschweißen unter dem Einfluss magnetischer und elektrischer Felder. Verlag Mainz, Aachen

Duarte M, Lasagni A, Giovanelli R, Narciso J, Louis E, Mücklich F (2008) Increasing lubricant lifetime by grooving periodical patterns using laser interference metallurgy. Adv Eng Mater 10(6):554–558

Duflou JR, Lauwers B, Verbert J, Tunckol Y, De Baerdemaeker H (2005) Achievable accuracy in single point incremental forming – case studies. In: Proceedings of the 8th esaform conference. Cluj-Napoca, Romania, pp 675–678

Duflou JR, Callebaut B, Verbert J, De Baerdemaeker H (2007) Laser assisted incre-mental forming: formability and accuracy improvement. CIRP Ann Manuf Technol 56(1):273–276

Duprè A (1869) Theorie Méchanique de la Chaleur. Gauthier-Villars, Paris, p 369

Ehrenstein GW (ed) (2004) Handbuch Kunststoff-Verbindungstechnik. Carl Hanser, München, p 710 ISBN 3446226680

Ehrlenspiel K (2009) Integrierte Produktentwicklung – Denkabläufe, Methodeneinsatz, Zusammenarbeit. 4th edn. Carl Hanser, München

Eifel S, Dohrn A, Gillner A (2010) Quality aspects in high power ultra short pulse laser ablation, In: Proceedings of the 11th international symposium on laser precision microfabrication (LPM), Stuttgart, ON-033

Emonts M (2010) Laserunterstütztes Scherschneiden von hochfesten Blechwerkstoffen. Dissertation, RWTH Aachen

Engel Austria GmbH (2007) Variotherme Werkzeugtemperierung neu entdeckt, Kunststoffe 97(7):64–65

Erhardt R, Schepp F, Schmoeckel D (1999) Micro forming with local part heating by laser irradiation in transparent tools. In: Proceedings of the 7th international conference on sheet metal. Erlangen, 27–28 Sept 1999. Meisenbach, Bamberg, p 497–504

Escuti MJ, Crawford GP (2004) Holographic photonic crystals. Opt Eng 43:1973

Esser M (2010) Stabilitätssimulation für das HPC-Fräsen. Apprimus, Aachen

Eversheim W (2003) Innovationsmanagement für technische Produkte. Springer, Berlin

Feess E (2004) Mikroökonomie: Eine spieltheoretisch- und anwendungsorientierte Einführung, 3rd edn. Metropolis, Marburg

Fink B (1997) Entwicklung einer Methode zur Simulation der thermischen Vorgänge in dynamisch temperierten Prägewalzen, Institut für Kunststoffverarbeitung, RWTH Aachen, unveröffentlichte Diplomarbeit, Betreuer: P. Blömer

Fowkes FM (1964) Dispersion force contributions to surface and interfacial tensions, contact angles, and heats of immersion. Adv Chem 43:99–111

Freyer C (2007) Schichtweises drahtbasiertes Laserauftragschweißen und Fräsen zum Aufbau metallischer Bauteile. Shaker, Aachen

Fritzsche G, Burger D, Burböck W (2002) Verfahren zum Entgraten von metal-lischen Werkstoffen Deutsches Patentamt. Offenlegungsschrift DE 33 44 709 A1, 19

Gausemeier J, Berger T (2004) Ideenmanagement in der strategischen Produktplanung – Identifikation der Produkte und Geschäftsfelder von morgen. Konstruktion, vol 9. VDI, p 64–68

Gebhardt A (2007) Generative Fertigungsverfahren. Rapid prototyping – rapid tooling – rapid manufacturing. Carl Hanser, München

Geiger M, Merklein M (2007) Sheet metal forming – a new kind of forge for the future? Key Eng Mater 344:9–20

Geyer F (2008) Konduktion: Der Heiztrick für hochfesten Stahl. Blech In Form 4:56–58

Giessauf J, Pillwein G, Steinbichler G (2008) Die variotherme Temperierung wird produktionstauglich. Kunststoffe 98(8):87–92

Gillner A, Dohrn A, Hartmann C (2008a) High quality laser machining for tool and part manufacturing using innovative machining systems and laser beam sources. In: Proceedings of the 3rd international CIRP high performance cutting conference, vol 1, pp 199–208

Gillner A, Hartmann C, Dohrn A (2008b) High quality micro machining with tailored short and ultra short laser pulses. In: Proceedings of the 3rd pacific international conference on application of lasers and optics 2008, pp 685–690

Gleich H (2004) Zusammenhang zwischen Oberflächenenergien und Adhäsionsvermögen von Polymerwerkstoffen am Beispiel von PP und PBT und deren Beeinflussung durch die Niederdruck-Plasmatechnologie. Dissertation, Universität Duisburg

Gornik C (2005) Einsatz von PVD-Beschichtungen in Plastifiziereinheiten. Kunststoffe 95(11):44–49

Göttmann A, Diettrich J, Bergweiler G, Bambach M, Hirt G, Loosen P, Poprawe R (2011) Laser-assisted asymteric incremental sheet forming (LAISF) of titanium sheet metal parts. Accepted Paper in Prod Eng Res Dev

Gottstein G (2004) Physical foundations of materials science, 1st edn. Springer, Berlin

Groche P, Dörr J, Erhardt R (2003) Halbwarmblechumformung – Werkzeugkonzepte und Verfahrensgrenzen. In: Tagungsband zum UKD, 8. Umformtechnischen Kolloquium Darmstadt. Darmstadt, 2–3 Apr 2003, pp 77–93

Grundke K, Jacobasch HJ, Simon F, Schneider S (1995) Polymer surface modifications relevance to adhesion. In: Mittal KL (eds) VSP Utrecht, The Netherlands, pp 431–454

Guosheng Z, Fauchet PM, Siegman AE, van Driel HM (1982) Growth of spontaneous periodic surface structures on solids during laser illumination. Phys Rev B 26:5366–5381

Hadoush A, van den Boogaard T (2008) Time reduction in implicit single point incremental forming simulation by domain dedcomposition. In: Proceedings of the 7th international conference and workshop on numerical simulation of 3D sheet metal forming processes. Interlaken, pp 411–414

Harary F (1969) Graph theory. Addison-Wesley, Reading

Harary F, Norman R, Cartwright D (1965) Structural models: an introduction to the theory of directed graphs. Wiley, New York

Hartmann C, Gillner A (2007) Investigation on laser micro ablation of steel using ps-IR pulse bursts. J Laser Micro/Nanoeng 2(1):44–48

Hartwig A, Albinsky K (1999) Qualitätssicherung der Oberflächenvorbehandlung von Kunststoffen in der Fertigung durch selektive Farbreaktion; Abschlussbericht AiF-no. 10900 N/1

Heinze M (1998) Wear resistance of hard coatings in plastics processing. Surf Coat Technol 105:38–44

Hennes J (2000) Einsatz der Mikrotechnik in der Extrusion, Institut für Kunststoffverarbeitung, RWTH Aachen, Pomotionsvortrag

Herzinger S (2004) Hochpräzises Heißprägen. Kunststoffe 94(10):151–154

Hinzpeter U (2008) Induktives Erwärmen von Spritzgießwerkzeugen. Kunststoffe 98(1):21–23

Hinzpeter U (2009) Heiße Werkzeugoberflächen lassen Formteile erstrahlen. Kunststoffe 99(1):17–20

Hirt G, Ames J, Bambach M, Kopp R (2004) Forming strategies and process modelling for CNC incremental sheet forming. Ann CIRP 52(1):203–206

Holtkamp J, Gillner A (2008) Laser-assisted micro sheet forming. In: Proceedings of FLAMN, fundamentals of laser assisted micro- and nanotechnologies. St. Petersburg, Russland, 25–28 June 2007. The International Society for Optical Engineering, vol 6985, pp 1–10

Höper L (2009) Kleben in Bremen – Personalqualifizierung im klebtechnischen Zentrum

Huang J, Beckemper S, Gillner A, Wang K (2010) Tunable surface texturing by polarization-controlled three-beam interference. J Micromech Microeng 20:095004

Hubka V (1976) Theorie der Konstruktionsprozesse – Analyse der Konstruktionstätigkeiten. Springer, Berlin

Hubka V, Eder E (1992) Einführung in die Konstruktionswissenschaft – Übersicht, Modell, Anleitungen. Springer, Berlin

Ion SE, Humphreys FJ, White SH (1982) Acta Metall 30:1909–1919

Ivanov T, Bührig-Polaczek A, Vroomen U (2010) Casting of microstructured shark skin surfaces and possible applications on aluminum cast parts. In: Proceedings of the 69th World Foundry Congress, Hangzhou, China

Jackson K (2008) The mechanics of incremental sheet forming. PhD thesis, University of Cambridge

Jenke K (2007) Konzept zur Lösung technischer Qualitätsprobleme in der Produktion durch Anwendung der Theorie des erfinderischen Problemlösens (TRIZ). Dissertation der Technischen Universität Kaiserslautern

Jeswiet J, Hagan E (2001) Rapid prototyping of a headlight with sheet metal. In: 9th international conference on sheet metal, Leuven, pp 165–170

Jeswiet J, Duflou JR, Szekeres A, Lefebvre P (2005a) Custom manufacture of a solar cooker – a case study. Adv Mater Res 6–8:487–492

Jeswiet J, Micari F, Hirt G et al. (2005b) Asymmetric single point incremental forming of sheet metal. Ann CIRP 54(2):623–649

Johannaber F, Michaeli W (2004) Handbuch Spritzgießen. Carl Hanser, München

JP 11171 592 (1999) Water repellant article and its manufacture. Japanese Patent Office, 29 June 1999

Junk S (2003) Inkrementelle Blechumformung mit CNC Werkzeugmaschinen: Verfahrensgrenzen und Umformstrategien. Schriftenreihe Produktionstechnik, vol 25. Dissertation, Saarbrücken

Junk S, Bambach M, Chouvalova I, Hirt G (2003) FEM modelling and optimisation of geometric accuracy in incremental CNC sheet forming. In: CIRP international conference on accuracy in forming technology (ICAFT/10). Chemnitz, pp 293–304

Kallien LH (2009) Druckgießen, Spezial 100 Jahre VDG – Teil 5. Giesserei Jahrgang 96:18–26

Keskitalo M, Mäntyjärvi K, Mäkikangas J, Karjalainen JA, Leiviskä A, Heikkala J (2007) Local laser heat treatments of ultrahigh-strength steel. In: Proceedings of 11th nordic conference in laser processing of materials. Lappeenranta, Finnland, 20–22 Aug 2007. Lappeenranta: Lappeenrannan Teknillinen Yliopisto, pp 536–543

Kief HB, Roschiwal HA (2007) NC/CNC Handbuch 2007/2008. Hanser, München

Kim YY, Kim SB, Park HH, Seo MD, Lee BC, Han MS, Kim TN, Cho SB (2008) Effect of cristobalite and quartz on the properties of gypsum bonded investment. J Mater Sci Technol 24(1):143–144

Kimerling TE, Liu W, Kim BH, Donggang Y (2006) Rapid hot embossing of polymer microfeatures. Microsyst Technol 12(7):730–735

Klaiber F (2010) Entwicklung einer Anlagen- und Prozesstechnik für die Herstellung superhydrophober Oberflächen im Spritzgießverfahren, ISBN 3–86130-972–6, Dissertation, RWTH Aachen

Klein B (2002) TRIZ/TIPS – Methodik des erfinderischen Problemlösens. Oldenbourg Wissenschaftsverlag, München

Klein-Wiele J, Wielea K, Simon P (2003) Fabrication of periodic nanostructures by phase-controlled multiple-beam Interference. Appl Phys Lett 83:4707–4709

Klocke F, König W (2006) Umformen. In: Fertigungsverfahren, vol 4, 5th edn. Springer, Berlin

Klocke F, Roderburg A, Zeppenfeld C (2008a) Design methodology for hybrid production processes. In: Proceedings of the 8th ETRIA world TRIZ-future conference, Enschede, Netherlands, pp 75–81

Klocke F, Zeppenfeld C, Pampus A, Mattfeld P (2008b) Fertigungsbedingte Produkteigenschaften – FePro. Apprimus, Aachen

Klocke F, Roderburg A, Wegner H (2009) Methodik zur inventiven Weiterentwicklung von Fertigungstechnologien. In: Gausemeier J (ed) Symposium für Vorausschau und Technologieplanung. HNI, Paderborn, pp 353–369

Klocke F, Heinen D, Liu Y, Arntz K, Ruset C (2010a) Kombinierte Oberflächenbehandlung von Werkzeugen. wt Werkstattstechnik online 6:508–512

Klocke F, Wegner H, Roderburg A, Nau B (2010b) Ramp-up of hybrid manufacturing technologies. In: Proceedings of 43rd CIRP conference on manufacturing systems, Vienna, pp 407–415

Köhler C, Conrad J, Wanke S, Weber C (2008) A matrix representation of the CPM/PDD approach as a means for change impact analysis. In: International design conference – Design 2008, Dubrovnik, pp 167–174

Koller R (1994) Konstruktionslehre für den Maschinenbau – Grundlage zur Neu- und Weiterentwicklung technischer Produkte, 3rd edn. Springer, Berlin

Koller R, Kastrup N (1994) Prinziplösungen zur Konstruktion technischer Produkte. Springer, Berlin

Kondo T, Matsuo S, Juodkazis S, Misawa H (2001) A novel femtosecond laser interference technique with diffractive beam splitter for fabrication of three-dimensional photonic crystals. Appl Phys Lett 79:725–727

Koshy P, Jain VK, Lal GK (1995) Mechanism of material removal in electrical discharge diamond grinding. Int J Machine Tools Manuf 36(10):1173–1185

Kozak J, Rajurkar KP (2000) Hybrid machining process evaluation and development. Proceedings of 2nd international conference on machining and measurements of sculptured surfaces, Keynote Paper, Krakow, pp 501–536

Kratky A (2009) Umformen von partiell mit Laserstrahlung behandeltem Halbzeug. Habilitationsschrift. TU Wien

Kuchling H (1996) Taschenbuch der Physik, 16th edn. Fachbuchverlag, Leipzig, pp 311

Kwiatkowski L, Urban M, Sebastiani G, Tekkaya AE (2010) Tooling concepts to speed up incremental sheet forming. Prod Eng Res Dev 4:57–64

Lang G (1973) Gießeigenschaften und Oberflächenspannung von Aluminium und binären Aluminiumlegierungen. Aluminium 49:231–238

Lange K (1990) Umformtechnik. Handbuch für Industrie und Wissenschaft. Blechbearbeitung, vol 3. Springer, Berlin

Lauwers B, Klocke F, Klink A (2010) Advanced manufacturing through the implementation of hybrid and media assisted processes. International Chemnitz Manufacturing Colloquium ICMC2010, Chemnitz

Lindemann U (2007) Methodische Entwicklung technischer Produkte – Methoden flexibel und situationsgerecht anwenden, 2nd edn. Springer, Berlin

Liu S-J, Lin K-Y, Li M-K (2008) Rapid fabrication of microstructures onto plastic substrates by infrared hot embossing. Int Polymer Process XXIII-3:323–330

Livatyali H, Terziakin M (2005) Forming of heat treatable-high strength steel sheets at elevated temperature. In: Tagungsband zur SCT, international conference on steels in cars and trucks. Wiesbaden, 5–10 June 2005. Stahleisen, Düsseldorf, pp 187–195

Lugscheider E (2002) Handbuch der thermischen Spritztechnik: Technologien-Werkstoffe-Fertigung. DVS, Düsseldorf

Lugscheider E, Bobzin K (2001) The influence on surface free energy of PVD-coatings. Surf Coat Technol 142–144:755–760

Lugscheider E, Bobzin K (2003) Wettability of PVD compound materials by lubricants. Surf Coat Technol 165(1):51–57

Mäkelä T, Haatainen T, Majander P, Ahopelto J (2007) Continuous roll to roll nanoimprinting of inherently conducting polyaniline. Microelectron Eng 84(5):877–879

Mäntyjärvi K, Leiviskä A, Karjalainen JA, Keskitalo M, Heikkala J, Mäkikangas J (2007) Local laser heat treatments in bending ultra-high strength steel. Proceedings of 11th nordic conference in laser processing of materials. Lappeenranta, Finnland, 20–22 Aug 2007. Lappeenrannan Teknillinen Yliopisto, Lappeenranta, pp 291–297

Mapleston P (2008) Turning up the heat: mould-temperature control. Plast Eng 64(9):10–16

Matchett E, Briggs AH (1966) Practical design based on method (fundamental design method). In: Gregory SA (ed) The design method. Butterworth, London, pp 183–199

Menges G, Michaeli W, Mohren P (1999) Anleitung zum Bau von Spritzgießwerkzeugen. Carl Hanser, München

Menz W, Dimov SS, Fillon B (2006) 4M2006: second international conference on multi-material micro manufacture, 20–22 Sept 2006, Grenoble, France. Elsevier, München. ISBN: 978-0-0804-5263-0

Menzies I, Koshy P (2008) Assessment of abrasion-assisted material removal in wire EDM. Ann CIRP 57(1):195–198

Micari F, Ambrogio G, Filice L (2007) Shape and dimensional accuracy in single point incremental forming: state of the art and future trends. J Mater Process Technol 191–1-3:390–395

Michaeli W, Klaiber F (2009) Investigation of laser-assisted moulding of micro- and nanostructures. In: Annual Technical Conference on Polymer Processing, Chicago, USA

Michaeli W, Scharf M (2009) Variothermal heating concepts for extrusion embossing of microstructured films, ANTEC 2009. In: Proceedings of the 67th annual technical conference and exhibition, Chicago, pp 1578–1582

Michaeli W, Blömer P, Scharf M, Brinksmeier E, Rickens K (2005) Mikrostrukturierung von Prägewalzen und Kunststofffolien, Institut für Kunststoffverarbeitung, RWTH Aachen, Abschlussbericht zum AiF-ZuTech-Forschungsvorhaben Nr. 105 ZN

Michaeli W, Gillner A, Klaiber F (2008) Analyses of laser-assisted moulding of micro- and nanostructures. In: The 1st international conference on nanomanufacturing, Singapur

Michaeli W, Ederleh L, Scharf M (2009) Simulation of the extrusion embossing process for microstructures. In: Tagungsband 25th annual meeting of the Polymer Processing Society (PPS)

Michaeli W, Klaiber F, Kremer C, Mäsing R (2010a) Kunststoffoberflächen – Optik, Haptik und Funktionalität/Superhydrophobe Kunststoffoberflächen – Urformen und Funktionalisieren in einem Prozessschritt, 25th Internationales Kunststofftechnisches Kolloquium, Aachen

Michaeli W, Scharf M, Eilbracht S (2010b) Variotherme Walzentemperierung beim Prägen mikrostrukturierter Folien, Tagungshandbuch zum 25th Internationalen Kunststofftechnischen Kolloquium des IKV, 3–4 Mar 2010, Aachen, Session 2: Extrusion – Anspruchsvolle Produkte sicher herstellen

Michaeli W, Schöngart M, Klaiber F, Beckemper S (2010c) Variothermal injection moulding of superhydrophobic surgfaces. Proceedings of the seventh international conference on multi-material micro manufacture (4M). Oyonnax, France

Michaeli W, Grönlund O, Neuß A, Wunderle J, Gründler M (2010d) Neuer Prozess für Kunststoff-Metall-Hybride. Kunststoffe 100(9). Carl Hanser, München

Michaeli W, Ederleh L, Scharf M, Eilbracht S (2010e) Variothermal heating concepts for embossing rolls. PPS-26 proceedings of the polymer processing society 26th annual meeting, Banff, Canada, 04–08 July 2010

Michel A, Rupprecht R, Harmening M, Bacher W (1993) Abformung von Mikrostrukturen auf prozessierten Wafern. KfK-Bericht, Karlsruhe

Mori K, Maki S, Tanaka Y (2007) Warm and hot stamping of ultra high tensile strength steel sheets using resistance heating. In: Tagungsband zur ICNFT, 2nd international conference on new forming technology. Bremen, BIAS, pp 15–29

Mori K, Saito S, Maki S (2008) Warm and hot punching of ultra high strength steel sheet. Ann CIRP Manuf Technol 57:321–324

Morioka M, Sakakibara S (2010) A new cell production assembly system with human-robot cooperation. CIRP Ann Manuf Technol 59:9–12

Mumme F (2002) Die Chemie muss stimmen. Plastverarbeiter 55(7):pp 52–53

Mumme F (2004) G(l)anz ohne Lack. Plastverarbeiter 55:22

Munzert U (2010) Bahnplanungsalgorithmen für das robotergestützte Remote-Laser- schweißen. Utz, München

Nachmani Z (2008) Randzonenbeeinflussung beim Schnellhubschleifen. Dissertation RWTH Aachen, Apprimus, Aachen

Neto VF, Vaz R, Oliveira MSA, Gràcio J (2009) CVD diamond coated steel inserts for thermoplastic mould tools – characterization and preliminary performance evaluation. J Mater Process Technol 209:1085–1091

Neugebauer R, Putz M, Bräunlich H, Kräusel V (2004) Schneiden und Lochen – ein entwicklungssorientierter Bereich der Blechbearbeitung. In: Tagungsband zur Internationalen Konferenz „Neuere Entwicklungen in der Blechumformung". Fellbach, 11–12 May 2004. Frankfurt, pp 255–276

Neugebauer R, Altan T, Geiger M, Kleiner M, Sterzing A (2006) Sheet metal forming at elevated temperatures. Ann CIRP 55(2):793–816

Neugebauer R, Lachmann L, Schönherr J, Rautenstrauch A (2009) Anforderungen generieren fertigungstechnische Innovationen. ZWF 104(9):725–729

Nezuka O, Yao D, Kim BH (2008) Replication of microstructures by roll-to-roll UV-curing embossing. Polymer-Plast Technol and Eng 47(9):pp 865–873

NN (2002) Eigenschaften und Anwendung von niedrigschmelzenden Metalllegierungen. Firmenschrift MCP-HEK Tooling GmbH, Lübeck

NN (2010a) Beheizte Sportbrille einstufig spritzgegossen. Kombination aus Metalldruckguss und Kunststoffspritzgießen am IKV, Kunststoffberater. Giesel, Hannover

NN (2010b) Polymer-Metall-Bauteile in einem „Guss". VDI Nachrichten, 26 Nov 2010

Nowotny S, Thieme S, Scharek S, Rönnefahrt T, Gnann RA (2008) FLEXILAS – laser precision technology for deposition welding with centric wire feeding. DVS-Berichte 250:318–322. ISBN 978–3-87155–256-4

Oehler G, Kaiser F (1993) Schnitt-. Stanz- und Ziehwege, 7th edn. Springer, Berlin

Oliver WC, Pharr GM (1992) An improved technique for determining hardness and elastic modulus using load and displacement sensing indentation experiments. J Mater Res 7(6):1564–1583

Orloff MA (2002) Grundlagen der klassischen TRIZ – Ein praktisches Lehrbuch des erfinderischen Denkens für Ingenieure. Springer, Berlin

Owens DK, Wendt RC (1969) Estimation of the surface free energy of polymers. J Appl Polym Sci 13:1741–1749

Pahl G, Beitz W, Feldhusen J, Grote K-H (2005) Konstruktionslehre – Grundlagen erfolgreicher Produktentwicklung, Methoden und Anwendung. Springer, Berlin

Pahl G, Beitz W, Feldhusen J, Grote K-H (2007) Engineering design – a systematic approach, 3rd edn. Springer, Berlin

Peng X, Lu X, Balendra R (2004a) FE simulation of the blanking of electrically heated engineering materials. J Mater Process Technol 145:224–232

Peng X, Qin Y, Balendra R (2004b) Analysis of laser-heating methods for micro-parts stamping applications. J Mater Process Technol 150:84–91

Petek A, Zaletelj V, Kuzman K (2009) Particularities of an incremental forming applica-tion in multi-layer construction elements. J Mech Eng 55(7–8):423–426

Peters M, Leyens C (2002) Titan und Titanlegierungen, DLR, Deutsches Zentrum für Luft- und Raumfahrt e. V. Wiley-VCH, Köln

Pires JP (2005) Interfacing industrial robotic systems with manufacturing tracking and control software: a choice for semi-autonomous manufacturing Abstract systems. Industrial Robot: Int J 32:214–219

Probst G, Raub S, Romhardt K (1997) Wissen managen – Wie Unternehmen ihre wertvollste Ressource optimal nutzen. FAZ, Frankfurt am Main

Pursche L (2002) Methoden zur technologieorientierten Programmierung für die 3D-Lasermikrobearbeitung. Meisenach, Bamberg

Rajurkar KP, Zhu D, McGeough JA, Kozak J, De Silva A (1999) New developments in ECM. Ann CIRP 48(2):567–579

Reisgen U, Stein L, Steiners M, Bleck W, Kucharczyk P (2010) Schwingverhalten von mit modifiziertem MSG-Kurzlichtbogenprozess gefügten Stahl-Aluminium-Mischverbindungen. Schweißen Schneiden 62(7/8):396–399

Ridder H, Schnieders J, Heim H-P, Jarka S (2009) Möglichkeiten und Grenzen variabler Werkzeugtemperierung. Kunststoffe 99(5):22–29

Roderburg A, Klocke F, Koshy P (2009) Principles of technology evolutions for manufacturing process design. In: Proceedings of the 9th ETRIA world TRIZ-future conference, Timisoara, Romania, pp 62–70

Rupprecht R, Hanemann T, Piotter V, Hausslet J (1997) Fertigung von Kunststoff-Mikroteilen für optische und fluidtechnische Anwendungen. Swiss-Plastics 19(4):5–8

Sauer H (1999) Untersuchung zur Haftung von Metallschichten auf Kunststoffen. Dissertation, Universität-Gesamthochschule Siegen

Schmidt R-A (2006) Umformen und Feinschneiden – Handbuch für Verfahren, Stahlwerkstoffe, Teilegestaltung. Hanser, München

Schmidt-Sandte T (2002) Laserstrahlbasierte Entgratverfahren für feinwerk-technische Anwendungen. IFS, Hannover

Schmitz GJ, Grohn M, Nominikat J (2004) Primary shaping method for a component comprising a microstructured functional element, International patent, publication no. WO 2004/087350 A2, RWTH Aachen University

Schmitz GJ, Grohn M, Bührig-Polaczek A (2007) Fabrication of micropatterned surfaces by improved investment casting. Adv Eng Mater 9(4):265–270

Schröter B (2008) Inertiale Positionserfassung zur Programmierung robotergestützter Handhabungsaufgaben. Apprimus, Aachen

Schuh G, Kreysa J, Orilski S (2009) Roadmap "Hybride Produktion" – Wie 1 + 1 = 3 Effekte in der Produktion maximiert werden können. ZWF 104:385–391

Schultz J, Nardin M (1999) Theories and mechanisms of adhesion. In: Mittal KL, Pizzi A (eds) Adhesion promotion technique. Technological applications, New York, pp 1–26

Seibel S (2005) Vielfalt am laufenden Meter. Kunststoffe 95(12):38–46

Seo S-M, Kim T-I, Lee HH (2007) Simple fabrication of nanostructure by continous rigiflex imprinting. Microelectron Eng 84(4):567–572

Sipe JE, Young JF, Preston JS (1983) Laser-induced periodic surface structure, 1st theory. Phys Rev B 27:1141–1154

Spur G, Stöferle T (1985) Umformen und Zerteilen. In: Handbuch der Fertigungstechnik, vol 2/3. Hanser, München

Stachowiak H (1973) Allgemeine Modelltheorie. Springer, Wien

Su H, Zhong Y, Wang X, Zheng X, Xu J, Wang H (2003) Effects of polarization on laser holography for microstructure fabrication. Phys Rev E 67:056619

Taleb Araghi B (2011) Eingereichte Dissertation am Institut für Bildsame Formgebung der RWTH Aachen

Taleb Araghi B, Manco GL, Bambach M, Hirt G (2009) Investigation into a new hybrid forming process: incremental sheet forming combined with stretch forming. Ann CIRP 58(1):225–228

Taleb Araghi B, Goettmann A, Bambach M, Biermann T, Hirt G, Weisheit A (2010) Development of hybrid incremental sheet forming processes. Steel Res Int 81(8):918–921

Taleb Araghi B, Göttmann A, Bergweiler G, Saeed-Akbari A, Bültmann J, Zettler J, Bambach M, Hirt G (2011) Investigation on incremental sheet forming combined with laser heating and stretch forming for the production of lightweight structures. In: Proceedings of the 14th international conference on sheet metal (SheMet2011), Leuven, Belgium

Tian L, Ren L, Liu Q, Han Z, Jiang X (2007) The Mechanism of drag reduction around bodies of revolution using bionic non-smooth surfaces. J Bionic Eng 4(6):109–116

Tillmann M (2009) Innovative Prozesskettenoptimierung (IPO). Dissertation der RWTH Aachen, Apprimus, Aachen

Tomiyama T, Gu P, Jin Y, Lutters D, Kind C, Kimura F (2009) Design methodologies: industrial and educational applications. Ann CIRP 58(2):681–700

Tönshoff HK, Peters J, Inasaki T, Paul T (1992) Modelling and simulation of grinding processes. Ann CIRP 41(2):677–688

Tuomi J, Lamminen L (2004) Incremental sheet forming as a method for sheet metal component prototyping and manufacturing. In: 10èmes Assises Européennes de Proto-typage Rapide, Paris

Ulrich KT, Eppinger SD (2004) Product design and development. McGraw Hill, Irwin

Unicam Software GmbH (2011) RobotMaster CAD/CAM for Robots. www.robotmaster.de. Accessed 4 Mar 2011

US PS 5599 489 (1997) preparing molded articles of fluorine-containing polymer with increased water-repellency. United States Patent, 04 Feb 1997

Van Stappen M, Vandierendonck K, Mol C, Beeckman E, De Clerq E (2001) Practice vs. laboratory tests for plastic injection moulding. Surf Coat Technol 142–144:143–145

VDI-Richtlinie 2221 (1993) Methodik zum Entwickeln und Konstruieren technischer Produkte. VDI, Düsseldorf

VDI-Richtlinie 2906, Blatt 2 (1994) Schnittflächenqualität beim Schneiden, Beschneiden und Lochen von Werkstücken aus Metall. Scherschneiden

Vita F, Lucchetta DE, Castagna R, Criante L, Simoni F (2007) Large-area photonic structures in freestanding films. Appl Phys Lett 91:1031141973

Vogler M, Wiedenberg S, Mühlberger M, Bergmair I, Glinsner T, Schmidt H, Kley E-B, Grützner G (2007) Development of a novel, low-viscosity UV-curable polymer system for UV-nanoimprint lithography. Microelectron Eng 84(5–8):984–988

Wagenknecht T, Collin H, Krajewsky P, Bloss P (2006) Heißprägen von Mikrostrukturen. Kunststoffe 96(3):55–58

Weber L, Ehrfeld W (1998) Mikroabformung – Verfahren, Werkzeuge, Anwendungen. Kunststoffe 88(10):1791–1802

Weck M, Brecher C (2006) Werkzeugmaschinen 4: Automatisierung von Maschinen und Anlagen. Springer, Berlin

Weck M, McKeown P, Bonse R (1995) Reduction and compensation of thermal errors in machine tools. CIRP Ann Manuf Technol 44(2):589–598

Weisheit A, Vitr G, Scheffler S, Wissenbach K (2005a) Local laser heat treatment of ultra high strength steels to improve formability. In: Proceedings of 1st international conference superhigh strength steels. Rom, Italien, 2–4 Nov 2005. Assoc. Italiana di Metallurgia, Mailand, pp 1–10

Weisheit A, Vitr G, Wissenbach K, Zajac J, Thoors H, Johansson B, Ribera E, Arino J, Sierra F (2005b) Local heat treatment of ultra high strength steels to improve formability. In: Proceedings of 1st international workshop on thermal forming. Bremen, 13–14 Apr 2005. BIAS, Bremen, pp 63–81

Weiss C (2004) Metallisierung von Folien auf der Basis von Polyetheretherketon (PEEK) für flexible Schaltungsträger. Dissertation, Universität Erlangen-Nürnberg

Wertheimer MR, Fozza AC, Hollander A (1999) Nuclear Instrum Methods Phys Res Sec B, Beam Interact Mater Atoms 151(1–4):65–75

Winkler R, Büttmann F, Hartmann S, Jerz A (2003) Thermal spraying of polymers: spraying processes, materials and new trends. In: Moreau C, Marple B (eds) Thermal spray 2003: advancing the science and applying the technology. ASM International, pp 1635–1638

Wöllmer H (2000) Untersuchung zum Präzisionsgießen metallischer Mikroteile. Dissertation, University of Freiburg, Germany

Yanagimoto J, Oyamada K (2005) Springback of high-strength steel after hot and warm sheet forming. Ann CIRP 54(1):213–216

Yang Y, Li Q, Wang G (2008) Design and fabrication of diverse metamaterial structures by holographic lithography. Opt Express 16:11275–11280

Yang C, Sheng Li B, Xing Ren M, Zhi Fu H (2010) Studies of microstructures made of Zn-Al alloys using microcasting. Int J Adv Manuf Technol 46(1–4):173–178

Yeo LP, Ng SH, Wang Z, Wang Z, de Rooij NF (2009) Micro-fabrication of polymeric devices using hot roller embossing. Microelectron Eng 86(4):933–936

Youn S-W, Ogiwara M, Goto H, Takahashi M, Maeda R (2008) Prototype development of a roller imprinting system and its application to large area polymer replication for a microstructured optical device. J Mater Process Technol 202(1–3):76–85

Young T (1805) Phil Trans R Soc Lond 9:255

Zhao G (2001) Spritzgegossene, tragende Kunststoff-Metall-Hybridstrukturen. Dissertation, Lehrstuhl für Kunststofftechnik (LKT), Universität Erlangen-Nürnberg

Zisman WA (1964) Relation of the equilibrium contact angle to liquid and solid constitution. Adv Chem 43:1–51 (Chapter 1)

Chapter 6
Self-optimising Production Systems

Robert Schmitt, Christian Brecher, Burkhard Corves, Thomas Gries,
Sabina Jeschke, Fritz Klocke, Peter Loosen, Walter Michaeli, Rainer Müller,
Reinhard Poprawe, Uwe Reisgen, Christopher M. Schlick, Günther Schuh,
Thomas Auerbach, Fabian Bauhoff, Marion Beckers, Daniel Behnen,
Tobias Brosze, Guido Buchholz, Christian Büscher, Urs Eppelt,
Martin Esser, Daniel Ewert, Kamil Fayzullin, Reinhard Freudenberg,
Peter Fritz, Sascha Fuchs, Yves-Simon Gloy, Sebastian Haag, Eckart Hauck,
Werner Herfs, Niklas Hering, Mathias Hüsing, Mario Isermann,
Markus Janßen, Bernhard Kausch, Tobias Kempf, Stephan Kratz,
Sinem Kuz, Matthis Laass, Juliane Lose, Adam Malik, Marcel Ph. Mayer,
Thomas Molitor, Simon Müller, Barbara Odenthal, Alberto Pavim,
Dirk Petring, Till Potente, Nicolas Pyschny, Axel Reßmann, Martin Riedel,
Simone Runge, Heiko Schenuit, Daniel Schilberg, Wolfgang Schulz,
Maik Schürmeyer, Jens Schüttler, Ulrich Thombansen, Dražen Veselovac,
Matthias Vette, Carsten Wagels and Konrad Willms

Contents

6.1	Research Program on Self-optimising Production Systems		698
6.2	Integrative Self-optimising Process Chains		702
	6.2.1	Abstract	702
	6.2.2	Motivation and Research Question	705
	6.2.3	State of the Art	711
	6.2.4	Results	717
	6.2.5	Industrial Relevance	737
	6.2.6	Future Research Topics	741
6.3	Integrative, High-resolution Supply Chain Management		743
	6.3.1	Abstract	744
	6.3.2	State of the Art	746
	6.3.3	Motivation and Research Question	751
	6.3.4	Results	753
	6.3.5	Industrial Relevance	785
	6.3.6	Future Research Topics	790
6.4	The Road to Self-optimising Production Technologies		793
	6.4.1	Abstract	793
	6.4.2	State of the Art	795

R. Schmitt (✉)
Werkzeugmaschinenlabor WZL der RWTH Aachen, Steinbachstr. 19,
52074 Aachen, Germany
e-mail: r.schmitt@wzl.rwth-aachen.de

	6.4.3	Motivation and Research Question	803
	6.4.4	Results	807
	6.4.5	Industrial Relevance	842
	6.4.6	Future Research Topics	847
6.5	Integrative Product and Process Design for Self-optimising Assembly		849
	6.5.1	Abstract	849
	6.5.2	State of the Art	850
	6.5.3	Motivation and Research Question	856
	6.5.4	Results	857
	6.5.5	Outlook	893
6.6	Self-optimising Assembly Systems Based on Cognitive Technologies		894
	6.6.1	Abstract	894
	6.6.2	State of the Art	897
	6.6.3	Motivation and Research Question	902
	6.6.4	Results	905
	6.6.5	Industrial Relevance	940
	6.6.6	Future Research Topics	942
6.7	Reconfigurable Assembly Systems for Handling Large Components		946
	6.7.1	Challenge	946
	6.7.2	State of the Art	947
	6.7.3	Research Questions	951
	6.7.4	Results	953
	6.7.5	Industrial Relevance	968
	6.7.6	Future Research Topics	970
References			971

6.1 Research Program on Self-optimising Production Systems

Robert Schmitt, Mario Isermann and Carsten Wagels

One of the central success factors for production in high-wage countries is the solution of the conflict that can be described with the term "planning efficiency". Planning efficiency describes the relationship between the expenditure of planning and the profit generated by these expenditures. From the viewpoint of a successful business management, the challenge is to dynamically find the optimum between detailed planning and the immediate arrangement of the value stream. Planning-oriented approaches try to model the production system with as many of its characteristics and parameters as possible in order to avoid uncertainties and to allow rational decisions based on these models. The success of a planning-oriented approach depends on the transparency of business and production processes and on the quality of the applied models. Even though planning-oriented approaches are supported by a multitude of systems in industrial practice, an effective realisation is very intricate, so these models with their inherent structures tend to be matched to a current stationary condition of an enterprise. Every change within this enterprise, whether inherently structural or driven by altered input parameters, thus requires continuous updating and adjustment. This process is very cost-intensive and time-consuming; a direct transfer onto other enterprises or even other processes within the same enterprise is often impossible. This is also a result of the fact that planning usually occurs a priori and not in real-time. Therefore it is hard for completely planning-oriented systems to react to spontaneous deviations because the knowledge about those naturally only comes a posteriori.

As a reaction to the observation that in a networked enterprise non-stationary, transient Ramp-Up and Ramp-Down procedures are the rule rather than the exception, with the change of the predominant paradigm of Taylorism successful approaches were developed that are oriented at the so-called value stream of an enterprise. In these approaches, central planning activities were reduced in favor of de-central activities and integrated into the value-adding process. The distinguishing mark of value-oriented production systems is their high degree of flexibility, as decisions are made depending on the situation. Still, the downside of purely value-oriented approaches is that much potential for optimization remains unused as it cannot be tapped without a holistic view and the according target orientation.

A successful enterprise that connects both planning-oriented and value-oriented approaches thus uses the advantages of both strategies. In order to achieve this, on the one hand it is necessary to exactly model the production process. On the other hand, by identifying the essential parameters to be influenced, a foundation has to be laid for the ability to autonomously and flexibly make decisions. However, the existing possibilities to optimize the behavior of one element within the whole system can be focused too tightly and bind resources even though in certain situations this could lead to an adverse behavior in other areas. This optimization task within a production system usually cannot be solved analytically.

The solution of this conflict becomes possible if a system is designed that can adjust its goals situatively. While in most cases the optimization of a system is controlled from the outside, e.g. by a person, in many cases the optimization by the technical system itself is a possible option. The developments in automatization technology show, however, that even in comparatively simple matters this has not yet been accomplished. Therefore in many cases people still play an important role. The implementation of self-optimizing abilities provides a substantial possibility to reduce the area of tension of planning efficiency.

The following research question thus is the focus of explorations within the research area Self-Optimizing Production Systems (Fig. 6.1):

> How can large and small production systems be harmonised through common models in terms of their control up to the point of self-optimisation?

Speaking of the term "self-optimizing systems" we understand systems that are able to effect independent ("endogenous") changes of their inner states or structure based on varying input conditions or interferences. In production processes, according target values can be e.g. capacities, number of pieces, quality, costs or processing times.

Self-optimizing systems are defined by the interaction of contained elements and the recurring execution of the actions (Adelt et al. 2009)

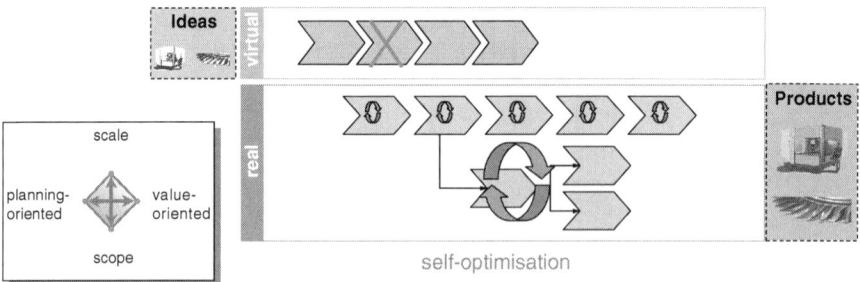

Fig. 6.1 Research question on self-optimizing production systems

- continuous analysis of the current situation,
- determination of targets, and
- adaptation of the system's behavior to achieve these targets.

From the perspective of a traditional control loop the behavior of the system is controlled by externally predetermined target parameters. If the control loop adapts control parameters to observed changes we speak of an adaptive system. The aspect of self-optimization puts the focus onto the dynamization of the target system. Compared to this traditional controlling a self-optimizing system is able to continuously determine the individual sub-targets based on internal decisions and to dynamically adapt the control path (Fig. 6.2).

In order to permit both the independent adaptation of targets and the change of the control path, e.g. of a production, assembly or even an organizational process (Fig. 6.3), self-optimizing production systems need an internal structure to register

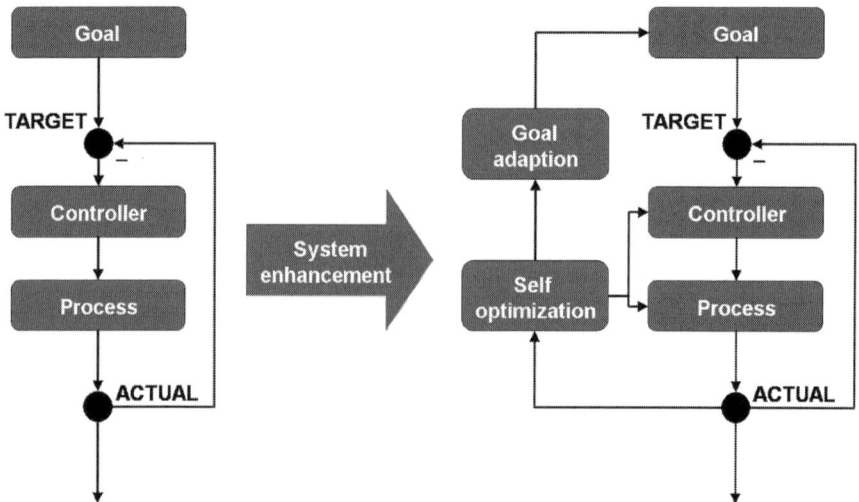

Fig. 6.2 From traditional controlling to self-optimization

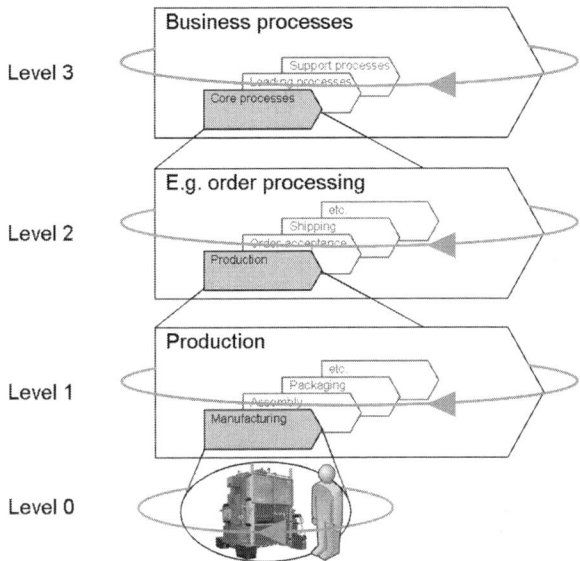

Fig. 6.3 Cross-level controlling in an enterprise

and analyze the current situation, to determine the system targets and to adapt system behavior. Self-optimizing systems are based on coupled sub-systems that can possess abilities of autonomy, versatility and cognition. Cognitive (sub-)systems are able to register information from their environment, to process it in a central processor and to convert it into behavior by being able to influence their environment.

Such a cognitive system can thus constitute the core of a self-optimizing production system. The transfer of cognitive mechanisms onto computer systems makes it possible to exactly process large quantities of data and thus to analyze even complex production processes with multi-level dependencies.

Therefore a self-optimizing system with its abilities to analyse data, to model and to make decisions offers an approach to master processes with no existing deterministic control function. It can basically be employed in various levels of an enterprise, from the direct control of machine parameters to assembly tasks and planning levels. Especially in cases where e.g. requirements of one-piece-flow or other non-analytical-deterministical cause-effect relationships prevail there are challenges to augment efficiency and cost-effectiveness through dynamic target synchronization of global and local target systems.

In production, a cognitive, self-optimizing system can intervene into a process and can concertedly vary individual values and tolerances in order to react to deviations in previous steps of the process. By doing this, scattering of the results of connected processes can be minimized and the quality of the end product can be augmented. Similarly, a cognitive system can act in the area of assembly that contains a high amount of manual labor, where the quality is decisively influenced by the experience-based knowledge of the workers and where aspects of assembly-compatibility of the construction also play a decisive role. Through specific matching of individual parts and through the variation of the according actuating variables from production resulting deviations can be compensated.

6.2 Integrative Self-optimising Process Chains

Robert Schmitt, Mario Isermann, Carsten Wagels and Matthis Laass

6.2.1 Abstract

6.2.1.1 Challenge

Only when a company can offer goods that satisfy the customer's requirements and for which a price can be obtained that will make a profit, can it be successful in the long run. The task of the management consists in optimising the product quality and thus the match of the customer's demands and the properties of the products produced. In addition to a highest possible liquidity ratio the other conditions to be taken into account in this optimisation task from a management standpoint include for example the pricing, production costs, the process through-put times or the flexibility of a complete production process as the basis for the ability to adapt rapidly and economically to changing governing conditions.

As a result consideration of Quality according to Garvin opens up further dimensions, which are to be taken into account when determining the target system. As a consequence a multi-dimensional target system is to be addressed during the optimisation process. In this case individual process optimisation alone is not expedient; in particular since the common optimum of the loss functions in the individual processes cannot be targeted in this way. Feedback of information from the individual processes to coordinate the whole process chain via overall control mechanisms offers a potentially successful approach. The challenge here consists in the developing the methodology and also specific applications for control and regulation.

6.2.1.2 Goal-Setting

The approach discussed here focuses upon the ability of production systems to adapt them-selves in self-optimising mode to changing conditions by using cognitive technologies. The main goal here is to ensure the product quality, wherein the process chains under consideration are aimed at fulfilling the required product functions by the design of pan-process closed-loop control systems. As a result the task of purely complying with tolerances of individual product components becomes a subordinate objective in the target system. This enables it (the target system) to permit deviations in the production system, dynamically during the production process, purposefully in favour of other targets (e.g. costs, quality, time) or to compensate these by targeted reactions, and increase its flexibility and hence also its competitiveness.

As a result it makes a contribution to the Cluster of Excellence "Integrated Production Technology for High Wage Countries to resolve the Polylemma of Production", while the dilemma between plan and value orientation by utilising self-optimising and cognitive technologies in the pan-process parameterisation of production systems is reduced.

The dichotomy Scale vs. Scope is being taken up by accelerating the identification and application of cause and effect relationships and the resulting associated faster deployment of these findings even with small quantities.

The influence of production parameters upon tolerances and consequently upon product quality will be considered along with the potential for making the production system more flexible by balancing the production system employing self-optimisation so that it will respond more rapidly to deviations in processes.

The aim is to fulfil the customers' requirements at lowest cost by taking into account possible constraints and at the same time safeguard the product quality economically in dynamic production environments.

Products and components are traditionally fabricated, geared towards dimensional, geometrical and positional tolerances and assessed in production, whereas customers primarily consider the production function.

The product function is determined essentially by geometric tolerances, but production is not considered explicitly. Function orientation is an approach that directs product features and hence the individual steps in the process chain towards the product function as the ultimate aim.

The product functions are awarded a higher priority than purely satisfying defined interim results of production steps, i.e. deviations and/or tolerances. Thus individual features may be varied as subordinate aims depending upon the given conditions. Deviations from previous processes can be compensated for in subsequent processes. This increases the flexibility of a production system and opens up additional potential for cost reduction, if costs for a tolerance can be set against the contribution to product function and consequently e.g. 'expensive' tolerances widened.

The objective is the ability of production to manufacture products based upon function and as a result to shift from production based upon tolerances to one based upon function.

If the target system is adapted and the derivation of control commands is carried out autonomously by continuous comparison with target conditions and the targets in the super-ordinate target system, this is described as self-optimisation.

With the combination of function orientation and self-optimisation a new approach for optimising production presents itself. It enables (the system) to respond autonomously to variations in processes and in the case of deviations in sub-targets of the target system to influence the remaining targets correctively so that attaining the primary target is still guaranteed.

This is however only possible if the processes are fully understood, how the processes integrate and which factors and production parameters influence the product parameters. The relevant information must be fed back to the correct points in the process as a basis for making closed-loop control decisions.

Accordingly the requirements are to identify factors, to make available all relevant production data, to analyse this data, to extract information from this data and finally to generate knowledge concerning the current cause and effect relationships from the information.

6.2.1.3 Results

With the aid of self-optimising mechanisms a production system may be optimised by considering the whole process chain. Self-optimisation offers the possibility of implementing improvements autonomously.

To determine production parameters in process chains, tools that can simulate technically the applicable procedure which mirror the human information process are suitable. To employ these tools to the greatest effect their specific strengths must be combined in operation with various types of tasks.

The tools can be combined by means of modular designed software architecture. The actual software implementation enables closed-loop control of production in production environments that are changing dynamically and controls dependencies between product and production parameters which cannot be described analytically.

It calculates solutions by means of efficient learning mechanisms and enables the closed-loop control of changing targets and disturbances to be adjusted rapidly in the cycle time. It implements the flexibility necessary for an individualised production scenario and at the same time helps to resolve the polylemma. Methodology and software are elucidated by means of application examples.

The transmission of torque by a shaft-hub connection is employed as a first application example, which permits a simple comprehensive explanation of the procedure. The second application example is the production of components for a car rear axle drive.

The focus here is on the acoustic emission of this drive during operation—a quality feature relevant to the customer that depends upon many components, component properties and in particular their tolerances. When the acoustic emission increases customers sense the generation of noise in the vehicle as annoying. If it lies below a vehicle-specific limiting value, then the drive noise will rarely be noticed.

The aim of optimisation is to meet the quality demands of the customer without additional cost and increased production costs. Increasing production flexibility to afford faster and efficient response to deviations will ensure the product quality and a relaxation of unnecessarily tight component tolerances favouring a homogeneous functional performance can reduce both fabrication costs and time.

This secures the competitiveness of the production. To achieve this, the interdependences between tolerances and product function fulfilment (here: low-noise power train) are analysed and fed back into the appropriate processes.

The application of cognitive methods in acquiring and evaluating production data and also generating decisions to optimise the production scenario is paramount in the application examples. Production data is evaluated with the aid of graphical data processing methods, modelling of the production processes is carried out by Artificial Neural Networks and the determination of process parameters is implemented by the cognitive architecture Soar as the core of optimisation software that can take decisions autonomously on the basis of control systems knowledge and practical knowledge.

In order to attain maximum production flexibility the capability of learning rapidly from available data is implemented in addition. Existing knowledge is updated via learning algorithms and the learning process when creating models is systematically supported and accelerated with quality management methods.

First the tasks and challenges of the developed cognitive software are derived. In Sect. 6.2.2 the motivation for the research projects and also the specific research question are derived from these, via which a contribution towards resolving the polylemma is evolved. Section 6.2.3 includes state-of-the-art techniques for self-optimising approaches to closed-loop control of process chains against the background of the polylemma to be resolved. The detailed work in the research projects is introduced in Sect. 6.2.4. This is divided according to the steps in human cognitive information processing. These comprise the cognitive extraction of information from production data, the analysis and interpretation of influences of specific feature characteristics upon the production results along with generating decisions to determine production parameters on the basis of knowledge and information, and also processes in order to learn from the combination of selected parameters and targeted results for future situations. Data mining processes are employed for analysing the data. The cognitive architecture by Soar is employed for decision making. Artificial Neural Networks are employed for illustrating cause-effect relationships and reinforcement learning is evaluated as learning process. In conclusion the industrial relevance of the solution concepts is evaluated in Sect. 6.2.5, before providing an overview upon the future research topics in Sect. 6.2.6.

6.2.2 Motivation and Research Question

The entrepreneurial task in design and production consists in optimising the liquidity ratio between the market demands and the properties of the products taking into account economic aspects. This means that a highest possible liquidity ratio is not to be striven for in principle, but that the constraints of the optimisation task, such as product pricing, throughput time in the process or the flexibility of a complete production on a company-specific basis are to be taken into account in all areas universally (Schmitt et al. 2007).

In the result a best possible quality of a product at best possible cost is to be delivered. Quality can be accessed from different points of view. Garvin differentiated the transcendental (subjective), the product-related, the customer-related and the value-based understanding of quality (Garvin 1984). The different views lead to various assessment results, since different criteria interact with each other, such as e.g. costs, functionality and service life. Objectives result from the various criteria for consideration and constraints which in combination form a multi-dimensional optimisation problem.

6.2.2.1 Tolerance Allocation in the Actual Process Chain

Tolerances in particular have a significant influence upon the perception of product quality, and upon the requirements and costs of production stemming from this. Quality demands determine and specify the subjective and objective functionalities of an overall product (Schmitt 2010).

Functionalities of a product arise due to the combination and/or interaction of its components, which in turn may be specified by their properties. A typical component property is its geometry that is defined by geometric dimensions and tolerances. Figure 6.4 shows the essential aspects of tolerances in the product life cycle: accordingly the product functions to be implemented are translated into geometries, nominal dimensions and tolerances in the design and construction phases of a product, which are the significant parameters in the assessment of the product components throughout the product development. These transformations are necessary since the fabrication according to the current state of art proceeds geometrically-based. The design of products or rather components is continued via the assembly of parts or components to form a sub-assembly through to the overall product (Jorden 2007). However the geometry is rarely the entire quality feature in the eyes of the customer, but the function that results from this. Geometries are however always prone to deviations in manufacture, that confronts designers with the task of specifying suitable dimensions and tolerances for components, so that the function of the sub-assembly or the overall product can be ensured and all customer requirements are at the same time taken into account optimally.

The area of tension in which the designer operates is characterised by two essential poles: firstly the functional capability must be ensured. Secondly the production costs rise with the requirements upon individual product properties (e.g. Accuracy of physical dimensions (Merget 2004; Jorden 2007)). A conflict of targets between production costs and a risk of dimensional deviations arises. Depending upon the

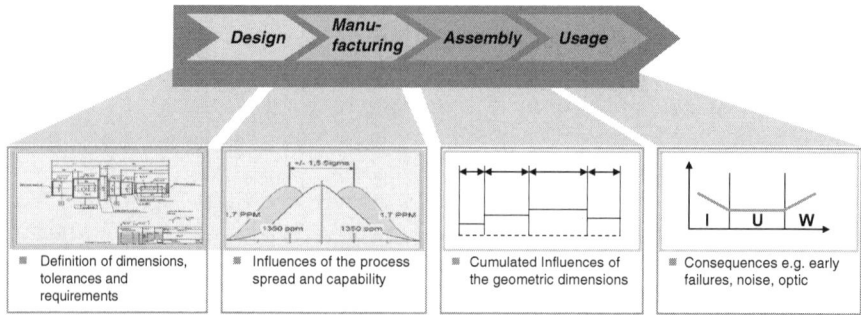

Fig. 6.4 Tolerances throughout the product life cycle

6 Self-optimising Production Systems

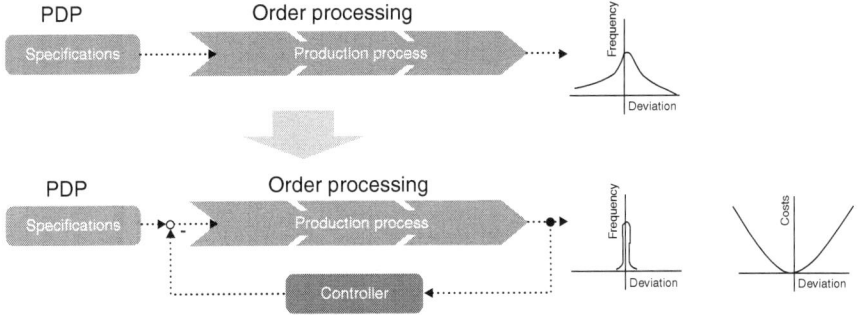

Fig. 6.5 Production control of a component

outcome of this conflict the compromise reached will lead either to quality problems on the product side or to cost problems on the manufacturing side. Geometrical product tolerances of product components are therefore one of the most frequent causes of quality problems in industrial manufacture (Scherms 1998).

Tolerance deviations occur with a distribution depending on the quality of the production process. Figure 6.5 shows the distribution of the tolerance deviations as an example of a production process for an individual component. With the aid of an effective production control (closed-loop) system the process quality was able to be improved and the probability of a tolerance deviation reduced. The precondition for this is the knowledge concerning the causes of deviations. The influence of tolerance deviations upon a product may be described by a loss function (Taguchi 1990). The loss function specifies the costs arising from a deviation from the optimum dimension. The loss function may be employed to decide for instance, at what dimensional deviation is re-work economic or the component must be treated as a reject. In most cases the production processes for individual parts of a product are controlled and optimised to the closest possible dimensional accuracy independently from one another.

A loss function example is shown in Fig. 6.5. The costs rise disproportionately to the deviation magnitude.

If the product or sub-assembly comprises several components, it is often possible for the deviations in down-stream process stages to cancel out, for example during assembly (Fig. 6.6).

For this reason, components typically from well controlled processes having a very small deviation from the target dimension are employed. Due to assembling these components based upon dimensions, dimensional deviations can be cancelled out and therefore rework prevented.

Due to the possibility of cancelling out deviations in individual components, the potential for optimisation arises. In order to exploit this however the loss functions can no longer be considered individually, but must be combined as multi-dimensional functions. This is for example practised in the assembly of roller bearings.

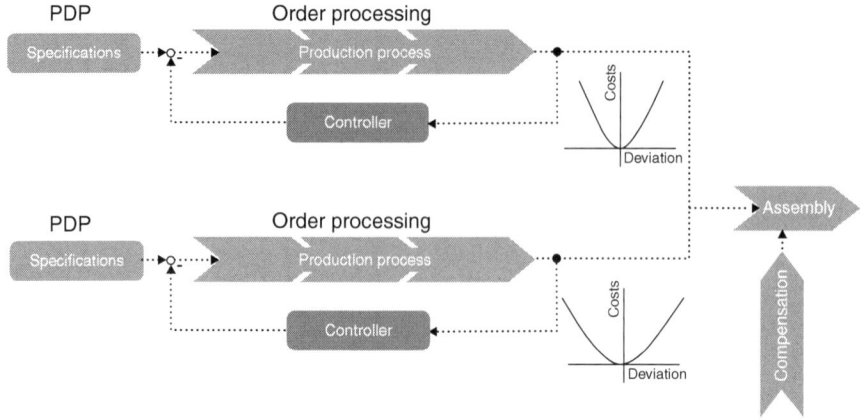

Fig. 6.6 Production control of several components

To achieve an optimum accuracy of fit between inner ring, rolling elements and outer ring a pairing by means of the dimensions achieved in the fabrication is carried out. Pairing is possible, since to fulfil the roller bearing function optimally the tolerancing of the individual components is not important in absolute terms, but is important relative to one another.

Figure 6.7 shows an example of the pairing between outer ring, rolling elements and inner ring carried out when manufacturing a motor car clutch release stop. The rolling elements must be manufactured with extreme dimensional accuracy and must enable the deviations arising from the pairing of outer ring and inner ring to be cancelled.

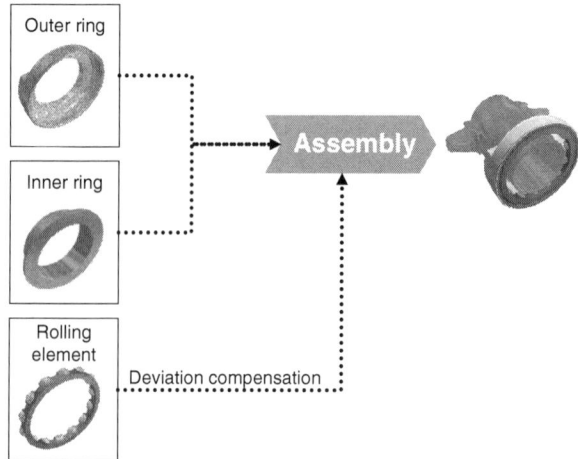

Fig. 6.7 Assembly of a roller bearing by pair-selection by means of rolling elements (e.g.: Clutch release stop). (Schmitt 2000)

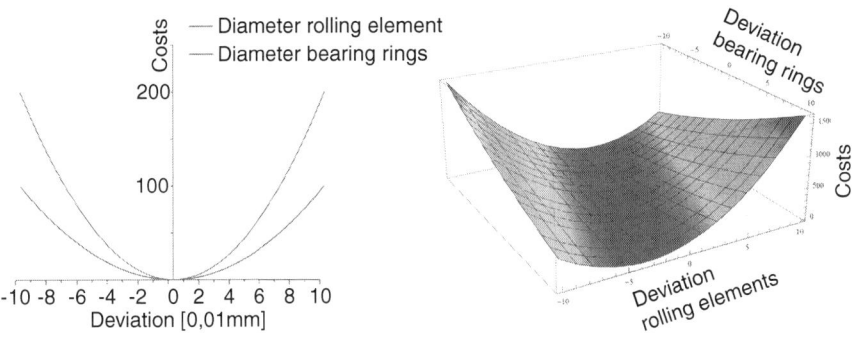

Fig. 6.8 Cost functions of bearing rings and rolling elements

This dimensional pairing of the individual parts permits cost-effective fabrication by simple component pairing, but nevertheless enables high precision bearings to be manufactured (Niemann et al. 2005).

Figure 6.8 shows the individual cost functions for the dimensional deviations in rolling elements and bearing rings on the left. The three-dimensional plane function shows the interactions of both cost functions. It clearly points out that there is a broad area for minimising the cost function, which may be attained by pairing individual components.

To establish the cost function it is necessary to be able to specify the relationships between individual deviations and the costs arising mathematically. Many production processes for demanding products however suffer from the lack of adequate knowledge of the relationships between production parameters and the functional performance of the component. Multi-faceted interdependences and a linking of several parameters are often present. An example of this is gearbox production, wherein the functional performance is characterised by dimensional stability, efficiency and acoustics. These dimensions may be influenced by numerous parameters in material choice, fabrication and assembly. In addition we are confronted by the challenge, that the functional parameters cannot be controlled individually, but in most cases stability, efficiency and acoustics are influenced interactively (Brecher et al. 2008d). It is not possible to establish an analytically minimising cost function.

Complex processes such as the gearbox production selected here as an application example consequently require methods that extend beyond simple pairing. A closed-loop control system is needed that can coordinate the various individual processes in parallel. Such a closed-loop controller is illustrated in Fig. 6.9 as an example for a bevelled gear wheel and ring gear pair. When the cause-effect relationships are unknown the closed-loop control system cannot however be reproduced analytically. In this case it is necessary to be able to simulate the cause-effect relationships in the production process via a model. A classical closed-loop controller is not possible. In this case it is necessary to be able to simulate the cause-effect relationships in the

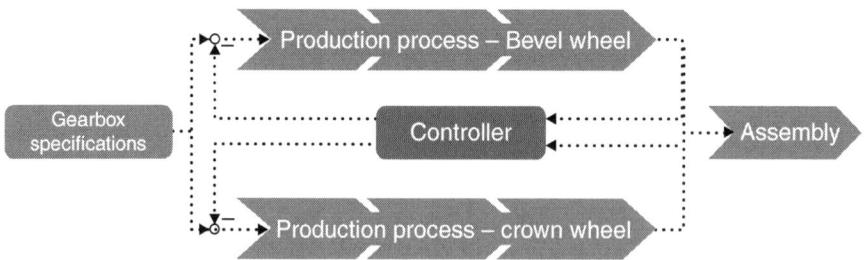

Fig. 6.9 Multiple target closed loop control in gearbox production

production process via a model. If an appropriate modelling mechanism is possible, an optimisation strategy is needed, which with the aid of the process model creates parameter suggestions for the production, with which the various defined goals can be fulfilled. As a result the targets that specify the product functional performance are of central significance. Subordinate aims such as for example tolerances of individual components, may be exceeded to a limited extent, if this can correct the result of another production stage in relation to the functional performance of the finished product.

Focusing upon the functional performance of the product beyond the features of individual components may be achieved by means of a self-optimising closed-loop controller. This is able to call into question also higher-level structures and interdependences. Classical approaches focus solely upon the optimisation of individual parts of the process (Pfeifer et al. 2003).

As part of this project a self-optimising system will be developed that enables processes to be controlled, whose analytical transfer functions are not known. The system is able to adapt itself flexibly to the process to be controlled and to adapt also to control targets, when this results in a higher level functional performance.

To realise this self-optimising system it is necessary to model the production system to be controlled and to take decisions for a closed-loop controller based upon the data generated by the model. Cognitive technologies are employed for modelling and decision-making. In particular this enables complex processes and ever-changing basic conditions to be optimised flexibly due to its inherent learning ability. The *research question* that arises from the starting position and the specified goal is formulated as follows:

> Can cognitive technologies control production processes having a number of inherent interdependences more flexibly, faster and using resources more efficiently, than planned production systems?

The work carried out is derived from this research question. This discusses, how and by which methods and techniques the possibilities of cognitive processes may be transferred to the challenges of optimising a complex production system.

6.2.3 *State of the Art*

Self-optimising control circuits can execute multi-target optimisation autonomously by open-loop and closed-loop applications that deploy cognitive technologies. Therefore an overview of existing approaches to self-optimisation in research is given below.

Existing technologies having cognitive character may be applied in data-recording and interpreting and also decision-making, while data-mining methods are being considered for recognition of cause-effect relationships. Artificial Neural Networks act as tools for modelling processes to evaluate Soar-generated decisions, prior to being utilised in the production system. The cognitive architecture by Soar is introduced at the application level for the optimisation. This is already employed in various application areas for generating decisions and implementing learning processes (Sutton and Barto 1999; Lehman et al. 2006).

6.2.3.1 Closed-Loop Control Systems in Production and Self-optimising Approaches

The application of closed-loop control systems is one of the value-based approaches and is utilised widely in industry at all levels of a company. Closed-loop control systems secure a corrective response when deviations from the target value occur in products and/or processes. Closed control-loops within the operative level aim at e.g. the condition of product features, by monitoring the maintenance of tolerances as closed-loop control systems near or inside machines (Conventional process control) The process concept of closed-loop control however must be broadened in the case of its application in quality management in contrast to the process concept of classical automatic control technology. All activities are described here by processes that influence the product development. Consequently a component of the control loop can also be people, machines and engineering processes, a method, products or materials and external influences in the work environment, such as the temperature or the humidity (Takeda 2009). The range and degree of difficulty of applications is increasing with an increasing complexity of modern products and heightened customer demand for a multiplicity of diverse variants. Companies are confronted with the challenge of reconfiguring their production systems into highly flexible, adaptable units, with which extremely individual products and the process variance resulting there from can be controlled.

Autonomous production systems pursue the goal of being able to adapt automatically to changing input conditions with the greatest possible precision (Among others: Pfeifer and Schmitt 2006; Chryssolouris and Mourtzis 2004). The monitoring scope of the closed-loop control circuit must be widened correspondingly in order to be able also to synchronise individual targets autonomously with one another across all processes. A *Self-optimising production system* pursues the goal of implementing the desired autonomy of the production systems, by carrying out the following steps in a recurring cycle, according to the definition by Gausemeier et al. (Gausemeier et al. 2009):

- Analyse the current situation
- Determine the (new) system targets
- Adjust the system behaviour

The Collaborative Research Centre 'Self-optimising mechanical engineering systems' (Adelt et al. 2009) as originator of this definition is admittedly concentrating upon mechatronic systems, thus primarily the product side, but offer among other things procedural models and tools along with practices, the use of which is also conceivable in production. In this way various methods of data analysis, a cognitive architecture having a 'Predictive algorithm', which is similar to an integrated model of the system to be controlled, possible future conditions are taken into account and various approaches for modelling and decision-making are employed. It is broken down into 'numeric processes based on a priori defined models of the physical system behaviour' as well as 'Learning methods and planning procedures based upon a black box specification of the system behaviour, which is arranged in part with the aid of accumulated experience'.

The Collaborative Research Centre 'Autonomous production cells' (Pfeifer and Schmitt 2006) along with further approaches by Mitsuishi et al. (2004) utilised an integration of physical process models in the machine control and installed model-based machine operation and monitoring as a constructional addition to this. Complex machining processes should be able to perform using optimal resources incurring minimum faults over longer time periods. Autonomous production cells do not have the possibilities of self-optimising production, since they do not have autonomous target adaption or decision-making.

The consideration of comprehensive influences during evaluation and decision-making requires a new approach, taking due account of this with the concept of a 'cognitive automation' (Onken and Schulte 2010). This concept is applied for example in the guidance of unmanned vehicles and also the control of virtual aircraft in flight simulation (Jones et al. 1999), which admittedly is comparable with an automated production in some aspects, but is only conditionally transferable.

With the so-called 'Cognitive factory' (Zaeh et al. 2009) there are approaches in research, to utilise cognitive mechanisms successfully in fabrication systems. All stages of the manufacturing processes are combined in the cognitive factory by means of EDP (e.g. Via Auto-ID-Systems such as RFID (Radio Frequency Identification)) and the information flow is implemented continuously from the customer requirements through to the delivered product, so that communication from component to production machine can be effected. Key areas of activity in the cognitive factory are:

Intelligent design: The CAD system provides knowledge regarding the capabilities of the production system employed. This avoids the need for long redesign-loops and for machines that are only used seldomly.

Artificial cognition in assembly scheduling: Production systems are equipped with their own planning capability. Thus new work sequences may also be generated and evaluated.

Cognitive assistance systems in assembly: Cognitive assistance systems permit the assembly robots to be adapted flexibly to the operators. By dint of this the operator can integrate flexibly with the robot. The robot detects the operator status and adapts its process steps correspondingly.

The individual modules in a cognitive factory are equipped with engineered systems. These comprise sensors and actuators for interacting with the environment, along with techniques for implementing perception, interpretation, learning and planning. The aim is to design engineered systems so that "they know what they are doing".

Technion (Israel Institute of Technology) is also grappling with the challenge of automated planning of assembly scheduling with the aid of cognitive technologies. Decisions for assembly scheduling are based upon a holistic model of the assembly process that takes into account both technological and economic aspects (Denkena and Shpitalni 2007).

The Fraunhofer-Institute for Optronics, Systems Technology and Image Assessment IOSB has developed a data mining tool for decision support in the production sector with ProDaMi (Assumed a company specialising in Data Mining). Data mining algorithms capable of learning enable intelligent evaluation of production and systems data. The goal is to generate knowledge capable of making decisions. The project makes it possible to recognise complex interdependencies in production processes, but does not enable systems to be modelled and is unable to make independent decisions.

To implement a self-optimising system for closed-loop control of complex production processes technologies for the following functional areas must be employed:

- Data analysis
- Modelling
- Optimisation

The state-of-art of the technologies applicable here are as follows.

6.2.3.2 Data Mining for Cause-Effect Relationships Analysis

Prior to the actual closed-loop controller the data to be processed is analysed. The aim of the analysis is to recognise interdependences within the data and also to simplify the issues to be addressed by reduction of data to information sources relevant to the issues. Methods borrowed from the data mining area may be employed for this analysis.

Data mining is known as the recognition and extraction of non-trivial and previously unknown patterns from huge amounts of data. In addition various statistical and mathematical methods are utilised. Data mining itself is divided into five sub-categories (Hand et al. 2001):

1. Explorative Data analysis
2. Pattern recognition
3. Content recognition
4. Descriptive modelling
5. Predictive modelling

In the case of explorative data analysis large amounts of data are browsed, whereby a priori no knowledge exists upon the data searched. In the context of pattern recognition the challenge is to recognise typical or atypical patterns in the data investigated. With content recognition structures within the data similar to patterns which are known and already identified, are sought. Descriptive modelling aims to find descriptions for the existing data sets, with predictive modelling additional attributes to the value are to be deduced with the aid of the existing attributes. Various methods such as classification or regression are employed for predictive modelling (Hand et al. 2001). One of these classification methods involves decision trees, which are seen to be especially attractive to categorise within the data mining for the following considerations:

1. Decision trees can be generated without additional inputs from the user
2. Compared with alternative methods decision trees can be generated in a very short time (Shafer et al. 2005; Gehrke et al. 2000)
3. The model quality of decision trees is very good (Michie et al. 1994; Murthy 1995; Lim et al. 1997)
4. Decision trees offer a very intuitive presentation format (Larose 2005)

6.2.3.3 Artificial Neural Networks as a Modelling Tool

Modelling of (production) systems may be carried out by various modelling types and tools according to the intended application. The requirements upon a model in the environment under consideration however include in particular the demand for calculability executed by software. This is necessary in order to be able to evaluate computer-based optimisation proposals, prior to being employed in the actual production. This renders the application of a mathematical model desirable, which consists of the equations for specifying the system behaviour (Unbehauen 2008).

This is not possible in the case of severely non-linear production processes, which are characterised by complex interdependences of many parameters and in addition must take account of various stochastical influences. The creation of mathematical models having explicit information of the system-specifying equations is not possible in this case or not possible with justifiable computing time. They are therefore unable to be employed for production control in real time. This also applies to the application example of the manufacture of rear axle drives which is considered here. Black-box models such as Artificial Neural Networks offer an alternative: although their behaviour follows mathematical laws, they do not need explicit information of the behaviour-specifying equations, since they can learn their behaviour numerically from sample data (Weinmann 1995). Artificial Neural Networks are therefore ideally qualified to be deployed for modelling purposes in the current application example.

Artificial Neural Networks are based upon the functionality of biological networks as information-processing systems. In 1943 McCulloch and Pitts described a neurological network for calculating logic and arithmetical functions that may be deployed also for pattern recognition. In 1949 Hebb created the Hebbsche learning rule for representing neural learning procedures. In 1960 a network was introduced with ADALINE (Adaptive linear Neuron) that uses delta rule learning for echo filtering in analogue telephones (Levine 2000; Golden 1996).

Following the introduction of Linear Associators by Kohonen in 1972 a non-linear neuron model was used in 1973 and Werbos defined the Back-Propagation method in 1974 for implementing a learning procedure. Mathematically based models of neural networks were developed by Grossenberg in 1976, prior to his developing an architecture for Artificial Neural Networks with Carpenter (Gurney 1997; Dören 2007; Müller and Reinhardt 1995; Zakharian et al. 1998)

The structure of an Artificial Neural Network in general comprises neurons arranged in layers, which exhibit links between layers. Artificial Neural Networks are characterised by these neurons, the architecture (structure of the layers, their nodes and the relevant links), the activation functions of the neurons and also the learning algorithms used.

Artificial Neural Networks have undergone specialised further development depending upon the challenges confronting various applications: (Dören 2007)

- Approximation of functions, e.g. Perceptron (Rosenblatt 1958)
- Image and speech recognition, e.g. The Kohonen Network (Kohonen 1984)
- Classification of patterns, e.g. Cooper's RCE (Zakharian et al. 1998)
- Prognosis, e.g. Perceptron (Rosenblatt 1958)
- Compressed storage, e.g. RAM
- Noise suppression, e.g. Adaline87 (Zakharian et al. 1998)

The Multi-Layer-Perceptron (MLP) was developed for approximating functions and for prognosis and is applied also in the software architecture introduced here. The sigmoid function is proven and widely used for activation (Kinnebrock 1994).

The network training function enables it firstly to anchor the interdependencies between input and output data to be mapped into the network. The Back propagation method (Error feedback method) is employed with MLP networks for this purpose. The Back propagation process is therefore defined as follows: "The term Back Propagation denotes the backwards propagation of an error signal through the network" (Nauck et al. 2002). The method attempts to reach a minimum deviation between calculated and actual trained results. Additionally the output pattern arising in the Artificial Neural Network is compared with a target and an error calculated from this. This error is then fed back into the network so that each internal unit is able to calculate its own error and thus to modify the individual weightings, so that the error is minimised (Nauck et al. 2002).

Neural networks can be found in many areas of production technology, in particular in process monitoring (Eversheim et al. 2006), such as e.g. in process monitoring and control of thermal spraying (Dören 2007) as well as monitoring tool wear (Fries 1999; Rehse 1999) and for decision support in grinding processes (Liao and Chen 1994).

6.2.3.4 Optimisation by Cognitive Technologies

Various algorithms are suitable for creating decision trees for deployment in production technology. An approach is chosen as optimisation technology in light of the present problem, which is designated as simulation-supported optimisation. In this case an optimiser and a process model are necessary for implementing the actual self-optimising closed-loop control. The optimiser generates parameter proposals, which are evaluated subsequently by the process model (among others in Fu 2002). In simulation supported optimisation simulation programs are employed for determining optimisation results (Sauer 2003). Here optimisation means seeking to achieve a desired process result by varying input parameters or variables. In contrast a conventional optimisation cycle achieves this by manual analysis of various process variants. For economic reasons however the process is not carried out physically at each optimisation stage, instead a numeric process analysis is brought into play, or the process is executed with the aid of simulation technology, or Artificial Neural Network, modelled in the computer. A number of strategies and algorithms are available for varying the parameters. With simple challenges a sensitivity analysis can be invoked (Assem 1986), and by these means the qualitative relationships between process parameter and result can be determined. Because of severe non-linearity and oscillations the target function is not practicable in the case of actual, complex production processes. Various investigations have revealed that it is sensible to separate optimising algorithm and process model completely. This separation also enables combinatorial optimisation procedures to be employed in which the relationships between process parameters and process result may be completely unknown (Becker 1991). Many modern optimisation systems are founded upon this concept, such as for example the Optimisation Shell by Grešovnik and Rodič. When aided by various combinatorial methods it offers the possibility of optimising diverse metal forming engineering problems automatically (Grešovnik and Rodič 1999; Posieek 2005).

Soar's cognitive architecture is suitable as an optimiser in the application examples described here. Soar is based on the KI systems GPS and OPS5, in which intelligence according to the principle of rationality is understood as the optimum achievement of targets. In Soar target-oriented problem-solving takes place as heuristic search in problem spaces. The search is affected through successive applications of operators until the target state is reached. In expanding classical planning systems the problem space search is built into a complex decision cycle. Soar is a condition-oriented program language, whose programs do not have a fixed sequential execution, but comprises rules that specify the space of possible conditions and the possible actions therein (Lehman et al. 2006; Laird and Bates Congdon 2008). The Soar program operates within the possibilities stipulated by the rules to reach a specified goal automatically. Initially the system acts like a random-based Monte-Carlo optimiser, but with advanced run-time the program learns from successes and failures and thus shortens the time required to solve the problem (Nason and Laird 2005).

6.2.4 Results

The challenge of solving the research question includes firstly the synchronisation of the technical dimensions in a company-wide closed-loop control system and secondly, an organisational dimension for coordinating all participants. This necessitates especially robust processes that are synchronised with one another. In particular the synchronising of processes with one another requires powerful approaches within a highly dynamic company environment. The implementation of closed-loop process control is effected by structured design of quality control systems in the company— and beyond the company's boundaries—by adopting analogies from closed-loop control techniques. Process-oriented closed-loop quality control supports the preparation and implementation of permanent competitive process landscapes, which adapt to the diverse and dynamic situations invoked by information and material flows (Schmitt et al. 2009a). This adaptation is effected by auxiliary tools from the cognitive sciences, which can recognise the information necessary for adaptation within all prevailing production data. As a result cognition provides the basis for the applicability of a closed-loop control system that will act upon this information. However while a classical closed-loop control system addresses the transfer function of the control loop, the closed-loop control process in the approach adopted here is carried out by a decision-making function by utilising self-optimising mechanisms, which constitute the various elements within the closed-loop quality control system.

6.2.4.1 Cascaded Closed-Loop Quality Control Systems

By designing closed-loop quality control systems to be deployed universally throughout the company relevant information can be distributed and used at the right time at the right place according to requirements. By intermeshing and linking closed-loop quality control systems at various levels in company unambiguous rules for decisions and escalation routines at engineering and organisational levels emerge. The continuous alignment between actual and target state enables continuous improvement to be institutionalised in the company.

By considering the processes at the various levels of the people and departments involved by means of cascaded closed-loop control systems, not only the improvement of individual partial aspects is ensured. On the contrary all the targets based upon the market demands from the company processes out to the finished products at the customer along with the extraction and recycling of user data will be achieved. Closed-loop quality control systems institutionalise this knowledge (Fig. 6.10).

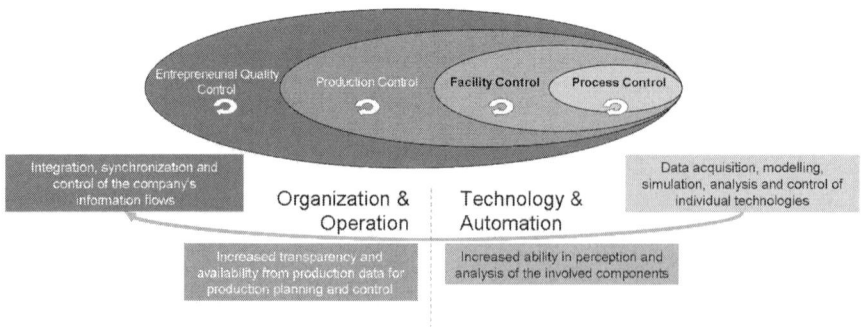

Fig. 6.10 Cascaded closed-loop quality control systems

The approach deployed makes use of the economic specification of tolerances against the background of the challenges discussed. Information from individual processes should be fed back to the correct points in the process chain with methodical support. Conclusions upon correlations in the manufacturing process are compiled; process and product parameters can then be adapted across all individual levels according to the situation. This adaptation allows individually specified parameters to be controlled and in this way the idea of self-optimisation with continuous alignment of target and actual states and the derivation of actions there from is realised.

6.2.4.2 Cognitive Information Processing

To execute closed-loop control across all levels, which adapts a target system automatically to the requirements of the various company levels, it needs intelligent coordination of the data flows, procedures for extracting exploitable information from the available data (knowledge extraction), and the preparation, making and validation of decisions. Conventional closed-loop control technology reaches its limits especially when making decisions autonomously. Making decisions to implement self-optimising production is comparable to human decision-making in its requirements. The engineering implementation of a self-optimising closed-loop production control in the style of human decision-making is the goal of the research. The engineering implementation of human decision-making is the cognitive science sphere of work and is explained in more detail as follows in the context of having transferred the procedures to the area of activity described.

The scientific definition denotes "Cognition" as a generic term for all processes and structures, which carry out the perception, classification and evaluation of facts and also decision-making (Strohner 1995). An example of this is human information processing: all actions that the human being executes in his environment are characterised by first perceiving the stimuli, then processing and subsequently acting upon these. This scheme is adapted and deployed here (Fig. 6.11).

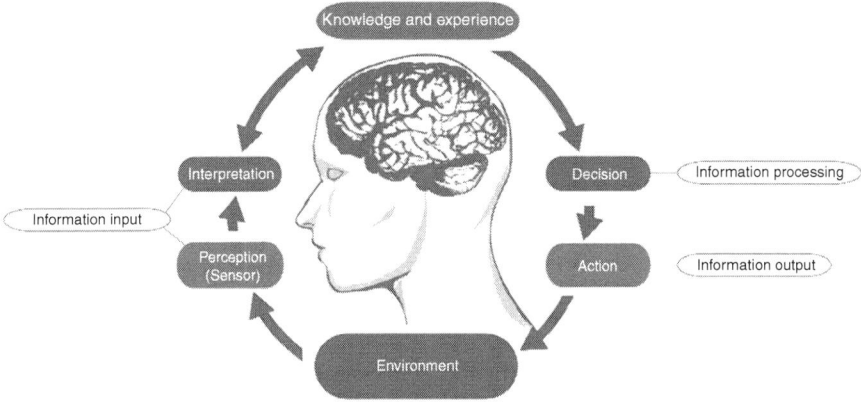

Fig. 6.11 Modelling cognitive processes

Immediately after acquiring the information, the perception, the information is processed. At this point the information acquired is first interpreted on the basis of knowledge and experience and a decision is then derived. Finally the information is output, i.e. decisions upon actions are implemented. The action executed represents a direct influence of the perceived environment, so that the cognitive closed circuit recycles.

Cognitive science pursues the aim of simulating this human information processing with the greatest possible precision by means of engineered systems. These engineered systems are called cognitive technologies. As part of the research work an architecture is being developed in which various cognitive technologies are employed. When developing the architecture the goal is to combine and shape the cognitive technologies, so enabling self-optimising closed-loop production control.

6.2.4.3 The Cognitive Tolerance Matching Architecture (CTM)

An essential requirement upon cognitive architecture for implementing a self-optimising production system is modelling the production processes to be optimised. The architecture must be able to generate production parameters, which can be evaluated by means of the process model. Also in order to adapt flexibly to varying basic conditions, the architecture must be equipped with an inherent learning capability.

The architecture developed enables self-optimised control of production processes by cognitive technologies. In addition it is being modularised according to the model of the processes and necessary functions of the human problem analysis and decision-making. It is composed of various problem-based modules, each representing parts of the overall process. The technologies of the individual modules may be selected individually with regard to the issues to be solved.

The architecture is designed as illustrated in Fig. 6.12. The perception and operation layer serve to exchange data and commands directly with the production

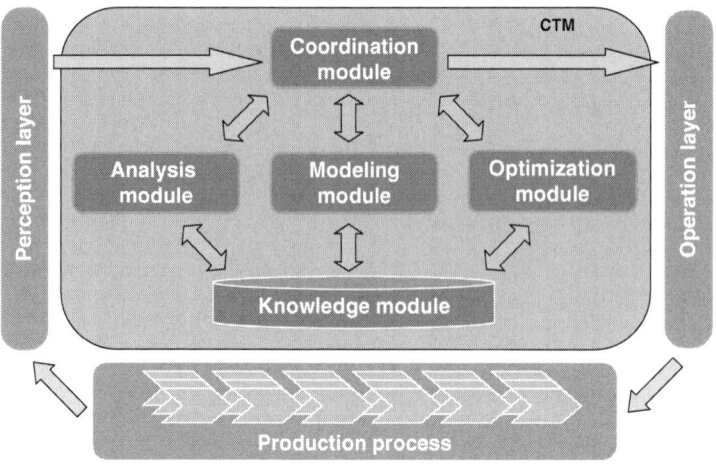

Fig. 6.12 The architecture of cognitive control software

system, while the central core of the architecture contains the elements necessary for the optimization. These provide the problem analysis and modelling, the decision-making (hence the actual optimization) and the mapping of knowledge with the aim of permanent storage and coordination of the individual elements. These tasks are represented respectively by their own modules or levels.

The *Perception layer* serves to acquire and forward sensor data and is employed as interface for the production system. Product and production data is acquired at this point. Implementation in engineering terms corresponds to a connection to the machines and measuring instruments. The perception layer makes the data available to the other modules with which these act, as a bundle and according to demand.

Coordination module communication between the various levels and modules. It provides the data employed and initiates the processes that are carried out by the other modules. In addition it forms the interface for direct interaction with the user.

The *Analysis module* provides the preliminary analysis of the data considered. It carries out e.g. correlation analyses and data reduction, in order to identify the essential data and keep the model complexity as low as possible, but nevertheless sufficiently accurate. The rules utilised for the optimisation are extracted on this basis.

The *Modelling module* is employed for mapping cause-effect relationships in production systems. It serves to predict results of individual production stages, which must be evaluated as a part of optimisation. In the modelling module both analytical and black-box models may be employed. Artificial Neural Networks are employed in the context of the previous implementations in light of a problem unable to be mapped analytically.

The *Optimisation module* creates the optimisation strategy autonomously and implements the flexible response to external influences. It deploys software by Soar in

6 Self-optimising Production Systems

essence, to make optimisation decisions on the basis of adaptively expandable closed-loop control knowledge, and to adapt the production system behaviour autonomously by learning processes.

The *Knowledge module* is responsible for saving and making available the relevant data, which maps the knowledge necessary for the optimisation. This comprises both the rules that are applied in Soar and also the data (input and results), which has already been deployed in the actual system and thus may be considered as substantiated data points.

The Operation layer serves to implement the optimisation decisions made and acts as interface for the production system. Parameters are transferred to the production system via this interface, so that the resources concerned are regulated to the target communicated to the closed-loop control system. The aim in a production (system) being deployed is to be able to implement the control commands via automated interfaces directly by means of the resources.

The interaction between the individual modules begins with the data acquisition in the perception layer. In this case it concerns both the process data from production machines and the measurement data from various sections of the production process. This data is distributed by the coordination module to the individual modules in the architecture. In the next stage the optimisation module generates a parameter proposal for a subsequent stage based upon the current data. This proposal is transferred together with the current process data to the modelling module by the coordination module. The modelling module then generates a prediction for the process result targeted in this case. The coordination module sends this prediction result back to the optimisation module, where it is evaluated in relation to target achievement. Further parameter proposals are then generated, until a result deemed satisfactory in accordance with the requirements for the optimisation module is reached. The parameters thus found are then output to the operation layer by the coordination module and from there accordingly deployed in the production system.

Production Mapping in the Modelling Module

The software generated here enables various tools to be integrated for process modelling by dint of its modular design. In light of the characteristics of the problems specified Artificial Neural Networks were chosen to evaluate the optimisation proposals, since although their behaviour follows mathematical laws, they do not require explicit information for the equations specifying behaviour, since they are able to learn their behaviour numerically from sample data (Weinmann 1995). Artificial Neural Networks therefore offer ideal conditions for deployment for modelling purposes in the present application example.

Artificial Neural Networks

Artificial Neural Networks have been developed to simulate the behaviour of biological nerve cells by engineering means. They are able to learn a behavioural pattern by established training data. As a result they are able to map technical facts such as for example the behaviour of a production process. They have been developed for various applications customised depending upon the problem (See Sect. 6.2.3), wherein the multi-layer perceptron (MLP) was developed for approximating functions and prognosis and also applied in the software architecture application presented. The multi-layer perceptron comprises an input and output layer with respectively n and m neurons. In addition it possesses intermediate layers. Each neuron is connected with all neurons in the surrounding layers. The sigmoid function is of particular significance as the activation function, since it is continuous and enables non-linear functions to be mapped in Artificial Neural Network applications (Dören 2007). Artificial Neural Networks are characterised by these neurons, the architecture (Composition of the layers by their nodes and associated links), and the activation functions of the neurons along with the learning procedures deployed.

The back propagation method (see e.g. Gurney 1997) is deployed for training the MLP network in the present context, which is ranked among supervised learning procedures. By feeding back errors of the training phase, it moves each inner entity to the layer to calculate its own error and to modify its own weighting so that the error is minimised (Nauck et al. 2002). This capability is used here to train networks. Training Artificial Neural Networks with the back propagation method in the application functions in such a way that they optimise their internal transfer functions against the target by adjusting weightings and activation values, based on data from input parameters and results there from provided for learning purposes, to produce output data having the least possible differences in relation to the given input values of the data. Consequently the prediction quality of the network is improved with the number of available substantiated data points, therefore operating points in the present case. As a result Artificial Neural Networks are also able to model non-linear interdependences.

Process Modelling with Artificial Neural Networks
Artificial Neural Networks are currently deployed successfully in many application areas, which may be both discrete event-based and continuous in nature. In the present application they are deployed to optimise process and product parameters in multi-stage process chains. This concerns producing components for rear axle drives, for which industry experts from the individual parts of the processes employed are working on a comprehensive deterministic model as the basis for the optimisation. The number of parameters considered and also the number of unknown cause-effect relationships is cause for restricting the optimising processes to individual aspects of production. Artificial Neural Networks have been identified as an approach, which dispenses with the expensive creation of process models and nevertheless is suitable for assessing results of production under known conditions, provided the possibility is provided to train the network by means of adequate data (Dittmar and Pfeiffer 2004). Comparison with regression analytical calculations also reveals the potential of Artificial Neural Networks to be able to map non-linear interdependences very sensitively also. An additional advantage of Artificial Neural Networks is the possibility of being able to map current changes by regular training.

Architectural Aspects When Modelling
Artificial Neural Networks can be deployed as black-box models, in which the interdependences and mapping rules do not have to exist provided in explicit mathematical terms, but are learned numerically. This reduces the model creation to taking into

account the essential characteristics of the present problem when creating basic structures of the neural networks. A mathematical specification of interdependences of the variables considered is not necessary, since the cause-effect relationships between the variables considered are learned automatically and deployed by the neural network. A "Supervised learning" procedure is deployed with back propagation in this case, with which an Artificial Neural Network can adapt its behaviour concerning the adjustment of the transfer functions. Consequently only its composition is relevant to the capabilities of Artificial Neural Networks, comprising the architecture of neurons and layers along with the links, the activation functions and the learning methods deployed. If multi-layer perceptron networks with sigmoid function and also back propagation as learning function are employed, only the neurons of the individual layers along with the number of layers remain to be determined Therefore the input layer consists of n neurons, whereby n is the number of input parameters, and the output layer of m neurons, where m represents the number of output parameters. Depending upon the problem complexity the formation of the middle layer(s) (called hidden layers) is still to be determined. This is affected in particular cases specific to the problem and is substantiated by appropriate studies, in which learning rates and deviations in results are compared (Pfeifer and Schmitt 2006).

Accelerating the Learning Process of Artificial Neural Networks by Means of Statistical Methods

To deploy Artificial Neural Networks as auxiliary resources when optimising production systems in real time, the speed of adapting to the given conditions is an essential aspect to consider. In addition to the required quality of results the demands upon deployment include the processing speed (Response time) as well as the time required to train a network with the actual conditions of the production to be optimised or to encounter deviations between calculated and actual result, as the internally mapped interactions of the model are updated.

In this connection the critical factor is "time", but not only dependent upon the processing speed of the network, but also of the production system to be mapped, the real time data for operating points to be learned must be made available. If this data must be generated only by trials in the production system, additional costs arise, which are to be minimised through reduction of the experimental expenditure by systematic design of experiments.

To accelerate the learning process of Artificial Neural Networks, aiming to set this up as rapidly and efficiently as possible, an efficient tool for reducing the experimental costs can be adopted from the area of six-sigma philosophy with the Design of Experiments (DoE). DoE includes various methods for systematic planning, execution and evaluation of trials for determining and mapping statistical parameters and their effects and interactions (Schmitt 2010). It helps to determine systematically the optimal parameters for processes subject to variation of the influences relevant to the process result. The aim in deploying this method is to obtain meaningful information and optimal process parameters with the lowest number of trials possible.

For application in the actual implementation experiments were planned and carried out systematically with the aim of comparing the learning process of neural networks (a) by applying DoE and (b) in conventional training. Artificial Neural Networks were deployed to determine parameters when assembling a shaft-hub joint in experiments conducted. A trial for calculating a transmittable torque has shown that the experimental expense was able to be reduced to 18 data sets taking as a starting point a conventional training run with 56 data sets from components actually constructed, whereas the prediction accuracy of the experimental results of individual parameter sets reduced by only 2% in the actual example (From 2.5–4.5% in the selected results range between 80 and 120 nm: 8 data sets determined systematically with DoE, 10 equidistant selected data points; the requested data points were not part of the training data).

In the experiment cited an Artificial Neural Network was first of all trained randomly with production data. The aim was to assess the ability to transmit torque on the basis of materials and tightening torques when screwing together a shaft-hub joint (an actual example is described in Sect. 6.2.4.3.2). A further Artificial Neural Network was trained by deploying DoE, wherein first the data on attribute and value ranges was investigated and then an experiment was designed. This experiment enabled known data points to be evaluated systematically as relevant. The data stemming from this then served as training data for the second Artificial Neural Network.

In a second experiment the whole spectrum of the results range of available data sets was utilised. The prediction accuracy of the Artificial Neural Network was compared (a) with training and validation with all 56 data sets available, (b) by deploying DoE with 8 data sets and subsequent post training with an availability of 16, 24 and 56 data sets. An Artificial Neural Network type MLP was deployed with one hidden layer, 3 input neurones and one output neurone, each fully connected with all neurons of each neighbouring layer, and by employing back propagation training each with 10,000 runs.

By means of the 56 data sets a prediction accuracy having a mean deviation of 4.2% was able to be achieved and after deploying DoE with only 8 data sets a mean deviation of 8.4%, (b) whereas the prediction accuracy with increasing number of training data sets (c) rose further likewise to 4.2% and thus to the level of experiment (a).

The learning curve for the prediction accuracy thus climbs at an early point substantially faster than when carrying out the conventional experiment.

If this system is deployed purposely in the training phase of Artificial Neural Networks, previous data is available that can be deployed for a training run (Fig. 6.13). This acceleration in the start-up phase is only detrimental to the prediction accuracy to a minor extent and incomparably better results can be achieved than with the random choice of training data far earlier than with conventional methods. Therefore the learning curve of the prediction accuracy climbs at an early point substantially faster than when carrying out the conventional experiment and nevertheless can be increased in further trials within the range possibilities, which conventional experiment additionally offers. Translating this to an application example of an actual production system time (reasonably good predictions available earlier) and money (experimental expense) are saved.

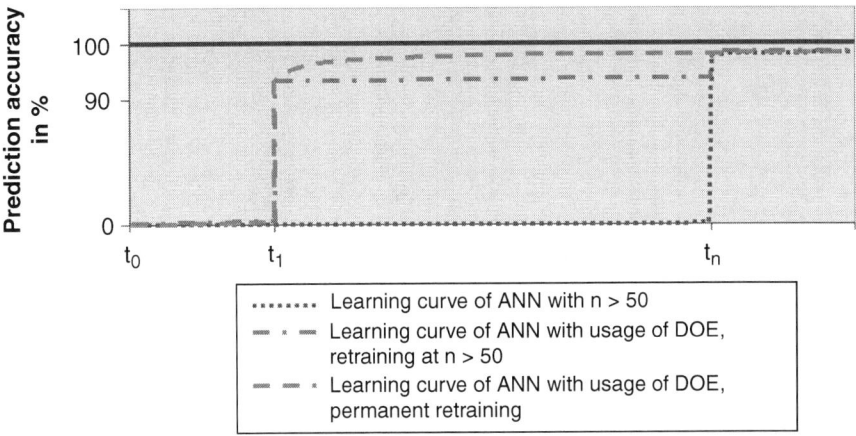

Fig. 6.13 Learning curves for artificial neural networks

Limits when deploying DoE however lie in the properties of the functions for mapping the interdependences. With a larger number of parameters and/or a non-linear function or a function able to approximate linearity only with large error, deployment must therefore be substantiated by studies. The DoE leads to a greater increase of the learning curve, but the prediction accuracy remains dependent upon the data employed.

Decision-making in the Optimisation Module
The scientific findings were established using, amongst other things, a specific shaft-hub model. This was based on a simplified problem suitable for developing and evaluating the cognitive architecture. In this respect, the ability to transfer a pre-determined torque from the shaft to the hub represents the required functionality (product function). Here, the use case exhibits the property of controlling the transmission of torque through various influencing variables which influence each other through interactions. The software controls the assembly parameters for the shaft-hub connection so that the connection is capable of transmitting a required torque, but slips at a defined maximum load. Hence, the connection fulfils the function of a slip clutch or predetermined breaking point, which helps protect other components from excessive stress and the damage this can cause.

During the assembly process for the shaft-hub connection, the individual assembly parameters are adapted dynamically by the CTM software so that the transmissible torque or the torque required for the connection to slip is achieved regardless of the dimensional stability of the shaft and hub used. In detail, the connection is comprised of two hubs fixed onto one shaft:

Figure 6.14 is a schematic of the connection setup. The lower hub (hub 1) is connected with the shaft via a tapered interference fit. In this respect, the strength of the connection is determined by the tightening torque of the screw used to fix the hub to the shaft as well as by the material and design parameters. The upper hub (hub 2) is connected with the shaft via a three-jaw chuck. Here, the strength is determined by the

Fig. 6.14 Shaft-hub demonstrator

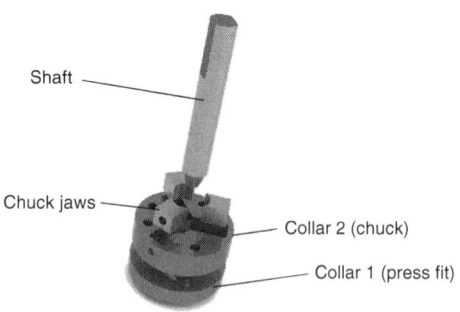

jaw material and the tightening torque of each screw. To simulate the predetermined breaking point, both hubs are fixed to one another and rotated against the shaft. The torque transmitted, i.e. the functional feature of the connection, can be measured.

The individual parameters for influencing the transmissible torque are dynamically calculated by the software. For this purpose, the assembly process for the shaft-hub connection is split into different subprocesses which are controlled by the software on a step-by-step basis. The individual steps are

- Hub 1 is predefined in each case,
- Selection of shaft,
- Assembly of tapered interference fit,
- Optional: measurement of the torque already achieved through hub 1,
- Selection of jaws and
- Assembly of the three-jaw chuck.

Thus, with respect to the assembly process, there is a wide variety of parameter variation possibilities that are calculated by the software and need to be checked. These variation possibilities allow a response to variations in the process, e.g. variations in the dimensional stability of the individual components. The Soar optimiser generates possible combinations of the individual assembly parameters, from which the modelling module extrapolates a resultant maximum torque. Next, the optimiser evaluates the torques provided by the modelling module with regard to the objectives to be fulfilled and varies the assembly parameters until the modelling module returns a result that meets the desired component function.

With reinforcement learning, the CTM system has an ability to learn that reduces the required queries of the modelling module and thus the computation time required for already known or similar starting situations (shaft-hub combinations).

Implementation of Reinforcement Learning in Assembly of the Shaft-Hub Connection

To be able to make an "intelligent" decision as to which of the actions or operators integrated in the Soar code should be selected, reinforcement learning is used. Reinforcement learning denotes a learning process based on reward and punishment. In Soar, positive or negative rewards are issued depending on whether an agent achieves its objectives and these rewards are allocated along the decision-making path that led to the corresponding result. The agent is thereby able to quantify the benefits of a decision and thus, step by step, learn the optimal way to achieve its objectives (Nason and Laird 2005; Sutton and Barto 1999).

At first, the agent makes random decisions about the subsequent operators. Once the agent has completed the search, the result is evaluated and reward points are issued for the decisions made depending on whether the objectives are met. The number of points awarded for fulfilling a particular objective can be fixed or freely selected by the programming. In the example of the shaft-hub production, dynamic point allocation was chosen. The nearer the operators bring the agent to the desired target torque, the higher the reward for the selected operators. Formula 1 shows the calculation for the rewards (r) as a function of the difference from the target torque (diff).

$$r = \frac{\max(r)}{(diff+1)*(diff+1)} \quad (6.1)$$

In further iterations of the agent, operators with high preferences are chosen more often. To avoid a situation where supposedly poor operators, or operators that have not yet been tried, are never selected, selection of a certain percentage of supposedly poor operators is always rewarded with a lower number of points. The frequency of selection of such operators can be freely determined. Experiments with 10 and 30% are presented below. This maintains the balance between exploration (undiscovered possibilities) and exploitation (of current knowledge).

Evaluation of Reinforcement Learning Using the Shaft-Hub Model

For evaluating reinforcement learning in respect of the shaft-hub production, a multi-step approach is selected. In the first step, there is no variation of hub 1 (tapered interference fit). On this basis, both the learning algorithms integrated in Soar are evaluated using Sarsa and Q-Learning. In the second series of experiments, hub 1 is varied too. The Sarsa and Q-Learning algorithms are analysed with a variable hub 1 as well.

Rewards points are distributed both after hub 1 assembly and after hub 2 assembly. For the transmission behaviour of the first hub, with 39–63 Nm transmissible torque, a comparatively greater range is considered able to fulfil the function. Ideally, the total torque should be 105 Nm; deviations of +/− 7 Nm are permissible. The agent receives positive rewards within these limits and negative rewards outside them. The points are issued according to the degree to which the objectives are met, as per formula 6.1.

In generating parameter proposals, it is important that new values are also generated in addition to the combinations already tried and tested. This is the only way to ensure that the optimiser does not just stick with in one local optimum, thereby leaving a potentially more favourable parameter combination undiscovered. Furthermore, a balance between tried and tested and new proposals facilitates a flexible reaction to variations in the production system. This deliberate selection of new parameter combinations is known as an indifferent selection strategy. In the experiments described here, the most widely-used procedure (Nason and Laird 2005) was selected using the epsilon-greedy method. This method provides the facility to specify an ε value of between 0 and 1 as the probability that new parameter combinations that deviate from the prioritisation will be selected. An ε of 0.3, therefore, means that in

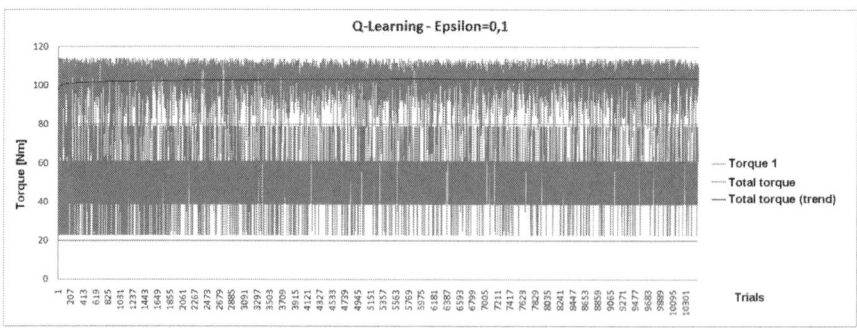

Fig. 6.15 Learning success without variation of hub 1

30% of cases, a new proposal will be selected (Nason and Laird 2005; Rummery and Niranjan 1994). Accordingly, when setting the value for ε, it is necessary to weigh up in each case whether the increasing flexibility produced by a higher ε is necessary, or whether the efficiency of the optimiser is reduced too much through more frequent testing of alternative parameter proposals. Frequently varying processes tend to require a higher value for ε whilst with very stable processes, the value for ε can be reduced.

Experiment 1: No Variation of Hub 1
With no variation of hub 1, the success of reinforcement learning can be measured at a very early stage. In carrying out the experiments, two learning algorithms were evaluated using Sarsa and Q-Learning. In addition, the learning rate ε was varied between 0.1 und 0.3. Figure 6.15 shows an example of this for Sarsa and an ε of 0.1. The diagram shows the transmissible torque resulting from the proposals of the optimisation module for hub 1 (torque 1) and the torque transmissible by both hubs together (total torque) across the total number of proposals executed. In all experiments, a significant improvement in the result quality is revealed after only approximately 100 proposals. With this learning success, the modelling module needs to be requested to find adequate assembly parameters less often, hence a result can be achieved in less time.

Figure 6.15 shows that over the entire length of the experiment, there are strong fluctuations in the predicted torque. These fluctuations are inevitable due to the use of the indifferent selection strategy and they enable a flexible response to changing boundary conditions. To compare the learning success of the various algorithms and the learning rates, the mean values and standard deviations of the calculated total torques are compared once at the beginning of the experiment and once at the end of the experiment. The study shows that after numerous experiments, a high epsilon also results in strong fluctuations of the predicted torque. This is reflected in an increased standard deviation of the values proposed by Soar (see Table 6.1). However, no significant differences due to the learning algorithm used are observed.

Experiment 2: Variation of Hub 1
If, in addition, the parameters of hub 1 are varied, this results in a significant increase in the complexity of the problem. The Soar program for manufacturing and

6 Self-optimising Production Systems

Table 6.1 Mean value and standard deviation without variation of hub 1

Learning algorithm	Epsilon	Tests	Mean value (Nm)	Standard deviation (Nm)
Q-Learning	0.1	1–30	89.67	17.23
Q-Learning	0.1	1,000–1,030	104.35	6.94
Q-Learning	0.3	1–30	97.58	16.73
Q-Learning	0.3	1,000–1,030	101.94	9.76
Sarsa	0.1	1–30	91.68	19.46
Sarsa	0.1	1,000–1,030	103.45	6.27
Sarsa	0.3	1–30	100.84	12.13
Sarsa	0.3	1,000–1,030	106.32	8.31

assembly of the shaft-hub connection contains approximately 100,000 rules. Due to the substantial increase in the number of parameter combinations, it takes longer to reduce the deviations of the total torque resulting from the proposals. A significant improvement in the result quality can be measured after approximately 2,000 proposals. Figure 6.16 shows an example of a learning curve for Q-Learning and an ε of 0.1.

Similar to the experiment without variation of hub 1, the graph shows strong fluctuations in the torque reached across the entire process. The fluctuations are much more pronounced over the first 1,000 tests. For a quantitative comparison of the learning success, as with the above tests, mean values and standard deviations are determined under variation of the algorithm and the ε. Compared with the permissible tolerance of ± 7 Nm, an ε of 0.1 generates very good values. With an ε of 0.3, the standard deviation of the total torque achieved increases significantly. In this experiment, which is characterised by a much higher problem complexity compared with the non-varying hub, Q-Learning exhibits significantly better results than Sarsa (see Table 6.2).

Rapidly identifiable learning success in relation to the possible parameter combinations is essentially connected with the distribution of reward points for interim results. In this case, the procedure makes it possible to evaluate unfavourable interim results, i.e. in this instance, the torque transmissible through hub 1, in advance. Thus,

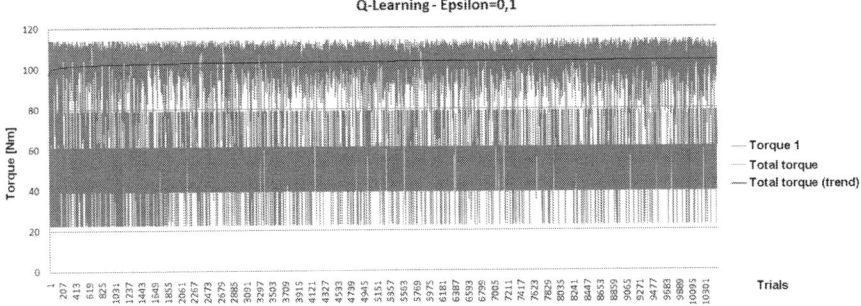

Fig. 6.16 Learning success with variation of hub 1

Table 6.2 Mean value and standard deviation with variation of hub 1

Learning algorithm	Epsilon	Tests	Mean value (Nm)	Standard deviation (Nm)
Q-Learning	0.1	1–30	92.85	18.41
Q-Learning	0.1	10,000–10,030	103.46	6.25
Q-Learning	0.3	1–30	97.42	15.23
Q-Learning	0.3	10,000–10,030	104.89	7.01
Sarsa	0.1	1–30	100.43	14.09
Sarsa	0.1	10,000–10,030	104.26	8.99
Sarsa	0.3	1–30	100.73	15.52
Sarsa	0.3	10,000–10,030	103.00	11.39

the potential, but not necessarily instrumental, degrees of freedom are effectively reduced without categorically excluding them through fixed rules or restricting the flexibility of the optimiser.

Generally, however, when using interim results, it is important to carefully check that, disconnected from the previous settings, they do not lead to interferences between the new and the previous settings.

Conclusion

Evaluation of reinforcement learning in the shaft-hub connection model is important for application of the CTM system in real industrial applications. These are usually characterised by greater complexity, i.e. a higher number of varying production parameters. Without an effective learning process, an optimiser with no physical model would not be able, within a practical period of time, to determine the production parameters that ensure the functional performance of the product to be manufactured.

However, the use of reinforcement learning in the shaft-hub model also demands a considerable amount of memory. At the beginning of optimisation, Soar initialises the learned value of each possible state with zero, i.e. the higher the number of possible states (in other words the number of possible parameter combinations), the greater the memory requirement of the optimisation program. The manufacturing software for the shaft-hub demonstrator requires up to one gigabyte of working memory depending on the degree of discretion of each parameter. The memory requirement increases exponentially with the concatenation of subprocesses. For this reason, preliminary analysis using data mining (see 6.2.4.3.3) is essential.

Another way to reduce the memory requirement, and thus also the computational complexity, is to split the Soar optimiser into several independent agents that process different sections of the production process. To create a working interface between the agents, it is important that only the information relevant for the next step is passed on. Since this can only happen reliably at certain points in the process, the use of multiple Soar agents must be tested in each case.

Data Analysis in the Analysis Module

The data recorded in the perception layer forms the basis for decision-making in the optimisation layer. Decision-making is implemented by the Soar cognitive architecture in cognitive tolerance matching. Both the effectiveness and the efficiency of decision-making decreases exponentially with an increasing number of parameters to be optimised ("curse of dimensionality", (Priddy and Keller 2005)). Hence for parameter optimisation that is useful in industrial practice, it is necessary to reduce the parameters to be optimised as much as possible.

In terms of parameter reduction, the challenge lies in identifying the parameters that have a significant influence on the product feature to be controlled. Often a large number of parameters, such as tolerances, dimensional measurements or machine settings, are available and their precise influence on the product feature to be controlled is not fully known. A prerequisite for controlling process chains with cognitive tolerance matching is to consider all relevant parameters that have an influence on the feature to be optimised. If a relevant parameter is not taken into account, the existing dependencies cannot be uncovered. With respect to data analysis, the challenge lies in reducing the scope of consideration by non-relevant parameters.

In the context of data mining, there are various statistical methods for interdependence analysis of large amounts of data. These make it possible to identify data or parameters that have a significant influence on the product feature to be controlled. As part of the research work, various data mining methods, including correlation analyses, principal component analyses and various classification methods, were examined for applicability in the cognitive tolerance matching system. A methodology was developed that can be established upstream of an optimiser in such a way that the optimiser only has to consider a greatly reduced number of parameters and hence the memory requirement and performance of the optimiser remain within practical limits, but the relevant area of the parameter space is covered.

Decision Trees as a Method of Data Mining
Based on the advantages described in Sect. 6.2.3, decision trees were identified in a preliminary study as the data mining method most suitable for CTM and thus were evaluated in detail.

Figure 6.17 shows an example of a decision tree for the quality of a shaft-hub connection. A decision tree is a directed cycle-free graph with a tree-like structure

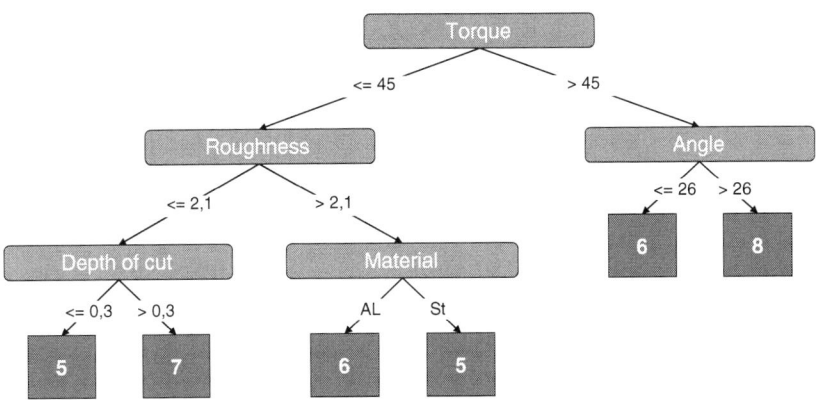

Fig. 6.17 Illustration of a decision tree

comprised of a root, edges, nodes and leaves. In the tree, each node denotes the test for an attribute, each edge indicates the result of such a test and each leaf represents a class. A decision tree makes it possible to intuitively visualise dependencies in relation to a product feature to be examined. At each node of a decision tree, all the available attributes are ranked. The attribute that is most useful for partitioning the tuple at this node into classes is selected as the test criterion. In the relevant literature, this test criterion is also known as the splitting criterion (Han and Kamber 2006; Agrawal et al. 2004).

In the leaves, the quality of the connection is rated using values between 5 and 8. The nodes and edges quantify the influencing factors that lead to achievement of a particular quality class. Hence a connection with a torque of 50 Nm, for example, and an angle of 31° is rated with an 8.

Evaluation of the Method

With respect to the applicability of data mining algorithms in general and for decision trees in particular, the effectiveness of the various algorithms was investigated using actual industrial data. The data used for this was from the industrial application of BMW rear axle drive production. In gathering the dataset, the aim was to use data mining to identify the parameters that have a significant influence on the acoustic behaviour of a rear axle drive. Therefore, ideal acoustics is the product feature to be optimised by the CTM system. In a quality control loop, the acoustics is the control variable.

To decide which parameters to include in the study, expert workshops were held at BMW in order to select all variables that, in the experts' opinion, could influence the acoustics. As a precautionary measure, all parameters whose influence could not be ruled out were also selected. In total, 193 different parameters were identified in the expert workshops. To gather the full database with respect to these parameters, 80 rear axle drives were produced and measured completely. The resulting data thus consists of 193 different measured and preference values which represent the process chain of production of a rear axle drive for the optimisation variable "acoustic behaviour". Figure 6.18 shows the distribution of the identified parameters along the process chain of rear axle drive production.

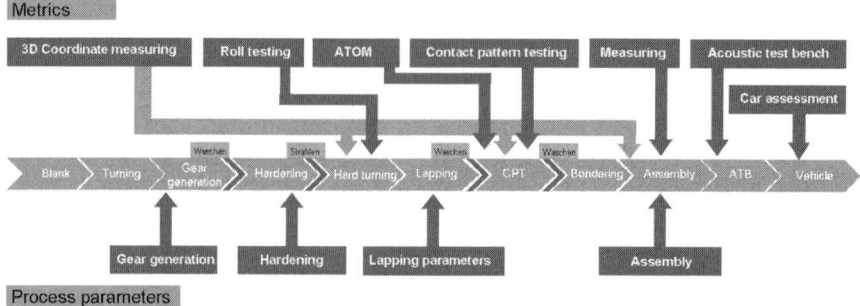

Fig. 6.18 Influencing parameters

6 Self-optimising Production Systems

Table 6.3 Assignment of acoustic test bed and evaluation indicator

KPI =>7	KPI <= 5			
	3,75	4	4,25	4,5
1,5	x	x	x	x
1,75	x	x	x	x
2	x	x	x	x
2,25	x	x	x	x
2,5	x	x	x	x

The acoustic behaviour of a drive is evaluated using a subjective key performance indicator (KPI) between 1 and 10. The aim is to achieve the highest possible value for the rear axle drive acoustics. A value greater than or equal to 7 is deemed exceeding requirements and a value smaller than or equal to 5 is generally perceived by customers as annoying and they would complain.

The KPI is heavily influenced by the subjectiveness of the test participant. For the sake of objectification, the measured values of the acoustics test bed were assigned to various KPIs. To this end, first the measured structure-borne sound signals of the acoustics test bed were standardised and then totalled for all measured axes. The measured values cannot be rigidly assigned to the KPIs because slight deviations may arise depending on the perception of the examiner making the evaluation. To compensate for subjective deviations, Table 6.3 was developed as an experimental design for variable assignment. The table shows 20 possible assignments for acoustically unobtrusive drives (KPI greater than or equal to 7) and acoustically obtrusive drives (KPI smaller than or equal to 5). The crosses each stand for one possible assignment of the limits for obtrusive and unobtrusive drives. The numbers represent the standardised and totalled measured values of the acoustic test bed. For example, the cross on the left on the first line means that a measured excitation smaller than 1.5 indicates an unobtrusive drive and a measured excitation greater than 3.75 indicates an obtrusive drive.

The evaluated measurement variable of the test bed is the integrated amplitude of individual frequency bands that can be perceived as annoying in the vehicle. To avoid subjective influence with respect to the evaluation indicator rating, decision trees were created for all 20 possibilities. For creating the decision trees, the three open-source methods "information gain", "gain ratio" and "Gini index" (Han and Kamber 2006) were evaluated, along with "Accuracy", a proprietary method of the "Rapid Miner" software used. Combining the four methods presented with the 20 possible evaluation indicator assignments produced, in 80 validation runs, decision trees which formed the basis for parameter classification independent of potential method influences.

The results from the analysis of the data mining process show that these vary considerably from method to method. Due to the significant differences in the data structures of various industrial applications, valid selection of a single method is not possible at this point in time. Automated use within a CTM architecture for prioritising the data gathered is therefore not feasible. The challenge in selecting the method suitable for the data to be examined lies in not knowing the relationships within the existing data. If it is not known whether, or in what way, dependencies exist between individual parameters, then precise selection of a suitable method of analysis is not possible. Hence, combining the individual methods was investigated as a possible solution. Adding up the results of the individual methods across all 20 evaluation indicator combinations clearly highlighted certain parameters so it can be assumed that these are of greater relevance. For example, the circumferential backlash and the block size of the gear set were identified as the most influential parameters with respect to the acoustics of the rear axle drive. A feasibility test by the experts at the BMW Group has confirmed the parameters obtained through data mining are appropriate. Industrial application of the methods therefore also seems thoroughly appropriate. However, this cannot be automated as yet; preparation with the help of the experts is needed.

Using data mining and decision trees in this way, complex process chains can be analysed with a variety of parameters that are considered relevant. Identifying subprocesses and individual parameters that have an increased influence on the product feature to be optimised only enables a cognitive optimiser to be used for those subprocesses.

Control Via the Coordination Module

To coordinate the individual tools, a comprehensive module has been installed that forms the interface to the user, controls execution of Soar programs as well as those of the modelling module and is responsible for the data exchange between individual modules. In real-time control of a production system, the coordination module also connects the production system, and thus the production machines, by transmitting the settings and control commands created for actions via the operation layer.

Soar Sequential Control

A Soar program is executed in "steps". Each step involves the execution of one Soar rule and thus the making of a decision. Between the executions of individual steps, it is possible to make Soar data available, or to read data from Soar. The coordination module controls the sequence of the individual Soar decisions and is responsible for the data exchange to and from Soar. For controlling a Soar agent, there is a C++ library which makes it possible to load Soar codes and execute them step-by-step as well as read and describe the individual variables of the Soar working memory.

Modelling Module Control

Similar to optimisation module control, the coordination module controls the data exchange between the modelling module and the other components. In the specific implementation of the software, the modelling module is represented by the artificial neural network software SNNS (Zell 1994). SNNS stands for "Stuttgart Neural Network Simulator" and simulates the components and processes of neural networks using technical means so that they may be recreated using a computer. SNNS allows artificial neural networks to be exported as a software library that can be linked directly to the coordination module and enables batch-based execution and querying of network actions.

The coordination module uses these functionalities and can access the relevant SNNS data and functions directly through specially designed interfaces in order to execute high-performance network queries. In addition to purely querying a network, the coordination module supports post-training of an existing network during run time. Hence in the event of deviations between a value predicted by the artificial neural network and a real measurement result, the network can be post-trained using the relevant data points (i.e. measurement results). The functionality provided by SNNS was extended accordingly for this purpose.

Storage of Knowledge and Experience in the Knowledge Module
Knowledge is provided in cognitive tolerance matching within Soar and by artificial neural networks. Soar provides two concepts for representing knowledge in the form of a long-term and a short-term memory:

The long-term memory stores operative knowledge for forming the problem spaces and control knowledge for controlling the search processes consistently in the form of productions.

The Soar memory model (Fig. 6.19) provides a more precise description of the long-term memory: it is divided into the semantic, the episodic and the procedural memory. Declarative knowledge, i.e. semantic and episodic knowledge, is organised in the form of so-called semantic networks. Short-term knowledge is represented by attribute-value lists which are grouped together to form objects. The short-term memory forms the working memory where all information processing takes place. The short-term memory is characterised in particular by its limited storage capacity.

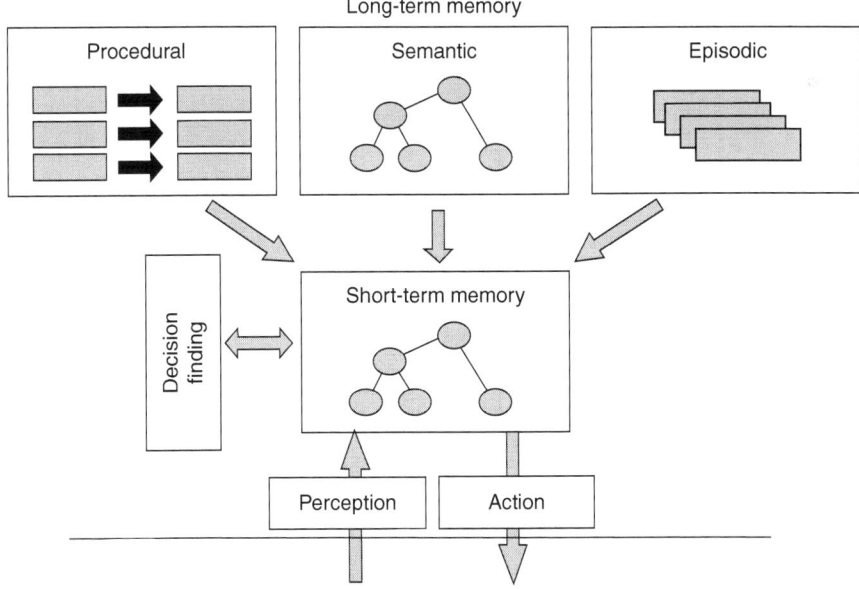

Fig. 6.19 Soar memory model

Long-term information must therefore be transferred into the long-term memory. Short-term memory is often referred to as working memory. The short-term memory processes the current to-do tasks.

The uniform representation and access mechanism as well as the ability to structure the working memory into areas means there is a strong similarity to blackboards. The open design of the working memory allows the addition of any modules that can use this memory, or an assigned segment thereof, for information exchange and coordination.

Information processing involves two phases. In the first phase of the knowledge search, applicable long-term memory productions which operate on the working memory fire. On the one hand, this process leads to the generation of new objects which can in turn allow other productions to apply. On the other hand, preferences emerge and vary, and these are then used in the second phase for control with respect to the further application of goals, problem spaces, states and operators.

In the second phase, based on the current knowledge in the short-term memory and with additional help from existing preferences, the decision procedure selects an operator and applies this to the associated problem space. Through successive application of operators, either the goal is met at some point or a dead end is encountered. In this case, a subordinate goal is generated and its task is to direct the search process out of the dead end. If a dead end cannot be resolved this way, then the problem space reverts to independent mechanisms such as backtracking (Laird and Bates Congdon 2008).

Through the use of reinforcement learning, the knowledge base is dynamically expanded during run time. The learned knowledge is taken into account when further decisions are made, thus reducing the likelihood of reaching a dead end (Nason and Laird 2005).

The knowledge about cause-effect relationships in the model is stored in artificial neural networks which serve as a black box model. To this end, using saved data, artificial neural networks are trained to calculate or estimate results for possible input parameters of the production system. If the input parameters are applied to a calculated result in real production, the real result obtained can be compared with the calculated result. If a significant difference is discovered, the model is updated with the corresponding newly acquired data through the learning process used, hence the newly acquired knowledge is added to the implicitly stored pool of experiences.

Data Acquisition by the Perception Layer

As a provider of information to the CTM architecture, the perception layer is dependent on the processes and machines to be controlled. The perception layer's task is to acquire all the characteristic variables of a production process that are considered relevant. Preliminary data analysis in the analysis module and optimisation by the control software requires the continuous recording and digital availability of all relevant production data. However, the challenges in this respect lie not only in the interfaces that must be created between the production means and the IT systems concerned, but also in interpreting the data in order to obtain useful information from it. Often, data is not available as concrete figures and thus not in a form that can be immediately analysed; instead, it often needs to be converted from a qualitative

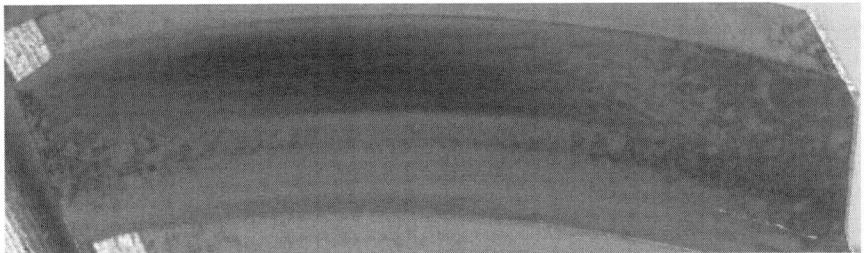

Fig. 6.20 Contact pattern of a ring gear

form into quantifiable values before the data, classified by the experts as potentially relevant, can be analysed in terms of its significance.

Hence when assessing a rear axle drive, for example, the contact pattern (Fig. 6.20) is an important evaluation indicator. The contact conditions of a gear set pair are visualised using contact pattern paste. Based on the quality of the tooth flank contact, conclusions may be drawn regarding, for example, the acoustics of a drive. However, to date, evaluation of the contact pattern has not been automated; instead, it has been conducted by trained staff and is therefore always subject to subjective influences and evaluation criteria. The evaluation results vary depending on the examiner and are therefore only suitable for further digital processing to a limited extent.

It is therefore necessary to evaluate this data deterministically and thus create reproducible evaluations. In the specific use case of the rear axle drive, an image processing system for automatic contact pattern recognition is developed. The measuring device is equipped with cameras to create colour images of the tooth flanks. Once the image has been captured, the tooth flanks are detected and extracted. Using a variety of graphic image processing algorithms, the contact patterns are identified and classified so they can then be converted into binary matrices. The relevant parameters of the contact pattern of rear axle drives are determined using these matrices. These parameters are optically recorded by the developed measuring device, they are captured by digital image processing and are thus used as input parameters for the optimisation software. In the next step, they are used by the analysis module which identifies relevant characteristic variables with the aid of data mining algorithms for the production process.

6.2.5 Industrial Relevance

The boundary conditions for the manufacturing industry in high-wage countries have changed immensely in recent years. Due to increased global competition, besides the demand for high-quality products, the demand for cost-effective production processes, in particular, has also grown. Many companies are no longer able to meet these challenges in Germany and are shifting their production to low-wage countries. There, lower wages and social security contributions as well as economies of scale through cost-effective mass production are leading to an apparently higher productivity (Pfeifer and Schmitt 2006; Beckmann 2009).

Another global trend is that customers are increasingly demanding individualised products available to them within a very short period of time. This is particularly evident in the automotive industry where customers are able to amend the configuration of their vehicle even shortly before its delivery (Lindemann 2006; Krause 2007).

So, successful companies must be able to adapt quickly to individual customer requirements which, in addition to short lead times, also requires an increased flexibility in production as well as stable processes (Pfeifer and Schmitt 2006).

Manufacturing processes for technically sophisticated products, in particular, increasingly have such complex dependencies between a high number of variable production parameters with an influence on the product to be produced that those relationships can no longer be fully comprehended. This leads to decreasing process stability. Production processes must be continuously optimised to achieve the desired product features and tolerances. However, these boundary conditions do not allow any last-minute product variations and, in addition, the processes are often not ideally designed because they can usually only be optimised in parts, especially in the case of multi-stage production processes. This results in the separate optimisation of individual elements of a system without the ability to fully assess the interaction of these changes with respect to the properties of the final product (Best 2010).

Function-oriented optimisation of the end product is more effective than optimising individual production steps. This is possible if individual targets, e.g. the dimensions to be obtained, can be adapted dynamically within the production process. That way, the superordinate goal, namely the functionality of the product, can be achieved.

For one thing, this kind of dynamisation of key process parameters will help reduce costs because individual tolerances can be selectively expanded without overlooking the required product features. For another thing, the flexibility of the production process will be greatly increased with respect to product modifications.

Self-optimising production systems are a crucial method of dynamisation. They allow value stream-oriented approaches to be adopted whilst at the same time enabling planning efficiency to be increased through the transfer of previously acquired knowledge into similar scenarios within production engineering. This facilitates brand new approaches for both production and assembly systems that constantly analyse and evaluate the current situation and dynamically adapt the system to changing goals.

Cognitive tolerance matching allows cross-level approaches to be developed in the field of coordination, planning, control and man-machine interaction. Hence, a more integrative framework of action is generated, enabling production systems to self-optimise with respect to different goals. This is achieved through the creation and implementation of cognitive mechanisms at planning and organisational levels across processes and process chains and through the creation of the ability to communicate in the areas concerned.

The business and technology case of rear axle drive production at the BMW Group in Dingolfing, Germany, is used to exemplify the industrial relevance. At this site, cognitive tolerance matching is used to optimise the production of rear axle drives with respect to the acoustics emitted during operation. The rear axle drive selected

Requirements for the vehicle:

- Sportiness   ⇒ Stiff bodywork
- Dynamic   ⇒ Less damping
- Light construction   ⇒ Low insulation
- Effectiveness   ⇒ High degree of efficiency

Requirements for the gear set:

- Minimal stimulated vibrations
- Optimal contact pattern
- Low gear loss
- Optimal adjustment specific to vehicle

Controlled parameters of the gear set:

→ Acoustics
→ Resistance
→ Efficiency

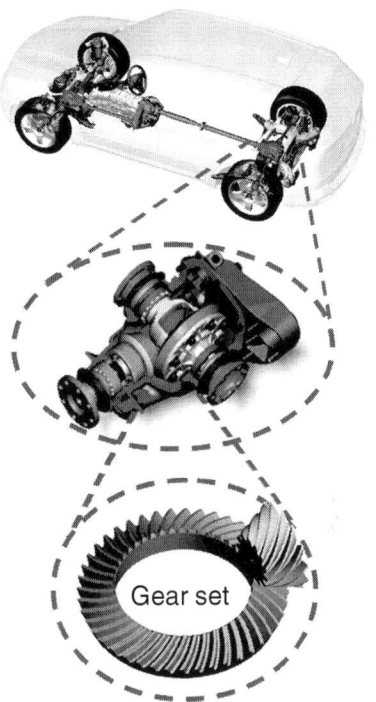

Fig. 6.21 List of requirements on the Hypoid gear sets for passenger vehicle axle drives of standard design

for the development of cognitive tolerance matching has a pronounced effect on the acoustics within the vehicle. It has very complex and multifaceted mechanisms that generate structure-borne sound which is then transferred to the interior of the vehicle via the car body. When the noise reaches a particular level inside the car, it may be perceived by the customers as annoying. The axle drive acoustics can sometimes be adjusted depending on the vehicle via the axle drive teeth (gear set). This presents a major challenge in connection with the occasionally contradictory demands for a high level of efficiency and strength. The vehicle acoustics, and therefore also the axle drive acoustics, are product features that are important to the customer (see Fig. 6.21).

Figure 6.22 illustrates the production process. It ranges from production of the gear set through to assembly and is characterised by many sensitive tolerances with complex dependencies.

The approach for optimising the acoustics is to optimise the specified tolerances throughout the entire production process. This means both expanding any unnecessarily tight tolerances to save costs and defining the critical tolerances more precisely to ensure the desired functionality of the end product.

The challenge lies in understanding the process, a pre-requisite for implementing the right optimisations at the crucial points.

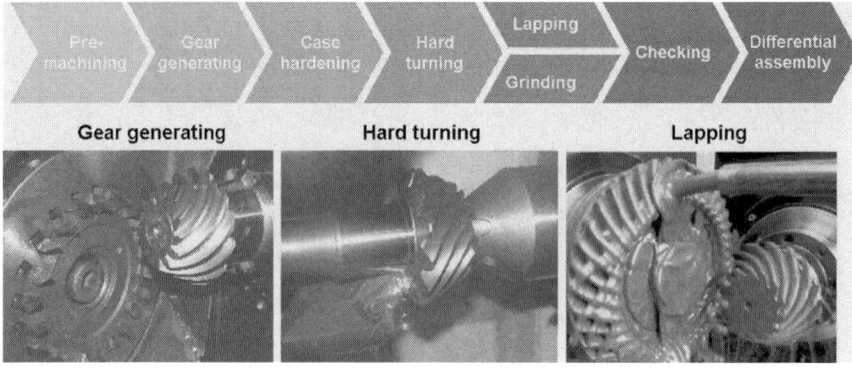

Fig. 6.22 Illustration of the process chain for manufacturing a hypoid gear set

Besides the objective of producing quieter vehicles, it is also important to identify the influencing parameters relevant for vehicle acoustics in design and production and to ensure that vehicles can be produced with constant acoustics without spread. The challenge here lies not just in merely identifying the acoustically relevant vehicle components and parameters, but, more specifically, in modifying them effectively. In this respect, the complex acoustic interactions between the individual vehicle components need to be taken into account.

Figure 6.23 illustrates the impact of cognitive tolerance matching throughout the entire production process. Firstly, the knowledge gained from production, assembly and use is fed back into development so that the production parameters can be adapted

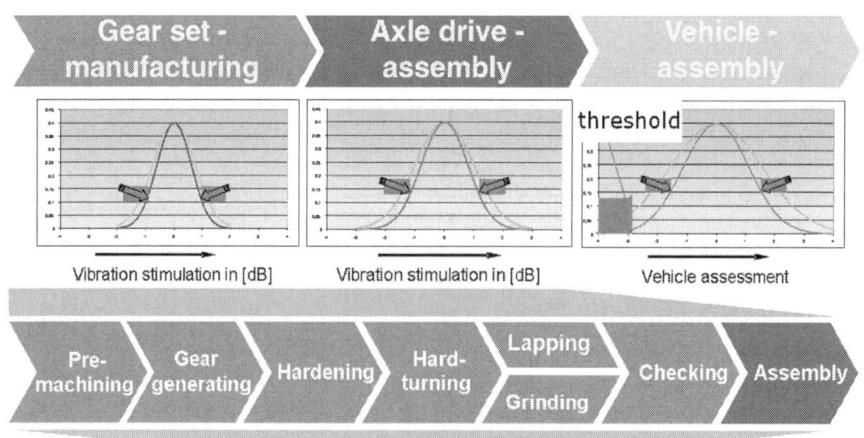

Fig. 6.23 Implementation of CTM in the production process

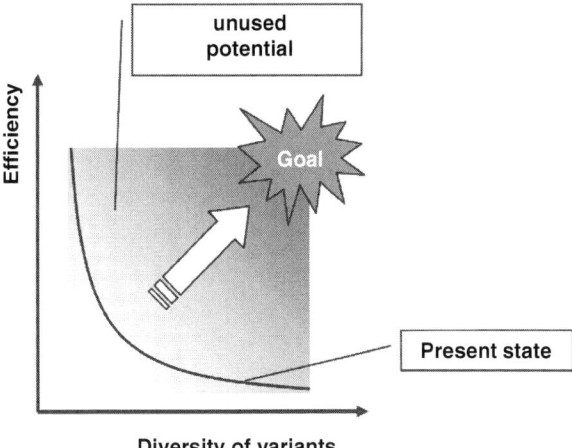

Fig. 6.24 Conflict between product variety and efficiency

accordingly. Secondly, through cognitive tolerance matching, individual process steps are interlinked, making it possible to respond immediately to deviations in the process. Hence, CTM contributes to the continuous improvement process in production. It also reacts specifically to deviations in upstream process steps, thereby ensuring the superordinate goal of an optimised drive in terms of acoustics, strength and level of efficiency can still be met. Thus the focus of deliberations is less on tolerance-oriented and much more on function-oriented production. Through these two mechanisms of action, the perceived quality can be both stabilised and increased.

In particular, the demand for increased product variety with increasingly complex products causes decreasing efficiency of the production process (see Fig. 6.24). For Germany to continue to be attractive as a production location, production efficiency must be increased as product variety increases.

The industry has recognised that in addition to focusing on how individual processes work, it is necessary to adopt a holistic view of the causal relationships within the value-adding chain.

6.2.6 Future Research Topics

Research and development in the field of self-optimisation in production and assembly already offers some promising approaches in the form of specialised solutions for individual problems. Technologies for implementing self-optimising applications in production engineering are being identified and successfully applied both in this and in other related research projects, e.g. in assembly scheduling (Sect. 6.6) or in production process parameterisation. Here, using self-optimisation approaches, significant improvements have been achieved in production and competitive advantages have been created. However, the research described herein presents just a small fraction of the possible applications that, in theory, can be improved through cognitive automated systems in production. Accordingly, there are other research tasks which may yield additional potential for these approaches. Deriving a systematic approach from the research presented makes it possible to transfer the methods to other areas of application. In this respect, their use is not limited to the operational level of a business, but can be applied at the planning and control levels too and, furthermore, these can be linked with the other levels. The following areas are therefore suggested as priorities for continuation of the research work:

- Development of a control model for the various production engineering applications within the levels of a company as well as on an inter-level basis: for planning the implementations of self-optimising control loops and the applications of cognitive components, it is necessary to model the use case as well as the solution components. For this purpose, analogous to classical control technology, mapping facilities need to be designed. In this respect, models need to be created in such a way as to be cascading and able to adapt to the granularity of the respective considerations.
- Development of modelling languages and standards: for modelling the applications of self-optimising control and regulation systems, it is necessary to develop modelling standards and languages that can serve as the basis for data exchange both between the models and between actual applications. The aim is to create standardised interfaces between the controlling and the controlled components. This includes creating semantic links between the individual cognitive architectures and technologies.
- Development of a software architecture as a modular framework: mapping the software technology in a modular framework allows standardised creation of control and regulation applications according to uniform guidelines. That way, different approaches can be concurrently evaluated quickly and easily so that the respective advantages and disadvantages can be compared instantly. The framework should be created using data structures based on the modelling languages to be developed.
- Further development of existing cognitive technologies: universal cognitive applications, such as SOAR, 3CAPS and ACT-R, offer significant potential for cognitive decision-making using technical means. To best exploit this potential, the capabilities need to be expanded with respect to the issues involved in production engineering.
- Evaluation of technologies for self-optimisation and cognition: the various technologies and cognitive tools used in the context of the Cluster of Excellence for implementing self-optimisation should be examined for their applicability in other cases. Capability profiles are useful in terms of selecting the best tool for the respective use case. Benchmarking in different applications provides support for such selection. A systematic selection methodology helps validate the decision for a specific technology.
- Evaluation and comparison of learning processes: learning processes provide the basis for generating knowledge from the available data and using it to positively

influence future situations. Different learning processes demonstrate different strengths and weaknesses depending on the application and these have to be balanced out against each other in accordance with the problem. The aim is to identify and implement stable and efficient learning processes depending on the application.

- Analysis of the effectiveness of cognitive and self-optimising technologies with respect to increasing the flexibility of production systems: individual studies within the problem area in question provide support for computer-based analysis of the effectiveness of various approaches as well as quantification of both the effectiveness and the efficiency prior to implementation.
- Cooperation of cognitive systems: cognitive systems must be capable of cooperation in order to be implemented on a cooperative or cascading basis at the same level. This includes strategy and sequence planning, coordination of a common component handling approach and the execution of transfers to the respective system boundaries. In addition to cooperation between machines, this also includes cooperation between man and machine and thus leads on to the task field of man-machine interaction.
- Man-machine interaction and cooperation: this includes the development of concepts and technologies for intuitive and safe interaction between man and machine in order to best combine the respective advantages of the possible scenarios. Amongst other things, this includes considering the challenges posed by the demographic change in production.

The afore-mentioned task fields address the question of how self-optimising control and regulation systems can be most effectively used throughout a production process in order to enhance the competitiveness of a production operation. This requires modelling the control loops to be depicted and modelling the specific use case in order to create a basis for identifying the right technologies for implementing the control loops. The optimal degree of automation with respect to throughput, flexibility and scalability needs to be identified through comparative assessment of possible approaches. In this connection, the optimal degree of integration of conventional systems and manual activities of human workers also needs to be determined. Changes in the boundary conditions (throughput, error rate) may also require adaptation of the production system configuration, even during the production, in order to enable a faster response to unforeseen interruptions. One example of this is an adaptive production line, e.g. in the automotive industry, which allows cognitive automation of customisable assemblies for multi-variant products with the involvement of humans.

6.3 Integrative, High-resolution Supply Chain Management

Günther Schuh, Tobias Brosze, Fabian Bauhoff, Niklas Hering, Simone Runge, Maik Schürmeyer, Sascha Fuchs and Till Potente

6.3.1 Abstract

6.3.1.1 Initial Situation and Problem

From the perspective of logistical planning and control, managing the increasing dynamics in customer-specific production and assembly is the most important challenge of the next few years (Zäh 2005). Causes of the increasing internal dynamics are shorter delivery times, greater variety of production and assembly processes (caused by the increasing product variety) and the use of more complex production systems (substitution of capital factor for work factor). The drastic reduction in delivery times has fundamentally changed the order situation and the capacity requirements of producing companies (Wiendahl 2006, p. 29 ff.). The necessary reductions in throughput time have only been made possible by the corresponding reduction of work in progress. This reduction in stock has necessarily led to more intensive coupling of the individual product resources. Capacity fluctuations and process instabilities in one resource thus affect the stability of the overall systems far more because stock can no longer be used as a buffer. At the same time, macroscopic industry-wide capacity fluctuations within the supply chain are increasing because there is no time buffer.

The increasing variation in process chains and times potentially intensify the effect of the capacity and throughput fluctuations described above. The "average value-based PPC" therefore no longer fits the requirements (Günther and Tempelmeier 2007). Conventional planning and control concepts which cannot respond to this variation necessarily impose further fluctuations on the system. The lack of a buffer and fluctuating requirements lead to loss of efficiency and increasing backlogs in production.

Absorption of the dynamics by means of stock levels and decoupling or maintaining reserve capacities for process synchronisation is no longer possible for reasons of cost pressure and customer-specification of products. In fact, new approaches for the planning and control of internal and inter-company production processes are required. These solutions need to cope with the dynamics of processes and capacity requirements and to incorporate these into the superordinate network and to engender further solutions to absorb the dynamics taking into account commercial and logistical goals.

6.3.1.2 Purpose of Research

High Resolution Supply Chain Management (HRSCM) describes the creation of a company-wide information transparency, which includes the interfaces with value chain partners, with the aim of ensuring product availability through decentralised, self-optimised control loops in industrial value chain networks. HRSCM pursues the aim of enabling organisational structures and processes to self-optimise, i.e. to adjust, according to consistent goals, to constantly changing framework requirements by means of decentralised production control mechanisms in the form of a cascaded control loop model. The designation "High Resolution Supply Chain" therefore relates to the almost unlimited transparency achieved at practically all levels of the industrial supply chains through the ubiquitous use of IT.

Information transparency plays a key role in the planning and control of a producing company. It is the basis for making decisions about the actual need for action and also provides support for actions to be undertaken. The central challenges are the complexity of information and their management along with the effective integration of human intuition and experience into the control loop of the supply chain management.

On the basis of adequate information transparency, this project pursues the scientific aim of radically improving the planning quality of planning and control processes while simultaneously reducing planning times and costs. This can only succeed when the currently prevailing average value-based planning logic of the MRP II concept is resolved by a decentralised actual data-based planning logic.

6.3.1.3 Results

To manage the initial situation and to implement the above mentioned aims, a comprehensive model of cybernetic production management was developed which represents the structural framework of the HRSCM approach. Furthermore, in order to enable the systems and subsystems to optimise dynamically and independently, logics were adopted from control theory and transferred to production planning and control and the optimisation of company processes.

To create a cybernetic production management model firstly a unified understanding of the production system processes to be controlled was required. For this purpose the processes and information flows involved in technical order processing and the tasks of the order-independent networking and cross-sectional functions were described The basis was the Aachen PPC model extended by Schmidt, in which the traditional processes of production planning and control were augmented by the processes of construction, assembly, dispatch and commissioning as well as project monitoring and control (Schmidt 2008; Schuh 2006). The result is a process-orientated illustration of the tasks and related information flows required to execute the operative order processing processes and thus a description of the production system to be controlled.

Thereafter the basic structures and mechanisms of a viable and changeable system were described from the perspective of management cybernetics. The principles, organisational structures, elements, information channels and mechanisms required by a cybernetic management system in order to manage the variety of complex system under dynamic influences were specified. The necessary expansion of the Aachen PPC model was then defined on the basis of these results with respect to the requirements of the cybernetic management theory. It was found that the model needed to be augmented by the principles of recursiveness, autonomy and synchronisation of target systems. On the other hand, it was also demonstrated that essential elements for the recording and use of high resolution information were as yet insufficiently defined.

The core of the cybernetic production management model is a reference model derived from the modules outlined above. Thereby the viable system model served as a structural template for the variety-orientated development and location of production management tasks, processes and information flows within an all-encompassing production management model. The architecture of the model is based on the order processing processes as the base units and is divided into four recursive structural levels. The autonomy principle was taken into account in the allocation of tasks in a way that in normal cases an order can proceed through the operative processes without intervention from the management system. To enable a cybernetic exertion of the management tasks, taking into account the manageability of the subtasks, the overall system was divided into five management units.

Based on the cybernetic production management model, control logics were applied to production management. Closed loop systems were modelled to permit automatic adjustment of the actual parameters. These dynamic adjustments of planning parameters require minimal planning time due to the high degree of automation but nevertheless permit high planning quality on the basis of data currency and granularity.

6.3.2 State of the Art

In the following, a summary of the state of the art in production planning and control is given. Furthermore approaches to the changeable design and control of production systems (Sect. 6.3.2.2) are presented.

6.3.2.1 State of the Art in Production Planning and Control

The operative planning and control of production is the core of production management. Current PPC systems are the result of various stages of development since the 1960s. The Order-Point System and Material-Requirements Planning (MRP) were the first of these development stages. Unlike the Order-Point concept, MRP is based on deterministic requirements in which part lists and associated gross secondary requirements are incorporated into the demand planning as influential values (Dangelmaier and Warnecke 1997, p. 258). The net demands which result in the order or the initiation of a production order are the result of balancing the gross secondary requirements with stock levels (Orlicky and Plossl 1994, p. 24). This material-requirement determination logic is still used today due to its simplicity and simple calculation algorithms, primarily for push-control of expensive and sporadically required parts (Ahsan et al. 1996; Askin and Goldberg 2002; Kiener 2006). Closed-Loop MRP was developed in 1969 as a refinement of the traditional MRP concept. At the heart of this refined concept is the assurance that all capacities required to implement the production plan are available (Kurbel and Endres 2005, p. 135 f.).

The term Manufacturing Resource Planning (MRP II) was coined in 1984 by Oliver Wight (Wight 1984, p. 51 ff.). Its most important addition is planning according to limited capacities. The integration of commercial and sales planning also enables analysis of implicit and financial aspects in PPC (Ahsan et al. 1996, p. 19).

ERP systems were first developed in the early 1990s, adding accounting-orientated modules, complex modules for operative production planning and control, maintenance, order management and human resources to the MRP-II approach (Kiener 2006, p. 267; Busch and Dangelmaier 2004b, p. 424 f.). The introduction of ERP systems and the use of Electronic Data Interchange (EDI) greatly improved communication and therefore also coordination within the supply chain. However, the MRP-II concept has some fundamental weaknesses. The centralised and push-orientated MRP-II planning logic, according to Pfohl, cannot adequately detect or plan the dynamic processes which frequently occur in a production environment as a result of problems (Pfohl 2004, p. 160). Weaknesses of MRP-II-based systems include lack of support for order release, a planning principle based on average values, the successive planning method and the use of limited partial models. The successive planning method breaks down PPC tasks into smaller task packages which prevents holistic analysis and achievement of optimum solutions (Hellmich 2003, p. 208). It is not possible to plan according to a higher-level commercial target system using the partial planning approach because of its inherently isolated partial analyses and reduced computation complexity (Finkler 2006, p. 23 ff.; Hufgard 2005, p. 152 f.; Kiener 2006, p. 269). Further weaknesses of MRP-II-based PPC systems according to Kletti are the inadequate analysis of the prevailing load horizon and the actual capacity exploitation, absence or delay of feedback about order progress, problems and poor information availability and transparency (Kletti 2006).

Since the 1960s there have been many attempts to expand the MRP-II approach with the aim of eliminating these weaknesses. At the end of the 1990s, the MRP-II concept was supplemented by so-called APS-Systems (Advanced Planning and Scheduling systems) to make them more useful by systematic improvements in planning (Gesatzki 2002, p. 1). APS systems are modular software systems which enable the integrated planning and control of companywide business processes (Albert and Fuchs 2007, p. 6). Exact mathematical optimisation processes are used for planning as well as heuristics for bottleneck-orientated planning (Schwindt and Trautmann 2004, p. 5). The use of APS systems makes it theoretically possible to incorporate all restrictions and available capacities simultaneously into the planning process and to continually update the plans (Albert and Fuchs 2007, p. 6 f.). Unlike ERP systems, APS technologies can be used to analyse alternative planning scenarios. In this case existing problems are not only resolved by increasing the efficiency of processes towards the customer but also optimizations at the interface to suppliers are taken into account (Gesatzki 2002, p. 4). APS systems are an attempt to obtain a more accurate prediction of the future using mathematical models with refined data. The

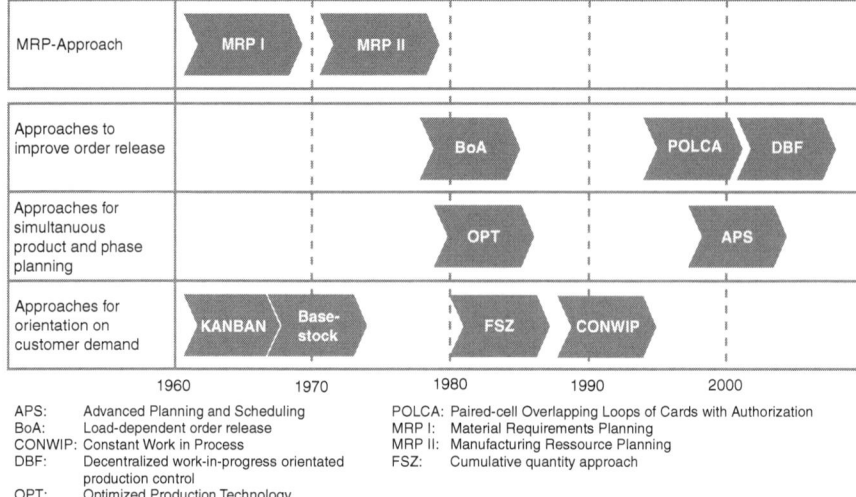

Fig. 6.25 The development of PPC concepts. (Meyer 2006)

computational results are nevertheless based on assumptions which do not fit the dynamic company environment, for example "an average production time per part".

Figure 6.25 provides an overview of the development of PPC systems from 1960 to the present day.

The following approaches represent the next development stage of the PPC whose basic characteristics are increasing decentralisation and self-organisation within PPC.

Scholz-Reiter et al. choose as the key issue the correct degree of use of self-control concepts for logistical processes (Scholz-Reiter et al. 2007, p. 1). The relationship between the degree of self-control of logistical systems and the stated degree of complexity to achieve an optimum target was analysed in an evaluation (Scholz-Reiter et al. 2007, p. 2 f.). The optimum degree of self-control that can contribute to the achievement of the desired targets can be identified with the help of the concept of positive emergence. Emergent characteristics of a system develop through the interaction of system components and not through the characteristics of individual system elements (Scholz-Reiter et al. 2007, p. 4.). These emergent characteristics may be quantifiable characteristic values (average throughput times, stock levels, use of resources, schedule-adherence etc.) or they may represent values which are not immediately quantifiable (flexibility, adaptability, robustness of the system etc). The supplementation "positive" in this case relates to the required characteristics and therefore to the achievement of the declared targets (Scholz-Reiter et al. 2007, p. 5 f.). The investigation of Scholz-Reiter et al. shows that a moderate use of self-control concepts has a demonstrably positive effect on production control.

SFB 614 at the University of Paderborn is involved with the use of self-optimising systems in mechanical engineering. The focus of the research work is on autonomously-acting technical systems which react flexibly to changing environmental conditions. Besides the establishment of self-optimisation within the scientific world, it is necessary to develop a method for the design of self-optimising systems (Gausemeier et al. 2006, p. 50 ff.).

The so called Hanover School has been looking at production control for some years. In this case control of the production system rather than the progress of individual orders is the focal point. Wiendahl is using basic principles of control theory for production planning and control to better engage with the deviation of production logistical processes from production schedules. The use of feedback mechanisms is designed to increase the speed of responses in the event of problems and therefore to reduce the gap between the actual and reference values (Wiendahl 2005, p. 348 ff.). The basic idea of the approach is to balance the planned values generated by the PPC for in-house and external production with actual production values fed back in the form of characteristic values. The analysis of the resulting discrepancies can be used to introduce suitable measures for their reduction with minimum delay. This type of production control enables target specifications to be achieved more quickly and more accurately (Wiendahl 2005, p. 348 ff.).

Nyhuis offers an approach wherein the expansion of production characteristics enables an increase in the degree of synchronisation of production goals. The characteristic curve theory is based on the progress elements which include the working processes and the throughput time of a production order. These throughput elements are the basis for different descriptive models, e.g. the hopper model developed by Wiendahl and the throughput diagram which is based thereon (Nyhuis 2008, p. 191 ff.).

In a general form, the production characteristics are valid for any production system. Specific characteristic curves for an analysed working system are possible however, taking into account additional framework conditions such as capacity, ongoing orders and the integration of the system into the material flow. Therefore a mathematical approach for the straightforward production of logistical characteristic curves for production processes is being developed at the Institut für Fabrikanlagen und Logistik (Nyhuis and Wiendahl 2003; Nyhuis 2008, p. 198). Determining production characteristics provides comprehensive assistance in the positioning of target systems in which relevant relationships between the stock and the throughput time are placed within a graphical context along with the use of resources in a normalised case.

6.3.2.2 Viable and Changeable Design of Production Systems

We now present actual approaches which particularly address the question of how cybernetic principles can be integrated into production systems to increase their changeability. Unlike conventional methods and general procedures in engineering or management sciences, system theory and cybernetics do not exclude complexity by restrictions and simplifying assumptions. It is in fact at the centre of all analyses.

Westkämper and Zahn present the Stuttgart company model, based on a holistic production system, which incorporates methods, tools and approaches for improving changeability in production companies with highly varied series production

(Westkämper and Zahn 2009, p. 25 ff.). The Stuttgart company model is a holistic model, which describes companies as complex systems which in turn comprise semi-autonomous performance units. In order to cope with highly dynamic production, decentralised, semi-autonomous organisms (performance units) form the basis of the Stuttgart company model. The performance unit comprises an organisational unit of a company of one or more employees, who pursue specific goals using resources (Westkämper and Zahn 2009, p. 49 ff.).

The investigations of Wiendahl, which were used within the context of SFB 467, are aimed at developing a situative configuration of order management in a turbulent environment. The term order management is used in order to differentiate it from PPC in terms of planning and control of orders (Wiendahl 2002, p. 18). Turbulence emphasises the fact that the environment is very unpredictable (Wiendahl 2002, p. 14), which leads to requirement fluctuations and unforeseeable events (Wiendahl 2002, p. 17).

Within SFB 467 Balve is developing a framework concept for the design of changeable order management systems (Balve 2002) on the basis of the Viable System Model (VSM). In this case an order management system is defined as an ideal and/or a real tool, supporting the economic and flexible processing of customer orders. The model described is incorporated into an overall procedural concept for changeable order management systems. This comprises the phases of team-building, situation analysis, target formulation, solution synthesis and subsequent evaluation (Balve 2002, p. 95 ff.)

Thiem provides another model for the socio-technical structuring of production systems based on the VSM. In his dissertation the author transfers the concepts of production control to the communication channels and systems of the VSM. Task steps and supporting tools (autonomy profile, information flow diagram) are assigned to stages of the procedural model (analysis of the actual situation, weakness analysis and development of reference structure). The design of the procedural model is followed by its application within a medium sized gearbox producer.

Haats proposed a solution for creating a lean computer-supported system for production planning and control based on VSM. Sub-concepts are developed from the overall concept of a lean PPC model. These sub-concepts are appropriated and concretised to form a catalogue of minimum requirements for informational processes. Haats demonstrates that with increasing decentralisation, while the requirements of communication support and integrated data access increase, the functional complexity of the communication systems overall can be reduced. The implementation and effectiveness of the concept is finally demonstrated in a case study (Haats 2000).

Espejo presents further application reports from various authors in his publication (Espejo and Harnden 1989, p. 103 ff.). The publisher opens the collection himself with his report on the application of the VSM as a diagnostic instrument within P.M. Manufacturers, a medium sized electrical engineering production company. The result of the project is a detailed performance specification. The application illustrates basic techniques for defining recursion levels and implementing the tasks of the VSM mechanisms within the organisational context.

6.3.2.3 Critical Evaluation of the State of the Art and Derivation of the Research Requirement

With regard to the PPC process and its development over time it is clear that many terms and organisational concepts for management cybernetics have also been established within operative processes. Therefore the development of the early deterministic successive planning models can be tracked to strategies and processes of self-control and optimisation in production and logistics. Existing approaches to production planning and control are inadequate for the dynamic processes often prevailing in today's production environments. The planning principle of today's ERP systems which is based on average values and the underlying successive planning prevents an integrated analysis of the PPC tasks and in most cases leads to less-than-optimum planning results. Planning, which is orientated towards an overall commercial target system, is not possible with today's PPC solutions because of their inherently isolated partial analyses and reduced computation complexity.

The approaches for self-control developed within various scientific disciplines show that in the analogy of naturally autonomous behaviour, forms of self-organisation can be transferred to social or logistics systems. The research finds that self-control strategies are suitable under certain circumstances for the management of dynamics and complexity in logistical production systems and exhibit a stable system behaviour. The analysis of the use of self-control strategies in the research we examined is often based on limiting assumptions with respect to the material flow structures and relates to simplified abstract models.

In conclusion, there are so far no models for defining in detail a structure for the decentralised, self-optimised planning and control of production systems and its incorporation into strategic and normative production management systems.

6.3.3 *Motivation and Research Question*

According to the state of the art, producing companies these days face an increasingly turbulent environment along with the associated increasing process complexity with rigid order processing coordination, which limits their response capability. The lack of reconciliation of elements within order processing due to the strict separation of planning and control means that a rigid and inefficient system behaviour has developed.

The planning and control of customer orders takes place in most companies within the context of operative production planning and control. Today, these tasks are supported by IT systems, so-called Enterprise Resource Planning (ERP) or production planning and control systems (PPC) in most cases.

Particularly planning based on static data (standard restocking times, standard throughput times etc.) is no longer appropriate in commercial practice as it is based on average values. The highly dynamic situation can therefore not be managed with the present planning systematic of many ERP/PPC systems.

High Resolution Supply Chain Management (HRSCM) however describes the creation of a company-wide information transparency, incorporating the interfaces with value chain partners, with the aim of ensuring product availability through decentralised, self optimised control loops in the PPC area. High Resolution Supply Chain Management pursues the aim of enabling organisational structures and processes to self-optimise, i.e. to adjust, according to consistent goals, to constantly changing framework requirements by means of decentralised production control mechanisms in the form of a cascaded control loop model. The potential of the High Resolution Supply Chain Management lies in the possibility of achieving a new level of capacity synchronisation through improved planning and control quality within the context of the PPC.

The designation "High Resolution Supply Chain" therefore relates to the almost unlimited transparency achieved at practically all levels of the industrial supply chains thanks to ubiquitous IT use. Information transparency plays a key role in the planning and control of a producing company. It is the basis for making decisions about the actual need for action and also provides support for actions to be undertaken. Companywide information availability of decision-relevant parameters accordingly serves as a fundamental basis for the operative planning and control of production value chains. The central challenges are the complexity of information and their management along with the effective integration of human intuition and experience into the control system of the High Resolution Supply Chain Management.

On the basis of adequate information transparency, the project pursues the scientific aim of radically improving the planning quality of planning and control processes while simultaneously reducing planning times and costs. This can only succeed when the current prevailing planning logic of the MRP II concept (Hellmich 2002, p. 27 ff.; Spath et al. 2002, p. 130 ff.; Wiendahl 2002) is resolved by a decentralised ACTUAL data-based planning logic.

The aim of the sub project "High Resolution Supply Chain Management" is accordingly limited to the significant increase of the planning quality with simultaneous reduction of the planning times and costs based on adequate information transparency in terms of a "High Resolution Supply Chain". This aim brings us to the following research question:

> How can we structure production planning and control considering the background of dynamic framework conditions to achieve a significant increase in planning quality while reducing planning times and costs?

In the future it will be possible to achieve the subgoal of reducing times and costs by increasing the rate of automation in data collection with the aid of auto ID technologies (e.g. with RFID), by situative aggregation and or disaggregation of information and by the direct availability of planning-relevant data.

The planning quality however is improved by the optimal use of ACTUAL values (information availability) and by the selection of an adequate level of detail (aggregation level) within the context of appropriate planning and control logics. For example, improved forecasting with the associated improved planning reliability can make a substantial contribution to innovative planning and control of a value chain.

6.3.4 Results

6.3.4.1 Principles

Production management is defined according to Kämpf as follows: *"Tasks, people, machinery and materials should be employed, controlled and co-ordinated such that products and services as a result of this activity are produced in the required quantity and quality at the specified time with the minimum outlay of costs and capital."* (Kämpf 2007, p. 5 ff.). In the following, production management should be understood as the planned and controlled employment of production factors in decentralised production systems (production network), in order to provide material products or services according to customer requirements with respect to quantity and quality at the required time with minimum costs and time.

At first, conventional production management is described in this section. It is followed by an explanation of the principles of cybernetic production management from which the preconditions for High Resolution Supply Chain Management (HRSCM) are derived.

Conventional Strategic and Normative Production Management The St. Galler Management Concept based on Ulrich's system approach differentiates between three levels of management: normative, strategic and operative. While normative and strategic management basically relate to design and development functions, operative management focuses on the steering function (Bleicher 2004, p. 80 ff.).

Normative management of a production company does not differ substantially from the normative management of any other company. The general company aims, principles, standards and company culture are defined and are intended to ensure the viability and development potential of the company. The overall company aim of a production company is typically to succeed in securing its existence.

The general aims of normative production management can be:

- the balanced fulfilment of the requirements of the various interested parties (partners e.g. neighbours, customers, shareholders and employees)
- obtaining or consolidating a significant position within
  - an industry,
  - a technology (e.g. laser technology) or
  - with respect to a material (e.g. special glass)
- concentration on the processes with the greatest added value irrespective of the traditional strengths of the company or of the industry in order to maintain the company sites and the company's size

Building on the aims of normative production management, we now address strategic production management.

Strategic production management in this case has two outstanding aims (Eversheim and Schuh 1996):

- Dynamic, market orientated development of the goods and services of the company
- Creation of long-term competitive advantages by establishing core competencies

To achieve long-term competitive advantages, core competencies, from which surprising products result, must be developed faster and more cost effective than those of competitors. The real sources of a strategic benefit therefore lie in the ability of the management to combine technologies and production capabilities (companywide) to form competences. Only then the business units will be strong enough to be able to respond rapidly to emerging opportunities in the future (Westkämper and Zahn 2009).

Conventional Operative and Tactical Production Management A comprehensive analysis of the task areas of operative and tactical production management was undertaken within the context of the Aachen PPC model (see Schuh 2006). The Aachen PPC model extended by Schuh serves firstly to describe various aspects of production planning and control (PPC) and secondly offers support in the definition of PPC aims and the application of various design and optimising methodologies. The main task is to divide the PPC into several smaller models, each taking a different perspective. This breakdown is required to identify the different aspect groups (personnel aspects, information technology aspects and commercial aspects) which have different influences on the aims and model requirements of the PPC. The four perspectives used in the Aachen PPC model are:

- the task perspective
- the process architecture perspective
- the process perspective and
- the function perspective.

The task perspective describes the tasks of the PPC on a generally applicable hierarchical level of abstraction. The process architecture perspective forms the interface between the task and process perspectives by creating connections between tasks at the network level and the company level. The process perspective identifies processes of the tasks in the task perspective in a chronological sequence in order to describe the order processing more precisely. Finally the function perspective describes the requirements for IT systems, which support the internal PPC (ERP-/PPC systems). The primary tasks of the task model can be broken down on the basis of the following structure (Fig. 6.26).

This structure divides tasks into industry-wide network tasks and business-wide core and cross-section tasks. The core tasks of production management in this case comprise production program planning, production requirements planning, procurement planning and scheduling and in-plant production planning and scheduling. Network tasks add industry-wide aspects at a strategic level to the task interpretation of the original PPC model (according to Luczak and Eversheim 1999) and include network configuration, network sales planning and network requirements planning. The cross-section tasks integrate the network tasks and the core tasks. These include order management, inventory management and financial controlling. All task areas

6 Self-optimising Production Systems

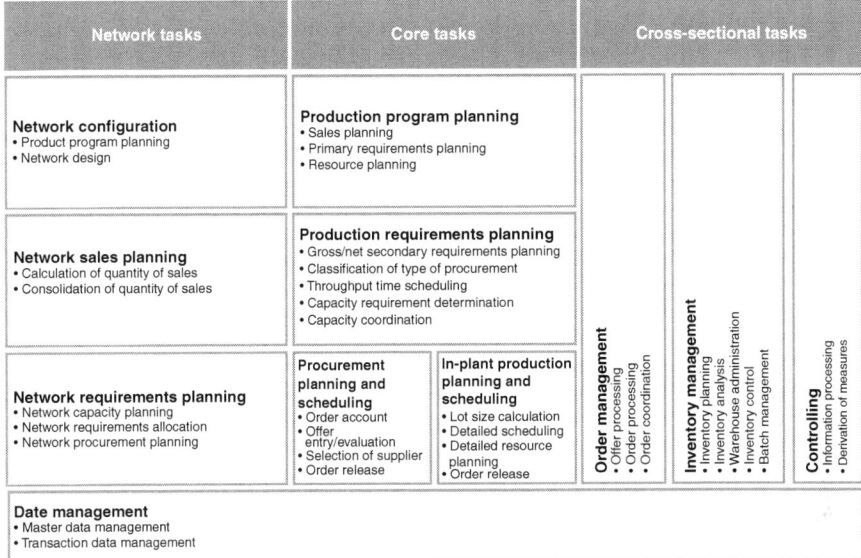

Fig. 6.26 Production management tasks. (From Schuh and Roesgen 2006, p. 28)

require access to the data management, which is therefore attributed to every task (Schuh 2006, p. 20 ff.)

Structure and Tasks of a Cybernetic Management System The basic ideas of cybernetics originated in the 1940s and 1950s. New methods were sought to solve complex problems. Central to cybernetics are the regarding of information as a central value and the use of closed systems and control loops. Already in the early phases of cybernetics, an analogy with living systems and natural organisms was sought. Stafford Beer adopted the principles of cybernetics into his Viable System Model (VSM), which serves as the reference model for describing, diagnosing and designing the management of organisations. The basis of the Viable System Model is the analogy with the human nervous system. Beer divides the organisational model into five subsystems, which can be assigned to three structural levels. The operative level of the whole system contains systems 1–3 and relates to the present and to the inner world of the system. At this system level, actions are initiated autonomously through routine behaviour and reflex-type adjustments to changes in the environment. System 4 aims at stabilising the system within the external world. System 5 represents the third structural level and is the normative instance. The following illustration (Fig. 6.27) is based on the cybernetic derivation of the model according to Beer, Malik, Gomez and Espejo (Beer 1979, p. 319; Malik 2006, p. 84; Gomez 1978, p. 24; Espejo and Harnden 1989, p. 99).

System 1 contains the management units (squares) and basic units (circles) in analogy with organs and muscles which implement the company processes. The steering units optimise the daily activities of the largely autonomous process units.

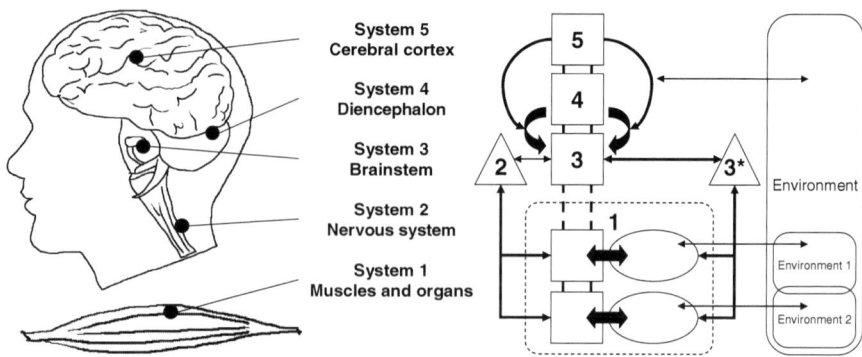

Fig. 6.27 The structure of the VSM. (From Beer 1979, p. 319; Malik 2006, p. 84; Gomez 1978, p. 24; Espejo and Harnden 1989, p. 99)

The operative units act autonomously within fixed limits at the horizontal level. The link between the individual systems 1 and the metasystem should be understood as a list and not as an hierarchical order (Beer 1979, p. 121).

System 2 coordinates and regulates the semi-autonomous systems in analogy with the nervous system. The core task of the system is to reinforce the self-regulating capacities through the supply of information and damping of oscillations between the operative units of system 1.

Like a brain stem, system 3 has an overall model which is superior to all systems 1 and their interactions. The operative overall management system ensures optimisation of the whole system. By its direct connections with all subsystems it detects simultaneously and in real time everything which occurs in system 1. It is also notified about activities in system 2. System 3 passes on instructions directly to system 1 via the central command axis. The monitoring and validation of information from the operative units are functions of system 3*.

System 4 represents the strategic system level in analogy with the brainstem, i.e. the external and future orientation. It detects and diagnoses its environment on a system-wide basis and constantly changes the orientation of the overall system on the basis of this information.

The normative level of system 5 (the "cerebral cortex") defines the identity of the system and specifies standards, values and strategies in order to maintain and develop that identity. It maintains a balance between the present and future perspectives.

Implementation of Cybernetic Principles in Production Management The production management system manages the production system by specifying control variables and identifying any intervention by actual comparisons with the control variables which are fed back. This control and monitoring can be described as a control loop (Zäpfel 2001, p. 3 f.). This interpretation is further supported in the control loop concept model in which the production system comprises the management system for planning and control (production management) and the performance

6 Self-optimising Production Systems

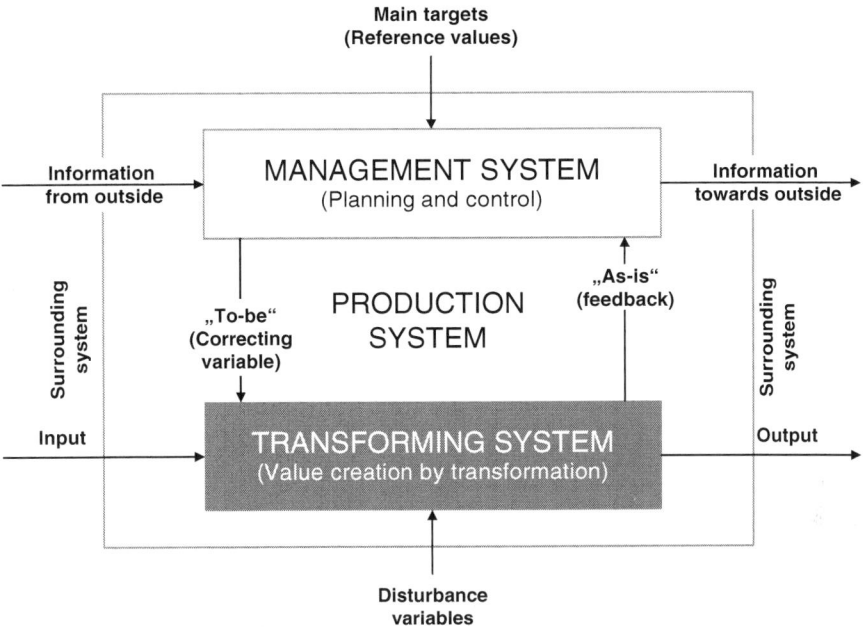

Fig. 6.28 Production management control loop. (From Dyckhoff and Spengler 2007)

system for adding value (see Fig. 6.28) The relationships between the management and transforming system (production system) are illustrated below.

In terms of the control loop, the management system transmits so-called reference values (control variables). These reference values define the quantitative and qualitative specifications for the performance system which in turn affect the economic and other target variables. External disturbance variables such as supply shortages, employee sickness etc. cause the actual results (control variables) of the production to deviate from the specified reference variables. This deviation is analysed with the help of further external information such as the development of procurement and sales markets and taking into account the global aims of the company (e.g. achieving maximum marginal income or profit). The results are transferred to the future reference variables. The central task of the production management system is therefore target-orientated planning and control of production taking into account organisation, human resources, information supply and monitoring (Dyckhoff 2003, p. 6 f., 29).

Approach and Preconditions of High Resolution Supply Chain Management
Thanks to technological innovations, real-time operative information is now available at a new level of granularity. Particularly new technologies of information collection such as Radio Frequency Identification (RFID) und and informational integration of companies have paved the way for this revolution (Fleisch 2008). This information is used for production management in HRSCM. The approach combines the

structuring of real-time data-based planning and control with an adequate decentralised organisational structure of production management.

Due to the interdisciplinary nature of the approach the preconditions of HRSCM are technological and organisational. The technological preconditions include the availability and usability of actual values for planning (reference value definition). These are the basic preconditions for enabling the transition from control to self-regulation within production management. Only realistic and achievable reference values permit the adjustment of regulated process variables. Therefore the generation of sufficiently detailed information, e.g. by sensors, and their availability through IT systems is a technological precondition for HRSCM.

From an organisational perspective, the complexity of companies as organisational units and their dynamic environment are the fundamental challenges. The organisational structure of the approach must overcome the complexity and be able to maintain internal stability as well as that of the external environment. As a basis for this, the structure must be able to satisfy the information requirements of the individual units which it comprises.

6.3.4.2 Structural Model of High Resolution Supply Chain Management

The aforementioned requirements determine the general requirements of the structure of a High Resolution Supply Chain Management (HRSCM) reference model. The organisation must be able to cope with the complexity of the company and the complexity of its dynamic environment. This can only be achieved by means of structural decentralisation and a high degree of response because decentralised interactions reduce the amount of coordination required (Frank et al. 2004).

Based on the Aachen PPC model and the VSM it is necessary (with regard to modularity) to define different reference perspectives to reflect the different perspectives of the overall model. The Aachen PPC model unifies four perspectives—the task-, process-, function- and process-architecture- perspectives. The function-perspective does not need to be considered as an actual perspective within the context of the model because it is aimed at the specific assignment of IT systems to implement the tasks. The process and task perspectives are adopted directly and extended for the newly established tasks and (management) processes within the metasystem.

Strategic and Normative Production Management System The mechanisms described next ultimately lead to the definition of strategic specifications for the operative production management system via the interface of the configuration channel. The core task of strategic and normative production management is to develop the business model and order processing according to changes in the market while preserving the actual company identity and values and to initiate targeted changes. The spectrum of these changes extends from organisational and structural changes to technical adjustments and innovations.

The external equilibrium pursues the ideal of effective orientation of the order processing with respect to the environment and environmental requirements. The

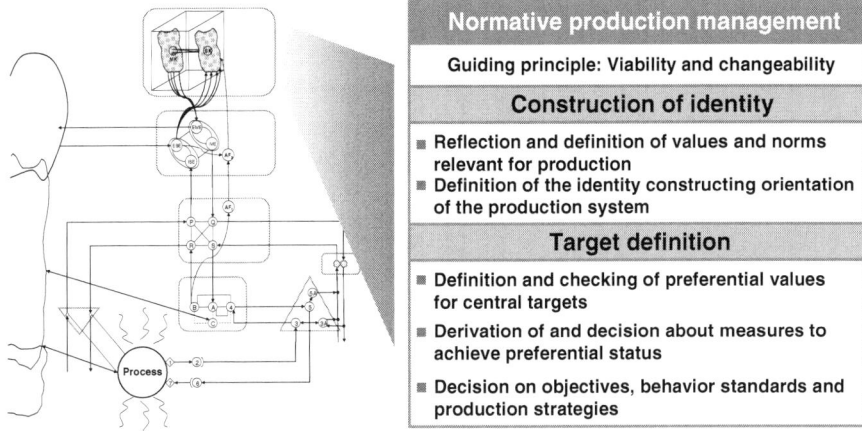

Fig. 6.29 Organisation and overview of the tasks of normative production management

equilibrium of the overall system ultimately involves balancing internal and external stability statuses. This overall equilibrium is monitored by the normative production management system which is responsible for the change, development and viability of order processing.

In the following sections we present the tasks, processes and information flows initially for normative production management. We then describe the tasks, processes and information flows of the strategic production management system.

The normative production management represents the highest level of the production system. Its aim is to ensure that the overall system remains viable in the long term by orientating the system both normatively and in terms of continuous change. As well as the aim of commercial viability, the target system thus engendered is also supplemented with normative ideals and values and requirements for its future scope of activity. The values and standards ensure the orientation of the target system within the identity of the company. Therefore normative production management is responsible, along with identity and values, for defining reference statuses and the initiation of measures to achieve these statuses.

The tasks of normative production management for securing the viability and flexibility of the production system are illustrated in Fig. 6.29.

The information flows for fulfilment of the tasks described above can be divided into the sensory (detecting) and motor function (instructing) information flows. The effectiveness of the system design depends on the interaction of the motor function and sensory mechanisms.

Sensory Components The sensors obtain all information about the internal and external stability statuses from the strategic production management system. The sensors also obtain all algedonic information which has been prepared by the alarm filters. If the actual status in the results varies from the desired status of the overall system,

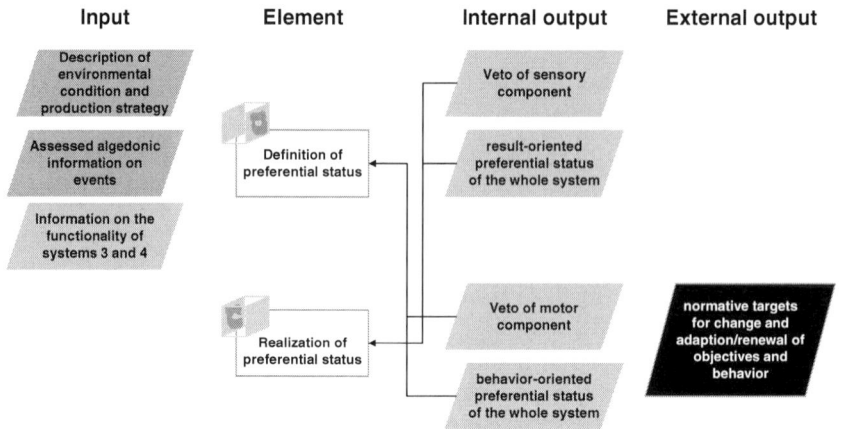

Fig. 6.30 Information flows of normative production management

measures are initiated. The sensors may in this case veto measures initiated by the motor functions.

Motor Function Components The normative instruction mechanism now checks the plausibility of the specifications from the sensors. Decisions are also based on preference statuses which can be assigned to them by the sensor system. The motor functions have corresponding knowledge of which results can be achieved with which responses. The statuses contain actual instructions which imply specific responses at the operative and/or strategic level.

Figure 6.30 illustrates the basic information flows of normative production management.

The task of the strategic production management centre is to anticipate potential futures, to initiate the adjustment of the organisation to the dynamic environment and thereby to adjust the strategic orientation of the production system so as to preserve existing success potentials and create further potentials (Gomez 1978, p. 24; Eversheim and Schuh 1996; p. 5 ff., Rüegg-Stürm 2003, p. 71; Malik 2006, p. 141).

Hill's concept is designed for the configuration of a differentiated production strategy which is particularly involved in the linking of the company and marketing strategy with the production strategy. Coupling is carried out by considering the qualification and differentiation features of the products (Hill 2000, p. 32 ff.). Qualification features make the company fit to compete for a market segment. Thus they enable the company to participate in the market. If qualification features are lost, the customer is no longer included in the decision to purchase the product. Differentiation features however have an impact on the purchasing decision of the customer when faced with comparable products. They differentiate the performances of the company in terms of satisfying a customer requirement from those of the competition (Hill 2000, p. 38).

6 Self-optimising Production Systems

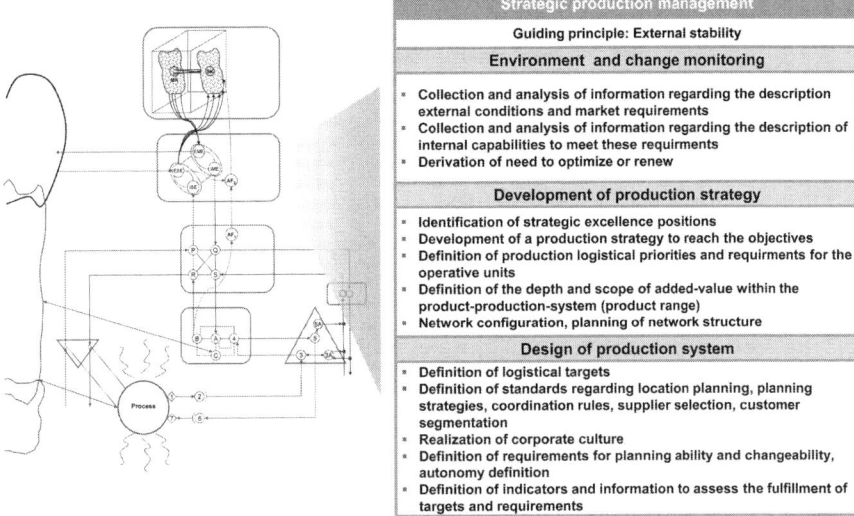

Fig. 6.31 Organisation and overview of the tasks of the strategic production management system

The tasks of the strategic production management system for ensuring the external equilibrium and effectiveness of the production system are illustrated in Fig. 6.31.

Like normative production management, the information flows also travel from the strategic level through a sensor and a motor function component. The essential difference in the structure as compared to the normative level is the presence of interfaces at higher levels of the production management.

Sensory Mechanism The central task of the sensory mechanism is to record and process information. In this case the stability detection mechanism (the internal reporting system) obtains information about the actual status of the equilibrium within the production system. This is information about the actual logistical parameters of use of resources, inventory, schedule-adherence and throughput time but also about the progress of structural and organisational changes.

Motor Function Mechanism The motor function mechanism is the executive organ of strategic production management. The mechanism obtains reference specifications about the external and resultant internal stability statuses in the form of the production strategy. In addition, normative production management is able to change strategic specifications at any time and define new goals for the system.

Alarm Filter 2 Alarm filter 2 passes on exception information to higher control levels. This is firstly the information from alarm filter 1 of the operative and tactical production management systems and additional information from the environment or from the programmes which implement the production strategy. Alarm filter 2 balances the information via internal disturbances with the market analyses of the strategic

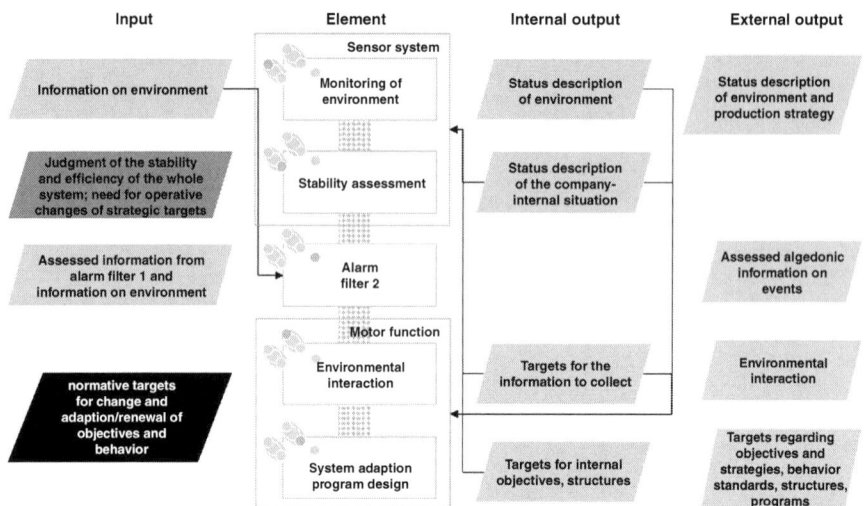

Fig. 6.32 Information flows of the strategic production management system

sensors. The interaction of the disturbed internal stability statuses and the market performance is the output of alarm filter 2. Figure 6.32 illustrates the information flows of the strategic production management system.

Operative and Tactical Production Management System In this section, in the same way as the previous section, we present the tasks, processes and information flows for operative and tactical production management.

Operative production management serves to steer the operative units with the aim of stabilising and optimising the system in the "here in now". This is carried out by designing the order processing in systems 1, 2 and 3 of the first recursion level (main order-processing processes, process coordination centre, tactical production management and by designing systems 1–5 of the second recursion level (main- and subprocesses, process control centre, process management).

The operative configuration and optimisation of the operative processes is the task of tactical production management (system 3 in VSM-terms) In this case, the primary issues are medium-term configuration according to strategic specifications from the strategic production management system and short-term optimisation of the processes without structural intervention. According to the basic assumption that the processes of order processing, when stable, proceed without intervention from higher management levels, unlike the previous research, order processing management tasks are largely decentralised. The tactical production management controllers can focus on stabilising interventions in the event of major unrest within the system and on continual improvement of the interplay of the subprocesses. To this end, the tactical production management system, unlike the process coordination centre, can intervene via the central control channel directly into the autonomy of the processes. The autonomy of the operative processes is thereby a gradual function and is limited or expanded according to the situation. Indirectly however, tactical production management can also influence the internal procedures via the process coordination centre (coordination channel) or via process monitoring (monitoring channel).

Configuration of the Production System to Ensure External Equilibrium A precondition for the short and medium-term configuration of operative processes is the understanding of external requirements of the logistical performance of the production system. These are identified in the strategic production management system and are input into tactical production management by a feed-forward mechanism. The dimensions of the requirements definition are primarily the reference delivery time (order throughput time), reference schedule-adherence and price level. The task of production management at this point is to check these external customer requirements for feasibility, translate them into internal process requirements and to simultaneously guarantee maximum efficiency of the overall process. This takes place by defining optimum operating points in the overall system and individual main processes. The target conflicts of the process planning in this case are minimised as far as possible by information transparency and organisational networking.

Operating characteristics according to Nyhuis and Wiendahl are a suitable means of choosing operating points. These have been established to illustrate the effective relationships between logistical target variables within production logistics, see for example (Gläßner 1995; Yu 2001; Lutz 2002; Nyhuis and Wiendahl 2003; Wiendahl 2005; Lödding 2008). Production characteristics are a condensed illustration of different operating statuses. Operating statuses are determined by analytical or stochastic monitoring modelling of real systems and by computation by means of simulation (Nyhuis and Wiendahl 2003, p. 36 f.). Figure 6.33 illustrates the target conflicts in production control on the basis of a characteristic curve diagram.

Based on the characteristic profile of the characteristic curves, the task of tactical production management is to position the reference operating points within the target conflicts of the PPC according to the strategic requirements. As a result, the reference variables of schedule-adherence, schedule deviation, throughput times, material and/or order levels, use of resources, quality of the process results and process costs per main process are defined and communicated as specifications.

If the desired reference values of the operating points cannot be achieved within an existing configuration, a series of tactical decisions must be made to determine the framework conditions for the operative processes and to configure the processes accordingly. In this way, the profile of the characteristics curves changes and enables new combinations of operating points. Decisions about the dimensioning of the use of resources are negotiated via process channels.

Depending on the targets in question and the resources available, the process channels then automatically define the effects of these targets in the form of control variables for the assigned subprocesses. This heredity ensures that the process

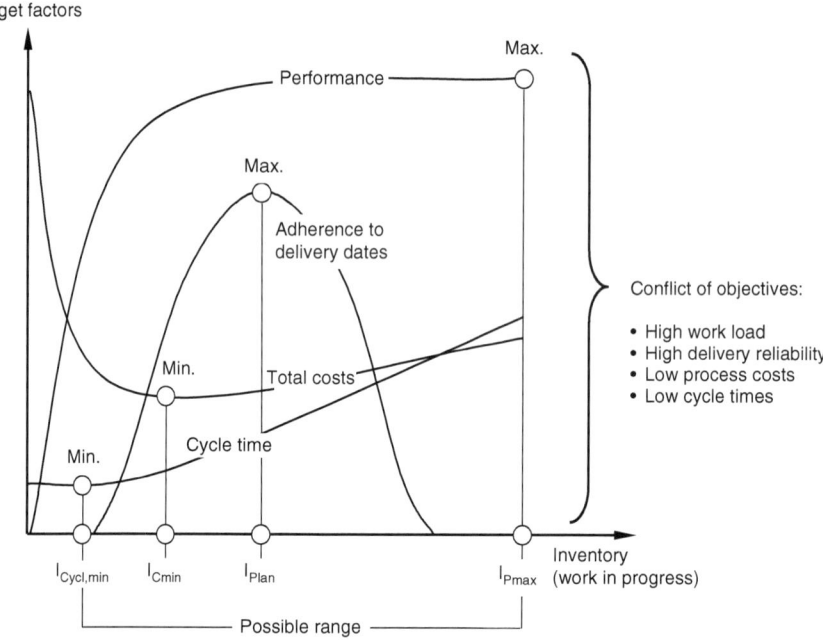

Fig. 6.33 Target conflicts in production control. (Wiendahl 2005, p. 35)

channels design the target corridors of the subprocesses in the recursion RL3 level such that the targets of the metasystem can be maintained at levels RL1 and RL2.

Monitoring and Optimising Horizontal Coordination to Ensure Internal Equilibrium As well as configuring the operative processes to ensure equilibrium with external requirements, the second task of tactical production management lies in monitoring and achieving internal equilibrium. Internal equilibrium is considered as having been achieved when the productive system fulfils the target values of schedule-adherence, throughput time, costs and synergy suggestions. Reference values are also defined based on the definition of escalation levels which describe alternative processes for handling discrepancies inside and outside the target corridors. Threshold values underpin these processes which determine when a particular disturbance should be balanced either locally by the process control centre or by the process coordination centre. If neither damping mechanism is able to find a solution for a problem, it is escalated to the tactical or even to the strategic production management system. The frequency and intensity of the discrepancy from reference values is the measure of the stability and therefore the internal equilibrium of the system or of subsystems.

As a reaction to these discrepancies, medium or short-term and one-off or recurring countermeasures can be instigated. One-off or sporadic measures are used to handle individual exceptions such as for instance major interventions in an individual order (machine use, order sequences, quality or cost compromises), the involvement of external advisers or downsizing or upsizing. This limitation of the autonomy of the decentralised units is implemented via the control channel and should be avoided if at all possible. Recurring countermeasures are incorporated as standards which so to speak pre-equip units with instructions, rules or response instructions. These are reflected amongst other things in changed reference values, changed rules for use of resources, changed procedures, changed decision-making mechanisms or physically changed resources, increased or decreased capacities or changes to responsibilities. Both types of intervention are aimed at ensuring and optimising the fundamental synergy pre-settings in the interaction of the main processes.

Process Monitoring for Ensuring Internal Equilibrium In the case of these processes, the tactical production management system is supplied with information from the process coordination centre and with unfiltered information from the processes. If deviations from reference values are symptoms of a lack of synchronisation, process monitoring is responsible for identifying the causes, in terms of which the actual situation is being interpreted (Herold 1991, p. 135 f.; Thiem 1998, p. 110; Malik 2006, p. 136 f.). Direct interfaces with the areas (IT systems or personal reporting on the part of group representatives) are used to request absolute information about the situation of orders and capacities (type, quantity, schedule, quality, progress etc) and suitable characteristic values for identification of deviations from the stable status of the units are disclosed. These overload manifestations are accompanied by changes in the performance values of use of resources, inventory, schedule-adherence, throughput time, costs and quality and can be expressed in the form of

- exhausted overtime accounts,
- high levels of sickness,
- continuously high order queues,
- low schedule-adherence in the case of subprocesses,
- urgent orders or "fire-fighting" and
- temporary capacity shortages.

Against this background the information described above is collected by the tactical production management system and used to improve the decision-making basis for all orders. In the case of a disturbance, corrective measures are generated to resolve the permanent or temporary shortages or short term responses to "unforeseen developments". These measures directly affect the procedures and are controlled by the monitoring channel (Herold 1991, p. 144). Examples of these kinds of immediate measures are the provision of additional capacities by approving overtime, involving external suppliers or hourly-paid labour as well as overriding locally-planned order sequences, batch sizes and resource planning.

Adequate Structuring of Planning Capacity by Implementation of High Resolution The success of the business processes as well as the coordination of the individual processes in the operative control is largely influenced by the planning capacity. This in turn depends on the availability of information and the quality of information flows (Wiendahl 2006, p. 16). This is particularly taken into account in the mathematical description with the time loss factor δi. Thus a core aspect in the tasks of the tactical production management system is to align the time delay for the transfer of information between the decision-making points with requirements. The appropriate interaction of the node points changes the decision-making quality of the controller. Poor availability, susceptibility to error, incompleteness, incomprehensibility and irrelevance of information reduce the decision-making quality of a node point and correspondingly the decision-making quality of the whole management mechanism. The task of the tactical production management system therefore is to define the requirements of granularity, real-time capacity and response time behaviour of the individual control loops and their associated management, control and disturbance variables. In principle, the resolution of information should be as high as necessary but not as high as possible.

The order processing target system must be expanded accordingly by optimising the planning capability. Planning capability (or planning reliability) is generated by a high level of information transparency and equally high planning data quality (Schmidt 2008, p. 63). The quality features of planning information thereby include completeness, currency, level of detail and absence of error of the informational content (Loeffelholz 1991, p. 27 f.; Schotten 1998, p. 44). As with the characteristic curves this target is inherited by the process channels and the coordination system in the form of reference operating points. The tactical production management system must test and optimise the fulfilment of the information quality criteria via the monitoring channel and the process coordination centre.

Adequate Flexibility of the Operative Systems The variability of individual process steps and of the overall production system has become a central aspect of production management (Blecker and Kaluza 2004, p. 4 ff.; Zäh 2005; Drabow 2006; ElMaraghy 2005, p. 264; Wiendahl and Hernàndez 2006, p. 147 f.). The aim of these endeavours is to create adequate product flexibility (to maximise variant variety with a specified infrastructure, process flexibility (to manufacture products using alternative processes, machinery, materials or process sequences) and capacity flexibility (scalability of the quantity of manufactured products) (Chryssolouris 2005). At the operative level, the focus is not the restructuring of the overall order processing or of the product mix because this is now a task for the strategic production management system. The aim here is the appropriate design of convert-ability, reconfigure-ability, flexibility and the versatility of the physical production system along with the logical processes and management mechanisms (Wiendahl et al. 2007, p. 787).

The tactical production management system now has the task of translating the strategic specifications for flexibility in the three main dimensions of product flexibility, process flexibility and capacity flexibility into operative specifications for the decentralised processes and to monitor their compliance. Specifications with respect to convert-ability, reconfigure-ability, flexibility and versatility are thus transferred to the design elements in the main processes where they are implemented. The essential enablers of the implementation are reconfigurable process planning, adaptive

6 Self-optimising Production Systems

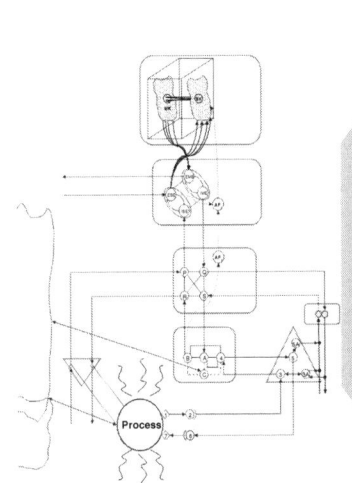

Tactical production management

Guiding principle: internal stability and efficiency

Process configuration
- Assessment and reconciliation concerning feasibility of strategic guidelines
- Adaption of topical structures based on strategic guidelines
- Definition of synergetic operation points for the total system
- Deduction of process-independent objectives, priorities and standards
- Definition of requirements of changing operative processes

Process direction and -control
- Forwarding and realization of strategic guidelines
- Definition and control of target operation points, resource availability and autonomy of all main processes
- Assessment and optimization of methods, operations and tools that should be used
- Definition of the order penetration point and planning mechanism
- Assessment and optimization of changeability and feasibility planning
- Support of operative processes on solving critical disturbances
- Control of middle-term supply chain design, supply chain demand and production program planning

Process coordination
- Definition of norms, priority rules and coordination routines
- Definition of escalation levels for the cooperation of coordinating units
- Assessment of process coordination center effectivity

Process monitoring
- Assessment of internal stability and efficiency of operative processes
- Sporadic interventions to save internal stability

Fig. 6.34 Organisation and overview of the tasks of the tactical production management system

production planning and control, reconfigurable production and assembly systems and flexible fabrication (Wiendahl et al. 2007, p. 789 ff.; ElMaraghy et al. 2009, p. 261 ff.). As with the characteristic curves this target is inherited by the process channels and the coordination system in the form of reference operating points. The operative production management system must test and optimise the fulfilment of flexibility criteria via the monitoring channel and the process coordination centre.

The tactical production management tasks of ensuring internal equilibrium and the efficiency of the production systems are illustrated in Fig. 6.34 and incorporated into the structure of the overall model.

Information about the external stability statuses reach the configuration channel via the connections between strategic and tactical production management. The configuration channel interprets the requirements of strategic production management and information about the current internal stability from the other channels. The balance of external and internal equilibrium results in a definition of reference operating points, target corridors in the characteristic curve field, characteristic values and escalation levels, standard behaviours and basic resource availability levels. These changes of the characteristic curve lead to the specification of new operating points in the main processes and therefore to the expansion of the internal stability via the control channel.

The configuration channel completes the information loop by providing information about the actual status of processes to the strategic production management system.

The Control Channel: Information Loop and Process Channels The tactical production management system intervenes into main processes directly via the central command axis of the control channel. The framework conditions for the reference operating points of the overall system from the configuration channel and information about the internal stability channel are translated via this channel into concrete instructions which are conveyed to the process channels of the main processes. These include particularly basic response instructions about core processes such as working times, maximum use of resources, capacity flexibilities or work organisation regulations and the definition of process related aims, e.g. the optimisation of process costs versus minimisation of throughput times. Thus the negotiation process over the allocation of resources takes place via this channel. The target system specified by the strategic production management system is therefore inherited by the next recursion level and appropriated per main process and/or negotiated with the process channels.

The Coordination Channel: Information Loop Incorporating Process Coordination Centre and Process Control Centre The basis of internal stabilisation is the reconciliation of the main- and subprocesses at synergetic operating points. The tactical production management system specifies the management values and target corridors within whose parameters the actual processes must adjust. The reference characteristic curves with the target corridors and operating points. The KPIs of the escalation model are transferred by the coordination channel to the process coordination centre. When selecting and defining the characteristic values, the mechanism takes particular account of the specifications for external stability on the part of the strategic production management system, feedback from the processes via the monitoring channel and the input via the profile of the actual operating points. The process coordination centre is a non-hierarchical instance. It controls the process operating points and the order status. In the event of a discrepancy it initiates compensating measures between the processes within the context of a predefined escalation model. The reporting (feedback) of the information is also carried out according to the specifications of the tactical production management system. In this case, only filtered information about discrepancies from plans, operating points or planning milestones is transferred to the tactical production management system. Thus no detailed image of the actual situation within the processes is permitted yet but it prevents information overload at this point.

The Monitoring Channel: Information Loop Incorporating Operative Main Processes The tactical production management system obtains only relative information about deviations from the reference state via the coordination and monitoring channels. However, the monitoring channel is designed to reflect the actual status of processes. It firstly identifies overload manifestations and new types of developments. Secondly, this channel provides the tactical production management system as required with detailed information about individual orders, load curves or quality data in order to permit corrective intervention in the case of shortages or problems. In addition to direct interventions, the knowledge obtained from the monitoring system is also used at other decision-making points within the tactical production management system. In order to identify new types of developments promptly, information about possible overloads is transmitted continuously whereas the specific order or capacity data is requested and fed back only as required.

The Alarm Filter: Direct Flow of Information Between All Levels So that neither the operative nor the tactical production management systems are flooded with information, information from process management and from the process coordination centre is filtered according to pre-settings. The set up of an integrated, direct channel is therefore required for the exception information flow (Malik 2006, p. 138 f.; Gomez 1978, p. 68 f.). The information flow here follows the principle of "Management by Exception" and is defined individually for each company and/or for each production system. It is transferred directly to the strategic and normative production management centres and is defined as sensitively as is required. Possible content might be serious breaches of schedules in the case of important customer orders, new major customers, specific quality problems, failure of a line or computer system etc.

Figure 6.35 summarises the information flows of the tactical production management system. The dotted line between the 5 channels symbolises the prevailing information transparency within the tactical production management system.

The tactical production management system, as the central switching point for operative processes and the interface with the strategic production management system, should incorporate representatives of all assigned processes for the exercise of its functions. This is particularly necessary to implement a synchronisation of the target systems. The representatives should participate regularly in the operative processes in order to be able to provide information to the monitoring channel independently from the system as well. The tactical production management system is supported in its activities by a control room which prepares transparent versions of the basic information from all four channels. Acting as a logistical control centre, it transports and visualises the characteristic values and status information of the monitoring, control, coordination and configuration channels together in appropriate time cycles.

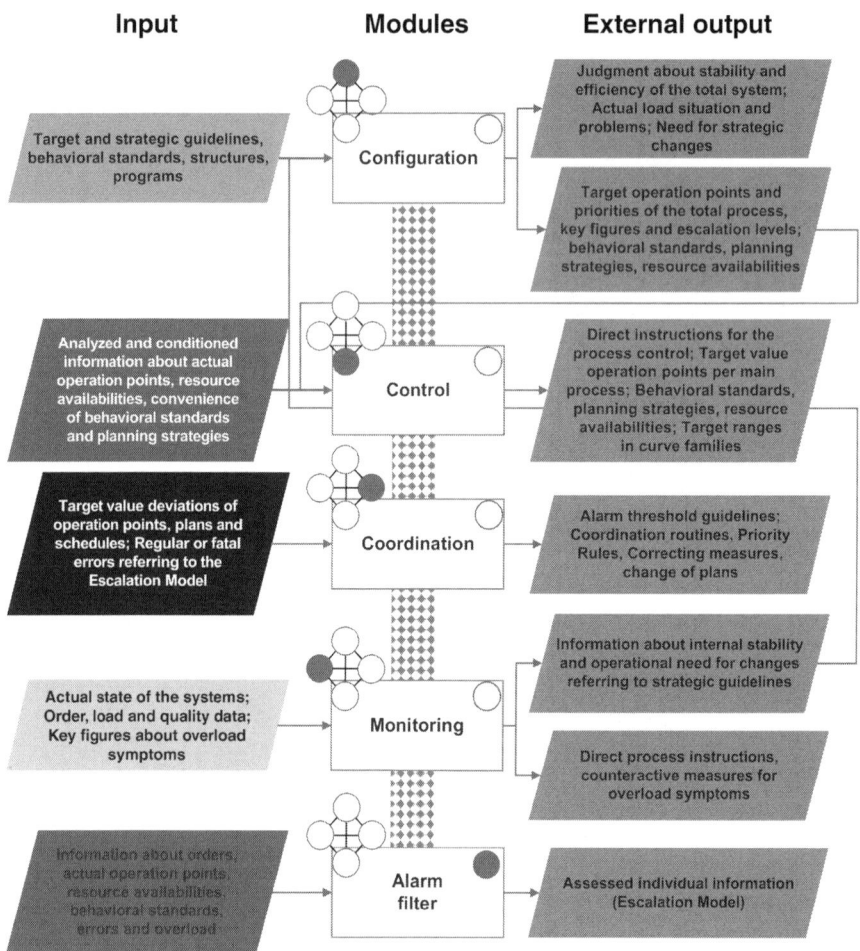

Fig. 6.35 Tactical production management information flows

While the VSM provides a conceptual model and the structural framework for the production management design, the model offers no methods for simulation, analysis and evaluation of the dynamic systems at the process and process control level. At the operative level, the management system can be supplemented by the transfer of specific control engineering components. Since both VSM and control engineering originated within the trans-disciplinary approach of system theory, these can be consistently integrated. However the description of the process control centre and process management defines abstract requirements for the control of the core processes of production management which must now be specified in more detail in terms of control engineering.

The process management systems organise the operative implementation of tasks and processes of technical order processing. The tasks are fulfilled according to the

6 Self-optimising Production Systems

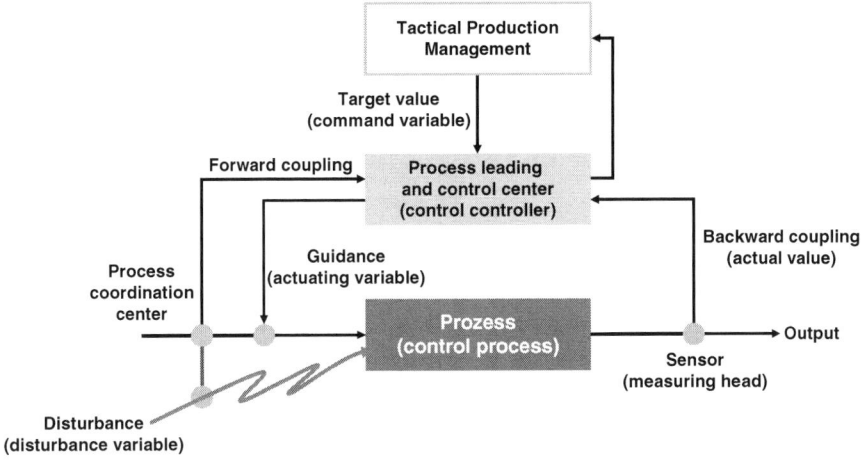

Fig. 6.36 Interaction of operative management units

requirements of viable systems by semi-autonomous value-adding process islands or performance-orientated capacity groups.

Up to four management instances along with the implementing resources are involved in the exercise of process management and control. The process management system of each process obtains reference values for the process management variables from the tactical production management system in the form of a servo mechanism. The process management system obtains information via the process coordination centre about possible and expected problems and the actual status of other systems and/or orders (feed-forward). The process management system receives the status of assigned processes via the process control manager (feedback). The reference variables from the tactical production management system and from the process coordination centre and the status information of the assigned process from the process control centre are translated into control variables (activities, programmes and schedules) for the management of the process. The implementation of these plans is once more initiated by the process control centre. Figure 6.36 illustrates the servo mechanism of the tactical and operative production management systems.

According to the basic logic of the maximum autonomy of the decentralised units, the process management system intervenes only if there is a discrepancy between the process and the reference value which cannot be counteracted by a standard routine. The known repertoire of routines for handling dynamic situations is saved in the process control centre and/or in the process coordination centre. These include particularly project monitoring, control tasks and countermeasures for the regulation of recurring problems. In accordance with the recursive structure of the overall model, the process management system and the process control centre jointly implement the tasks of systems 2–5 for the observed process. Figure 6.37 summarises the process management tasks.

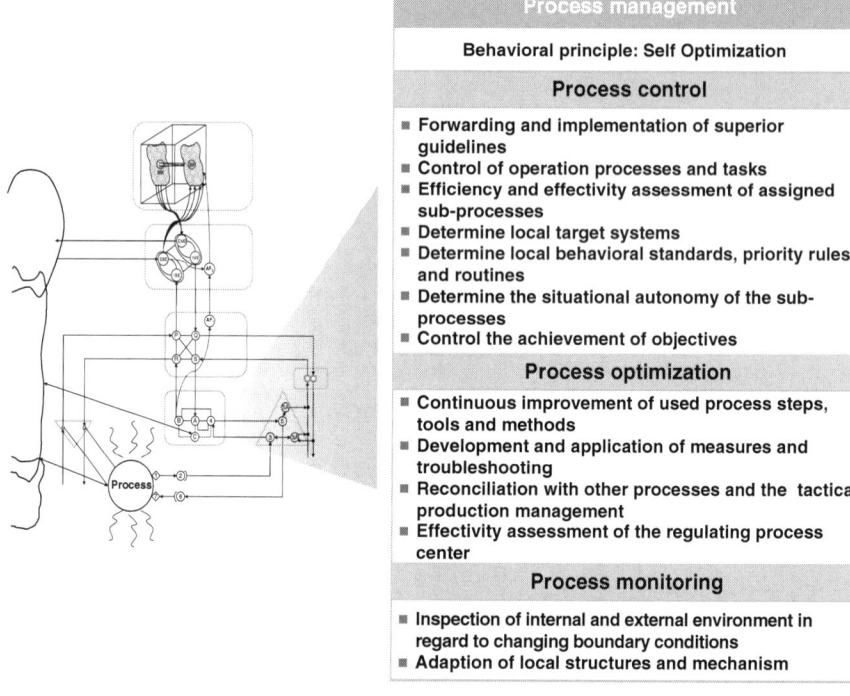

Fig. 6.37 Organisation and overview of process management tasks

The process management systems are supported by the process control centres, which carry out the tasks illustrated in Fig. 6.39.

The process coordination centre undertakes the process control centre tasks at the next recursion level up. It thereby regulates the interplay of the main processes and ensures implementation of the programmes selected by the tactical production management system. The process coordination centre does not play a hierarchal role in relation to the process management systems but rather acts as a link between the main processes in order to stabilise the overall operation. In doing so, the process coordination centre undertakes the following groups of tasks:

- Transfer and monitoring of reference values defined by the tactical production management system for internal stabilisation of the processes
- Construction and assurance of coordination rules and standards
- Construction and assurance of adequate communication channels.

The process coordination centre tasks for supporting self-optimisation and synchronisation of the main processes are summarised in Fig. 6.39.

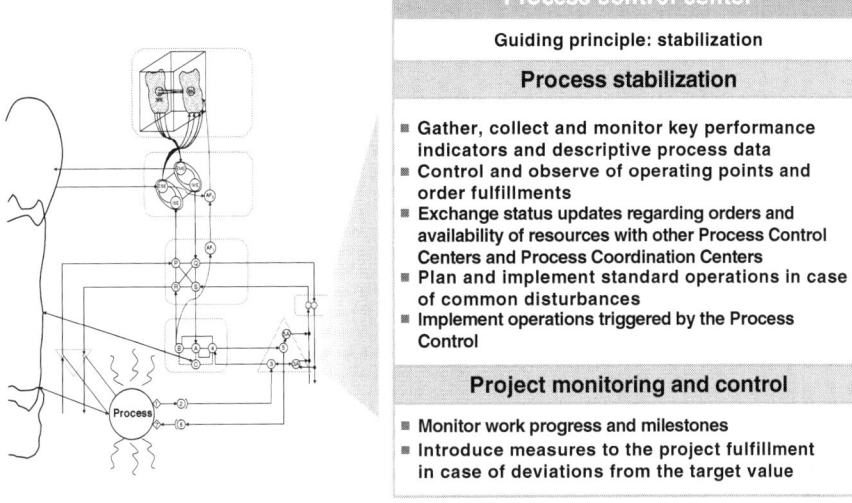

Fig. 6.38 Organisation and overview of the tasks of the process control centre

The tasks of the operative production management system also comprise guaranteeing the internal and external stability of the subprocess (process management system), monitoring and maintaining stability between the subprocesses (process control centre) and guaranteeing synchronisation between the main processes (process coordination centre).

To support the implementation of the aforementioned tasks, the associated management processes and information flows are described below.

The lines between the main processes represent the exchange of production factors which also include information (see Fig. 6.40). All information flows described in the model as internal output and internal input (Oi and Ii) are integrated ("2" in Fig. 6.40) as well as the material flow and access to the labour factor. The external information flows (Oe and Ie) exchange the processes via the connection with the environment ("1" in Fig. 6.40). All project monitoring and control information is fed into the process coordination system and can be requested in real time ("3" in Fig. 6.40). All main processes have interfaces with this data.

The logic of the VSM process images represent the framework for the design of the management processes of the tactical and operative production management system. They place the tasks of the individual controllers into a logical time sequence and structure the information flows through the channels between the controllers. To carry out the tasks described above, the most common interpretation of the VSM processes—according to Gomez—(see Gomez 1978, p. 48 ff.) must be supplemented with specific measures. The enhanced interpretation gives us the VSM process illustration in Fig. 6.41. The names of the node points correspond to the significance of their tasks.

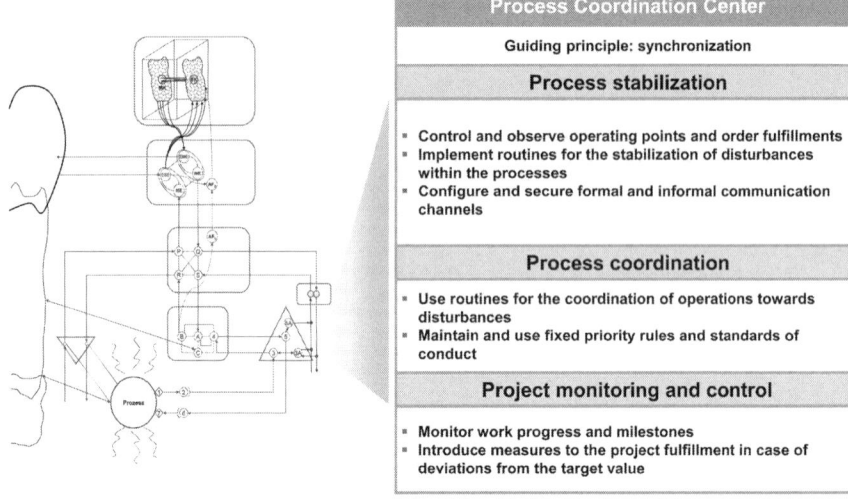

Fig. 6.39 Organisation and overview of the tasks of the process coordination centre

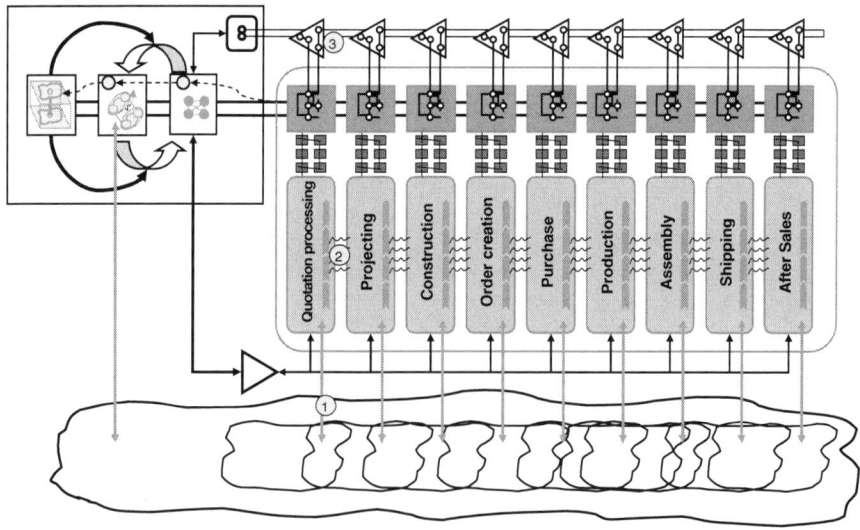

Fig. 6.40 Allocation of information types according to Schmidt (2008) into the structural model

Unlike nodes A, C and 4 the decision-making node B has only a forward-travelling information system. Nodes A, C and 4 reach decisions by iteration and coordination cycles in order to implement the specifications of the metasystem and to manage problems. While B is largely informed by coordination content, its central task is to

6 Self-optimising Production Systems

Tactical production management

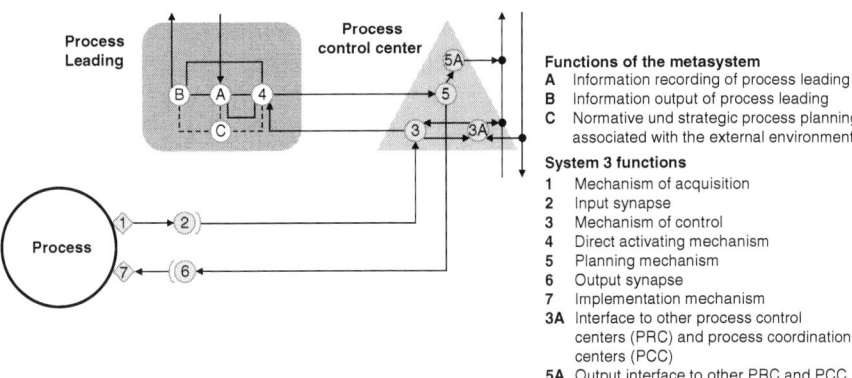

Fig. 6.41 Process illustration of the management mechanism of a main process

dispatch the decisions from A, C and 4 to the metasystem according to the quality criteria of the information.

Node 3A is not only an output but also an input for information into the control mechanism. With this adaptation, a process management system can see information from other processes or from the process coordination centre and can be taken into account in the decision-making process (feed-forward). The output of node 3A is now the relevant information of the monitoring mechanism, i.e. the status information for control of the order processing, availability and if necessary error messages and disturbance warnings. Node 5A however communicates the plans and programmes for process management as a response to problems in the process coordination centre cycle. This dual-differentiation means that status messages from a subprocess are detected by the overall system in real-time before a decision about intervention has been made. At this point it is necessary to decide where this increased transparency can add value and where it would simply cause unnecessary interference.

According to Beer, point C represents systems 4 and 5 of this recursion level. In order to correspond to the structure of the VSM and thereby to fulfil the conditions for viability, node point C allows interactive connection with the process environment. Only by this enhancement point C can undertake the strategic tasks of the local system 4.

The reference variables (reference values) which initiate these servo mechanisms are defined by the tactical production management system. In turn the process management system provides continuous information to the process coordination centre and the tactical production management system about the actual status of the reference values for internal stability. The process management system and the process control centre use the reference variables, available to them, to achieve the required status. Thus they interpret or transfer the global reference values into actual proposals, schedules and standard responses for the individual processes.

The modus operandi of the process coordination centre is based on the coordination measures as well as on an escalation model which is modelled by the tactical production management system and practised by the process control centres and the process coordination centre. In principle each process control centre attempts to compensate for any deviations from the planning milestones or reference variables itself before involving the process coordination centre in reconciliation with neighbouring or affected process units. If no solution can be found for an order or a central reference variable, the tactical production management system is involved to remedy the problems or prioritise and/or reschedule orders. The coordination mechanism is therefore able to absorb a high degree of variety and to induce only limited variety into the tactical production manager.

6.3.4.3 Implementation by Means of Control Logic Within Production Management

Generally, there is a differentiation between two types of control engineering system. In the open control system the incoming control signal is independent of the outgoing signal of the system. On the contrary, in a closed-loop control system, the input signal is influenced by the output signal in the form of feedback and control. The High Resolution Supply Chain Management (HRSCM) project focuses on closed-loop systems because these permit a feedback of information to the controller (e.g. time-dependent stock level, delivery times, restocking times etc) and can therefore react to external influencing factors. In this case the greatest challenge is in the control of dynamic systems which are defined as a functional unit whose input and output parameters change over time. Therefore they represent time-dependent functions. Special features of controllable dynamic systems include linearity, causality and temporal stability. The elements of a closed-loop control system are the dynamic system itself, the controller and a sensor integrated into the feedback loop. Figure 6.43 below illustrates the basic structure of a control system.

The influential variables on a system are its time-dependent input and output variables and equally time-dependent disturbances which can occur at various points along the entire control route and which influence the control variables. The input is manipulated directly via the controller. The measured deviation from the reference variable results from the difference between the reference variable and the fed back signal. The reference variable is the target value of the control system (e.g. target stock level or desired schedule-adherence). In order to achieve this, the control deviation is fed to the controller which in turn supplies the system with an input signal with the result that the reference variable is achieved.

In the context of Supply Chain Management (SCM), state-of-the-art approaches are usually based on optimisation problems. Some approaches are also involved with transferring control loops to the area of planning and controlling of logistical systems. Since the supply chain can be described as a dynamic system, the primary requirements are a flow-orientated and continuous analysis of the time-dependent input and output parameters. Specific requirements must be met in this case.

One of these requirements is expressed by the Nyquist-Shannon Theorem. It states that the measuring frequency must be at least twice the frequency of the input signal.

Fig. 6.42 Control system

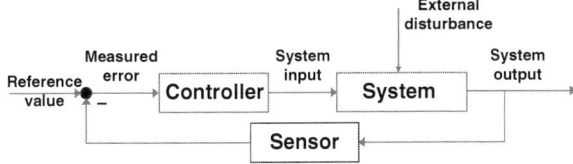

This is also referred to as the "Nyquist scan rate". Examples for the implementation of a high scan rate in practice include the employment of RFID technology and the application of additional real-time data. Along with the scan rate, the stability and the response behaviour are also critical elements of a control loop. Stability is a qualitative characteristic of the control loop and the essential requirement which is prioritised above all other criteria. A control loop is stable if, after changing any reference or disturbance variable, the control variable assumes a stable value or the status variables achieve equilibrium over time. In order to be able to appraise stability, there is a variety of very different stability criteria including for example the processes according to Nyquist, Hurwitz or Routh. The response of a second-order control system with a small damping factor on the increase of the service level is shown in Fig. 6.43 below.

Typical measured variables for stability and the response behaviour of a system are the adjustment time (T_s) and the rise time (T_r). While the rise time is defined as being the time required to increase from 10–90% of the target value, the adjustment time is the time taken for the fluctuations of the measured value to settle within a predefined corridor.

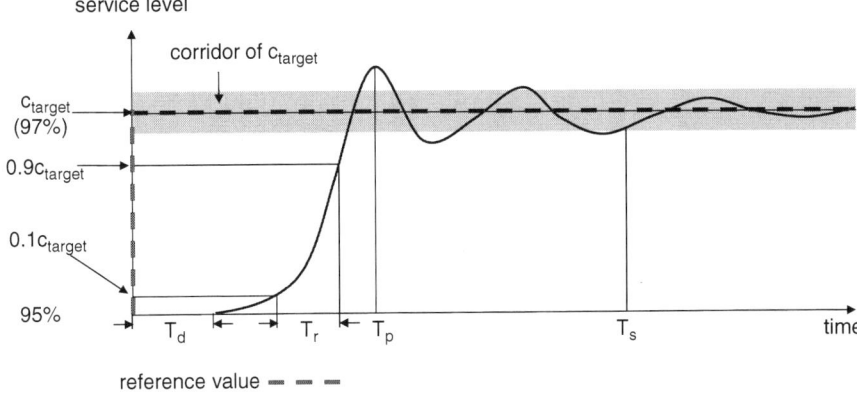

Fig. 6.43 System response in the event of an increase in the service level (92–97%)

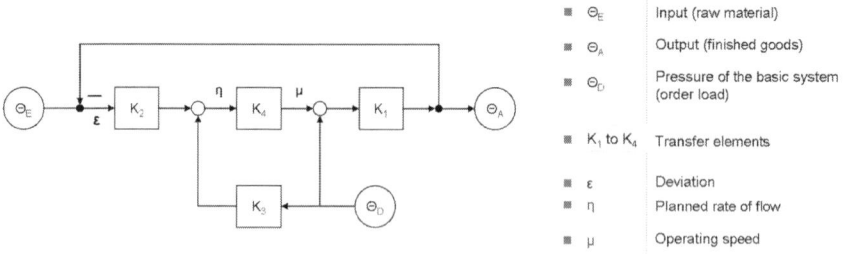

Fig. 6.44 Production control system according to Beer

As described above, the company processes can be considered as a control loop because they incorporate a sequence of target setting, monitoring the target achievement, discrepancy analysis, initiation of improvements and target correction as required.

In his book "Kybernetik und Management" Beer describes how control elements of systems, which do not comprise technical components, can be synthesised with the aid of control engineering. Therefore, the system of industrial production control can also be constructed on the basis of control theory and is described in brief below.

The input (Θ_E) of a production system is the raw material, the output (Θ_A) is the final product. The work between input and output proceeds at a specific speed (μ). The system is under a particular pressure (Θ_D) according to the ongoing orders. There is a primary feedback mechanism between Θ_E and Θ_A which measures the difference between the ideal status (output) and the initial status (input). This deviation is fed back into the system via an operator (K_2), which undertakes the necessary adjustment to the planned flow speed (η). This is transferred by a further operator (K_4) into the actual speed (μ). The order status (Θ_D) has an influence on the actual production speed and is also fed back by a third operator (K_3) in the form of the planned speed. The primary effect of the order load which affects production is immediately obvious at the output. It is represented by the most important operator (K_1) (Beer 1973, pp. 203–204). Figure 6.44 summarises the essential relationships based on a control loop for production control according to Beer.

6.3.4.4 Appropriation of the High Resolution Supply Chain Management Approach

Having described the basic structural model of HRSCM based on the VSM and control engineering, the practical implementation is addressed. For this, two typical processes are designed; "inventory management" and "production control", which represent two central processes of production management, according to the core process management mechanism described above.

Optimised Inventory Management Through High Resolution Supply Chain Management The sub-demonstrator "High Resolution Inventory Management Logic" exemplifies the improved planning quality with simultaneously reduced planning time and costs. Unlike conventional inventory management, the inventory management logic in the context of the above sub-demonstrator uses master data and historical item transaction data along with available real-time data. By now there is no inventory management process which is able to supplement the existing planning with real-time-compatible logic in order to plan order processing more efficiently and more cost-effectively and significantly increase the planning quality. It is expected that "high-resolution inventory management logic" will help inventory managers to achieve greater transparency over the entire range of articles, design and control processes more intelligently and improve the level of supply services to customers.

The central area of activity of inventory management on the part of the stock-based producer involves the three planning components; requirements, stock level and procurement planning. These are implemented in turn in each planning cycle. Requirement planning precedes the other two planning processes and is aimed at predicting as precisely as possible the expected market requirement in the subsequent period to ensure that customer orders can be filled completely and as fast as possible. Requirement forecasting depends on the ability to plan requirements either by programming means, i.e. based on deterministic assumptions of the consumption trends of the respective parts in the future or by consumption, i.e. based on known seasonal consumption levels.

The next component is inventory planning, i.e. defining the necessary safety stock levels and order preparation stock level (reorder level). This requires consistent inventory management and reliable forecasting or requirement values from the requirement planning stage. After forecasting the demand of future planning periods, the inventory management system determines the inventory required to achieve the necessary level of supply readiness. In this case a basic challenge is to integrate the ongoing orders into the forecast-based inventory planning. This indicates reconciling the reorder level (defined on the basis of the forecast) with incoming orders.

The next planning process—procurement planning—is used to define the optimum procurement quantities, taking into account all procurement costs (inventory management, purchasing, transport, etc.) along with storage and capital lock-up costs.

Inventory management faces various challenges in everyday company situations Guaranteeing production and supply while reducing stocks to minimise costs is a basic conflict of aims inherent in logistics. Furthermore, there is a necessity for improving the logistical efficiency through reducing procurement costs and capital commitment while increasing the liquidity of the company. The target system for the economic appraisal of procurement logistics can therefore be described by three quantifiable dimensions: the production supply, liquidity and procurement costs.

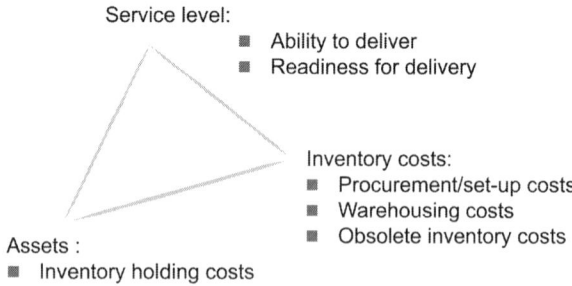

Fig. 6.45 Inventory management target system

The basic task of inventory management is to optimise these mutually-conflicting target dimensions. The aim of inventory management is to increase the production supply reliability (supply service) while simultaneously increasing liquidity and reducing procurement costs.

The supply service dimension comprises all factors required to ensure production with all necessary raw materials, semi-finished products and other materials. It is measured by the supplier's supply capacity in relation to the customer's desired schedule and by the supplier's supply readiness in relation to a delivery schedule previously agreed and confirmed with the customer. Liquidity describes the value-based commitment of capital to supplies and/or stocks. Procurement costs include all relevant ongoing costs, e.g. purchasing, inventory management, ordering and processing, obsolescence costs and the costs of storage (warehouse costs and capital lock-up costs). The operability of the target system variables is guaranteed by logistical parameters which are defined within a characteristic value system assigned to the target system. Figure 6.45 gives an overview of the most important target values of inventory management.

While the reference variables of many planning processes and planning logics of today's inventory management rely on the use of static average values, a key feature of high-resolution inventory management logic is the renunciation of average-based planning in favour of a *real-time-compatible adaptive planning method*. The planning method adapts average values periodically to the actual situation based on real time information. The resulting *real-time planning corridor* enables consideration of dynamic reference and control variables without however excessively increase the volatility of the downstream business processes. Planning complexity, time and costs are kept to a minimum while planning quality substantially increases. To prevent inefficiencies such as excessive stocks, integrated planning processes are supported within the context of high-resolution inventory management with a *differentiated procedure*. A set of procedures is deposited for each planning process in the context of demand, inventory and procurement planning. These are selected on the basis of decision-making logic to optimise the planning purposes of the articles with regard to the target system. The use of differentiated procedures enables more precise and also more efficient control of the company resources.

In order to use the real-time information appropriately within the context of inventory management, it must be prepared, filtered and aggregated previously.

6 Self-optimising Production Systems

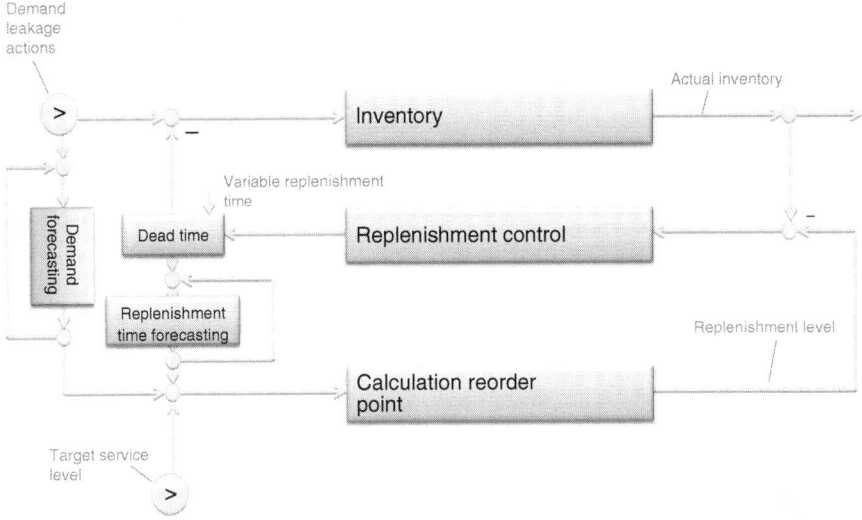

Fig. 6.46 Inventory management process control loop

Preparation and filtering processes ensure that the real-time information is free from error and only the respectively required data is used in inventory management. Hence information overload and increased complexity are avoided. *Aggregation of information* in this context means that the real-time data provided must be transferred to a different aggregation level according to quantity and time in order to be able to use it effectively for inventory management. For example it is not worth initiating procurement after every single sale of an item. For cost-effective dimensioning of the batch size and/or procurement quantity, the overall range of items must be taken into account over a specific period. Figure 6.46 shows a typical cascaded process control loop for inventory management.

In order to validate the added value and benefits of high-resolution inventory management logic for industrial companies, the potential of the developed planning system is demonstrated in a demonstrator the form of a test structure in comparison to the current state of the art. The test structure enables modelling and simulation based on real data. For this purpose, the inventory management process of an ERP system is modelled, in which as a basis products with its specific master and transaction data are provided.

The subsequent simulation analyses the characteristics of the planning systems based on different variation parameters and the formulation of two experiments: conventional inventory management logic and high-resolution inventory management logic. Variation parameters are selected in the way that they allow to simulate the situation as real as possible, e.g. requirement volatility, target available-to-promise level, replacement time, replacement time fluctuation, forecast errors, cost rates, warehousing model, consumption model. The object of the investigation of the simulation based on real data is the central target variables of inventory management illustrated in Fig. 6.45: stock costs, liquidity and service level.

Optimised Production Control Through High Resolution Supply Chain Management The challenges of production control are today not so much the development of tools to solve subtasks but much more due to managing the decision-making complexity. Production control problems have given rise to a variety of solution concepts. However there are no systematic approaches to guarantee the configuration of solution concepts while simultaneously being able to appraise the effects of this configuration on the logistical performance of the production.

Furthermore, current economic environments require a greater focus on the rapid adaptability of production control. Companies are being required to devote increasing time and energy to adjusting part list structures, changing inventory management logic or introducing software-supported fine tuning (e.g. MES systems). These examples can often take more than a year to implement. This shows that the adaptability of production planning and control can no longer keep up with the dynamics of economic change. As a result, targets have been reprioritised. The focus has shifted from use of resources and schedule-adherence to short throughput times and stock levels.

The approach to the configuration of production control that was developed in the context of HRSCM on the one hand comprises a configuration and a simulation model of production control. A four-level model as shown in Fig. 6.47 below is used as a basis.

The aim of the analysis layer is not the evaluation of traditional Key Performance Indicators (KPI's), but to illustrate the time-based processes with the opportunity of displaying data through the interaction with the user in aggregated or even fine grained form. This reveals interrelationships which cannot be illustrated by generating characteristic values which are usually based on averages. Figure 6.48 illustrates examples of an investigation of the process communality and throughput time variance.

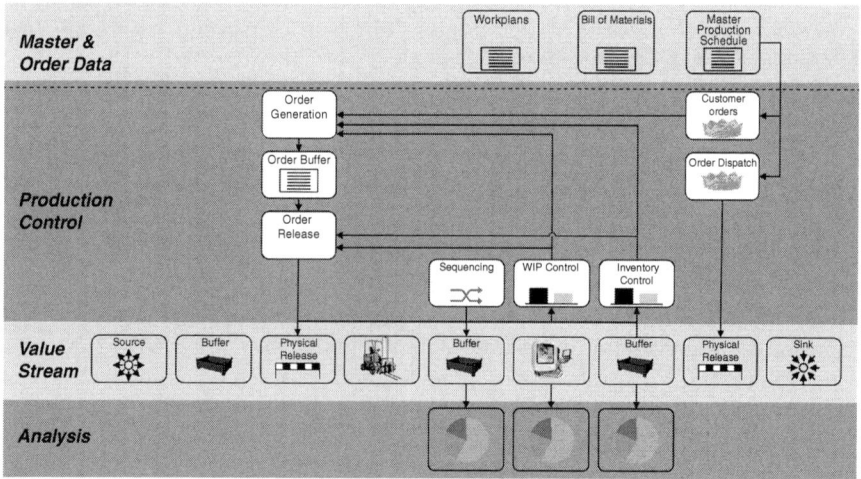

Fig. 6.47 Levels model of value flow-orientated production control

6 Self-optimising Production Systems

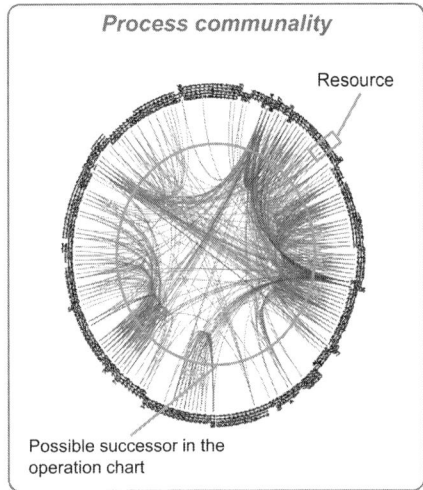
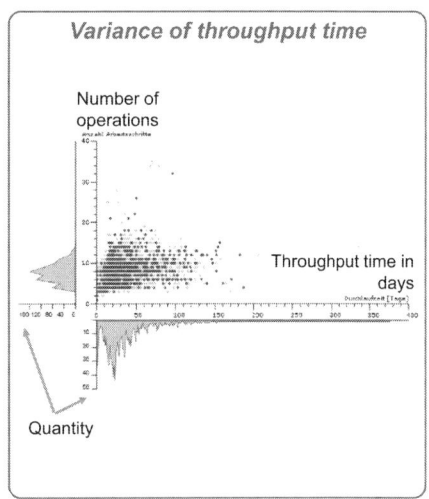

Fig. 6.48 Technical investigation results for process communality and throughput time variance

Process communality is a measurement of the variety of material flow within a production. Each production resource usually has one or more preceding or succeeding processes. If this kind of production system is exposed to a large number of production orders, there are often converging or diverging material flows. This means that there are few uniform material flow routes through the production resources. One measurement for this is process communality which is shown on the left-hand side of the figure. Within the context of a tool specially developed for the analysis, it is possible to select individual production resources and to investigate the number and or frequency of use of predecessors or successors.

In contrast to process communality, the throughput time is a traditional investigation variable. It is generally specified as an average in combination with its variance. In the right-hand part of the figure, the throughput variance is illustrated in the operational schedule depending on the number of operational steps. In the selected type of illustration, the frequency of distribution of the two variables was shown separately on the respective axes to gain additional information. It is possible to select individual periods and/or product groups for which the analysis should be displayed. By using such overviews, the information content increases sharply compared to separate KPIs. This provides a good starting point for analyses and reconfiguration of production control.

The value stream level represents the production process and is used to determine monitoring sections along the value stream. It is connected to the production control level which describes the control configuration.

The level model is in the following presented based on a value stream-orientated approach for the simulation of production control. The aim of the *value stream-orientated simulation* of production control is to enable systematic evaluation of these change processes within production control on the basis of the Enterprise

Dynamics simulation platform and to reveal the limits of existing configurations from the start to enable change processes to be introduced earlier. The basis of the simulation is a combination of the traditional value stream design and production control logic (Lödding 2005).

The work schedule data and the order data are imported into the master data layer (level 1) from files with standardised interfaces. The work schedule incorporates the item number, the workflow number, the resources employed, the processing and setup times, the batch size and the planned throughput time for evaluation. The order data must include the item number, a quantity and the required deadline.

The control layer (level 2) contains the interchangeable modules for order generation, order release and sequencing. They can be assembled in any combination by the user according to requirements. For example, LOR (*Load-Orientated Order Release*) can be combined with MRP order generation and a FiFo (*First In First Out*) rule for processing the queue upstream from a machine. The sequence rules are always applied at a buffer/store upstream from a production process.

The value stream of the control section to be investigated is reflected in the value stream layer (level 3). Here the material flow is modelled by use of source, sink, buffer/stock, release, production and transport elements. The source generates the item to be produced. These then awaits release in a buffer/stock. Once the order release is given according to the selected strategy (control layer in level 2), the item is released with the corresponding batch size for "processing". It is then forwarded via a transport module (facultative), according to the work schedule, to the processing components. Each processing step comprises a buffer/stock atom and the production process atom. The parts first reach the upstream buffer which if necessary initiates a sequential rule calculation (see level 2). They are then relocated to the processing station. After processing, the batches must be re-separated in the UnPack atom before new batches can be packed for the next processing cycle. These new batches then require a new release.

The analysis layer (level 4) contains analysis options. Wait times and inventory projections can be mapped. Strict separation of the information and material flows are required to ensure the modularity of the model. In this way the control modules can be exchanged without adjusting the value stream model. In most companies currently the order release rules ASAP (release the order as soon as it is recognized) or MRP (release the order on the release date) are used.

The value stream-orientated simulation of production control permits the systematic evaluation of the combination of control logics and a specific selection and recommendations for companies. Initial results from MRP control show that even slight turbulence in the order programme have major effects on the stock level. From a certain level of intensity of order exchanges in the order program on, stock levels no longer rise. By implication this means that minor improvements have hardly any effects on companies with high stock levels and MRP control. Only after a breakeven point results improve significantly.

6.3.5 Industrial Relevance

The industrial relevance of the research work is now illustrated on the basis of potential applications within industry. Applications in practice can be divided into two aspects. Firstly, the structural model developed offers support in the analysis and design of the production management system and of the business processes. Secondly, the resulting operative control loops of inventory management and production control enable optimisation of the fundamental components of production planning and control.

6.3.5.1 Analysis and Design of the Production Management System and of the Business Processes

The structural model described here can offer valuable support in projects for the analysis and design of the production management system and of business processes in practice. The structural model can already be used in the analysis phase. It provides help in structuring the as-is-analysis and accelerates the analysis of weaknesses. However, the developed structural model eliminates the open search for solutions because the solution approaches are already anchored in the model. It therefore requires a logic which assigns the potential solution approaches within the model to the analysed weaknesses. This takes place in the classification phase. The actions are then evaluated using standardised criteria and submitted to the project management which then decides on the implementation. The model application steps correspond to the design phase (see Fig. 6.49).

Phase 1—Analysis Phase The aim of this first phase is the substantiated analysis of the actual situation of company structures and processes. The reference model provides assistance in that it structures the process. Here the reference model functions as a template which is laid over the actual status for rapid identification of weaknesses. The template need not be known to the whole project team. Target-orientated moderation makes more sense here than overloading project participants with a partially abstract model.

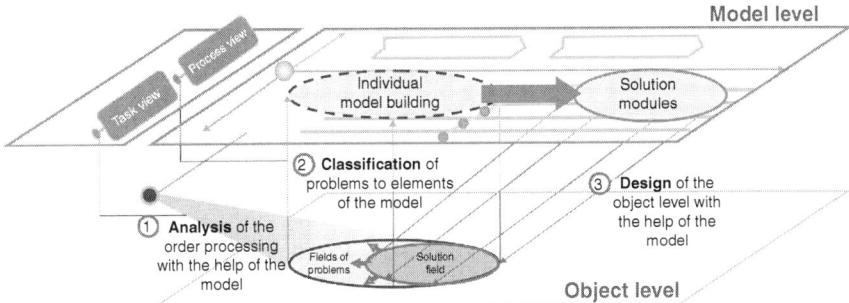

Fig. 6.49 Application logic of the structural model

Due to the different mechanisms and the variety of participants within the company, the process and structural analysis takes place within two analysis streams. The first part of the analysis is concerned with operative business processes and their coordination (part A). The target of the analysis is to improve the efficiency of the business processes. The second part of the analysis is aimed at investigating change mechanisms and at the integrated definition, transfer and monitoring of targets. It is involved with the effectiveness of the production management structures from a cybernetic perspective and takes into account management processes but not business processes (part B).

A—Analysis of the Efficiency of the Business Processes The process follows the logic of the process and structural analysis in the three-phases- concept developed by FIR at RWTH Aachen University. The aim of the module is to record the actual status of the order processing processes and the associated information flows. Methodological tools which can visualise value streams and IT-, process- and structural relationships can be used to document the organisational structure and the IT landscape and to investigate the workflow management and information flows. The record of the IT landscape involves analysing the data and the system architecture which support the order processing. The order processing processes are then collated into company-wide main- and subprocesses. Standard processes with no discrepancies are recorded first. In addition processes, which may or may not be standardised, are described in the case of disturbances or deviations.

B—Analysis of the Effectiveness of the Production Management System Decisions and changes within the company are seldom consciously perceived as structures and so it makes sense to implement this analysis phase with themes that are as concrete as possible. The financial controlling function is an ideal entry point. The detected characteristic values provide information about the control of the operative processes and about the control of the overall system. It also provides more or less structured forms of information about developments within the environment in terms of sales, production development and business management, Corresponding interviews can thus be used to create a map of the reference variables including their dimensions. In the next stage, the question of the availability of information forms the basis of decision-making. The recorded characteristic values are investigated to determine the extent to which they correspond to the criteria of currency, completeness, vulnerability to error, comprehensibility and relevance. If control elements are comprehensible, the individual decision-making processes and their links are then investigated. In this case, an analysis is made of which events can instigate or have instigated structural changes. In this case, particularly the two basic changes in direction from the operative to the strategic and vice versa should be analysed. The integration of management tasks is ultimately expressed in the extent to which changes are expressed over different hierarchical levels in successive characteristic value systems and change processes.

Phase 2—Classification Phase The starting point of this project phase is the assignment of the weaknesses analysed in phase 1 to causes which relate to lack of completeness or lack of integration of elements of the cybernetic production management system. This therefore transfers the reality of the situation to the model level of the cybernetic production management system. For this purpose, clusters of weaknesses with common causes are created and then the causes are assigned to problem-free elements from the reference model. Finally, the benefits and costs of the design of the individual elements are evaluated in order to prioritise the implementation sequence.

Phase 3—Design Phase Corresponding to the structure of the reference model and the approach to analysing weaknesses, actions are once again divided into two dimensions. The actions are detailed and implemented, their success evaluated and the project concluded in the design phase. In the process, the main features of the strategic actions (part B) are defined as they provide the basis for the target-orientated development of the operative processes and their efficiency (part A).

A—Actions for Increasing the Efficiency of the Business Processes The order processing process levels are now reorganised. In this case the project team starts with a reference design of the main processes. The subprocesses are then developed. The weaknesses and their causes are eliminated as far as possible by changing the processes, information flows and control mechanisms. For this purpose the structural model and the operative processes already developed provide a template which is used for the design process.

B—Actions for Increasing the Effectiveness of the Production Management System The actions for increasing the system effectiveness are implemented within the strategic production management system. The creation of a production system for implementing a production strategy through programmes should be the first step of the implementation. The requirements of the operative processes and their logistical performance, planning and flexibility are also defined at this level. In the process, the design of the actual production strategy is only the first step and must be reflected in the cybernetics-orientated design of the management mechanisms, information flows and physical value chain processes.

6.3.5.2 The Potential of Optimised Inventory Management Through High Resolution Supply Chain Management

The competitive pressure on companies is steadily increasing as they have to bring new and increasingly customer-specific products to the market in shorter lifecycle times. Customers are also demanding shorter delivery times. Implicit in these framework conditions amongst other things are greater demands on logistics systems. Particularly affected are the areas of procurement and distribution logistics along with inventory management at the material and also the finished product level.

The BVL study "Trends und Strategien in der Logistik" (2005) addresses this problem in its survey and proves that reducing stock levels and minimising capital

Table 6.4 Design of control loop modules in the case of classical logic and HRSCM logic

Controller	Classical logistics	Logic of high resolution supply chain management
Forecast of the replenishment time	Global average value	Moving average
Demand forecast	Simple exponential smoothing	Simple exponential smoothing
Calculation of the reorder level	Dynamic calculation	Dynamic calculation (current replenishment time considered)
Procurement control	Economic order quantity	Economic order quantity

lock-up costs are of great significance both for industry as well as for commerce. Against this background, the industrial relevance of support for inventory managers by situative automation of consumption-controlled scheduling is clear. In particular the flexible adjustment of control loop-based logic to changing framework conditions of procurement (for example, varying replenishment times and customer requirements) increases their usefulness and performance capacity as compared to classical processes.

In a study of potential an initial development stage of the HRSCM scheduling logic illustrated in Sect. 6.3.4.4 was compared with traditional methods currently used in industry. It displayed clear and considerable potential for optimisation. The investigation compared a traditional dynamic scheduling logic based on average values for replenishment times with the HRSCM logic described above (see Table 6.4). The HRSCM logic is distinguished in this expansion stage by the regular update of the replenishment time and its method of forecasting based on moving averages. The study accordingly reflects the potential of continuously recording the replenishment time and the use of actual values in planning in relation to the service and stock levels.

The investigation included determining the service level and the accumulated stocks achieved using the scheduling logic for different scenarios. The scenarios were formed by the combination of different requirement and replenishment time profiles (see Fig. 6.50) There was clear optimisation potential in the use of HRSCM logic in the form of an 8% reduction in stock level with the same service level and/or a 46% increase in the service level with a stock level of just 12% more. This represents a clear increase in the logistical performance capability.

6.3.5.3 Potential of Optimised Production Control Through High Resolution Supply Chain Management

Due to the variety of improvement actions in the area of production control, potentials are very difficult to quantify in general. In addition, the rate of improvement greatly depends on the starting point of the respective control system. Therefore it is possible only to indicate typical potentials. In the context of a case study,

6 Self-optimising Production Systems

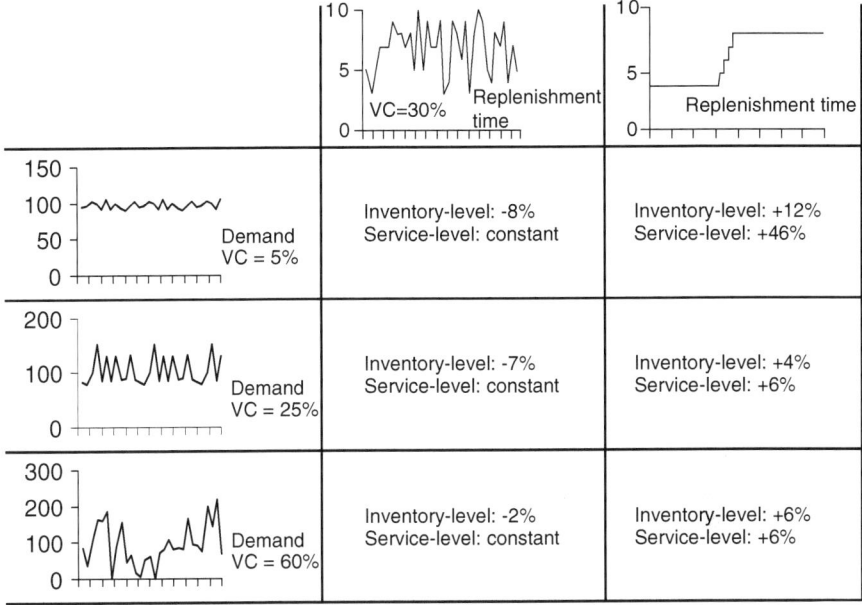

Legend:
VC = Variation coefficient
Inventory = cumulated inventory
Service-level = α-Service-level

Fig. 6.50 Optimisation potential of HRSCM logic in different scenarios

the proposed concept for the configuration of production control was applied to a job shop production system. This was characterised at the start by poor schedule adherence, long and particularly varied throughput times and high rotating stock levels. An Advanced Planning & Scheduling System (APS) was used to carry out all the tasks of production control. On the basis of the above analyses, the system was switched to FiFo (First in First out) control with shortage-orientated order release.

Measure variables were generated over the three operating statuses of starting status, transition phase and improved status in order to permit a comparison. Firstly, the rotating stock or WIP (Work in Process) was designated as a measurement for management of order releases which has a direct impact on the throughput time. Secondly the performance of the system was measured by totalling the processing minutes of the production orders leaving the system each day. Figure 6.51 illustrates the results of the measurements. It is obvious that there is an initial clear drop in performance just after the starting status. After this short fluctuation phase however a greater average performance is achieved along with a lower level of stock rotation.

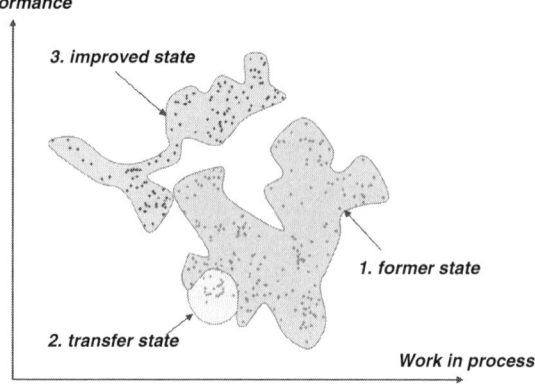

Fig. 6.51 Measurement results in the production control optimisation process

6.3.6 Future Research Topics

So far amongst other things the requirements of company viability of have been investigated in the "High Resolution Supply Chain Management" project. These include the following aspects:

- Real-time capacity of systems
- Vertical integration of information flows
- Horizontal integration of information flows
- Self-optimisation of systems

On the basis of the findings obtained, research into innovative implementation options emerges as a relevant future research theme. Therefore in the following possible future research themes related to this issue are presented.

Real-time capability in companies means having the right information at the right time at the right place for the right purpose (Martin 2003). Idealized real-time systems use information immediately after it has been generated. They automatically detect data as it is generated, avoid buffer and exhibit no medial or semantic breaks. Idealized real-time systems also select the most important information for a decision. They make and implement decisions automatically at the decision-making point (Fleisch and Osterle 2004, p. 3 ff.). Real-time capability requires a high degree of currency and an equally high granularity of data. This real-time data is generated partly within the actual company and partly externally.

So-called *vertical integration* is required to be able to use the real-time data generated within the actual company for the purposes of maximised information transparency. Vertical integration is the synchronisation and harmonisation of the different IT systems available within companies. In this way, for example, signals from operative hardware units are projected into the Operating Data Acquisition System (ODA), the Manufacturing Execution System (MES) and the Enterprise Resource Planning System (ERP). These operative hardware units include for example sensors and readers from auto ID technology e.g. 2D-Barcodes or RFID tags. It is also possible for manipulations in the ERP system to have direct effects on vertically integrated operative hardware units. For example, the entry of a customer order into the ERP system could lead directly to changes in an automated scheduling system.

As with vertical integration which relates to the reconciliation and synchronisation of internal information flows, *horizontal integration* offers potential for future research work. Horizontal information in this context is understood as the synchronisation and reconciliation of different IT systems located within the supply chain. The research theme of horizontal integration is highly innovative and of particular scientific interest because relevant information for the real-time compatible design of companies is not only generated within the company itself but can also come from cooperating concerns. Furthermore these cooperating concerns not only include direct customers and suppliers but also all the companies involved in the corresponding added value chains, from raw materials suppliers to consumers. Industry-wide order processing is already possible today with the use of standardised EDI interfaces. These interfaces simplify electronic order processing. It is possible to exchange further data in real time within a supply chain along with data about electronic order processing. In this way, logistical cooperation concepts can be implemented with minimum outlay and a high level of transparency. Future research themes might be concerned for example with the implementation of logistical corporation concepts as Web-based services. It is worth considering the IT implementation of the following logistical cooperation concepts and/or SCM concepts.

- Collaborative Planning, Forecasting and Replenishment
- Continuous Replenishment
- Efficient Consumer Response
- Industry-wide eKanBan
- Just-in-Time/Just-in-Sequence
- Quick Response
- Supplier Relationship Management
- Value Added Partnership
- Vendor Managed Inventory.

As well as the aspects described above, self-optimisation is a central requirement of the viability of companies. A system is self-optimising when the actual situation of the system is continually detected by the interaction of the system elements, when system aims are defined and prioritised within the context of a defined level of autonomy and when the system behaviour is adjusted accordingly (Frank et al. 2004, p. 22). In the context of the analysis of the actual situation of the system, the current internal system status and the systems environment are investigated and described in detail. The status description can also incorporate information from other systems and existing empirical values.

Changing the target system can be achieved by selecting targets from a pre-specified, discreet, finite number of possible aims, by changing existing aims or by generating new independent aims. Adjusting the system behaviour, which represents the feedback of the system-optimising control loop, can affect the parameters, structure or behaviour of the system (Frank et al. 2004, p. 22).

These self-optimisation findings lead to two further questions which relate to open research requirements:

- How can the self-optimisation of companies be achieved by the homogenous use of control logics?
- How can the reference value corridor of the operative unit be dynamically reconciled with a company's target system?

The *integral use of control logics for the self-optimising control of a company* follows the cascading principle. At the lowest hierarchical level, there are many control loops for continuous self-optimisation of the operative units which have already been intensively researched in the context of "system 1" within the project. These control loops work at a hierarchical level and are interconnected by defined interfaces. In case that it is not possible to control the operative units, pulses must be generated from a higher hierarchical level. These points were defined in the previous project as "system 3". From a control perspective, the pulse generator corresponds to a higher level control loop which can control subordinate control loops by varying additional control variables. This type of enmeshing of control loops can be organised according to the St. Galler Management hierarchical company structure. The highest cascade would accordingly represent the normative control of the company which was described in the previous process as "system 5".

Modelling this kind of complex system of enmeshed control loops must start at the lowest level to ensure that all relevant input and output variables are taken into account. The sales forecast and requirements planning are core tasks of production planning and control (PPC). They are particularly well suited to initial modelling and are analysed more closely below.

Amongst other things the following information flows must be taken into account in the control of sales forecasting:

- Historical trends of item- and customer-specific sales
- Internal company disturbance factors e.g. price fluctuations and sales promotion activities
- External disturbance factors e.g. sale of competing products and customers' promotional activities
- Sales forecasts of direct customers and indirect customers up to the consumer

Amongst other things the following information flows must be taken into account in the control of requirement planning.

- Item- and customer specific gross requirements
- Item- and customer specific stock levels
- Item- and customer specific sales forecasts
- Stock level status (booked/available)

The second question generated by the requirement of self-optimisation describes the *reconciliation of the reference value corridor* with the dynamically changing target system of a company. The target system of a self-optimising system plays a decisive role in the system optimisation and comprises internal targets which can adjust and change the system within the constraints of its autonomy based on external targets that are imposed by the direct environment or by other systems or inherent targets which results from the purpose of the design of the system (Frank et al. 2004, p. 30).

The need to adjust targets results from the localised analysis of preset targets prevailing within companies. In the case of non-uniform target setting, individual targets can cancel one another out with the result that preset priority targets cannot be met (Mälck 2001, p. 30 f.). Production economy target systems are the basis of financial production control whose task it is to reconcile production management with the production system (Hoitsch 1993, p. 27 f.). The literature contains a variety of approaches for company target-setting systems. These include target systems with a monetary focus (DuPont system, ZVEI system, RL system (Becker 2008, p. 166 ff.) or comprehensive target system concepts e.g. Balanced Scorecard according to Kaplan and Norton (2006)).

In order for a company to operate in accordance with its target system, a link must be created between this target system and the control mechanisms. Input variables for the system of interlinked control loops are specific parameters which in turn define reference value corridors within the control loops. The synchronisation of precisely these parameters with a previously-defined target system could be considered as a future research theme. To answer the research question mentioned above one might address the following secondary questions:

- How is the target system of a company structured?
- What interdependencies exist between the individual targets of the target system?
- At what points does the target system directly influence decision-making within the production management system? What mathematical relationship can be created?
- Is it possible to connect the target system with the enmeshed control loops of the production planning and control systems such that changes to the target system influence the reference value corridors of the PPC?

6.4 The Road to Self-optimising Production Technologies

Ulrich Thombansen, Thomas Auerbach, Jens Schüttler, Marion Beckers, Guido Buchholz, Urs Eppelt, Yves-Simon Gloy, Peter Fritz, Stephan Kratz, Juliane Lose, Thomas Molitor, Axel Reßmann, Heiko Schenuit, Konrad Willms, Thomas Gries, Walter Michaeli, Dirk Petring, Reinhard Poprawe, Uwe Reisgen, Robert Schmitt, Wolfgang Schulz, Dražen Veselovac and Fritz Klocke

6.4.1 Abstract

A product's success in its market is influenced by both technical and economic factors. If its fitness for purpose is given on the technical side, production costs come into focus. In an age of short-lived products and fast innovation cycles, manufacturing is coerced into being able to produce economically small production runs and individual items. While in mass production mostly costs relating to processing time dominate, the critical factor for production of individual items is often the time for production setup. In both cases, the resource-efficient achievement of product quality is a basic requirement.

Developments in the past have often aimed to increase performance of production systems by means of increasing processing speed. Advances in control technology as well as in drives have allowed the construction of machines where processes are pushed closer towards their physical performance limits. Processing at the limit of these process domains requires precise knowledge of the relationships between the setting parameters and their effects on process stability and processing result. The precondition for this kind of process control is knowledge about the conditions at the current operating point, in order to decide on the adaptation of values for setting parameters.

The described relationships clarify, that a multidimensional and highly non-linear task has to be mastered, which imposes genuinely new challenges to control systems. Existing approaches have only partly achieved industrial application so far. Many expert systems and Artificial Neural Networks, for example, do not yet have the stability and clarity of response needed for use on production machines running over multiple shifts.

Research in the area of self-optimisation aims to enable technical systems to adapt themselves to changing process conditions. To achieve this, research is needed in the area of process modelling and the development of reduced models to optimise setting parameters, the development of cognitive systems to determine the current operating point, and the creation of control and adjustment systems which, based on machine-readable models, decide and adjust setting parameters for themselves.

The concept of Model-Based Self-Optimisation opens the path to the integration of process models into technical systems which, by using cognitive components, are able to determine for themselves the current operating point within the parameter space, and use this to start carrying out process control. The basis for this is the methodical collection of information about the production process and the creation of reduced models, which can be used to define the target values and how they depend on the parameters with sufficient precision. On the basis of this process knowledge optimisation systems are created, in combination with decision algorithms, which enable an autonomous control of the production system achieving the defined target values, even when random disturbances occur during processing.

For this kind of functionality it is necessary to collect information continuously from the process and process it. The formulation of models to describe the product quality requires knowledge of quality relevant relationships between setting parameters. Even if the initial level of implementation does not provide autonomous process control, this knowledge can already be transferred into application. Validation of the research results on real life examples has led to the concept of information-processing sensor-actuator-systems (ISA-systems). These offer a way for Model-Based Self-Optimisation to realise flexible and re-configurable control platforms which can translate externally prescribed objectives into internal setting parameters using reduced models. This creates an abstraction layer between optimisation and machine control, which also permits implementation of stability and safety strategies.

The continuation of the research project addresses the extension of the parameter space under consideration. Where models are already based on multi-dimensional input parameters, a direct application as a control system is often rendered impossible by the complexity and the resulting scale of calculation resources required. As well as issues of process stability it is also necessary to model the process-machine interactions. But set-up assistants which exist already, and the first controls using scalar parameters are indicating today the potential and the sustainability of the concepts developed.

The solutions achieved in the area of self-optimisation of technical systems, and the documented progress in the area of production processes is based on work carried out by five research partners, who all contributed to the generic solution with diverse requirements, which in turn advanced the specific solution for each application. Many of these results have been transferred to industry, and a common platform was created as a basis for the second phase of the project. Under the title "self-optimising manufacturing systems in *verbund*"[1] open issues are addressed and the area impacted by self-optimisation is extended. With this approach, the whole production site comes into the design scope. This opens up the perspective onto an extended approach to manufacturing planning, in which the entire chain from product design through to manufacturing can be considered. Changes in specifications can hence be assessed with respect to their impact on the manufacturing process.

6.4.2 State of the Art

6.4.2.1 Artificial Intelligence

The concept of Artificial Intelligence (AI) was established in the 1950s by McCarthy (1959). Today, there is no distinct area of impact of the AI due to its diversification and transfer to a multitude of topics. Nevertheless, a certain set of AI methods can be named, which have been used in industry for process monitoring and control purposes.

Originally the first artificial systems aimed at the development of so called expert systems (Savory 1985). Expert systems are numeric systems, which collect and store information about a certain task, draw conclusions based on the gained knowledge and offer solutions to specific problems within the task (Engesser 1993). The stored knowledge, the rules and the input information that is to be processed—the facts—create the knowledge base of an expert system. This knowledge base contains two

---

[1] Verbund in this context describes the union of cooperating manufacturing systems which themselves react autonomously to changing requirements such that all entities within the union together achieve the global objective of the union.

types of knowledge: static, field-specific knowledge, which does not change during a consultation, and case-specific, dynamic knowledge, which is created using an inference engine or can be entered by an experienced user. This knowledge is structured in the form of relatively simple rules and facts, which as a whole represent the expert knowledge. Expert systems have been developed for the most diverse areas of interest. Examples are: informative and advising systems for design and synthesis in construction or fault diagnosis systems in assembly. (Mattke et al. 1993; Puppe et al. 1996; Cremers 1991). However, the use of expert systems often failed due their high acquisition and maintenance costs as well as due to the fact, that the heuristic knowledge bases and the programmed applications itself were error-prone. In most cases, the sensitivity of the overall system towards partly inherent knowledge gaps and faulty assumptions had a significant influence on the system's overall functionality. Therefore, such expert systems were not able to replace professional experts, but can be used to support their activities.

Artificial Neural Networks (ANNs) model the principle functionality of a brain, in that units which work in parallel, the neurons, are linked to each other via directional connectors. The structure of the network needs to be recreated for each individual problem, where it is impossible to predict its necessary topology in advance. The adaptation of an ANN to the problem is executed subsequently, for example by changing the values of a weight matrix using the back-propagation algorithm (Borgelt et al. 2003). The result of the learning process is an analytical function, which treats the given input information as a variable vector and the received output information as a result vector (Grauel 1992). One of the disadvantages of applying ANNs lies in the sensitivity of the network to the input data. Therefore, the training data and the system topology itself need to be selected in tight reference to the specific issue, in order to avoid overmatching or faulty decision bases. In addition, it is never certain that with an optimisation the global optimum is found.

Genetic algorithms (GA) belong to the class of evolutionary algorithms and are based on the principle of "survival of the fittest". This procedure uses populations of potential solutions and selects promising individuals based on their fitness. Based on the selection, new populations are created by combining the properties of the selected candidates. The candidates in the new population are changed by stochastic mutation, and an additional selection is performed. Evolutionary methods are suitable to handle any problems that are presentable as character strings or tree diagrams (Heidrich 2001).

6.4.2.2 Self-optimisation

Research in the area of self-optimisation covers the development of methods and systems for its implementation, as well as research on the processes to generate the required detail in process understanding. CoTeSys (Cognition for Technical Systems) for example is a project that investigates cognitive solutions for different technical systems such as factories, vehicles or robots. In the case of robots, the corresponding system is a human being and the cognition of these systems is implemented as a classic humanoid form. Hereby, the perception and orientation of the system is achieved using cameras for visual and microphones for audio input, in order to perceive the request that is imposed on the technical system. Alongside the technical aspect, Bannat et al. are researching the basis of humanoid coordination as part of this project. Regarding self-optimisation the project's contribution to self-optimisation consists of investigations on a robot's reaction to unforeseen events (Bannat et al. 2008; Laue and Röfer 2006).

In the field of communication networks, strategies for self-optimisation are already implemented (Litoiu et al. 2005). For example Witkowski et al. use self-optimising systems for the automatic creation of mobile ad-hoc communication networks between robots based on swarming behaviour (Witkowski et. al. 2008). Systems like these generate humanoid robot systems for disaster scenarios which improve their skills by learning.

Self-optimisation is also used for mechatronic systems such as the guidance module of rail-based vehicles. In the area of mechatronic systems, research was done on the needed system structure to achieve internal objectives with the assistance of classic control algorithms (Gausemeier 2005). This work was done in the context of the German collaborative research centre 614 "Self-optimising systems in mechanical engineering", which is currently being worked on at the University of Paderborn. Along with fundamental research on the general understanding of self-optimisation, methods and tools are developed to realise self-optimising products. The developed solutions employ sensors and actuators to control mechanical systems (Isermann 2002). In most cases, these have the advantage of being able to measure and evaluate their operating point directly in order to draw conclusions about their current status without having to build a model.

Production systems must meet the requirement to control machine components and a physical process simultaneously. Therefore, the realisation of self-optimisation for manufacturing systems often has to deal with the management of multi-dimensional parameter spaces and non-linear relations between setting parameters, process variables and process result. It requires solutions for handling complexity within the very short time-scales which are available for numerical calculation and optimisation algorithms during the production process. Nevertheless, the potential of self-optimisation for manufacturing processes in terms of reducing set-up time and improving a process's stability is large (Schmitt and Beaujean 2007). The expected added value from turning this potential into practical use in manufacturing scenarios in high-wage countries is therefore a strong driver for research activities in this field. The state of the art in technology shows gaps and starting points for the focussed manufacturing processes.

6.4.2.3 Meta-modelling and Model Evaluation

"Meta-models" in this context are numeric surrogate models that describe relationships between input variables and output variables on the basis of data sets. They therefore take the place of the complete physical description of the manufacturing process, also known as a white box model, and in this respect describe a process' behaviour, without having to be process-specific with respect to the modelling procedure. Suitable approaches to multi-dimensional data interpolation and data approximation are offered by regression models, Artificial Neural Networks (ANNs) or Radial Basis Functions (RBF), which from the modelling point of view are named black-box-models (Stork et al. 2007; Buhmann 2003). If process knowledge is included in the selection of the base functions or in the implementation of the model structure, then these models are considered to be grey-box-models. The process of creating a meta-model includes a series of steps which determine the quality of the resulting model. The decision which parameters are taken into account, the data sources, as well as the data structure and model structure, are just some of the possible sources of influence which require a high degree of expert knowledge when creating a meta-model and defining the configuration parameters (Coit et al. 1998). While these steps are usually carried out manually and case by case, there do exist common tools for the implementation of models, in the form of application programmes such as Matlab, SNNS (Zell et al. 1991), DesParO (Stork et al. 2007) or SUMO (Gorrisen et al. 2009). DesParO, programmed by the Fraunhofer Institute for Algorithms and Scientific Computing (SCAI) very rapidly creates surrogate models with the help of Radial Basis Functions (RBF). It is distinctive because it has a sensitivity analysis for input values and output values which means that irrelevant parameters can be identified and eliminated. This allows models to be downsized, meaning a reduction in the amount of numerical calculations to be executed. The SUMO product is sold by the Interdisciplinary Institute for BroadBand Technology (IBBT) at the University of Gent, as a toolbox for Matlab. Besides Radial Basis Functions, this platform also offers rational functions, splines, neural networks and support vector machines. In addition to standards like the squared error, algorithms such as pattern searching or particle swarm optimisation can be used to optimise models. SUMO offers an open system structure which allows the export of all model parameters and model functions. This allows the meta-models developed to be transferred to other platforms, such as real-time systems (Efron and Gong 1983).

Cross-validation is a procedure which allows the determination of the forecasting quality of a model with respect to process behaviour which can also be used to compare different models. For this, a data set with N data records is split into k subsamples ($k \leq N$), then, while removing one subsample each time, the single error rate from the training with the remaining samples is calculated. A special case in cross-validation is the leave-one-out, where each time one single data record is held back, so that if the data set consists of N data records then $k = N$. This also allows evaluation of the local error for the data point held back.

6.4.2.4 Manufacturing Processes

Milling Monitoring and control systems available in the market are limited in their field of application. Hereby, in most cases the recognition of a process disturbance is accomplished comparing measured process signals to reference curves acquired at stable process conditions. Therefore, recognition, evaluation and avoidance of disturbances are closely connected to specific operations. Due to their lack of transferability these strategies are not suitable for use in single item or small lot manufacturing. Here, system solutions are needed that enable a reliable process control at varying process conditions. This requirement applies at an even greater extent to machining freeforms, due to a continuously changing tool engagement (Reuber 2001).

The continuously changing boundary conditions and the special characteristics of hard to machine materials often require compromises in planning 5-axis milling processes. Therefore, in most of the cases conservative parameters are chosen, that lead to an overall drop in productivity (Li et al. 2004), increased costs and machining time, which are especially a constraint on competitiveness for production facilities in high-wage countries.

In the past, system-based approaches to control the manufacturing of complex free forms have been researched, but show substantial disadvantages in their industrial applicability. Ismail developed a system solution to suppress process vibrations when milling turbine blades, applying an offline adjustment of the feed-rate in combination with an online control of the cutting speed. Hereby, the feed-rate scheduling approach was more successful than the online control of the cutting speed, which at this point seemed as technically not feasible (Ismail and Ziaei 2002). The approach of controlling the feed-rate was also investigated by Erdim et al., where the feed-rate adjusted is non-proportional to the material removal rate. This approach can only be applied during the process planning steps, and not in process control, as the calculation required is too time consuming for online applications (Erdim et al. 2006). Dohner et al. developed an active control system which was able to manipulate the vibration behaviour of the tool and the spindle as a function of the dominant system vibrations. Using sensors and actuators, they eliminated process disturbances caused by vibration, such as chatter vibration. Although the implementation of a prototype was successful, the transfer of this solution in industrial applications has not been realized until today (Dohner et al. 2004). Mane et al. researched the control of process stability during finish milling thin-walled components by adjusting the cutting speed. Here, it was possible to determine the stability limits of the process offline, and to carry out an upstream process planning, reducing the number of cost- and time-intensive test runs. As this approach requires too much calculation time its applicability towards an online control was proven to be unsuitable for industrial practice (Mane et al. 2008). In 2008, Ferry et al. presented an approach to the optimisation of the feed rate where operating domains and the maximum allowable feed-rate for predefined operating points are calculated (Ferry and Altintas 2008). An approach that was also not applicable as an online solution, because of the required computing resources.

In summary, the current state of research covers certain approaches regarding process optimisation and process control, but did not reach the level of industrial applicability.

Plastics Injection Moulding The precise manufacture of complex and high-quality formed parts from thermoplastic substances produced by injection moulding, places high demands on the design of its tools, as well as on the process itself (Kudlik 1997; Menges et al. 1971). Conventional process control, now established for decades in injection moulding, is based on controlling the screw speed during the injection phase, together with the hydraulic pressure in the injection cylinder or the force acting on the screw during the holding pressure phase. The influence of the screw speed on the forming process in the mould cannot be clearly defined, because of various disturbances, such as variations in materials, pressure and temperature, as well as requirements on environmental conditions (Menges et al. 1971). Variations in quality of moulded parts from one run to the next are the result. These variations occur mainly after process set-up, since with conventional process control, at the start of production steady thermodynamic conditions need to be achieved.

After injection of the melt into the cavity, the holding pressure phase begins, where the material begins to shrink as it cools. By applying pressure, additional material is inserted into the cavity in order to compensate shrinkage. The pressure and temperature within the cavity have a particular importance here, as these values define the shaping of the moulded part. In an injection moulding machine, both electric motors and the hydraulic cylinders are used as actuators during the hold pressure phase. This means that by using the position signals on the servo-inverter or on the hydraulic vent, the pressure plot within the cavity, the so-called cavity pressure, can be adjusted. Based on the relationship between pressure, temperature and specific volume, or pvT ratio, which varies by material, by measuring the temperature a pressure can then be calculated to give a constant specific volume.

Both the direct adjustment of the cavity pressure and the adjustment of the target pressure course according to changes in boundary conditions of the process are included in some existing approaches in science and industry (Michaeli et al. 1992; Pramujati et al. 2006; Smud et al. 1991; Arburg 2010). Due to the behaviour of the controlled system, which is sharply non-linear, time-variable, and includes a time delay, the solutions provided so far which manage the cavity pressure settings and which are mainly based on PID controllers with an adaptation of the parameters, have not led to satisfactory results (Michaeli et al. 2004). The realisation of a freely selectable cavity pressure with a high level of precision and reliability, while being highly dynamic and with simple set-up and adjustment of the control system has not been solved so far.

To adjust the planned course of the cavity pressure to the boundary conditions of the process, the strategy of pvT-optimisation is well known, which allows a calculation of the necessary pressure course for the requested quality of moulding, depending on the melt's temperature course (Johannaber and Michaeli 2004; Michaeli and Lauterbach 1989; Menges and Thienel 1977). Based on the unsolved challenge of the control of the cavity pressure, the effects of pvT-optimisation on the quality of the mouldings could only be partially demonstrated.

Welding In the field of automated complex gas metal arc welding (GMAW) processes, the selection and setting of suitable welding parameters for specific welding tasks today is still done by the welding equipment operator. This implies well-founded technical experience and welding expertise of the operator in order to set the required welding parameters optimally for the given boundary conditions. The identification of optimal welding parameters for each application requires time-consuming, empirical welding tests. Since the mid-nineties some approaches have been developed for monitoring and controlling welding processes using Artificial Intelligence (AI). One application using Artificial Neural Networks (ANNs) aimed to predict welding seam geometry based on machine setting parameters that were entered off-line (Middle and Li 1993). Furthermore, the prediction of weld seam quality based on statistically processed transient data was also a focus in many developments in the field of on-line quality surveillance (Maqbool et al. 1998; Heidrich 2001; Kim et al. 2004, 2005). In addition, the use of Neural Networks for the detection of weld seam imperfections was investigated to determine weld seam quality. For this, process stability and the occurrence of spatter were determined using the transient process measurands welding current and welding voltage, wire stick-out lengths and torch angle (Kim et al. 2005). While Martin has used statistical data based on the electric process measurands for the detection of spatter and weld seam imperfections (Martin 1994), Gladkov et al. used the acoustic power density spectrum to identify weld seam imperfections (Gladkov et al. 1998). The optimal use of neural networks demands a sufficient quantity of training data to appropriately cover the process domain to be modelled, which results in enormous investments of time and economic expenditure to carry out the required welding tests (Matteson et al. 1992; Hichri 2005).

In another approach an expert system was used for on-line process optimisation of GMAW processes. For this, a dialogue-based, off-line expert system to determine parameters and optimise processes was modified, extended and linked to the welding process via a process computer. Several sensors were used to collect information about the welding process. To calculate the target seam geometry, the seam surface was measured by an optical sensor located behind the welding torch. By means of fuzzy logic the weld seam geometry, including the penetration depth was classified. At the same time, the electrical process measurands of the welding process were recorded which, together with the result of the evaluation, were used as input values to the on-line process optimisation by the expert system. The spatial distance between the sensor and the welding torch proved problematic in these works, as this spatial distance resulted in a time lag for the sensor information limiting the controller capabilities (Roosen 1997).

Laser Beam Cutting Sheet metal processing industry has, due to globalised markets, an increasing demand for cost efficient and flexible manufacturing systems. CO_2-Laser beam cutting machines enable the processing of a wide range of different types of materials, and thicknesses of material, which ranges from a few tenths of a millimetre thick stator sheets for electrical motors, to 30 mm thick stainless steel for power plant construction. With machines of this type, either individual items can be produced for medium-sized production companies or mass production parts for automobile construction. The local relevance of this is shown by the share of 5,200 laser flat-bed cutting machines sold in 2008, of which 37% were installed in Europe (Belforte 2009).

Along with the wide range of different materials, high cut quality and the ability to reproduce it are both significant factors for competitive production and machines. A smooth and dross-free cut saves expensive post-processing. Under laboratory conditions, the use of photo diodes or acoustic detectors was experimentally investigated to capture signals for controlling the cutting process (Keuster 2005). Along with the non-spatially-resolving sensors, cameras were used to capture geometric and thermal process emissions. For example, Haferkamp was able to use a heat-sensitive camera to show a correlation between temperature distribution in the interaction zone and the quality of the cut (Haferkamp 1999). Control of the setting parameters based on process changes such as, for example, variable properties between batches of material, or ageing and dirt on optical components which affect the laser beam, has so far only been accomplished in the laboratory (Keuster 2007; Schneider 2004; Zheng 1990).

In laser cutting systems today, devices such as penetration sensor systems, automatic nozzle changers, laser power controllers or the distance control of the cutting head have become state of the art. On the other hand, values for processing parameters are taken from empirically determined technology tables and adjusted by the user as necessary. The anticipated product quality is therefore dependent, among other things, on values such as the composition of alloys and the drift in focus position, which in an industrial setting leads to a loss of quality, or to increased costs for set-up and maintenance. Controlling the product quality today is still not state of the art.

Weaving The weaving process is noted for its very high efficiency. Once set up correctly for a certain fabric, a loom produces reliably for a long period. To further improve productivity, textile producers attempt to achieve this stable condition faster than before. Shortening set-up times after a change of fabric therefore allows a further increase in productivity.

Generally, when setting up a loom for a fabric, setting parameters are used from an in-house operations database. For new fabrics, setting parameters must be established experimentally. These tests involve additional costs for weaving mills. Costs are caused by the loss in productivity on the machine, plus the materials consumed during the set-up phase. Weaving mills in Germany in particular have high costs caused by set-up due to the intensive personnel time involved. In addition, the fabrics which are manufactured are generally technical fabrics. The raw material for technical fabrics, as for example aramide, glass and carbon, are expensive compared to raw materials for the clothing industry.

To shorten the time needed to find correct setting parameters for unknown fabrics, the company Picanol NV Ypres, Belgium introduced the EasyStyle-System in 2006. This system suggests machine settings based on a database of weaving parameters. The database contains setting parameters from many different weaving mills who have released their data for this system. The proposed machine settings in this system are not correlated to resulting characteristics in product quality.

At the Institut für Textiltechnik at RWTH Aachen University (ITA) an intelligent set-up assistant was developed to reduce the set-up time. The core of this set-up assistant uses the warp tension as the process measurand. Warp tension is a decisive process variable for weaving processes. It influences the machine's running behaviour, for example the number of warp breakages, or the cleanness of the front shed and therefore the weft breakages. The sum of these factors defines the economic efficiency of the weaving process.

By the combination of an Artificial Neural Network and an Evolutionary Algorithm, the set-up assistant defines the optimum course of the warp tension. This course is evaluated using defined performance criteria. A disadvantage of the set-up assistant is the extensive effort to train the Artificial Neural Network. In addition, no parameters for product quality are taken into account here.

6.4.3 Motivation and Research Question

The efficiency of production in high-wage countries has a significant impact on the total costs of a fit-for-purpose product. The further development of manufacturing systems therefore has to contribute to reducing costs per item through maximum productivity.

Classic controller structures are in most cases not suitable for controlling and optimising modern manufacturing processes. Processes are increasingly highly dynamic, and the interaction between input values and the processing result are mostly non-linear. Stability limits are therefore determined by a number of parameters and their correlations. Because of an increased number of materials to be processed, and because of batch variations, there is an increasing number of variations in the process input values which have a significant influence on the processing result. In addition, alongside geometric precision, in particular for safety critical components, requirements also apply for the creation of particular properties such as, for example, surface integrity. All this taken together shows clearly that for high performance processes a large number of functional criteria need to be produced within narrow tolerances. Consequently, this means that optimisation is also only possible in small process windows. Particularly in on piece flow production, this increases the production and testing effort because barely any consistent process models are available to describe the dependencies. The applied techniques for setting up a stable process are predominantly based on empirical methods.

It is from these circumstances that the research question is derived, whether technical systems can be enabled such that, under variable process conditions, they can autonomously steer the process at the performance limits, and reliably achieve the prescribed objectives. A pre-condition for this is that the system must know about the process at any time, and generate suitable values for setting parameters itself. This can also be stated more generically as follows:

Can complex and variable manufacturing processes be set up, monitored, controlled and adjusted by systems based on reduced physical models while taking the manufacturing history into account?

The research question addresses basically two areas. The creation of reduced physical models and in progressive steps the set-up, monitoring, as well as controlling and adjusting of a manufacturing system. The last part of the research question draws the production history, which contains information about relevant upstream manufacturing processes, into the design frame and optimisation frame. The necessary cognition of the technical system must be built such that all parameters which influence the processing result can be determined during processing, and the manufacturing system can make the required changes in the process behaviour in an adequate manner. What is principally needed for this, is that all relevant information about the current and the upstream manufacturing steps can be made available in a usable way.

As a first approach, the ability to model manufacturing processes is a significant task. Only once setting parameters and target values are combined in causal relationships and formally linked, the process can be modelled.

The following *research hypothesis* is derived from the research question: It is postulated that "Model-Based Self-Optimisation", consisting of a process model focused on product quality, and the necessary cognition, will be able to run a manufacturing process autonomously at the technical limits of the process domain, in such a way that consistent quality is achieved through all disturbances.

To verify the research hypothesis the following topics need to be studied:

- *Modelling:* The transfer of process knowledge into machine-readable models for product quality oriented process control and process adjustments
- *Cognition:* The development of suitable cognitive components for technical systems to determine the status of a process
- *Control and adjustment:* The control and adjustment of highly dynamic manufacturing systems, where the course of the process is unknown at the start of processing.

To create a generic solution in these areas, research will be carried out on five representative and practically relevant processes, while the model evaluation is viewed horizontally.

- *Gas metal arc welding* Welding and Joining Institute (ISF) of the RWTH Aachen University (ISF) with a manufacturing system from the industry partner CLOOS.
- *5-axis milling* Laboratory for Machine Tools and Production Engineering of RWTH Aachen University, Chair of Manufacturing Technology (WZL-TF) with a manufacturing system from the industry partner MAZAK
- *Laser beam cutting* Fraunhofer Institute for Laser Technology (ILT), Chair of Laser Technology of RWTH Aachen University (LLT)/Department Nonlinear Dynamics of Laser Processing (NLD) of RWTH Aachen University (NLD) with a manufacturing system from the industry partner TRUMPF
- *Plastic injection moulding* Institute of Plastics Processing at RWTH Aachen (IKV) with a manufacturing system from the industry partner ARBURG

- *Weaving* Institut für Textiltechnik der RWTH Aachen University (ITA) with a manufacturing system from the industry partner PICANOL
- *Model evaluation* Laboratory for Machine Tools and Production Engineering of RWTH Aachen University, Chair of Metrology and Quality Management (WZL-MQ)

6.4.3.1 Research Questions for Partners

Model Evaluation (WZL-MQ) Meta-models in the current context represent data records of measured values in the form of mathematical functions by means of interpolation and approximation. The creation of the models includes a series of steps which have an influence on the quality of the predictions. The choice of the setting parameters considered, the data sources, as well as the design of the model structure require apart from expert knowledge often also manual selection to achieve a good representation of the process by the model. The approach consists of visualising and evaluating the model quality and deriving strategies for efficient model creation, taking the influencing parameters into account. The derived question is as follows:

> How can surrogate models for diverse processes be efficiently created and evaluated with respect to the model quality, taking into account the influencing variables of model purpose, process robustness and complexity, choice of parameters and their density, spread and distribution?

Milling (WZL-TF) In 5-axis milling operations of free-form surfaces, the cutting conditions and therefore the process behaviour changes along the processing path. This results in a constant variation of the control-path behaviour of the production system itself, which prevents the implementation of a standard control system. The approach is the development of a model-based control. If it can be achieved to describe the control path behaviour in relation to the respective cutting conditions, then a stable process control can be realised that additionally includes the current cutting condition. This would be a substantial step on the path towards the overall goal of self-optimisation for a 5-axis milling process. From this context, the following research question for the 5-axis milling process is derived:

> Is it possible to realise a model-based, self-optimising control of the 5-axis milling process, that takes the current time- and position-dependent cutting conditions into account?

Welding (ISF) Automating welding systems that use the gas metal arc welding process (GMAW), often results in problems which negatively affect the required weld quality due to changes in process boundary conditions. The main causes of these problems are tolerances in component size, tolerances in positioning of components, as well as tolerances in previous finishing steps and tolerances due to thermal distortion caused by the welding process itself. Monitoring the welding process and applying an automatic correction of the welding parameters in case of deviations is often necessary to ensure the required weld quality. In the past, these problems have been approached in various ways, but these were always application specific solutions.

These required a time consuming collection of empiric data records by a welding expert, both for the initial set-up process and for carrying out the welding experiments. Therefore, the following research questions are posited for GMAW-welding:

> What optimisation methodology allows the generation of application-independent and model-based correction parameters, without having to carry out time-consuming and cost intensive welding tests?
> How, on the basis of information supplied by sensors during the welding process, can the current weld seam quality be determined using a model? What further requirements must be met by self-optimisation strategies in order to create optimum correction parameters based on the required weld seam quality characteristics?

Plastics Injection Moulding (IKV) For plastic injection moulding the product quality from a business point of view is determined by the resources required for set-up, and the cycle time, while from a manufacturing technology point of view, the weight and dimensions of the finished part are the major determinants of quality. The self-optimisation approach for plastic injection moulding consists of adapting the planned cavity pressure course, using the boundary conditions captured by metrology, such as for example the temperature in the cavity. The pvT-behaviour of the plastic can be used as a quality model, in order to enable the self-optimisation to compensate disturbances, such as temperature fluctuations. The research question therefore is:

> How can a self-optimising system be developed, built and applied, which is capable of ensuring constant product quality during injection moulding, by means of direct control of the cavity pressure, taking into account the pvT-behaviour?

Laser Beam Cutting (ILT/LLT/NLD) The laser beam cutting process is being investigated with respect to self-optimisation of steel sheet cutting, particularly focussing on stainless steel alloys. As the composition of the alloys itself has a major impact on product quality, the modelling activities in this area are supported by another sub-project (see Sect. 4.4) addressing metallurgical investigations. The quality of a laser cut product is determined economically by the time for set-up of the manfuacturing system, as well as the time and resources needed for the processing itself. In particular, the set-up of both the machine and the process for manufacturing of individual parts using special materials, have a significant impact on the economic efficiency of manufacturing. On the manufacturing technology side, the roughness of the cut edge and the amount of dross are the factors which determine quality.

For the laser beam cutting process there are challenges in the area of sensors and control, as well as process modelling. Measuring the values which characterise the process is currently an unsolved problem, due to the inaccessibility of the cut front. Due to the highly dynamic nature of the process, the question of how to execute control embraces the selection of setting parameters as well as suitable strategies to influence them a short time scale. Therefrom, three central questions arise for the laser cutting process:

> How can the laser cutting process be described such that machine-readable models for self-optimisation can be derived? How can process variables be measured which allow deducing product quality during laser beam cutting, so enabling self-optimising control of the laser

cutting machine? How can a laser cutting machine be controlled so that the cutting process itself can be optimised?

Weaving (ITA) The number of possible settings on a loom is very high. The individual components of a loom which are adjustable cannot be considered in isolation from each other. In most cases there is a complex interaction of the individual components.

The result of weaving is a product, the fabric, whose quality needs to be measured and evaluated. A direct transfer of the target value, product quality, to the multitude of setting parameters can currently not be expressed in closed form. The need for high-quality products and high-quality manufacturing in high-wage countries requires exactly this closed-form expression, and above all the feedback of target oriented setting parameters into the machine. A model-based approach is required to capture multi-parameter systems, to intelligently process and directly align to the target values. For the weaving process thus the following question exists:

> Is a multi-parameter system with the target product quality able to autonomously set multiple parameters on-line, by using a model-based evaluation algorithm in order to enable sustainable, intelligent production in high-wage countries?

6.4.4 Results

Increase in productivity entails not only the reduction of processing time, but also considerable challanges. In materials processing, higher performance of tools and faster drive technology with higher precision create new requirements for process control. If a process is to be run at its technical limits, then to ensure its stability it must be robust against disturbances. Static setting parameters meet this requirement through holding off the stability limit by experimentally determined safety margins. This limitation can be overcome if the manufacturing system is able to recognise critical operating areas within the process, and if a multiplicity of parameters can then be adjusted as needed within a very short time-frame. The prerequisite for this are technical systems which can determine their own operating point and can adjust their behaviour on-line. These are the fundamental characteristics of Model-Based Self-Optimisation.

Model-Based Self-Optimisation for manufacturing systems offers methods and strategies which can be used to run highly complex and highly dynamic manufacturing processes safely at their performance limits. In this area of design, the development and qualification of cognitive entities for the determination of operating points, and the application of reduced models to predict product quality is a major challenge. The high-level methodology for creating the model and the design for self-optimisation are stated generically. The process models developed, the cognitive components created, and the set-up assistants implemented demonstrate the approach to Model-Based Self-Optimisation.

The concept of self-optimisation in the current context is defined as: *Self-optimisation of a technical system consists of its adaptation to changing process inputs and boundary conditions based on embedded expert knowledge and direct process information without external intervention, so that the required output values are achieved optimally.*

The realisation of this concept is divided into two functional areas, the implementation of a system to optimise the process to match externally specified target values, and a system to transform internal objectives into control of setting parameters. Research in this area can be summarised under three main headings:

- Execution of Model-Based Self-Optimisation for manufacturing systems.
- Cognitive functions of technical systems for self-optimisation.
- Description of processes by models.

6.4.4.1 Model-based Self-optimisation for Manufacturing Systems

In product design customer needs are defined as functional and economic requirements for an end product. As part of the construction process these give rise to technical demands on components, which are each produced by a manufacturing system. Manufacturing systems fulfil the task of taking an input product and creating from it the requested output product.

In the simplest case, the required values of setting parameters for the processing task are set by an operator based on his knowledge and experience. Today, the values are taken from a technology table which is defined empirically from a large number of tests. These values define an operating point which generally offers adequate stability reserves against undetected and uncompensated disturbances. The actuators of the manufacturing system actualise the values of the setting parameters, while sensors check the work of the actuators and the result. In the event of a deviation between the predefined value of a setting parameter and the actual value, the control deviation is minimised using fixed classic controllers. This functionality can be described as a sensor-actuator (SA) system.

If all disturbances remain within the limits which also underlie the technology tables, the likelihood of achieving the product quality is high. In those cases where the scale of a disturbance exceeds the predefined limit, or if the input product does not meet the required characteristics, then it is likely that product quality will not be achieved. This needs to be precluded for mass production, and in the manufacturing of small batches or manufacturing of individual pieces it can lead to significant economic loss and delivery delays.

Figure 6.52 shows the principle of Model-Based Self-Optimisation for manufacturing processes. From suitable sensor signals, a meta model is used to establish the current operating point which is being compared to the operating point that would apply to optimally achieve the defined external objectives. From this, the optimisation potential is determined. This optimisation potential is used by the model-based optimisation system (MO-system) to autonomously generate a solution in the form of new internal objectives. The SA system is extended to include the ability to handle these internal objectives so that it becomes an information-processing sensor-actuator system

6 Self-optimising Production Systems

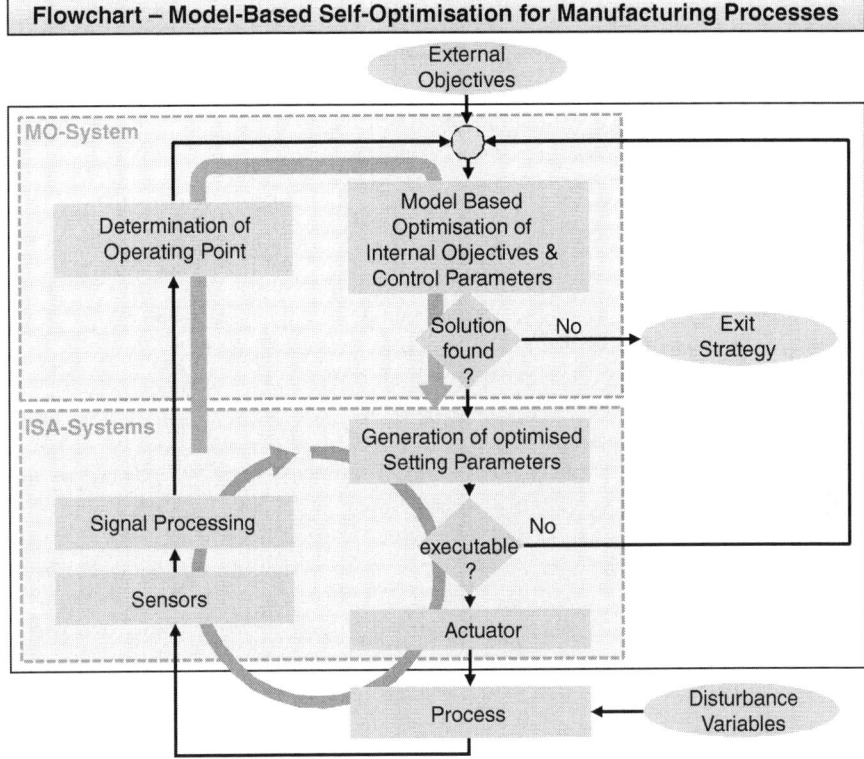

Fig. 6.52 Schematic presentation of the sequence of Model-Based Self-Optimisation for manufacturing processes

(ISA system). As well as translating the internal objectives into setting parameters, it also controls them and can be reconfigured in real-time during the process.

External Objectives of Manufacturing Systems The fitness for purpose of a product is achieved by matching the predefined characteristics for product quality (PQ). In most cases there are several quality characteristics to be taken into account. For example the product must meet predefined dimensions, have a certain level of roughness on its surfaces, or must meet requirements relating to rigidity and structure. This also applies to buisness requirements such as machining time and manufacturing costs. If resource efficiency is handled separately, then this parameter must also be included in the optimisation.

Process Variables Product quality generally correlates with a number of process variables. Model-based self-optimisation requires sufficient data, both in quality and distribution, about these variables, in order to be able to describe the conditions at the operating point with sufficient precision. Often suitable sensors are not available, or there is no access to the source of information. In many cases it is therefore necessary to develop new sensors, or to adapt existing systems to the current task.

An important process variable in the weaving process is the warp tension, whose temporal behaviour has a significant influence on product quality. To measure the warp tension with a high temporal resolution a new sensor has been developed which does not add any measurable friction to the process.

When finish milling thin-walled components the product quality is mainly defined by the surface quality and dimensional accuracy. In order to keep both targets in predescribed tolerances, the values of the significant process variables must also be kept within predefined limits. Here, the main variables are the behaviour of the component during processing, the tool used, the machine behaviour and the process with the significant setting parameters cutting speed and feed rate. Latter can be read directly from the machine control. This information is not sufficient for Model-Based Self-Optimisation, why further sensors have to measure additional process variables which are being integrated into the manufacturing system to enable self-optimisation. In the present example, further process variables which are relevant for quality are the vibration of the component and of the tool. It has been shown that neither force sensors, whose eigen frequency is too low, nor acoustic emission sensors, whose frequency range lies outside the observed frequency range, are applicable. Besides this criterion, other requirements need to be met by the sensor system, such as, for example, a sufficiently high signal-to-noise ratio, robustness against influence of coolant and material removal, as well as being independent of the mounting position. As tactile and optical sensors do not meet these boundary conditions, a multi-sensor system was developed. Two acceleration sensors were attached to the machining table and spindle housing, as well as an eddy current sensor facing the tool. The multi-sensor system delivers consistent information about the vibration behaviour of thin-walled components during finish milling. By using signal processing algorithms the measurement signals are converted into values for quality related process variables.

A further example is the selection of sensors for laser beam cutting. Most of the quality related process variables exist in the interaction zone of the laser beam and the material. Due to the high temperature of the molten material and the small width of the cut, of just a few hundred micrometres, it is difficult to acquire measurands to provide information about the roughness of the material at the solidified cutting surface. The process model however allows to derive the correlation between process measurands and process variables. Using the modeled relationship between the width of the cut and the focal position, the technical system determines the values of the inaccessible but quality relevant parameter, focal position, by measuring the width of the cut. In laser beam cutting an imaging sensor system observes the processing zone coaxially through the laser optics of the cutting head to acquire information from the interaction zone which has a causal relationship to product quality. The signal processing analyses the acquired sensor information and passes the data on for self-optimisation.

6 Self-optimising Production Systems

Determining the Operating Point: The operating point of a manufacturing process is given by the current values of the process variables. If measurands cannot be determined directly, then the dependencies of process variables on measurable variables must be known and described by models. In process models the functional relationships between setting parameters, process variables and process measurands are defined. Meta-models describe these relationships in a reduced mathematical form, and thus allow the determination of the current operating point during the manufacturing process.

In welding, the model to depict process variables and setting parameters with respect to product quality, for example the weld geometry, is also built up on fundamental physical relationships. The values of the dynamic process variables voltage, current and gap-width are determined by sensors. Other setting parameters, such as shielding gas quantity, which influences the shape and boundaries of the working areas, are predefined by the manufacturing task. The necessary calculations to determine the operating point are carried out numerically in real-time. This enables defining and tracking the operating point during the process. The process domain shown in Fig. 6.53 defines the parameter limits where the process has the same physical properties. Within this area the operating domain represents all values which are capable of delivering the pre-defined product quality.

Model-Based Optimisation The optimisation of the manufacturing process takes place in a number of steps. These include the determination of the process status at the current operating point, and the model-based assessment in relation to the external objectives. The externally defined objectives such as, for example, machining time, manufacturing costs, resource efficiency or surface roughness, dimensions, weight must be correlated to significant process variables in order to be transformed to internal objectives. Achieving the internal objectives is done by adjusting the setting parameters.

The models of the manufacturing processes enable prediction of the product quality characteristics from the setting parameters. The predictive use of the models

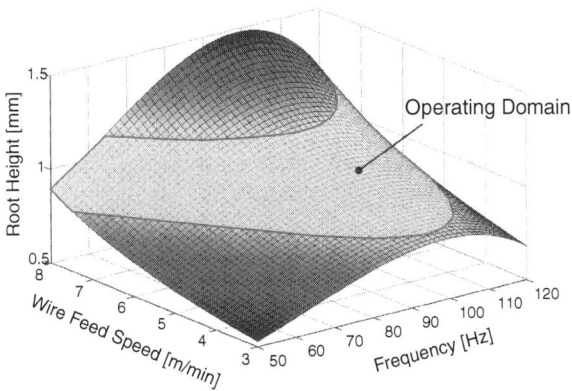

Fig. 6.53 Process domain of a gas metal arc welding process

supplies an unambiguous solution. If the models were bijective, then the inversion could be carried out using a mathematical operation. Because of the complexity of the processes, and the dimensionality of the parameter ranges, process models are in fact mostly not bijective. To determine the values for the setting parameters for a given product quality using a non-bijective model, an inverse solver is needed, which usually uses iterative methods.

A suitable solution generally includes meeting various objectives. Usually individual features of product quality, such as surface roughness, geometrical properties and process efficiency are competing with each other (Pareto optimisation). In addition, boundary conditions such as machine performance limits or process limits must be taken into account. For manufacturing processes, the dependencies between process variables which are part of the optimisation task and the process outcome are strongly non-linear. These types of problems are described as multi-criteria optimisation with boundary conditions. One approach for solving problems like these is transformation to a scalar optimisation function, with weightings for each of the targets and optionally the formulation of boundary conditions. The optimisation task that results from this can be solved with non-linear programming methods, such as gradient methods or Newton's method. Depending on the precision required, solutions can be created which are suitable for use in self-optimisation with regard to their calculation time.

In the context of Model-Based Self-Optimisation for manufacturing systems, there are two central goals. The first comprises achieving product quality by applying suitable operating points. If all operating points which deliver a required product quality are collected into the operating domain, then the minimum such domain is the optimum operating point.

The second target comprises the stability of the operating point. Meta-models are able to carry out a continuous calculation of target values within the parameter space. This enables optimisation solutions relating to stability to be assessed, by taking into account information about gradients and discontinuities.

Even if all operating points within the operating domain achieve the defined quality, some operating points will lie closer to instabilities than others. To evaluate robustness against disturbances, the area surrounding the operating point is also tested for instabilities. With this information, adjustments of the setting parameter values can be achieved so that the stability zone is increased.

Reaching an optimised operating point may require significant adjustments of setting parameters, which in turn can impact the stability of the process. In addition, because of the dynamic response of the process the adjustment can often not be made within one step of the control cycle, which makes it necessary to calculate intermediate steps. All the points on the trajectory, which are defined with the help of the meta-model, must work in a way that guarantees continuous quality and a sufficiently large operating domain to ensure robustness against disturbances.

The following condition could be interpreted as a safety risk or stability problem: If the system finds a operating point which meets the external objectives, but is in a operating domain which is not connected to the current operating domain, then no trajectory can be calculated for the transition. Therefore, this case must be handled separately, depending on the process and the application. But it is still possible that depending on the size of the new operating domain and the dynamic answer to the changes in values of the setting parameters, that this operating point is indeed the better choice for achieving the external objectives. The consideration of such decisions can be handled by adding additional terms into the target function for optimisation, or it can be solved with special decision algorithms. The result of optimisation is a new set of values for the internal objectives combined with sensitivities and validity areas.

Information-Processing Sensor-Actuator Systems Model-based optimisation, which is represented in the upper dashed box in Fig. 6.54 (MO-system), generates internal objectives and control parameters, which are used by the manufacturing system to achieve the external objectives. This information package is passed on to the system, which contains an information processing unit, a sensor and an actuator (and ISA system). The information-processing unit contains a static part with electronic interfaces to the sensor and the actuator as well as a variable part with programmable logic to reconfigure the control behaviour.

In a simple case, the information package handed over consists of a control value as an internal objective and the description of a simple proportional control element as a control strategy. If the relationships between the setting parameters and the process variables are non-linear, and if there are dynamic reactions to be taken into account, then the generation of the internal objectives and their control strategy becomes more complex. In addition, the information processing unit must even be able to be reconfigured at run-time if the internal objectives each depend on

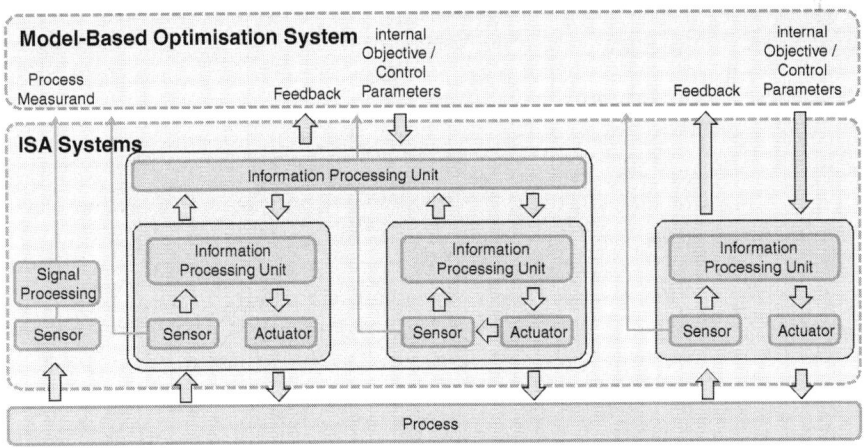

Fig. 6.54 ISA systems as a platform for controlling manufacturing systems with highly dynamic processes in variable configurations

the current operating point. On this point, the model-based approach offers precisely the possibility of generating dedicated meta-models which can be evaluated in the information processing unit in real time. This split between generating internal objectives and controlling the actuators introduces an abstraction layer between optimisation and control, which also enables decoupling the individual problems timewise.

The demands on the information processing unit increase with the complexity of the modelled relationships between process measurands and values for the setting parameters. For example, in weaving, the density of the fabric is an internal objective. Optimised values for the internal objective are transferred to the subordinate system and there translated into warp let-off and fabric take-off.

In welding, the external objective of the energy per unit length can be controlled via a system which consists of three ISA systems, combined as shown in Fig. 6.54. To meet the metallurgical demands of the product, the energy introduced into the product has to meet certain values over time and distance. Conversion of the external objective to internal objectives takes the form of electrical current and voltage values as the welding robot moves along its defined path. In this case, the values for welding current and voltage are managed by a superordinate information processing unit as an ISA system which controls the resulting arc power. This system receives instructions and constrains for the energy per unit length, as well as information about the pulse characteristics, like peak current or peak time, as well as pulse frequency. The ISA system for the arc power is parameterised by the characteristics of the welding power source, the welding current and voltage are measured continuously, while the ISA system for the robot controls the movement.

For high performance laser beam cutting, maintaining a given focal position is important, as this value has a significant impact on the quality of the cut. The control of the focal position is a very complex task as its position varies with the power of the beam and the diameter of the raw beam. In addition, the behaviour changes over time because of the optical system warming up. While processing, the focal position cannot be measured directly, but it has a strong correlation to the width of the cut. Therefore, a meta-model provides the width of the cut as an internal objective which needs to be met, in order to achieve the external objective. As this relationship is neither linear nor unambiguous, the information for the ISA system has to be extended to include control parameters and sensitivities before it can be transmitted. Using this information, the ISA system can pursue the internal objective. The coaxially mounted, imaging sensor observes the processing zone and determines the width of the cut using algorithms for image processing and signal analysis. This process measurand is used by the information-processing unit to activate the focussing unit in closed loop control. The transmission of the internal objective, the width of the cut, together with the control parameters such as dynamics and sensitivity, allow the ISA system to complete the task on a very short time-scale.

The generation of optimum values for the setting parameters in the injection moulding process, aims to achieve consistent geometric properties and weight of the moulded parts. The correlation between pressure, temperature and the specific volume of the plastic melt is used to calculate the optimum course of the pressure in the cavity, using the help of a model predictive controller. On the basis of a process model, which describes the process behaviour of the manufacturing system, the control input (control voltage in the servo-inverter or proportional vent) is compared to the controlled variables. The model predictive controller uses this model to predict the course of the pressure in the cavity, given the prescribed course for the control signal (cavity pressure course). The control signal is adjusted using an optimisation function, which is based on this model prediction. The injection moulding machine, the type of plastic granules, and the geometry of the cavity together define the system to be controlled. If one of these components changes then the process model is adapted accordingly to the new configuration.

The design of the ISA systems establishes an abstraction layer between optimisation and control, which decouples both areas in terms of time and content. The information-processing sensor-actuator systems (ISA systems) are able to track their internal objectives at a high control frequency while the model-based optimisation system (MO-system) monitors the existence of optimisation potential. Subsequently, new internal objectives are generated and applied to the manufacturing system using parameterisation or reconfiguration of the ISA systems.

6.4.4.2 Meta-modelling

Modelling Manufacturing Processes Manufacturing systems use a technical process to process materials, for example by joining, cutting or forming. If a physical process model exists for a manufacturing process, then the values for the setting parameters can be used to predict product quality. If such a model is also available in machine-readable form, then it can be used as the basis for self-optimisation. Models of this kind, often described as white box models, cannot be evaluated automatically and sufficiently fast in most cases. This means that, although they can be used to generate initial values for setting parameters, they cannot be used for process control.

An alternative approach is the use of meta-models, which in this context are seen as a reduced mathematical representation of the functional relationships between multiple input parameters and output parameters. These models are based on process knowledge which is gained from either experimental or simulated data and which describes the process in sufficient detail. Applying meta-models to the solution of a dedicated task allows the reduction of the complete system analysis to a lean numeric solution that can be automatically evaluated which enables control and adjustment of the manufacturing process. Possible implementations of meta-models are grey box models, where for example, the basic functions or the model structure are defined in advance, based on process knowledge. Black box models do not contain any process-specific information and only use data-based adjustment or training to adapt themselves to the behaviour of the process (Stork et al. 2007; Buhmann 2003).

The use of meta-models can refine or even generate process knowledge. Through the closed-form representation of the data, a direct comparison between experiment and theory is enabled, which allows examination of the sensitivity of setting parameters (Schüttler et al. 2009). A general disadvantage of approximation methods is their poor ability to describe discontinuities within a process domain. Hence, sudden changes in the behaviour of a process, as occur at the edge of process domains, usually have to be considered separately.

The creation of meta-models for manufacturing processes comprises several tasks that determine the quality of the model. Significant factors comprise the selection of the model structure, adjusting the model parameters to the data set and handling of discontinuities. Solutions of this task are supported by a process-independent methodology, which was developed as part of this study on model generation.

Methodology for the Development of Metamodels The methodology for creating models was developed as part of the project and evaluated against five different manufacturing processes. It includes, alongside the general approach, the necessary resources and tools as well as steps for optimising the models.

In Fig. 6.55 the column headed "B" describes the systematic process for creating a model. The first steps *(B1) Definition of process and model requirements*, *(B2) Selection of parameters*, *(B3) Determination of the process domain* and *(B4) Selection of data source* define the requirements and limits of the model. The second group of activities *(B5) Planning and setup of experiments* and *(B6) Generation of data* prepare the data for the calculations in the model. The last activities are for

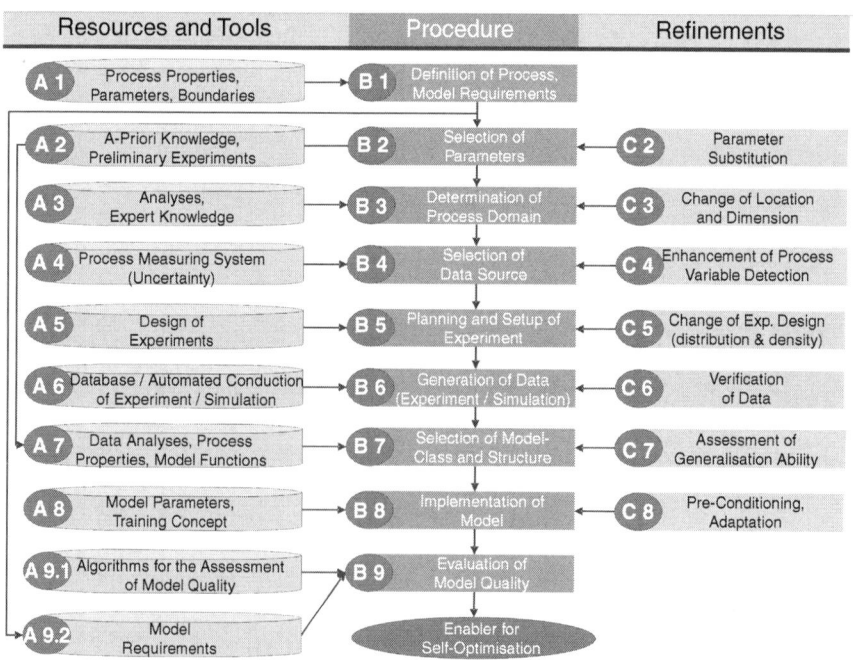

Fig. 6.55 Methodology for creating a meta model

the development of the model itself. Therefore, in *(B7) Selection of model class and structure* a suitable model type is selected, the model is created based on the acquired data in *(B8) Implementation of model* and finally the models are evaluated in *(B9) Evaluation of model quality*. The necessary supporting resources and tools are indicated with "A" and are assigned to the appropriate processes of creating the model. Since many different factors influence the overall model creation process, the required model quality is not always achieved after the first iteration of the model creation process. For this reason the method offers measures to optimise the model quality, which are shown with a "C" in Fig. 6.55.

Definition of Process and Model Requirements The generation of meta-models starts with the analysis of the process to be modelled. Input, output and influencing variables which affect product quality need to be identified and defined. In addition, on the basis of prior knowledge about the process, decisions are made, for example whether to use a dynamic or a static model, linear or non-linear correlations. The necessary requirements on the model, such as the maximum permitted computation time, the necessary model quality, or the differentiability of model functions must be taken into account. Adequate definition of the boundary conditions has a significant impact on achieving the required model quality, and minimises the number of necessary iterative steps right from the start.

Selection of Parameters The manufacturing process and the resulting product quality are influenced by multiple parameters, such as properties of the materials, machine settings and the prevailing environmental conditions. Existing process knowledge can be used to identify important parameters, or these can be identified by preliminary tests. Significances and correlations can be further detailed by experiments using design of experiments.

Determination of Process Domain After the selection of the parameters to be taken into account in the model their ranges of values are defined. This covers the process domain, the range of setting parameters and product quality in which the process has the same physical properties. A subrange of the process domain is the operating domain. This comprises all operating points at which the required product quality will be achieved, and does not necessarily have to be one continuous area. The self-optimisation of the manufacturing process therefore occurs within the overall process domain, taking into account the stability at the edges of the various operating domains.

Selection of Data Source Once the relevant parameters and their value ranges have been defined the data for the generation of the meta-model need to be acquired. In relation to the required model quality and the sensitivity of the setting parameters with respect to product quality, it is important to minimise noise, deviation and uncertainties in the input data and either adjust for or exclude the impact of disturbances. The selection and application of sensors, together with the conduction of experiments, have a significant influence on the quality of the model (Witt 2007). Particular attention must be paid here to the signal-to-noise ratio and the choice of a suitable time resolution for the acquisition of measured values and their average. In all cases, it must be taken into account that the sensors themselves have no significant influence on the process. For example, contactless distance sensors are suitable for measuring vibrations (Wibbeler 2002). The same considerations apply for data which is generated with the help of numeric simulations. Here, sensitivity and precision can usually be determined or estimated in advance.

Planning and Setup of Experiment If surrogate models are based on interpolation, then their quality depends substantially on the density and the distribution of the data. For weakly non-linear parameter correlations a suitable balance can be found between data density and model quality by having well distributed data points. This can be achieved using design of experiments. But in many areas, the modelling of highly dynamic and strongly non-linear manufacturing processes requires an individual distribution of data points within the parameter space in order to get reliable model predictions. A method which is often used to generate equal distributions is the Latin Hypercube Sampling (LHS) (Steinberg and Lin 2006). For an LHS-distribution first of all, the number of levels per parameter and the parameter range are defined and then, the parameter space is divided up using a hyper-grid into the corresponding levels. Then, test points are inserted so that exactly one point per row exists within each dimension. A more equal distribution in the upper dimensional space is provided by the Centroidal Voronoi Tessellation Method (CVT) (Romero et al. 2006). For this method, the parameter space is divided up around the randomly distributed initial data points into regions, such that each region contains exactly one data point. In addition, the distance between each data point and the centroid within a region is shorter than the distance to the centroid of any other region. A subsequent alignment of the data points provides that each of them is exactly the centroid of their region. The Latinised Centroidal Voronoi Tessellation Method (LCVT) is a combination of the two methods described above and of their advantages and was used successfully during the project to generate test points (Reisgen et al. 2010).

Generation of Data After defining the test plan, the necessary data for producing the meta-model is generated. Right from the beginning, the automated acquisition of measured data has to be a goal to minimise the manual experimental effort. The same applies for data that is generated using simulation. With the initial design of the simulation, interfaces should be implemented which support later parameterisation. In both cases, datasets are available after generation, from which the meta-model can be built up in order to predict the process outcome in a much shorter time or even in real-time.

Selection of Model Class and Structure Meta-models are defined by their class and structure. White box models in the current context describe the physical relationships of a manufacturing process using mathematical formulae, with the aim of providing an exact description of the real processes. While the precision of this class of model is very high, due to the use of very detailed process knowledge, the time required for implementation is also high, and its complexity often is associated with a high demand for computational resources. One approach to manage this problem is the use of reduced physical models that decrease the computational effort needed for a sufficiently precise solution, for example by applying an asymptotic expansion (Schulz et al. 2009). Grey box models combine process knowledge and adjustable structures. While the structure of the model is derived from the process knowledge, the coefficients are adjusted to the underlying data using a model fitting techniques such as linear regression with least squares method (Draper and Smith 1998). Other grey box implementations of this kind use symbolic regression with evolutionary algorithms. The most flexible and most general class of model generation is the black box model. They are adjusted to the underlying data records without predetermined structures based on process knowledge. One of the most widely used black box modelling tools is Artificial Neural Networks (ANNs) (Rojas 1996). ANNs consist of networked neurons which are defined as having inputs and outputs as well as an activating function. Several layers of such neurons can be used for the approximation of the underlying dataset, with the neurons being trained using the data. Another approach, which follows the paradigm of the black box model, is the use of Radial Basis Functions (RBF) (Jurecka 2007). These models carry out an interpolation into the underlying data records using a set of functions which are positioned at the sampling points and are weighted, and which are also scalable by means of a shape parameter. In areas which are highly dynamic and highly non-linear, the main risk is an under-adjustment. For self-optimising manufacturing processes, the importance of identifying and separately assessing the model quality in these areas is decisive. In grey box models this can be handled explicitly.

Implementation of Model After generating the data and choosing the type of model the meta-model is created. A series of application programmes such as SNNS (Zell et al. 1991), Matlab, DesParO (Stork et al. 2007) or SUMO (Gorrisen et al. 2009) offer modules to implement Radial Basis Functions, Artificial Neural Networks or a wide variety of regression models. When using Artificial Neural Networks, the main factor that influences the quality of the model is the number of training passes, when using Radial Basis Functions it is the definition of weightings and shape parameters. In many cases, there may be serious losses of model quality due to under-adjustments and over-adjustments. In addition, depending on the choice of model parameters and the number of underlying data points, the calculation time for creating the model can grow exponentially. In this case it needs to be checked whether a reduction of the model parameters or number of data records would compromise model quality.

Evaluation of Model Quality The model quality describes the usability of the created model. A suitable method is the determination of model error using cross-validation. In this, data records that were not used in creating the model are applied to validate the model by comparing them to the result of a prediction by the model. Various techniques of cross-validation are holdout, random sub-sampling, K-fold and leave-one-out (Efron and Gong 1983).

Classic error criteria are the mean absolute error or the mean square error. Neither of these take into account the global scaling of the model parameters, which means that after parameter transformation by pre-conditioning or scaling the relevant model error criteria are no longer comparable. The criterion of the coefficient of determination R^2, which defines the ratio between the mean square error and the standard deviation of the parameter is, for its part, not affected by scaling (Efron and Gong 1983). The values of R^2 range from 0 for a bad model to 1 for a "perfect" model and so enable comparison.

Models which describe the response to transitions between stationary process states require the application of a time horizon in order to evaluate their dynamic prediction quality. A k-step-ahead prediction (Ljung 1999; Nørgaard et al. 2000) allows looking at several discrete time steps, where each solution from the model is used as the input to the next iteration. By comparison of the prediction to a measured validation value, the predictive quality is obtained.

6.4.4.3 Results Model Evaluation (WZL-MQ)

The generated models for meta-modelling were tested for their model quality. The work on model evaluation contains additional research into the selection of a suitable evaluation criterion and the creation of software for model evaluation. For this, the AX-Workshop software was developed, which creates meta-models from the recorded process data, and reads in existing models in order to evaluate them.

Evaluation Criteria As first criterion to evaluate static models the criterion of the mean square error (MSE) was investigated. Starting from real output values y_i a comparison is made using the output value calculated by the model $\hat{y}_i$. The MSE is calculated by

$$MSE = \frac{1}{n}\sum_{i=1}^{n}(y_i - \hat{y}_i)^2 \qquad (6.2)$$

Over time, it appeared that the MSE-criterion only offers a relative comparison of models from the same manufacturing process. It offers neither a process-independent nor an absolute assessment of the model quality. Figure 6.56 shows as an example an evaluation of the laser cutting process using the MSE-criterion. It can be deduced that Artificial Neural Networks are best suited for the first two output parameters and reciprocal quadratic regression terms for the third output parameter.

To be able to make absolute statements about the quality of the model the suitability of the coefficient of determination R^2 was assessed. This is derived from the ratio of the square error

$$SS_{err} = \sum_{i=1}^{n}(y_i - \hat{y}_i)^2 \qquad (6.3)$$

and the absolute sum of the squares

$$SS_{err} = \sum_{i=1}^{n}(y_i - \bar{y})^2 \qquad (6.4)$$

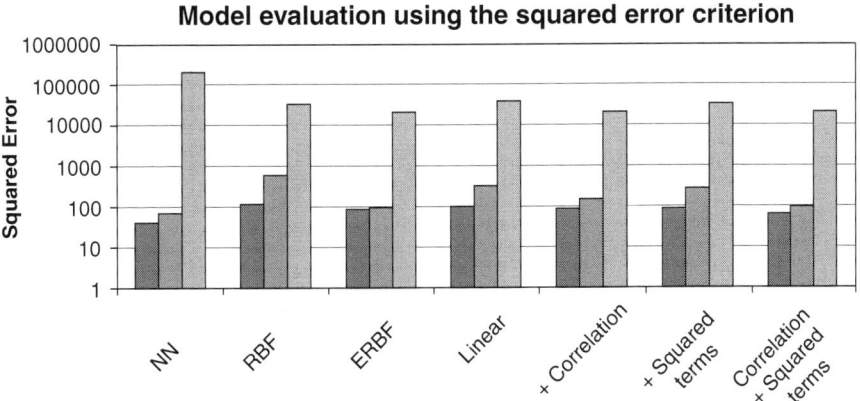

Fig. 6.56 Example of model evaluation on laser cutting using MSE-criterion

as a ratio to the mean value

$$\overline{y} = \frac{1}{n}\sum_{i=1}^{n} y_i. \quad (6.5)$$

R^2 is then defined as follows (Fig. 6.57 shows an R^2-evaluation of static models):

$$R^2 = 1 - \frac{SS_{err}}{SS_{tot}} \quad (6.6)$$

Particularly Artificial Neural Networks and the Radial Basis Functions generated by DesParO achieved overall a good quality of model (Buhmann 2003). The question as to which minimum R^2 value is necessary in order to obtain satisfactory rigging, monitoring and control of manufacturing processes within the area of self-optimisation still remains open, however.

Investigation of Distribution and Pre-conditioning of Setting Parameters An important criterion that determines the quality is the number of data points used to create a stable and precise meta-model. Response Surface Designs (Circumscribed, Inscribed, Faced) distributed in the parameter space were applied to the selection of data from the manufacturing process for meta-modelling. By the gradual reduction of the number of data points for the creation of the models and for their evaluation, the model quality was analysed for its dependency on the density of data and the data distribution. Using the circumscribed and inscribed distribution a model containing 593 data points could be reduced to just 25 data points, with only a minor loss of model quality (Fig. 6.58).

It was also examined how far pre-conditioning or normalising the data used contributes to improved model quality. The input data was transformed such that the

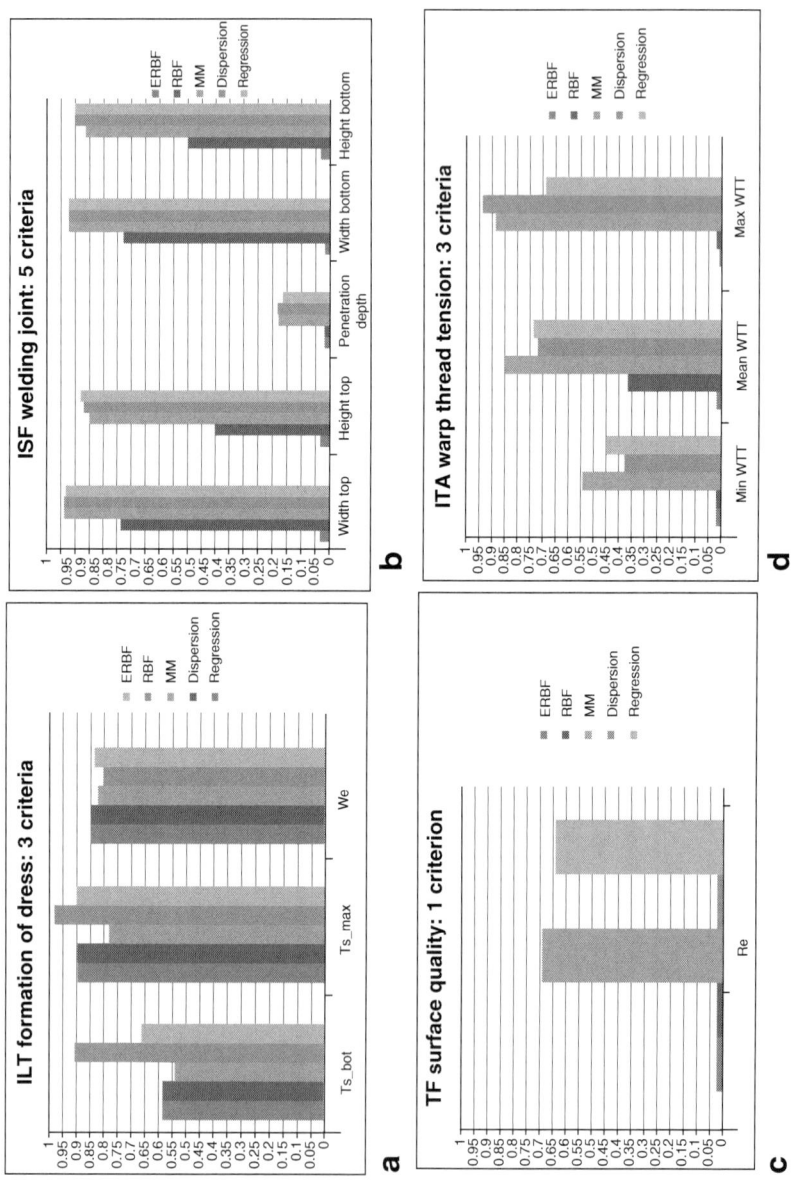

Fig. 6.57 R^2-evaluation of the models. **a** Laser cutting (ILT/LLT/NLD). **b** Welding (ISF). **c** Milling (WZL-TF). **d** Weaving (ITA)

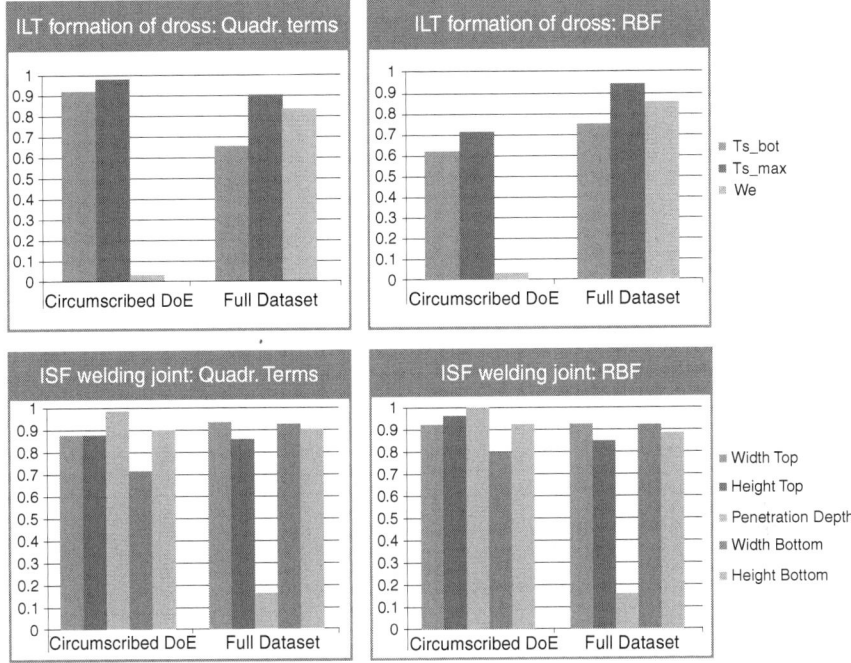

Fig. 6.58 Comparison of model quality of reduced test plans with full test scope

Fig. 6.59 Comparison of models with and without pre-treatment

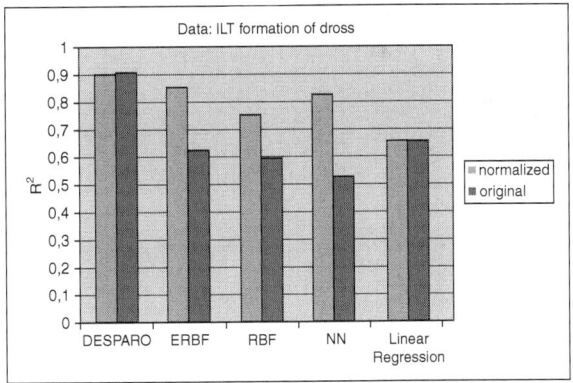

mean value μ for each dimension was μ = 0. Using a principal component axes transformation the standard deviations σ were scaled to σ = 1 and correlations between the different dimensions were removed. This delivered a clear improvement in quality of those models based on neural networks (Fig. 6.59).

Software for Creating and Evaluating Models As part of the research, a software system for meta-modelling was developed and further enhanced (Fig. 6.60). It

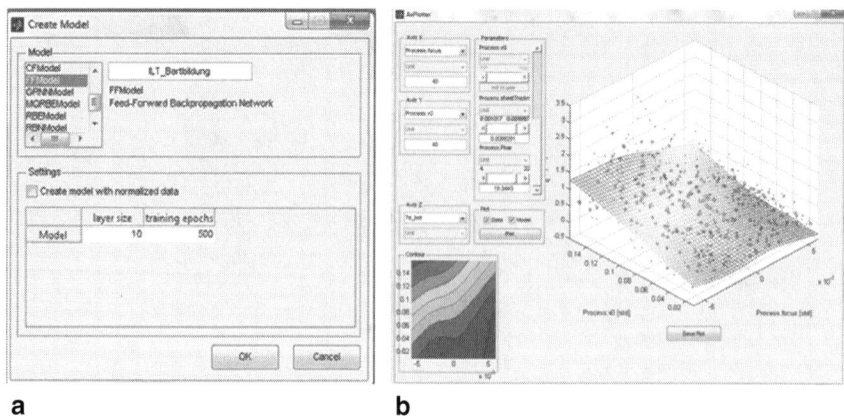

Fig. 6.60 Software for creation, visualisation and evaluation of meta-models. **a** User display. **b** Visualisation of the model

allows the efficient creation, visualisation and evaluation of different model structures. Radial Basis Functions, Neural Networks, regression functions, splines and other models are available as model structures. Alongside the visualisation, the models can be compared with the underlying data records, and can be evaluated using the implemented validation criteria (R^2, MSE).

A 20-fold cross-validation (Efron and Gong 1983) can be used as a good approximation for the global error using the coefficient of determination R^2. For conclusions about local errors in a meta-model, on the other hand, it is the Leave-One-Out cross-validation which is most useful, as each data point is verified (Efron and Gong 1983).

6.4.4.4 Results Milling Process (WZL-TF)

To ensure the reliability of a component's functionality and quality, from a manufacturing viewpoint the most important aspect in milling is keeping the geometrical tolerances. Here, dimensional and geometrical accuracy, roughness and surface quality were identified as the most important criteria of finishing operations. For roughing operations, the tool life and the removal rates are the most relevant quality measures.

For the analysis of dependencies between the setting parameter (SP), the process variables (PV) and the product quality (PQ) the three process measurands (PM) position, vibration and cutting force were selected. The selection of these process measurands is justified by the fact that the vibration behaviour of the part, and the mechanical load on the tool vary as a function of the position of cut. Based on this, there is a need for a position-oriented analysis and evaluation of the current operating point during the machining operation. This includes direct linking of sensor information to the position of the tool in the work piece coordinate system. Via this position a

6 Self-optimising Production Systems

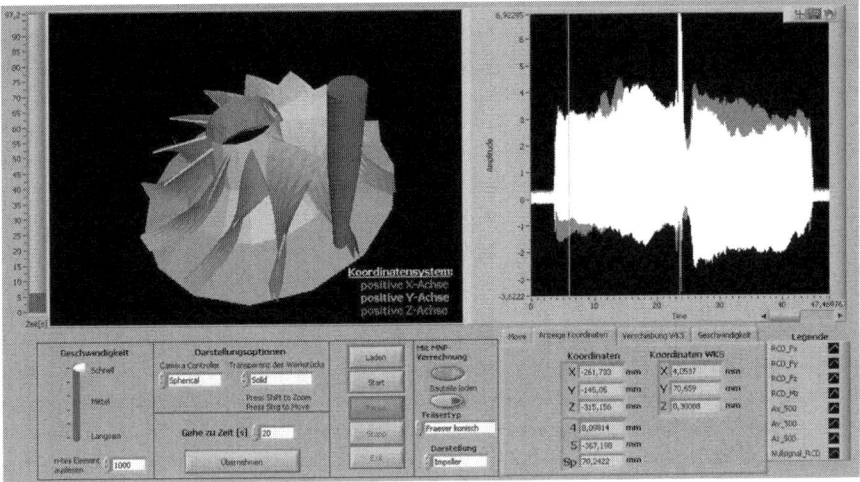

Fig. 6.61 Position-oriented monitoring during 5-axis milling

linkage of process and component parameters can be realised. This enables the analysis and evaluation of the process behaviour during changing engagement situations, as well as depending on the process disturbances and changes of material characteristics. Based on this information, suitable strategies can be derived for the design of a suitable control strategy. This approach can be applied to roughing and finishing operations. Due to this analysis, the whole machining sequence of 5-axis milling is structured more transparently and the general process understanding increases. The result of the position-oriented monitoring in 5-axis milling was demonstrated using a freeform geometry of an impeller and is shown in Fig. 6.61.

The appropriate use of a position-oriented monitoring system is based on the selection of suitable sensors to record the process measures. As an example of the finishing process, where according to Insperger the main causes of unsatisfactory process results are process instabilities, various sensors were tested in the laboratory for their useability (Insperger et al. 2003). Taking the position-oriented boundary conditions into account and the resultant requirements on the sensors, it was possible to demonstrate that the on the market available power measurement systems are not suitable for the investigation of finish milling operations. They offer too low eigen frequencies, so that the dominant process vibrations could not be resolved (Klocke et al. 2008). The measurement of the vibration behaviour with displacement sensors was done using only contactless operating sensor solutions. By doing so a direct linkage of the sensors to the machining operation system was established without affecting the system. Fibre optic sensors and eddy current sensors were used, both of which allowed the capture of dominant process vibrations. Although, looking for an industrial operable integration, both measurement systems are not suitable to be integrated in a manufacturing unit. In further investigations acceleration sensors were identified as suitable measurement systems for a position-oriented monitoring

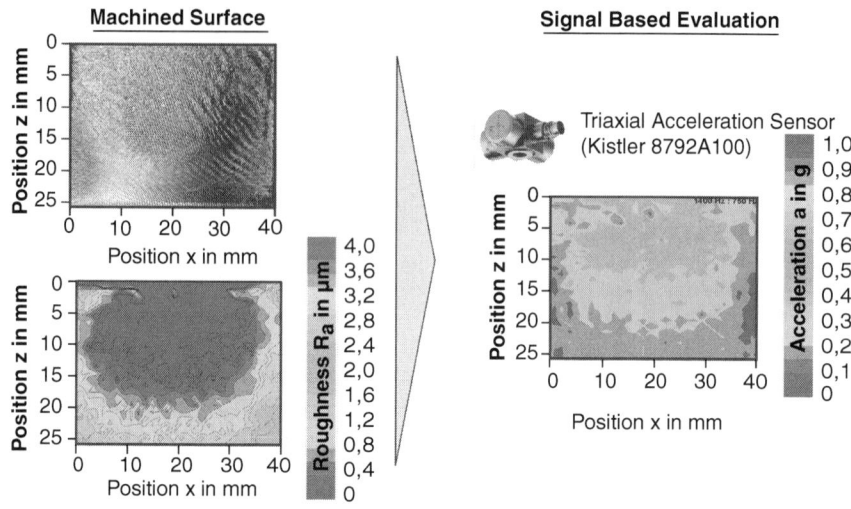

Fig. 6.62 Evaluation of finished surface based on acceleration signal

of the machining process stability. Regarding the finish milling process, the sensor signals were used to evaluate the surface quality manufactured by analysing the process stability. Hereby, the position-oriented analysis of the acceleration signals indicate to what extent the geometrical tolerance and surface quality meet the quality requirements, Fig. 6.62. In future work, the developed monitoring system will be used for rough milling operations, to enable the detection and control of instabilities in this area of the milling process too.

For the application of a model-based process control, solutions were developed for efficient model creation, to be able to provide the changing transfer behaviour in a machine-readable knowledge (Klocke et al. 2009). As the number of influencing factors during milling operations and therefore the time required to generate the data for a model creation, is very large, there was a need to develop a system which can be flexible configured. Furthermore, this system should allow an automatic evaluation of machining tests. In addition, the reproducibility of the test performance and data evaluation must be ensured, as this is the only way to generate reliable models. Knowing this, a system was designed and built which carries out automated machining tests, processes and analyses the signals received from the process-integrated measurement system (Klocke et al. 2010). Based on the analysis of the sensor signals, process-specific characteristics can be identified which correlate to the current engagement situation at any time, and are stored in a suitable database. Using this data, further work was done to develop process models to describe the transfer response behaviour in rough milling and finish milling operations and to integrate these into suitable process control concepts.

The basic models for a process behaviour description and those, which are to be developed for an application of process control, are a first step towards

6 Self-optimising Production Systems

Fig. 6.63 Self-optimisation during the plastic injection moulding process

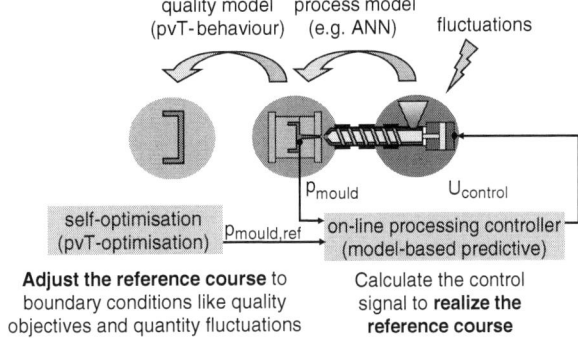

self-optimisation. A self-optimising 5-axis milling manufacturing system should be enabled by the use of these models in order to change its setting parameters autonomously and to meet the requested product quality demands at each position of operation.

6.4.4.5 Results Plastics Injection Moulding Process (IKV)

In injection moulding the assurance of a constant and high product quality requires a stable process control at an ideal operating point. The aim of such a process operation underlies the demand for a process which runs identically in each production cycle. The process is represented by different process variables, in particular pressure and the temperature of the melt in the cavity.

The effects of pressure and temperature on the specific volume (pvT-behaviour) of the material is used here as the basis for calculations to define the operating point and optimum process progress (Fig. 6.63). One requirement for the optimum operating point is a constant specific volume during cooling in the holding pressure phase. After measuring the temperature inside the cavity, it is possible to determine the set point for the control of the cavity pressure on the basis of the pvT-optimisation using the quality model. As actuating variable for process control a voltage is used, which enables the control of the servo inverter and hydraulic valve and thus the movement of the srew during the injection and holding pressure phase.

Typical disturbances in injection moulding are, for example, temperature variations of the melt and the tool, as well as fluctuations in the viscosity of the plastics material. Varying temperatures can occur during the manufacturing process, on the one hand during start-up or after breaks in production, when the stable thermal state has not yet been reached. On the other hand, there can also be control fluctuations or disturbances in the heating system which can lead to temperature variations. Viscosity fluctuations, which influence the pressure transfer to the tool cavity, typically occur due to variations in the batch of plastic or to the widespread use of recycled materials.

Table 6.5 Model prediction error and control error (Root Mean Square Error (bar) of linear and non-linear model structures with and without time information

	Linear (Regression)/bar	Non-linear (ANN)/bar
Model prediction error		
Without time input	5.2	3.2
With time input	5.1	2.7
Adjustment error		
Without time input	2.3	1.6
With time input	2.3	1.3

These disturbances should be compensated by a model predictive control as an ISA system, which has a cycle time of 8 ms using a dynamic process model. Here, the process model contains the correlation between the control voltage, which actuates the servo inverter or the proportional valve, and the cavity pressure as the control variable. Using identification tests, measured values are recorded which represent the non-linear and time-variable relationship between actuating and control variable. After calculating the time delay, which can be attributed to transporting the plastic melt when increasing and reducing the pressure, to the compressibility of the melt as well as to the inertia of the screw, the process models are created (Normeey-Rico and Camacho 2007). As the control path behaviour depends on the cavity used, the injection moulding machine and the material used, and thus often varies, the process modeling is automated as far as possible. This enables a simple identification of the control path behaviour. Using black box modelling, both linear and non-linear dynamic process models are created based on the measured data and evaluated by using k-step ahead prediction (Table 6.5). ANNs are used for non-linear modeling. These increase the quality of the model prediction. The quality of the model can be further improved by including the time as an input variable for the dynamic model. The prediction error of the process model then correlates with the control deviation of the tool cavity pressure control. In addition, the improvement in the model reduces the control effort, which helps to extend the machine's life.

A challenge when implementing the ISA-system is guaranteeing adherence to the control cycle of 8 ms on a micro-computer (Intel Core 2 Duo processor, 3 GHz with 3.25 GB RAM and Microsoft Windows XP operating system). The implementation of the ISA-system on a real-time system (NI 8353 RT from National Instruments Corporation, Austin, USA, running with an Intel Core 2 Quad Processor, 2.4 GHz with 2 GB RAM) with multi-processors even allowed reducing the control cycle to 4 ms. Future work will also include linking the real-time system to the MO-system.

Alongside the creation of the process model, a quality model is used to determine the internal objectives. The determination of the pvT-behaviour is being done today in laboratory devices. There are, however, limitations on how well the data derived in the laboratory could be transferred to an industrial injection moulding process. Determining the pvT-behaviour directly in the mould means that, although physically

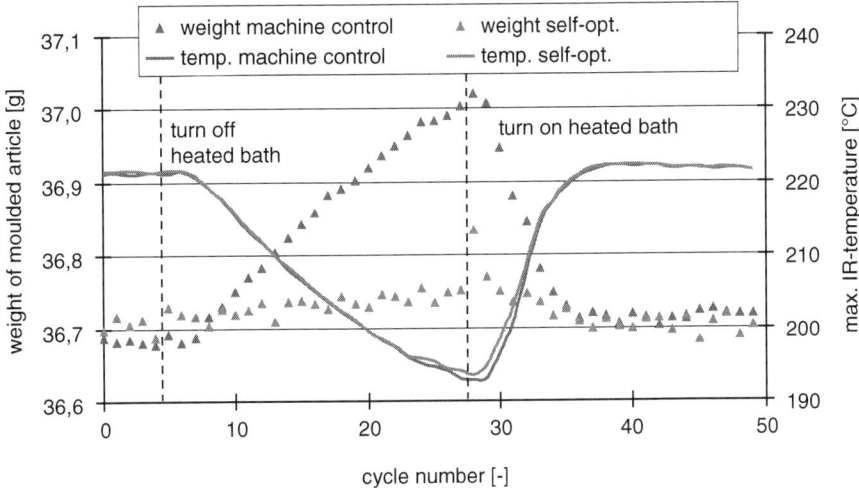

Fig. 6.64 Reaction of the self-optimising system to disturbances

correct pvT-diagrams cannot be obtained, instead characteristic diagrams of materials which are optimally aligned to the individual conditions can be gained (Michaeli and Schreiber 2010). The described process uses grey box modelling which is used to create the quality model. After capturing the temperature a reference course of the cavity pressure creates the internal objective which is implemented by an ISA-system. This cavity pressure set point is adapted by the quality model to the present temperature in the cavity. Within this project various options to acquire the temperature were researched. On the one hand, the melt temperature in the cavity was deduced indirectly using a mathematical approximation, and taking into account the temperature of the tool wall and the melt temperature at the tip of the nozzle. On the other hand, the melt temperature within the cavity can also be measured directly using an infrared temperature sensor. The expensive implementation of direct measurement competes with the inaccuracy of the mathematical approximation in the selection of method.

Model-Based Self-Optimisation should guarantee consistent, high product quality, even with disturbances in the process. Exemplarily introduced disturbances show the response of the Model-Based Self-Optimisation system. Figure 6.64 shows the response of the self-optimising system compared to conventional machine controls when both front heating zones in the plasticising unit are switched off. By measuring the quality values, such as the moulding weight or the actual moulding dimensions, it can be shown that the self-optimising system compensates the thermal fluctuations; the changes in temperature only have a negligible effect, and thus the quality consistency is increased. The active adjustment of the holding pressure course to the changed thermal conditions in the process therefore ensures a clearly improved consistency of quality of the moulded components. Thus, the system developed provides the basis for a complex self-optimising system that can work across cycles to intervene in the thermal balance of the injection moulding process. Additionally, the system ensures the compensation of thermodynamic fluctuations during process set-up and thus minimises set-up time. To prove its applicability in practice, the system was tested on the complex geometry of sports goggles, which are manufactured in a hybrid, multi-component injection moulding process.

6.4.4.6 Results Welding Process (ISF)

For monitoring and self-optimising on-line control of GMAW processes, it is absolutely necessary due to the dynamic of the process, that not only the unchanging initialisation values such as the base material, filler material, wire diameter, shielding gas, seam preparation, torch orientation etc. are taken into account, but also dynamic, sensor-based values. This is necessary since geometric changes on the component can occur due to the thermal influence of the welding process itself. Various sensors are used to gather this information from the current welding process. Electrical sensors measure the transient electrical current and voltage of the welding process. Using an optical sensor system the work piece geometry in the seam area is captured, allowing the determination of the width of the gap and contact tube distance.

In addition, a module of the simulation programme SimWeld was developed to calculate welded joint geometry as a stand-alone geometry solver, and optimised to be used during on-line operation. Using the initialisation values and the sensor-based data, the geometry solver calculates, on a continuous basis and dynamically, the current weld joint geometry, which is defined as the relevant product quality variable (Fig. 6.65). So even during the creation of the seam, it is already possible to react to any deviations from the prescribed seam geometry (seam height, seam width, root height, root width and weld depth). This information is used to monitor the weld seam geometry as well as to control the GMAW process, and therefore to guarantee the required product quality by selection of optimised setting parameters. This comprehensive provision of process and product information is a technological innovation for the welding process. It allows making the planning and control of a production line more transparent, and therefore more flexibly configurable.

When creating the models for self-optimisation the welding simulation programme SimWeld was a key tool. In cooperation with a research project on the production and processing of materials, SimWeld is used to simulate submerged arc welding processes as a component of the virtual process chain. The models used in this context are based on the developments needed for the simulation of GMAW processes. This makes it possible to amend the programme used for self-optimisation in such a way that it could be used to generate data and carry out on-line process simulation. The validation of models using real welding tests allowed further detail regarding the process behaviour to be added to the models used in SimWeld.

Generally, the SimWeld simulation system is a time-efficient way of providing process data to create models. This allows a considerable reduction of time and

6 Self-optimising Production Systems 831

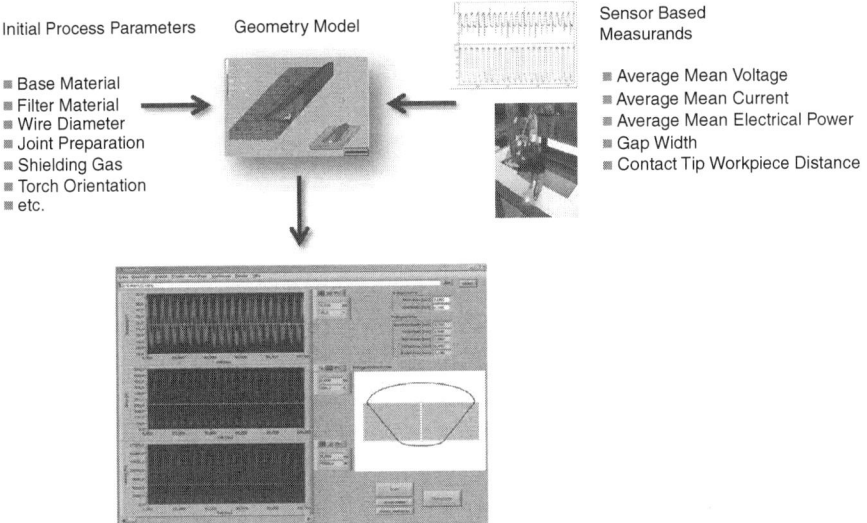

Fig. 6.65 On-line monitoring of weld seam geometry

technical efforts. Models to describe the behaviour of the overall process, and of each of the process domains for the different application areas, were created as the basis for implementing a method for Model-Based Self-Optimisation in the event of variations in the boundary conditions. Figure 6.66 documents exemplarily the welding process model created for two dependencies in the impulse welding process. In the left-hand part of Fig. 6.66 the height of the seam is displayed as a function of the wire feed speed and of the gap width for a square butt weld. The right-hand side shows the relationship between the gap width and the welding speed for a square butt weld.

With the help of the model relating to the overall process behaviour, it was possible to visualise in iso-lines the relationships between the setting parameters which led to an optimisation of the set-up procedure.

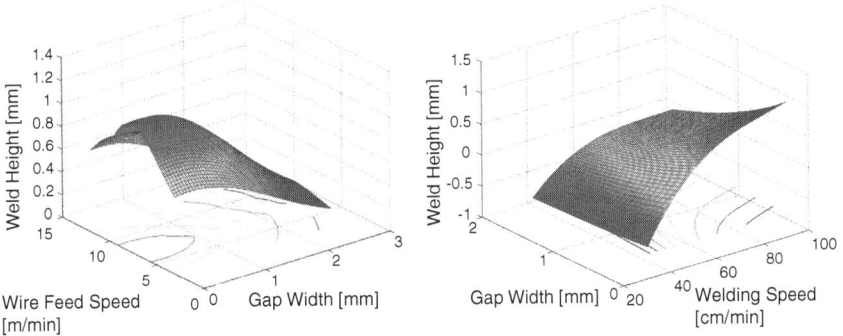

Fig. 6.66 Global surrogate model for impulse welding process

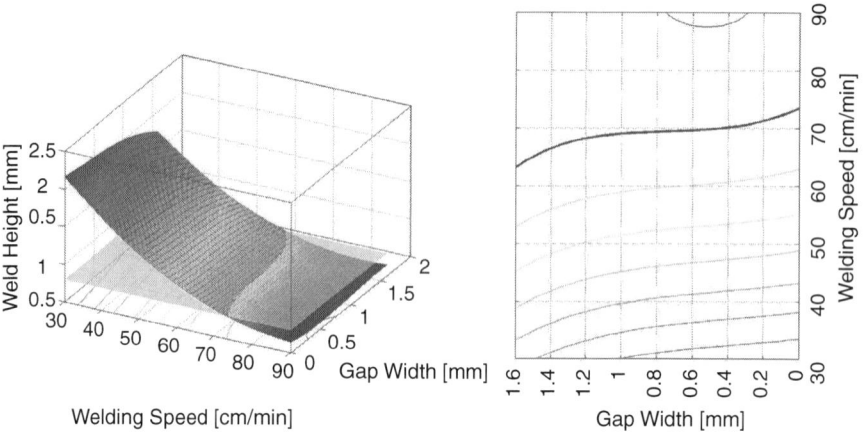

Fig. 6.67 Model-based description of root height

Figure 6.67 shows, for example, a model for the root height (left) and the derived relationship between the gap width and the welding speed with a constant seam height of 0.8 mm (on right). The other four setting parameters are kept constant. By analysing the iso-lines it was possible to derive the maximum possible weld speed at which a pre-defined product quality could be achieved. This functional relationship previously had to be determined by the operator applying his technical knowledge in time-consuming test welds. With the help of the model, this can now be done in the shortest possible time. This enables current on-line adaptation strategies, under which normally only the welding speed is adjusted during the welding process as a function of the measured gap width, a considerable reduction in both time and technical input.

A transfer of this approach to all six setting parameters is, however, not feasible, as the process models created do not show any strict monotonicity, so there is no functional relationship. Therefore, procedures had to be evaluated which enable inverse use of these models. In this context, various optimisers (stochastic and deterministic) were tested on the existing process models with regard to time efficiency and deviation from the optimum. For the set-up process this procedure did lead to an initial optimisation and selection of setting parameters. Future work will aim to qualify optimisation methods for on-line generation of setting parameters during the welding process.

6.4.4.7 Results Laser Cutting Process (ILT/LLT/NLD)

As with the other manufacturing processes studied, for laser beam cutting the key to self-optimisation is a model-based prediction of the processing result in relation to externally defined objectives. For laser beam cutting a large number of setting parameters need to be taken into account, which show strongly non-linear interactions.

If a meta-model supplies the relationships between setting parameters and processing results in order to control the process, then the cognitive component of a self-optimising manufacturing system can use the modelled process knowledge to determine the current operating point based on the acquired process measurands. In either case, the generation of a suitable meta-model can only succeed, if the parameter space is defined with high resolution. This requires large amounts of data, which can only be created at reasonable effort by using simulations.

Model Creation A particular challenge in the case of laser beam cutting consists in the fact that the actual "tool", the interaction zone between the laser beam and the work piece, changes depending on the process variables resulting from the process dynamics. For example, the absorption profile on the surface of the emerging melt creates a so-called free boundary, whose shape cannot be known in advance, but which is part of the equations describing the process. These free-boundary-value-problems can only be solved very poorly using conventional simulation tools or commercial FEM solvers. What is more, the relevant time-scales and length-scales cover multiple orders of magnitude, which very quickly makes the computational cost become uneconomic, if a conventional numeric procedure is used.

For this reason, the focus was set to the development of a numerically efficient process model, which is capable of describing the cut quality—and in particular the formation of striations on the cut surface—as a function of the process parameters. To do this, the cutting process was described using a reduced model where, by separation of the time-scales and length-scales involved in the process, a hierarchy of subordinated sub-processes is first created (for example, heat conduction, fluid dynamics of the melt and absorption of the laser beam) and this is then further simplified by reduction of dimensions (Schulz et al. 2009, 2010).

This procedure was then expanded to describe the non-linear dynamics of laser beam cutting at high-resolution in both space and time, such that the creation of striations, resulting from melting and recrystallisation, on the cut surface can be calculated. For this, the dynamics of cutting were reduced to two fundamental differential equations for the position of the melt surface $M(z, t)$ and the melt film thickness $h(z, t)$, which can be solved with minor computational cost (compare Fig. 6.68a):

$$\frac{\partial h}{\partial t} + 2h \frac{\partial h}{\partial z} = v_p, \quad \frac{\partial M}{\partial t} = v_p - 1, \quad v_p = \frac{1}{Pe \cdot h_m}(Q_A - Pe) \quad (6.7)$$

$$Q_A = \gamma \mu A(\mu) f, \quad \mu = \varepsilon \left(\frac{\partial h}{\partial z} - \frac{\partial M}{\partial z} \right), \quad A(\mu) = \frac{4\mu \cdot i}{2\mu^2 + 2\mu \cdot i + i^2} \quad (6.8)$$

for $z, t \geq 0$, where $v_p = v_p(z, t)$ describes the dimensionless speed of the phase boundary for melting. The dimensionless energy flow density of the laser beam at the area of absorption $Q_A = Q_A(z, t)$ is described by the Fresnel absorption $A = A(\mu)$ as a function of its polarisation and the cosine of the angle of incidence $\mu = \mu(z, t)$. The spatial distribution of the laser beam is indicated as f; the scaling value for

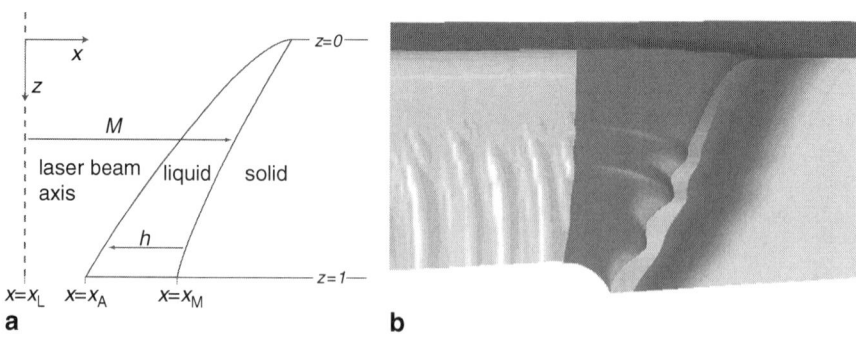

Fig. 6.68 Schematic 2D representation of the laser cutting process (**a**) and numeric simulation of the reduced model (**b**)

maximum intensity γ, the melt enthalpy h_m and the Peclet number Pe are explained in more detail in (Schulz et al. 1997).

The smallness parameter $\varepsilon = d_{m/d} \leq 1$ is the ratio of the typical melt film thickness d_m to the sheet thickness d. A detailed description of the reduced model can be found in (Vossen and Schüttler 2010). In the course of the studies it was demonstrated that the recrystallisation striations relevant for cut quality arise as the result of a self-organised melt dynamics.

Using the "QuCut" software which was developed (Vossen and Schüttler 2010), it is possible to calculate the virtual cut surface and quality criteria values for any parameter set, with strong correlation to the experimental results. Using the reduced model the calculation resources required for this could be reduced to the point where large-scale parameter variations are possible with high resolution.

These data form the basis for the meta-model, as they provide the sensitivities to parameter dependencies necessary for self-optimisation. The ability to calculate the relevant criteria values for the quality of a cut edge during laser beam cutting at this level of efficiency, with the help of simulations, has never been achieved before and is a significant result from the work carried out in this sub-project.

For one of the quality criteria, the virtually derived depth of roughness of the cut face R_z, an extract from the parameter space in Fig. 6.69 is shown as an example. Here, it becomes clear how the cut quality depends on the input parameters, and that significant local and global minima exist. The data produced through simulations are suitable for determining local and global optimum values, with the help of iterative optimisation algorithms, and for providing the corresponding internal objectives, whereas normally, in practice, both more parameters and also more elements (indicators) of PQ need to be taken into account than are shown in the example. For example, a global optimum value in relation to a single criterion may not be achieved, since infringement of another criterion or limits in the available ranges of individual setting parameters prohibit it.

Cognition The determination of the operating point of a manufacturing system for laser beam cutting is a significant step along the road to implementing

6 Self-optimising Production Systems

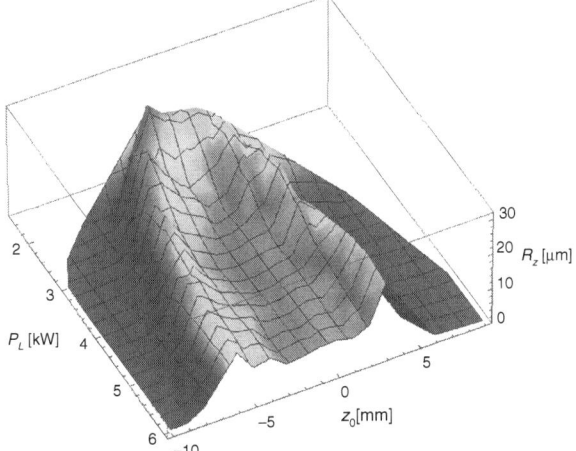

Fig. 6.69 Calculated quality indicators (depth of roughness R_z) as a function of two selected parameters (here: laser power P_L and focal position z_0)

self-optimisation. It is the pre-requisite for deriving decisions that can influence system behaviour. The focal position during laser cutting has a major influence on product quality but, particularly in systems which use CO_2 beam source, can only be measured with great effort. Special instruments to measure the beam enable determination of the energy distribution by using a scanning procedure, also in the focal position at energy densities of more than $5\,MW/cm^2$. But these measurements can only be made stationary within the machine, and generating measurement data during the process is not possible using this approach. Alternatively, the focal position can be determined experimentally using a so-called comb-cut. A cut is made in a thin sheet of metal at various focal positions, is taken out of the machine and measured with a feeler gauge. By determining the smallest width of all cutting kerfs the focal position can be determined. This value will be wrong if the optical components of the machine heat up differently in operation mode, or if the measuring equipment or manual evaluation were incorrect (Fig. 6.70).

If the manufacturing system is to become self-optimising, then it has to recognise the focal position itself and adjust it to the externally defined objectives and the requirements of the processing task, without external intervention. To achieve this, a cognitive unit is needed, which consists of a sensor to measure the width of the cut, a meta-model to translate the cutting kerf width into the focal position, and a controller to manage the focal position.

To determine the width of the cut when laser beam cutting, the heat conduction equation is discretised in one-dimensional space. This is done using two dotted stripes (see Fig. 6.71), in which the 1-dimensional heat conduction task is solved, in order to determine two data points for the melting isotherm. Because the melt film thickness next to the surface of the metal sheet is very small and the cut face in most relevant cases in practice can be assumed to be almost vertical, the melting isotherm on the sheet surface is a good measure for the expected cut width.

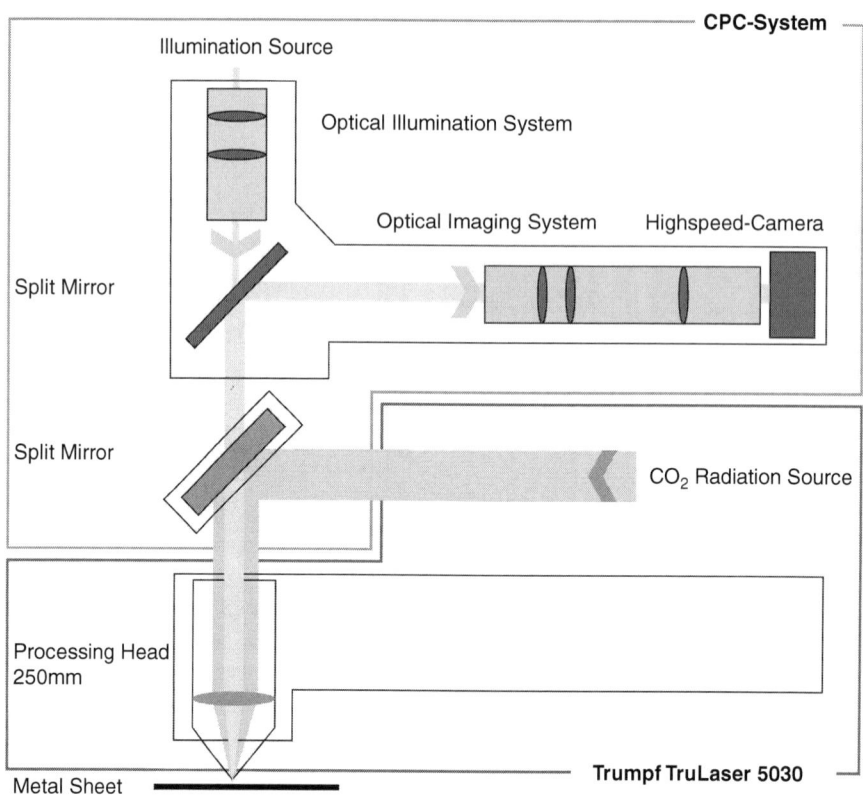

Fig. 6.70 System diagram of sensor placement in the TRULASER 5030

The point (x_0,y_0) or (x_1,y_1) within a dotted stripe where the melt temperature T_m is present as surface temperature T_S will be limited to the space between two data points and is approximated using linear interpolation. The surface temperature T_S and the depth of thermal penetration δ_{heat} are determined locally at the data point as a function of the local distribution of the intensity by using the following differential equations:

$$\frac{\partial T_S}{\partial t} = \frac{\kappa_T}{(1-b_1)\delta_{heat}} \left(\frac{Q_a}{\lambda_W} - b_1 \frac{T_S - T_0}{\delta_{heat}} \right) \qquad (6.9)$$

$$\frac{\partial \delta_{heat}}{\partial t} = \frac{1}{T_S - T_0} \left(\frac{Q_a}{\lambda_W} - \frac{\partial T_S}{\partial t} \delta_{heat} \right) \qquad (6.10)$$

with $b_1 = 3/5$, T_0 as ambient temperature, κ_T as temperature conductivity, λ_W as heat conductivity and $Q_a(r, t)$ as the heat flux due to absorption of the laser beam on the surface of the metal sheet.

Using the two points (x_0,y_0) or (x_1,y_1), which can be determined by resolving the above differential equations, and the x-axis as the axis of symmetry, a circle can be

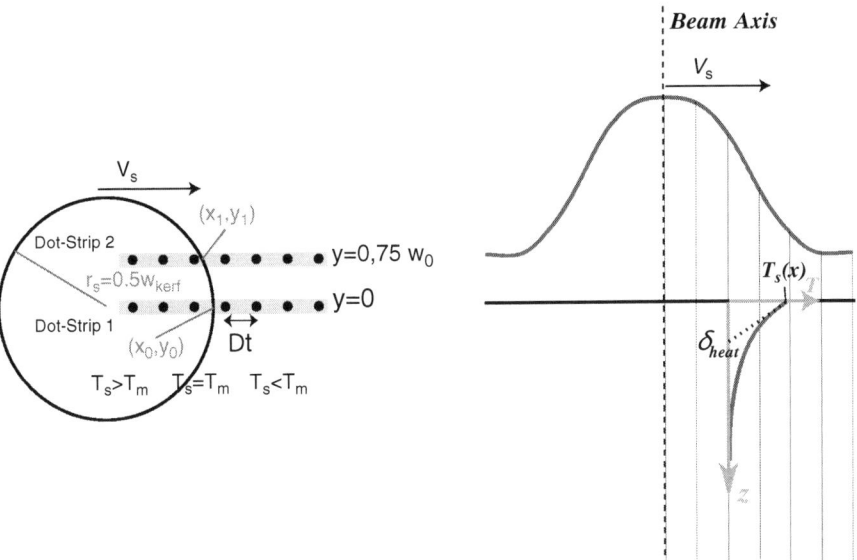

Fig. 6.71 Principle of the 1D pre-heating model to determine the cut width during laser cutting

defined, which is then used in the calculation of the cutting kerf width. This results in:

$$w_{Kerf} = 2 \cdot r_S = 2 \cdot \frac{(x_0 - x_1)^2 + 1.5 \cdot w_0}{2(x_0 - x_1)} \qquad (6.11)$$

With this calculation procedure to determine the cutting kerf width, and by varying the parameters, base data can be generated from which a meta-model can then be created to predict the cutting kerf width from the focal position.

In Fig. 6.70 the sensor system is shown which was developed for the coaxial process monitoring which enables the problem to be solved. The processing beam is combined via beam-splitters to the process monitoring, what allows monitoring of the working area during the process. Information about the process is imaged using optical elements on a 2-dimensional detector and this gives spatially resolved measurands. Signal processing allows extraction of process variables and makes these available in machine-readable form for optimisation.

The heating of the beam splitter has a major influence on the information content of the sensor data (Fig. 6.72b). Figure 6.72a shows an image of the cut without processing, so without the processing beam. In this case the beam splitter is at its nominal temperature and achieves the image quality calculated previously using optical simulation. In Fig. 6.72c the influence of heating can be seen with the resulting reduction in the information content of the sensor signals. The aberrations relating to power input can only partially be compensated in the algorithms which process the signals and which extract the width of the cut from the sensor signals. If this point had not already been covered in the design phase, this would have been a case for interating in the methodology entering the creation of a meta-model at Point C4 as shown in Fig. 6.55. In the case under consideration, a module was

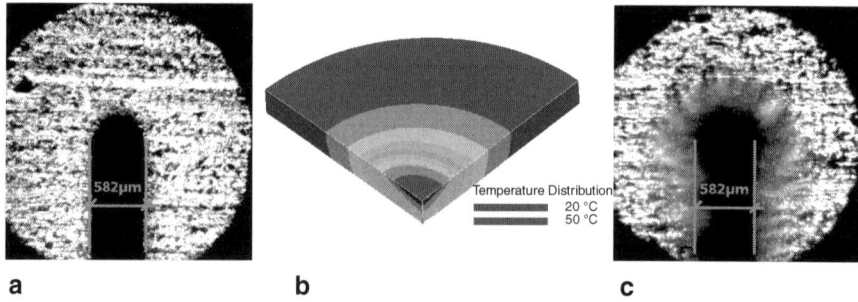

Fig. 6.72 Temperature behaviour of sensors in laser cutting. **a** Image quality at homogeneous temperature beam splitter **b** temperature distribution of beam splitter **c** Image quality at inhomogeneously heated beam splitter

developed to optimise the image quality, which receives data about the heating of the beam splitter from FEM data and makes this data available to the ZEMAX optical simulation. This allows imaging systems to be optimised based on power, energy distribution and cooling. The knowledge gained from these simulations, on the one hand feeds into the optimisation of the process monitoring, but on the other hand also allows the evaluation of signals collected from the process, and so the development of new algorithms. This module not only makes a contribution to the following steps to expand the cognition of laser cutting machines, but also makes a universal contribution to the simulation of thermally affected optical components.

In relation to the focal position, the signals for process monitoring are evaluated using algorithms which extract the characteristics of cut width. The image of the processing area shows the cutting kerf with its head and both side edges. The exact cut width is extracted using multi-stage image processing. The relationship between cut width and focal position is described by a meta-model. The input parameters for the model are the caustic, the power and power distribution of the laser beam, and the feed rate. Output parameter is the width of cut. By comparing the cut width calculated by the model to the measured value a control deviation is found. This is controlled by the ISA-system for the focal position.

The set-up assistant demonstrates the function of focal position measurement during processing, and therefore replaces the measurement by means of a comb-cut. This considerably reduces the time needed for rigging of the manufacturing system, and allows an instant, repeatable measurement, with the prospective possibility of closed loop control of the focus position during processing. The transfer of this parameter into self-optimisation is an example of gradual integration of self-optimising components in manufacturing systems.

6.4.4.8 Results Weaving Process (ITA)

The values used to set-up a loom are normally gained from the weaver's experience. In addition, tests are needed to check whether the required product quality is being

6 Self-optimising Production Systems

achieved. The aim of the study of the weaving processes is therefore self-optimisation in the loom set-up for a given product quality. With the help of a self-optimising set-up assistant, the test effort and its associated employee and materials costs are reduced. Important quality variables for fabrics include its mass per unit area and its tensile strength. The quality variables are depending on the characteristics of the used material, as well as the weave construction and the manufacturing parameters of the weaving process. An important factor is warp tension. It influences the crimp of the textile and hence the mass per unit area and the tensile strength. Basically the following context applies for the mass per unit area:

$$m_G = m_K + m_S \qquad (6.12)$$

$$m_K = Fe_K \cdot Fd_K \cdot \left(1 + \frac{E_K\%}{100}\right) \qquad (6.13)$$

$$m_S = Fe_S \cdot Fd_S \cdot \left(1 + \frac{E_S\%}{100}\right) \qquad (6.14)$$

(with a mass per unit area of m_G, mass of warp yarn m_K, mass of weft yarn m_S, warp density Fd_K, warp yarn count Fe_K, crimp in of the warp yarn E_K, weft density Fd_S, weft yarn count Fe_S and crimp of the weft yarn E_S). For the crimp, the relationship of

$$E\% = \frac{(l-r)}{r} \cdot 100. \qquad (6.15)$$

is applied where l is the length of the straight yarn and r is the length of the woven yarn (Latzke and Hesse 1974). The length of the woven yarn depends on the warp tension (Akgün et al. 2010).

For further research the minimum, maximum and mean traction of a warp during the weaving process were defined as process variables. The weft density, the loom speed, the warp tension of the entire warp system and the positioning of the warp stop motion (angle, vertical and horizontal positioning, Fig. 6.73) were selected as setting parameters.

The setting parameters have a significant influence on the time course of the warp yarn tension. The position of the warp stop motion changes the so-called rear-shed geometry of the loom. It has a determining influence on the course of the warp yarn traction during the weaving process. In comparison with other loom elements such as the back rest or the shafts, the effort needed to adjust the warp stop motion requires only a small manual effort.

To create the data base for modelling, the following parameters were varied using a factorial test plan:

- Loom speed [turns/min]
- Weft density [threads/cm]
- X-position warp stop motion [cm]
- Y-position warp stop motion [cm]
- Angle of warp stop motion [degrees]
- Warp tension of the entire warp system [kN]

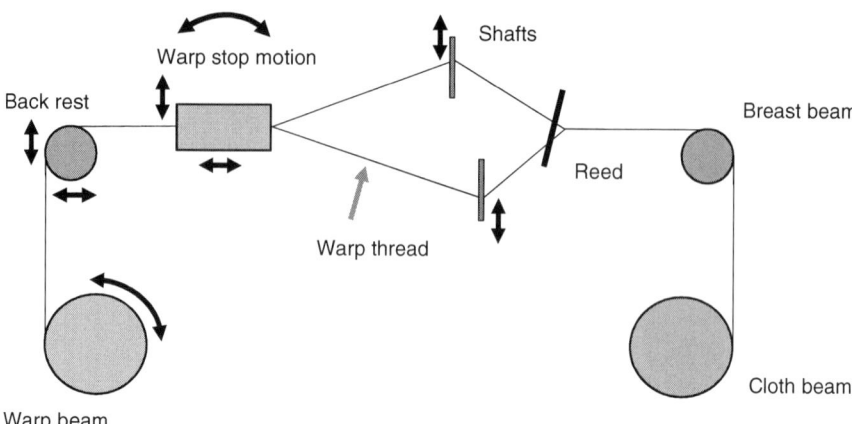

Fig. 6.73 Loom principle

The warp tension of the entire warp system, the loom's speed and the weft density are gained from the loom via a network interface using a specially created software. To capture the maximum, minimum and mean warp tension a sensor developed at ITA is used, which is shown in Fig. 6.74 (Gloy et al. 2009).

The sensor enables the acquisition of dynamic yarn tension. The data acquisition is implemented by a measurement amplifier, an analogue/digital converter and software to record the data. The values for the strength of the textile were obtained using tensile tests according to DIN EN ISO 13934-1. The mass per unit area is determined as defined by DIN EN 12127. Measuring the mass per unit area on the loom is currently not available industrially. In theory, there is an option to use radiometry sensors

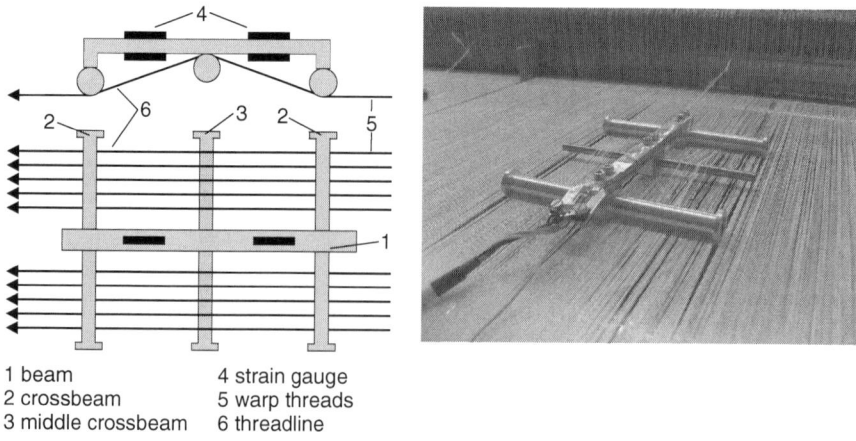

1 beam
2 crossbeam
3 middle crossbeam
4 strain gauge
5 warp threads
6 threadline

Fig. 6.74 Sensor to measure dynamic warp tension

6 Self-optimising Production Systems

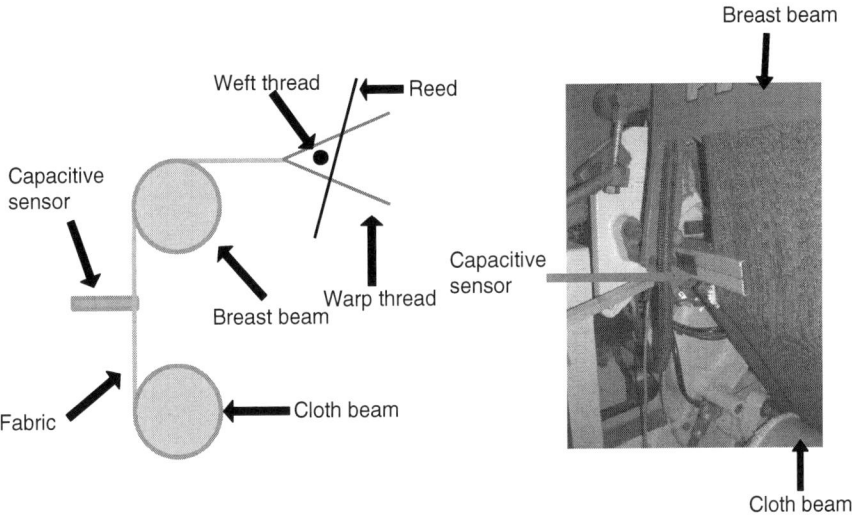

Fig. 6.75 Capacitive sensor on a loom

(Beta-back reflection method, ultrasound measurement and infrared measurement) to measure the weight of the textile on-line. But the investment required for this kind of equipment is higher than for the loom itself. Therefore, initially a capacitive sensor was inserted between the breast beam and the cloth beam on the loom (Fig. 6.75). The fabric is guided through the two capacitor plates. Any change in weight of the textile leads to a change in capacitance and so to a measurable change in the current output by the sensor.

Using the data base generated in this way, models are created to find the correlation between the setting parameters and the process variable of yarn tension. Here, it turns out that the positioning of the warp stop motion has the most impact on the course of warp tension, see Table 6.6. In contrast, the angle of the warp stop motion has only a minor impact on yarn tension and can be ignored for any further observations.

Analysing the correlation matrix of the process measurands in regard to the product quality, the mass per unit area, the minimum yarn traction has the greatest impact. Modelling of the correlation process variable to product quality regarding tensile strength is shown in Fig. 6.76. The bigger the box in an appropriate field is the

Table 6.6 Correlation matrix of settting parameters for the process variable of warp tension when weaving (from 0 minimum to 1 maximum influence)

Setting parameter	Minimum warp traction	Maximum warp traction	Mean warp traction
Revolutions	0.187	0.06	0.02
Weft density	0.195	0.129	0.133
X-position warp stop motion	0.06	0.05	0.01
Y-position warp stop motion	0.39	0.86	0.73
Angle warp stop motion	0.35	0.13	0.32

Fig. 6.76 Correlation matrix of the minimum, maximum and mean warp tension in relation to the elongation at break, maximum elongation, maximum tension and tension at break of a fabric

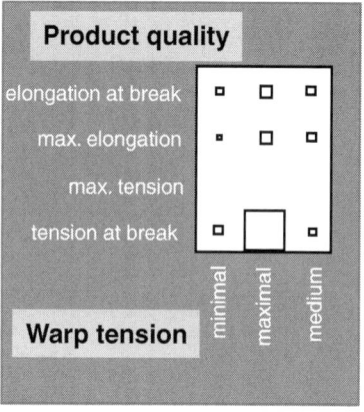

greater the influence of the parameter. It is notable that the maximum warp traction has the greatest impact on the strength of the textile.

Furthermore, a model was also investigated for the correlation between the setting parameters and product quality, using the "AX-Workshop" software developed as part of the project. The models used and their coefficient of determination are shown in Table 6.7. To improve the model quality, the base data needs to be expanded. Basically, it can be seen that the models are suitable for calculating product quality.

In ongoing work, inverse modelling is being researched, to enable the determination of the necessary setting parameters for a requested product quality. In collaboration with industry partner Picanol NV, Ieper, Belgium, a self-optimising set-up assistant is being developed as a next step.

6.4.5 Industrial Relevance

Increasing the competitiveness in an international scope using innovative and sustainable system solutions is a constant demand of manufacturing companies in high-wage countries. A distinct variable in meeting this general requirement is the manufacturing process itself. For companies, this creates the wanted result by the use of given social and technical production factors. The state of the art in technology in 2006 shows that manufacturing processes are not being run at their physical limits. To ensure product quality, mainly conservative manufacturing strategies are being selected, in order to avoid process instabilities from the outset. Depending on the level of complexity of the manufacturing process, and the dimensions of the determining

Table 6.7 Models for translation of SP to PQ and their absolute errors

Model for transference of SP to PQ	Coefficient of determination R^2
Regression model	0.89
Feed Forward neural networks	0.81
Exact RBF networks	0.2

6 Self-optimising Production Systems

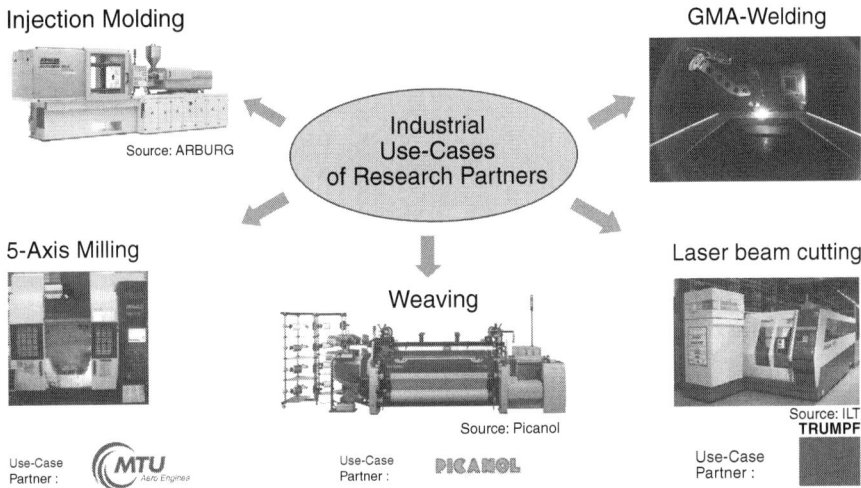

Fig. 6.77 Industrial use-cases with research partners

factor which are relevant for quality, this is achieved only with considerable set-up or inspection costs as well as re-work when deviations from predefined product quality occur. In addition, during the set-up phase after product changeovers, intensive test runs are needed in order to satisfy the required quality levels. This means, in particular for small batch sizes, a disproportionate amount of downtime in overall production. As a result, the described approach leads to high costs, in terms of both time and money.

The manufacturing industry seeks to optimise manufacturing processes in terms of processing speed and a reduction in downtime, always assuming product quality can be maintained. This goal is reflected in the basic research question, with the focus on rigging, monitoring and control of manufacturing processes. As a basis for ongoing validation of progress and growth in knowledge in close alignment with economic questions, each manufacturing process has an industrial use-case assigned to it. The industrial applications and the partner companies are shown in Fig. 6.77.

The joint work of the five research partners on the solution of the tasks ensures that solutions are generic in structure. Both the concept of Model-Based Self-Optimisation with its ISA-systems, and the process-independent methodology to generate meta-models, are tested by all five processes. This means that their application to further processes is possible, which guarantees the sustainability of the research outcomes in the conceptual domain. Sustainability in relation to gaining process-specific findings is demonstrated by the research partners' results.

6.4.5.1 Milling (WZL-TF)

The results achieved, if successfully transferred to industrial practice, will allow an improved process understanding and in a first instance enable a more efficient design and improved controllability of this manufacturing process. Integrated into the process design phase, the developed position-oriented monitoring system pinpoints unstable process behaviour in order to enable corrections in a targeted way. For industrial practice, this application would reduce the necessary number of preliminary test runs, resulting in a reduction of resource consumption and significant cost-savings for single and small lot production. Also, this optimization ensures the product's quality and therefore avoids, due to the acquired process information, instabilities by suitable parameter adjustments, In the simplest case the manufacturing process can be stopped at its current position, avoiding further damage to the component and therefore scrap costs. A reduction of both the manufacturing's scrap rate and required post-processing steps would significantly contribute to a company's competitiveness in the area of 5-axis milling.

Currently, the results achieved for machining thin, overhanging cantilevers are transferred to a demonstrator, optimizing the finishing operation of a blisk's thin-walled blades. A blisk (blade integrated disk), known as a safety-critical component in the aerospace industry, is subject to the highest quality requirements in order to avoid failures in the later use. At present, the expected product quality can only be achieved executing time consuming and expensive test runs. Integrating the results presented, a quality-oriented monitoring of the finish milling process will be established which reduces the number of necessary process design steps by far. For this purpose, the results of the process monitoring are combined with those of the virtual process design.

6.4.5.2 Model Evaluation (WZL-MQ)

The methods implemented in the AX-Workshop software for process-independent process modelling, visualisation and evaluation, enable the efficient creation of surrogate models for self-optimisation. An important part of this are the evaluation algorithms, by which the quality of the employed models can be evaluated in relation to various criteria, such as global and local representation quality, or robustness against disturbances, before they are put into use. In this way, evaluation provides the necessary criteria for the implementation of surrogate models for Model-Based Self-Optimisation in industrial practice. The AX-Workshop software and the common methodology for creating meta-models provide an important contribution to making model-generation systematic. The shift from an intuitive model-generation process, only open to implementation by experts, to a standardised procedure with software support, reduces modelling effort substantially, and allows a quality focussed generation of models.

6.4.5.3 Plastics Injection Moulding (IKV)

Injection moulding of thermoplastics is one of the most important processes in plastics processing. Thanks to the high level of flexibility displayed by mouldings' geometries and their materials, products can range from micro-components to car wings or complete telephone booths (Johannaber and Michaeli 2004; Michaeli 2006). In the automobile industry injection moulding is widespread, especially for technical parts. The injection moulding process is at the start of the process chain for creating complex assemblies. This means that fluctuations in mouldings' quality such as distortion propagate through the process chain. Therefore, it is important that the injection moulding process reacts robustly, even in the case of disturbances, as occur from the use of recycled materials in the mix. The works within the cluster of excellence focuses mainly on self-optimisation to stabilise the injection moulding process. In close cooperation with the machine manufacturer ARBURG, systems will be further developed in future to stabilise moulded part quality.

6.4.5.4 Welding (ISF)

The technological results achieved as part of the research work carried out in the area of automated gas metal arc welding, if adequately transferred to industrial manufacturing, allow to expect clear increases in efficiency in technical welding applications in the areas of cost-efficiency, quality assurance and flexibility.

By preparing and applying a global quality model, the rigging time required to prepare for manufacturing can be decisively reduced. The necessary setting parameters for the required product quality, in this case the weld seam geometry, are found almost instantaneously with the help of optimisation algorithms in the defined quality model, which means time-consuming and cost-intensive test runs are no longer needed. Depending on the application, the necessary time for setting up and running-in processes can be reduced by up to 50%. This is achieved by enabling expert staff to decide more rapidly on the adjustment of values for setting parameters, as they are supported by the technical welding expertise implicitly contained in the quality model.

The results presented in the area of on-line monitoring, particularly through reliable in situ prediction of the weld seam geometry, based on the welding current, welding voltage and sensor information such as gap width, workpiece to contact-tip distance and welding torch orientation, now for the first time enable a closed control loop to actively influence the weld seam geometry resulting from the weld pool, and to adjust the process to suit the required geometry. Thanks to this prediction of the weld seam geometry, the optimum process parameters can be determined and set in the light of the process information listed above, without extensive empirical tests, so that consistent weld seam geometry can be ensured. This leads to a sustainable reduction in rejects, and to a reduction in the necessary repair work of up to 30%.

The self-optimising gas metal arc welding manufacturing system is notable for its extensive flexibility, only restricted by the limits of the process model, and offers the possibility of reacting very swiftly, and therefore with greatly reduced economic effort, to changes in manufacturing conditions.

6.4.5.5 Laser Beam Cutting (ILT/LLT/NLD)

The process of laser beam cutting is an established industrial process. In many areas, flat-bed cutting machines are used to prepare steel sheets which are subsequently converted into switchgear boxes, façade claddings, construction components in the areas of mechanical engineering or plant engineering, or which find their use as sub-components in complex technical devices.

For the research in the area of self-optimisation, the company TRUMPF provided a flat-bed cutting machine to the cluster of excellence, equipped with a CO_2 laser beam source. This machine is sold in large quantities on the international market, and can manufacture continuously if automated accordingly. Optimisation potential remains in the area where the cutting process is not yet run at its technical limits, so where self-optimisation is capable of increasing the speed of manufacturing. This relates to both questions of process stability, as well as the achievable quality of the processing result.

The set-up assistant demonstrated as part of the project delivers a significant reduction in the rigging time of the focal position and increases the reliability of the result. The progress in modelling the cutting process already today enables the integration of information from simulation based on variable material data, which provides a way forward to its application for special alloy materials. On a second front, the prediction of the stability of the cutting process taking key parameters into account is now possible. The cognition which has been achieved in the manufacturing system, provided by sensor technology and model-based signal processing, has the potential to be extended to further parameters in the manufacturing process. This leverage has the potential for a significant increase in the processing rate, bringing it closer to the physical limits of the process. By replacing manual evaluation of a comb-cut to determine the position of the focal position with the use of focal position cognition, an important step has been taken in the direction of self-optimisation of a technical system. Parts of the solutions in this area are already finding their way into the development of a new generation of cutting machines.

It is expected that the application of self-optimisation to an extended set of parameters will enable a clear increase in the rate of manufacturing, and pave the way for expansion into other application areas, from the cutting of thicker materials to the rapid and precise processing of high-tensile steel for the construction of light-weight automobiles. Here, the research is making a contribution to current socio-political and economic success factors for high-wage countries.

6.4.5.6 Weaving (ITA)

The research in the area of weaving was carried out using looms from the loom manufacturer Picanol NV, Ypres, Belgium. Aim is the development of a set-up assistant for looms, which provides the necessary setting parameters to achieve a given quality of fabric. Picanol has provided a loom for the research work on which the sensors developed to monitor product quality on-loom are being evaluated. In addition, commands and interfaces for communication with the machine are being provided, which will allow direct transfer of results to an industrial application.

6.4.6 Future Research Topics

The research and development in the area of self-optimisation of manufacturing systems is focused on providing a generic answer to the question of how manufacturing processes can be described by models, so that they are robust against disturbances, and can themselves adjust parameters, so that the product which is produced meets its specifications. The strategy developed to implement self-optimisation, and the methods for generating meta-models, taken together with the description of processes in reduced machine-readable models, are the two main pillars of "Model-Based Self-Optimisation for manufacturing systems". The creation of set-up assistants shows the innate strength of the proposed solutions. In addition, the works in cooperation with industry partners show the possibilities for gradual implementation of self-optimisation through selective integration of ISA-systems in existing manufacturing systems. These results contribute to the reduction of rigging effort in product quality focused manufacturing.

In future studies it is necessary to further develop the concepts and methods, so that they can serve as a general basis for industrial engineering services. This is only feasible today in some selected cases. The validity of the methods for developing surrogate models has already been demonstrated, using the manufacturing processes which were studied. But to date it has not been possible to make any claims in relation to the stability arising from the interaction between meta-model, process and control system. Therefore, the expansion of self-optimisation from process monitoring and process control of isolated scalar parameters to dynamic multi-dimensional parameter spaces with the stability issues these entail, is a substantial focus for research.

The methods for the generation of models currently are targeting the manufacturing process. To successfully relocate within the parameter space, Model-Based Self-Optimisation has to take the machine behaviour into consideration. Therefore, methods need to be developed to combine algorithms for model-based optimisation with the machine-specific dynamic behaviour and performance of the machine. As it can already be predicted that, in most cases, this type of wide-ranging information is unlikely to be available in entirety and generically, methods also need to be developed to determine and describe the machine's behaviour under all relevant processing conditions.

The determination of the operating point within the manufacturing process has already been demonstrated for a reduced parameter space. Manufacturing processes which need to evaluate multiple process variables in order to reliably determine their operating point at times need to process a vast amount of data instantly. So that the

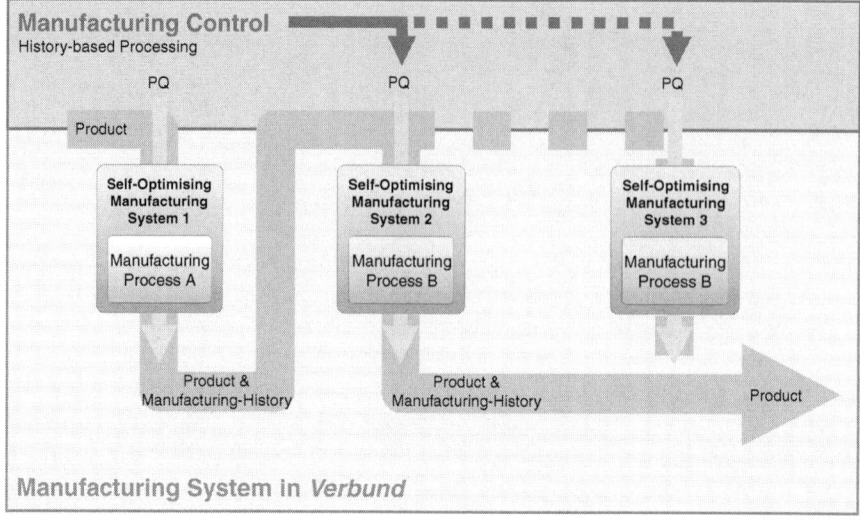

Fig. 6.78 Self-optimising manufacturing systems in *verbund*

analysis of this data can lead to an increase in cognition, new concepts for multi-sensor concentration are needed. Independent of its information content, acquired data needs to be used with respect to its context both for model-based optimisation systems in a superordinate scope, and also for subordinate systems within control loops.

Under the title "self-optimising manufacturing systems in *verbund*[2]" the aim of complete transparency of the manufacturing process will be pursued during the second phase of the cluster of excellence. This quality-focused, product-to-product transparency enables the realisation of quality-guided manufacturing Verbünde, where the final product defines the requirements which all manufacturing systems in the federation must comply with. Each manufacturing system also uses the manufacturing history of the product as an input variable, processes the product according to the externally defined objectives, and forwards the finished product accompanied by an extended history. Subsequent processes use this information to adjust their processing, so that they in turn achieve their externally defined objectives (Fig. 6.78).

"Self-optimising manufacturing systems in *verbund*" will demonstrate their ability to adapt to changed parameters on the input, and to changed requirements in the processing result. The gain in the area of stability, and the reduction in rigging, contribute to economic factors for the production of individual items, and address the scale-scope axis of the polylemma of production. The increase in flexibility and the

---

[2] Verbund in this context describes the union of cooperating manufacturing systems which themselves react autonomously to changing requirements such that all entities within the union together achieve the global objective of the union.

interaction with the planning level address the plan-value axis. By turning historical information from manufacturing steps into valuable input for the planning level, the processing parameters can be adjusted at an early stage for the subsequent manufacturing steps. This modularisation of the manufacturing site can also be applied using the model-based approach in virtual environments, as the quality-relevant behaviour exists in machine-readable form.

6.5 Integrative Product and Process Design for Self-optimising Assembly

Christian Brecher, Peter Loosen, Rainer Müller, Nicolas Pyschny, Alberto Pavim, Matthias Vette, Adam Malik, Kamil Fayzullin and Sebastian Haag

6.5.1 Abstract

In the scope of the Cluster of Excellence "Integrative Production Technology for High-Wage Countries" and more specifically within the sub-projects "Self-optimising Flexible Assembly Systems" and "Model-based Assembly Control" the application of self-optimisation principles in the automated assembly of complex, multi-variant products are under investigation. A miniaturised laser system developed in-house—the "MicroSlab" (see Sect. 7.3)—is used as demonstration scenario.

The motivation behind this research results in particular from the rapid development of optical technologies, which has caused a steady increase in the demand for products with optomechanical features. The technological development of such products is driven by a clear trend towards miniaturisation, which defines the highest demands on the precision of production and assembly technologies. Today, precision assembly is characterised by a high degree of manual or semi-automatic processes, which encourages production migration to low-wage countries. The automation of these processes is an essential prerequisite for manufacturing competitive products in high-wage countries. Due to high engineering efforts and high installation costs, a high degree of automation is currently only economically viable for the mass production of standardised products. A constant shortening of product life cycles and the demand for a bigger diversity of variants represent major impediments to extensive automation levels. Therefore, investigating the possibilities for increasing the flexibility of automation in optomechanical assembly is an essential task. Flexible automation enables the manufacturing of customised products at low production costs and contributes to the resolution of the "scale-scope" dilemma. Here, automation is studied in terms of job retention: a trained operator should be optimally supported and complemented, not replaced.

Turning automation processes more flexible induces increased organisational and planning effort. A detailed and hierarchical planning structure in form of a fixed assembly sequence for a wide range of products could only be implemented with very high effort and considering assembly components with high tolerance (and costs). The approach used in the sub-project for resolving the dilemma between planning and value orientation in assembly planning is based on the self-optimisation of assembly processes. In this respect, some of the relevant decision making is transferred to the assembly system during the execution of the assembly process. This implies reducing the planning efforts, because not all decision trees have to be fully defined in advance, but rather they can be created during the assembly process. According to the definition of self-optimisation, the assembly system must be able to derive new goals from observations of the assembly process and adapt itself to the dynamical boundary conditions. The required level of autonomy for such purposes can be achieved by means of implementing modular substructures (for example "software agents") that can perceive their environment and react to changes based on rules or models. Hence, system operators are made responsible only for coordination tasks, such as planning the product features or evaluating time, costs or quality of the assembly end product.

The following section introduces an insight into the state of the art in the field of micro and precision assembly of optical systems, from which the research purpose is derived.

The results section firstly develops the topic of self-optimising assembly systems on a conceptual basis. Based on a basic definition of self-optimisation in technical systems, the prerequisites, or building blocks, for self-optimising assembly systems are then derived and explained (see Sect. 6.5.4.1). These include:

- the flexibility and mutability of automated systems,
- the autonomy of assembly cells,
- automated planning in flexible assembly systems and
- cognition and learning ability in automation technology.

In the sequence, some specific solution approaches for the automated assembly of the MicroSlab are explored and the solutions developed for model-based assembly control are presented (see Sect. 6.5.4.2).

In addition to addressing the topic of self-optimising assembly systems, methods for the integrative design of products, processes and equipment in assembly were developed and these are presented in Sect. 6.5.4.3. Finally, future research in the individual topics is outlined.

6.5.2 State of the Art

Technological developments in many areas of the consumer and capital goods industry are driven by dynamic changes in customer requirements and a clear trend towards miniaturisation combined with functional integration (Lotter 2006). In various areas of consumer electronics, automobile manufacture and industrial production as well as in medico-technical applications, miniaturised approaches help to find better solutions to problems or to even make it possible to solve certain problems for the very first time (Pfirrmann and Astor 2006).

Based on rapid developments in microelectronics and semiconductor technology, a multitude of different manufacturing techniques from microtechnology have been established such as coating, lithographic and etching techniques, which, through parallel production (*batch processing*), are suitable for large lot sizes and thus achieve low unit costs.

Through monolithic integration, i.e. the integration of sensors and actuators using the structuring techniques mentioned above, various and sometimes highly complex technical solutions have been developed for actuation systems, sensors and analysis systems since the beginning of the 1980s under the general term *microsystems technology (MST)* (Völklein and Zetterer 2006). In a microsystem, in addition to the electrical and mechanical components, there are often optical functional elements too (Hankes 1999). Although there is no sharp definition of microsystems technology, there is generally a consensus that it describes the discipline of integrating as many different functions as possible in a small space, whereby at least one component of the microsystem is produced micromechanically.

In order to manufacture microsystems, driven by applications with micromechanical, microoptical and microfluidic components, attempts are being made to open up the third spatial dimension in production technology. Despite some progress, 3D capability in production processes remains extremely limited, becomes very costly as the complexity increases and is only viable where very large unit numbers are concerned. Against this background, the type of system integration (monolithic or hybrid) is increasingly determined by the economic boundary conditions and is following the clear trend towards hybrid integration from various production environments.

The area of production technology concerned with the hybrid integration of miniaturised products is usually referred to as microassembly. According to DIN 32564-2, the term "microassembly" is defined as follows: "microassembly is the assembly of microtechnical components, the fitting of microcomponents on mounting surfaces or their installation in housings, including electrical contacting and the creation of other connections (e.g. media)" (DIN 32564 2003). A more precise classification of microassembly compared with the related discipline of precision assembly as well as conventional assembly is shown in Fig. 6.79. According to this classification, microassembly is characterised primarily by small component dimensions, whereas precision assembly is characterised mainly by the required assembly precision of less than 25 μm.

6.5.2.1 Microassembly

For Germany as a production location, microsystems technology offers long-term employment prospects. At present, there are 680,000 jobs in the direct technological environment, plus the jobs in the application sectors. Economically, the market for

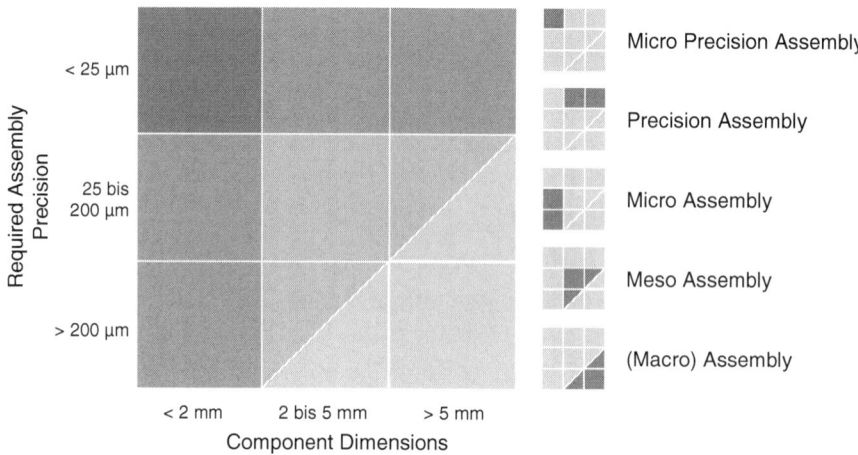

Fig. 6.79 Systemisation of assembly techniques. (Petersen 2003)

microsystems is already an above-average growth market with estimated average growth rates of 16%. These estimates do not include the value-adding portion of microsystems technology which results from integration into innovative products and of which experts expect a "leverage" factor of seven (Salomon 2006).

Due to the economies of scale through mass production, the unit costs of standard parts and components for microsystems are often very low, whereas the cost caused by microassembly is extremely high—up to 80% of the overall manufacturing costs of a microsystem. This is due to the predominantly low level of automation combined with high wage costs in Germany (Hesselbach et al. 2002). This could be interpreted as the reason why, in contrast to monolithic microsystems, a distinct increase in the unit numbers is still proving elusive in the field of series production of hybrid microsystems (Korb 2005).

In addition to high wage costs, there are other reasons for conducting intensive research into the flexible automation of hybrid microassembly. The small component dimensions cause difficulties with manual assembly because the highly precise positioning and focussing on small parts, e.g. under a microscope, are ergonomically suboptimal for human employees. If miniaturisation continues to progress, it will become more difficult for humans to carry out the required assembly operations safely and efficiently while maintaining the same quality.

6.5.2.2 Precision Assembly of Optical Systems

Optical technologies have now developed into a cross-sectional and key technology. This has resulted in a constantly growing demand for products with optical and, in particular, microoptical components in many technical fields (medical technology, information and communication technology, measuring technology, production

6 Self-optimising Production Systems

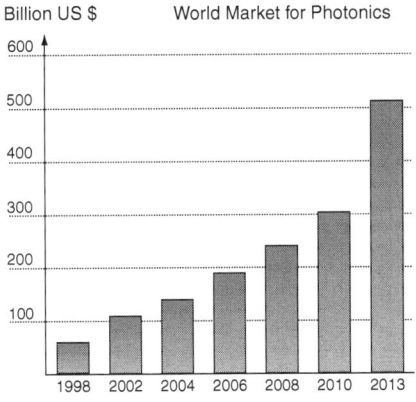
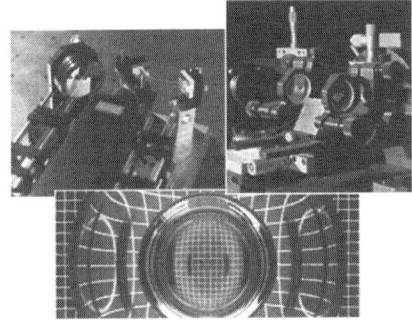

Fig. 6.80 Market forecast for optical technologies. (From von Witzleben 2006)

technology, etc.) and, therefore, these technologies are steadily gaining significance for the overall economic development of industrial countries (Fraunhofer IOF 2006) (Fig. 6.80).

Similar to the development of microsystems technology, through continuous miniaturisation and functional integration, optical systems are also opening up more and more new areas of application for non-optical components too. Essentially, these applications involve generating, manipulating or detecting light using optoelectronic components as well as components from electronics, actuator engineering and silicon micromachining (Beckert 2005).

From the perspective of the production technology polylemma, optical assemblies are characterised by evermore complex, multi-variant systems. At the same time, only small to medium unit numbers are usually required. Macrooptical systems are, for the most part, assembled and aligned in manual processes, whereby efficient execution requires many years of experience (Bauer 2008).

The classic system design is based on the so-called optical bench onto which the optical components are mounted in mechanical, manually adjustable manner. Optics, electronics and mechanics are generally treated as separate entities. These design principles cannot be transferred to miniaturised optical systems because they do not satisfy the requirements in terms of packing density, functional integration, robustness and unit costs (Beckert 2005).

As a result, there is a strong link between product design and the production process whereby the optical design of the assembly determines, in particular, the requirements on the assembly process. Miniaturisation restricts manual assembly to the extent that new solutions need to be developed especially in terms of flexible automation of the assembly of microoptical modules. In this respect, microoptical systems need integrative approaches that take an integrative view of product design, component manufacture and assembly.

To build complex hybrid microsystems with optical elements, in addition to an automation-friendly design, new and complementary technologies are required for joining and above all handling optical components (Bauer 2008). From a production engineering point of view, the joining methods used are decisive when it comes to different material pairings and, in particular, for opening up the third spatial dimension in assembly and alignment. Another key challenge is the precision required, which, depending on the application and component, is in the submicrometre to low two-digit micrometre range (or between low microrad and millirad) (Beckert 2005).

Hence, for further market penetration and to secure or recover jobs in high-wage countries, manual processes need to be replaced by flexible automation solutions for small and medium unit numbers (Photonics 21 2010). In this respect, based on their characteristics, the optical systems in question can be considered representative of a wide range of multi-variant, complex, miniaturised products.

6.5.2.3 Automation Solutions from Science and Industry

The relevant scope of automation technology for micro and precision assembly can be divided into the topics of handling and precision joining, which are explained in the following sections.

Handling The accuracy required for assembling optical systems is in many cases in the single-digit micrometre range and is often even in the submicrometre range. In the industrial world, huge efforts have been made to develop robots with similar positioning accuracy. Vertical articulated arm robots can achieve a positioning accuracy of approximately 50 μm (in some cases 20 μm) and SCARA robot kinematics, depending on frame size, can achieve down to 10 μm. The repeatability of Cartesian systems with linear axes, which are widely used in microassembly, is 1 μm for single axes with optical scales, but this is not necessarily the spatial positioning accuracy (Hesselbach et al. 2002).

Furthermore, manufacturers like PI and Micos produce high-precision single axes which are used in stacked systems for multi-axis alignment. There are also parallel kinematic concepts which allow multi-axis alignment with extremely high precision, but tend to only cover small working areas (e.g. Micos SpaceFab, PI HexAlign, PI M-810).

The high accuracy of precision robotics is associated with a high acquisition price. In many cases, for efficient assembly, large working areas are required to be able to feed the components and integrate the wide variety of sensors mentioned above. This increases the costs many times and sometimes is not even conceptually feasible.

Fully automated assembly systems are predominantly available for electronics production (pick-and-place, SMD, die bonding) or tailored to very specific applications (e.g. the Sysmelec SMS1000 for assembling microlenses and optical fibres). There are no flexible and cost-effective systems that allow integration of various modules from different manufacturers and therefore fast reconfiguration of peripheral equipment (Gramann 1999).

This fact reduces the availability of precise handling systems for laboratory automation, especially for young high-tech companies, which makes it impossible to exploit any potential for rationalisation in product development and series production.

In research, different approaches are used to achieve the required assembly precision (e.g. micromanipulators, microrobots). However, to date there is no consistent overall concept for modular systems. Prototype assembly systems developed in collaborative projects are always designed to facilitate assembly of a product.

The gripping of precision- and microcomponents has been investigated in various collaborative projects (e.g. German Collaborative Research Centre SFB 440 "Assembling hybrid microsystems") and various physical principles can be exploited to grip different components. For assembling microoptics, however, there are no solutions that address the specific requirements of these components: (1) to build a product, many different components need to be gripped and held, which means the gripping system needs to be flexible; (2) there is usually only minimal surface area available for gripping the component; (3) hybrid gripping, alignment and joining of components increase the demands on the robustness of the gripping principle and the process-safe holding of components in the aligned position.

Precision Joining of Microoptical Components To join microoptical components, a whole range of attempts have been made to make conventional joining methods from macroscopic production also suitable for microscopic applications. The feasibility of joining specific material pairings plays a central role in this respect. The German Collaborative Research Centres SFB440 and SFB356 focused in particular on the joining methods as micro-soldering, micro-electron beam welding and micro-bonding.

For optical components (non-metallic materials), soldering and bonding processes are most suitable. During assembly, it is crucial that these components can be aligned and fixed in the designated position. Using UV-cured adhesives, in principle, this kind of alignment is possible in the liquid adhesive provided a corresponding adhesive gap is created and maintained. In this regard and with respect to the state of the art, there has been no systematic investigations about the shrinkage behaviour of different adhesives and gap geometries, nor is it known for sure how alignment in the adhesive can be reliably and efficiently performed.

The company Coherent GmbH uses a patented soldering procedure (PermAlign) to assemble optics which requires relatively complex preparatory steps (Woods und Pflanz 2008). Various research projects are also striving to develop suitable soldering procedures.

Conclusion The state of the art is characterised by manual or semi-automatic solutions involving the use of highly specialised soldering or bonding technologies for individual components or assemblies. In terms of cost-efficient automation, the state of the art in micro and precision assembly should therefore be considered critical. While there are some fully automated systems already, most of them are designed for a specific purpose, i.e. product-bound, and are therefore inflexible and it would take considerable effort to convert them for use with other products. Automatic assembly units are associated with extremely high acquisition costs and the investment risk increases when sales figures are difficult to predict because this means it is not possible to guarantee that the amortisation point will be reached within the product life cycle.

Automated systems that guarantee improved cost-efficiency even with fluctuating unit numbers and a high variant-diversity are of significance, especially for small and medium-sized enterprises. Regarding the considerable growth potential of product technologies and applications, this is a promising starting point for new concepts and technologies which, based on the high demands optical assembly places on flexibility and precision, can also be transferred to other products and industries.

6.5.3 Motivation and Research Question

The Cluster of Excellence "Integrative Production Technology for High-wage Countries" addresses the challenges faced by manufacturing companies, especially those in high-wage countries. Given the complex competitive situation in the globalised market, many companies are struggling to devise a strategy for meeting the increasingly dynamic and complex demand. Unification of the strategies of differentiation and price leadership has not been achieved until today due to irreconcilable differences between economies of scope and economies of scale as well as between plan and value oriented production.

The Cluster of Excellence is developing contributions to production technology which, based on their integrativity, i.e. the integrative view and thus also the interdisciplinary solution of tasks, contribute to reducing both dichotomies mentioned above.

The key question in this chapter is how complex, multi-variant products can be efficiently manufactured in high-wage locations. The focus is set on assembly, which is increasingly becoming a key value-adding element and thus a core competence of the company. Assembly is also the last value-adding stage through which direct proximity to the market is influenced the most by a dynamic environment. In this respect, strategic positioning between the opposing corners of the production polylemma is possible. Figure 6.81 shows the corners on the relevant axes.

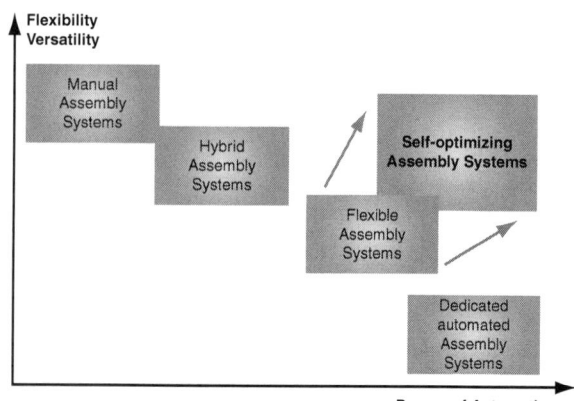

Fig. 6.81 Classification of assembly systems. According to (Weck 1998; Andreasen and Ahm 1988)

Automated assembly systems are efficient for high unit numbers and stand for economies of scale in mass production. Most processes have to be planned in advance and explicitly implemented because of the considerable adaptation work required, which means it is practically impossible to implement changes in the production process at short notice. In contrast, manual assembly is extremely flexible and copes better with variant diversity and fluctuations in demand. Value-oriented control mechanisms are increasingly being used in order to improve efficiency. The strategy of differentiation and the related economies of scope are, however, required because of the strong influence high wage costs have in international competition. As a compromise, hybrid or semi-automatic as well as flexible assembly systems are found between the extremes of the polylemma.

Concepts for self-optimisation should be implemented to create modern assembly systems that combine high productivity with high flexibility by reducing the effort required for planning, configuration and process ramp-up in flexibly automated systems. Self-optimisation offers solutions in which system adaptations are autonomously planned and implemented in order to respond to changed conditions.

The following section presents approaches for self-optimising assembly automation which help to reduce the production polylemma.

The technological focus of this chapter focuses on the assembly of complex optical systems which, as explained in the introductory sections, place the highest demands on the flexibility and accuracy of the assembly system whilst at the same time representing a wide variety of products that combine evermore and increasingly heterogeneous functional elements in the smallest of spaces in line with the trend towards miniaturisation. The basic research question is therefore:

How can high-tech products be assembled cost-efficiently in high-wage locations?

This chapter investigates how self-optimisation can help create flexibly automated systems that enable efficient production of complex, multi-variant products. The research aims to develop concepts for automated assembly of a miniaturised laser system (see also Sect. 7.3) and, on that basis, aims to demonstrate and study cognitive and self-optimising elements in industry-related applications.

6.5.4 Results

Based on a basic definition of the term self-optimisation, the following sections will first consider the technical and conceptual bases for self-optimisation in automated assembly systems. In light of the research question, the various aspects of flexibility and adaptability, autonomy as well as cognition and learning ability are brought into the context of self-optimising assembly systems.

Then Sect. 6.5.4.2 presents concrete solutions for the automated assembly of miniaturised laser systems which were developed in the project "Self-optimising Flexible Assembly Systems". One focus of the deliberations is autonomous planning in assembly cells (Project: "Model-based Assembly Control").

The concepts and methods developed to facilitate an integrative procedure for designing product, process and production equipment are presented in Sect. 6.5.4.3 using the example of a miniaturised laser system. Detailed information about the development of the laser system (MicroSlab) is given in Sect. 7.3, "Integrative Production of Micro-lasers".

6.5.4.1 Self-optimising Assembly Systems

With respect to the production polylemma, self-optimising assembly systems are seen as a means of alleviating the polylemma by allowing short-term adaptations to current conditions and objectives to be made with minimal planning, reconfiguration and change-over effort. In this respect, self-optimisation is to be understood as the ability of a system

- to continuously analyse of the current situation and
- to derive system objectives based on that analysis and then
- to autonomously adapt the system behaviour accordingly.

The first point ensures that the current state of the system and its environment is identified. The information obtained is then used to determine the new objectives of the system (point 2). These objectives should bring the system to an optimal operating point and they can be either selected, adapted or even newly generated. Once the new system objectives have been determined, the system behaviour is adapted to the new conditions (point 3). In layman's terms, self-optimisation means "thinking whilst doing" instead of "thinking before doing".

According to a definition in the German Collaborative Research Centre SFB614, self-optimising systems have the ability to respond autonomously and flexibly to changing environmental conditions, user intervention and actions of the system. Behaviour is optimised based on machine learning (Frank and Gausemeier 2004).

The basic prerequisites of an assembly system for implementing self-optimising mechanisms are therefore flexibility and mutability (as the basis for adapting system behaviour) and autonomy (for mastering complex processes robustly and without human intervention) (Fig. 6.82).

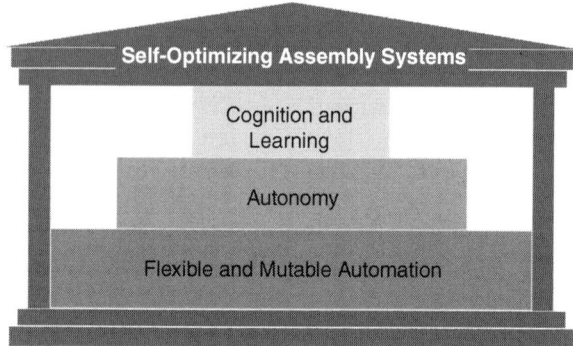

Fig. 6.82 The building blocks of self-optimising assembly systems

A self-optimising system is able to pursue new objectives based on dynamically changing boundary conditions through the use of automatic planning and optimisation functions or through the integration of cognitive structures without focussing on the planning effort of the system and without having to suffer corresponding losses in efficiency and profitability.

The next section first discusses the three main aspects of flexibility/mutability, autonomy and cognition as the fundamental basis for designing self-optimising assembly systems before presenting concrete implementation concepts for producing laser systems using an automated assembly cell.

Flexible and Mutable Automation In the context of assembly, flexibility refers to the ability to adapt an assembly system quickly and with little effort to changed influencing factors. With regard to flexibility, organisational and technical changes are limited at the time of planning by defined corridors (Abele et al. 2006). Hence a change in the number of units, for example, can be represented to a pre-defined extent within these corridors. Mutability, on the other hand, refers to the potential to implement possible changes on a reactive basis beyond the defined corridors (Reinhart et al. 2002). So, when planning mutable assembly systems, no explicit limits are set. Hence, the systems are largely solution-neutral; the necessary scope for possible changes is therefore taken into account (Nyhuis et al. 2008). Only by simultaneously increasing flexibility and mutability is it possible to meet the growing dynamics of turbulent market conditions (Spath and Scholtz 2007).

The influencing factors that lead to turbulence in the assembly system come from a number of external sources (e.g. the market, society, politics, technology, etc.) or from the production system itself. They affect assembly via a certain number of channels. The influencing factors, also known as change receptors, include:

- products and product variants,
- costs,
- time,
- number of units,
- quality and
- the elements of the assembly system (process technologies, system technology, tools, etc.).

All the requirements on flexibility and mutability can be traced back to changes of one or more of these receptors (Cisek et al. 2002). This results in various requirement categories which can be summarised as

- component and product flexibility,
- capacity and utilisation flexibility,
- process and technology flexibility and
- planning flexibility through connection to business data flows.

Component and product flexibility comprises the ability of an assembly system to adapt to frequent variant changeovers and to further/new product developments. Capacity and utilisation flexibility (or scalability) focuses on adaptability to varying unit numbers caused by fluctuations in demand or the evolvement of production volume within the product life cycle. Process and technology flexibility describes an assembly system's capability to apply different handling processes and, in particular, different joining processes depending on the component to be assembled as well as its capacity to use need-based measuring and testing technologies (Kurth 2005). A pre-requisite for planning flexibility is that the state of assembly is always transparent in the information systems, allowing the impact of short-term changes to be assessed. That way, the costs, time and quality parameters can be balanced out in direct market interaction by, e.g. by applying customer-specific pricing based on the variant and lead times (Westkämper et al. 2001).

To meet these requirements and to create a basis for reacting quickly to changing environmental conditions, an assembly system must have so-called change enablers which describe the necessary properties of the system elements. These include (Nyhuis et al. 2008):

- *Modularity*: modularity describes the ability of an assembly system to exchange standardised, functional units or elements (modules);
- *Compatibility*: compatibility between system elements allows individual elements to form networks with one another and interact in various configurations through uniform interfaces (mechanical, electronic, data-related);
- *Universality*: universality of system elements describes the dimensioning and design of these elements for various requirements and application areas;
- *Mobility*: mobility refers to the spatially variable usability of system elements;
- *Scalability*: scalability means the technical, spatial and personnel-related extensibility and reducibility of an assembly system.

Figure 6.83 summarises the information about the flexibility and mutability of assembly systems.

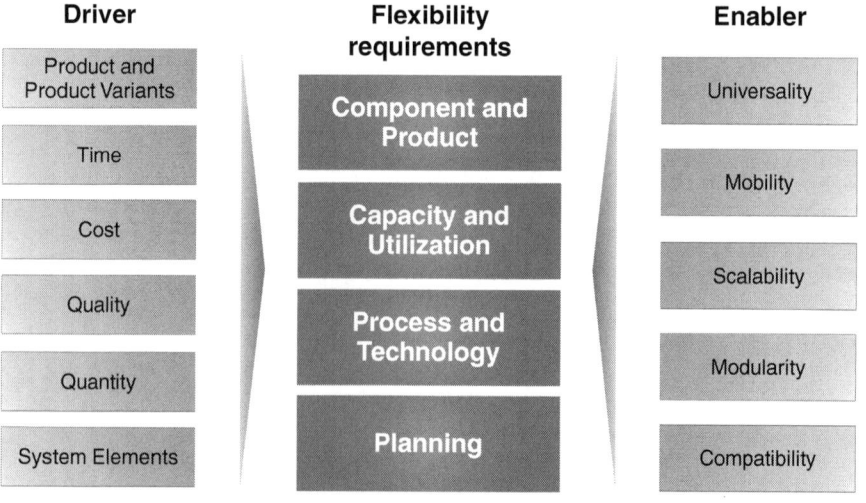

Fig. 6.83 Change drivers and change enablers

Flexibility and mutability create the degrees of freedom required for short-term re-planning and reconfiguration of assembly systems and therefore provide the basic foundations for self-optimising systems. Additional degrees of freedom lead first to increased system complexity and then to greater complexity in the planning and control of highly flexible systems. To cope with this complexity and get closer to the basic goal of self-optimisation with a view to reducing planning effort, additional components are also needed.

Autonomous Assembly Cells With respect to an assembly cell, autonomy means the capacity for reliable and error-free execution of complex assembly processes for a longer period and with a maximum degree of independence. Autonomy enables the system to respond proactively to environmental factors of influence (Pfeifer and Schmitt 2006). The need for an assembly system to offer the greatest level of autonomy possible is due to the fact that with the increasing depth of process automation, the number of process faults also tends to increase. In addition to producing a deterioration in process and product quality, this means that in order to remove these faults, time-intensive intervention by specialist staff is required and that staff must also be capable of coping with the increasing requirements caused by the growing complexity of processes and systems. For the transition from rigid automation solutions to flexible, fast and efficient assembly systems, the following three sub-aspects of autonomy can be identified:

- *Autonomy for the user through optimised support* The assembly system provides the user with a variety of functions and information and allows the user to intervene in the assembly process and in the planning of the next assembly steps;
- *Autonomy through fault tolerance* The assembly system has functionalities that enable it to react autonomously to faults caused by parameter fluctuations and other influences. Faults can be rectified autonomously if there is a high level of integration of the various system modules and fault rectification strategies can be planned;
- *Autonomy through closer integration of sub-functionalities* The assembly system benefits from an extended range of functions through the automation of all sub-operations of handling, joining, aligning, measuring and control as well as through additional services for process planning, monitoring and control.

Based on the key requirement of enhancing the responsiveness of automated assembly cells in the event of fluctuating influencing variables and making efficient use of the increased flexibility, there is a need for new functions along with new hardware and software concepts in the control system. One essential element in this respect is an information and communication technology infrastructure for connecting a variety of actuators and sensors in the form of a flexible and extendable sensor/actuator network. Furthermore, a self-optimising assembly cell must also have significantly expandable planning functions if it is to achieve the desired degree of autonomy.

Automated Planning in Flexible Assembly Systems The task of cross-process planning at the higher levels of production systems involves scheduling and coordinating the production and assembly orders that result from customer or planning orders and reacting to process-related changes caused by re-planning activities. To meet the needs of increasingly complex planning as result of dynamic influences, planning efforts have to be purposely decentralised, i.e. moving them to autonomous subsystems coordinated centrally (Pfeifer and Schmitt 2006).

Hence, flexible assembly systems are organisationally and technically divided into decentralised units (in automated systems: into autonomous assembly cells) which carry out process-related planning and control. If numerous cells have to work together to assemble a product, these cells are coordinated by a central entity. Coordination or even assembly planning includes allocating individual tasks to the subsystems. These tasks can be described in the form of complete programs that can be loaded into the cell control system for direct execution. This, however, involves major planning effort in the central unit and the result can usually only be used for a predefined cell configuration. Optimisation of process execution is usually done offline. A steady flow of new and changing tasks calls for frequent planning (diversity of variants, shortened product life cycles). Taking uncertainties into account when deliberating the plans hugely increases the workload and even hinders automation. If conditions vary, plans need to be changed and this delays execution. Program creation and optimisation (as part of planning) is time-consuming and requires a great deal of expertise, yet parts of programs can rarely be re-used.

What is considerably more flexible is an abstract, task-oriented interface between assembly planning and the executing cells, which are translated into device-related actions within the cells. Hence for an automated assembly cell, planning means determining the execution of a defined task. The result is a plan that describes a chronologically ordered sequence of actions. A complete plan contains the following aspects of the action description (Siegert and Bocionek 1996):

- Action planning (What?)
- Sequence planning (What is the temporal relationship?)
- Resource planning (Who?)
- Timing (When?)
- Execution/process planning (How? With which parameters?)

For automated planning, the cell control system requires a knowledge base with an environment model (cell model) and with rules regarding how a task is to be broken down into individual steps. At each planning level, only the section of the world relevant to the problem should be considered as this limits the number of states and therefore the planning effort.

In addition, synchronisation patterns are required for coordinating the activities and various algorithms are needed for planning specific device actions (grip planning, path planning, sensor integration). In planning, one distinguishes between the following procedures (Siegert and Bocionek 1996):

- *Goal-oriented planning*: In a perfect world, an overall plan would be created and executed without the actions being monitored.
- *Reactive planning*: A complete plan is created for an assumed world model. During execution, if action monitoring detects any deviations, a new plan is created or the goals are changed.

- *Opportunistic planning*: Plans are created for alternative scenarios and all goals are pursued simultaneously. In the next step, the plan promising the greatest success is selected based on the current state of the system.
- *Reflexive planning*: This involves an event-driven, step-by-step plan based on the actual data to hand with few forward-looking assumptions.

In conventional automated systems, planning is usually detached from execution. This basically corresponds with goal-oriented planning and results in rigid systems with a lack of responsiveness. During execution, the other types of planning mentioned above reflect the principle of self-optimisation: analysis of the current situation, determination of system goals, adaptation of system behaviour.

For self-optimising automation, it must be possible to move planning scopes, in the sense of reactive, opportunistic or reflexive planning, into the control system (depending on the planning task different alternatives should be preferred).

Automated planning in flexible assembly systems presents a challenge because of the special conditions. As the last link in the production chain, assembly is constantly exposed to changing quantities, variants and products. Thus, assembly depends on a lot of parameters of the prior production steps. As already mentioned, the difficulty is that not only faults and problems of the prior production steps have a negative impact on assembly precision or even the feasibility of assembly, but changing production tolerances and production processes also influence assembly. This reaffirms the need for reconfigurable and adaptive assembly systems.

Automation technology has been helping to make production processes faster, more precise and reproducible (Weck and Brecher 2006). At the same time, it has become a limiting factor. This is because the increasing complexity of the processes, and therefore also the ever-higher degree of subsystem networking, is confronted with the impermeability of different control domains (e.g. NC, RC, MC, PLC) combined with "weak" programming paradigms. Today's automation technology is well suited to reliably facilitating well-specified processes with short cycle times. Varying requirements or boundary conditions, however, are difficult to depict during process time and call for time-consuming planning processes or the direct intervention of human experts. Furthermore, the traditional approach hardly allows production and assembly processes to be viewed across multiple subsystems.

Cognition and Learning Ability in the Control Technology Cognition enables logical thinking about the relevance of the information acquired, allows planning and learning processes to be devised based on individual experience and facilitates intelligent decision-making as well as reliable and robust system behaviour (Beetz et al. 2007). The processes of logical thinking, learning and decision-making can be performed both on the basis of statistical data analysis and using the techniques of artificial intelligence.

Russell defines weak artificial intelligence as machines behaving as though they were thinking. In fact, they are not thinking, they are following a set of rules that can be as complex as desired. If these rules can be extended and adapted independently from the system during run time, then this is called machine learning (Russell and Norvig 2003).

Intelligent systems with the ability to learn are used in order to make decisions about their actions independently. Here we see a direct reference to the autonomy mentioned above. The better the decisions a system makes independently, the greater the degree of autonomy.

One difficulty in implementing intelligent systems is the wide variety of actual processes and process environments. This makes finding a solution highly complex. In most cases, not only is the solution path of interest, but also the associated costs. Variables, such as time duration, energy consumption, can be interpreted as costs in this respect. For example, efficiently programming complex automated systems is a major challenge. Only automatisms can drastically reduce programming effort. Such automatisms make use of the tools of artificial intelligence to determine optimal action plans, for example, or generate specific paths for robots. Using learning techniques such as neural networks, the automatisms can be refined independently of the system. Systems that able to achieve this to the full can presently only be found in the highly specialised domains with a very limited scope. The field of artificial intelligence remains an important research topic. Practical solutions designed to fulfil the requirements described with respect to flexibility and autonomy are of particular interest.

The means of cognition that an assembly system must possess for self-optimisation may, in the simplest case, be based on experience and the process data acquired may simply be statistically evaluated. In its most advanced state, however, cognition may take forms of artificial intelligence that are able to not only evaluate experiences, but also to make judgements and draw conclusions. In this respect, heuristics constitute another basis for self-optimisation whereby alternative decisions are specified and the system can chose between them.

Self-optimisation can be applied at various hierarchical levels within an assembly system. Planning and system configuration represent the top level while the lowest comprises the individual assembly processes such as joining a component. Ideally, self-optimisation should include all levels, allowing the entire assembly process to be adapted without human intervention. Should the boundary conditions or objectives change, the system autonomously re-configures itself, uses other joining technologies, measuring technology or handling units and changes the assembly and alignment process. At lower levels, individual processes optimise themselves: initial values for alignment tasks are adapted as the system learns in order to reach the desired result faster, join parameters are optimised in order to minimise deformation and misalignment and to allow faster assembly cycles. The changes made flow back to the user as information and the knowledge gained is fed back into the development process so that the necessary changes can be planned, assessed and implemented. In this respect, cognition presupposes the existence of flexibly deployable sensors which collect the necessary information, evaluate it and store it in a knowledge base so that situation-dependent decisions can be made. These decisions are converted within the system into corresponding control commands and fed back into the process via various actuator components.

6 Self-optimising Production Systems

Interim Conclusion The information above leads to the following understanding of the concept of self-optimising, automated assembly cells:

- Self-optimisation must mean that commands at task level can be transferred to the assembly cell control system either by the operator or by direct coupling to central planning systems, disregarding concrete hardware and solutions;
- The cell control system takes over the planning responsibilities, i.e. it independently breaks down the tasks into individual, device-related actions and generates the associated program code;
- The execution of actions is monitored and necessary changes to the plan are automatically implemented during run time (reactive/opportunistic/reflexive planning);
- Knowledge (learning effects) is gained from execution of the actions and taken into account in new planning operations (cognitive control).

From this it is possible to derive a variety of fields of research regarding reducing the production polylemma in assembly. The concepts and solution approaches presented in the following section are based on the results of the research projects "Self-optimising Flexible Assembly Systems" and "Model-based Assembly Control" conducted by the Cluster of Excellence "Integrative Production Technology for High-wage Countries" based at RWTH Aachen University. The work focuses on the following aspects:

- Increasing the flexibility and mutability of assembly systems along the entire process chain in order to improve efficiency and profitability when producing a broad product mix with low unit numbers;
- Increasing the degree of automation in flexible assembly through innovative, versatile assembly techniques and measuring equipment, by developing set-up and change-over procedures that can be automated and by designing open, integrated equipment control systems;
- Developing concepts for mapping and using planning systems and cognitive structures in assembly systems for self-optimising automation;
- Improving system transparency throughout the product creation process through systematic analysis of the dependencies within the process chain, from product development to process design to manufacturing equipment development.

6.5.4.2 Flexible Assembly Cells for Self-optimising Production

The automated assembly of complex products or diverse product variants requires conception of a flexible assembly system capable of continuously adapting to the production parameters such as quality, time and costs (Fig. 6.84). In some cases, various components and product variants can be assembled similarly, yet still have different features and this requires a change in the assembly system behaviour. Accordingly, the assembly system must demonstrate self-optimising behaviour by continually monitoring its current state and independently adapting to the new circumstances.

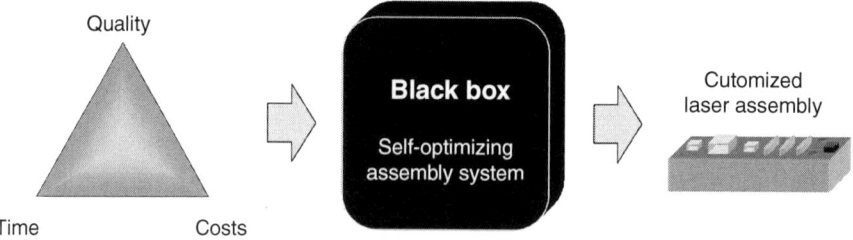

Fig. 6.84 Quality, time and costs as input parameters for self-optimising assembly

A highly flexible assembly system requires a completely modular equipment design with standardised interfaces for integrating various handling, joining and measuring technologies. Providing flexible assembly techniques in the form of interchangeable modules and developing automated change-over and calibration procedures as well as an open, integrated control system increases the degree of automation of flexible assembly systems and thus enhances the productivity and profitability of assembly. With respect to assembly control, solutions need to be developed to embed the decentralised control units of individual modules, especially of cooperating robots, in one central control system. This supports flexible order planning, thus delivering the prerequisite for self-optimising operation.

The following sections present solutions for designing a modular precision assembly system along with flexible and configurable equipment control systems as well as for automated planning of complex systems. The solution approaches focus on the self-optimising assembly of miniaturised laser systems.

Modular Multi-robot Cells for Precision Assembly As described in Sect. 6.5.2, as areas of assembly technology, micro and precision assembly are characterised by the component dimensions and precision requirements involved. Both of these factors play an important role when it comes to designing automation solutions for precision assembly. They produce the following overall set of requirements and tasks:

- Manipulation of objects in the micro[μm] and meso[mm] range;
- Setting up extremely precise spatial relationships between components (alignment);
- Sensor guidance of assembly processes;
- Monitoring and control of contact force;
- Execution of various physical and chemical processes (gluing, soldering, bonding, etc.);
- Linking/hybridisation of alignment and joining process.

It is in essence possible to derive design guidelines for precision assembly systems against the background of a high degree of product-specific operations and customer-specific adaptations to a wide variety of applications (Siciliano and Khatib 2008):

6 Self-optimising Production Systems

- Holism: a precision assembly system is composed of a wide variety of functional units (robots, manipulators, cameras, sensors) and their interaction must be considered integratively.
- Integrativity: although the handling technology in micro and precision assembly presents particular challenges with respect to robots and grippers, integration of the joining process is especially important for system design.
- Reconfigurability: as a basic requirement, reconfigurability can be achieved with a consistent, modularised system design that supports automated module changeover with respect to tools.
- Tolerance management: setting and complying with tolerances in the spatial relationship between components is the decisive factor for designing an automated system for precision assembly. One of the key aspects here is the purposeful shortening of tolerance chains in the system through referencing and closing local control loops. The precision that can be achieved then depends on the resolution of the sensors used and the movement resolution of the positioning units.

Fully automating the laser system assembly places highest demands on precision, but also on the flexibility and mutability of the assembly system. Hence, different component-dependent handling and joining processes, on the one hand, and need-based measuring and testing technologies, on the other, are used for building the MicroSlab. In order to implement the resulting variety of assembly tasks in one assembly cell, a modular system concept has been developed based on the use of multiple cooperating robots with standardised interfaces for interchangeable tools.

The basic idea of the system concept is to maximise the flexibility and mutability of an assembly cell by ensuring system modularity right down to tool level and thus reducing the configuration and set-up effort for the various process steps to a minimum.

During assembly, industrial robots take over positioning of the modular tools in the working area. These tools are used for gripping, fine-positioning, joining or measuring depending on the process step. For more complex assembly operations whereby components are actively aligned and a joining operation is performed at the same time or directly afterwards, several tools are used with three cooperating industrial robots (Fig. 6.85).

For joining optical components, a novel resistance soldering process is used which, based on the development of an automated joining module, can be transferred to a robot-based soldering process (see Sect. 7.3).

Since the robots do not provide satisfactory precision and step sizes for aligning certain optical components of the MicroSlab, an additional micromanipulator was developed providing the necessary precision. In line with the concept of a modular mounting head, this micromanipulator is designed for mobile use attached to conventional industrial robots to compensate for their positioning errors and to perform the subsequent high-precision alignment of components (Fig. 6.86).

This produces a seamless link between macro and micro working areas and therefore reduces the dichotomy, common in micro and precision assembly, between high precision and large working areas.

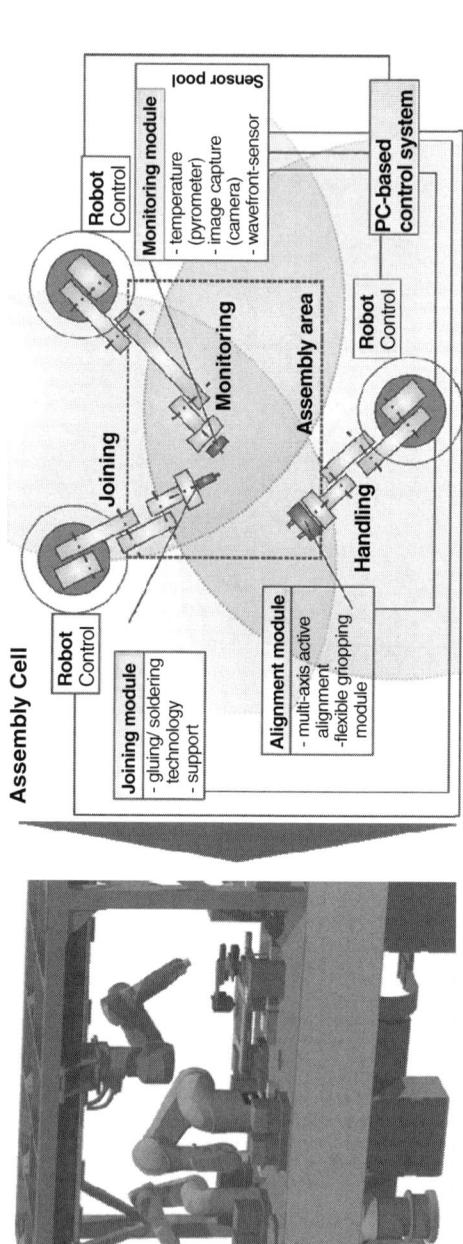

Fig. 6.85 Modular multi-robot cells

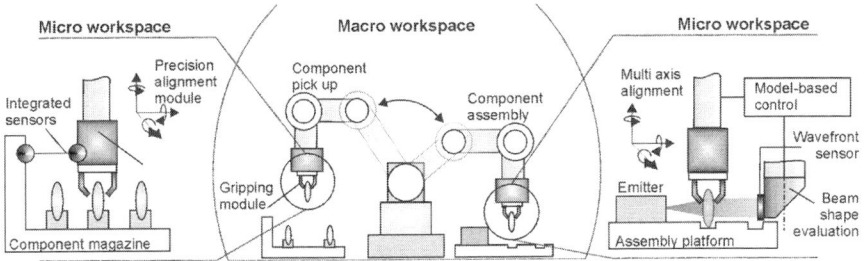

Fig. 6.86 Concept for integrating macro and micro working environments

Since the alignment of some components requires movement in all six degrees of freedom, the micromanipulator is based on a hybrid kinematic concept which meets these requirements while enabling the smallest increments in the micrometre range. The concept involves a parallel-kinematic structure whereby the movable end effector is connected to the frame via three struts of invariable length. The base points of the struts are mounted so as to allow movement in two translational degrees of freedom in the plane, resulting in six degrees of movement on the platform altogether.

All the passive joints of the kinematic structure are composed of special flexure-based joints which offer superior properties compared with conventional notch or leaf spring joints. Basically, with respect to the precision and movement resolution of a system, flexure-based joints offer advantages in that there is no friction or play and they are less susceptible to wear and tear. In theory, these advantages imply that the accuracy of the movement of the platform is ultimately determined solely by the movement behaviour of the drives, i.e. the smallest increment sizes correspond with the computational resolution of the system.

In the context of the metrological characterisation of the mounting head, it was possible to prove that the movement resolution of the drives can be transferred to the platform via the elastic structure without losses. Interferometer measurements show that the drive movements, including control oscillations in the single-digit nanometre range, are transferred to the platform and can be measured there. Unidirectional repeatability measurements with the same measuring setup yield a characteristic value of: 0.15 µm repeat accuracy (40 measurements with a standard deviation of 50 nm). The associated translational working area is approx. 4 mm × 4 mm × 4 mm (Fig. 6.87).

A vacuum gripper is used for handling components. For a highly accurate and reproducible grip, this gripper has multiple reference surfaces to which the components automatically align themselves when they are vacuum gripped.

With respect to assembly control, solutions need to be developed within the context of the system concept that will allow decentralised control units of individual modules of this type to be embedded in one central cell control system that supports hardware configurability.

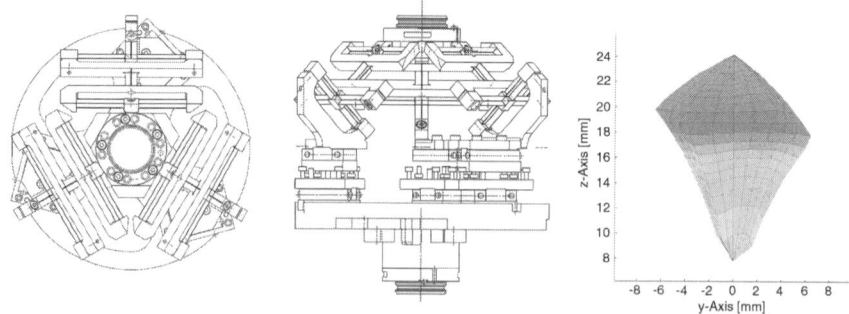

Fig. 6.87 Micromanipulator based on flexure-based joints (wireframe graphics and working area)

Agent-Based Control of the Assembly System To achieve the afore-mentioned flexibility, autonomy and cognition aspects desired in the assembly system, a concept for integrating distributed hardware and software platforms using agent-based technologies is used.

Modelling the control of an assembly system using an agent-based structure aims at the optimal utilisation of the assembly resources (Russell and Norvig 2003; Jennings et al. 1998). Although there are various definitions for the word "agent" in the literature, it is generally understood that an agent "is an entity that perceives its environment through sensors and can provide feedback to this environment by means of actuators" (Russell and Norvig 2003). The agent autonomously seeks one or more goals and usually remains in close contact with other agents in order to achieve these goals. Before the agent decides for a particular action, it takes into account its perception of the environment, its goals, its knowledge and its experience (Fig. 6.88).

The development of software agents focuses on the decomposition of a problem and the distribution of responsibilities to smaller, autonomous entities that follow

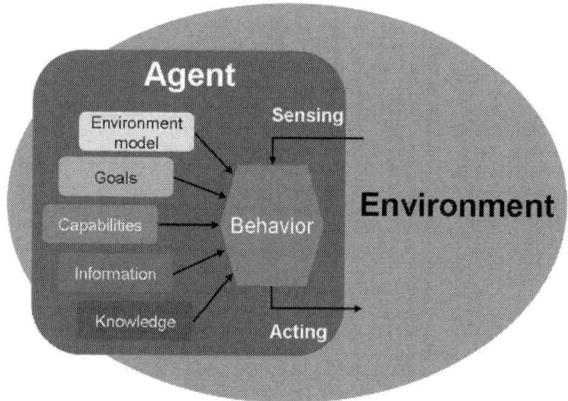

Fig. 6.88 Model of a software agent

certain principles, e.g. encapsulation, goal orientation, reactivity, autonomy, proactivity, interaction, persistence, adaptivity and intelligence or ability to learn, in order to accomplish their locally or globally distributed activities (Jennings et al. 1998; Göhner et al. 2004).

Within flexible production environments, at least three positive factors are introduced by using agent-based control structures (Brecher et al. 2009):

- The structure of an agent-based system allows an easy introduction of new agents (software or hardware) to the system without needing to re-programme the control logic;
- The application is based on a distributed system that allows the use of multiple operating system platforms and ensures communication between different hardware systems via a common medium according to a predefined communication protocol;
- Cooperation or competition between the various agents is the main cause for the desired operational autonomy of the production system.

The agents of a multi-agent system vary in terms of their abilities and responsibilities. There is, however, no specific command hierarchy between the agents from a functional point of view. In principle, all agents work on the same software level and exchange relevant information and services in order to achieve their goals. Nevertheless, from an organisational point of view, in terms of modelling and designing an agent-based assembly control structure, it is valuable and aids comprehension if the agents have a specific hierarchy. The hierarchy highlights the level of difficulty of implementation and the importance of the agent's functionality. These aspects are explained in the next section.

Agent-Based Control Structure of the Assembly System The self-optimising behaviour of the flexible assembly system can be divided into various levels or control loops. Figure 6.84 shows the highest level of self-optimisation whereby a dynamic strategy for laser assembly is derived directly from the production parameters of quality, time and costs defined by the customer.

The following extreme scenarios better illustrate this idea:

- *Assembly of a high-quality laser system:* In this case, the criterion quality is set as the highest priority while the time and costs of assembly are disregarded. Thus, the assembly sequence is carefully planned and the most suitable optical components as well as the most appropriate handling, positioning and joining processes are carefully selected. All necessary quality characteristics of the laser are metrologically assured and the optical components featuring critical positioning tolerances undergo active (in-process) alignment procedures;
- *Fast assembly of an inexpensive laser system:* By contrast, in this case the criteria time and costs are set as the priority while the quality is disregarded. Hence, the time allowed for planning the assembly, selecting the optical components and for handling, positioning and/or joining is reduced. The metrological systems are only used when necessary. To save on costs, fewer optical components are used where possible as long as the laser system remains operational.

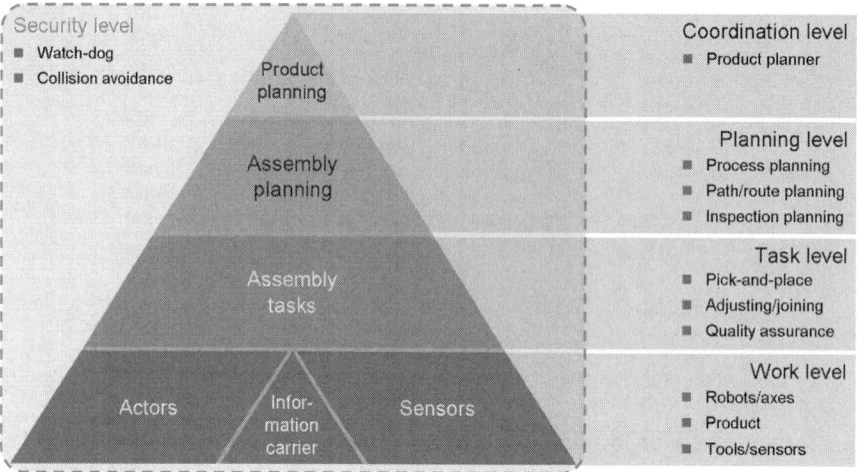

Fig. 6.89 Agent-based control structure of the self-optimising assembly system

Other assembly scenarios can also be depicted by choosing other concurrent configurations of the variables quality, time and costs. For the highest level of self-optimising assembly to take effect, the small, inner self-optimising control loops of the system must also be working. In this case, the black box element of assembly is conceived as a multi-agent system, whereby the cooperative/competitive exchange process between all agents involved creates the highest level of self-optimising system behaviour. Within the control structure, some agents are involved in reduced work groups in which smaller, self-optimising control loops operate. Figure 6.89 shows the overall agent-based assembly control structure, which is distributed across various levels (Brecher et al. 2009). Examples of tasks and agents that appear on different hierarchical structure levels of the system with different responsibilities and capabilities are as follows:

- *Coordination level:* Here, a product agent has high-level responsibility for product assembly planning. The desired production parameters quality, time and costs, as specified by an operator or user of the system, are incorporated directly here as input. Therefore, a strategy for assembling the product in accordance with the currently available resources is dynamically devised and orders are forwarded to the other lower-level agents in the control system. If no suitable strategy for assembly can be found because the combinations of production parameters are not viable, then a new, viable combination of parameters is suggested;
- *Planning level:* Here, several agents are responsible for optimised assembly planning based on the orders from the coordination level. Planner agents cooperatively define which and how assembly tasks should be carried out, in order to ensure that the expectations of the product agent are met. They translate the abstract assembly goals defined by the product agent into new, smaller and more specific orders for the task-level agents;

- *Task level:* Here, several agents are responsible for the interface between planning and working levels. Task agents translate the specific planned task orders into hardware and software commands that are to be executed by the work agents. They coordinate and assess whether the desired goals have been achieved after the task has been executed;
- *Working level:* Here, several agents are responsible for controlling the hardware and software modules (e.g. actuators and sensors). Some are also responsible for managing and providing information about the current state of assembly and of the product;
- *Safety level:* Safety agents are responsible for the integrity of the assembly work as well as of the agent-based system itself and avoid dangerous situations.

From this description of the responsibility assigned to each level of the control system, it is clear that the upper levels (task, planning and coordination levels) have a more complex level of design and implementation. These agents are characterised by intelligent capabilities (planning and coordination) and therefore need cognitive means in order to weigh up their actions and safely and robustly accomplish their tasks. The lower-level agents do not usually need these cognitive means and behave reactively (receive order, execute) to complete their activities.

The organisation of the various agents as well as their communication within the agent-based system is in accordance with FIPA[3] standards. Hence, FIPA-compatible tools are used for modelling (FIPA AUML[4]) and implementation (JADE[5]) of the system.

Top-Down and Bottom-up Approaches for Designing the Control Structure Based on these preliminary considerations, the control structure of the system for assembling the miniaturised laser was designed using a top-down approach. The behaviour of the entire system and the relevant smaller, inner assembly control loops were derived from the highest degree of self-optimisation via the coordination level. There are already important, inner, self-optimising loops in the planning level (e.g. assembly planning, path planning, test planning (Brecher et al. 2009)). Since the results generated during the assembly process do not always correspond with those that were predicted in the planning phase, these planning activities must be adapted based on the current state of assembly (Fig. 6.90).

Up to the planning level, the execution of agent activities is software-intensive and based on simulation and classification methods. From the task level down, these activities are closely related to the hardware control system, e.g. for component handling, positioning, aligning and joining.

Although the conception and rough modelling of the control architecture of the assembly system follows a top-down approach, the fine modelling and implementation of individual agents is developed using a bottom-up approach, because the lower-level agents constitute the basis of the assembly work and are essential for

---

[3] FIPA: Acronym for "Foundation for Intelligent Physical Agents" (http://www.fipa.org/).

[4] AUML: Acronym for "Agent UML" (http://www.auml.org/).

[5] JADE: Acronym for "Java Agent DEvelopment Framework" (http://jade.tilab.com/).

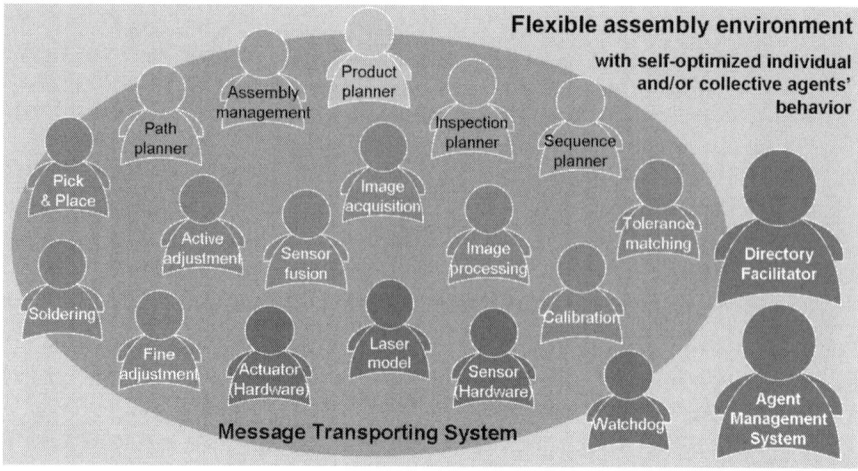

Fig. 6.90 Agent-based organisation for self-optimising assembly. (According to FIPA standards)

the minimal functionality of the system. This also includes their lower behavioural complexity in comparison with the planning agents.

The autonomy of the assembly system can only be achieved through the additional intelligence of machines (intelligent agents), which have the ability to perceive their environment using sensors and to process the obtained information in a goal-oriented way for fulfilling their task. Accordingly, sensor integration, intelligent controls and generic structures are essential for production systems of this type. On the basis of such formal descriptions, planning and optimisation components can determine action plans that are implemented by the control system or a management system (Fig. 6.91).

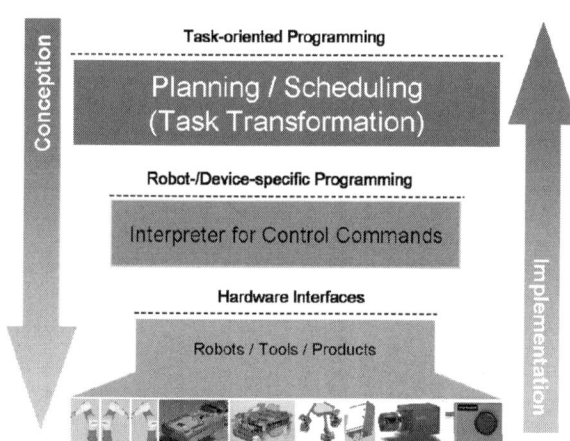

Fig. 6.91 Top-down and bottom-up approaches to the conception and implementation of the system

Agents for Dealing with Failure States The quality of the laser system depends not only on the quality of the individual optical components, but also on their positioning and correct pairing. In this respect, alignment precision has to meet the highest requirements (μm range). Deviations in component properties lead to new dynamic setpoint values for controling the assembly process. Hence, the entire assembly process must be flexible and supported by measuring technology for assessing the state of the system. Furthermore, the measuring technology should also be used for monitoring the entire assembly process, and active alignment of the optical components should be possible.

Depending on how the key performance indicator "quality" of assembly for a laser variant is defined, different laser system assembly steps must be metrologically assured. For a high-quality assembly, various laser characteristics have to be measured and tested:

- Geometric characteristics;
- Presence/absence and identification of components;
- Compliance with positioning tolerances;
- Characterisation and dynamic selection of components;
- Active (in-process) alignment of components with critical positioning tolerances;
- Characterisation of the laser beam profile and laser output power;
- Identification and resolution of failure states.

A number of measuring methods are used and integrated into the robot cell or the robots in order to ascertain the state of assembly (Fig. 6.92):

- robot-based, high-resolution image processing systems;
- a camera-based laser beam analysis system;

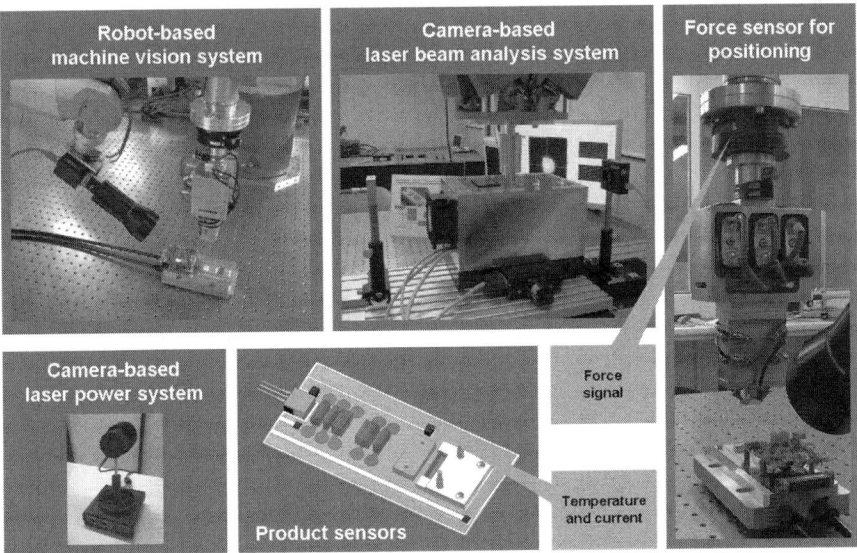

Fig. 6.92 Optical measuring methods for ascertaining the state of assembly

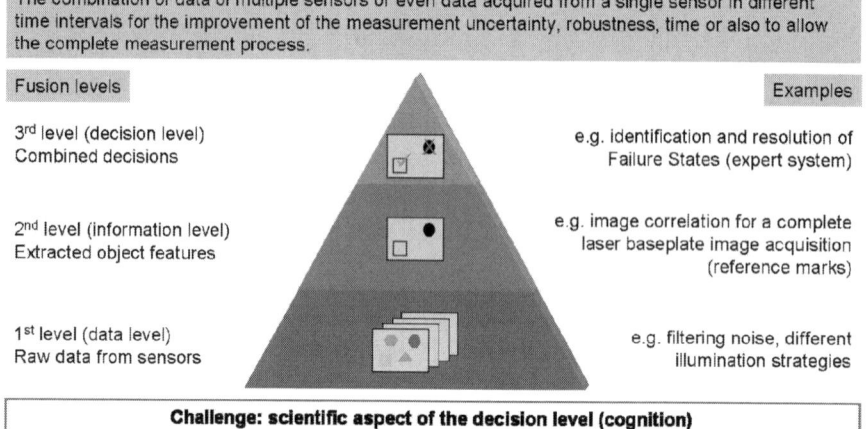

Fig. 6.93 Levels of the sensor data fusion strategy

- a camera-based laser power system; and
- electronic sensors for checking force, temperature and current.

Each sensor system is assigned a corresponding measurement and testing agent. Agents of robot-based measurement systems must constantly cooperate with robot agents to allow flexible measurement. The information gathered by the sensors is analysed by processing agents and stored as general knowledge. The complete state of the entire optical system can only be mapped by the fusion of data from all the measurement systems (Russer and León 2007). Combining the data from multiple sensors, or just the data gathered by a single sensor at different time intervals, improves measurement accuracy, robustness and time, or actually makes the measurement task possible.

The data gathered can be combined at three different levels (Brecher et al. 2009) (Fig. 6.93). At levels 1 and 2 (the data and information levels), images and features are combined to solve typical image processing problems. The capability of fault tolerance (one of the major challenges of laser assembly) is located on the 3rd level (the decision level). Fault-tolerant behaviour is coordinated and executed by an intelligent agent from the task level (failure states agent). The intelligence of this agent is based on an expert system which manages a model (part of the knowledge) of the laser assembly.

The aspects of cognition and decision-making ability of the failure states agent, which are required for the system to be fault tolerant, will now be briefly discussed.

The Intelligent Failure States Agent The use of cognitive aspects in technical systems focuses ensuring the capacity for intelligent decision-making. Modelling and implementing such cognitive abilities is supported by knowledge-based systems. These systems serve as the basis for knowledge representation as well as for inference and learning abilities that are required for dynamic and adaptive systems.

Compared with traditional programming approaches, knowledge-based systems differ by strictly separating the knowledge representation from the knowledge processing (Beierle and Kern-Isberner 2006). Hence, various knowledge-based approaches are based on a common core structure comprising a knowledge base for storing the data and an inference component for knowledge processing. The knowledge base can be created with various types of information which usually falls into two different classes (Beierle and Kern-Isberner 2006):

- *Case-based knowledge:* This is the specific kind of knowledge that relates only to the specific problem case. These are facts that exist because of observations or experimentations;
- *Rule-based knowledge:* This is the core of the knowledge base and may contain domain specific knowledge (theoretical knowledge or even practical knowledge) and general knowledge (problem-solving heuristics, optimisation rules or even knowledge about objects and relationships in the real world).

The failure states agent is supported by a special kind of knowledge-based system: an expert system. The crucial characteristic that differentiates the expert system from other knowledge-based systems is the origin of the knowledge contained in the knowledge base. In an expert system, the knowledge comes from experts who have a high degree of competence in the relevant field, both through appropriate training and through extensive practical experience (Beierle and Kern-Isberner 2006).

The failure states agent has a lot of "if-then" rules within its rule-based knowledge which depict a particular systematisation of the assembly of the laser system. Different information about the assembly is stored this way, e.g. the different components of the system, their setup and tolerances, the variety of metrological systems for checking the conformity of assembly tasks as well as the hardware resources for controlling and changing the system state. The intelligent agent requires further information from planning agents as input so that the inference machine can be initiated. The agent therefore obtains from the planning level a specific sequence or assembly plan for the components as well as the predefined quality level of the laser system. Thus, the agent can guarantee proactive and preventative intervention intermittently with the assembly tasks, in order to always assure the state of the assembly.

The agent then coordinates the measurement and test operations and delegates tasks to the metrological agents so as to support and assure the state of the assembly. This means first identifying the failure states (clear deviation between the desired and the actual assembly situation) by collecting and analysing metrological data. This process runs via the rule-based expert knowledge of the agent. The resulting state diagnosis (logic for deriving the failure state) can be presented by the agent in a comprehensible and transparent manner via an explanation component. Next, the causes of the failure need to be identified and interpreted. In some cases, additional metrological investigations can be used to initiate measures to correct the failure. The process of linking a failure with a solution usually applies the previous experiences of case-based knowledge.

Figure 6.94 shows a typical failure that occurs at the start of laser assembly. In this case, the pump beam is emitted from the diode laser, but instead of passing exactly

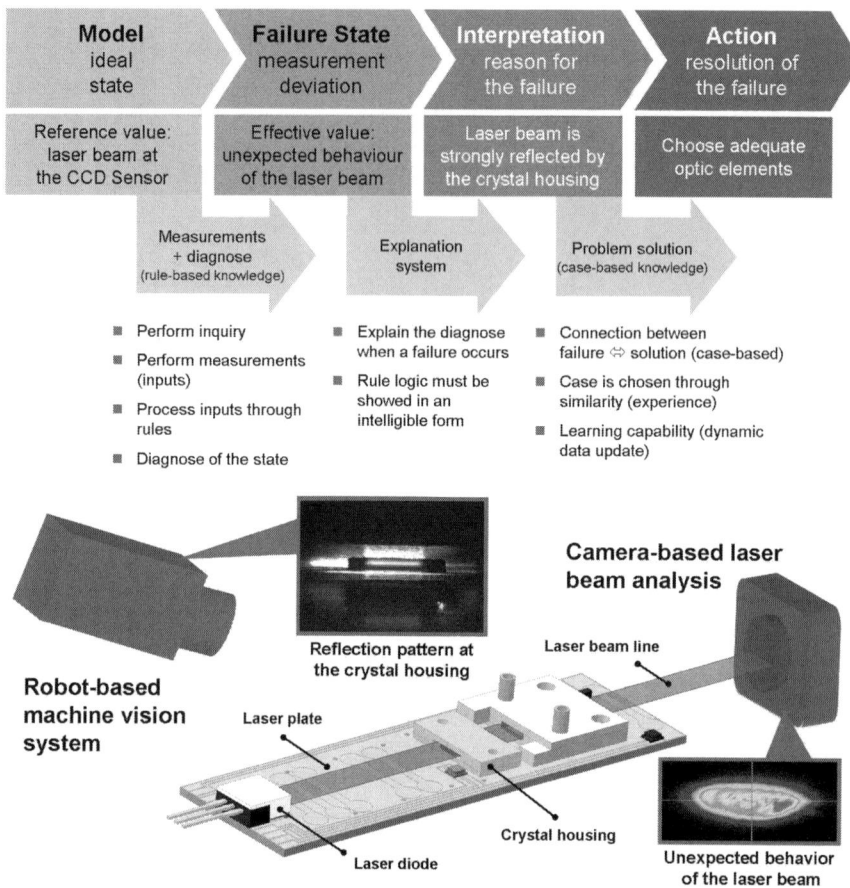

Fig. 6.94 System for identification and resolution of failure states

through the laser crystal, it hits the crystal housing. This failure can be identified by looking at a reflection pattern. At that moment, the diode laser and crystal or crystal housing are already firmly mounted or joined onto the laser mounting plate. A possible solution in this case is the dynamic selection of new optical components which are placed between the diode laser and the crystal, in order to correct the path of the pump beam. Planning agents run the dynamic selection of components as soon as the failure states agent provides the measurement data for the current state.

Model-Based Assembly Control In today's automation solutions, the knowledge about a particular task or system and the "solution competence" as expert know-how of the companies are usually inextricably intermingled. Corresponding subsystem-specific implementations are neither flexible (with respect to changes in the boundary conditions or objectives) nor transparent (with respect to global system behaviour) and thus offer no basis for self-optimising behaviour. Hence, separation into problem-specific modelling and a generic control architecture is presented as the basic scientific hypothesis.

With respect to self-optimising systems, it is still largely unheard of to have an interdisciplinary system and process modelling that covers the entire life cycle of a system ((Adelt et al. 2009) offer a concept geared mainly towards design). Hence, it was a particular scientific challenge to develop a metamodel based on general modelling languages, see for example (Brecher et al. 2007). The second scientific challenge was to develop control architectures capable of creating goal-directed behaviour based on the model information present, whereby current system states are incorporated into further planning during run time. Approaches from the fields of *artificial intelligence*, *operations research* or *cognition* that have been transferred into production engineering to date have promised little if any success (some approaches in production management engineering offer holonic production systems, cf. Schild and Bussmann 2007; Kempf 2010).

Integrative Vision The integrative vision for self-optimisation is based on continuous, model-based development, programming and implementation of self-optimising production cells. In this respect, the aspect of self-optimisation implies the assembly system's capacity for autonomous, in-process adaptation of system behaviour based on varying boundary conditions or objectives (Brecher et al. 2008b). If this takes place in the context of an automated solution, such a system can in principle be used for low-volume, multi-variant production processes as well as for producing large unit numbers (*reduction of scale/scope dilemma*). By using a consistent model description for the entire life cycle of a system, not only can engineering phases in the application be shortened but, above all, they can also be synchronised across all disciplines. In this respect, using the same model information for the runtime system too allows, in principle, global goal attainment (*reduction of plan/value-orientation dilemma*).

The potential offered by an integrative approach is obvious and will be demonstrated using a real assembly process for a complex product. Of particular interest is the possibility that the approach may be transferred to other systems and processes.

A self-optimising system that results from this will in the future be able to make essential, process-related decisions during run time both autonomously and with involvement of human experts and will be able to implement these decisions in different subsystems of the system without jeopardising the global, product-related goal variables (performance, costs). The specific goal of this subproject regarding self-optimisation was to prove the innovation potential of this integrative approach using the example of a realistic automated assembly process.

Results The problem of increasing individualisation and dynamisation of demand on the part of the customer and the resultant need for assembly systems that can be quickly and cost-effectively adapted to varying customer requirements and new products was already extensively investigated during exploratory work at the Chair of Machine Tools regarding operations control for flexible manufacturing systems with explicit focus on the shop floor level (Peters 2001; Possel-Dölken 2006; Fayzullin 2010). Although there are similarities, there are also many differences between the two domains of automated flexible assembly, on the one hand, and flexible manufacturing on the other, as per the flexible manufacturing systems usually found in industry. While, at shop floor level, the dependencies between the manufacturing system components involved are temporally and spatially encapsulated and can therefore be relatively easily organised and formalised in terms of function, at the cell and control levels the system components involved are closely interlinked. The assignment of functions to the individual components is not obvious here and can in fact change dynamically depending on the specific product and the application. A robot can, for example, be converted from a processing unit to a transport or handling unit by changing the gripper. Furthermore, the degree of abstraction of a control task varies: whilst at the shop floor level, clearly delineated work cycle elements, e.g. transport of an object from location A to location B, are referred to as elementary processes, a robot controller in an assembly cell can only run a linear interpolated movement between any space coordinates. In the latter case, there is no reference at all to the semantics of a process. When multiple robots are used in one assembly cell, the working areas of the individual robots may overlap, adding the problems of coordinate transformation and collision detection/avoidance. In terms of the desired self-optimisation, the operational planning task in the shop floor level refers to the allocation of the available degrees of freedom and the determination of an optimal sequence for the individual operations of multiple parallel-dispatched workflows. Planning tasks in cell and control levels are often based on few or even just one workflow instance which, however, has significantly more degrees of freedom. Obviously, the problem of operational control is characterised at cell and control levels by a large number of degrees of freedom, complex and not formally defined semantics as well as significantly higher demands on response time.

The concept of a possible control hierarchy for a self-optimising assembly cell includes a layer that is capable of optimally planning and controlling the assembly procedure during run time (Schmitt et al. 2008). The state of the art for such superordinate control systems is to date purely reactive, usually PLC (programmable logic controller)-based, master control systems. As demonstrated for the shop floor level, it is also assumed that the means for implementing a self-optimising system at cell level are the formal declarative model-based description of the problem domain and the development of algorithms for evaluating these models, together with implementation of the necessary optimisation runs and their execution during run time. The findings resulting from the development of a model-based control system for the automated assembly of the MicroSlab laser include four methodologies:

- Methodology for semantic modelling of flexible processes
- Methodology for model-based engineering of control systems
- Methodology for direct model execution (in contrast to imperative programming or model-to-code transformation)
- Methodology for optimising and planning the processes using models

These methodologies are described below.

Methodology for Semantic Modelling of Flexible Processes Different boundary conditions prevail at each level of the automation pyramid despite the general self-similarity of the control tasks. Quantitative characteristics vary dramatically from level to level, e.g. with regard to the guaranteed response time or the potential diversity of alternatives. Even in discrete manufacturing, processes continually running at the lowest levels, such as movement guidance, also have to be taken into account. In contrast, continuity at the higher levels of the automation pyramid is no longer a binding constraint because it is encapsulated within the limits of the respective manufacturing processes (Fayzullin 2010). The resultant diversity of variants at the individual levels calls for different approaches for implementing self-optimisation, taking into account the different boundary conditions.

The foundation for any process optimisation and planning is the formalisation of the relationships within the production engineering domain by means of adequate models. The need to create appropriate models also results from the demand for inexpensive and flexible development processes. In computer science, a new development paradigm has become established in recent years: model-driven development (Kühne 2005). In model-driven development, relationships within the specialist domain in question are mapped in appropriate models. This allows the problem specification to be separated from the implementation details. Furthermore, model-based development allows new approaches to be used with respect to validating the proposed solution (Fayzullin 2010).

A modelling approach for the semantic modelling of flexible process at shop floor level has already been developed at the Laboratory for Machines Tools and Production Engineering of RWTH Aachen University (Brecher et al. 2008a, c). Based on object-oriented modelling, an abstraction layer was designed in the form of a domain-specific language, which addresses the particularities of process descriptions in a flexible manufacturing system. However, because of the specific problems involved at cell control level, new modelling approaches and modelling input methods needed to be developed in order to engineer the model-based control systems. The significantly higher frequency of events occurring at cell control level and the associated need to map them leads to more detailed models and until now this required considerably increased effort in engineering the models. The purely graphical modelling tools developed thus far allowed the higher levels of the automation pyramid to be modelled with reasonable effort. For the cell control level, however, a new approach was required that would allow the existing meta-meta-model to be reused for faster engineering. The approach taken was to develop a modelling language that allowed graphical modelling to be supplemented with more textual input options.

Methodology for Model-Based Engineering of Control Systems As mentioned above, engineering model-based control systems is, to a considerable extent, comprised of creating the necessary semantic models. Since modelling is very time-intensive, it offers great potential for optimisation with a view to drastically shortening the overall engineering phase for model-based control systems and thus taking into operation of a control system faster. One option is to use template techniques. Templates can be used for modelling certain recurring parts of models. Modelling effort is reduced because actions are no longer elementarily modelled; instead submodels of specific, recurring processes, such as transporting a component from location A to location B, can be inserted directly into the model. To this end, a library of re-usable submodels was created. Another approach is the development of a text-based modelling language that provides a domain-specific scope. The limited scope of a domain-specific language offers the advantage of a shorter learning curve. The additional option of entering all input for the submodels via keyboard significantly reduces the modelling effort.

Methodology for Direct Model Execution The model-based approach is already being used in control technology. However, the state of the art is the transformation of the models created into rigid control logics processed by programmable logic controllers (PLCs). A much more flexible approach is the interpretation and direct execution of models because only this approach enables actions and processes to be planned online (Brecher et al. 2007). The interpretation of models offers some advantages over pure code generation. Change can be implemented faster because adaptations only affect the model and there is no need for renewed compilation and testing. The model is interpreted during run time and can in theory also be amended during run time. The interpreter can also be implemented for other platforms (cloud) and process the same models there.

Methodology for Optimising and Planning the Processes Using Models The duty cycle of the model-based control system includes perceiving information about the state of the assembly system, storing this information in the internal data model, planning actions online using the internal model and executing the plans created by triggering real actions. While these tasks are being processed, the model-based control system can have read access to the information stored in the semantic model regarding the types and instances of the system components and processes. If a state change is registered by the model-based control system or if an internal schedule changes, the control system modifies the content of the instances. The chronology of the tasks taking place in the decision cycle of the model-based control system is based on the preparatory work undertaken at the Laboratory for Machine Tools and Production Engineering (WZL) of RWTH Aachen University (Brecher et al. 2008b). However, unlike event processing at shop floor level, where querying the state of the system components is time-discrete, the time-critical nature of the cell control level requires a different approach.

Validation A flexible assembly cell was developed for the purpose of validating the methodology developed (Fig. 6.95).

Since the assembly cell consists of three cooperating robots that can be equipped with various grippers, tools or measuring instruments, a great deal of flexibility is offered in terms of performing assembly operations. The challenge was to create a close link between the system components at cell and control levels. As mentioned previously, each system component can be assigned different functions which may

6 Self-optimising Production Systems

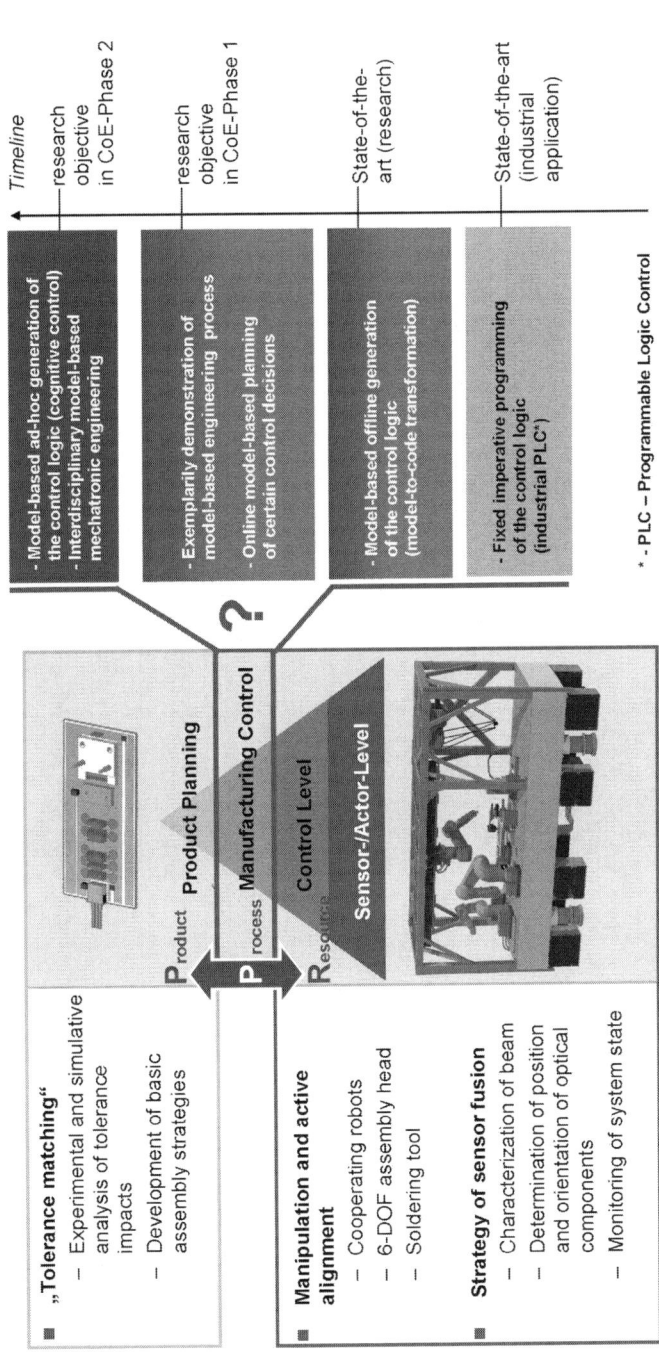

Fig. 6.95 Timeline of development of the cognitive control systems

be changed during the assembly process. Depending on the application or the product to be assembled, a robot can be reconfigured from a transport robot, for example, to a soldering robot merely by changing the gripper. This means the semantics of each element may change within the elementary processes compared with the superordinate process.

6.5.4.3 Integrative Product and Process Design

In assembly, products with defined functions are manufactured from a range of components or sub-assemblies. As part of this process, complex products have to be not only joined, but also put into operation, checked and aligned. In assembly, these steps are closely interlinked. Since assembly is usually at the end of the company's value-adding chain, it is the key to ensuring that product features are implemented at the desired quality level. It represents the last opportunity to influence features and their characteristics.

When planning an assembly system, the product to be assembled is usually taken as the starting point for deliberations. Taking this into account, first a rough process chain is created. Then the production equipment to be used in each process is selected based on that chain.

Given the objectives, some of which are fixed, with respect to quality and output quantity and in view of the goal of cost-optimal production, however, a strict unidirectional procedure would not make sense. There are mutual dependencies between product, process and production equipment that must be taken into consideration. Hence, the assemblability of a product should be taken into account during its design and development.

By integrating product and process design at an early stage, cost-intensive and time-consuming iterations can be reduced through integrative consideration of product, process and production equipment. Restrictions arising from the assembly process or the production equipment can therefore be taken into account during product design. This makes it possible to avoid costly changes to the product after ramp-up. Likewise, this avoids potential subsequent adjustments of the specified assembly process and the production equipment acquired.

With flexible assembly systems in particular, the process must be designed to be robust so that the system can be returned to operation quickly following a reconfiguration. This makes it possible to automate small series economically too.

A key factor for success in this respect is to collaborate with the company divisions involved, especially in the early planning stages. Traditional organisational structures with rigid, self-contained departments often have an entirely departmentalised way of thinking which causes significant friction losses in intersectoral cooperation. In this case, the departments usually only consider the part of the value-adding process that they are directly entrusted with. With the focus on departmental cost objectives, the individual departments tend to keep their areas of responsibility as small as possible and thus always choose the problem solution variant that is the least expensive. The aim of efficient business organisation, however, is to reach the overall optimum,

6 Self-optimising Production Systems

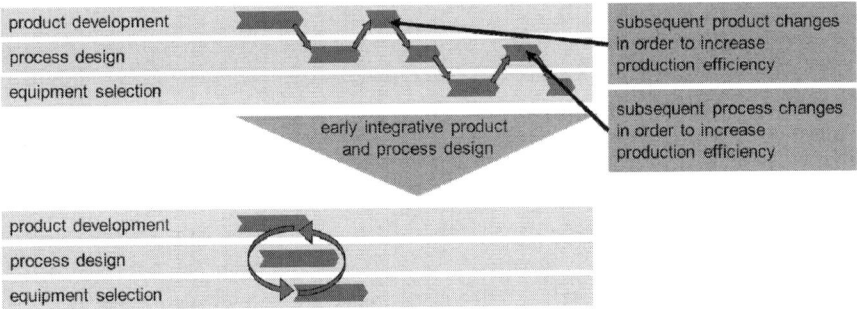

Fig. 6.96 Reduction of time-to-market and costs by early consideration of product, process and production equipment

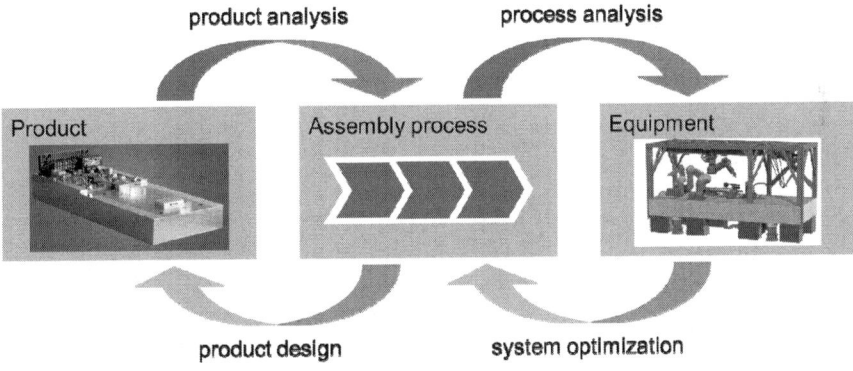

Fig. 6.97 Interactions between product, assembly process and production equipment

which in general cannot be achieved with localised optimisation measures. In fact, a procedure that is particularly inexpensive from one department's point of view can often lead to increased costs in the downstream business divisions. Accordingly, the best approach in terms of optimising production company-wide is to ensure the individual departments are closely interlinked or to even completely eliminate rigid departmental boundaries (Fig. 6.96). This facilitates, amongst other things, the integrativity of product design and production planning.

Due to the restrictions caused by the product, process and production equipment and due to economic influences, planning is becoming very complex, especially where flexible and adaptable assembly systems are concerned. In this respect, methods have been developed to assist the user in the design of flexible assembly systems. These methods can be assigned to the various stages of assembly system planning (Fig. 6.97).

Methodology for Product Analysis In assembly, to provide the required product features at the desired quality level, functional tolerances are defined during product design. Tolerance zones have to be defined because in no technical process can the target value of a property be achieved exactly and consistently (Trumpold et al. 1997). The objective of establishing tolerances is to find a compromise between the functionality of a component and the manufacturing costs. In this respect, the focus is often on the production of individual components and the features structure of the end product.

Often, products are designed which, due to their tolerances, can only be produced using the most complex assembly processes and the associated cost-intensive production equipment. Furthermore it is commonplace for assembly processes to be planned that only meet the required tolerances after the most laborious adaptation. To prevent this, it is essential that the tolerance problems are taken into account at an early stage in the product creation process. However, because of the large numbers involved, explicit consideration and control of all specified tolerances for complex products, e.g. the miniaturised solid-state laser, is unreasonable from an economic point of view and is often impossible from a technical point of view (Thornton 2004).

To facilitate early and efficient assembly planning despite the above challenge, the tolerances to be taken into account during planning are derived from the customer's requirements, which determine the product requirements. Whilst deriving this information, a fundamental understanding of the product is generated, which is essential for developing an efficient assembly process.

The tolerances to be taken into account are identified using the key characteristics method introduced in the 1980s in the USA and continuously further developed since then within the framework of scientific publications (Whitney 2004; Merget 2004; Müller et al. 2009a).

The term key characteristic (KC) refers to a quantifiable feature of a product, an assembly, an individual part or a process whose expected variation from the target value has an unacceptable impact on the costs, performance, quality (as perceived by the customer) or safety of a product (Merget 2004). Hence, to avoid such impact, KC variations must be limited by tolerances and these must be explicitly taken into account during assembly planning.

An essential element of the KC method is the key characteristic flowdown which shows the relationships between customer requirements and the KCs of individual parts (Merget 2004). In creating the KC flowdown according to the top-down approach, the customer requirements can be broken down into the KCs of the individual parts (Müller et al. 2009a). The tolerances to be taken into account can then be derived from the KCs of these parts (Merget 2004).

For a marking laser (see Sect. 7.3), beam quality, for example, is a customer requirement for marking on surfaces. The beam is influenced by the arrangement of the lenses and the crystal. For high beam quality, the beam must hit the centre point of the crystal exactly. To do this, the lenses must be aligned with the crystal (Fig. 6.98). Based on this, it is possible to define the functional tolerances that will ensure the beam offers the required quality.

Based on these functional tolerances, it is possible to derive the assembly sequence as well as the possible joining and handling technologies.

Methodology for Assembly-oriented Product Design Assembly processes and equipment are influenced significantly by the product design. Assembly-oriented

6 Self-optimising Production Systems

Fig. 6.98 Relationship between beam quality and arrangement of lenses

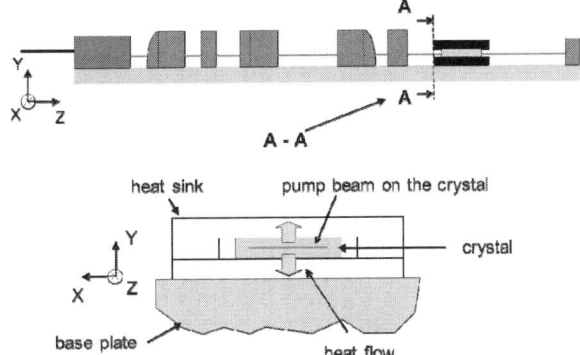

and handling-oriented product design is therefore always at the centre of the development process. The relevant literature contains a large number of guidelines to support the designer in this work. These guidelines provide a very good basis and offer valuable advice regarding assembly-oriented design. Due to the diversity and scope of the guidelines, however, it is seldom possible to fully incorporate all the recommendations.

When designing products, the restrictions caused by the process always have to be taken into account. Accordingly, it is technically impossible to, for example, assemble components with arbitrary precision. Hence, it makes sense to tailor product design to the assembly process.

Figure 6.99 shows an example of a conventional laser. Each lens is fixed in a separate carrier. The carriers have various settings as required for aligning and setting the beam. Especially for a complex product like the laser, whereby the interactions between the lenses, in particular, affects beam quality, alignment is very laborious and can only be performed by highly skilled and experienced personnel.

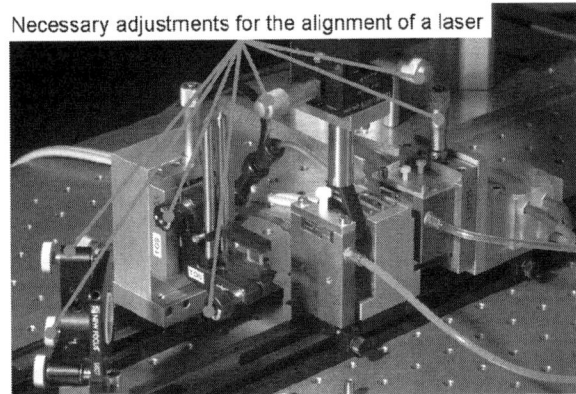

Fig. 6.99 Construction of a conventional laser

With assembly-oriented product design, the product can be modified so that it can be joined and aligned in an automated precision assembly process. In this case, the laser was designed so that the lenses could be joined from above using a robot. The new rectangular design of the lenses helps reduce the degrees of freedom of component positioning and orientation compared with the previous circular design. The lenses are joined to the motherboard by means of a soldering process. This allows the parts to be aligned during assembly whilst in the solder bath. When the solder cools, the lenses remain in the set position and orientation.

Furthermore, the laser was designed according to the so-called nest assembly, i.e. the individual parts or assemblies are arranged to a large extent side by side, as with an electronic circuit board for example (Lotter 2006). The order of installation is freely selectable. This results in a large number of acceptable process variants and optimisation approaches through variation of the assembly sequence.

Methodology for Process Analysis Process analysis considers the individual boundary conditions and objectives in the process. For the assembly process to be implemented, the functional tolerances must be maintained during assembly. The influences of the components and the production equipment on assembly tolerance must be identified in order to allow economic design of the production equipment. The aim is to compile the system so as to meet specific needs because excessive demands on the production equipment are usually associated with higher costs.

The tolerance analysis can be used in the production equipment design process. This analysis makes it possible to determine the assembly tolerance as a function of the production equipment. In this respect, possible sequencing options for the individual assembly steps are identified. In each of these steps two parts are assembled to form one assembly. These alternative sequences are then analysed to determine which of them make it possible to maintain the specified tolerances. The tolerance chain is visualised to facilitate identification and analysis of the sequencing options.

This step of the standardised approach includes, amongst other things, working out which assembly equipment and joining procedures can be used to assemble the individual parts. In addition, any component-induced or process-induced deviations arising during assembly are also revealed. To what extent these deviations need to be changed in order to efficiently meet the specified tolerances is also investigated. At the same time, implementation options for changing the deviations also have to be determined.

To help identify these options, the deviations that arise during each assembly step alternative are visualised in a graphical model (Fig. 6.100).

The nodes of the graph symbolise the characteristics of the assembly system (e.g. workpiece carrier stop). The edges of the graph describe the geometric deviations between the characteristics (e.g. the deviation in distance between the workpiece carrier stop and the edge of the laser mounting plate). This method of illustrating deviations is based on visualisation of the dimensional chains, the tolerance chains, the liaison diagram and the datum flow chain (Müller et al. 2009a). Since the graph visualises the greatest possible deviations between the actual and target variables and

6 Self-optimising Production Systems

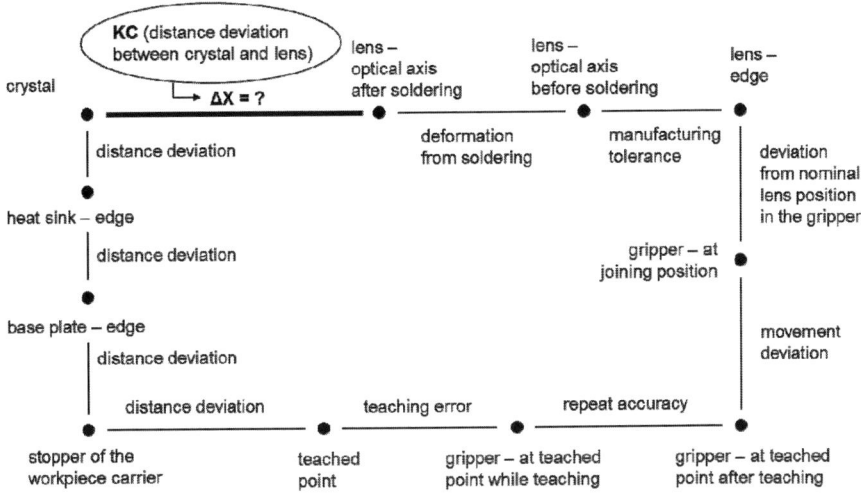

Fig. 6.100 Tolerance chain of laser assembly step alternatives

these determine the tolerances that can be satisfied, the graph is also referred to as a tolerance chain.

The deviation in distance between the crystal and the lens in the tolerance chain of the MicroSlab laser (Fig. 6.98) must be explicitly taken into account during assembly planning to ensure the laser can perform (i.e. this concerns a KC). The remaining elements of the tolerance chain symbolise the deviations that arise during the assembly step alternative in question and cause the deviation in distance between the crystal and the lens.

Methodology for System Optimisation The methods presented in Fig. 6.101 can be used in order create more efficient or shorter tolerance chains.

In the first method, the length of the tolerance chain is reduced by shortening one or more elements (Fig. 6.101, no. 1). So, in the tolerance chain from Fig. 6.100, for

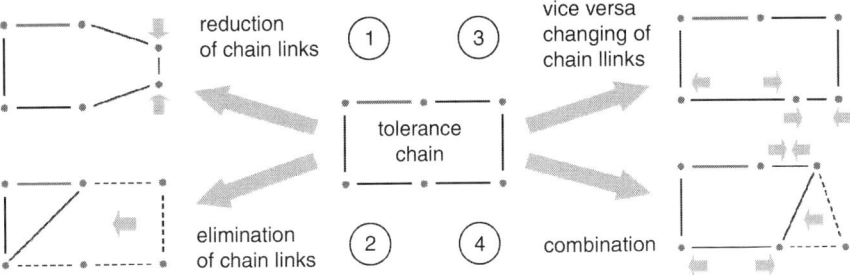

Fig. 6.101 Methods for changing the tolerance chain

example, one element can be shortened by using a robot with greater positioning accuracy. However, this is usually associated with higher costs.

Another method for shortening the tolerance chain is to reduce the number of chain links (Fig. 6.101, no. 2). Hence in the tolerance chain visualised in Fig. 6.100, the number of elements can be reduced by measuring the position of the lens with respect to the crystal because in doing so, the functionally relevant characteristic itself would be measured (distance between the lens and the crystal). Consequently, the repeat accuracy of the robot, for example, would no longer have any impact on the deviation in distance between the crystal and the lens.

To increase the efficiency of implementation of the tolerances while complying with the overall criterion of the tolerance chain (i.e. the specified tolerance), reciprocally changing individual elements can also be effective (Fig. 6.101, no. 3). This is the case, when, for example, shortening one element is associated with relatively low costs while lengthening another is associated with relatively high savings. Hence, instead of an articulated arm robot, a simple handling device could be used in conjunction with a fine-positioner. That way, the handling device puts the fine-positioner within the range required to join a lens. As already shown in Fig. 6.101, no. 2, the inaccuracies of the handling device are compensated by additional measuring technology.

Simultaneous application of the three methods specified shortens and/or increases the efficiency of the tolerance chain (Fig. 6.101, no. 4). These explanations clearly show that by examining the tolerance chain, it is possible to identify ways of shortening it and to uncover potential for making tolerance implementation more efficient. This potential is uncovered mainly through modelling and therefore focussing on the deviations and tolerances. The three methods of system optimisation are presented in detail below.

System Optimisation by Shortening Individual Elements The flexibility required for self-optimising assembly is usually achieved through robot systems. In this connection, storing a model of the robot and its environment in the control system allows robot programs to be automatically adapted when parts or processes are changed. The precision of execution of the new instructions is primarily determined by the model's consistency with the actual assembly system. However, significant deviations due to production tolerances mainly occur when nominal robot geometry is used to create the model (Elatta et al. 2004).

Within the context of self-optimisation, the assembly tolerances specified must be met without additional and complex automatable corrective measures, which is why highly accurate models are required. In this respect, through integrative system identification, it is possible to determine the relevant model parameters for the assembly system extremely accurately and thus significantly reduce the gap between simulation and reality. The following diagram (Fig. 6.102) illustrates the basic system identification process (Müller et al. 2010).

First, a nominal model is created for the assembly system. For the purpose of analysis and identification, the entire system is broken down into subsystems. The parameters of each subsystem are identified using established, state-of-the-art methods.

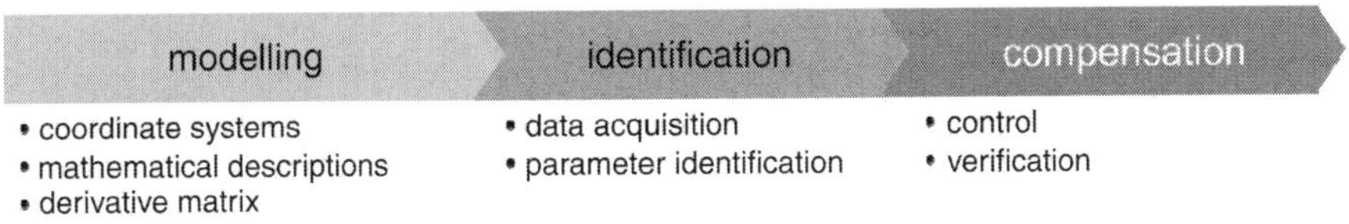

6 Self-optimising Production Systems

modelling	identification	compensation
• coordinate systems • mathematical descriptions • derivative matrix	• data acquisition • parameter identification	• control • verification

Fig. 6.102 System identification procedure

To this end, internal and external measuring systems are used to collect the data required to calculate the characteristic parameters needed to describe the model. Next, the control system and simulation environment are adapted accordingly. Finally, further measurements are taken to verify the model identified.

The first step of the procedure is to model the assembly cell using a parametrisable model. Based on the CAD data, the models are represented in a graphical simulation environment. For a complete model, not only the handling devices are included but also the product to be assembled, the necessary production equipment and the collision bodies (fencing, equipment housings, etc.). Next, all objects in the cell are provided with coordinate systems.

In offline programming, system behaviour must be precisely mapped in order to minimise the adaptation effort required in the actual cell. Complete and error-free mapping of the assembly system in a model, however, is impossible because some errors are of a stochastic nature and therefore cannot be explicitly modelled. On the other hand, not every single influence that can possibly be mapped has to be included in the model because identifying the model parameters based on the real system can be extremely laborious. In many cases involving robots, only the exact kinematic parameters are determined while the dynamic effects are disregarded because this simple calibration leads to a substantial and more than adequate improvement in accuracy.

The model to be identified must not contain uninfluential or linearly dependent parameters. In describing the robot using Denavit-Hartenberg parameters (DH parameters), some parameters may be interdependent. With a vertical articulated arm robot, for example, the movement parameters of the second and third axis are interdependent. Hence, for identification, one of the two movement parameters is fixed and not available for optimisation. Also, it is necessary to ensure that only those parameters that affect performance are made available for optimisation. If it is just the position that is determined with the measuring system, then parameters that only affect TCP (tool centre point) orientation cannot be identified because they cannot be monitored using the measuring system.

To allow numerical optimisation, it is important to use a reduced and non-redundant model. If this criterion of minimalism is not met, the system of equations cannot be solved explicitly. To be able to minimise the subsequent positional error between the model and the actual kinematics, therefore, a regular system of equations must be created for parameter identification. Then the parameters can be identified in the next step.

Fig. 6.103 Schema for parameter identification for industrial robots

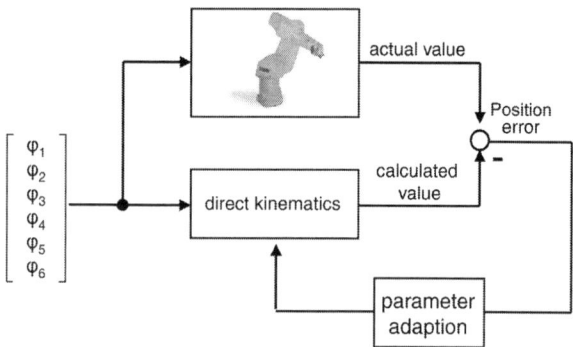

A numerical optimisation function is used to identify the parameters (see Fig. 6.103). To this end, the robot's TCP is measured for a particular set of joint angle combinations using an external measuring system. The TCP poses corresponding to these poses are calculated using a parameterised and non-redundant model. Based on the measured and calculated poses, an error function can be set up for each joint angle combination.

Using the least error squares method, the available model parameters are purposely changed until the positional errors for all joint angle combinations reach a particular termination criterion.

In addition to determining the exact model parameters, the kinematic parameters of the production equipment must also be identified. To this end, using an external measuring system, numerous significant edges, surfaces or points of the body are sampled in a reference coordinate system. By determining these characteristics, the body's own coordinate systems can be identified.

The final steps of the process are compensation and verification. In order for the identified model to be used, both the cell control system and the simulation environment have to be adapted. Problems can arise, especially with respect to adapting the robot controls, because not every control manufacturer provides the option for the user to adapt the relevant parameters. This is due to the fact that with the reverse transformations implemented, simplifications were made. Should the model deviate from these values, the stored reverse transformation can no longer be used and another way to adapt the control system must be found. This can be achieved by means of an upstream algorithm for specifying the Cartesian coordinates. In this respect, the target coordinates from path planning are converted into joint coordinates using the calibrated reverse transformation. With the nominal model used for the robot control system, the joint coordinates can be converted into Cartesian target coordinates. The manipulated Cartesian coordinates are then converted into the target coordinates by the robot control system and the actual kinematics (see Fig. 6.104) (Wiest 2001).

Once the models have been adapted in the simulation and to the control system, the result has to be validated. This step is essential, especially for numerically generated analogous models since these models have only been optimised for a specific working area.

Fig. 6.104 Algorithm for specifying Cartesian coordinates

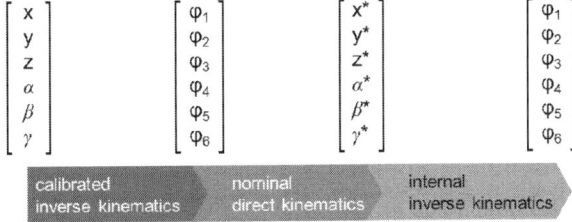

This approach to system identification makes it possible to improve the accuracy of models for robots and assembly systems and thus significantly shorten the corresponding tolerance chain.

System Optimisation by Eliminating Individual Elements The accuracy achieved by determining the exact kinematic parameters is sufficient for many applications, e.g. pick-and-place tasks. However, the accuracy usually remains insufficient for precision applications. Hence in many cases, additional measuring technology and sensors are used in order to improve the accuracy.

Figure 6.105 presents an example of camera-supported lens positioning. One robot holds a camera that is pointing towards the base plate. Another robot independently brings the lens to be mounted into the camera's image area. By measuring the base plate and the component, the camera incrementally guides the robot into the right joining position. This serves to eliminate many effects of the tolerance chain, e.g. the positioning accuracy of the robot.

6.5.5 Outlook

The theoretical foundations of self-optimising assembly systems presented in this paper are investigated and demonstrated with respect to various aspects based on the fully automated assembly of a miniaturised solid-state laser.

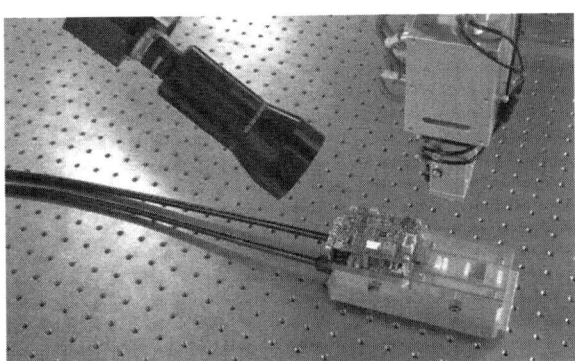

Fig. 6.105 Inline measuring technology for improving positional accuracy

Towards the vision of autonomous, self-optimising assembly systems, it is important to continue building on these results especially in the field of automated planning at cell and control levels. Due to the specific characteristics of multi-robot systems in particular, transferring the shop floor engineering solutions of flexible manufacturing systems to flexible assembly cells is only possible to a limited extent and requires new approaches. Through greater integration of system components at the lower control levels and through the configurability of components (mobile units, modular tools and sensors), functions and capabilities cannot often be assigned clearly and therefore considerably increase the degrees of freedom of planning activities.

When several robots are used in one assembly cell, the working areas of the individual robots may overlap, which is why additional aspects of coordinate transformation as well as collision detection and avoidance are added. Also, besides a large number of degrees of freedom plus complex and not formally defined semantics, the problem of assembly control is characterised at cell and control levels by significantly higher demands on response time, too. Only with appropriate real-time control solutions for modular, distributed systems along with new description models and planners for configurable and cooperative capabilities can the next step towards continuous self-optimisation in automated production be taken.

6.6 Self-optimising Assembly Systems Based on Cognitive Technologies

Marcel Ph. Mayer, Barbara Odenthal, Daniel Ewert, Tobias Kempf, Daniel Behnen, Christian Büscher, Sinem Kuz, Simon Müller, Eckart Hauck, Bernhard Kausch, Daniel Schilberg, Werner Herfs, Christopher M. Schlick, Sabina Jeschke and Christian Brecher

6.6.1 Abstract

Like business and production processes, entire production systems are often based on hypotheses that only provide a partial view of the value-adding chain or result in specific technological interactions. We rarely fully understand the way that processes, materials, production resources and humans interact or how this affects the product itself. It is impossible to predict how changes will impact on value creation as a whole. Because it is so difficult to gain an integrated view, optimisation processes often only focus on individual elements of a system. Existing ways of optimising the behaviour of a single element can become the focus of too much attention and can tie up resources even though, in certain situations, doing so can have a negative impact on behaviour in other areas of the system.

The solution to this conflict lies in designing a system that can adapt its goals according to the situation. While most system optimisation processes happen externally—i.e. a human controls them—having the system itself perform them can be an interesting option in many cases. However, developments in automation show that, even in comparatively simple situations, we still have work to do to achieve this goal. This means that humans often still play an important role in system optimisation. Implementing self-optimising capabilities offers great potential for resolving issues associated with planning efficiency.

Using cognitive functions as a basis, it is possible to take parts of the original range of tasks—like detailed algorithmic planning of an assembly process, and rule-based re-planning and adaptation planning—and transfer these from humans to the machines themselves. The first step in achieving this involves designing and developing a software architecture that is capable of fulfilling the specific requirements of a cognitive system, such as independently planning actions and executing them. This architecture is based on a three-layer model that is expanded to include modules for knowledge representation and human-machine interaction. The deliberative part of this architecture—the planning layer—is where the actual planning and decision-making happens, through targeted linking of production rules. A hybrid approach is used to achieve a flow-oriented, error-free assembly of randomly fed-in components. The approach combines traditional planning processes (an "offline planner") with reactive cognitive control (an "online planner"). Prior to execution, the machine's offline planner calculates the complex geometric analysis steps for assembly planning. This involves an assembly-by-disassembly strategy, which identifies all the assembly steps necessary to bring together individual components into a finished product. To do so, all possible pairs of subsets of components are disassembled and checked to establish whether this can be done collision-free. This process is repeated until the product has been completely disassembled. Because an assembly action can reverse every disassembly action, read backwards the disassembly steps produce all possible assembly steps. The results of this preliminary analysis are recorded as a state graph. The edges in the graph are weighted with cost values for the different work steps and also show information on any additional components needed. Weightings for the different interim states are also assigned to the nodes in the graph.

The online planner can evaluate this information during assembly to derive a suitable assembly sequence for the current production situation and a given target system. This involves updating the state graph on the basis of the available components and the current state of the partially assembled product. Next, the A* algorithm is used to calculate the best route from the current state to the target state (the end product). Under the prevailing boundary conditions, this path represents the optimal assembly sequence. The next work step involves investigating the assembly sequence to identify possible parallelisable assembly actions. The parallelisable actions are then combined in such a way that the assembly plan is made up of a sequence of sets of assembly actions that can be executed at the same time. This plan is then passed on to the decision-making component—a cognitive control unit (CCU) based on SOAR cognitive architecture—where it serves as a basis for decisions. The CCU checks the current situation in cycles and initiates appropriate actions. Under normal circumstances, from the set of parallelisable actions to be executed next, the CCU randomly selects an action and implements it. In unforeseen situations, either re-planning is initiated or the human operator is requested to solve the problem. This process continues in cycles until the desired product has been fully assembled.

A robot-supported assembly cell is presented to validate the system. In addition to covering all key aspects of an industrial application (relevance), the cell can easily illustrate the function and flexibility of a CCU (transparency). The assembly cell has a conveyor-belt system, which is used to feed in components. There are two jointed-arm robots, of which only one is initially connected to the CCU. This robot can be moved along a linear axis and, in addition to having a flexible multi-finger gripper, is equipped with a colour camera to identify components. The second robot separately feeds in components on a conveyor belt. In the centre of the assembly cell there is a surface that can be used as a workspace and as interim storage for parts that cannot be used immediately. The human operator's workspace is currently located directly next to the surface, separated by an optical safety barrier. The workspace has a human-machine interface that can be either stationary or mobile. By displaying process information in an ergonomic way, the interface provides details on the system status and can, if necessary, help human operators to interactively identify and resolve process errors.

To enable humans to safely and effectively manage and monitor the process, machine-made decisions (e.g. on establishing the assembly sequence) must be presented quickly and in a way that is easy to understand and will ensure reliable implementation. Furthermore, the assembly sequence must be planned in a way that conforms to human expectations. With this in mind, a concept was developed that enables system behaviour to adapt to operators' expectations by using a human-centred process logic based on the MTM-1 system. This concept was validated using the assembly cell. Two series of experiments under laboratory conditions sought to establish whether humans followed certain easy-to-generalise strategies when carrying out an assembly task. The experiments identified and validated three assembly strategies as rules: (1) People begin assembling from edge positions; (2) People prefer to assemble adjacent to existing objects; (3) People prefer to assemble in layers. A specially developed simulation environment shows to what extent taking the identified assembly strategies into account as production rules within the knowledge base of the CCU makes it easier to predict the actions of the controlled assembly robot, and to what degree the rules can be generalised.

The presentation of results concludes with a lab study designed to investigate visually presenting information to humans. The starting point was the following scenario: a robot plans the workflow for a production task previously defined by a human operator and then carries it out. However, an error occurs during processing. Because the cognitively automated system is unable to identify the error itself, the human operator must be put in a position to quickly and safely identify and resolve the error. The study compared different ways of displaying the information, in order to find the fastest and most reliable one for interactive error identification. A subsequent comparison looked at the differences between a TFT screen in a workroom and a head-mounted stereoscopic display. The results showed that a display fitted over the field of view improved error detection rates.

The research results made it possible, for the first time ever, to achieve an integrated demonstration of, and carry out scientific investigation into, the design, development and application of cognitive mechanisms in automation using a robot-supported assembly cell to control them. The work also identified the need for further research into whether the results can be transferred to products widely used in industry, and to the ergonomic design of human-machine cooperation.

The cognitive automation of production systems offers a technology that, with the same or even less planning effort, can efficiently and robustly automate product families with large numbers of variants—even in cases where only small numbers of each variant are produced. This effectively makes customer-oriented mass production possible. Cognitive mechanisms in automation offer high-wage countries in particular the chance to achieve considerable competitive advantages, and thereby to directly contribute to securing and expanding their own production locations.

6.6.2 State of the Art

Autonomous production cells can be considered the predecessors of cognitively automated production systems. They are mainly characterised by physical process models integrated directly into the machine control system and by machine operation and monitoring based on these models. Safe, effective human-machine interaction means that even complex handling processes can be carried out error-free over an extended period of time, and that the work system can be operated at the optimum level in terms of performance and operational requirements (Schlick 1999; Pfeifer and Schmitt 2006). However, automated production cells only have limited self-optimising planning functions; these functions are crucial to the concept of cognitive automation. The work of Onken and Schulte (2010) built on these functions and helped to shape the concept of cognitive automation and introduce it to the scientific community. But the original concept focuses heavily on using the technology in unmanned vehicles and is therefore only partially applicable to production systems. Although the corresponding concept of the "cognitive factory" (Zaeh et al. 2009) shows that cognitive mechanisms can be successfully integrated into manufacturing systems, the overriding subject of automation using cognitive models that incorporate the "human factor" for machine operation and monitoring has remained largely unexplored.

To present the current status of research into cognitive automation in production with a view to application in assembly contexts, we must look at a number of different aspects. In view of the research results presented in Sect. 6.6, what follows will focus on four central aspects: (1) Software architectures for cognitive systems; (2) Planning assembly sequences with formal methods; (3) Industrial automation; (4) Task allocation between humans and machines.

6.6.2.1 Software Architectures for Cognitive Systems

With regard to designing and developing automated robotic systems, numerous architectures have been proposed as basic structures for simulating cognitive functions

(Karim et al. 2006; Gat 1998). These software architectures combine a deliberative part for the actual planning process (planning level) with a reactive part for direct control (action level). A widely used approach here is the three-layer model that comprises cognitive, associative and reactive control layers (Russell and Norvig 2003; Paetzold 2006). The lowest layer (reactive) contains the components that control information processing, and is designed to influence system behaviour in such a way as to ensure that the required reference variables are achieved quickly and accurately. The associative layer monitors and controls the system. The majority of rule-based auxiliary functions for automation—like process control, monitoring processes and emergency processes, and adaptation routines for improving system behaviour—are all embedded here. In this top layer, the system can apply "reflexive" methods (e.g. planning and learning processes, model-oriented optimisation processes and knowledge-based systems) to use knowledge about itself and its environment to improve its own behaviour. The focus here is on the system's cognitive ability to carry out self-optimisation.

The software architecture used in Collaborative Research Centre (CRC) 614—"Self-optimizing concepts and structures in mechanical engineering"—picks up on this model (Gausemeier et al. 2009a, b). The Cognitive Controller from the Technische Universität München is also based on a multilayer model. The signals from the production system in question are prepared and processed by a standard controller and by a cognitive safety controller. Additional general studies of cognitive systems can be found in Onken and Schulte (2010) and, regarding the production environment in particular, in Ding et al. (2008). The latter focuses on implementing cognitive capabilities in security systems for plant control. As such, the study pays particular attention to safety in human-machine interaction and to safety in the workplace.

Complementary concepts and methods can be found for automobiles and air and space travel (Kammel et al. 2008; Putzer 2004). Particularly noteworthy here are the various architectures used by the winning teams in the DARPA Grand Challenge, a competition for driverless ground vehicles. These architectures provide regular insight into the latest research (Thrun et al. 2006; Kammel et al. 2008). Many of the architectures are modelled on a "relaxed layered system". In this system, every layer can use the services of the layers below. This makes the system more flexible and efficient by ensuring that every layer is fully supplied with information (Urmson et al. 2008).

This shows that modified architecture models are used to configure and develop cognitive systems. We can see that the most-used approach involves combining the multilayer model with other models. The final architecture must therefore be adapted to the specific application to ensure the highest level of system performance.

6.6.2.2 Planning Assembly Sequences using Formal Methods

"Planners" play a major role in cognitive systems. There are numerous formal approaches to solving planning tasks in different fields of application. Hoffmann (2001) developed the Fast-Forward Planner, which is capable of deriving actions for given problems in deterministic operational areas. By contrast, other planners can handle uncertainty (Hoffmann and Brafmann 2005; Castellini et al. 2001). All the planners mentioned are based on a symbolic knowledge representation. In the case of assembly planning, which requires the geometric relationship between conditions and their transitions to be adequately represented, this kind of knowledge representation becomes extremely complex, even for simple tasks. As a result, generic planners fail when it comes to calculating assembly sequences with even a short-term planning horizon.

Other planners have been specially designed for assembly planning. They work directly with geometric information to derive action sequences. A popular approach is the Archimedes system (Kaufman et al. 1996), which uses and/or graphs for formal representation, and the assembly-by-disassembly strategy to dismantle the end product into its individual components. Thomas (2008) picks up on this concept and develops it in such a way that the system requires no additional help from the user to deduce all possible disassembly steps for the end product—as was the case with the Archimedes system. Thomas only uses geometric information about the end product and the components it contains. This approach, however, is not capable of dealing with "uncertain" planning data. We can find another approach in the assembly planner developed by Zaeh and Wiesbeck (2008), which calculates action sequences almost independently. However, it is not designed to control a technical system, but as a support system for manual assembly. This means that the decision to carry out the sequence is made by a person, not a machine.

6.6.2.3 Industrial Automation

Industrial automation comprises many different controllers which, by working together, enable production plants to function. The components of an automation system used in production are usually assigned to different levels (DIN EN 62264 2008). The lowest level houses the sensors and actuators used to detect and change the state of a production plant. Sensors and actuators are linked, either directly or via fieldbuses, to device controllers. Depending on the machine type to be controlled, in manufacturing technology either a programmable logic controller (PLC), numerical control (NC), robotic control (RC) or motion control (MC) will be used. Each machine can then be controlled by one device controller or more. Groups of machines are combined to form cells, which are coordinated using device controllers. Depending on its size, an entire production plant comprises several cells and is controlled by a manufacturing execution system (MES).

The most commonly used device controllers are designed to meet the demands of "traditional" industrial automation. They are programmed using languages that are partially reminiscent of circuit diagrams, or using machine-like programming constructs that provide maximum control over the relevant device or group of devices (DIN EN 61131-3 2003).

The models and algorithms necessary for self-optimisation are not currently part of device controllers. It is only at the control level that "intelligent" planning algorithms come into play (Brecher et al. 2008a) and the concept of multi-agent systems is used (Brennan 2003). The prevailing programming paradigms in automation result in rigidly linked processes at the cell level and the device level.

In automation the term "intelligence", applied to the lower levels, is often taken to mean adaptive control. However, intelligent decisions on targeted action, which are necessary for assembly, are taken at the cell-control level and, in automated solutions, implemented in controlled movements. As a rule, using PC-based cell controllers (Possel-Dölken 2006; Upton 1992) in industrial environments also makes it possible to use the SOAR cognitive architecture employed in this project.

6.6.2.4 Task Allocation Between Humans and Machines

Looking at the role humans play in standard automated production, we see that their main task involves managing and monitoring the manufacturing system. In the event of a malfunction, they must be able to take over manual control and return the system to a safe, productive state. This concept, termed "supervisory control" by Sheridan (2002), involves five typical, separate subtasks that exist in a cascading relationship to one another: plan, teach, monitor, intervene, learn (see Fig. 6.106).

After receiving an (assembly) order, the human operator's first task usually involves planning the assembly process. To do so, he or she must first understand the functions in the relevant machine and the physical actions involved to be able to construct a mental model of the process. Using this basic understanding, the operator then develops a concrete plan that contains all specific sub-targets and tasks necessary. "Teaching" involves translating these targets and tasks into a format that can be used for machine control—e.g. NC or RC programs that facilitate a (partially)

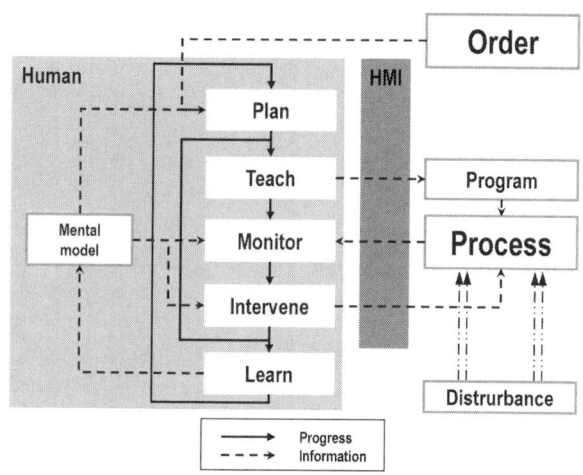

Fig. 6.106 Supervisory control. (Based on Sheridan 2002)

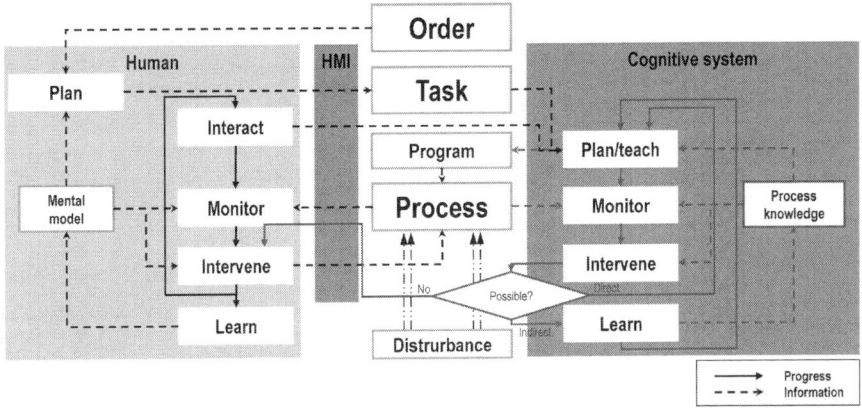

Fig. 6.107 Extended supervisory control approach for cognitive systems. (Based on Mayer et al. 2008)

automated process. This process must be monitored to ensure that it runs properly and produces products of the desired quality. The expectations for the process are drawn from the mental model the operator created at the start. In cases where reality deviates significantly from the model or where there are anomalies, the human operator can intervene by modifying the NC or RC program or by manually optimising the process parameters, for example. Ultimately, every intervention involves the human operator continually adapting his/her mental model, while existing process information, characteristic values and trend analyses help the operator better understand the process and develop a more detailed mental model.

With a cognitively automated system, the tasks change gradually, but in a conceptually relevant way (see Fig. 6.107). Because the cognitive control unit (CCU) can independently solve a certain class of rule-based production tasks, the human operator is relieved of performing repetitive, monotonous or very dangerous tasks. In a cognitive production system, the human operator defines the assembly tasks based on the status of the sub-product or end product, carries out adaptations or sets priorities as needed, compiles rough process plans, and sets initial and boundary conditions. The information-related pressure on the human operator is considerably reduced in the areas of detailed planning and teaching, because the cognitive system handles them. But shifting this load from the human to the machine can result in the human operator forming an insufficient mental model of the state variables and state transition functions in the assembly process. This is because knowledge relating to execution is already stored in the system's process knowledge. At the same time, however, humans must monitor system status and dynamics, and possibly make decisions based on this knowledge. Especially in the event of an error that the system cannot identify or solve, the human operator must receive all information relevant to the situation in an easily understandable form so that he/she can intervene correctly and enable system recovery.

6.6.3 Motivation and Research Question

In high-wage countries, automating production systems can cover over 70% of their functions. Given the law of diminishing marginal utility, raising the degree of automation even higher will not necessarily lead to a significant increase in productivity. Although automation can, as a rule, reduce the frequency of process errors, it also causes a disproportionate increase in the possible consequences of a single error (Kinkel et al. 2008). These relationships, which Lisanne Bainbridge incisively referred to as "ironies of automation" (Bainbridge 1987), are represented in Onken and Schulte (2010) as a negative feedback loop (vicious circle). To circumvent human shortcomings, a function that humans originally performed is automated. This increases the complexity of the system, which in turn places greater demands on the employee responsible for monitoring the automated function. The result is that the entire system potentially becomes less robust. The loop comes full circle when humans attempt to use automation again to compensate for these possible weaknesses. While it is not uncommon for an automated system's productivity to increase during the first iteration of the loop, humans often underestimate or even ignore the risks involved. Onken and Schulte (2010) believe that using mechanisms borrowed from human cognition presents an opportunity to "break the cycle" and design a flexible production system in which humans and machines work together safely and effectively—especially in process planning and monitoring, and in disturbance management.

A system that is capable of learning and of adapting to changing environmental conditions can increase planning efficiency by reusing acquired knowledge and transferring it to similar, new production cases. This can sometimes considerably reduce the number of iteration steps involved. Systems of this kind are known as self-optimising systems (Frank et al. 2004). Self-optimisation requires cognitive capabilities that, in current production processes, only humans possess.

In this context, the term "self-optimising production systems" describes a concept that can implement value-stream-oriented measures at the same time as increasing planning efficiency and improving process and product quality (Schmitt and Beaujean 2007). Transferring existing knowledge to similar or new production cases—the essence of self-optimisation—opens up new perspectives for production and assembly systems by enabling them to dynamically adapt system behaviour to keep pace with changing targets and situations. Incorporating humans' unique skills and experiences into the system is considered essential to self-optimisation. Innovative cognitive functions in the form of symbol-processing systems should support humans and, where necessary, relieve them of routine tasks.

Here, cognition is understood to mean processes such as perception, knowledge storage, reasoning and learning. Obviously software can only partially simulate the unique features of human cognition, but some models (Strohner 1995) can be partially transferred to technical systems and thereby provide a suitable basis for self-optimisation. Thus, cognition can largely be described as referring to the transfer and application of knowledge, and to the processing of information, either by a living being's central nervous system or in an artificial system (Strohner 1995). Within this context, the sub-project described here, which is concerned with self-optimising production systems, focuses on designing and realising a prototype of a CCU. This CCU can use symbolic knowledge representation to optimise itself according to predefined criteria. Most importantly, however, the CCU can be designed, developed, and operated safely and efficiently by highly qualified experts in a high-wage country like Germany.

Developing the design starts with the dilemma of planning orientation and value orientation in the polylemma of production (Schuh and Orilski 2007). In planning-oriented production, processes in a manufacturing system are centrally planned during operations planning and scheduling, in great detail, far in advance, and in line with the Taylorist principle of separating preparation and execution. Doing so makes it possible to closely align production steps with the overall target because all activities are analytically derived from the desired end result using a global target function. In value-oriented production, planning activities overlap with the actual value-adding process. In addition to carrying out the activity directly related to value creation, the person responsible for production defines sub-tasks, their sequence, and the use of production resources. The overall target is therefore generated in collaboration: organisational units independently define their sub-targets and sub-tasks, and these come together along the process chain. This approach has the benefit of allowing the production system to respond quickly to changing boundary conditions, which means that it can better handle the complexities and dynamics of its environment and the process itself. State-of-the-art value-adding chains are typically only partially capable of independently finding top-down solutions to specific problems on the basis of simplified models within a defined solution space. As a rule, these chains do not fully take into account interactions between processes, materials, production resources and the people working in the environment, meaning that knowledge of these interactions is usually incomplete. The same is true of how these interactions affect the design of the product structure.

Cognitively automated systems should pave the way for new concepts and technologies for production and assembly systems that should be able—through continuous data analysis, information fusion, interpretation and assessment of the actual situation—to dynamically adapt themselves to changing targets and boundary conditions. The research question that arises from this is: how can we achieve a highly dynamic system while at the same time ensuring that the targets of all activities are well-synchronised?

The solution posited in designing and using cognitive automation is as follows: a cognitively controlled production system reacts faster, more reliably and in a more resource-efficient way than a production system that uses traditional planning logic and methodology. Unlike a standard control system, a cognitively automated system can, on the basis of internal decisions, independently redefine reference variables in terms of targets, and adapt the control strategy accordingly. It would, however, be naïve to believe that such a system could function completely autonomously. Scientific research must focus on designing and producing prototypes of cognitively automated production systems that can be further developed and efficiently operated by highly qualified skilled workers in high-wage countries. These are no longer purely technical systems; they are complex human-machine systems that require an ergonomic design. Therefore, in defining the research question, we can identify two fields of activity:

> Firstly, this sub-project should address aspects of technical design and evaluating cognitive functions. Secondly, any design for this kind of future production system must focus squarely on humans and their superior cognitive skills.

The Cognitive Control Unit sub-project therefore explored the following research questions:

- Design and development of architecture for a cognitive control system: The requirements placed on a cognitive machine control system are reflected in the unique requirements placed on the software architecture of the system. The sub-project therefore investigates how this kind of architecture should be structured and designed. This involves taking into account the different time requirements for machine-oriented control systems (e.g. a robot cell) and planning for various data abstractions. The data flows and information shared between planning, control and human-machine interface must also be defined. A cognitive system works using a knowledge-based approach. This means that the software architecture must include components that allow the system to save and modify knowledge that machines can process and humans can understand.
- Design and development of a planning methodology for cognitive automation: Assembly planning presents a highly complex problem, even for current planning systems. If it also has to be flexible enough to adapt to unexpected events and ad-hoc changes in the process, we need to find special techniques for assembly planning and control that incorporate functions borrowed from human cognition.
- Usefulness of technical cognition in industrial automation: The numerous functions that have to be controlled in a production plant are predominantly designed to effectively and efficiently carry out a pre-defined process. Alongside the primary task (e.g. assembling a workpiece), supporting tasks also have to be controlled, such as transporting and handling the workpieces. To achieve end-to-end self-optimisation and control, all levels of industrial automation require cognitive functions. The sub-project therefore seeks to find out how to integrate cognitive functions into industrial automation, both conceptually and in terms of technical implementation.
- Human-centred design of a cognitive control system's knowledge base: A control system that has a knowledge representation which allows it to plan almost independently and according to the situation will have a considerable impact on the spectrum of tasks performed by human operators. Starting with the human role in this kind of cognitively automated production system, the sub-project uses robot-supported assembly to show how a human-centred CCU design can make system behaviour more compatible with the human operator's expectations. In doing so, the sub-project aims to develop a safe, productive and disturbance-free design for work processes in cognitively automated work systems.

6 Self-optimising Production Systems

- Ergonomic interface design: Another question relates to the way information is presented to human operators. One area that this research focuses on is ergonomically designing head-mounted displays and displays in workspaces using new methods of visualisation and interaction. In contrast to standard input and output media like keyboards and TFT screens, new technologies for user interfaces based on head-mounted, semi-transparent LCoS displays are being developed for use in production environments. They are also being ergonomically designed and evaluated with regard to their potential under operating conditions.

6.6.4 Results

6.6.4.1 Software Architecture

Russell and Norwig's three-layer model (2003), widely used in robotics, was chosen as the basic framework for the architecture. Compared with other architectures commonly used in this field, such as the blackboard model (Hayes-Roth 1985), a three-layer architecture has the advantage of a clear demarcation between abstraction levels and temporal demands. The planning layer operates on a high abstraction level with symbolic problem definitions and must satisfy only soft real-time demands. The reactive layer, on the other hand, has to monitor machine-related control loops in "hard real-time". The coordination layer mediates between these two layers. This is where abstract instructions from the planning layer are transformed into concrete machine control commands. In the reverse direction, the information from the various sensors is aggregated to form an overall picture of the situation and is transmitted to the planning layer as a basis for decisions.

In order to satisfy the demands of a holistic consideration of the human-machine system, the classic three-layer architecture was expanded to include further layers and modules (see Fig. 6.108)

The presentation layer forms the interface to the human operator. The interactive goal definition and description of the task, as well as the presentation of the current internal state of the CCU takes place by means of the human-machine interface. In addition, external data formats such as CAD data of the product to be assembled, are transmitted in internal representation forms and made available to the other components. The knowledge module contains the knowledge base of the CCU in the form of an ontology (Gruber 1993). The planning layer can send enquiries to the knowledge module by transmitting the system state received from the coordination layer. The knowledge module then analyses the objects contained in this state and can derive further information by means of reasoning via the ontology and transmit this to the planning layer. A further logging module forms the database for all the other components. All the data generated during system operation are persistently stored here so that in the event of an error, the cause can be reconstructed on the basis of the data if necessary. To support system transparency, all the data stored in the logging module during operation are accessible to the user via the presentation level.

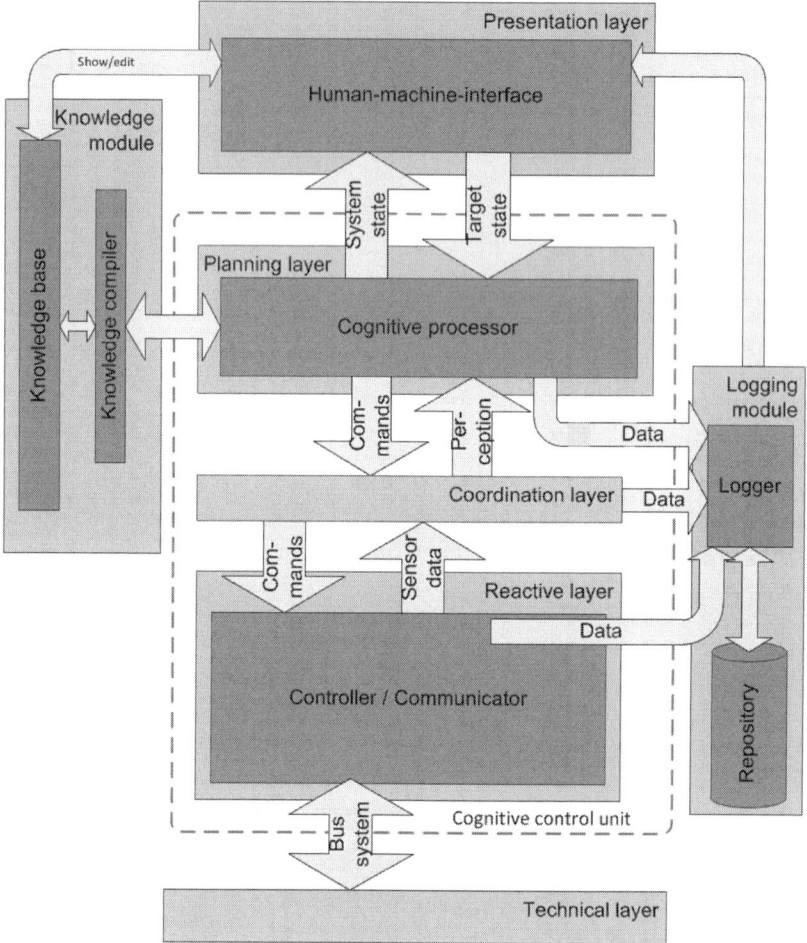

Fig. 6.108 Architecture for a cognitive planning and control unit

Furthermore, the data thus gathered can also be used for training measures that are integrated directly into work processes (embedded training, Odenthal et al. 2007).

The cognitive architecture SOAR, whose internal knowledge base is structured in the form of production rules (if-then rules), was chosen to simulate cognitive functions in the planning layer of the cognitive controller (Leiden et al. 2001; Langley et al. 2004). Compared with emergent systems such as artificial neuronal networks, a rule-based approach has the advantage of not needing time-consuming and potentially unreliable preconditioning. To a certain extent, SOAR is able to simulate rule-based human decisions and to take over repetitive and monotonous process steps (Hauck et al. 2008). It cannot, however, simulate genuine knowledge-based behaviour in the sense of reflecting on goals and their prioritisation (sensu Rasmussen 1986).

6 Self-optimising Production Systems

As SOAR was not designed for automation applications and has no interface to industrial control systems, a framework was designed and developed which makes it possible to model the knowledge and the corresponding algorithms necessary for controlling the assembly and the logistics (Kempf 2010). Furthermore, it provides the architecture required for system control and the necessary interfaces. The planning layer, coordination layer, reactive layer and technical layer were developed for this purpose (see Fig. 6.108). The technical layer essentially consists of the kind of control units used in today's manufacturing cells. The character of the control interfaces is thereby more or less predefined: a robot controller, for example, processes travel commands with the pattern "Move linearly from A to B at speed C"; a PLC expects switching commands with the pattern "If input A is true, then set output B"; a special image-processing software that is located in the reactive layer due to its complexity and the real-time demands imposed by the application expects an order such as "Start object detection and feed back result". The function of the technical layer is therefore to control the individual devices via a defined interface protocol with defined semantics. The main functions of the planning and coordination layers involve dynamically generating the individual device-specific control commands from a global and relatively abstract description of the task, and coordinating their execution in accordance with feedback from the sensors. One example of a component in the reactive layer is the image-processing software mentioned, which, due to its frame rates, has to be located in this layer.

A SOAR agent that takes control decisions independently consists of the model of a given problem space and the mechanisms for its processing provided by SOAR. Each agent has its own long-term and short-term memory, and its own input and output areas, where it operates cyclically, in a manner comparable with the function of a PLC. The planning layer contains SOAR agents which are responsible for selectively assigning control instructions for the production system. One such "dispatcher" thus represents an order. The coordination layer contains agents which execute the control instructions, whereby each operation is represented by a dedicated agent, the "executor". This split between several agents takes into consideration the aspects of hierarchisation and parallelisation (Kempf 2010, see Fig. 6.109).

However, a control architecture is characterised not only by layers and interfaces, but also by the way in which the internal temporal sequence is organised. The cognitive control architecture is generally based on a fixed processing cycle, comparable with the well-known behaviour of a cell controller in the form of a PLC. Similarly

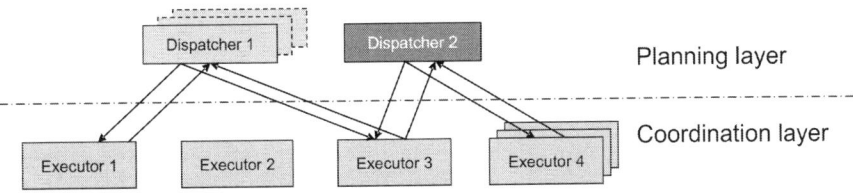

Fig. 6.109 Arrangement of the SOAR dispatcher and executor agents in the layer model

to that, the length of a processing cycle in this case also depends on the number and complexity of the active agents. Since there is no real-time operating system, no hard real-time can be achieved here. But this is generally not necessary for the execution and decision layers. The real-time-critical processing takes place in the reactive and technical layers. The SOAR agents are instantiated as independent objects of the agent class. First the active dispatchers are executed sequentially, then the executors. This means that decision and execution phases are always synchronous with one another. The same applies to the input and output or simulation. As a result, all the agents operate with the same image of the environment during a control cycle, and are therefore quasi-parallel; this also applies to the exchange of messages between the agents. A relatively simple but efficient scheduling policy is employed for the individual agents: an agent continues to receive computing time until it generates an output to the controller. This is frequently the case after one SOAR decision cycle, but in some situations the agent requires several cycles before it reaches a decision. All the agents run in one thread with the SOAR kernel and are processed in a fixed sequence. This process is flanked by two additional measures:

1. To avoid endangering the control functions at execution level, measures must be taken to ensure that an agent does not take too long (e.g. during virtual planning). This is achieved by ensuring that it is always the next agent's turn after a specified maximum number of decision cycles have elapsed. This also means that decision processes of individual agents can extend over several control cycles.
2. As only a limited number of resources are available at the decision level, a dispatcher at the front of the execution sequence tends to have an advantage. Users can prioritise production rules relative to one another. However, if they do not do this, the process follows the principle of "fairness" and the execution sequence in each SOAR cycle is defined at random.

Asynchronously to the actual processing, user inputs can be made for process control and visualisation, with semaphores ensuring data consistency.

6.6.4.2 Hybrid Method for Assembly Planning and Control

Starting from the general function of the CCU, which involves carrying out the planning and control of a robot-supported assembly process on the basis of production rules in conjunction with a formal-mathematical product model, this chapter initially deals with only the cognitive processor. As already mentioned, this is located in the planning layer. In the validation study presented here, the only input the CCU receives is a description of the finished product to be manufactured. "Action primitives" stored in the knowledge base of the CCU in the form of production rules serve as control commands for the industrial robot for component assembly, and are also linked in the processing cycles by the processor, depending on their state, to create a complete and efficient assembly process.

Simulation experiments investigated the influence of various factors on the dependent variables "CCU processor time" and "number of assembly cycles" in the whole process of assembling the target product (MTM-1 cycles, see Sect. 6.6.4.4). The target product consisted of identical parts. The independent variables of the simulation experiment are: (1) The size of the target product (six levels: 4, 8, 12, 16, 20 or 24 parts); (2) The number of parts fed in on the queue (seven levels: 1, 4, 8, 12, 16, 20 or 24 parts); (3) The type of feed-in (two steps: deterministic feed-in of the parts required or random feed-in, including parts not required). For the feed-in, a simple buffer in the form of a queuing model was used that is operated similarly to the FIFO principle. One hundred simulation runs were calculated for each of the $6 \times 7 \times 2$ combinations of factor steps. Given the high amount of computing time required, the simulations were performed on the RZ cluster in the computer centre at RWTH Aachen.

The simulation results show that the target product was correctly assembled in all 8,400 simulation runs. No assembly errors or blocking ("deadlocks") of the cognitive processor occurred.

The CCU processing time and the number of assembly cycles for the random feed-in of the parts are shown in Fig. 6.110. If we first consider only the time required, we can see that the processing time increases when more parts are used in the target object (Fig. 6.110, left). However, looking at the number of parts on the queue at the same time, it is surprising to note that processing time decreases as the number of parts increases. If we consider the number of simulated assembly cycles for a target product of a given size and for feed-in with given parts, we see that the anticipated number of assembly cycles is obtained (Fig. 6.110, right).

The corresponding results for the deterministic feed-in of parts are shown in Fig. 6.111. The simulation results clearly show a disproportionate increase in processing time as the size of the target product and the number of parts on the queue

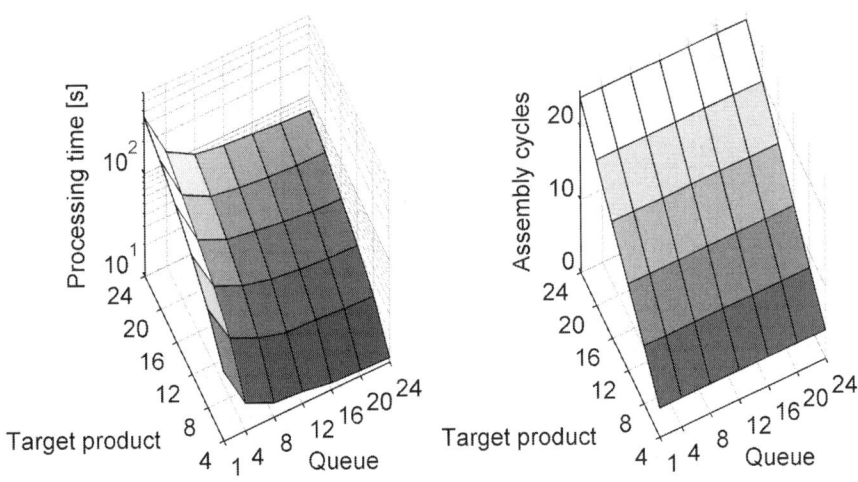

Fig. 6.110 CCU processing time (*left*) and number of assembly cycles required (*right*) as a function of the size of the target product and the number of parts fed in on the queue in random feed-in

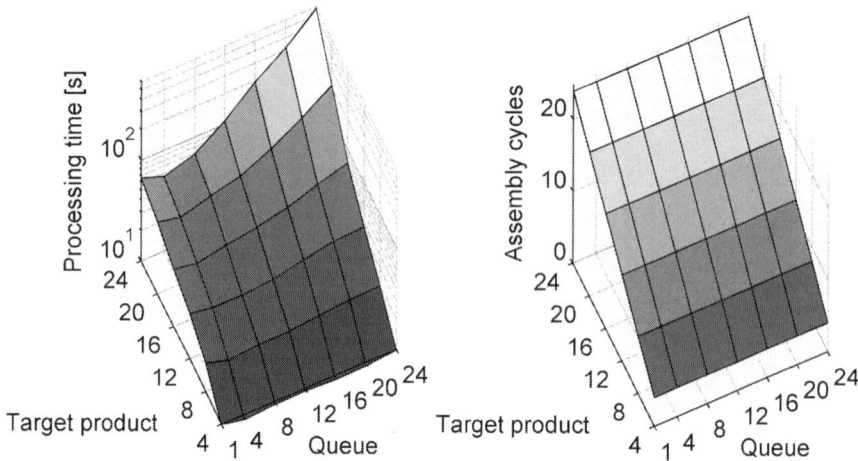

Fig. 6.111 CCU processing time (*left*) and number of assembly cycles required (*right*) as a function of the size of the target product and the number of parts fed in on the queue in deterministic feed-in

increase (Fig. 6.111, left). The number of assembly cycles, on the other hand, behaves as expected in relation to the size of the target product (Fig. 6.111, right).

The results of the simulation study show that a CCU based on SOAR cognitive architecture is able to reliably perform assembly planning. In the given application, SOAR proves to be particularly suitable for reactive planning under the boundary condition of random feed-in of parts. In the case of deterministic feed-in, however, the study showed exponential run-time behaviour. Although difficult to understand intuitively, this simulation result can be explained by the way SOAR functions during the decision-making process. It compares each required part on the queue with every possible position within the target product. These comparisons result in "proposals" that provide the basis for decisions on executing action primitives. Due to the combinatorics, this can lead to a non-polynomial increase in run-time behaviour (see, e.g. Barachini 1990).

If we want to achieve dynamic system adaptation to changing boundary conditions using—instead of SOAR—generic planning algorithms such as the Fast-Forward Planner (Hoffmann 2001), the only way to do so is by performing continuous re-planning or by completely forward planning all possible sequences of fed-in parts. Complete forward planning for products comprising just 15 parts would require drawing up over 3.6 billion plans. Forward planning can therefore no longer be controlled in normal cases by combinatorics. Continuous re-planning during assembly is also not possible, since combinatorics cannot satisfy the real-time demands for comparable reasons. By contrast, the concept presented in this paper of a CCU based on SOAR meets the demand for dynamic system adaptation very well, as the simulation results for random feed-in clearly show. However, the results also show that, unlike generic planning algorithms, the CCU has only limited suitability for classic deterministic planning tasks. The study observed, for example, exponential run-time behaviour with a strictly deterministic feed-in of required parts.

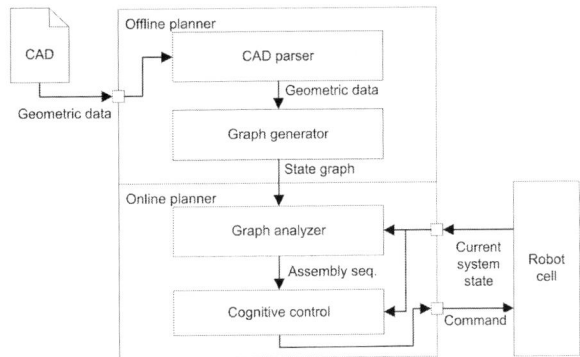

Fig. 6.112 Hybrid plan of the assembly planner

For this reason, a more advanced hybrid solution was developed (see Fig. 6.112). It involves making a distinction between an "offline planner", which computes complex planning steps based on geometric analyses prior to assembly, and an "online planner", which uses the results of the offline planner to generate assembly plans during the assembly process, depending on the current situation.

The offline planner consists of two components: the CAD parser and the graph generator. The parser receives the CAD description of the product to be manufactured and extracts from this the geometric information on the whole product and the individual components. This information is then forwarded to the graph generator. The generator follows an "assembly-by-disassembly" strategy (Lin and Chang 1993) to generate all the expedient assembly sequences for the given product. This involves starting from the fully assembled state and generating all possible disassembly steps recursively, until there is nothing but individual components. Read backwards, these disassembly steps show all the possible assembly sequences. The graph generator performs this process using the method presented by Thomas (2008). The system examines pairs of all subsets of components to establish whether these can be separated from each other, collision-free. Invalid disassembly steps are rejected, while valid steps are added to an and/or graph (Homem de Mello and Sanderson 1986) and further examined and evaluated on the basis of various criteria. An and/or graph for a simple tower made of four bricks is shown in Fig. 6.113.

Each hyperedge of the and/or graph represents one disassembly action, or, in the opposite direction, one assembly action. As some assembly actions have to be performed by different tools or by a person, different costs depending on the type of action are assigned as vectors to each hyperedge. The finished and/or graph is then transformed into a state graph (see Fig. 6.114) because the online planner uses a process—similar to those applied in assistant systems (Zaeh and Wiesebeck 2008)—that cannot be used for hyperedges.

During the transformation, each hyperedge is converted into a simple edge. The corresponding nodes, which represent the resulting disassembly steps, are merged into one node: if both nodes consist of more than one individual component, a new node is generated that contains both sub-products. If a node contains only one component, this node is removed and the respective component is added to the new

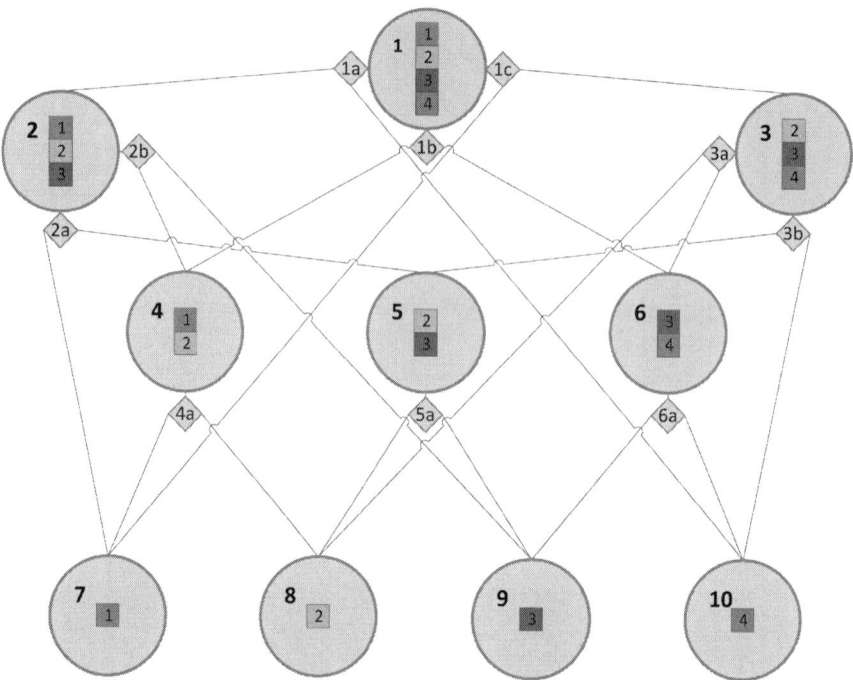

Fig. 6.113 And/or graph for a tower of bricks

edge as a requirement. The resulting state graph is then transferred to the online planner so that assembly can begin.

Within each assembly cycle, the current system state is transmitted to the online planner during assembly, the next action to be performed is calculated, and the robot cell is activated accordingly. Decision-making is carried out by three components: the graph analyzer, the paralleliser and the cognitive control (CC). The graph analyzer contains the system state and updates the state graph. First, the current state of the assembly cell is located in the diagram. Then all the following edges are updated. Edges whose required components are not contained in the current state receive additional "penalty costs". Edges further from the node describing the current state are assigned lower costs. The online planner thus gains the ability to speculate. Actions that are not possible at present, because the corresponding components are not available, are more likely to be integrated into a plan the further they lie in the future. The algorithm "hopes" that the required component will be delivered before it is required.

After updating the state graph, the least expensive path—according to the given target system—from the current state to the target node is calculated using the A* algorithm (Hart et al. 1968). When passing through the graph, A* selects as the next node to be examined the node x, for which the function $f(x) = h(x) + g(x)$ is minimal. The function $g(x)$ designates the costs of the path to be taken to arrive at node x.

6 Self-optimising Production Systems

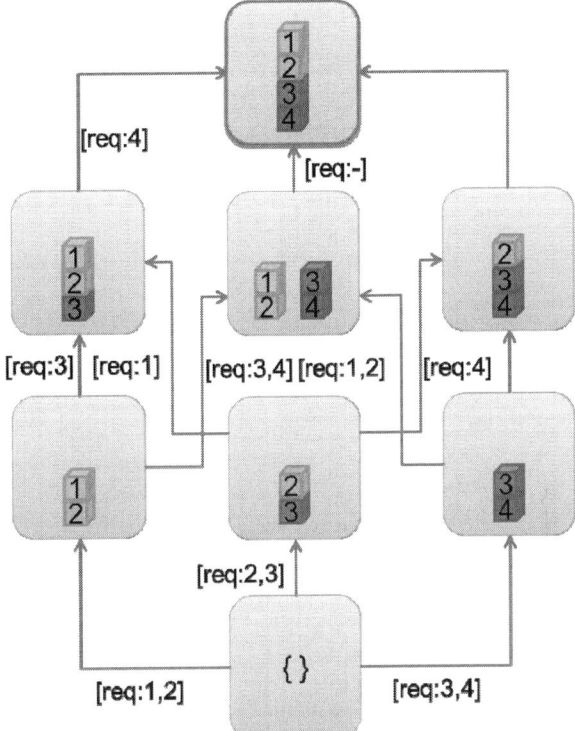

Fig. 6.114 State graph produced from the and/or graph

In addition to the costs of the individual edges of the graph, the cost calculation also takes into account any necessary tool changes. In view of prevailing safety regulations, higher costs are allocated to changes between robots and humans than to changes between tools. Paths with fewer changes therefore have lower costs. The function $h(x)$ is a heuristic which estimates the proximity of a node to the target node. The heuristic employed here uses the number of components already correctly assembled and the valuation assigned by the offline planner with respect to machine transparency. The path calculated in this way is passed on to the next component as the plan to be followed. The process described ensures that this is an optimal plan with respect to several criteria:

- High penalty costs for actions that cannot be carried out ensure that the online planner selects the most realistically probable assembly sequence.
- Additional costs for tool changes create a preference for choosing sequences where the same tool is used for longer periods or which can be carried out by a single human operator.
- Reductions in penalty costs for assembly steps that cannot be carried out, depending on their distance from the current state, create a preference for assembly sequences that can be started immediately, because actions that cannot be carried out are shifted to the end.

Fig. 6.115 Assembly plan as a sequence of sets of assembly actions—before and after parallelisation

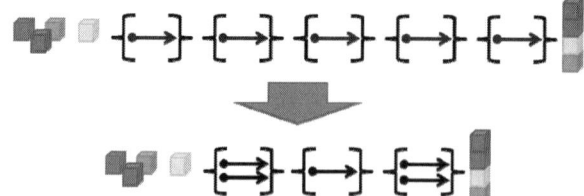

A further optimisation step to accelerate assembly is performed in the paralleliser component. The fixed sequence of assembly actions received is examined for actions that can be carried out in parallel. The sequence is thereby seen as a sequence of sets of assembly actions. Each of the sets is examined as to whether it can be carried out in parallel to the following set, i.e. whether all elements of both sets can be carried out in pairs at the same time. This parallelisation can be derived from the and/or graph: two actions, x and y, can be carried out exactly in parallel when their corresponding hyperedges—e_x and e_y—have a common predecessor hyperedge and when neither e_x nor e_y is a predecessor of the other. If two sets can be carried out in parallel, they are joined and examined to see whether they can be carried out in parallel with the next set. The result of this process is a sequence of sets of assembly actions (Fig. 6.115) which is then transferred to the decision-making component, the CC (Fig. 6.112).

Transferring a plan as sets of actions that can be performed in parallel has a number of advantages. On the one hand, assembly can be accelerated because several robots or even the human operator can work on the assembly at the same time. On the other hand, the CC has greater freedom for decisions because it is free to select the sequence in which the actions within a set are to be processed (if they cannot be performed in parallel).

As already mentioned, the CCU is based on SOAR, a cognitive architecture which attempts to simulate the human decision-taking process. The schematic sequence within this component is shown in Fig. 6.116. During assembly, the CC receives the current system state, transmits it to the online planner and receives back the assembly plan described above. On the basis of this plan and the current state, a decision on the next action is taken using the stored rule base. The CC can decide to follow the given assembly plan and to execute a suitable action from the first set of actions at its own preference. Alternatively, it can decide to ask the human operator for assistance, or to wait until the situation changes, for example through a new component being fed in. After executing the respective action, the resulting system state is queried and checked. If it corresponds to the target state, i.e. it contains the finished product, the assembly process ends. Otherwise the system examines whether the state has changed purely as expected, i.e. as a result of the action performed. If this is the case, it continues processing the assembly plan already received. If the situation has changed unexpectedly, the CC initiates re-planning by the online planner.

6   Self-optimising Production Systems

Fig. 6.116 Sequence of information processing in the CCU

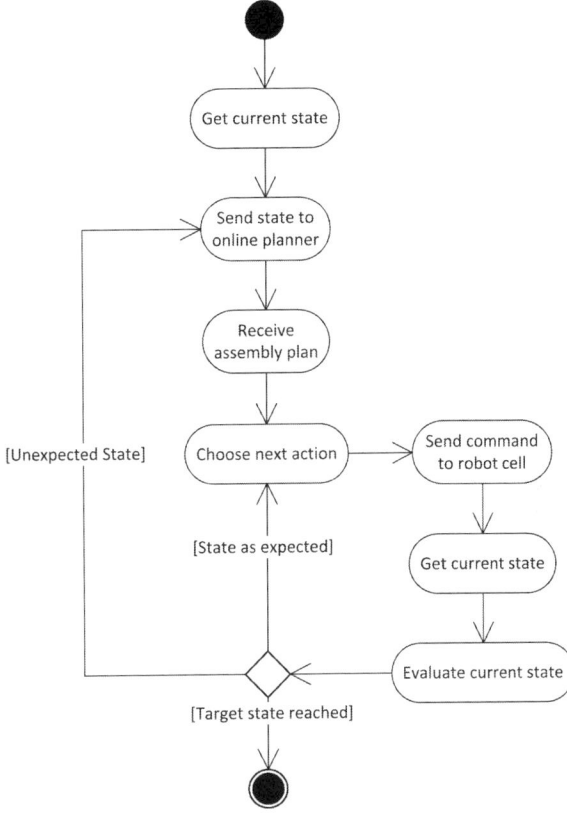

Fig. 6.117 Layout of the assembly cell

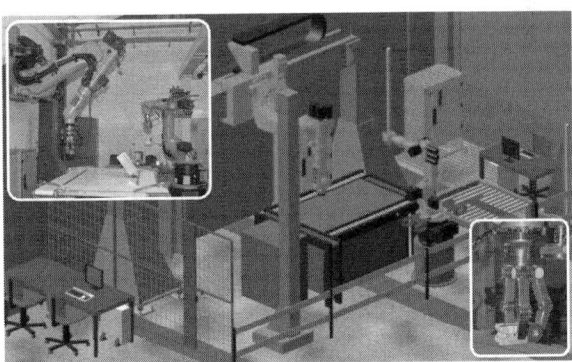

6.6.4.3   Prototype Realisation of a Cognitively Automated Assembly Cell

To test and develop a CCU in a near-reality production environment in a variety of different assembly operations, a robotic assembly cell was set up (see Kempf et al. 2008). The layout of this cell is shown in Fig. 6.117. The scenario was selected to

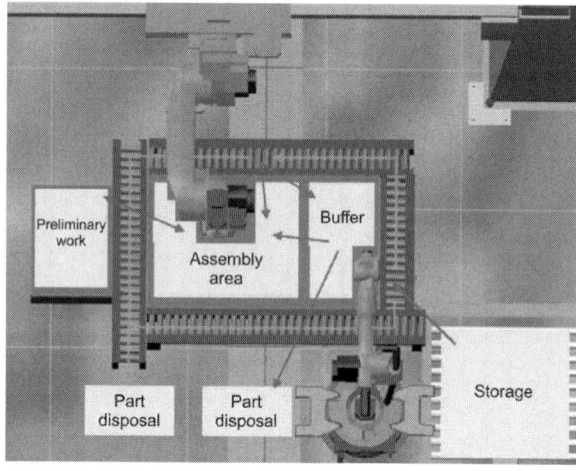

Fig. 6.118 Assembly and storage areas in the assembly cell

comprise major aspects of an industrial application (relevance), and at the same time to easily illustrate the potential of a cognitive control system (transparency).

The main function of the demonstrator cell is the assembly of known objects. Part of the cell is made up of a circulating conveyor system comprising six individually controllable linear belt sections. Several photoelectric sensors are arranged along the conveyor route for detection of components. Furthermore, two switches allow components to be diverted onto and from the conveyor route. Two robots are provided for handling the components, with one robot travelling on a linear axis and carrying a tool (a flexible multi-finger gripper) and a colour camera. Several areas were provided alongside the conveyor for demand-driven storage of components and as a defined location for the assembly (see Fig. 6.118). One area is provided for possible preliminary work by a human operator. This is currently separated from the working area by an optical safety barrier. The workstation has a multimodal human-machine interface that displays process information ergonomically, allowing it to provide information on the system state and to help solve problems, if necessary. Detailed information on the configuration of the multimodal interface can be found in, for example, Odenthal et al. (2008, 2009), and in Schlick et al. (2009). To simultaneously achieve a high level of transparency, variability and scalability in an (approximate) abstraction of the actual assembly process, building an assembly of LEGO Duplo bricks was selected as the assembly task. To take into account the criterion of flexibility for changing boundary conditions, the bricks are delivered at random (see Sect. 6.6.4.2). In terms of automation components, the system consists of two robot controllers, a motion controller and a higher-ranking sequencer. The latter takes the form of a CCU.

The initial state provides for a random delivery of required and non-required components on a pallet. One of the robots successively places the components onto the conveyor. The automatic-control task now consists in coordinating and executing the material flow, using all the technical components, in such a way that only the assembled product is on the assembly table at the end.

As has already been explained, the automatic generation of valid assembly sequences, e.g. directly from a CAD model, is an extremely complex problem for which a universal solution has still to be found. In the present case, however, it is possible to find valid sequences with reasonable computing time that can be generated using the hybrid planning process from Sect. 6.6.4.2. The result of this kind of upstream planning process is a set of sequences extracted from an assembly priority graph that serve the dispatcher as inputs. The assembly planner described in Sect. 6.6.4.2 is also involved in solving the assembly problem. Even for an assembly problem of average complexity, a number of additional boundary conditions have to be observed when planning the assembly graph in order to arrive at a valid assembly sequence. Such limitations exist particularly in the following basic operators, which are based on the MTM-1 taxonomy (Drumwright et al. 2006; see Sect. 6.6.4.4):

- REACH: Does the assembly situation, in combination with the gripper geometry, permit a valid approach trajectory?
- GRASP: Does the assembly situation, in combination with the gripper geometry, make it possible to grip the component during positioning (joining) (see Fig. 6.119, left)?
- POSITION: Does the assembly situation permit movement in joining direction (see Fig. 6.119, centre)? Does the assembly situation permit a stable force couple between the component to be positioned/joined and the assembly (see Fig. 6.119, right)?

The necessary call parameters have to be provided for commanding the functional units via the device interface. Important poses within the cell are preconfigured and are processed within the coordination layer. However, the exact target pose might only emerge at run-time, when a position is located on the pallet, for example. Operation always takes place in a discrete problem space. A component can be deposited on the pallet in, e.g. a given pattern that is stored as part of the plant model. The executor

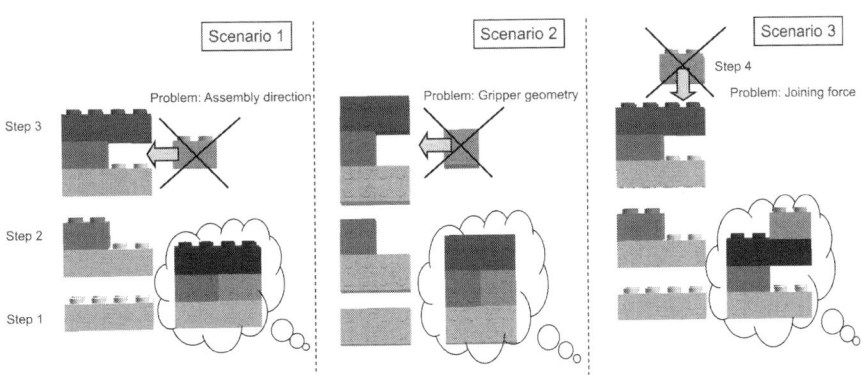

Fig. 6.119 Boundary conditions in positioning the demonstrator components

Fig. 6.120 Coordinate systems and transformations

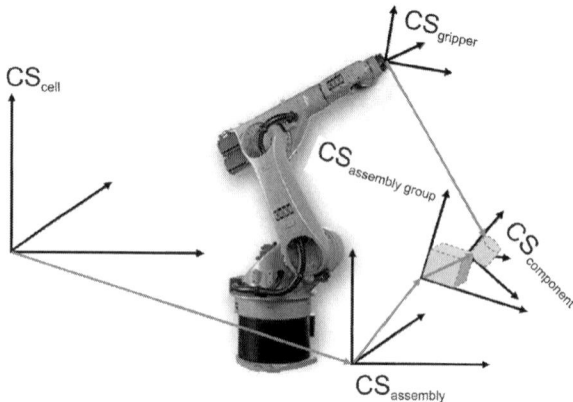

now supplies a discretised frame that corresponds to an actual (relative) component pose and has to be scaled to the modelled grid and offset against the reference pose. All the executors use this principle. Corresponding discretised poses are:

- Depositing poses of components or assemblies (in the cell coordinate system)
- Joining poses of components in the assembly (in the assembly coordinate system)
- Gripping poses on components (in the component coordinate system)

Frame and transformation operations are used to calculate the resulting target pose. This makes it possible to easily calculate random links between different coordinate systems (see Fig. 6.120).

Atomic commands are used in the technical layer to control the robots, grippers, conveyor belts and switches. The challenge here lies in linking the PC-based cognitive controller to the industrial device controllers.

Before a plant can begin production, it must be sufficiently tested. The advantages of using virtual commissioning for this purpose are increasingly being recognised. A cognitive control system probably has even greater need of simulative testing than a classic system does. This is because system behaviour in this case is not even known "on paper"—according to the requirement, it is not generated until run-time. First the question has to be answered as to how a control framework can make it possible to test the resulting processes in advance and—if possible—to visualise them. One possibility is by using a simulation tool of the kind used in connection with classic commissioning. These kinds of simulation tools generally have an OPC interface that allows them to connect to an external controller (generally in the form of a soft PLC). To avoid having to create several interfaces in the cognitive controller to device controllers and simulation systems, an abstracted interface was used that allows the different types of device controller and simulation software to be linked transparently. This abstracted interface communicates using the real-time-capable CORBA middleware (see Fig. 6.121).

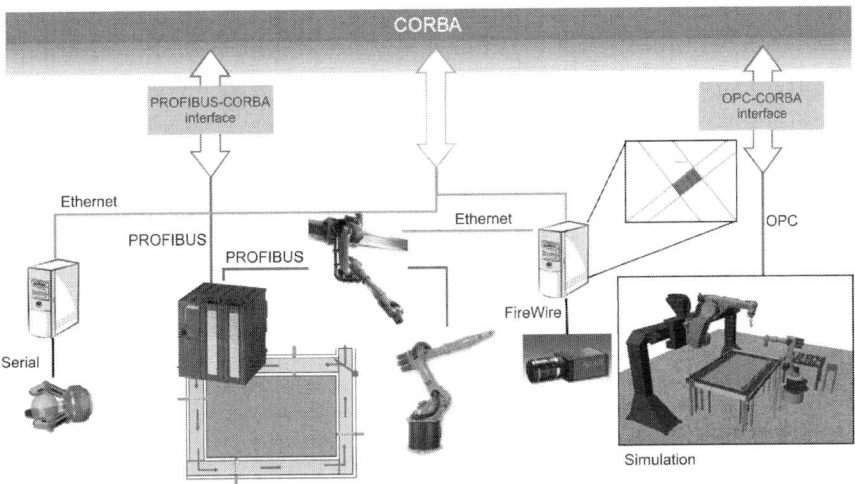

Fig. 6.121 Communication structure in the technical layer

In the technical layer, the function calls of the actuators and sensors with their parameters as control commands are either transmitted directly via the CORBA interface from the cognitive controller to the corresponding PC-based controller, or they are transferred to the fieldbus by a PC-based interface with a corresponding CORBA call. In the demonstrator cell, the following technical sub-systems are controlled via a PROFIBUS interface:

- Two KR C2 robot controllers made by KUKA
- One SIMOTION motion-control system made by Siemens
- Further actuators (two-jaw gripper, pneumatic switches) and sensors (photoelectric sensors) can also be reached as I/O systems via the PROFIBUS. These systems are linked to the cognitive controller via a CORBA PROFIBUS interface.

The following are linked directly via CORBA:

- An image processing system for identifying the individual components and detecting their position
- A gripper system for a flexible gripper, made by Schunk, with seven degrees of freedom

As the components have to be detected and gripped even when the conveyor belt is running, the robot must track them at a speed synchronous with that of the conveyor during the detection and gripping process. To achieve this, a direct control loop between robot controller and image processing system was created. At the technical level, the loop uses the KUKA RSI real-time interface to manipulate robot movements in the interpolation cycle. In the layer model, image processing in the control loop is located in the reactive layer due to its real-time character.

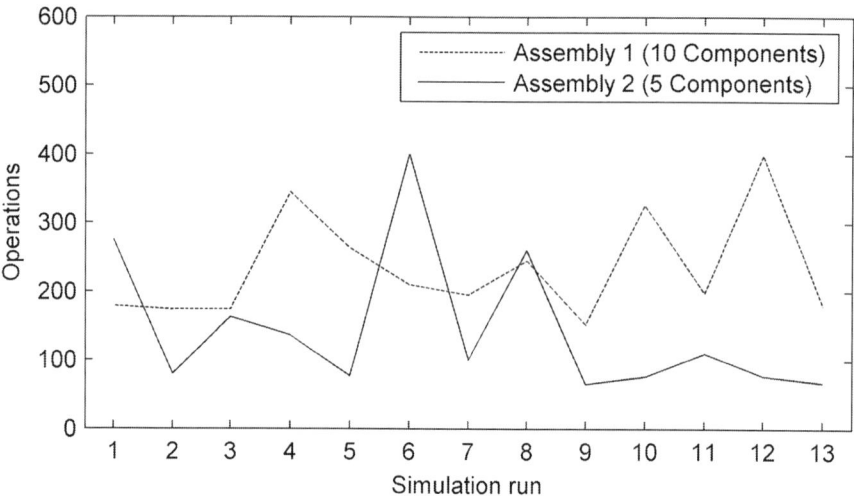

Fig. 6.122 Number of operations necessary for two parallel assembly operations

A truism for every production system is that a plant behaves correctly when it is appropriately designed, and using a cognitive control system does nothing to change this simple fact. However, the example of the demonstrator cell shows that the behaviour achieved with a correct model and self-optimisation completely met the expectations in every case (Kempf 2010). The duration of the process and the number of operations necessary depends to a large degree on the modelling and can be greatly influenced with just a few rules. Further factors are random influences—in this case the random feed-in of components, and non-deterministic decision-making within the SOAR agents. Figure 6.122 shows how the duration of an assembly process is affected by random influences alone.

Due to the semi-decidability of some planning tasks, no hard real-time behaviour can be expected from a cognitive controller at the planning level. Nevertheless, the run-time behaviour should still satisfy certain boundary conditions, particularly in the reactive layer.

As a rule, both the temporal behaviour and the necessary memory requirement are dominated almost exclusively by SOAR (planning processes are one exception; the planning module is responsible for these). The following times were measured on a normal desktop PC with a 2.5 GHz dual-core processor and 2 GB RAM. The control architecture means that there is only one control cycle that covers the planning, coordination and reactive layers. For the measurement of the cycle time it should be noted that the time resolution of the standard operating system (Windows XP) on the platform used was approx. 15 ms. To obtain a more precise statistical mean value, the total run-time was therefore divided by the number of necessary control cycles. It emerged, however, that the actual cycle time varied only very slightly from this mean value. This is also to be expected, because each SOAR agent usually performs exactly one decision cycle per control cycle.

If a normal industrial reaction time of 50 ms is assumed (see Fig. 6.123), the cognitive control system implemented always reacts within this period of time. The

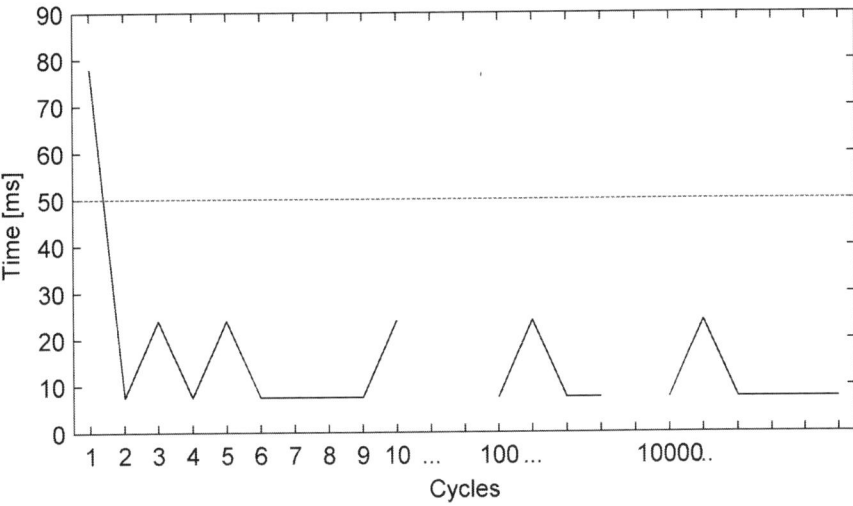

Fig. 6.123 Controller stability and cycle duration compared with normal industrial reaction time (50 ms)

coordination and reactive layers thus initially fulfil the speed demands of a suitable controller for the demonstration cell.

The current results are aimed solely at designing the technical "components" in the complex human-machine system of a cognitively automated assembly cell. From a purely technical point of view, the results demonstrated that robot-supported assembly processes can be cognitively automated, even though, with regard to planning, a hybrid approach had to be developed that combined classic planning elements with reactive, cognitive planning to meet the demands placed on reactive and adaptive planning.

Building on this, the following sections will focus on the human being in the overall system, and will examine how humans and the technical sub-systems interact.

6.6.4.4 Process Logic for a Cognitively Automated Assembly

As explained in Sect. 6.6.2.4, cognitive automation shifts the spectrum of tasks for the human being. Parts of the planning that originally lay in the human sphere of responsibility, because they involve the step-by-step transformation of a manufacturing strategy into an RC program, can be taken over by the cognitive control unit. This can lead to the abovementioned incompatibility between the human mental model and the process knowledge stored in the technical system. With the human and the machine, the working system has two totally different information-processing "systems", which either encode and process sub-symbolically as with the human, or merely have symbolically encoded information on the production process as with the machine (Fig. 6.107).

The work presented below aims to avoid such incompatibilities by developing a cognitive-ergonomic model of the process knowledge. The idea is to adapt the process knowledge stored in the cognitive control unit to human thought patterns, and to influence the behaviour of the cognitive system in such a way that humans can easily understand and reliably anticipate it.

Elementary components of the MTM-1 taxonomy were used for the cognitive-ergonomic design of the process logic for controlling the assembly robots. The hypothesis is that a sequence consisting of empirically validated basic elements/movements that conform to expectations can be quickly learned and, if necessary, optimised by humans, even if the executing instance is a robot gripper arm (see also Gazzola et al. 2007; cf. Tai et al. 2004). The MTM components transformed into production rules, or SOAR operators, are equal and therefore not defined in a sequence. They correspond to the MTM-1 basic movements REACH, GRASP, MOVE (with integrated TURN), POSITION and RELEASE (see Sect. 6.6.4.3) that are used to control the robots in the cell. In addition, further rules are stored which, depending on the basic elements used, contain the physical boundary conditions (e.g. joining direction or conditions for positioning an element) and assess whether a fed-in element can be directly fitted or has to be stored in a buffer until a later assembly step.

As the research work focuses on evaluating the concept—not on optimising the process purely in terms of time—no tabular time information is stored initially. This simplification is acceptable if all the bricks necessary for the assembly are available or if the component is made of identical bricks, since with fixed starting and finishing positions of the end effector, the sum of the paths and hence the overall time does not change, despite different assembly sequences.

A prototype implementation of the CCU described is carried out in a self-developed simulation environment, very similar to that of the assembly cell. For simplification, the conveyor belt was replaced by a panel similar to a chess board. The fed-in bricks are laid out on the panel's fields. The supply process can be varied from randomly feeding in a single brick through to delivering all the necessary bricks. The simulation included the workstation and buffer areas as independent areas. Regarding the simulation of the gripper, it is assumed that the geometry of the gripper means there are no restrictions in approaching or joining the elements. The prototype simulation environment is shown in Fig. 6.124.

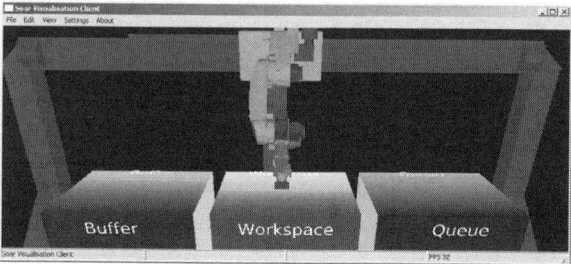

Fig. 6.124 The simulation environment

6 Self-optimising Production Systems

Table 6.8 Expected process sequence based on the MTM-1 taxonomy

	Step 1	Step 2	Step 3	Step 4	Step 5
Brick 1	Reach Start → Panel	Grip Brick 1	Move Box → Pos. 1	Position on Pos. 1	Release
Brick 2	Reach Pos. 1 → Panel	Grip Brick 2	Move Box → Pos. 2	Position on Pos. 2	Release
Brick 3	Reach Pos. 2 → Panel	Grip Brick 3	Move Box → Pos. 3	Position on Pos. 3	Release
...					

The prototype implementation of the CCU in the simulation environment should be considered as a reference model. The reference model was validated using several easy-to-assemble geometric objects made up of LEGO bricks. The objects differed in the number of bricks used, in colour and in shape (e.g. pyramid, cube, flat surface). The assembly results of a pyramid consisting of 30 identical bricks is explained here as an example. Identical bricks were chosen because this meant that each component could be installed in any position within the pyramid, thus resulting in a large number of possibilities ($\sim 10^{25}$) for the assembly sequence.

Without further defining the spatial dimensions within the workspace of the assembly cell, an assembly sequence like the one shown as an excerpt in Table 6.8 is expected. If we understand one complete cycle as running from the beginning of a REACH operator until the end of a RELEASE operator, we expect the number of cycles to be 30.

Repeated simulation runs ($n = 1,000$) with pre-picked sets of parts (all the necessary bricks were available) and with a random supply of parts (including bricks not required) show that the desired target object is always built, error-free and in the expected number of complete cycles. Given the simplified treatment of the gripper, no deadlocks occurred. This therefore replicated the simulation results presented in Sect. 6.6.4.2. It should be pointed out here, however, that the variance in the observable assembly sequences is immense, meaning that despite the use of an anthropocentric taxonomy in the form of MTM-1, the question arises as to whether the described approach is sufficient to ensure that system behaviour conforms to the expectations of the assembly cell user.

This question must be considered from different perspectives. If we look at the sequence of operators (from REACH to RELEASE) within a single cycle, the sequence conforms to expectations because it corresponds to the normal cycle of movements of the human hand/arm system. This does not apply, however, to the sequence of cycles, or assembly sequences. With SOAR these prove to be a statistical succession of possible sub-steps that is impossible for humans to understand because it happens with no recognisable heuristics.

From a technocentric point of view, the reference model provides valid results in the sense of a complete, error-free and expedient structure. From an anthropocentric perspective, however, the results appear to be inadequate since the high procedural variance means that they cannot be compatible with human expectations. Marshall's Schema Model (Marshall 2008) divides the types of knowledge for the human decision-taking process into four categories: (1) Identification knowledge (*What is happening at the moment?*); (2) Elaboration knowledge (*What has high priority and why?*); (3) Planning knowledge (*What has to be done and when?*); (4) Execution knowledge (*Who should do what?*). Regarding the discrepancy between the technocentric and anthropocentric approach, Mayer et al. (2009) conclude that the reference model significantly under-represents elaboration knowledge.

To examine the hypothesis that a stronger focus on elaboration knowledge will have a positive impact on the expectation conformity of the system behaviour, a series of experiments investigating assembly strategies was held under laboratory conditions and with 16 participants (13 male, 3 female). The participants had to assemble a complete assembly on the basis of a CAD drawing. To keep the results comparable with the assembly cell, limitations were imposed on the execution of the task. Participants were only permitted to use one hand for the assembly, and were not allowed to build sub-groups or pick up several bricks at once. The target object was a single-coloured pyramid of 30 LEGO bricks.

The analysis of the assembly strategies in the experiments produced three general rules:

- Rule 1: From the viewpoint of the participant, the first brick to be assembled is located in one of the left corner positions (87.5% of the sample cases).
- Rule 2: Preference is given to selecting bricks that can be positioned directly next to an adjacent brick during assembly (81%). This will be referred to as compliance with the adjacency relationship.
- Rule 3: The target object is built up in layers that lie parallel to the assembly surface (81%).

Validity of the Collected Data with Respect to the Defined Rules

A further series of experiments (ES2) was carried out to check the validity of the previously identified assembly rules. The series involved 25 people (14 male, 11 female) who are not involved in manual assembly during their day-to-day work. The subjects are therefore not classified as experienced workers in the sense of the MTM taxonomy. None of the participants had taken part in the first series (ES1). The average age was 26.9 years (SD = 3.4). For ES2 the task from ES1 was expanded to require the subjects to assemble ten identical pyramids in succession in a timely manner. This meant that, in spite of the laboratory conditions, the subjects gained a degree of experience that is quite comparable with that of small-series production. The reason for the expansion of the task was to avoid "unusual" methods of working due to the simple task. Subjects had to signal the start and end of each assembly sequence by double-clicking a pushbutton switch installed in the assembly area. The assembly process was subject to the same limitations as in ES1—one-handed assembly and no building sub-groups or gripping several bricks at once.

If the rules identified from ES1 are also applicable to ES2, at least equal relative frequency f_i for each individual rule should be recognisable in the empirical data.

6 Self-optimising Production Systems

Table 6.9 Results of the χ^2 goodness-of-fit test

	MV ES2 (EV ES1)	df	χ^2	p
ES2, Rule 1	80.4% (87.5%)	1	11.52	0.00
ES2, Rule 2	91.2% (81%)	1	16.25	0.00
ES2, Rule 3	97.2% (81%)	1	41.75	0.00

$\alpha = 0.05$; *MV* Mean value; *EV* Expected value; *ES* Experiment series

On the basis of this assumption, the following hypotheses can be formulated for the statistical test:

- H_1: The relative frequency of the position of the first brick in ES2 (f_{h2_Rule1}) is higher than the frequency in ES1 (f_{h1_Rule1}), or: H_{01}: $f_{h2\_Rule1} = f_{h1\_Rule1}$.
- H_2: In ES2, assembly that conforms to the adjacency relationship (f_{h2_Rule2}) occurs with a higher relative frequency than in ES1 (f_{h1_Rule2}), or: H_{02}: $f_{h2\_Rule2} = f_{h1\_Rule2}$.
- H_3: Assembly in layers in ES2 (f_{h2_Rule3}) occurs with a higher relative frequency than in ES1 (f_{h1_Rule3}), or: H_{03}: $f_{h2\_Rule3} = f_{h1\_Rule3}$.

To verify the null hypotheses, the χ^2 goodness-of-fit test is applied with a significance level of $\alpha = 0.05$.

The results of the χ^2 test for H_{01} (position of the first brick in left-hand corners) are shown in Table 6.9. This clearly shows that H_{01} has to be rejected. With a relative frequency of 80.4%, the observed distribution deviates significantly from that observed in ES1.

The results of the χ^2 test for H_{02} (taking account of adjacency relationships during assembly) are also shown in Table 6.9. The null hypothesis must also be rejected in this case. However, because the relative frequency of 91.2% is higher than the expected value from ES1 (81%), it can be said that the rule is followed more strictly than expected.

Finally, the results of the χ^2 test for H_{03} (assembly in layers) are also shown in Table 6.9. This null hypothesis must also be rejected. As the observed relative frequency is 97.2% (expected value from ES1: 81%), it is clear that this rule is also followed more strictly than expected.

Influence of the Rules on Prediction Quality

As already shown, the null hypotheses regarding the rules identified in ES1 had to be rejected. The results of the χ^2 goodness-of-fit test for the rules relating to compliance with adjacency relationships and to assembly in layers show that these are followed more strictly than expected. The only rule that could not be applied to the results of ES2 was the one relating to the position of the first brick. Nevertheless, in view of the relatively high satisfaction of the rule (80.4%), it will continue to be considered in the further course of the study.

Table 6.10 Overview of the simulation models compared in the study

	MTM-1 rules	Rule 1	Rule 2	Rule 3
Model 1	X			
Model 2	X	X		
Model 3	X		X	
Model 4	X			X
Model 5	X	X	X	
Model 6	X	X		X
Model 7	X		X	X
Model 8	X	X	X	X

Finally, we also wanted to investigate how the identified rules—individually and in combination—influence the CCU's prediction accuracy regarding assembly steps preferred by humans, and the generalisability of assembly steps carried out by humans. To do so, independent sets of rules, which consist of the rules of the reference model and the respective auxiliary rule, are transferred to the CCU and simulated several times in a simulation environment developed especially for the experiments. The resulting data give an indication of how each set of rules influences the prediction accuracy of the simulation, i.e. its ability to predict the next step in the human assembly activity, and provide information on the generalisability of the set of rules with respect to ES2.

The additional cognitive simulation models were systematically expanded to include the rules corresponding to the heuristics. An overview of the simulation models is shown in Table 6.10.

For space reasons, in what follows only one dependent variable is used for evaluating the simulation models. The variable is derived from the criteria provided in Langley et al. (2009) for evaluating cognitive architectures. This criterion is based on the "optimality" criterion and represents the simulation model's prediction quality regarding the assembly activity. The prediction quality of an observed model is defined as the probability of the simulation model positioning a given brick in conformity with the human action during simulated assembly. In other words, assuming a given state x_{i-1} and the stored process knowledge for reaching the next state X_i, the probability $p(x_i|x_{i-1})$ that this particularly state will be reached is evaluated. This observation is followed step-by-step for the assembly sequence until the target object is fully assembled.

As the simulated assembly is a Markov process, it is admissible to factor both the overall probability into the transition probabilities $p(x_i|x_{i-1})(2 \leq i \leq 30)$ described above, and the initial probability $p(x_i)$, where x_1 represents the initial state. The conditional probability of a sequence P_s is calculated as follows:

$$P_s = \prod_{i=2}^{30} p(x_i|x_{i-1}) \cdot p(x_1) \tag{6.16}$$

The logarithmic probability was calculated to simplify the interpretation of the resulting data. The prediction quality is therefore operationalised using "logarithmic conditional probability" (LCP):

$$LCP = \sum_{i=2}^{30} \log_{10} p(x_i|x_{i-1}) + \log_{10} p(x_1) \qquad (6.17)$$

In Eq. 6.17, $p(x_i|x_{i-1})$ is the conditional probability that the observed simulation model assigns to a brick that was originally positioned by the human. Thus, the LCP values vary between 0 (perfect prediction) and $-\infty$ (absolutely wrong prediction or behaviour that the simulation model cannot validly reproduce).

To examine the prediction quality of the different cognitive simulation models (see Table 6.10), the index of the simulation models is regarded as an independent variable. Expanding the reference model of the cognitive simulation (Model 1 in Table 6.10) is expected to increase the prediction quality when additional empirically identified rules are added to the knowledge base. The following null hypothesis is formulated on the basis of this expectation:

- H_{04}: The prediction quality of the cognitive simulation models exhibits no significant differences.

The Kruskal-Wallis test was used with a significance level of $\alpha = 0.05$ to test for differences in the prediction quality. This test can be regarded as a non-parametric form of the one-factor analysis of variance (ANOVA). This is necessary because the Lilliefors test for normal distribution rejects the normal distribution for all LCP values ($p < 0.01$ in each case). A post-hoc test with adaptations for multiple comparisons according to Bonferroni was also carried out.

Results

Table 6.11 shows the mean values for the dependent variable LCPs from Eq. 6.17, which describes the prediction quality for the 250 assembly sequences from the second empirical study.

On the basis of the simulation data underlying Table 6.11, it is clear that there is a significant difference ($p = 0.00$) in the LCP values. H_{04} must therefore be rejected. Figure 6.125 shows the simulation data as box plots of the observed cognitive simulation models. A higher LCP value means higher prediction quality with respect to human behaviour.

Table 6.11 Mean values for the logarithmic conditional probability (LCP) for the simulation models examined

Simulation model	LCP
Model 1	−24.595
Model 2	−24.422
Model 3	−20.579
Model 4	−20.279
Model 5	−20.430
Model 6	−20.192
Model 7	−15.848
Model 8	−15.671

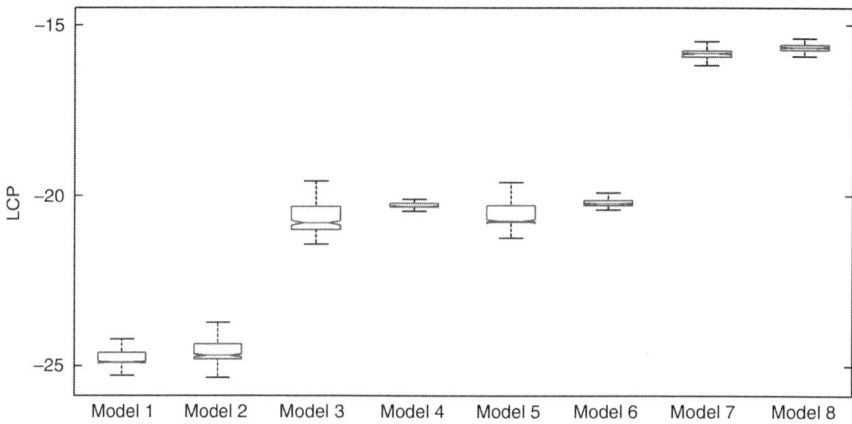

Fig. 6.125 Box plots of the logarithmic conditional probability (LCP) calculated for the simulation models examined

Table 6.12 Results of the multiple comparisons of the LCP values for the cognitive simulation models observed

	Model 1	Model 2	Model 3	Model 4	Model 5	Model 6	Model 7	Model 8
Model 1	-/-		X	X	X	X	X	X
Model 2		-/-	X	X	X	X	X	X
Model 3	X	X	-/-	X		X	X	X
Model 4	X	X	X	-/-	X		X	X
Model 5	X	X		X	-/-	X	X	X
Model 6	X	X	X		X	-/-	X	X
Model 7	X	X	X	X	X	X	-/-	
Model 8	X	X	X	X	X	X		-/-

Post-hoc pair comparisons were carried out to determine the differences between the cognitive simulation models. Table 6.12 shows the significant differences ($\alpha = 0.05$) as a cross-reference table, where X represents a significant difference ($\alpha = 0.05$).

If we compare all the cognitive simulation models, we can divide the models into three groups that are significantly different in terms of prediction quality. The first group has the poorest prediction quality and comprises Model 1 (the reference model for the cognitive simulation) and Model 2. Models 3, 4, 5 and 6 form the group with average prediction quality. Models 7 and 8 have the highest prediction quality.

The following conclusions can be drawn from the test results:

- Rule 1 has no significant effect when it is added to a cognitive simulation model.
- Rules 2 and 3 significantly influence the prediction quality of the cognitive simulation models, but they show significant differences in a direct comparison. Adding Rule 2 can improve the LCP value by between 16.32 and 22.39%. Rule 3 results in improvements ranging from 17.32 to 23.29%.

- The highest prediction quality is obtained when Rules 2 and 3 are combined. This increases quality by between 35.56 and 35.83%.

6.6.4.5 Human-Machine Interaction

As explained in Sect. 6.6.3, it is important to pay particular attention to the ergonomic design of the human-machine interface. Because the CCU can independently solve a certain class of rule-based production tasks, the human operator's duties mainly lie in cognitively demanding tasks. These include defining the production task, drafting rough process plans, defining the initial and boundary conditions and, in particular, monitoring the system state during the process and taking expedient action if disturbances or production errors arise. If the robot makes an assembly error, for example, the human operator must be able to quickly and efficiently intervene to enable system recovery. To achieve this, the sub-project designed, developed and evaluated a visualisation system that displays assembly information directly in the operator's field of view ("augmented reality", Ong et al. 2008).

A laboratory study was carried out to examine this visualisation system from the point of view of software ergonomics and to analyse how the relevant information should be displayed for the human operator. The task used in the experiment assumes that a robot has implemented the scheduling of an assembly task previously defined by the operator. During processing, however, a fault has occurred that the cognitively automated system cannot identify and remedy independently. Possible causes of assembly errors can be "noisy" sensor data regarding component detection, and flaws in the fed-in components themselves. Two separate series of experiments were performed to investigate the "optimum" ergonomic display of the assembly information with respect to the time necessary for fault detection and the corresponding accuracy. Two different displays (a head-mounted display—HMD—for visualisation directly in the field of view, and a table-mounted display—TMD—for visualisation in the workspace) and various types of visualisation were compared.

Experiment Scenario and Implementation

Different modes of visual representation (and interaction) were drafted and developed for the ergonomic system design and evaluation. These modes are based on known methods for designing written manufacturing instructions and are derived from guidelines for technical writing (see Alred et al. 2003). Exploded views and step-by-step instructions were used for ergonomically visualising assembly information in the field of view (Odenthal et al. 2008). To display assembly information in high quality and to permit quick and precise error detection with minimal mental effort, the assembly objects were accurately modelled as 3D objects and their colour and contrast were adapted to provide an ergonomic display on the HMD. LEGO bricks, which are easy to describe, were selected as assembly objects (see Sect. 6.6.1).

Table 6.13 Technical data of the HMD and TMD

	HMD	TMD
Resolution	1,280 × 1,024	1,280 × 1,024
Image refresh rate (Hz)	60	75
Monitor size	–	17″
See-through transmission	40%	–
Weight (kg)	1.3	7
Brightness	max. 102.79 Cd/m^2	Typical 230 Cd/m^2; min. 197 Cd/m^2
Monocular FoV	60° diagonal/100%	–
Technology	LCOS	TFT
Manufacturer	NVIS	ELO

Experiment Design

The experiment design distinguishes between four factors. The first is "display type" (DT). The three other factors represent the different modes of presentation of the synthetic assembly information. These factors are: augmented vision mode (AVM); a priori presentation of the target state of the fully assembled assembly (APP); the mode for interactive decomposition and composition of the assembly during the error detection phase (DCM). These factors with the corresponding factor levels and the associated experimental conditions are explained below. Table 6.14 provides an overview of the factors and factor levels.

- DT: The visualisation systems were designed on the basis of two different display technologies: (1) A high-resolution HMD based on liquid-crystal-on-silicon technology (LCoS) with two half-silvered mirrors in front of the user's left and right eye for stereoscopic display; (2) A high-resolution TMD using TFT technology, as is common in German industry. The TMD was selected and adjusted so that it corresponded as closely as possible to the specifications for the HMD. The light intensity was set to 500 lx at the point of the real assembly object in the working area. The technical data of the two displays are given in Table 6.13.
- AVM: The position and orientation of the HMD can be measured in real-time using an optical infrared tracking system. This makes it possible to calculate the user's viewing direction and adapt the presentation of the virtual information in the field of view. Furthermore, the system can determine the position and orientation of the real assembly object in front of the user. This means that virtual information can be precisely superimposed onto the real object when the HMD is in use. No comparable visual superimposition is possible when using the TMD since it does not measure the position and orientation of the user's head. Only the orientation (rotation of the real model around the vertical axis) was measured in the real model and adapted in the virtual model (rotation of the virtual image by 20° around the vertical axis). Because generating a "perspective view" is technically complex due to the tracking system, a company can only justify acquiring this if this presentation mode significantly improves performance or reliability. For this reason, a simple static information display was also investigated. The two factors are therefore "perspective view" and "static view". In the perspective view, the

6 Self-optimising Production Systems

Table 6.14 The different factors and factor levels

Factors	Characteristic	Examples
DT for augmented display	HMD	
	TMD	
AVM	Perspective view	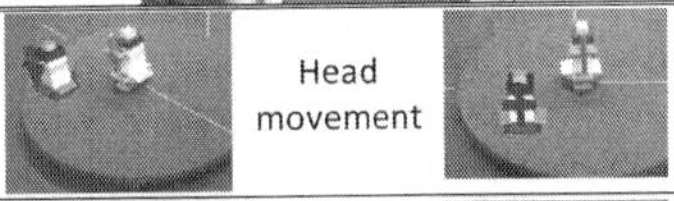 Head movement
	Static view	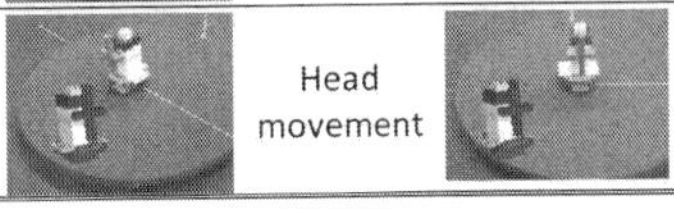 Head movement
APP	Rotation	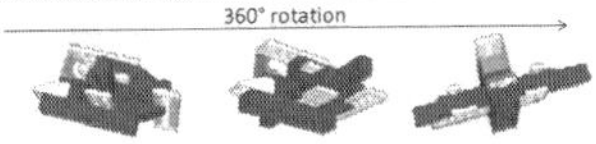 360° rotation
	Assembly	 Step-by-step
DCM	Step-by-step	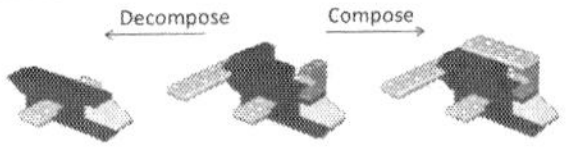 Decompose / Compose
	Exploded view	 Explode / Implode

display of the virtual assembly is adapted to the position and orientation of the real object and, when using the HMD, appears on the left at a distance of 4 cm from the real object. The position of the TMD was selected such that, as with the HMD, the virtual image appears to the left of the real model. The monitor was behind the turntable. This was intended to create as similar angle conditions as possible. In the static view, the virtual assembly is displayed in a fixed position relative to the monitor coordinates of the HMD or TMD. In the experiment the virtual object was displayed in the HMD, tilted at an angle of 20° to the vertical axis and in the third quadrant of the binocular display. In the TMD, the virtual object was displayed tilted at an angle of 20° to the vertical axis and in the centre of the monitor.

- APP: Before the actual error detection task, the subject was shown an a priori display of the target state (assembly). In this case a distinction was made between two factor levels: (1) Rotation: the complete assembly rotates 360° once, tilted at an angle of 20° to the vertical axis and at an angular velocity of 3.2 s.; (2)

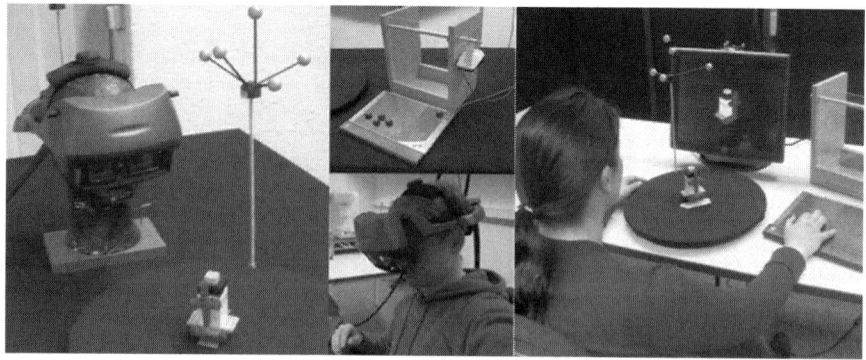

Fig. 6.126 Experiment configuration with turntable and HMD (*left*), experiment configuration with monitor (*right*), keypad for manipulating the virtual object (*top centre*), participant wearing the HMD (*bottom centre*)

Assembly: the virtual model is assembled virtually in steps with a cycle time of 1.2 s. per sub-element.

- DCM: Two factors were analysed with respect to the mode for interactive manipulation of the LEGO model during the error detection process: (1) Step-by-step: the subject can independently compose and decompose the virtual model in steps, using two buttons on a keypad (see Fig. 6.126); (2) Exploded view: the subject can interactively explode or implode the virtual model using another button on the keypad.

Table 6.15 also shows the types of errors that are generally relevant to error detection. Because detecting a colour error is simpler than detecting a position error or an error in shape or number, the experiment does not investigate colour errors. Eight different tasks with similar degrees of difficulty were developed to analyse the remaining error types. Two tasks contained the error "Position", three tasks contained the error "Type/shape", and three tasks contained the error "Number".

Table 6.15 Possible error types

	Colour (not investigated)	Position (investigated)	Shape (investigated)	Number (investigated)	
Target state	●	■	●	■	■
Actual state	●	■	●	■	■

Experiment Configuration

Figure 6.126 shows the main components of the visualisation systems that were used for the experiment. The LEGO model with the assembly error that had to be detected and identified was placed on the turntable, which was located on a conventional assembly bench in the participants' central field of view. Each person sat on a chair in a comfortable upright position during the experiment. The position and orientation of the HMD and the turntable were recorded with the smARTtrack real-time tracking system (made by ART GmbH). The participants used a coloured pen, which was located behind a transparent screen, to mark the incorrectly assembled LEGO brick. As soon as the person detected the error, he/she picked up the pen, which triggered a switch linked to the screen and recorded the time of the action. This point in time was taken as the moment of error detection for all participants. The keys for manipulating the virtual image were attached to a freely configurable keypad with the following key assignment: exploded view (left), decomposition/composition (top and bottom centre), repetition of the a priori presentation (right). The key at the top right has no function within this series of experiments.

Experiment Procedure

The laboratory studies for the HMD and TMD were performed separately, each with two groups of 24 participants. The same assemblies with identical errors were used in both studies. A full-factorial experiment design with measurement repetitions using three within-subject factors (AVM, APP, DCM) and one between-subject factor (DT) was selected. The task for the participants consisted of comparing the state of a real assembly object with the virtual representation on the screen (HMD, TMD) for possible errors. The experiment procedure was split into two main phases:

1. Pre-tests and training under experiment conditions

At the start of the study, general user data (age, profession, previous experience, etc.) were gathered using a questionnaire developed specifically for the study. Then the participants' visual acuity was recorded (tested in accordance with DIN 58220), as was their stereopsis and their colour vision (using the Ishiara colour test). Since wearing a HMD can quickly lead to visual fatigue and thus significantly influence factors like human performance and stress (Pfendler and Schlick 2007), participants' visual fatigue was recorded, based on Bangor (2000), using a questionnaire before and after performance of the task. Next, participants spent ten minutes practicing with the augmented vision system under the conditions that would be used in the experiment.

2. Data collection

Each participant performed the following sequence eight times:

- Starting a run using APP of the target state of the LEGO model. The virtual sequence was presented in the participant's field of view, without him/her being able to see the real object.
- At the end of the initial presentation, the test supervisor fastened the real assembly object to a turntable. Participants could use the keys on the keypad to manipulate the virtual object. They could call up the a priori presentation at any time, but could not cancel the displayed assembly sequence. The participants' task was to compare the real object (actual state) with the virtual object (target state) for differences, without knowing whether and/or how many possible assembly errors the component contained. If they identified differences, they had to mark them with a coloured pen. Each component had one assembly error.
- Each participant then completed the questionnaire on visual fatigue.

The total experiment, including the pre-tests, lasted roughly two hours for each participant.

Participants

A total of 48 people (16 female and 32 male) took part in the laboratory study. All participants satisfied the criteria of normal vision or corrected vision (visual acuity 0.8), stereopsis and colour vision. Apart from these physiological requirements, the groups were formed according to the following criteria: homogeneous age, comparable spatial perception (cube test, Liepmann et al. 2007), comparable experience with augmented or virtual reality (AR/VR) and comparable experience of assembly.

- Group 1—HMD. The participants were between 19 and 36-years-old (MV: 26.8 years; SD: 4.4 years). 95.8% used a computer every day. 58% stated that they had little or no experience with VR. 37.5% had experience with 3D computer games (on average 4 hours per week playing time). The average experience in LEGO assembly had a value of 3.0– on a scale of 0 (low) to 5 (high).
- Group 2—TMD. The participants were between 20 and 40-years-old (MV: 26.0 years; SD: 4.5). All participants used a computer every day. 64.6% stated that they had little or no experience with virtual or augmented reality systems. 50% had experience with 3D computer games (on average 4.7 h per week playing time). The average experience in LEGO assembly had a value of 2.9—on a scale of 0 (low) to 5 (high).

Dependent and Independent Variables

In accordance with the experiment plan, a distinction was made between four independent variables (see also overview in Table 6.14):

- Display type (head-mounted or table-mounted)
- Augmented vision mode (perspective or static view)

6 Self-optimising Production Systems

- A priori presentation of the target state of the complete assembly model (rotation or assembly)
- Decomposition/composition mode (step-by-step or exploded view)

The experiment measured the following dependent variables:

- Detection time: This was the time between the appearance of the real LEGO model and the detection of the difference by the participant (max. 15 min.). The experiment configuration measured the start and end point of this period.
- Error detection: Different cases could occur which could be represented by separate variables: (a) The participant detects the difference (error correctly detected); (b) The participant picks out a brick which is no different from the virtual (target) state (error wrongly detected); (c) The participant does not find the difference (error not detected).
- Visual fatigue: The study assumed that this sets in quite rapidly, particularly when using a HMD. This subjective variable was therefore also included in the study.

Null Hypotheses and Statistical Analysis

The following null hypotheses were formulated:

- The display type has no significant influence on detection time (H_{01}) or error detection (H_{02}).
- The augmented vision mode (H_{03}), the a priori presentation of the synthetic assembly information (H_{04}), and the decomposition/composition mode of the virtual model (H_{05}) have no significant influence on error detection time based on display type.
- The augmented vision mode (H_{06}), the a priori presentation of the synthetic assembly information (H_{07}), and the decomposition/composition mode of the virtual model (H_{08}) have no significant influence on the given cases of error detection based on display type.

On the basis of the data collected, inferential-statistical analyses were carried out using the program "Statistical Package for Social Science" (SPSS Version 17). First, the two data sets (Group 1—HMD; Group 2—TMD) were compared with respect to the dependent variables. A four-factorial repeated measures ANOVA was calculated with for the test of H_{01} (three within-subject factors: AVM, APP, DCM; one between-subject factor: DT). The groups were then analysed separately to identify any differences with respect to the different visualisation and interaction modes. A three-factorial repeated measures ANOVA was carried out for the test of hypotheses H_{03}, H_{04} and H_{05}. The detection time data were log-transformed before the ANOVAs were carried out (see Field 2005) to satisfy the qualitative requirements. The significance level was set at $\alpha = 0.05$. A Kolmogorov-Smirnov test was performed to examine the log-transformed data for normal distribution. Chi-square tests were carried out to verify the hypotheses H_{02}, H_{06}, H_{07} and H_{08} using the nominally scaled data from error detection. The significance level was again set at $\alpha = 0.05$.

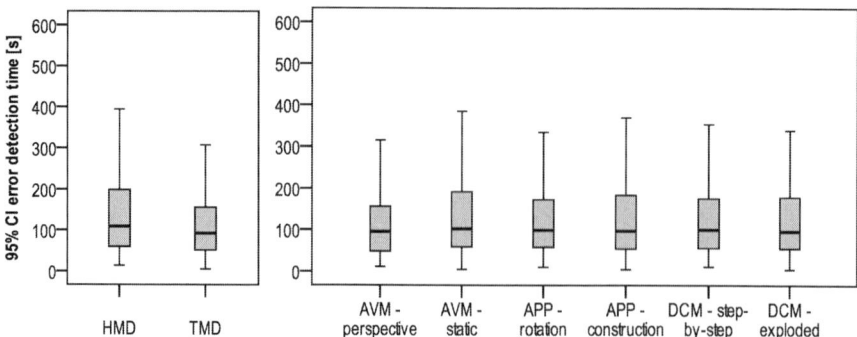

Fig. 6.127 *Left:* Error detection time for all participants, depending on the between-subject factor DT; *Right:* Error detection time for all participants, depending on the within-subject factors AVM, APP and DCM

Results and Interpretation

The log-transformed time data showed no significant deviation from the normal distribution. The mean values of the error detection time under the different experimental conditions are shown in Figs. 6.127 and 6.128 in the form of box plots.

Error-Detection-Time Results for All Participants: On average, the error detection time using the TMD was 27.64% shorter than with the HMD. However, this difference is not statistically significant ($F_{(1,45)} = 3.001$, $p = 0.090$).

A comparison of the perspective and static views in AVM showed that error detection times in the perspective view were on average 23.45% shorter than with the static view. This difference is statistically significant ($F_{(1,45)} = 8.854$, $p = 0.005$). In the case of the a priori presentation of the assembly information, the rotation condition led to an average error detection time that was 9.4% shorter than under the assembly condition. However, this difference is not significant ($F_{(1,45)} = 0.043$, $p = 0.837$).

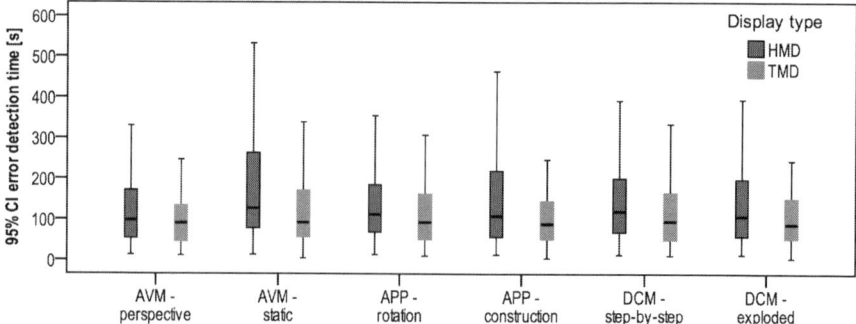

Fig. 6.128 Error detection time divided between the groups in the experiment (HMD, TMD) for the different conditions (AVM, APP, DCM)

When the participants worked with the exploded view, the average error detection time was only 2.2% shorter than with step-by-step decomposition/composition. Again, this is not a significant difference ($F_{(1.45)} = 0.103, p = 0.749$). No statistically significant interactions were identified.

Error-Detection-Time Results for Group 1 (HMD): For the perspective view in AVM, the study showed that error detection times were 29.61% shorter than with the static view. This difference is statistically significant ($F_{(1.22)} = 8.088, p = 0.009$). The rotation condition of the APP mode resulted in error detection times that were on average 18.19% shorter than the alternative assembly condition, but this difference is not significant ($F_{(1.22)} = 0.004, p = 0.929$). The average error detection times for both conditions in the decomposition/composition mode barely differed. With the exploded view, error detection times were an average of 2.78% shorter than with step-by-step decomposition/composition. This is not a significant difference ($F_{(1.22)} = 0.001, p = 0.971$). No statistically significant interactions were identified.

Error-Detection-Time Results for Group 2 (TMD): For the perspective view in AVM, error detection times were 13.07% shorter than with the static view. This difference is not statistically significant ($F_{(1.23)} = 1.526, p = 0.229$). The assembly condition in APP resulted in a 2.56% reduction in the average error detection time compared with the rotation condition. This difference is also not statistically significant ($F_{(1.23)} = 0.335, p = 0.568$). The average error detection times for both conditions in DCM barely differed. With the exploded view, error detection times were an average of 2.94% lower than with step-by-step decomposition/composition. This is not a significant difference ($F_{(1.23)} = 0.148, p = 0.704$). No statistically significant interactions were identified.

Error Detection Results for All Participants: The number of correctly detected errors was 36% higher with the HMD than with the TMD. The display type significantly influences all categories of error detection (correctly detected: $p = 0.019$; wrongly detected: $p = 0.023$; not detected: $p = 0.044$). With regard to AVM, the perspective view increased the number of correctly detected errors by an average of 18.5% compared to the static view. Compared with rotation, step-by-step assembly of the LEGO model with the APP mode enabled the user to correctly detect 14.5% more errors. Step-by-step decomposition/composition resulted in 10.7% more errors being correctly detected than with the exploded view. However, these differences are not significant. Error detection in the other two categories (error not detected or wrongly detected) is not significantly influenced by the levels of the independent variables.

Error Detection Results for Group 1 (HMD): A comparison of the two AVMs shows that, on average, the perspective view resulted in a 19% higher error detection probability (error correctly detected) than the static vision mode. With APP, a rotating LEGO model enabled the user to correctly detect 16% more errors than with step-by-step assembly. With regard to DCM, the difference in correct error detection between step-by-step decomposition/composition and the exploded view was far smaller (6%). The differences are not significant. Error detection in the other two categories (error not detected or wrongly detected) is not significantly influenced by the levels of the independent variables.

Error Detection Results for Group 2 (TMD): A comparison of the two AVMs shows that, on average, the perspective view resulted in a 17.4% higher error detection probability (error correctly detected) than the static view. With APP, step-by-step assembly of the LEGO model enabled the user to correctly detect 12.8% more errors than with a rotating model. With regard to DCM, correct error detection in step-by-step decomposition/composition was 17.4% higher than with the exploded view. The differences are not significant. Error detection in the other two categories (error not detected or wrongly detected) is not significantly influenced by the levels of the independent variables.

Visual Fatigue Questionnaire Results: As already mentioned, the participants' visual fatigue was recorded before the first and after the last cycle. A difference of 1 represented a subjectively perceived difference of 10% in visual fatigue. The largest average increase in visual fatigue for Group 1 (HMD) was 1.1 for the item "headache", followed by 0.65 for the item "mental fatigue". The largest average increase in Group 2 (TMD) was 0.37 for the item "mental fatigue". The other differences are smaller than 5% (0.5) and can therefore be ignored. The visual fatigue recorded was lower than expected, which indicates a good ergonomic design of the augmented vision systems. It should be pointed out, however, that the HMD was only worn for between 0.5 and 13.6 min. at a time during the experiment. The error detection phase was always followed by a 2–3 min. recovery phase, during which participants filled out the questionnaires (not wearing the HMD). It is expected that a longer cycle time would significantly increase visual fatigue (Pfendler and Schlick 2007).

Discussion

Display Type The results of the series of experiments carried out here confirm the research results of Tang et al. (2004) and Meyer et al. (2005). Using a HMD led to a significantly higher degree of precision in error detection, but not to a significantly shorter detection time. The lower error detection rate when using the TMD can be attributed to participants having to frequently shift their attention between the component and corresponding virtual model, and to the high mental strain this brings about. Furthermore, participants had to adapt to different light intensities between the real model and the representation on the screen. The participants consequently tended to overlook the error and thus arrived more quickly at a decision as to whether or not an error existed.

Augmented Vision Mode Compared to the static view, AVM with perspective view resulted in a shorter average error detection time for both groups. The perspective view did, however, have less influence on the error detection time when using the TMD. The difference between the two modes was not significant here. The perspective view resulted on average in more correctly and less wrongly detected errors.

Nevertheless, these differences were not significant in either group, which means the data analysis provided no statistically clear proof of a speed/accuracy trade-off.

The difference between the perspective and the static views—irrespective of the display used—is attributable to higher mental strain in the static view. In this view, the user frequently has to mentally rotate the virtual model during the error detection phase, which is a strenuous and time-consuming central process. Since Shepard and Metzler (1971) published their classic work, we know that people's reaction times in comparing and deciding whether or not two items are identical are proportional to the rotation angle between the two representations. The study confirmed this relationship. Reaction times increase as the objects presented become more complex (Funke and Frensch 2006). The perspective view allows users to make a direct perceptive comparison between the target and the actual assembly state, without drawing on important mental resources for rotation or translation.

As already mentioned, in the perspectively modified mode when using the HMD, the presentation of the virtual object is tracked if the turntable is rotated (vertical axis) or if the user moves his/her head (other axes). By contrast, the TMD only rotates the virtual object (vertical axis) if the turntable is rotated. These different degrees of perspective adaptation are probably the reason for the differences in performance and reliability between the perspective view and the static view in Group 2 (TMD). However, the differences are not significant.

A priori Presentation With APP, a shorter error detection time in rotation mode was observed for both groups. Using the HMD resulted in a shorter average error detection time, but correct error detection was lower than with step-by-step assembly. In other words, a tendency towards a speed/accuracy trade-off was observed here. The average error detection time when using the TMD was practically identical ($\sim 2\%$) in both rotation and step-by-step assembly. However, step-by-step assembly resulted in fewer correctly detected errors. Unlike the rotation sequence, the assembly sequence of the assembly object allowed participants to create a precise mental model of the product structure and the assembly procedure. Rotating the virtual model, which avoided the strain of mental rotation as already mentioned, resulted in faster but less reliable error detection.

Decomposition/Composition Mode In DCM, the exploded view allowed the user to switch quickly between a fully assembled product structure and an exploded view of a product structure, and to create a precise representation in their visual-spatial memory. However, in the step-by-step mode—particularly the assembly (bottom-up)—it took a certain time before participants detected the error. Nevertheless, and contrary to the expectation, only a very small time advantage of $\sim 3\%$ for the exploded view was observed in both groups. This appears to be due to the effects of overlapping, which make the exact localisation of the error more difficult. Correct error detection in the exploded view was lower than with step-by-step decomposition/composition. Consequently, the rate of wrongly detected errors and errors not detected was higher, but these differences are not significant.

Based on the results of this study, the head-mounted stereoscopic vision system was further developed to support the user in remedying the assembly error once it has been identified. Interactive graphic decomposition is possible here (Odenthal et al. 2011). A further study is in preparation (parts have already been conducted) that will compare different modes of visual assistance in the development of a cooperative human-robot decomposition strategy with regard to performance, reliability and mental strain.

6.6.5 Industrial Relevance

The concept of generic strategies first introduced by Porter in 1980 (Porter 2004) provides three strategies that a company can use to achieve a competitive advantage over its rivals. The cost-leadership strategy aims to gain a competitive advantage by keeping costs down. To do so, companies can adopt a number of different approaches (see Mintzberg et al. 2004), which may have opposing characteristics—e.g. economies of scale and scope. The differentiation strategy is about developing a unique selling point using factors like image, service or design. With the focus strategy, a company focuses on a very specific customer group or market segment, although it might also focus on differentiation or cost-leadership. A company in a high-wage country produces new products domestically and pursues a strategy of differentiation. Later in the product life cycle, however, it is very likely to move production to a low-wage country to pursue the cost-leadership strategy because other providers are flooding the market to benefit from the economic success of the product. The same applies to the focus strategy, which must undergo the same change.

The concepts and technologies for cognitive automation described in this paper make it possible—by taking targeted action on the polylemma of production—to adopt a position between Porter's strategies. A Business and Technology Case provided the opportunity to put the results into practice for the first time, as a prototype application. For this purpose concepts already tested with the demonstrator cell were simulated, in collaboration with connection technology manufacturer Phoenix Contact, in a real scenario involving switch-cabinet production.

In addition to final assembly in its mass-production-style manufacturing system, Phoenix Contact operates a customer-driven switch-cabinet assembly system. This involves assembling switch-cabinet components (control units, terminals, etc.) in configurations pre-planned by the customer, mounting them onto top-hat sections and passing them on as completed modules to switch-cabinet production (module assembly). Figure 6.129 shows an example of this system.

Fig. 6.129 Assembling switch-cabinet components on a top-hat section

Individual components must be mounted with additional plug combinations. The different mounting configurations planned are available as CAD data, as required by the CCU. The relevant mounting requirements are usually printed out, and employees then manually assemble the components on the top-hat sections. Next, even today, an image-processing-supported comparison is made between the mounting requirements and what has actually been mounted on the top-hat section.

Because the target is already available in electronic format, cognitively automating the process could bring considerable economic benefits in the future. The main challenges this involves are as follows:

- Building a continuous chain of information from the CAD system to the assembly system
- Making available robust and productive system components and joining processes for module assembly (handling devices, joining processes and aids for plugging, clamping and, if necessary, screwing)
- Applying logistical components and concepts for fitting the switch-cabinet components at the right time and in the right quantity
- Developing control concepts for establishing the ideal assembly sequence and for carrying out the assembly task itself
- Ensuring that sensors monitor the actual mounting situation and that human operators take corrective action in the case of errors

The switch-cabinet scenario bears strong similarities to the demonstrator-cell scenario. The following correlations are relevant to the transferability of the developed concepts:

- Defined workpieces: The LEGO scenario involves fitting a large number of building bricks from the whole range. The product specification means that all characteristics relevant to assembly—such as dimensions, tolerances, and colour—are known. Unfamiliar bricks or bricks that do not fit the boundary conditions for assembly are not fed in. The modules in the switch-cabinet assembly scenario correspond to the LEGO characteristics described. The modules are also precisely specified and are only fed in after quality control. Just like any LEGO bricks fed in by accident, any modules erroneously fed in can be separated out or, if they are to be assembled later or at another stage, they can be stored or fitted.
- Flexible handling technology: The flexible gripper hand installed in the demonstrator cell was chosen for two reasons: it could grasp a LEGO brick using a variety of grips depending on the assembly situation, and if the assembly scenario changed in any way, it could grip and join other workpieces with unfamiliar shapes. The gripper hand's seven axes and its push-button sensors mean that it could also assemble the switch-cabinet modules.
- Production logistics: As in the previously described scenario, the components in switch-cabinet production must be fed in individually. The demonstrator cell uses a robot to do this, along with a group of conveyor belts with photoelectric sensors and switches. A similar logistics system would also be suitable for feeding in modules for switch-cabinet assembly.

- Individual configuration according to objective: A particular challenge for the demonstrator cell involves planning and executing assembly processes that are unknown at the time of development. The definition of the target—i.e. the description of the assembly to be assembled—is provided for each assembly in the form of CAD data, which are used to establish the necessary assembly processes. Phoenix Contact produces a description of the components to be assembled for each customer-specific switch-cabinet. In the automated switch-cabinet assembly, the assembly system is responsible for feeding in the required components and for the assembly sequence. The case under discussion also showed that the system can be controlled using a cognitive controller.

6.6.6 Future Research Topics

Future research work should aim to achieve higher degrees and levels of cognition in production systems. In doing so, it should pay particular attention to topics relating to cognitive functions and technologies for fast set-up, fast start-up, flexible scalability of throughput, and to the possibility of cooperative cellular systems. Researchers must also address the question of how to construct cognitively automated production so that it is as versatile as possible and can be operated efficiently, safely and sustainably, even in the rapidly changing environment of a high-wage country like Germany. Within this context, it is especially important to integrate humans and their superior cognitive, perceptive and sensorimotor skills.

Industrial production often uses production lines with sequentially linked stations. Each station carries out the production-process steps assigned to it and then passes the (sub)product on to the next station. The only way to compensate for failures and delays at a station is by using buffers. In extreme cases, failures and delays can bring the entire system to a standstill. Using cooperative, cognitively automated cells and control structures can make the production system more error-tolerant and flexible. A shortage of materials, a broken tool, or delays would only minimally affect overall production, because the system would be able to dynamically adapt the production flow and have other CCUs take over certain production steps. Figure 6.130 shows a diagram of how this kind of process chain made up of cognitively automated cells might look.

The configuration of the process chain pictured here is based on the chaku-chaku principle (literally load-load in Japanese), also known as the "one-piece-flow system". In its original form, this system involves all stations producing more or less autonomously, and a human operator simply transports the parts from station to station. The principle can achieve a high level of flexibility when it comes to handling variants and fluctuations in production, and at the same time it can reduce processing times and space requirements. A classic chaku-chaku line is generally based on a high degree of automation and a linear-cyclical linkage, in which humans simply perform "residual tasks" that offer almost no scope for making decisions or taking

6 Self-optimising Production Systems

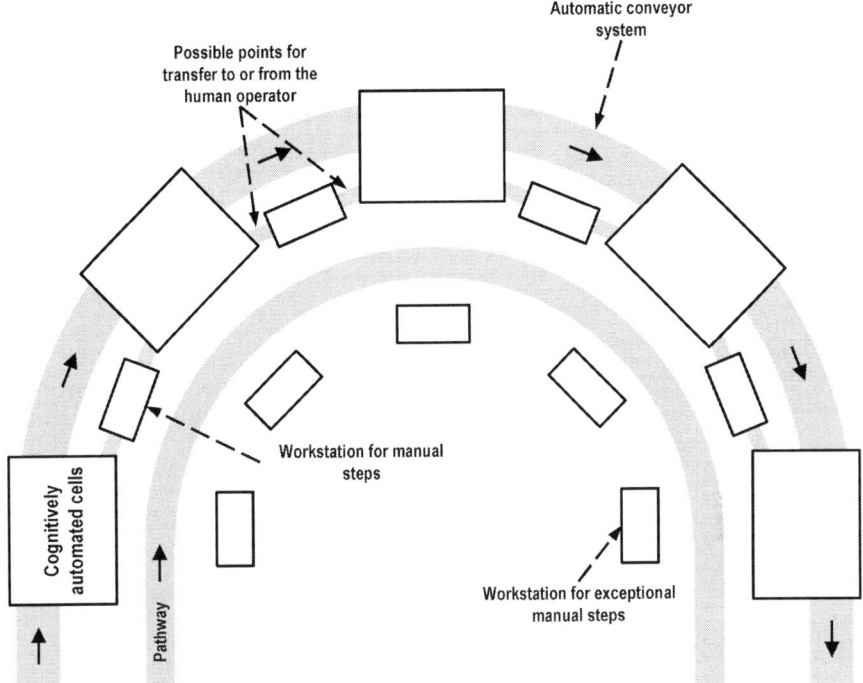

Fig. 6.130 Schematic construction of a cognitively automated process chain

action. Frieling and Sträter (2009) have investigated the productivity of such assembly systems and the health risks they pose to employees. Initial research findings from the samples they examined indicate that this kind of activity can have harmful effects on employees' mental workload (monotony and fatigue). This is presumably the result of highly repetitive assembly tasks and the lack of job rotation in the assembly concept (Enríquez Díaz et al. 2010)

To counteract the disadvantages of the chaku-chaku principle, future proposals should focus on parallelising and integrating tasks within the cells and on allowing humans to interact with cognitively automated systems in an ergonomic way. Looking at a single cognitive cell within the process chain, and based on Fig. 6.131, we can identify the following possibilities for processing a blank, a semi-finished product or a finished product:

- Fully-automated process: Feed-in, processing and feed-out happen automatically within the cell. Humans monitor and optimise the production process.
- Cooperative, partially automated process without removal: Feed-in and partial processing happen automatically. A human is required to manually perform some of the processing. To do so, the part in question is transported to the human operator's workspace, without leaving the secure, controlled area of the cell. While this is happening the cell can do nothing else. Once the human operator

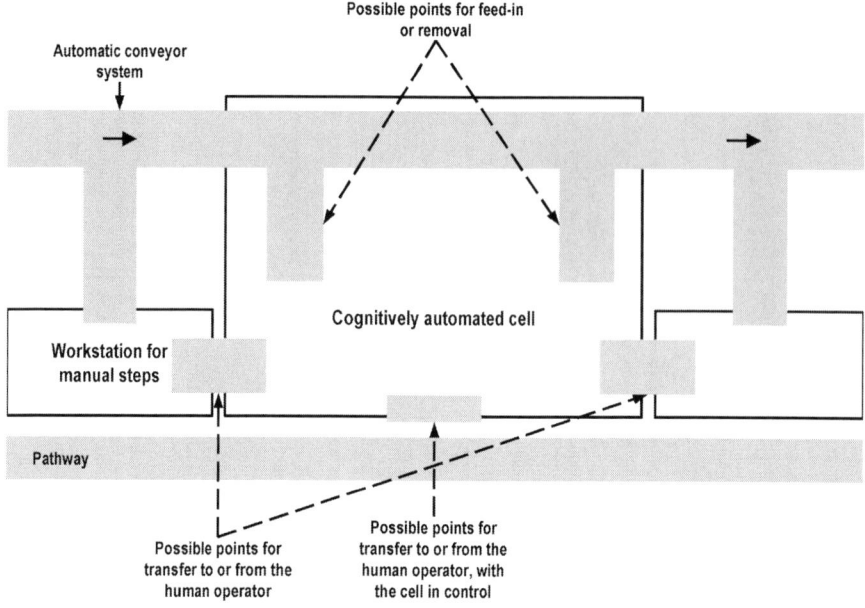

Fig. 6.131 Schematic construction of a cell in the cognitive process chain

has completed the manual work, the conveyor system takes the part through the remaining automated processing steps, the measuring and checking tasks, and on to feed-out.

- Cooperative, partially automated process with removal: Feed-in and part of the processing happen automatically. A human is required to manually perform some of the processing. To do so, the part in question is transported out of the cell to a manual workspace. The part is now outside the secure, controlled area of the cell, so the cell is free to carry out other tasks while the part is being manually processed. Once the human operator has completed the manual work, he/she feeds the part back into the cell, and the conveyor system takes it through the remaining automated processing steps and on to feed-out.

If the blank, semi-finished product or finished product does not require processing in a particular cell, the integrated automated conveyor system can make it skip that cell.

This concept requires research into the following:

- New concepts and technologies that allow cooperative CCUs to coordinate and communicate material requirements, process statuses and capacities.
- Cooperative planning algorithms that can dynamically respond to changes and adapt the behaviour of individual CCUs to keep the performance, reliability and safety of the overall system at an optimal level. Several concepts can achieve this: CCUs can coordinate directly with one another, or hierarchically structured

"meta CCUs" can perform higher-level planning tasks. The algorithms must be able to take into account limited availability of common resources, availability of materials, and the flow of products.
- This kind of system also makes it possible to integrate the specific skills of a human operator into the production process. Thus, using higher-level planning, work tasks that due to their complexity require a minimum level of experience can be assigned to the person best suited to the job. Furthermore, training and qualification processes can be carried out during periods when regular operations are slow (embedded training). This means that individualised production does not just apply to individualised products; it also refers to adapting production to the individual employees involved. Particularly in light of today's changing demographics, these kinds of systems could take physical impairments into account in production planning and execution and thereby offer staff support that is specifically targeted to their needs.

Given the many transfer points within the system, another field of research activities involves developing ergonomic designs for human-machine interaction. Because production is so varied and normally controlled by demand, one cannot assume that one fixed process model will emerge that humans can use for orientation. This underscores the necessity that human-machine interaction is safe and that it conforms to human expectations. Future research work should therefore investigate the following questions:

- To what extent can a cognitively ergonomic display of system status—e.g. a dynamic, multi-level flow diagram rather than the abstract, schematic displays that are common in today's control stations—increase system transparency, particularly given the fact that a human operator has to manage several cells and segments at once? Because the research focus in this case is not limited to intervening when errors arise, scientists need to develop additional concepts and methods for intuitively monitoring normal operation.
- How can targeted variations in a robot's motion sequences, in the sense of anthropomorphic kinematics and dynamics in the cell, inform the human operator—quickly, reliably and in a way that optimally responds to operational demands—that a transfer is coming up?
- Can anthropometric variables be integrated into overall system planning in such a way as to allow individual, ergonomic adaptation of things like transfer or processing points? How much flexibility do these points offer in terms of adapting to something like an unexpected staff absence?
- Can occupational safety aspects be taken into account during planning—e.g. by integrating an individual biomechanical model of the human operator—so that things like loads and torque are not set as standard, but are adapted to the specific human operator?

In conclusion, it is clear that a cognitive planning and control unit in the form of the detailed CCU concept described in this paper only provides a small example of what cognitively automated systems will be able to achieve in the field of production.

Looking at the process chain as a whole, it is already possible to use cognitive mechanisms to comprehensively optimise a system beyond the boundaries of existing tolerance areas. Using a targeted combination of components—known as cognitive tolerance matching (Schmitt et al. 2009b, see Sect. 6.2.4.3)—it is also possible to improve the quality of product. Taken together, the methods and systems developed so far offer companies in high wage countries in particular the chance to achieve considerable competitive advantages. By using cognitive mechanisms in automation, these companies can directly contribute to securing and expanding production locations in those countries.

6.7 Reconfigurable Assembly Systems for Handling Large Components

Burkhard Corves, Rainer Müller, Martin Esser, Mathias Hüsing, Markus Janßen, Martin Riedel and Matthias Vette

This chapter presents the results of the subproject "Reconfigurable self-optimising component handling", part of the Cluster of Excellence in the integrative cluster domain of "Self-optimising production systems". The Chair of Assembly Systems, Laboratory for Machine Tools and Production Engineering (WZL) is involved in the project as well as the Department of Mechanism Theory and Dynamics of Machines (IGM) of RWTH Aachen University.

6.7.1 Challenge

The project is motivated by fundamental changes in the operating framework for manufacturing companies in recent years. Reasons for increasing complexity and dynamics within these companies and in the industrial environment are not only increasing globalisation, but also the fast pace at which technology is developing and the altered resources situation (Möller 2008; Müller et al. 2009b). The consequences of this include the even greater shortening of product lifecycles, the persistent increase in the number of product variants and the constant pressure to reduce manufacturing costs (Lotter 2006). Assembly systems that are reconfigurable for the specific purpose are a highly promising approach in the conflict between achieving individualised production on the one hand and cost-reducing automation on the other.

Given this scenario, a current focus is the handling of large components that sometimes possess only little inherent rigidity. Such components are used in aerospace engineering for example, in shipbuilding and wind power installations. In aircraft construction in particular, we see increasingly large shell elements of CFRP materials being used for fuselages (Fig. 6.132) (Licha 2003).

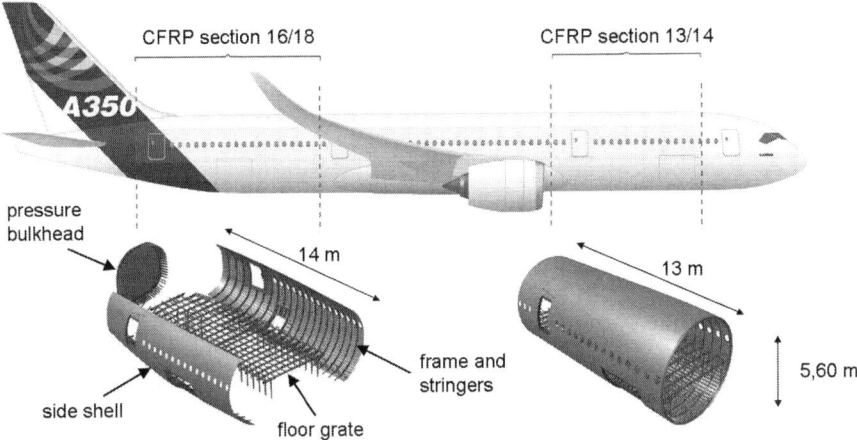

Fig. 6.132 Airbus A350 XWB sections 16/18 and 13/14. (Source: Premium AEROTEC GmbH 2011)

Especially for these components you have to consider that only limited forces may be applied to the component when it is handled. This also calls for high accuracy and an adequately large workspace. To satisfy these requirements, large jigs and fixtures are used for handling, specially matched to the particular component, and making a system highly inflexible and costly. A separate jig or fixture has to be constructed and provided in sufficient numbers for each component. The situation is complicated even more by the fact that the individual structural elements of an aircraft differ very much in shape, size and the position in which they are fitted, which will in most cases prohibit the use of a general-purpose jig or fixture.

The large variety of components and their application consequently necessitates an assembly system that can be expanded and reconfigured for the purpose. The assembly system must be scalable and universally applicable to respond to a changing economic and technical situation. To make automation economically attractive for assembling smaller production series too, it must be possible to ramp up a system again quickly after its reconfiguration to avoid the cost of long downtimes.

The approach taken here consists in composing an assembly platform (Fig. 6.133) to match the illustrated requirements. Different modules can be combined to implement the required assembly tasks and functions.

6.7.2 State of the Art

Industrial production shops are in many cases characterised by automated assembly processes. The degree of automation is relatively high in particular for small to medium-sized products in series production. However, this degree of automation becomes less as the size and complexity of assemblies and the scale of joining

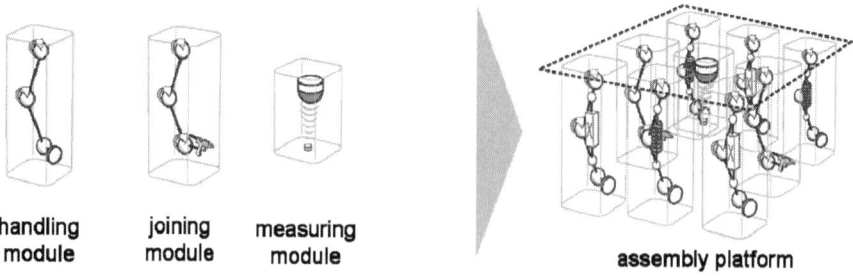

Fig. 6.133 Assembly platform

operations and processes increase. Large components in the aerospace industry for example, in shipbuilding and the construction of rail vehicles are for the most part assembled manually, supported by jigs and fixtures. The reasons for what seems to be a technical lag are to be found in the nature of the product to be assembled and high technical demands when it comes to accuracy and accessibility of the joints (Stepanek 2007).

Large components in particular frequently have little rigidity before they are assembled, so the force of gravity alone can cause inadmissible deformation, meaning that components must be picked up and supported at a number of points when handled. The task of the system includes joining components in addition to transporting and feeding them on a line. The joint tolerances may be just a few tenths of a millimetre while the components themselves are anything between ten and thirty meters (Wollnack and Stepanek 2004).

To complicate things more, the components are very different and are only manufactured in relatively small numbers, resulting in a whole number of high-investment stations and large space requirement when it comes to rigidly automated systems. This trend is especially noticeable in aircraft assembly where aircraft are sometimes manufactured for a number of decades using one assembly system. Here you generally find product-specific jigs and fixtures that not only ensure correct joining of the assembled objects by defined geometrical, functional elements but also support proper shaping of the components.

High investment costs and the long product life cycles of the assembly systems hinder the continuing development of new assembly technologies. In part, assembly systems are used for several decades. The introduction of new assembly technologies is frequently tied to the introduction of new products or variants, the result of which, in aircraft production, is long innovation cycles. But these, in most cases rigid systems are unsuitable for new market demands like individually matched and quickly available product variants. Instead a flexible system is needed that can respond to customer requirements without huge investment.

Motivated by the drawbacks of jig and fixture-oriented assembly of large components, new approaches are sought to make the assembly of large components more flexible and be able to automate it. This is a context in which the aerospace industry assumes a leading role.

Fig. 6.134 Positioning system in structural assembly of aircraft fuselage sections. (Wollnack and Stepanek 2004)

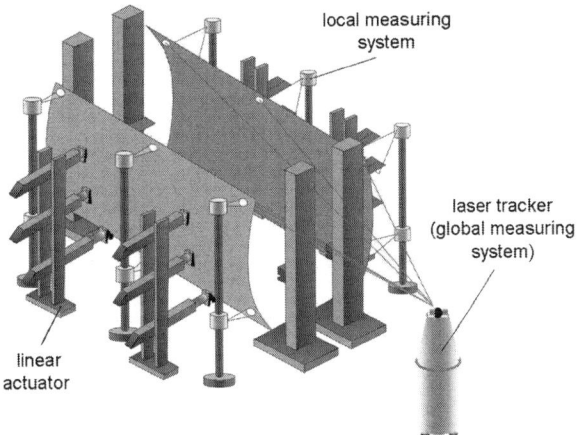

One approach to reducing product-specific jigs and fixtures, means of fastening, clamping and the like in aircraft structural assembly is the use of linear actuators (Fig. 6.134). The shell elements of the aircraft fuselage are taken up by a number of linear actuators and fed to the joining station. This increases flexibility compared to jig and fixture-oriented assembly systems. The configuration of the actuators is specified however, and workspace is limited.

The use of robot systems can further increase flexibility for automated assembly of large components in changing applications. Cooperating robots enable the implementation of highly flexible assembly cells. By resetting and reprogramming it is then possible to use cooperating industrial robots for different purposes. Inflexible jigs and fixtures to handle large components can be omitted because the component can be grasped and supported at different points by multiple robots.

But there are still a number of drawbacks associated with the use of cooperating robots. In addition to the high price of industrial robots there is, in particular, the time taken up by programming, or shutdowns needed for conversion and new programming.

Two concepts have emerged to date to control cooperating robots. Some manufacturers use a system in which a master controller consisting of a control computer and drive components is expanded by additional drive components for each extra robot (Fig. 6.135). The advantage of having just one control computer (i.e. single controller solution or SCS) is that the extra cost for each additional robot is very much less than if each robot had a fully featured controller with computer and drive modules (Bredin 2005). The disadvantage is that the system cannot be composed in any random way because a powerful central processor is always needed. There are also restrictions when it comes to achieving single operations of the handling modules, for feeding purposes for example.

Contrasting with this is a control concept in which each robot possesses a fully featured controller and there are consequently a number of computers available in

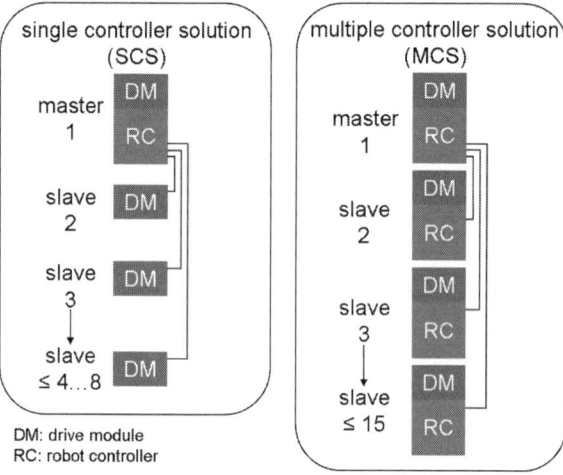

Fig. 6.135 Hardware concepts to control cooperating robots

the robot network (i.e. multiple controller solution or MCS). This concept can involve higher investment costs because each robot has its own complete controller. The additional computing power means that significantly more robots are able to cooperate. Despite a high-speed Ethernet connection the number of robots is still limited to 15 units (Kuka 2011a).

In an application with cooperating robots three operating modes are distinguished. In the simplest mode all robots act independently of one another but in overlapping workspace. One example is pick & place tasks where there is no direct interaction between the robots.

Other applications may require synchronised motion, i.e. where robots synchronise at certain points along a programmed trajectory. This makes sense, for instance, when one robot brings a component into a position for machining by a second robot. The situation requires coordination between the robots so that one can tell the other that a component has reached the position for machining, for example, or that machining is completed.

Precise path control of all robots is essential for coordinated motion, e.g. when they jointly move an object, because path deviations could result in deformation of or damage to a component. So the response of robots must be modelled as precisely as possible in the controller system. The single paths must also be generated at a fast interpolation rate to minimise path deviations between the computed interpolation points (Stoddard et al. 2004; Feldmann et al. 2007).

The programming of cooperating robots is independent of the hardware concept and similar for all operating modes. Each robot is invested with its own program, which is partly or entirely independent of the programs of the other robots. The robot can change between the operating modes within the program and coordinates with the other robots through the coordinate system of a master robot.

The disadvantage of this programming concept is that each robot possesses its own application-specific program that has to be configured for each new application. A high number of robots and frequent changes in work content mean a large programming effort, calling for more efficient concepts.

A further problem is the accuracy of robot systems. The position errors that appear are often kinematic in nature. They are caused by deviations between real system performance and the calculation model that is used.

The kinematics of a robot are frequently identified and modelled with inadequate precision. Instead of robot-specific calculation models, nominal models are used that do not correspond accurately to the genuine kinematics.

The length of robot components can vary as a result of production and measurement errors. A further error source is the position of rotary axes in relation to one another. There may be deviations in position if robot axes according to the model are not precisely parallel or at right angles to one another. In addition to axial position errors and deviations in length, another source of error is the zero positions of robot axes, which depending on kinematics may lead to a position and/or orientation error.

CAD data is used to model the other resources of the assembly system. Here too, assembly and manufacturing tolerances are not taken into consideration. Special importance attaches in reconfigurable systems to the position of the industrial robot, of the process station and of the handling object. When setting up heavy resources in particular, deviations of several millimeters may occur, as a result of which functional tolerances are no longer maintained.

Consequently these error effects must be measured and identified after assembly, a modification or repair, and taken into account appropriately in the model or for programming.

6.7.3 Research Questions

The aim is consequently to develop an assembly platform for reconfigurable self-optimising component handling that can be composed for differing handling purposes by a modular, versatile structure. It consists of multiple simple handling modules that are compact in shape and low in weight to allow fast reconfiguration. Use in a multiple-arm combination permits in particular the efficient handling of large-area components.

Two scientific questions, describing the individual tasks in more detail and allowing a target-oriented approach, were outlined for each of the four focal subjects presented in Fig. 6.136.

Under the heading "Design of versatile assembly units" those questions are resolved relating to the development of a reconfigurable handling system, covering kinematic reconfiguration of the hardware and the management of mechanical, electronic and software interfaces. To keep interface management as simple as possible, the approach consists in designing each module so that it is mechatronically self-contained, and in featuring it with ready defined interfaces so that it can be part of a mechatronic building set.

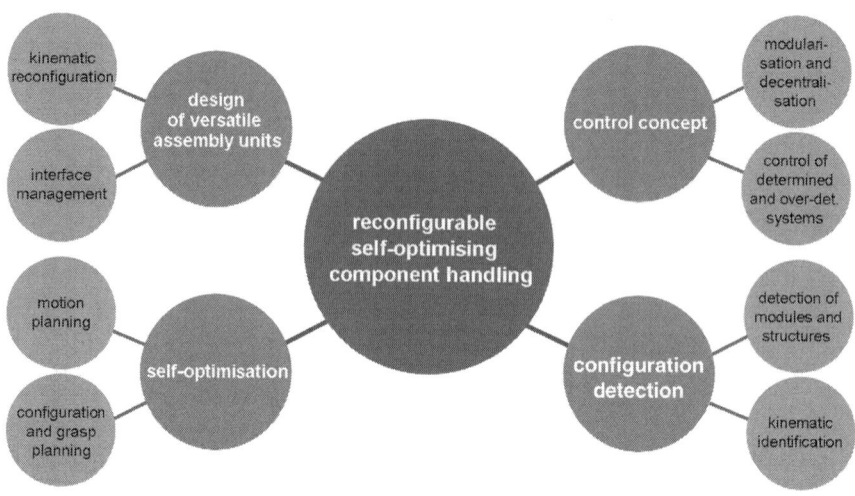

Fig. 6.136 Scientific questions in subproject

The subject "Control concept" deals with controlling and programming the modular handling system. This covers considerations of data storage within control, which must be managed decentrally at module level (e.g. transaction sets) on the one hand and centrally in a superordinate controlling instance (e.g. coordination of modules) on the other. Depending on the drive configuration the handling system can work both statically determined, i.e. with a total of six driven axes, and over-defined with further axes. A concept is devised by which both over-defined and exactly defined handling and multi-robot systems are simply programmed by specifying a path for the component from which the movements of every single handling module are indirectly derived.

The subject "Self-optimisation" looks at featuring the handling system with additional functions that support the user in planning the right movement and creating a suitable system configuration. To optimise movement, the trajectory of the component is analysed with computer support and subsequently adapted by the user interactively. Optimal system configurations can be calculated by an algorithm that optimises different weighting criteria, e.g. maximum drive speed, moment or accuracy, with reference to specified limits and thus determines the positions of the basic or reach coordinates for each handling unit.

"Configuration identification" means identifying the particular system configuration by means of measurement technology. After executing special movement patterns, the positions of the individual handling devices can be determined and deviations that appear in structuring the system can be compensated in control. Calibration of the handling devices is also provided to determine the exact kinematic parameters and match the calculation models. This is necessary to achieve maximum positioning accuracy of the robots.

6.7.4 Results

6.7.4.1 Idea of a New Handling System

Robot systems are often chosen to create the flexibility needed in an automated assembly used for changing applications. Cooperating robots are a solution for implementing flexible assembly cells. By reconfiguring and reprogramming it is possible to use industrial robots for different purposes. Drawbacks are the high purchase price of industrial robots and the complex programming. The need for synchronised path control, which also results in substantial setup time, means that such robot systems are often not an economical solution.

Against this background a new kind of handling strategy was developed that enables cooperative handling by simple handling modules and without elaborate setup procedures.

The handling method produced by this concept distinguishes three movement phases: coordinated movement, approaching movement and retracting movement. During coordinated movement, analogous to coordinated robots, multiple handling devices are joined to the component in a multiple-arm combination. Unlike conventional cooperating robots, the wrist axes are passive. During approaching and retracting movement the handling devices work autonomously and behave like separate mechanisms.

To implement this handling concept, and based on groundwork performed by the Department of Mechanism Theory and Dynamics of Machines (IGM), a suitable handling device was developed with six degrees of freedom (Fig. 6.137) (Müller et al. 2010).

The kinematic structure of the handling device corresponds to that of a vertical jointed-arm robot. A parallel double crank drives the third axis to apply the weight and inertia of the motor to the robot base. The wrist takes the form of a central hand and, unlike industrial robots, consists of three passive joints. In addition to the monetary saving by reducing the drive concept to only three active principal axes and three passive secondary axes, the handling device is lightweight and can be shifted by one operator without extra aids. The active degrees of freedom of the robot make it possible to join to the component by a certain grasp movement through a suitable contact element (magnetic or vacuum). The drives of the principal axes are actuated

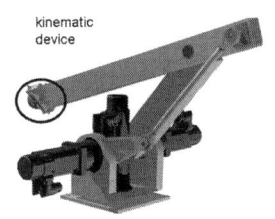
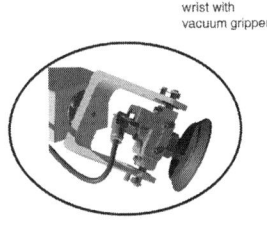

Fig. 6.137 Handling device with three active principal axes and three passive secondary axes

during the cooperating movement. The movement of the joint axes results from the kinematic structure of the multiple-arm combination, i.e. the wrists track passively.

In assembly platforms with multiple handling devices in particular, the economic advantage that results from reducing the active drives is considerable, because both the number of motors and the inertia plus complexity of the robot hand are very much reduced.

6.7.4.2 Steps of Reconfigurability

The wide-ranging field of use and the adaptability of the assembly platform to changing applications are achieved by multiple steps of system reconfigurability (Fig. 6.138). The usual boundaries such as workspace and choice of component or application are dissolved to produce an adaptable system that is reconfigurable and universally applicable.

The first step of reconfigurability is realised by altering the grasp-points on the object. This may be necessary if the planned object path leads to collisions between individual handling modules. Furthermore, rearrangement of the grasp-points is called for when the component changes. The second step of reconfiguration is relocation of the handling units within the assembly platform to design the workspace. The third step of reconfiguration involves enlarging or reducing the system by a number of handling devices. If more arm units are available in the layout than object handling requires, the object can be passed from one arm combination to the next. This re-grasping enables to re-design the workspace even during continuous object movement. Large translational object manipulations become possible, as well as unlimited object rotations about any arbitrary axis.

6.7.4.3 Mechatronic Modularisation of the Handling System

The prerequisite for these reconfiguration possibilities is seamless mechatronic modularisation of the handling system. Initially, module limits were defined to mark the

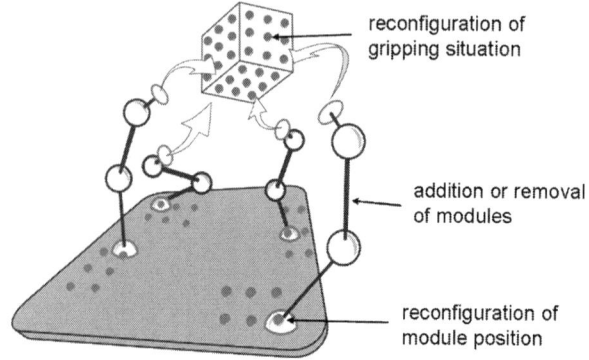

Fig. 6.138 Multiple steps of reconfigurability of an assembly platform

6 Self-optimising Production Systems

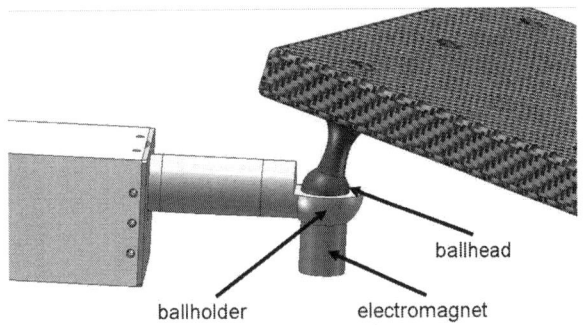

Fig. 6.139 Ball joint for handling non-magnetic large-area components

capabilities of a handling unit. In general terms, a module should exhibit a certain degree of autonomy, i.e. not be dependent on other modules, to prevent any meshed associations between modules and enable them to be exchanged as quickly as needed. Associated with this is the requirement to be able to assemble and start up a module in advance to ensure a fast restart upon replacement or reconfiguration. To satisfy these requirements it is consequently important that each handling device should be mechanically independent and also possess its own control, meaning mechatronic modularisation of the handling device. In addition to mechanical interfacing, simple control interfaces in particular are needed to ensure reconfigurability and scalability of the overall system.

To create appropriate mechanical interfacing of the handling module, the first three driving axes of the prototype were configured for simple conversion. The drives of the first axes were arranged coaxially to the particular rotary axis and the drive of the first principal axis positioned between the members of the handling device so that the latter can be set up on a baseplate. This is attached in the assembly cell by a drilling pattern. So the handling devices can be positioned in the same plane and extra degrees of freedom are provided for reconfiguration.

A further mechanical interface is the design of the wrist. Depending on the application, different solutions are conceivable for implementation of the handling module. Examples in the field of aircraft structure assembly are vacuum grippers (Fig. 6.139) or ball joints to handle large components.

The control of a handling module must possess functionality to control the drives, a local safety system, path interpolation and communication functions. Given the decentral control functions, the handling modules are designed so that they can be operated both autonomously and decentrally in association with a central controller. The assembly platform can thus be configured to requirement and reconfigured after completing an application.

Each handling module consists of a decentral control, a handling device with the mechanical structure, the servo drives and servo amplifiers.

The central module coordinates the assembly system and is responsible for crossboard functions such as communication and safety functions. Figure 6.140 is a schematic diagram of the related control system with the central module and the

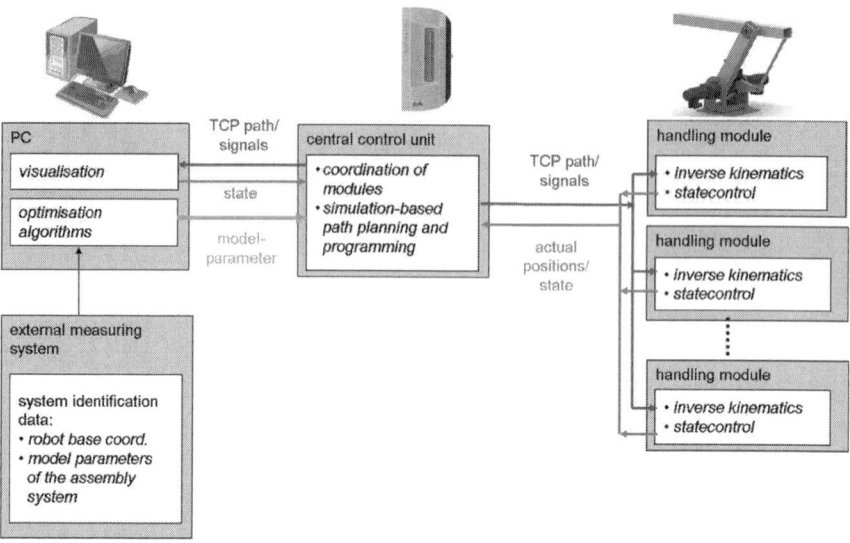

Fig. 6.140 Control architecture of the assembly platform

decentral handling devices. A simulation PC is also integrated that is responsible for optimisation functions in configuration planning and visualisation.

In addition to the mechanical interfaces of the handling device, the electrical and software interfaces were designed so that the system can be reconfigured fast without elaborate wiring (Fig. 6.141). The electric circuitry of each module was encapsulated

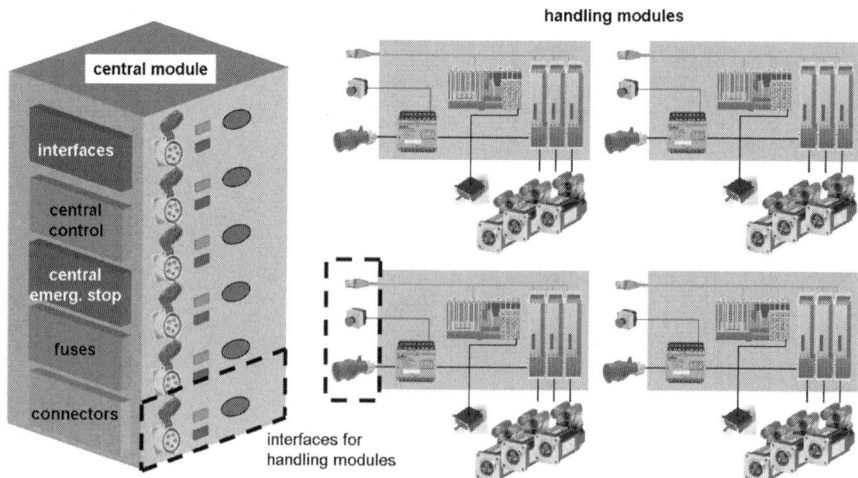

Fig. 6.141 Control concept of assembly platform with central module and decentral handling modules

6 Self-optimising Production Systems

in a separate cabinet and is self-contained. To connect the modules, different connectors were integrated in the central module and in the decentral handling modules to supply the decentral modules with power, enable communication with the decentral modules and connect the different emergency-off systems.

6.7.4.4 Grasp and Configuration Planning

Modularisation and reconfigurability generate new degrees of freedom. The reconfiguration steps give the user a variety of possibilities for configuring the assembly platform. But this major advantage compared to conventional *rigid* systems will only produce benefits in terms of efficiency and cost-effectiveness if both reconfiguration itself and the planning of an optimal configuration can be performed simply and fast. In addition to the accessibility of the grasp-points, stiffness, accuracy, load capacity and velocity transmission along the object motion must be considered as optimising criteria to find the optimal robot configuration.

A number of optimising tools were developed for this purpose, to automatically determine the additional adaptable degrees of freedom of the system, and thus to propose a possible configuration to suit the given application. In this way the assembly system is optimally reconfigured for a new application without prior testing.

Determination of grasp-points is automatically performed before each manipulation and has a large influence on the workspace of the object as well as the kinetostatic performances of the handling system, e.g. load capacity, stiffness or accuracy.

The handling devices used here possess a degree of freedom of $F = 6$, but unlike conventional industrial robots only three degrees of freedom are actuated. To be able to position and orient an object in the complete workspace, a handling system must have at least six actuated degrees of freedom. The handling concept is based on the combined working of multiple handling devices, so the system is even over-defined during cooperating component handling with three handling devices and a total of nine active driven joints. Further reduction of the number of drives at each individual handling device makes little sense because these three active degrees of freedom are needed to be able to grasp the object independently. The economic advantages compared to three cooperating robots with 18 active axes are nevertheless considerable because the concept used here saves nine drive units in this configuration.

This under-actuation of a separate handling device must be considered in grasp planning and performance however. The grasping pattern was consequently analysed in detail. The wrist features a spherical architecture with three serially arranged rotational joints whose axes intersect at one point. Figures 6.142 and 6.143 show a detailed view of the kinematic structure including terms and angles plus the prototype of a magnetic gripper.

Alignment of the contact element, e.g. an electromagnet or vacuum gripper, is passive upon contact with the surface of the object. Until contact with the object the joint is positioned centrally by spring forces. The sequence of a grasping operation with passive joint movement about the hand axes α_R and α_S is shown in Fig. 6.144.

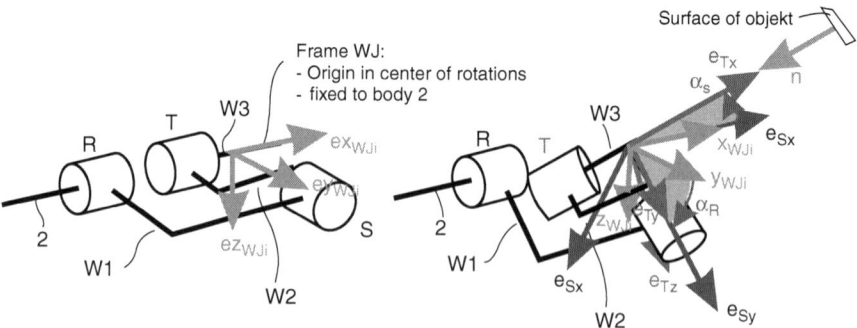

Fig. 6.142 Wrist structure: central position, deflected position

Fig. 6.143 Wrist with electromagnetic contact element

By experiments on a test bench for grasping movements and suitable mathematical models, it was possible to identify and optimise the parameters that are decisive for a good grasp pattern. In (Riedel et al. 2010a) it is shown that the area where the object could be placed on the ground and be grasped successfully can be multiplied almost tenfold by several optimisations concerning the grasp-process and the design of the wrist (Fig. 6.145).

Fig. 6.144 Passive alignment of wrist joint when grasping

Fig. 6.145 Improvement of the grasp process

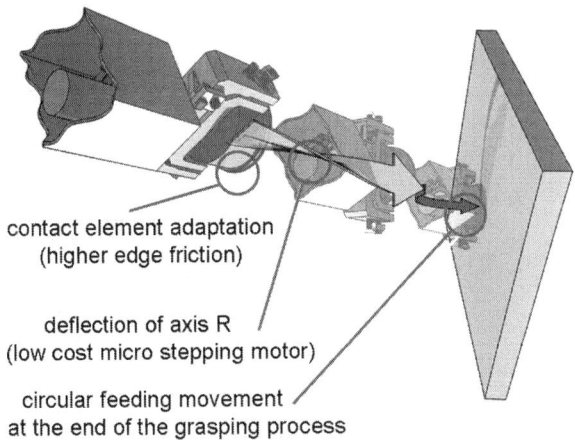

Taking a configuration for handling a cube-shaped object with magnetic grippers as an example, it was possible to expand the grasp area, which initially was only in the middle of the workspace (Fig. 6.146a), to the entire workspace at the level of the ground layer (Fig. 6.146b).

The major influencing variables are edge friction of the contact element and the grasp movement of the wrist. Spatial circular guidance of the wrist by the regional structure contributes very much to improving the complete grasp process. This circular path begins where the contact element sets down on the surface and is perpendicular to the joint axis e_{SY} of the wrist. The entire path of the wrists is shown in Fig. 6.147.

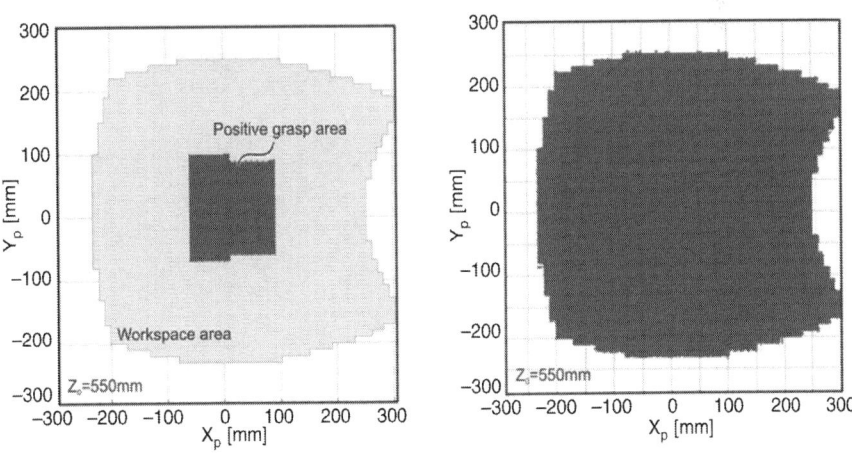

Fig. 6.146 Comparison of areas in which an object can be properly grasped. *Left:* grasp area before optimisation; *right:* grasp area after optimisation

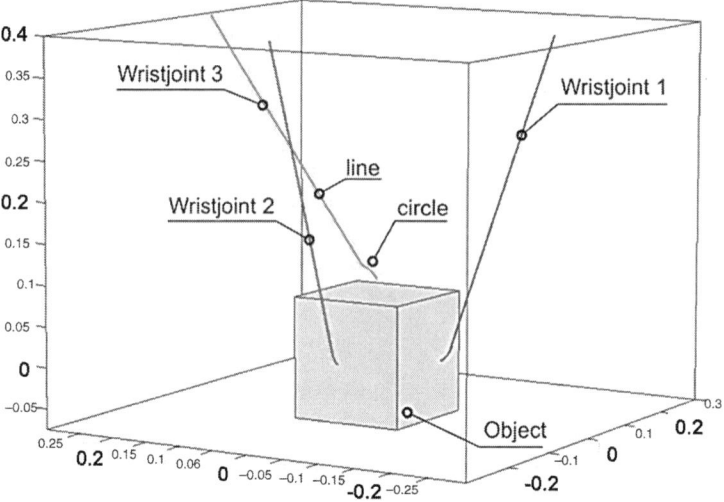

Fig. 6.147 Visualisation of grasp movement with circular track

Searching for and defining suitable grasp-points is performed by an optimising algorithm that is detailed in (Riedel et al. 2010b) and satisfies the following requirements:

- Optimisation of the grasp-points must consider the specific grasping pattern of the passive wrist.
- Grasp-points can only be selected and assessed correctly and meaningfully in a combination of the arms.
- Specification of a set of possible grasp-points or areas must be variable so that new objects can be integrated fast and simply.
- The user must be able to specify the objective of optimisation, e.g. minimal cycle time or maximum accuracy.

The possible grasp areas are given in the form of a set of discrete points or over continuous surfaces. Just nine points per face of a cuboid produce 148824 (54!/51!) possibilities of combination for the handling devices. Analysing each combination is not efficient. Consequently a smart and highly effective method of pre-selection was developed that first analyses only geometric information of the individual points. These include:

- It must be possible to grasp the point regarding the characteristics of the passive grasp process.
- The point must be within range of the handling device during the entire movement.
- All drive and joint angles must meet their permissible motion range during object manipulation.

Working through this sequence soon leads to exclusion of many of the points. Those points which fulfil the workspace conditions are comprised into a new set of possible combinations. The suitability of the combinations is examined at discrete points along the path, in that either the kinetostatic criteria allow a relative comparison between the solutions, or absolute performance data are evaluated by means of the dynamic model. If a combination of grasp-points satisfies all performance requirements, it goes into the final selection group. From this group the solution is then chosen that profiles best the user's needs. Figure 6.148 illustrates the flowchart for preselection for grasp-point optimisation.

If an object is mapped over continuous surfaces, a gradient-based method of optimisation can be used from dimensional synthesis in slightly modified form. The entire procedure of planning grasp-points is illustrated in Fig. 6.149.

The results of the different approaches to optimising are presented in Table 6.16. The first four solutions are selected manually, the next three (E through G) from a set of discrete points, and the last three (H through J) were optimised over a continuous surface.

The unsupported selection of grasp-points can easily produce unsatisfactory results. The cases show that grasp-points which were manually selected require too high driving torques (A), result in an undesirably large positioning error (B), can be grasped but violate the workspace condition during movement (C), or cannot even be grasped at all (D).

The automatic planning of grasp-points prevents these cases. In (E) all 148.824 combinations were evaluated, taking more than 12 min. of calculation. (F) uses smart preselection, so it considers all possible combinations and finds the best solution in the field after just half a second. Applications that are more time-critical can use search method (G), which halts at the first possible solution. Here the time to compute is negligible. But this approach only finds suitable solutions and not necessarily the best available. By its principle, optimising method (H) takes longer than a discrete search (E through G) because it takes at least 120 (6!/3!) optimizations if there is no preselection of surfaces. Furthermore, the optimising problem is not convex, so there is no assurance of finding a global optimum. That is why different starting values are given for surface optimisation, as a result of which the computing effort increases exponentially. Analogous to approaches (E through G) the time to compute is 4,757 s. for the complete run (H), 84 s. for optimisation with intelligent preselection, and 0.75 s. for optimisation with interrupt (J) (Table 6.17).

Despite the longer time to compute, these approaches are interesting if the performance of the handling system is to be improved further. In this way solutions are found that lie between the discrete points of the setting for (E through G), and further reduce the maximum driving torque from 8.6 Nm (F) to 7.5 Nm (I) or the maximum position error from 1.2 mm (F) to 1.0 mm (I). The choice of optimising approach here is user-specific or application-specific.

If a system embodies more than three handling devices, the combination of handling devices can also be optimised and selected for the movement application. Furthermore, the individual handling devices of the demonstrator can also be freely configured on the baseplate, resulting in a large number of possible combinations. If three handling devices can be set up at 100 base points, this alone means almost one million configuration possibilities–combined with the possible grasp-points the reconfigurability of the system is theoretically unlimited. For this purpose the approaches and algorithms of grasp-point selection were revised for efficiency and then

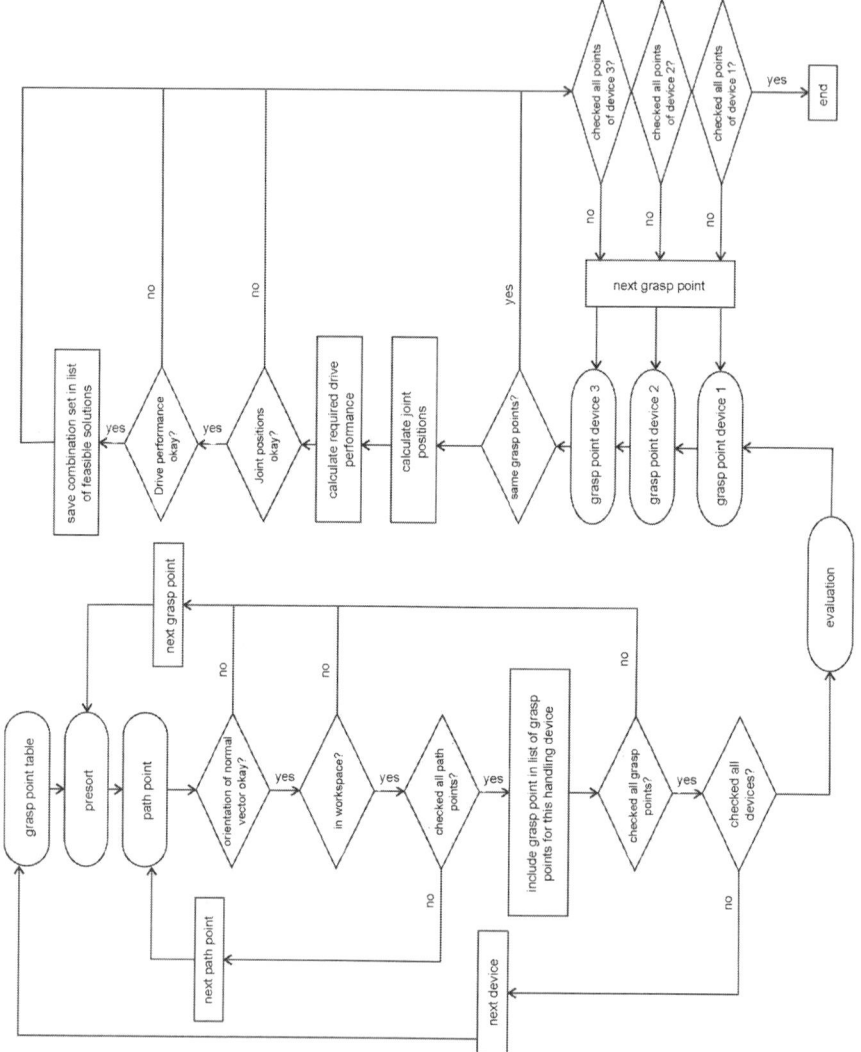

Fig. 6.148 Preselection to optimise grasp-points

6 Self-optimising Production Systems

Fig. 6.149 Optimising procedure

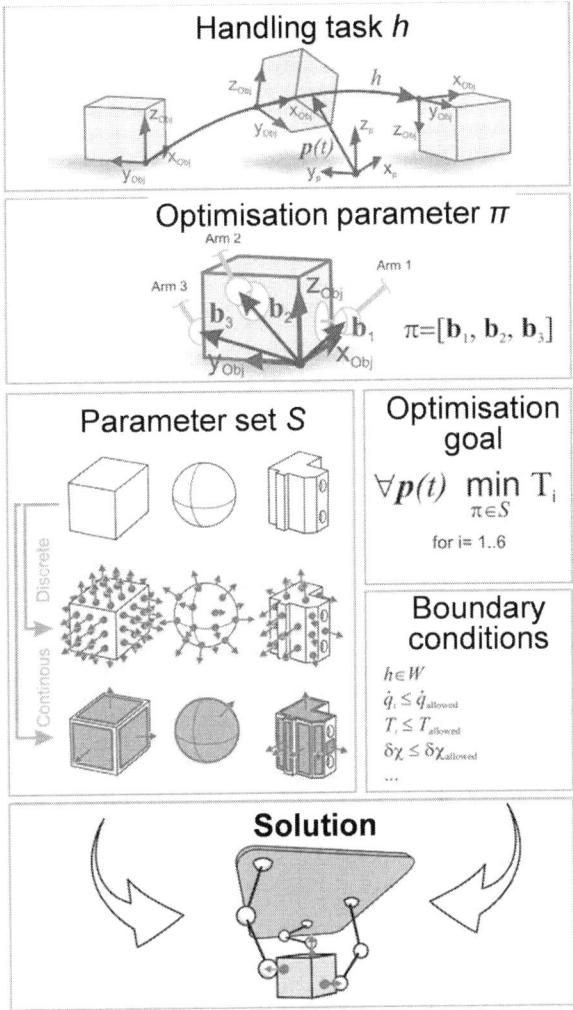

implemented. With reduced grasping possibilities the automatic planning algorithm finds an optimal solution in about five seconds.

In reduced terms, the search for base points works as depicted in Fig. 6.150. After specification of the optimising parameters, goals and limits, a first fast selection searches the specifications field and reduces it by the points or areas that cannot be reached because of the range of the handling devices. At the same time a second preselection examines whether the point or area underruns a certain direction angle to the grasp surface normal during the movement. If this is so, this surface cannot be managed from the particular frame point and is not examined further. Experience shows that the specifications field shrinks to 2–5% of the initial size, and it is then newly sorted and combined. The possible combinations of handling devices are subsequently analysed dynamically or kinetostatically and selected analogously to the planning of grasp-points.

Table 6.16 Specifications and boundary conditions for optimising task

	Task definition			
Path	cubic spline	P1	P2	P3
	$[x,y,z,]^T$ in [m]	$[-0{,}15;0{,}1;0]^T$	$[-0{,}05;0{,}15;0{,}1]^T$	$[0{,}1;0{,}15;0]^T$
	$[\varphi,\phi,\psi]^T$ in [deg]	$[0,0,0]^T$	$[0,45,0]^T$	$[0,90,0]^T$
Trajectory	law of motion	5th order polynomial		
Motion		duration	1	s
		length	0,48	m
		max. transl. velocity	0,91	m/s
		max. transl. acceleration	6,52	m/s^2
Object	Shape	Cube		
	Size	150x150x150	mm x mm x mm	
	Mass	1	kg	

Optimisation goal
minimize drive torque along path

	Boundary conditions		
Restrictions	Grasp criterion	true	
	Workspace criterion	true	
Limits	Maximal drive torque	15	Nm
	Maximal drive velocity	1000	rpm
	Maximal object error δX	3	mm

6.7.4.5 Path Planning and Programming

Programming of the assembly platform and/or handling system must also match the requirements of a decentral and reconfigurable control system. In conventional controllers for cooperating robots, application-specific programs are generated for each robot that map the particular movements and synchronise with the other robots through a network. The coordination effort is very high, for which reason online methods such as teaching and playback are unsuitable for programming multi-robot applications or for frequently changing purposes.

A more suitable method is offline programming where programs are created without using the real machine. The programs needed are generated and validated first in a simulation environment, and then transferred to the controller. For this purpose the simulation environment requires a model of the assembly system.

6 Self-optimising Production Systems

Table 6.17 Comparison of optimising approaches

Optimization method	Case	Grasp point vectors b_i [x;y;z] in respect to frame F_OBJ in [m] b_1	b_2	b_3	Optimization Time in [s] @2,66GHz	Grasp criterion	Workspace criterion	max T in [Nm]	max q in [rpm]	max δX in [mm]
non	A	-0,075 -0,060 0,030	0,030 0,075 0,060	0,000 0,030 0,075	0	true	true	31,0	784,4	2,2
non	B	-0,075 0,060 0,010	0,060 0,075 0,060	0,000 0,030 0,075	0	true	true	11,9	690,1	6,8
non	C	-0,075 0,060 -0,060	0,075 0,060 0,060	0,075 0,000 0,000	0	true	false	-	-	-
non	D	-0,075 0,060 -0,060	-0,075 0,060 0,060	0,075 0,000 0,000	0	false	false	-	-	-
discrete	E	-0,075 0,060 -0,060	0,060 0,075 0,060	0,000 0,000 0,075	734,59	true	true	8,6	647,1	1,2
discrete	F	-0,075 0,060 -0,060	0,060 0,075 0,060	0,000 0,000 0,075	0,51	true	true	8,6	647,1	1,2
discrete	G	-0,075 -0,060 -0,060	-0,060 0,075 -0,060	0,000 0,000 0,075	0,10	true	true	13,3	647,1	2,6
continuous	H	-0,075 -0,049 -0,060	0,043 0,075 -0,024	0,060 -0,035 0,075	4757	true	true	7,5	575,6	1,0
continuous	I	-0,075 -0,049 -0,060	0,043 0,075 -0,024	0,060 -0,035 0,075	84,50	true	true	7,5	575,6	1,0
continuous	J	-0,075 -0,060 -0,060	-0,060 0,075 -0,060	0,060 0,029 0,075	0,75	true	true	8,2	567,3	1,3

To shorten the setup time, the programming concept (Fig. 6.151) developed in this project implies that handling devices are controlled directly from the simulation environment. In this way no programs have to be exchanged on handling modules after a reconfiguration. The handling modules are simply reconfigured, the simulation model is matched appropriately, and handling modules are given setpoint values to match the new model.

The starting point for this programming is the component or the handling application. In a first step the handling system is modelled in the simulation environment. A collision-free component path is then defined by an initial pose, a final pose and the mode of movement. Proceeding from this component path, configuration and grasp

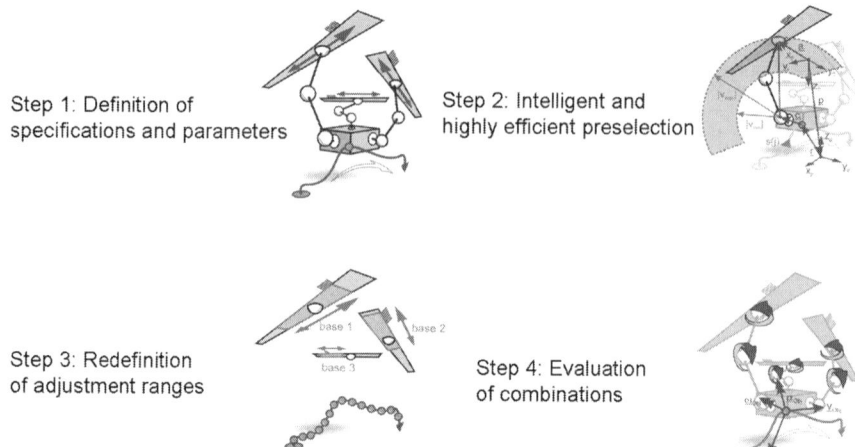

Step 1: Definition of specifications and parameters

Step 2: Intelligent and highly efficient preselection

Step 3: Redefinition of adjustment ranges

Step 4: Evaluation of combinations

Fig. 6.150 Schematic of search for frame points

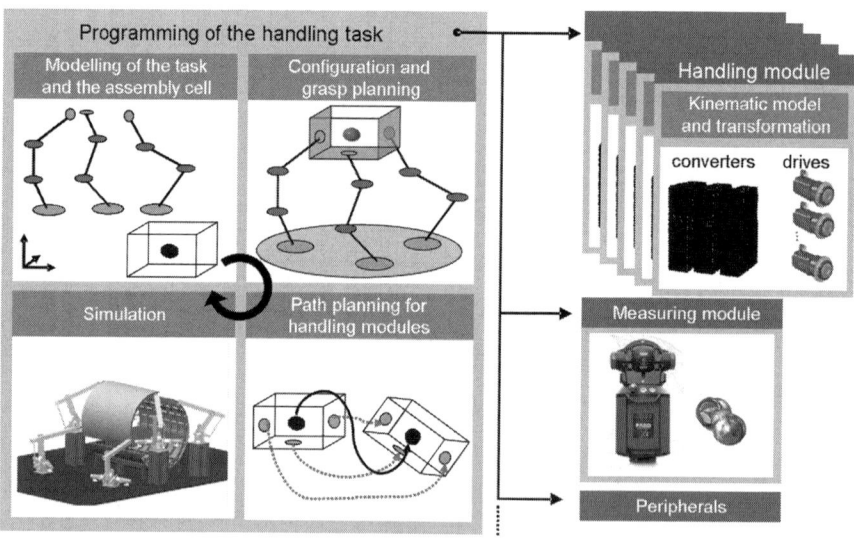

Fig. 6.151 Programming concept for reconfigurable assembly system

planning is performed, and the handling modules are matched appropriately in the simulation environment and the real assembly system.

The grasp-point paths of the handling modules can then be derived from the component path and the defined grasp-points on the component by simple transformation from the component coordinate system to the grasp-point coordinate system. Intermediate points of the component path are calculated in the interpolation cycle. Then the positions of the grasp-points are calculated by the fixed transforms to each intermediate point, and sent in the interpolation cycle on an interface as a setpoint to the handling modules (Fig. 6.152).

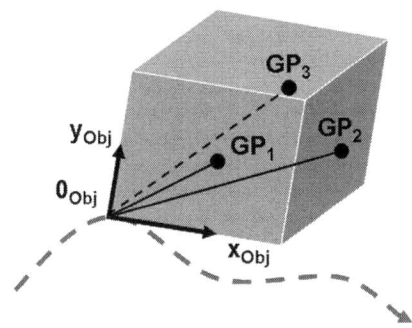

Fig. 6.152 Location of grasp-points (GP) and of component coordinate system as function of component path

The decentral handling modules read this interface in the interpolation cycle and use inverse kinematics to calculate the joint angles for the particular grasp-point, which are then set by the servo amplifiers. If necessary, finer interpolation of the setpoints is possible decentrally. This kind of setpoint generation very much simplifies the complex programming of cooperating robots.

Movement of the multiple-arm combination can be very complex, so a possibility was created of visualising the programs first. The setpoints on the controllers are calculated and then sent not to the servo amplifiers but instead to an interface for visualisation. The simulation cell is then updated by the real controllers in the interpolation cycle. A collision check supports the operator in judging movements.

By avoiding application-specific programs on the individual modules it is possible to exchange or add single modules. Thus, for example, industrial robots or other available handling devices could adopt the task of a handling module.

6.7.4.6 System Identification

An exact description of the system pattern is necessary for direct transfer of programs generated offline. Handling devices are calibrated after installation to significantly reduce the unavoidable gap between simulation and reality.

Here the real kinematic parameters of the handling devices are determined as accurately as possible. An external measuring system registers different poses and the associated joint angles. From the joint angles and the nominal model it is possible to calculate the set poses of the handling devices and compare them to the measured poses. Then, by a numerical method, the descriptive model parameters are altered with reference to gradients until the error between the measured pose and the calculated pose is minimal (Fig. 6.153). These parameters are subsequently returned to the kinematic model and stored on the controller. The real kinematics of the handling system documented in this way are then used for direct and inverse kinematic calculation. The additional setup time for matching simulated programs is thus minimised—so the possibility of real offline programming is given.

For further reduction of setup times, handling devices are automatically calibrated after reconfiguration. It is sufficient to position the units roughly in the assembly

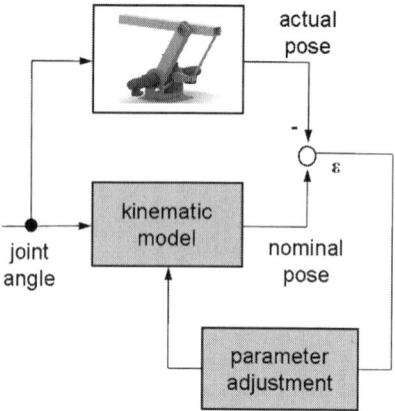

Fig. 6.153 Identification of optimal model parameters by recursive parameter matching

cell according to the configuration planning. They are then identified by short test movements and the appropriate transforms in the calculation model are matched to the real cell configuration.

By this system identification it is possible to determine the relevant model parameters of the assembly platform and of other objects in the assembly cell with high accuracy.

6.7.5 Industrial Relevance

Faced with constantly increasing cost pressure, a broader spectrum of types and shorter innovation cycles, there is greater demand for more flexible assembly and handling systems aimed at individualised production. As a cost-effective approach for adapting component-dependent applications, the concept of cooperating robots was further developed, based on grasping and moving objects with multiple handling devices. On this basis a reconfigurable, modular assembly platform is being developed which, beyond its handling devices, also allows the integration of measuring, testing and joining modules. In addition, the modular concept fulfils requirements for flexible and demand-driven configuration of multiple handling devices. An appropriate control system enables common handling of large components across several grasp-points. This also responds to a call from industry for harmonisation of production systems, i.e. the use of uniform systems for different components at all company locations.

One possible example of use of such an assembly platform is in aircraft structure assembly. The fuselage structure of a passenger aircraft usually consists of a number of sections (Fig. 6.132). The individual sections are composed of shell elements that are reinforced by ribs and stringers (Engmann 2008). The scope of work and the joining technologies for the sections are basically the same, but the sizes and shapes of the sections can differ. Today, to position the fuselage shells to join an aircraft

Fig. 6.154 Shell assembly

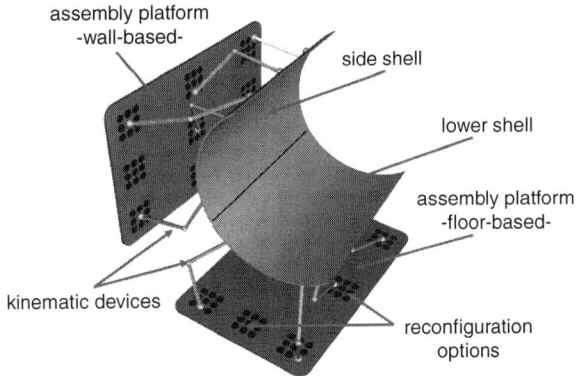

section, large rigid jigs and fixtures are used that map the component geometry. To maintain the required tolerances, the jigs and fixtures need a high degree of rigidity, which is usually only possible by a solid construction. As a result, the jigs and fixtures are large, heavy and costly.

An economically attractive alternative is reconfigurable assembly platforms, which are configured with handling devices to meet the demands of the specific application, and that will be able to replace the rigid jigs and fixtures in future. Here the assembly platform is configured according to the shell.

Since the workspace of the handling devices is very small compared to the dimensions of the shells, the shells must be conveyed by an additional handling system. While the transport system executes the macro movement of the component to the assembly station and roughly positions it there, the multiple-arm combination executes the smaller and more complex joining movement (Fig. 6.154). Shape and position corrections are possible by additional measuring means.

By separating the macro and micro movement and through the possibility of reconfiguring the multiple-arm combination, very different components can be handled and assembled.

For the purpose of combining large and smaller structures, there are different solutions such as wall and overhead gantries or Omnimove from KUKA (Kuka 2011b). The focus of the work is consequently on efficient, simple and cost-effective reconfiguration of the multiple-arm combination. The exact joining movement is executed by the handling devices. These position the two shell elements so that they can be joined manually or by automated means.

The advantage of this system is that the assembly platforms can always be configured to match the application, and the handling devices can be re-used for other purposes and products. This also makes the pilot series production of new aircraft types economical.

Another application example for this assembly platform is stringer integration. CFRP stringers of more than 20 m in length are laid in a shell and then pressed on for adhesion. Figure 6.155 shows a possible automated solution for this purpose. The

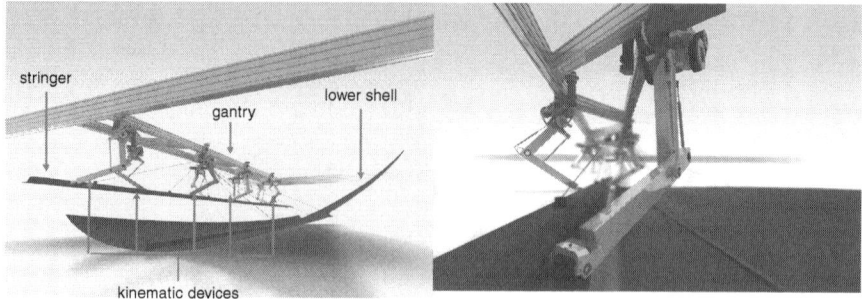

Fig. 6.155 Stringer integration

shell is positioned under an overhead gantry. Attached to the gantry are a number of handling modules that all engage with the stringer to prevent inadmissible sag.

Modularisation enables the number of handling devices to be matched to the length of the stringers. The low weight of the handling devices (approx. 25 kg) and of the stringer enables very simple and cost-attractive design of the gantry. In addition to laying stringers, the handling devices can lay further items of equipment needed for the autoclave process.

To achieve the previously mentioned scaling of the assembly system by addition and removal of handling modules with as little effort as possible, modularity must be consistent and interdisciplinary, from the mechanical system through the controller hardware to the controller software. The described controller concept can be implemented not only for handling devices but also for more extensive production systems that have to be reconfigurable. What is decisive is the modularisation and use of standard interfaces.

The presented, reconfigurable handling concept can be used to great advantage in other sectors of industry, wherever free configurability of large handling solutions can replace rigid jigs and fixtures or show the way to partial or full automation of small series.

6.7.6 Future Research Topics

Increasingly complex products, a growing number of variants and shorter product life cycles mean that automated production systems have less time to pay back.

As a result there is a continuing development in the automation of assembly towards flexible and adaptable systems accompanied by a variety of new applications. The greater functionality and configurability of assembly systems create new degrees of freedom. To enable optimum use of all this, self-optimising functions are aimed at reducing the effort and expense of reconfiguration as well as programming and planning complex applications.

A future research topic will be the integration of additional measurement and sensor technology. In the context of self-optimisation in particular, it is important to quickly identify system states. For this purpose, various methods are to be developed that not only determine the positions of robots but also the position of the component. The position data will then be used to monitor the component path and correct it if necessary. The correction data will be sent over a sensor input to the central controller, which automatically corrects the individual robots. In addition to monitoring of the component path, force/moment sensors can be integrated, to check and control the forces applied to a component. Forces could be intentionally introduced by tensioning of the robots to correct the shape of a component.

The description of a movement used in this project by reference to a component path very much simplifies the programming of cooperating robots. The programming concept, where the entire application is described in a central sequential program and the handling devices only possess simple standard programs to evaluate commands, also reduces the programming and modification effort for frequently changing tasks.

Programming of the handling system is nevertheless still elaborated in terms of the current state of development because individual movements of the handling devices outside the mode of cooperation have to be specified by the user. This is the case with the grasping movements for example, when every single handling device must be taken from the initial position to the grasp-point by a sequence of single movements.

Semi-automated path planning makes sense especially for these frequently recurring sequences, to support the user in planning and programming application-specific movements.

A future research topic is consequently the development of semi-automated path planning, by which a further function of self-optimisation can be integrated in the assembly platform, operation is simplified, and user acceptance can be enhanced. Existing methods of automated path planning will therefore be analysed, judged for suitability for the handling system and further developed.

Dynamic collision areas, which can be both machines, e.g. other robots, and humans, are a far greater challenge. In the case of machines and other controlled cell components, the position can be read from the controller and considered in path planning. Detecting the position of other dynamic obstacles requires external measuring technology (e.g. imaging), which means a much higher investment and stricter safety stipulations.

References

Abele E, Liebeck T, Wörn A (2006) Measuring flexibility in investment decisions for manufacturing systems. Ann CIRP 55(1):433–436

Adelt P, Donath J, Gausemeier J, Geisler J, Henkler S, Kahl S, Klöpper B, Krupp A, Münch E, Oberthür S, Paiz C, Porrmann M, Radkowski R, Romaus C, Schmidt A, Schulz B, Vöcking H, Witkowski U, Witting K, Znamenshchykov O (2009) Selbstoptimierende Systeme des Maschinenbaus. In: Gausemeier J, Rammig F, Schäfer W (eds) HNI-Verlagsschriftenreihe, vol 234. Westfalia Druck GmbH, Paderborn

Agrawal G, Jin R, Yang G (2004) Shared memory parallelization of data mining algorithms: techniques, programming interface, and performance. IEEE Trans Know Data Eng 16(10):1247–1262

Ahsan M, Hasin A, Pandey P (1996) MRP II: should its simplicity remain unchanged? Ind Manag 38(3):19–21

Akgün M, Alpay HR, Süle G Eren R (2010) Influence of warp tension on breking strength and strain of woven fabrics. Tekstil ve Konfeksiyon 1:30–36

Albert C, Fuchs C (2007) Durchblick im Begriffsdschungel der Business-Software. Universität Würzburg, Germany

Alred GJ, Brusaw CT, Oliu WE (2003) Handbook of technical writing. Bedford/St. Martin's, Boston

Andreasen M, Ahm T (1988) Flexible assembly systems. Springer, Berlin, p 44

Arburg (2010) http://www.arburg.com/com/COM/de/products/ac. Accessed 20 July 2010

Askin RG, Goldberg JB (2002) Design and analysis of lean production systems. Wiley, New York

Assem D (1986) Sensitivity analysis of linear systems. Springer, Berlin

Bainbridge L (1987) Ironies of automation. In: Rasmussen J, Duncan K, Leplat J (eds) New technology and human error. Wiley, Chichester

Balve P (2002) Ein Rahmenkonzept zur Gestaltung wandlungsfähiger Auftragsmanagementsysteme. Dissertation Universität Stuttgart. Jost-Jetter, Heimsheim

Bangor A (2000) Display Technology and Ambient Illumination Influences on Visual Fatigue at VDT Workstations. Published PhD thesis, Virginia Polytechnic Institute and State University, Virginia

Bannat A, Wallhoff F, Rigoll G, Friesdorf F, Bubb H, Stork S, Müller HJ, Schubö A, Wiesbeck M, Zäh MF (2008) Towards optimal worker assistance: a framework for adaptive selection and presentation of assembly instructions. In: Proceedings of the 1st international workshop on cognition for technical systems, Cotesys 2008, Technische Universität München

Barachini F (1990) Match-time predictability in real-time production systems. In: Proceedings of the 1st international Workshop on Expert Sys. Engineering, Lecture Notes in Computer Science, vol 462. pp 190–203

Bauer T (2008) Aufbau und Verbindungstechnik für Mikrooptische Systeme und Module, Jenoptik Polymer Systems GmbH, Jena, 6 Mai 2008

Becker M (1991) Anwendung von höheren Optimierungsmethoden in der Umformtechnik. Dissertation, Institut für Bildsame Formgebung, RWTH Aachen, Mainz, Aachen

Becker T (2008) Prozesse in der Produktion und Supply Chain optimieren. Springer, Berlin

Beckert E (2005) Ebene Keramiksubstrate und neue Montagetechnologien zum Aufbau hybridoptischerSysteme, Jena

Beckmann I (2009) Entrepreneurship-Politik: Neue Standortpolitik im politischen Spannungsfeldzwischen Arbeitsmarkt und Interessengruppen. VS Verlag für Sozialwissenschaften, Wiesbaden

Beer S (1973) Kybernetische Führungslehre. Herder und Herder, Frankfurt

Beer S (1979) The heart of enterprise. Wiley, Chichester

Beetz M, Buss M, Wollherr D (2007) Cognitive technical systems – what is the role of artificial intelligence? Adv Artific Intell 4667:19–42

Beierle C, Kern-Isberner G (2006) Methoden wissensbasierter Systeme Grundlagen – Algorithmen – Anwendungen. Vieweg, Wiesbaden

Belforte D (2009) Keeping the economy in perspective. Ind Laser Sol 24(1):4–11

Best E (2010) Process Excellence – Praxisleitfaden für erfolgreiches Prozessmanagement. Gabler, Wiesbaden

Blecker T, Kaluza B (2004) Produktionsstrategien – ein vernachlässigtes Forschungsgebiet? In: Braßler A, Corsten H (eds) Entwicklungen im Produktionsmanagement. Vahlen, München

Bleicher K (2004) Das Konzept Integriertes Management: Visionen- Missionen- Programme. Campus, Frankfurt a. M.

Borgelt C, Klawonn F, Kruse R, Nauck D (2003) Neuro-Fuzzy-Systeme: Von den Grundlagen künstlicher Neuronaler Netze zur Kopplung mit Fuzzy-Systemen. Vieweg, Wiesbaden. ISBN 3-528-25265-0

Brecher C, Fayzullin K, Possel-Dölken F, Valkyser B (2007) Optimierung der flexiblen Fertigung. Werkstattstechnik 97(6):464–470

Brecher C, Fayzullin K, Possel-Dölken F, Valkyser B (2008a) Optimierung flexibler Produktionsabläufe flexibel gestalten. ATP 5(50):60–69

Brecher C, Fayzullin K, Possel-Dölken F (2008b) Intelligent operations control. Prod Eng Res Dev 2(3):293–302

Brecher C, Herfs W, Fayzullin K (2008c) Operative Flexibilität – Ein workflow-basierter Ansatz für Fertigungsleittechnik, A&D Kompendium 2008/2009: Das Referenzbuch für industrielle Automation. Publish-Industry, München

Brecher C, Klocke F, Schmitt R, Schuh G (2008d) Wettbewerbsfaktor Produktionstechnik – Aachener Perspektiven. Apprimus, Aachen

Brecher C, Pyschny N, Loosen P, Funck M, Morasch V, Schmitt R, Pavim A (2009) Self-optimising flexible assembly systems. Workshop on self-X in mechatronics and other engineering applications, Paderborn, pp 23–38

Bredin C (2005) Team-mates – ABB multiMove functionality heralds a new era in robot applications. In: ABB Review 1/2005

Brennan RW (2003) From FMS to HMS. In: Deen SMH (eds) Agent-based manufacturing – advances in the holonic approach. Springer, Berlin

Buhmann MD (2003) Radial basis functions – theory and implementations. In: Cambridge monographs on applied and computational mathematics, no 12. ISBN 9780521633383

Busch A, Dangelmaier W (2004b) Integriertes Supply Chain Management – Ein koordinationsorientierter Überblick. Gabler, München

Castellini C, Giunchiglia E, Tacchella A (2001) Improvements to sat-based conformant planning. In: Proc. of the 6th European conf. on planning

Chryssolouris G (2005) Manufacturing systems – theory and practice. Springer, New York

Chryssolouris G, Mourtzis D (eds) (2004) Autonomous control of processes. In: IFAC conference manufacturing, modelling, management and control

Cisek R, Habicht C, Neise P (2002) Gestaltung wandlungsfähiger Produktionssysteme, ZWF – Zeitschrift für wirtschaftlichen. Fabrikbetrieb 97(9):441–445

Coit DW, Jackson BT, Smith AE (1998) Static neural network process models: considerations and case studies. Int J Prod Res 36(11):2953–2967

Cremers B (1991) Expertensysteme für die Planung der Produktion. Verlag TÜV Rheinland, Köln

Dangelmaier W, Warnecke H (1997) Fertigungslenkung – Planung und Steuerung des Ablaufs der diskreten Fertigung. Springer, Berlin

Denkena B, Shpitalni M (2007) Knowledge management in process planning. CIRP Ann Manuf Technol 56(1):175–180

DIN 32564 (2003) Fertigungsmittel für Mikrosysteme – Begriffe – Teil 1–3. Beuth, Berlin

DIN EN 61131-3 (2003) Speicherprogrammierbare Steuerungen. Teil 3: Programmiersprachen. Beuth, Berlin

DIN EN 62264 (2008) Integration von Unternehmensführungs- und Leitsystemen. Beuth, Berlin

Ding H, Kain S, Schiller F, Stursberg OA (2008) Control architecture for safe cognitive systems. In 10. Fachtagung Entwurf komplexer Automatisierungssysteme, Magdeburg

Dittmar R, Pfeiffer B-M (2004) Modellbasierte prädiktive Regelung: Eine Einführung für Ingenieure. Oldenbourg, München

Dohner JL, Lauffer JP, Hinnerichs TD, Shankar N, Regelbrugge M, Kwan C, Xu R, Winterbauer B, Bridger K (2004) Mitigation of chatter instabilities in milling by active structural control. J Sound Vib 269(1/2):197–211

Dören J (2007) Qualitätsmanagement und Neuronale Netze

Drabow G (2006) Modulare Gestaltung und ganzheitliche Bewertung wandlungsfähiger Fertigungssysteme. PZH Produktionstechnisches Zentrum, Garbsen

Draper NR, Smith H (1998) Applied regression analysis. Wiley series in probability and statistics. ISBN 978-0-471-17082-2

Drumwright E, Ng-Thow-Hing V, Mataric M (2006) Toward a vocabulary of primitive task programs for humanoid robots. In: Proceedings of the international conference on development and learning (ICDL). Bloomington, IN

Dyckhoff H (2003) Grundzüge der Produktionswirtschaft – Einführung in die Theorie betrieblicher Wertschöpfung. Springer, Berlin

Dyckhoff H, Spengler TS (2007) Produktionswirtschaft – Eine Einführung für Wirtschaftsingenieure. Springer, Berlin

Efron B, Gong G (1983) A leasurly look at the bootstrap, the jackknife, and cross-validation. Am Stat 37(1):36–48

Elatta AY, Gen LP, Zhi FL, Daoyuan Y, Fei L (2004) An overview of robot calibration. Inf Technol J 3(1):74–78

ElMaraghy HA (2005) Flexible and reconfigurable manufacturing systems paradigms. Int J Flex Manuf Syst 17(4):261–276

ElMaraghy HA, Azab A, Schuh G, Pulz C (2009) Managing variations in products, processes and manufacturing systems. CIRP Ann Manuf Technol 58(1)

Engesser H (1993) Eintragungen "Künstliche Intelligenz" und "Expertensystem". Duden "Informatik". Bibliographisches Institut und F.A. Brockhaus AG, Mannheim

Engmann K (2008) Technologie des Flugzeuges. Vogel, Würzburg

Enríquez Díaz JA, Frieling E, Thiemich J, Kreher S (2010) Auswirkung eines Chaku-Chaku-Montagesystems auf die älteren Beschäftigten am Beispiel der Abgasanlagen-Montage. In: GFA (ed) Neue Arbeits- und Lebenswelten gestalten. GfA-Press, Dortmund

Erdim H, Lazoglu I, Ozturk B (2006) Feedrate scheduling strategies for freeform surfaces. Int J Mach Tools Manuf 46(7/8):747–757

Espejo R (1996) Organizational transformation and learning – a cybernetic approach to management. Wiley, Chichester

Espejo R, Harnden R (1989) The viable system model – interpretations and applications of Stafford Beer's VSM. Wiley, Chichester

Eversheim W (1998) Organisation in der Produktionstechnik. Konstruktion. Springer, Berlin

Eversheim W, Schuh G (1996) Betriebshütte – Produktion und Management. Springer, Berlin

Eversheim W, Pfeifer T, Weck M (2006) Hundert Jahre Produktionstechnik. Springer, Berlin

Fayzullin K (2010) Autonome Planung und Entscheidungsoptimierung in der Ablaufsteuerung flexibel automatisierter Fertigungssysteme. Dissertation RWTH Aachen

Feldmann K, Ziegler C, Michl M (2007) Bewegungssteuerung für kooperierende Industrieroboter in der Montageautomatisierung. Werkstattstechnik 97(9):713

Ferry WB, Altintas Y (2008) Virtual five-axis flank milling of jet engine impellers part II: feed rate optimization of five-axis flank milling. ASME J Manuf Sci Eng 130:1

Field A (2005) Discovering statistics using SPSS. Sage, London

Finkler M (2006) Advanced planning and scheduling – die stille ERP-revolution. PPS Manag 3

Fleisch E (2008) High-resolution-management. Konsequenz der 3. IT-Revolution auf die Unternehmensführung. Schaeffer-Poeschel, Stuttgart

Fleisch E, Österle H (2004) Auf dem Weg zum Echtzeit-Unternehmen. In: Alt R, Österle H (eds) Retail-Time Business – Lösungen, Bausteine und Potenziale des Business Networkings. Springer, Berlin

Frank U, Gausemeier J (eds) (2004) Selbstoptimierende Systeme des Maschinenbaus: Definitionen und Konzepte/Sonderforschungsbereich 614. Heinz-Nixdorf-Institut, Univ. Paderborn, Paderborn

Frank U, Giese H, Klein F, Oberschelp O, Schmidt A, Schulz B, Vöcking H, Witting K (2004) Selbstoptimierende Systeme des Maschienenbaus. Definitionen und Konzepte. Sonderforschungsbereich 614. Bonifatius, Paderborn

Fraunhofer IOF (2006) Produktion von mikrooptischen Bauteilen und Systemen Status und Perspektiven, Fraunhofer Institut angewandte Optik und Feinwerktechnik, Jena

Frieling E, Sträter O (2009) Folgeantrag im Rahmen des Schwerpunktprogramms 1184 "Altersdifferenzierte Arbeitssysteme"

Fries E (1999) Anwendung neuronaler Netze zur Verschleißerkennung beim Fräsen. Dissertation, TU Berlin
Fu MC (2002) Optimization for simulation: theory vs. practice. INFORMS J Comput 14:3
Funke J, Frensch PA (2006) Handbuch der Allgemeinen Psychologie – Kognition. Hogrefe, Göttingen
Garvin DA (1984) What does "product quality" really mean? Sloan Manag Rev 1984(Fall)
Gat E (1998) On three-layer architectures. In: Kortenkamp D, Bonnasso R, Murphy R (eds) Artificial intelligence and mobile robots. AAAI Press, Menlo Park, pp 195–211
Gausemeier J (2005) Von der Mechatronik zur Selbstoptimierung – Herausforderungen an die domänenübergreifende Zusammenarbeit. Produkt Daten J, 1
Gausemeier J, Frank U, Giese H, Klein F, Schmidt A, Steffen D, Tichy M (2005) A design methodology for self-optimizing systems. Automation, Assistance and Embedded Real Time Platforms for Transportation (AAET2005), 16, 2005
Gausemeier J, Ramming FJ, Schäfer W (2006) Selbstoptimierung im Maschinenbau – Auf dem Weg zu den Maschinen von übermorgen. Forschungsforum Paderborn 10(1)
Gausemeier J, Adelt P, Donath J, Geisler J, Henkler S, Kahl S, Klöpper B, Krupp A, Münch E, Oberthür S, Paiz C, Porrmann M, Radkowski R, Romaus C, Schmidt A, Schulz B, Vöcking H, Witkowski U, Witting K, Znamenshchykov O (2009a) Selbstoptimierende Systeme des Maschinenbaus. HNI-Verlagsschriftenreihe
Gausemeier J, Rammig F, Schäfer W (eds) (2009b) Selbstoptimierende Systeme des Maschinenbaus. Definition, Anwendungen, Konzepte. HNI-Verlagsschriftenreihe, vol 234. Westfalia Druck GmbH, Paderborn
Gazzola V, Rizzolatti G, Wicker B, Keysers C (2007) The anthropomorphic brain: the mirror neuron system responds to human and robotic actions. NeuroImage 35:1674–1684
Gehrke J, Ramakrishnan R, Ganti V (2000) Rainforest – a framework for fast decision tree construction of large datasets. Data Min Know Dis 4(2/3):127–162
Gesatzki RP (2002) Elemente von advanced planning und scheduling-systemen zur optimierung der supply chain. SCM Newsletter
Gladkov EA, Maloletkov AV, Perkovski RA (1998) Predicting the quality of TIG butt welded joint using neural network models. Weld Int 3(12):215–219
Gläßner J (1995) Modellgestütztes Controlling der beschaffungslogistischen Prozesskette. VDI, Düsseldorf
Gloy Y-S, Muschong C, Schneuit H, Gries T (2009) Innovative Simulationstools zur Webmaschinenoptimierung. In: Küppers B (ed) Proceedings of the 3rd Aachen-Dresden International Textile Conference, Aachen, November 26–27, 2009, Aachen: DWI an der RWTH Aachen e.V.
Göhner P, Urbano PGA, Wagner T (2004) Softwareagenten – Einführung und Überblick über eine Alternative art der Softwareentwicklung. Teil 3: Agentensysteme in der Automatisierungstechnik. Automatisierungstech Prax 46(2):42–51
Golden RM (1996) Mathematical methods for neural network analysis and design. MIT Press, Cambridge
Gomez P (1978) Die kybernetische Gestaltung des Operations Managements – Eine Systemmethodik zur Entwicklung anpassungsfähiger Organisationsstrukturen. Haupt, Bern
Gorrisen D, De Tommasi L, Crombecq K, Dhaene T (2009) Sequential modeling of a low noise amplifier with neural networks and active learning. Spr Neural Comput Appl 18(5):485–494
Gramann W (1999) Industrielle Montage mikrooptischer Halbleiterbauelemente. In: Reinhart G (ed) Iwb-Seminarberichte Nr. 44: Automatisierte Mikromontage. Utz, München
Grauel A (1992) Neuronale Netze – Grundlagen und mathematische Modellierung. BI Wissenschaftsverlag, Mannheim
Grešovnik I, Rodič T (1999) A general-purpose shell tor solving inverse and optimization problems in material forming. In: Proceedings of the 2 ESAFORM conference on material forming guimararäes, Portugal

Gruber TR (1993) A translation approach to portable ontology specifications. Knowl Acquis 5(2):199–220 (Academic Press, New York)

Günther H, Tempelmeier H (2007) Produktion und Logistik. Springer, Berlin

Gurney K (1997) An Iintroduction to neural networks. UCL Press, London

Haats C (2000) Produktionsplanung und -steuerung für Unternehmen mit linearen Prozessketten. Dissertation Universität Stuttgart, Jost-Verlag, Heinsheim

Haferkamp H (1999) Thermographic system for process monitoring of laser beam cutting. Europto conference on optical measurement systems for industrial inspection, Munich, Germany, SPIE vol 3824, June 1999

Han J, Kamber M (2006) Data mining: concepts and techniques. Kaufmann, San Francisco

Hand D, Mannila H, Smyth P (2001) Principles of data mining. Massachusetts Institute of Technology, USA

Hankes J (1999) Sensoreinsatz in der automatisierten Mikromontage, Fortschritt-Berichte VDI 2 Nr. 459, Düsseldorf, VDI

Hart PE, Nilsson NJ, Raphael B (1968) A formal basis for the heuristic determination of minimum cost paths. IEEE Trans Syst Sci Cybernet SSC 4(2):100–107

Hauck E, Gramatke A, Henning K (2008) Cognitive technical systems in a production environment. In: Proceedings of the fifth international conference on informatics in control, automation and robotics. ICINCO: Madeira, Portugal

Hayes-Roth B (1985) A blackboard architecture for control. Artif Intell 26:251–321 (Elsevier, Oxford)

Heidrich J (2001) Optimierung und Überwachung des MSG-Schweißprozesses mit Hilfe klassischer Modelle und KI-Methoden. Aachener Berichte Fügetechnik, vol 3. Shaker, Aachen

Hellmich K (2002) Wettbewerbsvorteile durch realisierte Kundenorientierung der Wertschöpfung. Ind Manag 18(5):27–30

Hellmich K (2003) Kundenorientierte Auftragsabwicklung – Engpassorientierte Planung und Steuerung des Ressourceneinsatzes. DUV, Wiesbaden

Henrich P (2002) Strategische Gestaltung von Produktionssystemen in der Automobilindustrie. Shaker, Aachen

Herold C (1991) Ein Vorgehenskonzept zur Unternehmensstrukturierung – Eine heuristische Anwendung des Modells lebensfähiger Systeme. Dissertation Universität St. Gallen, Difo-Druck, Bamberg

Hesselbach J, Weule H, Klocke F, Weck M (2002) mikroPRO – Untersuchung zum internationalen Stand der Mikroproduktionstechnik. Essen, Vulkan

Hichri H (2005) Überwachung und Optimierung des MSG-Schweißprozesses mit Hilfe von KIMethoden und Spracherkennungsalgorithmen, vol 2. Aachener Berichte Fügetechnik, Shaker, Aachen

Hill T (2000) Manufacturing strategy. McGraw-Hill, Irwin

Hoffmann JFF (2001) The fast-forward planning system. AI Magaz 22;57–62

Hoffmann J, Brafman R (2005) Contingent planning via heuristic forward search with implicit belief states. In: Proceedings of the 15th international conference on automated planning and scheduling, Monterey, USA

Hoitsch H (1993) Produktionswirtschaft. Grundlagen einer industriellen Betriebswirtschaftslehre. 2., völlig überarbeitete und erweiterte Auflage. Vahlen, München

Homem de Mello LS, Sanderson AC (1986) And/Or graph representation of assembly plans. In: Proceedings of 1986 AAAI national conference on artificial intelligence, pp 1113–1119

Hufgard A (ed) (2005) Business Integration mit SAP-Lösungen – Potenziale, Geschäftsprozesse, Organisation und Einführung. Springer, Berlin

Insperger T, Mann BP, Stépàn G, Bayly PV (2003) Stability of up-milling and down-milling, part 1: alternative analytical methods. Int J Mach Tools Manuf 43(1):25–34

Isermann R (2002) Mechatronische systeme: grundlagen. Springer, New York

Ismail F, Ziaei R (2002) Chatter suppression in five-axis machining of flexible parts. Int J Mach Tools Manuf 42(1):115–122

Jennings NR, Sycara K, Wooldridge M (1998) A roadmap of agent research and development. Auton Agent Multi Agent Syst 1:275–306

Johannaber F, Michaeli W (2004) Handbuch Spritzgießen. Hanser, München

Jones RM, Laird JE, Nielsen PE, Coulter KJ, Kenny P, Koss, FV (1999) Automated intelligent pilots for combat flight simulation. AI Mag 20:27–41

Jorden W (2007) Handbuch für Studium und Praxis. Form- und Lagetoleranzen. Hanser, München

Jurecka F (2007) Robust design optimization based on metamodeling techniques. Dissertation Technische Universität München

Kammel S, Ziegler J, Pitzer B, Werling M, Gindele T, Jagzent D, Schroder J, Thuy M, Goebl M, von Hundelshausen F, Pink O, Frese C, Stiller C (2008) Team AnnieWAY's autonomous system for the 2007 DARPA Urban Challenge. J Field Robotics 25:615–639

Kämpf R (2007) Grundlagen des Produktionsmanagements. In: Gienke H, Kämpf R (eds) Handbuch Produktion: Innovatives Produktionsmanagement: Organisation, Konzepte, Controlling. Hanser, München

Kaplan RS, Norton DP (2006) The balanced scorecard. Translating strategy into action. Harvard Business School Press, Boston

Karim S, Sonenberg L, Tan AH (2006) A hybrid architecture combining reactive, plan execution and reactive learning. In: Proceedings of the 9th Biennial Pacific Rim International Conference on Artificial Intelligence (PRICAI), China

Kaufman S, Wilson R, Jones R, Calton T, Ames A (1996) The Archimedes 2 mechanical assembly planning system. Proc IEEE Int Conf Robotics Automat 4:3361–3368

Kempf T (2010) Ein kognitives Steuerungsframework für robotergestützte Handhabungsaufgaben. Apprimus, Aachen

Kempf T, Herfs W, Brecher C (2008) Cognitive control technology for a self-optimizing robot based assembly cell. In: Proceedings of the ASME 2008 international design engineering technical conferences and computers and information in engineering conference, American Society of Mechanical Engineers, USA

Keuster J De (2005) Methods for monitoring of laser cutting by means of acoustic and photodiode sensors. Adv Mater Res 6–8:809–816

Keuster J De (2007) Real-time adaptive control and optimisation of high-power CO_2 laser cutting using photodiodes. In: Proceedings of the 5th lane conference, laser assisted Net Shape Engineering pages, 2007, pp 979–992

Kiener S (2006) Produktions-Management. Oldenbourg, München

Kim IS, Jeong YJ, Lee CW, Yarlagadda V (2004) Prediction of welding parameters for pipe-line welding using an intelligent system. Int J Adv Manuf Technol 22(9–10):713–719

Kim IS, Son JS, Park CE, Kim IJ, Kim HH (2005) An investigation into an intelligent system for predicting bead geometry in GMA welding process. J Mater Process Technol 159(1):113–118

Kinkel S, Friedewald M, Hüsing B, Lay G, Lindner R (2008) Arbeiten in der Zukunft. Strukturen und Trends der Industriearbeit. Sigma, Berlin

Kinnebrock W (1994) Neuronale netze: Grundlagen, Anwendungen, Beispiele. Oldenbourg, München

Kletti J (2006) MES – Manufacturing Execution System – Moderne Informationstechnologie zur Prozessfähigkeit der Wertschöpfung. Springer, Berlin

Klocke F, Kratz S, Veselovac D, Mtz. de Aramaiona P, Arrazola P (2008) Investigation on forcesensor dynamics and their measurement characteristics. ASME international mechanical engineering congress and exposition. Boston, USA

Klocke F, Veselovac D, Auerbach T, Kamps S (2009) Controller parameter analysis during the milling of aerospace alloys. SAE 2009 aerotech congress und exhibition, Seattle, USA, Nov 2009

Klocke F, Veselovac D, Auerbach T, Kamps S (2010) Kennwertgenerator für die automatisierte Versuchsdurchführung von Zerspanversuchen. In: Jamal R, Heinze R (eds) Virtuelle Instrumente in der Praxis -Begleitband zum 15. VIP-Kongress 2010. Hüthig, Heidelberg, pp 75–81. ISBN 978-3-8007-3235-7

Kobayashi I, Panskus G (eds) 20 Keys® Die 20 Schlüssel zum Erfolg im internationalen Wettbewerb. Adapt-Media-Verlag, Bochum

Kohonen T (1984) Self-organization and assoziative memory. Springer, Berlin

Korb W (2005) Mit Pepp zur wirtschaftlichen Mikromontage, Mikroproduktion 1/2005. Hanser, München, pp 21–22

Kötzle A, Göbel E (1997) Strategisches Management – Theoretische Ansätze, Instrumente und Anwendungskonzepte für Dienstleistungsunternehmen. Lucius und Lucius, Stuttgart

Krause F-L (2007) Innovationspotenziale in der Produktentwicklung. Hanser, München

Kudlik N (1997) Reproduzierbarkeit des Kunststoff-Spritzgießprozesses. Dissertation RWTH Aachen, ISBN: 3-89653-412-4

Kühne T (2005) What is a Model? In: Proceedings of Language Engineering for Model-Driven Software Development, Internationales Begegnungs- und Forschungszentrum fuer Informatik (IBFI). Dagstuhl, pp 1–10

KUKA Robotics GmbH (2011a) http://www.kuka-robotics.com/germany/de/products/software/hub_technologies/. Stand: March 2011

KUKA Robotics GmbH (2011b) http://www.kuka-omnimove.com/en/ Stand: March 2011

Kurbel K, Endres A (2005) Produktionsplanung und -steuerung: Methodische Grundlagen von PPS-Systemen und Erweiterungen. Oldenbourg, München

Kurth J (2005) Flexible Produktionssysteme durch kooperierende Roboter. wt Werkstatttechnik online 3

Ljung L (1999) System identification. Prentice-Hall, Upper Saddle River

Laird J, Bates Congdon C (2008) The soar user's manual, 9 edn. The Regents of the University of Michigan, USA

Langley P, Cummings K, Shapiro D (2004) Hierarchical skills and cognitive architectures. In: Proceedings of the twenty-sixth annual conference of the cognitive science society. Chicago, IL

Langley P, Laird JE, Rogers S (2009) Cognitive architectures: research issues and challenges. Cogn Syst Res 10:141–160

Larose DT (2005) Discovering knowledge in data. Wiley, New Jersey

Latzke PM, Hesse R (1974) Textilien Untersuchen-Prüfen-Auswerten, Hrsg. Fach-verlag Schiele und Sohn GmbH, pp 123–125. ISBN 3 7949 0228 9

Laue T, Röfer T (2006) Getting upright: migrating concepts and software from four-legged to humanoid soccer robots. In: Pagello E, Zhou C, Menegatti E (eds) Proceedings of the workshop on humanoid soccer robots in conjunction with the 2006. IEEE International Conference on Humanoid Robots, Genoa

Lehman JF, Laird J, Rosenbloom P (2006) A gentle introduction to soar, an architecture for human cognition. Technical Report, University of Michigan, USA

Leiden K, Laughery KR, Keller J, French J, Warwick W, Wood SD (2001) A review of human performancer models for the prediction of human error. Prepared for: national aeronautics and space administration system-wide accident prevention program. Ames Research Center, Moffet Filed, CA

Levine DS (2000) Introduction to neural and cognitive modeling. Psychology Press, New York

Li ZZ, Zhang ZH, Zheng L (2004) Feedrate optimization for variant milling processes based on cutting force prediction. Int J Adv Manuf Technol 24(7/8):541–545

Liao TW, Chen LJ (1994) Neural network approach for grindind processes: modeling and optimization. Int J Mach Tools Manuf 37(7):919–937

Licha A (2003) Flexible Montageautomatisierung zur Komplettmontage flächenhafter Produktstrukturen durch kooperierende Industrieroboter. Meisenbach, Bamberg

Liepmann D, Beauducel A, Brocke B, Amthauer R (2007) I-S-T 2000 R Intelligenz-Struktur-Test 2000 R, 2., erweiterte und überarbeitete Auflage. Hogrefe, Göttingen

Lim TS, Loh WY, Shih YS (1997) An empirical comparison of decision trees and other classification methods. Technical Report 979, Department of Statistics, University of Wisconsin, Madison

Lin AC, Chang TC (1993) An integrated approach to automated assembly planning for three-dimensional mechanical products. Int J Prod Res 31(5):1201–1227 (Taylor & Francis)

Lindemann U (2006) Individualisierte Produkte – Komplexität beherrschen in Entwicklung und Produktion. Springer, Berlin

Litoiu M, Woodside M, Zheng T (2005) Hierarchical model-based autonomic control of software systems. ACM SIGSOFT Softw Eng Notes 30(4)

Lödding H (2008) Verfahren der Fertigungssteuerung – Grundlagen, Beschreibung, Konfiguration. Springer, Berlin

Loeffelholz FV (1991) Qualität von PPS-Systemen – Ein Verfahren zur Analyse des Informationsgehaltes. Dissertation RWTH Aachen, Springer, Berlin

Lopitzsch JR (2005) Segmentierte adaptive Fertigungssteuerung. Dissertation der Universität Hannover. PZH Produktionstechnisches Zentrum GmbH. Garbsen

Lotter B (2006) Montage in der industriellen Produktion. Ein Handbuch für die Praxis. Springer, Berlin

Luczak H, Eversheim W (1999) Produktionsplanung und -steuerung: Grundlagen, Gestaltung und Konzepte. Springer, Berlin

Lunze J (2008) Regelungstechnik 1, 7 edn. Springer, Berlin

Lutz S (2002) Kennliniengestütztes Lagermanagement. VDI, Düsseldorf

Nørgaard M, Ravn O, Poulsen NK, Hansen LK (2000) Neural networks for modelling and control of dynamic systems. Springer, London

Mälck H (2001) Zielsysteme in der Produktion: Das Führen mit Zielen als Grundlage für optimierte Technik-, Organisations- und Personalentwicklungsprozesse. RKW, Eschborn

Malik F (2006) Strategie des Managements komplexer Systeme – Ein Beitrag zur Management-Kybernetik evolutionärer Systeme. Haupt, Bern

Mane I, Gagnol V, Bouzgarrou BC, Ray P (2008) Stability-based spindle speed control during flexible work piece high-speed milling. Int J Mach Tools Manuf 48(2):184–194

Maqbool S, Smith JS, Lucas J (1998) Neural networks for the control of welding systems. 8th international conference on computer technology in welding, Maastricht, pp 125–133

Marshall SP (2008) Cognitive models of tactical decision making. In: Karowski W, Salvendy G (eds) Proceedings of the 2nd international conference on applied human factors and ergonomic (AHFE) 14–17 July 2008. Las Vegas, Nevada

Martin PJ (1994) Artificial neural networks in welding. Join Mater 6(2):62–67

Martin W (2003) Kundenbeziehungsmanagement im Echtzeit-Unternehmen. Strategic Bulletin CRM 2004. itresearch, Sauerlach bei München

Matteson A, Morris R, Tate R (1992) Real-time GMAW quality classification using an artificial neural network with air born acoustic signals as input. In: International conference on computerization of welding information IV, USA, pp 189–197

Mattke F, Berlekamp CG, Queren-Lieth W (1993) Die Darstellung von schweißtechnischem Wissen in Expertensystemen. DVS-Berichte, vol 156, DVS-Verlag Düsseldorf, pp 147–150

Mayer M, Odenthal B, Grandt M, Schlick C (2008) Anforderungen an die benutzerzentrierte Gestaltung einer Kognitiven Steuerung für Selbstoptimierende Produktionssysteme. In: Gesellschaft für Arbeitswissenschaft eV (ed) Produkt- und Produktions-Ergonomie – Aufgabe für Entwickler und Planer. GfA-Press, Dortmund

Mayer M, Odenthal B, Faber M, Kabuß W, Kausch B, Schlick C (2009) Simulation of human cognition in self-optimizing assembly systems. In: Proceedings of 17th world congress on ergonomics IEA 2009, Beijing

McCarthy J (1959) Programs with common sense. In: Proceedings of the teddington conference on the mechanization of thought processes, Her Majesty's Stationery Office London, pp 756–791

Menges G, Stitz S, Vargel J (1971) Grundlagen der Prozeßsteuerung beim Spritzgießen. In: Kunststoffe, Februar, pp 74–80

Menges G, Thienel P (1977) Pressure-specific volume-temperature behavior of thermoplastics under normal processing conditions. Polymer Eng Sci 17(10):758–763

Merget M (2004) Kostenoptimierung durch Toleranzvariation im Simultaneous Engineering. Shaker, Aachen

Meyer D, Steil T, Müller S (2005) Shared Augmented Reality zur Unterstützung mehrerer Benutzer bei kooperativen Montagearbeiten im verdeckten Bereich. In: Kuhlen T, Kobbelt L, Müller S (eds) Virtuelle und Erweiterte Realität. Shaker, Aachen

Meyer M (2006) Logistisches Störungsmanagement in Kundenverbrauchsorientierten Wertschöpfungsketten. Dissertation RWTH Aachen, Shaker, Aachen

Michaeli W (2006) Einführung in die Kunststoffverarbeitung. Hanser, München

Michaeli W, Lauterbach M (1989) Die pmT-Optimierung "Konsequenzen aus dem pvT-Konzept zur Nachdruckführung". Kunststoffe, Sept 1989, pp 852–855

Michaeli W, Schreiber A (2010) Der Weg zum geregelten Prozess. Swiss Plastics, Nov, pp 21–26, Aarau, AZ Fachverlage AG

Michaeli W, Breuer P, Hohenauer K, Von Oepen R, Philipp M, Pötsch G, Recker H, Robers T, Vaculik R (1992) Qualitätsgesichertes Spritzgießen. Kunststoffe, Dezember, pp 1167–1171

Michaeli W, Hopmann C, Gruber J (2004) Prozessregelung beim Spritzgießen. Kunststoffe, Januar, pp 20–24

Michie D, Spiegelhalter DJ, Taylor CC (1994) Machine learning: neural and statistical classification. Horwood, London

Middle JE, Li J (1993) Artificial neural networks applied to process modelling for robotic arc welding, modelling of casting, welding and advanced solidification processes, 6th international conference on modelling and welding process, USA, pp 127–134

Mintzberg H, Quinn BJ, Ghoshal S (2004) The strategy process. Prentice-Hall, Hemel Hempstead

Mitsuishi M, Ueda K, Kimura F (2004) Manufacturing systems and technologies for the new frontier, 41st CIRP conference

Möller N (2008) Bestimmung der Wirtschaftlichkeit wandlungsfähiger Produktionssysteme, Forschungsbericht N IWB, vol 212. Utz, München

Müller B, Reinhardt J (1995) Neural networks. An introduction. Springer, Berlin

Müller R, Esser M, Janßen C, Brecher C, Pyschny N (2009a) Flexibel automatisierte Montagesysteme – Toleranzoptimierte Montage von miniaturisierten Produkten. Proceedings der VDITagung Mechatronik 2009, Wiesloch, 12–13. Mai

Müller R, Buchner T, Gottschalk S, Fayzullin K, Herfs W, Hilchner R, Pyschny N (2009b) Montagetechnik und – organisation. Apprimus, Aachen

Müller R, Corves B, Hüsing M, Esser M, Riedel M, Vette M (2010a) Reconfigurable self-optimising handling system. In: Fifth international precision assembly seminar IPAS 2010, Chamonix

Müller R, Esser M, Janßen C, Vette M (2010b) System-Identifikation für Montagezellen – Erhöhte Genauigkeit und bedarfsgerechte Rekonfiguration. In: wt Werkstattstechnik online 100, 9, S 687–691. ISSN 1436-4980, Internet: www.werkstattstechnik.de. Springer-VDIVerlag, Düsseldorf

Murthy SK (1995) On growing better decision trees from data. Ph. D. thesis, Department of Computer Science, Johns Hopkins University, Baltimore

Nason S, Laird J (2005) Soar-RL: integrating reinforcement learning with soar. In: Cognitive systems research, special issue of cognitive systems research – the best papers from ICCM2004

Nauck D, Klawonn F, Kruse R (2002) Neuronale Netze und Fuzzy Systeme. Vieweg, Wiesbaden

Niemann G, Winter H, Höhn BR (2005) Maschinenelemente, Band 1: Konstruktion und Berechnung von Verbindungen, Lagern, Wellen. Springer, Berlin

Nise N (2009) Control systems engineering, 4 edn. Wiley, Hoboken

Normeey-Rico JE, Camacho EF (2007) Control of dead-time processes. Springer, London

Nyhuis P (2008) Produktionskennlinien: Grundlagen und Anwendungsmöglichkeiten. In: Nyhuis P (ed) Beiträge zu einer Theorie der Logistik. Springer, Berlin

Nyhuis P, Wiendahl H (2003) Logistische Kennlinien – Grundlagen, Werkzeuge und Anwendungen. Springer, Berlin

Nyhuis P, Heinen T, Reinhart G, Rimpau C, Abele E, Wörn A (2008) Wandlungsfähige Produktionssysteme – Theoretischer Hintergrund zur Wandlungsfähigkeit von Pproduktionssystemen, wt Werkstatttechnik Online pp 1–2

o.V. (2006) Gibt es eine verschwenundungsfreie Logistik? In: 3. Lean management summit. Aachener Management Tage, Aachen, pp 29–45

Odenthal B, Mayer M, Grandt M, Schlick C (2007) Concept of an adaptive training system for production. In: Hinneburg A (ed) Lernen – Wissen – Adaption, Workshop Proceedings. Martin- Luther-University, Halle-Wittenberg

Odenthal B, Mayer M, Grandt M, Schlick C (2008) Konzept eines Lehr-/Lernsystems einer kognitiven Steuerung für Selbstoptimierende Produktionssysteme. In: Gesellschaft für Arbeitswissenschaft e. V (ed) Produkt- und Produktions-Ergonomie – Aufgabe für Entwickler und Planer. GfA-Press, Dortmund

Odenthal B, Mayer M, Kabuß W, Kausch B, Schlick C (2009) Error detection in an assembly object using an augmented vision system. In: Proceedings of 17th world congress on ergonomics IEA 2009, Beijing

Odenthal B, Mayer M, Kabuß W, Kausch B, Schlick C (2011) An empirical study of disassembling using an augmented vision system. In: Proceedings of the HCII 2011

Ohno T (1993) Das Toyota-Produktionssystem. Campus, New York

Ong SK, Yuan ML, Nee AY (2008) Augmented reality in manufacturing: a survey. Int J Prod Res 46(10):2702–2742

Onken R, Schulte A (2010) System ergonomic design of automation. Dual-mode cognitive design of vehicle guidance and control work systems. Springer, Berlin

Orlicky J, Plossl G (1994) Orlicky's material requirements planning. McGraw-Hill, New York

Paetzold K (2006) On the importance of a functional description for the development of cognitive technical systems. In: International design conference 2006, Dubrovnik

Palani S (2010) Control systems engineering. McGraw-Hill, New Delhi

Peters A (2001) Benutzerorientierte Leitsoftware für automatisierte Fertigungssysteme in Komponentenstruktur. Dissertation RWTH Aachen

Petersen B (2003) Flexible Handhabungstechnik für die automatisierte Mikromontage. Aachen, Shaker

Pfeifer T, Schmitt R (eds) (2006) Autonome Produktionszellen: Komplexe Produktionsprozesse flexibel automatisieren. Springer, Berlin

Pfeifer T, Tillmann M, Wimmer M (2003) Prozesskettenoptimierung – Mit der IPO-Systematik innovativ zu ganzheitlichen Lösungen. In: REFA-Nachrichten 6/2003

Pfendler C, Schlick C (2007) A comparative study of mobile map displays in a geographic orientation task. Behav Inf Technol 26(6):455–463

Pfirrmann O, Astor M (2006) Trendreport MST 2020. Basel, Berlin

Pfohl H (2004) Logistiksysteme: Betriebswirtschaftliche Grundlagen. Springer, Berlin

Photonics 21 (2010) Lighting the way ahead. Düsseldorf

Porter ME (2004) Competitive strategy. Free Press, New York

Posieek S (2005) Einsatz kombinatorischer Optimierungsmethoden bei automatischer Optimierung von Umformprozessen. Shaker, Aachen

Possel-Dölken F (2006) Projektierbares Multiagentensystem für die Ablaufsteuerung in der flexibel automatisierten Fertigung. Shaker, Aachen

Pramujati B, Dubay R, Samaan C (2006) Cavity pressure control during cooling in plastic injection molding. Adv Polym Technol 25(3):170–181

Premium AEROTEC GmbH: http://www.premium-aerotec.com/Binaries/Binary5687/A350-Grafik_work_packages_DE.jpg

Priddy KL, Keller PE (2005) Artificial neural networks. SPIE – The International Society for Optical Engineering, Bellingham

Puppe F, Gappa U, Poeck K, Bamberger S (1996) Wissensbasierte Diagnose- und Informationssysteme Mit Anwendungen des Expertensystem-Shell-Baukastens D3. Springer, Berlin

Putzer HJ (2004) Ein uniformer Architekturansatz für Kognitive Systeme und seine Umsetzung in ein operatives Framework. Verlag Dr. Köster, Berlin

Rasmussen J (1986) Information processing and human-machine interaction. An approach to cognitive engineering. North-Holland, New York

Rehse M (1999) Flexible Prozessüberwachung bei der Bohr- und Fräsbearbeitung in einer autonomen Produktionszelle. RWTH Aachen, Dissertation

Reinhart G, Berlak J, Effert C, Selke C (2002) Wandlungsfähige Fabrikgestaltung, wirtschaftlichen Fabrikbetrieb 97(1/2):18–23

Reisgen U, Beckers M, Willms K, Buchholz G (2010) Einsatz und Vorgehensweise bei der Ersatzmodellierung beim Impulslichtbogenschweißverfahren. VS-Berichte 268:79–84

Reuber M (2001) Prozessüberwachung beim Schlichtfräsen von Freiformflächen. Dissertation RWTH Aachen

Riedel M, Nefzi M, Corves B (2010a) Grasp planning for a reconfigurable parallel robot with an underactuated arm structure. In: Proceedings of the 1st international workshop on underactuated grasping. Montreal, Canada, 19. Aug

Riedel M, Nefzi M, Corves B (2010b) Performance Analysis and Dimensional Synthesis of a Six DOF Reconfigurable Parallel Manipulator. In: Proceedings of the IFToMM symposium on mechanism design for robotics. Mexico City, Mexico, 28.–30. Sept , 2010

Rojas R (1996) Neural networks. Springer, Berlin

Romero V, Burkardt J, Gunzburger M, Peterson J (2006) Comparison of pure and "Latinized" centroidal Voronoi tessellation against various other statistical sampling methods. J Reliabil Eng Syst Safety 91(10/11):1266–1280

Roosen S (1997) Online-Prozessoptimierung beim MAG-Schweißen mit Hilfe eines Expertensystems, vol 4. Aachener Berichte Fügetechnik, Shaker

Rosenblatt F (1958) The perceptron. A probabilistic model for information storage and organization in the brain. Psychol Rev 65:386–408

Rüegg-Stürm J (2003) Das neue St. Galler Management-Modell: Grundkategorien einer integrierten Managementlehre; der HSG-Ansatz. Haupt, Bern

Rummery GA, Niranjan N (1994) On-line Q-learning using connectionist systems, Technical Report CUED/F-INFENG/TR 166, Cambridge University, GB

Russell SJ, Norvig P (2003) Artificial intelligence: a modern approach. Pearson Education, Upper Saddle River

Russer H, León FP (2007) Informationsfusion – eine übersicht. Tech Mess 74(3):93–102

Salomon P (2006) Micro sensors and microsystems. Eurosensors Göteborg, Sept 2006

Sauer W (2003) Prozesstechnologie der Elektronik, Modellierung, Simulation und Optimierung der Fertigung. Hanser, München

Savory S (1985) Künstliche Intelligenz und Expertensysteme. Oldenburg, München

Scherms E (1998) Optimierte Tolerierung durch Qualitätsdatenanalyse. Konstruktion 50: 31–36

Schild K, Bussmann S (2007) Self-organization in manufacturing operations. Commun ACM 50(12):74–79

Schlick C (1999) Modellbasierte Gestaltung der Benutzungsschnittstelle autonomer Produktionszellen. In: Luczak H (ed) Schriftenreihe Rationalisierung und Humanisierung, vol 61 (zugl. Dissertation RWTH Aachen). Shaker, Aachen

Schlick C, Odenthal B, Mayer M, Neuhöfer J, Grandt M, Kausch B, Mütze-Niewöhner S (2009) Design and evaluation of an augmented vision system for self-optimizing assembly cells. In: Schlick C (ed) Industrial engineering and ergonomics. Springer, Berlin

Schlick C, Mayer M, Odenthal, B (2010) MTM als Prozesslogik für die kognitiv automatisierte Montage. In: Britzke B (ed) MTM in einer globalisierten Wirtschaft – Arbeitsprozesse systematisch gestalten und optimieren. Wirtschaftsbuch, München

Schmidt C (2008) Konfiguration überbetrieblicher Koordinationsprozesse in der Auftragsabwicklung des Maschinen- und Anlagenbaus. Dissertation RWTH Aachen, Shaker, Aachen

Schmitt R (2000) Aufbau flexibler Mess- und Prüfstationen für die automatisierte Montage. Shaker, Aachen

Schmitt R (2010) Qualitätsmanagement. Strategien, Methoden, Techniken, 4 edn. Hanser, München

Schmitt R, Beaujean P (2007) Selbstoptimierende Produktionssysteme. Zeitschrift für wirtschaftlichen Fabrikbetrieb 9:520–524

Schmitt R, Betzold M, Hense K (2007) Das Aachener Qualitätsmanagementmodell. In: Schmitt R, Pfeifer T (eds) Masing Handbuch Qualitätsmanagement. Hanser, München, pp 38–41

Schmitt R, Pavim A, Brecher C, Pyschny N, Loosen, P, Funck, M, Dolkemeyer J, Morasch, V (2008) Flexibel automatisierte Montage von Festkörperlasern: Auf dem Weg zur flexiblen Montage mittels kooperierender Roboter und Sensorfusion. In: wt – Werkstatttechnik, Ausgabe 11/12, 2008, Jahrgang 98, pp 955–960

Schmitt R, Stiller S, Beaujean P (2009a) The aachen quality management model – normative framework of the entrepreneurial quality philosophy

Schmitt R, Wagels C, Isermann M, Matuschek N (2009b) Cognitive optimization of an automotive rear-axle drive production process. J Mach Eng 9(1):78–90

Schneider F (2004) Überwachung, Regelung und Automatisierung beim Hochgeschwindigkeitsschneiden von Elektroblechen mit Laserstrahlung. Dissertation RWTH Aachen

Scholz-Reiter B, de Beer C, Bose F, Windt K (2007) Evolution in der Logistik Selbststeuerung logistischer Prozesse. In: VDI Berichte

Schotten M (1998) Beurteilung von EDV-gestützten Koordinationsinstrumentarien in der Fertigung. Dissertation RWTH Aachen, Shaker, Aachen

Schuh G (2005) Produktkomplexität managen: Sstrategien – Methoden – Tools. Hanser, München

Schuh G (2006) Produktionsplanung und -steuerung. Springer, Berlin

Schuh G, Orilski S (2007) Roadmapping for competitiveness of high wage countries. In: Proceedings of the XVIII. ISPIM conference. Warschau, Polen

Schuh G, Roesgen R (2006) Aufgaben. In: Schuh G (ed) Produktionsplanung und -steuerung. Springer, Berlin

Schulz W, Kostrykin V, Zefferer H, Petring D, Poprawe R (1997) A free boundary problem related to laser beam fusion cutting: ODE approximation. Int J Heat Mass Trans 40(12):2913–2928

Schulz W, Niessen M, Eppelt U, Kowalick K (2009) Simulation of laser cutting. In: Dowden JM (ed) The theory of laser materials processing: heat and mass transfer in modern technology. Springer Series in Materials Science, pp 21–69

Schulz W, Nießen M, Vossen G, Schüttler J (2010) Modelling and process monitoring of laser processing. In: Erscheint in proceedings of NUSIM multiphysics 2010 (und in The International Journal of Multiphysics)

Schüttler J, Lose J, Schmitt R, Schulz W (2009) Exploring process domain boundaries of complex production processes using a metamodeling approach, 12th CIRP conference on modelling of machining operations, Donostia-San Sebastiàn – Spain, vol 2. In: Arrazola P (ed) Conference on modelling of machining operations, vol II, Mondragon, 7.–8. Mai 2009, pp 835–841

Schwindt C, Trautmann N (2004) Advanced- planning- systeme zum supply – chain- management, Universität Karlsruhe, Karlsruhe

Shafer J, Agrawal R, Mehta M (2005) SPRINT: a scalable parallel classifier for data mining. In: Readings in database systems. Massachusetts Institute of Technology, USA, pp 668–679

Shepard RN, Metzler J (1971) Mental rotation of three-dimensional objects. Science 171(972):701–703

Sheridan T (2002) Humans and automation: system design and reserch issues. Wiley, New York

Shingo S (1992) Das Erfolgsgeheimnis der Toyota-Produktion: eine Studie über das Toyota-Produktionssystem – genannt die "Schlanke Produktion". Verlag Moderne Industrie, Landsberg/Lech

Siciliano B, Khatib O (eds) (2008) Springer handbook of robotics. Springer, Berlin

Siegert H-J, Bocionek S (1996) Robotik: Programmierung Intelligenter Roboter. Springer, Berlin

Smud SM, Harper DO, Deshpande PB, Leffew KW (1991) Advanced process control for injection molding. Polymer Eng Sci 31(15):1081–1085

Spath D, Scholtz O (2007) Ideen gegen Verlagerung der Montage ins Ausland. Werkstattstechnik 97(1/2):2–4

Spath D, Klinkel S, Barrho T (2002) Auftragsabwicklung in dezentralen Strukturen: Erfolgsfaktoren und Probleme – Ergebnisse einer Studie. Zeitschrift für Wirtschaftlichen Fabrikbetrieb 97(3)

Steinberg DV, Lin DKJ (2006) A construction method for orthogonal Latin hypercube designs. Biometrika 93(2):279–288

Stepanek P (2007) Flexibel automatisierte Montage von leicht verformbaren großvolumigen Bauteilen. Shaker, Aachen

Stoddard K, Kneifel W, Martin D, Mirza K, Chafee M, Hagenauer A, Graf S (2004) Method and control system for controlling a plurality of robots. Patent No.: US 6,804,580 B1 (12 Oct 2004)

Stork A, Thole CA, Klimenko S, Nikitin I, Nikitina L, Astakhov Y (2007) Simulated reality in automotive design. International conference on cyberworlds, pp 23–27

Strohner H (1995) Kognitive Systeme. Eine Einführung in die Kognitionswissenschaft. Westdeutscher, Opladen

Sutton RS, Barto AG (1999) Reinforcement learning, Journal of Cognitive Neuroscience, Massachusetts Institute of Technology, USA

Taguchi G (1990) Quality Engineering

Tai YF, Scherfler C, Brooks DJ, Sawamoto N, Castiello U (2004) The human premotor cortex is "mirror" only for biological actions. Curr Biol 14:117–120

Takeda H (2009) QiP – Qualität im Prozess: Leitfaden zur Qualitätssteigerung in der Produktion. Finanz Buch Verlag GmbH, München

Tang A, Owen C, Biocca F, Weimin M (2004) Performance evaluation of augmented reality for directed assembly. In: Ong SK, Nee AYC (eds) Virtual and augmented reality applications in manufacturing. Springer, New York

Thiele M (1997) Kernkompetenzorientierte Unternehmensstrukturen – Ansätze zur Neugestaltung von Geschäftsbereichsorganisationen. DUV, Wiesbaden

Thiem I (1998) Ein Strukturmodell des Fertigungsmanagements. Dissertation Universität Bochum, Shaker, Aachen

Thomas U (2008) Automatisierte Programmierung von Robotern für Montageaufgaben. Shaker, Braunschweig

Thornton AC (2004) Variation risk management. Focusing quality improvements in product development and production. Wiley, Hoboken

Thrun S, Montemerlo M, Dahlkamp H, Stavens A, Aron J, Diebel P, Fong J, Gale M, Halpenny G, Hoffmann K, Lau C, Oakley M, Palatucci V, Pratt P, Stang S, Strohband C, Dupont LE, Jendrossek C, Koelen C, Markey C, Rummel J, van Niekerk E, Jensen P, Alessandrini G, Bradski B, Davies S, Ettinger A, Kaehler A, Nefian A, Mahoney P (2006) The robot that won the DARPA Grand Challenge. J Field Robot 23(9):661–692

Trumpold H, Beck CH, Richter G (1997) Toleranzsysteme und Toleranzdesign. Qualität im Austauschbau. Hanser, München

Unbehauen H (2008) Regelungstechnik I: Klassische Verfahren zur Analyse und Synthese linearer kontinuierlicher Regelsysteme, 15 edn. Vieweg+Teubner, Wiesbaden

Upton D (1992) A flexible structure for computer-controlled manufacturing systems. Manuf Rev 5(1):58–74

Urmson C, Anhalt J, Bae H, Bagnell JAD, Baker C, Bittner RE, Brown T, Clark MN, Darms M, Demitrish D, Dolan J, Duggins D, Ferguson D, Galatali T, Geyer C, M, Gittleman M, Harbaugh S, Hebert M, Howard T, Kolski S, Likhachev M, Litkouhi B, Kelly A, McNaughton M, Miller N, Nickolaou J, Peterson K, Pilnick B, Rajkumar R, Rybski P, Sadekar V, Salesky B, Seo YW, Singh S, Snider JM, Struble JC, Stentz AT, Taylor M, Whittaker WRL, Wolkowicki Z, Zhang W, Ziglar J (2008) Autonomous driving in urban environments: boss and the urban challenge. J Field Robot – Special Issue on the 2007 DARPA Urban Challenge 25:425–466

Völklein F, Zetterer T (2006) Mikrosystemtechnik. Vieweg und Sohn, Wiesbaden

Vossen G, Schüttler J (2010) Modelling and stability analysis for laser cutting. Submitted to Mathematical and Computer Modelling od Dynamical Systems

Vossen G, Nießen M, Schüttler J (2010) Optimization of partial differential equations for minimizing the roughness of laser cutting surfaces. In: Diehl M, Glineur F, Michiels W (eds) Recent advances in optimization and its applications in engineering. Springer, New York

Weck M (1998) Maschinenarten und Anwendungsbereiche, Werkzeugmaschinen Fertigungssysteme, vol 1. Springer, Berlin, p 20

Weck M, Brecher C (2006) Werkzeugmaschinen 4 – Automatisierung von Maschinen und Anlagen. 6 edn. Springer, Berlin

Weinmann A (1995) Regelungen II: Analyse und technischer Entwurf, vol 2. Springer, Wien

Westkämper E, Zahn E (2009) Wandlungsfähige Produktionsunternehmen – das Stuttgarter Unternehmensmodell. Springer, Berlin

Westkämper E, Bullinger H-J, Horvàth P, Zahn E (eds) (2001) Montageplanung – effizient und marktgerecht, Stuttgart. Springer, New York

Whitney DE (2004) Mechanical assemblies. Their design, manufacture, and role in product development. Oxford University Press, New York

Wibbeler J (2002) Frequenzselektive Vibrationssensoren mit spannungsgesteuerter Resonanzabstimmung in Oberflächenmikromechanik. Dissertation Technische Universität Chemnitz

Wiendahl H (2002) Situative Konfiguration des Auftragsmanagements im turbulenten Umfeld. Dissertation an der Universität Stuttgart, Stuttgart

Wiendahl H (2005) Betriebsorganisation für Ingenieure. Hanser, München

Wiendahl H (2006) Methodische grundlagen. In: Schuh G, Westkämper E (eds) Liefertreue im Maschinen-und Anlagenbau: Stand-Potenziale-Trends. Aachen und Stuttgart

Wiendahl H, Hernàndez R (2006) The transformable factory – strategies, methods and examples. In: Dashchenko AI (ed) Reconfigurable manufacturing systems and transformable factories. Springer, Berlin

Wiendahl H, ElMaraghy H, Nyhuis P, Zäh M, Duffie N, Brieke M (2007) Changeable manufacturing – classification, design and operation. CIRP Ann Manuf Technol 56(2)

Wiest U (2001) Kinematische Kalibrierung von Industrierobotern. Dissertation Universität Karlsruhe, Aachen, Shaker

Wight OW (1984) Production and inventory management in the computer age. Reinhold, New York

Witt S (2007) Integrierte Simulation von Maschine, Werkstück und spanendem Fertigungsprozess. Dissertation RWTH Aachen

Woods S, Pflanz T (2008) DPSS lasers rival fiber lasers for marking applications. LaserFocusWorld, Januar

von Witzleben A (2006) Workshop des Thüringer Ministeriums für Wirtschaft, Technologie und Arbeit Entwicklungsschwerpunkte Thüringens im Bereich Optische Technologien, 17. Juli 2006, am Fraunhofer IOF Jena

Wollnack J, Stepanek P (2004) Formkorrektur und Lageführung für eine flexible und automatisierte Großbauteilmontage. Werkstattstechnik Jahrgang 94(9):414–421

Womack JP, Jones DT (1996) Lean thinking. Simon und Schuster, New York

Yu K-W (2001) Terminkennlinie: Eine Beschreibungsmethodik für die Terminabweichung im Produktionsbereich. VDI, Düsseldorf

Zäh M (2005) A holistic framework for enhancing the changeability of production systems. In: 1st conference on changeable, agile and virtual procduction (CARV 2005b), München

Zaeh M, Wiesbeck M (2008) A model for adaptively generating assembly instructions using state-based graphs. In: Manufacturing systems and technologies for the new frontier. Springer, London

Zaeh MF, Beetz M, Shea K, Reinhart G, Bender K, Lau C, Ostgathe M, Vogl W, Wiesbeck M, Engelhard M, Ertelt C, Ruehr T, Friedrich M, Herle S (2009) The cognitive factory. In: ElMaraghy HA (ed) Changeable and reconfigurable manufacturing systems. Springer, Berlin

Zakharian S, Ladewig-Riebler P, Thoer S (1998) Neuronale Netze für Ingenieure. Vieweg Verlagsgesellschaft

Zäpfel G (2001) Grundzüge des Produktions- und Logistikmanagement. Oldenbourg, München

Zell A (1994) SNNS: stuttgart neural network simulator, Universität Stuttgart, Stuttgart
Zell A, Mache N, Sommer T, Korb T (1991) Design of the SNNS neural network simulator. Österreichische Artificial Intelligence Tagung, pp 93–102
Zheng HY (1990) An experimental study of the relationship between in-process signals and cut quality in gas assisted laser cutting. SPIE – CO_2 Lasers and Applications II 1276:218

Chapter 7
Integrative Business and Technology Cases

Christian Brecher, Achim Kampker, Fritz Klocke, Peter Loosen,
Walter Michaeli, Robert Schmitt, Günther Schuh, Thomas Auerbach,
Arne Bohl, Peter Burggräf, Sascha Fuchs, Max Funck, Alexander Gatej,
Lothar Glasmacher, Julio Aguilar, Robert Guntlin, Ulrike Hecht,
Rick Hilchner, Mario Isermann, Stephan Kratz, Matthis Laass,
Meysam Minoufekr, Valentin Morasch, Andreas Neuß, Christian Niggemann,
Jan Noecker, Till Potente, André Schievenbusch, Georg J. Schmitz,
Stephan Schmitz, Jochen Stollenwerk, Dražen Veselovac,
Cathrin Wesch-Potente and Johannes Wunderle

Contents

7.1	Objective of the Business and Technology Cases	988
7.2	Process Planning and Monitoring of Blisk Production	988
	7.2.1 Abstract	989
	7.2.2 State of the Art	989
	7.2.3 Motivation and Research Question	990
	7.2.4 Procedure to Create a Continuous CAx Process Chain with Integrated Process Evaluation	991
	7.2.5 Industrial Relevance	994
	7.2.6 Future Research Topics	994
7.3	Integrative Production of Micro-lasers	995
	7.3.1 Abstract	995
	7.3.2 State of the Art	998
	7.3.3 Motivation and Research Question	1003
	7.3.4 Results	1005
	7.3.5 Industrial Relevance	1030
	7.3.6 Future Research Topics	1035
7.4	Integrative Process and Product Development for Hybrid Plastic-Metal Structural Components	1038
	7.4.1 Industrial Task	1038
	7.4.2 Objective	1039
	7.4.3 Solution	1040
7.5	Integrated Rear Axle Drive Production for Cars	1042
	7.5.1 Abstract	1042
	7.5.2 Starting Point for Rear Axle Drive Production	1042
	7.5.3 Motivation and Research Question	1044
	7.5.4 Cognitive Methods to Improve Rear Axle Drive Production	1045
	7.5.5 Future Research Topics	1052

G. Schuh (✉)
Werkzeugmaschinenlabor WZL der RWTH Aachen, Steinbachstr. 19,
52074 Aachen, Germany
e-mail: g.schuh@wzl.rwth-aachen.de

7.6	Integrative Levelling of Production		1053
	7.6.1	Abstract	1053
	7.6.2	State of the Art	1054
	7.6.3	Motivation and Research Question	1057
	7.6.4	Results	1058
	7.6.5	Industrial Relevance	1060
	7.6.6	Future Research Topics	1062
7.7	Integrative Production of TiAl-Products		1062
	7.7.1	Abstract	1062
	7.7.2	State of the Art	1063
	7.7.3	Motivation and Research Question	1066
	7.7.4	Results	1067
	7.7.5	Industrial Relevance	1070
	7.7.6	Future Research Topics	1071
References			1071

7.1 Objective of the Business and Technology Cases

In order to strengthen the relevance and integrativity of research in the Cluster of Excellence, current best practice "business and technology cases" were selected. Hereby the theories, hypotheses, predictions and technology projects developed in the Cluster of Excellence are evaluated and advanced in close collaboration with leading production companies in Germany and Europe. To make the work more transparent, different application scenarios will be developed.

Objectives of the business and technology cases as well as the selected demonstrators are the monitoring of research subjects and the corresponding adjustment, evaluation and validation, as well as the transfer of research results into industrial context. The cases feature scientific content, which has especially been derived for actual applications and generates concrete benefit, to the partners in the industrial advisory board.

The set of business and technology cases in the Cluster of Excellence is defined in a way that all relevant aspects of research from the individual ICDs are covered. The following chapter provides detailed information on the business and technology cases conducted in the Cluster of Excellence.

7.2 Process Planning and Monitoring of Blisk Production

Fritz Klocke, Thomas Auerbach, Lothar Glasmacher, Stephan Kratz, Meysam Minoufekr and Dražen Veselovac

7.2.1 Abstract

The aim of the business and technology case study "Blisk Production" is the implementation of a hardware and software solution which allows a position-oriented analysis and design of 5-axis machining processes. For this purpose an interactive software system is realised to evaluate and optimise the process sequence before and after the manufacturing. The current case study focuses on the development and integration of a more efficient optimisation strategy including the correction of process parameters after the initial machining. Thus, by taking into consideration the process-specific signal measurements, which refer to the tool position in work piece coordinates, the process sequence can be analysed and optimised based on recent approaches in online-process monitoring and NC data analysis.

7.2.2 State of the Art

Process planning and design with the scope of identifying stable process areas are associated with time-intensive correction loops consuming a high amount of resources. The number of iterations and hence the required effort mainly depends on the practical knowledge of the person in charge of the process. This is because know-how acquired during many years and a profound understanding of the process are required to design the machining strategy for complex production tasks. Another issue is the gap of information about the process behaviour which is relevant for understanding the machining process with regard to process parameters. Hence, only time- and cost-intensive inspections can determine the targeted product quality for the selected machining strategy. Modern CAx systems consider the kinematic behaviour of the machine to optimise the tool path before the first machining of the parts takes place (Klocke et al. 2003) and to correct NC data sets with regard to the machine dynamic. Klocke, Altmüller and Glasmacher have suggested ways of analysing NC programs by taking into account the machine dynamic (Klocke et al. 2001, 2003; Altmüller 2001). The generated and optimised NC data sets consider, amongst other aspects, the dynamic behaviour of the machine tool and its rotational and translational axes. The resulting tool paths are corrected according to the maximum possible acceleration of the individual machine axes. Further work was carried out based on a continuous CAM-NC simulation chain to investigate the harmonic design of NC tool paths (Klocke et al. 2003). This simulation chain was used by Klocke et al. to present a CAM-based system to model process forces (Klocke et al. 2007). In Klocke et al. [2008] and Klocke et al. (2009a, b) methods based on geometrical models were developed to calculate and analyse machining processes and carry out modifications in advance. Also, the use of simulation systems to analyse process behaviour is more and more in the focus of recent research activities. These simulation techniques are based on a discretisation of the work piece to be machined and vary according to its geometric design (Glaeser 1997; Jerard et al. 1989; Robert et al. 1987). Milling simulation methods based on the Voxel or Dexel model are presented in Ayasse et al.

(2001) and Stautner (2005), and are partially deployed in commercially available systems (Cimatron 2011; Open Mind 2011; Machine 2011).

Production systems may use monitoring systems to maximise a tool's operation time by guaranteeing the functionality of the manufactured product throughout the whole process. In industrial use, the objective of such systems is to reliably detect disturbances such as process instabilities in the process, and to provide an opportunity to prohibit them. As a consequence monitoring systems make a major contribution to the safety and quality of the process. Expensive damage and outage of the production system caused by disturbances can therefore be significantly reduced. A reliable functioning monitoring system is a major prerequisite to increase the level of production automation.

Industrial monitoring systems which are available on the market are primarily used for large-scale manufacturing and mass production. This is due to relative simple and inflexible monitoring strategies used to evaluate the process behaviour. Hereby, disturbances are detected by referencing the process to signals measured during a disturbance-free process. Evaluation and disturbance detection are hence compared to a specific behaviour within the process. Such a strategy's lack of flexibility to cope with process variations does not allow its application to single piece and small-scale manufacturing. Therefore, a monitoring system is required that during all stages of the process considers the frequently changing machining sequences. This applies to a greater extent to the machining of free-form surfaces. Here, meshing and machining conditions are continually changing and occurring monitoring strategies are referring to the required process changes (Markworth 2005).

Besides thermal causes of disturbance, mechanical process disturbances may also lead to an exceedance of process and/or machine tolerances and hence undesired process deviations (Witt 2007). Occurring process disturbances, which are a result of mechanical process loads depend on large extent on the selected process parameters and the work piece's and machine tool's stiffness and damping properties. Hence, a prerequisite for monitoring a process is the ability to determine the actual status of the machine and the process at any time. First of all a suitable sensor system must be chosen or, if necessary, developed to reliably detect in-process disturbances. Starting point for this case study is the detailed analysis of the technical challenge presented by the finish milling of thin-walled structures developed in the D2 sub-project entitled "Technology Enablers for Embedding Cognition and Self-optimisation into Production Systems", which has yet to be solved.

7.2.3 *Motivation and Research Question*

The primary goal of modern production technology is to guarantee the long-term competitiveness of manufacturing companies in high-wage countries. This requires the ability to control and modify process systems based on the specified and dynamically changing demands and constraints. A key sector in German industry currently facing these challenges is the jet engine manufacturing industry which, within the scope of this project, is a cooperating partner in the investigations.

The business and technology case (B&T case) for the "process planning and monitoring of blisk production" focuses on the manufacture of safety-critical components such as blisks (blade integrated disks) which are machined using 5-axis roughing and finishing operations. These machining operations have to guarantee high quality standards to ensure the safe operation of the manufactured product when put into operation. At the same time, the manufacturing system has to be flexible to meet the changing requirements of the market, i.e., the capability to produce different component types for the jet engines and to cope with the frequent change of the product portfolio. This makes it difficult to control the resulting complexity of such production processes and their susceptibility to disturbances as their conditions are changing permanently. Consequently, process design and production are subject to the highest technological demands.

Within the scope of the B&T case presented here, the goal is to introduce new system solutions, which reduce current shortcomings in process planning and production helping companies to save resources.

The process design stage requires multiple optimisation steps to repeatedly ensure disturbance-free manufacturing. This iterative optimisation prevents a productive use of a machine tool. Caused by these situations the aim of this case study is to carry out a position-oriented evaluation of the process. This will allow quicker and better optimisations and reduce the necessary resources. In this context the following research question is investigated:

> How can a position-oriented evaluation strategy be integrated into the CAx process chain to reduce the number of iteration steps required to identify an optimal machining process within the process design stage?

Answering this question involves the interdisciplinary cooperation of the process monitoring and CAx research area. In each research area different questions have to be addressed in order to meet the primary goal:

- How can CAx technologies be used to create a continuous CAx process chain?
- How can the existing strategy for position-oriented evaluation be transferred to 5-axis free form machining operations?
- How can the interface between the CAx process chain and process monitoring be realised to allow a continuous data exchange?

7.2.4 Procedure to Create a Continuous CAx Process Chain with Integrated Process Evaluation

7.2.4.1 NC Data Analysis and Optimisation of the Traverse Path

In the context of efficient process planning, the developed CAx process chain is applied to the machining task presented for manufacturing a blisk geometry. This software module allows the design of the process by process-related parameters, such as feed and cutting parameters. Thus, modifications and optimisation measures can be applied to the virtual process model during the planning stage (Klocke et al. 2008). In this way, optimised machining strategies for the tool path can be identified and implemented in the CAM and NC process design. Depending on the blisk feature to be machined the right machining strategies have to be selected. The first step is to evaluate the complexity based on criteria such as blade height, chord length, blade thickness and distortion of the blade profile. Afterwards the integrated machine simulation is used to verify the favourite finish and rough milling strategies for possible collisions between tool, work piece and machine tool. Furthermore, technological useful process parameters such as cutting speed, feed and depths of cut are determined. In addition, an NC data analysis carries out a final optimisation of the NC tool paths before the machining of prototypes is processed on the manufacturing system.

7.2.4.2 Position-Oriented Evaluation of the Process Behaviour

In addition to the process planning, a monitoring system specially developed for the finish milling of thin-walled parts is integrated into the blisk machining. Both the measuring system and the components required for the online recording of position data must be integrated into the machining system. This has an important effect on the implementation of the evaluation strategy. Reading the individual machine axes means the position of the tool centre points can be calculated in the machine's coordinate system. Axis transformation is then used to determine the position in the work piece's coordinate system. This information is needed to analyse process signals with respect to its current state. Downstream process evaluation involves the recording of measurement parameters that reflect the true state of the process. Here fore sensors, which measure the process behaviour according to the physical principles, are used to convert process information into electrical, evaluable signals.

Within the context of machining thin-walled free-form surfaces, an intensive analysis yields two measurands, which are representative of the process behaviour and hence could be subsequently used for monitoring purposes. The measurands are the position and the vibration. The choice is justified by the fact that the vibration behaviour is proven to be a major source of disturbance and that in the process it is characterised by its position-related variability. To correlate recorded signals with the actual values of the prevailing meshing conditions, the position of the tool must be determined relative to the work piece. The position data also provide a suitable interface to the process planning as these data can be used to synchronise both system solutions—i.e., the CAx system and the process monitoring system.

Earlier work focused on the development and implementation of methods to record and analyse data for monitoring purposes. In this context and as an analogy to blisk machining, the first step was to machine and simultaneously evaluate 2 mm thin, 36 mm overhanging beams for milling operations. Acceleration signals recorded in process at the spindle housing and on the machine table were evaluated. During the course of this case study, the developed algorithm must be enhanced to the

5-axis machining operations and the signal evaluation has to be linked to the process planning module.

7.2.4.3 Coupling of Process Planning and Monitoring

Using an integrated approach during the preparation of the process involves the coupling of planning and monitoring methods with one another. The result is a more efficient analysis and identification of suitable machining strategies. In this case, knowledge extracted during on-line monitoring of the process behaviour is transferred into the planning system. This allows a faster identification of critical settings regarding each individual machining strategy. Here "critical" means the product quality is not achieved and set tolerances are not met. Due to the increased density of information, NC analysis can be carried out at an early stage of process planning to modify the process parameters at the virtual level of production. Modifying machining strategies in this manner means that disturbance-free manufacturing processes and set quality goals are achieved in a faster way, so that the costs of the required planning is reduced.

Suitable data formats and data synchronisation methods are needed to integrate information into the virtual planning environment that was acquired by the process monitoring system. Hereby, data synchronisation ensures effective process planning coupling with process monitoring information in a viable manner. The evaluation of the process referring to the tool's position delivers a suitable interface for data synchronisation as it simultaneously exists at the virtual level. The result of such a conjunction is shown in Fig. 7.1 where a free-formed test geometry is exemplary displayed.

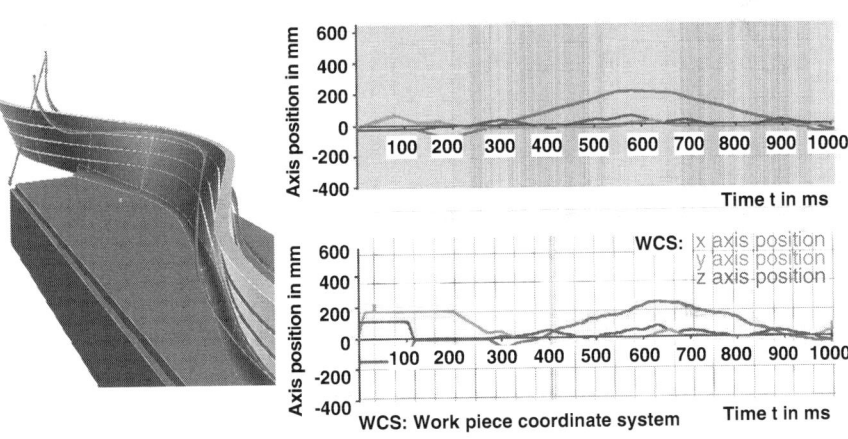

Fig. 7.1 Verification with the help of a use case: WCS positions of the target data (*top right*). WCS positions following PM import (*bottom right*): x: red, y: green, z: blue

7.2.5 Industrial Relevance

The manufacturing of safety-critical components requires a 100% guarantee of functionality at the final stage of production, which prohibit a failure of the product during operation later on. During the phase of process design at all times the product quality requirements have to be taken into account to ensure that the quality goals are met at all stages of the process. At the same time, however, the cost of identifying an optimal machining strategy must be minimised. In this context "optimal" means that the machining time to produce the component is minimized.

The turbine engine manufacturing industry produces safety-critical components for the aviation industry and is therefore confronted with the described challenges. For this reason, a use case was defined in cooperation with a turbine engine supplier and contains the representative blisk component. This component is used to firstly demonstrate how methods developed in the areas of process planning and process monitoring can be transferred to industry. Secondly it demonstrates the potentials created when both areas are combined at an early stage of the process design phase.

If the planned procedure is successful, an integrated system platform, which couples CAM-based process design with in-process monitoring solutions, will be available to a large number of manufacturing companies. Complex tool paths can then be analysed and evaluated on the basis of real process behaviour. It will be possible to optimally exploit the potential performance of machining centres and to machine at the process limits. This results in significantly more efficient machining operations.

The process design stage is also speeded up as the gathered information significantly reduces the number of correction loops required in the CAx process chain. This would directly reduce the use of resources as a result of fewer test runs and shorter machine times. Another important benefit arising from the integration of process monitoring and process planning is the lower cost in quality management.

CAD/CAM manufacturers in the component development sector will also benefit. The implementation of interfaces for process monitoring will allow the integration of adaptive system methods and simulation algorithms, which can be implemented into existing modules. This improves their competitiveness in the international market.

7.2.6 Future Research Topics

The scientific challenge is the continuous mapping of in-process results and their transfer into the CAx process chain (CAD/CAM, NC data generation, process simulation). All partial simulations have to be tightly coupled to the overall process design system, including process monitoring solutions. The result is the optimisation of production processes.

The development of an adaptive control system is also investigated. One objective is to perform position-oriented process analysis using acceleration signals to reduce the number of dynamic disturbances during machining (Fig. 7.2).

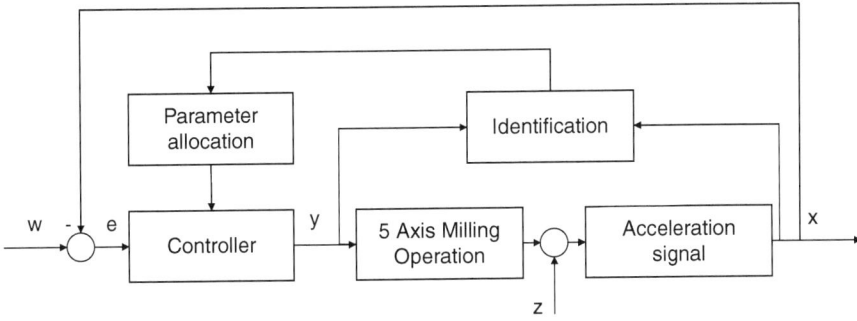

Fig. 7.2 Controlled adaptation in the control loop

Cutting speed is the parameter to be controlled in the process as feed, axial and radial depth are more or less kept constant during finish milling process as a result of the required finish quality and the kinematic roughness. The controller parameters are modified by identifying the transmission behaviour between the cutting speed (actuating variable y) and the system's vibration behaviour (control variable x) which is passed on to the controller in accordance to set rules of assignment.

7.3 Integrative Production of Micro-lasers

Peter Loosen, Max Funck, Alexander Gatej, Valentin Morasch and Jochen Stollenwerk

7.3.1 Abstract

As a non-contact, flexible and precise tool, lasers are playing an increasingly important role in modern manufacturing. Its main area of application with just under 50% of the world market comprises the fields of marking and micro machining. Modern lasers are optical systems characterised by a strong interplay of mechanics, optics and electronics enabling lasers to work with variable parameters and under the most adverse conditions. Due to their versatility, lasers are not only used in a great variety of areas; a diverse range of lasers must also be provided and manufactured.

Due to the increasing trend for miniaturisation, which has also reached the production of laser systems, industry is facing increasing demands with regard to mobility, flexibility and costs. The average power and dimensions place the micro-laser at the lower specification end of typical material processing systems. With just up to 10 W average output, high-intensity pulses ($>10^6$ W/cm^2) capable of marking plastics and metals by removing coatings, inducing colour changes, or engraving can be generated. Small dimensions, low weight and low costs make the micro-laser a suitable product for the mass production of individual applications in material processing. Through flexible, automated manufacture customised solutions can be provided.

Solid state lasers contain doped crystalline structures as their active laser medium and provide many advantages for use in compact laser markers as they allow variable pulse parameters and are also principally well suited for automated micro assembly. Nevertheless, the automated production of lasers is in many respects an extremely demanding task, and requires the collaboration of design and development, assembly, metrology and systems engineering. In addition to the pure collaboration involved, the development of such an integrative approach to production also requires answering the multidisciplinary issues which arise from such an integration.

This project has produced significant findings in the fields of product design, joining technology and assembly strategies. Conventional designs, which are primarily based on mechanically adjustable supports, are virtually unsuitable for automation.

The product, taken as the starting point for the design of the assembly, must therefore undergo significant changes as part of the automation and miniaturisation processes in order to resolve the conflicts between automation, systems engineering and product design. The developed strategy is based on surface mount technology with parts mounted on a planar ceramic substrate. This type of structure provides a great deal of flexibility with regard to the positioning of components while allowing vertical positioning by automatic units. Difficulties caused by poor accessibility are thus avoided and standardised component dimensions allow comparably simple handling. Electrical connections can be realised by coating the ceramic carrier with conducting material and subsequent structuring so that sensors may be added to record the system's status. Due to the lasers internal heat production, the mandatory thermal control may likewise be achieved by selecting a suitable ceramic material. Aluminium nitride, with a thermal conductivity of $180 \, W/(m \cdot K)$, has proven to be particularly suitable and enables heat dissipation via the mounting plate. Individual components are chosen and designed according to the planar design and the associated constraints, such as the limited number of available degrees of freedom for the alignment of components.

In order to avoid complex supports and alignment processes, optical components are attached directly to the ceramic substrate. Soldering is an excellent solution for joining as it offers a high level of thermo-mechanical stability. A resistance soldering process was specifically developed for planar assembly taking advantage of the heat induced by electrical current in the layer of solder. Structured solder pads, 10-20 µm in thickness, are applied to the ceramic carrier and molten by applying an electrical current. The geometry of the solder pads as well as positioning and alignment of the optical components play an important role in the process sequence. The joining process may be controlled by observing the voltage drop due to a change of electrical resistance of the molten solder. This allows accurate control of the process that is completed successfully within a few seconds. Shear tests of components joined to the substrate have recorded a loading capacity of up to 25 kg and samples have successfully passed environmental tests such as thermocycling and vibration tests. Thanks to the small thickness of the layer of solder absolute volume contraction of the solder, a major cause of misalignment is drastically reduced. Soldering components to the basic substrate requires precisely aligned contact areas that are parallel to one another. As a consequence it is only possible to carry out alignments within those three degrees of freedom which do not influence the parallelism of the alignment. An additional soldering process has been developed for components demanding up to six degrees of freedom using thicker layers of solder and incremental solidification. Depending on the needs, the more suitable of the two processes may be selected.

Design adaptation is an integral step towards the viability of automated and flexible assembly as the required levels of accuracy need to be maintained. In particular, the limitations resistance soldering is placing on alignment must be addressed. In order to resolve these, the design is laid out to be insensitive to component movements along the axis vertical to the mounting plate. Furthermore components are selected and paired by a procedure called tolerance matching in order to compensate tolerance effects. It is particularly beneficial to select cylindrical components for elements of the pump optics and the resonator for planar assembly as the functions may be decoupled. In addition, components with a vertical cylinder axis can always be centred by alignment. Tolerance matching makes use of tolerance deviations to compensate for errors. For example, the centration of lenses is characterised and suitable lenses combined during assembly. Omitting classification as common during selective assembly according to tolerance classes makes the procedure attractive for smaller quantities. Trough tolerance matching the centration of the pump radiation within the laser crystal can for example be raised by a factor of four for a series of six lasers.

Integration of sensors, actuators and an electronic control system increases the range of product functions and assists assembly, placing an emphasis on the communication between the product and the assembly system. This makes it possible to adjust product parameters during assembly. In addition to providing the laser with power, the electronic control can also deal with many other tasks such as temperature control, electro-optical light modulation for pulse generation and monitoring of the laser operation. The integration of purely electrical functions is bundled together in an electronic control system of modular design and the electronic power board is assembled on the cooling plate common to the laser mounting plate in accordance with the planar design. The basic functions required for the operation of the laser, such as the regulation of laser diode current, availability of safety functions and temperature control of the laser diode are all realised within one module. A second module, the controller board, controls the electronic power board and contains a microcontroller which allows both manual control by the operator as well as automated laser operation. There are also signal processing circuits for evaluating the integrated sensors. Expanding the functions through additional sensors, new types of sensors or, for example, signals or image processors aiding the manufacture of customer-specific products can be achieved by exchanging the controller board. The selected electronics concept grants the assembly system online access to the laser control and the sensors integrated in the product, which is necessary for various alignment processes.

An active compensation, the control parameters of which are adjusted during assembly, is used to correct thermally induced changes to the beam parameters when changing the laser output power. A module for the compensation of the thermal lens is used for this purpose, which identifies and corrects the output-based changes in the geometry of the laser spot.

Despite the adapted design, the execution of the assembly and in particular the assembly sequence is crucial for the assembly results. Interactions arising from component positions and parameter deviations must be taken into consideration during assembly and the detection of alignment signals preclude some assembly sequences from the outset. A comprehensive tolerance analysis, which provides a precise forecast of the expected effects of component deviations, is therefore indispensable. The selected sequence is adjusted according to the tolerance deviations of the components, measurement uncertainty and alignment precision in order to meet the various requirements and reduce the effort involved in assembly. The requirements are thus examined separately in this respect and are mapped as functions of the key factors. This process may be reasonably supported by self-optimisation strategies addressing the resolution of planning efforts, a major aspect of the polylemma of production.

7.3.2 State of the Art

7.3.2.1 Compact Solid State Lasers for Marking and Micro Machining

Marking is one of the most common laser technology applications and consequently a large number of possibilities exist to achieve the laser parameters specific to each of the applications. Marking metallic surfaces by engraving or ablation requires high energy densities which are usually achieved by pulsed operation in compact systems. Depending on the pulse parameters, it is possible to reach peak pulse power levels of multiple-10 kW at an average laser output power of less than 10 W. Several approaches will be presented below and their suitability discussed. When selecting the presented concepts, particular attention is paid not only to compactness and efficiency, but also the ability to automate or, if necessary, the degree of automation of the assembly technology of the respective solutions, whereby commercially available products are introduced in addition to laboratory prototypes.

Classic laser markers are CO_2 and solid state lasers with a wave length of around 10 μm or 1 μm, respectively. Due to the lower wavelength, the latter may be better focused by a factor of 10 and therefore provide advantages in terms of resolution. The following designs are therefore limited to beam production in the infrared spectral range around 1 μm.

The laser concept with the greatest potential for growth is the fibre laser concept. Conventional laser sources are currently being replaced by fibre lasers in ever more fields of application. Due to the high level of cycle amplification achieved, typical applications of this laser lie in the generation of high average outputs in continuous operation. However, current fibre lasers are either unable to directly reach the peak pulse power levels of solid state lasers due to numerous non-linear effects (self-focusing, self-phase modulation, scatter effect) or only with considerable effort. This requires components and assemblies outside the fibre, or multi-stage MOPA- (Master

Oscillator Power Amplifier) or MOFPA- (Master Oscillator Fibre Power Amplifier) systems, which have far more complex designs. Industrially available examples of this include the IPG Photonics Corporation YPL series, the Rofin-Sinar Laser GmbH PowerLine F series, the TRUMPF TruMark Series 5000 and the Keopsys Kult 1 μm series.

The Optically Pumped Semiconductor Laser (OPSL) is an innovative laser concept with great potential for development. The integral feature of such systems is the option of wavelength tuning instead of choosing between discrete emission wave lengths as is the case with conventional solid state lasers. OPSL systems are capable of covering the entire visible wave length spectrum of light with average outputs of less than 10 watts in continuous operation. The main fields of application of the OPSLs therefore include medical, biological and entertainment technologies. OPSL systems are available commercially from Coherent.

Another type of laser with a large market share is the diode laser, which is able to achieve high power densities in continuous operation and is used in material processing for applications which require high laser output power with good beam quality. Industrial products are available from Dilas GmbH, Laserline GmbH, TRUMPF or Jenoptik AG, for example. Although diode lasers may be used to generate pulses by modulating the diode current, they are unable to generate peak pulse power levels of the same magnitude as those generated by Q-switched solid state lasers, either in direct operation or in pump operation for solid state lasers. Pulsed diode lasers with short (ns and shorter) pulses are used increasingly in communications and medical technology (Coherent GmbH, Lumics GmbH, etc.) and do not provide the peak pulse power required for engraving applications.

Diode lasers are also particularly appropriate as pump sources for solid state lasers, whereby end- or side-pumped rods are particularly suitable and are often used for resonator configurations due to their high level of cycle amplification. Disc lasers are yet another wide-spread resonator configuration whereby, due to the low single cycle amplification, high peak pulse power levels are only reached at above average outputs and the systems are therefore less compact. Disc lasers are available commercially from TRUMPF and Rofin-Sinar, among others.

An established method for generating pulses of higher peak outputs in solid state lasers is Q-switching, whereby the resonator losses (or resonator Q-factor) may be varied by active or passive elements in the resonator. Thereby the "storing" of energy in the laser crystal and subsequent decoupling of short (in the nanosecond range), energy-rich pulses are being enabled. Another advantage arising from the generation of high peak outputs is the particular suitability for the generation of shortwave radiation by means of frequency conversion, provided the conversion efficiency increases as the pulse output increases. Shortwave radiation in the green spectral range, which may be generated by frequency-doubling of solid state lasers, exhibits considerably better absorption behaviour in many plastics in particular and other materials, e.g. copper, and can also be focussed better. Laser marking systems are therefore commercially available from numerous manufacturers, with frequency conversion as an optional extra.

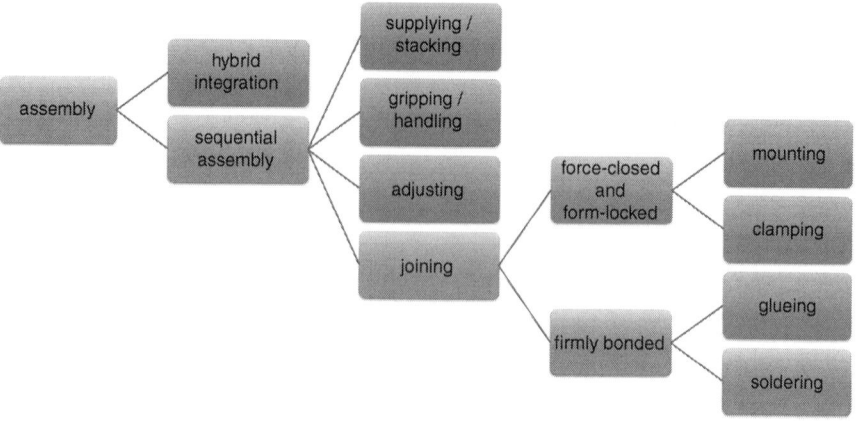

Fig. 7.3 Assembly sub-processes according to (Banse 2005)

7.3.2.2 Structural Design Technology and Joining Technology for Optical Systems

The assembly of optical systems may be subdivided into two main categories (cf. Fig. 7.3). While the aim of hybrid integration is to combine individual modular systems to form a super system, the alternative is based on the sequential integration of individual components in the entire system by means of suitable joining processes (Banse 2005). Despite this strong theoretical division, the majority of systems usually involve modular integration into a hybrid system as the final stage.

Laser systems are firmly established in current industrial manufacturing, biotechnology and medicine and new fields of application are constantly being researched and opened up. The classic structural design technology as illustrated in Fig. 7.4 is based on the principle of mounting individually adjusted lenses and fixing them to a common alignment platform. The advantage of this procedure is the option of aligning each component separately and exchanging them due to wear and tear. However, poor mechanical and thermal long-term stabilities, complex and time-consuming manual alignment procedures and the considerable structural size due to the use of indirect parts such as holders and mountings prove to be disadvantageous (Beckert 2005).

The desire for a greater integration density in production processes or the requirement for compactness and ease of handling of the laser sources justify the clear trend towards miniaturisation. In the classic type of construction, the limits of miniaturisation are usually set according to the space requirement of the adjustable mountings. This results in a diverse range of options for replacing the adjustable holders with material-bonded joints with long-term stability on a common mounting plate. The wide variety of laser components and the associated high number of requirements placed on their mounting and alignment demand joining processes which support

7 Integrative Business and Technology Cases

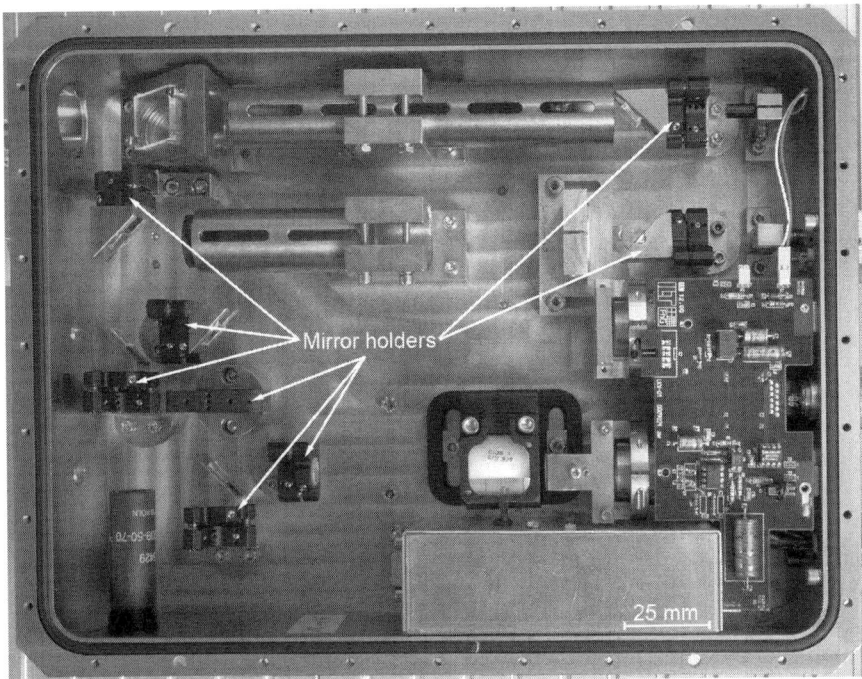

Fig. 7.4 Classic structural design technology with lens- and mirror holders

these. Some of the most well-known design concepts and implementations in the field of miniaturised laser assembly are explained in brief below.

Options for miniaturising by means of a direct construction on a common substrate are discussed in a patent specification (Halldorsson et al. 1989) and subsequent patents of Messerschmitt-Bölkow-Blohm GmbH. The structure represents a significant miniaturisation of an optical bench with the option to integrate optical and electro-optical components. In this example, a common mounting plate of a silicon substrate is used on which miniaturised laser systems may be mounted in a planar type of construction by means of precise, component-specific adjustment aids consisting of guide and fixture groves produced by anisotropic etching processes. This technology is principally similar to the SMD technology of electronic manufacturing, whereby universal application is limited by component-specific adjustment aids. A similar approach is taken by (Beckert 2005) with geometrically stable, smooth ceramic substrates used for the mounting plate material, whereby the thesis focuses on the analysis of producing mechanical structures suitable for mounting optical components.

Various other miniaturisation processes are not based on the positioning of the lenses by high precision mechanical stops, but are optimally adjusted using high-precision handling devices and joined with long-term stability. These options are classified according to the joining process and discussed in greater detail in the following chapter.

The need of joining different materials in a material-bonded manner limits the choice of joining processes to glueing and soldering. Other requirements are also placed on the joint: high position stability even in the curing process, thermo-mechanical stability over the entire operating range, high thermal conductivity for the compensation of the reduced contact area of the components, and minimal gas emission for the protection of the optical surfaces. Despite the high loading capacity of both types of joint, soldering has the advantage of speeding up the process by omitting UV light curing and provides reduced volume changes compared to glueing (Beckert 2005). Further analysis is therefore limited to solder-based technologies.

Soldering of optical parts requires a heat source which heats the components to the soldering temperature and thus controls the process. Electro-thermal heating resistors, laser radiation and heating gases are being used as heat sources.

The use of a heating gas stream in particular is an established method in surface mount technology. The hot gas stream, which is directed onto the solder, evenly heats the components to be soldered and joins them when the soldering temperature is reached. The main disadvantage of this process is that it is not possible to guarantee the heating of a specific localised soldering point due to the dispersing gases.

The good focusing ability of laser radiation is making it ever more significant for localised heating. A large number of processes have therefore been developed on this basis in recent years, whereby these processes may be subdivided into direct and indirect radiation processes. In the case of direct application, the beam is focussed through a transparent component or mounting plate onto the solder, while in the case of indirect application, the beam heats an adjacent part and melts the solder by thermal conduction. Processes based on heating by laser radiation in combination with thin solder layers have been developed by the Fraunhofer Institute for Applied Optics and Precision Engineering (IOF) (Banse et al. 2005) and by Choi et al. (2003). The main drawback of both processes is the limited number of options available for adjusting the components within thin solder layers, which is why other processes have attempted to compensate for this disadvantage by using thicker layers of solder.

Leica Geosystems' "TRIMO-SMD" therefore uses a vertical mount with a pool of solder on the underside which may be taken hold of by a robot, aligned and joined from the underside by laser radiation (Scussat et al. 2000). Furthermore, the IOF has developed a process called "solder bumping" in which a ball of solder is fed into a capillary and melted by laser radiation. The molten solder is pushed out of the capillary by the emitted inert gas and comes into contact with the laser-heated wetting surface (Beckert et al. 2007).

In addition to being able to heat the solder using a laser source, it is also possible to generate the heat using a thermal or electrical contact. Patents of Adlas Lasertechnik (Seelert and Elm 1997) and Zeiss (Tayebati and Holderer 2007) use the effect of resistance heating to melt solder between a carrier ceramic and an optical component. The use of thin layers of solder has the same drawbacks as seen in the laser processes. Optical components may therefore only be positioned in two translatory axes and one rotational axis thus making it impossible to carry out alignments in all six degrees

7 Integrative Business and Technology Cases

of freedom. The "PermAlign" process developed by Coherent extends the spatial flexibility of the process by using a thicker solder pool in which an alignment can be made in the melt (Woods and Pflanz 2008). The solder is let into a dish located in the mounting plate and heated from the underside by an electrical resistance heating element.

7.3.3 Motivation and Research Question

While diode lasers are manufactured mostly automatically thanks to wafer level production, solid state lasers are predominantly manually assembled. This places limitations not only on miniaturisation, but also on the reduction of manufacturing costs, making it impossible to meet the demand for compact low power systems. Such small and low cost systems are predominantly demanded for marking and micro machining and a large future market is forecasted for these applications due to quality assurance measures, part traceability, miniaturisation and solar cell production.

The increasing trend towards product customisation is driving companies to offer tailor-made solutions suiting specific applications as part of their marketing strategies. This in turn leads to an increased number of products with only small production runs or even single batches requiring a greater degree of flexibility in the manufacturing process. Flexibility and customisation traditionally stand in opposition to automation, which is necessary to increase productivity and maintain competitiveness in high wage countries. Existing automation solutions are usually designed for the manufacture of a single product and allow only a small degree of flexible adaptation. An increase in flexibility places high demands on the robustness of the automation as well as on associated planning of processes. In order to address these factors, both, product design and assembly must be adapted to meet the increased level of flexibility.

The production of laser systems places great demands on the technical qualification of staff and is therefore an excellent example of a production process in high-wage countries. Component handling, assembly and alignment all require particular care and high precision along with a trained approach and understanding. Errors are often only pinpointed by experts due to the complex interrelationships between the laser components. An increase in productivity as a result of automation must take this fact into account and permit the system operator suitable opportunities to carry out corrections.

Flexible automation and the transfer of planning tasks to the system itself provide substantial benefits to manufacturing in high-wage countries. Repetitive and time-consuming manual tasks which tie up the capacity of skilled employees may be performed by the system thus making better use of human skills. While the planning and implementation is performed by the machine, the machine operator makes the decisions. This prevents possible errors and significantly increases productivity.

It is possible to reduce the automation of the production of solid state lasers to assembly, being the production process with the highest proportion of added value.

In the case of manual assembly, a large number of mechanical, optical and even electronic components are assembled and aligned. The alignment process is particularly characteristic for the production of laser systems and optical devices in general. Individually developed supports and mounts with a high degree of mechanical and thermal stability optimised for manual assembly are used but are unsuitable for automation. Automation therefore requires alternative product concepts oriented towards robot-based component handling. Supporting assembly by an integrated electronic control of product functionalities is another crucial factor in implementing automation. The communication between the assembly system and the product guarantees that functions can be met during operation. Analysing possible product control architectures are therefore just as necessary as the integration of suitable sensors and actuators for assessing the system's state and adapting it to the current assembly status.

Joining and fastening technologies such as mounting and clamping must be replaced by technologies which enable simple automation and which have the smallest number of additional elements possible in order to reduce the number of operations. This quick processing simultaneously significantly reduces the cycle times and helps increase cost effectiveness. Due to the high level of demands placed on positioning and alignment, the robustness of the processes is crucial, particularly with regard to mechanical and thermal stresses, as the absorbed radiation adds heat to the system. The joining of optical components using soldering technology represents a promising solution, the suitability of which must be examined with respect to flexibility, adaptation to the design, robustness and long-term stability.

The inherent sensitivity of the laser to disturbances and tolerances requires a particular handling of component- and position-deviations. On the one hand, the joining technology must allow component alignment; while on the other hand, strategies to avoid alignment must be implemented in order to reduce assembly complexity thereby simplifying automation. Depending on the assembly concept, sensitivities within the optical system must be limited to mainly moveable or adjustable parameters and critical tolerances avoided as far as possible. For parameters which may not be adjusted, selective assembly or tolerance matching may be a way of increasing quality by reducing sensitivities. In order to prevent rejects and increase the precision of a component paring, a classification of components into tolerance classes should be avoided.

The developments focus on the automation of an adapted product concept which enables the planning of product variants whilst retaining flexibility and the ability to automate. In addition, an automated joining process and a tolerance compensation concept are devised. The investigations may be summarised by the question how small production quantities of a wide variety of laser systems may be automatically assembled with a high degree of robustness and how design, system and assembly processes must be adapted for this purpose.

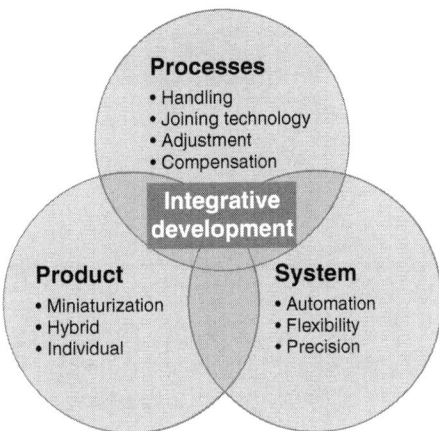

Fig. 7.5 Development requirements

7.3.4 Results

7.3.4.1 Assembly-oriented Product Development

The diverse range of requirements placed on micro-lasers and their assembly make it necessary to carry out a comprehensive examination of all technological disciplines involved in the development of products and production systems. These requirements and constraints are collected and systemised in a first step (cf. Fig. 7.5). It is then helpful to subdivide these requirements according to the separate development steps in the production process. This allows the classification of individual tasks, thereby simplifying the planning of detailed solutions. It is also important to identify contradictory or even conflicting requirements which may only be resolved by an integrated examination due to the common nature of the tasks (cf. Brecher et al. 2008).

Development requirements consist of both contributions to the resolution of the polylemma of production as well as technical constraints of the product and assembly system. A customised production process which allows the manufacture of bespoke products must be flexible, technologically high developed and automated. As far as the product is concerned, a high level of miniaturisation is desired in addition to the mechanical, thermal, electrical and optical functions required due to the hybrid character of the micro-laser. This leads to the following basic requirements:

- Miniaturisation of the product
- Automation of the assembly
- Customisation of the product and the manufacturing process.

In order to achieve a generic solution, these requirements must first be addressed and continually mirrored by generalised product demands in order to meet the typical requirements of the micro-laser. Table 7.1 presents a qualitative overview of the mechanical, optical, electrical and thermal requirements.

Table 7.1 Component requirements: (+++) very high, (++) high, (+) low, (O) none

Components	Mechanical requirement	Optical requirement	Electrical requirement	Thermal requirement
Pump source	++	++	+++	+++
Laser medium	+++	++	O	++
Lenses	++	++	O	+
Mirrors	++	+++	O	+
Control electronics	++	O	+++	+
Pulse generation	+++	++	++	+

If miniaturisation is considered during the design and development of the product, it is possible to reduce the number and dimensions of optical and mechanical components. A reduction in the number of optical components equates to a greater integration of functionality in the individual parts. Complexity is therefore transferred to the production of components. This is definitely of benefit to the assembly process. However, highly-integrated functional components are often difficult to adjust since several functions are simultaneously influenced during alignment processes. Functional integration is thus particularly useful when separate alignment parameters are available for the different functions of a part. This is of considerable importance for the automation of the assembly process since the dependencies of functions may often cause errors due to their interactions. Avoiding the use of mechanical elements to mount optical components has great advantages for both miniaturisation and automation. However, proven techniques, such as clamping and mounting of optical elements, must be rejected and replaced by joining processes which allow the optical components to be connected directly to the mounting plate. The use of joining processes (e.g. soldering) may considerably reduce the number of interfaces between components, thereby making better use of tolerances and achieving a higher level of robustness (Brecher et al. 2008).

Customisation in the sense of flexible product design and production systems is essentially possible due to an increased level of flexibility with regard to products and assembly as well as a standardisation of processes. It must be possible to freely position parts and components and they must subject to as few restrictions as possible with regard to their applicability, number and type. If this is achieved, product variants may be designed freely and in real-time. A large product variety is achieved in many industrial applications by means of platform concepts (Schäppi et al. 2005). This allows the exchange of components and subassemblies and the platform serves as a common base for the assembly. Using a standardised joining technology is helpful for achieving a high level of customisation as it allows components or modules to be attached to a platform as required. The customisation of the assembly means implementing flexible and adaptable sequences and processes that allow assembly sequences which meet the requirements. The use of exchangeable tools in the assembly system and the application of a flexible control architecture can additionally support a customised assembly.

Assembly with automatic units primarily requires that sequences may be repeated and that there is good accessibility with regard to the positioning of components and modules. In principle, these requirements combine well with customisation which also relies on free positioning in order to increase the freedom of design. A standardisation of sequences and processes is also helpful for automation in order to reduce the level of complexity. Standardisation is particularly advantageous with regard to joining and fastening technologies and component handling. The standardisation of measurement and adjustment processes across product variants is another aspect which may enable flexible automation. Due to the level of complexity and precision, an automation of the assembly of laser systems also requires the use of flexible measurement and adjustment devices as well as a suitable method for planning flexible and customised assembly processes.

Key points that need to be addressed come to light if the options for realising the conceptual requirements are compared to the actual product requirements. For example, the requirements placed on miniaturisation and handling during the production of micro-lasers may be achieved without the need for adjustable supports, but this means the requirements placed on the precision and adjustability of the component position may no longer be fulfilled to the same degree. However, a comprehensive examination shows that the adaptation of the product and the use of suitable tolerances and compensation principles may be exploited in order to do without adjustable supports. In this case, the expertise gathered from the fields of design, measuring technology, assembly and joining technology, and systems engineering may be combined to resolve the issue.

Likewise, synergies arise when using soldering technologies for mounting components, which may fulfil the thermal, mechanical and system requirements placed on the ability to automate, thereby simultaneously resolving several problems.

Design Using Planar Technology As a design option, planar technology (Fig. 7.6) offers clear advantages in terms of miniaturisation and automation. Components may be arranged relatively freely and the basic design therefore offers a high degree of freedom with regard to the product design. The excellent level of accessibility simplifies the handling of components, which may be fastened using standardised joining processes. Designing parts of similar geometry makes it possible to carry out similar joining processes, thereby increasing the level of robustness and reliability. There is no need for additional elements for supporting and adjusting components, substantially reducing the number of components. This saves costs, allows a high level of integration and simplifies the assembly process. The only limitation of planar assembly is the predetermined alignment of components due to the planar mounting plate. Tilt and vertical deviation of the components directly depend on this. Alignment may be aided by joining processes which allow adjustments to be made within tolerance deviations and which only marginally impede the basic planar principles. The individual selection and allocation of components (so-called tolerance matching) may be used to exploit compensation effects and play out deviations of several components against each other (Brecher et al. 2008; Funck and Loosen 2009).

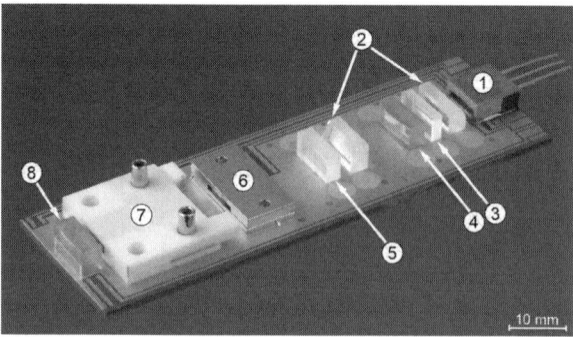

Fig. 7.6 Planar assembly of the MicroSlab micro-laser with pump diode (*1*), imaging optics (*2*), micro lens array (*3*), focusing lens (*4*), coupling-in mirror (*5*), crystal package (*6*), Pockels cells (*7*) and end mirror (*8*)

The electrical and thermal requirements are achieved using a ceramic carrier plate. Aluminium nitride distinguishes itself in terms of a high level of electrical isolation and good thermal conductivity. Electrically conducting paths may therefore be realised by structuring metal layers. Moreover an active or passive temperature control of the ceramic is possible. Soldering as a joining technology is particularly suitable for meeting electrical, thermal and mechanical requirements in equal measure. The creation of a defined thermal contact zone with a high level of thermal conductivity and a high melting point allows the use of high performance components with substantial thermal loads. However, the use of an irreversible joining and fastening technology also introduces significant limitations. Assembly may only be performed sequentially as it is impossible to carry out subsequent corrections to already assembled components. This places specific requirements on assembly planning and assembly implementation as well as product design.

Figure 7.6 illustrates the set-up of a slab-laser using planar technology. The ceramic mounting plate is plated and structured with conducting paths and reference points. Solderable coatings in the form of pads are also provided for mounting optical elements of glass.

Selection and Design of Components for Planar Assembly The micro-laser consists of various functional modules, which may be examined separately. These are: the pump source, the beam shaping optics, the laser crystal, the resonator, the pulse generator and the sensors and electronics. Depending on the laser, additional functional modules such as an amplifier, frequency converter or wave length stabiliser may be added. Each of these functional modules consists of a different number of components which are selected and designed for the concept of planar assembly.

The best choice of pump source is diode lasers in the form of single emitters, broad stripe emitters and diode laser bars. There are considerable differences in their emitted output and radiation characteristics. Single emitters exhibit a proportionally higher output than diode laser bars which can be attributed to thermal factors. However, the achievable total output of diode laser bars is considerably higher and they have a smaller divergence along the slow axis. Diode lasers are also available in different installation formats. Concepts which are encased by the manufacturer and already equipped with a fast-axis collimator are particularly suitable for automated planar

7 Integrative Business and Technology Cases

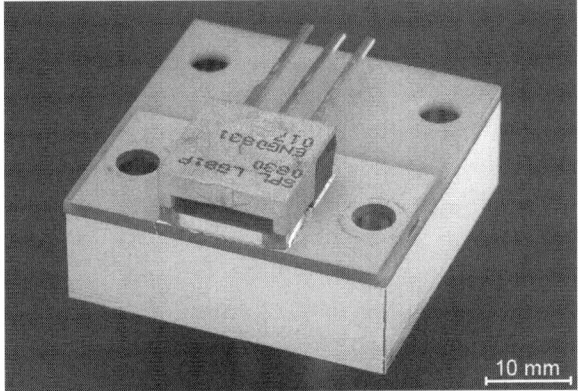

Fig. 7.7 Module with soldered diode laser bar

assembly using soldering technology. This makes it possible to do without the highly precise alignment of the micro lens and mount a finished package instead. The higher outputs available allow the use of diode laser bars in the implemented micro-laser design, which emit an output of up to 30 W. Figure 7.7 illustrates the diode laser bars used as a soldered package on a submount.

The active laser medium in a solid state laser is a doped crystal which is brought to an excited state by the absorption of light. The return of excited electrons to a lower energy level causes heat loss in addition to the desired laser radiation (Siegman 1986). Stable operation of a laser requires systematic heat dissipation even at moderated outputs of a few watts. In order to guarantee this, the crystal is soldered into a casing with good thermal conductivity and this is presented as a package for assembly. The manufacture and assembly of this package is particularly simple if the crystal geometries are rectangular. In this case, these are known as slab crystals.

Figure 7.8 illustrates a slab crystal soldered into a gold-plated copper housing for the purpose of better heat dissipation. The thermo-mechanical design gives the crystal an even and symmetrical temperature distribution despite the dissipation of heat via

Fig. 7.8 Crystal soldered into a copper heat sink

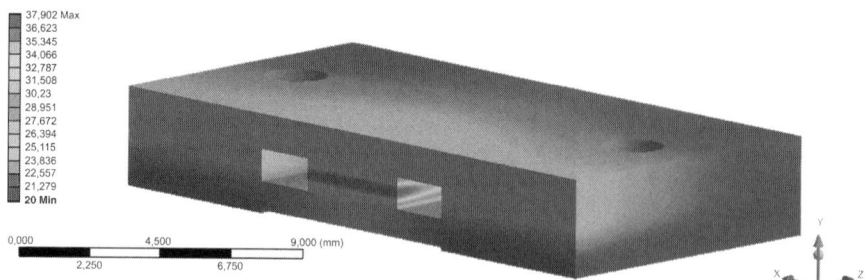

Fig. 7.9 Simulated temperature distribution (°C) of the connected Nd:YAG with a heat sink at 30 W pump output

the laser mounting plate. Choosing a mounting plate of aluminium nitride guarantees electrical isolation and a good level of thermal conductivity (180 W/(m · K)) which is sufficient to prevent the crystal overheating. At higher outputs it is also possible to implement active cooling via this interface.

The thermal influence of the pump radiation on the crystal also plays a significant role in the functionality of the laser. Along with the temperature profile, absorption of the pump radiation also creates a refractive index profile along the vertical axis. This causes a delay in the wave front near the centre of the crystal, thereby focusing the radiation, which must be taken into account when designing the resonator (Koechner 2006).

Figure 7.9 illustrates the temperature profiles calculated using the finite element method based on a linear absorption distribution over the width of the crystal. As the temperature changes with the output of the pump laser, a thermal lens also develops which in turn leads to a displacement of the laser focus relative to the workpiece being machined. This effect may be adjusted by a suitably active compensation lens.

The purpose of the beam shaping optics is to align the radiation of the pump source to the laser crystal and the resonator. The radiation emitted by the laser diode bars is already collimated vertically to the mounting plate (fast axis) and exhibits an approximately Gaussian intensity profile along this axis. The distribution along the horizontal axis (slow axis) is non-uniform caused by the interference of the individual bar emitters. However, optical pumping of the laser crystal requires the homogeneous radiation of a portion of the laser crystal over the entire width of the laser. The required height of the pump distribution is determined approximately by the beam height of the generated Eigenmode of the resonator at the position of the laser crystal (Koechner 2006). This gives the key functions of the beam-forming or pump optics: the homogenisation and adjustment of the line width along the horizontal axis and the adjustment of the beam height.

The specifications of the pump optics also determine the positioning of the pump line in both coordinate directions as well as the maximum tilt. Particularly, the vertically centred alignment is critical due to the developing temperature distribution and the associated thermal lens effect.

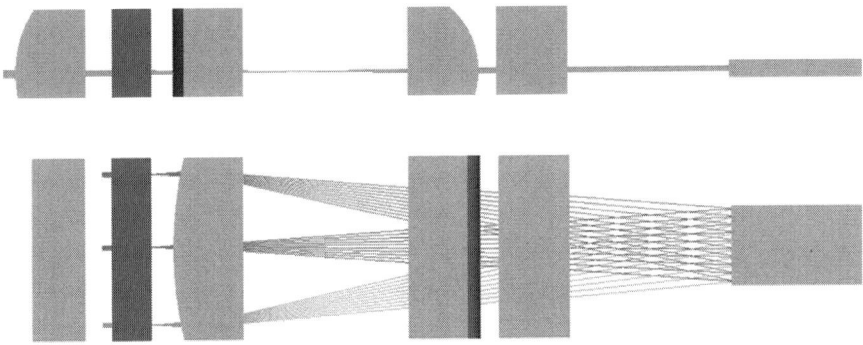

Fig. 7.10 Pump optics: Optical imaging (*top*) and homogenisation (*bottom*)

The distribution in vertical and horizontal coordinate directions predetermined by the laser diode may also be used to design the pump optics. While it is entirely possible to reduce the number of parts by grouping together components, the possibilities for compensation and the complexity of the requirements placed on the position and location of the pump line argue against it.

In principle, the radiation may be homogenised by integrated elements based on the total internal reflection in glass rods and micro-optical components. In one instance, the modelling of the end face of the integrator, and in another, the focussing of the far field distribution, are required to guarantee a homogeneous distribution along the horizontal axis where the crystal is located. In both instances, the coherence effects should be taken into consideration although they may be disregarded due to the number of emitters in the bar.

Two components are modelled for the adjustment of the beam height along the vertical axis (Fig. 7.10). Theoretically, this may also be achieved using only one component, although this does have some disadvantages. Due to the tolerance related deviation of the emission level of the laser diodes and the position of the laser crystal within the package it is difficult to maintain the position of the pump line within the crystal. The centering of the imaging lenses therefore also exerts a substantial influence on the position of the pump line and may thus be used as a compensator. Due to the planar assembly, this may be achieved by exchanging components since any alignment is not feasible in the conventional sense. The use of a single lens as compensator is generally suitable, although it provides a lower degree of freedom with regard to correction and does not allow the separate control of the angle of incidence of the radiation and the vertical position. Figure 7.10 illustrates the side view and top view of the pump optics of the micro-laser.

The laser resonator is one of the most sensitive parts of the laser to adjust. During the design process it is therefore particularly important to avoid any occurring sensitivities if possible or control them by alignment. As with the pump optics, cylindrical components are recommended for planar assembly. Since the laser crystal only has a relevant thermal lens in the vertical axis, it is possible to implement a resonator of plane surfaces along this axis. This has the advantage that assembly on the mounting plate does not cause a centering error and that only angle deviations are critical. Along the horizontal axis, it is not possible to avoid a curvature of the resonator mirror for realising a stable resonator, although it is possible to carry out an alignment here. With this configuration it is only necessary to separately examine the alignment of the angle.

In order to be able to generate high intensity pulses with a laser, many applications use Q-switching which interrupts the feedback of the resonator so that more energy may be stored within the laser medium and a more powerful pulse may be generated when the resonator is disengaged again (Paschotta 2008). Typical pulse lengths for Q-switching lie in the nano-second range. Therefore peak pulse power levels of several 10 kW are achievable at average outputs of around 10 watts. A distinction is made between active and passive Q-switching technologies. Passive Q-switches are optical elements which possess the characteristic of being able to absorb a certain energy density before abruptly becoming transparent. The frequency of the laser pulses generated by passive technologies may only be modulated by output control in a narrow window, and the triggering of the pulse required for a precise marking is inherently impossible. With active Q-switches, each pulse emission is initiated by an external parameter whereby a variation of the peak pulse power is enabled by an output and frequency modulation. This variation substantially extends the spectrum of materials that can be laser-processed and in addition to triggering the pulses, represents another advantage for its use in laser marking applications. Established technologies for active Q-switching are based on the use of electro-optical or acousto-optical modulators. Acousto-optical modulators use the principle of a spatial deflection of the beam by bending at an optical mesh generated by a sound wave (Davis 2000). Based on this principle, acousto-optical modulators combined with a high resonator cycle amplification and a short resonator possess only a limited ability of extinction. The electro-optical modulators influence the polarisation of the laser beam. Depolarised portions of radiation are decoupled from the resonator by a polariser and are not sensitive in terms of the length of the resonator. There are no commercially available modulators suitable for integration in the design of lasers for miniaturised planar assemblies in a linear beam geometry. A comparison of both types of modulator in terms of their functionality and ease of integration in the planar laser design is conducted during the construction of another prototype and is described in more detail in (Funck et al. 2010). This paper compares the in-house development of an electro-optical modulator with a miniaturised special design of a commercial manufacturer of acousto-optical modulators (Fig. 7.11). The electro-optical modulation is characterised by a better extinction ratio and may therefore also achieve higher peak pulse power levels. From an assembly point of view, this is more favourable to implement as it allows a planar implementation and the high voltage required for control may be reduced. The specific requirements of the parts are relevant for assembly. While passive elements are relatively simple to introduce to the beam path without excessive precision, the applied electro-optical modulators must be electrically insulated due to the high voltage.

The set-up of a micro-laser with its corresponding complexity and variety of components requires the integration of electrical or electro-optical functionalities.

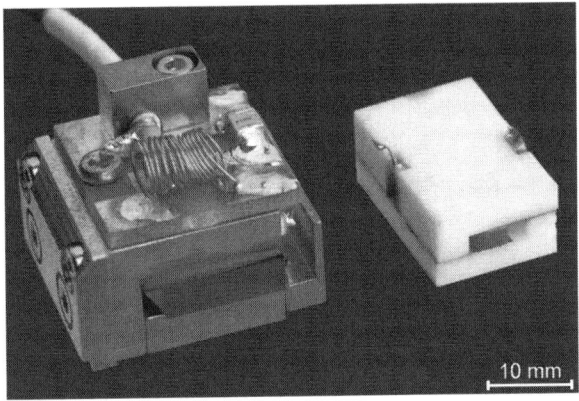

Fig. 7.11 Comparison of Q-switches; acousto-optical switch (*left*) and electro-optical switch (*right*)

In addition to providing the laser with power using the laser diode (pump diode), it is also necessary to deal with many other functions such as heat dissipation and temperature control, electro-optical light modulation for pulse generation in a Q-switch and monitoring the laser operation, etc. The integration of purely electrical functions is bundled together with an electronic control system of modular design, the electronic power board of which is assembled true to the planar design on the cooling plate common to the laser mounting plate. This allows the basic functions of the laser operation, such as regulation of the laser diode current, availability of the safety functions and tempering of the laser diode. A second module, the controller board, controls the electronic power board and in addition to a microcontroller, which allows both manual control by the operator as well as automated laser operation, also includes signal processing circuits for analysing the integrated sensors. Temperature sensors and photodiodes are also integrated in order to determine the laser output power. Exchanging the controller board expands the functions by means of additional sensors, types of sensor or signals or image processors, for example, which the manufacture of customer-specific products requires. The selected electronics concept allows the assembly robot to access the laser control and use the sensors integrated in the product even during assembly, which is necessary for various adjustment processes.

Components with an electro-optical function (laser diodes, Q-switches, photo-diodes) combine the optical requirements in terms of the precise positioning with the requirements for realising the appropriate electrical interface. In the case of the laser diode, which is aligned precisely with and joined to the laser mounting plate during pre-assembly, this interface is realised by means of a suitable arrangement of the electronic control system to the laser assembly plate in order to enable direct contact. The next stage of integration is an adaptation of the explored handling and joining technologies involved in the assembly of lenses to the electro-optical system components. When designing individual modules, such as the self-developed Q-switch, particular attention is paid to the practicability of the already tested planar assembly technologies.

It is also possible to integrate hybrids, e.g. electro-opto-mechanical expansion modules, which may be provided as an optional extra for improving or adjusting the radiation properties. A module for the compensation of the thermal lens is used for this purpose, which identifies and corrects the performance-based changes in the geometry of the laser spot.

7.3.4.2 Joining Technologies

Joining technologies may also be regarded as an essential component of the design process. The basic ideas discussed in Sect. 7.3.2.2 and the implementation strategies for thermal energy input in the soldering process underline the interest by both science and industry in making a robust joining process available. The resistance soldering thus developed offers the opportunity to automate the assembly using robot technology, increase long-term stability and considerably reducing gas emissions compared to glueing (Schmitt et al. 2009a).

The joining technology is essentially based on planar technology as this is similarly advantageous for the key assembly steps. The stackability and storage of components is simplified, the different parts may be handled using the same gripper and the components may be adjusted and joined accordingly without exchanging the robot tool. The aim is to successively join several components on a carrier substrate by means of resistance soldering, without affecting the joints between existing parts and moving the lenses out of adjustment. This requires a local energy input and therefore a suitable soldering geometry design. The structure and development of resistance soldering is explained in greater detail below.

Assembly on a Carrier Ceramic Depending on the selected process, it is possible to attach an optical component directly to an insulated material in order to direct the current across the joined components. The electrically insulating aluminium nitride (AlN) is used as the carrier material as it has a high thermal conductivity and is therefore capable of dissipating the temperatures generated in the system. With thermal expansion coefficients of $4.63 \cdot 10^{-6}$ K^{-1} it also provides a good compromise between the majority of glass components such as quartz, BK7 and SF6 ($0.5 \cdot 10^{-6}$ $K^{-1} - 9.0 \cdot 10^{-6}$ K^{-1}). The PVD process is used to apply solder to the ceramic, whereby a bonding agent and a diffusion barrier are used to improve the adhesion of the solder on both the base material and the components. The design of the solder pad is based on the need for electrical contact and is structured so that this is able to occur on the component side.

The structure in the schematic in Fig. 7.12 illustrates the gold-layered AlN mounting plate and the optical component, as well as the vapour-deposited solder on the carrier plate and the electrical contacts formed by a DC source.

A current passing through the solder pad causes a power loss due to the pad's resistance and brings the system to the soldering temperature. As a result, the pad begins to melt locally at the point of greatest resistance and joins the components

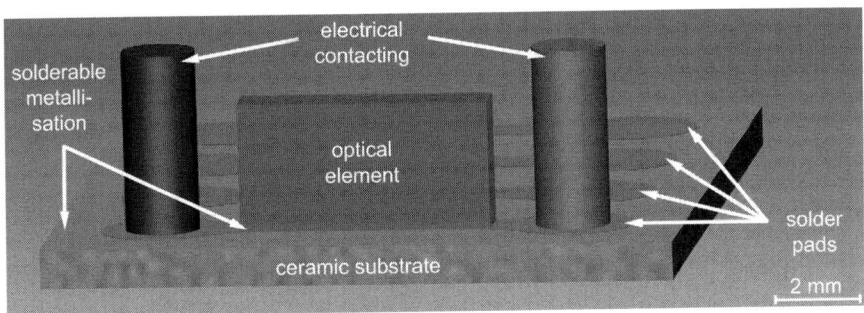

Fig. 7.12 Schematic of the "pick & join" structure

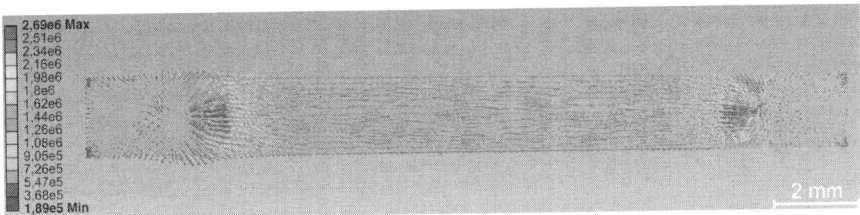

Fig. 7.13 Current density distribution (mA/mm^2) of a solder pad without structuring

two-dimensionally to the substrate. At the same time, the heating of the entire system improves the wetting ability of all components giving an optimum soldering result.

The "Pick & Join" Process In the "pick & join" process, the vapour-deposited solder has a thickness of 10–20 µm in order to guarantee a good degree of wettability while causing only a minimal amount of change in volume during the phase change. This process thus does not allow active alignment in all spatial axes and is therefore only used for insensitive components. The total surface soldering of the lens requires a homogeneous current density distribution in the joining zone which should fuse the solder locally under the components. The generation of melt at other points impairs or interrupts the electrical conductivity of the solder and prevents a uniform soldering. This phenomenon is examined using FEM simulations and compared to experiments. Figure 7.13 therefore illustrates that there is a homogeneous current density distribution across the entire area in the rectangular solder geometry of the basic shape when an electrical voltage is applied, whereby maximums are identifiable in the areas around the electrical contacts.

In order to prevent the pad melting on one of the contact pins, it is made smaller in the area around the optical components whereby a radius is introduced in order to reduce voltage surges at sharp junctions. The result, as illustrated in Fig. 7.14, exhibits the highest current density in the bridge and leads to a targeted fusion underneath the lens. To reduce the excessive temperatures at the contact points, the contact pins are made of copper in order to dissipate the heat generated in the contact area by the high thermal conductivity.

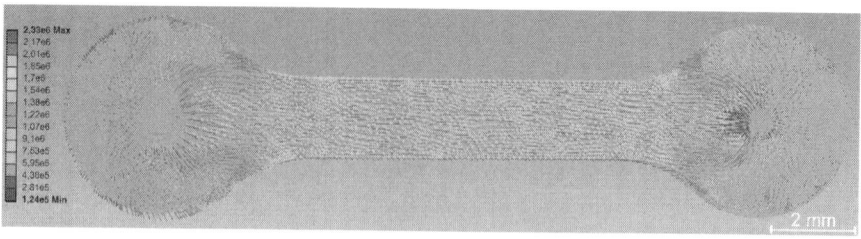

Fig. 7.14 Current density distribution (mA/mm^2) on the tin layer

A large radius may be integrated in the bridge area in order to optimise the geometry, thus defining the melting point with greater accuracy and compensating for tolerances in the evaporation process. Tin is selected for the solder as it is comparably cheap compared to tin-precious metal alloys and with a melting point of 232°C lies in the same range as them. The latter point is important as the circuit board is fitted with temperature- and photodiodes, among others, by reflow soldering with tin-based alloys before the components are soldered thus preventing the dissipated heat from melting the existing soldered connections during the resistance soldering.

Once the process design has been defined, the process control parameters are determined. Two options examined involve a specified current or a specified electrical voltage under variable settings of each of the other parameters. The regulation of both parameters, and therefore the output, is also conceivable whereby the choice of process will be discussed in greater detail below.

If the current profile is specified and the voltage changes, the increasing solder resistance at the phase transition causes the voltage and therefore the power fed into the system to increase and the solder thus overheats (Fig. 7.15).

In contrast, with increasing resistance a predetermined voltage curve ensures a fall in the current profile and therefore a self-limitation of the process. The inhibitive feedback effect ensures that even less energy is fed into the system the more material

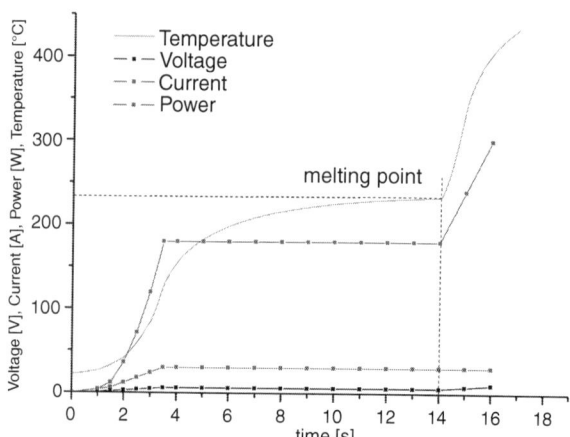

Fig. 7.15 Example process sequence under current regulation

Fig. 7.16 An example of a process sequence under voltage regulation

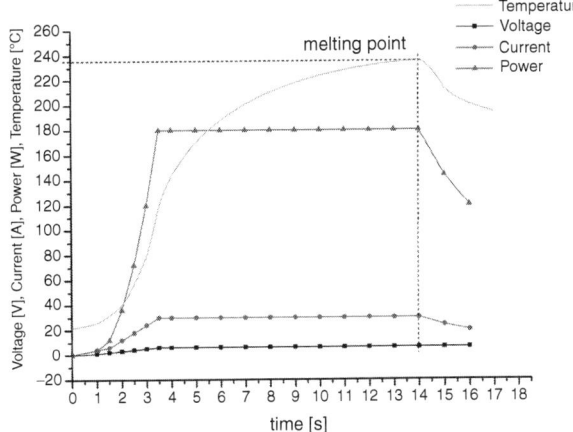

melts. It is therefore conceivable, as the example illustrated in Fig. 7.16 clarifies, that the given output is insufficient to keep the system at the soldering temperature and the solder therefore immediately solidifies again. It is possible to prevent this effect by selecting appropriate parameters which compensate for the heat loss.

The current profile may be used to identify the melting point in both instances and thus for process monitoring and control. The advantages of voltage regulation mean this type of energy input is selected and there is no need for a detailed evaluation of a power control system as this produces the same disadvantages as current control. A shut-off criterion which takes into account the gradient of the current profile is used to automate process control.

Planar Alignment As a consequence of the small thickness of solder layers together with a change in volume during phase change, the solder wets the lenses only at the points which were already in contact with the solder prior to melting. Accordingly, it must therefore be ensured that the lens is aligned in a planar arrangement on the substrate when soldering in order to maximise the wetting surface. Due to the unavoidable production tolerance of the components, each part must be individually aligned in a planar manner on the solder pad prior to soldering. The alignment may be performed by the system itself, whereby a force sensor integrated in the gripper tool measures the forces and moments when attaching and the moments are successively reduced by corresponding angle corrections.

Stress Analyses In addition to the robustness of the process, analysis of whether the joint offers long-term stability should be carried out in order to ensure that the optical properties of the system do not change over time. Due to the choice of different materials being joined it is impossible to avoid stresses on the components. The example FEM simulation in Fig. 7.17 qualitatively illustrates the deformation of the components during the cooling process.

Suhir's model allows the analytical calculation of the shear stresses and normal stresses of a three-component system along the soldered breadth (Suhir 1987),

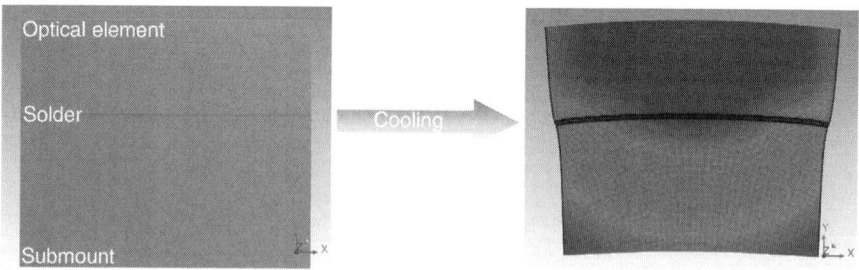

Fig. 7.17 Comparison between an unstressed (*left*) and deformed (*right*) soldered joint

whereby the simulation results confirm these values. A system consisting of quartz glass, tin and aluminium nitride (5 mm, 20 μm and 1.5 mm) is calculated for three sizes of solder in order to estimate the maximum stress. The shear stress profile along the underside of the lens exhibits an increase in the maximum stress with an increase in the size of solder, whereby the gradient flattens out around the axis of symmetry (Fig. 7.18).

If the shear stress of a specific material is exceeded, the dislocation movements within the material cause plastic deformations to occur in it. Analysis indicates that the boundary zone may be particularly affected, resulting in the formation of cracks and material fatigue. At less than 30 MPa, the calculated stresses lie below the permissible stress for quartz glass, whereby the deformations lie in the elastic range and are therefore uncritical for the material.

The "Pick & Align" Process Another process which ought to expand the limited alignment options of the "pick & join" process is the so-called "pick & align" process. A thicker solder is required to enable alignment, which is able to absorb the misalignment of the components. To that effect, the structure differs in that instead

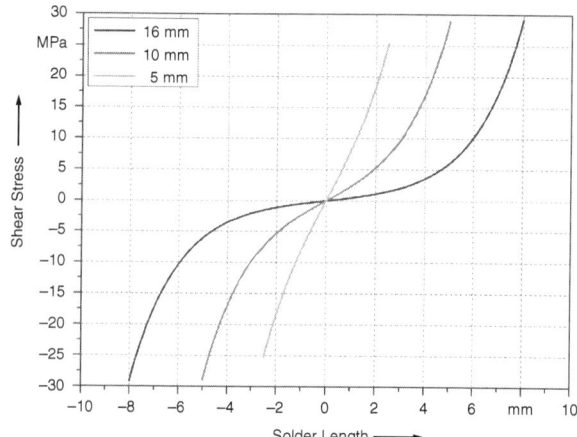

Fig. 7.18 Shear stress according to Suhir for different sizes of solder

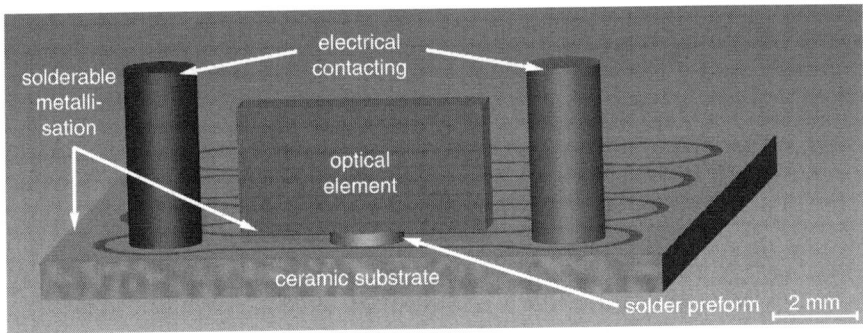

Fig. 7.19 Schematic of the "pick & align" structure

of a solder in thin-film technology, a preform such as a stamped part is used and laid between the plated ceramic and the lens (Fig. 7.19). Electrical contact is formed directly via the gold-plating, whereby in order to achieve a systematic current feed on the basis of the pick & join solder pads, a structure is created by laser structuring.

The selected solder alloy Bi58-Sn42 based on the Bi-Sn eutectic has a low melting point of 138 °C (Fig. 7.20) and a volume change at the phase transition from liquid to solid of 0.77% (Bauer et al. 2007)–an optimum condition for this process. It is particularly important that during the time-consuming alignment, the solder must remain molten, whereby the low melting point of the solder limits the heat which enters the system. As an advantage compared to glueing, it should be mentioned that once the adhesive has hardened, no further adjustment is possible, while with a soldered connection, not only is it possible to re-adjust the components by repeatedly

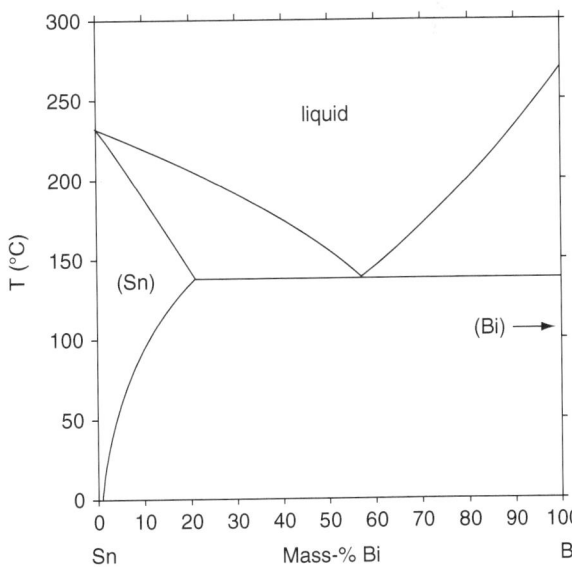

Fig. 7.20 Bismuth-tin phase diagram (Bauer et al. 2007)

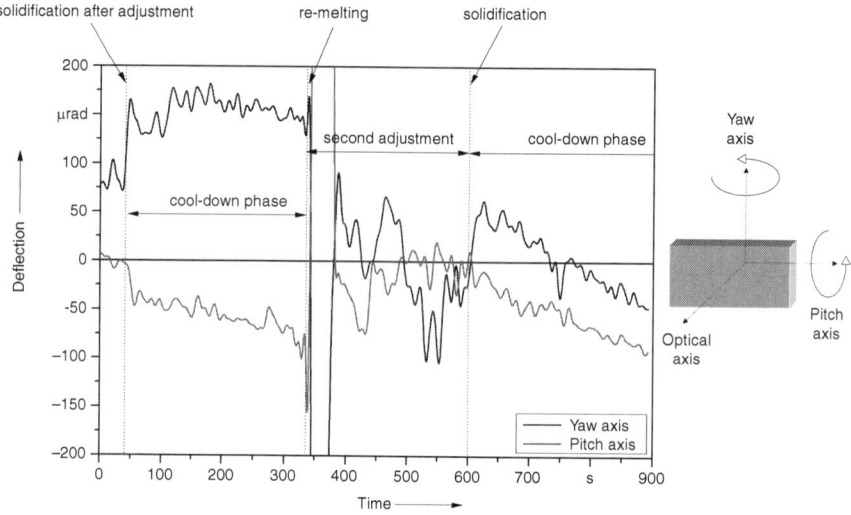

Fig. 7.21 Example of an alignment profile with repeatedly melting the solder in a "pick & align" process

melting the solder, it is also comparatively easy to disassemble a part and replace it with a new one as required.

Alignment Process The success of alignment may be determined by directly observing laser properties or by optical measuring systems determining component position. Due to the change in volume at the phase change, misalignment is unavoidable and may be compensated for by an iterative adjustment. Figure 7.21 illustrates an example of the evaluation of an autocollimator measurement of the yaw and pitch angle during an alignment process. Solidification causes a sudden change in position, which continues to change during the cooling process. Re-melting the solder allows the position to be corrected, which has reduced deflection characteristics following solidification. The need for multiple adjustments means the time expended for the "pick & align" process is greater than for the "pick & join" process, whereby this process is only used for sensitive components.

If a free deformation of the solder is used, the change in position is minimised by selecting a solder with low material parameters (volume change, melt temperature and coefficient of thermal expansion), keeping the joint clearance small and only adjusting the angle a minimal amount (cf. Eq. 7.1), whereby the last of these may only be influenced to a limited extent.

$$\frac{\Delta \beta_V + \Delta \beta_T}{\beta} = \frac{\Delta h_V + \Delta h_T}{h_0} = \varsigma - \alpha \cdot \Delta T \qquad (7.1)$$

$\Delta \beta_V$ Change in angle due to volume change

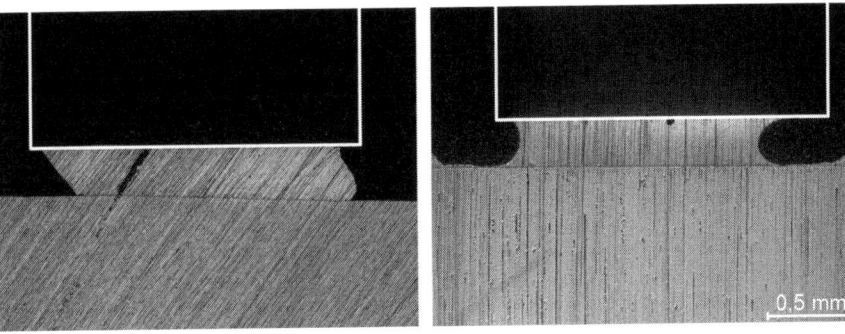

Fig. 7.22 Cross-section of solder geometries

$\Delta\beta_T$	Change in angle due to thermal expansion
β	Adjusted angle
Δh_V	Translational shift due to volume change
Δh_T	Translational shift due to thermal expansion
h_0	Joint clearance
ς	Volume change at phase change
α	Coefficient of thermal expansion
ΔT	Temperature difference

An estimation of the angular misalignment with an alignment angle of 0.1° gives a change of up to 13 µrad, whereby the actual changes in angle measured are larger by up to a factor of 10. Accordingly, the calculated angle deviations may not be calculated by this formula alone.

The uneven distribution of the solder may be regarded as the cause of the increased deviations. The manufactured cross-sections of soldered specimens with angular solder exhibit variable wetting angles which are caused by the displaced soldering material. Figure 7.22 illustrates an example of the different wetting properties of two soldered connections.

FEM models of the area around the cross-section are constructed to compare simulation and experiment. Depending on the situation modelled at room temperature, a mechanical simulation based on a change in temperature from −40 °C to 60 °C gives a tilt of the mirror of up to 156 µrad. The simulation thus confirms that the final geometry of the solder and the material combination are relevant to the direction and magnitude of misalignments.

7.3.4.3 Assembly and Alignment Strategies

Alignment and assembly strategies play an important role in the assembly of micro-lasers. The sensitivity of the system to disturbances, be they environmental conditions, component tolerances or deviations in component positioning, requires the laser to be aligned precisely to the intended operating point. In this context, the alignment of the pump optics to the laser crystal and the alignment of the laser resonator are particularly significant.

In order to make a thorough selection of the alignment or compensation elements it is necessary to investigate the effects of alignment on multiple requirements, for example the laser output power and beam quality. Alignment elements which mainly influence one requirement and are as independent of other parameters as possible offer particular advantage. In time-critical alignment processes, one element with influence over several requirements may also be beneficial, although independent alignment parameters are also an advantage in this regard. The subdivision of the pump optics into elements for the beam formation along the slow and fast axes already takes this into account during the design phase.

Developing assembly strategies in a targeted manner requires a tolerance analysis to uncover the interrelationships between the system parameters. The alignment constraints must also be considered. This affects the limitations placed on practicability due to the available measurement signals, for example, or the measurement accuracy and the associated precision of the alignment.

For the tolerance analysis of the pump optics it is advisable to use ray-tracing software to simulate the optical performance. This makes it possible to determine both the sensitivities to individual parameter deviations as well as statistical analyses (Monte Carlo Method). In this type of tolerance analysis, all the system parameters are varied randomly according to predetermined probability distributions of the tolerances, thus generating a series of random systems. The experiments performed on the micro-laser indicate that the position of the pump line on both the horizontal and vertical axes is particularly critical. While the deviation in the vertical position may be attributed to centering errors of the modelled components, the pump diode and the crystal, the horizontal deviations are caused by the alignment of the micro lens array and the associated focusing lens.

Compensation for the deviations created by an alignment requires the measurement of the position of the pump line in the crystal, which may be performed using a camera. However, a useful measurement is only possible when both the micro lens array and the focusing lens generating the pump line are brought into the beam path. In this respect there is therefore a considerable limitation which must be taken into account when selecting the assembly sequence. Depending on the available systems engineering and position accuracy, it may be more useful to first position the micro lens array and then align the focusing lens, or vice versa.

In order to reduce the complexity of assembly planning and to allow flexible solutions depending on the constraints and specific requirements, it may be possible to support the process with automation. The use of multi-agents using an approximated functional description of the system makes it possible to devise suitable compensators and possible alternative courses of action for each specification. The search for an assembly sequence which describes the sequence of positioning, measurement and alignment, and maximises the fulfilment of the objective with minimal effort, may be supported by the self-optimising processes of the software agents (Morasch et al. 2009).

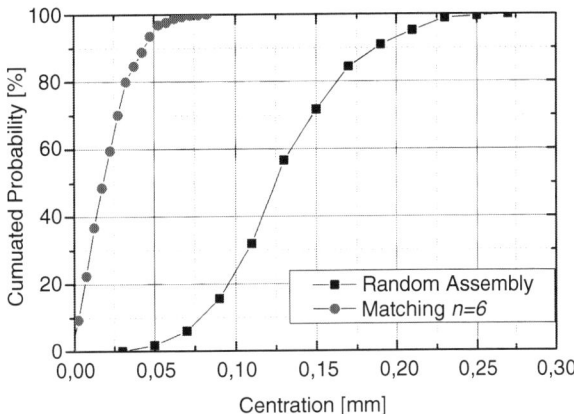

Fig. 7.23 Simulation of tolerance matching (Funck and Loosen 2009)

Tolerance Matching Apart from the significant advantages of planar technology, planar structural design technology and the "pick & join" process cause some limitations which must be taken into consideration when designing the system and the assembly strategy. Attaching the components to the substrate and the subsequent soldering using thin layers reduce the degrees of freedom. The remaining ones are the two tilt angles and the vertical direction to the substrate. Centric incidence of pump light in the crystal, crucial for the formation of a symmetrical laser beam, may only be realised using tight component tolerances. An alternative would be to apply tolerance matching, a type of selective assembly. Selective assembly is a method widely used in other industries to raise quality and lower manufacturing costs in large volume production by avoiding tight tolerances (Warnecke 1996). In contrast to selective assembly, tolerance matching does not sort components into tolerance classes. It requires characterisation of components, selection of components which suit one another as well as precise handling and assembly. The diverse tasks require the close collaboration of the participating research institutes in order to be able to implement tolerance matching in the automation cell. The positive effect of tolerance matching and selective assembly in general is based on the compensation of errors caused by deviations of component properties.

The selection and classification of suitably decentred lenses with regard to the laser diode and the crystal may lead to a significant reduction in the centering deviation of the pump line to the crystal, even for small quantities.

Simulations of this so-called matching process are based on measured or assumed probability distributions of the component tolerances. In this instance these are: the vertical location of the laser diode's emission, the centering of both imaging lenses and the distance between the centre of the crystal and the ceramic mounting plate of the laser. Figure 7.23 clearly illustrates that even during the assembly of a small series of six laser systems, a significant increase in quality may be expected (Funck and Loosen 2009). The assembly of larger quantities would increase this positive effect further. The small series assembly of six systems is performed in simulations

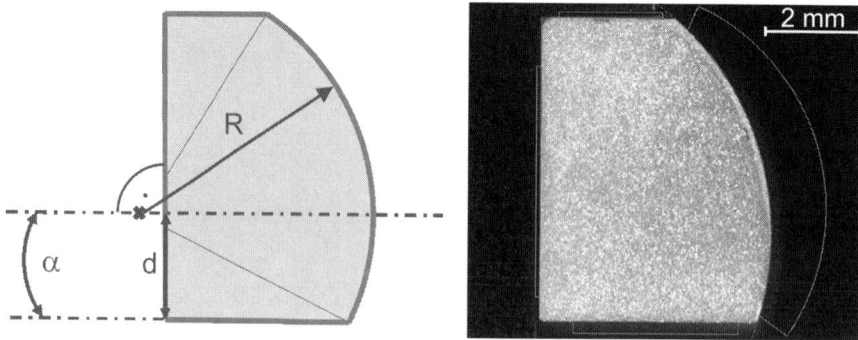

Fig. 7.24 Ideal lens with parameters and camera image

and repeated 100 times in order to generate a statistical statement. In the Figure, the cumulated probability of the worst case of every series is plotted and random assembly given as a benchmark.

Tolerance matching may be used to make expensive components—such as the laser diode and crystal—usable without tight tolerances by selecting suitable components of lower price that can compensate for tolerance deviations. The simulation is based on random systems obtained from Monte Carlo analyses, taking component classification into account. A software tool which may be universally applied to optical systems is developed to calculate the increase in quality. The simulation enables forecasts of the expected assembly results by means of statistical analysis and makes it possible to analyse arbitrary quality criteria.

In order to compensate for the fluctuations in the vertical position of the pump light on the laser crystal, it is necessary to determine the height of the laser crystal, the height of the laser diode and the centering of the cylinder lenses. After characterization suitable components are selected and assembled. The centration of lenses is defined in DIN ISO 10110 as the deviation of the cylinder axis from the mechanical axis of symmetry. The lenses used in the laser are plano-convex but do not have the typical geometry of a cylinder lens as they are not symmetrically centred.

The schematic in Fig. 7.24 illustrates the parameters of the ideal cylinder lens. The optical axis is vertical to the planar side and goes through the centre of curvature of the second surface. The distance d corresponds to the centering of the lens. It is given by the distance from the point of intersection of the optical axis on the planar surface to the bottom surface of the lens.

Characterizing the centering of the cylinder lenses is accomplished by an image processing measuring station recording the cross-section of the components. The side surfaces are observed with a CCD camera equipped with a telecentric measurement lens and the contour of the lens is determined by edge detection. The recorded images may be evaluated using scripts with a comparably small amount of effort and the data thus obtained may be used to calculate the centering. The scripts record the entire contour of the lens and map the optical axis as a connection between

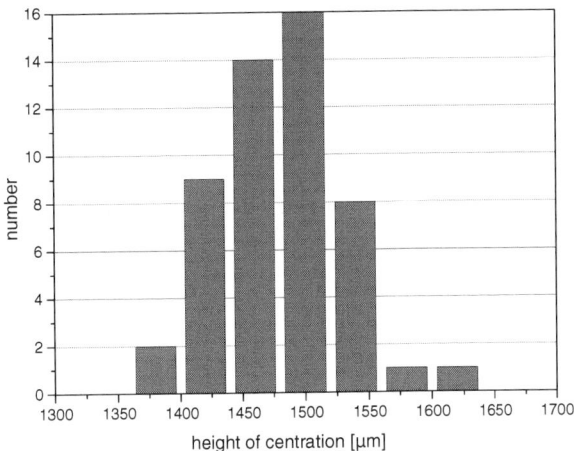

Fig. 7.25 Histogram for the centering of a MicroSlab lens

the centre of curvature of the convex surface and the normal to the planar side of the lens. Measurement uncertainty may be determined during the calibration of the measuring station. As it is not possible to guarantee the exact positioning of the lens in the measurement setup, calibration is indispensable and is performed using a distortion target in the test setup.

The coating on the underside of the component required to solder lenses to the ceramic carrier as well as the roughness of that surface influence measurement uncertainty as they influence edge detection. Measurement accuracies of ± 5 μm are determined for the centering of the lens with a greater curvature.

Measurement results of a series of components can be used to construct a histogram which illustrates the distribution of the centering and forms the starting point for the simulation of tolerance matching. The histogram of the centration of lens B (Fig. 7.25), has a mean value of 1480 μm and a variance of 47 μm. At 1500 μm, the maximum of the distribution lies on the nominal specification value. More than 90% lie within a tolerance of 100 μm deviation.

In order to be able to perform tolerance matching, the height of the laser crystal and the laser diode must be known in addition to the centering of lenses. These parameters are therefore also determined prior to assembly. The distance between the centre of the crystal and the ceramic carrier plate, and the emission level of the laser diode must be determined. This is also performed by digital image processing.

Figure 7.26 illustrates images of the crystal and the switched-on diode. It is possible to identify the radiation of the individual emitters.

Component selection is based on simulations of different component combinations. In doing so, it is important to determine the optimal selection for an entire production series (or small batch), taking the quality of all the resulting products into consideration. Experimental verification of the positive effect is only possible if a statistically significant number of production series are assembled. To avoid this,

 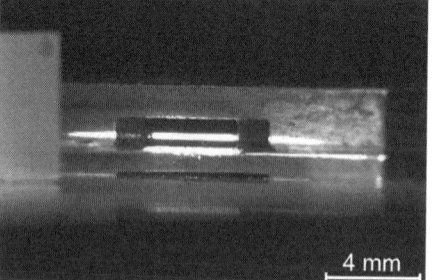

Fig. 7.26 Images for determining the mean height of the laser crystal (*right*) and the emission level of the diode (*left*)

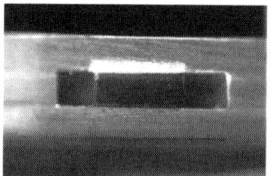 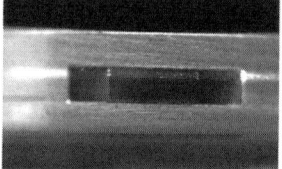 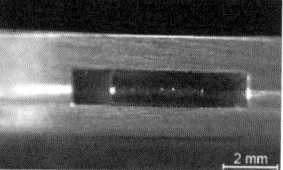

Fig. 7.27 The effect of different lens pairings on the position of the pump line relative to the laser crystal

the modelling accuracy and forecasts made by the simulation are examined for a variety of combinations, in order to be able to draw conclusions regarding the simulation results and to identify any possible influencing factors. The position of the pump line in the crystal is conveniently observed via the resulting fluorescence. Figure 7.27 illustrates the pump line as fluorescence in the crystal and the reflection of the stray radiation on the crystal housing. A heavily exaggerated decentering is depicted for better visibility.

Figure 7.28 provides a graphical comparison of measured pump line centrations and predicted values from simulation. For the depicted combinations it may be concluded that measured and calculated centration show good coincidence. However, the difference between measured and calculated values may be up to 50 μm. The measurement uncertainty of the calculated decentering of lenses may be statistically determined and amounts to ± 9 μm.

The remaining deviation may either be attributed to errors in the measurement of the emission level of the laser diode or the centre of the laser crystal, but also to insufficient edge detection when evaluating the pump line on the crystal. This therefore requires further analysis. The results indicate nonetheless that it is possible to achieve a clear reduction in the tolerance errors.

Tolerance matching may be universally applied to optical systems and may be simulated for any optically calculable lens. Further application possibilities include the classification of micro lens arrays, for example, and focusing lenses for setting a more constant pump line width or an alignment to the crystal breadth for optimising

7 Integrative Business and Technology Cases

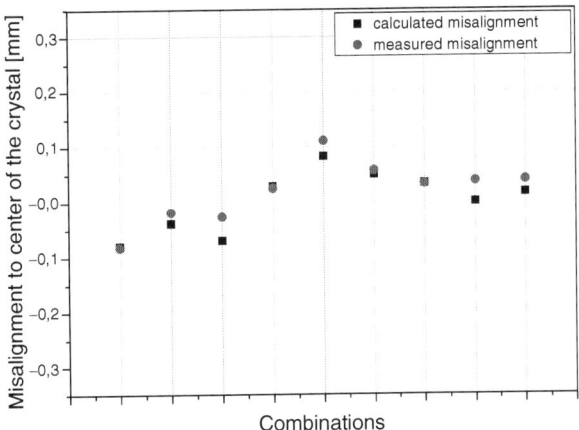

Fig. 7.28 Pump line to laser crystal centration for different component combinations

the degree of efficiency of the laser. Minimising required alignment of resonator mirrors is also conceivable. The compensation of centering errors is not limited to correcting the image position. Aberrations may also be controlled and the quality of images increased.

Resonator Assembly Assembly and alignment of the resonator are among the final stages of product creation and are also the most crucial for the beam quality and output of the laser. The alignment of the resonator is also sometimes strongly influenced by preceding assembly steps. The setup of the pump optics, the direction and position of the pump line and the laser crystal all significantly determine the subsequent alignment status of the resonator. When assembling the resonator, it is not possible to assume a constant starting position, despite maintaining the tolerances of the pump optics. The detailed setup of the resonator, i.e. in the context of the alignment adjustments, therefore always differs from one laser to the next. The absolute positioning of the resonator mirror at the ideal position by the robot in the final assembly step is therefore unable to guarantee that the laser will be brought to resonant amplification so that the laser initially remains off. This means there is no control signal required for an alignment. This situation may be addressed by one of various strategies. Either the alignment search area is extended by presetting new target positions for the resonator mirror until a control signal is found, or an alternative control signal is generated, which represents the optimum alignment of the resonator, or at least approximately. Aligning the resonator mirror to a common optical axis is a good starting point for further fine adjustments. The alignment of the mirror based on the reflected radiation of a collimated laser beam serves well and achieves a good preliminary alignment. It is then possible to use the laser output power and beam quality as control variables for optimising the alignment status (Dolkemeyer et al. 2010).

The sensitivity of the resonator is indicated by the angle positions of the mirror in particular. Deviations in the range of μrad must be set by the alignment and also maintained following the joining process. In order to achieve such precision, a fine

Fig. 7.29 Fine adjustment tool for connecting the robot system for the assembly

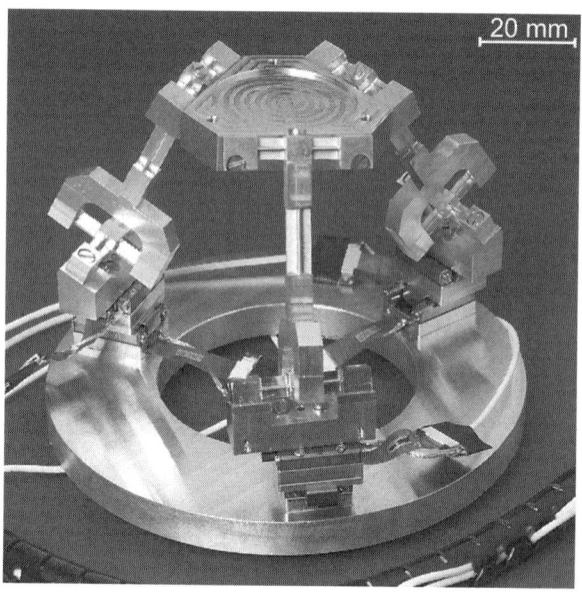

adjustment tool is used, which is guided by a jointed-arm robot. The conflict between positioning the components in the macro range and an alignment in the micro range are overcome by the combined used of the robot and the fine adjustment tool. While the jointed-arm robot may be used to position within an operating range of several 100 mm, it is possible to achieve an alignment with sub-micrometre precision in an operating range of just one cubic millimetre. With the resonator alignment in particular, the components are initially positioned using the jointed-arm robot, the position of which is then determined in order to perform a subsequent fine adjustment. The fine adjustment tool illustrated in Fig. 7.29 achieves a high level of precision by using solid joints and precise piezo motors.

The active resonator alignment is performed at the end mirror and, due to the changing baseline conditions and the self-adjusting control, is a self-optimising process which speeds up the search for the optimum alignment. Figure 7.30 illustrates the experimentally determined sensitivity of the resonator compared to a misalignment of the mirror and clarifies the required high level of precision of the alignment and joining technology. It illustrates the variation in output depending on the tilt angle.

The values of an example alignment presented indicate the start position and an interpolated performance distribution based on 30 measured values depending on the two alignment angles. It is clear that due to the crystal geometry of the MicroSlab, which has a width-to-height ratio of 4, the alignment of the pitch angle is more sensitive to a similar order of magnitude. The algorithm correspondingly corrects successively the critical pitch angle to the local maximum output and afterwards the yaw angle to the global maximum.

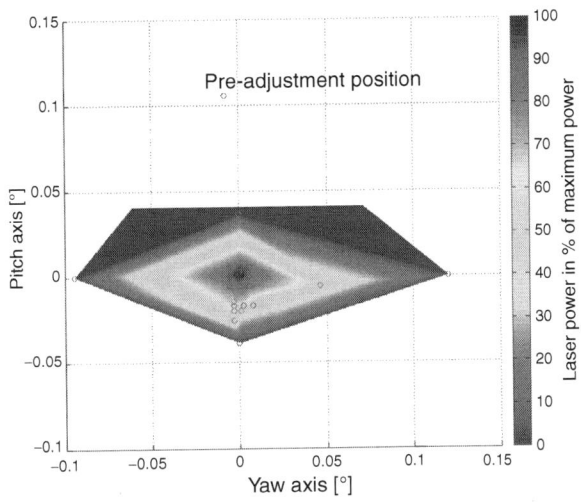

Fig. 7.30 Interpolated performance data from the resonator alignment

Dynamic Compensation Dynamic changes in the product properties due to modified environmental conditions or operating parameters, for example, may not be absorbed by a robust design in all instances, and must therefore be compensated during operation. The emitted wave length of the laser diode, for example, may therefore depend on the operating temperature, and the absorption of the crystals on the pump wave length. Although the operating temperature may be dynamically adjusted by the respective cooling control circuit, this results in strong variations in the temperature of the crystal in the initial phase and this effect is superimposed over the effect of the thermal gradient in the lens. Thermally induced changes also occur during the user-defined modulation of the power output, whereby the resulting effect of the thermal lens leads to a dynamic change in the length and radius of the midsection. During the marking process, however, the invariance of the laser spot size is a prerequisite for a good quality marking. A static compensation of the thermal lens, for example, is possible by using a suitable laser crystal shape or an adaptive lens. However, this only applies to one operating point and is therefore not a solution to the problem. The dynamic influence of the thermal lens may be minimised by a good and symmetrical heat dissipation from the laser crystal, although it is not eliminated completely. This limits the range of operation. In laser marking, this means that the range of materials which may be marked with the same system is extended, for example. An active compensation removes this limitation. An active compensation module, consisting of a three-component telescope and a beam characterisation unit (Fig. 7.31), may be used to keep both the focal position and the spot diameter almost constant. A small proportion (< 1%) of the operating beam is therefore deflected onto the CCD sensor. Suitable image processing algorithms may be used to extract the beam dimensions. This information is used to construct a control circuit which determines the position of the central telescope lens by means of a real-time computer. This also keeps the

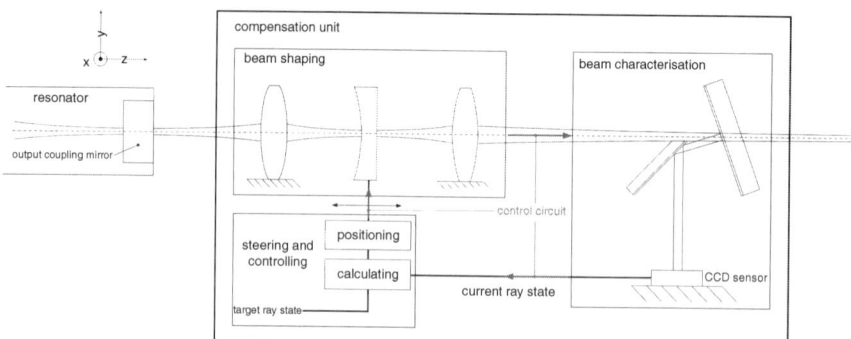

Fig. 7.31 Compensation unit thermal lens

Fig. 7.32 Prototype of the compensation unit thermal lens

size of the spot on the workpiece constant, so that a high quality marking is achieved on different materials at different outputs.

The structure of the compensation unit is therefore designed according to planar technology and the optical components may be assembled according to the planar structural design technology developed for the laser assembly (Fig. 7.32).

7.3.5 Industrial Relevance

The economic environment of industrial companies has undergone a dynamic change. Current trends, highly characterised by customisation, are overriding the strictly separate generic competition strategies of (Porter 1980) and are attempting to lead

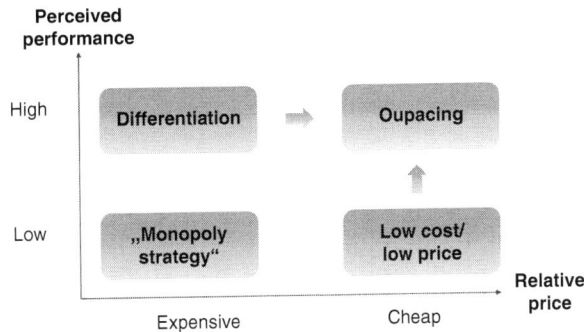

Fig. 7.33 The competitor strategy system according to (Hungenberg and Wolf 2007)

on price while following a differentiation strategy (Fig. 7.33). Such hybrid strategies, referred to in the literature as outpacing strategies (Gilbert and Strebel 1987), provide, on the one hand, the benefit that the offered pricing advantage leads to an increase in demand and thus the use of learning curves and the benefits of scale in production. On the other hand, the broad range of products is strengthening the market position thus creating a barrier to substitutes and potential competitors (Hungenberg and Wolf 2007).

Current buzzwords such as mass customisation and open innovation serve to underline the need for industry to follow these trends. However, implementing the desired market position is inevitably linked to overcoming the barriers. Leading on cost therefore requires an unavoidable standardisation of production and its environment, while differentiation requires a customised product range tailor made for the customer. A consequence of these developments is that numerous companies will be confronted with the problem of overcoming the incongruity of simultaneous mass and customised production and therefore competitively managing the balancing act between large quantities and a high level of product variety. Resolving the resulting scale-scope dilemma may only be achieved by a high level of flexibility, which although relocated centrally in production, also includes upstream steps, such as product design and scheduling.

The continuous increase in possible applications of the laser as a tool and the ever growing laser technology market share means these trends and issues are being encountered more often in this industry. Almost half the globally produced industrial lasers are used for marking and engraving tasks in particular (Fig. 7.34). This therefore provides a good starting point for the customisation and automation of what have up until now largely been manually produced lasers.

This increasing number of applications is also mirrored in the product range of the manufacturers, who currently offer a highly fragmented range of products covering marking and microprocessing lasers. Since the development of the laser, an exceptionally large number of potential realisations and types of lasers have also been designed which are all justified for their respective fields of application. There is no such thing as a universal laser, which is why there is such a diverse range of

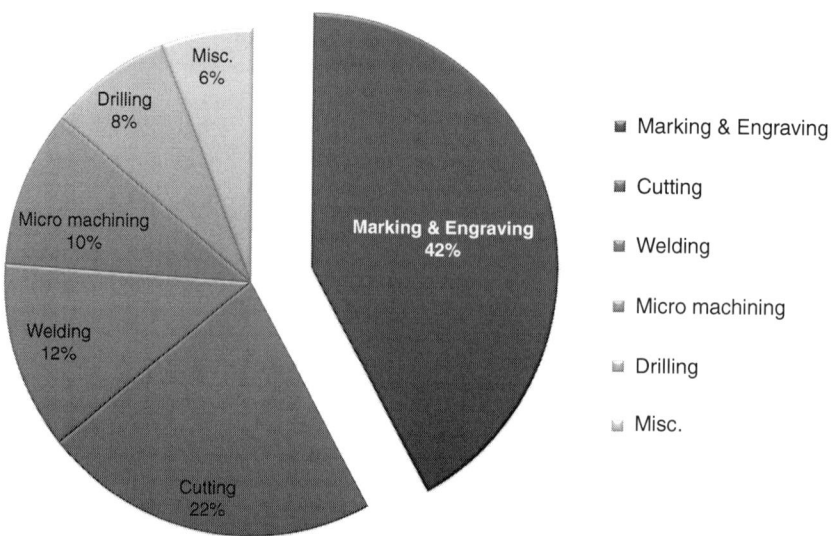

Fig. 7.34 The global industrial application of lasers in 2008 by the number of units (Belforte 2009)

laser types available today, from solid state lasers to gas and dye lasers and fibre lasers.

In recent years therefore, miniaturised systems have undergone particular advancement which is having an effect on all types of laser systems. The comparison of low output solid state laser systems shows that there has been a continuous decrease in the size of the products in recent years. Assembly technologies which rely less on mechanical supports or direct joining processes, such as the one used on the laser system in Fig. 7.35, are being used increasingly more often.

Bonding of optical components in particular is a commonly used technology for components under low thermal stress. There are various reasons for this trend towards miniaturisation. A reduction in costs may also be observed due to the availability of value for money fibre lasers, which are particularly cost effective to manufacture due to their product architecture. This is based on the fact that the alignment and assembly tasks are transferred to the production process. Despite these advantages, the fibre laser is unable to cover all applications, particularly since its peak pulse power is limited. An additional benefit of miniaturisation and driver of the development is the integration density of lasers in machining facilities. Many lasers are thus combined in an overall system in order to increase the throughput when marking products, for example.

Mobility and flexibility are thereby crucial factors in this development and the saving of installation space not least offers the opportunity to save costs and resources. Miniaturisation increases the level of complexity involved in assembly, whereby this is becoming increasingly important in terms of being a value-adding process.

Fig. 7.35 Laser marker based on MicroSlab technology

As a significant driver in increasing productivity and reducing manufacturing costs, the industrial relevance of automation should not be underestimated. In high-wage countries, competitiveness based on pricing and differentiation in terms of high and consistent quality primarily depends on automation (VDI 2010). Therefore, in the manufacturing of laser systems, it is useful to focus automation on assembly as assembly represents the greatest proportion of added value in industrial practice along with the greatest level of complexity. While automated solutions have been successfully implemented to all intents and purposes in other areas of the optical industry, such as in the millionfold manufacture of semiconductor lasers or LEDs, the assembly and alignment of more complex and more sensitive systems is less well developed. The demand for suitable solutions is not least spurred on by the increasing level of international competition, but also depends on the requirements placed on miniaturisation of systems by automated and partially automated solutions. The production of optical systems as surface mountings, as occurs in the electronics industry, could not least also help optical technologies achieve a clear advance in the length of time required for development and therefore the capacity for innovation.

The development of a suitable product architecture for implementing automated strategies and the associated joining technologies is of great importance in an industrial sense. Mechanically and thermally robust joining processes, which may be fully and partially automated, may be integrated into today's products in a considerably more direct manner. This is why precisely such developments are of particular importance. Adjustable supports, which are prominent in many of today's applications, suffer from thermo-mechanical stresses and under extreme requirements, such as in aerospace applications, fail to provide the necessary level of reliability. Soldering technology is also suitable for such challenging applications and is very versatile. Thermal stresses and the dissipation of heat via the soldered connection are particularly useful for laser systems. Conventional mounting technologies are often undefined with respect to their thermal contact and do not really allow precise forecasts by calculations.

There has always been a close connection between the assembly of optical systems and the alignment of optical components, which is particularly unavoidable in lasers. The development of new joining technologies inevitably leads to a fundamental change in alignment processes. Components, once attached, may not be subsequently adjusted depending on the process and may also not be positioned in all degrees of freedom even prior to soldering. Further use of joining technologies will inevitably lead to the application of alternative strategies such as tolerance matching and the systematic design and desensitisation of component positioning. In precisely those areas in which a qualification of the process used has the greatest priority, standardised joining process with ever more similar process parameters will be preferred and classic alignment procedures will no longer be generally applicable. This renunciation of the repositioning of those components already attached for the sake of greater durability will have a corresponding effect on the assembly sequence, the planning of which will be particularly significant.

Modular expansion will extend the laser system by compensating for the effect of the thermal lens. This compensation is necessary due to the requirements placed on spot geometries when labelling and marking and the adjustment of the output to different materials. In addition to the expansion around this module, other functions may also be realised by using additional components (e.g. frequency conversion) on one and the same basic structure. This increases the number of product variants and reduces the effort required for assembling the variants.

Active elements for expanding the product function are already being used in laser systems to stabilise the wave length. When planning the overall laser power supply concept, particular attention should be paid to cost minimisation. By installing the electronic control system directly in the laser, cheap industrial power supplies may be used to provide the electrical power, which place only low requirements on voltage stability. The lasers may also be cooled using economical water coolers from the PC industry: these only dissipate the heat, the precise temperature control of the pump diode being performed using Peltier elements, which are also controlled by the electronic control system.

The chosen control concept with its central unit in which all sensor information converges and which controls the laser, makes the automated and self-optimising implementation of the electronic control system possible. In this respect, the control specifications are varied and statements about the functionality of individual modules

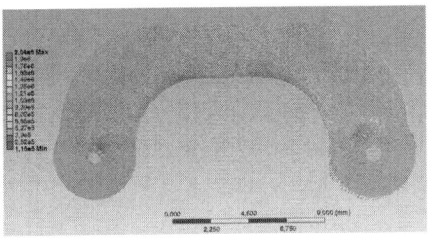 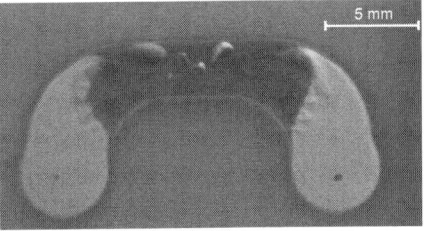

Fig. 7.36 Simulation of a curved solder pad (*left*) and a practical test (*right*)

(e.g. pump diode) or individual sensors and about the success of individual pre-assembly steps (e.g. thermal connection) are made according to the plausibility of the sensor responses. The increase in additional sensor information also makes it possible to conduct causal research into implausible sensor responses and localise the error. Calibration tables are also compiled on installation, which guarantee an optimum level of efficiency for each individual system for a broad operating range. These tables are stored individually for each system on the respective electronic control system and provide, for example, the target value for the temperature control of each operating point at which the absorption in the laser crystal is at its maximum. In the majority of conventional systems, these procedures are performed by the human operator and are therefore associated with substantial costs.

7.3.6 Future Research Topics

The automated assembly of micro-lasers has taken a crucial step forward due to integrative examination, the adaptation of products and production systems, and measurement and systems engineering. Nevertheless, some issues remain open and new issues have arisen which offer an exceedingly fascinating field of activity for future research projects.

The developed structure and the joining process are very well suited to the mounting of glass optical components on the ceramic carrier material used. Other laser components such as the crystal package, diodes and sensors have so far been soldered in successive reflow processes in advance. An expansion of the resistance soldering process to include other types of component is desirable to reduce the necessary process steps. This requires a geometrical arrangement of further solder pads on the existing substrate along with analyses of the need to adapt the packaging materials. Initial simulations have been performed to minimise the amount of lateral space required and to preserve existing ceramic geometries, which take into consideration any possible geometrical changes (Fig. 7.36). It is therefore possible to use curved structures, for example, to shift the space required for contact along the axis, thus saving lateral space. A comparison with practical tests shows that it is possible to robustly implement this in practice using existing means and requiring only a minimal adjustment of the soldering parameters.

There is a need for further research into the development of the soldering process chain. The current process is based on coating technologies carried out in vacuum processes and which do not allow much flexibility as all the solder pads are applied to the substrate using masks prior to assembly. The manufacture of the necessary coating masks in particular is both time and cost intensive and does not allow any dynamic adjustment during the laser assembly process. A greater degree of flexibility and a faster process chain from the design to the finished coating would be desirable with particular respect to the customisation of a product.

The application of layers of solder using printing technologies, as in the printing of nanoparticle layers, presents a possible solution as it allows the flexible alignment of the solder geometry and the arrangement of the optical components on the ceramic substrate. It also makes it possible to process the substrate more quickly. Printing technologies print molten dispersions with metallic particles on the basic substrate before sintering them. Lasers, among others, are also used for sintering whereby they melt the particles together to form an enclosed, electrically conductive layer of solder. Apart from the flexible positioning of the layers, the free choice of material combinations provides other freedoms with regard to the design and use of the joining technologies.

In addition to the development of the process chain, extending the soldering process to include component packages represents another useful expansion. Up until now, only materials suitable for both the joining process and for fulfilling thermal and electrical requirements could replace the packaging materials used. The package may also be electrically passified in order to be able to use resistance soldering. If such passivations are sufficiently good at conducting heat, laser diodes and crystal packages may be soldered directly onto the ceramic mounting plate and the entire laser system therefore assembled automatically.

In assembly planning, the implementation of tolerance matching and the use of desensitisation strategies are only the beginning. The adjustment of the design to the assembly strategy is given here as a further area of research. If it is possible to optimise the product design together with the assembly strategy, the level of quality may be increased further still. The dynamic use of tolerance matching and alignment can also have a positive influence on the assembly time depending on the requirement. If tolerance matching is extended yet further to include all components and modules, it is possible to achieve a completely customised and function-oriented assembly. The systematic measuring of relevant parameters may be replaced by test results from production so that only existing information is used. While today's measurement results are almost exclusively used for inspection, quality assurance and production control, tolerance matching allows them to be used in assembly as well.

The self-optimising planning of the assembly sequence, for example, by the interaction between software agents and the optical simulation, and the available means of production could only be used rudimentally up until now. The subsequent continuation of these activities should lead to an integrated solution of the design task and assembly implementation. It is conceivable that both the planning of the assembly and alignment steps, and tolerance matching may be implemented by the intervention of software agents on the optical systems specified and simulated in the design

7 Integrative Business and Technology Cases

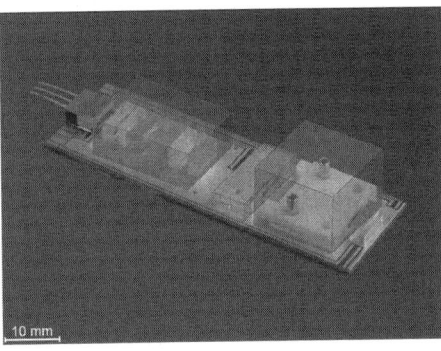 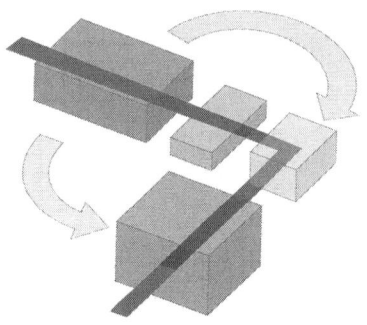

Fig. 7.37 Modularisation and self-optimising assembly

software. If the agent system measures and records the requirements and limitations on the unit, such a tool provides the opportunity to register the effects of the assembly during design and consider them accordingly.

Flexible assembly may be supported by the additional allocation of functions among subassemblies and modules, which may be assembled with less effort. In this way, complexity may be shifted to the assembly of individual subassemblies while making it possible to flexibly automate the final assembly. In order to enable such an approach, the product must be subdivided into suitable groups and the interface issues must be addressed in particular. It is not yet possible to answer the question how optical, mechanical and electrical interfaces may be realised. One interesting possibility is the use of active components in the product which allow a co-ordinated alignment after or during assembly. This idea has already been realised in terms of approaches towards the active compensation of the thermal lens, although the notion may be explored much further. Setting up the product in a self-optimising manner using individually co-ordinated active elements is therefore conceivable. This would make it possible to reduce the final assembly to the purely mechanical combination of subassemblies and components and avoid the issue of optical interfaces altogether. The extent to which such approaches may be deployed economically remains unexamined.

Tolerance matching may be extended by modularising the product according to a classification of all modules and take into consideration not only the absolute increase in quality but also the assembly sequence. The common issue is therefore the optimum configuration of modules and components in terms of time, cost and quality.

A flexible configuration of the modules could be automated according to an abstracted product description connecting the modules in a similar manner to routing in electronic manufacturing. The self-optimising arrangement of the components and modules on a platform may therefore enable an adaptation to a standard platform (Fig. 7.37).

Apart from the pure assembly, further research is needed with regard to the examination of the entire production chain, starting from material production and component manufacture. The interlinking of product design and production is reflected in both the assembly and also much earlier on in the manufacture of components and production of materials. Complex optical components in particular may be rendered suitable for subsequent use by means of an integrative examination of the manufacturing and design process. The reciprocal adjustment of design and production in the sense of a common optimisation offers the potential for reducing tolerances, thereby creating a more robust assembly. This can also lead to cost savings.

The integration of a greater degree of functionality in individual parts by means of holographic elements, gradient lenses or free-form surfaces, for example, may ultimately make it possible to combine the functions of a module in a single component, thereby reducing the effort required for assembly. The laser may also be extended for the purpose of material processing by using components for deflection and focussing. Particularly innovative concepts for small and economical deflection units are of special interest for making laser radiation suitable for flexible applications.

A micro-controller based approach is currently used for the real-time monitoring and control of the product for adapting to changing environmental conditions. Further research is required regarding the development of real-time control which also allows the self-optimising configuration of monitoring and control algorithms. This would allow the adaptation of the products to changing parameters.

7.4 Integrative Process and Product Development for Hybrid Plastic-Metal Structural Components

Walter Michaeli, Andreas Neuß and Johannes Wunderle

7.4.1 Industrial Task

In recent years, the number of complex applications demanding various integrated functionalities within one injection moulded part has been growing continuously. The increasing requirements often cannot be fulfilled by a single material. Therefore, the combination of different materials, such as plastics and metals is a promising approach for future applications. Hence, the hybrid technology has grown in importance during the last decade. The hybrid technology combines the advantages of both materials, plastics and metals, in a single product.

Traditionally, hybrid parts are manufactured in a multi-stage process. Firstly, the raw materials are pre-processed in separated processing steps. Secondly, the preforms are assembled by welding, bonding, clamping, screwing or similar to each other. Another option to manufacture hybrid parts is the overmoulding of metal inserts by an injection moulding process. A well-known example is the so called hybrid technique. Hybrid technique means the manufacturing of a hybrid composite, by overmoulding a profiled sheet metal part with a thermoplastic resin using injection moulding. In this process an interlock between metal and plastic is created through the injection moulding step. Actual trends strive for the replacement of the interlock through an adhesive bond that is enabled by a primer on the sheet metal profile. Besides to the metal profile in the hybrid composites the plastic component can support the mechanical properties. For example flat areas of the sheet metal can be reinforced by plastic ribbings. This allows the manufacturing of more resilient and highly functional hybrid composites. Typical applications can be found in the automotive sector. Automotive is representing a branch with high industrial and economic relevance. Possible examples are frontends, instrument panel reinforcements, engine enclosures, door/roof modules or roof beams. However, the described process chain for the production of such hybrid composites has disadvantages regarding the necessary number of manufacturing steps, the productivity and the achievable complexity of the manufactured parts. Regarding the achievable geometry, the process chain includes expensive and limiting process steps, for example punching, bending, stamping, positioning, assembling, and so on. Especially the combined use of punching-/bending machines implies the focus on high volumes to guarantee an economic production. Hence, it is nearly impossible to align the production for different variants, in the sense of a diversified product range (Economies of Scale vs. Economies of Scope). In addition, the production is characterised by a high planning effort. Due to the sequential order of the different production steps a precise timing is required and the integrated buffer stations need to be synchronised. (plan- vs. value oriented production). In the existing production processes, the material groups (metal or plastic) are generally considered separately from each other. Especially the processes plastic injection moulding and metal pressure die casting are handled completely isolated from each other, although the process course, the machines and the moulds show many similarities. In summary, hybrid plastic metal parts can be found in different applications in particular in the automotive industry. But there are no hybrid processes that allow a flexible manufacturing using short process chains.

7.4.2 Objective

The goal of this Cluster Business and Technology Case is the development of a guideline for the holistic and integrative production of hybrid plastic-/metal structural parts for future applications in the automotive sector. In cooperation with the industrial partner LANXESS Deutschland GmbH, Dormagen, and based on the results of the Cluster of Excellence the guideline will be developed. Thereby especially the research areas "Hybrid production systems", "Individualised production" and "Virtual production systems" are involved. As The main manufacturing concept for the production of the structural components should be the developed hybrid multi-component pressure die casting, which enables to create short process chains for plastic-/metal hybrids based on the combination of the primary forming processes metal pressure die casting and plastic injection moulding. The notion of a variant manufacturing is planned to be realised through modular concepts of hybrid multi-component moulds that can be transferred from the research area "Individualised production". Key notes for an increasing predictability and planning quality of the new manufacturing concept using simulation tools can be gained from the research area of "Virtual production systems". The documented information in the guideline for process- and product development of hybrid plastic-/metal structural parts will provide direct hints for capabilities in the development as well as particular production scenarios.

7.4.3 Solution

In the research area "hybrid production systems" of the Cluster of Excellence a new, efficient production chain for the manufacturing of plastic-/metal parts has been developed. By combining the master shaping processes metal pressure die casting and plastic injection moulding, hybrid parts can be produced in a multi-component process using one machine and one mould. By this way of production, hybrid structural components with a high level of integrated functions can be manufactured in significantly shortened process chains compared to state of the art technologies. In a first step a structure is produced by metal pressure die casting using an aluminium alloy. In a second step the structure is overmoulded with plastic elements. In the experiments conducted, the feasibility of this new hybrid process has been demonstrated. Moreover, within this research area basic knowledge is gained how to produce high performance bond strength between metal and plastic using just thermal processing, so without the exertion of macroscopic undercuts. This information is supposed to be included in the process control of the hybrid multi-component die casting in order to realise a mechanically resilient bond between plastic and metal. Furthermore, a method for the modularisation of moulds for pressure die casting has been developed in the scope of the research area "Individualised production" it will be integrated into the advanced technologies of the modularisation of injection moulds. Technical solutions for the qualification of the new hybrid multi-component process as a bulk processing method for the production of small volumes of products with varying geometry can be found. Thus, for example, the increasing variety of hybrid plastic-/metal structural parts can be taken into account. The identification of potential applications that can be produced with the hybrid multi-component process, especially for the automotive sector as representative of a branch with industrial and economic relevance, is provided by the participating industrial partners. This selection of potentially feasible components is supported through methods of the systematic identification of hybridisation potentials, which have been developed in the research area "Individualised production". To improve the planning efficiency and quality for the process and, in particular, for the properties of the manufactured hybrid plastic-/metal parts, conclusions of the research area "Virtual production systems" will be embedded. At this point, possible scenarios are evaluated at the process level in order to implement a coupled simulation using the developed simulation platform based on existing simulation tools for the separated single processes. On microscopic level, phenomena in the interface between plastic and metal are regarded. Here, for example, the influence of the still warm pressure die casted part on the later overmoulded plastic is of interest. In particular, the changes of the inner properties of the plastic are analysed (e.g. crystallisation behaviour). The influence of changing thermal boundary conditions on the inner properties of the plastic is already research topic in research area "Virtual production systems".

The integrative involvement of specific know-how from the research areas and the industry for finding solutions can be divided into four steps (Fig. 7.38). Based on the delphi-method, hybrid technology experts from the industrial partner Lanxess, who have established knowledge of current possibilities and limitations of hybrid components in industrial applications, are interviewed by an open questioning about an operationalised problem. The objective of the survey is on the one hand the determination of the basic requirement profiles for plastic-/metal parts, focused on automotive. On the other hand, the survey identifies potential future applications, which can be manufactured using the hybrid multi-component technology. Following the results of the market analysis are transferred into a standardised list of requirements.

Based on the list of requirements, subsequently the experts are interviewed in a second step.

In this step it is planned to discuss and to render more precociously the list of requirements with a total of five commonly recognised experts in the areas automotive, plastic injection moulding, metal die-casting, moulding technology for primary forming processes and simulation. The results of these interviews will be evaluated and reflected during a workshop to the respondents. In addition, the workshop is inspired by presentations from the participating research areas and by the industrial partners to gain an intensive discussion regarding the development of new potential applications of the hybrid multi-component die casting and for the derivation of specific production scenarios. Finally, the results of the workshop will be edited. Based on this, the guideline for process- and product development of plastic-/metal structural parts is generated. The guideline gives essential hints for a holistic and

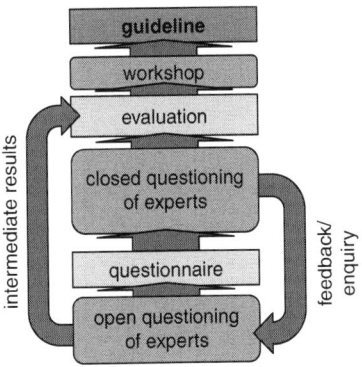

Fig. 7.38 Procedure for creating the guideline

integrative approach of the newly developed hybrid multi-component die casting for future applications. Key areas of the guideline are on the one hand knowledge about hybrid production systems and on the other hand practical approaches for implementing a production for individualised products using the hybrid multi-component technology as a mass production process. In addition to that, knowledge for verifying and optimising the simulation platform developed in the research area "Virtual production systems" is generated by using these practical hybrid production scenarios. An essential component of the guideline are the future areas of applications for hybrid plastics/metal products which gathered in cooperation with the industrial partner, respectively its interviewed experts. These future areas illustrate the relevant industrial aspects for the implementation of the new technology in the industrial manufacturing environment.

7.5 Integrated Rear Axle Drive Production for Cars

Robert Schmitt, Mario Isermann, Matthis Laass and Christian Niggemann

7.5.1 Abstract

Rear axle drives for cars are characterised by a large number of variants and a complex variety of production stages. As a result, the process must be frequently modified and the cost of planning is high. This article shows how, using cognitive methods, rear axle drive production can be made flexible, planning can be dynamically adapted to new situations and costs can be reduced. The focus is on a function-oriented product optimisation based on a flexible modification of the individual process steps during production.

7.5.2 Starting Point for Rear Axle Drive Production

A global trend shows that customers are increasingly demanding a high level of variability in products and that the products must be available quickly. This applies in particular to the car industry where customers are still permitted to change the car configuration shortly before delivery. Any modification indirectly affects other components though. For example, a vehicle with a tow bar requires a different type of rear axle drive. This means successful companies must be able to respond quickly to the individual wishes of a customer which in turn means short cycle times, flexible product planning and robust processes (Pfeifer and Schmitt 2006).

It is possible to split the customer's requirements for the vehicle into individual modules and components (Fig. 7.39). The attributes of each component needed to implement the requested vehicle functionality can then be defined.

Requirements for the vehicle:

- Sportiness   ⇒ Stiff bodywork
- Dynamic   ⇒ Less damping
- Light construction   ⇒ Low insulation
- Effectiveness   ⇒ High degree of efficiency

Requirements for the gear set:

- Minimal stimulated vibrations
- Optimal contact pattern
- Low gear loss
- Optimal adjustment specific to vehicle

Controlled parameters of the gear set:

→ Acoustics
→ Resistance
→ Efficiency

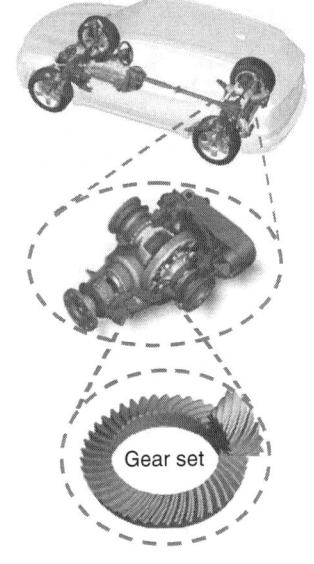

Fig. 7.39 Attributes of the rear axle drive. (Source: BMW Group)

The rear axle drive has a major influence on the acoustics inside the vehicle. It typically has a large number of variants because of its dependency on the possible vehicle configurations (Schmitt et al. 2009b). If the noise inside the car caused by the rear axle drive exceeds a certain limit, the customer may feel disturbed by it. The mechanisms which give rise to this structure-related noise are multi-layered and complex. A large number of cause and effects play a role here. A study by the BMW Group in Dingolfing discovered 193 factors in the gear set product development process which influence the acoustics (Schmitt et al. 2010a,b). The quantitative impact of the individual influencing factors on the acoustics is largely unknown.

Production control or optimisation of modern rear axle drive production processes faces two major challenges. Firstly it must be possible to control the complexity of the high level of process, and secondly a high level of flexibility must exist to cope with the large number of variants. These two challenges explain why planning methods for production control and optimisation are increasingly reaching their limits.

A large number of control variables and disturbances in integrated process chains cause a high level of complexity. With each successive process step there is an exponential increase in the number of states which the process can be in. The computing time needed to control the process likewise increases exponentially. A general lack of analytically deterministic models for the individual process steps, which describe for example distortions during tempering, means a planning system must consider every possible state of the production system and carry out the corresponding model calculations for these states. This takes an unacceptably long time with currently available computer hardware. Real-time control by planning systems is not viable in the foreseeable future.

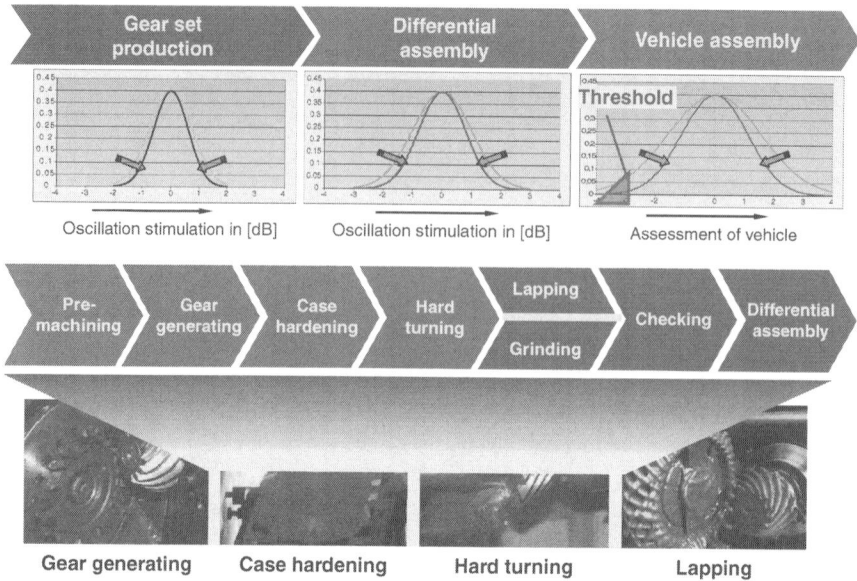

Fig. 7.40 Production of acoustically-optimised hypoid gear sets. (Source: BMW Group)

7.5.3 Motivation and Research Question

Manufacturing processes of highly technical products in particular are increasingly subject to the multi-layered dependencies of a large number of variable production parameters and the influence these have on the product being manufactured. This specifically applies to rear axle drive production as well (Fig. 7.40). The large number of strong inter-dependencies means the influence of production and assembly parameters on the acoustics cannot be fully determined (Schmitt et al. 2010a,b). This lowers process stability, and production processes must be continually optimised to achieve the required product characteristics and tolerances. The effect of these constraints is two-fold. Firstly short-notice product variations are not permitted. Secondly process designs are still not ideal as multi-level production processes in particular usually need to be optimised on a step by step basis. This leads to separate optimisations of the individual system elements, which means there is no way of fully estimating the interactive effect of these changes on the characteristics of the end product. In the case of the rear axle drive, lap process optimisation alone cannot improve the acoustics of the hypoid gear set.

A function-oriented optimisation of the end product is more effective than an optimisation of the individual production stages. This is possible if individual specifications such as final dimension can be dynamically modified during the production process. The higher level goal of product functionality can thus be achieved by considering the specifications.

Rear axle drive production should be optimised to eliminate disturbing acoustical effects inside the vehicle. The durability and efficiency of the drive should not suffer as a result. It is difficult to transfer the parameter set for a drive variant with good acoustical properties to a new variant. This gives rise to the following research question:

> How can rear axle drive production become more flexible (individualised production) while still remaining efficient, and less dependent on the human experts, who currently carry out the acoustical optimisation (self-optimisation)?

7.5.4 Cognitive Methods to Improve Rear Axle Drive Production

Self-optimising production systems are a key factor in guaranteeing competitiveness. They combine a value stream-oriented approach with a simultaneous increase in planning efficiency (Scholz-Reiter and Höhns 2007). This is achieved by transferring existing knowledge to comparable use cases. In self-optimising production systems, the actual state of the process is continually recorded and compared with its normal state (Frank et al. 2004). Depending on the calculated deviation of the actual state, a dynamic decision is made about how to modify the process, taking into consideration the specified and, to some extent, contradictory product requirements.

One way to face the challenge described in sect. 7.5.2 is to use adaptive systems based on cognitive technologies as they enjoy a high degree of freedom in finding solutions to problems. Compared with planning systems, this implies that cognitive systems do not explicitly consider and evaluate all possible alternatives. Instead they independently seek an adequate solution to a functional product requirement using a defined set of rules and acquired knowledge. Cognitive technologies allow more efficient searches in multi-dimensional and large problem spaces. Cognitive systems are capable of collecting information about their environment, processing it and converting it again into actions which change the environment (Strohner 1995). Such systems may be the core of self-optimising production (Zäh et al. 2007, 2008).

Cognitive Tolerance Matching (CTM, cf. sect. 6.2), (Schmitt et al. 2010a,b) can be used to create self-optimising production systems. The objective of such a system is to optimise production parameters for the whole production process. The result is a reduction in costs achieved, for example, by widening unnecessarily narrow tolerance bands and implementing flexible responses to production deviations which lower the rejection rate and increase production quality. For example, a different mounting dimension can compensate for a small deviation in contact pattern. In particular, production processes with a large number of variants, such as rear axle drive production, should be optimised. Where there are a large number of drive variants, a cognitive system must exhibit maximum flexibility in order to respond quickly and effectively to changes in the general framework. These include changes to both the production environment and the product configuration. In addition, the system must extract knowledge acquired from previously manufactured drives and apply it to new drive variants and process states.

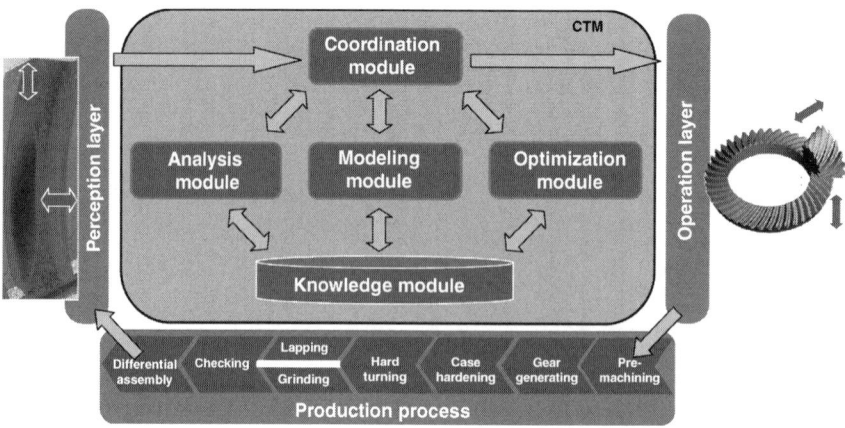

Fig. 7.41 Cognitive tolerance matching architecture for manufacturing hypoid gear sets. (Source: BMW Group)

Cognitive Tolerance Matching can be used to develop cognitive systems for process control and optimisation. Such systems deploy intelligent parameterising, optimising and matching methods in order to respond more quickly and deploy resources more efficiently than planned production systems. The quick response time is the result of continuous monitoring of production processes. Efficient resource use implies that deviations in the production process are dealt with step by step instead of modifying the whole process. In addition, planning costs can be reduced by looking for similarities in other drive variants. The idea of CTM is to intervene in the individual steps of the production process and correct tolerance deviations before the next process step is started. This applies throughout the entire production. The focus is on the final functional specification of the product and not individual process tolerances.

Figure 7.41 shows the CTM architecture in rear axle drive production using the gear tooth contact pattern a major factor in drive acoustics as an example. To interact with the production process, CTM is embedded between a perception layer and an operation layer. The perception layer comprises mainly sensors and measuring technology, while the action layer contains actuators for intervening in the production process where necessary. The example shows how a contact pattern can be determined by a single flank test machine. This involves spraying paint on the engaged gear set. The paint is forced onto the contact points giving a clear picture of contact pattern. Once a picture of the tooth profile has been created, digital image processing techniques are used to assign parameters to the contact pattern such as the core area of the contact pattern or the distances to tooth edges. These parameter values may indicate that preventative measures are necessary. For example, the mounting dimension for the relevant gear set may need to be adjusted in the axial or radial direction during the assembly process.

7 Integrative Business and Technology Cases   1047

The defined CTM architecture is located between the perceptive layer and the action layer, and is characterised by three main modules: the analysis module, the modelling module and the optimisation module. These three main modules access a knowledge module and are controlled by a coordination module.

In its current form, CTM is designed to control and optimise assembly line processes which manufacture products in large quantities. The structure of the optimisation and modelling modules in particular take this fact into consideration.

The modelling module can be implemented using artificial neural networks (ANN). ANNs also allow complex processes to be simulated which simple analytical modelling cannot cope with. ANNs should make it easier to predict the impact of parameter modification more correctly. It should be kept in mind that there are some processes which do not allow valid statements to be made using the known parameters, for example about the concavity of ring gears following case hardening. Distortions which occur during hardening are extremely complex so that discrepancies between the expected values calculated using ANN methods and the actual measured values may be very large.

To use an ANN reliably, it must be trained from the beginning with meaningful production data. Advanced training such as this however is impossible if the production process has a high level of variability. The large number of drive variants often means an inadequate sample size is available for all the relevant parameters of each variant. In this case other models must be used to do the work of the ANNs, or at least to provide them with support during the learning phase in the shape of special adaptive techniques so that the ANNs can be quickly put to use.

The optimisation module can be implemented using Soar cognitive architecture (Soar 2010). This can be used to realise a highly flexible optimiser which is controlled by both defined rules and self-acquired knowledge. At the moment, each modification to the production process requires a time-consuming adjustment of the Soar optimiser. As a result, it may not be possible to pass over the acquired knowledge as it is explicitly stored for specifically defined states. If the Soar optimiser is used in a production system with a large number of variants, its design must be correspondingly flexible to quickly adapt to a new general framework.

Soar is an open architecture used to generate programs which operate cognitively. It mainly uses a rule-based approach and independently looks through all defined rules to find a solution to a specific problem. In doing so, Soar can learn from its successes and failures by using different mechanisms to develop preferences during decision-making and to generate new rules for problem-solving.

An optimiser developed on the basis of Soar has the potential to control a production or assembly line. This optimiser can use artificial neural networks to search for the optimal parameters of individual process steps. As a result it can control the production or assembly process in real time. The developed Soar program uses reinforcement learning to benefit from the acquired knowledge. Reinforcement learning grades the decisions made during problem solving depending on whether they were successful or not. This facilitates a quick search of the problem space. Based on the successes and failures, the program learns to repeat decisions which achieved their goal and avoid decisions which did not.

Such an optimiser is tightly coupled however with the production process which is to be observed. Process changes cannot be carried out without costly modifications to the core of the optimiser. This includes any changes to both the observed product parameters and the process parameters being controlled.

A modular optimiser is needed to control and optimise production with the large number of variants, or even individualised production (one piece flow). Each individual process step requires entities which can be arbitrarily combined to create an overall process. A constitutive element is also needed to connect the individual modules together and manage the logical organisation of the overall process operation. Particular attention must be paid to clearly define and design highly flexible interfaces between all relevant modules.

Optimisation modules for individual process steps should also be implemented using Soar methodology. A process step here does not mean an individual machining job or operation but covers all the actions carried out by a machine or during a certain type of machining such as, for example, gear set cutting. If an optimisation module is developed for each type of machining, individual modules can be combined together to create individual process chains. For rear axle drive production this means, for example, the different process steps can be adapted to the gear set in question i.e., lapping can be used instead of grinding, and face hobbing used instead of face milling, welding or screwing together the differential.

Preference should be given to use an identical basic structure for each described module as this ensures the same set of rules applies to them. Individual module design subsequently takes place according to three principles:

- Each production stage is modelled
- Individual Soar set of rules
- Acquired knowledge is evaluated using reinforcement learning

Within the individual modules modelling can be based on different model types. Attention should be paid to implement a common interface. The use of individually optimised models is the responsibility of the modelling module.

The Soar set of rules is a representation of the available expert knowledge of the individual process steps. It defines the possible actions within each module and stores previous knowledge about the technical and economic context. Knowledge can be added to this set of rules at any time. Adding new knowledge narrows down the solution space and hence improves the efficiency of the optimisation. A solution can then be found more quickly during the next run.

Besides the pre-defined rules, the modules are also capable of acquiring knowledge themselves. For this purpose, reinforcement learning is directly implemented in the modules. An additional knowledge base is thus created whose contents depend on the used modules and the achieved production results. The optimiser can use the knowledge base to learn from past decisions.

A higher-level Soar program coordinates the individual optimisation modules. Its main task is to detect process deviations and faults, and make sure these deviations are compensated for in downstream modules.

7 Integrative Business and Technology Cases

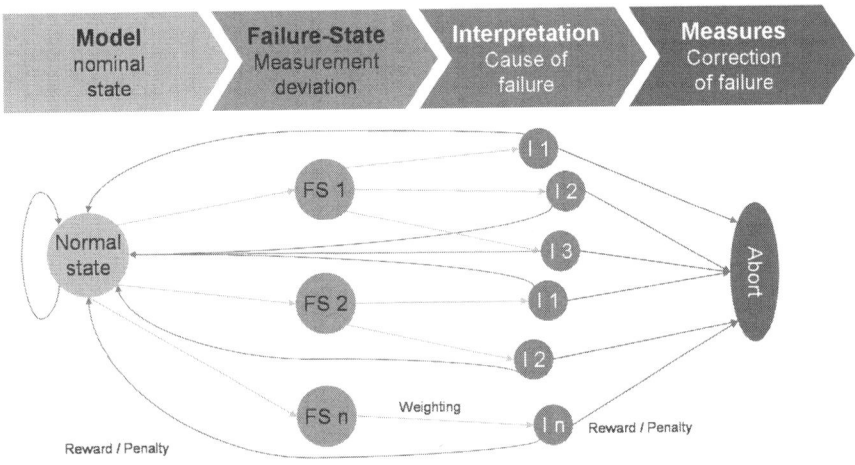

Fig. 7.42 Information flow in failure state system

Deviations and faults in the individual process steps are recorded using so-called failure state agents (cf. sect. 6.3, (Pfeifer et al. 2010)). These agents use multiple sensors to continuously record information about the production environment. When deviations from the ideal normal state are too large, they can implement the necessary corrections to the production process. The normal state is a specified point between the conflicting interests of time, cost and quality. This point can shift with new drive variants.

Figure 7.42 shows the system which identifies and solves fault conditions in the production process.

Using previous knowledge about the production process, sensors only record actual states which are of relevance. It then compares them with the normal state. Previous knowledge about the production process means only information about the relevant production stage needs to be recorded and evaluated. This knowledge is based on expert knowledge or is generated by data mining techniques. If the fixed set of rules detects an unacceptable deviation, the FS_i failure state is diagnosed. The diagnosis is explained and a search is carried out to find the cause of the deviation. The fault is solved by comparing the state with similar failure states in the knowledge base. The normal state can then be restored. Newly detected faults are stored in the database so that cognitive production system can learn about them and use them in future comparisons. The failure state agent provides a system which is similar to a feedback control system with the required flexibility. This is achieved for example by the ability to add extra components depending on the fault. Available knowledge about the process means possible failure states can be inferred for every process step. Knowledge is available about the causes and solution of known failure states. Firstly this means knowledge about drive variants can be accessed. Secondly it

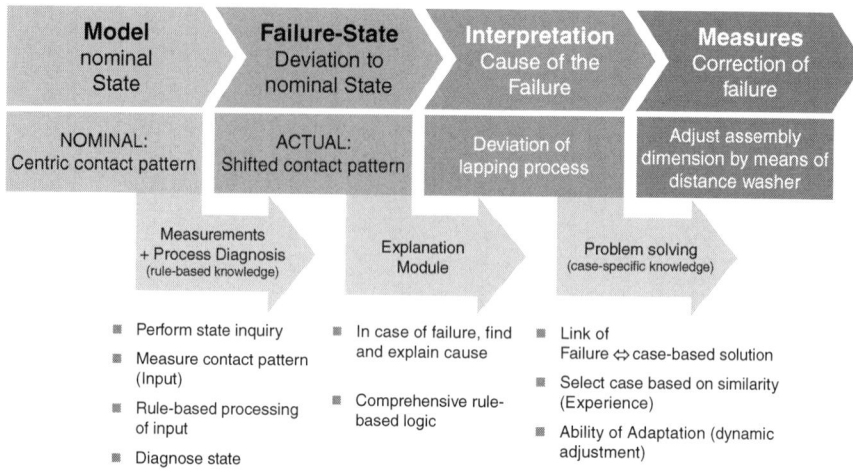

Fig. 7.43 Failure state system for rear axle drive production

means knowledge about possible failure states, causes and preventative measures is continually growing due to self-learning.

Figure 7.43 shows the event sequence of the failure state system during rear axle drive production. The contact pattern displacement which has occurred is used as an example. A normal contact pattern is specified by various parameters such as core area of contact pattern, edge distances etc. Any contact pattern displacement which occurs during single flank testing is recorded in the profile picture. If used to calculate the actual values of the parameters, it will be seen that that the edge distances are no longer correct. A failure state is thus diagnosed. Contact pattern displacement can be due to a variety of factors in the upstream process steps, for example, a change in the grinding process. As well as dealing with the process change, it should be checked if the faulty component can still be used. Once again this uses the concept of CTM. In the given example, the displaced contact pattern can be dynamically compensated for in the installation situation by using a suitable pinion and ring gear combination and correspondingly modifying the mounting dimension during rear axle drive assembly by using a distance spacer for example. This solution does not have any negative effect on the acoustics.

This approach to production control and optimisation combines the concepts of individualised and self-optimising production. This ensures self-optimisation can be implemented when there are a large number of variants so that the cost of rear axle drive production remains comparable with that of mass production despite its high dynamism and large number of variants.

The production must use a model suitably adjusted to the situation as this allows production processes to be flexibly adapted to cognitive methods and the results based on these methods to be evaluated using Soar.

As meta-modelling represents the essential features of products and processes, it should be used to transfer specific process models of a particular product to general models of complex processes, for example to the model of a manufacturing machine or an assembly process. The ideal actual state of the process must be specified. In addition, knowledge about relevant factors influencing the process must be available. This can be existing expert knowledge about the process, or information extracted from the CTM analysis module. It must be represented using IT methods so that it can be processed and interpreted by the optimiser. Knowledge about the influencing factors is also needed to choose suitable sensors for the failure state system. If required, the sensors then record the state of the relevant influencing factors. The methods mentioned above such as data mining are applied to the recorded measurement data to identify it, consolidate it into parameters and compare it with the normal state. If a deviation is unacceptable, it is cognitively interpreted and a search for its cause begins. When the cause of the deviation is found, it is corrected by the relevant actuator integrated in the process.

This procedure is effectively a control loop which compares the data of a model with the measured data thus giving the CTM software (for decision-making and optimisation) the chance to learn and hence extend the production rules used in the control loop according to requirements.

Depending on the complexity of the observed processes, a decision is made about whether to model the whole process chain as a single model, to model each process as separate sub-models or to use different modelling for different parts. If only one model is used for the modelling, it is recommended selecting a type of modelling suitable for the whole process chain; if several models are used, different types of modelling can be selected. The individual modules must be linked together via a common interface to allow evaluation of each process chain in terms of its contribution to product quality and its interaction with other process chains. Processes can be represented using both analytically deterministic models and statistic models.

Depending on the complexity and available understanding of the process, the requirements may not be fulfilled for consequent use of analytical models by every process in the use case under consideration due to the various interactions and partly unknown disturbances. As a result it may be necessary to re-generate the representation functions or re-calculate them, at some expense, when processes are modified. In view of the conflicting goals of "computing time" and "quality of results", a suitable compromise must be found for representing the production in the model. Another compromise relates to the quantity of data required to create the model. A model based on a small set of production data is not very accurate, whereby a large set of sample data is very costly and may involve elaborate testing. Added to this is the importance of testing how well intelligent mechanisms update existing models in real time (post-trained) so that the model basis can be adaptively adjusted to the real production.

The described cognitive methods should thus be used to develop and implement a select-and-evaluate logic which can be used to select the best model for any situation and the best method for updating and maintaining a sufficiently accurate model at run-time. For this purpose, a system is being developed which evaluates the respective advantages and disadvantages of the various types of models used in the sub-projects. The evaluation must be validated for rear axle drive production in terms of suitability and self-learning.

In this specific use case analytically deterministic models or artificial intelligence techniques such as artificial neural networks are considered alternatives to the model representation of production. It is important here to consider possibilities which combine different models and the specification of the interfaces.

To summarise, this solution is important to rear axle drive production because it combines CTM and failure state methodology at the meta-modelling level. This combination adds value to the knowledge base and hence allows dynamic and production-oriented modifications of the production process to be carried out more efficiently. This in turn helps to improve processes which need a large amount of planning and have a large number of variants.

7.5.5 Future Research Topics

The vision of integrated rear axle drive production technology can only be realised if the level of self-optimisation is increased compared to its current level. This should be achieved by extending the interaction between the cognitive production system and the failure state system to include a closed control loop in which knowledge acquired in one domain is exchanged with the other complementary domain. The starting point for pattern recognition is the successive execution of the actions of analysis and storage of the newly detected failure states. In this way regularities can be detected in the failure states, for example, whether a state occurs particularly often, what its causes are and which preventative measures have proven to be the best in terms of component functionality. Patterns detected in the failure states should be exported to a rule and used in the cognitive production system, as this narrows down the search and solution space. The result is a significant reduction in the time taken to find a solution to a production process based on simulation. The path from individual observations of the failure states to a generalised theory is based on an inductive approach. If the derived rule is implemented for example in the Soar simulation program by the cognitive production system itself (automatic rule derivation), the result is a self-optimising system. The rule which has been exported to the source code is then available as another option when the program runs again.

The significance of the generated knowledge can only be evaluated if it is categorised into semantic, episodic and procedural knowledge. The first category is the most valuable and generally valid form of knowledge. Using this categorisation and reinforcement learning methods, possible preventative measures can be weighted. The newly generated and implemented rule helps to improve the quality of the recommended preventative measures generated by the cognitive production system with the aim of restoring the desired normal state. This closes the control loop between the optimisation tool of the cognitive production system and the failure state agents.

The research questions summarise the work to be done in future i.e., can complex production processes use cognition to derive rules from failure state patterns, and can they automatically implement these rules themselves to achieve self-optimisation?

7.6 Integrative Levelling of Production

Günther Schuh, Arne Bohl, Sascha Fuchs, Till Potente and Stephan Schmitz

The Business- & Technology-Case, called B&T-Case, "Integrative Levelling of Production" is based on the results of the topics "Integrative Assessment and Configuration Logic for Product and Production Systems" (see sect. 3.1) and "Integrative High Resolution Supply-Chain-Management" (see sect. 6.2). The scientific questioning concerning "Integrative High Resolution Supply-Chain-Management" is about how planning and controlling quality of MRP can be increased, whereas expenditure in stochastic and individualized production systems is decreased. The "integrative benchmarking and configuration of production systems" creates a configuration logic that enables a balancing of standardization of products and production.

The B&T-Case, "integrative levelling of production" represents the transformation of the developed approach to optimize scale effects in production, industrial engineering and engineering design department. These solution components and methods were tested and validate within practice cases. Vice versa the gained knowledge finds one's way into theoretical development. For example marginal conditions for the use of several procedures of the "levelling of production" topic, which are not theoretical deducible, were derived from simulation and analysis of real corporate data. Therefore the B&T-Case is a classical research activity where practical experiences were used for theory construction by coincidentally integrating the gained knowledge into the several concepts and methods. Again this output was used for practical application.

7.6.1 Abstract

The purpose of the B&T-Cases is to configure a product-production system, which allows a high diversity of variants for the customers and at the same time broad standardized and levelled production processes with the aid of control-loops. On a basis of constitutive product features and manufacturing workload technology sequences as well as resources and tools were configured. Despite similar products and processes practical knowledge based working schedules are the reason, why potential economies of scale are not realized sufficiently. The integrative benchmarking and configuration logic enables the industrial engineering division to increase economies of scale with regard to technology, process and products. The integrative high resolution supply-chain-management raises the value stream orientation within the manufacturing control by assisting the setting of working schedules and the definition of standard times. With the aid of control-loops between the production data acquisition within the production, the industrial engineering division as well as the engineering design department the efficiency of the integrative technology sequence is increased. Figure 7.44 shows the result of the originated working schedule process.

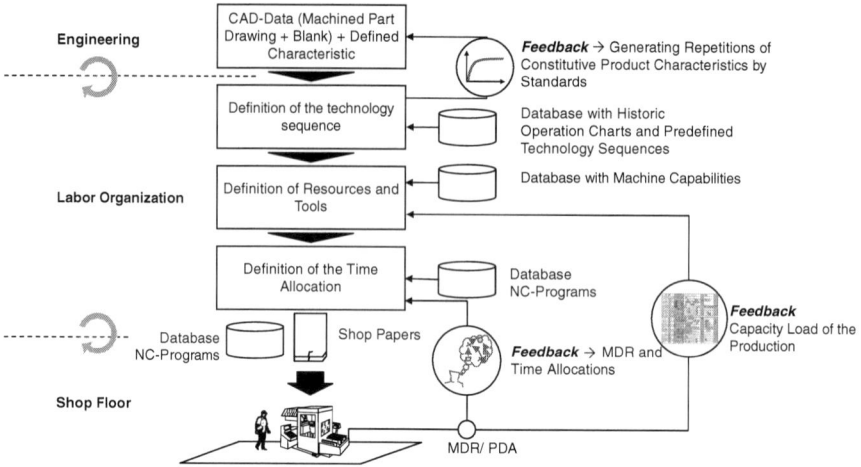

Fig. 7.44 Course of a "Learning Labor Organization"

7.6.2 State of the Art

As one can see in current practice a high degree of labor division seperates planning, realizing and controlling departments. As a consequence, single divisions within a company emerged, which are called engineering design department, industrial engineering and manufacturing. As a result of this the order process is characterized by a strong fragmentation within and between the various divisions. The handling of those orders is determined by a strong sequential method (Wiendahl 2010). The flexibility of a company is noticeable reduced because of these many interfaces between the divisions. Being settled between the engineering design department and manufacturing the industrial engineering has a central position (Eversheim 2002). Beside the engineering design department the industrial engineering owns the most influence to set up product cots. In general the industrial engineering consists of the work scheduling and the manufacturing control department. On the one hand the work scheduling defines the production sequences and creates the required working documents and on the other hand the manufacturing control monitors the progress of the production process. The work scheduling advises the engineering design department on manufacturing features and shows up technological limits of the production resources. (Eversheim 1996)

Creating working plans is the essential task of the working schedule division. Its output depends on constructive product information of the engineering design department (DIN Deutsches Institut für Normung e. V. 1984). The quality of the working schedule is based on the correctness of that constructive information. The workers make use of this working schedule to fulfil their job as well as other divisions in a company that apply this working document (REFA 1985). Within this context the working schedule faces a big challenge concerning reliability and correctness (König

1994). Due to rapidly growing needs regarding product individualization especially batch production are affected by this development (Wiendahl 2010).

From experience there are a lot of tools, which support the employees to manage the several different processes. These tools are typically using computerised support to model the operational processes. A frequently used IT-tool is called the enterprise resource planning (ERP) which connects the different divisions in a comprehensive way. In general the term ERP is used to describe the entrepreneurial task of efficiently managing the operational processes (Thome 2007). Within the context of enterprise resource planning the term manufacturing resource planning is often used. This is basically a system, which is developed from the Manufacturing planning II concept (MRP II) (Eversheim et al. 1999). The MRP II-concept is being the most common manufacturing planning system (Hellmich 2002; Hufgard and Hecht 2005; Wiendahl 2002). As an advancement of the MRP II-concept, ERP-systems not just include MRP II-features rather than integrate all operational planning tasks within a company. Nevertheless the production planning feature within ERP is based on the MRP-logic. Insufficiently supporting the interface of the engineering design department and the industrial engineering division is the key issue of ERP. Within this context especially the lack of detection of the actual manufacturing situation and the progress of production orders have to be mentioned. To solve this problem operating and machine data logging are being used. Unfortunately these systems are often implemented as isolated application, which is why integration between other IT-tools exemplarily from the working schedule division is not realized (Beckert and Hudetz 2002). Thus actuality of the working schedule is reduced and interface problems as well as a confusingly amount of data arise. To resolve this issue closed-loop control of the production connected to ERP is necessary (Schuh et al. 2011a). The organisational weak point of many MRP-systems is often described by the term black box, because it is not clear to the workers, wherefrom the planning input emerges. The system's credibility decreases due to unreliability and all too frequent changing planning inputs. As a result of this a lack of willingness to stick to planning input rises as well as relaxation of discipline to feedback (Wiendahl 2005). As a consequence of realizing a MRP-concept usually increased processing times have to be accepted (Gronau 2010). In addition to that Fig. 7.45 shows the various IT-tools mentioned and class these systems in the product engineering process.

For the design phase constructors use computer-aided design (CAD) as well as product data management (PDM) and ERP for managing numerous material masters (Spur and Krause 1997; Scheer 1990). Due to the variety of product characteristics and in view of the fact that specific product details and features are not known at the beginning of an order these IT-tools support the user insufficient. Constructors usually face the decision either to do an elaborate search for existing product models or to generate a complete new one (Kölle 1990). As long as there are precisely predefined product families present IT-solutions do perform well. However in case of reutilizing existing product models proper IT-support is rather limited.

Today's IT-tools such as CAD or ERP just offer solutions for creating, saving and managing product models (Schneewind 1994). Search facilities for existing product

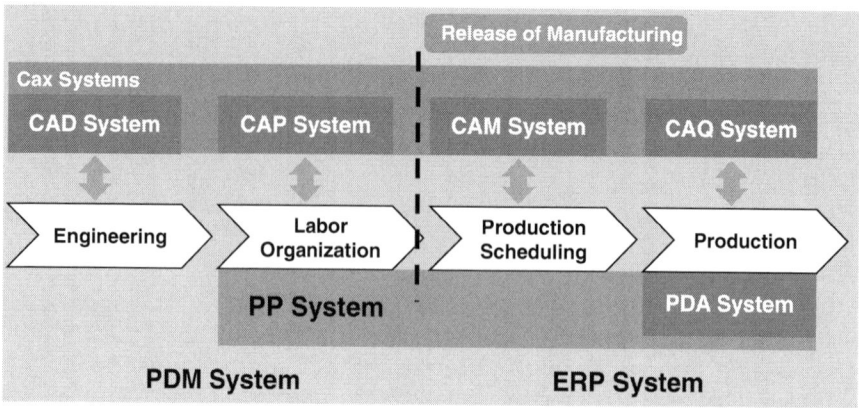

Fig. 7.45 Relationship between the DP systems in the product emergence process. (Vajna et al. 2009)

models are restricted to single IT-tools not allowing a common search within the IT-support (Kölle 1990). For example ERP allows classifying product models but only for manually entered data. Especially against the backdrop of huge product model databases this searching method is very complex. Because of operators sometimes not knowing how to exactly classify the quality and definiteness of the result is variably sufficient. Due to that fact the operator relies on his personal practical knowledge. Comfortable search capabilities within the IT-Support (ERP and CAx) for exact model geometries or particular product model features are not developed yet. In summary, it can be stated that there is a certain discharge of routine jobs through the use of current IT-support within the order procedure process. Nevertheless these IT-solutions represent isolated applications. The sequential order procedure process favours a solid structure of the company's divisions. Although single IT-solutions work fairly well, today's IT-support with its interface problems amplify these isolated applications. Looking at the process chain there is no continuous flow of information to support the order procedure process. The architecture and digital presentation of the IT-tools do not entirely fit to the operators' habits and way of thinking. Therefore many operators work with standard software such as Microsoft Excel by using excel tables handle their daily tasks. At the end the operators do work for a certain time with that tool without synchronizing it with the remaining IT-support. Doing this afterwards means a lot more administrative work for the users. In conclusion it can be stated, that operational processes are characterized by a minor transparency and a lot of interfaces between the different divisions of a company. In addition to that there is a lack of communication combined with large feedback loops. As a consequence, companies face organizational friction loss, additional expenditure in coordination and a poor system availability and utilization (Strina 1996).

7.6.3 Motivation and Research Question

Especially in mechanical and plant engineering and the small- and medium-sized batch production are confronted with the problem to handle orders with different lot sizes and a variety of different scales. In particular, the volatility of customer demands produces a high variance and dynamics in the processes. Therefore only a few procedures have been able to be standardized yet. Although the challenges in the design of the production lies less within the development of supportive tools for solving subtasks as much more in the creation of structures for the control of decision complexity in the overall context. For the problems in the production design lots of solutions already exist. However, there is still a lack of systematic approaches for an integrated design of operation charts and investment decisions considering the product complexity and an integrated configuration of a production system.

The task of process planning is to assign the right machine for each process step in terms of cost and material flow. As a connection between design and production, it thus has a central position and has a great impact on the manufacturability. As a connection between design and production process planning has a central position and great impact on the manufacturability. Downstream in production caused by such a pre-determined resource allocation, however, major performance losses can occur. The analysis of a company from the machinery and equipment industry showed that for example 8,000 manufacturing orders were produced in 4,000 different routes. This led to strongly intersecting material flows. As a result the future capacity impact of newly released orders is not predictable. Any form of planning and controlling becomes unstable. The following Fig. 7.46 shows the high number of possible routes within a production system and the caused spread of lead times.

Besides production, design and development have a significant influence on process planning and thus the production. The use of identical parts, the transfer of

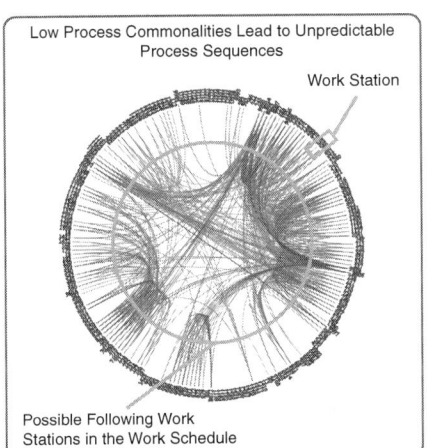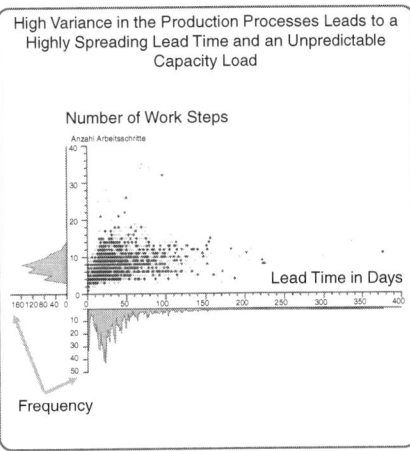

Fig. 7.46 Process commonalities and spread of lead time of a production system

variant formation in later production process steps with the use of standardized semi-finished products and the implementation of functional requirements not only have a significant impact on the component variance, but also on the number of different work schedules and technological sequences. There is still potential for rationalization, which has not been sufficiently exploited. The target is therefore to create structures which allow long-term improvement measurements and the stabilization of processes.

The success of the described Business & Technology Cases can be measured via different ways. On the one hand the potential of this concept can be seen on a lead time reduction in order processing through standardized processes in design, work planning and production. Furthermore, there are reductions in inventory, as work in process can be reduced drastically by improving production processes. This also leads to lower throughput times in production while on-time delivery can also be increased. Declining stocks by the standardization of parts and a simultaneous reduction of delivery times are the long-term potentials.

7.6.4 Results

To achieve the former described goal a concept has been implemented, which considers the previously identified departments (engineering, process planning and production) within a holistic approach. This concept is divided into several control loops with different viewing and repetition horizon. The following control loops have been built:

7.6.4.1 Short Term Production Control Loop

Especially in small and medium-sized companies with high product variance the processing times are often poorly maintained within the operation charts. The often by REFA recordings determined or estimated values differ from reality. To still have any practical usability the feedback from the production of possible deviations is an important feedback loop. This is still an open issue in reality. Modern machine and production data acquisition systems can close this loop through the feedback from work stations and the monitoring of machine parameters.

7.6.4.2 Mid Term Production Control Loop

Up to now specifications made within process planning are usually taken for granted. Changes to bills of material are often carried out only once in serious cases, since they involve high efforts. So there is no direct feedback between production and process planning which allows a dynamic adjustment of process plans based on the capacitive load situation on the shop floor to suggest alternative routes for the order through production. Similar to modern navigation systems process relevant "traffic reports" could (congestion, construction, closures, etc.) be taken into account in planning a route and suggest alternatives. These alternatives are evaluated against the possible lead time, production costs and delivery date, so that the process planner is able to make decisions. The traditional push principles, in which a pre-established process plan is followed, can be replaced by a pull-principle which generates load-dependent process plans.

7.6.4.3 Long Term Production Control Loop

Modified operational concepts in production, such as segmentation, lead to fixed predecessor and successor relationships and therefore a leveling and harmonization of production. For segmentation two main approaches can be distinguished. For complete processing the former individually carried out process steps are integrated on one machine. On the other hand the process is divided into multiple stages, which are simpler to carry out and can be processed on cheaper machines. At the same time such segmentation approaches achieve a standardization work plans, as material flows are bundled. Also approaches for leveling or harmonization purposes of process times can only then reach their full potential.

7.6.4.4 Short Term Engineering and Development Control Loop

Engineering after cost-optimal criteria is based mostly on experience of personnel and does not follow a systematic approach. Within CAD a new free-form is created quickly but later has considerable cost implications although this could have been avoided. A short control loop between engineering and the knowledge within process planning enables a cause-oriented allocation of the costs already during the engineering process.

7.6.4.5 Mid Term Engineering and Development Control Loop

The use of identical parts through standardization can be problematic for engineering. The poor comparability of existing CAD models, the lack of classification characteristics by which existing components could be found again and the beginning of a new design model based on basic bodies lead to the variety of components found today. The aim is therefore based on standardized semi-finished parts to influence actively the point of variance generation. Furthermore component classes are defined with great similarities in processing. Both leads to a significant reduction in effort in process planning and thus reduce the processing time of an order in both the direct and indirect areas.

7.6.4.6 Long Term Engineering and Development Control Loop

In the long term the goal of engineering and development is not to find a new solution for each customer demand but customize prefabricated parts. A modular product design provides the basis for this. This means that individual parts, components and pre-assemblies are incorporated into multiple end products. The central question here is how products and components have to be designed so that they match dynamically changing customer requirements. A high degree of common parts leads to a shift of the customer decoupling point "downstream" and shortens the delivery time. The upstream processes are standardized and allow a much easier run through production. Also process planning can access standards. A higher degree of automation in the whole process cycle from customer inquiry through to the start of production would be possible, which saves a considerable part of the otherwise typically long lead times.

7.6.4.7 Implementation of the Control Loops

The solution of the B&T Case "levelling of production" was divided into the following three steps: In an analysis phase the current structure of production and products was mapped. The focus was set particularly on the resources, the process plans and product variety. The aim of the analysis phase was to establish a system to visualize the complexity and the controllability of the production.

In the further course there was a segmentation of the production in control segments, which allows a self-contained approach each. In addition technology chains were investigated at the level of process plans, based on which process segments can later be identified in production (Schuh et al. 2010). The classification of products by determining the production process characteristics (e.g. diameter, tolerances etc.) also gave indications of a possible potential for standardization and the summary of product classes.

The last step was the systematic adjustment of process plans, which now divides into several stages. In the selection of technological processing sequence it has to be distinguished between standard and customer specific sequences. The know-how has to be stored systematically and provide support. The assignment of the machines based on the processing steps is oriented to the capacitive situation. Therefore the capacitive load of the resources had to be verified simulative.

7.6.5 Industrial Relevance

The previously described relationships are particularly evident in manufacturing companies that represent a wide range of products as in small batch manufacturing. Besides standard products the features offered by those companies mainly allow the customization. These adjustments to the customer requirements lead to significant adoptions of existing products or new designs (ElMaraghy 2009). For the realization of individual customer orders these companies often have productions with a high degree of vertical integration and often more than 100 production resources. Typically the resulting diverse manufacturing needs are met with a performing-oriented production structure. The processed parts are transported between the different groups and are manufactured. However, the differences in material flow prevent a simple and clear determination of the current situation in production. The organization of the job shop production is undertaken mainly by the production masters, which are responsible for the information flow within and between the workshops. A continuous flow of information between the employees of the manufacturing sectors is thus not completely guaranteed, resulting in a difficult communication of disturbances, capacity conditions, order confirmations and production know-how.

Due to an intransparent situation in manufacturing the fluctuation in load becomes unpredictable for the workers. The need of a continuous production status acquisition becomes clearly combined with the transparent display of the capacitive situation. In addition each machine operator works individually and locally separated in his field and carries out limited processes of the overall manufacturing process. Therefore exchange of employees along the value chain in production is not possible, so that existing process-specific and individual knowledge of the ma-chine operators cannot be exchanged between each other.

Between the components to be manufactured there are huge differences in the geometric dimensions. Nevertheless similarities can be found between some parts. Despite these similarities new constructions and process plans are generated due to the absence of adequate search capabilities within the ERP and PDM systems of drawings and planning documents. Because of these circumstances the designers as well as the planners usually rely on their personal experiences.

The non-continuous flow of information between engineering and process planning also manifests itself in the type of documents or media used. Digital information of CAD models from engineering are given to process planning in form of drawings and bill of materials. The planner has the task to interpret the information from these plans with the help of his personal experience and transfer it into manufacturing instructions. Figure 7.47 illustrates the function of the personnel as an interface between the computer systems.

Fig. 7.47 The human as an interface between the DP systems. (Hartmann 1992)

A two-way communication between engineering and programming based on an integrated interface is denied. Therefore there is a large difference between functional and technological knowledge between the functional areas.

7.6.6 Future Research Topics

Parts of the results have already demonstrated a clear potential for improvement in practice. However, since they could only be realized mainly through the development of the previously described short-term control loops, the future research is needed for the further development of the long-term control loops. Here, the production-driven definition of product standards is a priority. The creation of a product modularization concept based on standardized routings and process sequences allows fulfilling the individual customer requirements and compensates the load over all functional areas (Schuh et al. 2009).

On the one hand the communication and knowledge transfer within the indirect areas has to be improved. On the other hand, this must be a systematic analysis of the active relations between product and production process, which allows an insight into the impact of the variability of a product feature to the related operations of manufacturing (Schuh et al. 2011b). Based on this the necessary standards within the production and the resulting degrees of freedom defined to product variation could be derived which enables the fulfillment of individual customer requests while maintaining economies of scale in production.

7.7 Integrative Production of TiAl-Products

Georg J. Schmitz, Julio Aguilar, Ulrike Hecht, André Schievenbusch, Robert Guntlin, Günther Schuh, Achim Kampker, Bastian Franzkoch, Peter Burggraef, Jan Noecker, Cathrin Wesch-Potente and Rick Hilchner

7.7.1 Abstract

Based on lasting developments toward the production of components made of titaniumaluminides and the related application of various technologies—particularly numerical approaches being partially developed in the Cluster of Excellence—present activities aim at the optimisation of the manufacturing processes and of the component properties toward commercial applications. The actual target of the present Cluster-B&T-case is the development and the application of process planning methodologies with a view to designing and establishing a foundry for investment casting of titaniumaluminide components in Aachen, Germany.

The actual task of this B&T-Case is the application of an interdisciplinary and integrative planning concept being developed within the Cluster of Excellence for (i) planning an investment foundry for TiAl-components, for (ii) the verification of the planning results via simulation and for (iii) commercial viability studies.

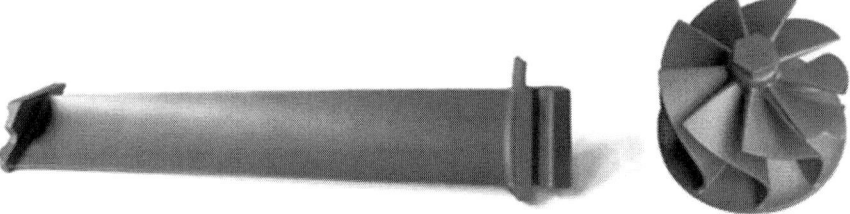

Fig. 7.48 Example of high-performance titaniumaluminide components for the commercially attractive aviation industry (low-pressure turbine blade, *left*) and for the automotive industry (turbocharger wheel, *right*)

7.7.2 State of the Art

Manufacture of high-precision components represents a key-technology in high wage countries especially for the automotive and aviation sectors. Examples are exhaust gas-turbo-chargers for the automotive sector or turbine components for airplane engines (Fig. 7.48). As a prospective material class for such components destined for the future, titanium aluminides (TiAl) have proven their potential especially with respect to weight reduction. Due to their reactivity in the molten state and their difficult machinability, the production of near-net shape components made from titaniumaluminides represents a challenging engineering task.

By now, most production steps for manufacturing and processing titanium aluminides have been developed at the laboratory scale. The extremely adverse processing properties require highly developed and sophisticated production systems comprising an elaborated process control, which, until recently, had only been available for producing a few prototypes. The market volume for TiAl turbochargers is estimated at 250,000 units per annum. The annual market for TiAl turbine blades for aircraft engines is predicted to be approximately 100,000 units.

Considering a pricing as defined by the market, new casting technologies have been developed for the production of respective quantities at target costs. Simultaneously with and independent of the activities of the Cluster of Excellence the degree of automation for single processes has been signficantly increased. Investigating a production system being suitable for mass production of TiAl components in the frame of a business & technology case represented one of the most interesting topics for all development sectors participating in the Cluster of Excellence along the value-adding chain (material sciences, plant engineering, control systems, and production management). The development of such a production system itself has been widely realised in the framework of several cluster-external projects. In many cases, however, methodologies being developed or being improved in the Cluster of Excellence provided significant contributions.

Examples are the simulation of the distortion of wax-models for investment casting toward well-defined model geometries by simulation of injection moulding (Fig. 7.49), microstructure simulations during solidification of TiAl-materials by multi-phase-field methods coupled to thermodynamic databases (Fig. 7.50),

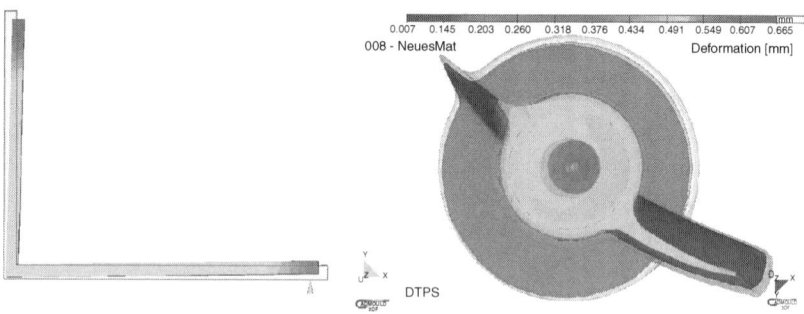

Fig. 7.49 Shrinkage and distortion simulation during injection moulding of wax-models for TiAl-turbine blades. This simulation provided a valuable insight into the selection of suitable model waxes as well as about the geometrical design of the gating system

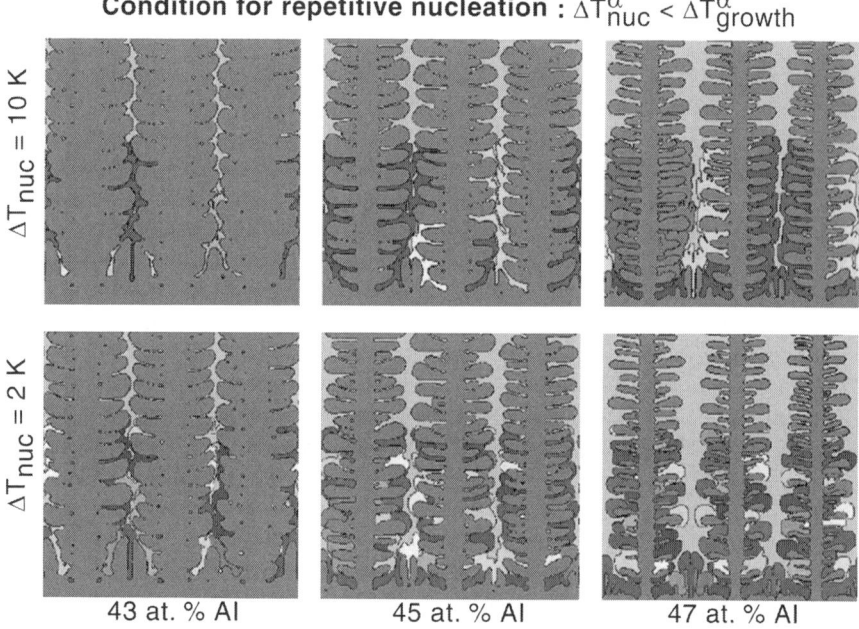

Fig. 7.50 Multi-phase field simulation of nucleation and solidification in titanium aluminides. Alloy composition, nucleation conditions and the process parameters all influence the microstructure and accordingly the properties of the component significantly

the simulation of mould filling and solidification during centrifugal casting of TiAl-components (Fig. 7.51), simulations and experimental studies of mechanical processing of TiAl-components (Fig. 7.52), the development of innovative joining technologies for TiAl-components based on gallium containing solders as well as studies of cost targeting for investment-casting of TiAl-components.

In April 2010, the interplay of all these different activities has led to the inauguration of a production scale centrifugal casting plant for the production of TiAl-turbine blades providing a capacity of up to 30,000 turbine blades per annum. (Fig. 7.53)

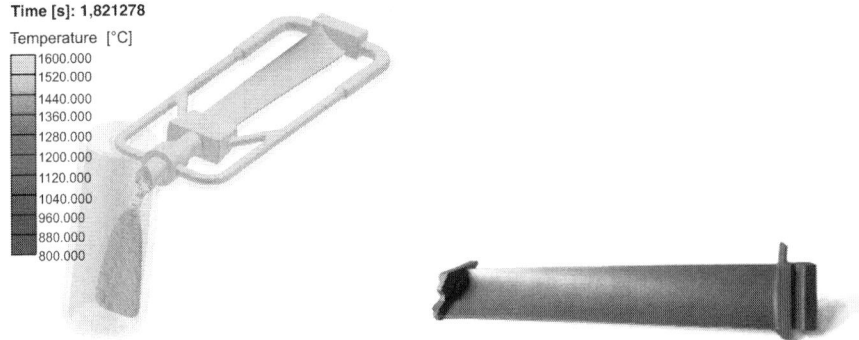

Fig. 7.51 The simulation of mould filling and solidification during centrifugal casting of TiAl-turbine blades allowed for the identification of optimised geometries for the casting system as well as a parameter optimisation for centrifugal casting processes

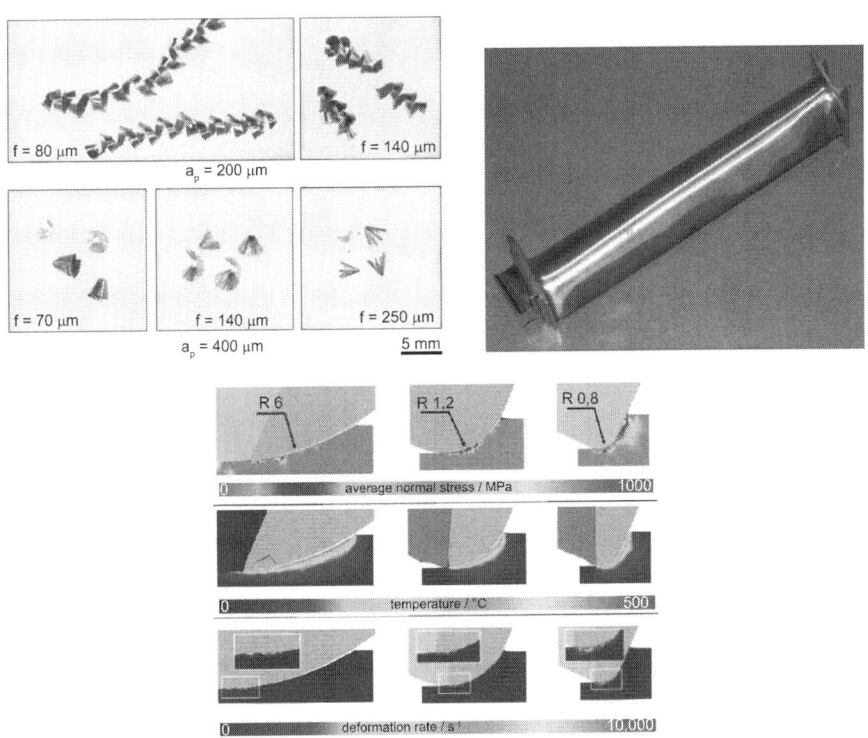

Fig. 7.52 Experimental and numerical studies of machining of titanium alloys and titanium aluminides and machined TiAl-turbine blades (courtesy Leistritz)

Fig. 7.53 The production scale centrifugal casting plant with a cold-wall-induction-crucible and a double chamber system for the production of TiAl-turbine blades being installed at Access e. V. (Techcenter, Jülicher Str., Aachen) allows manufacturing up to 30,000 turbine blades per annum

7.7.3 Motivation and Research Question

The actual task of this B&T-Case is therefore the application of interdisciplinary and integrative methodologies, which have been developed in the Cluster of Excellence for designing an investment-casting foundry for TiAl-components, as well as the verification of planning results via simulation and commercial viability studies. It provides the option to develop advanced competencies against competition, which supports Germany's Hi-Tech position and creates new working places.

The B&T-Case "Integrative factory planning for a TiAl-component investment casting foundry" in the Cluster of Excellence focuses on the "Scale-Scope" as well as on the "Plan-value orientation-dilemma"; hence, it is specifically relevant for the CoE. Such processing is possible only with interdisciplinary cooperation, based on the combined know how of manufacturing planners, material engineers and process engineers.

The objective of that B&T-Case is the concept development for industrial production of TiAl-components. The process of integrative planning shall take into account the essential requirements of manufacturability, quality, and costs. Overall, the manufacturability of TiAl-components shall be proven and the results shall be marketed as a demonstrator. Therefore, the following intermediate objectives will be pursued:

Intermediate Objective 1: Feasibility study and evaluation of current research results

The first objective of this case is evaluating the existing research results in order to transfer them to high volume production processes. Along that procedure, relevant side conditions will be recorded and analysed for potential risks.

Intermediate Objective 2: Development of a virtual factory for TiAl-components

For certifying the results, a virtual factory will be created, which contains the relevant properties and links between the particular objects of planning. The objective of the digital verification is a significantly higher quality of planning and reduction of planning efforts particularly in times of crisis.

Intermediate Objective 3: Creation of a target image and the necessary development of needs for mass production

The concept developed for this case contains along with the concept of planning, also the actual planning results, but on the other side, it also analyses the implementation efforts. This research comprises various scenarios for long-term realisation and concept advancement.

Intermediate Objective 4: Demonstration of integrative development via exchange of simulation data

The material flow will be simulated for achieving most accurate planning results concerning the capacities required and the processing times to be expected. The entry information for the material flow simulation derives from the timing that was preset by the technological study. With that measure, the use of information beyond model levels will be used to increase the planning quality significantly.

7.7.4 Results

The results of the research project derive from the following four consecutive phases:

AP 1: Analysis phase
For planning a factory, an information basis situation has to be created where the requirements and peripheral conditions for the planned production have to be defined.

AP 2: Simulation
The simulation model structure is based on the analysis results, which allows testing of different alternatives.

AP 3: Conception
Based on the results from analysis and aided by the simulation model, the processes, resources, and capacities, as well as the layout are conceptualised.

AP 4: Migration Planning and Economical Feasibility Evaluation
A migration plan has to be created for transferring the planning results, which will be correlated to a detailed financial business plan.

The relevant results per phase are outlined below:

AP1: Analysis Phase
Product Analysis The product program (product types) and the production program (quantities produced, call-off situation, and prognoses), as well as anticipated changes are analysed during a coarse product analysis. The required input is provided by Access e. V. and/or TITAL GmbH. Product clusters are formed based on the

product and production programs for structuring the production segments. Moreover, planning scenarios are created for the segments.

Process Analysis The material and information flow in the manufacturing process is analysed using reference products. To achieve this, the technical and time sequences in the production process are recorded. This includes the study of different manufacturing process chains, including the supply and delivery chains. Moreover, the throughput times (process and transfer times) are checked on basis of existing data. The above-mentioned analyses are used to assess the performance of the production sector and logistics.

Production Structure Analysis Goal of this processing step is to record in detail the current production structure. This comprises recording the production layout (layout structures) with material flows and area occupancy, as well as the existing resource capacity (operating supplies, workforce). Besides detailing the restrictions for additional planning, this study reveals the relationship between bottlenecks and structural restrictions of process design. Parallel to this, the current capacity and area availability is determined.

Demand for action In this processing step, the analysis results are collated and the need for action derived. This comprises a strength-weakness analysis and optimisation goals for the existing process organisation as well as for the production structure. In a production structure matrix, the requirements for changes, flexibility, and intervals with respect to products, production processes, and production technologies are compared, and the need for action with respect to space and capacity requirements versus availability (operating supplies, layout, and infrastructure) are determined. The goal is to determine relevant restrictions and structural potential, which then form the basis for the design phase.

AP2: Simulation

Based on the controlling requirements of production processes, direct optimisation measures have been determined and evaluated, and a concept for production control and processes of the future processing landscape were created. To prove the feasibility and advantage of a largely interlinked production, various flexibility models have been tested via process simulation whereby the effects on the process efficiency have been determined. The simulation has been executed according to the following segments:

- Definition of the relevant product and program spectrum for simulation
- Preparation of relevant production data: The WZL-project team will create an exact data definition during the analysis phase. Access e. V. will supply the relevant data.
- Definition of processing times, layer models, respectively qualification matrixes
- Creation of a regulation matrix for order inflow
- Creation of a simulation model for areas of interest and execution of process concept simulation studies for validating the process concept.

- Change of planning parameters for concept optimisation (e.g. process decoupling, layer models, etc.) for concept optimisation.
- Simulation of different flexibility models (variable daily production volume, variety of variants, process interruptions, etc.).

Different design scenarios will be created (order mix, batch sizes, order sequence, and tooling sequences). The above-mentioned control parameters are adjusted via simulation studies in such a way that the production target sizes will be optimised.

AP3: Conception

Rough Planning

- Based on the analysis results, during this processing step, a rough planning of production elements will be created. With the aid of the WZL-factory planning-ppt, the layout of particular resources as well as the related processing steps, material provisions and material flows are prepared in form of 2D area planning (block layout): Creation of different rough layout variants in conjunction with the physical layout structure of particular sectors; variant evaluation in a planning workshop conducted with management and workers of the relevant areas; selection of layout alternatives based on quality criteria.
- The material flow planning studies especially the bottleneck areas and relevant conflicts of resources, for example, due to converging/diverging material flows, will be identified.

Creation of a Digital model

- Incorporation of planning objects: Incorporation of relevant planning objects, machine drawings, buildings and process restrictions, machine sequences, as well as additional machines for expansion scenarios. As far as available, the relevant data are provided and jointly processed by Access e. V.
- Creation of a digital production model: Digital modelling of relevant objects of particular segments with consideration of restrictions; Creation of 2D and 3D-models; preparation of the factory planning table and layout workshop (creation of planned layout and first alternatives).

Fine Planning

- For fine planning, workshops will be laid out with the responsible experts. In this way, the optimised layout structures and sequences will be dimensioned along with the capacity, resource requirements, and processes.
- To visualise different layout variants for evaluation, an "Open Factory 3D" factory planning table will be used. Moreover, with aid of the factory-planning table, the operative workers can be included in the early planning stages to clear existing resentments and concerns.
- The fine layout planning will be documented in form of a CAD-layout and a 3D-film (AVI) (optional), as well as with a list of images.

AP 4: Migration

The migration planning follows the conceptualisation: Results of the previous work-packages will be integrated and migrated to a realisation (action) plan. On one hand, this plan contains all building measures for realising the planned hall layout. On the other hand, alternative factory planning does have to be created during migration planning when certain planning premises change over the time (especially product and production program). In addition, the migration planning has to cater for production commencement in the new factory as well as the necessary interim steps for start-up have to be defined for securing the scheduled production quantities.

The economic assessment is continuously conducted within the project. The objective of economic assessment is a transparent illustration of the cost-benefit ratio. The benefit of planned changes is evaluated according to cost saving potentials to be realised (for example, non-recurring effects, and reduction of operating costs). The expenditures are recorded with aid of technical investment planning. Therefore, and based on the results of the conceptualisation phase, tender documentation has to be prepared for the required construction projects, machines and systems. For the most important variables, manufacturers have to be identified and supplier discussions have to be completed. Based in this information, an investment plan will be created, which will be constantly updated and amended with the estimations in line with standard procedures.

7.7.5 Industrial Relevance

Aviation industry expects a doubling of passenger and freight volume in air traffic within the next twenty years according to Boeing and Airbus. The aviation sector, growing at rates of 5% annually since years, estimates the value of all airplanes to be manufactured up to 2026 to be 2,600 to 2,800 Billion US$. In this context, the aviation industry has committed itself to significantly reduce sound and exhaust emissions of future jet-engine generations. Significant weight reductions by using titanium-aluminide turbine components are considered as one of the most promising pathways to meet this objective. A market volume of at least 100,000 cast TiAl-turbine blades per annum is expected according to statements of leading turbine manufacturers.

Present developments in the automotive industry are targeting especially on energy saving measures and CO_2 reduction. In this context, the next generation of turbo chargers requires light and high temperature resistant materials. Titaniumaluminides have emerged as an interesting alternative to current solutions, as they allow for entirely new turbocharger designs due to their small masses. Such designs lead to an increased functionality in the sense of improved response qualities ("time to torque") and further foster the realisation of downsizing concepts being characterized by smaller engines with higher performance. For combustion engines comprising TiAl-turbo charger concepts, fuel savings of 0.4 l per 100 km are estimated. For an individual car with a total mileage of 120,000 miles (approx. 200,000 km) during its life cycle cost savings totalling to approx of 1,200 € can be realized (considering a typical fuel price of 1.50 €/l). Even if only a part of Germany's fleet of vehicles were fitted with TiAl-products, 280 million litres of fuel could be saved and CO_2-emmisions could be reduced by 6,800 tonnes. Thus, a productive and cost efficient manufacturing process for app. 250,000 turbocharger wheels per annum at a price of approx. 35 € per wheel has to be specified and qualified.

Based on the market potential for TiAl-components described above, the strong interest of all partners along the value chain to exploit this market, and the start-up of the first production size plant, the TITAL Company expressed significant interest in establishing an investment cast house for manufacturing turbine blades for aero-engines within the next few years in Aachen, Germany. Similar developments are expected for turbochargers in the automotive sector.

7.7.6 Future Research Topics

Titaniumaluminides have a high potential to become the future material of choice. The industrialisation of this material will make it an economical and technological alternative for present engine manufacture. The success of the TiAl-technology in the market largely depends on an integrative design of the entire product-production system. Within the domain "Virtual production systems" of the Cluster of Excellence "Integrative production technologies for high wage countries" respective methods and instruments have been developed allowing for a holistic view of a new technology with high reliability. Respective methods already have found applications in some externally funded projects for developing TiAl-turbine blades. As one of the major results of the Cluster of Excellence, the activities now culminate in the planning of an investment-casting foundry for TiAl-components eventually creating/securing employment in a high-wage country.

In case of a successful industrialisation of TiAl-turbine blades using the outlined methods for factory-planning, other research fields and requirements for investigating the industrialisation of this material for further products like e.g. above turbo chargers for the automotive industry, will arise.

References

Altmüller S (2001) Simultanes Fünf-Achs-Fräsen von Freiformflächen aus Titan. Diss. RWTH Aachen

Ayasse et al (2001) Interactive manipulation of voxel volumes with free-formed voxel tools. VM 2001, Stuttgart

Banse H (2005) Laserstrahllöten – Technologie zum Aufbau optischer Systeme. Dissertation, Jena

Banse H, Beckert E, Eberhardt RS, Vogel J (2005) Laser beam soldering – a new assembly technology for microoptical systems. Microsyst Technol 11:186–196

Bauer P, Bruzzone P, Portone A, Roth F, Vogel M, Vostner A, Weiss K (2007) Review of material properties, past experiences, procedures, issues and results for a possible solder filled cable as plan B conductor for the EFDA dipole magnet. infoscience | ecole polytechnique federale de lausanne

Beckert B, Hudetz W (2002) Stand und Potenzial produktnaher Datenverarbeitung. PPS Manag 7(2):35–39

Beckert E (2005) Ebene Keramiksubstrate und neue Montagetechnologien zum Aufbau hybridoptischer Systeme. Dissertation, Jena

Beckert E, Burkhardt T, Oppert T, Azdasht G (2007) Solder bumping – ein neues, flexibles AVTVerfahren. Fraunhofer IOF Jahresbericht

Belforte DA (2009) Keeping the economy in perspective. Ind Laser Solutions Manu 24(1):4–11

Brecher C, Loosen P, Schmitt R, Funck M, Morasch V, Dolkemeyer J, Pyschny N, Pavim A (2008) Flexibel automatisierte Montage von Festkörperlasern. wt online, 2009

Choi YB, Park SJ, Jeong KT, Park TS (2003) Patentnr. US20030116547

Cimatron Homepage [Online] Cimatron GmbH, (2011), http://www.cimatron.de

Davis CC (2000) Lasers and electro-optics: fundamentals and engineering. Cambridge University Press, Cambridge

DIN Deutsches Institut für Normung e V (1984) Begriffe im Zeichnungs- und Stücklistenwesen: Zeichnungen, Berlin(DIN 199-1)

Dolkemeyer J, Funck M, Morasch V (2010) Flexible Automatisierung in der Produktion der Zukunft. RWTH Themen, ISSN-Nr. 0179-079X (1/2010)

ElMaraghy H, Azab A, Schuh G, Pulz C (2009) Managing variations in products, processes and manufacturing systems. CIRP Ann 58(1):441–446

Eversheim W (1996) Grundlagen. Organisation in der Produktionstechnik Band 1. Studium und Praxis, 3rd edn. VDI, Düsseldorf

Eversheim W (2002) Arbeitsvorbereitung. Organisation in der Produktionstechnik Band 3. Organisation in der Produktionstechnik, Walter Eversheim, vol 3, 4th edn. Springer, Berlin

Eversheim W, Bleicher K, Brankamp K, Bender K (1999) Produktion und Management. Springer, Berlin

Frank U, Giese H, Klein F, Oberschelp O, Schmidt A, Schulz B, Vöcking H, Witting K (2004) Selbstoptimierende Systeme im Maschinenbau. Definitionen und Konzepte. HNI Verlag, Paderborn

Funck M, Loosen P (2009) Statistical simulation of selectively assembled optical systems. Proceedings SPIE

Funck M, Morasch V, Dolkemeyer J, Loosen P (2010) Design of a minaturized solid state laser for automated assembly. Proceedings SPIE

Gilbert X, Strebel P (1987) Strategies to outpace the competition. J Bus Strat 8(1):28–36

Glaeser G (1997) Efficient volume-generation during the simulation of NC-milling. In: Proceedings of the international workshop on visualization and mathematics'97. Springer, Heidelberg, pp 89–106

Gronau N (2010) Enterprise resource planning. Architektur, Funktionen und Management von ERP-Systemen. Lehrbücher Wirtschaftsinformatik, 2nd edn. Oldenbourg, München

Halldorsson T, Kroy W, Peuser P, Seidel H, Zeller P (1989) Patentnr. DE 3925201 A1

Hartmann D (1992) Anwendung wissensbasierter Informationstechnologie in Konstruktion und Planung – dargestellt an prototypischen Beispielen aus dem CAD CAP-Bereich der Bauindustrie und des Maschinenbaus. Fortschrittberichte VDI@Reihe 4, Bauingenieurwesen, vol 114. VDI, Düsseldorf

Hellmich K (2002) Wettbewerbsvorteile durch realisierte Kundenorientierung der Wertschöpfung. Ind Manag 18(5):27–30

Hufgard A, Hecht H (2005) Business integration mit SAP-Lösungen. Potenziale, Geschäftsprozesse, Organisation und Einführung. SAP Kompetent, 1st edn. Springer, Berlin

Hungenberg H, Wolf T (2007) Grundlagen der Unternehmensführung. Springer, Berlin

Jerard RB et al (1989) Approximate methods for simulation and verification of numerically controlled machining programs. Visual Comput 5(6):329–348

Klocke F et al (2001) Dynamikorientierte NC-Programme für die simultane 5-Achs-Hochgeschwindigkeitsbearbeitung, Fraunhofer-Institut für Produktionstechnologie IPT, Aachen

Klocke F et al (2003) Technologisch-optimale Prozesskette zwischen CAD-Geometrie und Bauteilfertigung. Tagungsband zum Seminar Wirtschaftliches Zerspanen – Intelligente Systeme für die Prozessauslegung und -simulation

Klocke F, Markworth L, Glasmacher L (2003) Simultaneous five-axis-machining of complex shaped parts. In: Proceedings of the 3rd International conference on machining and measurements of sculptured surfaces, pp 55–65

Klocke F et al (2007) Modelling of milling forces in a CAM-based system, CIRP conference on modelling of machining

Klocke F et al (2008) Model based optimization trochoidal roughing of titanium, 11th CIRP international conference on modeling of machining operations

Klocke F et al (2009a) Advanced simulation for rough machining titanium alloys. J Mach Sci Technol

Klocke F et al (2009b) CAM based material removal simulation and force extraction strategies, 12th CIRP international conference on modeling of machining operations

Koechner W (2006) Solid-state laser engineering. Springer, New York

Kölle J (1990) CAD/PPS-Integration. Konzepte und Erfahrungen. Hanser, München

König D (1994) Wissensbasierte Techniken zur automatisierten Arbeitsplanerstellung. Diss. Universität Dortmund, 1994. Fortschritt-Berichte VDIReihe 20, Rechnerunterstützte Verfahren, vol 118. VDI, Düsseldorf

Machine Works. Online (2011) http://www.machineworks.com

Markworth L (2005) Fünfachsige Schlichtfräsbearbeitung von Strömungsflächen aus Nickelbasislegierungen; Aachen, Dissertation, WZL der RWTH Aachen

Minoufekr M (2010) CAx-Prozessketten zur Fertigung und Reparatur von Turbomaschinenkomponenten. Tagungsband zum Seminar Potenziale und Trends im Bereich der CAD/CAM/NC Verfahrenskette

Morasch V, Funck M, Pyschny N, Pavim A (2009) Self-optimising flexible assembly systems. In: Proceedings of the selfx in engineering conference

Open Mind – The CAM Company (2011) http://www.openmind-tech.com/de

Paschotta R (2008) Field guide to laser pulse generation. SPIE Press, Bellington

Pfeifer T, Schmitt R (2006) Autonome Produktionszellen. Komplexe Produktionsprozesse flexibel automatisieren. Springer, Berlin

Pfeifer T, Schmitt R, Pavim A, Stemmer M, Roloff M, Schneider C, Doro M (2010) Cognitive production metrology: a new concept for flexibly attending the inspection requirements of small series production. In: Hinduja S, Li L (eds) Proceedings of the 36th international MATADOR conference. Springer, London, pp 359–362

Porter M (1980) Competitive strategy: techniques for analyzing industries and competitors. Free Press, New York

REFA (1985) Methodenlehre der Planung und Steuerung. Teil 1: Grundbegriffe, 4th edn. Hanser, München

Robert L et al (1987) Discrete simulation of NC machining. In: Proceedings of the third annual symposium on Computational geometry. Ontario, CA, pp 126–135. ISBN 0-89791-231-4

Schäppi B, Anreasen MM, Kirchgeorg M, Radermacher F-J (2005) Handbuch Produktentwicklung. Hanser, München

Scheer A (1990) CIM, computer integrated manufacturing. Der computergesteuerte Industriebetrieb, 4th edn. Springer, Berlin

Schmitt R, Pavim A, Brecher C, Pyschny N, Loosen P, Funck M, Dolkemeyer J, Morasch V (2009a) Flexibel automatisierte Montage von Festkörperlasern. Werkstatttechnik Online

Schmitt R, Wagels C, Isermann M (2009b) Production optimization by cognitive technologies. J Mach Eng 9(1):78–90. ISSN 1895-7595

Schmitt R, Laass M, Wagels C, Isermann M, Matuschek N (2010) Selbstoptimierende Produktionssysteme durch kognitive Technologien – Produktionsmanagement, Fertigungssteuerung, Arbeitsgestaltung. In: wt Werkstattstechnik online 100, 1/2, ISSN 1436-4980, pp 28–29

Schmitt R, Wagels C, Isermann M (2010a) Production optimization by cognitive controlling systems. In: Dimitrov D (Hrsg) Proceedings of the international conference on competitive manufacturing – COMA' 10. Stellenbosch University Stellenbosch, Südafrika, pp 153–158. ISBN 978-0-7972-1322-7

Schmitt R, Wagels C, Isermann M, Matuschek N (2010b) Cognitive optimization of an automotive rear-axle drive production process. J Mach Eng 9(4):71–80. ISSN 1895-7595

Schneewind J (1994) Entwicklung eines systems zur integrierten Arbeitsplanerstellung und Fertigungsfeinplanung und -steuerung für die spanende Fertigung. Diss. RWTH Aachen, 1994. Berichte aus der Produktionstechnik, vol 94,8. Shaker, Aachen

Scholz-Reiter B, Höhns H (2007) Selbststeuerung logistischer Prozesse mit Agentensystemen. In: Schuh G (ed) Produktionsplanung und -steuerung, 3rd edn. Springer, Berlin

Schuh G, Lenders M, Arnoscht J (2009) Focussing product innovation and fostering economies of scale based on adaptive product platforms. CIRP Ann 58(1):131–134

Schuh G, Kampker A, Franzkoch B et al (2010) High resolution information management decentralized and self optimizing manufacturing control in a multi machine operating environment. Adv Intell Soft Comput 66:725–734

Schuh G, Stich V, Brosze T, Fuchs S, Pulz C, Quick J, Schürmeyer M, Bauhoff F (2011a) High resolution supply chain management – optimised processes based on self-optimizing control loops and real time data. Production Engineering, Sonderheft 1

Schuh G, Arnoscht J, Bohl A, Quick J, Kupke D, Nußbaum C, Vorspel-Rüter M (2011b) Assessment of the scale-scope dilemma in production systems. An integrative approach. In: Production engineering, 5 Special Issue

Scussat M, Würsch AC, Salathé RP (2000) An innovative flexible and accurate packaging technique suited to fabricate low cost micro optoelectronic modules. 50th Electronic components and technology conference

Seelert W, von Elm R (1997) Patentnr. DE19540140, 2000

Siegman AE (1986) Lasers. University Science Books, Sausalito

Soar Technology (2010). http://sitemaker.umich.edu/soar/home 15. Nov. 2010

Spur G, Krause F (1997) Das virtuelle produkt. Management der CAD-Technik. Hanser, München

Stautner M (2005) Simulation und Optimierung der mehrachsigen Fräsbearbeitung. Vulkan Verlag Essen, Dortmund

Strina G (1996) Anwendung von Prinzipien der Selbstähnlichkeit und Selbsterneuerung auf das Innovationsmanagement in kleinen und mittleren Produktionsbetrieben. Diss. RWTH Aachen, 1995. Fortschritt-Berichte VDIReihe 16, Technik und Wirtschaft, vol 83. VDI, Düsseldorf

Strohner H (1995) Kognitive systeme. Eine Einführung in die Kognitionswissenschaft. Westdeutscher Verlag, Opladen

Suhir E (1987) Die attachment design and its influence on thermal stresses in the die and the attachment. In: Proceedings of the 37th electronics components conference

Tayebati P, Holderer H (2007) Patentnr. DE102007004185

Thome R (2007) Business software. ERP, SCM, APS, MES – was steckt hinter dem Begriffsdschungel der Business-Software-Lösungen. Lehrstuhl für BWL und Wirtschaftinformatik, Mainfränkisches Electronic Commerce Kompetenzzentrum, Würzburg

Vajna S, Bley H, Hehenberger P, Weber C, Zeman K (2009) CAx für Ingenieure. Eine praxisbezogene Einführung, 2nd edn. Springer, Berlin

VDI (2010) Technologiezentrum European Technology Platform Photonics21. Lighting the way ahead

Warnecke H-J (1996) Die montage im flexiblen Produktionsbetrieb. Springer, Berlin

Wiendahl H (2002) Situative Konfiguration des Auftragsmanagements im turbulenten Umfeld. Diss. Universität Stuttgart, 2002. IPA-IAO-Forschung und Praxis, vol 358, Jost-Jetter, Heimsheim

Wiendahl H (2005) Stolpersteine der PPS. Symptome – Ursachen – Lösungsansätze. Institut für Industrielle Fertigung und Fabrikbetrieb (IFF), WI 2670/1 (DFG), Stuttgart

Wiendahl H (2010) Betriebsorganisation für Ingenieure, 7th edn. Hanser, München

Witt S (2007) Integrierte simulation von maschine, Werkstück und spanendem Fertigungsprozess. Dissertation, RWTH Aachen University, pp 40–41

Woods S, Pflanz T (2008) DPSS lasers rival – fiber lasers for marking applications. Laser Focus World

Zäh MF et al (2007) Kognitive Produktionssysteme – auf dem Weg zur intelligenten Fabrik der Zukunft. ZWF-Z Wirtsch Fabrikbetr 102(9):525–530

Zäh, MF et al (2008) The cognitive factory. In: ElMaraghy HA (ed) Changeable and reconfigurable manufacturing systems. Springer, Berlin

Subject Index

A

Aachen (Aix) Virtual Platform for Materials Processing concept (AixViPMaP), 6, 320, 332, 372, 419
Aachen factory planning model, 306
Aachen House of Integrative Production Technology, 2
Aachen PPC model, 745, 758
Abaqus software, 334, 345, 369
abrasive energy, 476
academic network, 70
acoustic behaviour, 732
acoustics test bed, 733
acousto-optical modulator, 1012
active sum, 182
actual process chain, 706
ACTUAL value, 752
ADALINE (adaptive linear Neuron), 715
adaptability, 299
adaptive project organisation, 308
added value distribution, 283
additive manufacturing, 136
additive production process, 136
    powder bed-based, 138
adhesion
    energy, 649
    mechanism, 486
    model, 486
    strength, 544
adhesive bonding, 481, 482, 542
adsorption theory, 488
advanced planning and scheduling system (APS system), 747, 789
aggregation of information, 781
AIIuS framework, 257, 259
air plasma spraying (APS), 502
air restrictor, 682

AixViPMaP, see Aachen (Aix) Virtual Platform for Materials Processing concept
alarm filter, 761, 769
alignment process, 1020
analysis module, 720
anti-reflective surface, 629
application programming interface (API), 258, 328
arc ion plating (AIP), 646
Archimedes system, 899
area and space planning, 317
artificial intelligence (AI), 795, 801, 879
artificial neural networks (ANN), 705, 711, 715, 721, 796, 798, 819, 1046
    as a modelling tool, 714
    process modelling, 722
assembled assembly, 930
assembly
    agent-based control, 870
    error, 909, 933
    on a carrier ceramic, 1014
    planning, hybrid method, 908
    platform, control architecture, 956
    platform, programming, 964
    process, 1006
    strategy, 1021
assembly-by-disassembly strategy, 895, 899, 911
assembly-oriented product
    design, 886, 888
    development, 1005
atmospheric pressure plasma, 504, 515, 541
atomic force microscopy (AFM), 642
attainability matrix, 464, 466
augmented reality, 929
augmented vision mode (AVM), 930, 934
automatch, 260

1078 — Subject Index

automated
  assembly system, 34, 857
  micro assembly, 996
  planning, 862
  production, 25
  robotic system, 897
automatic
  geometry optimisation, 184
  rule derivation, 1052
automating laser process, 625
automation engineering, 33
automation markup language (AutomationML), 214
automatism, 864
autonomous
  assembly cells, 861
  production cells, 897
autonomy, 861
availability, 295
AX-workshop software, 820, 842
axial stress, 662
axis-aligned bounding box (AABB), 408

B

back-propagation method, 715, 796
Balanced Scorecard, 69, 71, 85, 793
balancing
  production, 124
  sequences per segment, 125
ball screw drive (BSD), 226
Barker's solution, 676
batch processing, 851
Beer's viable system model, 312
benchmark of bond strengths, 542
benchmarking process, 507
best-cost country, 18
black box model, 445, 454, 714, 722, 736, 815, 828
blackboard model, 905
blade integrated disk, see blisk
blisk (blade integrated disk), 12, 416, 844
  geometry, 991
  production, 12, 988
blue gene system, 199
BOBYQA method, 202
bolt test, 675
bond strength, 527, 542
bonding agent, 138
Boolean operation, 407
bottleneck, 102
boundary
  deformation, 200
  representation, 410

box constraint, 202
braiding process, 354, 360
brittle plastics, 684
brownfield planning, 282
Brundtland report, 292
build-to-order (BTO), 209
build-to-stock (BTS), 209
build-up rate, 145, 165
building design, 317
business, 281
  profit-oriented, 282
  segments, 179
business and technology case, 988, 1053
business intelligence, 277
business management, 39
  input/output analysis based on Kloock, 44
business process, 785
  analysis of the efficiency, 786
  execution language (BPEL), 220
buy-to-fly ratio, 138

C

CABRUN, 354
CAD, see computer-aided design
CAE, see computer-aided engineering
calculation
  model, 39, 347
  on demand, 372
  procedure, 837
calibration table, 1035
CALPHAD method, 324
CAM, see computer-aided manufacturing, 565
canonical data model, 258, 259
capability maturity model integration (CMMI), 214
capacity
  flexibility, 766
  utilisation, 295, 296
capital equipment, 43
carbonyl, 503
Carreau model, 199
Cartesian
  coordinates, 892
  system, 854
  velocity, 388
cascaded closed-loop quality control system, 717
casting
  pressure, 675
  process, 521, 685
  simulation, 364
  temperature, 525, 675
CASTS program, 333, 335, 342

Subject Index

cause-effect
  chain, 456, 460, 470
  diagram, 461, 480
  relationships analysis, 713
cavity pressure, 800
CAx
  data, 215, 224
  framework, 7, 343, 374, 412
  process chain, 566, 994
  system, 210, 211, 989, 992
CCU processor time, 908
cell control system, 865
cell model, 862
centroidal voronoi tessellation method (CVT), 818
chaku-chaku principle, 942
change enabler, 860
CHEOPS framework, 246
Chi-square test, 935
chip geometry, 395, 396
classic controller structure, 803
classical
  cost function, 43
  production theory, 42
Clio, 260
closed node index, 264
closed-loop control, 704, 709, 711, 717, 776, 1052
cloud computing, 247
Cluster of Excellence (CoE), 2
CMB-process (Controlled Metal Build-up), 138
coating
  development, 646
  system, 629, 633, 638
Cobb-Douglas production function, 43, 112
cognition, 34, 718, 834, 863
cognitive
  architecture, 712, 914
  assistance system, 713
  automated system, 742
  automation, 712, 897, 904
  control, 912
  control unit (CCU), 11, 895, 901, 904
  factory, 712, 897
  function, 895
  information processing, 718
  simulation model, 927
cognitive system, 34
  software architectures, 897
cognitive technology, 702, 716, 894
cognitive tolerance matching (CTM), 10, 719, 731, 735, 738, 1045

cognitive-ergonomic model, 922
cognitively automated assembly cell
  process logic, 921
  prototype realisation, 915
cognitively automated system, 901, 903
cold chamber pressure die casting process, 521
cold forming, 581
cold Metal Transfer welding process (CMT), 550
collision test, 376
COMA++ 249
comb-cut, 835
combined process steps, 61
common object request broker architecture (CORBA), 245
commonality, 95
  index, 99
  of product architecture, 97, 98
compact solid state laser, 998
compatibility, 860
competence
  orientation, 291
  production-based, 291
competitive strategy, 80
compile-cycle, 613
complementarity, 297
complexity
  control, 35
  management, 131
  reduction, 35
component
  geometry, 551
  library, 225
composite casting mould, 484
computation time, 557
computational
  fluid dynamics (CFD), 324
  steering, 251
computer simulation, 176
computer-aided automated validation, 230
computer-aided control engineering (CACE), 378
computer-aided design (CAD), 565, 1055
  CAD-CAM-NC-chain, 7
  data, 79, 141
  geometry, 376
  model, 192, 552, 562
  programme, 172, 177
computer-aided engineering (CAE), 565
computer-aided manufacturing (CAM), 375
  conventional planning, 375
  analysis of milling processes, 389

computer-aided manufacturing (contd.)
    extensions, 594
    planning, 374
    system, 388
computing time, 1051
Comsol multiphysics simulation software, 539
concatenation of services, 262
Condor, 246
conductivity value, 537
conductor track, 538, 546
cone-like protrusion (CLP), 654
configuration
    identification, 952
    logic, 27, 133, 308
    model, 219
    planning, 957
configure-to-order (CTO) strategy, 209, 229
consistency matrix, 459
constitutive product feature, 84, 95
constructive information, 1055
consumption function, 43, 46
consumption-controlled scheduling, 788
continuous
    system development, 444
    time basis, 275
contour joining, 516
control
    architecture, 907
    channel, 768
    circuit model, 386
    computer, 949
    concept, 952
    engineering system, 770, 776
    error, 828
    fluctuations, 827
    logic, 776
    of complexity, 27
    simulation, 374
    structure, 873
control loop, 33, 129, 386, 595, 611, 700, 776, 1051
    adaptive, 33
    implementation, 1060
control technology, 610, 863
    cognition, 863
    learning ability, 863
controllability, 24, 31
controller
    board, 997
    software, 229
conventional mass production, 56
coolant lubrication system, 404
cooling, 361, 581, 636

capacity, 168
medium temperature, 659
process, 507
cooperative planning algorithm, 944
coordination
    channel, 768
    module, 720, 734
coordination channel, 763
CORBA middleware, 918
core box, 678
core competency, 291
corporate
    culture, 302
    identity, 303, 305
    strategy, sustainability-oriented, 293
cost
    analysis, 184
    calculation, 189, 913
    degression effect, 296
    effectiveness, 1004
    leadership strategy, 289
    optimisation, 293
    planning, 316
    reduction, 295
    theory, 47
cost-effective fabrication, 709
cost-effectiveness of system automation, 100
cost-efficient manufacturing, 172
cost-leadership strategy, 940
CoTeSys (Cognition forTechnical Systems), 796
coupled simulation, 404, 412, 420, 539
    loop, 399
    of grinding processes, 405, 410
coupling, 244, 277
    mechanism, 380
    of monitoring, 993
    of process planning, 993
COVISE, 251
Craig-Bampton method, 390
criticality, 182
cross process technology planning, 412
cross sectional process, 65, 67, 73
cross-CAM-planning, 377
cross-technology solution, 439
crystal plasticity, 330
crystallinity, 510
crystallisation, 350
cupping test, 585
CURARE, 593
curve theory, 749
customer

Subject Index 1081

relationship management (CRM), 248
requirement, 290
satisfaction, 37
customer-specific
　product variant, 126
　production, 744
customisation, 178, 1030
　of mass production, 207
　of products, 55
cybernetic production management, 745, 753, 755, 787
cyclic stress, 362

D
damping effect, 419
data
　analysis in the analysis module, 730
　conversion process, 258
　element tool-based analysis (DELTA), 260
　exploration, 278
　mining, 713, 731
　transformation, 259
　visualization, 279
database intensional knowledge extractor (DIKE), 260
deadlocks, 909
deburring process, 609
decimation method, 267, 268
decision tree, 714, 731
decision-making, 42
　in the optimisation layer, 730
decoupling of the product architecture, 121
degree
　of commonality, 113
　of efficiency, 111
delivery reliability, 106
Delphi method, 13
DELTA, see Data Element Tool-based Analysis, 260
demand for action, 1068
demonstrator, 71
demouldability, 673
demoulding process, 634, 671
Denavit-Hartenberg parameter, 891
dependency, 188
deposition welding process, 604
descriptive model, 49
desensitisation strategy, 1036
design
　logic, 132
　methodology, 443
　of experiments (DoE), 723
　process, 443

specification, 57
structure matrix (DSM), 121, 314
theory, 461
Dexel model, 406, 408, 409, 989
diagnosis-based process development, 144
die
　casting, 685
　categorisation, 185
　design, standardised process, 193
　temperature, 528
die-based production technology, 4, 170
differential thermo analysis (DTA), 156
differentiation strategy, 289, 296, 940, 1031
diffusion theory, 487
DIGIMAT software, 360
digital direct fabrication, 136
digital model, 1069
DIKE, see Database Intensional Knowledge Extractor, 260
diode laser, 138, 684
dipolar coupling forces, 488
direct
　digital manufacturing, 136
　metal deposition, 136
　metal laser remelting (dmlr), 140
　model execution, 882
directed acyclic graph manager (DAGMAN), 254
dismantling of factories, 282
dominant effect, 665
double-loop learning, 72
dual-beam process, 499
dynamic
　compensation, 1029
　management of visualization, 274
　recrystallization (DRX), 577, 580
　tensile test, 587
dynamisation, 738

E
ease of explanation, 132
eclipse-based functionality, 413
economic
　analysis, 35
　research, 2
　success function, 46
economies
　of scale, 3, 21, 81, 296
　of scope, 3, 21, 80, 296
edge weighting, 465
EDX, 363
efficiency of systems, 111
efficient order processing, 209

elastic mesh update method (EMUM), 201
elasticity
 equation, 201
 parameter, 113
electrical
 conductivity, 538
 construction (ECAD), 218
electromagnetic
 energy, 516
 levitation, 155
electronic data interchange (EDI), 747
element topology, 260
embossing
 process, 635, 664
 roller, 636, 663
energy
 consumption, 403
 demand simulation, 380
 requirement simulation, 402
engineer-to-order (ETO) strategy, 127, 209, 229
engineering
 design theory, 180, 439, 468
 process, 29
 production function, 47
enterprise
 dynamics simulation, 784
 resource planning (ERP), 129, 751, 790, 1055
 service bus (ESB), 219, 249
ergonomic interface design, 905
Erlangener bearing, 529
error
 detection phase, 930
 detection time, 936
 feedback method, 715
European Industrial Research Management Association (ERIMA), 51
evaluation model, 87, 88, 110, 113
 fundamental structure, 88
excellence initiative, 65
executable UML (xUML), 212
explainability at the point of sale, 82, 90, 93, 94, 132
extension point, 413
external variety, 223
extract, transform and load (ETL) process, 250, 262
extraction process, 260
extrusion
 embossing process, 630, 635, 657
 velocity, 203

F

factory design
 strategic orientation, 288
factory dismantling, 282
factory planning, 6, 58, 280
 competence orientation, 291
 goal system, 287
 hierarchy levels, 284
 of the future, 280
 Porter's five forces, 289
 project nature, 284
 resource orientation, 290
 sustainability orientation, 292
 synchronisation, 310
factory productivity, 299
factory redesigning, 283
failure state system, 1049
false solution, 468
fast-forward planner, 898, 910
fatigue test, 587
feature clustering, 83
 in the product programme, 119
feature modelling, 221, 230
feature tree, 119, 120
feature-oriented domain analysis (FODA), 213
feedback loop, 130
feeding back error, 722
fibre laser system, 32
fibre-optics, 147
FiFo (first in first out) rule, 784
filter algorithm, 402
financial indicator, 132
fine planning, 1069
finite element method (FEM), 324, 369, 377, 378, 557
 calculation, 191
 simulation, 1015
fit of variety, 82, 90, 91, 132
fixed cost
 degression, 295
 reduction, 122
fixture-oriented assembly system, 949
flame spraying, 502
flexibility, 294
 of the product architecture, 96
flexible
 assembly cells, 865
 assembly system, 34, 861, 865
 fabrication, 767
 handling technology, 941
 link concept, 286
 product design, 1006
 production system, 294

Subject Index

floor skirting, 204
flow channel, 176, 194
    cross-section, 194
    geometry, 172, 193
    optimisation, 194
flow solver, 198
FOAM framework, 249
forming forces, 562
Fourier transformation infrared spectroscopy (FT-IR), 503
Fowkes division, 648
framework ontology, 266
free-formed test geometry, 993
Fresnel absorption, 833
function structure, 95
function-oriented optimisation of the end product, 1044
functional material design, 322
functional surface, 630

G

g-lite, 246
gain ratio, 733
Galerkin finite element method, 201
Galerkin/least-squares stabilisation method, 199
gamma machine, 561
gas metal arc welding (GMAW) process, 800, 804
Gaussian
    intensity profile, 147, 1010
    output density distribution, 150
gel-permeation chromatography (GPC) analysis, 511
genetic algorithms (GA), 796
geometrical
    accuracy, 589
    element, 196
    module interface, 121
geometry
    Deformation, 199
    optimisation framework, 208
German national economy, 19
German Research Foundation (DFG), 2
Giesekus model, 208
Gini index, 733
global competitive, 18
global user data (GUD), 615
global variant production system (GVP), 84
globalisation, 22, 56, 135, 490
globus, 246
goal system

    for the plant, 301
    of planning, 297
goal triangle, 299
gradient method, 812
graph theory, 462
graphical
    control element, 275
    effect, 273
    modelling language, 212
grasp-point, 961, 966
grasping operation, 957
gravity die casting, 685
greenfield planning, 282
grid
    computing, 245, 328
    middleware, 246, 252, 255
grinding process, 380, 406, 441
    simulation, 419
grounded theory, 70
guided expert interview, 70
gypsum, 674

H

Hall-Petch relation, 365
handling strategy, 953
Hanover School, 749
harmonisation of product structure, 28
hatches, 161
hatching, 160
head-mounted display (HMD), 929
health promotion, 304
heat transfer coefficient, 365
heating, 477, 658
heuristic model, 456, 463
high power pulse magnetron sputtering (HPPMS) technology, 683
high resolution supply chain management (HRSCM) 10, 126–128, 743, 752
high velocity oxygen flame (HVOF), 502
high-quality laser system, 871
high-wage country, 1
holism, 867
holographic
    interference, 651
    lithography, 632, 641
HoMat mathematical homogenisation software, 336
homogenisation, 348
Hooke matrix, 337, 349
Hoppe's algorithm, 270
hopper model, 749
horizontal integration, 791
hot chamber pressure die casting process, 521
hot embossing, 634

hot forming, 581
hot isostatic pressing (HIP), 137
hot rolling simulation, 342
hot working steel, 606
hot-forming simulation, 342
human-machine interaction, 897, 929, 945
hybrid
    die technique development, 523
    incremental sheet forming, 552
    kinematic concept, 869
    machine tools, 62, 592
    machining centre, 598, 600, 617, 626, 628
    manufacturing center, 227
    manufacturing process, 449
    manufacturing technology, 439, 453
    material combination, 481
    mechanical engineering, 522
    microassembly, 852
    multi-component injection moulding, 529, 548
    multi-component pressure die casting (M-HPDC), 521
    plastic-metal structural component, 1038
    process, 624
    production system, 7, 31, 60, 61, 435, 1040
    technique, 484
    welding process, 441
hybrid laser system, 615
    machining head, 624
    machining tool, 599
hybridisation, 31, 32, 436, 599
    and interconnection point (HIP) 462, 468
hybridised control architecture, 625
hydrogen
    atom, 545
    bridge connection, 487
hydrophobic surface, 632

I

iMap, 260
implementation planning, 316
improvisation, 282
in-mould assembly process, 482
incremental sheet forming (ISF) 8, 551
    laser heating, 574
    laser-assisted, 551, 558, 559
    geometric accuracy, 555
index plate technology, 547
individual production, scope-orientated, 143
individual simulation tool, 247
individualisation, 78
    of industrial production, 28
individualised production, 3, 27, 55, 77, 1040

research programme, 78
technology radar, 55
tool-less production technology, 135
induction
    heating, 668
    joining, 483, 512
inductor power, 667
industrial
    automation, 899
    engineering, 1054
    mass production, 169
    monitoring system, 990
    production, 19, 26
influence matrix, 466
information
    flow, 760
    gain, 733
    loop, 768
    planning, 317
    transparency, 745, 752, 763
information-processing sensor-actuator-systems (ISA-systems) 794, 808, 813
initial graphics exchange specification (IGES), 248
injection compression moulding, 634
injection embossing, 637
injection moulding, 634, 652, 665
    machine, 530, 531
input object, 40
integrated manufacturing technology, 438
integration system, 256
    architecture, 257
integrative
    balancing, 84
    cluster domains (ICD), 26
    computational material engineering (ICME) 6, 320
    configuration logic, 83
    evaluation model, 82, 114
    high resolution supply-chain-management, 1053
    levelling of production, 1053
    material simulation, 325
    process chain simulation, 318
    product, 884
    research, 3, 64
    self-optimising process chains, 702
    vision, 879
    visualization, 267
integrative production
    of TiAl products, 1062
    research program, 25

Subject Index   1085

technology, 135
theory, 36
integrative simulation, 243
   of machine tool, 373
   solution, 344
integrativity, 867
intelligent
   agent, 874
   design, 712
   failure states agent, 876
   production system, 34
   system, 864
interdisciplinary engineering, 218
interface, 188
   bond strength, 482
internal
   strength, 482
   variety, 223
international competition, 20
interpersonal communication, 304
inventory
   management, 778, 779
   planning, 779
inverse modelling, 327
investment
   casting process, 672
   cost, 298, 948
   planning, 316
Ishiara colour test, 933
ISO, 15504 214
isoquants, 112
IT support, 1056

J

Java
   Business Integration (JBI), 219
   code, 247
joining
   pressure, 512
   technology, 1014
jointed-arm robot, 1028
just in time (JIT) delivery, 289

K

kanban
   control loop, 37
   implementation, 37
KAON2 264
Kelvin mechanism, 354
key characteristic (KC), 886
key performance indicator (KPI) 733, 875
Kienzle model, 393
kinetic energy, 504

Kloock theory, 44
knock-out requirement, 470
knowledge, 453
   case-based, 877
   module, 721
   rule-based, 877
Kolmogorov-Smirnov test, 935
KombiMasch project, 593
Kruskal-Wallis test, 927
KUKARSI real-time interface, 919

L

labor costs, 25
Ladyzhenskaya-Babuska-Brezzi (LBB)
   compatibility condition, 199
Lamé parameters, 201
LARSTRAN, 342
LARSTRAN/-Shape software, 335
laser
   ablation, 608
   cutting process, 832
   deburring, 609
   deposition welding, 603, 606, 617, 624
   engineered net shaping (LENS), 137
   hardening, 603
   heating, 574, 590
   induced periodic surface structures
      (LIPSS), 653
   joining, 483
   machining head, 625
   metal deposition (LMD) 136, 142
   microstructuring, 496, 541, 548, 638
   movement, 607
   output power, 497, 517, 607, 612
   power, 158, 608
   processing unit, 601
   production process, 596
   protection system, 620
   radiation, 1002
   source controls system, 620
   spot oscillation, 627
   structuring, 518, 633, 640, 650, 680
   system technology, 595
   transmission joining, 516, 517
   welding process, 617
laser beam, 79, 645
   cutting, 801, 804, 806, 846
   radiation, 620
laser-assisted
   machining, 31, 468
   process, 61
   shearing, 472
laser-based heating, 661, 669
laser-machining process, 437

LaserSpy sensor, 620
LaserWeld3D, 343
Latin hypercube sampling (LHS), 818
latinised centroidal voronoi tessellation method (LCVT), 818
law of diminishing returns, 41, 42, 46
layer thickness, 159, 163
layer-based manufacturing, 136
layout planning, 316
lean production, 289
learning
    curve, 100, 104, 724
    effect, 296, 301, 865
    process, 723, 742
    success, 729
Leblond model, 334
length scale, 322
Leontieff production function, 46
levelling of production, 1053, 1060
levels of detail (LOD), 273
lifecycle
    calculation, 327
    model, 327
Lilliefors test, 927
linear associator, 715
local cost optimisation, 40
logarithmic conditional probability (LCP), 926, 927
logistic postponement, 126, 127
logistics system, 787
long-term
    control loop, 1059, 1062
    memory, 736
loop system, 14
LOR (load-orientated order release), 784
lotus effect, 631, 664
low melting metal alloy, 530
low-wage country, 18

M

M-function, 228, 618
machine
    concept, 144
    control system, 610, 620
machine tool
    engineering, 212
    manufacture, 29, 79
machine-oriented control system, 904
machined surface, 644
machinery and control simulation, 58
machining process, 437
macroscopic simulation, 331
macrosegregation, 363

macrosimulation of the operating load, 369
macrostructure simulation, 350
magnetron sputter ion plating (MSIP), 647
make-or-buy decision, 283
make-to-order manufacturer, 109
make-to-stock manufacturer, 109
man-machine interaction, 743
management by exception, 769
manufacturing execution system (MES), 375, 790, 899
manufacturing functional surface, 656
    extrusion process, 656
manufacturing method
    laser-assisted, 442
    vibration-assisted, 442
manufacturing planning II concept (MRP II), 1055
manufacturing process, 38
    levelling, 38
    model-based optimisation, 811
    of high-precision casting, 362
    operating point, 811
    technological boundary, 459
manufacturing resource planning (MRP), 317, 747
manufacturing technology, 29
Marangoni
    convection, 152, 158
    effect, 153
marching cubes algorithm, 271
market dynamics, 56
market orientation, 288
Markov process, 926
Marshall's schema model, 923
martensitic transformation, 370
mass customisation, 27, 28, 57, 78, 143, 1031
mass production, 1, 55, 135, 629, 857
    customisation, 207
    die, 179
    scale-orientated, 143
master control system, 880
master oscillator fibre power amplifier (MOFPA), 999
master oscillator power amplifier (MOPA), 641, 999
MatCalc, 342
material
    and process simulation, 58
    connection mechanism, 484
    engineering, 322
    flow planning, 317
    science, 322
    suitability data, 471

Subject Index 1087

material-requirements planning (MRP), 746
mathematical
    homogenisation, 330, 348
    optimisation process, 747
Matlab, 214
Matlab/Simulink, 387
maximum capacity utilisation, 293
maximum utilisation of machines, 102
MCAD data, 222
mean square error (MSE), 820
mechanical
    construction (MCAD), 218
    joining process, 482
mechatronic
    modularisation of the handling system, 954
    system, 797
mechatronic-oriented modular system, 211
media planning, 317
medical engineering, 168
melt pool
    dynamic, 152
    geometry, 154
mental fatigue, 938
MES system, 782
message-based communication, 261
meta-modelling, 797, 815,
metal alloy, 493
    flow capacity, 534
metal processing, 481
metallic joining partner, 507
methodology development, 446
MICRESS
    phase field program, 339
    software, 334, 342, 364
micro injection moulding, 530
micro-controller, 1038
micro-laser, 12
    Integrative Production, 995
microassembly, 851
microcontroller, 1013
microfeature geometry, 678
micromanipulator, 855, 869
microoptical components, 855
microrobot, 855
microsimulation of the transformation, 367
MicroSlab, 850, 858, 867, 889, 1028
microstructural simulation, 345
microstructure, 9, 632
    grid, 272
    simulation, 329
microstructured
    surface, 629
    wax model, 672

microsystems technology (MST), 851
mid-term production control loop, 1058
MIG welding process, 441
migration planning, 1070
milestone trend analysis, 313
milling, 805, 843
minimum inventories, 293
mirror-prism, 564
model
    creation, 833
    evaluation, 797, 805, 844
    prediction error, 828
    predictive controller, 815
    quality, 819
model-based
    assembly control, 849, 587, 865, 878
    control system, 882
    engineering of control systems, 881
    flow control, 11
    optimisation system (MO-system), 808, 815
    process control, 826
    self-optimisation, 794, 804, 807
Modelica, 214
modeling cognitive process, 719
modeling language, 214
    Matlab, 214
    Modelica, 214
modeling manufacturing process, 815
modeling module, 720
    control, 734
modeling standard, 742
modular framework, 742
modular function deployment (MFD), 85
modular multi-robot cells, 866
modular product, 55
modular design system, 171
    mechatronic-orientated, 29, 79
modularisation, 55, 171, 174, 206, 970
    Balanced Scorecard, 85
    for die-based production technologies, 170
modularity, 27, 860
module
    definition, 183, 184, 190
    design, 192
    drivers, 85
    evaluation, 192
    indication matrix, 85
molecule orientation, 350
monitoring
    channel, 763, 769
    horizontal coordination, 764

Monte Carlo
    method, 1022, 1024
    optimiser, 717
Moore's law, 212
morphological box, 472, 476
motion control (MC), 899
motor function mechanism, 761
mould
    production, 167
    technology, 533
    temperature, 534
moulded interconnect devices (MID), 485
moulding
    geometry, 534
    process, 671
    temperature, 675
    test, 665
MRP system, 129
MTM taxonomy, 917, 924
multi-beam lithography, 632
multi-body simulation models (MBS models), 377, 389, 411, 419
multi-layer perceptron (MLP), 715, 721
multi-phase field simulation, 1064
multi-robot system, 894
multi-technology platform, 32, 625, 628
multiple
    controller solution, 950
    dataset, 273
    simulation, 273
mutable automation, 859

N

nano-imprinting, 634
nanostructure, 9, 632, 683
nanostructured surface, 629
nanostructuring, 641, 644
natural strengthen alloy, 493
Nd:YAG laser, 137
neoclassical production theory, 41, 42
nest assembly, 888
network
    actor, 66
    development, 66
    effectiveness condition, 66
    efficiency condition, 66
    project-orientated, 67
    service, 67
    support, 45
    theory, 65
networked models of the digital factory, 286
neural network, 715
Newton-Raphson method, 389

Newton's method, 812
non-uniform rational B-splines (NURBS), 200
normative instruction mechanism, 760
normative production management system, 753, 758
ns-laser, 644
null hypothesis, 935
numeric material removal model, 392
numerical control (NC), 228, 374, 386, 625, 899
    core extension, 613
    machining process, 375
numerical design, 176
numerical program, 383
numerical simulation, 557
NURBS (Non-Uniform Rational B-Splines), 177
Nyquist scan rate, 777
Nyquist-Shannon theorem, 776

O

objective function, 176
octree model, 406
offline planner, 895, 911
ohmic heating, 539
Omnimove, 969
one-piece-flow, 28, 31, 79, 436
online planner, 895, 911
open grid services architecture (OGSA), 245
open innovation, 143
open-loop control, 711
OpenDX, 250
operating data acquisition system (ODA), 790
operation layer, 721
operative production management, 754
opportunistic planning, 863
optical bench, 853
optical fibre, 148
    V parameter, 148
optical sensor system, 830
optical system, 565
optically pumped semiconductor laser (OPSL), 999
optimal utilisation of resources, 293
optimisation
    algorithm, 193, 845
    framework, 172, 198, 207
    module, 720, 725
    of fluid dynamics, 682
    of production systems, 48
    parameter, 200
    Shell, 716
    simulation program, 716
optimised production control, 788

Subject Index

optimiser, 201
optimising
  algorithm, 960
  Horizontal Coordination, 764
optimum level of standardisation, 86
OptoRep machine, 592
order processing, 185
order-point system, 746
organisational networking, 763
oriented bounding box (OBB), 408
orthotropic material model, 351
oscillating
  drop, 155
  laser, 605
outpacing strategy, 1031
output
  intensity distribution, 149, 150
  object, 40
outsert technique, 484
overbending, 555
overmoulding process, 484, 536
oxygen, 154

P
PA66 509
  GF30 509
PAM-Crash finite elements software, 354
parameterisation, 177
parametric constraint evaluation (PCE), 227
ParaView, 250
Pareto optimisation, 109, 812
participative factory planning, 303
passive sum, 182
path planning, 964
payment function, 46
pearlite, 338
penalty costs, 912
pentaho data integrator (PDI), 250, 262
perception layer, 720
  data acquisition, 736
performance
  indicator, 69
  measurement system, 69, 71
PermAlign, 1003
phase-manipulated multi-beam lithography, 632
physical
  model, 455
  vapour deposition (PVD), 630, 646
physical-analytic model, 40
pick & align process, 1018
pick & join process, 1015
pilot mould, 533, 536

pipeline, 333
piston rod material, 358
plan orientation, 287
planar
  alignment, 1017
  Assembly, 1008
  technology, 1007
planning
  agent, 874
  assembly sequence, 898
  costs, 416
  domain definition language (PDDL), 267
  economy, 21
  effectiveness, 298
  efficiency, 22, 30, 298, 698, 895
  expenditure, 27
  framework, 306
  goal-oriented, 862
  intensity, 298
  module, 286, 306
  object, 52
  participants, 309
  process, 287, 312
  scope, 298
  stakeholders, 309
  utility function, 298
plasma pretreatment, 505, 515
plastic
  carrier, 536
  injection moulding, 799, 804, 806, 817, 744
  processing, 481
plastic-to-plastic welding, 520
plasticisation, 513
plausibility check, 470, 471
point-of-interest (POI), 278
polarisation theory, 487
polyamide, 492, 514, 547
polybutene terephthalate (PBT), 492, 541
polycarbonate (PC), 492, 518, 541
polycrystalline, 323
polyflow programme, 196
polylemma of production, 20, 903
polymer
  material, 670
  shrinkage, 545
Porter's five forces, 289
post-mould assembly process, 482
postponement, 84
power consumption, 404
precision
  assembly of optical systems, 852
  optical manufacturing, 136
prediction quality, 925

predictive algorithm, 712
preparation process, 259
pressure
    control, 513
    die casting, 174, 178, 184, 186, 521
    sensor, 531
price
    leadership, 296
    limit, 43
primary shaping process, 323
printing, three-dimensional, 136
prioritising planning orders, 313
process
    analysis, 888, 1068
    behaviour, 992
    channel, 768
    commonality, 82, 102–104, 123, 132, 783
    control centre, 768, 773
    coordination centre, 771
    design, 884
    flexibility, 766
    force simulation, 389
    integration into machine tools, 591
    management system, 773
    model of the methodology, 451
    monitoring, 278, 765
    morphology, 117
    optimisation, 881
    planning, 218, 570, 1057
    sequences, 104, 124
    structure, 101
    variables, 809
process chain, 123
    shortening, 591
    simulation, 319
process-machine interaction, 596
processing location, 147
    intensity distribution, 148
procurement planning, 779
product
    analysis, 885, 1068
    complexity, 57, 131
    customisation, 1003
    customisation, 55
    data management (PDM), 1055
    design, 887, 1007
    family, 218
    features, 93
    flexibility, 766
    function, 703
    geometry, 29, 38
    information, 58
    lifecycle management (PLM), 214
    modularly-designed, 55
    morphology, 116
    planning, 58
    production system, 78
    quality, 809
    reconfigurability, 57
    segment, 23
    structure, 95
    variation, 57
product architecture, 81, 88, 94, 181
    commonality, 82, 97, 98
    decoupling, 84, 120
    flexibility, 82, 96
    operationalisation, 90
product programme, 81, 88
    configuration logic, 92
    operationalisation, 90
product-production system, 87, 130
    integrated construction methodology, 134
    integrative design methodology, 134
production, 281
    balancing, 122
    chain, 479
    costs, 589
    die-based, 4
    economics, 42
    economy, 21, 793
    efficiency, 47
    engineering, 2, 286
    flexibility, 22
    history, 480
    individualised, 27, 55
    integrative levelling, 14
    logistics, 941
    mapping in the modelling module, 721
    morphology, 116
    network, 283, 753
    plan, 213
    planning and control (PPC), 36, 248, 751, 754, 792
    polylemma, 858
    self-optimising, 10
    start, 299
    task, 458, 929
    times, 555
    tool-less, 4
production control, 778
    value stream-orientated simulation, 783
production function, 40, 41
    based on Cobb-Douglas, 43
    based on Leontieff, 46
    S-shaped course, 42
    Type A, 42

Subject Index
1091

Type B, 43
Type C, 44
Type D, 44
Type E, 45
Type F, 45
production management, 42, 746
    analysis of the effectiveness, 786
    design, 770
    implementation of cybernetic principles, 756
production process
    chain, 362
    die-related, 144
    standardisation, 81
production structure, 81, 100
    analysis, 1068
    operationalisation, 90
production system
    classification, 24
    configuration logic, 87
    constitutive features, 129
    degree of efficiency, 111
    economic efficiency, 48
    evaluation model, 87
    integrative configuration, 115, 118
    performance, 47
    self-optimising, 62
production technology, 490
    as a product, 23
    as an instrument for manufacturing of products, 23
    die-based, 170
production theory, 34, 36
    based on Dyckhoff, 46
    based on Gutenberg, 43
    based on Heinen, 44
    based on Küpper, 45
    based on Matthes, 45
    classical, 42
    deficiency, 46
    development, 34, 41
    evaluation, 46
    industrial relevance, 36
    neoclassical, 41, 42
productivity, 131
profile extrusion, 175
    die, 178, 206
profile geometry, 172, 175
profit-oriented business, 282
programmable logic controller (PLC), 228, 377, 882, 899
programming technique, 616
project
    configuration, 306
    execution, 285
    management, 285
    network, 66
property-optimised manufacture, 325
ps-laser, 642, 644
psychological inertia vector, 451
pulsed laser radiation, 608

Q

Q-factor, 999
Q-Learning, 729
Q-switching, 99, 1012
qualification feature, 760
quality control, 549
quality function, 198
    deployment (QFD) valuation, 188
QuCut software, 834
quenching process, 367
questionnaire, 70

R

radial basis functions (RBF), 798, 819
radio frequency identification (RFID), 712, 757
ramp-down procedure, 699
ramp-up, 299, 699
rapid manufacturing, 136
rapid miner software, 733
re-indexing transformation, 266
REACH operator, 923
reactive planning, 862
real-time
    capability, 790
    compatible adaptive planning method, 780
    planning corridor, 780
rear axle drive production, 1042, 1050
    failure state system, 1050
    production control, 1043
rear-shed geometry, 839
reconciliation of the reference value corridor, 792
reconfigurability, 867, 954, 957
reconfigurable assembly system, 946
recrystallization, 576
recrystallized microstructure, 579
reduction of complexity, 27
redundancy, 297
REFA, 302, 1058
reference values, 757
    definition, 758
reflexive planning, 863
reinforcement learning, 726, 730, 736
relational data model, 262
relaxed layered system, 898

RELEASE operator, 923
reliability, 299
remote procedure call (RPC), 245
repetition function, 44
replication process, 656
representative volume element (RVE), 328, 330, 335, 348, 358
requirements tree, 119
research
    program, 25
    requirement, 53
residual cooling time, 665
residual stress, 361
    solidification-induced, 363
resistance soldering process, 996, 11016, 035
resonator assembly, 1027
resource
    costs, 25
    orientation, 290
    structure, 101
    suitability, 302
    utilisation, 82, 102, 124, 132
response surface design, 821
Reynolds number, 357
rheological test, 526
rigid body, 355, 356
rival competitor, 289
roadmapping, 50
robot
    control (RC), 610, 613, 896
    controller, 880, 907
    kinematics, 611, 615
    mounting, 622
    network, 950
    programme, 594
    system, 666, 953
    technology, 512
robot-based
    cutting, 596
    processing system, 604
robot-machine cooperation, 612, 626
robot-specific calculation model, 951
robot-supported assembly cell, 896
robotic assembly cell, 11
RootAsset, 221
rough planning, 1069

S
sand moulds, 673
sandblasting, 495, 509, 511
Sarsa and Q-learning, 727
scalability, 860
    of the visualization, 275

scale-scope-dichotomy, 4
scan line
    spacing, 160
    stability, 152
    width, 149
scan strategy, 161
scanning
    electron microscope (SEM), 514, 642, 647, 670
    velocity, 158, 159
SCARA robot kinematics, 854
schwarzschild optics, 642
selective laser melting (SLM), 4, 29, 135, 140, 167
    process efficiency, 141
selective laser sintering (SLS), 138
self-control, 748
self-learning system, 64
self-optimising
    automated assembly, 11
    component handling, 12
    control loop, 742, 872
    flexible assembly system, 849, 857, 865
    manufacturing systems in verbund, 848
    production technology, 793
    systems in mechanical engineering, 797
self-optimising assembly system, 858, 893
    based on cognitive technologies, 894
self-optimising production system, 10, 33, 62, 697, 902
    hardware, 63
    human and organisation, 63
    software and control, 63
semantic
    modelling of flexible processes, 881
    network, 735
    transformation, 259, 266
    web rule language (SWRL), 263
semi-automated path planning, 971
sensor system, 876
sensor-actuator (SA) system, 808
sensory mechanism, 761
separation process, 440
service-oriented architecture (SOA), 245
shaft-hub combinations, 726
shark skin effect, 631
shear
    force calculation, 393
    resistance, 474
    strength, 474
    test, 996
shearing process, 472, 476
sheet

Subject Index   1093

geometry, 571
metal forming, 588
thinning, 572
shell assembly, 969
shell-core interface, 166
short-term
control loop, 1058, 1062
memory, 735
Sigmasoft processor simulation software, 344, 345
signal processing algorithm, 810
simulation
chain, CAM-integrated
coupling, 244, 245, 250
graph, 252
loop, 386
model portability (SMP), 246
of machining, 367
of the Heat Treatment, 364
software, 251, 326
system, 30
tool, 243, 251, 256, 595
simultaneous joining, 517
SimVis, 246
SimWeld simulation system, 338, 830
sine law, 554
single dexel ray casting procedure, 407
single-point incremental forming (SPIF), 553
Sinumerik, 840D, 600
skin effect, 512, 513
skin-core strategy, 145, 158, 164, 165
slit profile, 203
small and medium-sized enterprises (SMEs), 215
small batch production, 553, 558, 588, 1060
smooth sheared portion, 475
snap decision, 282
Soar
agent, 907, 920
cognitive architecture, 716, 895, 906, 910, 1047
memory model, 735
program, 728
Sequential Control, 734
social goals, 303
employee-related, 303
society and environment-related, 303
software
agent model, 850, 870
architecture, 905
product line (SPL), 213
tool, 60, 286
solder bumping, 1002

solder-based technology, 1002
soldering technology, 1009, 1034
solid freeform fabrication, 136
solid state transformation, 329
solidification, 329, 353, 634
solution competence, 878
special machines, 209
SphaeroSim, 345
microstructure, 348
splitting criterion, 732
St. Galler Management Concept, 753
hierarchical company structure, 792
stakeholder value, 303
Standard d'Exchange et deTransfer (SET), 248
standard machines, 209
standardisation in production process, 81
static calculation, 39
statistical
analysis, 935
package for social science (SPSS), 935
step effect, 159
step-by-step assembly, 939
STEP-NC, 627
stiffness matrix, 390
stochastic homogenisation, 330
Stokes equation, 198
strategic production
management, 753, 758
planning, 174
strategy
of cost leadership, 80
of differentiation, 80
stress
analysis, 1017
field, 113
stretch
forming, 558, 562, 566
StrucSim software, 335
structural design technology, 1000
structuring and assessment of modular product platforms, 84
Stuttgart company model, 749
Stuttgart Neural Network Simulator (SNNS), 734
submission file, 252
substitute product, 290
success level, 46
successive planning method, 747
SUMO product, 798
superhydrophobic surface, 630, 681
supervised learning procedure, 723
supervisory control, 900
supplier structure, 289

supply chain, 81, 83, 88, 105
    capital efficiency, 83, 105, 107
    configuration, 110
    effectiveness, 83, 105, 106
    high-resolution management, 126
    management (SCM), 10, 776
    operationalisation, 90
    postponement, 126
surface
    coating, 633
    energy, 488, 504, 649
    geometry, 248
    imperfection, 663
    mount technology, 996
    optimisation, 961
    pretreatment, 495, 515, 549
    quality, 152, 157, 826
    tension, 153
    topography, 495, 509, 545
sustainability, 303
    orientation, 292
switch-cabinet production, 940
synchronousactions, 612
SysML modelling tool, 215
system
    behavior, 35
    capacity utilisation, 295
    identification procedure, 891
    modelling tool, 218
system optimisation, 889
    by eliminating individual elements, 893
    by shortening individual elements, 890
system-side interaction, 621
systems diagnostics, 64
systems engineering, 217
    model-based, 213
systems modeling language (SysML), 5, 211, 212, 217
SYSWELD, 338

T

table-mounted display (TMD), 929
tactical production management, 754, 762, 766
tailored hybrid blanks, 558, 560, 583
takt review, 313
tangential stretch forming, 570
target sector, 23
task allocation between humans and machines, 900
Taylor wear formula, 378
Taylorism, 699
technical thermoplastics, 494
technical-scientific law, 47
technological boundaries, 456
technology, 453
    flexibility, 860
    innovation process, 54, 450
    intelligence process, 54
    of optical systems (TOS), 563
    planning, 58, 59
    pool, 451
    radar, 52, 53
technology roadmapping, 3, 50
    for integrative research, 51
    integrative development, 54
temperature, 508
    profile, 659
tensile shear test, 350, 520, 524, 541
tensile strength, 475, 518, 527
test case
    line pipe, 333
    plastic component in automotive interior, 344
    stainless steel casting, 360
    textile reinforced piston rod, 352
    transmission component, 339
test geometry, 585
testing agent, 876
text-based modelling language, 882
textile reinforced metals, 353
T grid, 357
thermal
    conduction, 584, 635
    efficiency, 685
    expansion coefficient, 507
    joining process, 481, 507, 523, 544, 550
    spraying, 541, 501, 541
    stress, 511
    transfer coefficient, 512
thermal conductivity, 522, 534, 672, 1015
    joining, 483, 508
thermo-calc, 324
thermocouple, 525
thermodynamic
    efficiency, 114
    simulation, 341
thermography, 538
thinning, 554
three pillars model, 293
three-layer model, 905
TiAl
    components, 1066
    product, 14, 1062
titanium alloy, 14
tolerance
    allocation, 706

Subject Index

analysis, 888, 998, 1022
chain, 890
deviation, 707
management, 867
matching, 13, 997, 1004, 1007, 1023, 1037
tool
    centre point (TCP), 386, 401, 891
    geometry, 680
    production, 167, 638
tool-less production, 4, 135
top box, 352
top-hat intensity profile, 146
total cost, 184
total surface energy, 630
track editor, 354
trade-off matrix, 459
transfer technology, 529
transformation, 281
    calculation, 612
    function, 45
    process, 40
    service, 261
triangulation process, 71
tribological pairing, 682
TRIZ methods (theory of inventive problem solving), 444
two-point incremental forming (TPIF), 553

U

ultra-lightweight construction, 169
ultra-short pulse laser, 633, 684
ultrasonic joining, 483, 519
ultrasonic-assisted process, 61
UNICORE, 246
uniform pre-processing of datasets, 267
Unigraphics NX5 CAM system, 568
universality, 860
utilisation flexibility, 859
utility function of planning, 298

V

V model, 212
validation of system properties, 226
value creation, 33, 35, 301
    of a business, 281
value orientation, 287
value-adding network, 283
van der Waals forces, 488
variation point, 224
variothermal
    extrusion embossing, 639, 658
    heating, 629, 636, 656, 658
    injection moulding test, 639
    process, 635, 639
    temperature control, 660
    tool temperature, 681
velocity distribution, 196, 202, 204
versatile assembly unit, 951
vertical integration, 790
viable system model (VSM), 750, 755
vibration test, 586
virtual control system, 388
virtual machine (VM), 247
virtual manufacturing system (VMS), 7, 374, 383, 398, 619
    cost-effective application, 417
    data flow, 385
    during machine development, 417
    during product development, 418
    in production, 416
virtual material testing, 348
virtual planning, 993
virtual process chain, 258
virtual product development, 436
virtual production, 5, 30, 58, 242, 244, 1041
    intelligence, 277, 279
    technology radar, 59
    tools, 59
virtual programmable logic control system, 418
virtual reality for scientific and technical applications (ViSTA), 276
virtual tensile test, 329
virtual test, 330
virtual workpiece, 398
virtualisation of production system, 30
virtualised ontology, 264
viscoelastic model, 208
viscosity, 154
visual fatigue questionnaire, 938
visualisation, 250, 618, 619
    framework, 246
    scalability, 275
    Tool Kit (VTK), 328
Voronoi cells, 270
Voxel model, 989
vtkQuadric Decimation, 272

W

warp system, 840
waste product, 43
weak boundary layer, 488, 516

weaving, 802, 805, 807, 838, 846
web
   interface, 252
   ontology language (OWL), 262
weld seam geometry, 607, 845
welding, 800, 805, 830, 845
wetting process, 488
white light interference microscopy (WIM), 642
Wiegers' relative weighting method, 313
WIP (work in process), 789
Wolter's time model, 275
work piece agent, 611
working memory, 735
workload balancing, 301
workpiece
   boundary zone, 454
   coordinate system, 394
Worst Case Exposure Scenario, 620

X
XML data model, 248, 258
XPS analysis, 506

Y
Young equation, 488, 649
Young-Dupré equation, 649
Young's modulus, 647

Z
ZaKo software, 344
ZEMAX optical simulation, 838

Printed by Publishers' Graphics LLC